High-Speed Digital System Design

Anatoly Belous • Vitali Saladukha

High-Speed Digital System Design

Art, Science and Experience

Springer

Anatoly Belous
Integral
Minsk, Belarus

Vitali Saladukha
Integral
Minsk, Belarus

ISBN 978-3-030-25408-7 ISBN 978-3-030-25409-4 (eBook)
https://doi.org/10.1007/978-3-030-25409-4

This Springer imprint is published by the registered company Springer Nature Switzerland AG
The registered company address is: Gewerbestrasse 11, 6330 Cham, Switzerland

Preface

The book, proposed for the readers' attention, is unique—as for the structure of composition and the contents of materials, it so far has no analogues in the world scientific-technical literature. This book is meant for a wide circle of readers: students, postgraduates, professorial staff of technical universities, and engineers, specializing in the sphere of design development and production arrangement of the various electronic devices and systems, whose common and specific feature is the high rate of data processing and transfer. First and foremost, it refers to the contemporary digital communications (navigation, mobile communications, Internet of Things, etc.), which are characterized by the high rates of signal processing and transfer, while these speeds keep on increasing, the number of the accompanying problem multiplies.

Here, one should differentiate the problems of ensuring the rates of processing and the problems of ensuring data transfer, as the causes of the origination and the methods of their solution bear an entirely different nature. While the data processing rate primarily depends on the technical solutions of the hardware of digital communications, the rate of the data transfer is determined by the technical solutions of the involved transmission channels. In the fast response digital circuits and on the printed circuit boards of the modern high-speed electronic devices, the short fronts of the processed and transferred digital signals prompt the emergence of multitude of adverse effects, as the designers of these devices encounter in practice the distorted beyond recognition digital signals, aggravated by the background of the plethora of disturbances in transmission.

The design engineers often treat such effects as something of a mystery in the professional stratum nicknaming them as "black magic manifestations," implying that at the high frequencies, the high response devices run amok inexplicably. Specifically due to this reason, some authors of the books, dedicated to the problems of designing the high-speed digital devices, embellished the captions of the books with the "thematic" headings of the type: *High-Speed Digital Design: A Handbook of Black Magic* and *High-Speed Signal Propagation: Advanced Black Magic*. These books enjoy a great popularity in the consortiums of the design engineers of the high-speed digital devices.

There exist quite many publications, including monographs, dedicated to these problems. However, they usually resort to the detailed description of the individual concrete methods and technical solutions, applicable to the various electronic devices, lacking the description of interaction among themselves and the sequence of their application at the system level.

Figuratively speaking, this is equal to the presentation of a detailed description of solely stand-alone trees, which makes a reader to exert strenuous efforts in the attempts to create a vision of a general picture of the forest. Therefore, the authors endeavored to systematize all known methods and create a comprehensive picture of such a "forest" (a type electronic device), consisting of the "trees" (major stages of design development and production arrangement of the fast response electronic devices).

Here, the authors deem it expedient to say a few words about the methodology for creating this book.

It is common knowledge that in order to stand ground on the secured and amass the new segments of the world market, one needs the innovative products with the advanced quality features. Such products can be created by those engineers only, who have mastered the most advanced design methods. These methods can be accessed by means of two approaches—to gain the individual practical experience of the highly qualified engineers and to study the contemporary technical literature.

In the process of their major activities for a great number of years, the authors were science directors of a variety of large teams of specialists involved in the design development and production arrangement of a multitude of integrated circuits of the commercial, space, and military applications, as well as finished high-tech functional fast response radio-electronic devices and systems, designed on the basis of these integrated circuits. More often than not in the process of the design development and implementation into the industrial production of the similar devices, appliances, and systems, there emerged situations which were reported to the superiors by the designers as the abovementioned "effects of black magic." A most trivial and comprehensible practical example is as follows: why does one encounter the situations when the designed and fabricated integrated circuits of the central processor and RAM operate flawlessly at the frequencies above 100 MHz and the developed electronic devices and systems with application of the very same integrated circuits can reliably function only at the cycle frequencies not in excess of 50 MHz?

In each such case, all these "effects of black magic" were, indeed, pinpointed and taken care of, which every single time commanded time and means and which delayed the due dates of the appearance of the new products on the market.

As a rule, no mystic transformations happened with the integrated circuits and circuit boards in these devices and fixtures; just an in-depth comprehension was required of the physical mechanisms of emergence of the similar effects to preclude the causes of their manifestation in the process of designing the devices (circuit boards, transmission lines).

It is worthwhile mentioning that some of these physical mechanisms and methods of their neutralization received an ample and comprehensive consideration, for instance, in the 1993 classic textbook *High-Speed Digital Design: A Handbook of*

Black Magic by Howard Johnson and Martin Graham, as well as in the 2003 book, published by "Prentice Hall," *High-Speed Signal Propagation: Advanced Black Magic* by the same authors. Meanwhile, for the years passed, there appeared quite a number of new data, innovative technical solutions, as well as new physical mechanisms of origination of the similar "effects of black magic," which were published in various scientific articles and monographs.

Therefore, the authors strived for an ambitious task—to systematize the main known theoretical and practical aspects of designing the high-speed device along the "entire chain"—starting from the top level of hierarchy of the high-speed systems to the bottom most level, the base elements and production arrangement.

The top level (Chap. 1) covers the contemporary informational-communicational technologies and telecommunications systems (TCS) on their basis. The next level comprises the data transmission (exchange) channels among the TCS segments (wire interfaces, Chap. 2; and wireless communications, Chap. 3).

Nowadays, there exist a variety of configurations and architectures of the high-speed electronic devices, aimed at solving the concrete tasks, while it is advisable to have readily available a list of the optimal base elements ("bricks"), on whose basis these devices are constructed. Thus, the chapters dedicated to the description of the composition, technical parameters, and specific features of the applications of the passive and active base components (Chap. 4) and special purpose integrated circuits (Chap. 5) emerged.

Furthermore, it was necessary to generalize the available effective technical solutions on the problems and methods and means of ensuring the noise resistance of such fast response devices—resulting in the creation of Chap. 6.

It is impossible to circumvent the consideration of the design development problems of the printed circuit boards of the fast response devices—bringing in print Chap. 7 "Design Fundamentals of Printed Circuit Board Structures."

The design engineer is supposed to select the optimum technology of packaging the designed microelectronic device, and this is one of the most complex and expensive operations. Under the packaging, the specialists understand the integration process of the various electronic elements (integrated circuits, sensors, capacitors, etc.) into one finished functional device by means of formation among them of the system of conductive connections at different levels. It is on the correct selection of the used packaging technologies that the reliability, cost, and functional efficiency of the electronic device depend. Chapter 8 is dedicated specifically to these problems.

However, during designing of the high-speed electronic devices on the basis of these active and passive elements, it is necessary to apply the special measures for their protection from the various kinds of interferences and electromagnetic irradiations (Chap. 9).

At any rate, even the "correctly" designed electronic device with consideration of all recommendations in the previous chapters is not certain to function reliably if the system of its power supply management is "incorrectly" arranged. Certainly, this problem did not surface yesterday, but an acute actuality it has acquired lately is due to the dramatic rise of power consumption in the contemporary high-speed telecommunications systems. These are the origins of Chap. 10, focused on the specific features of designing power management systems, selection of the optimum nomenclature of the special purpose integrated circuits, and methods of noise suppression in the power supply circuitries crucial for practical application peculiarities of arranging the grounding systems of the electronic equipment (convection systems, "floating ground," protective earthing, protective filters, etc.)

Quite recently, during the creation of the high-speed devices, they use ever more often the various microwave devices, operating in the centimeter, millimeter, and even sub-millimeter wave bands. Here, the major problem is posed in the peculiarities of arranging the process of their analysis and measurements, depicted in the conclusive Chap. 11.

In the long run, the authors were compelled to tackle the maximum task and to describe the major stages of creation of the high-speed systems as per the following algorithmic progression: "systems → interfaces → methods of ensuring noise resistance → methods of designing printed circuit boards → base components of the fast response systems → special purpose sets of integrated circuits for them → methods and fixtures of their protection from the electromagnetic noises → assembly technologies → arrangement of the correct system of power management → organization of analysis and measurements of the high-speed electronic devices."

When writing this book, the authors have been deriving guidance from the following principles, which hardly presented difficulties in unveiling, but were quite intricate in their further realization in practice:

1. The design engineers of the high-speed electronic devices, students, and their professorial staff have always to keep "handy" a certain systematized reference book of collected materials on the methods of design development and production arrangement of such devices.
2. For the sake of acquiring a pronounced popularity among a wide circle of scientists, engineers, and students, this book should manifest simultaneously the integral functions and be a classic textbook alias a concise reference book for the readers to be engrossed in just the captivating reading.
3. While presenting a huge volume of the reference information, one should endeavor by means of an utmost simple language to recount both the theoretical and practical aspects of estimation and design development of the electronic devices, sidestepping whenever possible the complex mathematical expressions and physics formulae. The book should comprise only those methods and technical and technological solutions whose effectiveness was first verified by the practice of their application.
4. In the text, it is required to use the maximum possible amount of graphic material, reflecting effectiveness of the various working scenarios.

Whether the authors have succeeded in applying these principles, the judgment is with the reader.

Minsk, Belarus — Anatoly Belous
Minsk, Belarus — Vitali Saladukha

List of Authors of the Major Works Whose Graphic and Textual Materials are Referred to in this Publication, to Whom the Authors Would Like to Express Their Deepest Gratitude

Chapter 1:
Bogachev L, Glasskott-Jones A, Korchak V, Stackler M, Tuzhikov E, Frass S. Sombe, Chantie N, as well as: Bodea M, Brimley Sh, Dan C, Flelaoui M, Garcia Oya Jose Ft, Ghannouchi FM, King Z, Kwan A, Lopez-Morillo E, Marquez Lasso F, Munoz Chavero F, O'Rourke R, Pellerin Ch, Perry WJ, Richmond J, Roper WB, Spiridon S, Spiridon F, Torralba, Silgado A.
Chapter 2:
Blinkov OE, Drobyshev A, Ponomar VN, Senyakin Yu, Silin AV, Smirnov M, Merdanov MK, Wissel F, Shcheglov A, Yarzhembitsky VB, as well as: Bhagwath N, Conchas J, Gupta N, Kinnaird C, Kugelstadt T, Patel A, Ritter D, Taylor A.
Chapter 3:
Blinkov OV, Bochechka G, Vishnevsky V, Emelyanov VA, Efimenko SA, Zhurba VM, Makushin MV, Martynov V, Merdanov MK, Pirpoint M, Podrubny OV, Silin AV, Starkloff E, Sukhoroslova Yu, Syakersky VS, Tikhvinsky V, Turtsevich AS, Frolov S, Chachin P, Shakhnovich I, Yarzhembitsky VB, as well as: Amitabha G, Chih-Lin I, Ghosh A, Corbett R, Cudak M, Curbuz OD, Ding J, Gang L, Karroy A, Khan F, Kishiyama Y, Kovarik T, Li G, Li J, Miao Q, Nakamura T, Pi Z, Raman S, Rappaport TS, Rebeiz GM, Shuangfeng H, Thomas TA, Wang T, Wang Y, Wonil Roh, Zhang J, Zhengang P, Zhikun X, Zihir S.
Chapter 4:
Bagdasaryan AS, Bagdasaryan SA, Belov L, Belyanin AF, Bessonov LA, Blank T, Blackwood M, Butenko VV, Bykhovsky MA, Varava N, Gulyaev Yu V, Goldberg Yu, Gudkov V, De Los Santos G, Jurinsky KB, Dneprovsky VG, Egorov RV, Efimov IE, Zaitsev VE, Zemlyakov V, Zlotnikova EA, Zhitkovsky VD, Zhitnikova M, Karapetyan G. Ya, Kozhevnikov V, Kondratyev SN, Kruglikov D, Kuznetsov R, Kukk KI, Lebedev A, Lobanov A, Lokshin MG, Lvov V, Maiskaya V, Mashinin OV, Merdanov MK, Mnatsakanov T, Mokeev A, Neretina SP, Nikitov SA, Nikolaeva SO, Nikonorov M, Novikov S, Orlov MM, Ostankovich GA, Pomortseva L, Posse E, Praporshchikov VV, Prokimov A, Pronin S, Rctnyuk V, Richards R, Sabitov RA, Samoilov AI, Samoilovich MI, Sbruev S, Sevastyanov DI, Semenov VV, Sergeev VI, Sigov AS, Sinitsyna TV, Soldatenkov F, Smolyaninov I, Taube LM, Tumachek A, Khazarchiev Yu D, Khlebnikov VI, Chernykh M, Tsotsorin A, Shestakov SP, Yurkov S, as well as: Agarwal A, Chen W, Delisle J-J, Domingue F, Drews J, Ehwald KE, Fall EM, Fang W, Fedder GK, Fouladi S, Fuchs J, Johnson P, Karstensen H, Kaynak M, Koufogiannis I, Kress G, Li S, Mansour RR, Mattes M, Moyer B, Mukherjee T, Qureshi SUH, Reinke J, Rupflin M, Thibeault B,

(continued)

Tornatta P, Ward-Foxton S, Wettstein K, Saddow SE, Sorolla E, Su W, Yu M, Zaghloul M, Zahirovic N, Zhang S.	
Chapter 5:	
Afonsky VP, Balakirev A, Verkhulevsky K, Vikulov I, Goloshchapov S, Golyshev V, Demidov D, Dyakonov VP, Zolina KG, Ioffe D, Kertis D, Kilpatrik D, Kichaeva N, Kishchinsky AA, Krasnobrov D, Kudryashov VE, Merdanov MK, Mikhalev P, Odnolko A, Smirnov M, Sinha Manish, Turkin AN, Fedorov Yu, Shakhnovich I, Sherill Robby, Yunovich AE, Farmikoun G, Fedorov Yu, as well as: Betten J, Green DS, Kollman R, Lidow A, Narendra Mehta, Palczynski J, Reusch D, Rosker Mark J, Strydom J.	
Chapter 6:	
Avdeev VB, Avdeeva DV, Alybin VG, Ashkenazi D. Ya, Basalaev K, Belov LA, Belyaev VP, Berdyshev AV, Bogdanov Yu, Brigidan AM, Brodulenko GI, Budzinsky YuA, Buevich EA, Bykovsky SV, Vildanov SA, Golenitsky II, Dzhurinsky K, Drozdovsky NV, Drozdovskaya LM, Dukhanin ND, Dyakov IV, Zakharov D, Zie S, Zhadnov VV, Kantyuk SP, Karapuzov MA, Karchevsky EV, Kirillov AV, Kirshchov VM, Kondratov AS, Korotkov S, Kochemasov V, Krutov AV, Lagovier BB, Lazunin YuA, Lebedev IV, Liberman LS, Mandelshtam M. Ya, Mastryukov MA, Merdanov MK, Mitlin VA, Murashkina ZN, Peresadin A, Perets RI, Polessky SN, Polzunov VV, Prokhorov RA, Rebrov AS, Rebrov SI, Ropiy AI, Rozhkov VM, Romashchenko KV, Sestroretsky BV, Skorobogatov DB, Smirnov A, Starik AM, Strigin I, Stroganova EP, Titovich NA, Ugnichev LV, Ugnichev DV, Fefelov AG, Filippov A, Khasyanova E, Shnitnikov AS, Shpirt VA, Shur M, Shutov KK, Yakuben LM, as well as: Armstrong AL, Bakeman PE, Bell T, Booth JC, Byrd D, Cannon N, Card KC, Collado C, Collins S, Cox NW, DeGroot DC, Flynn K, Forrez M, Gawronsri M, Hartman R, Hill CN, Jargon JA, Kugelstadt T, Lam PC, Mardiguian MA, Mateu J, Milazzo C, O'Callaghan JM, Orloff ND, Padilla A, Ramay RL, Rocas E, Rucken CT, Soushworth GC, Steer MB, Wamoss JW, Wilkerson JR.	
Chapter 7:	
Bodnar D, Vladimirov S, Gakhariya V, Gakhariya L, Dorokhov V, Emelyanov VA, Zakharov D, Kondratyev V, Korotkov S, Krasnov S, Medvedev AM, Merdanov MK, Peresadin A, Savenko S, Yarzhembitsky VB, as well as: Bhyrav M. Mutnury, Bogatin E, Brooks D, Chao FL, Coondor J, Digiacomo JJ, Eblen M, Feller A, Chang RW-Y, Graham M, Guo Wei-De, Johnson H, Kaupp HR, Kerinami MH, Kim J, Lee H, Michalka T, Morant K, Olney B, Oswal M, Ramahi OM, Rubin BJ, See K-Y, Shin J, Soh W-S, Wang L-B, Warner J, Wu RB, Zhang Pi.	
Chapter 8:	
Alekseev SA, Bogatyrev Yu, Bodnar D, Vasilenkov N, Vikulov I, Glebko AS, Grabchikov S, Gulbin V, Dedovskikh IA,Dzyubanenko SV, Didilev S, Epifanovsky I, Efremov G, Zabolotny V, Ivanov NN, Ivin VD, Kochetkov A, Lastovsky S, Maksimov A, Medvedev A, Merdanov MK, Nisan A, Petrov EV, Raikunov GG, Starostin E, Tikhomirov K, Fedorov SS, Khokhlun A, Shavlovsky IV, Shiryaev A, as well as: Adams L, Andrae ASG, Beacour J, Boeske S, Bonnet B, Brizoux M, Carriere T, Chang-Chien P, Chang Y-W, Chen C, Dandu P, Diao C, Drumm C, Eloy J, Engelmaier W, Fan XJ, Fan W, Grivon A, Gusak AM, Hsiao H-Y, Huang AT, Ingram B, Kelleher A, Lai YS, Lau JH, Leadon R, Lee JK, Leidecker H, Liang YH, Liu Y, Maia Filho WC, Mangeret R, Mao H, Millward D, Monier-Vinard E, Morianz M, Nickson R, Ouyang F, Passenheim B, Soret C, Spratt J, Stahr J, Strobel D, Subramanian KN, Tu KN, Yajima M, Yan YG.	
Chapter 9:	
Batishchev AG, Veksler GS, Galper AM, Grabchikov SS, Grishin SA, Gurevich VI, Goreev A, Glazenko TA, Dmitrenko VV, Zamesov AYu, Lantsov V, Merdanov MK, Nakauchi Ed, Nasyrov II, Naumov PYu, Nedochetov VS, Petyuk AL, Senkov VI, Sosnovskaya LB, Cheledyuk AV, Sharapa TE, Eranosyan S, as well as: Altamura F, Christoph J, Goldman SJ, Nalhorczyk AJ, Sgrigna V.	

(continued)

Chapter 10:
Gurevich VI, Kuznetsov MB, Makarenko V, Matveev MV, Merdanov MK, Noikhoizler Yu, Sadler S,as well as: Armstrong K, Barreto RM, Bridges JE, Duff WG, Joffe EB, Lock KS, Longoria SN, Miletta J, Ricketts LW, Tuite D, Whitaker JC.
Chapter 11:
Abubakirov BA, Andreev VS, Andriyanov AV, Bondarenko IK, Valitov RA, Vikulov IK, Gasanova LG, Glebovich GV, Goldenberg LM, Gonorovsky IS, Gudkov KG, Gusinsky AV, Deinega GA, Diyakonov VP, Elizarov AS, Emelyanov AV, Efimenko SA, Kalinin O, Kostrikin AM, Lanne AA, Magrachev V, Maks Zh, Nechaev AV, Petrovsky AA, Rimskogo GV, Turtsevich AS, Chigir GG, Sharov GA, Shvedov SV, Shermina GA, Shpak II, as well as: Gartner UF, Hillstrom, Jargon JA, Marks RB, Rytting, Schick B, Schneider I, Vanderstein G, Yyes Rolain

Acknowledgments

The main distinguishing feature of this book, as compared to all works previously published by the authors, is the fact that the authors herein acted only as a kind of "system integrators" of very extensive and yet sketchy technical and scientific data, currently available on various aspects of the complex problem of designing modern high-speed electronic devices. On a practical level, their role was to systematize, critically analyze, and summarize the core of scientific, technical, and practical problems arising from design and production of such systems and devices, relying, among other things, on their own professional experience in this industry.

Thus, the most relevant materials from more than 1200 sources of information, published in Russian and in English languages, have been selected for inclusion into this book as a result of a huge amount of work carried out by the authors together with great many experts involved, which are authoritative in this field of knowledge.

The authors of these publications were both ordinary RnD and process engineers, not "overburdened" with high academic titles, and also professors, doctors of science, academicians, and managers of electronic industry enterprises.

As a result, a technical encyclopedia on the most effective design methods used at all successive stages of creating high-speed electronic devices was virtually presented to the reader. Here, the authors consider it appropriate to quote a famous writer and philosopher Denis Diderot, the creator of the Encyclopedia of Sciences, Arts and Crafts: "Indeed, the purpose of an encyclopedia is to collect knowledge disseminated around the globe; to set forth its general system to the men with whom we live, and transmit it to those who will come after us, so that the work of preceding centuries will not become useless to the centuries to come; and so that our offspring, becoming better instructed, will at the same time become more virtuous and happy."

The authors express their gratitude and appreciation to the academician of the National Academy of Sciences of Belarus and elected foreign member of the Russian Academy of Sciences, Professor Vladimir A. Labounov, for his critical feedback and helpful advice which contributed greatly to producing the book in this particular format, as well as to the following colleagues, Yuri Sizov, Vladimir

Ilyenkov, Viktor Chikilev, Marina Biryukova, Lyudmila Ignatovich, and Olga Antipenko, for their assistance in preparing the materials for the book.

Introduction

Materials of the book are formulated as 11 chapters.

Chapter 1 considers the tendencies of the development of the informational-communications technologies, the basic principles of arrangement, structure and base components of the telecommunications systems, standard protocols of functioning, modes and media of data transfer, layout of the high-speed data transmission, access arrangement methods, and commutation technologies of the high-speed channels. Here are under coverage the peculiar features of integrated circuits and semiconductor devices, used in the telecommunications systems of the new generation.

Chapter 2 is dedicated to the specific features of the arrangement of the wire interfaces of the high-efficiency digital devices for information processing: disclosed are the basic principles of arrangement and methods of computation, as well as peculiar features of designing and application of the most common interfaces (RS-485, SerDes, LPVECL), DDR-type high-efficiency memory buses, recommendations, and schematic technical solutions of their protection from the electrical overvoltages.

Chapter 3 reviews the basic tendencies, problems, and development prospects of the wireless telecommunications systems.

Under coverage are the basic technical parameters and specific features of all generations of advancement of this scientific technical trend: 1G, 2G, 2,5G, 3G, 3,5G, 4G, and 5G.

Under perusal are the tendencies, problems, and development prospects of such directions, as "Internet of Things" (IoT) and "Internet of Everything," and the arrangement peculiarities of the inter-machine (M2M) interactions in the similar networks.

A special attention is attributed to the status, problems, and development prospects of technologies of the wireless communications of the fifth generation (5G); under focus are the concrete technical solutions of one of the ways of transition to 5G on the example of the specific features of creation of communications channels with the phased aerial lattices, examples of using the millimeter wave band.

Chapter 4 concentrates on the description of structures and technical parameters of the base components of the high-speed electronic devices—from the passive components (wave guides, coaxial communications lines, phase stable cables of the microwave band, adapters, attenuators, absorbent filters, directional couplers, power dividers and adders, encoders) to the intricacies of designing the radio-frequency circuits, surface-acoustic-wave filters, MEMS items, and software configured radio systems.

Chapter 5 is dedicated to consideration of the special purpose integrated circuits for the high-speed telecommunications devices—principles of construction, basic technical parameters, and application peculiarities of the integrated circuits of series SerDes, SEPIC, base sets of integrated circuits of transceivers for the range of millimeter waves, special purpose integrated circuits for the standard interface RS-485, and high-voltage interfaces LVDS and DV1.

A separate paragraph is dedicated to the innovative dynamically developing in the world direction of microwave devices on the basis of the heterogeneous integration technology.

Chapter 6 dwells in detail on the basic methods of ensuring noise resistance and reliability of the contemporary high-speed electronic devices. Here in depth are described the mechanisms of emergence and methods of counteracting the relatively new for the telecommunications systems types of interferences, in particular, the passive intermodulation (PIM) (mechanisms of emergence, peculiarities of manifestation in the materials of the printed circuit board, in the band, coaxial and wave guide lines, and others). There were proposed effective methods for reduction of the level of intermodulation distortions in the high-speed electronic devices.

Here under meticulous coverage are the various protective microwave devices for the high-speed electronic devices (semiconductor, gas discharge, vacuum), as well as special components for filtration and suppression of the various electromagnetic interferences.

Chapter 7 is dedicated to the peculiarities of developing designs of the modern circuit boards for the high-speed electronic devices. Usually, the designers of electronic devices use a circuit board solely as a certain physical medium for the design realization of their instrument under development (device, system) and select and use the numerous available on the contemporary market options of this class of products.

The aspects of the design analysis of a circuit board or the technology of its fabrication are usually approached by the designers of the electronic devices only when in practice they encounter problems—everything is designed and simulated apparently correctly, but during the tests, a device behaves erroneously. The matter is that the producers of the numerous varieties of circuit boards are not obliged to notify their consumers without a special query of the numerous design and technological peculiarities of their products, although quite often these differences and peculiarities are the causes of emergence of similar parasitic effects.

This chapter describes a number of design and technological features of contemporary printed circuit boards, which are directly related to functional reliability of high-speed devices in which these boards are used. The features are presented on the basis of comprehensive consideration of analysis and generalization of existing positive and negative experience available to designers.

Besides, under perusal are the various methods of ensuring integrity of signals on the circuit boards of the high-speed transmission lines; the methods, expressions, and formulae are provided for computations of the optimal structures and layouts of semiconductor, including indication of the empiric rules for design development of layouts for the circuit boards with the high-speed signals.

Chapter 8 is dedicated to the problems of packaging electronic device—one of the most complex and expensive operations of the multistage process of creation of electronic devices.

Packaging specialists understand the integration process of various electronic elements (microcircuits, sensors, capacitors, etc.) into one functional device by means of forming between them systems of conductive interconnections at different levels. The correct selection of the used technology of packaging determines reliability, cost, and functional efficiency of the electronic device. Therefore here, one considers the basic development directions of IC packaging technologies, the basic types and peculiarities of the packaging technologies of the high-speed electronic devices, especially, advantages and drawbacks of the technological solutions of the type "system in package" and "system on chip." A number of sections of the chapter are dedicated to description of the concrete technological processes of chips mount on the circuit board, their assembly on the wafer, peculiarities of packaging microwave integrated circuits on the wafer, multichip modules, and 3D assembly technologies.

Also, under consideration here are the basic peculiarities of the assembly technologies of the electronic devices, intended for the military and space applications, as well as the technology peculiarities of the soldered joints fabrication, substantially influencing reliability of the modern electronic devices.

Chapter 9 is dedicated to the practical recommendations on the application of the various design-technological and hardware methods for protection of the high-speed electronic devices from the various kinds of electromagnetic interferences with consideration of the concrete methods of application of the film electromagnetic screens for protection on the on-board electronic equipment of the space and military hardware from the influence of radiation and quasi-stationary magnetic fields.

Chapter 10 is dedicated to the issues of designing one of the vital systems—power supply systems—of the high-efficiency electronic devices.

Under consideration are the basic requirements to these systems, as well as the composition, technical parameters, and peculiarities of application of the special purpose integrated circuits—controllers and power supply drivers, pulse voltage stabilizers, integrated circuits of supervisors, sequence formers of the control signals in the electric supply circuits, special purpose integrated circuits of load switchovers, integrated circuits of ensuring the "hot mode" of enabling the electronic equipment, etc.

Under coverage are the concrete technical solutions of the various effective methods of noise suppression in the supply circuits of the high-speed devices, as well as the arrangement peculiarities of the grounding systems of the electronic equipment of the electric energy objects, integrated into the composition of the telecommunications systems.

The materials of the conclusive Chap. 11 are completely dedicated to the problems of measurements and analysis of the microwave devices, which are widely used in the creation of the contemporary high-speed devices and systems. The studies of the parameters of the microwave devices during their creation and tests of their compliance with the specification requirements when arranging production demand application of the specific software and hardware means. A list of such measuring tools is quite expansive, as it should ensure measurements of the microwave devices in the different frequency bands and for the various types of transmission lines. Therefore, here in detail are considered the scalar and vector analyzers, including the tasks, methods, and tools of the instrument analysis of the microwave circuits in the different bands of the wavelengths. Significant attention is attributed here to the methods and hardware means for measurements of parameters of microwave integrated circuits (amplifiers, switchovers, phase shifters, attenuators, etc.) directly on the semiconductor wafers.

Contents

1 Information and Communication Technologies and Systems Based on Them 1
1.1 Main Trends in the Development of Information and Communication Technologies 2
1.2 Basic Principles of the Organization of Telecommunication Systems 4
1.3 Extracts from the History of Development of High-Speed Data Networks 13
1.4 Standard Protocols of Functioning of Telecommunication Systems 16
1.5 Basic Standards of Telecommunication Systems 21
1.6 Practical Basics of Physical and Logical Data Coding 24
1.7 Narrowband and Broadband Systems with Data Multiplexing 31
1.8 Data Transmission Modes and Media 35
1.9 Structured Cable Systems 43
1.10 Topology of High-Speed Data Transmission Systems 46
1.11 Methods of Organizing Access to High-Speed Channels 49
1.12 Switching Technologies 51
1.13 Features of Organizing Communication Between Segments of Telecommunication Networks 54
1.14 Semiconductor Element Base for Telecommunication Systems of the New Generation 61
References 77

2 Wired Interfaces of High-Speed Electronic Devices 79
2.1 Communication Lines as a Component of the Interface 80
2.2 Organization Methods and Basic Parameters of Interface Buses of High-Speed Electronic Devices 81

2.3 Basic Principles of Application and Methods for Calculating the Characteristics of Wired Communication Lines 86
2.3.1 Calculation of the Electrical Parameters of Wired Communication Lines 86
2.3.2 Calculation of the Dynamical Parameters of Wired Communication Lines 90
2.3.3 Energy Parameters of Wired Communication Lines 92
2.3.4 Design Options for Implementation of Wired Communication Lines 93
2.4 Organization of the Process of High-Speed Data Transmission Based on Transceivers with SerDes Type Topology 94
2.4.1 Principles of Operation of the Transceivers with SerDes Type Topology 94
2.4.2 Methods for Signal Correction During Transmission Over Video Cable 98
2.5 Design Features of High-Speed DDR Memory Buses 101
2.5.1 Principle of the DDR Module Operation 101
2.5.2 Methodological Features of the Stage of Preparation for Modeling 105
2.5.3 Certain Methodological Features of the Modelling Stage 108
2.6 Features of the Organization of High-Speed Data Transmission Channels with RS-485 Interface 111
2.6.1 Synchronous Data Transmission Channels Based on RS-485 111
2.6.2 Asynchronous Data Transmission Channels Based on RS-485 114
2.7 Circuit-Based Methods of Protection of RS-485 Interface Buses from Electrical Overvoltages 116
2.8 Practical Recommendations for the Design of the Bus Node of the High-Speed Interface RS-485 121
2.9 Design Features of High-Speed Electronic Devices Based on Programmable Logic Chips 126
2.9.1 Selection of a Specific Type Standard of the FPGA 126
2.9.2 General Recommendations on the Selection of the Structure of Power Supply System for High-Speed Electronic Devices 129
2.9.3 General Recommendations for Choosing the Optimal Clock Signal Tree Structure 130
2.9.4 General Recommendations for Designing Topologies of High-Speed Printed Circuit Boards . . . 132

2.10 Circuit-Based Options of Coordination of Transmission Lines of the LVPECL Standard 135
References 142

3 Wireless Telecommunication Systems 145
3.1 Trends, Prospects, and Problems of Development of the Element Base of Wireless Telecommunication Systems 145
3.2 Basic Methodological and Terminological Problems of the Introduction of Fifth-Generation Wireless Telecommunication Systems 152
3.3 State, Prospects, and Problems of the Development of Fifth-Generation Mobile Communication Technologies (5G) 157
3.4 Major Trends and Challenges in the Development of Wireless Technologies for the Internet of Things (IoT) 165
3.4.1 Technical Background Pf the Emergence of the Internet of Things 165
3.5 Communication Channels of the 5G-Generation Telecommunication Systems Based on Phased Antenna Arrays 168
3.6 Features of Using the LPWAN Technology for the Internet of Things 171
References 173

4 Basic Components of Telecommunication Systems 177
4.1 Features of Using Passive Components in HF and SHF Devices 178
4.1.1 Necessity of Using Special Materials for Operation in Extreme Conditions 178
4.1.2 Waveguides and Coaxial Communication Lines 179
4.1.3 Adapters and Terminal Equipment 181
4.1.4 Attenuators and Absorbing Filters 181
4.1.5 Directional Couplers and Power Splitters/Adders . . . 182
4.1.6 Basic Mechanisms of Degradation of SHF Components 184
4.2 New Generation of Encoders, Features of Their Use in Radioelectronic Equipment 185
4.3 Basic Technologies of Radio Frequency Systems 191
4.3.1 Qualitative Analysis of the Technologies of Production of Radio Frequency Components 191
4.3.2 Features of Selection of Design and Technological Parameters for High-Power SHF MESFET Transistors Based on Silicon Carbide 197
4.4 Structural and Technological Features and Applications of Impedance SAW Filters 202

4.4.1 Design Features of SAW-Based Chips 202
4.5 Structural and Technological Features of Impedance SAW Filters 207
4.5.1 Fields of Application of Impedance SAW Filters . . . 212
4.6 SHF Applications of MEMS Technologies 218
4.6.1 Features of Implementation of Radio Frequency MEMS/CMOS Devices 218
4.6.2 Radio Frequency MEMS Switches 223
4.6.3 Radio Frequency MEMS Capacitors of Variable Capacity 226
4.6.4 Integrated MEMS/CMOS Resonators 228
4.6.5 MEMS Technologies in the Tasks of System Integration of High-Speed Electronic Devices 230
4.7 Software-Configured Radio Systems 238
4.8 Cable Assemblies of the SHF Range 244
4.8.1 Purpose, Classification, and Features of the Use of Cable Assemblies 244
4.8.2 Basic Specifications of Radio Frequency Cable Assemblies 246
4.9 Basic Specifications and Design Solutions for Phase-Stable SHF Cables 253
4.9.1 Temp: Flex Radio Frequency Cables by Molex 253
4.9.2 Radio Frequency Cables with Silicon Dioxide Insulation 260
4.9.3 Phase Track Radio Frequency Cables 264
4.9.4 Huber + Suhner SucoPearl Radio Frequency Cables 269
4.10 FOL in LAN and High-Speed Information Collection Systems: Selection of Physical Environment and Active Component Base 273
4.10.1 Main Trends in the Development of Fiber-Optic Technology 273
4.10.2 Standard Specifications of the Basic Set for FOL 280
References 285

5 Specialized Chips for Telecommunication Systems 291
5.1 Principles of Construction and Main Specifications of Telecommunication Chips of the SerDes Series 292
5.1.1 Basic Provisions of the Basic Standard IEEE 802.3 for Ethernet Telecommunication Network 292
5.1.2 Basic Specifications of SerDes Series Chips by Semtech 293
5.1.3 Structure and Operation Principle of SerDes Series Chips 296

5.2 Methods for Designing Pulse Voltage Converters with SEPIC Topology 299
5.2.1 Analysis and Calculation of Parameters of the Standard SEPIC Converter 299
5.2.2 Analysis and Calculation of Bipolar Voltage Source Based on the SEPIC Converter 304
5.2.3 Selection of the Converter Chip 306
5.2.4 Determining Parameters of Other Components of the Circuit 308
5.2.5 Brief Description of the Modeling Process of the Converter SEPIC 310
5.3 Principles of Organization and Application of Chips of Converters with DCS-Control Topology 314
5.3.1 Basic Principles of Operation of the Chips of DCS-Control Converters 314
5.3.2 Features of Operation of Converters in Energy-Saving Mode 317
5.3.3 Organization of Seamless Transition Between Controller Modes 320
5.3.4 Examples of Practical Application of DC/DC Converters with DCS-Control Topology 326
5.4 Use of System-on-Chip (SoC) for Creation of Bidirectional Communication Lines in the 60 GHz Millimeter Wave Band 333
5.4.1 Millimeter Wave Band Modem 333
5.4.2 Basic Set of Transceiver Chips for the Millimeter Wave Band 337
5.5 Specialized Chips for Construction of the RS-485 Bus 340
5.5.1 Functional Features of the Transceivers of the Interface RS-485 340
5.5.2 Practical Recommendations on the Use of Interface Transceivers 342
5.6 Specialized Chips for High-Voltage LVDS and DVI Interfaces 347
5.6.1 Basic Technical Solutions for Building of a LVDS Interface 347
5.6.2 Basic Technical Solutions for Building of a DVI Interface 349
5.7 SHF Chips for High-Speed Electronic Devices 356
5.7.1 Development of Gallium Nitride Technology and the Prospects for Its Application in SHF Electronics 356
5.7.2 GaN-Based SHF Chips by RFHIC for Wireless Communication Systems 357

5.7.3 SHF Chips by Hittite Microwave 369
5.7.4 Powerful Gallium Nitride Transistors 407
5.7.5 SHF Devices Based on Heterogeneous Integration Technology 425
5.8 Methodological Features of Design of Radio Frequency Chips 435
References 443

6 Methods and Means of Ensuring Interference Resistance of High-Speed Electronic Devices 447
6.1 Electromagnetic Compatibility of Electronic Devices: Terms, Definitions, Classification 448
6.1.1 The Nature of Electromagnetic Interference 448
6.1.2 Interference Signal Representation Forms 449
6.1.3 Measurement of Electromagnetic Radiation 451
6.1.4 EMC Terms, Definitions, and Classification 452
6.1.5 Regulations and Standards for Electromagnetic Compatibility 456
6.2 Assessment of the Tolerance of the Semiconductor Element Base of High-Speed Electronic Devices to the Effects of Electromagnetic Interference 466
6.3 Ensuring Electromagnetic Compatibility of Microprocessor Control Units of High-Speed Electronic Devices 473
6.3.1 Design Features of Printed Circuit Boards Optimized in Terms of Electromagnetic Compatibility 473
6.3.2 Features of Measurement of the Level of Interference Emitted by Microcontrollers 478
6.3.3 Ensuring Electromagnetic Compatibility in Wired Lines of Telecommunication Networks 487
6.3.4 Design Features of the Topology of Printed Circuit Boards for High-Speed Communication Systems 490
6.4 Methods of Reducing the Level of Intermodulation Distortion in High-Speed Electronic Devices 501
6.4.1 Mechanisms of the Occurrence of Passive Intermodulation Events 501
6.4.2 Methods of Reducting the Level of Passive Intermodulation in Radio Frequency Connectors 503
6.4.3 PIM in the Printed Circuit Board Material 504
6.4.4 PIM in Band, Coaxial and Waveguide Transmission Lines 504
6.4.5 PIM in Directional Couplers, Frequency Duplexers, and Transformers 505
6.4.6 External PIM Sources 506

6.4.7 PIM Level Evaluation Methods 507
6.4.8 Specialized Equipment for PIM Level Testing 510
6.5 Protective SHF Devices for Telecommunication Systems 512
6.5.1 Classification and Application Features of Protective SHF Devices for Communications Electronics Equipment 512
6.5.2 Gas Discharge Protective Devices 514
6.5.3 Semiconductor Protective Devices 516
6.5.4 Vacuum Protective Devices 522
6.6 Special Components for Filtering and Suppressing Electromagnetic Interference in Telecommunications Systems 523
6.6.1 Ceramic Condensers 523
6.6.2 Coaxial Low-Pass Filters 525
6.6.3 Surface Mount Filters 530
6.7 Methods for Assessment of the Resource of High-Speed Electronic Considering the Reliability of Mechanical Components 532
References 539

7 Basics of Designing Structures of Printed Circuit Boards of High-Speed Electronic Devices 543
7.1 Printed Circuit Board Design and Manufacturing Technology 544
7.1.1 History of Printed Circuit Boards 544
7.1.2 Designs of Printed Circuit Boards 546
7.1.3 Basic Technological Features of the Production of Printed Circuit Boards 552
7.1.4 Problems of Microminiaturization of the Elements of Printed Circuit Boards 556
7.1.5 Printed Circuit Boards for Components with High Heat Dissipation 558
7.2 Design Features of the Printed Circuit Boards for High-Speed Transmission Lines 563
7.2.1 Classification of the Main Types of Transmission Lines of High-Speed Electronic Devices 563
7.2.2 Constructive Features of Designing High-Speed Signal Transmission Lines 568
7.2.3 Features of Calculation of Characteristics of Conductors and Transmission Lines on Printed Circuit Boards of SHF Devices 569
7.3 Design and Technological Features of the Design of Printed Circuit Boards for High-Speed Electronic Devices 572

7.3.1 Methodological Aspects of the Design of Printed Circuit Boards of High-Speed Devices with Consideration of the Insertion Loss 572
7.3.2 Finishing Coating for HF Boards 578
7.4 Main Trends in the Development of Printed Circuit Board Design for High-Speed Electronic Devices 581
7.4.1 Main Trends in the Development of Innovative Technologies of Printed Circuit Boards 581
7.4.2 Certain Technological Features of the Implementation of Embedded Components of Printed Circuit Boards 583
7.4.3 Some New Applications of Copper in Interconnects of Printed Circuit Boards of High-Speed Electronic Devices 585
7.4.4 Methods of Designing of Special Substrates for Microassemblies of Electronic Devices 587
7.4.5 Features of Placement of Optic Networks on Boards 589
7.5 Methods for Ensuring Integrity of Signals in High-Speed Communication Lines 592
7.5.1 What Is Signal Integrity? 592
7.5.2 Signal Integrity 595
7.5.3 Protocol Selection as a Means of Increasing the Signal Integrity 597
7.6 Electromagnetic Interference and Methods for Ensuring Integrity of Signals on Printed Circuit Boards of High-Speed Electronic Devices 600
7.7 Methods for Ensuring Integrity of the Signal During All Stages of Designing of Printed Circuit Board 605
7.8 Method for Calculation of Zigzag Delay Lines for High-Speed Printed Circuit Boards 616
7.8.1 Calculation of Crosstalk in Connected Transmission Lines 616
7.8.2 Mechanism of Interference Formation in the Loop . . 620
7.8.3 Equations of the Loop Output Signal 622
7.8.4 Method for Determining the Distance S and Coefficient of Communication Between Neighboring Sections 625
7.8.5 Analytical Expressions for Determination of the Distance Between the Loop Sections 627
7.8.6 Practical Recommendations for Selection of the Section Length 628

7.8.7 Method for Calculation of the Magnitude of the Lead Time of the Output Signal Due to Interference 631
7.8.8 Method for Final Calculation of the Delay Line 632
7.9 Practical Recommendations for Eliminating Undesirable Effects Caused by Hanging Vias 633
7.10 Basic Rules of Thumb for Designing Topology of Boards with High-Speed Signals 642
References 653

8 Packaging Technologies for Electronic Devices 655
8.1 Trends in the Development of Chip Packaging Technology 655
8.2 Main Trends in the Development of Packaging Technology for High-Speed Electronic Devices 658
8.3 Systems in Package and Systems on Chip 662
8.4 Technology of Installation of Chips on the Board (Chip-on-Board (COB) Technology of Mounting) 666
8.5 Wafer Level Packaging (WLP) Technology 668
8.6 Features of Packaging of SHF Chips on Wafers 671
8.7 BGA Technology of Chip Assembly a Ball Grid Array (BGA) Chip Assembly Technology 676
8.8 Multi-Chip Modules and Printed Boards 678
8.9 Assembly of 3D Products Using Flip-Chip Technology 681
8.10 Special Features of Designing Electronic Devices for Spacecraft Applications 682
8.10.1 Technologies of Assembly of Microcircuits with Integrated Elements of Radiation Protection 682
8.10.2 Technologies of TSV Assembly of Chips for Aerospace Applications 688
8.11 Problems of Ensuring Reliability of Soldered Connections 692
References 700

9 Protection of High-Speed Electronic Devices from Electromagnetic Interference 703
9.1 Constructive and Technological Methods of Protecting Electronic Devices from Exposure to Electromagnetic Radiation 704
9.2 Information on Filters for EMI Protection Required by the Developer 709

9.2.1 Practical Recommendations on Selecting Reliable Filters for Protection of SHF Devices from Electromagnetic Pulse . . . 709
9.3 Design Features of Protective Devices for Radioelectronic Equipment with Frequency-Adjustable Drives . . . 716
9.4 Practical Recommendations for Developers of Radioelectronic Equipment on Solution of Problems of Protection from Conductive Interference . . . 720
9.5 Application of Multilayered Film Electromagnetic Screens for Protection of Onboard Equipment of Spacecrafts from the Effects of Quasistationary Magnetic Fields . . . 730
References . . . 739

10 Power Supply Systems of High-Speed Electronic Devices . . . 741
10.1 Design Features of Power Supply Systems for High-Speed Electronic Devices . . . 741
10.2 Specialized Chips for Power Supply Systems of Electronic Devices . . . 749
10.2.1 Generalized Structure of the Power Supply Circuit of Telecommunication Devices . . . 749
10.2.2 Controllers and Drivers in Power Supply Control Circuits . . . 749
10.2.3 Pulse Voltage Stabilizers . . . 753
10.2.4 LDO-Stabilizers for Voltage . . . 757
10.2.5 Supervisor Chips . . . 761
10.2.6 Chips for Formation of Sequences of Control Signals in the Power Supply Circuits of Electronic Devices . . . 766
10.2.7 Specialized Chips for Provision of the Hot Mode for Connecting Electronic Equipment . . . 769
10.2.8 Specialized Chips of Load Switches . . . 777
10.3 Step-Up and Step-Down DC–DC Converters for High-Speed Battery-Powered Electronics . . . 783
10.4 Circuit-Based Methods for Suppressing Interference in the Power Circuits of High-Speed Electronic Devices . . . 792
10.4.1 Analysis of the Typical Wiring Diagram of the Power Supply Circuit of SHF Device Board . . . 792
10.4.2 Basic Circuit-Based Methods for Interference Suppression in Power Supply Circuits . . . 797
10.5 Features of the Organization of Grounding Systems for Electronic Equipment of Power Facilities . . . 802
10.5.1 Grounding of Electronic Equipment: Analysis of Unevident Contradictions . . . 802
10.5.2 Problems of Conventional Grounding Systems . . . 804

10.5.3 Topical Issues of the Use of Protective Filters for Grounding of Electronic Devices 807
10.5.4 Floating-Type Grounding 809
10.5.5 Optimal Solution for the Problem of Organization of Protective Grounding of Electronic Devices 812
References 814

11 Measurement and Analysis of SHF Devices 817
11.1 Features of Measurement and Analysis of Circuits in the Millimeter Wave Band 818
11.2 Instrumental Analysis of SHF Circuits in the Millimeter Wave Range (Tasks, Methods, Tools) 821
11.2.1 Measured Circuit Parameters of SHF Devices 821
11.2.2 Measurement Goals of the Analysis of SHF Circuits 823
11.2.3 Certain General Requirements for Devices for Measurement of the Parameters of Millimeter Wave Band Circuits 827
11.3 Retrospective Analysis of Methods and Means of Measurement of SHF Circuits 828
11.3.1 Probe Methods 828
11.4 Reflectometric Methods for Measuring Parameters of SHF Circuits. Reflectometers 837
11.5 Trends of Further Development and New Fields of Application of Vector Circuit Analyzers 840
11.5.1 Main Trends in the Development of VNAs 840
11.5.2 New Frequency Ranges and Measurement Media 841
11.6 Improvement of Precision Characteristics of VNAs and Further Improvement of the Methods and Means of Their Metrological Support 844
11.7 Extension of Functional Possibilities and Application of Circuit Analyzers for Measurement of Parameters of Active and Non-linear SHF Devices 847
11.7.1 Frequency-Time Transformations 847
11.7.2 Modulation Vector Analysis of Circuits 847
11.7.3 Features of Measurement of Parameters of Non-linear Microwave Circuits 849
11.8 Automated Hardware for Measuring the Parameters of Integrated Circuits on Semiconductor Wafers in the Millimeter Wave Range 850
11.8.1 Testing of SHF Integral Circuits on Semiconductor Wafers 850
11.8.2 Coplanar Probes 852

11.8.3 Automated Complexes of Testing of ICs on Semiconductor Wafers 853
11.8.4 Features of Calibration of Measuring Systems during Testing of ICs on Semiconductor Wafers 858
11.8.5 On Advantages on RF and Microwave Testing of ICs on Semiconductor Wafers 860
11.9 Vector Analyzers of Millimeter Wave Networks. Classification and Building Principles 861
11.9.1 Types and Classifications of Vector Network Analyzers 861
11.9.2 Heterodyne VNAs 863
11.9.3 Homodyne VNAs. Measurement Methods Depending on the Modulation of the Reference and Measuring Signals 866
11.9.4 Circuit with Amplitude Modulation in the Measuring Channel 867
11.10 Homodyne Circuits with Simultaneous Modulation of Reference and Support Signals 874
11.11 Block Diagrams of Automatic Analyzers of Homodyne Networks 878
11.11.1 Block Diagrams of Network Analyzers with 14-Pole Measuring SHF Path 878
11.11.2 Block Diagram of the Network Analyzer with 10-Pole Measuring HF Path 881
11.11.3 Block Diagram of the Vector Network Analyzer with 8-Pole Measuring SHF Path 883
11.12 Network Analyzers by Keysight 885
11.12.1 PNA-X Series Network Analyzers 885
11.12.2 Features of Testing of Amplifiers with PNA-X Network Analyzers 889
11.12.3 Features of Testing of Non-linear Devices with PNA-X Network Analyzers 892
11.12.4 Means of Calibration of Network Analyzers 895
11.12.5 Measuring Systems Based on Network Analyzers . . . 897
11.12.6 PNA-L Network Analyzers 901
11.12.7 PNA Series Network Analyzers 902
11.12.8 Non-linear Vector Network Analyzers (NVNA) 907
11.12.9 Signal Generator Keysight UXG 908
References 914

Index 917

Abbreviations

A2/AD	Anti-access/area denial
ADC	Analog-to-digital converter
ADSL	Asymmetric digital subscriber line
AESA	Active electronically scanned array
AFC	Amplitude-frequency characteristic
AFD	Antenna-feeder device
AGC	Automatic gain control
APAA	Active phased-array antenna
APAR	Active phased-array radars
ARBF	All-route broadcast frame
ASCII	American Standard Code for Information Interchange
ASIC	Application-specific integrated circuit
ASL	Active service life
ASMT	Automated surface-mount technology
ATM	Asynchronous transfer mode
BAW	Bulk acoustic wave
BER	Bit error rate
BGA	Ball grid array
BiCMOS	Bipolar complementary metal-oxide-semiconductor
B-ISDN	Broadband-integrated services digital network
BLS	Board level shielding
BPS	Bits per second
CADS	Computer-aided design system
CBGA	Ceramic ball grid array
CCGA	Ceramic column grid array
CDM	Code division multiplexing
CDMA	Code division multiple access
CDR	Clock and data recovery
CL	Communication lines
CML	Current-mode logic

CMOS	Complementary metal-oxide-semiconductor
CMOS	Metal-oxide-semiconductor complementary structure
COB	Chip-on-board
COSMOS	Compound semiconductor materials on silicon
CP	Central processor
CPR	Common public radio interface
CPU	Central processing unit
CRC	Cyclic redundancy check
CSD	Circuit switched data
CSMA	Carrier sense multiple access
CSP	Chip scale/size package
CTE	Coefficient of thermal expansion
CTLE	Continuous time linear equalization
CWDM	Coarse wavelength division multiplexing
D2D	Device-to-device
DA	Destination address
DAC	Digital-to-analog converter
DCA	Direct chip attach
DCD	Duty-cycle distortion
DCE	Data circuit-terminating equipment
DCP	Digital control potentiometer
DDR	Double data rate
DFE	Digital front end
DLL	Delay-locked loop
DPR	*Digital* (microprocessor-based) *protection relay*
DRAM	Dynamic random access memory
DRC	Design rule checkers
DS	Data strobing
DSP	Digital signal processor
DTE	Data terminal equipment
DWDM	Dense wavelength division multiplexing
EBCDIC	Extended binary coded decimal interchange code
ECL	Emitter-coupled logic
ED	Electrodeposited
EDGE	Enhanced data rates for GSM evolution
EDVAC	Electronic discrete variable computer
EF	Exposure factor
EFT	Electrical fast transient
EMC	Electromagnetic compatibility
EME	Electromagnetic environment
EMI	Electromagnetic interference
EN	End node
ENIAC	Electronic numerical integrator and computer
ENIG	Electroless nickel immersion gold

EPROM	Electrically programmable read-only memory
ERC	Electrical rule checks
ESD	Electrostatic discharge
ESR	Equivalent series resistance
EV-DO	Evolution-data optimized
EW	Electronic warfare
FBGA	Fine ball grid array
FDM	Frequency-division multiplexing
FEC	Forward error correction
FEP	Fluorinated ethylene-propylene
FFE	Feed forward equalization
FFT	Fast Fourier transform
FOC	Field-oriented control
FOCL	Fiber-optic communications lines
FPGA	Field-programmable gate array
FT	Fast transient
FTP	Foil twisted pair
GCN	Global computer network
GPRS	General packet radio service
GSM	Global system for mobile communications
GSM/DCS	Global system for mobile communications/digital cellular system
HASL	Hot air solder leveling
HPF	High pass filter
HSPA	High-speed packet access
HSTL	High-speed transceiver logic
HTCC	High-temperature cofired ceramic
IAB	Internet architecture board
IBA	Intermediate bus architecture
IC	Integrated microcircuits (интегральные микросхемы)
ICEM	Integrated circuit electromagnetic emission model
ICIC	Intra-cavity interconnection
IDT	Inter-digital transducer
IMD	Intermodulation distortion
IMTS	Improved mobile telephone service
IoE	Internet of Everything
IoI	Internet of Things
IP	Internet protocol
IPX	Internetwork packet exchange
IR	Infrared (инфракрасный)
IRS	Integrated radiation shielding
IRTF	Internet Research Task Force
ISDN	Integrated Services Digital Network
ITE	Information Technology Equipment
JTRS	Joint Tactical Radio System

LAN	Local area network
LDMOS	Laterally diffused metal oxide semiconductor
LDO	Low dropout
LE	Logic element
LNA	Low noise amplifier
LoRaWAN	Long-range wide-area networks
LPISM	Liquid photoimageable solder mask
LPWAN	Low-power wide-area network
LTCC	Low-temperature cofired ceramic
LTE	Long-term evolution
LVDS	Low-voltage differential signaling
LVPECL	Low-voltage positive emitter-coupled logic
LVTTL	Low-voltage transistor-transistor logic
M2M	Machine-to-machine
MAC	Address media access control address
MBA	Mobile broadband access service
MBE	Molecular beam epitaxy
MCM	Multi-chip module
MDI	Medium-dependent interface
MDI-X	Media-dependent interface crossover
MEM	Microelectromechanical
MEMS	Microelectromechanical systems
MESFET	Metal-semiconductor field effect transistor
MFS	Multilayer film shield
MHS	Message handle systems
MIC	Monolithic integrated circuit
MIMO	Multiple input multiple output
MLT	Multilevel transmission
MM	Machine model
MMIC	Monolithic microwave integrated circuit
mMTC	Massive machine-type communications
MS	Message switching
MU	Memory unit
NEMP	Nuclear electromagnetic pulse
NRZ	Nonreturn to zero
NRZI	Nonreturn to zero with ones inverted
OA	Operational amplifier
OFDMA	Orthogonal frequency-division multiple access
OMPAC	Overmolded plastic array carrier
OSI	Open system interconnection
OSI/RM	Reference Model for Open Systems Interconnection
OSP	Organic solderability preservative
OSPF	Open shortest path first
OTN	Optical transport network

OVP	Overvoltage protection
PAA	Phased-array antenna
PBGA	Plastic ball grid array
PCB	Printed circuit board
PD	Power divider
PDN	Power distribution network
PDU	Protocol data unit
PFR	Phase-frequency response
PIM	Passive intermodulation
PLL	Phase-locked loop
PMT	Photomultiplier tube
POL	Point of load
PON	Passive optical network
PoP	Package on package
PPGA	Plastic pin grid array
PRCS	Pseudorandom clock search
PS	Packet switching
PSM	Power surface mount
PSRR	Power supply rejection ratio
PTF	Phase thermal factor
PTFE	Polytetrafluoroethylene
QA	Quality assurance
QFP	Quad flat package
RA	Reflective array
RA	Rolled annealed
RAM	Random access memory
RC	Root complex
RCD	Residual current devices
REE	Radio electronic equipment
RES	Radio electronic system
RFC	Request for comment
RFID	Radio-frequency identification
RFR	Radio-frequency resource
RIP	Routing Information Protocol
RMS	Root mean square
ROM	Read-only memory
RPC	Remote procedure calls
RS	Recommended standards
RT	Rise time
RTF	Reverse-treated foil
S/W	Software
SA	Source address
SATA	Serial advanced technology attachment
SAW	Surface acoustic wave

SCS	Structured cabling system
SCSI	Small computer systems interface
SDN	Software-defined networks
SDR	Software-defined radio
SDRAM	Synchronous dynamic random access memory
SEMI	Semiconductor device
SEPIC	Single-ended primary inductance converter
SER	Secondary electron resonance
SerDes	Serializer/deserializer
SHC	Sample-and-hold circuit
SHF	Super high frequency/microwaves
SI	Signal integrity
SiP	System in package
SJ	Solder joint
SLC	Surface laminar circuit
SLICC	Slightly larger than an IC carrier
SMOBC	Solder mask over bare copper
SMPS	Switch-mode power supply
SMS	Short message service
SMT	Surface-mount technology
SNA	Systems network architecture
SNR	Signal to noise ratio
SoC	System on chip
SPI	Serial peripheral interface
SPM	Scratchpad memory
SPX	Sequenced packet exchange
SR	Source routing
SRBF	Single-route broadcast frame
SSN	Simultaneous switching noise
STP	Shielded twisted pair
STTL	Schottky transistor-transistor logic
SWR	Standing wave ratio
TAB	Tape-automated bonding
TACS	Total access communications system
TCLE	Thermal coefficient of linear expansion
TCP	Transmission control protocol
TDA	Technical design assignment
TDM	Time division multiplexing
TDMA	Time division multiple access
TLM	Transmission line model
TR	Transparent
TS	Technical specifications
TS	Telecommunication system
TSV	Through silicon vials

TTL	Transistor-transistor logic
TVS	Transient voltage suppressors
UEMI	Unintentional electromagnetic interference
UI	Unit interval
uMTC	Ultrareliable machine-type communication
UMTS	Universal Mobile Telecommunications System
UNIVAC	Universal Automatic Computer
UTP	Unshielded twisted pair
VAC	Volt-ampere characteristic/I-V curve
VBR	Breakdown voltage
VCG	Voltage controlled generator
VCL	Current limit voltage
VLSI	Very large-scale integrated circuit
VoIP	Voice over IP
VRM	Voltage regulator module
WAN	Wide area
WCDMA	Wideband code division multiple access
WDM	Wavelength division multiplexing
WSA	Wafer-scale assembly

Chapter 1
Information and Communication Technologies and Systems Based on Them

At the end of the 90s of the last century, the process of combining the two branches of science and technology, computer science and telecommunications, began, resulting in the emergence of information and communication technologies combining information technology and communication technologies.

In the future, these technologies were called "telecommunication technologies" and on their basis telecommunication systems (TCS) were created, which are a combination of various software-compatible hardware integrated into the system in order to organize data transfer from one place to another. Here "data" means any text, graphic, voice, or video information.

The main requirement for up-to-date TCSs is to ensure high performance. Here it is necessary to separate the problems of ensuring the processing speed and the problems of ensuring the speed of data transmission, since the causes of their occurrence and the ways of their solution are of a completely different nature. If the data processing rate primarily depends on the technical solutions of the hardware of the TCSs, then the data transfer rate is determined by the technical solutions of the transmission channels used in the TCSs.

Therefore, in order to study and correctly apply current methods of increasing the performance, it is necessary to clearly understand the structure and characteristics of the organization of TCSs. This is considered in this chapter.

Here, in Sects. 1.1, 1.2, 1.3, 1.4, 1.5, 1.6, and 1.7, the main technical solutions of TCS hardware, basic standards, standard protocols of narrowband and wideband TCSs are considered, and Sects. 1.8, 1.9, 1.10, 1.11, 1.12, and 1.13 consider technical solutions of data transmission channels, various data switching technologies, topology of high-speed data transmission systems, features of organization of communication between various TCS segments. Section 1.14 gives a detailed overview of the microcircuits intended for use in up-to-date TCSs.

A. Belous, V. Saladukha, *High-Speed Digital System Design*,
https://doi.org/10.1007/978-3-030-25409-4_1

1.1 Main Trends in the Development of Information and Communication Technologies

Today, without information and communication technologies (ICT) and systems based on them—telecommunication systems (TCS) it is impossible to imagine both the life of an ordinary person and functioning of a variety of modern infrastructures responsible for the socio-economic well-being of citizens and for a whole range of problems related to the national security of the state. Information and communication technologies have penetrated into all spheres of human activity, but the most active results of their development are used in the interests of military and special departments.

First of all, the military departments of the world's leading powers, who know well the real state of affairs and possible unique prospects for the development of this direction, finance in sufficient volumes a number of both individual projects and special comprehensive programs.

For example, as far as it is known from open information sources [1–9], in 2014, to counter threats to national security in the military sphere and ensure absolute technological superiority, the US Department of Defense launched another set of measures for the innovative development of armed forces called the Defense Innovation Initiative (DII), aimed at *implementing* the strategy of complete military superiority of the USA over all potential adversaries, principally—Russia and China. This is the third such strategy.

The first strategy to ensure the military superiority (offset strategy) of the USA was based on nuclear weapons and means of its delivery, and, according to American experts [4, 5], it was successfully implemented during the Cold War. The basis of this strategy and measures for its implementation included the first theoretical substantiation of technologies providing military superiority (technology offset strategy), developed by Admiral William Perry (William J. Perry) during the period when he was the US Deputy Defense Secretary for Research and Development [6]. It can be considered that periodic publications in the open press of individual elements of such strategies are a kind of "invitation" to potential opponents of the USA to participate in the "Arms Race".

The second strategy was based on the synergistic effect of the use of precision weapons, space reconnaissance systems, missile and air defense systems (MD/AD), and technologies to reduce the visibility of weapons and military equipment.

The main objective of the third strategy implemented in 2014 (third offset strategy) is aimed at achieving absolute US military superiority over all potential adversaries possessing modern countermeasures (lock) of access to their own or controlled territories (Anti-access/Area Denial, A2/AD). In particular, such A2/AD means includes a complex of weapons, including precision-guided munition (PGM), defense systems (anti-air, anti-space, anti-missile, anti-ship, anti-submarine), and electronic warfare (EW).

At the same time, absolute superiority in this strategy is understood as the unconditional achievement of military success in all areas of the struggle—in space, in the air, at sea, on land, and in cyberspace.

The choice of the working title of the program "quickness" is far from being accidental, as the main objectives of the strategy are the following [1, 6–9]:

- military operations based on large-scale and integrated use of the capabilities of robotic systems;
- air operations with the use of long range nearly invisible aircraft;
- submarine warfare with the use of autonomous complexes consisting of various technical equipment;
- design of integrated systems of weapons and military equipment and their quick integration into an integrated weapon system.

As part of the implementation of the third strategy, five integrated R&D areas were identified:

- autonomous machines and systems capable of continuous self-learning;
- human–machine interaction technologies that provide effective decision support;
- new technical equipment to improve the efficiency of human activities (military serviceman);
- technologies of interaction of groups of weapons and military equipment and robots;
- semi-autonomous weapons systems (arms), effectively functioning in conditions of large-scale use of electronic warfare (EW) by the enemy.

Obviously, for the technical implementation of both the above-listed main tasks and these complex R&D areas, it is required to use the most high-speed telecommunication systems, the main technical problems of designing and production engineering of which this book is devoted to.

For these reasons, it is impossible to find out, in most cases, the specific technical solutions of these systems and their individual components from published open sources of scientific and technical information.

But for ordinary technical developers of modern telecommunication systems, all these "blackouts" have no common sense.

As will be shown below, it is enough for them to simply analyze the product lines of the leading global manufacturers of the element base for these systems, and it is possible to "decrypt" the key technical points of all these secret "concepts" and "strategies" quite easily.

But before considering the key technical problems of designing high-speed electronic devices that underlie the construction of various modern telecommunication systems, it is necessary to at least briefly consider the theoretical and practical bases of their organization and functioning presented below (http://pdst.narod.ru/_20_el_uch/ost_wpd_01).

1.2 Basic Principles of the Organization of Telecommunication Systems

At the end of the 90s of the past century, the process of convergence of two branches of science and technology began—informatics and communications (telecommunications), which resulted in the so-called information and communication technologies, combining information and communication technologies and which were so named as telecommunication technologies (http://pdst.narod.ru/_20_el_uch/ost_wpd_01). Hardware telecommunication technologies are implemented by *telecommunication systems*.

The most common general definition of the term telecommunication system (TCS) is a combination of hardware and software compatible equipment interconnected in a single system for the purpose of data transfer from one place to another. The flow diagram of the simplest TCS is presented in Fig. 1.1 (http://pdst.narod.ru/_20_el_uch/ost_wpd_01).

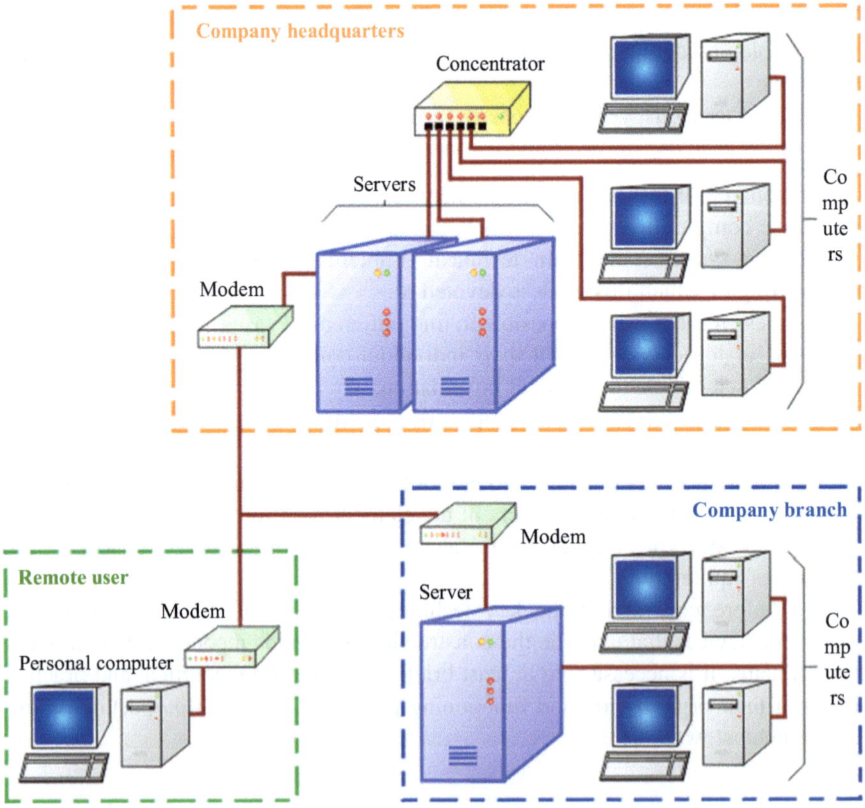

Fig. 1.1 Block diagram of the elementary telecommunication system

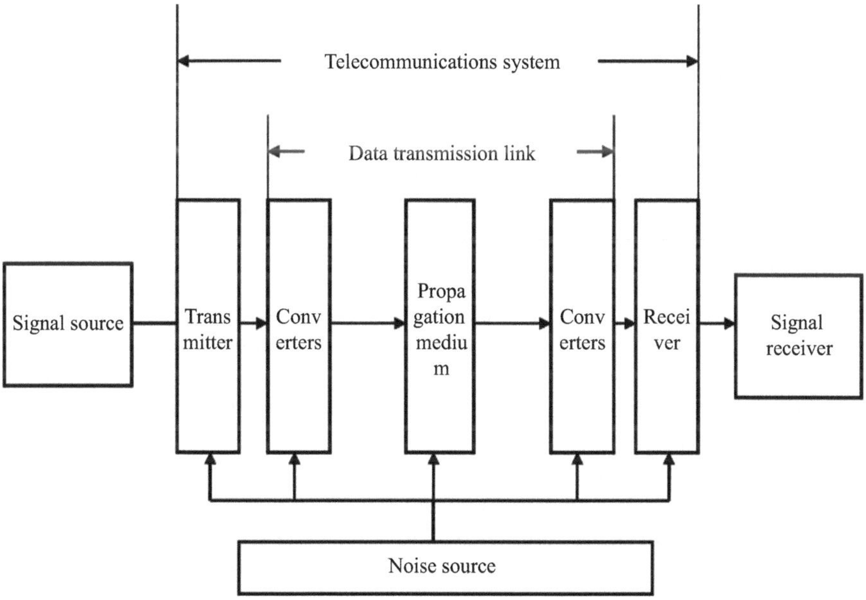

Fig. 1.2 Simplified block diagram of the data transmission channel

Any TCS shall be capable of transmitting text, graphic, voice, or video information; therefore, the basis of any TCS is data transmission channels, a simplified flow diagram of which is shown in Fig. 1.2 (http://pdst.narod.ru/_20_el_uch/ost_wpd_01). On the transmitting side, the user's computer generates the original data signal, which is converted by a converter (modem) into a form (analog or digital), convenient for transmission through the transmission medium. On the receiver side, the same modem converts the received signals into understandable to the other user computer signals. According to a similar pattern, data is transferred in the opposite direction to the computer of the user who sent the original data order.

Let's consider the main components of the standard commercial TCS, as well as in plain language, explain how these individual components work together with each other, forming as a result different types of TCS.

The structure of such a typical TCS includes host equipment (servers, user computers, communication equipment (communication channels)), as well as the so-called active equipment (various modems, hubs, and a number of other equipment).

As can be seen from Fig. 1.1 the following basic components form the elementary TCS:

- *servers* storing and processing information;
- *work stations* and user *PCs* designed for entering database queries, receiving and processing query results, and performing other tasks of end users of information systems;

- *communication channels*—communication lines through which data is transmitted between the sender and receiver of information. It is important to understand that communication channels use different types of data transmission medium: telephone lines, fiber-optic cable, coaxial cable, wireless and other communication channels;
- "*active*" *equipment* is various modems, network adapters, hubs, switches, routers, and so on. These devices are necessary for data transmission and receipt;
- *network software* that controls the process of data transmission and receipt and the operation of individual parts of the TCS.

So, the main functions of any TCS are to implement the seemingly simplest function—it is just necessarily to quickly transfer information from one point A and get it in another point B. For this, the TCS needs to perform a number of operations that are usually hidden from users. Firstly, before the TCS starts transmitting information, it needs to establish a reliable connection between the *sender* and *receiver* sides, then quickly calculate the optimal route for transmitting this data, perform the initial processing of the transmitted information (for example, check that your message is transmitted to the person to whom you sent it). But, most importantly, it is necessary to convert the transfer speed of the computer into the speed actually supported by the communication line.

TCS controls the flow of information transmitted (traffic), therefore, to implement these functions, any TCS shall contain a variety of hardware and software components that need to work together to transmit this information to its intended purpose. These network components shall actively "communicate" with each other, following a set of special rules. This set of standard rules controlling the entire process of data preparation and transmission between two points of the network is called a *protocol.* Each device in the network shall correctly "understand" and, of course, follow the protocol of the other device.

The main functions of network protocols seem to be very simple: to identify each device involved in the data transmission process, quickly check whether this data does not need to be retransmitted, retransmit quickly if an error has occurred, as well as a number of additional functions, which will be discussed below.

The bad "old news" for TCS developers is the fact that numerous commercial, government, and computer institutions are aware of the need to introduce common standards for data transmission; in reality, there are no universal standards in the industry at the time this book was published.

A special issue is the signal transmission over communication lines, where the information flow in any TCS is transmitted as electronic signals. Signals are just of two types: analog and digital. An analog signal is a continuous sinusoidal oscillation. Such signals are used mainly for the transmission of voice messages.

The digital signal, in contrast with analog signal, is discrete and pulse-type. Using digital signals, information is transmitted, previously encoded by two discrete values of the signal: 0 and 1. This form of data transmission is very convenient when using computers that understand binary information. But in most communication channels it is impossible to transmit digital data without some prior conversion—all digital

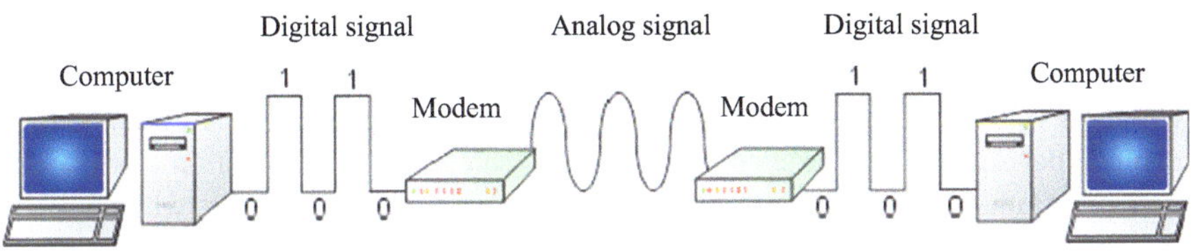

Fig. 1.3 Simplified block diagram of data transmission using modems

signals shall be converted to analog signals before being transmitted over a communication channel.

One of the devices used to convert signals is a modem (modem-MODulation/DEModulation, modulation/demodulation).

A *modem* is a device that converts digital signals transmitted by a computer into analog form. On the receiver side, the same modem performs the reverse conversion of the signal from analog to digital form.

Simplified operation of the elementary modem is shown in Fig. 1.3 (http://pdst.narod.ru/_20_el_uch/ost_wpd_01). On the sender side, the modem converts digital signals from the computer into analog signals that can be further transmitted via a telephone line. On the receiver side, the same modem converts the analog signals into digital signals understandable to the other user computer signals.

Usually, when transferring data using a modem, ordinary telephone lines are used as communication channels.

Modem not only connects a computer to a telephone line, but also performs another important function—converts computer signals into a form in which they can be transmitted over a telephone line.

On the sender side, the modem converts digital signals to analog form, then they are transmitted via a telephone line, and on the receiver side the same modem performs the inverse conversion of signals from analog to digital form, understandable to the computer.

By design, modems can be *external* and *internal*. The external modem is usually a small box, which has a block of device operation indicators on the front panel. The rear panel contains two slots:

- for connection of the cable connecting the modem to the computer (RS-232 Interface);
- for connection to the telephone line (RJ-11 Interface).

A standard modem is usually connected to the computer serial port. RS-232 standard cable is used for connection to the computer. Cable with RJ-11 connectors is used for connection to the telephone network.

The internal modem is inserted into the free expansion slot like any other card. After that, it is only necessary to connect the modem to the telephone network with an RJ-11 connector.

When constructing TCS various types of communication channels are used. Let's consider briefly the types of TCS communication channels. It is important to note that, in general, the *communication channels* are the communication lines through which one network device transmits data to another one. Such communication channels can use different types of data transmission medium: twisted pair, coaxial cable, fiber optics, radio and infrared waves, satellite communication lines. Each type of communication channels has its advantages and disadvantages, which will be discussed more detailed in subsequent chapters of this book. Usually, high-speed channels are more expensive, but they can quickly transfer large amounts of data (which reduces the value of the price/bit index). For example, the price/performance ratio may be better for the user in the case of using a satellite communication channel than using a dedicated line, but provided that the user constantly (100% of the time) uses satellite communication. It is also necessary to take into account that when using any transmission medium, the data exchange rate strongly depends on the configuration of the used hardware and software.

Telephone lines are the most common type of communication channels. In telephone lines, a two-core copper cable is used as the transmission medium. There are two types of telephone lines by which modem communication can be performed. The first type is the public *dial-up* telephone communication. They are slow, not very reliable, and require the user to manually establish connections for each session. The transmission speed in such ordinary telephone lines is low, from 14 to 56 Kbps. Nevertheless, it is the telephone lines, thanks to their global presence, that still form the basis of many networks, including the global network Internet.

Another type of telephone lines—*leased, or* so-called, *dedicated lines.* They are distinguished from ordinary telephone lines by the fact that such a channel is used only for the organization of communication between two specific points. In the case of using such a dedicated line, there is usually no need to dial the number of the subscriber with whom you want to contact—you have a direct line of communication with them, which is always at your service. The quality of communication via such a dedicated line is always higher than the communication via a dial-up telephone line. At the same time, the transmission speed in dedicated lines is slightly higher—from 64 to 512 Kbps. The economic aspect—usually the owner of a dedicated line pays a fixed amount for its use. But keep in mind that it's usually possible to use such a dedicated channel only for digital data transmission—for other purposes (for example, for voice communication) it is not suitable.

When using a dedicated line to dial the number is not necessary, since there is a direct channel of communication between subscribers.

Other types of communication channels include various types of coaxial cables, twisted-pair cable, various types of fiber-optic cables, microstrip lines, etc. In more detail these and other types of communication channels will be discussed in other sections. Here only the basic types will be briefly considered.

Coaxial cable used in telecommunication networks is very similar to that used in television. Its design contains one copper conductor in insulation and metal braid, which serves as a shield. This shield is necessary to protect against interference, which allows the use of this type of signal transmission cable over long distances than a twisted pair cable. This type of cable has become widely used in Ethernet-type local networks, although since 2010 coaxial cable has gradually got into out of use, mainly due to its high cost and relatively low data transfer rate (10–20 Mbps).

The *twisted pair cable* contains no longer one but several pairs of copper wires and is most common in local networks, since many new buildings in the Russian Federation are built with a prefabricated cable system that uses the twisted pair cable. It should be noted that the available communication lines in the design can be successfully applied for other purposes (for example, voice communication or signaling system). The data transmission speed in networks where twisted pair cable is used is from 10 to 100 Mbps. The most commonly used cable is UTP5 (Unscrewed Twisted Pair—unshielded twisted pair, category 5). In cases where increased data protection from interference is required (for example, from strong electromagnetic fields), the developers use shielded twisted pair. Twisted pair cable is widely used in local networks based on Ethernet and Fast Ethernet technologies.

A *fiber-optic cable* contains thousands of glass fibers, each thinner than a human hair. For transmission the data is converted not into electrical signals that are familiar to us, but into special light pulses that are transmitted over fiber optics using a laser device with a speed from 500 Kbps to several gigabits per second. On the one hand, fiber-optic communication lines are much faster, lighter, and much stronger than copper wires; therefore, fiber optics are used as communication channels in those systems where large amounts of data need to be transmitted quickly. But on the other hand, fiber-optic cable requires a large financial cost when laying, in addition it is more expensive. It is best used for laying backbones, and for connection to a network of user computers it is better to use a cheaper twisted pair. Fiber-optic cable is used in local networks where a high degree of data protection against unauthorized use (for example, in banks) is required, since it is impossible to connect to the fiber optics for the purpose of data interception.

Wireless communication channels using radio or infrared waves as a medium do not make direct physical contact with sending and receiving devices, therefore such communication channels are the main alternative to contact data transmission methods based on telephone lines, twisted pair and fiber optics. The most commonly used wireless data transmission devices today are cell phones, smart phones, cordless phones, satellite TV systems, and microwave communication systems. Wireless communication devices also include remote controls for various household devices and other infrared (IR) communication systems. Below we consider only those technologies that are used in telecommunications data networks, namely radio communication systems, satellite communication, as well as infrared devices.

Microwave systems are mainly terrestrial radio-relay lines, which in telecommunication systems are used for transmission of large amounts of information between two points. Here, the microwave signals spread over a distance in a straight line, which limits the transmission distance to 40–50 km due to the curvature of the earth. Another significant disadvantage of these systems is their dependence on weather conditions.

Satellite communication systems are free from the disadvantages specific for radio-relay lines, because here the signals are transmitted from the ground station to the satellite, which serves as a repeater. A conventional parabolic antenna is used as a receiving device. Communication satellites are very cost effective in transmitting large amounts of data over long distances, so communication systems of this type are used in large, geographically distributed organizations, as well as in places where cable or radio relay links cannot be used. A typical example, the Rite Aid pharmacy network uses satellite communications systems to organize fast two-way communications between regional warehouses and the corporate mainframe located in Camp Hill, Pennsylvania. Each warehouse has a special server that serves cash registers, a local network of computers of pharmacy workers and managers, and also checks for the availability of necessary drugs and equipment. This server can quickly connect to the mainframe via satellite for sending sales reports, for accessing the company database, for scheduling, etc.

Of special note are Internet access systems based on the use of satellite communications systems. Since the transmitting equipment is quite expensive, many companies use DirecPC systems, where satellite antennas are used as receiving devices, and transmit information from users to the Internet via regular terrestrial communication channels.

It should be noted that if conventional communication satellites are in stationary orbits at an altitude of about 35 km above the Earth, then new low-orbit satellites operate in orbits as close as possible to the earth surface, which allows them to receive signals even from low-power transmitters. At the same time, such satellites (micro and nanosatellites) consume less energy, and their launch and operation is much cheaper than traditional satellite communication systems. The use of low-orbit satellites makes it possible to use communication services from anywhere in the world.

In terms of *infrared* (*IR*) *communication* systems (*infrared/IR-/systems*), it is used, as a rule, in small offices to organize wireless communication between various devices, for example, to connect a computer to a printer or to connect to a local network from two up to six computers at a short distance from each other (1–8 m). These systems are used in mobile offices when there is no need or ability to lay a cable. Networks built on the basis of IR-systems are quickly deployed and configured, but have a low data transmission speed and are not very reliable.

The characteristics of the communication channel chosen by the TCS developer determine the effectiveness and capabilities of the TCS. The main characteristics include the transmission speed, the direction of the signals, and the transmission mode.

For the signal transmission the above-mentioned international organizations have developed certain standards of transmission modes, called *conventions*, which are necessary for all devices operating on the network to adhere to. For example, during *asynchronous transmission*, one character (8 bits) is transmitted over the communication channel at a time. The start bit is placed at the beginning of each such character, and the stop bit and parity bit are placed at the end (http://pdst.narod.ru/_20_el_uch/ost_wpd_01). Asynchronous transmission is used when operating low-speed communication channels, for example, ordinary telephone lines.

By the asynchronous method, data is transmitted by a serial stream. During such transmission, each character (letter, number or symbol) is decomposed into a bit sequence. Each of these sequences is separated from the others by the start bit and stop bit.

When receiving data during an asynchronous transmission, the receiving computer uses the start and stop bits to control synchronization, thereby preparing to receive the next data byte. Communication of this type is not synchronized. The sending computer simply sends the data. The receiving computer accepts the data and checks it to make sure that the data was received without errors.

Of course, the error probability during such transmission is not excluded, therefore, asynchronous communication always carries out the so-called parity check of information, which uses a special bit—the parity bit. It is used to control errors. For parity check, the number of individual bits sent and received shall match. If this number matches, then the data is received correctly. If it does not match, a request for retransmission of data automatically follows.

Synchronous communication is based on a synchronization scheme agreed between two devices, in which the sending modem selects special bits from the data group and sends them in blocks called *frames*. Special symbols are used for setting synchronization and periodically checking its correctness. In a synchronous transmission, several characters are transmitted over a certain time interval, but at the beginning of a communication session, the devices that receive and send data synchronize the operation of their electronic circuits. Since the start and stop bits for synchronous communication are not needed, in this case the transmission is faster. Synchronous communication is used in the operation of fast telecommunication channels—dedicated lines, etc.

The transmission ends at the end of one frame and begins again at the next frame. Of course, this method is more efficient than asynchronous transmission. At the same time, synchronous communication requires less verification information and therefore works faster.

If an error is detected, the synchronous recognition and error correction scheme simply repeats the frame transmission.

The *transmission speed* describes how quickly the information bits were transmitted to the communication channel. In this case, the total amount of information that can be transmitted through the communication channel is measured in *bits per second* (*BPS*).

Sometimes another measuring unit for the transmission speed is *baud rate*. A baud rate is a binary event that reflects a signal change from plus to minus or vice versa. It should be understood that the baud rate is not always equal to the speed in bits per second. At high speeds, as well as in the case of the use of modern data compression methods, more than one bit of information can be transmitted during the signal change from plus to minus. Therefore, the speed in bits per second is usually higher than the baud rate.

The *data compression* procedure significantly reduces the time required for data transmission (by redundant element or empty area removal).

It should be noted that since during the phase of the signal change (or per cycle) one or several bits per second can be transmitted, the performance of any communication channel depends on the frequency at which the channel allows data to be transmitted (frequency is measured in Hertz). The frequency range that a particular communication channel has is called *bandwidth*. Therefore, the bandwidth is the

Table 1.1 Transmission speed and cost of channels for standard transmission lines

Communication channel	Speed	Cost
Phone line	14.4 KBPS–56.6 KBPS	Low
Dedicated line	28.8 KBPS–256 KBPS	Low
Coaxial cable	10 MBPS–20 MBPS	Low
Radio relay line	256 KBPS–100 MBPS	Low
Twisted pair	10 MBPS–100 MBPS	High
Satellite channel	256 KBPS–100 MBPS	High
Fiber optics	500 KBPS–10 GBPS	High

difference between the highest and lowest frequencies at which a communication channel can transmit data. The higher the frequency range, the greater the bandwidth, the more efficient the communication channel. Table 1.1 shows a comparative analysis of indicators such as transmission speed and cost for the most common telecommunication channels (http://pdst.narod.ru/_20_el_uch/ost_wpd_01).

Note to the Table 1.1:

- KBPS—kilobits per second;
- MBPS—megabits per second;
- GBPS—gigabits per second.

When transmitting data, the direction in which a data stream may follow in a communication channel is usually taken into account. In *simplex transmission* systems, data can always be transmitted in one direction only. *Half-duplex transmission* allows data transmission in two directions; but at any time, devices can only transmit or only receive data. Full-duplex transmission systems can simultaneously send and receive data.

The *network equipment* includes devices that provide transmission and reception of network traffic (data traffic transmitted through a communication channel (network traffic)), as well as the connection of one TCS to others. The main devices in this group are modems, network adapters, hubs, repeaters, bridges, routers, and gateways. You are already familiar with modems, so let's briefly review the purpose of the other devices.

A *network adapter* is a device with the help of which a computer in a network can send and receive data. There are several types of network adapters, the most common of which are *network cards*.

When creating large networks, it is often necessary to lay cables over long distances. It should be considered that in a long cable segment, *signal damping* may occur, i.e., distortion of the waveform, which leads to distortion of the transmitted information. To prevent this, *repeaters* are used—special repeaters that amplify the signal and restore its shape.

In some types of signal transmission lines (for example, based on twisted pair), it is necessary to use special matching devices that connect all computers to network adapters in one node and transmit data within this node. These devices include the so-called *hubs*.

It is often necessary to connect two or more networks together. Bridges, routers, and gateways are used for these purposes.

A bridge is a device connecting two or more networks of the same type. By their effect, bridges are similar to repeaters, but unlike the latter, they can filter data, i.e., transmit to other segments or networks only part of the traffic, which reduces the load on communication channels.

Routers are used to organize communication between networks and transfer data through the optimized routes. Most often, a router is a specialized computer that has its own processor for calculating data transmission paths, as well as memory for storing a database of routes and characteristics of communication channels. Compared to the bridge, the router provides greater control over the paths and greater protection of the transmitted data, but it is much more difficult to install, configure, and operate it.

A gateway is a device that performs the conversion of transmitted data from one format to another. The gateway can be software (for example, an e-mail gateway) or hardware, for example, to connect a twisted pair network with a fiber optics network.

A host is any computer connected to the TCS.

Telecommunications software is a special software used to manage and support the operation of a telecommunications network. This software is installed on the hosts and active network equipment. The main functions of telecommunications software are network management, data access control, data transmission control, transmission error detection and correction, and data protection. Computer operating systems perform some network management and monitoring functions.

1.3 Extracts from the History of Development of High-Speed Data Networks

Analyzing the historical experience of the creation and development of network technologies for high-speed information transmission, it should be noted that the main factor that led to the emergence of these technologies is the creation and development of means of computer technology (http://pdst.narod.ru/_20_el_uch/ost_wpd_01). In turn, the Second World War became an incentive for the creation of computer equipment (electronic computers). To decode the coded messages of the German agents, the English specialists needed a huge amount of calculations, and they had to be made immediately after receiving the radio interception. Therefore, the British government at that time established a secret laboratory to create a special electronic computer called COLOSSUS, in the creation of which the famous British mathematician Alan Turing took part. As a result of this work, the world's first electronic digital computer was created.

It can be said that the Second World War influenced the development of computer technology in the USA. The army needed shooting tables that were used to target heavy artillery. So, in 1943, John Mooshley and his student J. Presper Eckert began to design an electronic computer, which they called ENIAC (Electronic Numerical Integrator and Computer), an electronic digital integrator and calculator. It consisted of 18,000 vacuum tubes and 1500 relays. This monster weighed about 30 tons and consumed 140 kW of electricity. The unit had only 20 registers, each of which could contain only a 10-bit decimal number.

After the war, Mooshley and Eckert were allowed to organize a school, where they shared their work with fellow scientists, after that other researchers began to work on designing such electronic computers. The second computer was EDS AS (1949), which was designed by Maurice Wilks at the University of Cambridge. Then came JOHNIAC—at the Rand corporation, ILLIAC—at the University of Illinois, MANIAC—at the Los Alamos laboratory, and WEIZAC—at the Weizmann Institute in Israel and many others.

Inspired by the result, Eckert and Mooshley soon began work on the EDVAC machine (Electronic Discrete Variable Computer), followed by the development of UNIVAC (the first electronic serial computer). In 1945, John von Neumann was involved in their work, creating, as we know, the basic principles of modern computing technology. Since Von Neumann realized that creating computers with a large number of switches and cables takes a long time and is very tiring, he came to the idea that the program should be represented in computer memory in digital form along with data. He also noted that the decimal arithmetic used in the ENIAC machine, where each bit was represented by 10 vacuum tubes (1 tube was on, 9 were off), should be replaced by binary arithmetic. As a result, the new Von Neumann machine consisted of five main parts: memory—RAM, processor—CPU, secondary memory—magnetic drums, tapes, magnetic disks, input devices—reading from punched cards, information output devices—printer. The need to transmit data between parts of such a computer subsequently served as an incentive for the development of high-speed data transmission and organization of modern computer networks.

The official “date of birth” of Soviet computer engineering should probably be considered the end of 1948 [3]. It was then that in the secret laboratory in Feofania, near Kiev, under the leadership of Sergey Aleksandrovich Lebedev (at that time the director of the Institute of Electrical Engineering of the Academy of Sciences of Ukraine and part-time Head of the Laboratory of the Institute of Precise Mechanics and Computer Engineering of the USSR Academy of Sciences) the work on the creation of a Small Electronic Accounting Machine (SEAM) began (http://pdst.narod.ru/_20_el_uch/ost_wpd_01). Lebedev put forward, substantiated, and implemented (independently of John von Neumann) the principles of a computer with program kept in memory. In 1953, the team under the leadership of S.A. Lebedev created the first large computer—BEAM-1 (Big Electronic Accounting Machine), released in one item. It was already created in Moscow, at the Institute of Precision Mechanics (abbreviated IPM) and the Computing Center of the USSR Academy of Sciences, whose director was S.A. Lebedev, and it was assembled at the Moscow Plant of Calculating Analytical Machines (abbreviated as CAM). At the same time, work was carried out in the Special Design Bureau No. 245 situated near Moscow, which was under the leadership of M.A. Lesechko, also founded in

December 1948 by the order of J.V. Stalin. In 1950–1953 the team of this design bureau, but already under the leadership of Y.Y. Bazilevsky, developed a digital computer “Strela” with a speed of 2 thousand operations per second. This computer was produced until 1956, and only 7 copies were made. Thus, “Strela” was the first industrial computer,—MEAM, BEAM existed at that time only in one item. As a result, in 1953 serious computational problems for the needs of the country’s defense, science and national economy could be solved on three types of computers—BEAM, “Strela”, and M-2.

Originally, punched tapes and punched cards were used to transmit data between computers, then magnetic tapes and removable magnetic disks were used. Later, special software (soft) appeared—operating systems that allow many users from different terminals to use one processor, and one printer. In this case, the terminals of a large machine (mainframe) could be removed from it for a very limited distance (up to 300–800 m). With the development of operating systems, it became possible to connect terminals to mainframes using public telephone networks with an increase in the number of terminals and corresponding distances. However, no common standards existed. Each manufacturer of large computers developed their own rules (protocols) of interconnection and, thus, the choice of manufacturer and data transmission technology for the user became lifelong.

The introduction of integrated circuits with low cost has led to the fact that computers have become smaller in size, more affordable, more powerful, and more specialized. Companies could already afford to have several computers designed for different departments and tasks and manufactured by different manufacturers. As a result, a new task appeared: interconnection of computer groups. The very first companies that connected these “islands” were IBM and DEC. The DEC protocol of the data transmission system was DECNET, which is no longer used today, and for IBM it is SNA (System Network Architecture—the first network data transmission architecture for IBM 360 series computers). However, computers from the same manufacturer were still limited to connecting with their own kind. When connecting computers from another manufacturer, software emulation was used to simulate the operation of the desired system.

In the 60s of the last century, the US government set the task of ensuring the transmission of information between computers of various organizations and made significant for that time financing of the development of standards and protocols for information communication. The ARPA, the Research Agency of the US Department of Defense, took up the task. As a result, it was possible to develop and implement the first computer network ARPANET, with the help of which the US federal organizations were eventually connected. In this network, TCP/IP protocols and the first communication technology of network segments of the US Department of Defense (DoD), known today as the Internet, were introduced.

Personal computers that appeared in the 80s began to merge into local area networks (LAN).

Gradually, more and more manufacturers of equipment and, accordingly, software appear, active developments are being carried out in the field of interaction of equipment from different manufacturers. At present, networks that include equipment and software of various manufacturers are called heterogeneous networks. The need to "understand" each other leads to the need to create not corporate data transmission rules (for example, SNA), but common to all. There are organizations that create data transmission standards, define the rules by which private clients can work, telecommunication companies, rules for combining heterogeneous networks. Such international standardizing organizations include, for example:

- ITU-T (telecommunication standardization sector of the International Telecommunication Union, successor of CCITT);
- IEEE (Institute of Electrical and Electronics Engineers);
- ISO (International Organization for Standardization);
- EIA (Electronic Industries Association);
- TIA (Telecommunications Industry Association).

In parallel, individual private companies keep up the developments (for example, Xerox developed Ethernet technology, and CISCO developed 1000Base-LH and MPLS technology).

With the cheapening of technology, organizations and companies have been able to integrate their computer islands, which are located at different distances (in different cities and even continents), into their private—*corporate network*. Such corporate network can be built on the basis of international standards (ITU-T) or standards of a single manufacturer (IBM SNA).

Finally, with the further development of high-speed data transmission, it became possible to combine various organizations into one network and connect not only members of a single company, but any person who fulfills certain access rules. Such networks began to be called global. Note that if a corporate network is a network that is not open to any user, then the *global network*, on the contrary, is open to any user.

Currently, almost all networks are heterogeneous, although basic information is usually born in corporate networks. The main volumes of information circulate there. Hence there is the obvious necessity to study them and the ability to create such networks.

1.4 Standard Protocols of Functioning of Telecommunication Systems

As is known in 1977, the International Standardization Organization (ISO), consisting of representatives of the information and telecommunication technology industry, created an ad hoc committee to develop communication standards in order to ensure universal interaction of software and hardware of many manufacturers. The result of its work was the so-called open systems interconnection basic reference model OSI. This model determines the levels of interaction in computer networks, describes the functions that are performed by each level, as shown in Table 1.2 (http://pdst.narod.ru/_20_el_uch/ost_wpd_01).

Table 1.2 Basic levels of network interaction in accordance with OSI

7	Application
6	Representative
5	Session
4	Transport
3	Network
2	Data link
1	Physical

Table 1.3 OSI levels and protocol data blocks

7	Application	Message
6	Presentation	Packet
5	Session	Packet
4	Transport	Packet Segment
3	Network	Packet Datagram
2	Data link	Frame
1	Physical	Bit

Since different modern computers have different processing and transmission speeds, different data formats, different types of connectors, different ways of storing data, accessing data (access methods), different operating systems, and different organization of fixed and random access memory, any TCS developer has a lot of problems connecting them into one system. All these issues were classified and distributed into functional groups—the so-called OSI levels (Table 1.3).

As a rule, these levels are organized as a separate vertical stack (Fig. 1.4). Here, each level performs its specific group of similar functions required for organizing communication of computers. In the implementation of more primitive functions, it relies on the lower level (uses its services) and is not interested in the details of this implementation. In addition, each such lower level offers its services to a higher level.

Let us consider in more detail the operation algorithm at the user level. Let the user application process, which runs in the end system "A", requests the application layer, for example, the file service. Based on this request, the application level software generates a standard format message, which usually consists of a header and a data field. The header contains the service information that needs to be transmitted through the network to the application layer of another computer (end system "B") in order to inform it what actions are required to be performed. For example, the header should contain information about the location of the file and the type of operation that shall be performed. The data field can be empty or contain any data, for example, those that need to be written to a remote file. In order to deliver this information to its destination, many tasks have to be solved. But other underlying levels are responsible for them.

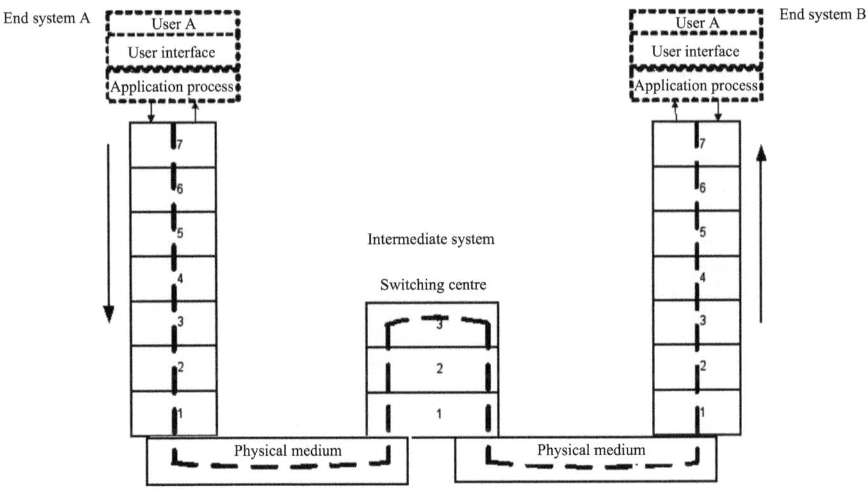

Fig. 1.4 Network processes architecture in accordance with OSI

This generated message is directed to the application layer down the stack to the *representative layer* (*Presentation*). Based on the information obtained from the application level header, the program module of the representative level performs the required actions and adds its service information to the message—the header of the representative level, which contains instructions for the module of the representative level of the receiver computer. The formed data block is transmitted down the stack to the *session level*, which in turn adds its own header, etc. When the message reaches the lower *physical level*, it "overlaps" with headings at all levels. The physical layer provides for the transmission of a message over a communication line, that is, via a physical transmission medium.

When the message finally arrives at the receiver computer, it is received by the physical layer and sequentially moves up the stack from level to level. In addition, each level analyzes and processes its own header, performs its functions, then deletes this header and transmits the remaining data block to the adjacent upstream layer.

So, these rules (specifications), according to which the components of the systems interact, are called protocols. In the above OSI model, there are two main types of protocols. In *connection-oriented network service protocols*, before data exchange, the sender and receiver (network components of the same level in remote systems) shall first establish a logical connection and, possibly, select the protocol they will use. After completion of the dialogue, they shall break the connection. In protocols without a *connectionless network service*, the sender simply transmits the data. These protocols are also called *datagram*.

A hierarchically organized set of protocols sufficient for organizing the interaction of nodes in a network is called a "stack of communication protocols".

To designate a data block with which modules of a certain level deal, the OSI model uses the common name protocol data unit (PDU). At the same time, a data block of a certain level has a special name (Table 1.3) (http://pdst.narod.ru/_20_el_uch/ost_wpd_01).

Briefly consider the main functions related to different levels of OSI, given in Table 1.3.

Thus, the *physical layer* provides a stream of bits to the physical transmission medium. Basically, it only defines a specific specification for cable and connectors, i.e. only the mechanical, electrical, and functional characteristics of the network medium and interfaces.

At this level, the total number of devices involved is determined, the main mechanical parameters are the number of pins (connector type), electrical parameters (voltage, single signal pulse duration), functional parameters (what each pin of the network connector is used for, how the initial physical connection is established, and how it is torn down).

Typical examples of the practical implementation of such physical layer protocols are the well-known RS-232, RS-485, RS-449, RS-530, and many ITU-T V and X series specifications (for example, V.35, V.24, X.21).

If we talk about the "*data link*" *level*, here the bits are already organized in groups (frames). A frame is a block of information that has a logical meaning for transferring from one computer to another. Each frame is supplied with the addresses of the physical devices (source and destination) between which it is sent.

The protocol of the data link layer of the local network provides a procedure for the delivery of a frame between any nodes of this network. If a local network uses a shared transmission medium, the data link layer protocol performs a check on the availability of this transmission medium, i.e. it implements a certain method of access to the data transmission channel.

In global networks that rarely have a "regular" topology, the data link layer provides for the exchange of frames only between adjacent nodes in a network connected by an individual communication link.

In addition to forwarding frames with the necessary synchronization, the data link layer also performs error control, connection control, and data flow control. The beginning and end of each frame are indicated by a special bit sequence (for example, flag—01111110). Each frame contains a control sequence that allows the receiving party to detect possible errors. The link layer can not only detect, but also correct damaged frames due to retransmission.

The "header" of the data link layer usually contains detailed information about addresses of interacting devices, frame type, frame length, information for data flow control, and information about higher level protocols that receive a packet placed in a frame.

The *network layer* is one of the most difficult to implement. The main task of this layer is to transmit information over a complex network consisting of a set of islands (segments), where completely different message transmission principles between end nodes and other subscribers, computers, can be used within segments. This network, consisting of many segments, we today call the Internet.

Finally, the data (packets) transmission between individual network segments is performed using *routers* (Fig. 1.5). We can imagine a router as a device in which two

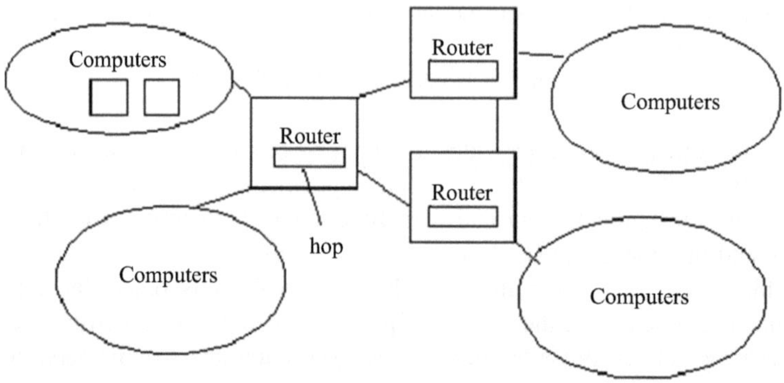

Fig. 1.5 Block diagram of the transmission of information between segments of a complex network using routers (http://pdst.narod.ru/_20_el_uch/ost_wpd_01)

processes function. One of them processes incoming packets and selects an outgoing line for the routing table for them. The second process is responsible for filling in and updating the routing tables and is determined by the route selection algorithm. In turn, the route selection algorithms can be divided into two main classes: adaptive and nonadaptive. Nonadaptive algorithms (static routing) do not take into account the specific topology and current state of the network and do not measure traffic on the communication lines. The list of routes is loaded into the memory of the router in advance and does not change with any change of network status. Adaptive algorithms (the so-called dynamic routing) change the decision on the choice of routes when changing the network topology and, most importantly, depending on the current line load.

Note that in modern networks there are two methods of dynamic routing: distance vector routing (RIP protocol that minimizes the number of hops through intermediate routers—the number of hops) and routing taking into account the state of the channels (OSPF protocol, which minimizes the time to reach the desired network segment).

The evidence from practice shows, at the network layer it may be necessary to split the received frame into a number of smaller fragments (datagrams) before their following transmission.

Well-known examples of the most widely used network layer protocols are the TCP/IP stack IP interworking protocol and Novell IPX/SPX stack IP interworking protocol. With regard to the transport layer, it should be noted that the *transport layer* is the core of the protocol hierarchy. It is designed to optimize data transmission from the sender to the recipient, control the data flow, organize the application or the upper layers of the stack for the required degree of reliability of data transmission, regardless of the physical characteristics of the network or networks used. Starting from the transport layer, all upstream protocols are implemented by software, usually included in the network operating system.

Session layer allows users of different computers to establish communication sessions with each other. This provides for opening a session, managing the device dialog (for example, allocating space for a file on the disk of the receiving device) and ending the interaction. This is done using special software libraries (for example, RPC-remote procedure calls from Sun Microsystems). In practice, few applications use the session layer.

The *presentation layer* is a very important component of the TCS, because it performs data conversion between computers of a network with different formats of character codes, for example, ASCII and EBCDIC, i.e., it overcomes the syntactical differences in data representation. At this level, data can be encrypted and decrypted and compressed, thereby ensuring the secrecy of data exchange for all application services at once.

And, finally, the last in the list, but not least, is the application layer (the layer of specific applications).

The *application layer* is a set of various protocols (standards) through which any network users can access shared resources, such as files, email, hypertext WEB-pages, and printers.

At this layer, interaction usually occurs no longer between computers, but between specific applications: a model by which files will be exchanged is determined, rules by which we will send mail are established, a virtual terminal, network management, directories, etc. are organized.

Typical examples of protocols at this level are: Telnet, X.400, FTP, HTTP.

1.5 Basic Standards of Telecommunication Systems

In the general case, the standard (protocol) refers to the methods and rules by which data is exchanged between TCS computers. However, some actually used protocols from the point of view of OSI perform functions of only one layer, and other protocols—of several layers. Vice versa, the protocol can perform the functions of some layer partially. While the ISO was developing its own model of partitioning of data communication functioning, various large manufacturing companies created their own rules, while some of them followed the guidelines of the ISO, some did not. Table 1.4 shows the main practical stacks of communication protocols and levels of OSI X (http://pdst.narod.ru/_20_el_uch/ost_wpd_01).

It can be said that a standard is a legalized version of the protocol implementation in hardware or software.

In general, the whole variety of standards can be divided into three large groups:

- *Legal standards*. They are confirmed by laws adopted by the state (for example, the FSS in Russia requires compliance with certain safety standards for equipment used by this organization).
- *Actual standards*. These standards exist, but their use is not defined by law. For example, the TCP/IP stack (the most popular protocol stack used in both global and local networks) and its associated protocols are the de facto industry standard for connecting data network segments.

Table 1.4 Conformity of industrial stacks of telecommunication protocols of OSI model

OSI layers		Network architecture protocols		
		IBM	Novell/NetWare	TCP/IP
7	Application	SNA		Telnet FTP HTTP SMTP
6	Presentation			
5	Session			
4	Transport		SPX	TCP UDP
3	Network		IPX	IP ICMP
2	Data link		Ethernet, token ring, FDDI	
1	Physical			

- *Corporate standards*. These are the standards that private commercial companies develop and implement for their products (for example, Novell's original IPX/SPX protocol stack, which is developed for its NetWare operating system in the early 1980s).

It should be understood that the standards developed by non-private standardizing organizations are not corporate, they may become an industrial standard de facto, and at the same time they may not. Examples are the IEEE 802.3 standards and the OSI protocol stack.

It is important for developers of TCS to know the main international standardizing organizations in the field of data transmission:

ITU—International Telecommunication Union is a structural unit of the UN, formed in 1865 as the International Telegraph Union. The main working bodies of the ITU are:

- The Telecommunication Standardization Sector (ITU-T), which is the successor (CCITT);
- Radiocommunication Sector (ITU-R);
- Telecommunication Development Sector (ITU-D).

ITU-T developments are of advisory nature in the field of traditional telecommunications, data transmission, and information networks. Nevertheless, ITU-T recommendations are in fact international standards in their respective engineering fields. A series of similar ITU-T recommendations are usually denoted by Latin letters, for example, Q—switching and signaling, X—data networks and open systems interconnection, V—data transmission over the telephone network, Y—Global information infrastructure and aspects of Internet protocols.

A well-known example of such standards is the X.25 standard.

ISO—The International Organization for Standardization (also the International Standards Organization). It is a non-profit organization with headquarters in Geneva (http://www.iso.ch/), which develops international standards in many areas, including computer technologies and communications. It was founded in 1946 as the world federation of standardization bodies. ISO members are more than 130 national standardization institutes (for example, ANSI—American National Standards Institute). The name ISO is not an abbreviation—it comes from the ancient Greek word isos, meaning "equal, equivalent". The ISO consists of many working groups in various fields. OSI protocols are an example of ISO standards.

IEEE—The Institute of Electrical and Electronics Engineers, Inc. (USA). This largest professional organization in the world (www.ieee.org) was founded in 1963, it unites more than 300 thousand technical specialists from 147 countries, at the time of publishing the book it is the leading standardization organization responsible also for network standards. IEEE conducts a great deal of publishing and educational activities, subsidizes the development of standards for computers and, from the point of view of data transmission, is responsible for the specifications of the 802 series of standards that are essential for high-speed data transmission.

EIA—Electronics Industries Alliance (EIA Alliance). Until October 1997, the abbreviation was decoded as the Electronics Industries Association. Based in the USA, the organization specializes in physical layer standards (interfaces); it develops electrical and functional standards with an RS (Recommended Standards) identifier. An example of such a standard is the well-known RS-232C serial interface.

TIA—Telecommunication Industry Association—US Telecommunications Industry Association, TIA Association. It is an association of communication equipment manufacturers that specializes in the development of standards for cable systems.

ETSI—European Telecommunications Standards Institute—was established in 1988, is a non-governmental organization that develops European standards. Examples of the most well-known standards are the GSM standard for digital mobile communications (Global System for Mobile Telecommunications), DECT, and TETRA.

And finally, it is impossible not to include in this list *IAB*—Internet Architecture Board—Coordination Council on Internet Architecture. IAB includes:

- IETF—Internet Engineering Task Force, in fact, this is the Internet Technical Commission, formed in 1986 and is still engaged in solving current problems. For example, its functions include the standardization of the TCP/IP protocol stack. Within the framework of this commission, separate working groups are quickly created for a short period of time in order to quickly solve the particular task that has arisen. Additional IETF functions include preparing RFC documents (Request For Comment). Of course, not all RFC documents are Internet standards, but many contain useful comments on a standard, or just a description of any Internet problem.
- IRTF—Internet Research Task Force. This so-called Internet research group is engaged in promising long-term research on TCP/IP stack protocols and the standardization of new information technologies that are being developed.

Russian developers need to remember that Russian Federation laws are in force in the Russian Federation (for example, Federal Law on Information, Information Technology, and Data Protection), GOSTs of Federal Agency on Technical

Regulating and Metrology, SNiP—building standards and rules of State Committee of the Russian Federation for Construction and the Housing and Utilities Complex, PUE—rules for electrical installations, VNTP and RD—departmental norms and rules and working documents of the Ministry of Information Technologies and Communications. For example, RD.45.128–2000, network and data services. All these legal documents may, for objective reasons, differ from the above standards.

In conclusion of this section, it should be noted that protocols and standards are extremely important for the organization of data transmission. However, although the number of protocols and standards at the time of this book's publication is very large, the development of data transmission facilities is going so fast that it is necessary to monitor this important segment of the TCS.

1.6 Practical Basics of Physical and Logical Data Coding

Each type of computer has its own internal form of coding for representing processed and transmitted data—symbolic and textual information (http://pdst.narod.ru/_20_el_uch/ost_wpd_01). The most commonly used codes are ASCII (American Standard Code for Information Interchange) and EBCDIC (Extended Binary Coded Decimal Interchange Code).

So, ASCII is an 8-bit encoding for representing decimal digits, Latin and national alphabets, punctuation marks, and control characters. The first half of the so-called code table (0–127) is occupied by US-ASCII characters, which include 95 printable and 33 control characters (developed by ANSI—American National Standards Institute). The second half of the table (128–255) contains national fonts (Cyrillic) and pseudographic characters. This code is used in personal computers and in large IBM incompatible machines. On large computers (mainframes), the 8-bit EBCDIC code developed by IBM is used. When transferring data from one computer to another, character transcoding may be required, which is performed by the system software of the transmitting or receiving computers. These actions are functions of the presentation layer of the OSI model. Consider next the most commonly used encoding methods at the physical level.

Most computers for the representation of "0" and "1" usually operate with standard logic layers of signals, which are determined by the specific type of chips used. The TTL logic represents 0.5 V as "0", and 5 V as "1". ECL and CMOS logic represent −1.75 V as "0", and − 0.9 V as "1". For data transmission, for example, in fiber-optic systems, a special chip is installed in the transceiver, processing any logic and issuing a control signal to a light source with conversion of 0.5 and 5 V TTL to 0 and 50 mA, respectively (turn on the light, turn off the light).

In most modern computer networks, all digital data is transmitted with a digital signal, i.e. pulse chain. More than two signal levels can be used for data transmission, while a single signal pulse can represent not one bit, but a whole group of bits. The reverse situation is also possible, when two signal pulses can be used for highly reliable transmission of only one bit.

In digital transmission, both potential and impulse codes are used. In *potential codes*, only the value of the signal during the bit interval is used to represent logical ones and zeros, and the signal edges that form the completed pulses are not taken into account. *Pulse codes* are a logical "zero" and a logical "one" by voltage drop of a certain direction. The value of the pulse code includes the entire pulse along with its fronts.

Signals in the form of pulse sequences have an infinite spectrum. The main signal energy is concentrated in the frequency range from zero to the frequency $f = 1/t_0$ (the first lobe of the signal energy spectrum), where t_0 is the baud interval, that is, the duration of a single pulse of a linear signal.

Theoretically, in accordance with the known Nyquist limit, the maximum permissible rate of change of values of a discrete signal ($B = 1/t_0$, baud rate) when transmitting a sequence of rectangular pulses over a communication channel, the equivalent of which is an ideal LF-filter with a rectangular amplitude frequency response and linear circuit phase response and cutoff frequency f_{gr} is equal to $B_{\max} = 2f_{gr}$. This limitation is due to the presence of transients at the output of the LF-filter, while the rise/fall time of the signal edge is defined as:

$$t_H = \frac{1}{2f_p} \tag{1.1}$$

At the maximum allowable transmission rate of the signal, $t_0 = t_н$. If the interval $t_o < t_н$, there is an unacceptable distortion of the waveform at the channel output and, as a result, reception errors. Thus, the information transfer rate N (Bps) depends on the B (baud) signal transmission rate and the chosen method of signal coding at the physical layer.

In any case, the physical digital coding method used by TCS developers should achieve several goals:

- to provide the smallest width of the signal spectrum at a given information transmission rate N (bps). To minimize the value of the constant component in the spectrum of a linear signal.
- to provide the receiver the possibility of clocking synchronization. The so-called self-synchronizing codes allow the receiver to extract a clock frequency from the received digital stream and then generate clock pulses from it for any statistics of the bit stream at the input of the transmitter.
- to have the ability to recognize errors.
- to have a low implementation cost.

Consider next the methods of physical digital coding of signals.

Potential Code Without Returning to Zero NRZ (Non-Return to Zero) The signal in the transmission line always has two levels: the lower level corresponds to zero, the upper one to one. Transitions occur at the boundary of the bit interval. When transmitting a sequence of ones, the signal does not return to zero during the bit interval.

Fig. 1.6 Features of coding by NRZ method

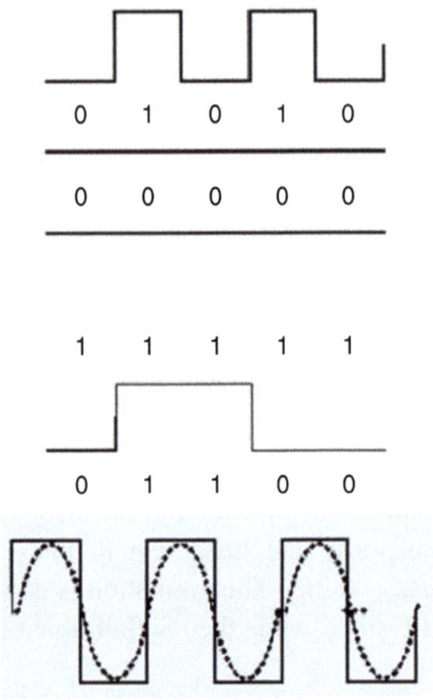

Let us consider in more detail some particular cases of data transmission by the NRZ code (Fig. 1.6): alternating sequence of zeros and ones, a sequence of zeros, and a sequence of ones. Let us determine the frequency of the main harmonic of the signal spectrum in each of these cases.

With the alternation of ones and zeros and the transfer rate N (bps), the period of the main harmonic in the signal spectrum is $T = 2t_0 = 2/N$ (s). The frequency of the fundamental harmonic f_0 is equal to $f_0 = N/2$ (Hz).

When transmitting only ones, or only zeros, the signal in the line is a direct current.

The spectrum of the real signal is constantly changing depending on what is the structure of data transmitted over the communication line. When transmitting long sequences of zeros or ones, the signal spectrum shifts towards lower frequencies. The NRZ linear signal usually contains a constant component and does not always provide the receiver with the ability to synchronize with the incoming signal. On the other hand, the NRZ code is simple to implement, has good noise immunity (due to two sharply different signal levels). The main signal energy in the NRZ code is concentrated at frequencies from 0 to $N/2$ (Hz).

NRZ code is not used in networks alone. Nevertheless, its various modifications are used, in which they successfully eliminate both poor self-synchronization and presence of a constant component.

Non-Return to Zero with Ones Inverted, NRZI With this encoding method, the transmission of zero corresponds to the signal level that was set in the previous bit interval (the signal level does not change), and when one is transmitted, the level changes to the opposite (Fig. 1.7) (http://pdst.narod.ru/_20_el_uch/ost_wpd_01).

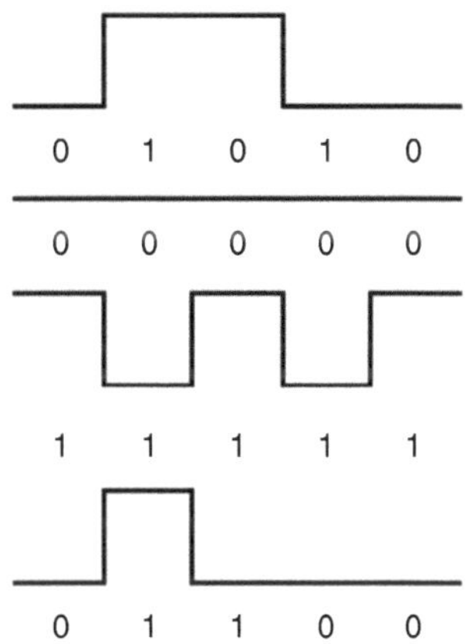

Fig. 1.7 Features of coding by NRZI method

This code is used when transmitting via fiber-optic cables, where the receiver consistently recognizes two signal states—light and dark.

Let us determine the frequencies of the main harmonics of a linear signal for particular cases of bit sequences.

So, for a sequence of alternating ones and zeros, the period of the signal is $T = 4t_0$ (s), the main frequency of the signal is $f_0 = N/4$ (Hz), with a sequence of ones—$f_0 = N/2$ (Hz), when transmitting a sequence of zeros $f_0 = 0$—direct current in the line (or absence of light).

The NRZI code usually uses only two signal levels and therefore has good noise immunity. The spectral components of the signal about frequency $N/4$ (Hz) have the maximum energy. It should be noted that the NRZI code has become central to the development of improved coding methods for high-speed data transmission systems.

Bipolar Alternate Mark Inversion, AMI, Quasi-Triple Code In this method, not two, but three signal values are used—"−1", "0", and "+1". To distinguish between these three levels, it is necessary to provide the best signal/noise ratio at the receiver input. An additional level requires an increase in transmitter power of about 3 dB to ensure the same bit-count integrity on the line. It should be noted that this is a general disadvantage of codes with several signal states compared to two-level codes.

Zero potential is always used to encode a logical zero, the logical one is usually encoded alternately with either a positive potential or a negative one (Fig. 1.8) (http://pdst.narod.ru/_20_el_uch/ost_wpd_01).

Note that when transmitting any bit sequence, the signal in the line does not contain a constant component. When transmitting ones, the main harmonic of the signal is at the frequency $f_0 = N/2$ (Hz). In the case of an alternating set of ones and zeros, the main harmonic is at the frequency $f_0 = N/4$ (Hz), which is two times less than that of the NRZ code. But there is a synchronization problem when transmitting a sequence of zeros.

In general, using different AMI codes for different bit combinations results in a narrower signal spectrum than for the NRZ code. The AMI code also provides some

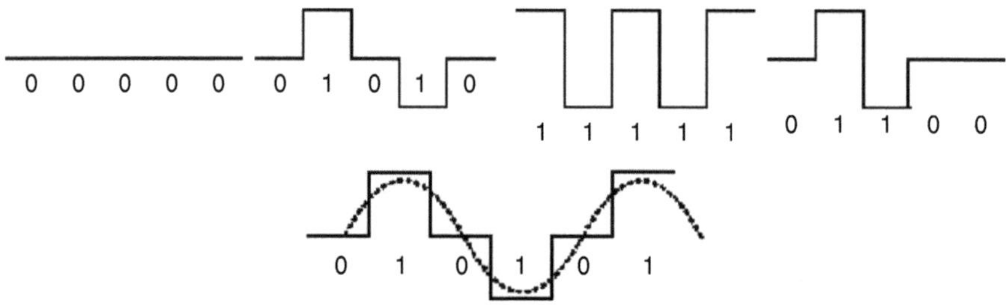

Fig. 1.8 Principles of quasi-triple coding (AMI)

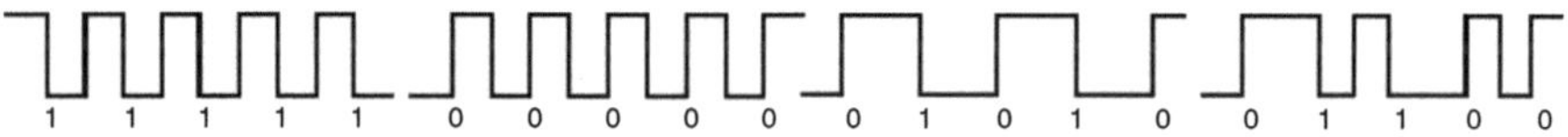

Fig. 1.9 Manchester coding principle

possibilities for recognizing false signals. Thus, the violation of the strict alternation of the polarity of the signals indicates a false pulse or disappearance of the correct pulse.

Manchester Code The Manchester code belongs to self-synchronizing codes and has two levels, which provides good noise immunity. Here, each bit interval is divided into two parts. Information is encoded by a level difference occurring in the middle of each interval.

The one is encoded by the difference from the high voltage level of the transmitted signal to the low, and zero by the reverse difference (Fig. 1.8). At the beginning of each bit interval, the so-called service signal difference can occur (when transmitting several ones or zeros in a row) (Fig. 1.9).

Note that when transmitting any bit sequence, the signal does not contain a constant component. The duration of a single pulse of a linear signal t_0 is equal to half of the bit interval, i.e. $B = 2$ N. The frequency of the main harmonic of the signal depends on the nature of the bit sequence and is in the range $f_0 = N/2 - N$ (Hz).

The Manchester code is widely used in Ethernet networks with a transmission rate of 10 Mbps (specification 10Base-T).

Currently, the developers have come to the conclusion that in many cases it is more rational to use potential coding, eliminating its disadvantages with the help of the so-called logical coding.

Potential Code 2B1Q This is a potential code with four signal levels for encoding data. The name reflects the essence of coding—every two bits (2B) are transmitted in one clock by a signal of a certain level (1Q). In this case, the linear signal has four states (Fig. 1.10).

Dibit "00" corresponds to a potential of −2.5 V (−3), "01"—potential −0.833 V (−1), "11"—potential +0.833 V (+1), "10"—potential +2,5 V (+3). The signal transmission rate B with this encoding is two times less than the information transmission rate *N*.

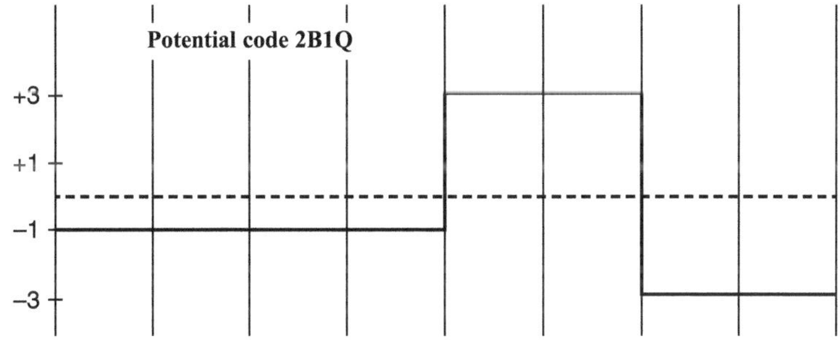

Fig. 1.10 The time curve of signal in 2B1Q code

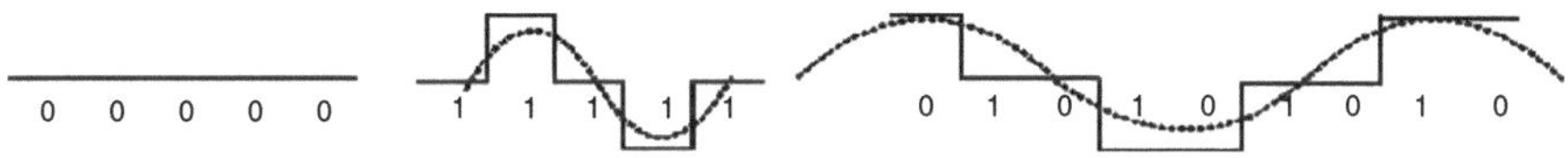

Fig. 1.11 Typical waveform in MLT-3 code

The basic frequency of the signal usually does not exceed $f_0 = N/4$ Hz. However, to implement this coding method, the transmitter power shall be higher so that all four potential values are clearly distinguished by the receiver against interference.

MLT3 Code (Multi Level Transmission—3) Here three levels of a linear signal are used: "−1", "0", "+1" (Fig. 1.11). A logical one corresponds to a mandatory transition from one signal level to another. When transmitting a logical zero, the change in the level of the linear signal does not occur.

When transmitting a sequence of ones, the period of change in the signal level includes four bits. In this case, $f_0 = N/4$ (Hz). This is the maximum fundamental frequency of the signal in the MLT-3 code. In the case of an alternating sequence of zeros and ones, the main harmonic of the signal is located at the frequency $f_0 = N/8$ (Hz).

Logical encoding is performed by the transmitter prior to the physical encoding discussed above, usually by means of the physical layer. At the stage of logical encoding, they are struggling with the disadvantages of the methods of physical digital encoding—lack of synchronization, the presence of a constant component. Thus, at first, corrected data sequences are formed using the means of logical encoding, which are then transmitted via communication lines using physical encoding methods.

Here, logical encoding implies replacing the bits of the original information sequence with a new sequence of bits carrying the same information, but possessing, in addition, additional properties.

There are two methods of logical encoding:

– redundant codes;
– scrambling.

Redundant codes are based on splitting the original sequence of bits into groups and replacing each source group in accordance with a given table with a code word that contains more bits.

The 4B/5B logic code replaces the original 4-bit groups (words) with words of 5 bits in length. As a result, the total number of possible bit combinations 25 = 32 is greater than for the original groups 24 = 14. The code table includes 16 code words that do not contain more than two zeros in a row, and use them for data transmission. The code ensures that with any combination of code words on the line cannot meet more than three zeros in a row.

The remaining code combinations are used for signaling (transmission synchronization, the beginning of the data block, the end of the data block, transmission control). Unused code words can be used by the receiver to detect errors in the data stream. The price for the obtained advantages with this method of data encoding is the reduction of the transmission speed of useful information by 25%.

There are also codes with three signal states, for example, in the 8B/6T code, for coding 8 bits of the source information, code words of 6 elements are used, each of which can take one of three values. The redundancy of the 8B/6T code is higher than of the 4B/5B code, since there are 36 = 729 resulting combinations for the 28 = 256 initial combinations.

In code 8B/10B, every 8 bits of the original sequence are replaced with ten bits of the code word. In this case, there are 1024 resulting combinations for the 256 original combinations. When replacing in accordance with the code table, the following rules are observed:

- none of the resulting combination shall have more than 4 identical bits in a row;
- none of the resulting combination shall contain more than 6 zeros or 6 ones.

All the considered redundancy codes are applied on Ethernet networks that are widely used. Thus, 4B/5B code is used in 100Base-TX/FX standards, and 8B/6T code is used in the 100Base-T4 standard, which is currently practically not used. 8B/10B code is used in the 1000Base-X standard, the 64/66 code in the 10 GbE standard (when fiber is used as the transmission medium).

Network adapters perform logic encoding. Since the use of the conversion table is a very simple operation, the logical encoding method with redundant codes does not complicate the functional requirements for this equipment.

To ensure a given line capacity, a transmitter using a redundant code shall operate at an increased rate (clock frequency). To ensure the information transmission rate of 100 Mbps using 4 V/5 V + NRZI encoding, the transmitter shall operate at 125 MBaud. In this case, the spectrum of the linear signal is expanded. However, the spectrum of the signal of the redundant potential code that is already the spectrum of the signal in the Manchester code, which justifies the additional logical encoding step, as well as the operation of the receiver and transmitter at an increased speed.

The *scrambling* is the "mixing" of the original data sequence so that the probability of occurrence of ones and zeros on the line becomes close to 0.5. Devices (or software modules) that perform such an operation are called scramblers (scramble—dump, random assembly).

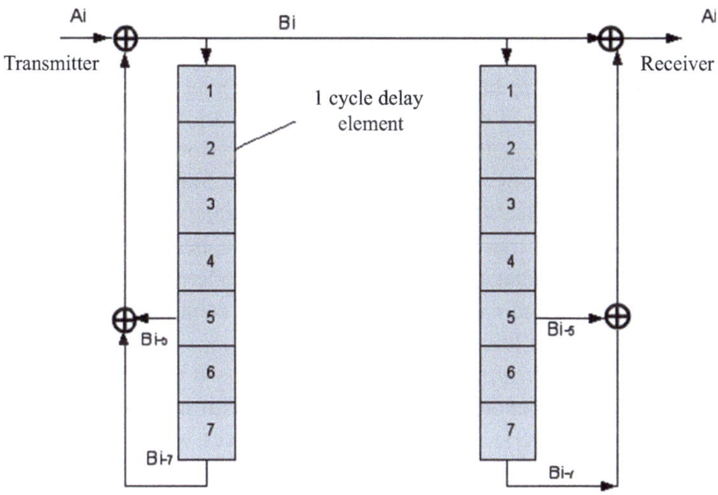

Fig. 1.12 Scrambling implementation variations (http://pdst.narod.ru/_20_el_uch/ost_wpd_01)

The scrambler in the transmitter performs the transformation of the structure of the original digital stream. The descrambler in the receiver restores the original bit sequence. Practically the only operation used in scramblers and descramblers is XOR—"bitwise exclusive OR" (2 modulo addition).

Let, for example, a scrambler implement the relationship Bi = Ai + Bi-5 + Bi-7 (Fig. 1.12).

Here Bi is the bit of the result code obtained at the i-th operating cycle of the scrambler; Ai is the bit of the source code that arrives at the transmitter to the input of the scrambler on the *i*-th cycle; Bi-5 and Bi-7 are the bits of the result code obtained at the previous operating cycles of the scrambler, respectively, at the "i-5" and "i-7" cycles.

The descrambler in the receiver restores the original bit sequence using the ratio

$$\text{Ci} = \text{Bi} + \text{Bi-5} + \text{Bi-7} = (\text{Ai} + \text{Bi-5} + \text{Bi-7}) + \text{Bi-5} + \text{Bi-7} = \text{Ai}.$$

1.7 Narrowband and Broadband Systems with Data Multiplexing

The *baseband system* uses a digital signal transmission method. Although the digital signal has a wide spectrum and theoretically occupies an infinite frequency band, in practice the width of the spectrum of the transmitted signal is determined by the frequencies of its fundamental harmonics. They give the main energy contribution to the formation of the signal. In a narrowband system, the transmission is carried out in the original frequency band, there is no transfer of the signal spectrum to other frequency domains. It is in this sense the system is called baseband. The signal occupies almost the entire bandwidth of the line. Special devices—repeaters—are used for the regeneration of the signal and its amplification in data networks.

Examples of implementation of baseband transmission are local networks and relevant IEEE specifications (for example, 802.3 or 802.5) (http://pdst.narod.ru/_20_el_uch/ost_wpd_01).

Previously, baseband transmission was used at distances of about 1–2 km over coaxial cables due to attenuation of signals, but in modern systems, due to various types of signal coding and multiplexing and types of cable systems, the limitations are shifted to 40 or more kilometers.

The term *broadband transmission* was originally used in telephone communication systems, where it denoted an analog channel with a frequency range (bandwidth) of more than 4 KHz. In order to save resources when transmitting a large number of telephone signals with a frequency band of 0.3–3.4 kHz, various schemes of multiplexing of these signals were developed, ensuring their transmission over a single cable.

In high-speed network applications, broadband transmission means that an analog not pulsed, carrier is used for data transmission. By analogy, the term "broadband Internet" means that you use a channel with a bandwidth greater than 128 Kbps (in Europe) or 200 Kbps (in the USA). A broadband system has high bandwidth, provides high-speed data, and multimedia information (voice, video, data). The examples include ATM, B-ISDN, Frame Relay, CATV cable networks.

The term "multiplexing" is used in computer technology in many aspects. By this we mean the integration of several communication channels in one data channel.

Here are the main multiplexing techniques: frequency division multiplexing (FDM), time division multiplexing (TDM), and wavelength division multiplexing (WDM).

WDM is used only in fiber-optic systems. Cable television, for example, uses FDM.

FDM At frequency multiplexing, its own analog carrier is dedicated to each channel. In this case, any kind of modulation or a combination of them can be used in FDM. For example, in cable television over a coaxial cable with a bandwidth of 500 MHz, 80 channels of 6 MHz each are transmitted. Each of these channels, in turn, is obtained by multiplexing of subchannels for the transmission of sound and video.

TDM In this type of multiplexing, low-speed channels are combined (merged) into one high-speed channel, through which a mixed data stream formed as a result of aggregation of the original streams is transmitted. Each low-speed channel is assigned its own time slot (time interval) within a cycle of a certain duration. Data is represented as bits, bytes, or blocks of bits or bytes. For example, the first 10 bits within a time interval of a given duration (frame, frame) are allocated to channel A, the next 10 bits— to channel B, etc. In addition to the data bits, the frame includes service bits for transmission synchronization and other purposes. The frame has a

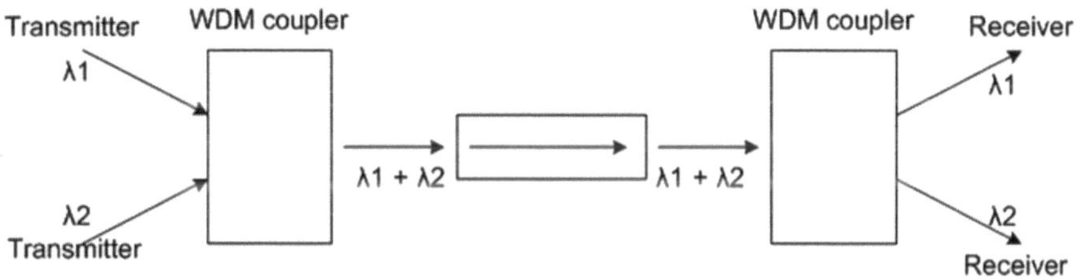

Fig. 1.13 WDM multiplexing implementation diagram

strictly defined length, which is usually expressed in bits (for example, 193 bits) and structure.

Network devices that multiplex data streams of low-speed channels (tributary, component streams) into a common aggregate stream for transmission over a single physical channel are called *multiplexers* (*mux*). Devices that divide the aggregate stream into component streams are called demultiplexers. Synchronous multiplexers use fixed time slots division. The data belonging to a particular component stream has the same length and is transmitted in the same time slot in each frame of a multiplexed channel. If information is not transmitted from a device, its time slot remains empty. Statistical multiplexers (stat muxes) solve this problem by dynamically assigning an empty time slot to the active device.

WDM WDM uses different wavelengths of the light signal to organize each channel. In fact, this is a special kind of frequency multiplexing at very high frequencies. In this type of multiplexing (Fig. 1.13) (http://pdst.narod.ru/_20_el_uch/ost_wpd_01), transmitters operate at different wavelengths (for example, 820 and 1300 nm). Then the rays are combined and transmitted over a single fiber-optic cable. The receiving device divides the transmission according to the wavelength and directs the beams to different receivers. Special devices—couplers—are used for merging/dividing the channels by wavelengths. The following is an example of such multiplexing.

Among the main constructions of couplers distinguish transmitting, reflecting couplers, and centrally symmetric reflecting couplers (SCR). Transmitting and reflecting couplers are tiny "twisted" in the center pieces of glass in the form of a star (Figs. 1.14 and 1.15). The number of output beams corresponds to the number of ports of the coupler. And the number of ports determines the number of devices transmitting at different wavelengths. And the number of ports determines the number of devices transmitting at different wavelengths.

A centrally symmetric reflecting coupler (Fig. 1.16) uses the reflection of light from a spherical mirror. In this case, the incoming beam is divided into two beams symmetrically with the center of bending of the mirror sphere. When the mirror is rotated, the position of the bending of the sphere and, accordingly, the path of the reflected beam change. A third fiber-optic cable can be added and it can redirect the reflected beam to another port. The implementation of WDM—multiplexers and fiber-optic switches is based on this idea.

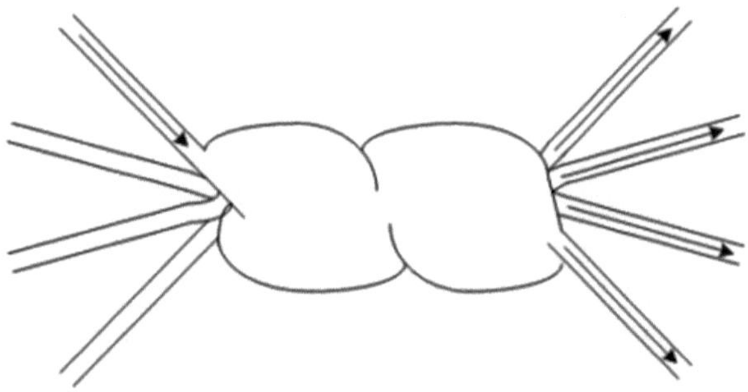

Fig. 1.14 Sketch of a "transmitting star" system

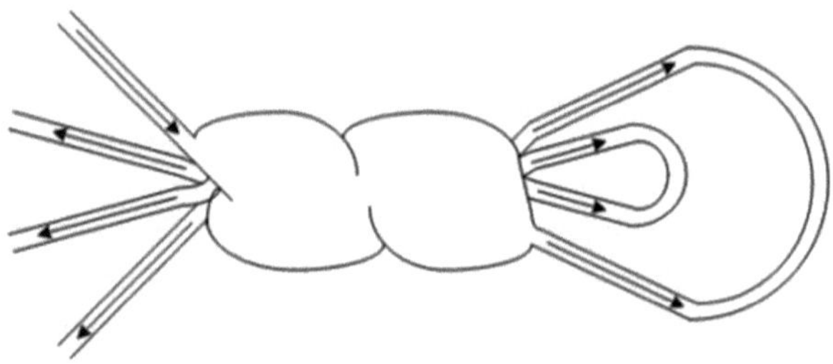

Fig. 1.15 Sketch of the "reflective star" system

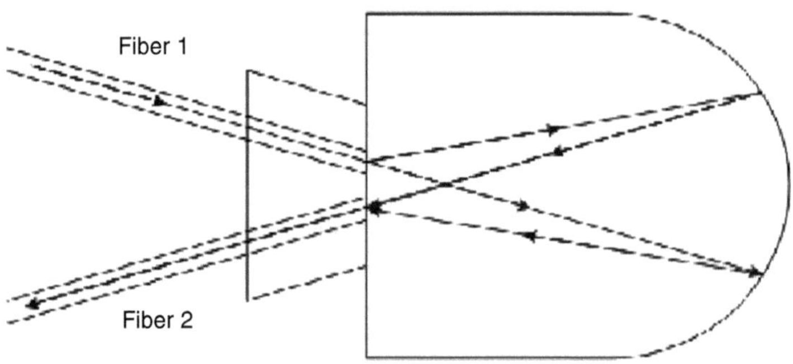

Fig. 1.16 The principle of operation of the centrally symmetric reflecting coupler

Optical multiplexers can be realized not only with the help of CSR-couplers, but also with the help of reflecting filters and diffraction gratings.

The main factors determining the capabilities of various implementations are interfering crosstalks and channel separation. The amount of crosstalk determines how well the channels are separated, and, for example, shows the part of the power of the 820-nm beam was on the 1300-nm port. A target of 20 dB means that 1% of the signal appeared on an undesignated port. To ensure reliable separation of signals, wavelengths must be "widely" spaced. It is difficult to recognize close wavelengths, for example, 1290 and 1310 nm. Usually 4 multiplexing schemes are used: 850/1300, 1300/1550, 1480/1550, and 985/1550 nm. CSR-couplers with a system of mirrors, for example, two, has the best characteristics (Fig. 1.17).

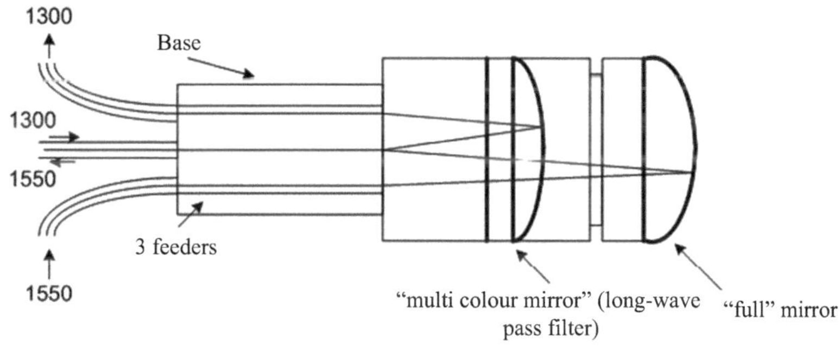

Fig. 1.17 The principle of construction of the SCR-coupler with two mirrors

WDM technology, which is one of three types of spectral compaction, occupies a middle position in terms of spectrum efficiency. In WDM systems, spectral channels are combined, the wavelengths of which differ from one another by 10 nm. The most productive technology is DWDM (Dense WDM). It provides for the integration of channels separated by a spectrum of not more than 1 nm, and in some systems even 0.1 nm. Due to this dense distribution of signals along the spectrum, the cost of DWDM equipment is usually very high. The least efficient spectral resources are used in new systems based on CWDM technology (coarse WDM, sparse WDM systems). Here, the spectral channels are separated by not less than 20 nm (in some cases this value reaches 35 nm). CWDM systems are commonly used in metropolitan area networks and LAN, where low-cost equipment is an important factor and requires the organization of 8–16 WDM channels. CWDM equipment is not limited to one part of the spectrum and can operate in the range from 1300 to 1600 nm, while the DWDM equipment is tied to a narrower range 1530–1565 nm.

1.8 Data Transmission Modes and Media

A standard communication line between remote computers usually consists of a physical medium through which signals are transmitted, data transmission equipment of the DTE (or data circuit terminating equipment DCE) and intermediate equipment of the communication network (Fig. 1.18) (http://pdst.narod.ru/_20_el_uch/ost_wpd_01). Intermediate equipment may include multiplexers, demultiplexers, switches, routers, repeaters, etc.

In the general case, the following elements can be distinguished in the complex of hardware and software connected to the transmission medium (http://pdst.narod.ru/_20_el_uch/ost_wpd_01):

- DTE (data terminal equipment)—terminal equipment for input, output, processing, and storage of data. Most often this is the computer where the application is running.

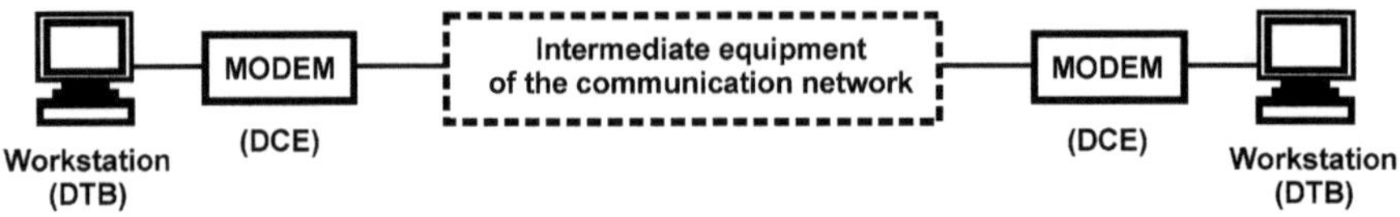

Fig. 1.18 Simplified structure of the communication line between remote computers of TCS

- DCE (data circuit-terminating equipment)—data transmission equipment (modem, network adapter), physical and data link devices, connecting the DTE with the communication line. In fact, DCE serves as an interface between a computer and the transmission medium (for example, between a computer and a fiber-optic cable, or between a computer and a telephone line). The DCE has the function of converting the signal from the DTE to a form suitable for transmission over a communication line—the medium format. The transmission medium determines the ways and possibilities of data transmission. There are special protocols for interaction between DTE and DCE. An example of a DTE-to-DCE interface is an RS-232 serial interface.

A *modem* is a special type of DCE device used to transfer data from a DTE to an analog transmission medium (for example, from a computer to a telephone line or a microwave transmitter). The modem modulates the signal with an analog carrier data sequence in transmission and performs the inverse transformation on reception. Modems are used when it is necessary to transfer over sufficiently large distances. Another example of the use of modems may be the need to have several transmission channels in the same medium (for example, over a single cable).

To connect two computers at a short distance, it is possible to use their connection with a null modem cable. It will make an electrical and mechanical connection between the transmitting port of one computer and the receiving port of another. There is no need to convert a digital signal to analog one.

When computers interact, simplex, duplex, asynchronous, and synchronous transmissions are possible. Simplex transmission means the ability to transfer data over a communication line in one direction only. Half duplex transmission is the transfer of data by devices in both directions, but in turn. Full duplex transmission implies the possibility of simultaneous transmission of interacting devices. In modern high-speed systems, duplex transmission is used.

From the point of view of organizing data transmission synchronization, there are: *asynchronous and synchronous data transmission protocols*. With asynchronous (start-stop) transmission (Fig. 1.19) (http://pdst.narod.ru/_20_el_uch/ost_wpd_01), the combination of the bits of each character of the alphabet is preceded by the start bit and ends with a stop bit (or two stop bits). Start and stop bits synchronize the transmission of bytes. Error control is performed by adding a parity (or imparity) check of the number of ones in a byte.

The controller in the receiving device performs the following main functions:

- detects the corresponding start bit on the line (the start/stop transition is a signal indicating the beginning of reception of the bit combination);

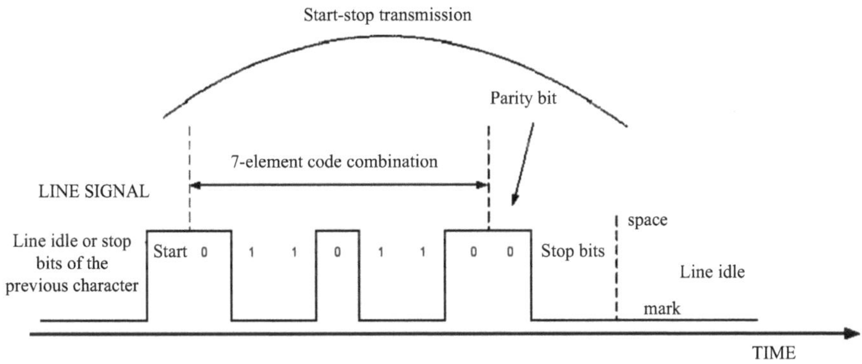

Fig. 1.19 Typical timing waveform diagram of the signal in the line with the start-stop transmission of the controller

- starts to register at certain intervals the incoming information bits;
- generates a byte in the receive buffer, performs validation checks.

Such synchronization is fairly simple to implement, but slows down the signal transmission.

In modern high-speed networks, synchronous transmission protocols are used, in which blocks of transmitted data are synchronized. Before the block, a special signal shall be transmitted, which will ensure that the receiving device will enter the synchronous mode. The transmitted data block has an established format. Error control is carried out by adding to the data block a control sequence obtained as a result of packet encoding with a noise-immune cyclic code (CRC). The length of the control sequence can be 16 bits (CRC—16) or 32 bits (CRC—32). The receiving device calculates the value of the control sequence in accordance with a known algorithm from the contents of the received block. In case of coincidence of the calculated and received control sequence, it is considered that the received data block contains no errors. The error control in the received data block is performed by the data link layer protocol (mas-sublayer). An example of synchronous transmission (USB interface) is shown in Fig. 1.20.

Next, let us take a brief look at the *transmission media* used in data networks. Each medium has its own advantages and disadvantages. Usually the price, ease of implementation, the possible transmission rate and noise immunity, the ability to protect transmitted data from unauthorized access are taken into account.

All media are divided into *bounded* and *wireless*.

Bounded media are cables (twisted pair, coaxial cable, fiber-optic cable) that transmit electrical and light signals. Data transfer capabilities are limited by cable capabilities: At the same time, various manufacturers of computer equipment have different requirements for the implementation of cable systems. For example, cable systems for connecting terminals to the IBM AS/400 are different from cable systems used to connect personal computers to the AS/400. AMP and RID cable systems have different capabilities.

Synchronization sample	Packet (up to 1024 bytes): address fields, data packet type fields, CRC check sequence	End-of-packet

Fig. 1.20 The structure of the synchronous packet data transmission

Fig. 1.21 Appearance of RJ-45 connector

Wireless media provide microwave, laser, infrared, and radio transmissions.

Usually bounded media are used in high-speed data transmission over limited distances. When constructing mobile networks, large corporate networks, or global networks, a combination of bounded and wireless media is used.

Below, we will more specifically consider the types of restricted environments, the typical representative of which is a twisted pair.

A *twisted pair* is formed by a pair of insulated intertwined copper conductors (cores). These conductors are combined in a single cable with an insulating sheath. To connect network devices via twisted pair, RJ-11 connectors (4 pins), RJ-45 (8 pins—4 pairs) (Fig. 1.21), and multi-pin RS-232, RS-449 connectors are used. Twisted pair can be shielded (Shielded Twisted Pair—STP, Foil Twisted Pair—FTP) and unshielded (Unshielded Twisted Pair—UTP).

Shielded twisted pair has an additional screen in the form of foil or metal mesh. STP was developed by IBM for Token Ring networks (for example, STP IBM Type 1). The screen standard has not yet been approved due to the complexity of grounding and high cost.

Since the last century, unshielded twisted pair is used in data transmission networks. EIA/TIA published a TSB-36 bulletin, where UTP categories according to the frequency characteristics of the passband and the measurement parameters of these cables are described (Table 1.5) (http://pdst.narod.ru/_20_el_uch/ost_wpd_01). The UTP cables used in high-speed data transmission, according to the EIA/TIA 568 standards, have 8 cores (4 pairs) and certain characteristics. Certification of cable systems of manufacturers for compliance with these characteristics has been conducted since 1991 by a special laboratory—Underwriter's Laboratories.

Table 1.5 Standard unshielded twisted pair categories

UTP category	Passband, MHz
3-cat3	16
4-cat4	20
5-cat6	100
5e-cat5e	125
6-cat6	250

Table 1.6 Main technical characteristics of twisted pair cable of UTP type

	Cat3		Cat4		Cat5	
Frequency, MHz	Attenuation, dB	Next, dB	Attenuation, dB	Next, dB	Attenuation, dB	Next, dB
10	30	26	20	41	20	47
16	46	23	21	33	25	44
100	–	–	–	–	67	32

All cables usually have the same design and are distinguished by the density and quality of winding. It should be noted that measurements of cable characteristics are carried out on 70 parameters, at certain frequencies and at a certain temperature. TCS developers need to focus on the measured characteristics of an unshielded twisted pair, such as:

- Attenuation;
- NEXT- near end crosstalk;
- Impedance; 100 Ohms for all categories ±15% at all frequencies (Table 1.6) (http://pdst.narod.ru/_20_el_uch/ost_wpd_01).

The generally accepted advantages of an unshielded twisted pair cable of UTP type are low price, compatibility with existing telephone cable systems, availability of multiple standards, relative ease of installation, and relatively low cost of diagnostic equipment.

The disadvantage of UTP is susceptibility to electromagnetic effects, which leads to the need to use a variety of encoding and scrambling tools to ensure high-speed transmission.

In the bounded media category, two more categories need to be considered.

A *coaxial cable* consists of two conductors located on the same axis ("co", "axis"—axis) and separated by an insulating sheath. The data transmission systems of large computers also use cables consisting of three conductors—twinax. According to their characteristics (passband, maximum distances), these cables are in the middle between UTP and fiber optics. A 75 Ohm RG-59 (PK-75) cable is used for cable television. For the "old" Ethernet-networks, designed for a transmission rate of 10 Mbps, cables RG-11 and RG-58 were used. In modern high-speed systems, coaxial cables are not used, since they are more expensive and heavier than UTP, and on the other hand, they are closer to the cost of fiber optics.

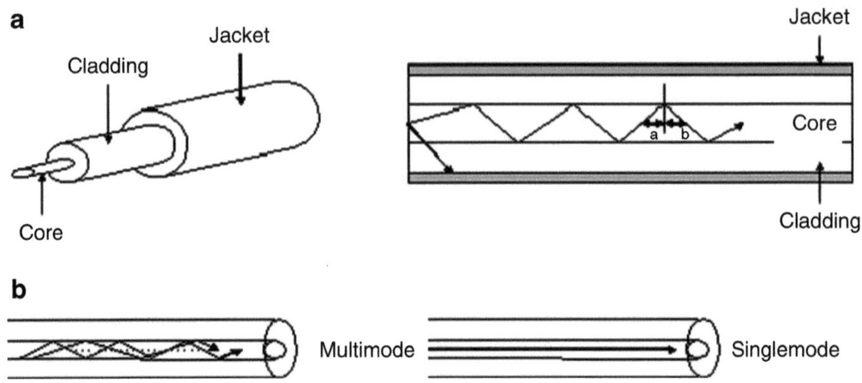

Fig. 1.22 Sketch of the design of fiber-optic cable of type MM (**a**) and SM (**b**)

Fiber-optic cable is a thin light-conducting glass or plastic cores in the glass reflective cladding, enclosed in a protective jacket. There are many designs of fiber-optic cable depending on the type of installation and transmission speed requirements. Unlike previous cable systems, fiber optics is immune to electromagnetic effects.

There are two main types of optical fiber, depending on the diameter of the glass core and glass reflective cladding:

- multimode fiber (MM)—(62.5/125 microns and 50/125 microns) (Fig. 1.22a);
- single-mode fiber (SM)—(8–10/125 microns) (Fig. 1.22b).
- At short distances and relatively low transmission rates, multimode MM type cables are used, at long distances and high rates single-mode SM type cables are used. The light beam is transmitted through different types of fiber at different wavelengths:
- multimode fiber—850 and 1310 nm with attenuation of 2.4 and 0.5 dB/km, respectively;
- single-mode fiber—1310 and 1550 nm with attenuation of 0.17–0.5 dB/km.

A single-mode SM type cable can transmit signals for hundreds or even thousands of kilometers, depending on the type of radiation source, wavelength, and data transmission rate. Since it is almost impossible to visually distinguish multimode fiber from single-mode fiber, most manufacturers tend to facilitate this task. Multimode cables usually have an orange cladding, and single-mode cables—yellow cladding.

Fiber-optic cables have a very large passband and, accordingly, allow for a high signal transmission rate. Single-mode fiber passes frequencies up to 50–100 GHz. Light is transmitted through it by a thin beam (one mode), the light source is a laser. Potentially, lasers can generate a light carrier with a frequency of up to 100 THz, and optical fiber can transmit a signal with a frequency of up to 1 THz, the overlap distance without regeneration can reach 300 km in actual conditions and 10,000 km in the laboratory conditions.

Fig. 1.23 Appearance of fiber-optic connectors

Plastic fiber-optic cables have several other designs, use a wavelength of 660 nm and red light sources, provide transmission at a maximum speed of 50 Mbps per 100 m. They are not applied for high-speed data transmission.

The advantages of modern fiber-optic cables are: low cost (glass components are much cheaper than copper), lightness of the cable, very high transmission rate compared to copper cables, insensitivity to interference, and high security against unauthorized access.

Disadvantages are the still high cost of the corresponding network and diagnostic equipment, qualification requirements for installation personnel.

It should be noted that modern fiber-optic transmission systems, in addition to cables, include a number of important elements:

- transmitters, transceivers—devices that convert electrical signals into light signals. The light source can be a LED or a laser.
- receivers, transceivers—devices that convert a light signal into an electrical one. Its main elements are usually a photodiode and a chip that regenerates and amplifies the signal.

Connectors and splice connectors, which provide the connection of fiber-optic cables to each other, connection to transmitters and receivers. Connectors are of various types depending on the signal power loss that occurs on them, the invariance of these losses in time, cost, possibility of reinstallation, and types of fiber optics.

Currently, *ST-connectors* are widely used (Fig. 1.23) (http://pdst.narod.ru/_20_el_uch/ost_wpd_01). The ST connector was first developed by AT&T in the mid-80s and was most widely used in optical subsystems of local networks. It is used to connect all types of multimode and single-mode fiber, connecting old network equipment. The connector is simple, relatively cheap, and easy to install. The main disadvantage of the ST-connector is the need for rotational movement when connecting to a connector socket. To overcome this drawback, a SC-type connector (NTT Corporation) was developed. The *SC-connector* has the mechanical reverse isolation of the terminal, locking element, and cable. Connection and disconnection is performed linearly (push-pull). SC connectors are widely used in single-mode and multimode networks with data transmission at rates of 1995 Mbps. New optical active equipment developed after 1995 is available only in versions with SC-ports.

FC connectors (NTT Corporation) are focused on the use of single-mode long-distance lines and specialized systems, as well as cable television networks. FC connectors withstand vibration and shock.

A broader classification is adopted in the wireless media.

Let us take a brief look at the main representatives.

Microwave data transmission systems exist in two versions—satellite and terrestrial. The latter are organized, for example, with the help of two parabolic antennas on the roof of buildings, they work in the lower part of the gigahertz range and under conditions of direct visibility. Microwave systems are cheap and high speed. They are sensitive to interference, audio interception, and atmospheric phenomena.

Laser transmission is carried out using a narrow beam of light generated by a laser. The system operates at higher frequencies than the microwave transmission, and is more narrow-band. Lasers are used as emitters, and photodiodes are used as receivers. Laser transmission is resistant to interference, audio interception, but is highly dependent on atmospheric phenomena and works at short distances in direct visibility conditions.

For *infrared data transmission*, infrared diodes and photodiodes with a frequency higher than 1000 GHz are used. Data transmission rates are close to fiber-optic systems, but the overlapping distance does not exceed 25 m at direct visibility.

Radio transmission is data transmission in the frequency range from 3 MHz to 3 GHz. Radio systems are widely distributed, have low cost, and are used for mobile technological applications, radio (FM is the transmitted frequency range up to 15 KHz), television (transmitted frequency range up to 6 MHz). Data transmission rates are relatively low, and transmission is sensitive to interference and audio interception.

In *wireless networks*—a rapidly developing class of the "unbounded media" category—data is transmitted in the ISM-range. The ISM-range (ISM—Industrial, Scientific and Medical) are frequencies that were reserved by developers (by US government in 1985 for civilian purposes and in 1950 for military purposes) for industrial (902–928 MHz), scientific (2400–2483 MHz), and medical (5725–5850 MHz) purposes. This data transmission is carried out using a broadband, noise-like signal. Each network has its own unique code for such a signal (implemented using a chip), allowing working in this region only to those who have the same chip. This code (extending pseudorandom sequence) is added to any message. Each message can be transmitted on the whole frequency spectrum of the range (spreading spectrum), which can be carriers (spread sequence). Transmitters operate with hopping pseudorandom frequency tuning (hops). When receiving a signal from other such networks in this network, it is perceived as noise. For wireless LAN and last-mile solutions, IEEE 802.11 and 802.14 standards are approved. The range of frequencies used was expanded to 11 GHz, the speed to 135 Mbps, and the range to 50 km.

To summarize, it can be said that for bounded media the most promising and very rapidly developing technology is fiber-optic transmission. The bounded media are used with the inefficiency of laying cable systems (mobile systems, high cost of cable laying). The following is possible in any media: simplex, duplex, asynchronous, and synchronous transmission.

1.9 Structured Cable Systems

For the organization of high-speed data transmission, two main types of specialized cables—twisted pair and optical fiber—are used. In the 80s of the last century, it was found that with decreasing costs of such cables and connectors, the cost of creating cable systems and, especially, their modification when employees move, increases and dramatically exceeds the cost of telephone systems. The solution was found by the well-known consulting company Gartner Group. They proposed to build data transmission systems by analogy with telephone systems—star topology, structure of the system (partitioning into functional modules), and cabling of buildings not by the number of existing employees, but according to ergonomic requirements (in Russia it is 6 square meters per person). Then, the number of working places is determined by dividing the area of the building intended for work by 4. At the same time, the concept of structured cabling systems and, accordingly, EIA/TIA standards appeared. Since data is transmitted through cable systems of buildings, and they are connected to computers, strict fire safety requirements and special tests that are performed by the above-mentioned UL (Underwriters Laboratories) have arisen. These are tests for meeting the following requirements:

- prevention of burning (insulation and jacket of the cable system shall be non-flammable);
- absence of smoke emission in case of fire;
- absence of toxic emissions in case of fire (halogens).

Cables that pass this test are labeled as L(ow)S (moke)Z(ero)H(alogen). There are labels for communication cables that have partially passed the tests (for example, CMR or OFNR).

Subsystems of the building cable system. A standard structured cabling system (SCS) (http://pdst.narod.ru/_20_el_uch/ost_wpd_01) consists of a set of subsystems, each of which represents a set of cables, terminals, connectors, and other products necessary for a cost-effective solution of the problem of data transmission in a given territory. According to the TIA/EIA 568 cable construction standard, SCS has the following characteristics:

1. Topology of any subsystems: “star”.
2. Types of devices connecting cable subsystems: horizontal cross (HC), intermediate cross (IC), main cross (MC), and equipment room (ER)—room for active network equipment (Fig. 1.24).
3. No more than one cross shall be between the main and horizontal cross. Between any two horizontal crosses there should be no more than three crosses.
4. The maximum length of the horizontal segment for a twisted pair does not depend on the type of cable and is 90 m.
5. The maximum length of the trunk segment depends on the cable type and is presented in Table 1.7 (http://pdst.narod.ru/_20_el_uch/ost_wpd_01):

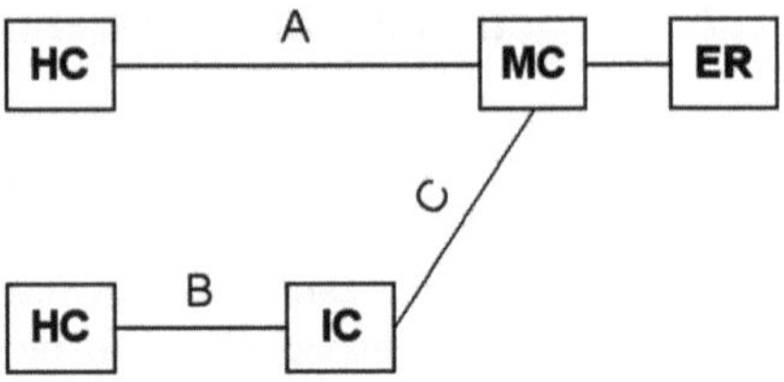

Fig. 1.24 The main types of devices connecting cable subsystems

Table 1.7 Maximum length of the trunk segment

	Maximum allowable distance		
Cable type	A (HC-MC)	B (HC-MC)	C (IC-MC)
Twisted pair	90 m	90 m	90 m
Multimode fiber	2000 m	500 m	1500 m
Single-mode fiber	3000 m	500 m	2500 m

When using intermediate cross IC, the total distance from HC to MC should not exceed the maximum length of the trunk segment.

The TIA/EIA 568A standard defines the following subsystems of structured cabling systems:

- Building backbone;
- Campus backbone; campus is a system of buildings separated by distances not exceeding those shown in Table 1.7 (http://pdst.narod.ru/_20_el_uch/ost_wpd_01);
- Horizontal subsystem of the building;
- Administrative subsystem;
- Workplace subsystem.

The rooms (HC, IC, MC) where the cable connecting devices are located are called the telecommunication closet—TC, and the rooms where the network equipment is located are called the equipment room—ER (in small systems they are combined with a telecommunication cabinet). We show these subsystems in the diagram and describe in more detail.

The workplace subsystem (Fig. 1.25) serves to connect a terminal, computer, or telephone to the horizontal subsystem. The transmission medium is UTP/STP/Coaxial cable. Connection is carried out with the help of a workplace socket. The socket may contain a special adapter, matching the resistance of various cable systems (balun).

The horizontal subsystem is the horizontal part of the cable system. It connects the telecommunication socket in the workplace zone with the administrative subsystem of the floor in the telecommunication closet. The transmission medium—STP/UTP/Coaxial/Fiber.

The administrative subsystem consists of a set of switching cables (patch cords), devices (patch panels), connectors, and blocks that connect the horizontal subsystem to a vertical system of the building. The administrative subsystem is located in telecommunication cabinets.

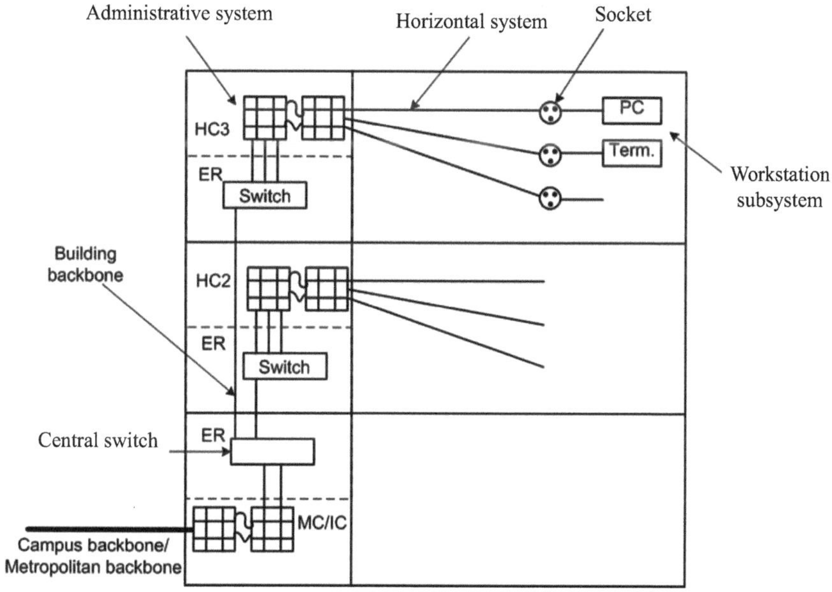

Fig. 1.25 Typical SCS subsystem

Building backbone is the vertical backbone of the building. It provides a connection between the nodes of the administrative subsystem. The transmission medium is UTP/Coaxial/Fiber. The subsystem has a star topology in which each horizontal closet (HC) is connected by cable to the main (MC) or intermediate (IC) closet.

Campus backbone (metropolitan backbone) is the campus highway connects various buildings in a limited territory. Table 1.7 shows that according to the standards, the length of this trunk is determined by the type of fiber and is 2500 m for multimode fiber and 3000 m for single-mode fiber. In general, such a limited territory corresponds to the territory of the local network. The transmission medium is usually fiber optics. The subsystem topology is a star, the main cross is in the central building. Connection to the city backbone or global wide area network (WAN) is carried out in the main closet of the building or campus. If there are several buildings on campus, then the main closet is arranged in the building to which the city backbone approaches, and in each of the other buildings an intermediate closet is arranged.

EIA/TIA Standards The creation of cable systems is based on a variety of standards. Here we will describe only the basic standards necessary for the implementation of high-speed data transmission (http://pdst.narod.ru/_20_el_uch/ost_wpd_01):

- EIA/TIA 568 is the standard for the creation of telecommunications of office and industrial buildings, planning of cable systems for buildings, methods of constructing a telecommunication system for office and industrial buildings;

- EIA/TIA 569 is the standard that describes the requirements for premises in which a structured cabling system and communication equipment are installed;
- EIA/TIA 606 is the standard for administering telecommunication infrastructure in office and industrial buildings;
- EIA/TIA 607 describes the requirements for the infrastructure of a telecommunication system for grounding and potential balancing in service and industrial buildings.

The transmission medium is very important for high-speed data transmission. Groups of computers shall be interconnected by structured cabling systems. Such systems consist of modules-subsystems (horizontal, administrative, building trunk, campus trunk, workplace subsystem) and are built according to EIA/TIA standards (Table 1.8).

1.10 Topology of High-Speed Data Transmission Systems

In general, there are two types of system topologies: physical and logical. The physical topology is the location of the physical devices and connection between them. The logical topology reflects the path of data transmission over the network.

The data networks use the "point-to-point" concept (direct interaction of two devices via a communication line) and "point-to-multipoint" concept (interaction of three or more devices, one of which acts as the root node). In the first case, device addressing is not required (it is already clear), in the second case, it is required to divide the data transmission channel between devices, specify the device addresses and determine the method of access to the data transmission channel.

The following classes of network topologies can be distinguished (http://pdst.narod.ru/_20_el_uch/ost_wpd_01): directly connected, mixed, radial-node (star, hierarchical star), ring, bus, hybrid. It should be noted that there is no standard for these concepts.

When using a *directly connected topology*, devices in the network (nodes) are connected according to the principle of "each with each". This topology is not used in modern networks, since if there are N nodes and each node shall have an (N-1) interface, which is absolutely impossible.

In a network based on *a mixed topology* (Fig. 1.26), each node is connected to several adjacent nodes, so that loops are formed in the network. Mixed topology is used in the construction of large, geographically distributed networks.

All network devices with *a "star" topology* (Fig. 1.27) are connected according to the "point-to-point" topology to the central device. The central device in such a network is a hub (repeater) or switch. Hubs and switches can be connected to each other according to "hierarchical star" topology.

A passive hub makes a simple connection without regeneration and signal amplification (in fiber-optic systems, these are splitters). Active hubs, in addition to connecting all devices, regenerate and amplify the signal (physical layer).

Table 1.8 The layout of TIA-568A/B standard (http://pdst.narod.ru/_20_el_uch/ost_wpd_01)

The layout of T568A/B RJ-45

Pin	T568A Pair	T568B Pair	T568A colour	T568B colcur	Pins on the front side of the connector (in the socket—vice versa)
1	3	2	White/green stripe	White/orange stripe	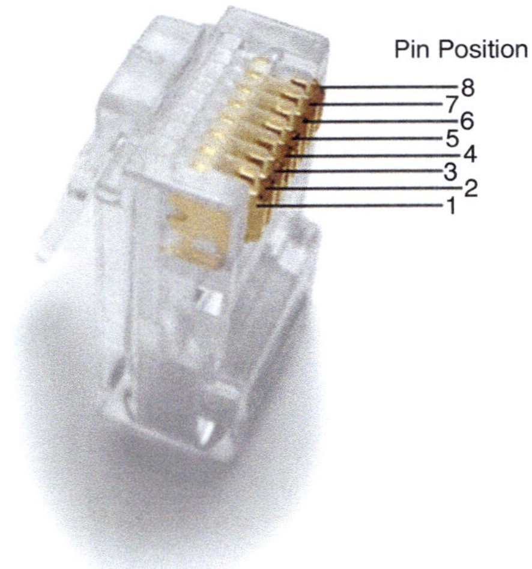
2	3	2	Whole green	Whole orange	
3	2	3	White/orange stripe	White/green stripe	
4	1	1	Whole blue	Whole blue	
5	1	1	White/blue stripe	White/blue stripe	
6	2	3	Whole orange	Whole green	
7	4	4	White/brown stripe	White/brown stripe	
8	4	4	Whole	Whole	

Fig. 1.26 The example of mixed topology

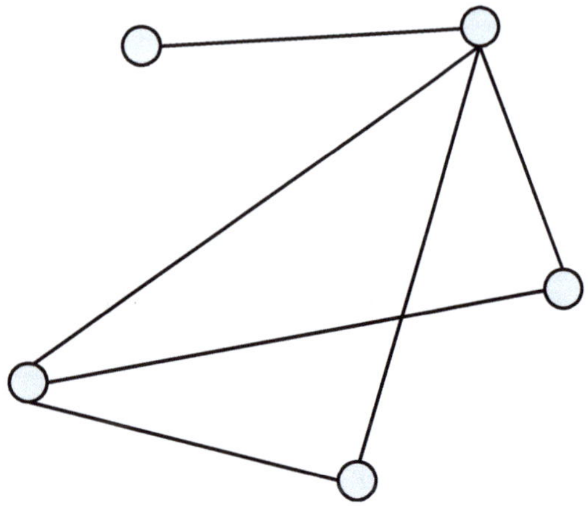

Fig. 1.27 Network with physical "star" topology

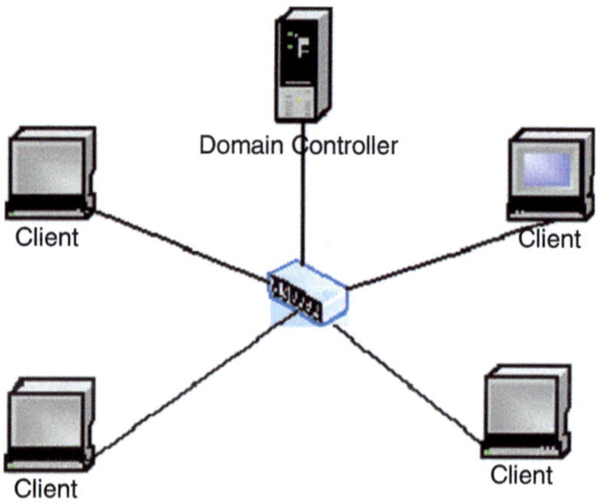

Modern switches perform functions up to the third and fourth layer of the OSI in the network. In "star" topology networks, it is easy to detect errors (this is why it is used in all modern Ethernet variants). On the other hand, such a network requires a large amount of cable, and therefore tend to use a relatively cheap twisted pair UTP. In case of failure of the central device, the entire network stops functioning.

In the network with the "*bus*" *topology*, a linear connection of devices to the cable segment is used (Fig. 1.28). It is very economical and simple topology, but it is difficult to localize equipment malfunctions and software failure. The network is inefficient in terms of system modification (expansion). An example is older Ethernet versions on a coaxial cable (10Base-5 and 10Base-2).

"*Ring*" *topology* is a ring of repeaters or switches. This network provides a "point-to-point" connection between two adjacent nodes. Let's emphasize that this is not a ring of workstations. An example is the FDDI network.

Finally, *hybrid topologies* combine star, bus, ring topologies. This is the most commonly used topology. An example is the Token Ring network, which generally

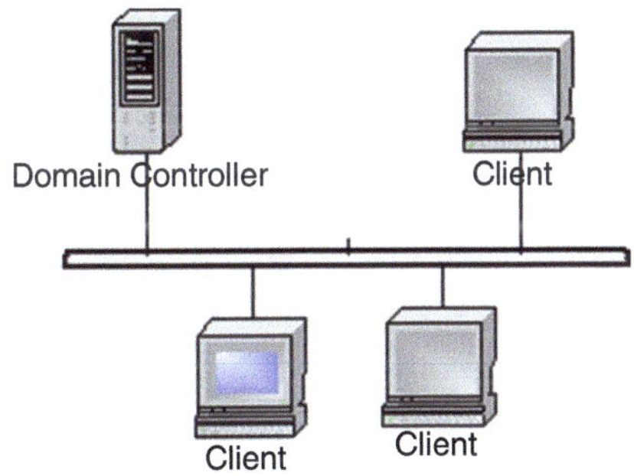

Fig. 1.28 Network with physical "bus" topology

has a combined star-ring configuration. Hybrid topologies are most common in modern data transmission systems.

1.11 Methods of Organizing Access to High-Speed Channels

If two devices are connected to each other in a point-to-point version, then there is no need to use complex data transmission rules. The transmitting device transmits data at any desired time. But, if several devices share a transmission channel (a network with "bus" topology), a method that will allow each device to transmit information without imposing on the transmission of another device is required. The *channel access method* is the rules that describe how devices divide the communication channel, access the channel, and release it. There are three main methods of channel access: competition, polling, and token transfer. Theoretically, the channel access method does not depend on the network topology. But almost every topology implements a specific access method. In this case, the access method affects the effective data transmission rate, that is, the actual network bandwidth.

Let us briefly consider these methods (http://pdst.narod.ru/_20_el_uch/ost_wpd_01).

With a *competitive access method*, any device tries to start the transmission whenever it wants. There is no arbiter which would allow or prohibit work. Of course, if two or more devices transmit simultaneously, then there is a distortion of the transmitted frames—a collision. There are protocols in which devices listen to a signal in the communication channel (carrier) before transmitting, and determine whether the frame can be transmitted at the moment or should be postponed in order to avoid collision. Such protocols are called CSMA (carrier sense multiple access). There are two types of multiple access: CSMA/CD collision detection (for example, IEEE 802.3 Ethernet) and CSMA/CA collision warning (IEEE 802.11 Wi-Fi). In modern applications, CSMA/CA is typically used in unbounded media, and CSMA/CD in bounded media. The network topology is a star. The advantage of these access methods is their simplicity, and the disadvantage is that access time is predictable

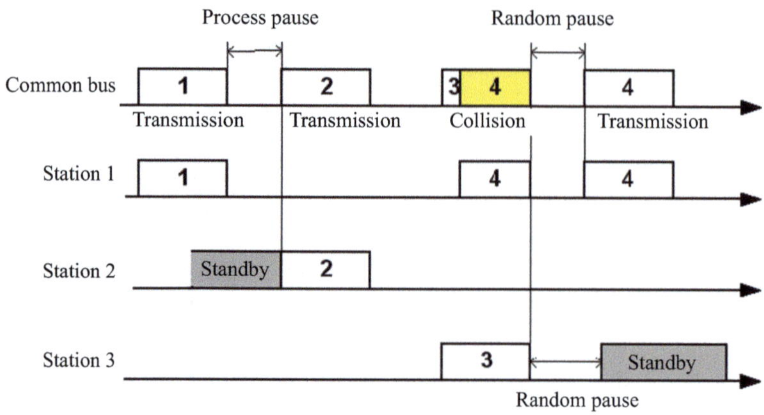

Fig. 1.29 Access method CSMA/CD

only on the basis of statistical considerations, and the difficulty of prioritizing devices to speed up the access. Figure 1.29 shows a timing diagram illustrating the access method CSMA/CD in the peer-to-peer network. Consider the concept of collisions for Ethernet networks with a "bus" topology.

All connected devices (workstations) "listen" to the common communication channel (bus) in order to determine its state—busy or free, and copy the frame transmitted via the bus into the receiving buffer if the frame is addressed to this network device.

The station may begin transmitting a frame if the channel is free. If the channel is busy, frame transmission is postponed. Having started the transmission, the station continues to monitor the channel status by comparing the type of the sent signal with the one present in the communication line. If a conflict (collision) is detected, the stations involved in the collision should stop the transmission. The transmission may be resumed at the expiration of a random period of time. The conflicts are a normal, albeit undesirable, phenomenon in networks with multiple (collective) access to a common transmission medium.

With the "*polling*" *access method*, one device polls and controls the other devices of the network, acting as an access administrator (master, controller). Typically, the network topology is a star or bus. An example of an implementation is a mainframe controller polling terminals, or a server cluster controller. In unlimited environments, the method partially uses the 802.16d family of protocols (Wi-Max). In this case, the base station allocates to each of the devices a frequency band for operation at the entrance to the network (device competition takes place only at the initial entrance to the network) and can prioritize the devices according to the dedicated band. The advantage of polling access is the predictability of the minimum and maximum access times, the ability to prioritize traffic.

With a *token data transmission method*, it is possible to start upon the receipt of a token—a specific small packet of a certain format. Usually it is generated by the network device with the lowest address. The device that received the token controls the channel while it owns the token. In turn, the channel is monitored by each of the devices in the network according to the rules established by the protocol. An example of the implementation of this access method is the IEEE 802.5 Token Ring network. Typically, the logical network topology is a ring. The advantage of token access methods is the predictability of the transmission time and a lot of traffic management capabilities. The disadvantage is their complexity, since every device on the network shall be able to do everything, to have full functionality. That is why the network devices have complex software.

Modern high-speed networks use a combination of the capabilities of the described access methods (for example, IEEE 802.3 Ethernet—CSMA/CD networks with additional 802.1 management functions).

Summarizing the results of the above-mentioned, it should be noted that the methods of access to the transmission medium can be divided into *probabilistic* (CSMA) and *deterministic* (polling, token access). A common disadvantage of probabilistic access methods is the indefinite frame passing time through the network, which increases sharply with increasing load on the network. With the deterministic method, nodes access the medium in a predetermined order. This order is determined by the network controller, which can be centralized (its functions can be performed, for example, by the server) or/and distributed (functions are performed by the equipment of all nodes).

1.12 Switching Technologies

Three main switching technologies are used in the telecommunication networks: circuit switching (CS), message switching (MS), and packet switching (PS) (http://pdst.narod.ru/_20_el_uch/ost_wpd_01).

At "circuit switching", a physical connection is established between the sending and receiving devices (A and M) (Fig. 1.30). An example is a connection in a TF-OP telephone network or an ISDN network. The connection established in the circuit-switched network is maintained until the end of the communication session, regardless of whether information is being transmitted or not, and is destroyed by the initiative of one of the terminals. The advantages of this switching method are its simplicity and absence of delays in the transmission of information after the connection is established. The disadvantages include the inefficient use of bandwidth due to the presence of temporary pauses in the information flow between terminals

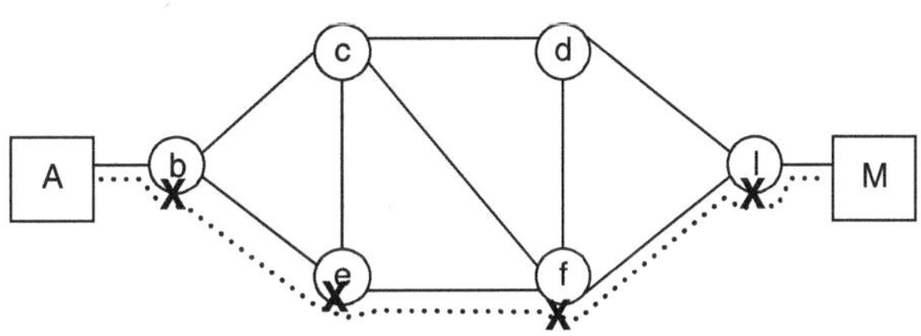

Fig. 1.30 The scheme of information transmission in the network with CS

Fig. 1.31 The scheme of information transmission in the network with MS

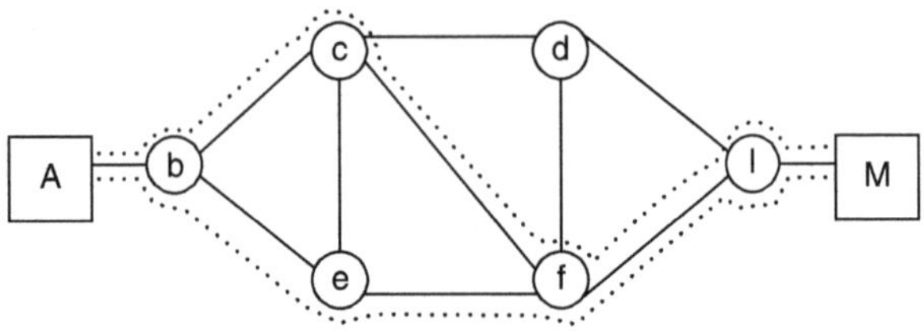

and possible network failures to establish a connection. During the construction of modern high-speed networks, such switching is practically not used.

Message switching is a set of operations for a network node from a terminal device or another node to receive the entire message (file, data block), to store the received message in the node's memory and then transmit it according to the address it contains. Thus, the message step-by-step with a re-reception at each node is transmitted through a number of nodes to the destination. The transmitting station (source) (Fig. 1.31) (http://pdst.narod.ru/_20_el_uch/ost_wpd_01) supplies the message with the destination address (DA) and source address (SA). Different messages between the sender and the recipient (A and M) can take place in the network in different ways.

A telegraph network can serve as an example of the implementation of this switching method. In computer networks in its pure form, this type of switching does not apply, although the very idea of "store and forward" is used in mail transmission systems (MHS-message handle systems).

Message-switched network nodes should have buffer memory of an indefinite size and large disk space for temporary storage of data. Difficulties arise when working in real time due to unpredictable and large message delays in the network. The advantages of such a switching method include more efficient, than at switching channels, using network channels (messages are transmitted in turn, using channels reaches 95–98%), wide traffic management capabilities, and the ability of broadcast messages.

Packet switching is the main method in computer networks and was first proposed by Rand Corporation, and then widely applied in the implementation of the ARPANET network (1964–1967). The transmitted message (Fig. 1.32) (http://pdst.narod.ru/_20_el_uch/ost_wpd_01) is divided into relatively short parts (packets), each of which is supplied with a header (service information). It is assumed that such a packet has a source address and a destination address (SA and DA). Since the packet has a fixed maximum length, it does not require disk memory to store it, there is enough RAM, which significantly reduces transmission delays.

There are two methods of packet switching: *datagram* transmission and *virtual circuit* transmission. Consider the difference between these methods.

When transmitted in datagram mode, each packet contains the sender and recipient addresses, service information, and the sequence number of the packet in the message. Datagrams that belong to the same message move independently from each other in the network. The receiving device collects the message from the packets

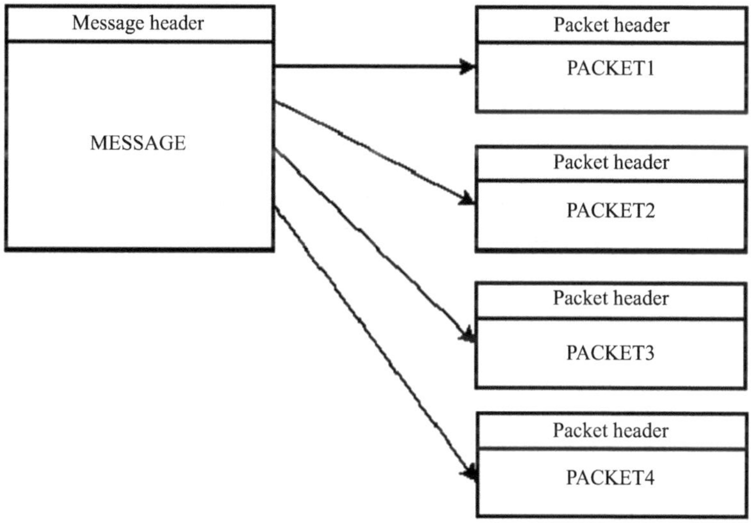

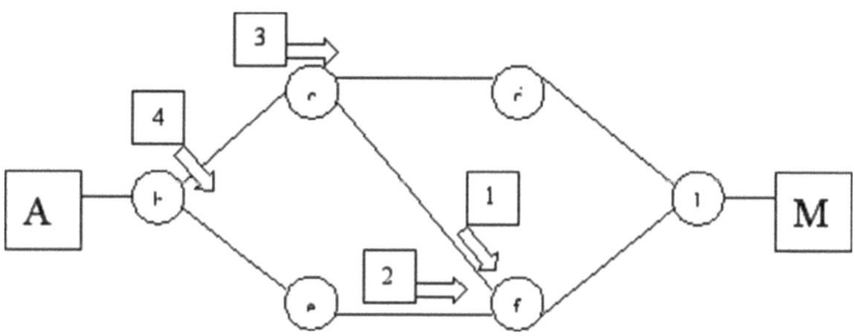

Fig. 1.32 The scheme of information transmission in the network with PS

according to their numbers. Package length is quite significant. The transmission can be carried out in long packets (used in Ethernet networks) or short packets (ATM networks). Too long packets bring the network closer to the message-switched network. Network time increases, and network efficiency drops. Too short packets significantly increase the share of service information (overhead), since each packet has a fixed-length header.

When transmitting via virtual channels, a logical connection between devices is created (logical connection), i.e. a route is established in the network for transmitting packets of a certain information flow (message). A connection is established before data transmission starts by exchanging service packets between the sender and receiver. They contain the transmission parameters (maximum packet size with data, transmission path, transmission rate, the need for confirmation of delivery (acknowledgment), coordination of error control procedures, and connection control procedures). In the memory of each node through which a service packet passes, a buffer zone is reserved for intermediate accumulation of data packets that will be transmitted over this virtual connection. In the routing table of each node, the service packet leaves an order that has the following meaning: packets that have a logical number (virtual channel number) K in the header, coming on the incoming physical line a, should be sent to the outgoing physical line b and assigned the virtual channel number L. Each network segment (between nodes or between a node and an attached computer) has its own virtual channel number, which is set by the sender of the packet in this segment. Thus, a virtual, i.e., conditional logical connection exists only in the memory of the switching node. Practically several hundreds and even thousands of virtual channels can be organized in one physical channel.

A logical connection can be temporary (SVC, established for one session) or permanent (PVC, persists for a long time, usually up to several months).

Transmission using datagrams is more flexible, requires less administration costs, but is less reliable than transmission over virtual channels.

Packet switching requires complex software solutions and fast switching devices, and this has hindered its development earlier. At packet switching, small delays of packets within the network are provided, since there are no long messages and, therefore, there are no delays in the buffer memory of the nodes. There are wide opportunities to manage traffic (it is possible to have different algorithms for finding routes and bypass the loaded sections of the network). Therefore, packet switching is implemented in all modern high-speed data transmission systems.

So, summing up this section, it should be noted that there are three main methods of switching: circuit switching, message switching, and packet switching. Switching channels is characterized by minimal delays after allocating a channel, but inefficient use of bandwidth. Message switching uses message forwarding at each node. Delays can be significant, although network links are much more efficient. Packet switching implies a balance between delay parameters and channel efficiency. With the advent of high-speed and managed switching devices, packet switching is the main switching method in networks.

1.13 Features of Organizing Communication Between Segments of Telecommunication Networks

After solving the problem of combining individual computers in the network (the 80s of the last century), it became necessary to connect computer networks with each other (http://pdst.narod.ru/_20_el_uch/ost_wpd_01). This connection is made with the help of special devices—switches, routers, and other devices. The term "network segment"—a part of a network that does not contain connecting devices—has emerged. Devices connecting networks (segments of one large network) are divided into types depending on the layer of OSI on which they operate (Table 1.9).

Table 1.9 Functional layer of presentation of network devices

Level 1	Amplifiers/repeaters/hubs
Level 1	Amplifiers/repeaters/hubs
Level 3	Routers (routers) all levels gateways

As is known, a signal passing through a cable system changes under the influence of various interferences and attenuates, due to which the cable system limits the data transmission over a distance. Therefore, in networks, devices that can amplify and restore the waveform are used.

A structure known as a hub does not analyze information. It instantly memorizes the values of the signal “0” or “1”, regenerates them accordingly, amplifies, and sends them to all connected network segments. Theoretically, these functions should be performed on the way from the source to the recipient as many times as necessary. In practice, many networks limit the number of hubs. For example, in versions 10Base Ethernet on a coaxial cable, the number of hubs should not exceed 4 (5 network segments).

If there is a lot of traffic in the network, it is advisable to divide it into segments in which computers most often work with each other. For this purpose, the so-called *bridges* are used, which forward the message not to all network devices, but only to that network segment in which the recipient is located. Bridges are devices that divide networks. They work with the physical addresses of stations at the data link layer of the OSI protocols. Unlike a hub, a bridge can allow access to physical devices, or deny it, that means, it can regulate traffic. In the process of operation, the bridge polls the network devices, writing down the addresses of new devices in a special hash table in the RAM.

Working in segment 1 (Fig. 1.33) (http://pdst.narod.ru/_20_el_uch/ost_wpd_01), each such bridge receives all frames of this segment, ignores frames addressed to the stations of segment 1, and frames addressed to the stations of segment 2 transmits to the corresponding port.

When connecting several network segments (Fig. 1.34), bridges should route frames. There are three types of bridge routing protocols (http://pdst.narod.ru/_20_el_uch/ost_wpd_01):

- TR or STA (transparent or learning)—uses the STA algorithm (Spanning Tree Algorithm), which is used, for example, in all versions of switched Ethernet;
- SR (source routing)—information about the route is contained in each packet transmitted; used in the IBM Token Ring network;
- SRT (source routing transparent)—a combination of the two types listed above.

TR or STA bridges do not require any initial programming when included (initialization). They analyze the traffic and “learn” how the addresses of devices belong to network segments (each port is associated with one segment). Such a bridge creates a dynamic database of device addresses and determines whether to transmit or delete a frame according to the recipient’s address. Typically, these bridges are used in networks that have a *hierarchical star* topology: the central bridge and each ray of the star is a tree of bridges. Such a combination is called a

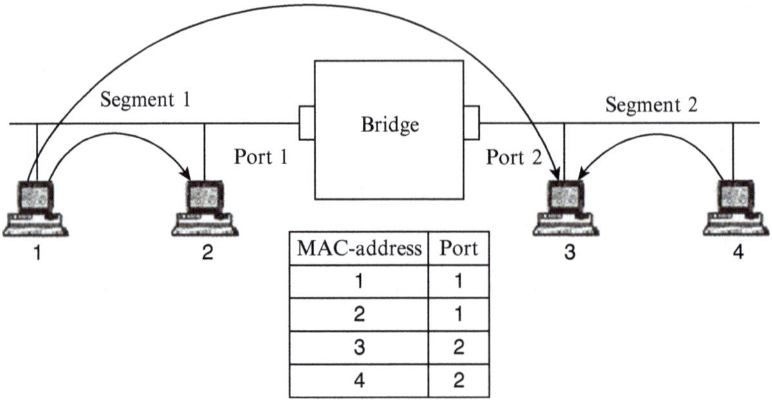

Fig. 1.33 Connection scheme of two segments 1, 2 of the network using a bridge

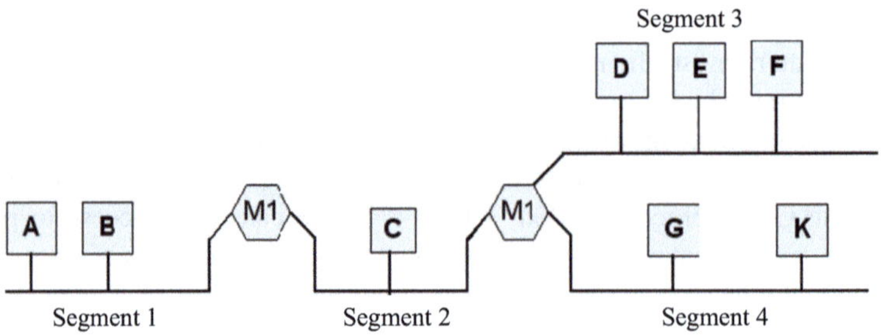

Fig. 1.34 The structure of four segments (Segment 1–4) and two bridges (M1 and M2)

"collapsed backbone". The protocol is called transparent because it is transparent for network stations (each station can communicate with any other as in one large segment without thinking about the route) and, on the other hand, is transparent for protocols starting from the network layer and above.

Bridges with routing from the source SR (Fig. 1.35) (http://pdst.narod.ru/_20_el_uch/ost_wpd_01) are used to connect the Token Ring rings and FDDI. The bridges combine into a ring, and each ring in turn is joined by another ring of stations. Such a combination is called a "token backbone". The sender places into each sent frame all address information about the intermediate bridges and rings that the frame shall pass before it enters the ring to which the recipient station is connected. The bridges and rings have identifiers for setting the route.

SR bridges when moving frames use information from the corresponding fields of the data frame. For the operation of the source routing algorithm, two additional frame types are used: SRBF (single-route broadcast frame) and ARBF (all-route broadcast frame). SR bridges are configured by the administrator to transmit ARBF frames to all ports except the source port of the frame, and for SRBF frames some bridge ports need to be blocked so that there are no loops in the network. The SRBF frame-researcher is sent by the source station, if the recipient station is in another

Fig. 1.35 Connection circuit of the ring segments with SR bridges M1

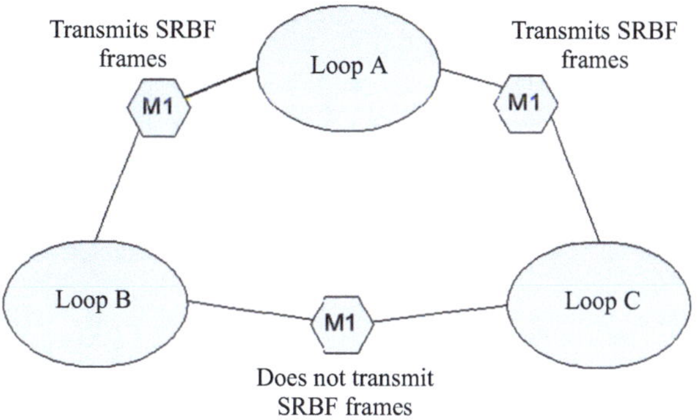

ring and it is unknown through which bridges and rings the path goes to this station. SRBF, spreading through all the rings of the network, reaches the recipient station. In return, the recipient station sends all route broadcast frame (ARBF). This frame is transmitted by bridges through all ports. The source station generally receives several response frames that have passed along all possible routes of the composite network, and selects the best route (usually by the number of intersections of intermediate bridges).

SRT bridges operate as TR if there is no routing information in the frame, and as SR if there are such information.

Bridges are very fast and cheap devices, but in order to ensure transparency, they are forced to transmit all common service (broadcast, multicast) traffic to all network segments. Moreover, when working with segments where there is a different data transmission rate, bridges become a bottleneck because they can't manage it. They are applicable only in small networks (up to 50 users).

A switch is a multiport bridge. It provides frame transmission (cells in the ATM network) from station to station in point-to-point mode. In this case, the stations in the network operate in parallel, that is, the transmission can be conducted simultaneously between all pairs of ports. Switching is carried out by the physical addresses of devices (MAC-addresses). In this case, with the help of special protocols of the third layer of OSI, a lot of network traffic control functions are performed.

There are two types of switching (http://pdst.narod.ru/_20_el_uch/ost_wpd_01):

- store and forward, if the frame transmission rate over the channel exceeds the maximum processing rate, the buffer may overflow, and frames that continue to arrive will be rejected;
- cut-through; switches using this form of switching are called cut-through, and they begin to broadcast a frame to the output port immediately upon receipt of the header, without waiting for the frame to be received. Let us give an example of the switch operation.

Figure 1.36 (http://pdst.narod.ru/_20_el_uch/ost_wpd_01) shows a functional diagram of an EtherSwitch switch proposed by Kalpana in 1990.

Fig. 1.36 Functional diagram of the switch of Kalpana company

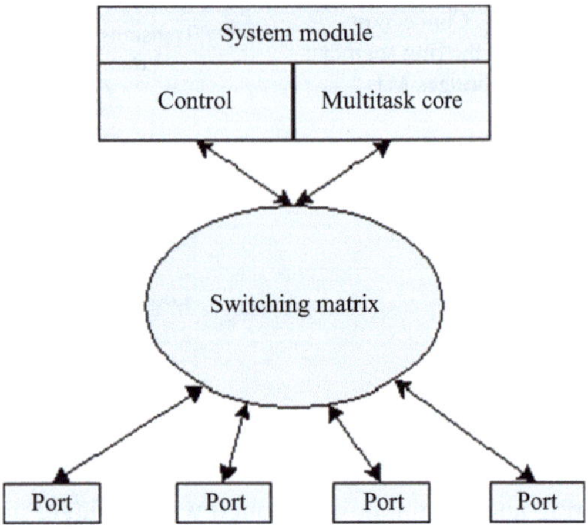

In this switch the system module supports the general address table of the switch. The switch matrix is responsible for forwarding frames between ports. Each port has its own frame processor. When a frame arrives at one of the ports, its processor sends the first few bytes to the buffer in order to read the destination address. After determining the address, the processor makes a frame transmission decision without analyzing the remaining bytes. According to the address table, the corresponding output port is selected. The switching matrix forms the connection of the input and output ports. If the received address is not in the address table, it is recorded in a new line, and the frame is transmitted by broadcasting through all ports, except for the receiving one.

The routers demand separate consideration. A router is a device operating at the OSI network layer and using one or more metrics to determine the optimal path for transmitting network traffic based on the information (packet header) of the network layer. The path is estimated by quantitative indicators, which are called metrics. The best path is the path with the lowest metric.

Based on information about devices in the network (routing tables) and certain rules (routing protocols), the router selects the output physical port for forwarding the packet from the network layer to the receiver or router next on the route.

At the network level, a logical network address is created. This address is assigned by the operating system or by the system administrator simply to identify a group of computers. This group is otherwise called subnet. A subnet may or may not coincide with a physical segment. It shall be remembered that the physical addresses of the devices are set by the hardware manufacturer (using hardware or software). For example, the physical address of the workstation is the unique address of the network adapter, which is flashed by the manufacturer. The physical address database is maintained by Xerox. There cannot be two devices with one physical address in the network. Routers do not see the physical addresses, they send information to the logical addresses of the subnets.

Determining the route of data transmission occurs programmably. Relevant software are called routing protocols. The logic of their work is based on routing algorithms. The simplest routing algorithms determine the route based on the smallest number of transit nodes on the way to the destination. More complex algorithms take into account several indicators, for example, the delay in the transmission of packets, the capacity of communication channels, or the monetary value of communications. The main result of the routing algorithm is to create and maintain a routing table, in which all the routing information is written.

The most well-known dynamic routing protocols that are commonly found on all routers are *RIP* (*Routing Internet Protocol*) and *OSPF* (*Open Shortest Path First*).

RIP uses the hop count through the routers as the metric of the path. The maximum allowed number of hops is 15. The router with a certain periodicity (by default every 30 s) retrieves the addresses of the recipients of information and metrics from its routing table and places this data in update messages sent to neighboring routers. Neighboring routers compare the received data with their own routing tables and make the necessary changes. After that, they themselves send update messages. Thus, each router receives information about the routes of the entire network. RIP protocol can only work efficiently in small networks.

OSPF is a more complex protocol, it refers to the channel state protocols and is oriented to use in large heterogeneous networks. To determine the status of connections, neighboring OSPF routers often exchange short HELLO messages. To distribute link state data, routers use a broadcast message of a different type called *router links advertisement*—an advertisement about router links (more precisely, link state). In this way, OSPF routers ultimately receive status information on all network connections. This information is used to build a network connection graph, which, naturally, is the same for all network routers. In addition to information about the neighbors, the router lists the subnets with which it is directly connected. The calculation of the route with the minimum metric to each subnet is made directly according to this graph using the Dijkstra's algorithm.

Channel state protocols are difficult to implement and need a large amount of memory to store channel state information. The examples of these protocols are OSPF, IS–IS, Nowell NLSP, and Cisco EIGRP. There are many other routing protocols: BGP, IGRP, SNA APPN, NLSP, RSVP, SMRP. Routing protocols are constantly being improved.

Routers allow reaching the recipient using multiple paths, managing traffic based on metrics, filtering traffic for security reasons, working with network segments of various technologies (for example, simultaneously with Ethernet, ATM, and FDDI). But they are not fast devices and are difficult to maintain. Routers in large corporate and global networks are used to connect segments.

Gateway is a device for connecting subnetworks by protocols above the 3D OSI layer. They are used in complex heterogeneous networks. For example, if we want to attach a segment with personal computers representing characters in ASCII code to the mainframe that represents characters in EBCDIC code. There are gateways that convert all seven layers of OSI protocols (usually the hardware at the first two layers and the software at the other layers). However, they can be dedicated only for the

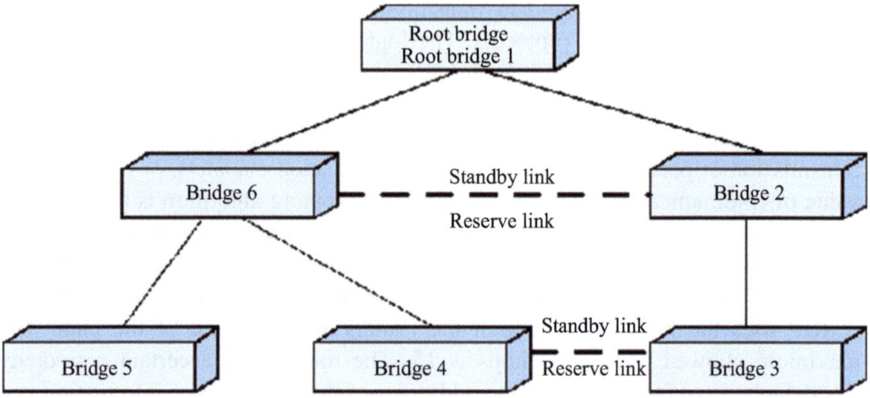

Fig. 1.37 Bridge tree in accordance with the STA algorithm (http://pdst.narod.ru/_20_el_uch/ost_wpd_01)

implementation of the functions of compliance of various protocols to each other, and can also perform other functions. For example, the gateway and simultaneously network file server.

The *Spanning Tree Algorithm* is used to connect network segments.

In order to prevent the loss of network performance in case of failure of the device connecting the network segments, it is necessary to organize backup links between the segments. For example, the same network segment can be reached via three bridges/switches. But then an algorithm that would prevent the presence of closed loop paths should be provided. This algorithm is standardized by IEEE for TR bridges and switches and is called the IEEE 802.1d STA (Spanning Tree Algorithm).

The STA algorithm formalizes the network in the form of a graph, the vertices of which are switches in the network segment (Fig. 1.37). The algorithm provides a search for a tree-like topology of connections with a single path from each switch and from each segment to some dedicated switch (Root Switch) at the minimum possible distance. The switch with the minimum address is selected as the root switch, it is assigned the highest priority. The distance in the STA is metric—a value inversely proportional to the bandwidth of the segment. The metric is the transmission time of one bit, measured in 10-nanosecond units (conditional segment time). For example, for 10 Mbps Ethernet segment, the metric is 10 arbitrary units.

The root switch sends a special "hello packet" to the other switches. The switches relay this packet so that each switch determines the minimum distances from all its ports to the root switch. The algorithm determines which switch or communication between switches is the main and which is backup. The main ones are marked as forward, and the backup ones—as standby. If the primary link or switch fails, they will be replaced by the backup ones. That is why, there is one path for each segment, and the backup paths wait in the wings if the switch fails or the connection between switches occurs.

The algorithm allocates root ports on the switches (this is the port with the shortest distance to the root switch) and limits the number of switches (hops) to 7, and the segments, respectively, to 8.

So, to connect network segments or split a network into segments, specialized devices are used: hubs, bridges, switches, routers, and gateways. Switches are used in high-speed data transmission, and gateways are used to connect networks with different protocols above the third and fourth layers of the OSI. The switches work much faster than routers and, performing the functions of protocols 2 and 3 of OSI, they are comparable in functionality, easier to manage, and are used in networks of all sizes.

1.14 Semiconductor Element Base for Telecommunication Systems of the New Generation

Below we present the results of the analysis of the composition and main technical characteristics of the basic elements of promising TCS on the example of one of the world leaders in this field—the company with the short name "E2V" (France, Gronoble) [10].

This company has been cooperating for many years with technical specialists from the military departments of the USA, England, and Canada, participating in promising R&D and its products are widely used in a variety of telecommunications systems for military and commercial purposes.

A few words about the company itself, more precisely about the "semiconductor" direction.

The e2v semiconductor component division designs and supplies an ADC and DAC with a very high speed of the order of a few GSa/s. This division also produces high reliability processors and memory IS for mission-critical applications, including for avionics, space, and military equipment.

The traditional market for e2v products includes a segment of defense equipment, incl. radar and electronic countermeasures, which require a high sampling rate and a wide bandwidth.

Some other, much more capacious markets require data converters operating at a frequency of several GHz to provide signaling for the implementation of new broadband services. E2v company supplies high-speed broadband data converters for equipment designed to test telecommunication systems. These devices are also used in the network equipment of base stations, 3G/4G, WiMax networks, and in Giga-Ethernet fiber-optic networks.

In many of these markets, powerful QorlQ communication processors from Freescale, Micron, and EVERSPIN TeCHNOLOGIES compete with e2v company in applications that require high processing power or work with large amounts of data—routers, switches, network gateways, controllers for base stations and radio networks, access gateways for LTE networks.

The article [10] describes several new telecommunication applications in which the semiconductor components of e2v semiconductors, which provide a qualitatively new level of solutions, allow expanding the capabilities of projected advanced systems.

If analyze the composition of both the already manufactured and the developed products of this company, we can only formulate global trends in the development of telecommunication systems for the near-term (2017–2020) and medium-term (until 2030) prospects using this example. These trends are graphically presented in Fig. 1.38.

First of all, it is a deep modification of military systems (radar, radar combat equipment), standard radio receivers and radio transmitters, modernization of satellite communications using the principles of direct conversion of HF signals, replacement of standard frequency converters with high-performance broadband DAC, development of the so-called software–hardware systems (Software Defined Radio—SDR), and much more.

The e2v semiconductor components division provides a complete set of military-grade electronic components for signal processing systems, starting from input analog blocks, data converters, data processing blocks, and ending with interfaces.

Figure 1.39 [10] shows the entire range of high-speed data converters and microprocessors for highly reliable systems, which produces a semiconductor branch e2v.

A few words should be said about the features of the use of direct frequency conversion in modern high-speed electronic devices.

So, for example, in modern modems of telecommunication systems an analog-digital architecture, which requires frequency converters to be implemented, is used. However, one high-performance broadband DAC in principle allows replacing a large number of these converters.

The ability of a new generation DAC to directly convert a digital signal in the L, S, and even C (6 GHz) frequency bands can be considered a huge step forward in increasing the system's functional flexibility and reducing power consumption, since these DAC eliminate the need for frequency converters [10] and therefore, they allow to get rid of PLL circuits and VCO jitter (voltage controlled oscillator), which occurs at every frequency rise.

When using a DAC in the gigahertz range to convert with increasing frequency, the effect of spectral overlap (aliasing) gives a positive result. Figure 1.40 [10] illustrates the effect of aliasing of a 100-MHz signal at the output of a DAC with a sampling frequency of 3 GSa/s.

Here, the solid blue line corresponds to the sincX DAC type characteristic, which operates in Non-Return to Zero, NRZ, coding mode with zero at the sampling rate. The 400 MHz signal is clearly visible in the first Nyquist zone. In the second, third, and fourth Nyquist zones, the signal is also reflected. To eliminate these reflections (responses), special filtering of the signal is applied near the specified frequencies.

Table 1.10 [10] shows the data for comparison of a typical DAC operating in the gigahertz range with a standard wideband IQ modulator.

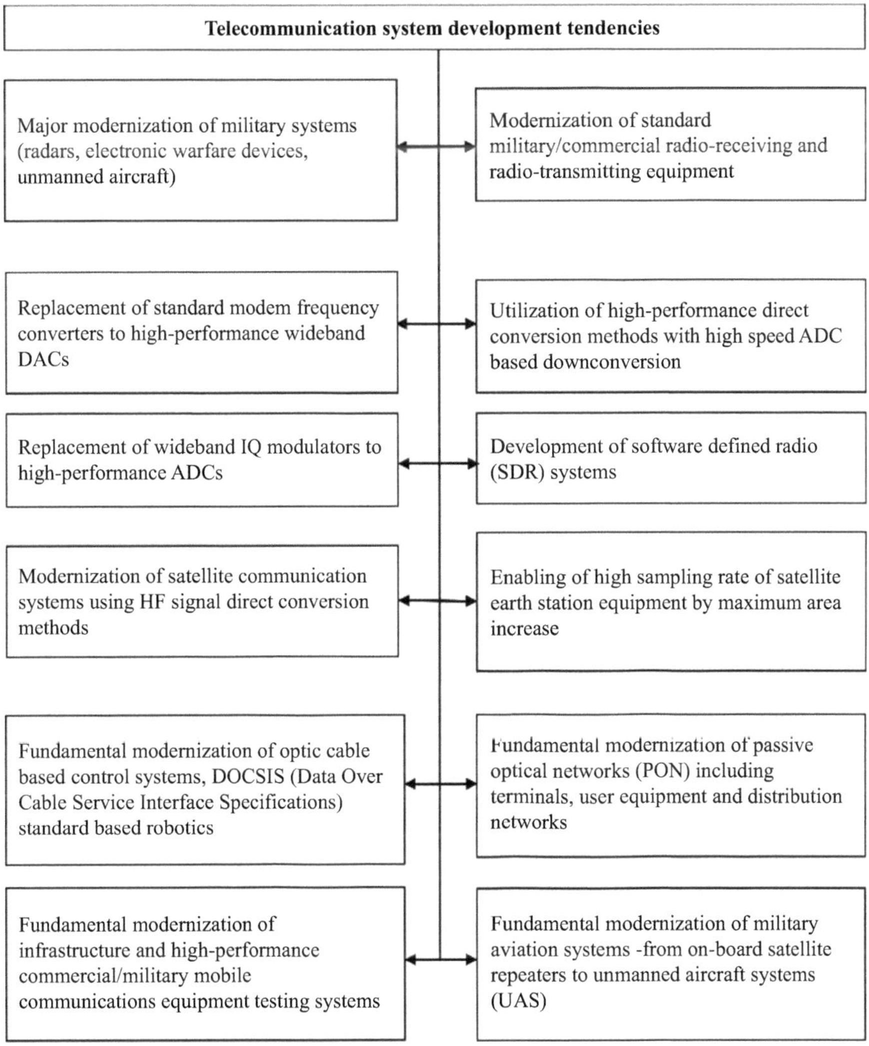

Fig. 1.38 The main trends in the development of telecommunications systems are the near (2017–2020) and long-term perspective

It can be seen that in many cases the DAC works better than the fully analog component, and also has a compact case of 15 × 15 mm.

The replacement method shown in Fig. 1.41 [10] is also used for down conversion. In this case, it is necessary for the ADC to have sufficient performance to lower the required frequencies.

As it is well known in the traditional scheme of building standard radio receivers, one or several frequency reduction cascades using superheterodyne are usually used. In the case under consideration [10], a local oscillator and a frequency reduction

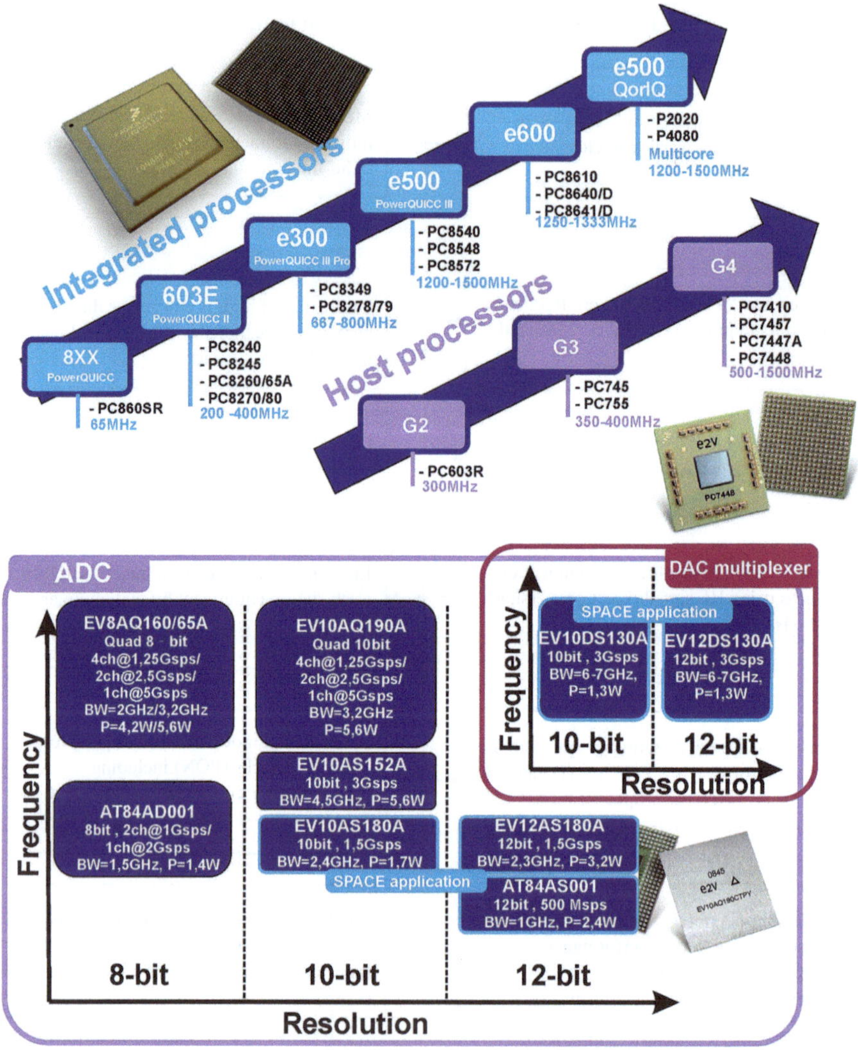

Fig. 1.39 Nomenclature of data converters and microprocessors of the e2v company [10]

circuit are used (see Fig. 1.42). As a result of the conversion, signals appear at the frequencies $F_{LO}\,{}^{-}F_{IN}$ and $F_{LO}\,{}^{+}F_{IN}$. The signal with a higher frequency (and higher harmonics) is removed by filtering, which is facilitated by the ADC unit at an intermediate frequency.

Developers of the element base in recent years carry out the active work in the direction of direct signal conversion with frequency dip.

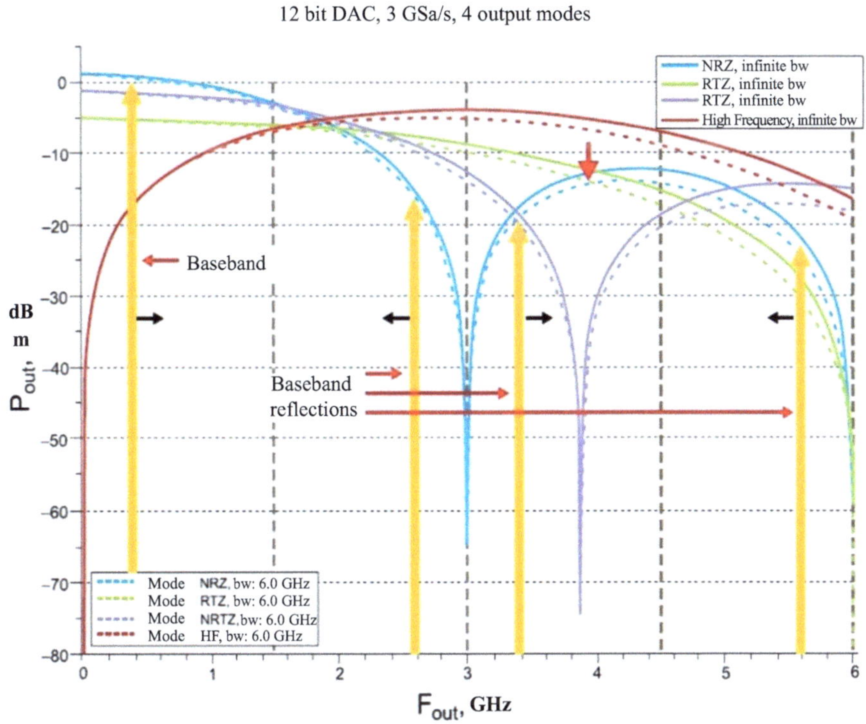

Fig. 1.40 Typical reflections (replicas) that occur at high frequencies

Table 1.10 Comparison of the characteristics of the DAC with high sampling rate and IQ-modulator

Output signal with 3 GHz frequency	EV12DS130A	IQ modulator
Harmonic distortions	−66 dBm (SFDR)	−55 dBm (third harmonic)
Adjacent Channel Power Reduction (ACPR)	59 dB	60 dB
Inherent noise level	−160 dBm/Hz[a]	−157 dBm/Hz[a]
Gain change	0.02%/°C	n/a
Power dissipation	1.3 W	1.5 W[b]

[a]Only constant bias voltage was measured. The actual value for the EV12DS130A DAC, obtained by measuring the noise power ratio, is −142 dBm/Hz

[b]1.5 W—for one modulator only; in most cases, a DAC is also used to generate a signal in the main frequency band

According to the Kotelnikov theorem, for sampling an analog signal, it is required that the sampling frequency should be at least twice the maximum frequency component in the signal. This requirement eliminates spectral overlap, ensuring full recovery of the input signal. Figure 1.43 shows that in the process of sampling signals with a higher frequency can contribute to the main band, since the ADC works as a down converter.

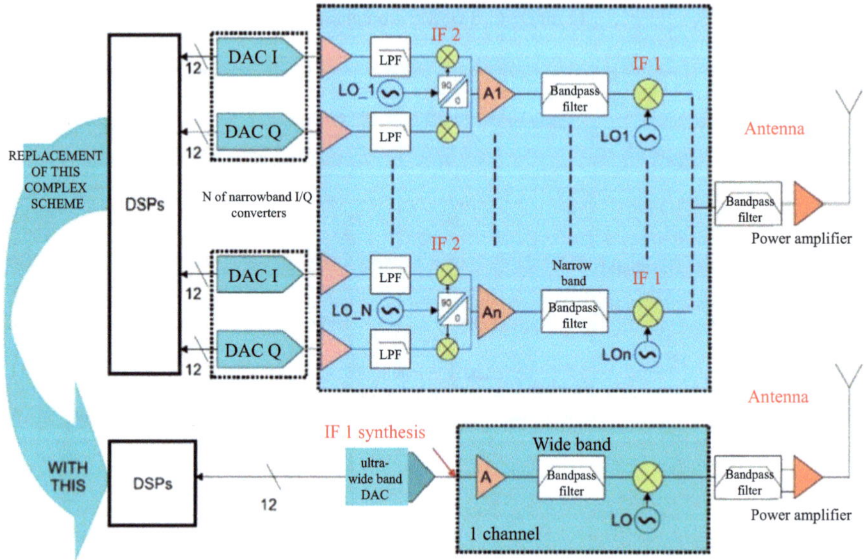

Fig. 1.41 Replacing multiple DACs with one broadband DAC

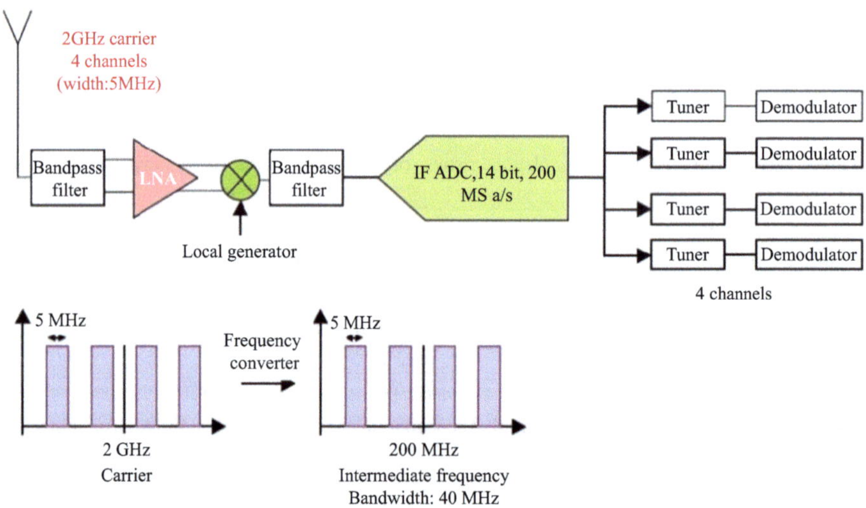

Fig. 1.42 Conventional equivalent frequency dip circuit

As the sampling frequency increases, the distance between the Nyquist zones increases, which greatly simplifies signal filtering. It is known that with a very small sampling frequency between zones, interference occurs. In practice, this effect is used to directly convert the signal with a decrease in frequency in receivers

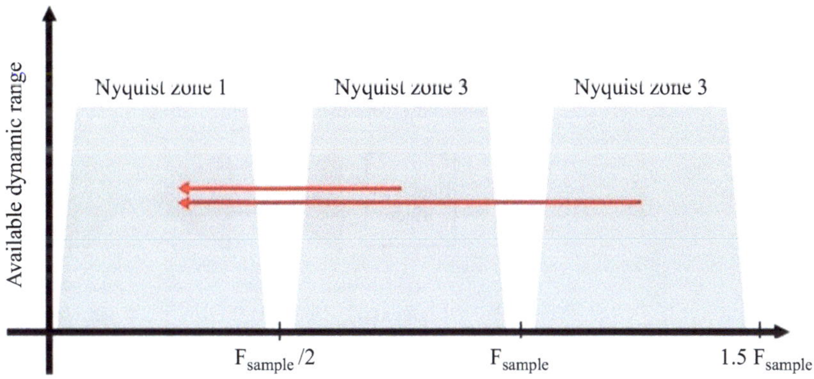

Fig. 1.43 Nyquist zones

and transmitters, where higher frequencies are intentionally introduced into the main band.

Software defined radio systems (SDR) use transceivers that perform up-and-down conversion between baseband and high frequency carrier exclusively in the digital domain. As a result, the HF interface is reduced to a power amplifier in the transmit path, a low-noise amplifier (LNA) in the receive path, and a small analog filtering circuit (which may be missing).

In such a promising radio system, it is possible to update or completely change functions only by updating the software. Due to the recent increase in the number of different modulation standards for TCS customers, there is a growing interest in the ability to change the functions of a radio system using only special software, without changing the hardware of the system. For example, the US military program JTRS (Joint Tactical Radio System) is designed to provide 33 radio carrier modulation methods in the 2 MHz–55 GHz range within one radio system.

The possibility of creating such a radio system previously was not feasible due to the lack of appropriate HF-digital devices that would convert the signals on carrier frequencies directly into digital data. However, with the advent of new methods and high-speed semiconductor chips, the creation of a software-defined radio system becomes more realistic. A simplified structure of such a software-defined radio system is presented in Fig. 1.44 [10].

The obvious advantages of such an SDR system include: greater flexibility due to software channel separation; the absence of component drift, as well as the reduction of their number, which improves the cost indicators of the system and increases its reliability. Moreover, the need in the inverter operating by a specific frequency scheme, and use multiple external interfaces to different frequency bands are excluded. All that is required to work with a new carrier is to select the appropriate sampling rate and use the appropriate bandpass filter.

Other advantages include a high frequency hopping rate due to the absence of a local PLL generator. In addition, due to the software definition of functionality, it is

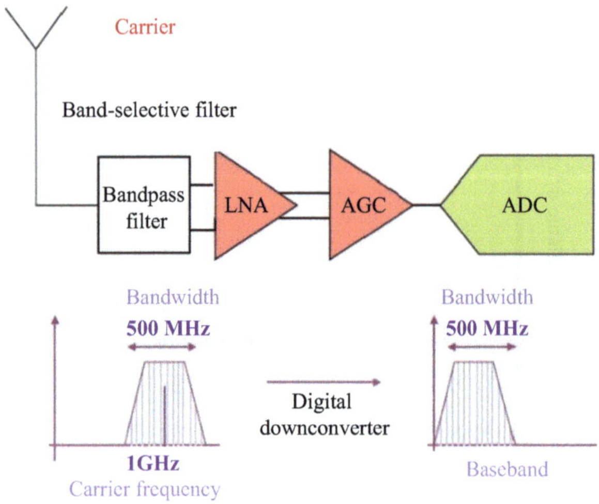

Fig. 1.44 Simplified block diagram of a software-defined radio system

possible to perform the update in the field. However, the use of high-speed ADC entails the need for certain trade-offs related to throughput, sampling frequency, dynamic range, resolution of the ADC, and power consumption. In particular, since high-speed analog-to-digital converters consume more power, SDR systems are mainly used in base station applications.

One direction, the use of direct conversion—is a practical application in the field of satellite communications.

In modern satellite telecommunications systems, a number of transducer stages are commonly used to process HF signals. Previously, the input signal in such systems passed through at least two down-stages installed before the ADC. However, thanks to the use of a high-performance data converter, which ensures the appropriate quality of digital signal processing, it has recently become possible to get rid of one or more conversion stages and, therefore, to get a cheaper telecommunications satellite system with high flexibility.

Figure 1.45 [10] shows simplified structural diagrams of the organization of the satellite signal processing system and the basic communications equipment of high-performance satellite repeaters.

It should be noted that the output signals of the DAC EV12DS130A from e2v are of high quality up to the lower limit of the C-band (5 GHz). These transducers are also available for various space communications systems. Usually, DAC EV12DS130A, in combination with the ADC EV10AS180, is used in high-quality transceivers, where there are no down/up conversion stages, which significantly reduces the cost of such satellite communication systems.

In addition, in space communication systems with high computational power, which include applications for communication routing, weather monitoring, and synthetic aperture radar, recently the processors P7448, P8548 from e2v have been used. These first space processors operating at gigahertz clock frequencies will

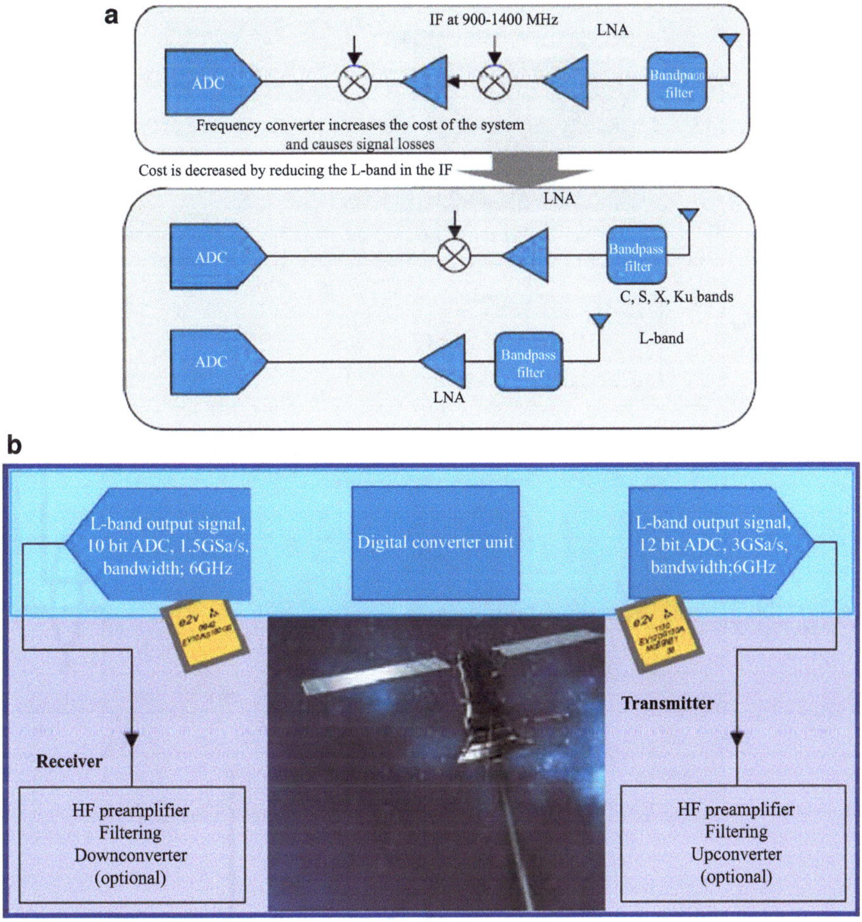

Fig. 1.45 Block diagrams of satellite signal processing system (**a**) and onboard communication equipment (**b**) satellite repeaters

significantly transform the design process of satellite advanced communication systems.

To make the equipment of ground stations of satellite communication work with a high sampling rate, it is necessary to maximize the Nyquist zone (instantaneous bandwidth) and to provide a fairly wide bandwidth of the analog input signal for down-conversion from the L-band.

In addition, it is necessary that the parasitic harmonics were small, because for this connection, multi-stage averaging is required to reveal signals from, for example, space probes for the study of near and far space. The network interface also requires a high-performance processor that is capable of operating at extreme temperatures.

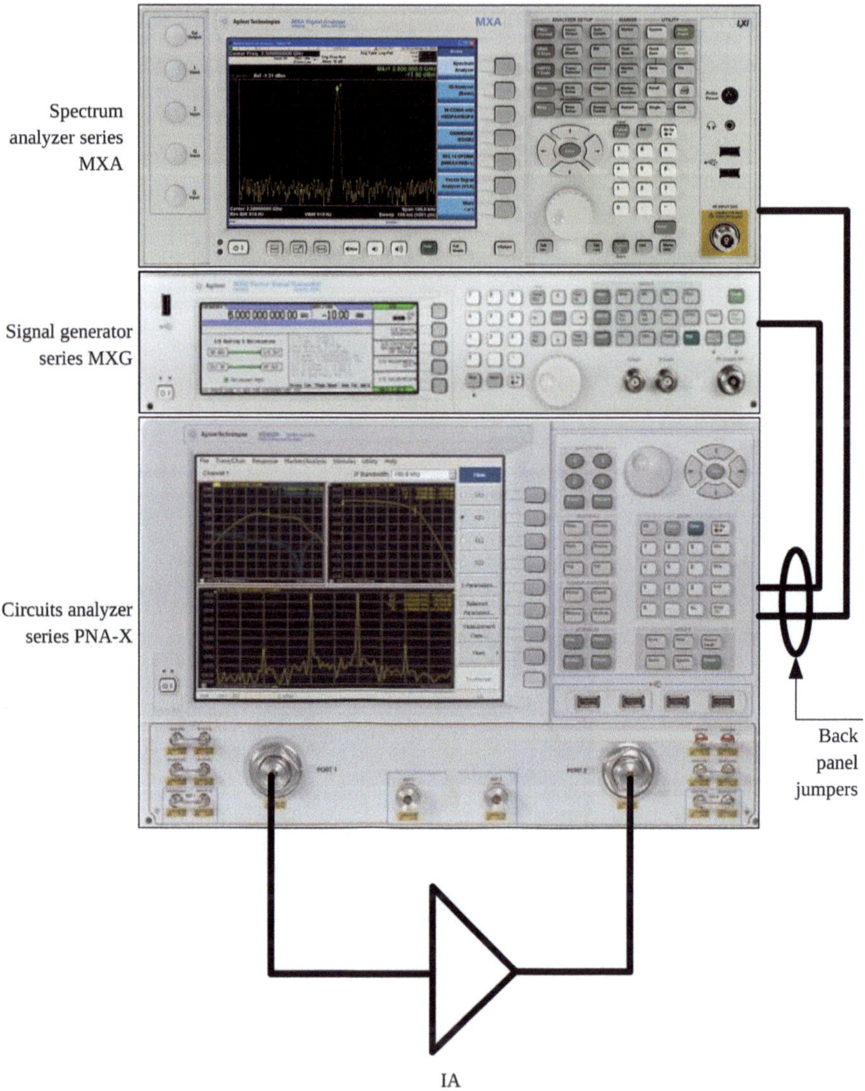

Fig. 1.46 Typical distribution of power consumed by a terrestrial satellite communications station

Currently, the equipment of terrestrial satellite communications stations uses such e2v products as EV8AQ160, EV10AS150, and EV12AS200, whose noise floor is −152 dBm/Hz (see Fig. 1.46) [10]. In network routers that operate at low temperatures (−55 °C), PC8548 and P4080 processors are widely used.

One of the important components of telecommunication systems is their cable control systems and passive optical circuits.

- Developers of telecommunication systems and devices are well aware of such a document as DOCSIS (Data Over Cable Service Interface Specifications), a standard for transmitting data over a coaxial (television) cable, which very precisely defines the requirements for noise characteristics. For the design of such DOCSIS systems, modulators with the following characteristics are required:
- Nyquist zones far exceed 1 GHz;
- linear and flat frequency response in the frequency range of the output signal 50 MHz–1 GHz;
- large dynamic range of 12 bits or more;
- small harmonics H2, H3 and low intrinsic noise;
- high isolation from adjacent channels at central frequencies $F_{out} = 50–950$ MHz.

The DOCSIS specification has very strict requirements for supporting multiple channels of analog video signal. In particular—the power of the adjacent channel should be better—73 dBc.

It should be noted that the extended version 3.1 of the DOCSIS standard requires that the downlink output signals have a frequency of 1700 MHz (see Table 1.11) [10]. For this, it is necessary that the sampling rate be at least 4 GSa/s, taking into account the "tails" remaining after filtering, unless a special architecture with a boost converter is used. Figure 1.47 in schematic form shows the algorithm of DOCSIS.

Table 1.11 Comparison of DOCSIS 3.0 and 3.1 specifications

Parameter name	DOCSIS 3.0	DOCSIS 3.1, Phase 3
Downlink frequency range	108–1002 MHz	500–1700 MHz
QAM-modulation level in the downlink	256	1024 and more
Downlink capacity	1 Gbps	10 Gbps

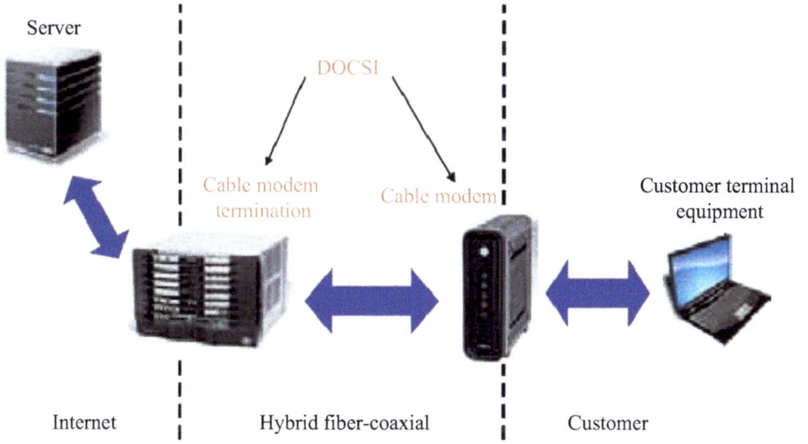

Fig. 1.47 DOCSIS network operation principle

Table 1.12 DOCSIS 3.0 specification for parasitic interference and noise

Band	Number of channels			
	1 (dBc)	2 (dBc)	3 (dBc)	4 (dBc)
Adjacent channel	−58	−58	−58	−58
1st bypass	−62	−60	−60	−59
2nd bypass	−65	−64	−63	−62
3rd bypass	−73	−70	−65	−64
Other channels	−73	−70	−67	−64

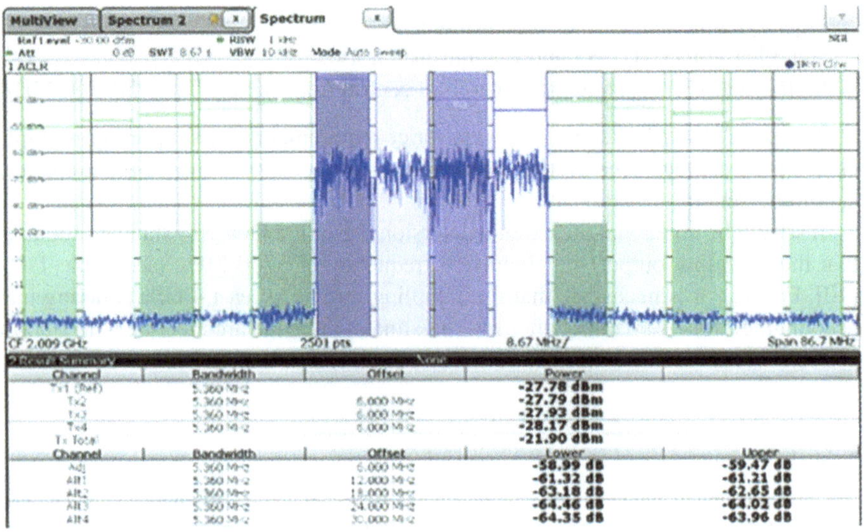

Fig. 1.48 EV12DS400 generated four-channel QAM signal at 2 GHz

Another widely used communication standard, IEEE 802.ad Gigabit WiFi, which is an addition to the 802.11 standard, provides for wireless communication in the frequency band up to 60 GHz. This specification defines a channel bandwidth of 50 GHz (50 times wider than the existing standard), allowing for increased bandwidth.

It should be noted that in order to comply with these requirements, a DAC with a sampling frequency higher than 4 GSa/s is required. DOCSIS 3.0 requirements for parasitic interference and noise are presented in Table 1.12 along with a four-channel QAM-modulation scheme, generated taking into account the fact that the sampling frequency is 4.5 GSa/s, and the output central frequency is 2 GHz. It can be seen that this feature is in line with the requirements of DOCSIS 3.0. Figure 1.48 [11] shows the picture of the EV12DS400-generated four-channel QAM signal at 2 GHz.

It should be noted that the EV12DS400 DAC chip can generate a signal without using internal interpolation filters. As a result, the attenuation becomes uneven in the passband, it increases the phase shift, and is limited by the amplitude frequency response of the system.

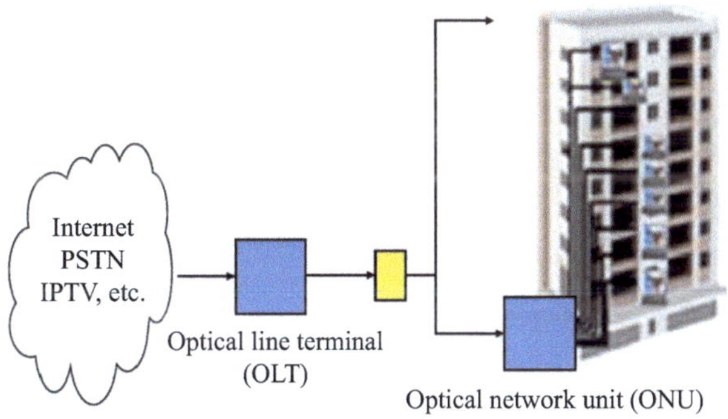

Fig. 1.49 Typical GPON networking

As is known, a passive optical network (PON) is a node in a multipoint network that uses passive optical splitters to service different premises. The existing systems operate at frequencies of 10 and 40 GSa/s, and to ensure optimal performance, domestic and foreign developers use different multiplexing methods—WDM, CDM, TDM.

A typical passive optical circuit usually consists of a terminal, a subscriber unit, and corresponding distribution network (see Fig. 1.49) [11].

There are two standard methods for converting data from these devices: using a high-speed converter with a low resolution or increasing channel efficiency with QAM-modulation; and using a converter with a higher resolution but with lower speed.

The method of using the EV10AQ190 converter in systems with QAM and OFDM modulation, in which the sampling rate reaches 13 GSa/s, is described in [11]. Since only the 2 GHz band is used in these systems, the data transmission rate is easily increased. Systems are flexible enough to adapt to changes in the subscriber line.

These systems require a high sampling rate and good resolution to generate and decode signals with a small difference in amplitude and phase, which are used in the modulation scheme. In these systems, a quad 10-bit AD10 EV10AQ190 with a frequency of 5 GSa/s and a 12-bit DAC EV12DS130 with a frequency of 3 GSa/s are used.

Another obvious problem should be noted: as the number of standards supported by mobile subscriber equipment increases, the production tests of such devices and related components become more and more complex. To test these systems and work with different modulation schemes, the test equipment used shall have sufficiently high flexibility. In addition, since such new standards as 5G LTE are not yet fully defined, even at the development stage of these systems it is required to ensure their high degree of flexibility.

Today, the frequency spectrum of wireless communications is determined by many effective standards, which include the following most widely used: WCDMA (Wideband Code Division Multiple Access) (V) for a frequency of 880 MHz; GSMDCS (Global System for Mobile Communications, Digital Cellular System) for the frequency of 1.82 GHz; WCDMA (I) for the frequency of 2.12 GHz;

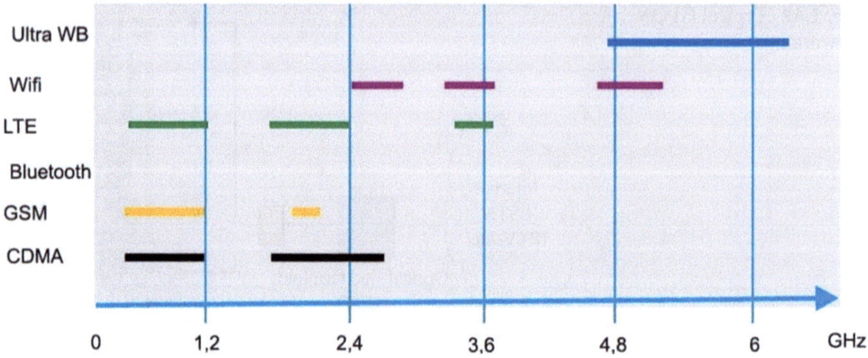

Fig. 1.50 Standard frequency spectrum allocation for various communication protocols

Bluetooth for the frequency of 2.4 GHz; WiMAX for the frequency of 3.5 and 5.8 GHz; and IEEE 802.11a for the frequency of 5.2 GHz.

Figure 1.50 shows the scheme with the distribution of the frequency spectrum.

Under actual operating conditions, when multiple signal re-reflections occur, MIMO (Multiple Input Multiple Output) systems are often used. These systems increase the signal-to-noise ratio using a number of antennas, transmitting one message over all paths or simultaneously separate parts of the message with different antennas, which increases the bandwidth of MIMO applications. It is expected that this method will be included in the standard being developed for 5G applications.

The many antennas used can increase system capacity and service quality, eliminating the need for increased transmission power or bandwidth. However, these devices require appropriate test systems to simulate different operating conditions.

The ability of wireless test equipment to generate or digitize signals in wide frequency bands becomes very attractive in terms of flexibility and compliance with system performance requirements. It is expected that the 5G standard will provide low latency, primarily in signal processing systems for solving critical tasks. For this purpose, a high decoding speed at the receiver side and a small delay in the Rx circuit will be required if we are talking about systems with automatic gain control (AGC). In addition, due to the large number of processing stages that make up the converter, problems may arise with synchronization between several converters, which adversely affect the operation of the M1M0 systems.

Requirements to provide high flexibility and work with different frequencies lead to the need to use multiband systems for direct conversion as a basis for creating universal test equipment. For example, a data converter with a large instantaneous bandwidth can operate with frequencies of different magnitudes at the input or output of the system. If we use a multi-standard receiver/transmitter, installing several different receivers with support for different standards in one receiver, the occupied area and power consumption of such a system will increase significantly [12].

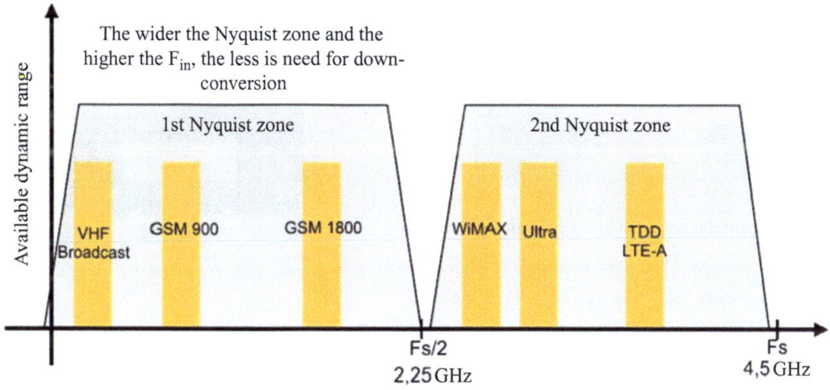

Fig. 1.51 The case when one data converter supports different communication standards

Table 1.13 Frequency ranges and sensitivity values defined by different communication standards [13]

Standard	Frequencies	Bandwidth	Sensitivity, dBm
GSM	869–1990 MHz	200 kHz	−102
UMTS	1710–2110 MHz	5 MHz	−117
WLAN 802.11b	2400–2485 MHz	14 MHz	−104

Obviously, the use of a data converter with a high sampling rate (in this case equal to 4.5 GSa/s) ensures operation with several frequency bands within one Nyquist zone (see Fig. 1.51). This means that replicable equipment is required for each communication standard (external HF interfaces, down converters, and demodulators), since all the relevant functions can be performed in the digital domain. It should be noted that this conclusion applies to the sending and receiving part of the equipment. Table 1.13 shows the frequency ranges and sensitivity values defined by different communication standards.

Due to the fact that the signal bandwidth is small, usually the effective SNR value is significantly reduced at the receiver with digital filtering. Figure 1.52 shows that the self-noise level of the AD12 EV12AS200A is 152 dBm/Hz; this value corresponds to SNR = 80 dB (13 ENOB) for a bandwidth of 10 MHz.

Thus, a modern high sampling rate data converter is capable of operating in many bands due to the wide instantaneous bandwidths, while ensuring the appropriate SNR value when the signal is restored.

Soon, wireless test systems will support different communication standards using data converters with a large instantaneous bandwidth (see Fig. 1.53) [10].

Here it is necessary to note the existing specific problems of the development of military aircraft systems.

As is known, communication systems used in severe environments require highly reliable processors and storage devices. Among such systems are those that provide data transmission channels for unmanned aerial vehicles (UAV) (see Fig. 1.54) [10].

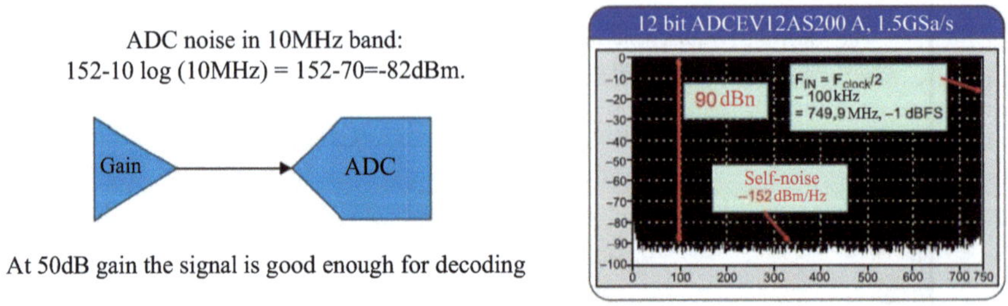

Fig. 1.52 Noise characteristic of EV12AS200A

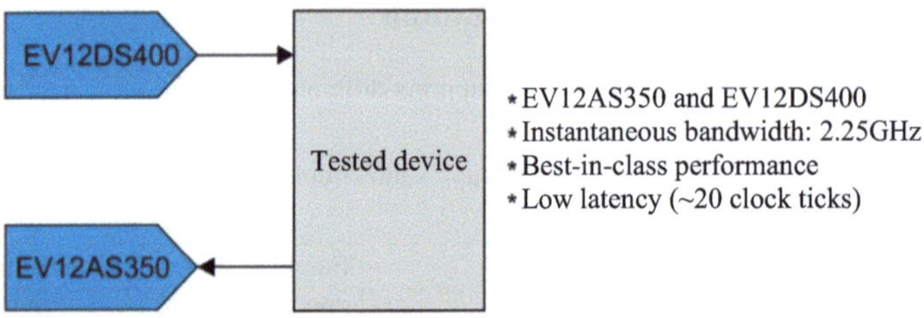

Fig. 1.53 An example of building a universal wireless test system

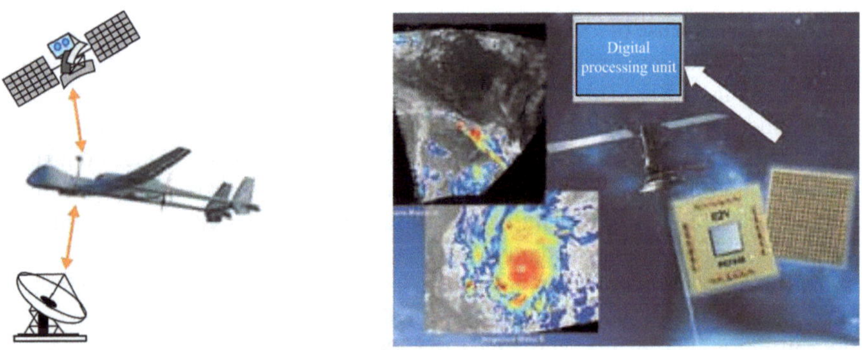

Fig. 1.54 The use of highly reliable processors in the communication systems of UAVs and satellites

In 2016, e2v launched the PCS-7448 and PC8548 microprocessor, designed for operations requiring high computing power. In particular, these processors are used in the onboard communication equipment of satellite repeaters (see Fig. 1.54).

The use of these processors will allow processing and compressing data directly onboard the satellite, which will greatly simplify the problem of transmitting signals from space to ground stations. In addition, the processors will allow creating highly efficient space communication networks between satellites.

References

1. Bogachev, L., Korchak, V., & Tuzhikov E. (2014). Initiative of the US defense ministry in the field of defense innovations. *Electronics NTB* 6, 136–144.
2. Pellerin, C. (2014, November 15). Hagel announces new defense innovation, reform efforts. *DOD News*.
3. Richmond, J. (2014, November 24). Work explains strategy behind innovation initiative. *DOD News*.
4. Memorandum dated November 15, 2015, from Secretary of defense chuck hagel to the deputy secretary of defense and other DOD recipients on the defense innovation initiative. http://www.defense.gov/pubs/OSD013411-14.pdf
5. Brimley, S. Offset Strategies L Warfghting Regimes. *War on the Rocks.com*. October 15, 2014, http://warontherocks.com/2014/10/offset-strategies-warfighting-regimes
6. Perry, W. J. (Fall 1991). Desert storm and deterrence. *Foreign Affairs 70*(4), 66–82.
7. Statement by Dr. William B. Roper. Jr. Director, Strategic Capabilities Office. (2014, April 12). *Strategy and implementation of the Department of Defense's technology offsets initiative in review of the Defense authorization request for fiscal year 2017*.
8. Keller, J. (2015, January 14). BAE systems to develop MAD ASW drone to help Navy P-8A find submarines from high altitudes. *Military Aerospace Electronics Unmanned Vehicles*.
9. O'Rourke, R. *Navy Virginia (SSN-774) class attack submarine procurement: Background and Issues for Congress*.
10. Glaskott-Jones, A., Chantier, N., Stuckler M., & Frass Sombe, S. (2012). Future communication systems based on E2V semiconductors. *Electronic Components*, 1, 17–24.
11. King, Z., et al. *Demonstration of software-defined flexible-PON with adaptive data rates between 13.8 Gb/s and 5.2 Gb/s supporting link loss budgets between 15dB and 35dB*. Huawei Presented at ECOC2014 Paris.
12. Garcia Oya, J. R., Kwan, A., Chavero F. M., Ghannouchi, F. M., Flelaoui, M., Lasso, F. M., et al. (2017). *Subsampling receivers with applications to software defined radio systems*. Intech Publication.
13. Spiridon, S., Spiridon, F., Dan, C., & Bodea, M. (2011). Deriving the key electrical specifications for a multi-standard radio receiver. In *Conference: COCORA 2011*.

Chapter 2
Wired Interfaces of High-Speed Electronic Devices

Most up-to-date telecommunication systems are built on the bus-modular principle: the main units of the system are formed as complete electronic modules with available functions that are combined into the necessary configurations using a system of interface communication channels. Communication channels are divided into two types—wired and wireless. This chapter describes only wired interfaces.

In Sects. 2.1 and 2.2, the organization methods and the basic parameters of such interfaces are discussed.

Section 2.3 considers the methods of calculating the electric, dynamic, and energy parameters, as well as various design options for the implementation of wired interfaces.

Section 2.4 is briefly discussing the features of the organization of the process of high-speed data transmission using transceivers with a topology of SerDes type. The need to use high-speed SerDes chips in TCS (serializers/deserializers or parallel-serial or serial-parallel conversion devices) is due to the actions of the well-known Moore's law. As you know, the easiest way to transfer data is to directly connect each bit of one chip to the similar input/output bit of another chip. However, the density of placement of the pins (contact pads of the chip) did not increase at the same pace as the density of the transistors on the chip. Therefore, today's SerDes chips allow transferring serial data between chips on a printed circuit board with a speed of up to 10 Gbit/s and above.

Section 2.5 discusses the design features of high-speed buses of standard DDR memory modules.

Sections 2.6, 2.7, and 2.8 contain the analysis of the features of the organization of high-speed data transmission channels using standard RS-485 interface protocols, both synchronous and asynchronous channels, and describe the schematic methods for protecting front-line buses from electrical overloads (overvoltages).

Section 2.9 discusses the features of designing high-speed channels based on programmable logic ICs.

In the final Sect. 2.10, practical examples of the organization of the data transfer interface based on the LVPECL standard are considered.

A. Belous, V. Saladukha, *High-Speed Digital System Design*,
https://doi.org/10.1007/978-3-030-25409-4_2

2.1 Communication Lines as a Component of the Interface

Most modern high-speed digital electronic information processing and transmission systems are built according to the modular backbone principle of organization, in which individual units of the system are formed as complete modules with their specific functions that can be combined on a printed circuit board into the configurations needed by the developer using a wired or wireless communication system The basis of such a trunk-modular organization is aimed at achieving the ultimate characteristics of the highest possible level of standardization of the elemental base of the internal blocks of modules and individual components of "intra-module" and "inter-module" links, as well as an extensive system of information, address and control buses. The high performance of such systems is provided by the specific characteristics of the used circuit-based basis of the components that make up the complete modules of the electronic system. At the same time, with regard to the choice of the type of element base used, for "intramodular" components with short lines of communication, relatively low output load capacities, a small level of generated noise are characteristic, it is more efficient to use a CMOS circuit-based design, which provides rather high functionality of such modules. "Intermodular" components are characterized by "long" communication lines and a high level of interference; therefore, to ensure high load capacity, the most effective is the use of a bipolar circuit-based design (already "outdated" ECL, still existing transistor–transistor logic Schottky, I^2L-circuitry), or a combined technology BiCMOS—depending on the required data transmission rate.

In modern high-speed digital information processing and transmission systems, communication between various functional modules is carried out using special communication lines, called "buses." Hereinafter, the following terms are used to characterize the interface bus [1].

Interchange circuit is a physical medium designed to transmit information between items of equipment that take part in information exchange, including data, control signals, and synchronization.

Data Transmission Channel is a set of physical media and hardware, including equipment for converting signals involved in the process of transferring information between equipment.

(Driver):

(a) an electronic circuit or a relay contact (source) on the transmitting side of the interchange circuit through which binary digital signals are transmitted to the terminal load through a connecting cable;
(b) a binary digital signal transmitter.

Terminator:

(a) an electronic circuit (a consumer at the receiving side of the exchange circuit), by means of which binary digital signals are received from the builder via a connecting cable;
(b) a receiver of binary digital signals.

The communication line is a component of any interface. Indeed, any telecommunication system is a set of complete functional hardware units and individual elements interconnected in accordance with the required architecture of the electronic system through a system of communication lines (buses) designed for high-speed transmission of information between the various units of this system. If the bus interconnects the internal functional units of the system, then it is called the *internal* bus. An example of such internal bus is the communication bus between controller and internal registers of the central processor. The electrical and functional characteristics of such a bus are important to chip developers, but they are of absolutely no interest to users.

The bus interconnecting the main components of the computing system, such as, for example, the central processing unit (CPU) and the RAM, is called *external*. Knowledge of the basic technical characteristics of external buses is necessary for the proper construction of the information and communication system of the architecture given by the customer. Usually, external buses are divided into three groups: *data bus* (for information transmission), *address bus* (for specifying the source and direction of information transmission), and *control bus* (regulating the sequence of actions on other buses). A collection of data, address, and control buses is called a *system bus*. As it follows from its definition, the system bus connects the central processor with memory and I/O devices.

In a broad sense, the interface of a telecommunication system is understood as the combination of a system bus and other circuits necessary for connecting various components to it, including the bus control logic. The presence of various interface chips simplifies the task of interface design and advanced bus control logic, especially since modern processors and microcontroller chips are characterized by partial (sometimes even complete) placement of interface functions and bus control logic directly in the crystals of these chips.

I/O interfaces, for example, shall receive data from the system bus, store and output it to the system bus, exchange information with the CPU, receive and execute commands from the CPU, and transmit information to the CPU from external connected devices.

External memory interfaces interact directly with the memory by controlling the system bus.

Consider further the main parameters of such buses and the principles of the organization of their work on the example of some standard technical solutions [1].

2.2 Organization Methods and Basic Parameters of Interface Buses of High-Speed Electronic Devices

A variety of applications of modern high-speed electronic systems—from household and industrial devices to space systems—cause a variety of modifications to the system bus. At the same time, the degree of applicability of these system buses is connected not only with the areas of their application and their technical parameters, but also with some purely economic factors. Taking into account that consideration of the principles of organization and technical characteristics of specific system buses is not the main purpose of this chapter, let us confine to a brief listing of them and take a brief excursion into their history—let's evaluate the applicability of the buses of the previous generation of digital devices [1].

Thus, AT-96 and Industrial PC buses were not widely used in the past (last century), they were distributed only in Germany, the Futurebus, despite its advanced technical characteristics, was also not widely used, many buses of brands NuBus (Apple), Micro channel (IBM), and Q-bus (DEC) are also not widely used due to limited use in industrial systems.

The STD bus is one of the oldest microcomputer buses; however, it was fairly widespread, especially in the USA (STD-32).

The ISA bus was the most common bus in the world and is still used in industrial computers in some places. The PC-104 bus is a modified version of the ISA bus, although their electrical characteristics vary significantly (the load capacity of the PC-104 is reduced to 4 mA). Despite the fact that the PC-104 bus market was already quite narrow, the PC-104 is sometimes still used as an expansion card for single-board industrial computers.

The VL-Bus quickly lost its popularity at the same time as Intel-484 processor, which was well known at the time, left the market. The EISA bus was occasionally used in high-performance equipment, but then it was completely excluded out by the PCI bus.

The Multibus II bus, considered at one time as a competitor to the VME bus, also lost popularity due to Intel's failure to support it.

The VME bus in the middle of the last century was widely used for industrial and other special applications, but the basic principles of the VME bus, introduced by Motorola during the active development of TTL-IC type circuits in the 1970s, were a brake for its development. Despite the emergence of an enhanced version of VME-64, the share of products with a VME bus on the market of high-speed electronic systems by the end of the last century actually approached zero.

The PICMG (Association of Manufacturers of Equipment with PCI bus) has developed the specification for Compact PCI bus.

The PCI architecture also had a major impact on the Card Bus, a further development of the PCMCIA standard. It should be noted that the applicability of the newest high-speed buses up to 2000 was often not necessary and in 90% of cases, a bus with a throughput of just up to 1 MB was enough to exchange with the analog and digital input equipment.

Data Width, Address Range Central processors (microprocessors), depending on the target destination, usually process data with width of 8, 16, 24, 32, or 64 bits. Obviously, it is not economically feasible to choose a bus that is optimized for 32–64-bit processors, if the system is intended mainly for work with 8-bit devices. Conversely, if the system is currently working with 8-bit devices, but in the future it can be transferred to work with 32-bit processors, it is necessary to take measures to ensure that the bus used will provide adequate support for 32-bit devices.

Component-level buses typically contain only 16–20 address lines, which provides a total amount of address space in the range from 64 KB to 1 MB. Although this range may be sufficient for the implementation of small process control systems (for example, machinery equipment), it is not enough for many modern applications, especially those related to the use of graphical tools or the maintenance of several users. Therefore, for any multi-user general-purpose system, it is advisable to allocate at least 24 address lines. And for high-performance systems processing large amounts of data, required, in particular, when working with graphics or for image analysis, it may be necessary to use 32 address lines.

Data Transmission Rate and Protocols To ensure reliable data transmission between two functional modules, it is necessary to follow a certain set of rules, called an exchange protocol, which should provide a solution to three main problems.

The first problem is that when data or an address is transmitted over several parallel lines, a single signal (strobe) of the system's main device is passed to the bus, the slave device does not receive all signals addressed to it simultaneously. Such a state in practice is called "skew." The second problem is that in practice there is a finite duration of the rise time of various signals and the achievement of the switching threshold of the bus receivers. The third problem is the need for coordination in time (synchronization) of the work of the master and slave devices. These include, for example, such questions: how to ensure the interaction of the fast master device, first with the fast and then with the slow slave; what situation occurs if the call is made to a non-existent slave device or other errors occur?

There are two fundamentally different approaches to solving the above problems. The first involves the use of *synchronous*, and the second—*asynchronous* exchange protocol. The easiest way is to use the synchronous protocol (Fig. 2.1a), providing for the commit and wait cycles.

In this case, the data arrive on the parallel bus lines, and then after a delay of sufficient magnitude to eliminate the maximum expected bias that can occur in the system, a control (strobe) pulse is generated. The slave responds to it by performing the requested operation. If the slave device has less speed than the master one, it generates a commit impulse. Thus, the slave device sends a request to the master to form the wait cycles until the slave device completes the operations it performs.

However, the performance of all synchronous systems is limited by two factors. Firstly, after a certain value of the period of clock pulses is established, it can no longer be changed. This means that it is impossible to realize the advantages of faster schemes appearing on the market in the system, as a result of which the maximum performance of the system for the whole period of its operation remains fixed at some fixed level. Secondly, slow devices shall be synchronized. This means that a commit pulse shall be generated very quickly (which requires high-speed logic), so

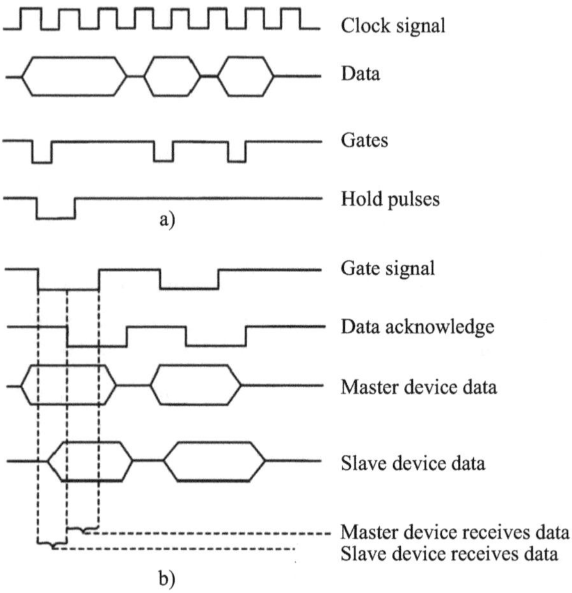

Fig. 2.1 Interface bus exchange protocols: (**a**) synchronous, (**b**) asynchronous [1]

that the master device has time to decide whether to proceed to the next data processing cycle or be interrupted by inserting the idle cycle (wait cycle).

All these problems can be resolved by using an asynchronous protocol. In this case, as shown in Fig. 2.1b, all signals are fully "acknowledged." Acknowledgment is provided by using data strobing (DS) signals and acknowledgment of data reception (DR) signals shown on the first two lines.

The *master device* places its data on the data transmission lines (information lines), waits for transmission errors of its own transmitter to be corrected, and only then issues the DS signals.

The slave device distinguishes DS signals, waits for confirmation of the fact of elimination of "skew" errors of its own receiver, and then during all the time required for this (which may be different in each particular case) receives the data addressed to it or exposes new stable data on the corresponding transmission lines. After that, the *slave device* issues DR signals. The master device stably fixes its data as long as it distinguishes the DR signals. Then, it simultaneously issues data and DS signals or first receives data from the slave device and only after that it issues DS signals. The slave device stably fixes its data as long as it distinguishes the flow of DS signals. When this process ends, it deletes its data and issues DR signals, thereby preparing for the next cycle. In this way, the value of the data transmission rate is automatically smoothly set to the maximum possible value for the two modules participating in the process of information exchange.

For the block-by-block transfer, the considered protocol can be modified, providing a doubling of the exchange rate. To this end, the developers of electronic systems use both fronts of DS and DR signals.

Bus Transfer Protocol, Priority Compliance, and Preemption In any multiprocessor system, a method should be provided for the orderly transfer of the bus to maintain its processors. In general, the bus is transferred to the disposal of the next processor for the time interval that is needed in each particular case. For obtaining access to the bus any number of processors in the system can compete. If we talk about the organization of data transfers in a computer system, then it should be emphasized that there is no one bus master in it. However, this is precisely why it is important to have some mechanism for controlling access to the common bus (arbitration mechanism) to decide which particular processor out of several available in the system should be given access to the bus. Such arbitration logic can be concentrated in one block or be completely distributed.

One of these simplest arbitration schemes is the so-called daisy chain. With this scheme, processors are connected sequentially in order of decreasing priorities. When the processor with the highest priority does not function, it transfers control to the next processor, etc. in the chain: from higher priority processors to lower priority processors.

The main disadvantage of this approach is that when a single card is removed from this chain (failure), *all the cards with lower priorities that follow it stop working* and that the bus priority is fixed and depends on the function of the boards.

It turned out that the most successful approach to solving this problem is to use parallel arbitration with separate lines for arbitration code signals. In such systems, arbitration can be performed in parallel with the usual data transfers. At the same time, when one main device—the current bus "master"—releases it for the next "master," the time delay required for it to go to work with this bus is minimal.

Above, in discussing the problem, it was assumed that each processor of a high-speed electronic system has a fixed priority and that all priorities are different. With regard to the requirements for the organization of the bus operation, this situation should always occur. However, it can make it very difficult to really access the bus for low priority modules. A rational method for resolving this problem is to implement the so-called fairness algorithm. However, if the condition of "fairness" shall be strictly observed; in a system with many processors, the waiting time for accessing the bus by each specific processor ("access delay") may be unacceptably long. To eliminate the problem, it can be possible to dynamically switch the actual priority levels of each processor between the level of "fairness" (in accordance with the above algorithm) and the level of fixed priority (in the sense discussed above).

Interrupt Mode Most modern high-speed electronic systems operating in real time should respond as quickly as possible to interruptions from a control object or peripheral controllers. If the central processor and the interrupt source are located on different boards, the bus should be provided with the means of quick transmission of interruption signals. However, this requirement will not affect larger multiprocessor systems, since each separate processor–memory combination will be located on each board, and such systems will be able to perform initial interruption processing only locally. Therefore, such a high-speed multiplex bus is not required to be equipped with any special means to provide interrupts.

In a multiprocessor system, resources are shared among several users or tasks. To reserve these resources for a specific task, each resource is supplied with a so-called "flag" (sign) of employment. The flag is read by the requesting processor, and if the resource is free, then this is fixed by setting the flag in the "busy" status. Such an operation is known as "test and install." It includes two bus operation cycles: reading and subsequent record. In order to avoid violating the required order, the operation shall be carried out in such a way that no device can change the state of the busy flag in the period between reading and recording operations.

Reliability, Detection, and Elimination of Errors As systems become more complex, the probability of errors in them increases. These errors can be caused by failures of both software and hardware. When an error is detected, it is necessary to find the corresponding defective module in which it occurred, and thereafter provide an automatic system reconfiguration, in order to bypass (block) the detected failure. In the event of a failure occurring in the logical control unit of a bus that is part of a module, certain alternative means should be provided for accessing this module.

Alternative handling can also be provided using a second subsequent bus. However, if the failure is caused by a short circuit in one of the bus lines, data transmission to all bus modules will still not be possible. A possible solution to the problem in this case is to use dual buses and, in critical situations, triplex buses. Then the system will be able to resume work on the additional ("alternative") bus. This can only lead to an insignificant change in system performance. Such an approach implies the observance of a number of certain limitations, the entire bus shall in this case be connected to one or several connectors, which could be duplicated within the printed circuit board of a standard format.

2.3 Basic Principles of Application and Methods for Calculating the Characteristics of Wired Communication Lines

2.3.1 Calculation of the Electrical Parameters of Wired Communication Lines

The parameters of communication lines of modern information and switching systems largely determine their most important characteristics such as speed, power consumption, overall dimensions, and reliability.

I/O interfaces, for example, shall receive data from the system bus, store and output it to the system bus, exchange information with the CPU, receive and execute commands from the CPU, and transmit information to the CPU from external connected devices.

In the general case, a wired communication line of any high-speed electronic system can be represented in a simplified form (Fig. 2.2), where the transmitter and

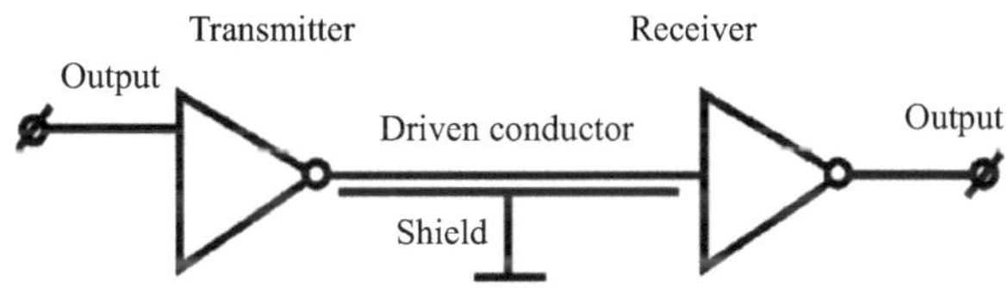

Fig. 2.2 Generalized diagram of a wired communication circuit of a microprocessor system

receiver perform the functions of generating levels of electrical signals transmitted from one functional unit to another via an actuated conductor. The functions of the transmitter and receiver can be performed by both special interface chips (in the case of large lengths of communication lines) and their own buffer (input and output) stages of microprocessor circuits.

Figure 2.3 shows the organization of interface lines of communication of the most widely used standards of single-wire communication circuit RS-232C (Fig. 2.3a) and two-wire communication circuit of standards RS-423A (Fig. 2.3b), RS-422A (Fig. 2.3c), RS-485 (Fig. 2.3d), MIL-STD 1553B (Fig. 2.3e). As a rule, the transmission and conversion of signals are carried out in standard TTL levels at the input and output of the communication circuit.

Naturally, for each standard, their own types of interface chips (receivers and transmitters) are used, the nomenclature of which currently exceeds several 100 names.

Each standard, except for the mandatory internal organization of the communication circuit and the receiver and transmitter switch-on circuits, provides a specified level of the main technical parameters of the interface bus: the length of the communication line l of the guaranteed transmission and the rate D of information transmission (bps) for this length of the line.

Let us consider in more detail the basic parameters of standard communication line interfaces.

The main characteristics of communication line are their electrical, dynamic, and energy parameters.

When calculating usually the following basic limitations are used: Electromagnetic energy propagates through the link channel in the form of a TEM-wave, when the active conductivity of the dielectric tends to zero (an ideal dielectric).

Electrical capacitance C depends on the dimensions of the structural elements, methods of shielding, and properties of the electrical insulating material. Provided the loss of distribution of capacitance C is neglected, the capacitance C is identical to the electric capacity of the line; for conductors with circular sections, capacitance C is easily determined using a cylindrical coordinate system, for rectangular sections, as well as for combinations of circular and rectangular sections, conformal mapping methods are used.

The inductance L is related to the electric capacity by an inverse relationship and can be calculated for the case of a homogeneous medium using the formula [1]:

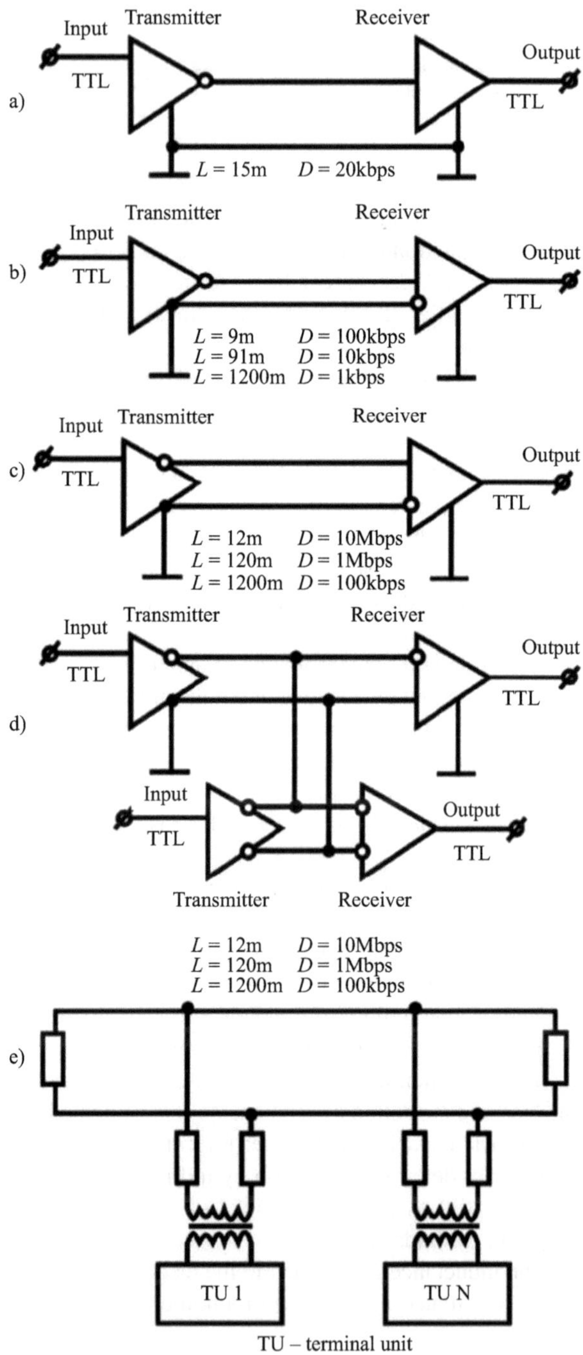

Fig. 2.3 Diagrams of interface communication lines [1]: (**a**) RS232C, (**b**) RS-423A, (**c**) RS-422A, (**d**) RS-485, (**e**) M1L-STD 1553B

$$L = \frac{1}{C^2 C_0}, \tag{2.1}$$

where $C = 3 \cdot 10^8$ m/s is the speed of light in vacuum; C_0 is the linear capacity of the communication line.

The resistance of the communication line R_f is calculated for two cases: low and high operating frequency. If the known Ohm's law is applied at low frequencies, then at high frequencies it is necessary to take into account the "skin"—the effect leading to an increase in the active resistance of the communication line R_f. So, for the case of a circular section of a communication line conductor, the following expressions are used [1]:

$$R_f = R_0(X + 0.25) \text{ at } X > 1;$$

$$R_f = R_0\left(X + \frac{X^3}{3}\right) \text{ at } X < 1;$$

where $X = \frac{d}{4}\sqrt{\pi f \sigma \mu_0}$.

Here d is the diameter of the conductor; σ is the conductivity of the material; $m_0 = 47\pi \cdot 10^{-7}$ GN/m; f is the operating frequency, Hz; R_0 is the resistance at a constant current, Ohm. For practical calculations of the R_f value, taking into account the "skin" effect for circular section:

$$R_f = 7.975 \cdot 10^{-3} \frac{\sqrt{f}}{d}; \text{Ом}$$

and for rectangular section:

$$R_f = 3.94 \cdot 10^{-3} \frac{\sqrt{nf}}{ab}; \text{Ом}$$

where f is the operating frequency of information transmission in the communication line, MHz; a and b are the width and thickness of a rectangular conductor, mm.

The characteristic impedance Z_0 of the communication line is calculated from the expression:

$$Z_0 = \sqrt{\frac{R + j\omega L}{G + j\omega C}}. \tag{2.2}$$

When transmitting high-frequency signals over a communication line, the active components R and G can be neglected and then:

$$Z_0 = \frac{120\pi \cdot \varepsilon_0}{C_0} = \frac{33.33 \cdot 10^{-14}}{C_0}.$$

2.3.2 Calculation of the Dynamical Parameters of Wired Communication Lines

The minimum theoretical value of the propagation delay of a communication line signal with a length l is equal to [1]:

$$\tau_{\text{лсmin}} = \tau_0 \cdot l \approx 3.315 \cdot l\sqrt{\varepsilon_1 \cdot \mu_1},$$

where τ_0 is the specific delay in propagation of the signal in the communication line; l is the length of the communication line; ε_1 and μ_1 are the relative values of the dielectric and magnetic permeability of the insulation material of the communication circuit (usually $\mu_1 = 1.0$; $\varepsilon_1 = 2$–8).

In practice, the actual signal delay in the communication line is $\tau_{\text{cl}} > \tau_{\text{clmin}}$, which is due to the nature and completeness of the communication line matching, the presence of losses and discontinuities in the lines, taps from the line, etc.

Communication line immunity is determined by two main factors: the mutual influence of communication lines (causing the appearance of crosstalk) and mismatching.

Line matching is provided by means of matching resistors connected to the communication line according to certain rules. The connection rules for these resistors are determined by the type of the communication line, the type of logic gates located at the transmitting and receiving ends of the communication line (ECL, TTL, TTLS, CMDS, etc.).

There are short and long lines of communication. As is known from the general theory of electrical circuits, the mismatch of the communication line between the receiver and the transmitter can lead to phenomena of reflection of signals from the end of the line. For the worst case of mismatch (a long line is connected to the output of an integrated circuit with a low output impedance), the reflected noise signal can reach a value comparable to the amplitude of the useful transmitted signal. Reducing the length of the line leads to a decrease in the duration of interference and, starting from a certain critical length of l_{Gy}, to a linear decrease in the amplitude of the interference. The boundary length of l_{Gy} corresponds to the propagation delay τ_{Gy}, equal to half the duration of the front of the transmitted useful signal:

$$\tau_{\text{гр}} = \frac{1}{2}\tau_{\phi} \text{ or } l_{\text{гр}} = \frac{\tau_{\text{гр}}}{\tau_0},$$

It is considered that for a communication line with a length of no more than l_{Gy} no agreement is required.

A similar situation is the case for the analysis of crosstalk in communication lines: in the system of closely located long-range communication lines, the originating crosstalk can reach a significant value, up to and including the switching threshold of the input logic element, receiver, and spurious reception. As the length of the interaction region decreases, the crosstalk decreases in duration, and starting from the critical length l_{Gy} and in amplitude. Therefore, communication lines with less length have good resistance to crosstalk and do not need shielding.

For the coordination of long communication lines ($l_{cl} > l_{Gy}$) subsequent and parallel methods of coordination are used. For the correct choice of the first or second method, it is necessary to take into account the energy characteristics of a communication line—the power dissipated by the line and the loss resistance.

Loss in communication lines is mainly determined by the ohmic resistance of their conductors:

$$R_{ЛС} = \rho \frac{l_{ЛС}}{S_{ЛС}}; \tag{2.3}$$

where ρ is the resistivity of the conductor; l_{cl} is the conductor length; S_{cl} is the cross-sectional area.

Short lines are characterized by low loss resistance: usually it is assumed that $R_{cl}{}^{K} = 0$. Then the delay in signal propagation in an unmatched short communication line will depend on the output impedance of the transmitter's R_{out} chip and for the case of $R_{out} < Z_0$ will be determined by the expression:

$$\tau_{ЛС} = \tau_{ЛС\min} = \tau_0 l; \tag{2.4}$$

for the case $R_{out} > Z_0$,

$$\tau_{ЛС} = R_{вых} C = \frac{U_{ЛС}}{I_{вых}} \cdot \frac{\tau_0 l}{Z_0};$$

where C is the line capacity; R_{out} is the output impedance of the transmitter chip; U_{cl} is the amplitude of the signal switched to the communication line; I_{out} is the value of the output current of the chip during switching; Z is the wave resistance of the communication line.

Damping of the communication line characterizes a decrease in the magnitude of the voltage of each signal and an increase in the voltage times and decay of the signals at the line output. The attenuation coefficient of the communication line is determined by the well-known expression:

$$\alpha = \frac{1}{2}\left(\frac{R_f}{Z_0} + Z_0 \cdot G\right); \tag{2.5}$$

Transient attenuation characterizes the level of protection of the interface lines from internal interference.

2.3.3 Energy Parameters of Wired Communication Lines

During the propagation of a useful signal over a communication line, power is dissipated, which is actually a pulsed output power of a microcircuit—a signal transmitter. The power value can be determined from simple expressions [1]:

$$P_{\text{КЛ1}} = \frac{C\,U^2_{\text{ЛС}}}{2\tau_{\phi}}\,(\text{для}\,R_{\text{вых}} \leq Z_0); \tag{2.6}$$

$$P_{\text{КЛ2}} = \frac{C\,U^2_{\text{ЛС}}}{2\tau_{\phi}}\,(\text{для}\,R_{\text{вых}} < Z_0). \tag{2.7}$$

Parallel matching of the line is characterized by the minimum value of the signal propagation delay over communication line; however, the power dissipated at the same time is the largest compared to sequential matching. This power is of a static nature and is dissipated on the terminating resistor $R_c = Z_0$ included at the end of the line, preventing the reflection effect of the signal from the end of the line $P = U^2/Z_0$.

The delay of signal propagation to any point of the X line for this case is determined by a simple expression:

$$\tau_{\text{ЛС}\,\text{X}} = \tau_{0\text{X}}. \tag{2.8}$$

With a consistent matching of the communication line, the matching resistor R_c is switched on in series at the beginning of the line so that the condition $R_c = R_{\text{out}} = Z_0$, where R_{out} is the output impedance of the transmitter chip. Then the signal transmitted over the communication line is reflected from the end of the line (receiver input) with a double amplitude, returns again to the beginning of the line, and is "absorbed" by the equivalent resistance $R_c + R_{\text{out}}$ without reflection. It is obvious that the condition of constancy of the value of $R_c + R_{\text{out}}$ is feasible for the transmitted signals of logical levels "O" and "1" only at very low values of R_{OUT} and Z_n.

The delay of signal propagation along such a line to an arbitrary intermediate point X is determined from the expression [1]:

$$\tau_{\text{ЛС}\,\text{X}} = \tau_0(21 - x). \tag{2.9}$$

which is more than for the case of parallel coordination of communication lines.

At the same time, the power of dispersion of a communication line has a dynamic nature and, in fact, is equal to the output pulse power of the transmitter chip. For the case

$$R_{\text{вых}} = R_{\text{лС}} = 0, P_{\text{имп}} = \frac{U^2}{2Z_0}.$$

In practice, the concept of reflection coefficient is often used [1]:

$$\rho = \frac{R_{\text{вх}} - Z_0}{R_{\text{вх}} + Z_0}. \quad (2.10)$$

where R_{in} is the input resistance value of the receiver circuit.

2.3.4 Design Options for Implementation of Wired Communication Lines

The main design options for the implementation of communication line of external and internal interface of high-speed systems are two-wire coaxial cable, twisted pair, conductor over a grounded plate, microstrip line, strip line, multi-core ribbon cable, and multichannel three-wire ribbon cable.

The main parameters that determine the criteria for selecting the type of cable are:

- exchange rate, the value of which determines the duration of the transmitted bit of information;
- minimum signal level: at the receiver input, necessary to recognize the transmitted binary states;
- the maximum permissible level of signal distortion;
- the maximum required length of the communication line.

The duration of the information bit (Tb) is determined by the minimum allowable time interval between transitions of transmitted binary states. If the signal voltage in the line does not have time to reach the level corresponding to the transmitted binary state before the next transition appears, the specified transition will appear at the receiver input with some time shift that leads to intersymbol distortions. When choosing a cable, the ratio of the length of the leading edge to the duration of the information bit (tr/Tb) at the connection point of the most remote receiver should be taken into account.

The signal level present at the receiver input shall be at least its sensitivity threshold. In this case, the minimum value of the input voltage should be selected with a margin depending on the intensity of interference affecting the communication line and the receiver, the permissible probability of error, as well as the permissible level of distortion at the receiver input. To determine the cable parameters, it is necessary to set the minimum signal level at the input of the most remote receiver, taking into account the listed factors.

Signal distortions are determined by its time shift relative to the position during the transmission in ideal conditions. Quantitatively, distortions are expressed as a percentage of the total duration of the information bit. When choosing a cable, the permissible level of distortion at the input of the receiver located at the most distant point of the communication line shall be taken into account.

In addition, the design of the communication line includes such means of combining devices as connecting and terminating resistors.

And finally, wired communication lines can work both in an inconsistent mode and in a coordinated mode, and their electrical and time coordination can be carried out both sequentially and in parallel.

2.4 Organization of the Process of High-Speed Data Transmission Based on Transceivers with SerDes Type Topology

2.4.1 Principles of Operation of the Transceivers with SerDes Type Topology

This section discusses the features of organizing the transmission of multimedia data at high speed using SerDes type transceivers. These devices have a number of significant advantages for high-speed transmission of large amounts of data, providing acceptable power consumption, performance, and cost of the system. In addition, modern analog-to-digital methods of signal correction are widely used here when it is distributed over cables with a length of more than 300 m.

So, a SerDes type transceiver (serializers/deserializers) [2] is a parallel-serial or serial-parallel conversion circuit that allows switching from the transmitter from a multi-bit (wide) parallel bus with single-ended signals to a serial bus with a balanced signal. The receiver of the same transceiver produces an inverse transformation. SerDes type transceivers provide the transmission of large amounts of data “by bits,” which can significantly reduce the complexity of the electronic system, its cost, power consumption, and space occupied on the board. The advantages of using SerDes type devices are especially high as soon as the frequency of parallel data buses begins to exceed 500 MHz (1000 Mbps).

The advantages of serial data buses are obvious [1, 2], and we will not dwell on them.

The principle of operation of SerDes type transceivers is relatively simple (see Fig. 2.4). The data from the parallel bus, which switches with a certain frequency, goes to the parallel input bus of the SerDes device. This data is synchronized on the internal parallel SerDes registers based on the back or leading edge of the reference signal, simultaneously arriving with the parallel data from the MAC controller or system processor. Synchronization is also provided by an appropriate reference signal source (the simplest quartz oscillator or clock oscillator chip). Data loaded on parallel registers is encoded using standard schemes, for example 8V/10V.

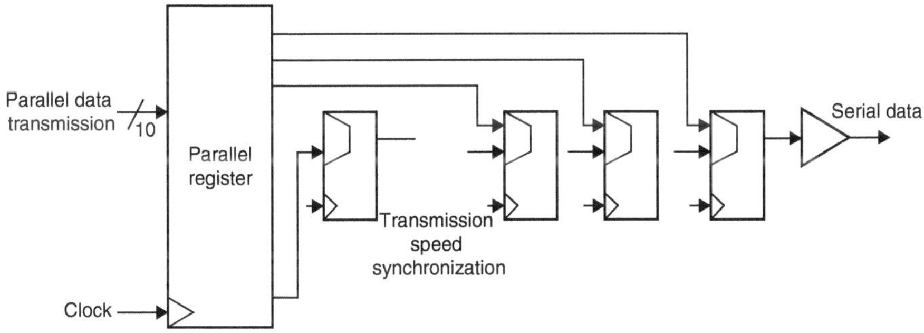

Fig. 2.4 Block diagram of the process of converting parallel data into a serial form for a transceiver [2]

SerDes transceiver encoder/decoder (ENDEC) performs several important functions. First of all, the incoming data stream shall be definitely prepared for the subsequent automatic parallel-serial conversion. For example, the original raw data streams may include long continuous combinations of zeros (0000000) or ones (1111111). In these cases, SerDes has to independently and very quickly find the end of such a long sequence of identical bits.

The CDR (Clock and Data Recovery) scheme of data synchronization of the next SerDes device determines those specific sections of the continuous stream of transmitted data, where the data flow density changes and implements some special functions to avoid missing bits. When encoding, the incoming word of parallel data (a predetermined number of bits of parallel data) is automatically converted to some standard combination that best corresponds to the parallel-serial conversion mechanism.

Another problem associated with uncoded data is the possibility of losing the boundaries of the transmitted frame in the process of its conversion into a sequential form. In order to simplify the work of the receiver, the ENDEC device provides special code words (so-called *preambles*) denoting the end limits of the transmitted word. The receiver can then automatically use these preambles to delimit bytes. For example, in the mentioned 8B/10B scheme, a special comma is used to establish the initial code limit in the data stream.

The coded data comes from the parallel bus to the serial bus for subsequent transmission over a single differential channel. Often such a transformation is implemented using the shift registers shown in Fig. 2.4 [2]. Data in parallel registers shall be synchronized.

The serialized data from the bus then goes to the line driver, or so-called a *differential signal buffer* that directs the flow of data through a copper cable or other backplane conductors.

On the receiver side of the SerDes transceiver, the stream of transmitted data already serialized is fed into the input buffer of the differential signal. The input buffer inside the receiving circuit of the SerDes transceiver automatically turns sequentially incoming bits (in the form of the corresponding voltage amplitude) into a digital parallel data stream.

Another important function of this receiver is to "recover" the clock synchronization of the incoming serial data stream. A special *clock recovery device* extracts a clock signal from the serial data stream arriving at the receiver. The recovered signal is used to synchronize the incoming data stream, which is then converted back into a parallel form.

In the process of converting, serial data is collected into parallel words, which can then be decoded into initial words. This process depends on the CDR circuit, which uses the recovered clock signal to control the synchronous operation of the embedded shift registers, which automatically collect the parallel coded data into the right words.

The parallel bit stream is decoded into the original data sequence, which is then fed to the parallel output registers and clocked by the output buffers of the parallel signals. Typically, the output buffers are designed to receive single-ended signals. The clock signal is parallelized and transmitted along with parallel data. It is aligned in phase with the data transmission clock signal. Often the receiving and transmitting paths of the SerDes transceiver operate on the same frequency, forming the so-called *full-duplex serial channel.*

The *phase-locked loop* (PLL) is an important component of any SerDes type device, because it generates a high-speed clock signal to control both the serial transmitter and the receiving path of the device. Depending on the SerDes architecture, this device may have one PLL circuit for both the transmission and reception paths, or two separate circuits for each of them.

The PLL circuit is the main unit of the CDR circuit, which is controlled by the incoming reference signal. Therefore, the incoming reference signals shall meet strict electrical requirements and jitter, otherwise the "bad" reference signal will generally have a negative impact on the performance of the SerDes transceiver. The recovered clock signal, as a rule, is parallelized and transmitted further along with parallel output data.

Separately, it is necessary to consider the peculiarities of the formation of reference clock pulses.

The *reference clock signal* is fed to the input of the SerDes transceiver to control the PLL of this device. There is a definite relationship between the reference clock and the serial data rate at which the SerDes device should operate. For example, SerDes devices with 10-bit parallel interfaces use a 125-MHz reference frequency so that the serial data transmission rate is 1.25 Gbit/s. In this case, the internal PLL circuit shall provide a tenfold multiplication of the frequency of the reference signal, provided that it is set by the leading edges of the signal.

As mentioned above, the quality of the reference signal is the main factor determining the operation of the SerDes transceiver. If the "jitter" in the reference signal is large, in the stream of serial data, most likely, the content of this interference will also be unacceptably high. If the reference signal is unstable and "wanders," the same undesirable characteristics appear in the data stream. Thus, the reference signal of the SerDes device shall simultaneously meet a set of sufficiently stringent requirements for jitter, pulse offset in time (phase-pulse modulation), and pulse rise/droop time.

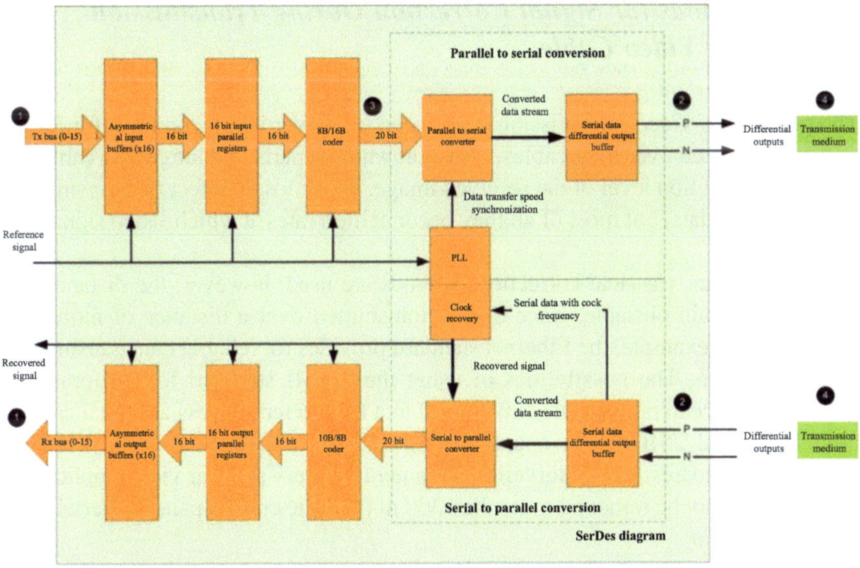

Fig. 2.5 The block diagram of a SerDes type transceiver [3]

The specific requirements for the reference signal and the supported frequency range of the SerDes transceiver vary depending on the specific vendor of these devices. For example, such solutions as Texas Instruments' SerDes transceivers, such as the TLK1501, TLK3131, TLK3134, and TLK6002, support fairly wide input frequency ranges and therefore provide a high data transmission rate according to a wide range of standards: with Gigabit Ethernet, fiber-optic, 10G Ethernet, and public radio interface standards.

Devices such as SerDes have a number of significant advantages in transferring large amounts of data, ensuring acceptable system power consumption, ease of use, performance, and reasonable cost. Modern solutions of SerDes transceivers are devices with a high degree of integration (see Fig. 2.5) [3].

Here, the parallel input-output (1), as a rule, are asymmetrical. They support signal transfer levels such as HSTL, LVTTL, and TTL. In applications with high data transmission rate, the parallel interface also uses differential LVDS transmission levels. In differential signaling, two lines are used (2) in accordance with the LVPECL, CML, and VML standards. The stream of serial data (3), as a rule, is encoded according to the 8B/10B scheme. The transmission medium (4) allows the flow of data from one point to another. Copper or fiber-optic cables are commonly used as this medium.

2.4.2 Methods for Signal Correction During Transmission Over Video Cable

The problem of video transmission is the need to deliver large amounts of information at high speed over long cables. As is known to experts, the longer the cable, the lower the resolution level of the received image. These losses affect the transmission of any type of data, but most of all they occur at high rates at which video signals are transmitted.

In such cases, classical correction methods are used; however, the distance still remains the main obstacle, since data is transmitted over a distance of more than 300 m [4]. For example, the Ethernet standard provides for reliable data transmission only for 100 m. The possibilities of using the HDMI standard for equipment of entertainment centers, as a rule, are limited to a few meters, unless a special cable is used. Peripheral USB devices operate at a distance of no more than 5 m from a PC in order to avoid losses. Video surveillance camera is a very popular video application, which should not be removed more than 300 m from the control panel, otherwise the *image will suffer.*

Such distances are not large enough in the case of video transmission at airports, shopping centers, or stadiums, when cables with a length of more than 300 m are required. Therefore, developers use a variety of analog and digital signal correction methods as it propagates over long cables.

Typically, devices designed to compensate for losses in the system and that are called *equalizers* provide the maximum identity of the output and input signals. Loss compensation is not the only way to transmit video over long distances, but it is the best method. Another well-known method is to apply a high voltage signal to the input of a communication line. For example, to transmit a video signal over a distance of 1500 m over a Cat 5 cable (24AWG twisted pair) with a 70 dB signal attenuation at the extreme subcarrier frequency of 3.85 MHz, the input voltage should be 3000 V. In this case, the amount of current flowing through a very thin twisted pair will be 30 A, and too much heat will be generated for such wires.

The use of the correction method is preferable because it is easily implemented at the receiving end of the line and does not require the generation of a high voltage signal. For example, at high frequencies, it is required to amplify the signal by 70 dB (in full, the video signal occupies a band of about 5 MHz) in order to provide the corresponding image. In this case, the signal is amplified 3000 times, not the voltage. In this case, the thermal load on the cable is absent, and there is no potential threat of destruction of the wires. A single-stage equalizer boosts the signal by 14 dB, and five of these devices provide 70 dB amplifying.

This is usually a fairly reliable solution, but it never completely eliminates the threat of unwanted interference from electronic amplifiers, which can completely or partially suppress the useful video signal. The noise *at the input of the circuit is amplified with the same coefficient as the useful transmitted signal. In order for it to be acceptable on the receiver side, the equalizer noise shall be at least 30 dB less than the video signal at its input. It should be said that usually the dynamic range of a typical equalizer is 100 dB.*

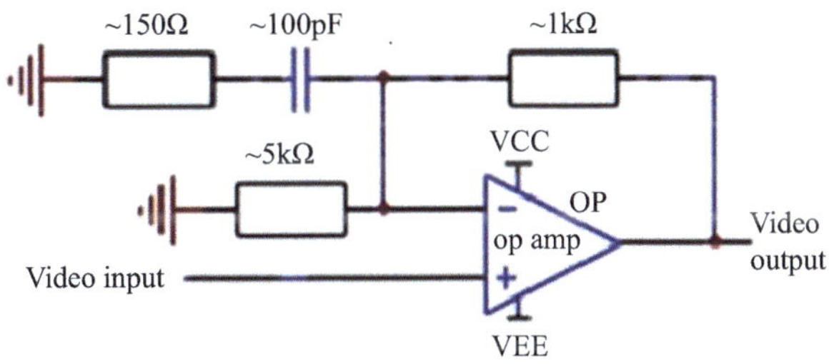

Fig. 2.6 Diagram of a single stage equalizer for signal correction in the 300-m Cat 5 cable

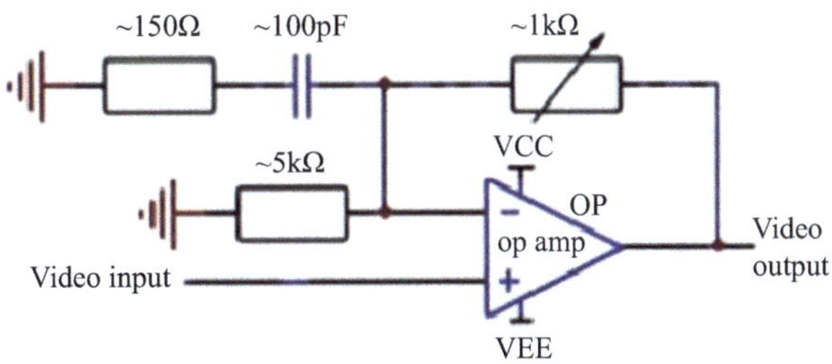

Fig. 2.7 Scheme of equalizer with variable amplification factor for cables of up to 300 m

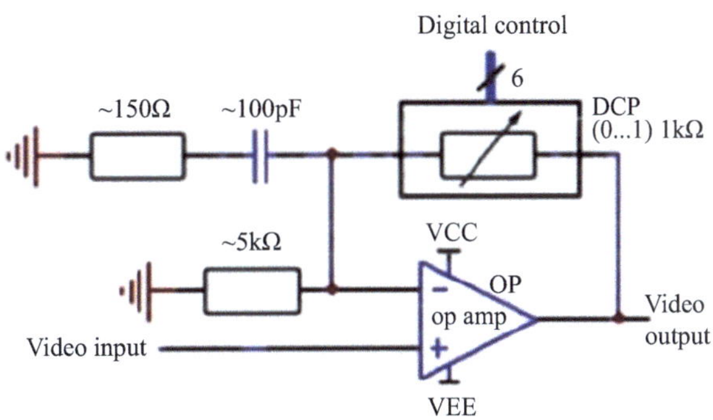

Fig. 2.8 Variable gain cascade implemented by DCP

There are various technical solutions scheme of *equalizers*.

Figure 2.6 [5] presents the amplification circuit for signal compensation in the 300-m Cat 5 cable. This scheme is a low-noise controlled object with a sufficiently large product of the amplifying per bandwidth, providing the required amplifying of 14 dB.

When transmitting over shorter distances, less amplification is required (Fig. 2.7) [4]. Here in the feedback circuit, the resistor acts as a potentiometer.

To control the correction using a microcontroller or a digital logic circuit, it is necessary to equip this variable resistor with a digital input. For example, it is possible to use a digitally controlled potentiometer (DCP) (see Fig. 2.8). The ability to build this circuit on a single chip provides significant advantages in cost, functionality, and programmability.

Figure 2.9 shows a simplified diagram of the automatic equalizer [3], which consists of eight blocks. The feedback control circuit and user interface are

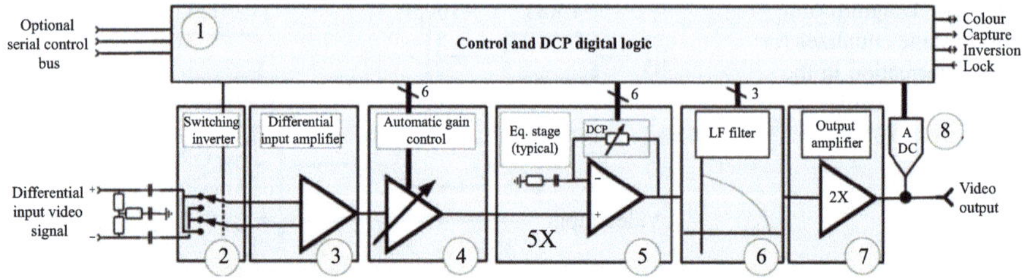

Fig. 2.9 Block diagram of the combined correction circuit

implemented using a sufficiently small number of elements of a digital logic block (*Block 1*). The logic determines to what extent the signal needs to be corrected and how to adjust the correction. Equalizer and filter codes provide the same behavior for each crystal.

Block 2 is an additional block. Since the video cable in this case is a twisted pair, it is possible to switch the conductors by inverting the signal. This circuit detects switching and inverts the signal for compensation. In addition, since this video cable is a two-wire twisted pair, the input stage shall be differential.

Block 3 is a differential input amplifier that receives a signal from a twisted pair. This amplifier removes common-mode noise, leaving only the video signal.

The voltage swing of this video signal is 1 V. *Block 4* provides automatic gain control (AGC), correcting the variations in video signal level.

The main block of the equalizer is responsible for amplifying high frequencies, which disappear when transmitting over long distances. The cascades of this unit shall be low-noise and suppress the noise of the power supply. *Block 5* consists of five cascades, which we have already discussed.

Block 6 is a noise filter. If the equalizer is configured to amplify, for example, a 5-MHz signal, it can amplify it to about 10 MHz, after which the signal drops. As a result, noise is amplified, for cutoff of which a low-frequency steep-sided filter is applied.

The signal that passes through the noise filter enters another amplifier unit (*Block 7*). This amplifier provides sufficient gain and control current to interface with other video equipment.

Finally, *Block 8* is an ADC that digitizes the output signal and feeds it to the digital control circuit. In case of an error of any of the equalizer blocks, the circuit performs debugging of this block. For example, if the output signal declines, AGC is turned on to compensate.

Figure 2.10 shows the original high-quality video signals received after passing 1500 m over a Cat 5 cable *with and without correction.* It can be seen that at such large distances the color is lost, the image is blurred, and the text becomes illegible.

The single-crystal equalizer described above allows fully restoring the image transmitted over a cable with a length of more than 1500 m.

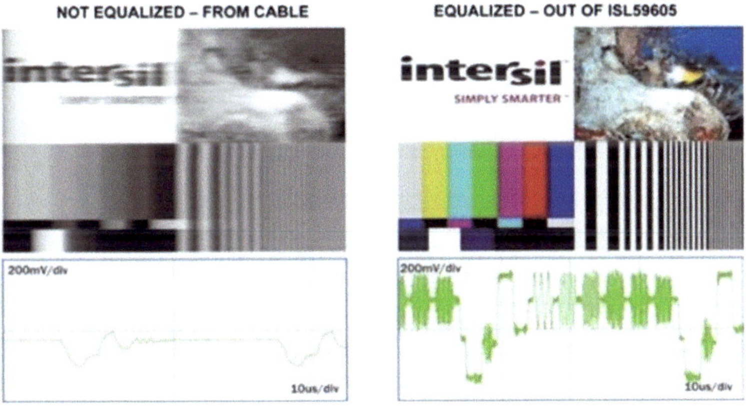

Fig. 2.10 The original signal after passing through the 1.5-km cable Cat 5 with correction (right) and without it (left)

2.5 Design Features of High-Speed DDR Memory Buses

2.5.1 Principle of the DDR Module Operation

The abbreviation DDR (Double Data Rate) means double data transmission rate without increasing the frequency of the memory bus clock signal [5]. DDR-type modules are a further development of the widely known SDRAM (Synchronous DRAM) memory modules. The main difference between DDR and SDRAM is that instead of doubling the clock frequency to speed up the electronic device, DDR modules transmit data twice in one clock cycle. DDR1 (Double Data Rate 1) modification is a newer version of DDR; unlike DDR, it provides the ability to work at a much higher clock frequency thanks to an improved design, as well as a large number of contacts: it increases from 184 (for DDR) to 240 (for DDR2 modification). The next generation of modules (DDR3 and DDR4) are distinguished by improved technical parameters; they are also designed in the 240-pin version, but are not electrically combined with DDR and DDR2.

The DDR bus is one of the most common and widely used high-speed memory buses. Despite the fact that its design is quite simple, the design of DDR boards has a number of specific features that shall be well known and taken into account by the developers of such boards and electronic systems based on them. Here, first of all, it is necessary to clearly understand the purpose of the basic structural elements of the bus.

Such a bus usually consists of one controller at one end and one or more DRAM memory chips at the other. In general, the more memory required by the system being developed, the greater the number of DRAM chips that shall be used. The main operations performed by the controller are recording data to DRAM chips and reading data from DRAM chips.

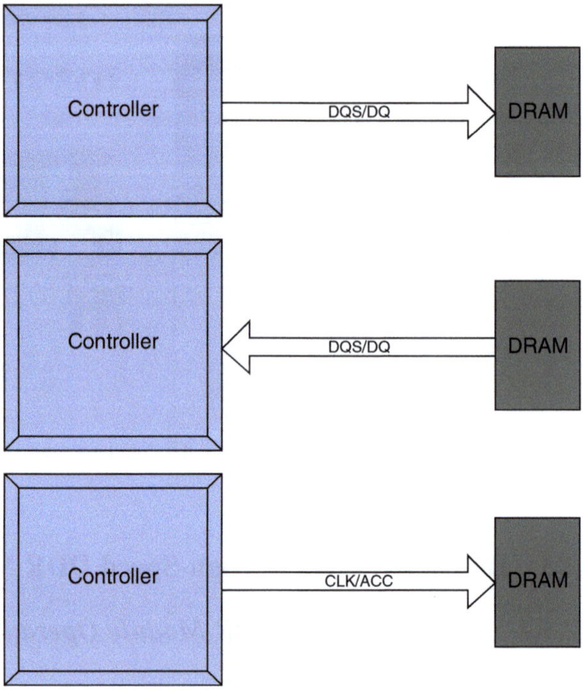

Fig. 2.11 DDR bus uses two sets of signals

Let us consider these operations in more detail. The main thing a developer needs to know is that all memory management operations are performed using two synchronized sets of signals: via the data bus and the address/command bus, as shown in Fig. 2.11 [5].

Bidirectional data bus is the main element of the system, which is characterized by the number of lines for data transmission. Each such line consists of a unique DQS (strobe) signal and the corresponding data signals. Normally, a line consists of only 8 bits, but there are DRAM modules with only 4 bits of data, so some lines can consist of only 4 bits. Regardless of the number of bits in the line, every bit is fixed during the transmission of the DQS signal. Data signals are all one-way. DQS signals for DDR3 and DDR4 buses are always differential. For DDR2 buses, DQS signals can be either one-way or differential.

During recording mode, the controller sends a signal approximately midway between two transmissions of the DQS signal (Fig. 2.12). Moreover, during the current transmission of the DQS signal, the data signal should be stable. Next, the DRAM chip captures the data signal during the DQS transmission. This is the first important element: it is necessary to meet the requirements for installation time and retention time on DRAM. Since the controller supplies data signals approximately midway between two DQS transmissions, in order to center the DQS signals with respect to stable data signals, it is necessary that the propagation delay of the DQS signal and the signal delay of this line are not very different from each other. The

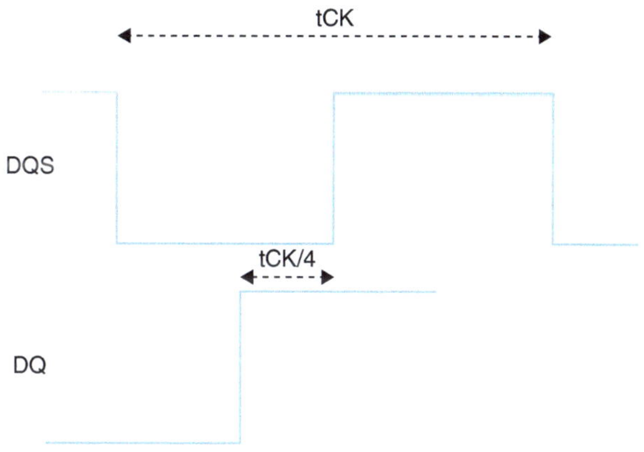

Fig. 2.12 For recording, the controller generates a signal approximately midway between two DQS signal transmissions

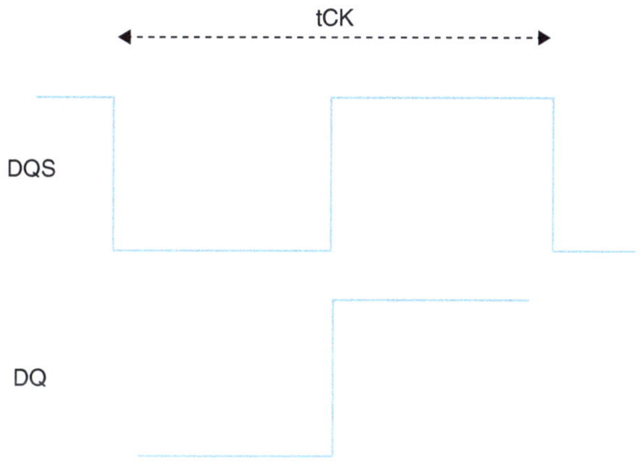

Fig. 2.13 While reading, as opposed to recording, the data signals are attached to the DQS

magnitude of the signal propagation delay between the lines may vary slightly, since the strobe signal is used only for data bits on the line under consideration.

In reading mode, the DRAM module provides data signals synchronously with the DQS signals (Fig. 2.13). The controller then needs to delay the data signal and/or the strobe signal for the right time in order to fix it with DQS.

In a one-bit system, each line is connected to only one DRAM chip. In multi-bit systems, a line (data bit and strobe signal) can be connected to multiple DRAM chips. During the transmission, only one of the DRAM chips connected to the line is always active. The ability to access a memory chip is determined using the memory element select signal (CS signal); for other DRAM connected to this line, the selection signal will be inactive. Usually, the number of "bits" in the system is equal to the number of active CS signals, which is equal to the number of DRAM chips connected to a given line.

The address/command bus consists of several address bits (the exact number depends on the amount of DRAM memory to be accessed), several bits for the command transmitted on the DRAM, and the clock signal. Unidirectional bus; commands are sent only from the controller to DRAM.

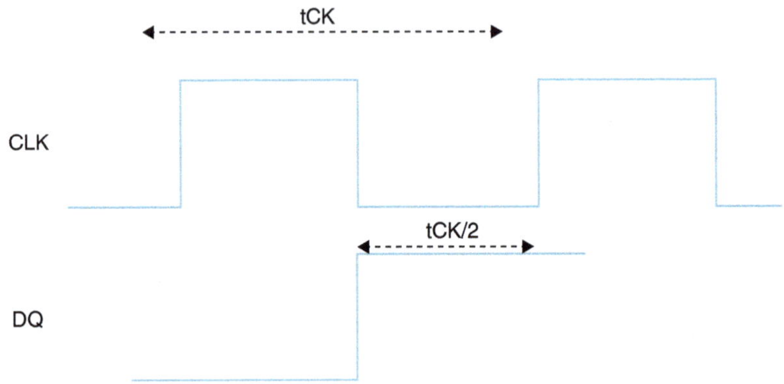

Fig. 2.14 For commands, command and address signals are generated by the controller during a clock drop

Address signals and command signals are unbalanced. The clock signal is differential, synchronization occurs at the intersection of the main and inverted clock signal.

To execute a command, the controller gives the corresponding address signal and the command signal at the moment when the clock signal goes down (Fig. 2.14). The signals remain at the same level as the clock signal begins to rise, and then during the subsequent clock drop, the transition to the new command is performed. Therefore, the command signal is stable when the clock signal is at the top level.

This leads to another important issue: it is necessary to coordinate the installation time and the retention time of address signals and command signals on DRAM. Since the signals are sent by the controller in such a way as to ensure the stability of the address/command bits during the rise of the clock signal, it is necessary to accurately delay the address/command signals by an amount equal to the clock. Since the address/command bus is distributed across different DRAM chips, the signal delay from the controller to each DRAM shall be the same for all address bits and clock signals. Therefore, even if the delay on DRAM chips can vary, the delays for the address and the clock signal shall be synchronized on any given DRAM.

Finally, there are also certain requirements that shall necessarily be met for DRAM modules between DQS signals and clock signals (CLK), namely, synchronization of the data bus and address bus. DQS and CLK signals need to be synchronized on each DRAM.

For example, to ensure this process, the DDR2 bus needs that the time delays from each controller to each DRAM module for each of the DQS signals are equal to each other, and the total value equals the value of the time delay from controller to DRAM for the clock signal.

In the buses of DDR3 modifications, a new concept of tracing is already presented, known to specialists as "fly-by" [6]. This is when the address and clock signals are given by the controller and reach each DRAM in turn, one after another. This also implies that a clock signal reaches each DRAM chip at a different time. If

DQS signals are produced of equal length, then there is no guarantee that the requirement for the DQS/CLK signals mentioned above will be fulfilled.

In this case, the controller shall be able to independently delay the DQS signal so that it reaches the clock signal at about the same time.

It should be taken into account that for DDR2 buses and for controllers that do not support this method, the clock signals and DQS signals shall be consistent with each other.

2.5.2 Methodological Features of the Stage of Preparation for Modeling

At the beginning of this section, some of the critical tasks of ensuring the integrity of the signal that engineers face in the process of designing DDR buses [5, 7] were listed. It is best to organize their work before organizing the production of such boards, with which it will be possible to detect possible errors in the board's functionality and also reduce the number of iterations of the design process. This means that the time required to install, run, and analyze the simulation results should also be reduced, otherwise the simulation process becomes ineffective.

Therefore, the developer's goal is to use the necessary tools to quickly conduct simulations and obtain data that is easy to understand and analyze, in order to focus on critical points then. Ideally, such a tool should do all of the above.

In order to reduce installation time, it is necessary to carefully study the input data and determine their values by the time of the start of the simulation. The developer usually receives most of the required information from the customer (for example, a board file or a terms of reference) and from a supplier (specifications for materials and components).

The first thing that is needed is the driver and receiver models for the memory controller and DRAM chips. The IBIS-controller model can always be obtained from the supplier or downloaded from the manufacturer's website. The IBIS model for DRAM can also be requested from DRAM suppliers such as companies like Micron, Samsung, or Hynix. Models should contain all the information for various parameters and signal termination options (ODT method) available for the respective chips.

Some vendors mostly offer Spice-models instead of IBIS. It should be understood that usually Spice-modeling can be slightly more accurate and often slower processes larger orders of magnitude than IBIS. The publication [7] shows that the setup and installation to launch the Spice driver from the developer took only 221 h, whereas it took only about 3 h to launch the IBIS driver. Thus, if the main task is to save time, the use of the IBIS model can significantly speed up the process. However, for the sake of fairness, it should be said that there are very poorly created IBIS models, so it is best if the developer requests comparative data from the supplier they choose.

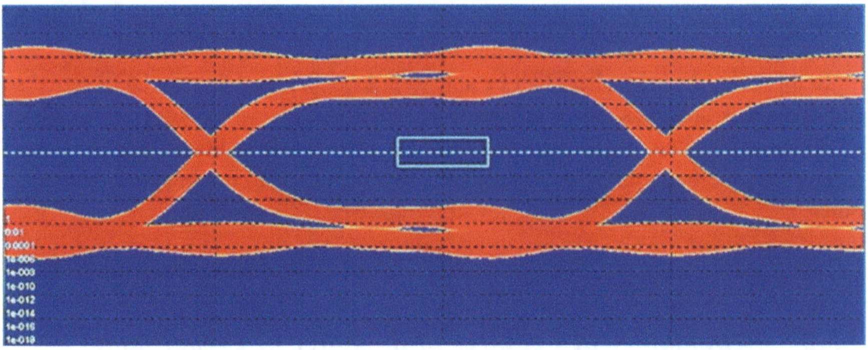

Fig. 2.15 The eye diagram displays a large number of signals above the mask (central rectangle) in order to show the margin. In this example [5] the margin is very good

The next condition is the creation of the "fit/unfit" criterion for various integrated circuits. Modeling tools can generate signal fluctuations, but only the chip supplier can provide calculation criteria for classifying the resulting signals as "fit" or "unfit." As a simple example, let us give the eyewidth for the controller signals, which is necessary when reading incoming DQ signals. For DRAM, this information is provided by the Jedec organization; it is already available in DRAM specifications and on the Jedec website (jedec.org). Accordingly, the requirements for controllers are given in the specifications for the controllers.

But it happens that the suppliers of controllers do not publish such information in their specifications. In such cases, it is necessary to request information from the process engineer of the supplier. It should be emphasized that these requirements do not apply to the modeling itself, but are necessary to characterize and describe the parameters of the designed bus, if suddenly the board will be returned for updating. After all this, if an eye diagram is plotted on an oscilloscope, as shown in Fig. 2.15, and there is no mask with which to compare the diagram, it is impossible to assess the suitability of the system and find out how large margin in terms of the parameters the design will have. Therefore, it is better to request from a supplier's process engineer the data related to checking the integrity of the system's signal, rather than checking the simulation results.

General parameters that in this case need to be requested from the supplier are:

- the ratio of the input time between the DQ/DQS signals during reading operation.
- deviation of the output DQ/DQS signals during recording operation.
- deviation of the output DQS/CLK signals during recording operation.
- deviation of output address/CLK signals during the transmission of an address/command.

For standard DRAM, this information is usually easily available on the Jedec website, and the specifications contain information concerning not only the integrity of the signal, but also the reliability of the DDR channels (Fig. 2.16 shows an example of the information submission) [5].

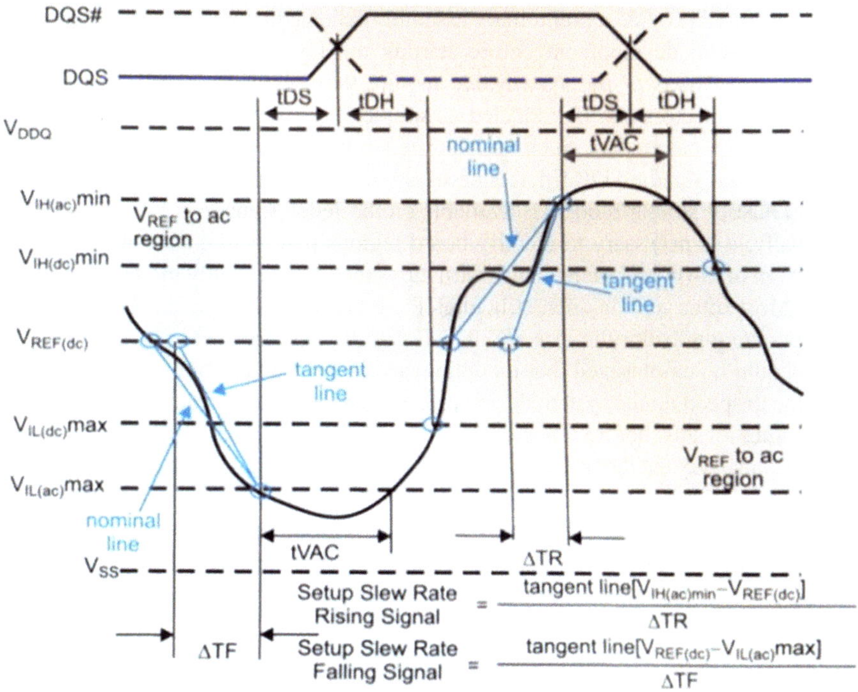

Fig. 2.16 An example of information required for modeling

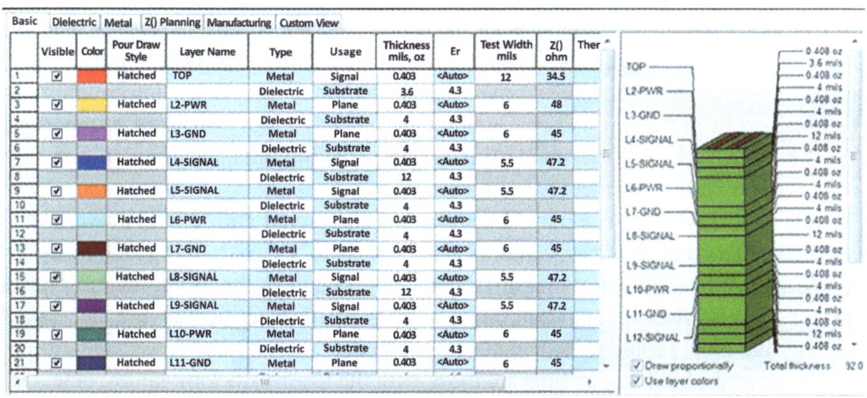
Basic | Dielectric | Metal | Z() Planning | Manufacturing | Custom View

	Visible	Color	Pour Draw Style	Layer Name	Type	Usage	Thickness mils, oz	Er	Test Width mils	Z() ohm	Ther
1	☑		Hatched	TOP	Metal	Signal	0.403	<Auto>	12	34.5	
2					Dielectric	Substrate	3.6	4.3			
3	☑		Hatched	L2-PWR	Metal	Plane	0.403	<Auto>	6	48	
4					Dielectric	Substrate	4	4.3			
5	☑		Hatched	L3-GND	Metal	Plane	0.403	<Auto>	6	45	
6					Dielectric	Substrate	4	4.3			
7	☑		Hatched	L4-SIGNAL	Metal	Signal	0.403	<Auto>	5.5	47.2	
8					Dielectric	Substrate	12	4.3			
9	☑		Hatched	L5-SIGNAL	Metal	Signal	0.403	<Auto>	5.5	47.2	
10					Dielectric	Substrate	4	4.3			
11	☑		Hatched	L6-PWR	Metal	Plane	0.403	<Auto>	6	45	
12					Dielectric	Substrate	4	4.3			
13	☑		Hatched	L7-GND	Metal	Plane	0.403	<Auto>	6	45	
14					Dielectric	Substrate	4	4.3			
15	☑		Hatched	L8-SIGNAL	Metal	Signal	0.403	<Auto>	5.5	47.2	
16					Dielectric	Substrate	12	4.3			
17	☑		Hatched	L9-SIGNAL	Metal	Signal	0.403	<Auto>	5.5	47.2	
18					Dielectric	Substrate	4	4.3			
19	☑		Hatched	L10-PWR	Metal	Plane	0.403	<Auto>	6	45	
20					Dielectric	Substrate	4	4.3			
21	☑		Hatched	L11-GND	Metal	Plane	0.403	<Auto>	6	45	

Fig. 2.17 An example of the presentation of information about the assembly of all layers of the board, including the width of the tracks and properties of the dielectric

Information on the assembly of the board shall be very accurate (Fig. 2.17). These data are usually given by the developers or manufacturers of the boards, and they include detailed information about the purpose and location of the layers, the width of the tracks (for a given impedance), and the properties of the dielectric. The board developer shall provide a preliminary assembly scheme that will be used at the stage of experimental development before starting the layout. Then it is necessary to identify some parts of the system that require the use of third-party models. For example, connectors can be modeled as simple RLC-circuits, or as more complex models with S-parameters. Such options need to be discussed with the developer. If the system has standard DIMM, it is necessary to use the appropriate card models for them. A list of similar models is available on the Jedec website.

Finally, it is necessary to identify board regions that require special analysis. In the design of the board, there may be structures that are preferably modeled using 3D tools. Most often among such structures there are vias. The higher the data transmission rate, the better the vias are modeled as 3D structures.

It should be emphasized that modeling should be used as a tool with which it is possible to speed up the system design process. The modeling itself should not take a lot of time; for this, before it begins, it is necessary to carry out appropriate training. That is, analyze the board, contact, and request technical information from the suppliers of DRAM chips, controllers, and connectors. With such information, development and design verification will become easier and faster.

2.5.3 Certain Methodological Features of the Modelling Stage

The first step after performing the DDR simulation is to study the bus requirements. In fact, these are the requirements of the receiver to the controller (during reading) and DRAM (during recording and transmission of the address/command). Such information for the controller is given in its specification. DRAM information can be obtained from the supplier or found on the Jedec website. This step is important, since it is possible to understand all the nuances of the obtained simulation results only if the basic requirements are known.

The only way to ensure that the results are interpreted correctly is to work with modeling tools that give easily understandable results. Often, based on the results of modeling, a huge amount of measurement data is obtained; therefore, for subsequent analysis, their filtering and ranking according to their criticality are necessary. Usually the most problematic signals are the first.

The most problematic circuits are those with the lowest stock of values, at least in one of the measured parameters. For example, a signal that has the shortest installation time among all circuits can be considered as the most problematic circuit. In general, it is a good tradition to analyze problem chains, even if they do not violate the performance. Even when the simulation did not show a malfunction of the structure, the results can be quite close to the limit, which indicates a trouble spot in the project and manifests itself in the form of failures in work, interruptions often occur in real systems, which further forces expending significant efforts to eliminate deficiencies. This is especially true when one signal has an abnormally low margin for each of the measured parameters. For example, if the DQ0 signal has a time margin of only 5 ps, while other signals have at least 40 ps, then it is necessary to probably pay attention to the problematic DQ0 signal (and the corresponding strobe signal).

Such incorrect behavior of signals may indicate several phenomena, and their causes may be different. It often happens that either the margin on installation time, or the margin on holding time is very low. If one of the indicators is low (for example, the installation time margin) and the other indicator (in this case, the hold time) is high, then the strobe signal (or clock signal) is not synchronized with the DQ signal (or address). In such cases, it is necessary to make corrections to the parameters of the controller and, if possible, calibrate each bit individually. After that, the simulation can be performed again (automatically, if the simulator has such a function) with the corresponding delays for the signal bits. If the controller does not allow calibration for each bit, then for wiring some of the signal bits a higher or lower delay value should be set.

The next element to analyze is undesirable oscillations that can cause unnecessary changes in several parameters, including an increase in the maximum overshoot value or a change in set/hold time. Some deviations, for example in overshooting, lead to chip failure. Such failures do not always occur in the first inclusion and are more likely to be observed after some time. So the sooner you learn about it, the less rework and debugging will be needed in the subsequent stages, which in turn will reduce costs. Undesirable oscillations are often caused by incorrect termination of signals. This is usually corrected by selecting different ODT values on the receiver until the undesirable oscillation is reduced to an acceptable level. Figure 2.18 shows how the signal falls outside the threshold value Vdd (1.5 V). Significant overshoot of the signal does not occur (for this it is necessary to cross the 1.9 V threshold), but the values of the voltage/time region shown in red in Fig. 2.18 [6] may be enough to violate the Jedec specification requirement for the area of maximum overshoot.

The third problem—excessive noise and crosstalk affecting the signal—is solved by spatially separating the signals from each other.

Another problem is the presence of loops in the circuit (Fig. 2.19) [6], which causes the appearance of signal reflections, often looking like "steps" in signal oscillations (Fig. 2.20). If the loop can be shortened, then the signal integrity is improved. Often, such "steps" in the form of oscillations appear because measurements were taken during reading operation at the controller output. This can be quickly corrected by raising the measurement point to another area. Similar problems appear because quite often the IBI5 file for the controller takes measurements on the controller outputs.

After identifying the main problems and the causes of their occurrence, the next step is to eliminate them on the board and then simulate them. To save time, it is better to first simulate the work of only those areas where changes were made. And only after local fixes, it is possible to proceed to the modeling of the entire board/system.

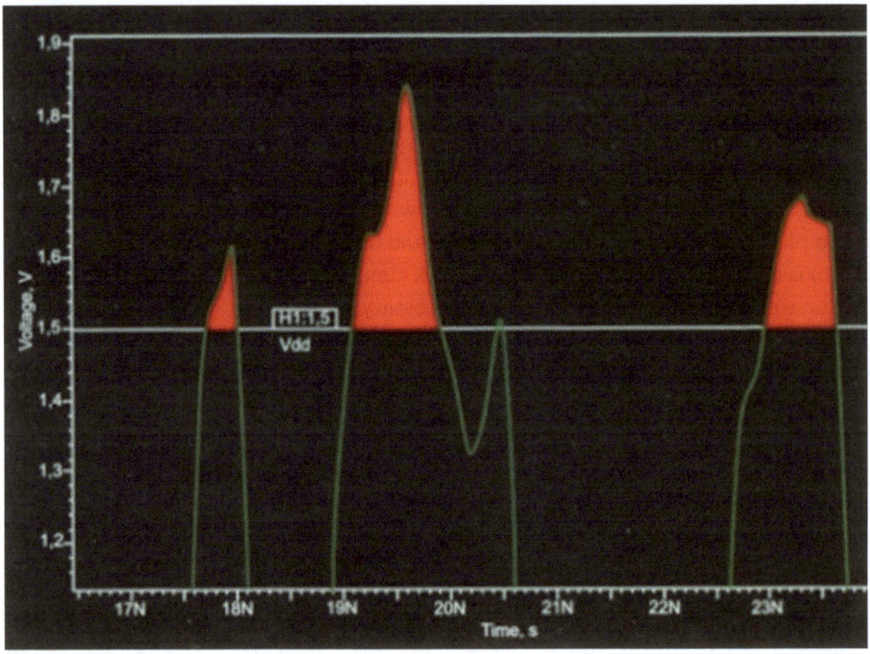

Fig. 2.18 Overshoot scope

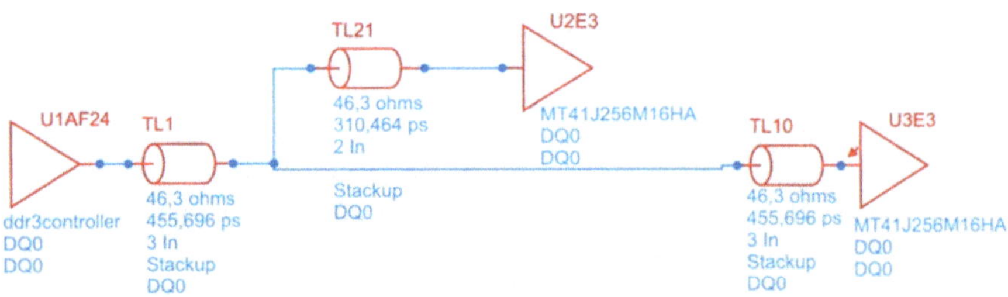

Fig. 2.19 Typical parameters of the loop in the designed electronic system

Finally, after the printed circuit board is manufactured, the simulated waveforms can be compared with the signals obtained with an oscilloscope. And there is no need to measure each signal on the bus with an oscilloscope—it is enough to measure several signals and verify them with the results obtained during the simulation, which will give confidence in the successful implementation of the simulation.

Correct simulation settings and correct analysis of the results will help to reduce the total time required for designing high-speed DDR subsystems.

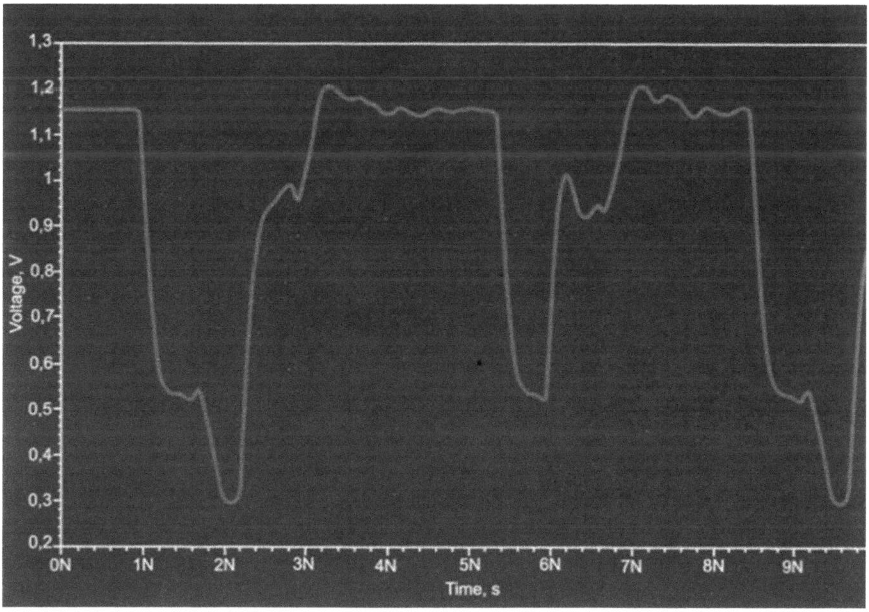

Fig. 2.20 Reflection effect due to loop

2.6 Features of the Organization of High-Speed Data Transmission Channels with RS-485 Interface

2.6.1 Synchronous Data Transmission Channels Based on RS-485

Data transmission channels based on high-speed modems with an R5-485 interface are widely used in applications such as GPS transceivers, traffic monitoring systems, and seismic prospecting complexes. Communication channels in such systems should provide data transmission with a speed of no less than 8–20 Mbps over a distance of 70–300 m and more. The development of such communication channels is a rather complicated technical task, since it requires the developer to take special technical solutions to compensate for the growing time delays in signal propagation through the cable and through the transceiver, as well as taking into account the frequency-dependent losses and distortions in the cable that are inevitable [8].

Here, first of all, it is necessary to conduct a qualitative analysis of such a very important factor in practice, as the ratio of data transmission rate and cable length.

Figure 2.21 presents a generalized data transmission rate—cable length graph [8]. As can be seen from this figure, with a cable length of about 1220 m, a voltage divider formed by the resistance of a twisted pair cable with a wire diameter of 24 AWG (0.51 mm) and a load resistor resistance corresponding to the characteristic impedance of the line attenuates the signal by almost two times (6 dB). Typically,

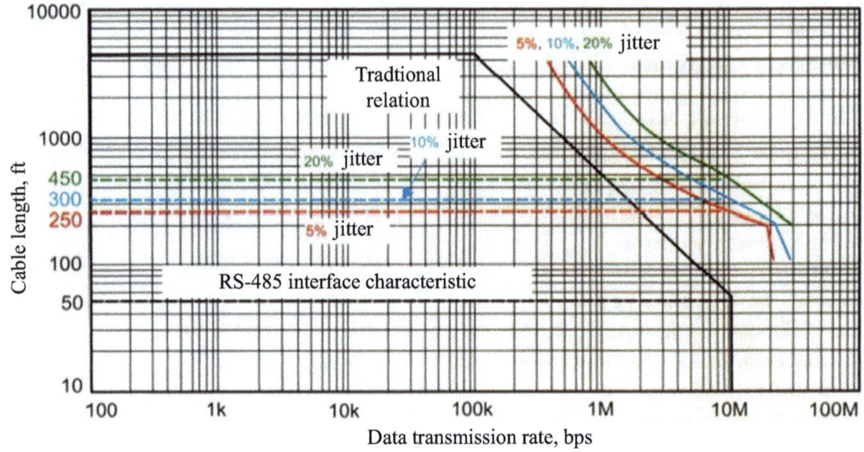

Fig. 2.21 Generalized dependence graphs of data transmission rate on cable transmission line length [8]

this signal attenuation is the reference value by which the system designer determines the maximum cable length of the transmission line. Due to the signal loss, most modern data transmission lines over long distances mainly operate at speeds of 10–20 Kbps, or 100 Kbps. As is known [1], at higher data rates, the inductive and capacitive characteristics of the line begin to affect, which limits the maximum cable length between traditional transceivers. With an even greater increase in the transmission rate, the quality of the transmitted signal begins to be determined by the characteristics of the transceiver itself. The speed range of 1–10 Mbps is the main operating range for most modem applications during data transmission over longer cables.

The use of traditional high-speed transceivers often leads to significant phase distortions of the signal (jitter) during data transmission, which is usually associated with a number of factors: with delays in the propagation of signals through the transmission channel; with losses; with the decline of the characteristics at high frequencies; there are also a number of problems associated with the occurrence of intersymbol interference, depending on the adopted method of data encoding and the irregularity of delays (temporal delay skew) while using for the transmission of combined line signals—shielded cables with several pairs of wires and unshielded twisted pair cables (UTP).

As it was noted above, in general, any data channel can be represented as consisting of a signal source (transmitter), a transmission medium (cable and connectors), and a signal receiver. In addition, each of these sections of the transmission channel has its own delay in signal propagation and its own temporary "delay skew," of which the common delay in signal propagation and the total temporal skew of delays in such a transmission channel are usually added.

In turn, standard synchronous interfaces usually consist of two parallel transmission channels: one to transmit synchronizing clocks, and the other to transmit the

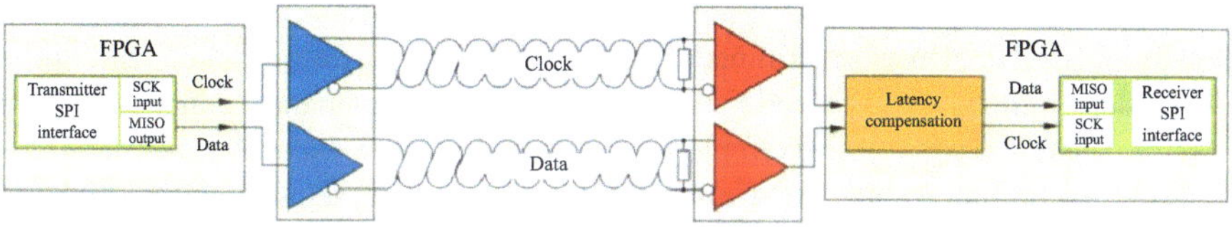

Fig. 2.22 The structure of the organization of standard channels

data itself (see Fig. 2.22). It is clear that for correct data exchange through the interface it is necessary to ensure full synchronization of signal transmission on both channels in order to avoid such problems, for example, the occurrence of bit errors at the receiving side of the information channel. Obviously, with this approach, the factor of temporary "delay skew" between these two channels will play a more significant role than the factor of the overall delay in signal propagation.

Here it is necessary to clarify once again the meanings of the terms "signal skew" and "temporal delay skew." As is known, the time delay of signal propagation exists in all types of components of the transmission medium; this is the difference between the start of transmission of time (the signal enters the cable channel) and the time when this signal appears at the opposite end of the channel. The numerical value of the signal skew for a twisted pair cable directly depends on the speed of the signal passing through the cable, its length, and the frequency of the signal.

In turn, the rate at which a signal travels through a cable depends on the type of dielectric materials used in this cable and is usually measured as a percentage of the light speed. For example, as follows from the analysis of statistical data, most of the widely used cables of category 5 (CAT-5) have signal propagation speeds in the range of 65–70% of the light speed, and such cables that are widely used in modern telecommunication cables for the RS-485 interface, as Belden 3105A (with one pair of wires) and 3109A (with several pairs of wires), are characterized by increased rates in the range up to 78% of the light speed.

So, to ensure the conditional correct transmission of signals in international standards for local area networks (LAN), specialists determined that this delay in signal propagation and temporal skew of all delays should not exceed a value that is only a few tens of nanoseconds per 100 m of cable length. For example, for cables of CAT-5 and CAT-5e categories, such maximum nominal values for signal skew and temporal skew delay should not exceed 550 ns/100 m and 45 ns/100 m, respectively. For cables designed directly for the EIA-485 interface, even lower signal propagation delays and time skew delays are typical. For example, the delay skew for this standard can be no more than 15 ns/100 m.

Increasing the cable length leads to a proportional increase in the numerical value of the signal propagation delay.

As is known [1], different signal pairs of wires in the same cable, as a rule, have different time delays for signal propagation. The difference between the values of these delays, called by the developers as "time delay skew," is mainly influenced by the difference in the number of wire twists between different pairs of wires used to minimize crosstalk between pairs of signal wires.

So, to ensure the correct signal transmission in the standards for local area networks (LAN), it was determined that the "signal propagation delay" and "time delay skew" should not exceed a few tens or hundreds of nanoseconds for every 100 m of cable length. For example, for cables of CAT-5 and CAT-5e categories maximum nominal values for signal skew and temporal skew delay should not exceed 550 ns/100 m and 45 ns/100 m, respectively. For cables designed for the EIA-485 interface, even lower signal propagation delays and time skew delays are typical. For example, the delay skew in this cases can be no more than 15 ns/100 m.

Of course, the temporal delay skew between the synchronization and data transmission channels shall be eliminated, taking into account the signal transmission rate, cable length, and transceiver synchronization mode. As a rule, this "skew" is eliminated by compensating for the signal delay on the receiver side, for which an algorithm is built into the FPGA program for measuring the time difference between the first incoming signal (either a clock pulse or data) and the next incoming signal. To do this, in logic circuits of transmission channels, a sampling frequency is often used that at least eight times exceeds the working frequency of data transmission synchronization, which, in turn, is always twice as high as the actual data transmission rate.

For example, when transmitting a data stream through a communication channel with a speed of only 5 Mbps, the frequency of transmission clock pulses is 5 MHz (or 10 Mbps). In this case, the sampling frequency of the FPGA used should be at least 40 MHz. To limit the delay compensation time with the duration of one transmission synchronization cycle, in order to simplify the data transmission process, it is very important to correctly select the data transmission rate, as well as the cable and transceiver parameters. That is why at high data transmission rates it is recommended to use high-quality cables with minimal time delay skew.

Consideration of the time delay skews between components of transceivers, sometimes called intercomponent skews, is usually important only for "short" data transmission channels, which have a chance to fully compensate for delays. However, in cases of using "long" cables, the problem of time delay skews in them becomes dominant, since, due to the large cable lengths, the numerical values of these "skews" may already exceed several times the value of the numerical value of intercomponent skews.

2.6.2 Asynchronous Data Transmission Channels Based on RS-485

Developers of modern data transmission channels are well aware that all the above features also exist when organizing asynchronous channels.

A significant capacitive load on the cable of an asynchronous transmission line also significantly "rounds" the rectangular edges of the pulses of the original signal, which can lead to the effect of signal jitter. Phenomena of charge and discharge of the accumulated capacity of the cable can be the cause of the loss of whole bits of

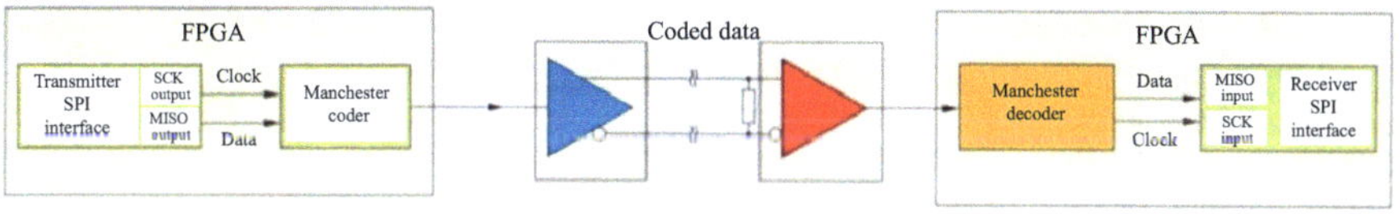

Fig. 2.23 An example of the organization of cable high-speed single-channel data transmission interface

data at those moments when the "long" sequence of "ones" of bits in the transmitted binary code changes to a single "zero" bit, and vice versa.

To prevent the effect of irregularities in the processes of discharges and charges of capacitive components, it is recommended when organizing such a transmission line, all data signals be encoded into a self-synchronizing stream. In applications with an R5-485 interface, the Manchester code is usually used as an encoding Scheme [9, 10].

The simplest equivalent logic circuit for "Manchester" coding usually consists of an "EXCLUSIVE OR" logic circuit connected to a serial peripheral interface (SPI) port. Transitions of signals from the lower level to the upper level and from the upper level to the lower level at the output of the coding scheme correspond to "1" and "O" on the data line; at the same time, long sequences of bits of the same polarity turn into a series of sync pulses. The encoded signal is fed to the input of the generator and transmitted over the communication channel. Decoding occurs at the output of the receiver, which is quite simply implemented using FPGA (see Fig. 2.23).

It should be noted that the integrity of the transmitted signal can be significantly improved with the help of transceivers, the receivers of which are equipped with equalizers. Such transceivers allow amplifying not only the high-frequency components at the receiver input, but also the signals themselves. To do this, an active high-pass filter (HPF) is usually built in between the input of the comparator and the cable.

This approach makes it easy to optimize the characteristics of an internal filter selected by the developer for the needs of a particular application, which often effectively solves the problem of noise reduction and allows using lower order filters in the lower megahertz range to adjust cable attenuation parameters. It is clear that in the range of several tens of megahertz to obtain the same accuracy, the designer of the transmission channel shall apply filters of a higher order.

It should be noted that equipping the receiver with an equalizer eliminates the undesirable influence of high-frequency radiation and allows protecting the transmission channel from electromagnetic interference discussed in detail in Chap. 4, while providing only a slight increase in electricity consumption due to the use of the appropriate high-pass filter.

In support of the above-mentioned, it is possible to bring the data in Fig. 2.24 [8], which compares the dependence of the data transmission rate on the line length for a standard EIA-485 transceiver and upgraded transceivers, in which the receivers are equipped with built-in equalizers. As can be seen from this figure, device "a" transmits data at a rate of 10 Mbps over a cable with the length of 250 m. If the developer can mitigate the requirements and allow the presence of minimal jitter,

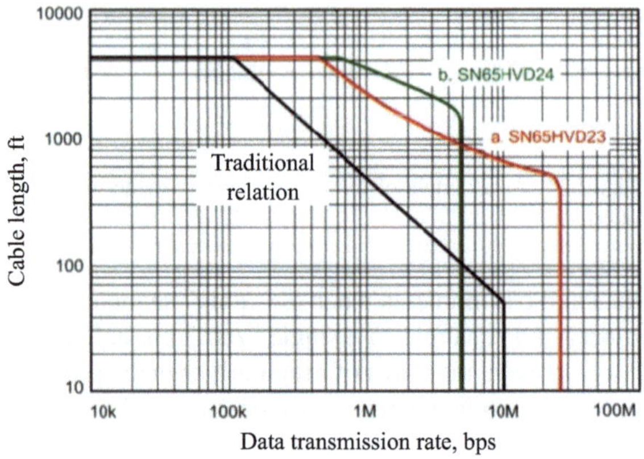

Fig. 2.24 Dependence of the data transmission rate on the transmission line length for a traditional RS-485 transceiver and receivers with equalizers

then this device will be able to transmit data even with a cable with the length of 300 m. Device "b" transmits data already at a distance of almost twice as large (more than 500 m). At the same time the transmission rate in the channel is also reduced by half—to about 5 Mbps.

Thus, by combining Manchester coding and a modern transceiver, the receiver of which is equipped with an equalizer, it is possible to design a highly efficient single-channel asynchronous data line, which, unlike its synchronous analog, does not require an additional channel for transmission of clock pulses and any requirements for the quality of the cable used.

Then taking into account the above-mentioned recommendations, for high-speed reliable data transmission over long distances, developers need neither the high-precision additional synchronization diagrams nor the use of high-quality expensive cables nor expensive transceivers. A reliable and inexpensive data transmission channel can be implemented using proven coding schemes and a transceiver with small electromagnetic radiation and a receiver having an equalizer circuit in its design.

Here it is appropriate to provide specific examples—for the organization of reliable operation of high-speed data transmission channels with a large difference in earth potential, Texas Instruments manufactures 5N65HVD23 and 5N65HVD24 transceivers with a built-in equalizer, which provide a wide dynamic range for in-phase signals of −20 to 25 V, which clearly demonstrates dependency diagrams (Fig. 2.24).

2.7 Circuit-Based Methods of Protection of RS-485 Interface Buses from Electrical Overvoltages

The RS-485 bus is widely used due to the fact that the minimum amplitude of the differential signal is 1.5 V and the bus operates steadily over a wide range of common-mode voltage of −7 to 12 V. Initially, the RS-485 bus was used only in

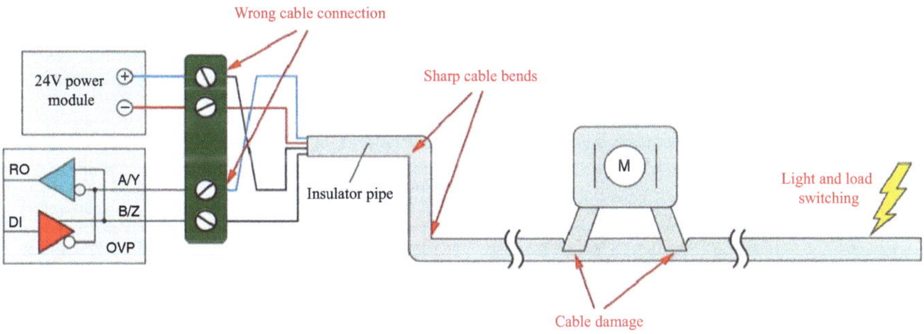

Fig. 2.25 Typical channel structure for the case when the transmission line is laid in the same cable as the DC power line

communication networks of laboratory instrumentation, and then it became widespread in industrial automation and building automation systems, production PLC networks, in the management of production processes, and in alarm systems of oil towers, coal mines, and the petrochemical industry and so on [1].

With the growing use of RS-485 buses, consumers have come up with requirements to increase the amplitude of the output voltage in order to guarantee a greater noise immunity margin; and also to expand the range of common-mode voltage, which allows to increase the difference between the earth potentials of remote bus nodes; and provide protection against periodic voltage surges (overvoltage protection, OVP), which in practice far exceed the maximum voltage level of the transceiver, as defined in the technical description.

Using the work [8] let us consider some features of ensuring the protection of electronic devices with this bus from overvoltages and fast transient processes.

Typically, 24 and 48 V DC power lines for industrial and telecommunication systems are laid in the same cables and channels as data transmission lines in RS-485 networks. Moreover, situations that contribute to the appearance of voltage surges in such systems often arise.

Figure 2.25 [8] a typical structure of the organization of the DC power line is presented in a simplified form.

This figure lists the main causes of possible failures in data transmission over communication lines.

For example, if a DC power supply uses the same connector or screw terminal block as the data lines of the adjacent bus node, sometimes there are failures due to improper installation of one or more power wires with the bus terminals of the transceiver.

Another reason for the failure is the method of laying the cable itself. The sharp bends violate the requirements for the minimum allowable bending radius of power and communication cables. Over time, the increased mechanical pressure on the cable can lead to a breakdown of its insulation up to the appearance of a short circuit between power and information lines. This phenomenon is also observed when production or telecommunication equipment is installed directly on an insulating

Fig. 2.26 The dependence of the peak power absorbed by the 600-W transient voltage suppressor on the duration of the pulses

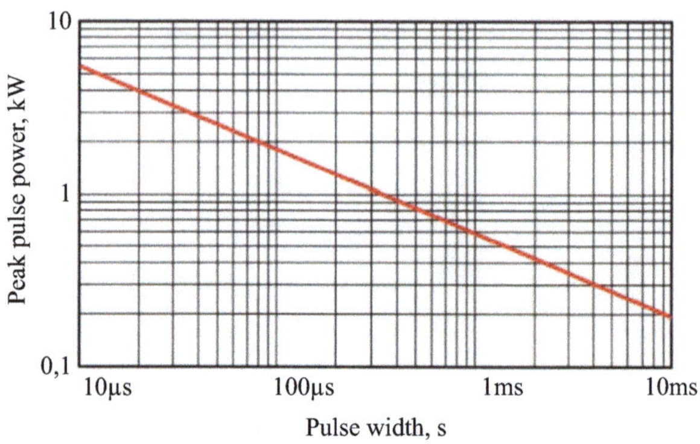

pipe, which leads to its flattening. In this case, there will also be chaotic voltage surges.

The developers of communication lines for high-speed electronic systems often consider that installing an external transient voltage suppressor (TVS) into a standard transceiver provides effective protection against short bursts and prolonged voltage surges. However, this is a wrong point of view, since the maximum power that a limiter can absorb significantly decreases with increasing duration of the transient process.

Figure 2.26 shows the typical dependence of the power absorbed by the standard 600-W transient voltage suppressor [9]. In the above case, the absorbed power of 600 W is normalized for a pulse with a width of 1 ms. The range of change of values along the time axis is 1 μs–10 ms, and the power level varies from 6000 to 2000 W. From this characteristic it follows that as a result of prolonged exposure to electrical overvoltage, the device (transceiver) may well become out of order. To protect bus nodes from overvoltage in a wide range of values, it is required to use special transceivers protected from such failures, for example, ISL3245xE family from Intersil. These transceivers are equipped with special protection against direct current overvoltage up to ±60 V and against voltage surges during the transient process up to ±80 V.

As is known [8], there is built-in and discrete protection against such failures.

Discrete solutions that provide protection for a standard transceiver against overvoltage increase the cost of the system, and also constructively take up more space than a transceiver with its own built-in protection.

Suppose that the full protection function of a half-duplex ISL32452 standard transceiver (Fig. 2.27a) is performed using a discrete circuit (Fig. 2.27b). First, the paths for transmitting and receiving the signal shall be separated in order to realize an output stage with a high maximum voltage during continuous operation. In this case, a full-duplex transceiver is required. The output stage is implemented using four discrete transistors or a built-in half-bridge, the control inputs of which need to convert bus signals from the RS-485 format to the voltage levels of logic signals TTL or CMOS. This requires a control logic circuit between the transceiver and the discrete output stage.

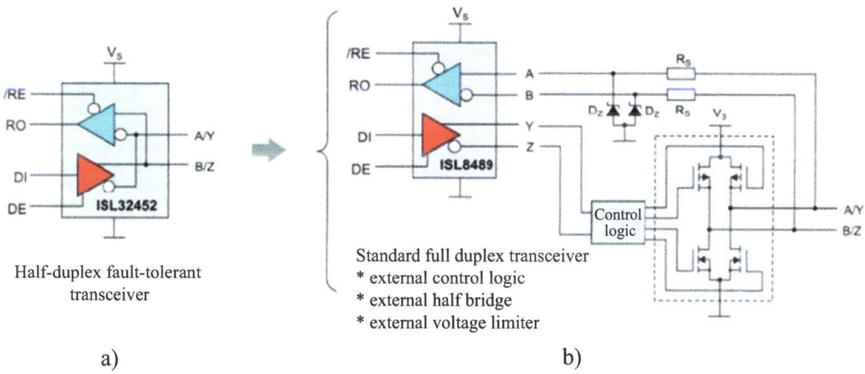

Fig. 2.27 (**a**) a half-duplex transceiver with protection against failures and (**b**) a standard full-duplex transceiver

A special discrete voltage suppressor consisting of voltage-reference diodes and series resistors is installed in the receiving path in order to limit the bus voltage at voltage surges.

Figure 2.27 shows that the discrete circuit is already overloaded, although it is equipped with only basic functions for overvoltage protection. At this, the current suppressor, which is a very important component of electrical protection, is missing in the circuit.

Current limitation is the most important function that is used when protecting against voltage surges, when the driver is actively controlling the bus. Due to the fact that a driver-to-ground connection usually has a low impedance, the bus currents flowing through this driver become huge and can damage the device if they are not limited.

Protected transceivers, in which the common-mode voltage range is wider than that defined by the RS-48S standard, require twice the limitation of the reverse current in the driver circuit. Figure 2.28 characterizes the current-limiting function of a protected transceiver from the ISL3245x family, which operates in a wide range of common-mode voltage of ±20 V [11].

The first level of 63 mA reverse current ensures that the driver never experiences overloads due to reverse current when controlling loads in the whole common-mode voltage range of 40 V. A very small second value of the reverse current of 13 mA allows minimizing power dissipation in the active state of the driver. This circuit does not allow the current to exceed the value specified in the RS-485 specification, even in the event of a failure condition.

In the event of a short circuit danger, the protective shutdown function in case of overheating disconnects the driver from the line as soon as the crystal temperature exceeds a predetermined value. As a result, power dissipation is reduced, and the crystal is cooled. The driver automatically starts after the temperature drops by 15 °C. If the condition for the occurrence of a failure persists, the protective shutdown cycle in case of overheating and start-up are repeated until this condition

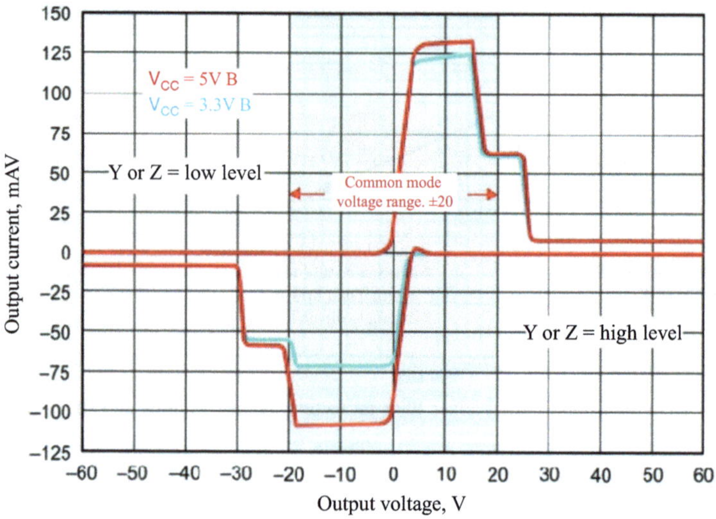

Fig. 2.28 Output current limitation of the driver depending on overvoltage

disappears. The receiver continues to function during a protective shutdown in case of overheating, and the protection against failures remains active regardless of the driver state or the disconnection of this chip.

The final overvoltage during an extreme transient, for example, caused by a lightning strike, can easily exceed the protective level of the transceiver. In order for its operation not to be disturbed, it is necessary to ensure the absorption of excess energy by external diodes TVS-suppressor. When installing external TVS devices in a fail-safe transceiver, the following basic conditions shall be met [11–14]:

1. The breakdown voltage of the TVS suppressor should be 1 V higher than the maximum common-mode application voltage or maximum DC supply voltage if it is higher than the common-mode voltage. In applications with a standard range of common-mode voltage −7 to 12 V with a minimum breakdown voltage of $V_{BR-min} \geq 13$ V, the value of V_{BR-min} for bus lines adjacent to the power lines with a nominal voltage of 24 V should not be lower than 31 V, since 24-V systems deviate from the nominal value up to 30 V.
2. The peak response voltage of the TVS suppressor should not exceed the maximum protective level of the transceiver.

Figure 2.29 shows the corresponding protection circuit (a), as well as a comparison of the breakdown voltage and response voltage (V_{BR} and V_{CL}) of the TVS suppressor with maximum common-mode voltage, DC voltage, and levels of protection against failures (b).

The transceivers with protection against failures and a wide range of supply voltage can be used in 5-V and 3.3-V systems, which simplifies the delivery of the required components and reduces the cost of the order with large volumes.

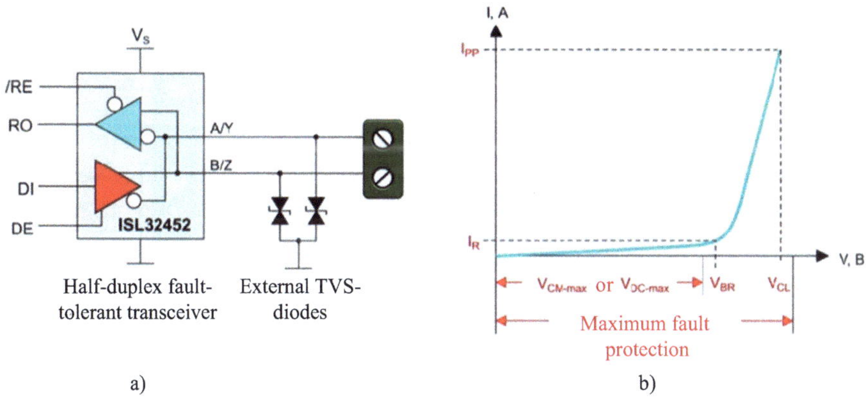

Fig. 2.29 (**a**) a half-duplex transceiver with protection against failures; (**b**) maximum protection against failures

2.8 Practical Recommendations for the Design of the Bus Node of the High-Speed Interface RS-485

Let us take a brief look at the basic requirements that shall be used when developing high-speed data transmission systems based on the RS-485 standard [15–19].

In high-speed and ultra-speed data channels of the RS-485 standard, as a rule, the main configurations shown in Fig. 2.30 (two-point, full- or half-duplex, or multipoint) are applied.

These circuit-based options do not reflect the whole possible variety of design solutions, starting with the choice of a transceiver and ending with the topology of printed circuit boards for each bus node, as well as the choice of cable and configuration of interconnections. Developers use the term "bus node" to refer to an electronic device, the design of which includes a transceiver, a controller (usually an FPGA-based one), capacitors and resistors for protection against voltage surges and threshold discharges, and interference suppressors for elimination of noise caused by transients in buses. All of these requirements need to be taken into account in order to ensure higher data transmission rates and the use of longer cables, while ensuring minimal data transmission errors.

Signal degradation due to jitter is a major factor that limits the choice of cable length in practice. Jitter appears as a result of the desynchronization of the transmission processes between the signal generator and the receiver, as well as in the cable [16, 17]. The temporary asynchronous behavior between the generator and the receiver is the difference between the propagation delays for the rising and falling edges of the signals of these devices.

The second type is cable asynchronous behavior, which depends on the configuration of the transmitted bits, caused by the variation of the rise and fall times of the signal pulses on the bus, which is associated with a change in the composition of the bit sequences. Because of this asynchronous behavior, the amplitude of the data

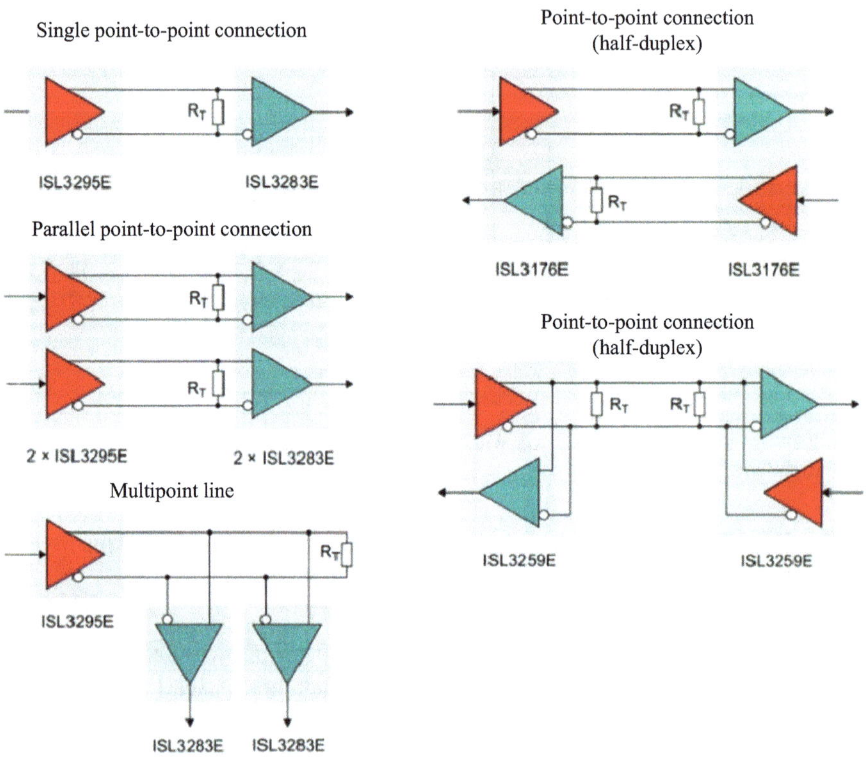

Fig. 2.30 Basic configurations of the construction of a transmission line of standard RS-485 based on Intersil microcircuits

signals is reduced, their edges are "blurred", a time shift occurs, and the pulses are also "smeared" in adjacent transmitted bit intervals.

The use of the standard Manchester 8b/10b or 33hex coding diagram makes it possible to significantly reduce jitter, since time intervals appear in encoding the data stream, which makes it possible to evenly charge and discharge the parasitic cable capacitance, and also provides the required signal amplitudes. However, due to the fact that data encoding shortens the charging and discharging time of the cable capacity, the amplitude of the bus signal decreases.

Therefore, one of the priorities of the high-speed electronic system developer is the choice of a specific type of transceiver.

To provide a reliable signal at the receiver input, it is recommended to use high-speed transceivers with a large differential output voltage V_{OD}. Due to the large value of this parameter, the reduction of the signal amplitude due to its weakening in the cable, data encoding, and the common-mode load is compensated, and a sufficient margin of noise immunity is provided at the remote inputs of the receiver.

At the same time, in low-voltage transmission systems, as far as possible, those 3-V transceivers with output stages that provide the output voltage required by the

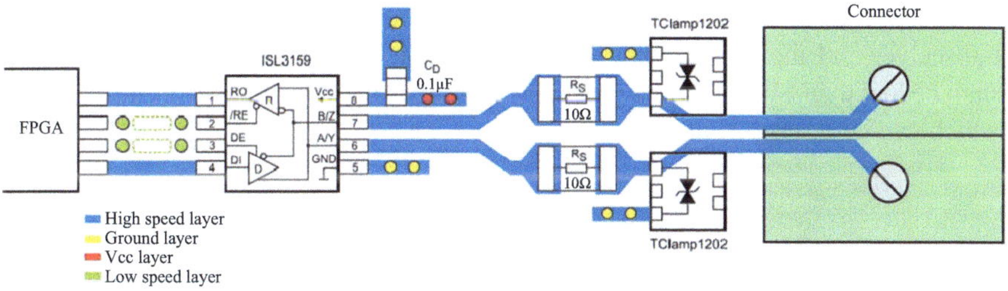

Fig. 2.31 Topological draft of the location of the main components and bus node interconnection

R5-485 standard should be avoided only at supply voltages above 4 V. At lower supply voltages, the efficiency of the output transistors of the microcircuit drops noticeably, with the result that the V_{OD} value becomes 40% less than the 1.5 V minimum that is defined by the RS-485 standard. Such a small output voltage usually does not provide enough noise margin to trigger the remote receiver at the second end of the line.

Thus, the minimum V_{OD} value of the Intersil high-speed transceiver family is in the range of 160% of the required minimum of 1.5 V at a supply voltage of 4.5 V to 100% at a supply of 3.0 V, which exactly corresponds to the value specified by the R5-485 standard. Typical V_{OD} values measured at maximum operating temperatures exceed the requirements of R5-485 by 27% with a supply of 3.0 V and by 70–93% in the case of transceivers with a supply voltage of 5 V.

Moreover, due to less signal asynchronous behavior in the transceiver, the total jitter of the data channel is also reduced. All Intersil high-speed transceivers have a maximum signal mismatch of 1.5 ns, and the mismatch between the units of these devices does not exceed 4 ns.

Let us consider some practical aspects of designing a bus node of RS-485 interface, which were analyzed in detail in [15, 18, 19].

Figure 2.31 shows a typical bus node, which includes a transceiver, an FPGA controller, resistors for protection against voltage surges, and suppressors of interferences caused by a transient process for protection against lightning. The specified location of the transceiver terminals allows simplifying the design procedure due to the fact that the A/Y and B/Z bus pins are on the same side of this IS, and the one-way DI, RO data lines, and DE/RE control lines are on the opposite side of the ISL3159 chip.

The use of a data transmission line with controlled impedance minimizes EMI. On the bus side, the differential impedance of its conductors shall be matched to the impedance of the transmission medium (100 or 120 Ohm). On the control side, the impedance of conductors on a one-way line is typically 50 Ohm.

In a standard bus node, which consists of a receiver and a controller, the lines with controlled impedance are realized quite simply, if to carefully choose the length, width, and height of conductors and the distance between them, as well as provide electrical connection to the low-impedance layer—ground or power. It is necessary to correctly calculate the length of the conductors connecting the controller to the transceiver, as well as the length of the differential signal conductors between the transceiver and the cable connector. In each case, the electrical connection length shall be less than 1/10 of the length of the leading edge of the generator output signal, which can be described as follows [19]:

$$L_{\mathrm{COND.}} = 30.48 \cdot (t_{\mathrm{HAP.}}/10) \cdot v \cdot c,$$

where $L_{\mathrm{COND.}}$—conductor length, cm; t_{NAR}—the duration of the leading edge of the generator signal; v—signal rate as a percentage of s; c—the speed of light.

For a one-way line located near and above the earth layer, as recommended when connecting the controller and transceiver, the relative signal speed (v) is usually 56%. Since the length of the leading edge of the ISLZ159E device used here is 2 ns, the maximum conductor length is 33 mm.

The length of the conductor between the transceiver and the cable connector is calculated from the rate of the signal propagating along the conductors on a standard woven-glass reinforced substrate. Since this speed is 40%, the maximum conductor length is 24 mm.

Using the standard field analyzer, it is possible to get the exact location and geometrical dimensions of the conductors for any board design. This software calculates the characteristic and differential impedances, signal speed, crosstalk, and also allows analyzing different topologies. In addition to the line width, thickness of the dielectric layer and the dielectric constant, this tool can also be used for calculation of thickness of the conductors, parameters of the solder mask and etching.

To minimize the electromagnetic interference on the board, it is recommended to use four layers, which are placed in the order shown in Fig. 2.32. At the top of the board is the signal layer, under which the earth and power layers are placed. The control layer is at the bottom of the whole structure.

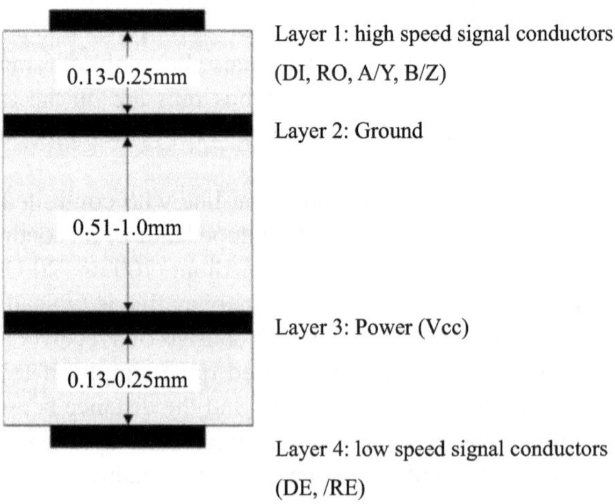

Fig. 2.32 Recommended design of the four-layer printed circuit board

Placing high-speed signal conductors at the top of the board avoids unnecessary complexity and increasing inductance, which is associated with the addition of through holes. This simplifies the interconnects between the bus connector and bus terminals of the transceiver, as well as between one-way high-speed data lines and the node controller.

Placing a solid earth layer under the high-speed signal level allows controlling the impedances of the transmission line interconnections and provides a low-impedance path for the return current. Placing the power layer under the earth layer leads to the appearance of an additional high-frequency shunt capacitance.

Placing a layer with low-speed signals at the very bottom of the board increases the versatility of the solution. As a rule, these communication channels are resistant to such irregularities of the structure as through holes, and the separation with a high-speed layer makes it possible to almost completely eliminate crosstalk.

When designing boards for high-speed interfaces, developers need to take into account a number of other equally important practical recommendations:

1. The size of through vias for signal conductors should be as small as possible to minimize the effect of differential impedance.
2. Continuous power and earth layers should be used to control and minimize noise in the power supply lines of the transceiver.
3. The electrical length of the conductors between the RS-485 connector and the transceiver should be as short as possible to minimize attenuation and reflection.
4. Smoothing capacitors (for example, with a capacity of 10 μF) are recommended to be installed as close as possible to the voltage regulators or near the power supply connector on the printed circuit board.
5. Next to the V_{CC} terminal of the transceiver, it is required to install decoupling capacitors of a small capacity of 0.1 and 0.01 μF to provide local power sources for chips that require current for internal switching.
6. If necessary, through vias are used to connect V_{CC}, ground, decoupling capacitors, and TVS diodes to high-speed integrated circuits.

Moreover, when designing a system, it is possible to use the following recommendations for choosing a cable, bus termination resistance and creating a branch from the main signal line.

So the RS-485 standard recommends using an unshielded twisted pair (UTP) with a characteristic impedance of Zo = 120 Ohm. As a rule, either single-pair RS-485 cables with Zo = 120 Ohms, or CAT-5 cables with four signal pairs and Zo = 100 Ohms are used.

When using CAT-5 cable in single-pair applications, three signal pairs that are not used are grounded through a resistor Rr = 100 at both ends of the cable to prevent data wires from penetrating noise from unused wires.

Note that when choosing a cable other than CAT-5 with several pairs of wires, it is impossible to use the so-called unbalanced cable, as it is designed to transmit analog signals with low or zero harmonic content. This type of cable produces large crosstalk and errors in high-speed digital data transmission.

Since the RS-485 standard allows the use of two terminating resistors Rr, these resistors should always be installed at both ends of the high-speed data link for optimal bus coordination. The resistance value Rr shall be matched with the characteristic impedance of the cable Zo or with the controlled impedance of the transmission lines on the printed circuit board: Rr = Zo = 100 Ohms.

Taking into account the above recommendations when designing printed circuit boards that provide high-speed data transmission with minimal asynchronous behavior and supply voltage of 3.3 and 5 Vcan overcome the problems of signal quality degradation that inevitably arise in high-speed RS-485 communication lines.

2.9 Design Features of High-Speed Electronic Devices Based on Programmable Logic Chips

2.9.1 Selection of a Specific Type Standard of the FPGA

When designing a modern high-speed electronic system using field-programmable gate array (FPGA), the developer first needs to solve a number of priorities, the main ones are the formulation of requirements for performing specific necessary functions in this programmable device and the organization of an efficient synchronization system [20–24].

The initial stage of development includes the determination of the overall architecture of the electronic system, which shall meet the specified requirements, as well as the specification of the way to implement this architecture at the hardware layer.

Figure 2.33 shows the typical architecture of such a system, built on the basis of a standard FPGA.

The following generally accepted and understandable designations are used here:

Config Metogu—configuration memory; Config—configuration data; Rx Data—input data; Tx Data—output; CAN IF—CAN interface; NVRAM—non-volatile RAM; Address; Data; Control; FPGA; Syncs—synchronization; Red—red pixel data; Green—green pixel data; Blue—blue pixel data; Display IF—display interface; JTAG—JTAG-interface; JTAG Connector; Power Supplies; Core Voltage; 1/0 Voltages—input/output port supply voltage; Misc. Voltage—auxiliary supply voltage; Reset; Clock—clock signal generator; Clock [7:0]—clock signal [7:0]; Pixel Clock—pixel synchronization.

Figure 2.34 shows an example of the structure of the power supply system of some typical electronic device. The following designations are taken here:

IP Voltage—input voltage; Isolating DC/DC—Isolating DC/DC converter; Intermediate Voltage—intermediate supply voltage; Switching Regulator—switching regulator; Linear Regulator—linear regulator; Core Voltage—core voltage; 1/0 Voltage—the voltage of the input–output ports; Misc. Voltage—auxiliary supply voltage; Reset—a unit for generating a reset signal; Reset—a reset signal.

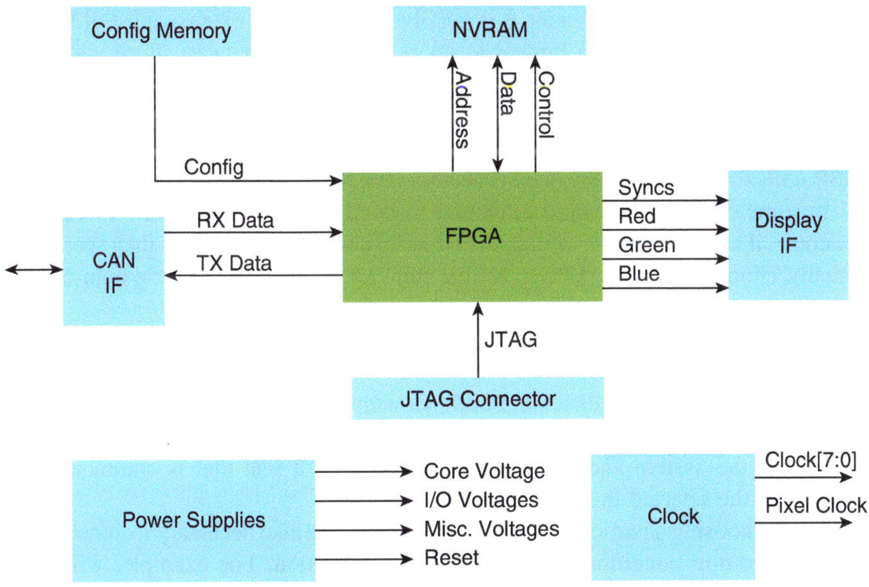

Fig. 2.33 Typical system architecture based on FPGA [20]

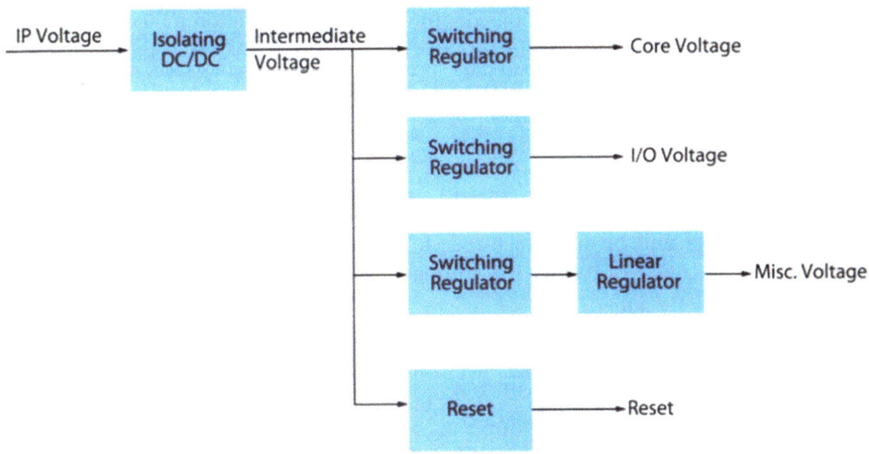

Fig. 2.34 Typical structure of a high-speed electronic device power supply system

The most important decision at the beginning of development is the choice of the most suitable FPGA from a sufficiently large number of available options. This choice is influenced by many criteria. The first and most important criterion is the amount of resources available in the FPGA.

As a result of this assessment, it is possible to reduce the extensive list of FPGAs offered on the market to a limited to about 10 types of devices, from which the optimal FPGA is then selected for a specific project under development.

Another defining criterion is the additional resources that may be needed in a designed high-performance electronic system, such as digital processing devices—DSP units, multipliers, embedded processors, or high-speed serial channels.

In some cases it is required to choose a specific "subfamily" of the FPGA. For example, if additional logic or high-speed serial channels are needed, then specialists usually choose today the FPGA Xilinx chips of Srapan-6 LX or Spartan-6 LXT series.

The required number of inputs and outputs of the FPGA determines not only the choice of the type of device, but also its case design, since today it is possible to order the same device in different buildings with a different number of user inputs/outputs. In this regard, it is necessary to take into account the possibility of further upgrading the system and choose a case with such a seat that is common to all devices of the selected family.

When choosing a particular type of FPGA chip, it is also necessary to consider the future operating conditions for a particular application. For example, will it be enough to use the "commercial" version of the device, or should we consider the possibility of acquiring the "industrial" version of the device? In some cases, it will only be necessary to set up a military or space quality category.

In addition, it is necessary to consider the device configuration method. Will it be an acceptable option for this application FPGA based on RAM, or is the non-volatile version of this device better suited? When choosing an FPGA, it is also necessary to consider the possibility of implementing a proactive reading of the newly recorded information and encoding the data stream.

As a rule, the cost of the device is also one of the most important determining factors, regardless of whether serial production of the product will be established or only one sample will be manufactured, for example, for a system launched into space. It is necessary to envisage the volume of maximum development costs and in the future to strictly adhere to this plan.

A number of these parameters are also important when choosing other devices for use in the developed high-speed electronic system. In particular, an important criterion for the choice of components is the possibility of their operation from the same supply voltage, which simplifies the architecture of the power supply system and reduces its cost.

It is also necessary to reduce the nominal values of the operating parameters of the components. In technical documentation for components, manufacturers usually provide only the maximum allowable values of parameters and basic operating modes. If build an electronic system in such a way that, for example, the device operates with an electrical load just below an acceptable level, then the reliability of this system will drop dramatically. To ensure an acceptable level of reliability, it is necessary to reduce the electrical load on the system components, taking into account the actual electrical modes in which the designed device will operate.

2.9.2 General Recommendations on the Selection of the Structure of Power Supply System for High-Speed Electronic Devices

After the developer chooses a specific FPGA part type, the next most critical factor is the structure of the power supply system. In modern FPGAs, the core voltage is 0.9–1.5 V. Often these devices, especially high-performance FPGAs, consume a fairly significant current in all modes of operation of the device.

For these reasons, a correct estimate of the total consumed electronic power is required already in the early design stages in order for the power supply system to be optimal. In this case, the energy consumption assessment should be performed necessarily both in the static and dynamic FPGA operation modes, including at the maximum operating frequency and at the extreme operating temperature ranges for the entire spectrum of specific applications. Having determined the power consumed by the FPGA, it is necessary to correlate the obtained result with the overall energy budget of the system in order to determine the final structure of the power supply system.

In computing systems with a high level of current consumption, it is recommended to use pulsed DC/DC converters in order to ensure a sufficiently high overall efficiency and not complicate the subsystem to ensure the removal of excess generated heat. In applications with low current consumption and low noise, such as systems with high-speed serial data transmission channels or ADC and DAC that are sensitive to interference, linear voltage regulators, and additional power filtering may be required. In all cases, it is necessary to carefully examine the technical documentation for the component to be selected.

A more detailed assessment of FPGA energy consumption at the last design stages is recommended to be performed using well-known developers of special CAD tools, for example, using XPower Analyzer in the Xilinx ISE development environment.

In addition, developers of high-speed electronic devices should be very attentive to the interchange in power supply for devices placed on the printed circuit board, using the recommendations detailed in subsequent chapters.

Since modern electronic devices switch many times faster than their power sources, including those that stabilize the supply voltage, without the use of decoupling capacitors, this would lead to voltage drops in the power supply buses, despite the fact that the stabilizer would in all cases maintain the nominal voltage on them. Therefore, it is necessary to choose these decoupling capacitors in such a way that the nominal value of their impedance in combination with the capacity of the interlayer bus of the printed circuit board does not exceed 0.1 Ohm in the frequency range of 100 kHz–1 GHz.

To achieve such parameter values, the developer should use not one type, but probably a whole series of decoupling capacitors of different capacities with different parameters, with their own resonant frequency.

It should also be possible to minimize the number of additional power buses in the boards of the designed high-speed electronic system, it is necessary to check the fulfillment of the system requirements for the required power supply sequence and the voltage rise rate on them to make sure that the designed circuit meets them. If any deviations from the specified requirements are identified, it is necessary to make the appropriate changes and carry out the simulation rerun.

2.9.3 General Recommendations for Choosing the Optimal Clock Signal Tree Structure

As is known, in any designed electronic system, at least one clock (synchronizing) signal is required. Usually, a standard highly stabilized generator of the required frequency is used to generate this clock signal. The key parameters of such a generator are usually its output frequency and stability. Stability is usually expressed in parts per million (ppm). Typical stability values for such generators are $\pm$ 50 or $\pm$ 100 ppm.

Clock generators controlling high-speed serial channels or ADC/DAC shall also have low phase noise and, accordingly, low jitter, which is objectively necessary to reduce the bit error rate of the high-speed data transmission channel and also to reduce the signal-to-noise ratio of the ADC and DAC.

Let us explain this problem in more detail. The problem of choosing a specific standard output signal deserves high attention. The transmission of differential signals, such as LVDS or LVPECL, provides a higher noise immunity of the system than the use of unbalanced signals—LVCMOS or LVTTL. Differential signals also reduce electromagnetic interference and provide shorter signal rise and fall times. However, the transmission of differential signals may be accompanied by a higher phase noise than the transmission of unbalanced signals. To reduce phase noise and jitter, high-quality systems use sinusoidal signal generators as the source of the primary clock signal; first of all, it concerns clocking ADC and DAC.

Since several autonomous synchronization areas are usually supported within the FPGA, currently high-speed electronic systems usually contain more than one clock signal. Often in such systems, not only high-frequency, but also low-frequency clock signals are used at the same time, if the frequency division cannot be ensured only with the help of its own internal resources, for example, such as the well-known delay-locked loop (DLL), DCM-module, and PLL. In addition, different protocols or algorithms sometimes require different clock frequencies, which leads to the need to generate a multitude of clock signals and create several parallel synchronization regions in an electronic system. Figure 2.35 shows such a standard typical clock signal tree. The following designations are taken here:

Prime Clock—the main clock signal; Redundant Clock—backup clock signal; Clock Selection—selection of a clock signal; Clock Fan·Out—a splitter of clock signals; Clock [0:7]—clock signals [0:7]; Pixel Clock—pixel synchronization.

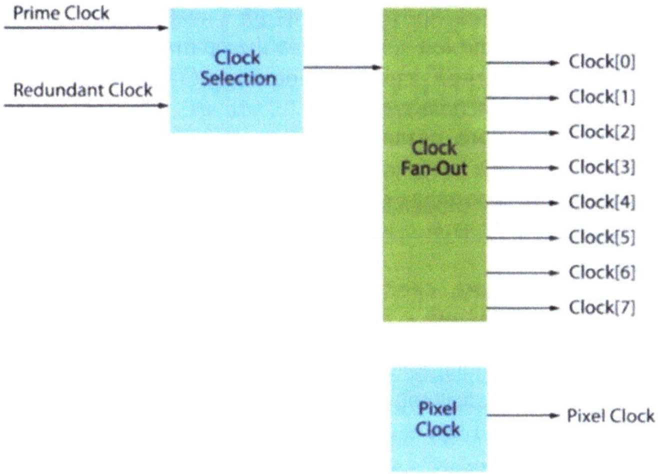

Fig. 2.35 Standard clock signal distribution system

In many projects based on FPGA, it becomes necessary to form special reset signals for a number of peripheral devices in order to ensure their correct initialization immediately after turning on the power of the electronic system or by a special command from an external device, for example, from the "button" of the control panel. Reset signals can sometimes cause unplanned problems in the operation of the system, since usually these signals are generated randomly. As soon as the operating voltage is established on the supply buses, the clock signal generators are usually started immediately, and only after the completion of all transients they start working in a stable mode. Therefore, it is desirable that this reset signal on the printed circuit board of the electronic system be formed in advance—even before the appearance of voltage on the power buses and the transition of this generator to the standard operating mode. Since this signal inside the FPGA, as a rule, is strongly branched, it should be carefully filtered so that spurious emissions and electromagnetic interference do not lead to unauthorized system reset.

The developer of the electronic system should also ensure that, in the FPGA-based system, the reset signal is connected only to those resources where it is really needed (the FPGA and RAM registers can be initialized during the configuration of the system from SRAM).

In addition, the developer should also ensure that turning off the reset signal does not lead to unstable states inside the FPGA, i.e., so that turning off this signal near the clock signal front does not cause any transients.

It is also necessary to accurately determine the nature of clock signals and reset signals at the input of the FPGA, namely, the clock signal is defined as a global or local resource, and the system developer shall work closely with the FPGA specialist to correctly use these resources.

Since FPGA support a range of standards for unbalanced and differential I/O signals, the main task of the developer is also the optimal allocation of device I/O ports and compliance with strict rules for allocating I/O banks.

For assigning I/O ports, developers usually use special software tools, such as PlanAhead from Xilinx, which check compliance with the rules for assigning and placing the FPGA terminals. This is especially useful when using high-speed serial channels, since in such cases it is necessary that the user terminals are not used near the power terminals. Otherwise, interference may be caused to the power bus, which will worsen the quality of data transmission in a high-speed channel.

It is also necessary to check the integrity of the signals (including the rise and fall times of the signals), which determines the maximum length of the conductors before connecting the terminal load. Most FPGAs support built-in load balancing, reducing the need for discrete components. This approach works correctly only if the controlled impedance of all layers of the printed circuit board is ensured.

The developer of a high-speed electronic system should also work very closely with the design engineer involved in this project to determine the specific physical dimensions of the printed circuit board, the position of all the required connectors, etc. Modern FPGAs can heat quite strongly, that is why the developer also has to solve the difficult task of correct heat removal. Solving this problem also affects the mechanical design of the product, since it may require the use of forced cooling or even a special radiator.

If the application of the designed electronic device is intended for operation in harsh environments, the mechanical design will also impose additional restrictions on the location of the printed circuit board, since it will be necessary to take into account the impact and vibration strength of the product.

To calculate and simulate various thermal conditions of a designed high-speed electronic device printed circuit board, all design engineers and system designers involved in the project shall exchange information in the form of DXF files with each other in order to coordinate and verify the correctness of their chosen technical solution.

2.9.4 General Recommendations for Designing Topologies of High-Speed Printed Circuit Boards

The topology of the printed circuit board is the most critical factor ensuring the required characteristics of the designed electronic devices. In any high-quality system, a special multilayer printed circuit board with controlled impedance layers is always used, for which the number and order of the layers of the printed circuit board, as well as the width of the conductors and the gap between the layers, are crucial. To ensure the best performance in high-performance electronic systems, it is necessary to provide an accurate calculation of all the main technical characteristics of microstrip lines in order to provide the required 50-Ohm impedance for

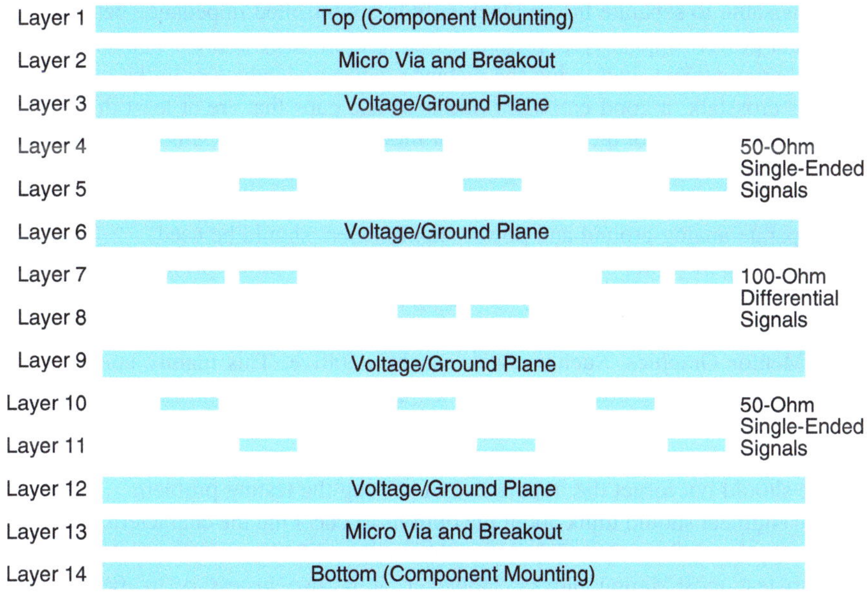

Fig. 2.36 Typical set and order of manufacturing layers in a 14-layer printed circuit board

asymmetric signals and 100-Ohm impedance for differential signals. This will allow for correct termination of the transmission lines.

In addition, a sufficiently close arrangement of the metallized layers of the power supply buses and the metallized layers of ensuring the flow of reverse currents in such a printed circuit board to provide all the necessary capacity couplings are of great importance. In this case, the printed circuit board is used as a place to house at least one additional decoupling capacitor. Figure 2.36 shows a typical set and order of manufacture of layers of printed circuit boards with controlled impedance. The following designations are taken here:

Layer—layer; Thor (Component Mounting)—top layer (component mounting); Micro Via and Breakout—transient pinholes and breaks; 50 Ohm Single-Ended Signals—50 Ohm single-ended signals; Voltage/Ground Plane—power/ground layer; 100 Ohm Differential Signals—100-Ohm differential signals; Bottom (Component Mounting)—bottom layer (component mounting).

At this stage, it is also necessary to define a set of additional requirements for the developed printed circuit board topology, which should include restrictions on the placement of components, taking into account the division of the board into analog and digital parts. It is also necessary to determine the requirements for matching the lengths of buses and signal conductors on the board, taking into account the effect of crosstalk between signals.

It is necessary as far as possible to reduce the number of signal conductors in one layer and dissolve the conductors so that the signals in the layers above and below do not run parallel to each other. Otherwise, additional sources of crosstalk may appear.

It is advisable to separate the conductors with a controlled impedance between the earth and power supply layers perpendicular to each other in order to exclude their parallel arrangement. Increasing the distance between conductors in the same layer reduces crosstalk; a good empirical rule is to use gaps that are at least three times larger than the width of the conductor. For more critical signals, this distance should be increased five or even seven times relative to the conductor width.

Analog and digital signals should be separated as far from each other as possible, and separate analog ground and power supply layers should be used.

One of the most important final stages of the design of the printed circuit board topology is to check the integrity of the signals of the entire system, as well as the integrity of the power supply using development tools, such as NurerLynx SI and PI from Mentor Graphics. Such a check can be iterative. This mainly concerns the integrity check of the power supply, since high FPGA core consumption currents may require correction of the wiring of several layers of the printed circuit board at once.

We should not forget the importance of solving the testing problem.

The engineer should think about the options of checking the characteristics of the designed printed circuit board. It is possible to use a JTAG controller and peripheral scan to test most digital blocks. However, it is also necessary to evaluate the functioning of the power supply system, clock signal generation circuits, and reset signals. Test probes connected to the power bus outputs will allow quickly checking the power supply system. Measurements should use series resistance to limit the current. To measure the consumption current of the FPGA and other components, resistors of small nominal values (of the order of mOhm) are used. If clock signal buffers are used in the project and there is a spare output, it should be ensured that it is correctly matched to the terminating load in order to be able to test the clock signals.

LEDs, which clearly reflect the state of the system, can provide a great help when debugging. In addition, in some cases, the boards provide for the possibility of placing connectors for connecting a logic analyzer or test signal generator. It is necessary to make sure that access to them is carried out simply not only when measuring the open board, but also as part of the system when the product is enclosed in a case.

For debugging FPGAs, special tools are used, for example, ChipScope, which allows determining the bit error rate in high-speed serial channels.

Designing systems using FPGA can be associated with the need to solve a number of other interesting tasks. Today development tools are available to the engineer that can assist in solving them, technical documentation and a variety of manuals are available. However, the best resource is a perfect knowledge of the functionality and characteristics of the basic components that make up the system under development.

2.10 Circuit-Based Options of Coordination of Transmission Lines of the LVPECL Standard

First of all, it should be noted that LVPECL is an "old" technology for building interface lines that goes back to the times when microelectronic technologies did not allow the production of high-speed bipolar PNP transistors in the same process with high-speed NPN transistors. Accordingly, external passive elements had to perform the role of the first. This requirement was used in LVPECL (as well as in the previous ECL technology) by combining the functions of passive binding to earth and matching the transmission line. However, this technology is still used in a number of applications, and often the interface device designers have difficulty choosing electrical circuits for matching circuits for LVPECL transmission lines, because the role of the matching chain as an integral part of the output stage scheme is usually not considered in the literature.

The purpose of considering the above circuit-based variants of matching LVPECL [25] communication lines in this section is to show what the main factors (output transistors currents, number of components, power consumption) determine the choice of matching network circuit and its component ratings. Modern technologies of semiconductor production make it possible to manufacture high-speed PNP transistors, which has led to the emergence of such standards for differential signal transmission with earth level readings like HCSL and LVDS. These standards completely eliminate the need for the developer to develop a voltage VTT.

Let us consider in more detail the features of circuit-based solutions of high-speed differential interface LVPECL, including recommendations on the choice of components of the entire transmission line.

LVPECL Driver The simplified LVPECL valve is usually implemented on the basis of an open-emitter driver, as shown in Fig. 2.37 [25] and is characterized by the following main features:

1. Output transistors are closed, so the driver will not produce an output signal:
   (a) to switch Q and nQ, each of these lines shall be DC loaded on an electrical circuit connected to earth;
   (b) if the output is not used, the Q and nQ contacts can be left floating to ensure the closed state of the output transistors, minimizing power consumption and noise level.
2. The emitter Q and nQ currents are determined by the choice of external earth-binding resistors. The resistance of these resistors is selected based on the value of the voltage VCC0, the dissipated power, and the magnitude of the signal voltage.
3. VCC0 = VCC. Unlike LVCMOS, VCC0 cannot be lower than VCC, since it is necessary to provide interfacing with clock receivers having other supply voltages. For example, with VCC = 3.3 V and VCC0 = 2.5 V, the output transistors will go into saturation mode, as a result of which the switching time will drastically decrease.

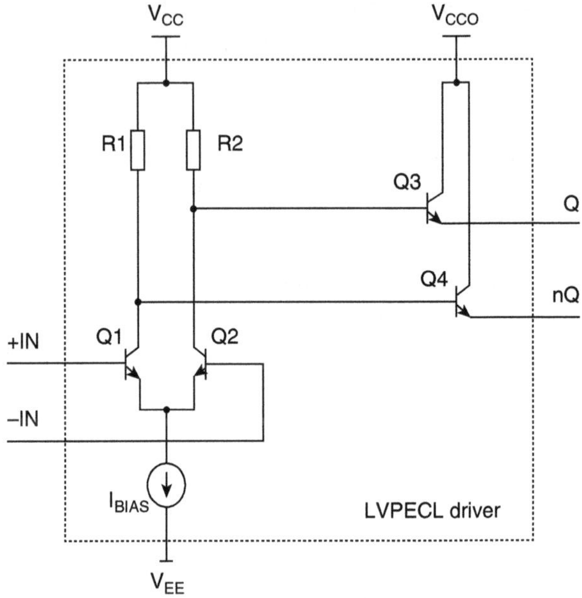

Fig. 2.37 Equivalent circuit of the LVPECL open emitter driver

Differential LVPECL drivers are capable of operating at frequencies of the order of gigahertz. Such frequencies require proper matching of transmission lines with LVPECL receivers. Designers can choose matching circuit diagrams based on the ease of component layout and the trade-off between power consumption and rate, but it is always necessary to check the signal integrity by using circuit simulation taking into account the length of the tracks and vias.

Let us further consider circuit-based solutions to the problem of matching LVPECL transmission lines with DC coupling.

Standard Matching Chain Figure 2.38 [25] shows the standard matching chain for LVPECL transmission lines with DC coupling. Based on Fig. 2.37, the VTT rated voltage was set according to Kirchhoff's law for a circuit consisting of VCC (=VCC0), R1, Vbe Q4, and Rterm, with a low level at the Q4 output and an emitter current of 2 mA. VTT is the voltage applied at the connection point of the terminating resistors, which satisfies the Kirchhoff's law for voltages and varies together with VCC and VCC0.

The advantage of VTT is that the fluctuations in the supply voltage do not affect the output currents of the driver. This is especially important for a low-level transistor: by preventing its jumping to the cut-off mode, we prevent the device switching rate from decreasing. An obvious disadvantage of this matching circuit is that it requires a stabilizer with an output voltage of 2 V, relative to VCC0.

The Agreement on the Thevenin Equivalent Circuit In practice, the VTT source is often disposed of by agreement with the scheme shown in Fig. 2.39 [25], where the

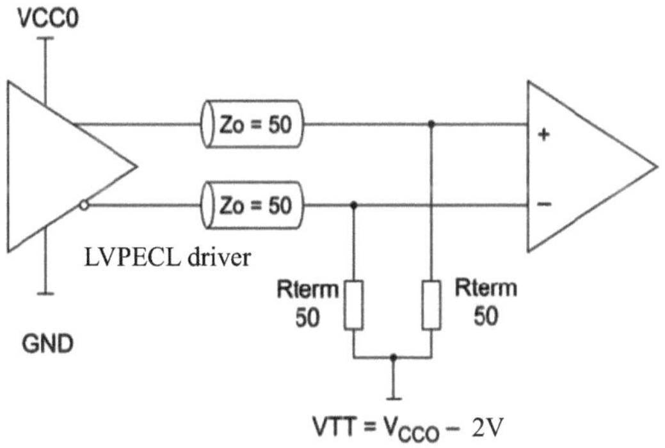

Fig. 2.38 Standard matching chain of LVPECL

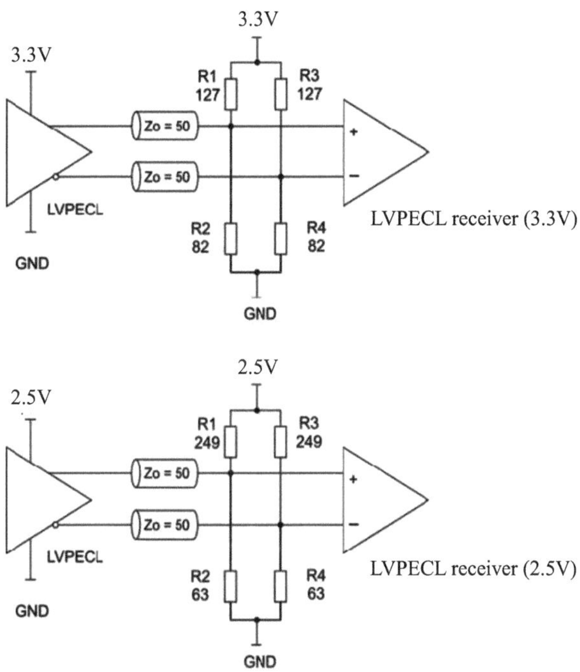

Fig. 2.39 Approval of LVPECL transmission lines according to Thevenin equivalent circuit for supply voltages of 3.3 and 2.5 V

voltage VTT and the matching resistance of 50 Ohms are realized according to the Thevenin equivalent circuit using two resistive bias chains. The separate coordination of Q and nQ allows these lines to be laid separately in unrelated transmission lines to different asymmetrical loads.

The agreement with the Thevenin circuit has five disadvantages [25]:

1. The bias chain is another consumer in addition to the emitter circuits of the output driver (about 80 mW per pair of outputs at a supply voltage of 3.3 V).

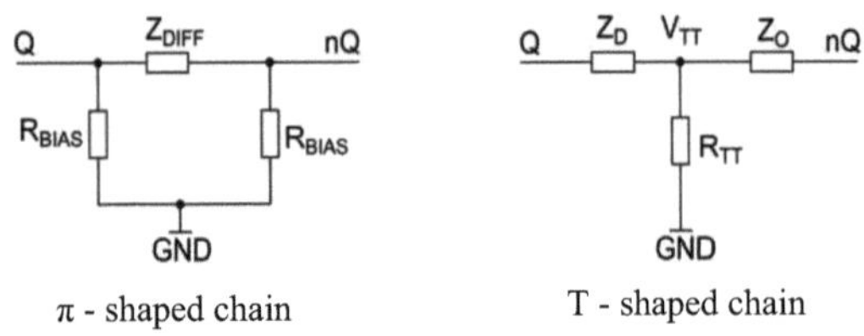

Fig. 2.40 T- and π-shaped matching chains for LVPECL transmission lines

2. There are no degrees of freedom left to take into account the inconstancy of VCC0, temperature, and device response.
3. VCC0 oscillations in accordance with the Thevenin equivalent circuit are transformed by a bias voltage divider into smaller voltage fluctuations VTT (proportion 39% when VCC0 = 3.3 V and 20% when VCC0 = 2.5 V). Therefore, emitter currents, especially at a low output level, are not as effectively controlled as with a standard matching circuit.
4. With differential matching, three resistors are sufficient to obtain a VTT voltage and two matching resistances of 50 Ohms each.
5. Two additional resistors at the input of the clock signal receiver for Q and nQ signals complicate the board wiring and reduce the operation rate compared to a single resistor due to the additional parasitic capacitance.

Shunt Matching Chains (π- and T-shaped) It is possible to get rid of the disadvantages of the Thevenin circuit in whole or in part, using connected transmission lines with π- and T-shaped resistive chains, as shown in Fig. 2.40 [25]. The impedance of the π-shaped chain for the common-mode signal is Rbias/2, and the T-shaped one is Zo/2+ RTT. The matching resistance of a connected transmission line with a π-shaped chain is equal to Zdiff (which, in turn, is 2Zo), and with a T-shaped chain it is directly equal to 2Zo. Common-mode voltage at RTT is designated VTT by analogy with VTT = VCC—2.0 V.

In these chains, the disadvantages in the agreement with the Thevenin circuit are overcome as follows [25]:

1. The loss of power in the displacement chain disappears, since the matching chain is fed directly by the driver currents.
2. The common-mode impedance (and, therefore, the negative feedback ratio) can be adjusted to achieve an acceptable compromise between driver current and variability of VCC0, temperature, and device response, as shown in Table 2.1.
3. T- and π-shaped matching chains allow better control of the emitter currents due to in-phase current feedback. For example, in the case of a T-shaped chain with RTT = 50 Ohms and VCC0 = 3.3 V, the common-mode feedback between the VCC0 and VTT nodes is 66% (actually slightly lower due to differences in the operating points of the output transistors)—this is 60% higher than the standard straight matching chain. Although the difference between the two circuits may seem insignificant at first glance, it is significant if consider the relatively small voltage drop across the 50-Ohm emitter resistor with a low logic level at the output.

Table 2.1 Bias resistors π- and T-shaped matching chains depending on the output current of the transistors [25]

Case	Vcco	Vout_h	Vout_I	Ibias (min.), mA	Ibias (med.), mA	Rbias	R9
Nominal value	3.3	2.35	1.65	2	10.9	183	67
− 5% Vcco	3.164	2.214	1.514	2	11.1	165	58
− 5% Vcco the worst response of the device,—40–85 °C	3.164	2.074	1.289	1	11.6	142	46
Nominal value	2.5	1.55	0.850	2	12.7	94	22
− 5% Vcco	2.375	1.425	0.725	1	11.9	91	20
− 5% Vcco the worst response of the device,—40–85 °C	2.375	1.285	0.5	0	14	64	7

4. Only three transistors are required—the minimum necessary for biasing two outputs and matching the transmission line. Transmission lines Q and nQ for each chain shall be connected, since each of them implements a shunt matching circuit.
5. At each input terminal of the LVPECL receiver there is only one terminating resistor (and the corresponding parasitic capacitance).

Table 2.1 shows how it is possible to correct the resistance of the common-mode bias resistors of the π- and T-chains to compensate for the variability of VCC0 and the response of the device. The IDT 853S006 chip was chosen as a representative device, for which the largest and smallest voltage values of logical "0" and logical "1" at the LVPECL output in the industrial temperature range were determined. The calculation was made for three cases: nominal voltage VCC0, extreme case (VCC0 deviation—5% and minimum common-mode output voltage over the entire temperature range and at the worst device response), and intermediate case (VCC0 deviation only—5%).

The minimum bias current at the logic low output, which is desirable for achieving the maximum switching speed, is reduced proportionally to compensate for the decreasing probability of two cases with a deviation of VCC0—5%. The resistance of the bias resistors shall be reduced in order to compensate for the decreasing voltages on the emitters of the output transistors, as the spread of these voltages increases.

Decreasing the resistance of the bias resistors increases the average bias current. In contrast to this trend, with a decrease in the minimum bias current Ibias, the average current decreases, as it should be expected. This effect is visible if we compare the bias currents at a nominal VCC0 = 2.5 V and with a deviation of −5% from this voltage.

The average bias currents for the differential pair are calculated with the VCC0 that is taken in each case. In all cases, the bias currents increase when VCC0 deviates by +5%. The designer should choose the resistance of the resistor, based on these cases, taking into account the tolerances of the design under development.

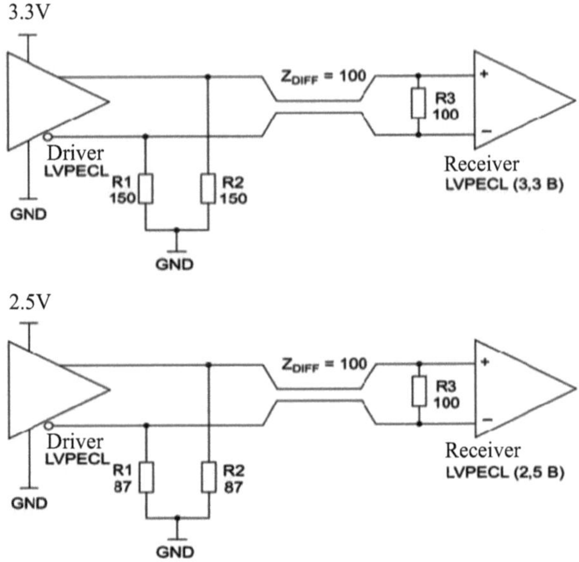

Fig. 2.41 External π-shaped matching chains for LVPECL transmission lines of 3.3 and 2.5 V power supply voltage

π-Shaped Matching Chain and Associated Transmission Lines When using the π-shaped matching chain shown in Fig. 2.40 [25], with the associated transmission line, Zdiff is replaced by an equivalent impedance associated with the transmission line, loaded on Zdiff. This circuit retains the advantages associated with the possibility of using Rbias for independent installation of bias currents Q and nQ of the driver, as well as the presence of parasitic capacitance from only one resistor at each input terminal of the receiver. Specific forms of the π-shaped chain are shown in Fig. 2.41 [25].

Such coordination facilitates printed circuit board layout in cases where a single component (for example, in a BGA package) serves as a receiver for several LVPECL signals. In fact, this is the same advantage that the LVDS standard provides on the receiver side. Each transmission line at the Q and nQ outputs has only one Rbias resistor, which facilitates their installation on one side of the printed circuit board with the LVPECL driver, and this provides a significant gain in terms of signal integrity.

In some devices, the LVPECL input cascade contains a built-in differential circuit of 50-Ohm terminating resistors with an in-phase offtake to the contact of the case, which is usually referred to as VT. When implementing a π-shaped matching chain, VT contact is left floating.

T-Shaped Matching Chain and Associated Transmission Lines When using the T-shaped matching chain shown in Fig. 2.40 [25], with the associated transmission line, this entire chain is located at the end of the line at the location of the LVPECL receiver. Each input terminal has one resistor contributing parasitic capacitance, and the parasitic capacitance associated with the RTT does not affect the integrity of the signal in the differential mode, since it is located in the common-mode VTT node. A

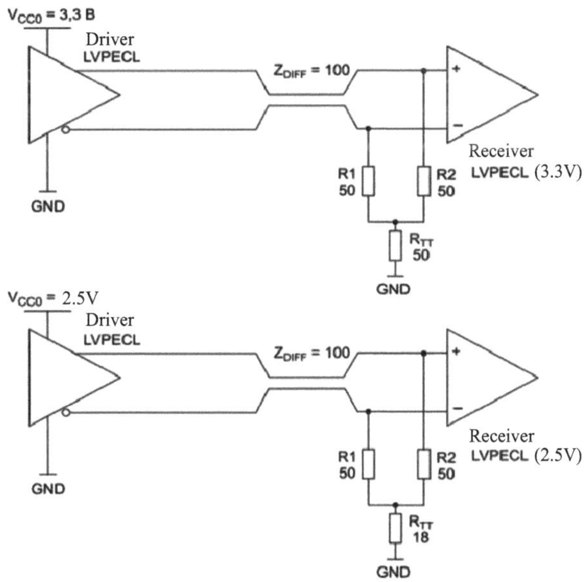

Fig. 2.42 External T-shaped matching chains for LVPECL transmission lines of 3.3 and 2.5 V power supply voltage

T-chain is preferable in LVPECL devices with a high degree of branching, for example, in a buffer with a degree of branching of 1:22, where while using a π-chain, it would be difficult to place all the ground-reference resistors of the clock signal driver.

Specific forms of the T-shaped chain are shown in Fig. 2.42. If the LVPECL device has a built-in differential circuit of 50-Ohm terminating resistors and an in-phase offtake to the VT body contact, this contact is connected to earth through the external RTT resistor shown in these examples.

LVPECL Driver Output Offset When Matching Transmission Lines with AC Coupling It is often necessary to interface the clock signal LVPECL drivers with AC current with receivers based on a different type of logic with different switching threshold voltages. The output of the LVPECL driver will still have a constant offset, but due to the use of coupling capacitors, the offset shall be within the common-mode range, and the matching resistance of the transmission line shall be located at the place of the LVPECL receiver. The overall configuration of the outer part (only for the LVPECL driver) is shown in Fig. 2.43 [25].

The values of the resistors Rbias should be taken from the Table 2.1. They were calculated at the emitter current of the output transistor corresponding to the switching threshold. The CC value is calculated from the clock frequency and the impedance module at that frequency.

The coupling capacitor CC is selected based on the maximum permissible parasitic reactive impedance that the capacitor can bring into the signal path. This value was arbitrarily assumed to be 0.2 Ohm when compiling the Table 2.2 [25], in

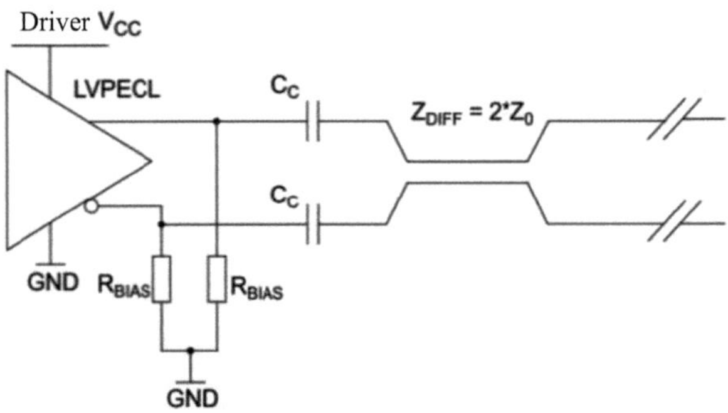

Fig. 2.43 General circuit of the LVPECL driver bias with AC coupling

Table 2.2 Standard separation capacitor ratings and corresponding frequency ranges

Clock frequency range, MHz			Clock frequency range, MHz			Clock frequency range, MHz		
Min.	Max.	CC, pF	Min.	Max.	CC, nF	Min.	Max.	CC, μF
3700	5400	220	370	600	2.2	38	54	0.022
2500	3700	330	250	370	3.3	26	38	0.033
1700	2500	470	170	250	4.7	18	26	0.047
1200	1700	680	120	170	6.8	12	18	0.068
800	1200	1000	80	120	10.0	8	12	0.1

which the standard capacitor values and the corresponding frequency ranges are given.

Therefore, taking into account the above recommendations will allow users of this standard to avoid undesirable effects in the design and manufacture of high-speed electronic systems.

References

1. Belous, A., & Yarzhembitsky, V. (2001). *Digital IC engineering for information processing and transmission systems: Education book*. Minsk: Technoprint Unitary Enterprise.
2. Smirnov, M. (2010). Modern data transfer methods. *Electronic Components*, 11, 84–87.
3. Patel, A. *The basics of SerDes (serializers/deserializers) for interfacing*. www.eetimes.com
4. Ritter, D. *Applying advanced analog and digital techniques to long video-cable equalization*. www.eetimes.com.
5. Nitin, B. (2014). About DDR PCB design. *Technologies in the Electronics Industry*, 7, 28–31.
6. Bhagwath, N. (2015, January) *DDR4 board design and SI challenges*. DesignCon.
7. Bhagwath, N. (2014, June). Designing PCBs for DDR busses. *PCD&F/CirCuits Assembly*.
8. Kugelstadt, T. (2014). High speed data transmission over long distances. *Electronic Components*, 11, 14–14.
9. EnDat 2.2. Bidirectional interface for position encoders. *Technical Information*. www.heidenhain.us.
10. *Manchester coding basics*. Atmel. www.yumpu.com.
11. *Reference data for SN65HVD23 and SN65HVD24*. www.ti.com.

12. Kinnaird, C. *Use receiver equalization to extend RS-485 data communications* (Application Report (SLLA169)). Texas.
13. Belous, A., Blinkov, O., & Silin A. (1990). *Bipolar circuits for automatic control systems interfaces* (p. 272). Mechanical Engineering, Leningrad Division.
14. Belous, A., Ponomar, V., & Silin A. (1998). *Bipolar circuit design for high-performance information processing systems* (p. 162). Minsk: Polyfakt.
15. Senyakin, Yu. (2014). Overvoltage protection for RS-485 bus. *Electronic Components*, 12, 38–41.
16. Kugelstadt, T. *High-speed connections require.*
17. Belous, A., Merdanov, M., & Shvedau, S. (2014). Microwave electronics in radar systems and communications. In *Technical encyclopedia* (2 Vols., p. 728). Moscow: Technosphere.
18. Drobyshev, A. (2014). High-speed RS-485 interface. *Electronic Components*, 12, 42–44.
19. Kugelstadt, T. *High-speed connections require RS-485*. www.edn.com.
20. Shcheglov, A. (2013). Fundamentals of system design using FPGA. *Electronic Components*, 12, 95–98.
21. Taylor, A. (2013). Nuts and bolts of designing an FPGA into your hardware. *Xcell Journal.*
22. Gupta, N. *Successfully designing FPGA-based systems*. www.cadence.com
23. *Power considerations in FPGA design*. www.latticesemi.com
24. Conchas, J. *Simplifying power supply design in FPGA-based systems*. www.silabs.com
25. Wissel, P. (2015). Schematic options for LVPECL standard transmission line conditioning. *Electronic Components*, 3, 42–45.

Chapter 3
Wireless Telecommunication Systems

In the general case, as shown in Chap. 1, all up-to-date telecommunication systems consist of only two components: a high-speed data *processing* device and data *transmission* channels.

In Sect. 3.1, trends and problems of the development of microelectronic technologies used to build wireless data transmission channels are considered. Section 3.2 discusses the main methodological and terminological problems of the creation of wireless telecommunication systems, in Sect. 3.3, the state, prospects, and problems of the development of fifth-generation mobile communication technologies (5G) are considered in more detail. Section 3.4 discusses the main trends and problems of the development of wireless technologies for the "Internet of Things" (IoT). Section 3.5 shows specific examples of the organization of communication channels of 5G telecommunication systems based on phased antenna arrays, and Sect. 3.6 considers specific examples of the use of LP WAN technology for the design of communication channels of the "Internet of Things".

3.1 Trends, Prospects, and Problems of Development of the Element Base of Wireless Telecommunication Systems

The main subject of this book is the methods of designing high-speed electronic devices, briefly referred to as "High-Speed Design" in English.

In the general case, as shown in the first chapter, all modern information and communication systems consist of only two main complex functional components: devices for high-speed information *processing* and channels (buses, lines) for data *exchange* and control signals between similar electronic devices.

A. Belous, V. Saladukha, *High-Speed Digital System Design*,
https://doi.org/10.1007/978-3-030-25409-4_3

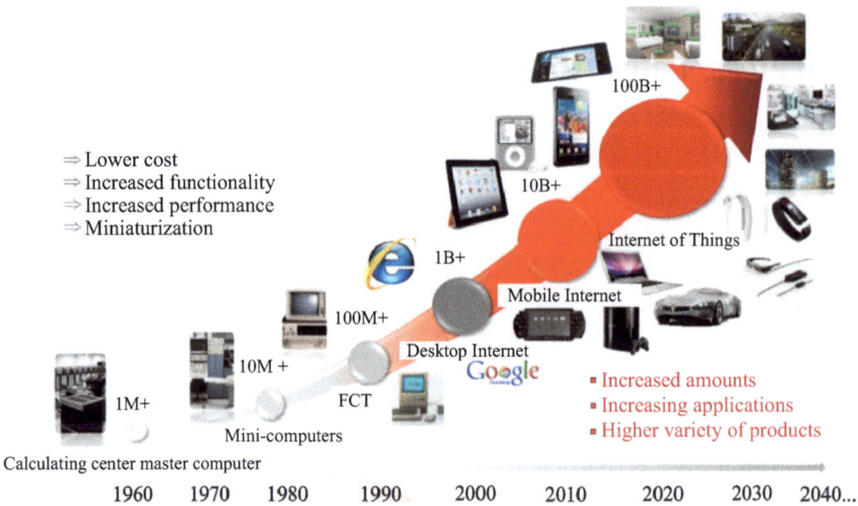

Fig. 3.1 The main trends of the global increase in the range of users of wireless telecommunication systems

Methods and ways to solve the problem of improving the performance of electronic devices and information *processing* systems in the world are described in many works, including most of the works of one of the authors of this book [1–8].

So, Fig. 3.1 [1] presents in simplified illustrative graphic form the main stages of the history of creation and development prospects of this direction—from the moment of the appearance of the first boards in 1966, mini-computers in 1970–1980, personal computers in the mid-1990s, Desktop Internet in 2000, the mobile Internet in 2010, and the expected appearance of the "Internet of Things" after 2020.

As can be seen from this popular in the Internet image, the main "criteria for the survival" of each new direction were: minimum price, maximum functionality, more advanced technical (electrical and dynamic) parameters, and microminiaturization.

Figure 3.2, 2015 shows the same simplified illustrative graphic form of a "road map" of the development of the basic elements of telecommunication systems—computers—of various architecture, starting from 2015. As can be seen from this figure, for each stage of development of TCS basic computers it is possible to distinguish their main features.

If for the basic computers of TCS in 2015, the main trends were 3D integration, the use of constructions such as interposer and separate integrated photonic links, then by 2020 it will be: new architectures and new materials for the implementation of storage devices (new memory materials and architectures), highly integrated 3D constructive design of microelectronic devices (3D VLSI and high density 3D), numerous integrated elements of photonics (integrated silicon photonic links), and wider use of the principles of neural networks (neuromorphic for advanced chipset architecture).

Figure 3.3 [1] presents global trends in the area of crystals of silicon processors, which form the basis of the hardware of TCS. In accordance with the well-known

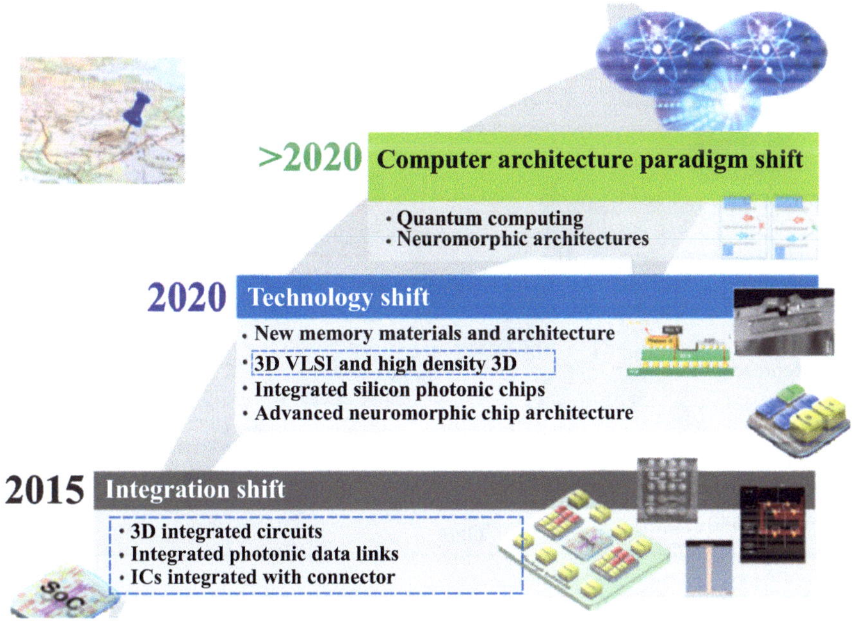

Fig. 3.2 "Road map" for computers

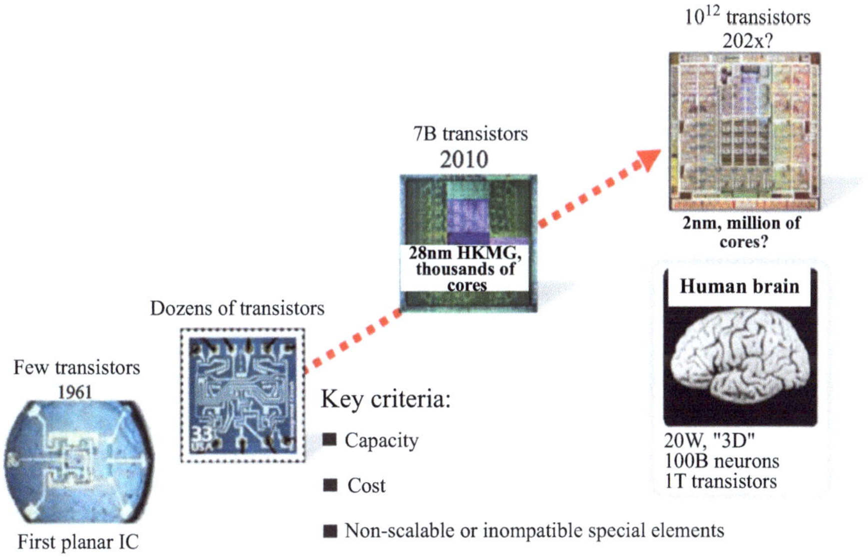

Fig. 3.3 Moore's Law in use

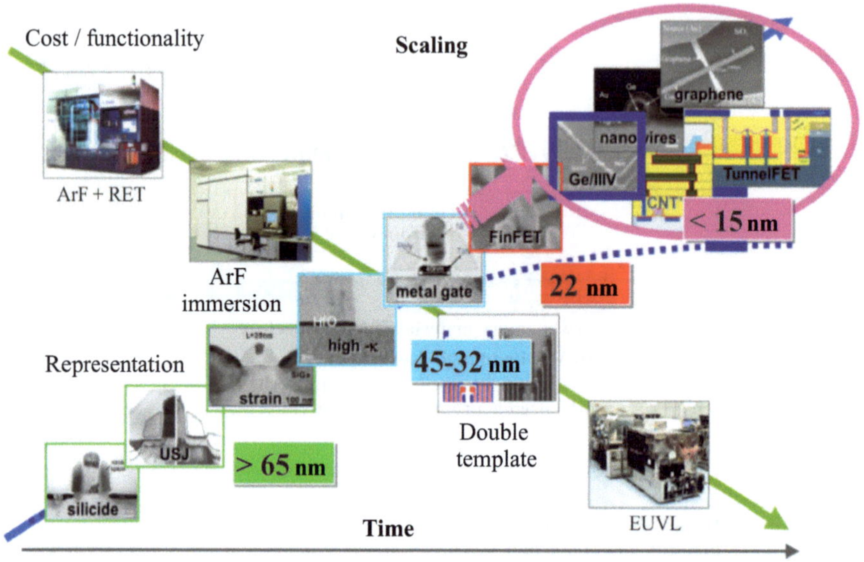

Fig. 3.4 Roadmap of CMOS chip scaling

Moor's Law, the number of transistors on a chip since 1961 (the appearance of the first integral transistor) doubles every 24 months. If in 2010 a typical processor manufactured according to the technology with design standards of 28 nm allowed to place a thousand cores on a single chip (which is seven million transistors), then in the middle of the 20s of this century it is predicted that they can be manufactured according to 2-nm design standards on a single processor crystal that has over a million cores (this is at least a trillion (!) transistors).

And what kind of transistors will they be? Figure 3.4 [1] shows the roadmap of decreasing of the CMOS LSI area by scaling the linear dimensions of the elements of these transistors tied to the trend of changes in design standards (from 65 nm to 15 nm or less). As can be seen from this figure, experts predict simultaneous development of two differently directed trends, indicated by arrows. The first and the main trend is the improvement of the technical characteristics of these transistors—the "bricks" from which these multi-core processors are built. These characteristics will steadily improve in the foreseeable future due to the emergence of new and more advanced designs of transistors and the corresponding new technologies for their manufacture based on the use of more and more new materials: from silicide, used successfully for many years, to metalgate and then to new designs like FinFET, to the use of new materials (nanotubes, graphene, etc.).

The second obvious global trend is a continuous decrease in the cost/function ratio of microcircuits, a very rigid concept of microelectronic component manufacturing is predicted for this parameter.

Leading semiconductor companies in the world are already fighting hard in the market of semiconductor component manufacturers for high-speed electronic

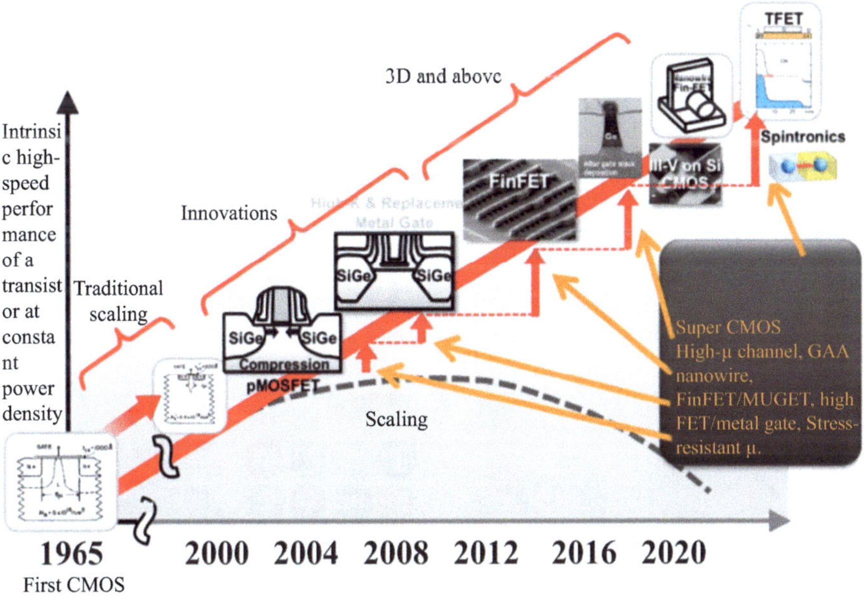

Fig. 3.5 Evolution of base transistor designs

devices, investing dozens of billions of dollars in relevant R & D and new technologies.

Figure 3.5 shows, in simplified form, the main results achieved and trends in the development of the designs of microelectronic device basic transistors listed above in Fig. 3.4 over the period from the appearance of the first CMOS chip (1965) to 2020.

Here we see three main stages: classical scaling [9], innovative constructive-technological solutions [10], and three-dimensional microelectronic devices [1].

What is "beyond the horizon" of the 2020s—experts cannot yet formulate specific forecasts, but it is obvious that this trend is linear, so inevitably new designs as well as new technologies will appear, allowing to produce super-fast, low-loss transistors, smaller and smaller in geometrical size and with lower energy consumption.

Naturally, the developers of information and communication technologies and various telecommunication systems based on them have worked both to develop technical solutions and to expand the scope of their application, various types of new telecommunication devices and systems were created and developed, and their application has been constantly expanding.

Thus, Fig. 3.6 in simplified symbolic and graphical form shows the evolution of functions of the well-known Internet—the World Wide Network (or the World Wide Web).

All independent experts unanimously agree on one thing—with the evolutionary process of development of this area of information and communication technology,

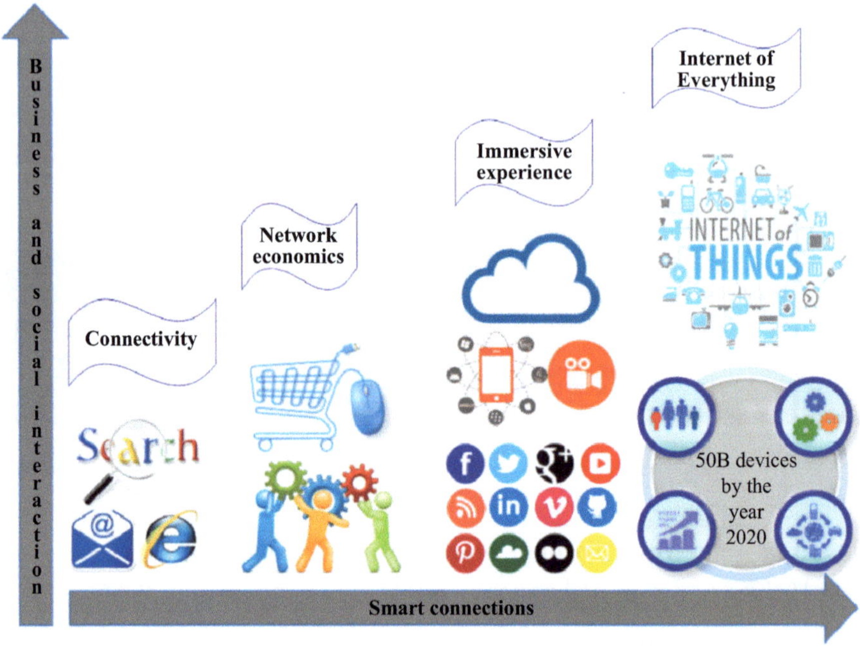

Fig. 3.6 Evolution of the World Wide Web

the mankind today is in the brink of the next stage (form) of its development—the Internet of Things, which will be discussed in more detail below. Moreover, analysts and experts in the field of information and communication technologies, technical specialists and managers of large companies are already actively working on the issues of specific technical solutions to create and use the system that is to follow the "Internet of Things"—the "Internet of Everything".

What is the "Internet of Things" and "Internet of Everything", as well as the main problems and prospects of their development will be considered below, but first the reader needs to pay attention to Fig. 3.7, from which we can conclude that the "Internet of Things" is in fact the basis of the "Internet of Everything".

Indeed, the collective authors of this picture, professional analysts of the well-known Beecham Research Limited group, tried to solve this seemingly unsolvable task—to show the diversity of the process of this information and communication "monster" penetrating into the daily life of planet Earth and each person in just one picture. Nevertheless, to a certain extent, they succeeded—analysis of the information presented in this visually readable figure leaves no doubt—the "Internet of Things" is, as the classics of Marxism–Leninism used to say, an inevitable near future "objective reality that will exist regardless of our consciousness."

Well, since it will be an inevitable "objective reality", specialists need to understand the relevant material (technical) problems of design and manufacturing of both suitable TCSs and their basic elements, and also well understand the problems of their future operation, the prospects for their long-term and short-term development.

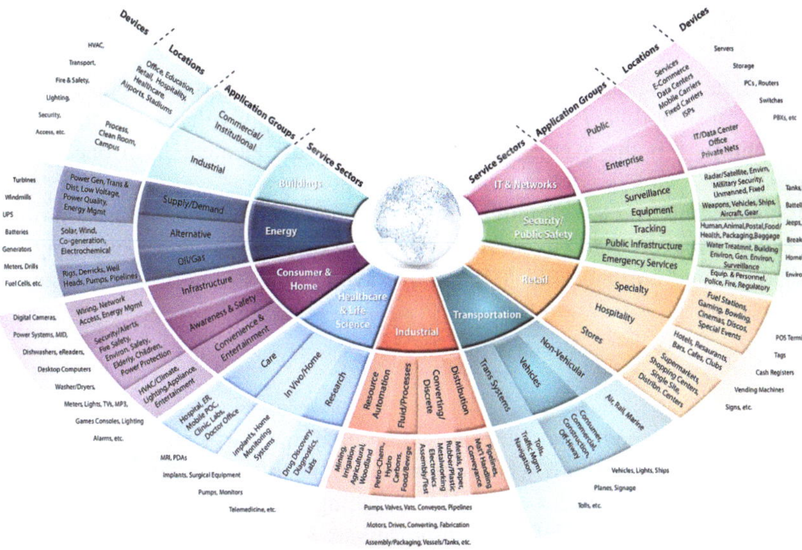

Fig. 3.7 The global scope of the Internet of Things (source: Beecham Research Limited)

The subsequent materials of this chapter, according to the intention of the authors, are aimed at helping specialists who are engaged in solving this difficult task.

But before proceeding to these subsequent materials in the chapter, the authors consider it necessary to explain why this chapter of the book includes the materials of paragraph Sect. 3.1.

If we simplify the content of the materials of this subsection to the utmost, we can formulate the following important points.

First of all, at the heart of the global development trend of information and communication technologies and their applications (the Internet, the Internet of Things, the Internet of Everything) are very simple laws: processing speeds and information transfer speeds will be constantly increasing each year, but the laws of physics still apply (despite the postulates of "White Magic")—the product of the numerical value of the average delay time of signal transmission by the base element (or its switching time) and the average numerical value of its power consumption from the supplying element (power supply network) will always remain a constant value, regardless of the design of any basic elements of the TCS (such as transistors), and manufacturing technologies.

In the general case, this simple but fundamental law can be written in a very simplified form as follows [2]:

$$A = P \cdot \tau = \text{const}, \tag{3.1}$$

where A is the switching operation of the TCS base element; P is the average value of its power consumption ($P = U \cdot I$, where U is the supply voltage, I is the current consumption); τ is the average time of switching the element from one state to another.

Of course, this is a too simplified formula, since the power consumption of the base element may be static (constant) or dynamic (at the moment of switching into the state "1" and into the state "0"), and the average switching time τ is generally determined by the expression [2]:

$$\tau = \tau_{10} + \tau_{01}/2, \tag{3.2}$$

where τ_{10} is the element (transistor) switch-on time, τ_{01} is its switch-off time.

Its main theme of the book which contributes to its title is the basics of designing *high-speed* electronic devices dedicated to finding technological design solutions with which the increase in information processing speed achieved by reducing the average switching delay of various basic elements should not reach the threshold of violation of this fundamental limitation—corresponding value of the maximum power consumed by the element.

If for the first basic component of TCSs (computer information processing hardware) the ways to solve this problem are sufficiently developed [1–10], then for another basic component of TCSs (information transfer channels) there is not yet any single technical solution.

In the following chapters, some particular technical solutions to this difficult problem will be presented, formulated by the authors, both on the basis of our own research and on the basis of the existing diverse (positive and negative) experience of domestic and foreign development of such channels and TCS information transmission lines.

3.2 Basic Methodological and Terminological Problems of the Introduction of Fifth-Generation Wireless Telecommunication Systems

Before considering the main problems and prospects for development of fifth-generation TCSs (5G), it is nevertheless necessary to make a brief overview of the complicated history of development of wireless technologies and understand what these (G) mean [11]. First it should be noted that G (Generation) is the so-called next "generation" of network technology. The development of these generations originates the distant 1980s, when several of the first such network technologies appeared, in the USA it was AMPS technology, and in Europe TACS, NMT, and their combinations.

Although it must be said that several variants of the simplest mobile networks existed before, but the first three (AMPS, TACS, and NMT) are considered by the experts to be the first generation (1G), because they allowed the widely used mobile phones of today to become truly mass products.

It should also be noted that "at the time of 1G" no one thought about the services of data transmission, those were just purely analog systems designed exclusively for voice calls that were in demand at the time, although they provided some additional modest opportunities. The cost of a minute of conversation at that time was such that only a wealthy person could afford a mobile phone, since this type of product was considered an unaffordable luxury.

Surprisingly, the first mobile communication system in the world was the Soviet "Altai"—an automatic mobile communication system that had actually worked in Moscow since 1963. The "Altai" phones were called "mobile" because they were installed mainly on vehicles of managers and major specialists of law enforcement agencies, employees of special services, government officials, and managers of large industrial enterprises.

It was a truly full-fledged *telephone*—one could talk on it as if it were a regular "landline wired" telephone, in duplex mode. To call another Altai or a regular landline phone, it was enough to simply dial the number without any additional switching of channels and conversations with dispatch operators.

A similar American IMTS (Improved Mobile Telephone Service) was tested in the United States in a special pilot zone only a year later, and actually began to be used in various states since 1965. In fact, mass production of these first American mobile phones began only in 1969 (4 years later) than in the USSR.

But in the USSR, by the next year—1970, the Altai system worked effectively in more than 30 cities.

A decade later—in the early 1990s, the first digital "cellular" telephone networks of the second generation (2G) appeared, which had a number of very significant advantages over analogue ones, namely improved sound quality and greater noise immunity.

While mobile communication systems based on Qualcomm's CDMA and D-AMPS standards were actively developing in the United States at that time [11], in Europe the GSM triumphal march was already beginning. These communication standards already supported the transfer of short text messages (SMS), as well as the pioneering for that time technology CSD, which allowed transmitting data to the base station in digital form. Moreover, the speed of such transmission reached 14.4 kbps, which was comparable with the speed of transmission of standard modems of the mid-90s.

In order to initiate data transfer using CSD technology, the user had to make a special "call". It was like a telephone modem—you were either connected to the network or not. Given that the tariff plans at that time were measured in dozens of minutes, and the CSD technology was akin to an ordinary telephone call, there was almost no practical benefit from this technology at that time.

Now, a few words should be said about the "intermediate" generation (2.5G) [11].

The emergence of the General Packet Radio Service (GPRS) in 1997 actually became a turning point in the history of cellular communication, because it proposed using *continuous data transmission* technology for GSM networks. With this technology, for the first time it became possible to use data transfer only when it is needed—there is no more "stupid" CSD which is more like a telephone modem. In addition, the GPRS system could work with a higher speed than CSD—theoretically up to 100 kbit/s, and the carriers received the long-welcome opportunity to rate the *traffic* itself, not the *time* on the line.

It should be noted here that this protocol (GPRS) appeared at a very opportune moment—when people began to continually check their email inboxes.

However, this innovation still did not allow adding a full-weight unit (3G) to this new generation of mobile communication. While GPRS technology was already on the market, the International Telecommunication Union (ITU) approved a new standard—IMT-2000—describing the specifications of "real" 3G. The key point here was to ensure the data transfer rate of 2 Mbit/s for fixed terminals and 384 kbit/s for mobile ones, which was beyond the power of the GPRS protocol. Thus, GPRS was stuck between the 2G generations, which it exceeded, and 3G, which it did not reach, and therefore this generation received the conditional 2.5G status. By the way, this situation will be repeated many times over, we also know intermediate generations (2.75G, 3.5G, 3.75G) [11].

In addition to the aforementioned data transfer requirements, 3G specifications declared a relatively easy transition from second-generation networks, for which the UMTS standard became basic for all GSM carriers, and another standard, CDMA2000, provided backward compatibility. It must be said that after the precedent with GPRS, the developers of the CDMA2000 standard offered their own technology of continuous data transfer, called 1xRTT. Note that, although officially the CDMA2000 standard is a 3G level standard, it provides a data transfer rate only slightly higher than the GPRS standard—about 100 kbit/s.

The next "intermediate" standard EDGE (Enhanced Data rates for GSM Evolution) was developed as a way for GSM network carriers to "squeeze out additional juices" from 2.5G installations without investing substantial funds in upgrading equipment [11]. So, using a phone that supports EDGE, one would get a speed twice the speed of GPRS, which was quite good for that time. However, many European carriers did not bother with EDGE, because they were committed to implementing the UMTS standard, which was provided with the most powerful advertising support in social networks.

Discussions on the classification of "generations" arose again in the expert community. So to which generation does the EDGE standard belong? It is not as fast as UMTS or EV-DO, so it can be argued that it is *not 3G*. But at the same time it is clearly faster than the GPRS standard, therefore, it should be better than *2.5G*. Therefore, many experts have called EDGE the intermediate *2.75G* technology.

After another decade, CDMA2000 networks have received a new upgrade to the EV-DO Revision A level, which offered a slightly higher *incoming speed* and a much higher *outgoing speed*. In the original specification, which was then called EV-DO Revision 0, the outgoing speed was limited to 150 kbit/s, and the new version allowed it to be done ten times faster, which could already be attributed to the *3.5G generation*. About the same thing happened for UMTS technology: HSDPA and HSUPA technologies allowed to significantly add speed for incoming and outgoing traffic.

From numerous literary sources today it is known that further improvements in UMTS use the HSPA + (dual-carrier HSPA +) and HSPA + Evolution standards, which theoretically can provide bandwidth from modest 14 Mbps to very impressive 600 Mbps [11].

In terms of attributing them to specific generations, experts are not unanimous: Some experts still believe that they belong to *the new generation 4G*, and some propose to call them *3.75G* (by analogy with EDGE and 2.75G).

The situation with *the 4G generation* also deserves separate consideration. Just as it happened in its time with the 3G standard, ITU took control of the 4G standard, tying it to a specification known as IMT-Advanced. This document calls for incoming data rates of 1 Gbps for fixed terminals and 100 Mbps for mobile ones. These are, respectively, 500 and 250 times faster than IMT-2000, and these are really huge speeds that can outrun any ordinary DSL modem or even a direct connection to a broadband channel.

As we know, wireless technologies play a special role in providing broadband access, especially in rural areas. This is the most cost-effective way—to build only one 4G station, which will provide communication at a distance of tens of kilometers, rather than to cover the farmland with a "blanket" of expensive fiber-optic lines.

But, in the opinion of a significant part of experts, these specifications are so aggressive that no commercial standard in the world yet complies with them. Historically, WiMAX and Long-Term Evolution (LTE) technologies, which were designed to achieve the same success as CDMA2000 and GSM, are only considered fourth-generation technologies, but this is only partly true: they both use new, extremely efficient multiplexing circuits (OFDMA, as opposed to old CDMA or TDMA, which have been used for the past twenty years), and both lack a special separate channel for voice transmission. 100% of their bandwidth is used exclusively for data services, and this means that voice transmission will be considered as VoIP. Considering how strongly the modern mobile society is focused on the transfer of mainly mass data, this can be considered as a fairly good solution.

The disadvantage of WiMAX and LTE is the data transfer speed, which is 40 and 100 Mbit/s, but in practice the real speeds of commercial networks do not exceed 4 and 30 Mbit/s accordingly, which is very good in itself, but does not meet the high goals of the IMT-Advanced standard. In the process of updating these standards (WiMAX 2 and LTE Advanced), the developers promise to do this work, but it has not yet been completed and the real networks that use them do not exist at this time [11].

Nevertheless, it can be argued that the original WiMAX and LTE standards are significantly different from the "classic" 3G standards, in this sense, it is not entirely correct to talk about the change of generations. Indeed, most carriers around the world who have deployed such networks call them 4G. According to independent experts, this is used as marketing because the parent organization ITU does not have the appropriate authority to counteract this. Both technologies (LTE in particular) are already being installed at many telecom carriers around the world under the name "4G".

Another well-known example comes from the recent history of the development of telecommunication systems. The American carrier T-Mobile, which had never previously announced its intention to upgrade its HSPA network to LTE, decided to brand an upgrade to HSPA + as 4G. In principle, this step makes sense: 3G technology can ultimately reach speeds greater than just LTE, approaching the requirements of IMT-Advanced. Indeed, there are many markets where the HSPA + T-Mobile network is faster than WiMAX from Sprint. And neither Sprint, nor Verizon, nor MetroPCS—the three US carriers with a live WiMAX/LTE network—at the time of this book's release, are not yet offering VoIP services. They continue to use their 3G frequencies for voice and will continue to do so for a long time.

As a result, it must be said that the carriers won this terminological battle: ITU "retreated", saying that the term 4G "can be applied to the predecessors of this technology, LTE and WiMAX, as well as other evolved 3G technologies that provide a significant increase in performance and capabilities compared to the initial third-generation system" [11]. And in a sense, this is true—no one will argue that the so-called "4G" networks today resemble 3G networks of the 2001 period: We can stream video of very high quality, download large files very quickly and even, under certain conditions, use some of these networks as a replacement for DSL.

Experts point out that perhaps the WiMAX 2 and LTE-Advanced standards will indeed be called "4G" by the time they are widely available, although the capabilities of these networks will already be very different from the 4G networks that exist today.

Summarizing this section, we can formulate the main distinctive features of the generations listed above [11]:

- 1G is pure analog signal processing;
- 2G is digital signal processing, circuit switching, both in the radio channel and in the core network;
- 2.5G, 2.75G is digital signal processing, packet switching, both in the radio station and in the core network;
- 3G is a transitional generation that ensures compatibility of the "junior" and "subsequent" generations;
- 4G is only packet transmission of signals, both in the radio and in the core network. Further development of this technology is PSTN and PLMN.

As can be seen from the above, on the evolutionary path of development of wireless telecommunication systems from 1G through 3G to 4G, many issues arose in terms of terminology—to which generation to assign a particular communication standard.

What then can be said about the prospects of the fifth generation (5G)—what are the contours of this future telecommunications system? What technical problems need to be solved on the path "from theory to practice"?

We will try to briefly examine these questions in the next section of this chapter, using the materials from [12–35].

3.3 State, Prospects, and Problems of the Development of Fifth-Generation Mobile Communication Technologies (5G)

As follows from the analysis of the above material, the 5G standard is the next stage in the development of information and communication technologies, which should provide *unrestricted access* to network for individual users and devices. The development of the 5G standard took into account the improved capabilities of LTE and HSPA, as well as other technologies of the so-called radio access, aimed at solving a number of specific tasks.

From the moment of their appearance up to the moment of publication of this book, the mobile communication network has come a long way of development; there are new types of user devices—smartphones and tablets. The opportunities that mobile technologies open today have long gone beyond voice services only, creating new ways of communicating, data sharing, and business models. The proliferation of these devices has led to an exponential increase in traffic across networks around the world. However, this is only the beginning of the technical revolution that contributes to the active development of technologies that connect society.

With the development of mobile networks and their adaptation to new types of devices and services—from smart electricity meters, cars and connected household appliances to industrial facilities—new diverse requirements are being put forward. In this regard, the classic one-for-all technology approach can hardly be effective in meeting the growing number of diverse needs from both business and individual users.

All these technologies will continue to evolve towards *higher productivity* and an increasing number of opportunities. As shown in Fig. 3.8 in addition to the existing radio access technologies, new technologies will appear (hitherto unknown) that will allow to solve tasks that cannot be solved using 3G/4G.

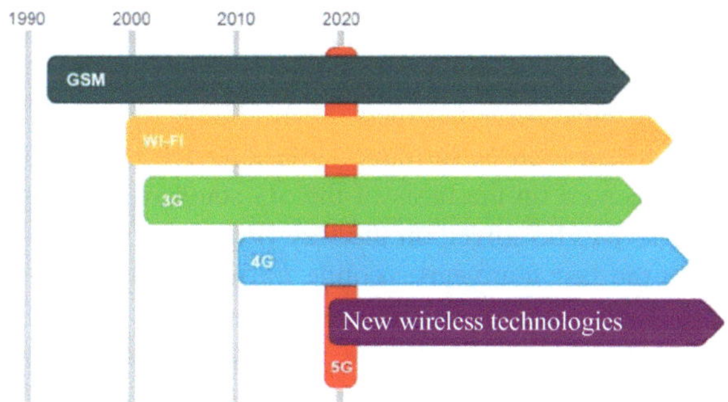

Fig. 3.8 Dynamics of development of generations of wireless technologies [4]

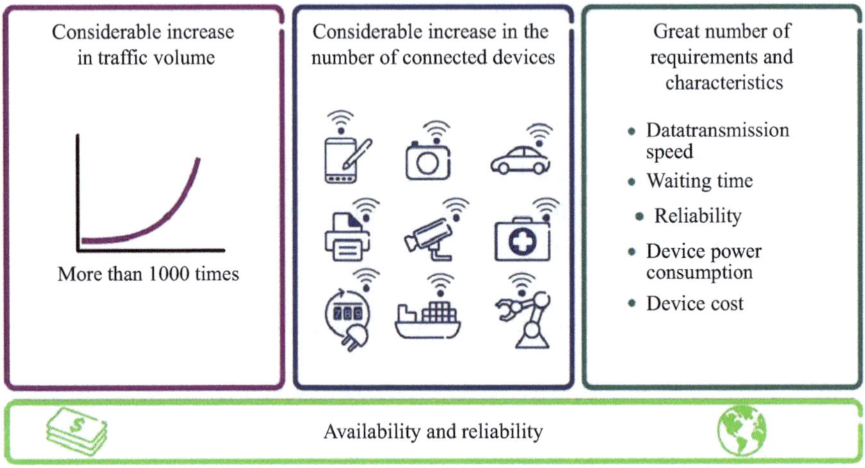

Fig. 3.9 The main problems of radio access system implementation

In the long run of such development, what we call today 5G will finally appear, i.e. a set of organically integrated radio access technologies (Fig. 3.8) [12, 14, 19, 26]. First of all, it is the LTE-evolutionary step in the development of mobile communication technologies, as well as the development of HSPA and Wi-Fi. Therefore, this is not about replacing existing technologies with 5G, but about their development and addition of new radio access technologies designed for specific scenarios and specific goals.

The main requirements imposed by users on mobile broadband access (MBBA) networks, first of all, relate to *performance*. The development of modern networks is committed to ensuring that in the future access to data is instantaneous, and the provision of services occurs without delay and is not interrupted due to unreliable communication. The role of MBAN for society and business is steadily becoming larger and, often, reliable communication can be a matter of life and death, especially when it comes to providing a reliable connection for medical and rescue services.

The number of connected devices is increasing rapidly. The general trend is that in the end, everything will be connected that can benefit from network connectivity, ranging from traffic lights, home appliances to cars, medical equipment, and power supply systems (Fig. 3.9). This opens up unlimited opportunities for people, business, and society. Providing such connections is the task to be solved by 5G technologies.

Efficient high-performance solutions needed to provide such connections can be implemented on the basis of the 5G standard, although a number of difficult tasks will have to be solved on the way (Fig. 3.9).

In the last few years, mobile traffic has been showing steady growth, and this trend will continue in the future. Based on various forecasts [11–14], it can be concluded that after 2020, the capacity of the systems will have to handle traffic that exceeds the current value by more than 1000 times.

Today, more than five billion mobile devices [11] are operating in networks around the world, most of which are mobile terminals or devices that provide mobile broadband access and are integrated into portable computers and tablets. In the future, it is expected that the number of connected devices involved in smart cities, smart homes, and smart grids will exceed the number of user devices by 10–100 times.

To ensure the smooth operation of 50 billion (and possibly 500 billion) devices is not an easy task. Along with the increase in the number of connected devices, the network requirements will also increase significantly.

Video streaming, file sharing networks, and cloud services will continue to be the most popular applications, *requiring ever higher speeds*.

In offices and urban environments, where the density of network calls is the highest, it is necessary, first of all, to ensure a data transfer rate of several Gb/s. Only such a speed will allow to synchronize local storage with cloud and network drives, transmit ultra-high-definition video, and support the work of applications of virtual and augmented reality.

To solve the problem of providing unrestricted access to information, data transmission speeds in hundreds of Mbps should be available to users.

In addition, in order to ensure the "gigabit" speeds required for the *operation of virtual or augmented reality applications*, it will be necessary to further reduce the response time to a few milliseconds, which puts in the first place the search for ways to solve the problem of improving the performance of the data transmission system.

For some purposes, for example, for remote video surveillance (both on Earth and in the large space), it is necessary to transmit significant amounts of information in a very short period of time, while for other applications, for example, tracking cargo traffic by shipping or railway companies, the amount of data transfer can be quite small.

Thus, in order to successfully cope with the tasks described above, the development of new, more advanced and efficient wireless access solutions is required.

In this context, 5G can be viewed as the result of development of existing radio access technologies. The resulting system, which includes various second-, third-, and fourth- generation radio access technologies, fully integrated and used as a whole, can be called the fifth-generation system (5G), as it will allow you to drive the quality of user network experience and performance to a level that is unavailable to fourth-generation systems (Fig. 3.10).

Below we list only the most obvious basic technical tasks that developers will need to solve on the path to implementing 5G technology [14, 36, 37]:

1. *Providing high-quality access to mobile broadband access (MBBA) services.* Existing mobile broadband access technologies, such as HSPA and LTE, will continue to evolve and provide the foundation for a new fifth-generation network standard. Ensuring high-quality access to high-speed mobile services will be possible due to the introduction of intelligent antenna technology with a large number of controlled elements, the development of new band ranges of a higher degree of coordination between base stations.

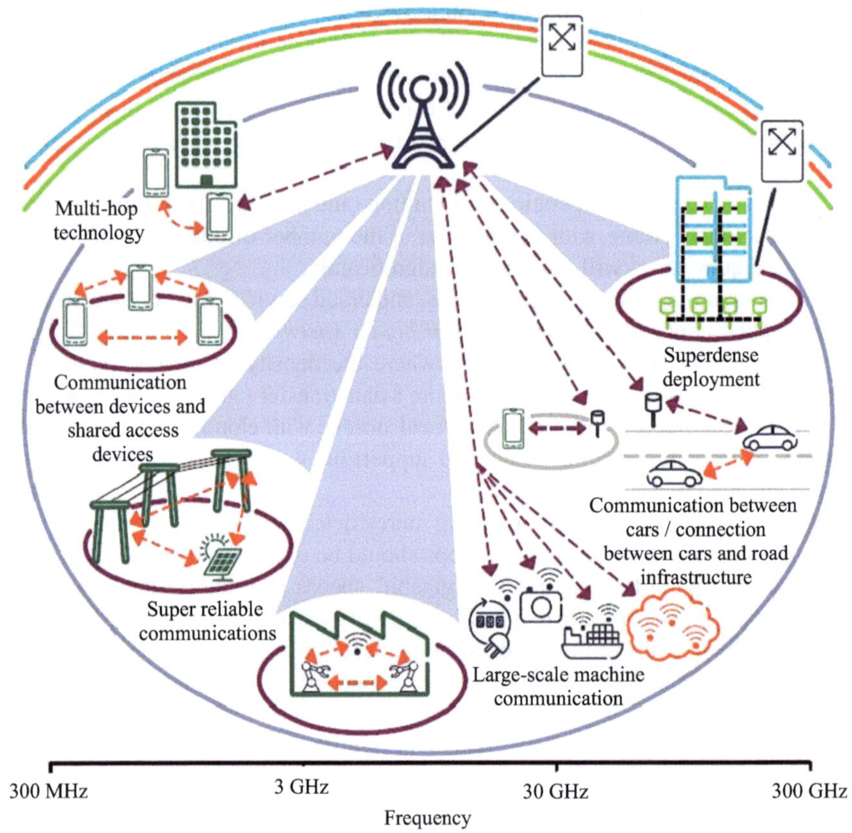

Fig. 3.10 Main areas of application of 5G standard

2. *Providing ultra-high bandwidth and network capacity*. In order to ensure the processing of ultra-high user traffic and data rates, measured in several Gb/s, specific scenarios foresee the introduction of super-dense network deployment in which base stations will use a very wide bandwidth in the upper frequency bands using new radio access technologies.
3. *Support of a huge number of low-power M2M devices*. Activities aimed at expanding existing technologies (mainly LTE) in order to support the huge number of connected M2M devices are already being developed within the framework of 3GPP standards. However, LTE standards will not be able to meet the highest requirements of some applications. Therefore, it is necessary to introduce alternative technologies, for example, to support devices with extremely high energy consumption limits. Such technologies will be fully integrated into cellular technologies, which will provide users with the possibility of continuous access. For example, capillary networks using alternative technologies will connect to the rest of the world through cellular networks.

4. *Application of proximal communication.* In situations where users are close to each other, and especially when information is specific to a particular place of use (traffic safety, state security, National Security Personnel System (NSPS), general short-range services, when the user interacts and exchanges information with the immediate environment), it makes sense to organize data exchange directly between devices using the device-to-device protocol (D2D), rather than through the network infrastructure.
5. *Ensuring super-reliable communications.* In military and space technology, in industrial communications and social services, for example, while ensuring traffic safety, in e-health services and in managing smart cities, modern radio networks cannot always meet reliability requirements. In some cases, associated with the use of communication tools for intelligent power systems or traffic safety, the delay time requirements, for example, may be higher than modern systems can offer, no more than a few milliseconds with end-to-end data transfer. This will require an overhaul of the interaction of elements of today's mobile broadband access systems, which mainly focus on bandwidth, coverage, and data rates.
6. *Ensuring energy efficiency and sustainability.* As mentioned above, energy efficiency will play an even more important role in the future, which should be the main goal in the design of all 5G solutions.
7. *Mastering new frequency spectra.* 5G technologies by 2020 and beyond will require a significantly larger frequency range and wider carriers in order to support the expected traffic growth and even higher data rates. Until 2020 and beyond, new frequency spectra will be needed, both in bands already used today and in the upper bands.

Obviously, to realize this perspective, it will be necessary to solve the problem of the increasing number of devices connecting to the network. The variety of connected devices, applications, and tasks suggests that 5G will be the result of integrating radio access technologies, both advanced LTE and HSPA, and more specialized technologies, serving specific goals.

It can be assumed that the concept of a 5G system, developed, for example, in the METIS (Mobile and wireless communications) project, may be proposed for standardization in a few years. In this case, according to most competent experts, a gradual transition to 5G networks can begin as early as 2020.

Another reason that requires a change in the paradigm of wireless access networks—the "Internet of Things", as well as its subsequent development—the Universal Internet (or the Internet of Everything). From telecommunication point of view, this means, first of all, the explosive growth of active transceiver devices, and wireless ones. They are all connected into one global network—the Internet. And, after all, we are talking not only about low-speed telemetry and control streams, but also about high-speed traffic, for example, video. The structure of the channels itself is changing—the concept of machine-to-machine (M2M) communication via the network becomes relevant. We are talking about global connectivity, the explosive growth of not only the speed in the channel, but also the number of such channels.

Another trend that is becoming more pronounced today is the concept of cloud computing. Its essence is: procedures for complex processing, storage of large data arrays strive to be transferred "to the cloud", that is, to data processing centers, distributed or local. And this is an industrial trend, a global paradigm shift of industrial management (one of the names of which is Industry 4.0). Games and office applications, industrial object management systems, and databases from personal devices and local servers go "into the cloud." Moreover, this trend has radically affected the telecommunication technologies: The concept of software-defined networks SDN is nothing more than the transfer of complex processing functions from distributed network devices (routers) to a certain center. In general, cloud technologies mean that the era of "big analog data" is coming—there is a huge variety of sources of raw data that should be transmitted via telecommunication channels to the processing centers. And, again, the vast majority of sources of such data are wireless devices.

The emergence of a multitude of data sources, the *ever-increasing speed demands in a single transmission channel*, and an increase in the number of base stations lead to an explosive increase in power consumption by wireless networks. In communication systems, a problem emerges that has already been encountered in the field of super high-performance computing (supercomputers)—*energy efficiency*. And this is not just an environmental struggle—it is a real need. For example, the world's largest mobile operator China Mobile in 2012 consumed 14 billion kWh of electricity in its network of 1.1 million base stations. And by 2014, another 800 thousand base stations were added to China Mobile networks, that is, power consumption grew by more than 70% [36, 37].

In addition to speed, the *transmission delay time plays an increasingly important role*. Whereas earlier this was most acutely affected by exchange operations and network gamers, now, with the advent of the Universal Internet, the era of global management through the network, connection establishment, data transfer, and response delivery become more and more critical.

So, *super-speed (1–10 Gbit/s), ultra-small delays (less than 1 ms), ultra-large number of active physical connections, and their ultra-high energy and spectral density are the main tasks* that cannot be solved within the concept of 4G wireless networks, such as LTE Advanced. The 5G network should provide access to data everywhere and always, to all people and to each device.

There are not many known methods that can potentially solve the problems described. At the level of physics, it all comes down to the following methods [36, 37]:

- operation in the millimeter wavelength range (30–300 GHz);
- use of spatial channel separation and spatial modulation (which means formation of narrow radiation patterns in the receiving and transmitting devices, the use of multichannel antenna systems MIMO);
- application of new signal and code constructions, for example, multiplexing with "non-orthogonal" frequency division of channels, which leads to complication of signal processing algorithms in devices.

The principle of organizing the communication channels themselves needs to change as well. If 2G and 3G systems allowed only the "base station - subscriber device" channel, and in 4G systems the possibility of direct communication between base stations appeared, then in the context of 5G people speak about direct device-to-device communication. And this is fundamental—as part of a globally connected network. We are talking about the so-called mesh-networks, which are mobile, dynamically emerging, and reconfigurable.

Here, the main principle is the so-called *heterogeneous network principle*, where heterogeneity is possible in a variety of parameters, including the technologies used, the frequency ranges, and the division of functions between macro and microcellular BSs. For example, data is exchanged via a micro-BS, and control information—via a macro-BS. Or, the downlink can be implemented in the millimeter range in the narrow beam formed by the BS, and the uplink—in the centimeter range.

In the case of super-dense BS location (with a grid spacing of hundreds of meters), the supporting networks connecting them will play a fundamentally different, key role. These can be high-speed networks of the millimeter range, radio-relay systems with the capabilities of mesh networks, etc.

Of course, all this is not a complete list of all technical problems and conceptual approaches to the organization of 5G networks.

Today, all experts are unanimous in that 5G networks will be based on operation in *two frequency domains—"below 6 GHz" (i.e., in the cm-band) and "above 6 GHz"*. The latter means millimeter range, in the range from 28 to 95 GHz. Moreover, while communication technologies in the bands below 6 GHz are, in fact, an evolutionary development of 4G technologies (for example, the transition from OFDM modulation to non-orthogonal FDM), operation in the millimeter range is a fundamental new approach to the organization of wireless communication systems. Let's note a few points that characterize this new paradigm of building mobile communication networks.

First of all, the millimeter range is a huge unoccupied frequency resource (formally from 30 to 300 GHz, actually—up to 100 GHz). It allows you to operate with bands above 1 GHz, which provides a transmission rate of 1–10 Gbit/s and above. However, communication in the mm-range requires directional antenna systems. This achieves a significant energy gain compared with traditional wireless communication systems [37].

Due to the small wavelength, the physical dimensions of the mm-band antenna systems at high gain are very small. Since the spacing between the elements of the antenna array is half the wavelength, that is, a few millimeters (5 mm for 30 GHz, 2.5 mm for 60 GHz), active grids can be formed on the substrates of the microcircuits, in their cases, etc. This significantly simplifies and cheapens the element base and the development of communication systems as a whole. Due to the small physical size of the antenna arrays, they can be embedded in subscriber devices (for example, in telephones).

Due to the high directivity of mm-range antenna systems, high spatial selectivity is achieved—neighboring devices can operate at the same frequency without interfering with each other, that is, the problem of highly efficient use of the frequency resource is solved. High attenuation of electromagnetic radiation in the range of 60 GHz due to resonant absorption on oxygen molecules naturally solves the problem of interference, which is especially important for the organization of a wireless core network. Compact, tunable devices provide the flexibility required for the 5G concept.

In addition, the millimeter-wave communication systems are also safer compared to today's cell phones—due to the high attenuation in water and greater energy efficiency, the effect of mm-band electromagnetic radiation on vital human organs is significantly lower than in cm-band communication systems (almost all radiation attenuates at the level of the outer skin).

A few years ago, the vast majority of specialists were very skeptical about the possibility of organizing communication systems in the mm-range. Indeed, this is "practically optics", "only direct visibility", "a huge attenuation of free space". Millimeters are for radio-relay systems, fixed-line communications with narrowly directional parabolic antennas, and radar systems with liquid cooling. And most importantly, how much will the element base cost, what mass and dimensional characteristics will the end devices have, what will their power consumption be?

All these problems have recently seemed irresolvable. But it is precisely for this reason that we talk about the "new generation", because it means a change in the technological paradigm, a transition to a new ideology, which a generation ago seemed an unscientific fiction. A fundamentally new element base and new equipment based on it make this fiction a reality. Thus, for example—not only discrete microwave components have been developed, but also monolithic integrated circuits (MIS) operating in the millimeter range. And then the silicon VLSI technologies, such as BiGMOS (SiGe) and even CMOS, went into millimeter range. The new element base is already actively used in E-band radio relay communication systems (71–76, 81–86, 92–95 GHz), in the equipment of local wireless networks in the 60 GHz band (WiGig, IEEE 802.11ad, WirelessHD, IEEE 802.15.3c) [35]. And, very importantly, due to the small wavelength, antenna arrays are often formed directly on the substrates of integrated circuits. A new category of the element base appears, where all microwave processing, analog and digital, as well as the antenna system allowing the formation of a narrow tunable radiation pattern, are integrated on a single chip or in a single package. The most complex problems of designing microwave devices are solved at the VLSI/MIS level and are hidden from the hardware developers. At their disposal are real components with a high-speed digital interface and an integrated antenna array.

Summarizing this section, we can draw two obvious conclusions:

- It's not yet possible to talk about the real appearance of 5G networks until 2020–2025.
- The hardware basis of these networks will be exclusively high-speed electronic devices, to the theoretical, technical, and practical bases of the design of which this book is devoted.

3.4 Major Trends and Challenges in the Development of Wireless Technologies for the Internet of Things (IoT)

3.4.1 Technical Background Pf the Emergence of the Internet of Things

The term "Internet of Things" (IoT), in a broad sense, means the concept of a new computer network uniting various physical objects of the real world (the world of things) equipped with built-in means of identification, measurement, reception, and transmission of data. Examples of such physical things are machines and sensors, and virtual things are computer programs.

It is believed that the very idea of building an IoT type system was formed in 1999 at the Massachusetts Institute of Technology (MIT) [36], while the real IoT birth moment is the period from 2008 to 2009, when the number of devices connected to the global network surpassed the population of the Earth. Experts predict that by 2020 the number of devices on the web will increase to 34 billion.

Some IoT applications will generate very large data transfer rates with data requiring to be analyzed in real time. In some systems, such as telecommunication, telemetry systems, millions of events are transmitted per second. To meet these requirements, various distributed stream computing platforms (DSCPs) have been created. For processing streams with a very high data transfer rate, they usually use parallel architectures, which allows solving problems such as real-time analytics and pattern recognition.

The rapid development of IoT will have a significant impact on the development of future wireless standards, such as 5G. Consider below the main trends in the development of wireless technologies used in applications for the so-called "Internet of Things" (IoT) [36–38]. First of all, remember that experts in the history of the development of communications distinguish three historical periods ("eras"). At the time of the book's release, we are entering a third era of communications that will change the way we use wireless communications. In the first era, wire telephony and Internet technologies were the main methods of connecting houses and businesses. Over the past couple of decades, there has been a transition to a second era with its wireless technologies.

The connection of things is the third era of communications, the onset of which is approaching much faster than many of us believe. Today, there are more than seven billion mobile devices in the world, connecting more than 3.8 billion people. According to the Ericsson Mobility report presented in November 2015 [39], the number of machine-to-machine (M2M) connections will increase to almost 12 billion over 5 years. And by 2030, according to Ericsson's forecasts, there will be ten times as many interconnected machines than people who use wireless communications.

The main feature of this new era of communication, as already noted above, is the transition to the microwave range.

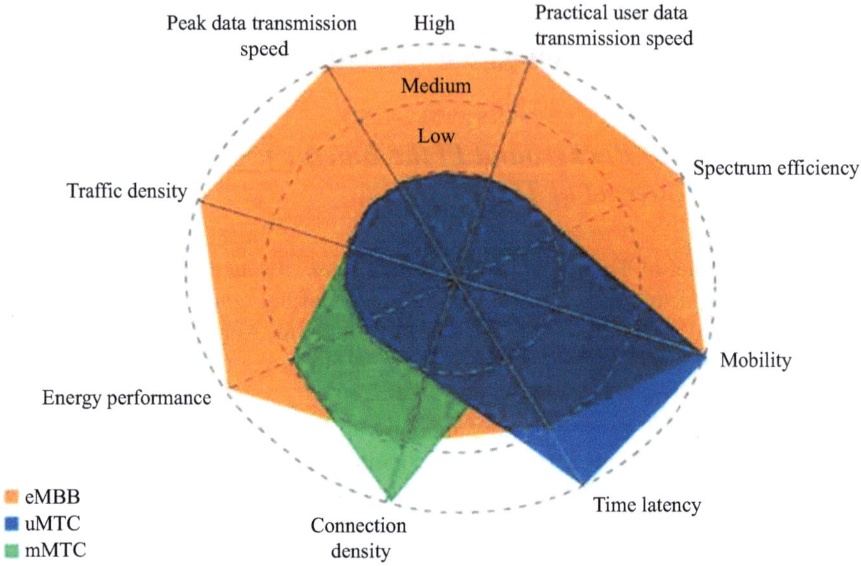

Fig. 3.11 The main areas of use of 5G-systems in 2020, suggested by ITU (source—ITU-R M. (IMT.VISION)) [41]

In the past, the requirements to new standards for wireless communications were dictated mostly by users. In the future, these requirements will be determined primarily by the need to connect the machines. The differences in the requirements for communication between people and between things are clearly stated in the proposed use cases of the fifth generation of telecommunication systems (5G). The International Telecommunication Union (ITU) has established three different scenarios for the use of wireless communications [40], each of which is focused on achieving different performance in terms of delays, bandwidth, connection density, and mobility (Fig. 3.11) [39].

The second feature is the emergence of a new mobile network with advanced features. The first variant of the use of SG-systems are enhanced mobile broadband networks, eMBB. They can be considered as a natural evolution of LTE and LTE-Advanced standards. Just like the transition from GPRS to HSPA and then to LTE provides consumers with ever higher data transfer rates, the 5G standard will continue this trend. The usage scenario of eMBB defines technical requirements for achieving peak data transfer rates of 10 Gbit/s—at the same time, the total network traffic can increase 10 thousand times.

Another distinctive feature of the new era of communication *is machine-to-machine interaction in telecommunication systems*.

Two other scenarios for the use of 5G, defined by ITU, are designed to *solve the problem of connecting things together, not people*.

Not surprisingly, these scenarios—massive machine-type communications (mMTC) and ultra-reliable machine type communication (uMTC)—require quite different requirements than eMBB.

For the TMTC scenarios, the basic requirements for radio modules include ultra-low power consumption and ultra-low cost. In these radio devices, one battery must provide power for up to ten years or more. On the other hand, uMTC scripts are targeted at critical applications. The main requirements for uMTC systems are delays <1 ms and reliability of at least 99.999%.

Although the requirements of 5G are clear evidence of how IoT shapes the development of future standards, this is not the only example. Wireless systems should be designed to connect things more often than people, and this already has an impact on the development of expected wireless technologies, such as 802.11ah, and mobile communication technologies such as LTE-M.

As consumers in an increasingly interconnected world, many of us can easily imagine how wireless technology will fit into consumer electronics. These applications will be able to have a significant positive impact on society. As for industrial application scenarios, a significant part of them is not yet completely obvious, but over time their number will constantly increase.

One of the families of applications where wireless technologies provide innovation is a car with network capabilities (connected saga). Residents of North America may remember how General Motors introduced the first OnStar system on several 1997 Cadillac models. The original system used AMPS mobile phone networks [42] (the predecessor of CDMA and GSM) and allowed vehicles to contact the call center in the event of an emergency. Today, this service uses LTE-modems and can even provide a Wi-Fi connection for passengers in a car.

Although OnStar technology has been available for nearly 20 years, the use of cellular modems inside a car is still relatively uncommon. In 2015, only 20% of cars sold worldwide had the ability to connect to the network infrastructure. In the future, this figure will approach 100% [43]. Thus, the European Union in 2015 resolved that after 2018 all new cars would have to be equipped with e-call technology—the European equivalent of OnStar, which provides notification of an accident in case of an emergency [44, 45]. Focusing on the development of technologies such as e-call, traditional providers of information and entertainment devices, such as Harman, are increasing the production of modules supporting both Wi-Fi and LTE. This requires investment in new production test methods, such as parallel testing.

Thus, the further spread of the Internet of Things depends largely on the successful development of wireless technologies. The joint efforts of researchers, developers, test engineers, and other specialists will allow the creation of wireless devices and systems ensuring the widespread adoption of the Internet of Things in everyday life and in industry.

3.5 Communication Channels of the 5G-Generation Telecommunication Systems Based on Phased Antenna Arrays

As was shown in a number of papers [46–49], transmitters and receivers with phased antenna arrays operating in the millimeter wavelength range are supposed to be used to create promising generations of telecommunication systems. The main features of construction of such communication channels for 5G systems with phased antenna arrays were first considered in detail in [47].

The key element of such a solution is an antenna, the physical dimensions of which for narrowly targeted transmission should be about eight times the length of the transmitted wave. But in order for the antenna to maintain a size that is acceptable for commercial use, it is necessary to use shorter wavelengths or higher frequencies—this explains the huge interest in the unloaded 60 GHz band. The most commonly standard used in this range is the IEEE 802.Had (WiGig) standard, which provides up to 7 Gbit/s transmission speed due to the use of transmitters and receivers with phased antenna arrays that increase channel efficiency, improve system parameters, and help optimize the transmission path.

Although the benefits of implementing 5G and multi-gigabit transmission channels are very attractive, the possibility of their implementation is determined by two key factors: the creation of inexpensive phased antenna arrays and their adaptation for use in communication channels 5G. Scientists at the University of California at San Diego (UCSD) in close collaboration with Keysight. Technologies developed in 2015 the world's first integrated transmitter with a 64-element (8×8) and 256-element (16×16) phased antenna array. Its highly efficient built-in antennas provide connectivity at gigabit speeds within a radius of several hundred meters. During the demonstration, the operating capacity of such a channel was proved and record performance was shown.

The preparation of this demonstration required tremendous effort and relied on two important factors. One of them is creation of the industry's first integrated 64-element (8×8) [46] and 256-element (16×16) phased antenna arrays operating in the 60 GHz band (Fig. 3.12). This development was based on early research conducted by UCSD and Tower Jazz, with financial support from the Defense Advanced Research Projects Agency (DARPA). As a result, on-chip integrated systems were created, each of which contained a 60 GHz signal source, distribution circuits, phase shifters, variable gain amplifiers, and highly efficient integrated antennas.

At about the same period, after the development of integrated phased antenna arrays was completed, the UCSD University was ready to switch to the implementation of these crystals in 5G communications equipment. For this purpose, the university scientists working on the project needed a high-speed hardware and software measuring system that *would provide the necessary characteristics and accuracy.*

Fig. 3.12 Integrated 5G transmitters with 256-element 60 GHz integrated phased antenna arrays created at UCSD University [46]

Such a system was created, a 60 GHz communication channel was created stably operating at a distance of 4 m, then 30 m and, finally, about 100 m.

To create this channel, researchers from UCSD used a full-featured instrument kit from Keysight (Fig. 3.13). The kit included an M8190A arbitrary waveform oscillator, an E8267D PSC vector waveform oscillator, and an Infiniium C Series high-resolution oscilloscope with an l0-bit ADC (Fig. 3.14). An analog signal oscillator E8257D PSC was used as a heterodyne.

The test results showed that the 60 GHz communication channel was working steadily. At a distance of 4 m (full channel 802.Had), a transmitter with a 64-element phased antenna array created a complex communication channel at all scan angles (±45° in all planes) and with almost no additional distortion, providing a transmission speed of 3.85 Gbit/s with 16-QAM modulation. Almost the same results were obtained at a distance of 30 m. At a distance of more than 100 m, the communication channel provided a transmission rate of 1.54 Gbit/s with QPSK modulation at scanning angles of ±45° in all planes.

Successful demonstration of the world's first 5G communication channel based on a transmitter with a beam forming function with a 64- and 256-element phased antenna array is a big step towards the introduction of 5G technologies.

The example given here shows tremendous potential opportunities contributing to the transition to a narrowly focused 5G connection. It is equally important that this work paves the way for future research and development in the field of millimeter-wave communication systems.

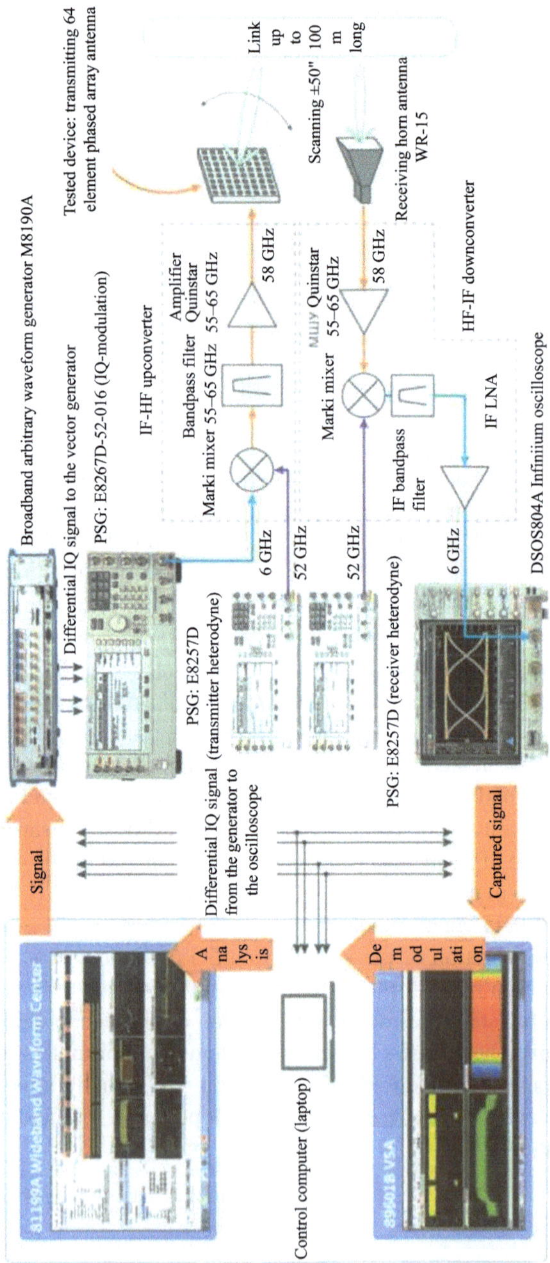

Fig. 3.13 Chart of the measurement system used by UCSD to measure the characteristics of a 60 GHz communication channel [46]

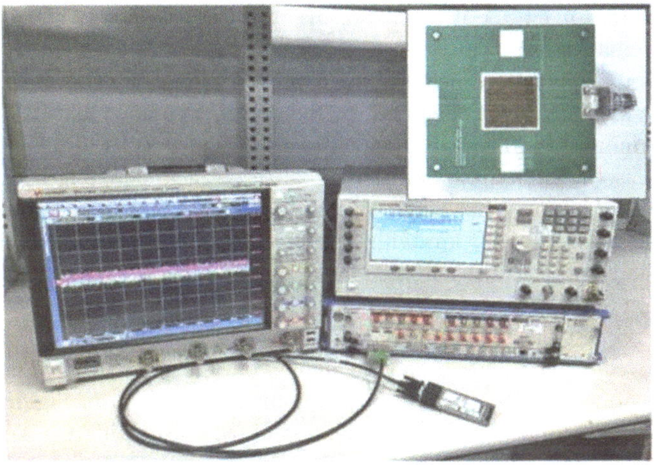

Fig. 3.14 General view of Keysight equipment for measuring the characteristics of the 60 GHz 802.had channel using a 64-element phased array antenna [46]

3.6 Features of Using the LPWAN Technology for the Internet of Things

As was shown above, for the implementation of the Internet of Things it is important for carriers to have a set of various information and communication technologies as an environment allowing them to connect their devices to the World Wide Web using various communication technologies—from expensive high-speed to cheap low-speed ones.

For example, "fast" channels can be used for self-driving (unmanned) cars, and cheap low-speed solutions are needed for millions of sensors and counters.

LPWAN (low-power wide-area network) technology, an energy-efficient long-range network, is intended to provide this wide enough field of application. In other words, LPWAN is a group of wireless technologies for transferring small amounts of data over long distances, developed for distributed telemetry systems, machine-to-machine interaction, and IoT [49].

LPWAN technologies are designed for use in M2M/IoT modules, which do not require high-speed data transmission, provide a wide geographic coverage and a long (more than 10 years) battery life.

In fact, LPWAN complements the above-described traditional 2G, 3G, and 4G technologies and is widely used in applications such as intelligent household and industrial meters (water, electricity, gas), street and industrial lighting, etc.

Obviously, the approach used to build LPWAN networks is very close to the principles of operation of mobile communication networks, but their structures are much simpler, since they require an order of magnitude fewer base stations and switches. One of the main disadvantages of this technology is the lack of a single standard that defines the physical layer and control access to the environment.

At the time of this book's release, the following standards are used: LoRa, SigFox, Weightless [49].

LoRa is the abbreviated name of the LoRaWAN open network protocol (long-range wide area networks) developed by Semtech Corporation and IBM Research. At the beginning of 2016, LoRa appeared under the brand LoRaAlliance (IBM, Semtech, Cisco, Actility, etc.) with the goal of de facto adopting and promoting the LoRaWAN protocol as a single standard for LPWAN. For such a solution, the coverage area is up to 20 km with a data transfer rate of up to 11 Kbps. The undoubted advantage is the minimum energy consumption—more than 3 years of operation on one AA battery.

SigFox was established in France in 2009. By the end of 2015, SigFox networks were deployed in 20 countries, and the number of officially registered devices with SigFox support exceeded eight million. The main advantages of SigFox are ultra-low consumption (up to 20 years of operation of the sensor with two AA elements) and high penetrating power in the conditions of a modern city.

Weightless is a group of open standards of the non-profit organization Weightless SIG.

In Russia, deployed LPWAN networks use Strizh technology developed by Strizh Telematika.

Unlike most traditional low-power LPWAN systems, it is a fast-growing alternative technology group based on the 3GPP (Third-Generation Partnership Project) consortium protocols, which are in fact the basis of modern cellular communication systems. Their distinctive feature is redundant for most existing applications of the Internet of Things, namely the data transfer rate in the LTE-M standard, leading to high power consumption. But this fact is naturally due to the difference between the GSM and LTE protocols.

According to Cisco forecasts (Fig. 3.15) [12, 14, 49], the share of LPWAN technologies in mobile M2M connections in the world will grow from 4% in 2015

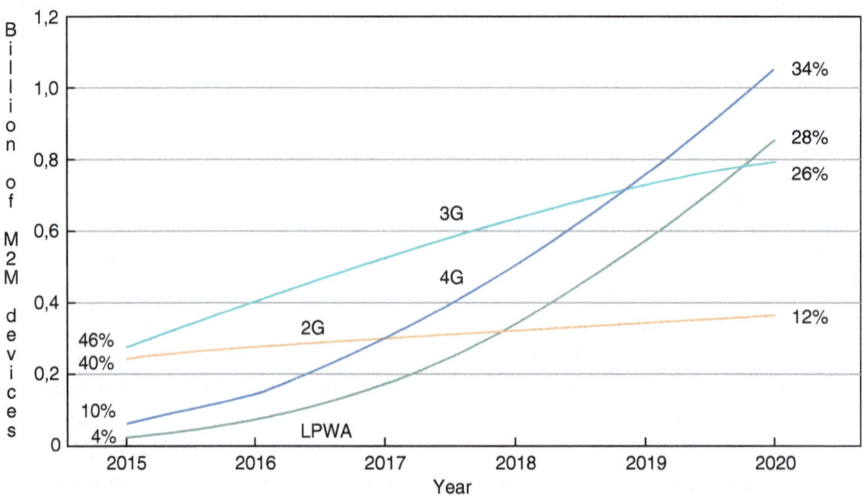

Fig. 3.15 State and forecast of the global market for M2M connections (source: Cisco)

to 28% in 2020. According to the calculations of Strategy Analytics, more than 2020 billion devices will be connected to LPWAN networks by 2020.

References

1. Belous, A., Saladukha, V., & Shvedau, S. (2017). *Space microelectronics*. Artech House, Inc.
2. Belous, A., Emelyanov, V., & Turtsevich, A. (2012). *Fundamentals of microelectronic device circuitry* (p. 472). Moscow: Technosphere.
3. Belous, A., Shvedau, S., & Merdanov, M. (2016). Microwave electronics in radar systems and communications. In *Technical encyclopedia* (2 Vols., p. 1416). Moscow: Technosphere.
4. Belous, A., Efimenko, S., & Turtsevich, A. (2013). *Semiconductor power electronics* (p. 216 + colour attachment on 12 pages). Moscow: Tekhnosfera.
5. Belous, A., & Yarzhembitsky, V. (2001). *Circuitry of digital microcircuits for information processing and transmission systems* (p. 116). Minsk: Tekhnoprint UE.
6. Belous, A., Emelyanov, V., & Syakersky, V. (2009). *Designing integrated circuits with low power consumption* (p. 320). Minsk: Integralpoligraf.
7. Belous, A., Blinkov, O., & Silin, A. (1990). *Bipolar IC for automatic control system interface* (p. 272). Leningrad: Mashinostroenie.
8. Belous, A., Podubny, O., & Zhurba, V. (1992). *K1815 LSI Microprocessor Kit for digital signal processing* (p. 256). Moscow: Radio and Communications.
9. Belous, A., Saladukha, V., & Shvedau, S. (2015). *Space electronics* (Vol. 1, p. 696). Moscow: Tekhnosfera.
10. Belous, A., Saladukha, V., & Shvedau, S. (2015). *Space electronics* (Vol. 2, p. 488). Moscow: Tekhnosfera.
11. *2G, 3G, 4G, and everything in between: An Engadget wireless primer*. https://habrahabr.ru/post/112535/
12. Ericsson. (2013). *Networked society essentials* [pdf]. Stockholm: Ericsson. Website: http://www.ericsson.com/res/docs/2013/networked-society-essentials-booklet.pdf
13. Ericsson. (2013). *Ericsson mobility report – On the pulse of the networked society* [pdf]. Stockholm: Ericsson. Website: http://www.ericsson.com/res/docs/2013/ericsson-mobilityreport-june-2013.pdf
14. Cisco. (2013). *Cisco visual networking index: Global mobile data traffic forecast update, 2012–2017* [pdf]. USA: Cisco. Website: http://www.cisco.com/en/US/solutions/collateral/ns341/ns525/ns537/ns705/ns827/white_paper_c11-520862.pdf
15. Ericsson. (2013). *Technology for good – Ericsson sustainability and corporate responsibility report 2012* [pdf]. Website: http://www.ericsson.com/res/thecompany/docs/corporate-responsi bility/2012/2012_corporate_responsibility_and_sustainability_report.pdf
16. METIS. (2013). *Mobile and wireless communications Enablers for the Twenty-twenty information Society* [pdf]. Website: https://www.metis2020.com/wp-content/uploads/2012/10/METiS_factSheet_2013.pdf
17. Pi, Z., & Khan, F. (2011). An introduction to millimeter-wave mobile broadband systems. *IEEE Communications Magazine*, 101–107.
18. Pi, Z., Khan, F., & Zhang, J. (2010). *Techniques for millimeter wave mobile communication*. US Patent Application Publication No. 61/299304, priority October 29 2010.
19. Vishnevsky, V., Frolov, S., & Shakhnovich, I. (2010). The millimeter range as an industrial reality. 802.15.3c standard and WirelessHD specification. *Electronics: NTB*, 3, 70–79.
20. Vishnevsky, V., Frolov, S., & Shakhnovich, I. (2011). Radio-relay communication lines in the millimeter-wave range: New speed horizons. *Electronics: NTB*, 1, 90–97.
21. Recommendation ITU-R M. 1645. (2010). *Framework and overall objectives of the future development of IMT-2000 and systems beyond IMT-2000*. ITU.

22. *Cisco visual networking index: Forecast and methodology, 2014–2019*. www.cisco.com/c/en/us/solutions/collateral/service-provider/ip-ngn-ip-next-generation-network/white_paper_c11–481360.html
23. Chih-Lin, I., Rowell, C., Han, S., Xu, Z., Li, G., & Pan, Z. (2014). Toward green and soft: A 5G perspective. *IEEE Communications Magazine*, 2, 66–72.
24. Wang, T., Li, G., Ding, J., Miao, Q., Li, J., & Wang, Y. (2015). 5G spectrum: Is China ready? *IEEE Communications Magazine*, 7, 58–65.
25. ICT-317669-METIS/D1.1. (2013). *Scenarios, requirements and KPIs for 5G mobile and wireless system*. Project METIS.
26. Shakhnovich, I. (2014). Myth about free space attenuation: What G.T. Friis did not write. *First Mile*, 2, 40–45.
27. Rappaport, T. S., Sun, S., Mayzus, R., Zhao, H., Azar, Y., Wang, K., et al. (2013). Millimeter wave mobile communications for 5G cellular: It will work! *IEEE Xplore Digital Library*, 1, 335–349.
28. Rappaport, T. S. et al. (2013). Broadband millimeter-wave propagation measurements and models using adaptive-beam antennas for outdoor urban cellular communications. *IEEE Transactions on Antennas and Propagation*, *61*(4), 1830–1839.
29. Rappaport, T. S. et al. (2014). *73 GHz millimeter-wave indoor and foliage propagation channel measurements and results* (Technical Report 2014–003). NYU WIRELESS: Department of Electrical and Computer Engineering, NYU Polytechnic School of Engineering .
30. Ghosh, A. (2013). Can Mmwave Wireless Technology meet the future capacity crunch. In *IEEE ICC*.
31. Roh, W. (2013). Performances and feasibility of mmWave beamforming prototype for 5G cellular communications. In *IEEE ICC*.
32. Tikhvinsky, V., & Bochechka, G. (2014). The prospects of the millimeter range for 5G in Russia. *The First Mile*, 2, 36–39.
33. Cudak, M., Kovarik, T., Thomas, T. A., Ghosh, A., Kishiyama, Y., & Nakamura, T. (2014). Experimental mmWave 5G cellular system. In *Globecom 2014 Workshop Mobile Communications in Higher Frequency Bands* (pp. 377–381).
34. 4G Americas. (2015, August). *5G spectrum recommendations*. www.4gamericas.org.
35. 5G Candidate Band Study. (2015). *Study on the suitability of potential candidate frequency bands above 6GHz for future 5G mobile broadband systems* (Final report to Ofcom). www.quotientassociates.com.
36. Martynov, V., Makushin, M., & Sukhoroslova, Y. (2016). Paradigm of paradigms or the internet of things. *Electronics: Science, Technology, Business*, 10.
37. Makushin, M. Ten main directions of development and problems of IoT for 2017–2018. In *Express information on foreign electronic equipment* (Vol. 9, No. 6597, pp. 7–11). Moscow: JSC Central Research Institute Electronics.
38. Starkloff, E. (2014). Wireless technologies for IoT - Incentives and development trends. *Electronics: NTB*, 6, 118–121.
39. *Ericsson Mobility Report*, November 2015, www.ericsson.com/res/docs/2015/mobility-report/ericsson-mobility-report-nov-2015.pdf
40. *IMT Vision - framework and overall objectives of the future development of IMT for 2020 and Beyond*. International Telecommunication Union, September 2015, www.itu.int/dms_pubrec/itu-r/rec/m/R-REC-M.2083-0-201509-1!!PDF-E.pdf
41. LTE-M - optimizing LTE for the internet of things. *Nokia Networks*. https://gsacom.com/paper/nokia-lte-m2m-optimizing-lte-for-the-internet-of-things/
42. *Most analog cellular to fade away next week*. PC World, February 2008, www.washingtonpost.com/wp-dyn/content/article/2008/02/15/AR2008021500034.html
43. *2025 every car connected: Forecasting the growth and opportunity*. SBD, February 2012, www.gsma.com/connectedliving/wp-content/uploads/2012/03/gsma2025everycarcon nected.pdf
44. *eCall Whitepaper Version 1.5. QUALCOMM*, March 2009.
45. RF Power Amplifier and transceiver market tracker. *Databeans*, Q42015.

46. Pierpoint, M. (2014). Transition to 5G telecommunication systems - creation of communication channels with phased antenna arrays. *Electronics: NTB*, 6, 114–117.
47. Zihir, S., Curbuz, O. D., Karroy, A., Raman, S., & Rebeiz, G. M. (2015). A 60 GHz 64-element wafer-scale phased-array with full-reticle design. In *2015 IEEE MTT-S International Microwave Symposium*, Phoenix, AZ.
48. Zihir, S., Gurbuz, O. D., Karroy, A., Raman, S., & Rebeiz, G. M. (2015). A 60 GHz single-chip 256-element wafer-scale phased array with EIRP of 45 dBm using sub-reticle stitching. In *2015 IEEE Radio Frequency Integrated Circuits Symposium (RFIC)*, Phoenix, AZ.
49. Chachin, P. (2017).Use of LPWAN radio technologies for the IoT market. *Electronics: NTB*, 1, 140–144.

Chapter 4
Basic Components of Telecommunication Systems

This chapter presents the results of the analysis of the basic components of up-to-date high-speed telecommunication systems. Figuratively speaking, here is a detailed description of the main building blocks of which the architecture of up-to-date telecommunication systems is being built: waveguides, adapters, attenuators, filters, tappers, dividers and power combiners, encoders, coaxial communication lines, etc.

In order to design a really high-speed TKS, the designer should be well aware of all the technical characteristics and features of the application of these basic elements in actual operating conditions.

So, in Sects. 4.1, 4.2, and 4.3 the features of the use of waveguides, coaxial communication lines, adapters, attenuators, and absorbing filters are considered. It should be borne in mind that at ultra-high frequencies (SHF) high-power radiation is capable of generating a huge amount of heat in small volumes of materials of basic element structure. Therefore, any elements of the TCS, including semiconductor devices and microcircuits that are critical to the effects of high-power radiation, should be designed to ensure maximum stability with respect to such phenomena.

Section 4.4 considers the features of the design and application of filters on surface acoustic waves (SAW filters), and Sect. 4.5 considers various options for the structure of elements designed on the basis of microelectromechanical systems (MEMS).

Sections 4.6, 4.7, 4.8, and 4.9 discuss other basic elements necessary for creating high-speed TCSs: software reconfigurable radio systems, microwave cable assemblies, radio frequency cables, fiber-optic communication lines, etc.

A. Belous, V. Saladukha, *High-Speed Digital System Design*,
https://doi.org/10.1007/978-3-030-25409-4_4

4.1 Features of Using Passive Components in HF and SHF Devices

In this section, we consider a number of specific requirements for passive components most frequently used in high-speed systems. The development and production of HF and microwave passive components are associated with numerous design and technological limitations, and the methods for analyzing these components differ significantly from those used in most other electronics applications [1].

This applies both to the use of special materials in high-speed electronic systems (microwave systems) operating under extreme conditions, as well as to all the basic passive components of these systems and devices.

4.1.1 Necessity of Using Special Materials for Operation in Extreme Conditions

With the increase in power consumption characteristic of modern microwave systems, the requirements for structural materials and their production technologies increase significantly. For example, in high-power telecommunication systems and military radars, along with huge output powers, it is necessary to ensure a high level of performance. Since many of the well-known technologies and materials cannot be used in high-power microwave installations, they use special structural materials and unique technological processes.

As we know [1], high-power HF and microwave radiation can generate huge amounts of heat in small volumes of structural materials. Often the results of such an impact are detected only after the failure of individual elements or the failure of the entire system. Such scenarios are characteristic of powerful military, aerospace, and telecommunication equipment.

High-power HF and microwave radiation throughout the signal propagation path can activate damage associated with the accelerated material aging processes, design errors, and other similar effects. Besides, such damage can be caused by the so-called deliberate external powerful microwave pulses (microwave weapons) aimed at microwave systems [1]. Given these circumstances, any structural elements, including electronic components that are critical to the effects of high-power RF and microwave radiation, should be designed to ensure maximum resistance to this kind of phenomena.

In addition, under conditions of high-power HF and microwave radiation, the effects of passive intermodulation and harmonic distortion that disrupt equipment operation are more pronounced. These phenomena must also be taken into account in the design and production of structural materials and electronic components intended for high-power HF and microwave equipment.

Any connection of a transmission line or components through which high-frequency and microwave radiation propagates is characterized by the so-called insertion losses due to the fact that part of the energy is absorbed by the construction materials. Since these elements are most susceptible to destruction under the action of HF and microwave radiation, maximum allowable values of power are limited. Normally, the same HF and microwave equipment operates in different modes of radiated power, depending on the current required task. Specifically, the power level restrictions are set for pulsed or for average allowable values. Additional destructive influence is exerted by fluctuations in operating temperature, humidity, vibration level, as well as accumulated time of operation in extreme conditions. However, not all manufacturers specify in detail the maximum permissible values of the parameters listed above, which also increases the risk of an accident in the field. Special requirements are imposed on HF and microwave electronic components the failure of which can lead to a cascade and complete destruction of the entire system. In particular, elements of the communication lines of microwave devices can be attributed to such components [2].

4.1.2 Waveguides and Coaxial Communication Lines

Depending on the used wavelength of the HF and microwave radiation, its power, as well as the specific application, various wave equipment uses either waveguides or coaxial cables.

As we know [2, 3], coaxial lines are usually used for wavelengths greater than 2 cm. At higher frequencies, the loss in the coaxial line increases significantly, due to the fact that the effective surface area of the internal wire is small compared to its large resistance.

Also a big role in coaxial lines is played by the loss in material that isolates the outer wire from the inner wire.

When the wavelength is less than 1 cm, waveguides in the form of a metal tube of circular or rectangular cross section are usually used, inside which the electromagnetic wave propagates.

The developers of microwave devices use waveguides only when the used wavelength is shorter than the "critical" value, which is approximately two times the numerical value of the transverse size of the waveguide. For example, if a waveguide has a diameter of 1 cm, then the "critical" wavelength is 2 cm, longer waves cannot propagate through such a waveguide.

Two types of electromagnetic waves can propagate in waveguides. Waves (indicated as "E") have an electric field located both in the transverse and in the longitudinal directions, as well as a transverse magnetic field. In magnetic waves ("H"), the magnetic field is aligned in the transverse and longitudinal directions, and the electric field is only aligned transversely.

Two types of adapters are used to connect the coaxial line with the waveguide. In the first case, an electrical connection is used, using a metallic conductor probe, which is installed inside the waveguide along the electric field lines at the point with the highest electric field strength.

In adapters of the other type, a so-called magnetic loop is used, located in a plane perpendicular to the magnetic force lines, at the point where the magnetic field is maximum (see Fig. 4.1) [2].

Waveguides can transmit high power in the HF and microwave bands. It should be noted that this technology is much more expensive than coaxial lines.

In areas in which small capacities and ranges of more than 2 cm are required, where the price of the product and installation is critical, it is preferable to use coaxial lines. For these lines, a very wide range of standard adapters, connectors, and attenuators is available on the market today. Since coaxial lines are designed for low power and breakdown voltages, components for them, as a rule, have small dimensions and weight. This fact is an undoubted advantage when using HF and microwave technologies in aerospace equipment.

In applications that require increased reliability, high power, and microwave range, waveguides are mainly used. In some cases, both technologies are used in the same equipment operating in different modes with tunable frequency and power. In such devices, the powerful signal arriving from the waveguide is attenuated with special attenuators and fed through micro-connectors to the coaxial line (see Fig. 4.2) [3].

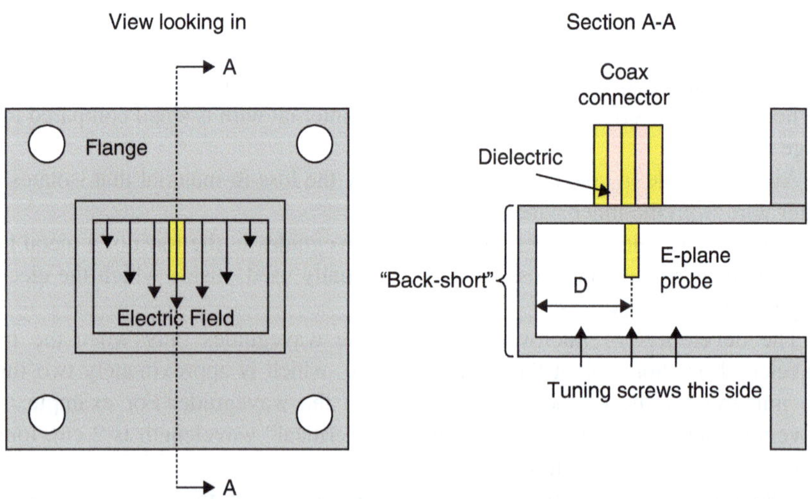

Fig. 4.1 Two types of devices are used to adapt the coaxial line and waveguide: magnetic loop and electric connection with a metal probe

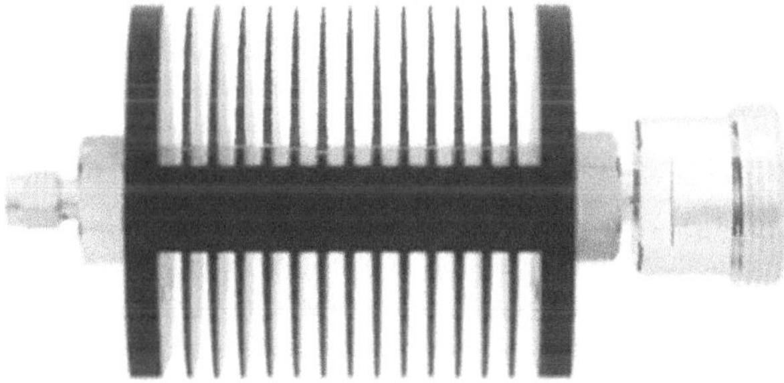

Fig. 4.2 Attenuators attenuate the waveguide signal to a level acceptable for switching to miniature coaxial connectors

4.1.3 Adapters and Terminal Equipment

Adapters, connectors, and other terminal equipment can cause overheating and failure of electronics due to the known effects of distortion, reflection, and absorption of microwave radiation. A very careful approach is required when choosing these components for each specific high-frequency installation. There is a fairly wide range of models of adapters and connectors of different designs. For example, these include straight, angular, triple, matching, attenuating, etc. models. Each model is designed to work with certain powers, frequencies, and maximum values. Each type of waveguide is designed for a strictly specified frequency and power. Each type of coaxial line has its own cutoff frequency.

Presently, a wide range of adapters are available matching coaxial lines with different parameters, for example, in frequency, capacitance, insertion loss, active and reactive resistances. Since terminal devices can dissipate significant amounts of energy in high-power HF and microwave installations, they are usually equipped with forced cooling systems.

One of the main parameters characterizing the terminal device of HF and microwave equipment is the voltage standing-wave ratio (SWR). Excess of SWR and impedance can cause irreversible damage to active electronics. For example, inconsistency of the SWR of a terminal device with the parameters of the power amplifier is one of the most frequent causes of accidents at HF and microwave installations.

4.1.4 Attenuators and Absorbing Filters

Attenuators are designed to dissipate radio frequency energy inside the device, without causing unwanted distortion and reflections of the signal. Constant power, adjustable and variable attenuators of small and medium power are produced. There are also numerous families of coaxial and waveguide attenuators [1, 3].

Attenuator models are available on the market that adapt coaxial lines to waveguides. As a rule, powerful attenuators are equipped with radiators with passive or active cooling systems.

High-frequency absorbing filters are nothing more than frequency-dependent attenuators. These devices absorb signals outside a certain frequency band and dissipate their energy as heat.

Reflecting frequency filters emit signals at a specific frequency and return them to the source. Filters of this type can damage the output stages of the transmitting device in cases of power and voltage overload. It should be noted that threshold power values depend on the frequency of the signal—the higher it is, the lower the power threshold. In addition, when choosing a filter, it is also necessary to take into account the average characteristics given for the recommended frequency band.

4.1.5 Directional Couplers and Power Splitters/Adders

A directional coupler (DC) is designed to divert part of the electromagnetic energy from the main channel to the auxiliary one. The main line is called the primary, and the auxiliary line—the secondary. A well-known example of a coupler is a television antenna splitter. The working arm of the secondary line is selected depending on which wave should be diverted from the primary line—the incident or reflected. By design, directional couplers are similar to adapters. The difference is that DCs can also be used as an adder. In this case, when power is applied to the auxiliary line, the resulting power is summed in the main channel.

A very important feature of multichannel DCs is that the transmitted power on individual channels can be reduced by orders of magnitude compared with the case when it is transmitted on one channel. This property is used in combined DCs, in which a powerful HF signal from a waveguide forks into several coaxial lines equipped with miniature HF connectors. The appearance of such a directional waveguide power divider is shown in Fig. 4.3 [2].

Fig. 4.3 Directional waveguide power divider with coaxial pins on auxiliary channels

It should be noted that there are DCs, the principle of operation of which is somewhat different from that described above. For example, hybrid-type directional splitters and hybrid matrix splitters provide uniform separation or summation of signals in HF and microwave networks. This family includes 3-dB hybrid splitters with a signal phase shift of 90°, which evenly distributes the HF signal between two equivalent channels.

One of the important problems arising in the design and use of DCs is the insertion loss and distortion caused by the reflection phenomena. Since these effects are significantly dependent on the frequency and power of the signal, incorrect design solutions and the use of DCs in lines that do not correspond to their specification can cause huge insertion losses and noticeable distortions.

Another technical problem associated with the use of DCs is related to the correct choice of the load of the primary and auxiliary lines. To transmit small power, you can use standard terminals. However, when transmitting powerful microwave signals with an unmatched load, even small pickups and reflections can cause significant energy surges in the main channel. In this regard, the terminal devices of the main channel must be designed for capacities that significantly exceed the loads of the auxiliary channels, and also have a reliable system of protection against overloads of various kinds.

Power splitters (PSs) are designed to separate the signal passing through the main line into several identical ones. There are in-phase splitters and quadrature splitters with a signal phase shift of 90°. The principle of operation of microwave power splitters (PSs) is similar to that used in the DC. However, the main requirements for the PSs are to harmonize the wave impedances and preserve the waveform in the main and auxiliary channels. Thus, when designing a PS, special attention is paid to the problems of phase distortions, which lead to insertion losses, distortions, and reflections. In the development of microwave PS, different methods of signal separation are used. Table 4.1 presents the parameters of microwave power splitters for different types of structures [3].

Table 4.1 Impedances of microwave power splitters for different types of structures

Splitter	$m = 1$	$m = 2$	$m = 3$	$m = 10$	$m = 100$
Wilkinson	$Z_1 = Z_2 = 2^{1/2}$ $R_2 = 2$	$Z_1 = 2060$ $Z_2 = 0.612$ $R_2 = 2121$	$Z_1 = 2630$ (hard to implement) $Z_2 = 0.385$ $R_2 = 2.309$	Not available	Not available
Ring type	$Z_1 = Z_2 = 2^{1/2}$	$Z_1 = (3/2)^{1/2}$ $Z_2 = 3^{1/2}$	$Z_1 = 2/3^{1/2}$ $Z_2 = 2$	Not available	Not available
Two-branch	$Z_1 = 1$ $Z_2 = 1/2^{1/2}$	$Z_1 = 2^{1/2}$ $Z_2 = (3/2)^{1/2}$	$Z_1 = 3^{1/2}$ $Z_2 = (3/2)^{1/2}$	Not available	Not available
On linked lines	$Z_{01} = 2414$ $Z_{02} = 0.414$	$Z_{01} = 1931$ $Z_{02} = 0.518$	$Z_{01} = 1732$ $Z_{02} = 0.577$	$Z_{01} = 1365$ $Z_{02} = 0.732$	$Z_{01} = 1060$ $Z_{02} = 0.905$

Complex power splitters, as a rule, are adjustable in power and frequency, and also allow to connect waveguide and coaxial lines of different types. This approach makes it possible to choose the best variant of the terminal device for a specific microwave line.

4.1.6 Basic Mechanisms of Degradation of SHF Components

Probably, firstly we should note the failures of microwave components due to the effects activated by the so-called passive intermodulation interference [2].

Passive intermodulation interference (PII) occurs when two or more frequencies pass through elements with non-linear characteristics, which causes a lot of additional spurious frequencies, called "intermodulation products."

Effects such as increased temperature and humidity, vibration, mechanical damage, contact oxidation, electrolysis at the junction of various metals, pollution in the connector, and other similar phenomena lead to the appearance of intermodulation products. At high powers, the intermodulation effect in microwave devices can cause energy dissipation and destruction of materials of structures. For example, moisture penetration is often one of the reasons for changing the electrical characteristics of a material, increasing the power of dispersion and, as a result, destruction of a structure (see Fig. 4.4) [1].

Considering the danger and consequences caused by PII, we can talk about special requirements imposed on powerful passive components used in such equipment. There is a separate class of passive microwave components designed specifically for operation in harsh environments and resistant to the onset of PII. When designing PII-critical microwave devices, first of all, one should very carefully select the appropriate components that are resistant to the emergence of third-order and higher intermodulation products. Besides, it is recommended to test all equipment for PII resistance prior to its commissioning.

Fig. 4.4 An example of destruction of a rotary microwave connector due to ingress of moisture into the structure [2]

High levels of microwave power are also inevitably accompanied by dissipation of energy in structure materials that have a heterogeneous structure. As a result, at peak loads, overheating occurs at local points of the structure, which can lead to a change in the structure of the material and its destruction during long-term operation. Thus, power in different operating modes is selected so that the working temperatures are within reasonable limits.

Most manufacturers of passive electronic components determine the operating capacity and operating temperature ranges of finished products based on permissible material operation parameters. This is especially important for critical applications in which equipment downtime caused by any accidents during thermal overheating is unacceptable.

Along with temperature effects, increased humidity and vibrations contribute to degradation and destruction of structure material. The influence of these factors makes it necessary to further reduce the recommended power to levels at which safe operation of the microwave components is guaranteed. Considering the importance of safe operation with a long service life, the leading manufacturers of passive microwave components conduct complex and expensive tests of finished products under conditions of high temperature, moisture and salt content in the air, smog, vibrations, shock loads. As a rule, the results of such studies are provided to authorized distributors, which allows for optimal selection of components for each specific case.

4.2 New Generation of Encoders, Features of Their Use in Radioelectronic Equipment

In the last decade, there has been a constant "boom" in the field of encoder designs—one of the main components of modern radioelectronic systems. Therefore, the results of development of a new generation of encoders [4]—encoders with output switching signals, which are in demand in almost all drive control applications, will be listed below. The need to use these devices is growing due to the wider distribution of brushless DC drives (BLDC), which provide significant advantages in control, accuracy, and efficiency.

As an introduction to this section, remember that the principle of capacitive measurement is used in touch switches; the user's finger acts as the second capacitor plate. Data on changes in its capacity is read by the interface circuit. Thus, the function of a traditional electromechanical button is reproduced. Touch switches are protected from dust, water, and improper operation due to the absence of moving parts and small metal plates that fit tightly to the mounting surface.

The use of capacitive sensors in digital systems instead of optical or magnetic encoders has several advantages, including that it provides the implementation of intellectual functions in the development of encoders with output switching signals.

Fig. 4.5 Encoders with output switching signals determine the rotational speed, direction, and acceleration of the drive shaft [4]

The main task of an encoder is to determine the final position of the drive shaft and transfer this data to the system controller (see Fig. 4.5) [4]. Using this data, the controller automatically switches the drive windings, and also determines its necessary rotational speed, direction, and acceleration, which are used to maintain the specified drive parameters.

The principle of operation of encoders is based on different methods, which ultimately allow to obtain at the system output certain digital quadrature signals A and B, as well as the so-called output index signal for determining the direction of rotation (see Fig. 4.6a). Standard U, V, and W phase switching signals are also emitted from the encoder output (see Fig. 4.6b) [4].

Standard encoders, as a rule, produce 48–2048 pulses per revolution (p/rev), whereas in most applications, values of 800–1024 p/rev are demanded. It should be understood that the use of encoders with higher accuracy increases the cost and complexity of the solution, and also imposes an additional computing burden on the system controller or digital processor used. Excessive accuracy can even lead to increased noise levels in the system, vibrations, and to the infamous jitter.

Although in some systems encoders are not used, these devices are actually a necessary link in the vast majority of control and feedback loops used in practice.

It should be noted that, along with the use of brushless DC motors, sensorless circuits that do not require encoders to determine the position of the shaft are actively used today. In this case, electric drives are controlled using different algorithms, including the vector method (field-oriented control, FOC) [4].

Although excluding an encoder from the system is an economically attractive enough opportunity, there are several serious drawbacks to vector control: it does not provide the same accuracy as the sensor method; the control system skips certain items and periodically requires a reset; control is complicated at certain values of torques; the system processor needs significant computational resources. As a result, today, vector control is used in applications that do not require high accuracy, continuous tracking of shaft position and speed. Such applications include, for example, household appliances—washing and drying machines. At the same time, for most military industrial systems, the cost of encoders is not as critical compared with stringent performance requirements.

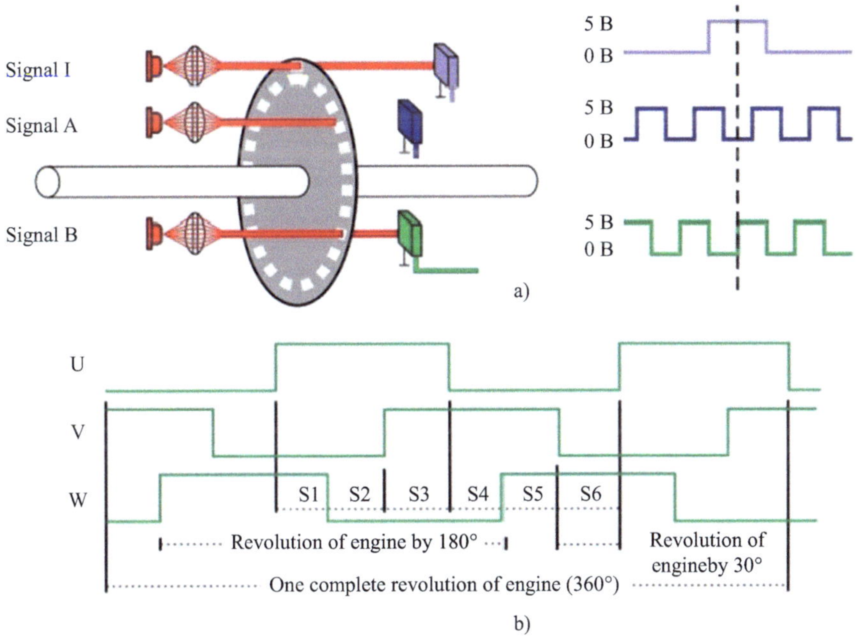

Fig. 4.6 (**a**) Standard quadrature signals A and B, as well as an index signal at the output of the optical encoder; (**b**) U, V, and W signals at the encoder output

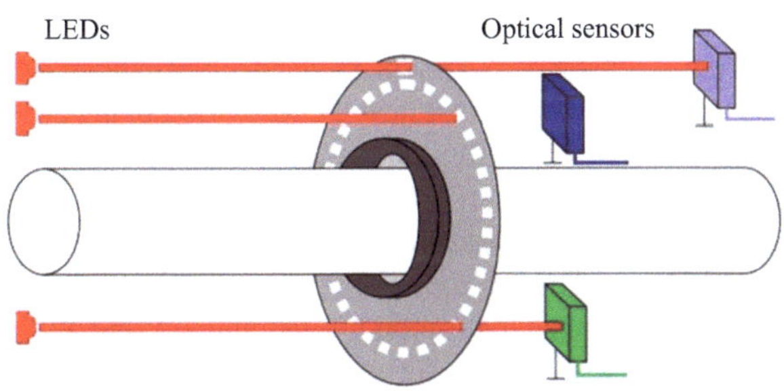

Fig. 4.7 An optical encoder works by reading the information about the shaft turns using light that passes through slits in the disk

Optical, magnetic, and capacitive encoders have been widely used to date. The optical method uses a slotted disc, where LEDs and phototransistors are installed opposite each other (see Fig. 4.7). When this disk rotates, the optical path is periodically interrupted, and the speed and direction of rotation of the shaft is determined by the incoming pulses. Despite the low cost and high efficiency of the optical encoder, its reliability depends on the effects of dust, dirt, or oil trapped in the light path. In addition, as we know, LEDs fade over time—their brightness, as a rule, decreases by half in just a few years of operation.

By design, the magnetic encoder is similar to an optical sensor, but it involves a magnetic field, not a light beam. Instead of an optical wheel with cuts, in a magnetic encoder, a magnetized disk is used, which rotates over an array of magnetoresistive sensors. Any rotation of the wheel causes the sensors to respond, which enters the input cascade to convert the signal, allowing to set the position of the drive shaft. The undoubted advantages of a magnetic encoder include a very long service life, and the disadvantages are the relatively low accuracy and susceptibility to magnetic interference from electric drives.

In addition to optical and magnetic encoders to encode the shaft position, Hall sensors are commonly used. These efficient and reliable devices, however, have relatively low resolutions.

Considering the need for an accurate and reliable method of coding the position of a rotating shaft, most of the encoder developers use the principle of capacitive measurement, implemented in a standard encoder, which converts the angular position of the shaft to linear, which is also used for many years in the above-mentioned digital caliper.

It is important to remember that the capacitive encoder has all the indicated advantages of optical and magnetic sensors, but does not have their disadvantages. Such an encoder consists of two groups of strips or lines, one of which is mounted on a stationary element of the structure, and the other—on a moving element. Both of these elements form a variable capacitance, configured as a transmitter/receiver pair (see Fig. 4.8).

The attractiveness of this capacitive technology lies in the fact that when it is used, the elements of the system are not subject to wear, and the encoder is protected from dust, dirt, and oil, which ensures higher reliability of operation than optical devices. In addition, capacitive encoders have the advantages of digital control functions. These include the ability to adjust the resolution of the device (the number of pulses per revolution), which eliminates the need to replace it with an encoder with a different resolution (higher or lower).

Developers of modern high-speed systems should pay attention to a number of devices of the AMT31 series from the CUI company, which in their own way are examples of a capacitive encoder providing quadrature A and B signals, an index signal, and also U, V, and W signals for switching windings. This encoder, which has 20 resolution values in the range of 48–4096 pulses/rev, supports seven values

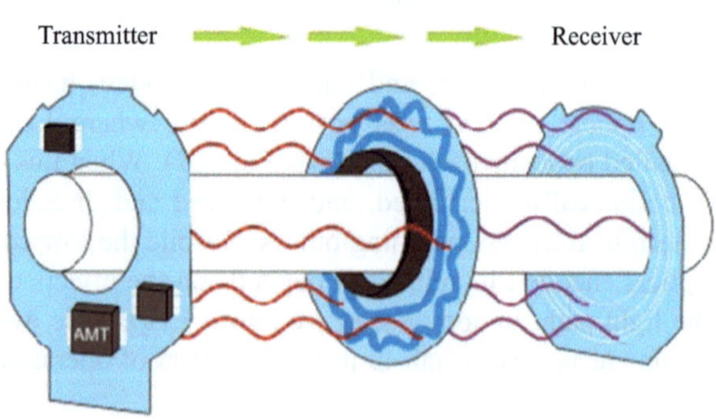

Fig. 4.8 A capacitive encoder counts simulated pulses when a signal passes from the rotor on the drive shaft

of drive pole pairs in the range of 2–20. In addition, the AMT31 series encoder also has a locking bolt, which facilitates its installation. Encoder supply voltage: 5 V, supply power current: 16 mA.

The advantages of a capacitive encoder are not limited only to high technical characteristics, versatility, and reliability. In contrast to the optical and magnetic encoders, the digital functions of a capacitive encoder provide the system with many unique advantages at all stages of use, from product development to installation and even maintenance.

Obviously, the usual output signal of an optical or magnetic encoder is quite functional, but, unfortunately, it does not provide universality, the possibility of presenting more complete information. On the contrary, the capacitive encoder is equipped with various digital functions, uses its own ASIC and a microcontroller, which provide additional capabilities. The intelligent encoder largely changes the operating scenario while maintaining absolute compatibility with standard encoders.

Taking into account all of the above, we will discuss in more detail the functionality expansion provided by the use of ASIC and a microcontroller, using the example of the AMT31 series encoder from CUI [4]:

- The digital functions of the capacitive encoder make it easy to implement the reset function. To do this, it is enough to block the shaft in the required position using the appropriate phases of the electric drive and give the command to zero. All these operations will not take more than 1 or 2 min and don't require special tools.

It should be noted here that the process of resetting an optical or mechanical encoder is multistage, complex, and not always inefficient. This is because at first it is required to block the rotor, change the position of the encoder, then proceed to control the drive with feedback signals using an oscilloscope in order to achieve a reset using the back-EMF and the encoder. Since in many cases this iterative process requires a very precise fit and verification, it takes at least 15–20 min to complete.

- The digital functions of the AMT series encoder can significantly accelerate the design process. In particular, since one capacitive encoder operates in a wide resolution range and with a large number of pole pairs, a programmable resolution function can be used to dynamically adjust the characteristics of the PID control loop when designing a controller and creating an algorithm, eliminating the need to purchase another encoder.

The AMT series encoder provides autonomous diagnostics, which speeds up the analysis of types and causes of failures. In addition, it is possible to check the correct operation of the encoder or to determine the cause of failure due to mechanical misalignment of the shaft. Moreover, the developer can use this feature to take precautionary measures—for example, to conduct an encoder test before starting the application.

- Since the system parameters (rotational speed, the number of pole pairs, and the direction of switching) which use the capacitive encoder are set using software, there is no need for other models of encoders for multi-drive engines or several different engines.

The advantages of a capacitive encoder are especially evident in the production and installation stages. Its mounting holes coincide match holes of other types of encoders (see Fig. 4.9). Thus, one such encoder can be installed on shafts of different diameters with the adapters, which allows reducing the number of assortment positions at the production warehouse and reducing the repair fund.

Structurally, this universal AMT11 encoder, consisting of a capacitive transducer and an electrical interface based on a customized ASIC, is a small module with a diameter of 37 mm and a thickness of 10.34 mm, which operates from a + 5 V power supply. The output signals of the encoder can be of two types: asymmetric quadrature signal of CMOS levels with a phase difference of 90° and a differential signal from the output of the line driver. Both of these signals are electrically compatible with standard optical and magnetic encoder signals. The connection with axial and radial orientation is carried out depending on the requirements of the application. The operating temperature range of the encoder is −40 to 105 °C.

A few words should be said about the user interface. Thus, the Viewpoint software for AMT encoders, running on a Windows PC, can accelerate the design process and reduce the time to perform routine tasks, such as determining the model number and version. To connect the encoder to a PC, only a USB cable is required.

Figure 4.10 shows the standard graphical user interface used to configure the encoder to the needs of a particular application [4].

The user interface window with parameters allows to set the main characteristics of the encoder and the time mode of operation. The device is programmed using this interface in a few mouse clicks, and it takes only a few seconds to align and zero the A, B signals, the index mark or the switching mode, which is significantly faster compared to same task in a non-programmable encoder.

Importantly, in the demo mode, users always have the opportunity to get acquainted with the graphical interface and perform several operations associated with the encoder. Finally, this interface also supports creation of component numbers for ordering specific models of encoders, taking into account the output signal format, adapter type, mounting support, etc.

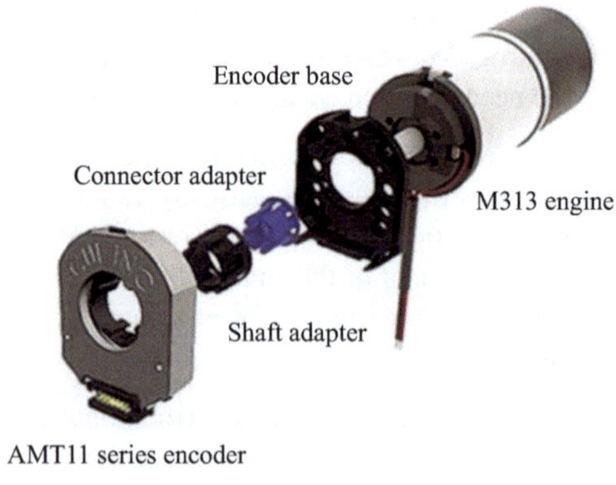

Fig. 4.9 The mounting holes of AMT encoders are compatible with non-capacitive encoders [4]

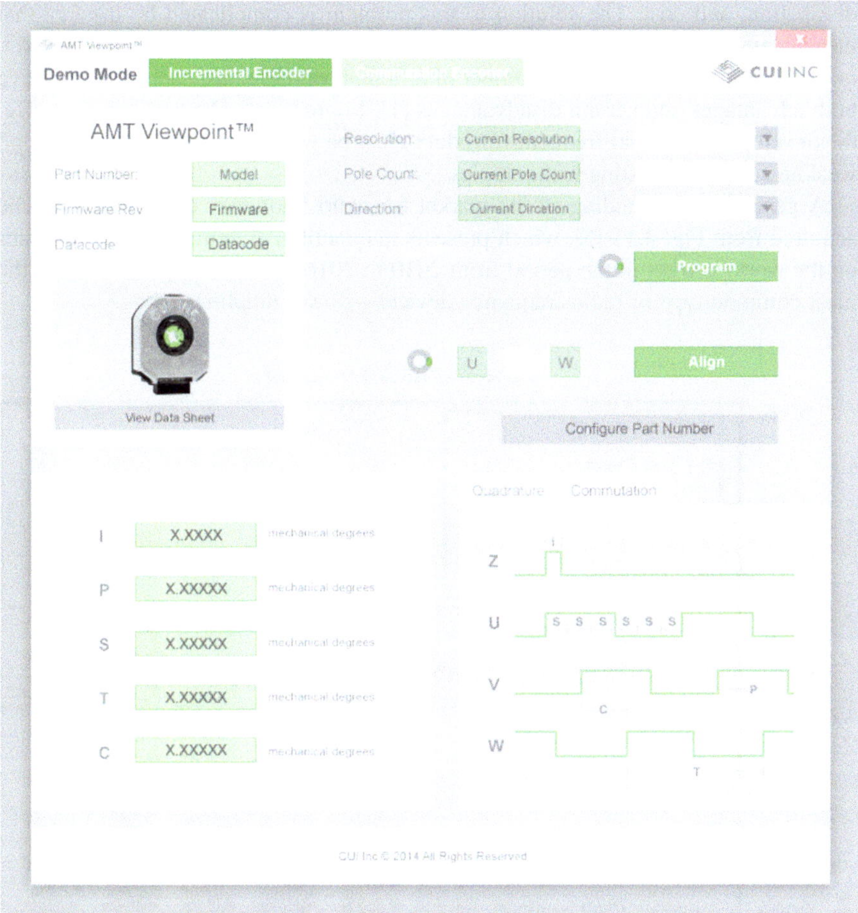

Fig. 4.10 The Viewpoint software for AMT encoders provides an easy-to-use interface [4]

Thus, on a particular example of only one type of the AMT31 series encoder, from one company CUI, we examined the main features of a new and rapidly developing class, the so-called intelligent encoders with output switching signals.

4.3 Basic Technologies of Radio Frequency Systems

4.3.1 Qualitative Analysis of the Technologies of Production of Radio Frequency Components

For developers of high-speed radio frequency systems, a very topical issue is the choice of a technology for manufacturing radio frequency components that would meet to the greatest extent all the requirements for data transfer speed, power dissipation, and cost. The solution to this problem is complicated by the fact that there is already a very wide range of radio frequency components on the market, manufactured on the basis of a number of different technologies, each of which have both advantages and certain disadvantages [1]. In order to make the correct choice, the developer of a radio frequency product must be well aware of the strengths and weaknesses of all existing technologies.

A general understanding of the market for radio frequency components can be acquired from Fig. 4.11 [5], which presents in quantitative terms the sales volumes on the world market in the period from 2010 to 2016 (predicted up to 2021) of the most common type of radio frequency devices—power amplifiers.

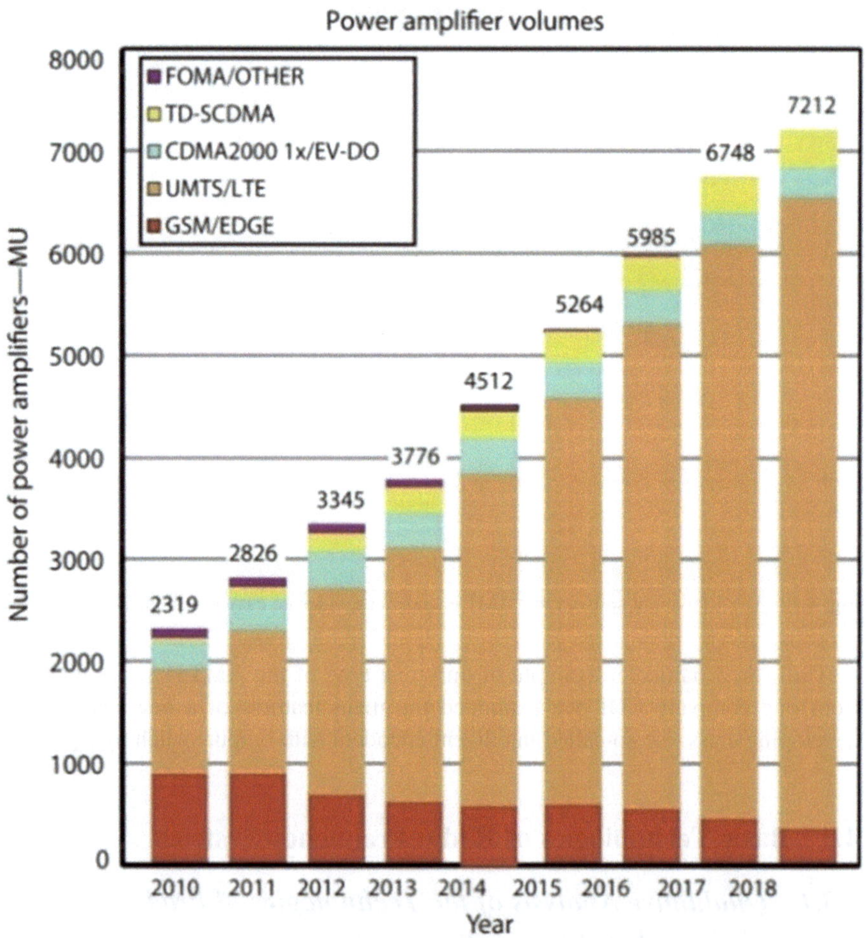

Fig. 4.11 The fastest growing segment of the power amplifier market are devices for UMTS/LTE systems [5]

Notably, for each year of production, the segments characterizing the number of consumed components in the context of specific stationary communication systems are highlighted: GSM/FDGE, UMTS/LTE, CDMA2000, 1x/EV-DO, TD-SCDMA, and FOMA.

Among the technologies competing in this market, one should mention both silicon-based technologies (LDMOS and CMOS/SOI) and technologies based on complex semiconductor compounds (GaAs, GaN, SiC).

The technologies most common today are LDMOS and GaAs, which have been developing for several decades. They provide rather high performance in many radio frequency applications. For example, LDMOS crystals are most often used in power amplifiers for wireless infrastructure. These devices are formed on the basis of an epitaxial layer of silicon grown on a heavily alloyed silicon substrate. LDMOS devices provide a fairly high output power of the order of hundreds of watts and a drain-source breakdown voltage of 60 V.

Usually LDMOS devices operate in the frequency range up to 3 GHz. However, companies such as Freescale and NXP have developed an LDMOS technology that effectively operates at frequencies up to 3.5 GHz for wireless broadband applications. Since we know that LDMOS technology is inherently highly reliable, it is often used in adverse environmental conditions, for example, in wireless base stations. When technology develops over many years, the cost of production tends to decrease, which makes this technology also attractive in terms of price.

While LDMOS devices are high-voltage wide-gap semiconductors, GaAs-based devices operate at lower voltages: 10–12 V. GaAs is a semiconductor material with a wide band gap of the AZV5 group with a high electron saturation rate and high electron mobility. In this regard, GaAs transistors can potentially operate at frequencies up to 250 GHz. Since GaAs is an unalloyed material, it provides insulation between microwave components. GaAs is widely used in low- and high-frequency systems, as well as in low and medium power applications such as cell phones.

Figure 4.12 [6] presents the state and prediction of development of the market of power amplifiers intended specifically for use in cellular phones—GaAs leads. As a rule, GaAs devices operate at powers well below 50 W. Their advantages are a small crystal size in combination with a high level of efficiency of the amplifier.

The GaN-based technology that has emerged in recent years makes it possible to achieve the characteristics of HF power amplifiers that were not available for GaAs and LDMOS devices. GaN is a semiconductor material with a wide band gap, which provides high levels of mechanical stability, heat capacity, and thermal conductivity; it also has very low sensitivity to ionizing radiation. This opens the way for GaN devices to operate in space and in military systems. This technology is well suited for use in high-output systems operating at high frequencies, which allows them to be used in the range of millimeter waves. As for power, GaN devices can reach hundreds of watts.

In most cases, silicon carbide (SiC) substrates are used to form GaN devices, since SiC exceeds GaN in thermal conductivity by a factor of three. GaN-on-SiC technology provides higher reliability and durability; therefore, it is used in pulse applications with high output power, for example, in S-band radars in military and

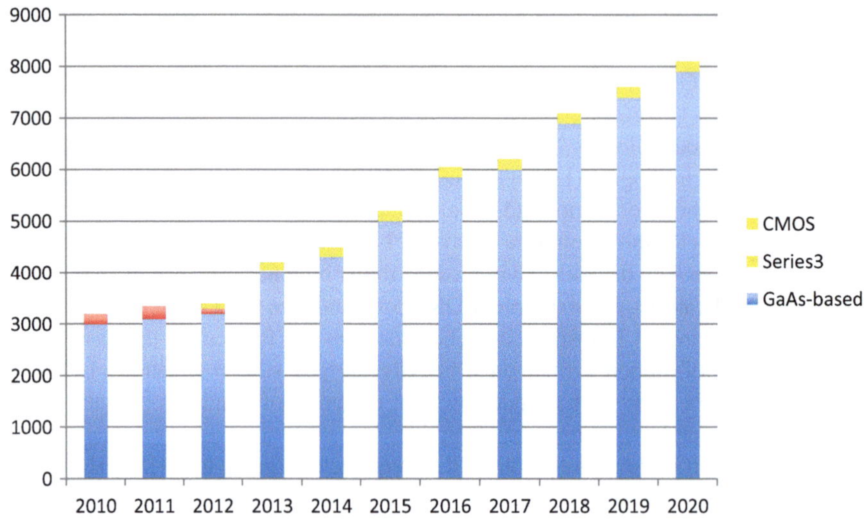

Fig. 4.12 Prediction for the market of power amplifiers for cellular terminals [6]

aviation systems. Although radio tubes have traditionally been dominant in this market, GaN-on-SiC technology provides much shorter pulses. There are also many other applications in which GaN-on-SiC based devices can be successfully used.

GaN technology is advisable to use in high-frequency (over 3 GHz) and high-power systems, but in other applications its use is limited for reasons of price. Nevertheless, GaN technology begins to enter some elements of the wireless infrastructure, in particular, it begins to be used when creating remote radio heads and active antennas. However, LDMOS technology dominates this area. For GaN devices, it will be difficult to challenge LDMOS in the low-frequency applications market due to the higher cost. GaN crystals are still several times as expensive as LDMOS crystals.

As the operating frequency increases, when the cost of buildings and assemblies increases, the total costs begin to level off. For example, an LDMOS device for S-band radars costs about the same as a GaN device of the same purpose.

Can GaN technology replace GaAs in terms of radio frequency applications, and at what power level lies the line between GaN and GaAs applications? Some experts believe that such a line is about 10 W of power, while others call the range of 35–40 W. However, it seems that a completely reasonable amount of power separating GaN and GaAs applications should be considered 20–25 W, i.e., applications with an output power of up to 25 W will be left to GaAs devices [6, 7].

The main direction of GaN technology development is GaN-on-SiC, which is aimed at serving the market of high-power and high-frequency applications, for example, wireless transport networks, military systems, satellites, etc. However, the cost of SiC material itself and the small size of the plates limit the entrance of GaN-on-SiC to this market. Despite the fact that GaN technology is characterized by high reliability and radiation resistance, the problem also lies in the substantial mismatch of the thermal properties of GaN and Si, which complicates their integration. However, since many companies are developing this technology, it is likely that the most pressing problems will be resolved, and this will expand the capabilities of radio frequency systems.

For many years, the technical characteristics of radio frequency devices made on the basis of CMOS technology have been significantly inferior to GaAs devices. The technological process developed in 2016 by Peregrine Semiconductor allowed the creation of silicon-based power amplifiers that are not inferior to their GaAs analogs [6].

Figure 4.13 [5] shows power summation efficiency for typical power amplifiers manufactured on the basis of standard CMOS and GaAs technologies and on the basis of Peregrine's new CMOS technology. (The power summation efficiency is the main characteristic of the industry for a power amplifier.) As can be seen from this figure, power summation efficiency for WCDMA signals with a permeability coefficient in the adjacent channel at the level of −38 dBc for power amplifiers manufactured on the basis of Peregrine's GaAs and CMOS technologies is about 50%. The closest competing CMOS device has a 10% lower efficiency value. Figure 4.13 reflects data obtained without tracking the envelope or digital predistortion. With the introduction of the envelope tracking parameter, the power efficiency of Peregrine increases by about 20%.

Peregrine's UltraCMOS 10 technology is based on a SOI process (silicon-on-insulator technology). Sapphire is used as a substrate, since it has almost perfect radio frequency properties. The UltraCMOS 10 technology uses 130 nm standards.

We know that while the main problem of the standard CMOS process is low breakdown voltage of the field-effect transistor. Most companies solve this problem by combining the output voltage to minimize the voltage on the field-effect transistors.

However, this method is associated with increased losses and leads to a significant decrease in efficiency. Peregrine's approach differs from this method and consists in creating a virtual heterojunction bipolar transistor, which allows a configuration similar to that used in GaAs-based power amplifiers.

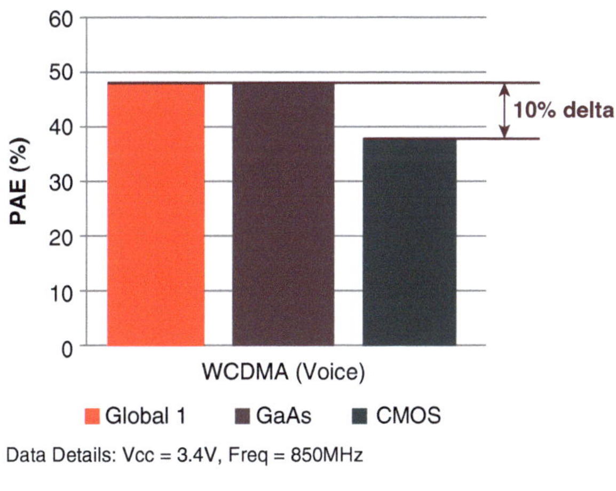

Fig. 4.13 Comparison of power summation efficiency for a power amplifier made on the basis of CMOS and GaAs technologies and on the basis of the new CMOS/SOI technology from Peregrine (Global 1 platform) [5]

HF transistors created using UltraCMOS 10 technology are characterized by a low $R_{ON} \times C_{OFF}$ value, a widely used indicator of radio frequency key quality.

Figure 4.14 shows data on the decline in $R_{ON} \times C_{OFF}$3a over the last 8 years. This decrease is mainly due to the successive transition to the 130 nm process [5, 7].

In particular, power amplifiers based on UltraCMOS 10 technology are part of Peregrine's HF input path Global 1. The Global 1 platform includes a three-channel multimode multiband power amplifier, output keys, an antenna key, and an antenna tuner. Since all the constituent blocks of the platform are made using CMOS technology, they can be integrated on a single chip.

The implementation of this approach promises a number of advantages. First of all, the rejection of GaAs devices means that you can use a reliable supply chain of CMOS crystals in large batches for a lower price with guaranteed reproducibility of parameters. In addition, a long-term plan was developed to improve the technological process and increase the diameter of the plates.

When all blocks of the radio frequency path are made using CMOS technology, higher levels of integration are possible. Previously, in HF input paths that contained GaAs-based power amplifiers, multi-chip modules were used. A higher level of integration provides, in turn, a higher degree of configurability—a huge advantage in meeting new requirements for the input circuits of radio frequency systems. For example, LTE uses a rather complex transmission modulation circuit (SCFDMA) to increase data throughput, which requires higher linearity to optimize the transmission channel [5].

While LTE does not have a single international frequency range, it is therefore necessary to dramatically increase the number of frequency bands to support the global LTE service area. Besides, an increase in the data rate means that even at the regional level, an increase in the frequency spectrum is required to support this growth. In many cases, the deployed frequency spectrum provides a more complex interference environment in combination with carrier aggregation.

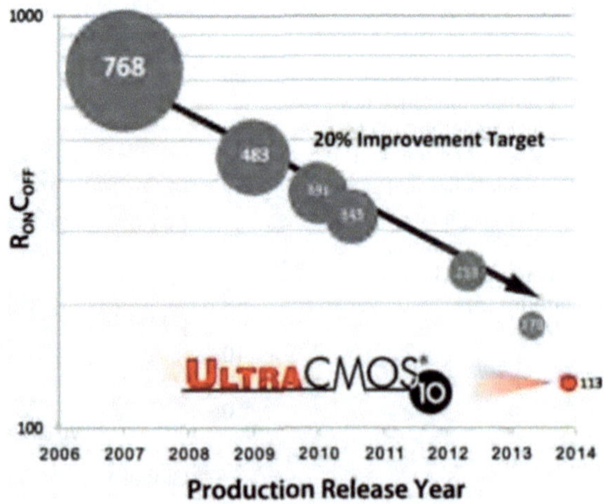

Fig. 4.14 Dynamics of changes in $R_{ON} \times C_{OFF}$ due to the improvement of the technological process of component manufacturing

The problem of the growing data may be solved using more than 40 frequency bands worldwide for LTE, and over the next five years more than 5000-fold increase in the number of operating states is predicted (this indicator depends on the number of frequency bands, modulation circuits, power amplifier modes, antenna settings, and carrier aggregation in the downlink). This means that in order to meet these requirements, the RF input paths should become increasingly complex. Reconfigurable input RF paths make it easier to solve this problem.

Input RF paths of the Global 1 platform can be configured electrically (rather than manually adjusting the force to a specific frequency range), which provides quick and easy modification of supported frequencies and optimization of antenna switch interface and the power amplifier. This, in turn, eliminates any mismatch due to different lengths of conductors and losses.

In addition, the ability to reconfigure the input path provides tremendous benefits to smartphone makers in terms of reducing material costs. A typical example—Apple's iPhone 5S was released in five versions for different regions with differing radio input paths. If the input radio path for all models were common, then it would be possible to save money at all stages of product creation—in the design process, at acceptance tests, in production, in supply management and material and technical supplies. When truly international cellular phones appear, they will be configured during the production process depending on the supported frequency range, and then adjusted in operating conditions to the appropriate operating mode (i.e., WCDMA, LTE, or carrier aggregation on the downlink, etc.).

4.3.2 Features of Selection of Design and Technological Parameters for High-Power SHF MESFET Transistors Based on Silicon Carbide

The unique properties of silicon carbide (SiC) provide characteristics unattainable for silicon devices, namely: high breakdown voltage (more than 1 kV), resistance to ionizing radiation, a wide range of operating temperatures. Since silicon carbide is chemically inert, this allows the use of caseless SiC devices in various corrosive environments. The development of the technology of SiC monocrystal growing using the modified Lely method and chemical vapor deposition processes to form epitaxial structures on the SiC substrate made it possible to achieve significant success in creating high-performance power and microwave devices based on SiC, in particular, powerful MESFET transistors [8–10]. Below we consider some key factors that determine the choice of structural and technological parameters of high-power microwave MESFET transistors based on SiC [8].

Today, there are a large number of structural and technological variants of the basic Schottky barrier field transistor, or metal-semiconductor field-effect transistor (MESFET), the most common type of high-power microwave transistors based on silicon carbide. To create such devices, semi-insulating silicon carbide substrates are used, on which the p-type buffer layer and the n-type channel layer are formed using the epitaxial growth method. Sometimes the epitaxial structure for creating MESFET transistors contains an additional n + layer for forming ohmic contacts to the drain and source areas. The method of obtaining ohmic contacts is alloying the channel layer under the contact metallization, nickel or titanium [11, 12]. The n-type gate and channel layer form a Schottky junction. The space charge region from the reverse biased Schottky contact overlaps the channel n-layer up to the buffer p-layer, thereby closing the transistor. Nickel, titanium, or molybdenum is often used as the gate material.

In particular, the influence of channel and buffer layer parameters on the static characteristics of the MESFET transistor was considered in a number of Russian studies [13, 14]. In these papers, it was shown that the behavior of the MESFET transistor is mainly determined by the ratio of the alloying level to the thickness of the channel and barrier layers, which actually determines the effective channel thickness of the transistor and the presence or absence of leaks through the buffer layer. It is depending on this ratio that four main variants of structural and technological performance of this type of devices can be distinguished [8].

A sufficiently high level of alloying in the channel area with small thickness and/or low alloying level of the buffer layer. The key feature of this structural and technological version of the transistor is the leakage through the buffer layer up to the semi-insulating substrate. This variant is characterized by the maximum conductivity of the MESFET channel of the transistor due to deep depletion of the buffer region and the flow of current in it. In the channel cutoff mode, there is a significant dependence of the transistor drain current on the current applied between the drain and the voltage source, which negatively affects the linearity of the transistor characteristics. Bear in mind that with this ratio of transistor parameters, the threshold voltage will also depend on the supply voltage. In addition, quality of the semi-insulating substrate and transition layers in the epitaxial structure affects operation of the device. If there are any defects in these elements, there may be leaks when the transistor channel is closed, as well as premature breakdown of the transistor structure. Using compensated (rather than particularly pure) silicon carbide as a substrate under conditions of elevated temperatures or exposure to ionizing radiation can lead to conductivity in the semi-insulating layer, which will cause the transistor to malfunction due to the departure of the threshold voltage values and the transfer characteristic offset. Therefore, this design is not recommended for use in devices and systems with an increased requirement for reliability (space and military applications).

A high level of alloying in the channel area with a ratio of the alloying level to the thickness of the buffer layer prevents leakage current from flowing to the semi-insulating substrate at a given supply voltage. With this design, leakage through the buffer layer of the MESFET transistor still occurs, but negative effects due to the influence of the semi-insulating substrate and the transition between the substrate and the buffer epitaxial layer are excluded. As in the first case, this design is characterized by dependence of the threshold voltage on the supply voltage of the transistor. The drain current is influenced by the voltage between the drain and the source in the channel cutoff mode, but this dependence is less pronounced and decreases as the level of alloying of the buffer layer increases with a constant level of alloying of the channel layer. This constructive-technological variant of the MESFET transistor can be called a compromise. On the one hand, saturation currents of the transistor are smaller than in the first case, but rather large, and on the other hand, using this approach eliminates the influence of the semi-insulating substrate on the properties of the MESFET transistor.

The alloying level and the thickness of the channel and buffer regions are such that there are no leaks through the buffer layer. This design version of the MESFET transistor is characterized by lower saturation currents than the previous two, the threshold voltage does not depend on the supply voltage of the transistor and there is no dependence of the drain current on the voltage between the drain and source in transistor cutoff mode. This provides a better linearity and stability of the transistor when operating with a large signal. Using this variant when designing a MESFET transistor, it is possible to achieve breakdown voltages above 500 V.

A low alloying level of the channel region and/or small thickness of the channel layer and high alloying level of the buffer layer. With this type of design, the MESFET transistor will be a normally closed device due to the propagation of the space charge region from the p–n junction "channel layer–buffer layer" into the depth of the channel layer and its complete overlap. Depending on the final application of the MESFET transistor, the developer can choose the optimal combination of technological parameters of the channel layer to achieve minimum resistance and mobility in the channel, slope or breakdown voltage between the drain and source, and threshold voltage. In this case, as the channel resistance decreases due to an increase in the channel layer alloying level, the breakdown voltage decreases and the mobility of the main carriers in the transistor channel decreases.

One of the most important tasks when designing high-power microwave MESFET transistors is to minimize parasitic capacitances of the device. In order to optimally select the parameters of the channel and buffer layers [15], computer simulations of parasitic capacitances of a MESFET transistor designed for use in microwave technology were carried out. Epitaxial layer parameters were chosen on the condition of absence of leaks. The calculation was carried out at the alloying level of the channel layer, which is about 1017 cm^3, since under such conditions a relatively low resistance of the transistor channel can be obtained with a small thickness of the channel epitaxial layer. It is known that the mobility of the main carriers in silicon carbide decreases sharply at high alloying levels [15]. An alloying mix concentration of 1017 cm ~ 3 allows to maintain sufficient carrier mobility for use of such structures in microwave electronics. The thickness of the channel layer in [15] was chosen so as to ensure the threshold voltage of the transistor of about −10 V.

The simulation was performed using Sentaurus TCAD. In electrophysical modeling, it is also necessary to take into account such features of silicon carbide as anisotropy of properties and incomplete ionization of the impurity. Figures 4.15, 4.16, 4.17, and 4.18 present the simulation results, dependence of the parasitic capacitances of the transistor on the alloying level of the channel and buffer

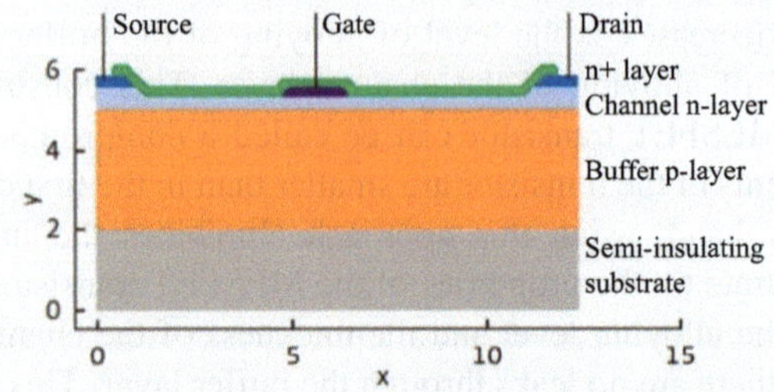

Fig. 4.15 Two-dimensional model of silicon carbide-based MESFET transistor

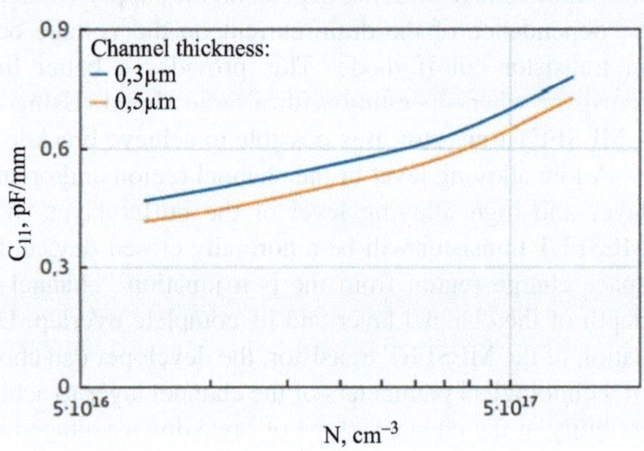

Fig. 4.16 Dependence of the input capacitance on the alloying level of the channel region

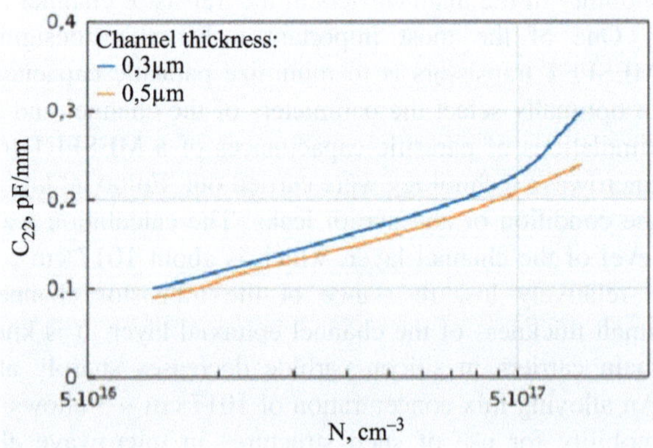

Fig. 4.17 Dependence of the output capacitance on the alloying level of the channel region

regions [8]. It shows that with increasing alloying level, the main component of the input capacitance—the Schottky junction's own capacitance—increases due to an increase in the number of charge carriers in the MESFET channel of the transistor, and also due to a decrease in the thickness of the space charge region under the Schottky gate (Fig. 4.16). As the channel thickness increases, the capacitive coupling of the barrier metal with the buffer layer decreases, which, in turn, affects the value of the input capacitance.

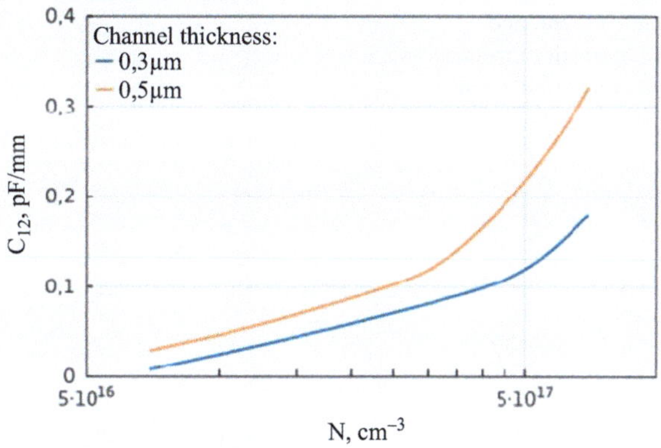

Fig. 4.18 Dependence of the feedthrough capacitance on the alloying level of the channel region

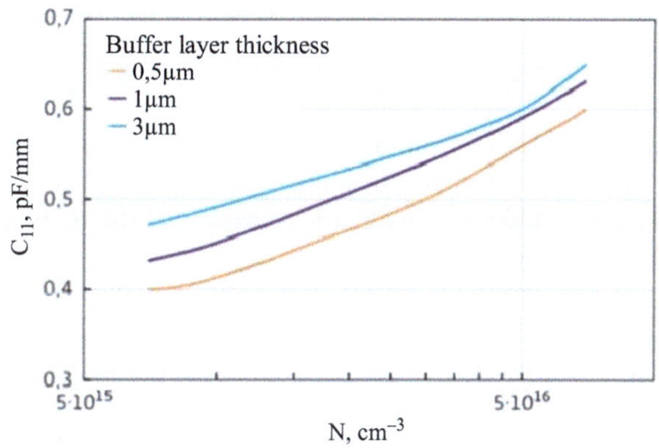

Fig. 4.19 Dependence of the input capacitance on the alloying level of the buffer region

The dependencies of the output and feedthrough capacities on the alloying level of the channel layer are also of increasing nature (Figs. 4.17 and 4.18). Increasing the capacity in both cases is due to an increase in the number of free carriers in the volume of the channel layer.

While the change in the alloying level of the buffer region negligibly affects the output and feedthrough capacitance of the transistor, the input capacitance has a pronounced dependence on the parameters of the buffer layer—it increases noticeably with increasing alloying level and increasing the buffer region thickness of the MESFET transistor (Fig. 4.19).

During the simulation, the dependence of the input, output, and feedthrough capacitances on the drain voltage with a DC-grounded source was also calculated (Fig. 4.20). To estimate the performance of the MESFET transistor, the response of the drain current at a supply voltage of 50 V in a circuit with a common source to a sinusoidal change in the gate voltage with an amplitude of 1 V and frequency of 1 GHz was simulated. A bias voltage of −2 V was applied to the gate, which corresponds to the operation of the transistor in amplification mode according to

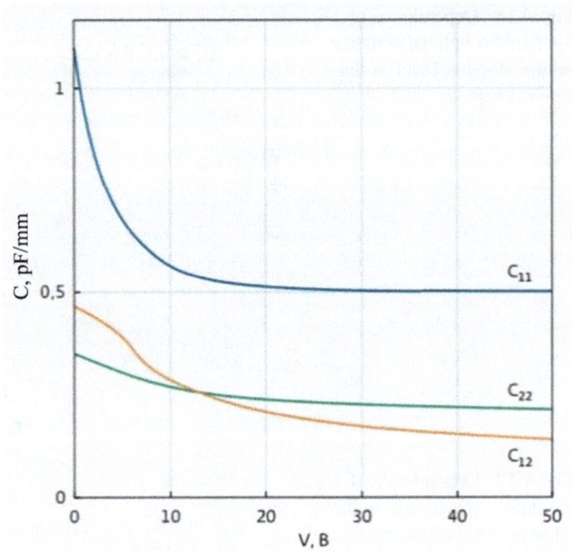

Fig. 4.20 Dependencies of input, output, and feedthrough capacitances on the drain voltage with a grounded source

class A. The drain current delay from the control signal was $2 \cdot 10^{-11}$ s (Fig. 4.21). It should be noted that the model used by the authors [15] does not take into account parasitic inductances of the wire leads and the transistor case, which significantly distort the signal level.

According to the results of the above, general recommendations can be formulated: to minimize the parasitic capacitances of high-power microwave MESFET transistors, it is necessary to reduce alloying levels in the channel and buffer regions, as well as to reduce the thickness of the buffer layer while maintaining the optimal ratios of these parameters [8]. The selection of structural and technological parameters of epitaxial layers of high-power microwave MESFET transistors should be determined based on specific design problems, taking into account the recommendations given.

4.4 Structural and Technological Features and Applications of Impedance SAW Filters

4.4.1 Design Features of SAW-Based Chips

Normally the task of creating miniature stable microwave oscillators is solved by using high-quality solid-state microwave resonators in the operating frequency range of the oscillator, technologically compatible with hybrid integrated circuits. Such devices with operating frequencies of 0.3–1.5 GHz provide a loaded Q-factor of the resonator of at least 3 · 103 in the upper part of the range and over 5103 in the lower one, as well as good temperature stability and long-term frequency stability.

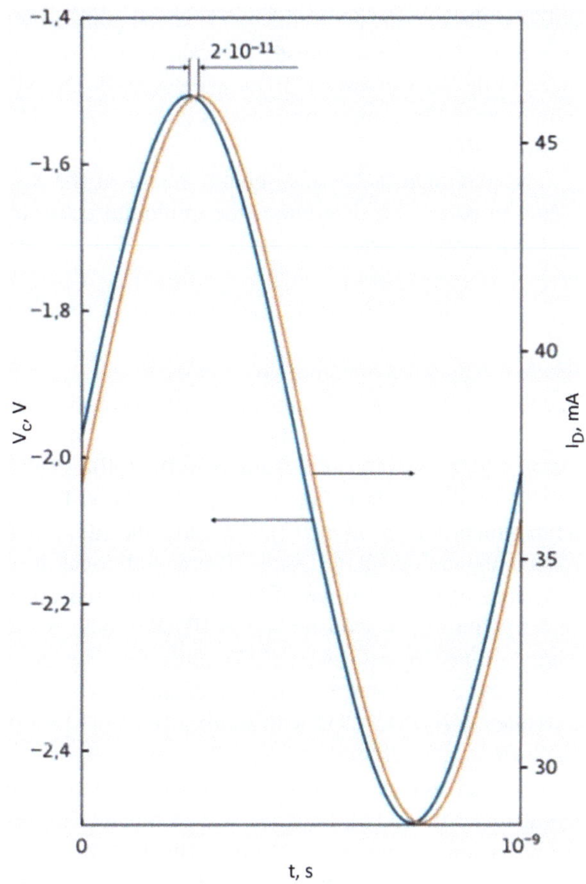

Fig. 4.21 The drain current delay from the sinusoidal voltage at the gate

This allows to create a microwave oscillator with an operating frequency of at least 800 MHz, having a phase noise level below −80 dB/Hz at a frequency of 100 Hz.

One of the directions of microwave oscillator development is the use of surface acoustic wave (SAW) devices [16]. In the USSR, at the end of the 1970s, the Pulsar Research and Development Institute began to solve the problem of creating mass-produced frequency-selective SAW devices and microwave oscillators based on them.

Let's briefly consider the special aspects of development of this class of chips on the example of experience in the Pulsar Research and Development Institute.

The designs of SAW resonators can be divided into two main types: *resonators with reflective gratings and resonators in which the delay line (DL) is an element of frequency stabilization*.

DLs allow to make higher-frequency resonators, but they have large insertion losses (IL) than resonators with reflective gratings (RG) and with them, it is more difficult to ensure single-frequency operation. The high slope of the phase-frequency characteristic obtained on a long SAW DL (in principle, this leads to a decrease in noise of the microwave oscillator) is accompanied by large insertion losses, which as a result increase the noise. The calculations showed that, in the ideal case, the noise of the microwave oscillator using a SAW DL will be determined by the same amount of SAW propagation loss as in the case of SAW resonators with reflecting gratings (RG).

The maximum achievable quality factor of the SAW resonators with RG (this value, in particular, determines the minimum achievable noise level) is expressed by the following formula:

$$Q_m = \pi \cdot 1000 \cdot 20 \cdot lg(e)/(/(a_1 + a_2 \cdot F) \approx 27300/(a_1 + a_2 \cdot F), \qquad (4.1)$$

where e is the base of natural logarithms; a_1 and a_2 are SAW attenuation coefficients due to the external atmosphere and SAW absorption into the substrate, respectively; F—frequency in gigahertz. For ST quartz, X-cut, and under air load, these coefficients, found experimentally, are: $a_1 = 0.47$ dB/cm and $a_2 = 2.62$ dB/cm/GHz.

From the formula it follows that $Q_m = 8800$ for $F = 1$ GHz. A chart of Q_m as a function of F is given in [17]. Naturally, the additional losses in the SAW resonator reduce the unloaded Q-factor. These additional losses are associated with the resistance of the electrodes of the interdigital transducer (IDT), with the conversion of SAW into bulk acoustic waves (BAW), with SAW leakage from the resonant cavity formed by RG, with SAW absorption by the metal film forming the IDT electrodes, etc. At the same time, reducing losses increases the quality factor, so at a frequency $F = 0.5$ GHz with an air load $Q_m = 15300$, and for a vacuumed device ($a_1 = 0$) – $Q_m = 20800$.

A big advantage when creating miniature stable microwave oscillators on SAW resonators with RG is the small size of this device, because the reflecting gratings form a resonant cavity for SAWs. The IDT, which excites SAWs, is located in this resonant cavity between two RGs. The input electrical resistance of the IDT decreases, and the devices have small losses when working without matching in the 50-Ohm path even for such a weak piezodielectric as quartz ST, X-cut. Substrates from this material are suitable in terms of temperature and technological parameters for creating surfactant resonators (for example, these substrates are acoustically and technologically in good agreement with aluminum films used as electrodes of the microcircuit).

RG resonators can also be divided into two types: *single input and double input.* In a single-input resonator there is only one IDT, it is described as a two-port network and is a functional analogue of a widely used resonator based on bulk acoustic waves (BAW). In the meter range of wavelengths for self-oscillators on single-pass resonators, well-known circuit solutions developed for quartz oscillators are used, as a rule. The advantage of this type of RG resonators is a simpler design. In particular, in this design it is easier to ensure low input electrical resistance of the device for operation without coordination in the 50-Ohm path, since only one IDT should be placed in the resonant cavity. Consequently, it can be made with a larger number of electrodes than a two-input resonator, in which there are two IDTs—an input and an output, and which is a four-port network. According to

Hewlett-Packard, in 1981, this type of resonator became the world's first commercially available SAW resonator (its center frequency was 1981 MHz).

When using a single-input resonator in a self-oscillator, the main disadvantage is the relatively large amount of the static capacity of the device passing through, mainly due to the capacity of the IDT. With an increase in the operating frequency of the oscillator, the use of a single-input resonator in the microwave range becomes difficult, since inductance is required to compensate for the static capacitance, and this complicates the design and tuning of the oscillator circuit.

A SAW device that is free from the drawbacks of a single-input resonator and delay line is a two-input resonator. In the traditional design of a two-input resonator, two IDTs are used in the resonant cavity located near the reflective gratings [18], made in the form of shallow grooves. The first prototypes of such resonators at frequencies of ~1.2 GHz were fabricated using electron-beam lithography. However, this technology in the manufacture of serial samples had to be abandoned due to high percentage of rejected devices. There are two reasons for the high reject percentage: poor alignment of the image fields by the electron beam, which leads to distortion of the characteristics of the SAW resonator; the second is a large technological variation in the size of submicron elements, which led to a large variation in the center frequency of the instruments.

However, the traditional topology of a two-input SAW resonator with an RG used in the 321FE1x chip did not allow to obtain small insertion losses without matching at a load of 50 Ohm (path impedance). To obtain small insertion losses (~6 dB) without coordination, a 321Exxx chip was developed, the topology of which had 3 IDTs.

Low insertion loss and a large number of IDT electrodes, in the above-described design, led to a significant influence of reflections from IDT (both electrical and mechanical) on the parameters of the resonator. To account for these dependencies, they were investigated, in particular, using the multichannel resonator method. The following are typical parameters of SAW resonators made on ST quartz, X-cut. The electrical parameters of the devices were measured in the 50-Ohm path without coordination (see Table 4.2).

A wide range of microwave transistors produced by the Pulsar plant made it possible to analyze the dependence of the spectral power density of the phase noise self-oscillators using a large sample of devices (Fig. 4.22).

Studies have shown that self-oscillators on silicon bipolar devices (KT640, KT642, KTZ115, KTZ132) have a noise level 15–20 dB lower than self-oscillators on gallium-arsenide FETs with Schottky barrier (AP604, AP602, AP307). Similar dependencies were observed by Shea, Sullivan, and Kindell from Sawtek when

Table 4.2 Typical parameters of SAW resonators

Center frequency, MHz	333.33	532.00	715.00
Loaded Q-factor	5000–7000	4000–5500	3000–4000
Insertion loss, dB	5–6	5–6	7–8
Size of metal-ceramic case (height × diameter), mm	6.5 × 17.0	2.5 × 9.5	2.5 × 9.5

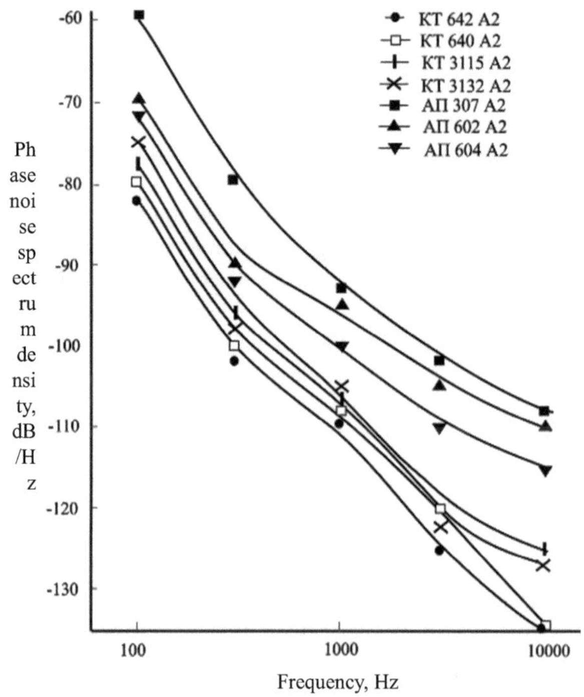

Fig. 4.22 Spectral density of phase noise of self-oscillators with SAW resonators, the amplifier stages of which are made using different transistors

studying microwave oscillators with SAW resonators at an operating frequency of 500 MHz [19]. The spectral density of phase noise at a detuning frequency of 100 Hz in their measurements was −60 dB/Hz for field GaAs transistors and −80 dB/Hz for silicon bipolar transistors.

The phase noise level of self-oscillators on medium power transistors (KT640, KT642) is 10–15 dB lower than the oscillators on low-noise low-power transistors (KT3 115, KT3 132), which correlates well with the spectral density of the low-frequency noise of transistors.

The minimization of phase noise spectral density of a self-oscillator can be done not only by selecting the operating mode of the transistor, but also by optimizing the coupling of the SAW resonator with the load. Optimal matching is achieved at such a coupling value of the resonator in the oscillator at which a further increase in Q-factor leads to an increase in losses in the resonator, worsening the oscillator noise. In practically relevant cases, the optimum is achieved at an instrument insertion loss of ~6 dB. Such losses can be obtained by matching for the frequency-selective chip 321FE1x or without coordination for the chip 321FEEEh (as a rule, additional elements in the oscillator increase its noise and complicate the design and configuration of the microwave device).

In terms of phase noise, microwave oscillators with SAW devices are inferior to quartz oscillators with frequency multipliers under normal conditions. However, in real equipment, self-oscillators often operate under the influence of vibrations, shocks, and acoustic effects. Under these conditions, the advantages of self-oscillators with SAW resonators over quartz oscillators become obvious (Fig. 4.23).

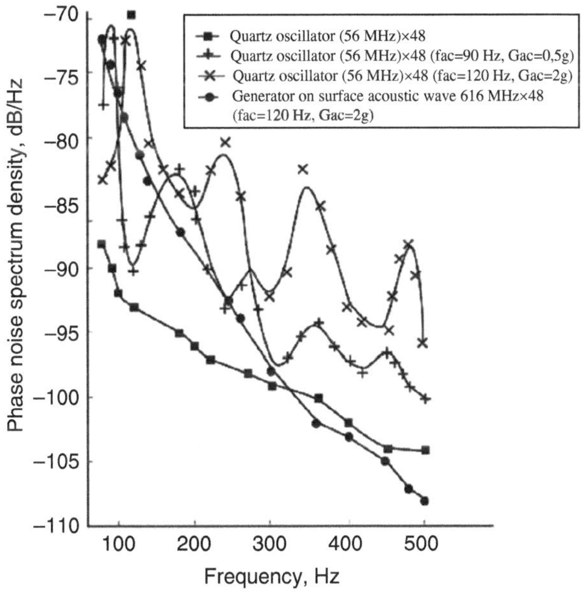

Fig. 4.23 Spectral density of phase noise of various types of microwave oscillators when affected by vibration

The results of studies of microwave oscillators with an operating frequency of 2.5 GHz on various types of devices have shown that even at minor vibration loads, the parameters of quartz oscillators are significantly worsened. At vibrations with an acceleration of ~ 2 g, phase noise level of quartz oscillators is 10–15 dB worse than that of oscillators with SAW resonators.

4.5 Structural and Technological Features of Impedance SAW Filters

This section is based on the cycle of papers [16–46] by Russian scientists who proposed a new class of filters on surface acoustic waves (SAWs)—impedance SAW filters, which combine the advantages of classical LC filters (low insertion attenuation, small frequency response pulsations and group delay in the passband, large power dissipation) with known advantages of SAW filters (high manufacturability, using the capabilities of planar micro- and nanoelectronics, high reproducibility of parameters, small weight and size, low cost). Currently, impedance filters are used in mobile and cellular communication systems, mobile phones, digital television, high-definition television, radio frequency identification systems, etc.

Employees of the following Russian institutions and enterprises took part in the creation and development of this elemental base of telecommunication systems (TCSs): FGUP Scientific Research Institute of Radio, Institute of Radio Engineering and Electronics, V.A. Kotelnikova of the Russian Academy of Sciences, OOO NPP TRIIS, LLC Piezotron, UFU, LLC BUTIS.

Creation of mass low-cost telecommunication systems (TCSs) based on a unified series of functional devices for frequency selection base on surface acoustic waves (SAWs) is one of the priority tasks of microwave electronics, which is solved by creating a unified technological platform of the "system in case" type as the basic functional design of frequency selection, amplification, and signal processing devices for developed and modernized electronic systems and control complexes, television, and communication. Today, SAWs are considered as a unique technology that provides great opportunities compared with its analogs [17–19, 21–24]. Table 4.3 [20] summarizes the place and role of SAW in modern TCSs.

The use of such impedance SAW filters in transceiver nodes and units can ensure the achievement of high characteristics of the hardware functions of TCS.

Table 4.3 Place and role of SAW in telecommunication systems

Types of telecommunication systems	Basic hardware functions implemented with SAW devices	Typical SAW devices
Radar, navigation tools, and systems. Cellular communication. Cell phones	Insertion attenuation, frequency selection, power dissipation, reliability, size, cost	Bandpass filters, capacitive weighting filters, delay lines, dispersion delay lines, Fourier processor Impedance bandpass filters, multichannel impedance filters for parallel Fourier processors
Satellite, tropospheric and radio-relay communication lines. Mobile communication systems	Frequency selection, fast Fourier transform, noise-like signal, pseudo-random search for operating frequencies, reliability, mass dimensions	Bandpass filters, capacitive weighting filters, delay lines, dispersion delay lines, Fourier processor Impedance bandpass filters, multichannel impedance filters for parallel Fourier processors
Automated means of radio control, electronic equipment	Frequency selection, fast Fourier transform, reliability, weight, and size	Bandpass filters, capacitive weighting filters, delay lines, dispersion delay lines, Fourier processor Impedance bandpass filters, multichannel filters
Television—satellite, digital, cable, high definition	Insertion attenuation, frequency selection, intersymbol interference, power dissipation, reliability, size, cost	Bandpass filters for TV receivers, transmitters, repeaters, cable television networks, high-definition television, satellite television, mobile television, filters with capacitive weighting. Impedance bandpass and notch filters of television, including For digital television

The structural and technological difference between the considered impedance filters and the classical ones is that they contain only interdigital transducers (IDTs), with no SAW reflectors, which significantly simplifies the design [20].

As known [1], classical bandpass and notch LC filters usually contain sequential and parallel LC circuits, combined into G, P, T, and bridge circuits (Fig. 4.24). As was shown [26–30], SAW impedance filters have equivalent circuits that are identical to LC. Therefore, the diagrams shown in Fig. 4.24a, b (serial and parallel LC circuits) correspond to the topology of the impedance SAW filters (Fig. 4.25a, b).

Impedance SAW filters corresponding to P and LC circuits are constructed in a similar manner (Fig. 4.24c, d). In these cases, a bandpass impedance filter will contain three IDTs, one of which, connected in series, with a small aperture, and two connected in parallel, with a large one (Fig. 4.26a). In a notch SAW filter, on the contrary, there is one IDT with a large aperture, connected in parallel, and two—with a small aperture, connected in series (Fig. 4.26b).

A T-type bandpass impedance SAW filter (Fig. 4.24d) contains, like the P-type filter, three IDTs. In this case, there are two converters with a small aperture, and one with a large aperture (Fig. 4.27a). As in G- and P-filters, in these filters, in the

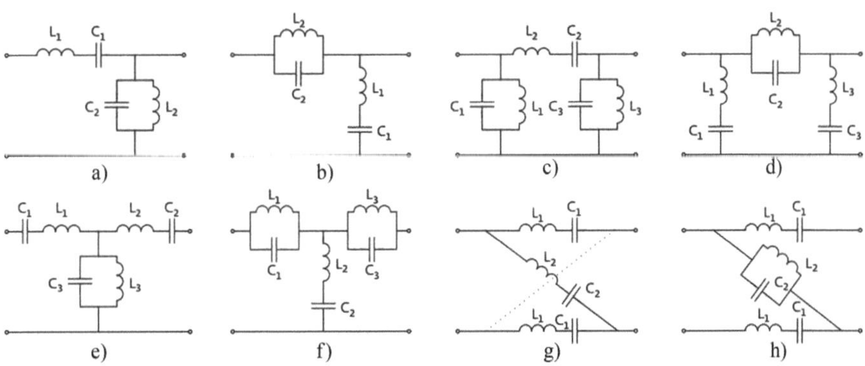

Fig. 4.24 Bandpass and notch LC filter diagrams: G-type (**a**, **b**); P-type (**c**, **d**); T-type (**e**, **f**); bridge (**g**, **h**) [20]

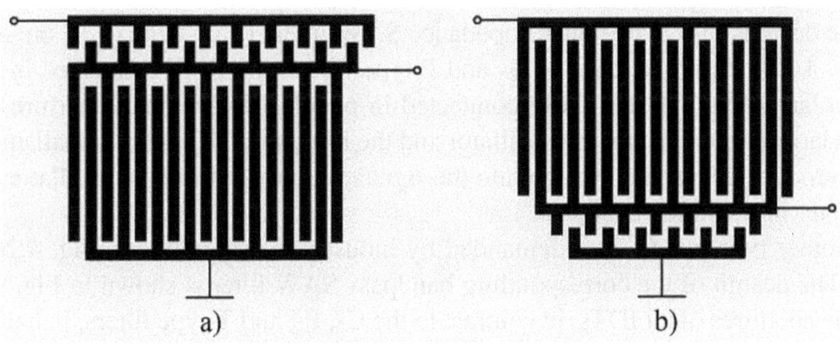

Fig. 4.25 G-type impedance SAW filters: band (**a**) and notch (**b**)

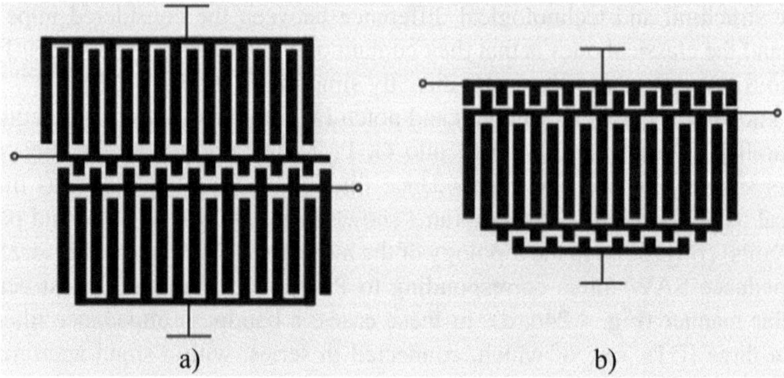

Fig. 4.26 P-type impedance SAW filters: band (**a**) and notch (**b**)

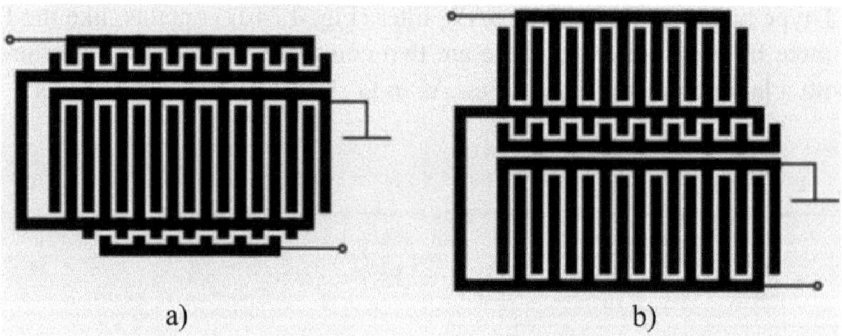

Fig. 4.27 T-type impedance SAW filters: band (**a**) and notch (**b**)

passband, impedances of the IDT with a large aperture are large and do not shunt the oscillator and the load. The impedances of IDTs with a small aperture are much smaller and introduce a small attenuation when the signal passes from the oscillator to the load.

Outside the bandwidth, large-aperture impedances of IDTs have small values, and small-aperture impedances of IDTs have large ones, which provides shunting of the generator and loads of IDTs with a large aperture, as well as a significant attenuation of the signal as it passes through IDTs with a small aperture.

The designs of T-type notch impedance SAW filters (Fig. 4.24g) [20] are shown in Fig. 4.27b. As in the case of G- and P-type filters, the IDTs connected in series have a large aperture, and those connected in parallel have a small aperture. IDTs with a large aperture shunt the oscillator and the load, and IDTs with a small aperture will introduce a large attenuation into the signal as it passes from the oscillator to the load only in the notch band.

Another type of LC filter demanded by industry is bridge filters (Fig. 4.24g, h) [20]. The design of the corresponding bandpass SAW filter is shown in Fig. 4.28a, and the apertures of all IDTs, in contrast to the G-, P-, and T-type filters, in bandpass filters can be very close. In the case of a close or equal number of electrode pairs

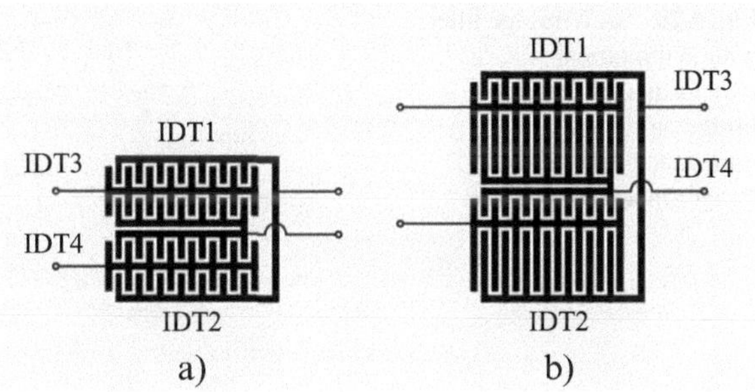

Fig. 4.28 Bridge-type impedance filters: band (**a**) and notch (**b**)

in all IDTs far from f_0 (center frequency, acoustic synchronism frequency), where the transducer impedance is capacitive, it is easy to obtain a bridge balance and, therefore, a large insertion attenuation.

Typically, a chip contains four IDTs connected by a bridge circuit. The frequencies of acoustic synchronism of IDT1 and IDT2 are set to be the same, but different from the frequencies of acoustic synchronism of IDT3 and IDT4, which are also the same. The ratio between the periods of the IDT 1, IDT 2 and IDT 3, IDT 4 depends on the passband of the filter. In the passband of the filter, the impedances of IDTs located in different arms of the bridge (IDT1, IDT2 and IDT3, IDT4) can vary significantly, the bridge will be unbalanced, and the losses in the filters will decrease significantly.

In such filters, as in G-, P-, and T-types, extended IDTs are used, for which $Nk^2 \gg 1$ (the number of N electrodes is so large that the SAW does not go beyond the IDT [28]). For notch filters (Fig. 4.28b), the apertures of IDT1 and IDT2 should differ greatly from the apertures of IDT3 and IDT4, and the period of IDT with a large aperture should be chosen in such a way that $Y_{\mathrm{IDT1,\ 2}} = Y_{\mathrm{IDT3,\ 4}}$($Y_{\mathrm{IDT1,\ 2}}$ and $Y_{\mathrm{IDT3,\ 4}}$ are conductances of IDT1, 2 and IDT3, 4, respectively) at the notch frequency. Then the bridge will be unbalanced at all frequencies, except for a small frequency band in the region of the notch frequency.

The number of IDTs in a bridge circuit can be reduced using a symmetric transformer (Fig. 4.29a). In this case, the impedance filter circuit will contain only two IDTs with apertures half as large (Fig. 4.29b).

It should be noted that the impedance filters contain only equidistant IDTs, in which the width of the electrodes and the gaps are the same and equal to a quarter of the period of IDT. These filters do not contain additional elements, in contrast to low-loss SAW filters of other types (reflectors, couplers, phase-shifting circuits, IDTs with internal reflectors), and therefore these are low-loss SAW filters of the simplest design. This allows to significantly reduce the size of the piezoelectric substrate, increase the technological reproducibility of electrical parameters, and therefore reduce the cost of producing such filters, which is important when they are mass-produced.

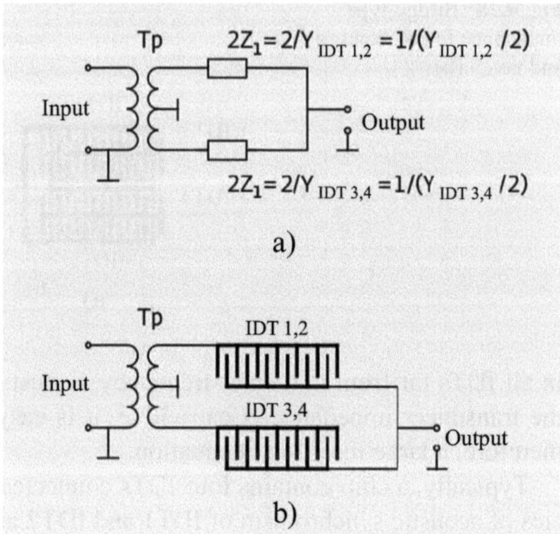

Fig. 4.29 SAW bridge filter with symmetrical transformer: differential bridge circuit (**a**); impedance filter with transformer (**b**)

4.5.1 Fields of Application of Impedance SAW Filters

The main areas of application of SAW filters in microwave devices are discussed in detail in [1]. Below we consider several main applications and technological features of impedance SAW filters. It is television—channel filters for broadcast television systems and their opposite—notch impedance SAW filters for closing television channels. To prevent unauthorized viewing of television channels, today various systems of channels closing are used. The system considered to be the simplest and most reliable one is based on interference at the input of a television receiver, which is contained in a signal transmitted from a television studio. The interference is a narrowband signal modulated by an audio frequency signal (the carrier frequency is quite close to the carrier frequency of the television signal image). When entering the television receiver's input stages, this signal is mixed with the main television signal, which distorts the audio and video signals and, consequently, renders it impossible to watch television programs. If you put a notch filter on the input of a television receiver that passes the signals of all television channels and does not pass the interference signal, then the interference will not get into the input circuits of the television receiver. In this case, the video and audio signals will not be distorted, and watching a television program will be possible.

It is important to note that the frequency of rejection (and, therefore, interference) cannot be chosen arbitrarily within one television channel (8 MHz), but should be located in the range of the television channel where the frequency components of the signal are minimal [20]. Otherwise, the notch filter will cut a significant part of the signal, which will lead to image distortion. And the closer to the carrier frequency is the notch band, the more distorted the image.

As the measurements showed, image distortions become hardly noticeable if rejection occurs at frequencies 2.8–3 MHz above the image carrier frequency. In this case, the image is very slightly distorted when viewing video materials (2 MHz signal bandwidth), its clarity is slightly deteriorated when transmitting a full television signal (6 MHz). For the effect of interference to not affect the image, the rejection should be of at least 30 dB. Thus, a notch filter should simultaneously provide a notch of at least 30 dB at a frequency 2.8–3 MHz from the image carrier frequency and a loss at the carrier frequency of no more than 3 dB, which practically does not degrade the quality of image reception.

Due to temperature fluctuations in the room (10–15 °C), the rejection filter should be either temperature-stable or provide such a notch band at which the frequency of the interference would not go beyond its limits.

The first option is practically difficult to realize because of the small value of electromechanical coupling coefficient of a thermostable ST cut quartz used for these purposes in SAW devices. Indeed, to satisfy the condition $4k^2N/\pi^2 \gg 1$ for a quartz substrate, the number of pairs of electrodes must reach 1000 or more. To reduce the number of electrodes, additional LC circuits can be used [20]. However, this complicates the design of the filter and increases its cost for mass production.

The second option can be implemented on substrates with a high value of the square of the electromechanical coupling coefficient. Thus, for substrates of YX/128° cut of lithium niobate, the $4k^2N/\pi^2 \gg 1$ condition is easily fulfilled when $N = 200–250$, and rejection of at least 30 dB can be obtained using only one T-link in the design of the impedance filter, which is quite acceptable.

At the same time, the temperature coefficient of this cut of lithium niobate is $80 \cdot 106$ 1/deg. Since the temperature in the room, as a rule, is in the range of 15–30 °C, the relative temperature departure of the notch frequency is 1.2×10^3. For example, in the Russian eighth television channel, the interference frequency should be 198.25 MHz. Then, when the temperature changes from 15 to 30 °C, the notch frequency will shift by about 200 kHz, that is, the notch band should exceed 200 kHz.

SAW velocity in lithium niobate substrates YX/128° cut is in the range of 3960–4000 m/s on the free surface. Then, at $k^2 = 0.058$, SAW velocity under the IDT V_{SAW} will be equal to 3924 m/s with $V_{SAW} = 3980$ m/s, the period of the IDT with a small aperture will be equal to 3924/196.25 = 19.99 microns. IDT apertures will be 40 and 400 μm, respectively, and the period of the IDT with a large aperture will be $19.99\ (1 + 4 \cdot 0.058/3.142) = 21.47$ μm [28]. For lithium niobate, the capacitance of a single-wave of the IDT section is $C_S = 450$ pF/m. As calculations [28] show, with the use of T-link, it is possible to get a rejection of more than 30 dB in the notch band, with the number of electrode pairs in an IDT with a large aperture of $N_1 = 250$ and the number of pairs in an IDT with a small aperture of $N_2 = 200$.

With a small change in the number of electrodes, the IDT should also slightly shift the frequency of antiresonance [28]. This circumstance was used by the authors to tune the notch filters by trimming the IDT electrodes, as during their mass production, the notch frequency shifts due to different process errors (discontinuities of IDT electrodes, variation in the SAW speeds under the IDT, variation in the SAW speeds on the disk surface).

Filters were manufactured using the group method on a lithium niobate disk with a YX/128° cut. After cutting the disc into individual filters, they were placed in the housing, which is a metal tube, inside which a substrate is attached to a plastic base (Fig. 4.30).

At the ends of the tube there are openings through which the packaged filter is inserted into the antenna connector of the TV, and the antenna is in turn inserted into it. In order for an IDT with a small aperture to be connected to a body (tube) in a plastic body, a special copper conductor is provided in the plastic base (Fig. 4.30), which is output to the outside of the tube and is welded to it by spot welding. At high frequencies, the inductance of this conductor is sufficient to shift the notch frequency, as calculations show, by more than 200 kHz in the direction of lower frequencies. To compensate for this shift, the IDT periods are changed so that the notch frequency is increased by 200 kHz.

A separate area is television channel filters for broadcast television systems. Television channel filters are designed to highlight the signal of one television channel and suppress the signals of the other channels [33]. Based on this, their characteristics are set: passband of 8 MHz, loss within passband of no more than 3.5 dB, irregularity of frequency response in the passband no more than 1 dB, suppression of nearby non-adjacent channels at the carrier frequency of the image of at least 20 dB, suppression away from the bandwidth not less than 40 dB. The nomenclature of a set of channel filters depends on the channels broadcasted in the area. Particular attention in the design and production of television channel filters is given to ensuring a small non-uniformity of the amplitude–frequency characteristic (AFC) in the passband. Unfortunately, in the types of SAW filters used today, it is difficult to combine small losses in the passband and low unevenness of the frequency response. But in impedance filters based on the change in impedances of the IDT with frequency, it is easy to obtain small insertion loss, the necessary out-of-band suppression, and small unevenness in the passband.

As an example, consider a channel filter built on a differential bridge circuit (Fig. 4.31). Such a circuit allows halving the number of IDTs (as compared with a bridge circuit) in the presence of a symmetric transformer. A YX/128° cut of lithium niobate was selected as a piezo-substrate; to reduce the non-uniformity in the passband, transducers were apodized using the Hamming function with a pedestal

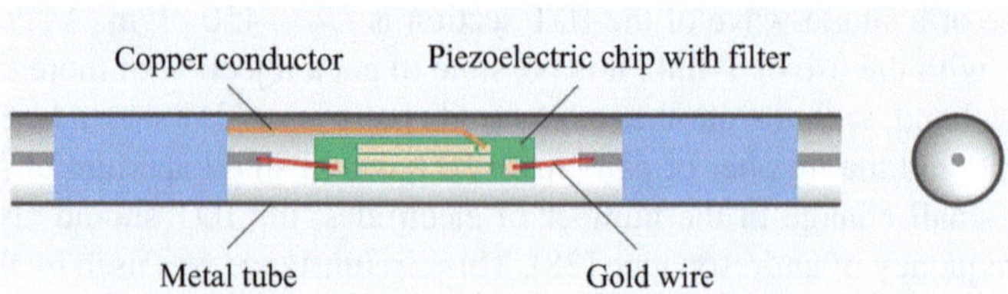

Fig. 4.30 Notch impedance filter in the antenna connector housing [28]

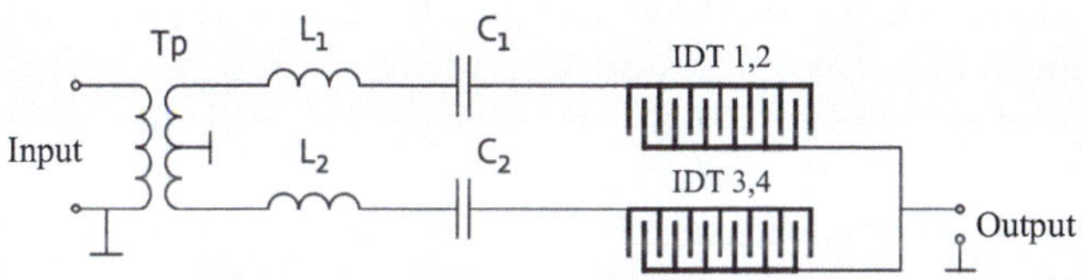

Fig. 4.31 Diagram of the channel filter on the second special channel

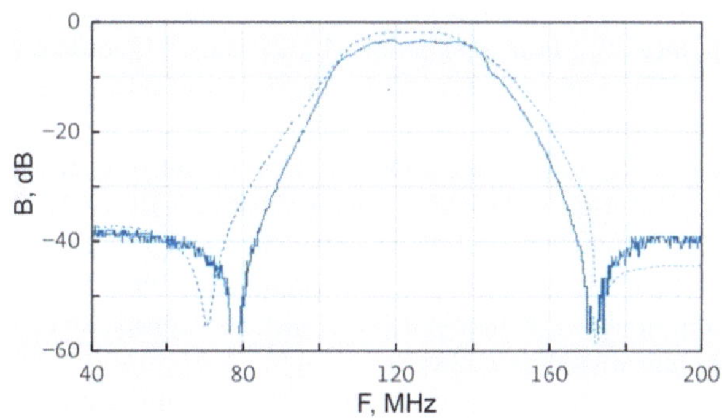

Fig. 4.32 Frequency response of a channel TV filter for the second special channel, dotted line—calculation, continuous line—experiment

of 0.1. To increase the slope of the frequency response fronts, expansion inductances were used [25, 28]. The latter allowed to expand the bandwidth to the required value of 8 MHz and provide the necessary slope of the frequency response fronts. The calculated insertion loss was no more than 2 dB. One of the IDTs was designed for a frequency of $f0_1 = 118$ MHz, and the second one—for a frequency of $f0_2 = 126$ MHz. The first IDT had an aperture of 80 microns, a period of 32.95 microns, and the number of pairs of electrodes was $N_1 = 250$, and the second IDT had the same aperture and the same number of electrodes as the first IDT, but its period was 30.86 microns. Capacities are $C_1 = C_2 = 9$ pF, inductance—$L_1 = L_2 = 20$ μH. The substrate with the IDT on it was placed in a hermetic case, which, in turn, was placed in another case with 75 Ohm cable connectors.

The measured and calculated frequency dependence of the channel filter insertion loss is shown in Fig. 4.32. Filter insertion loss is 3 dB.

Filters for other TV channels are calculated in the same way. In this case, capacitances and inductances were selected in such a way that, with the necessary slope of the frequency response fronts, the necessary relative band would be provided (together with the expansion inductances). Such channel filters and amplifiers based on them (large-scale production) are used for cable TV [32–36].

Among other applications of impedance filters, the use of oscillators as frequency-setting stabilizing elements, narrowband and ultra-narrowband filters, wideband filters as part of telecommunication systems (TCSs), dispersive delay lines for fast Fourier transform (FFT), and pseudo-random frequency search (PRFS) should be noted. Impedance filters can also be successfully used in self-oscillators [28, 37], in terminal stages of TCS transceiver devices, including cellular communication systems [28, 38], narrowband and ultra-narrowband filters, broadband filtering [28, 39], in multichannel filters for parallel Fourier transducers (FT) and other devices, nodes and blocks of different TCSs (Table 4.3).

This allows to achieve such limiting characteristics of hardware functions as part of the TCS as insertion attenuation, frequency selection, NTS, PRFS, FFT, power dissipation, intersymbol interference, group-delay ripple, reliability, size, cost, to significantly reduce insertion attenuation (by at least 1–2 orders of magnitude) and weight and size characteristics (approximately twofold). Let's show it on the example of a self-oscillator.

Indeed, SAW delay line on the substrate of a ST cut of quartz with a relative bandwidth of 0.1% at f_0 = 380 MHz has a substrate size of 10 mm (including acoustic absorbers at its ends), while an impedance bridge filter with a symmetric transformer can be made on a ST quartz substrate cut with a length of only 3 mm [37]. In addition, no space is required for acoustic absorbers, since this filter does not have a traveling wave mode [28].

For creation of a self-oscillator as a frequency-setting stabilizing element, an ultra-narrowband impedance filter was developed (Fig. 4.33a). This circuit allows

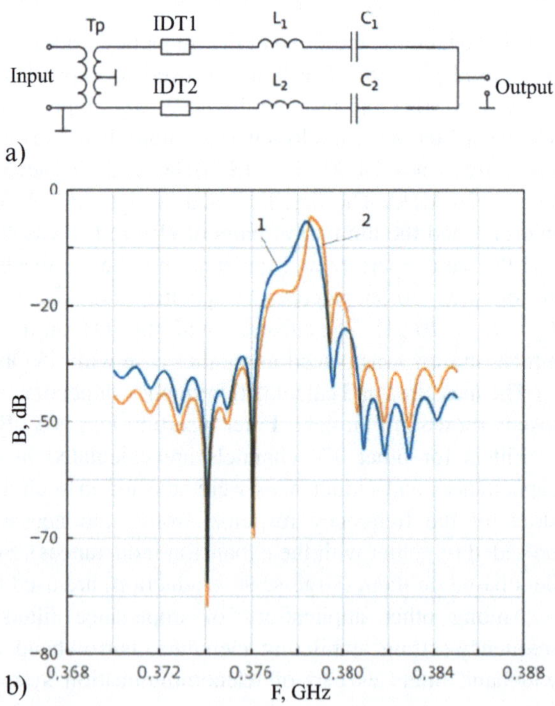

Fig. 4.33 Ultra-narrow differential bridge filter. Circuit with trimmers (**a**) and dependence of the frequency response on external circuits (**b**): without trimmers (curve 1), with trimmers (curve 2)

using external inductances and capacitances to make adjustments within the passband. For example, it is possible to shift the frequency up by half the bandwidth (Fig. 4.33b, curve 2).

The phase-frequency response (PFR) of this filter is similar to the PFR of an LC filter, i.e., it passes through zero (satisfies the phase balance) only once within the main lobe of the frequency response. Therefore, an oscillator with an impedance filter will work only in one frequency, unless amplifier gain exceeds the suppression of the side lobes of the frequency response.

The use of expansion inductances allows using a bridge circuit to obtain broadband filters [28]. Figure 4.34a, b shows the diagram and frequency response of this filter. The filter has a center frequency of 160 MHz and a bandwidth of 50 MHz, that is, a relative bandwidth of 31.25%, with an insertion loss of 3 dB and irregularities in the passband of not more than 1 dB.

The use of substrates in which near-surface waves are excited makes it possible to obtain frequency response with low irregularity in the passband due to the fact that the zeroes of the active and reactive components of the radiation conductivity "float in." Figure 4.35 shows the frequency response of a SAW filter on a substrate of lithium tantalate YX/42° cut ($k^2 = 0.064$). The filter consists of two identical G-circuits connected in series. The IDT with a small aperture has an aperture of 10 microns, and the number of pairs of electrodes is 250, and the IDT with a large aperture has an aperture of 260 microns and 250 pairs of electrodes.

Fig. 4.34 Broadband filter: differential bridge circuit (**a**) and frequency dependence of insertion loss (**b**)

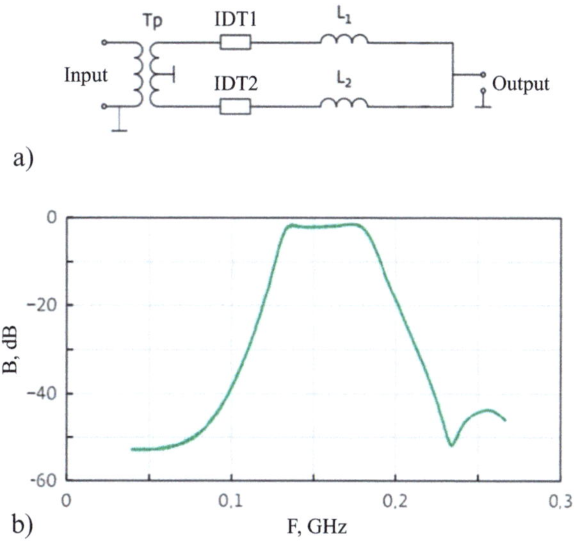

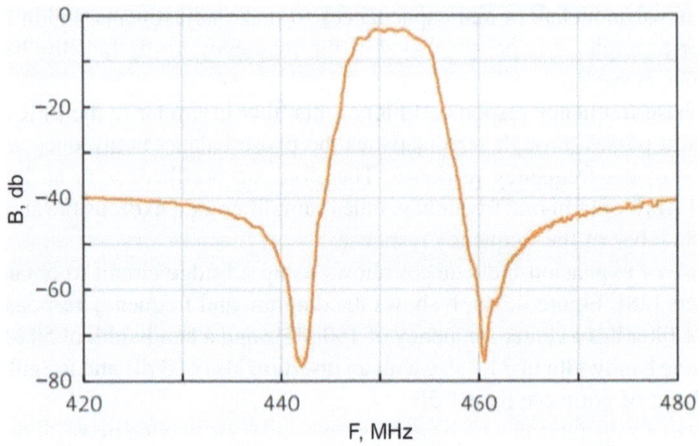

Fig. 4.35 Frequency dependence of the insertion loss of a lithium tantalate YX/42° cut filter

4.6 SHF Applications of MEMS Technologies

4.6.1 Features of Implementation of Radio Frequency MEMS/CMOS Devices

The creation of modern HF and microwave devices and systems based on them requires numerous reconfigurable RF input units that can work with a wide variety of communication protocols. In such blocks, you need to use tunable modules, not channel switches. These modules should be small-sized, have low losses, a large frequency tuning range, and low power consumption. As shown in [1], modern CMOS circuits can no longer fully meet these requirements; therefore, developers of microwave systems often use reconfigurable MEMS devices running classical CMOS chips to solve their problems. Below is a brief look at recently created MEMS/CMOS switches, switch capacitors, resonators with electrostatic, electro-thermal, piezoelectric control, which in the near future will replace classical switches in the RF systems based on standard field-effect transistors and high electron mobility transistors (HEMT) [47–62].

The first MEMS switches were successfully demonstrated by IBM (in 2004) and NTT (in 2006). Thus, NTT's MEMS/CMOS chip was a multichannel single-pole switch for eight directions with a voltage of 3.3 V. The MEMS switch was manufactured using low temperature processing. According to the company, the device withstood 10^9 switching cycles without failures [47].

RF MEMS until recently were supplied to the market in a small volume mainly for test equipment. Interest in their use in smartphones appeared in the middle of 2010, when difficulties arose with the reception of signals by the iPhone 4 when changing its position. It turned out that the so-called death grip problem can be eliminated with the help of RF MEMS, providing frequency adjustment of the antenna.

WiSpry, a company which develops tunable RF semiconductor devices for wireless systems, in early 2010 received a patent for MEMS capacitors with digital capacity tuning, actuators, and methods for implementing them. In the same year, the company entered into an agreement with IBM for the integration of new MEMS-capacitive switches with 0.18-μm multilayer metallization CMOS chips. MEMS capacitors functioned as passive RF components and were controlled independently of each other. This allowed the developers to combine them to produce tunable RF devices used in filters, power amplifiers, antenna matching devices, matching circuits, universal capacitors.

An example is a two-digit tunable MEMS capacitor containing four components: a primary and a secondary actuator, two RF capacitors, and an RF signal transmission line. The actuators can be connected to each other or connected to separate master nodes. The RF signal transmission line and actuators are separated by an air gap and silica. Insulation of the fixed electrode of the MEMS capacitor and dielectric coating of the first metallization layer of the CMOS circuit make it possible to minimize the parasitic shunt capacitance. Additional parasitic capacitance was reduced by reducing the size of the pads for mounting the device using the flip-chip method. A cell containing four tunable WiSpry capacitors had the following characteristics [49]:

Capacity when off, C_{min}, pF	0.13
Capacity when on, C_{max}, pF	1.27
Tuning range	10:1
Quality index:	
at 1 GHz	160
at 2 GHz	87
Permissible continuous power, W	2
Console pull voltage, V	27
Self-operating voltage, V	40
Working voltage, V	3.5–5.0

In 2011, WiSpry released two MEMS/CMOS microchips for tuning the impedance of a mobile phone antenna—WS2017 and WS2018 for the frequency range 824–2170 MHz [49]. WS2018 circuit can work in the LTE-frequency range 700/2600 MHz. Thus, these antenna impedance tuning chips could be used in any smartphone chipsets on the market at the time. The circuits contained a low-loss inductance unit and MEMS/CMOS capacitors with digital capacitance tuning in 0.125 pF steps and allowed to dynamically compensate for a VSWR equal to 20:1 and more. The WS2017 chip setting to compensate for changes in antenna impedance was controlled by a modem via an SPI serial interface, and the WS2018 chips were controlled via SPI and RFFE interfaces.

In November 2013, the RF MEMS circuit of a WS1050 capacitive switch with three individually controlled and configurable capacitors appeared on the market (Fig. 4.36). The scheme provides a broadband antenna tuning, the natural frequency of which exceeds 3.5 GHz. Tests using National Instruments' PXI vector network analyzer showed that the capacitive tuning time of the WS1050 is 20 μs.

Each capacitor of the MEMS circuit has a separate output. The circuit may be connected in series or in parallel with the transmission line. Eight configurable source identifiers support up to eight separate tuners per mobile device bus. WS1050 meets the requirements for 2C, 3G, and LTE mobile phone tuning blocks. It can be used in instrumentation, phase shifters, tunable filters, and amplifiers.

At the Mobile World Congress 2014 WiSpry presented three mobile phone antenna tuning chips with four individually controlled MEMS capacitors—WS1040, WS1041, and WS1042 (Fig. 4.37) (www.gnuradio.ru). The included DC converter ensured operation with a single power bus. The use of unified source identifiers and sequential or parallel system configurations provide flexibility to the design of the telephone system.

The total capacity of the antenna impedance matching circuit and the WSJ040 mobile phone input unit is 12 pF. The circuit was designed for such currently developed systems as wearable mobile devices. Such areas of application as smartphones, phase shifters, and power amplifier matching devices are also promising.

Fig. 4.36 MEMS circuit WS1050 in a standard LGA type package

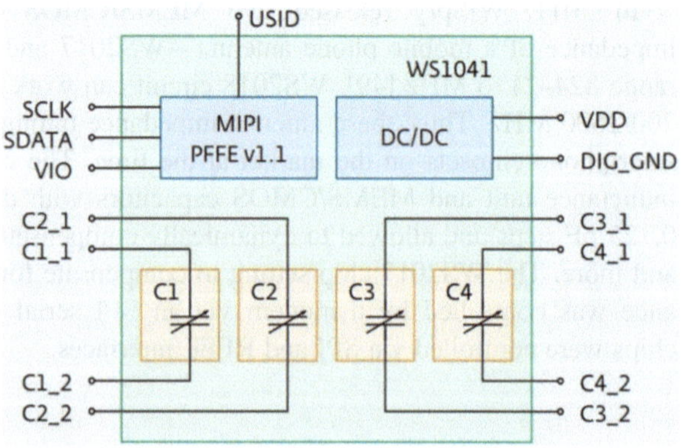

Fig. 4.37 A flow chart of a radio frequency MEMS circuit type WS1040, WS1041, and WS1042 containing four tunable capacitors

The radio frequency circuit WS1041 is designed to adjust antenna frequency. The total capacity of the frequency tuning circuits is usually less than that of the antenna impedance matching circuits and the input blocks. They must provide highly accurate tuning and withstand high operating voltages. At the same time, such circuits designed for LTE-standard mobile phones and supporting carrier aggregation technology (CA) should not generate harmonics or degrade the antenna quality factor. The WS1041 circuit that can withstand voltage up to 80 V satisfies these requirements. The generated third harmonic is—150 dBc, and the quality factor is 159 at the maximum value of capacitance and low frequencies. The total capacity of the circuit is 6 pF. The WS1041 matches well with the WS1050 circuit, providing a wide range of impedance matching and frequency tuning.

The WSI042 circuit is designed for use in custom filters. Due to its high linearity, it can be used in receiving and transmitting channels. The total capacitance of the circuit is 3 pF, its increment does not exceed 25 pF, the speed of adjustment of the filter frequency is no more than 20 μs. The quality factor of the circuit is 200 at maximum capacity. The operating voltage is up to 100 V. Thus, WiSpry offers a software-controlled digital matrix of RF capacitors based on MEMS structures with extremely high quality, low insertion loss, and high linearity.

In 2010 Cavendish Kinetics was the first company to report the creation of a standard CMOS circuit of a MEMS switch with metal welded contacts embedded in the metallization [50]. Reliability tests of the switch have shown its ability to withstand up to 10^9 switching cycles with little to no change in the voltage of the actuation. The actuation voltage remained constant in the temperature range from −150 to 300 °C, although the contact resistance slightly altered, possibly due to the temperature dependence of the metal resistivity and adhesion force.

The switch was designed to create a non-volatile memory circuit capable of operating in adverse environmental conditions, as well as a capacitive switch. A constant voltage applied to its upper electrode causes a cantilever beam to rise, accordingly leading to an increase in the gap between the electrodes and a decrease in capacitance. When voltage is applied to the lower electrode, the gap becomes smaller and the capacitance increases (Fig. 4.38). Reliability tests of the MEMS/CMOS circuit also showed trouble-free operation with 10^9 switching cycles. The switch was designed for non-volatile MEMS memory capable of operating in adverse conditions. As a result, a third-generation RF MEMS capacitor with tunable capacitance was created for mobile phone antenna frequency tuning in the 700 MHz to 2.5 GHz frequency range [51].

The chip uses MEMS capacitors of the smallest size. The insertion loss is small. In addition, the switch has very small values of minimum capacitance C_{min}. The switch frequency tuning range is 5:1. Deviations from the nominal value do not exceed 0.25 PRM. Developers managed to get low losses by reducing the equivalent series resistance (ESR) of the circuit. The greater the ESR, the greater the losses in the circuit: an increase in the equivalent resistance of 1 Ohm leads to an increase in antenna losses of 2–3 dB. Therefore, the device's ESR settings should be as small as possible.

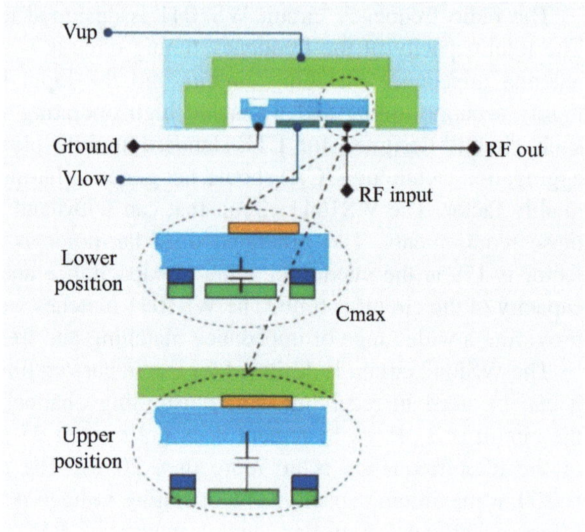

Fig. 4.38 Cavendish's radio frequency MEMS-capacitive switch

The developers have solved this problem by forming tunable capacitors in a multilayer metallization structure and using a single dielectric layer. This reduced the ESRh to provide a maximum operating voltage of 650 V. As was already mentioned, the MEMS capacitance depends on the voltage applied to the upper or lower electrode. In the structure of the tuning device, mechanical stops are provided, fixing two positions of the cantilever beam. Therefore, the applied voltage does not exactly determine the position of the beam, i.e., the capacitance value is most likely determined by the speed of its movement from one position to another. In the developed frequency tuning circuit, a voltage in the range of 22–24 V generated by the intracircuit pump charge unit with a voltage of 30 V is applied to the cantilever beam. For each particular MEMS structure, you can set the required actuation voltage, and the lower the voltage, the less the insertion loss. However, if the voltage is too low, contact with the limiter deteriorates, and if the voltage is too high, a collision with it is possible.

Even with such a "digital" solution to the problem of frequency tuning, the switch capacity may vary. Therefore, the developers use one "capacitor" consisting of small capacitors connected in parallel, which allows smoothing the value variation of the device capacitance. This matrix allows to create different combinations of capacitors and thereby obtain different values of capacitance. The created MEMS/CMOS circuit of the antenna frequency switch includes a matrix which contains ~ 700 small capacitors, some of which are connected to obtain 32 different capacitance values. The capacitive switch is mounted in a housing that is comparable in size to a crystal and is slightly larger than a conventional capacitor. The ESR of the antenna frequency tuning MEMS/CMOS circuit is 0.3 Ohm, which corresponds to a loss of 1 dB. The ESR of a switch which uses traditional tunable capacitors with a dielectric based on strontium-barium titanate is 3 Ohms or a loss of 3.5 dB.

However, the use of high-quality discrete capacitor allows to get an ESR of 0.25 Ohms.

In the summer of 2013, the company provided samples of the MEMS circuit of a switched capacitor to its strategic customers to determine their compliance with technical requirements and reliability. Thus, the MEMS circuit of a capacitive switch for mobile phone antenna tuning has “launched,” and perhaps in a year or two we will encounter such tuning circuits in next-generation cell phones [50].

4.6.2 Radio Frequency MEMS Switches

The main advantages of MEMS switches are lower insertion loss, higher isolation coefficient, operating voltage, and linearity compared to pin diodes and CMOS switches.

Among the RF MEMS/CMOS chips created by post-processing in the BEOL-metallization layers of a CMOS circuit, note the thermo-controlled switch [52] connected in series to the transmission line. The switch was created jointly by specialists from Hughes Network Systems, John Washington University, and the University of California, Santa Barbara. The MEMS switch structure is implemented in the BEOL-metallization of a CMOS chip manufactured using the 0.6-micron technology from AMI Semiconductor. The cantilever beam and the teeth of the comb structure form three layers of aluminum and silicon dioxide, and the beam is made using two RIT operations. As the material of the heater leads to the activation of the MEMS switch, polysilicon was used. Contact pads are provided for supplying both a constant bias and an RF signal. For proper separation of the cantilever beam and thermal actuator between them, a small gap is made (Fig. 4.39).

The capacitance adjustment of the MEMS switch is ensured by its comb structure and thermal actuator. Heating the thermal actuator when a constant voltage is applied to it causes the cantilever beam to be lowered and the comb teeth located on its two sides to come in contact, which corresponds to the open state of the switch. The teeth overlap determines the capacitance value. In the absence of voltage on

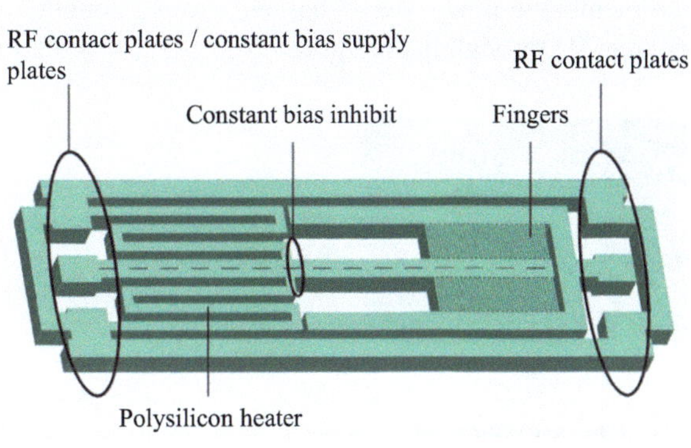

Fig. 4.39 Thermally controlled MEMS-capacitive switch

the thermoactuator, the internal voltage of the cantilever beam leads to its rise, breaking the teeth contact. The switch returns to the disabled state.

The ratio of capacitance values in the on and off states was 15:1. The insertion loss and isolation at a frequency of 5.4 GHz are less than 1.6 and 33 dB, respectively. The switch is designed for RF input circuit diagrams (voltage controlled generators, filters, matching circuits) of industrial, scientific, and medical systems, as well as wireless local area networks.

Researchers at the University of Waterloo, in order to obtain a large ratio of capacitance values in the off and on states, as well as to reduce insertion loss, increase the response speed and operating voltage, suggested a MEMS/CMOS-capacitive switch with electrostatic actuation connected in parallel to the transmission line [53]. The circuit is made using the above-considered processes of post-processing of BEOL-metallization layers of 0.34 μm CMOS circuit manufactured by contractor TSMC. The switch structure contains a section of a coplanar waveguide and two upward bent membranes made on the basis of aluminum/silicon films. To reduce the insertion loss and eliminate the parasitic effects introduced by the low-resistance silicon substrate, an air gap of 65 microns depth is made under the signal conductor.

The membranes are mounted on the center conductor and hang over the ground wire of the waveguide (Fig. 4.40). The bending of the membranes occurs after the final manufacturing of the MEMS switch due to the difference in the values of residual voltage forming the MEMS structure of the metal and dielectric layers. The capacitance between the signal and grounding conductors of the switch is minimal when bent diaphragms are in the up position, and maximal when they are in the down position. The bending of the membranes, which, as a rule, is considered undesirable, is useful in this structure because it leads to a decrease in the minimum capacitance and an increase in the maximum capacitance and thereby improves the isolation of the switch.

High impedance polysilicon actuation electrodes are located in the gap between the signal and ground electrodes of the waveguide and are separated from the RF ports. Applying a constant bias V_b to the actuating electrode, located under the membrane, triggers the switch. In order for the actuation voltage to be low, the air gap between the actuation electrodes and the membrane must be small. To bring the membrane down, the signal and ground conductors are supplied with constant voltage V_h through the RF ports.

a)

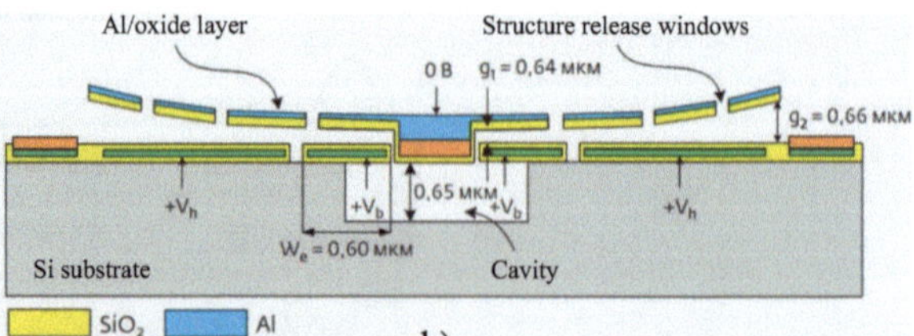

b)

Fig. 4.40 MEMS/CMOS-capacitive switch circuit developed at the University of Waterloo (**a**); switch cross section (**b**)

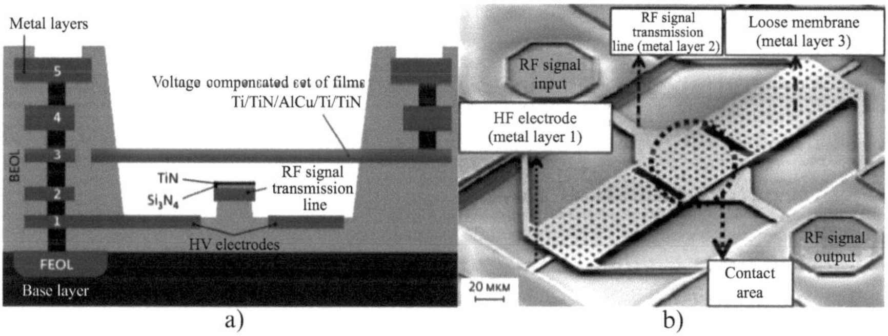

Fig. 4.41 Integrated RF MEMS-capacitive switch: structure (**a**); switch view obtained with a scanning electron microscope (**b**)

The RF MEMS circuit of a curved membrane capacitive switch was used by the University of Waterloo specialists to create a MEMS/CMOS circuit for matching antenna impedances and telephone input blocks that is tunable in the 13–24 GHz range The circuit is created on the basis of a distributed MEMS transmission line (DMTL). Suspended slow-wave (SSW) coplanar waveguide with eight MEMS switches connected to it is located on a 1.3 × 3.2 mm crystal, which allows to obtain 256 different impedance matching nodes on a Wolpert diagram with VSWN up to 11.5:1. At a frequency of 24 GHz, the impedance matching exceeds 10 dB, the power transfer coefficient is—2.84 dB [54].

The joint development of specialists from IHP, the Research Institute of Electronic Nanosystems of Fraunhofer (Germany), and the University of Sabanci (Turkey) is interesting. With the help of BEOL-metallization layer post-processing technology of 0.25-micron SiCe: C BiCMOS-chip with five-layer metallization, they created a high-voltage MEMS/CMOS tuner. The MEMS-capacitive switch (Fig. 4.41a, b) is formed by the second (M2) and third (M3) metallization layers. The first layer (Ml) was used to form high-voltage electrodes, while the signal lines with a thin Si_3N_4/TiN coating were located in the second metallization layer. The distance between the high-voltage electrode and the signal line, which also serves as a brake for the switch membrane, is large enough. The multilayer membrane of the MEMS switch is created on the basis of a voltage compensated set of Ti/TiN/AlCu/ Ti/TiN films of the third metallization layer (voltage compensation does not lead to disruption of the functional characteristics of the M3 layer). An important advantage of this MEMS/CMOS chip is only one additional process of masking and drying in the manufacture of the BiCMOS circuit on a 200 mm plate.

Measurements of the characteristics of a capacitive switch on the plate showed that the capacitance in the off state (the membrane in the upper position) is 25–30 fF, in the on state (the membrane in the lower position) 210–250 fF. Thus, the ratio of capacitance values in the off and on states is on average 1:10, which ensures good isolation in the millimeter wavelength range. Thus, the isolation and insertion loss in the range 60–110 GHz was more than 15 dB and less than 1.65 dB, respectively.

The technical capabilities of a MEMO-capacitance switch integrated into a CMOS circuit are demonstrated by the example of a voltage controlled dual-range oscillator. It was found that the oscillator frequency can be changed from 3.55 to 4.15 GHz. When performing 10^9 cycles of frequency switching, no degradation of characteristics was observed [55].

4.6.3 *Radio Frequency MEMS Capacitors of Variable Capacity*

Consider as an example the MEMS/CMOS tunable counter-combed capacitors developed at Carnegie Mellon University (USA), in particular, the MEMS/CMOS three-digit digital capacitor chip with a 60:1 tuning range. The circuit contains a set of connected in parallel switched capacitors, the capacitance of which is regulated by an operating electrothermal actuator located above the structure and a horizontal electrothermal actuator, located in the plane of the structure of the electrode with electrostatic activation. In the initial state, the working electrodes of the capacitor due to thermal voltage are located above the plane of the structure (Fig. 4.42a). During electrothermal activation, they descend to the plane of the structure containing stationary electrodes (Fig. 4.42b). Electrostatic activation of the horizontal electrode leads to the attraction of both electrodes and their mechanical contact. Since during the formation of the electrodes, a passivation film was applied to their side planes, there is no electrical contact when the electrodes are closed, and the capacitance of the capacitor is maximum. At this time, DC power consumption is absent.

Measurements of the characteristics of a switched MEMS/CMOS capacitor showed that its Q-factor at a frequency of 1 GHz is 150, and the natural generation frequency exceeds 10 GHz. At voltages of electrothermal and electrostatic activation

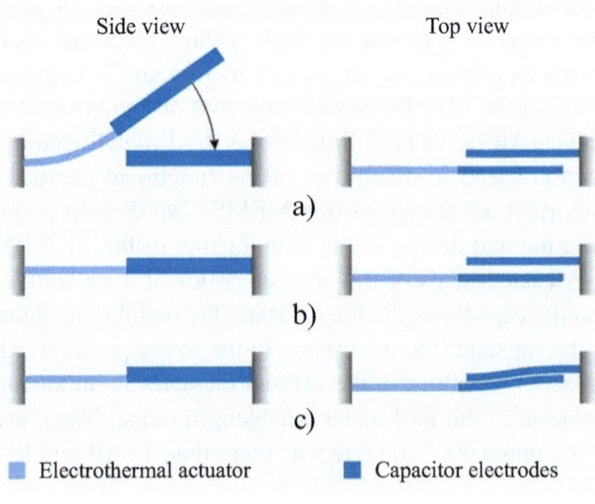

Fig. 4.42 The principle of operation of the three-digit digital MEMS capacitor at Carnet-Mellon University: (**a**) after the final implementation of the MEMS capacitor; (**b**) after electrothermal actuation in the vertical plane; (**c**) after electrostatic actuation in the horizontal plane

of 4 and 20 V, respectively, the switching time of the capacitance does not exceed 1 ms [56].

In recent years, interest in batteries of tunable capacitors has increased due to their reliability and the possibility of continuous (analog) control of many capacitors. However, during electrostatic activation of a capacitor, the bistable hysteresis of the volt-capacitance characteristic makes it difficult to adjust analog MEMS tuning devices. Moreover, the reliability and adjustment of electrostatically activated MEMS devices are influenced by the effects of charging a dielectric, leading to a "blurring" of the volt-capacitance characteristic (Fig. 4.43) and to failures during testing for long-term reliability. To adjust the characteristics of a MEMS device and ensure their consistency, it is necessary to accurately control the position of its mechanical elements.

Researchers at the University of Waterloo have proposed using the piezoresistive properties of the polysilicon layer located under the MEMS structure. To demonstrate the capabilities of an integrated means of controlling the position of elements of a structure, a MEMS element with a curved cantilever beam was selected. A change in the bend of the beam leads to a change in the resistance of the piezoresistor, which can be read using the built-in Wheatstone bridge. In order to evaluate the advantages of using the sensor mechanism for controlling the position of mechanical elements, the dependence of the capacitance of a MEMS tunable capacitor on voltage and resistance in the absence and presence of a piezoresistive sensor was measured (Fig. 4.43a, b). As can be seen from Fig. 4.43a, at a given voltage due to hysteresis and charging of the dielectric, there are various possible values of capacitance, while dependence of the normalized capacitance of the capacitor on the resistance of the beam is formed by three straight sections corresponding to the operating modes at fixed values of the control voltage, which ensures a stable position of the structure elements. The suggested method allows to solve the problem of possible instability of capacitance values.

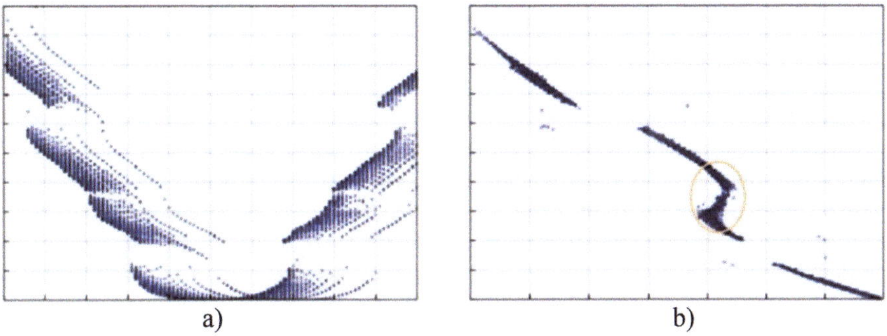

Fig. 4.43 Dependence of capacitance of an alternating capacitor on the voltage in the absence of a piezoresistive sensor (**a**) and dependence of capacitance on resistance in the presence of a sensor (**b**)

4.6.4 Integrated MEMS/CMOS Resonators

MEMS oscillators with mechanical oscillating resonators appeared on the market as early as the 1980s, but were not widely used. This was due to their high temperature coefficient, the need to use complex temperature compensation circuits, as well as expensive metal or ceramic bodies. For compact low-cost MEMS resonators with high performance characteristics to appear on the market, the combined efforts of manufacturers of various CMOS devices were required. At the seventeenth International Conference on Micro Electro Mechanical Systems in 2004, IBM presented the first MEMS resonators manufactured in BEOL-metallization of BiCMOS circuits. The cantilever beam of the resonator had a dielectric layer of copper metallization. As a result, its resonant frequency was ~4 MHz, and the quality factor—1200. The size of the MEMS resonator was 6 × 15 microns.

But the first real success was achieved by a new company that did not have its own production facilities, SiTime, which reported in 2006 the creation of SiTllxx (for fixed frequency values) and SiT8002 (with programmable frequency) models of MEMS/CMOS oscillators manufactured using silicon technology. MEMS oscillators for frequency from 1 to 125 MHz were supplied in standard QFN type 2.0 × 2.5; 2.5 × 3.2; and 3.2 × 5.0 mm cases (case thickness for all variants—0.85 mm) [60].

The emergence of high-quality factor resonators based on low-cost MEMS/CMOS platforms, which ensure a fast production cycle, attracted the attention of the creators of signaling units and RF communication systems. MEMS/CMOS resonators with a high level of integration, good quality factor, measured in thousands of units, and lack of DC power consumption have drastically changed the technical capabilities of not only conventional electronic devices, but also integrated microsystems.

The problem of creating MEMS/CMOS resonators with a sufficiently large gap between the electrodes, which does not affect the mechanical boundary conditions of the beam, with low mobility-induced impedance and high quality, was solved by researchers at Tsinghua National University (Taiwan). Using CMOS amplifier chip BEOL-metallization post-processing, a structure was created with a free-free beam located between two retractable frames with four thin supporting rods fixed in the functional nodes of the structure (Fig. 4.44). The rods are designed to reduce the oscillation energy flow to the fixed substrate, which prevents the resonator from depending on the mechanical boundary conditions of the movable beam, preserves the resonance energy of the structure, and ensures a high quality factor of the resonator. In order to reduce the voltage required for the frames, they are connected to adjacent latches using a zigzag spring. The manufacture of a MEMS/CMOS resonator structure in the metallization layers made it possible to significantly increase its thermal stability: the frequency in the temperature range from −40 to 80 °C changed by less than 1%. The placement of tuning electrodes under the retractable frames provided a "quasilinear tuning" of the frequency at which the dependence of frequency on voltage is linear.

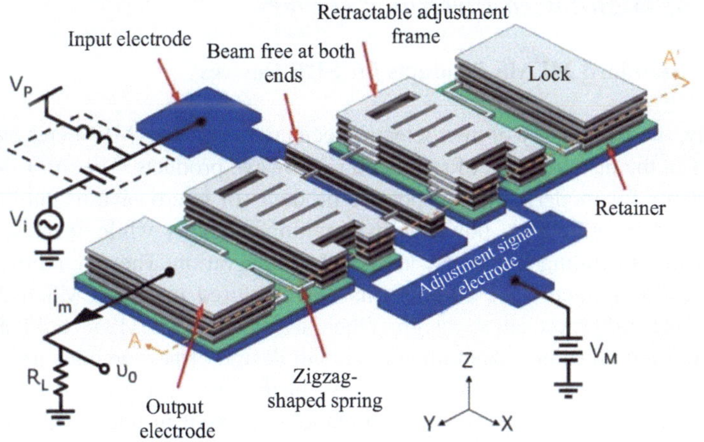

Fig. 4.44 Structure of the MEMS/CMOS resonator with a free–free cantilever beam and a large submicron gap

The developers of this university created and tested MEMS/CMOS resonators with free–free cantilevered beams and gaps of 110, 210, and 210 microns. The measured resonant frequency of the device with a gap of 275 microns was 11.56 MHz, the Q-factor exceeded 2000, the total resistance due to beam mobility was 112 kOhm. When using a resonator array, it is possible to further reduce the impedance by promoting the use of the MEMS resonator in oscillator circuits. The size of a MEMS resonator integrated in a CMOS amplifier chip is 300×130 μm [61].

Thus, we can summarize this section as follows:

1. With the miniaturization of RF chips and the growing need to improve their functions, the trend towards integrating MEMS and CMOS devices for RF applications will increase.
2. MEMS/CMOS technology is already quite well mastered, and the choice of a particulate solution depends on the manufacturing cost and quality of the MEMS device. Of the three methods of MEMS/CMOS integration discussed above, the post-processing method is the most promising. The fourth method of CMOS circuit BEOL-metallization post-processing is of interest only for university researchers, because it is cheap and allows the use of chips manufactured by various contract manufacturers.

4.6.5 MEMS Technologies in the Tasks of System Integration of High-Speed Electronic Devices

4.6.5.1 Standard MEMS Products for SHF Devices

Currently, the technology of microelectromechanical systems (MEMS) is one of the priorities in the development of radio and microwave products. The most important requirements for modern and perspective products of microwave technology is to reduce their weight, size, power consumption, and cost while increasing their functionality, operating frequency, and level of integration. These requirements are fulfilled by new microwave components manufactured using MEMS technology using a systematic approach to design of their architecture. MEMS technology opens up new opportunities in system integration and design of new architectures.

The interest in using MEMS technology in the microwave range is due to its flexibility. It allows to overcome the limitations inherent in integrated microwave devices, and implement higher-level circuits. Therefore, the ultimate goal of using MEMS in microwave technology is to extend the advantages of the "device" level to the "system" level and achieve record characteristics of the final product, as shown in Fig. 4.45. Below we briefly consider examples of a number of already produced MEMS circuits [62].

MEMS Circuits on Inductance Coils Inductance coils are key elements of resonant circuits, providing, in particular, matching of network impedance, ensuring operation of low-noise amplifiers and voltage controlled oscillators (VCOs).

An increase in the gain, power dissipation, or the presence of phase noise in such circuits leads, in turn, to the need to combine MEMS coils inductance on a single chip. Perhaps the most striking example of the advances resulting from the use of

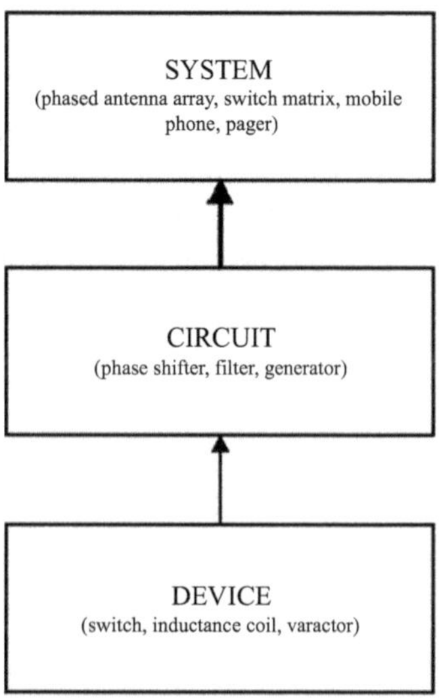

Fig. 4.45 Strategy for system integration of radar devices based on MEMS technologies

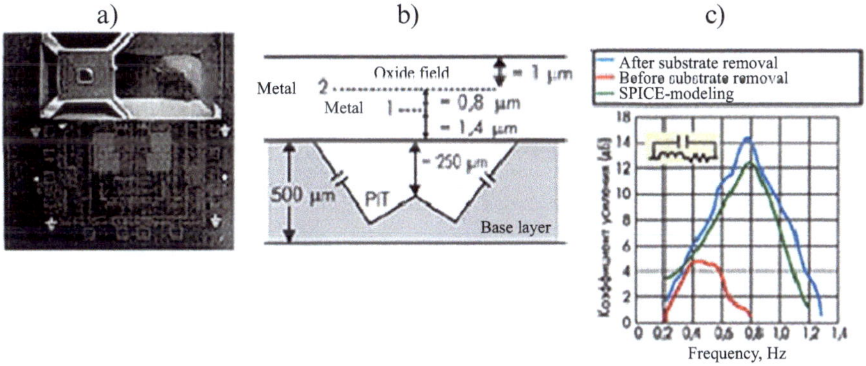

Fig. 4.46 Microwave CMOS amplifier with "suspended" MEMS inductance coil (**a**), cross section of the inductor (**b**) and its characteristics (**c**)

MEMS inductance coils is the first demonstration of a resonant microwave amplifier with a complementary metal-oxide transistor by Chang, Abidi, and Gaitan.

Figure 4.46 shows the comparative characteristics of amplifiers with detached and non-detached inductance coil substrates.

The use of a MEMS inductance coil gives a gain increase of 12 dB and a double increase in the resonant frequency.

MEMS Circuits Based on a Varactor Varactors are essential components wherever circuitry is required (for example, in various matching circuits and VCOs).

Since the downsizing trend continues in this case, and good semiconductor varactors, unfortunately, cannot be manufactured using traditional technologies used in the manufacture of integrated circuits, only MEMS-varactors possess the necessary characteristics so far. An example is the technical solution of the researchers Dec and Suyama, who, using a parallel flat capacitor 1.4 pF ($Q = 14$ at 2 GHz), made using standard technology of micro-processing of polycrystalline silicon surface, demonstrated the operation of a VCO at 2.4 GHz with a phase noise of 122 dBc/Hz with a detuning from the carrier frequency of 1 MHz and the possibility of tuning the band by 3.4% relative to a changing voltage of 5 V.

Switch-Based MEMS Circuits The prototype MEMS switch in the frequency range from 0 to 4 GHz provides an insertion loss of about 0.1 dB and isolation at 50 dB.

This shows their great potential for replacing semiconductor switches incurring significantly higher losses and consuming more power. This includes T/R switches, phase shifters, switchable filters, cross-over multiplexers, tunable antennas, and phased arrays.

In this area, a multiband 4-bit phase shifter has been created with averaged insertion loss of only 1.4 dB and return loss of more than 11 dB (Fig. 4.47).

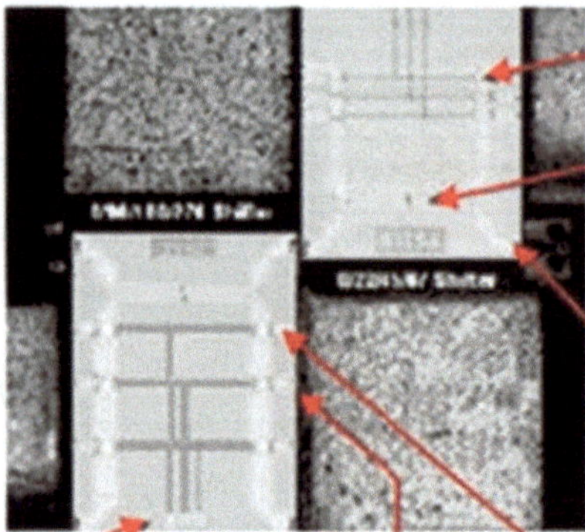

Fig. 4.47 4-bit X-band phase shifter

MEMS Circuits on Micro-Processed Cavity Resonators It is well-known that the quality factor Q of a cavity resonator is proportional to its volume. Therefore, it is natural to consider the use of volume resonators in traditional circuits and applications in which the level of frequencies and speed cannot be realized differently.

These are circuits of oscillators, VCOs, and filters. Example: MMIC (monolithic microwave integrated circuit), a 33.2 GHz frequency oscillator, stabilized by a micro-processed volume resonator shown in Fig. 4.48, provides 113 dBc/Hz phase noise at 1 MHz tuning, which is 18 dB better than MMIC without stabilization.

Micromechanical Resonator Circuits At low frequencies, cavity resonators are excessively large. Micromechanical resonators in this regard are very attractive, since their resonant frequency is proportional to the square root of the stiffness-to-mass ratio.

Thus, the development of MEM resonators has a significant effect, especially for creating filters.

Currently, the resonant frequency is well below 1 GHz (at 156 MHz), and the Q-factor, measured under vacuum conditions, is approaching 9400.

Filters operating at frequencies from several kHz to several MHz have been shown. Figure 4.49 shows the equivalent circuit (a) and the response (b) of a two-cavity filter operating at 7.8 MHz.

Transmission Line MEMS Circuits Transmission lines are widely used in microwave electronics, since they are key elements of many circuits and systems, complementing the circuits enclosed in them. A number of circuits used by different types of lines (namely filters, diplexers, and antennas) are shown in Fig. 4.50.

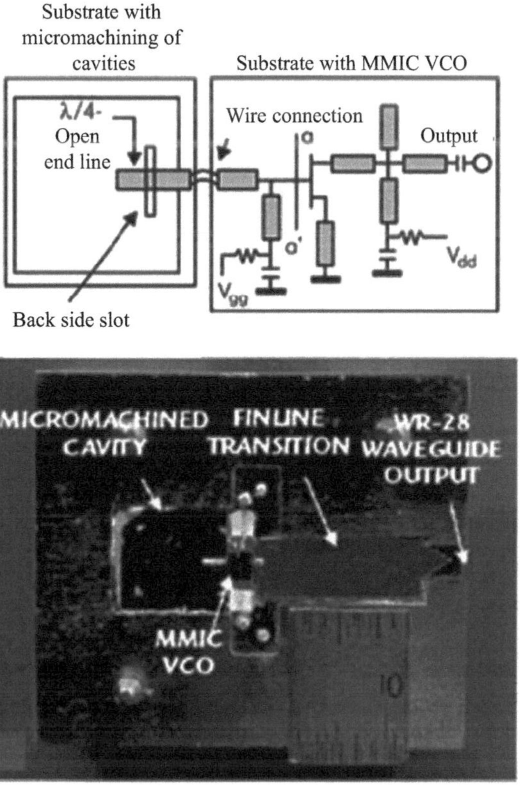

Fig. 4.48 VCO on micromachined resonator cavities; diagram (**a**) and top view (**b**) of a volume generator mounted in a WR-28 waveguide package [63]

4.6.5.2 Installation Technology of Microwave MEMS Devices

The requirement to increase the power of communications and radiolocation extremely exacerbates the problem of integration, since in such products as cell phones, radio networks, radars, LMDS systems, MMDS, controlled pattern antennas, and satellite communications, cost reduction, power consumption, and weight remain paramount. One of the most promising innovations of MEMS is the possibility of increasing the degree of integration of microwave components, which allows to take the next steps in system integration and create new architectures aimed at fulfilling modern requirements.

There are two possible approaches to microwave MEMS installation—"bottom-up" and "top-down."

The bottom-up approach involves simply replacing microwave elements with new MEMS devices according to a pre-designed architecture. In the top-down scenario, it is necessary to begin with the development of a system architecture that is not subject to prejudice of usual constraints imposed by conventional microwave components.

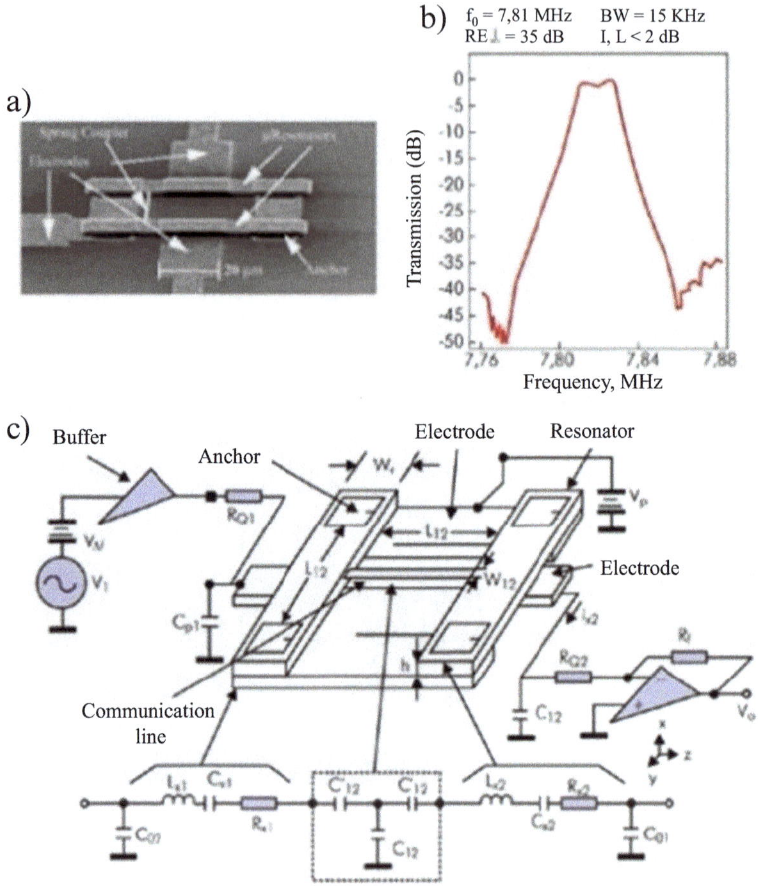

Fig. 4.49 Tico-resonator MEMS filter (**a**), specifications (**b**), and circuit model (**c**) [63]

A typical example of a bottom-up approach is shown in Fig. 4.51 with conventional transceiver architecture. In this context, it is clear that passive components outside the microcircuit, switches, filters, VCOs, mixers, generators, and diplexers are candidates for direct replacement with MEMS duplicates. Another example of a bottom-up construction could be a phased array antenna (PAA).

Since MEM switches require nanojoule switching energy and permanently "virtual zero" energy, it is easy to imagine PAAs containing several orders of magnitude more elements than can be provided by replacing conventional switches.

At the same time, a certain number of reports on developments based on the top-down approach have already appeared in the literature, especially with regard to transceivers. Two such receiver concepts are shown in Fig. 4.52.

The option in Fig. 4.52a uses a tunable input filter that improves circuit integration by drastically reducing the intermediate frequency. The second option (Fig. 4.52b) uses a switched set of acoustic resonant filters, which simplifies the system due to the fixed first local heterodyne used for this.

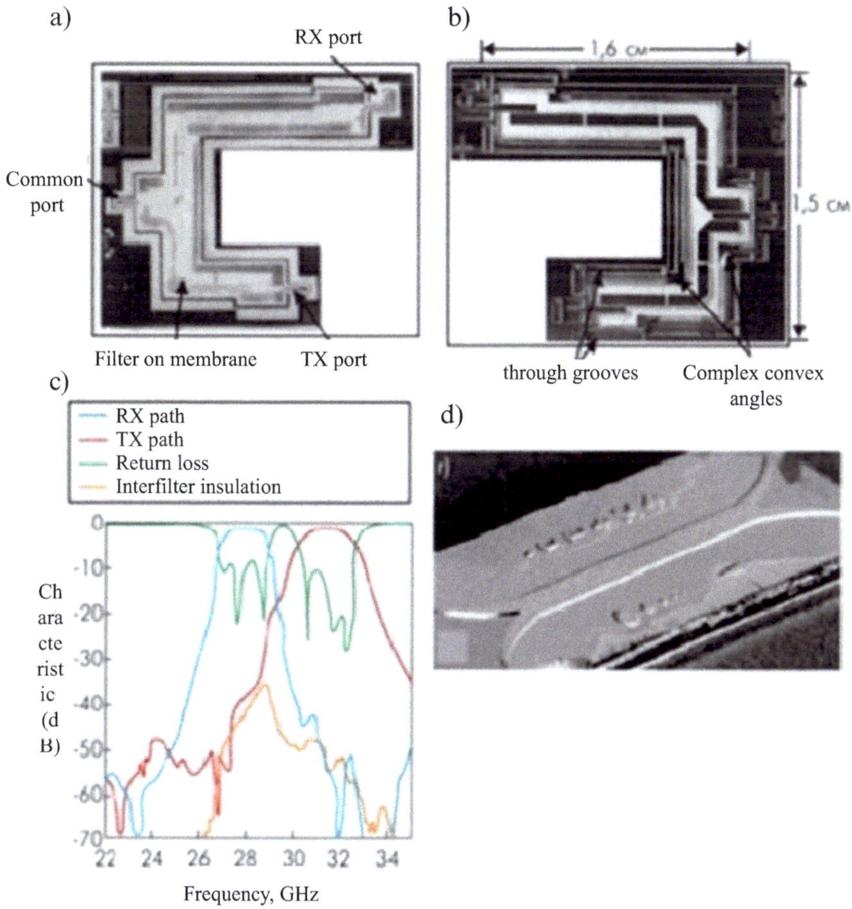

Fig. 4.50 Top view (**a**) and front view (**b**) of a MEMS chip on microstrip lines, its frequency response (**c**) and fragment (**d**) of appearance of a waveguide antenna of 140–220 GHz [62]

A similar approach is shown in Fig. 4.53, where the choice of the reception band is carried out by a matrix of switched filters with matching circuits.

Without creating an appropriate environment, development, as well as rapid deployment of production systems based on MEMS technologies and products, is impossible due to the scale of the projects, as well as the complex physical and multi-domain nature of MEMS devices.

Thus, the creation of appropriate tools for designing MEMS products becomes a very essential condition for the emergence of systems based on microwave MEMS.

CAD (computer aided design) tools provide an environment with all aspects of design and production of a MEMS device: from the principles of physical modeling to the system level of device models.

A typical design cycle for such a microwave MEMS device is shown in Fig. 4.54.

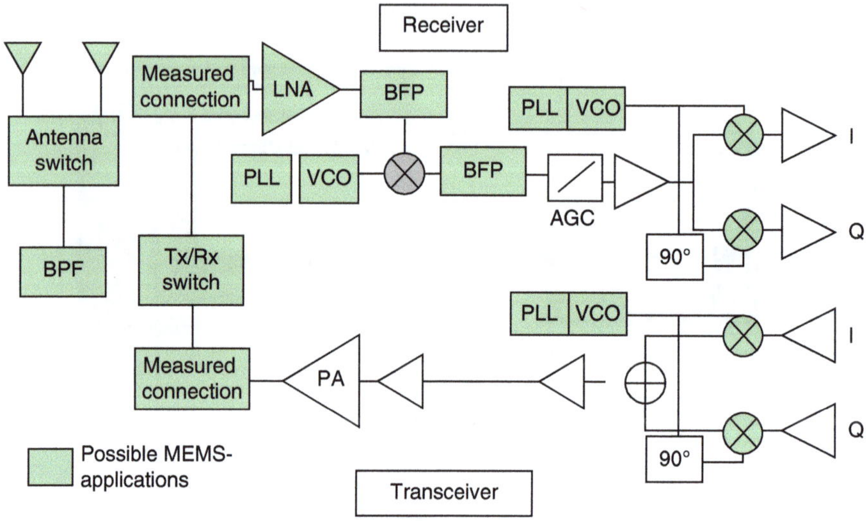

Fig. 4.51 Simplified architecture of a conventional transceiver [62]

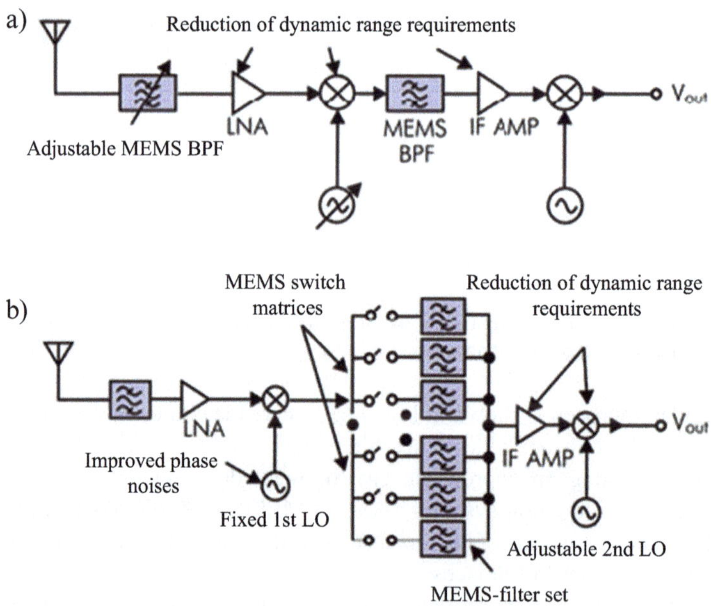

Fig. 4.52 Block diagrams of MEMS transceivers based on a tunable input MEMS filter (**a**) and matrices of MEMS switches and MEMS filters with a set of acoustic resonant filters IF AMP (**b**) [62]

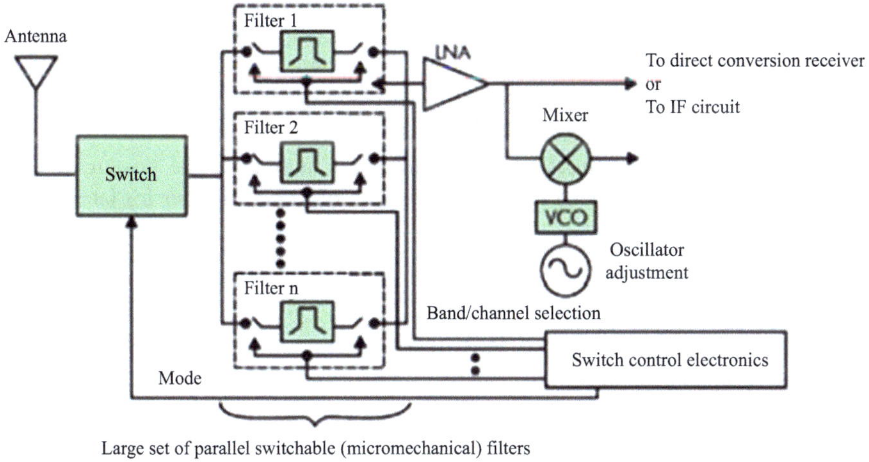

Fig. 4.53 MEMS receiver structure with adjustable receive band [62]

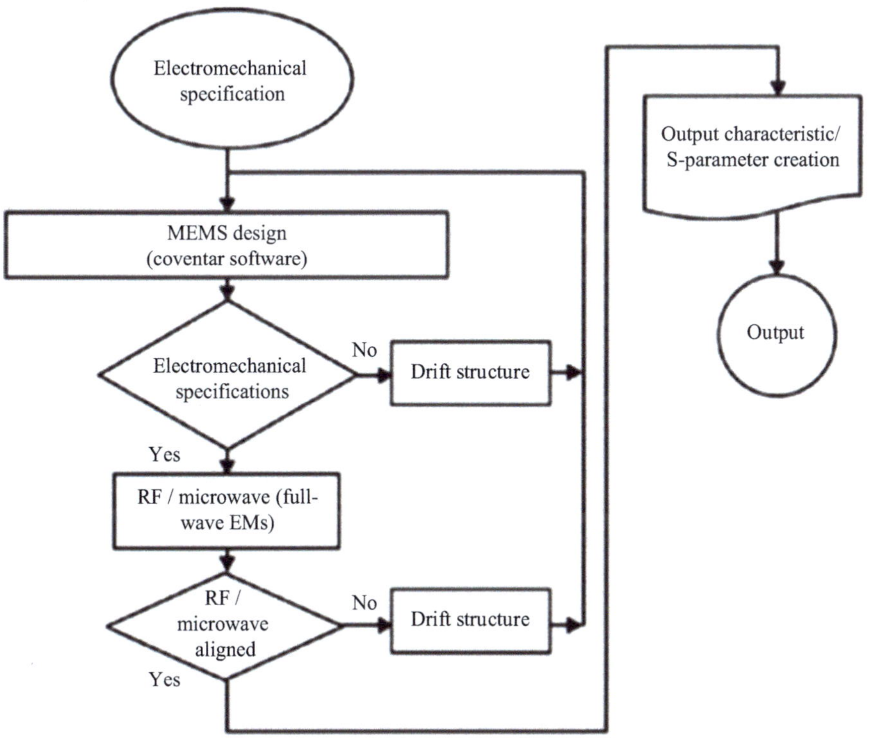

Fig. 4.54 Algorithm for creating microwave MEMS devices

Using mechanical and electrical specifications, the design process is interactive as long as both specifications are not installed in the right place in the project. Within the iterative feedback loops, special software can be used to adjust the project in terms of testing some potential external influences and device production, as well as the devices themselves. These can be, for example, mechanical effects during production or changes in temperature and humidity. The design process ends with a short S-parameter file and/or a model of a certain order for later use in a simulator.

4.7 Software-Configured Radio Systems

Software-defined radio systems (SDR) are usually designed for military and special applications: both when organizing communication in various networks (GSM, Tetra, Wi-Fi, Wi-Max), and for providing wireless services, as they allow to serve a large number of protocols [64–67] (www.ru.wikipedia.org; www.rtl-sdr.ru; www.gnuradio.ru). In addition, this technology is gaining popularity among amateur radio engineers due to the introduction of cheap and affordable SDR receivers (SDR-RTL) on the market.

Already in the mid-1970s, methods of digital signal processing (DSP) were developed, which allowed performing almost any mathematical operations on a signal represented by samples in digital form. The operations of filtering, extrapolation, interpolation, fast Fourier transform, and others have become easily realizable. Digital methods allowed the use of complex algorithms for digital signal processing, and the technology of manufacturing integrated circuits with a high degree of integration made it possible to create cheap and reliable devices.

Any signal could be represented by samples in digital form, then processed in accordance with the implemented algorithm in a program on a computing device—a processor or a microcontroller.

Thus, a new stage of evolution began, digital radio receivers appeared—in fact, specialized electronic computers, in which an additive mix of signal and interference at a radio or intermediate frequency was subjected to analog-to-digital conversion (ADC), and all subsequent processing (filtering, demodulation, decoding, estimation of signal parameters, etc.) was performed digitally [65]. The new generation of digital devices was much more convenient and efficient than analog processing systems. They allowed to solve important practical problems that were previously difficult to implement in analog technology. The stability of parameters, accuracy of digital signal operations, ability to reconfigure the circuits, and unification of processing units have become indisputable advantages of digital devices over analog technology, determining the future direction of development of this industry.

The rapid development of telecommunication technologies, communication equipment, microelectronic technologies and devices led to emergence of a large number of types of protocols and equipment standards. Often, different manufacturers produced equipment that operates according to the same standards and protocols, but is not compatible with each other. As a rule, manufacturers with this approach pursued the goal of market monopolization. The corporate standard required branded equipment, and large companies actively promoted their standards in the global market. The introduction of digital technologies made it possible to appropriately reconfigure the existing receiving equipment, and the use of universal digital platforms eliminated the need to adapt the system to a specific manufacturer. Thus, universal modems appeared in the world market, in which a simple change in the firmware could ensure compatibility with one or another proprietary protocol or equipment. The modems provided not only equipment compatibility, but also performed a more complex function—adaptive correction of channel distortions (www.ru.wikipedia.org). Modems could adapt to different data transmission channels with different characteristics and quality of performance. The task of the modems was to ensure the guaranteed transfer of data over the maximum distance at maximum speeds with minimal energy consumption, while the infrastructure of data transmission networks was far from perfect. Copper twisted pair has remained the main type of wire connection since 1900. Without adaptive correction of distortions in the communication line, working at high speeds under the conditions of "copper" infrastructure would simply be unrealizable, and the setting up for the line could not be implemented without a reconfigurable receiving system.

The term software defined radio (SDR) was coined in 1984 by the development team of the E-Systems company in Garland, Texas (later the company became Raytheon). The first article was published about a digital receiver with implemented software-defined blocks for compensation of intersymbol interference and signal demodulation, with a digital filter with a thousand taps in the delay line. Such a receiver was implemented based on a massively parallel processor with shared memory [66].

Today, the term SDR means a software-defined system—a radio transmitter and/or a radio receiver, which allows using software to change the following system operating parameters:

- type of modulation;
- operating frequency range;
- width of the received/transmitted signal spectrum;
- output power.

The parameters set by the manufacturer according to the device architecture remain unchanged: ADC bit width, the limiting bands and modes of operation of the digital device, its presets.

In the simplest case, an SDR is composed of:

- Analog part, consisting of an antenna, amplifiers, and filters;
- high-speed ADC and digital-to-analog converter (DAC);
- digital signal processor (DSP) and/or computer.

See a diagram of such a device in Fig. 4.55.

The reception path is usually organized as follows: the signal from the output of the antenna-feeder path enters the amplifier, then is filtered in a low-frequency filter, then the signal is digitized on the ADC and goes to the signal processor for final processing.

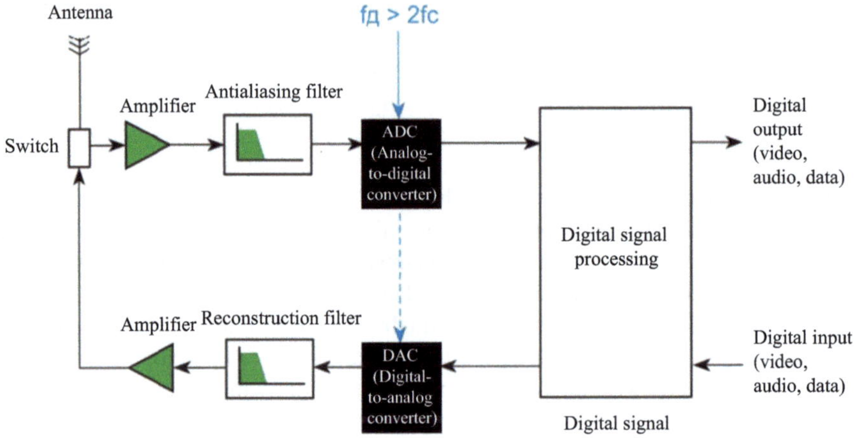

Fig. 4.55 Structure of a simplest SDR

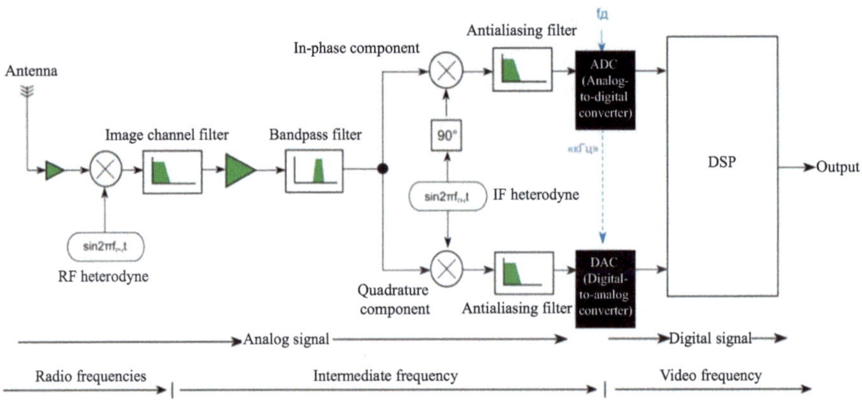

Fig. 4.56 Block diagram of the first generation of SDR receivers

The transmission path is organized as follows: The digital signal from the signal processor is fed to the DAC, where it is converted into an analog signal, which is processed on a recovery filter, after which the signal is amplified and fed through switches to the antenna-feeder path.

Consider the three evolution stages of SDR receivers [64].

The first generation of SDR receivers appeared in the early 1990s. Figure 4.56 presents its structural diagram. As can be seen from the figure, the analog part of the receiver lowers the frequency of the signal from radio frequency to intermediate frequency using a reference oscillator. Since the most optimal for DSP is signal processing at a video frequency—when the sampling rate, and, accordingly, the data flow rate to the signal processor, is the lowest possible, a quadrature signal processing circuit is used [64]. To do this, using the second reference oscillator

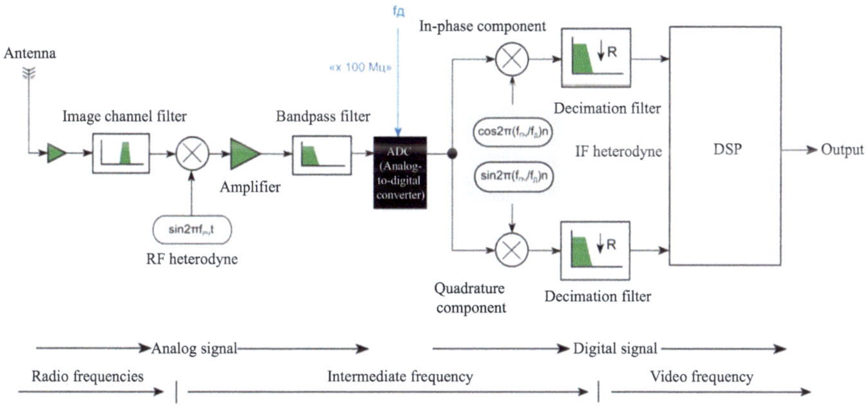

Fig. 4.57 Block diagram of the second generation of SDR receivers

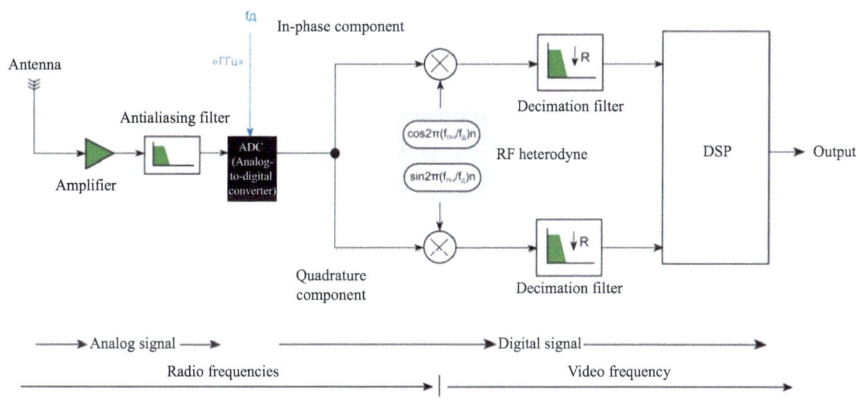

Fig. 4.58 Block diagram of the third generation of SDR receivers

and a phase shifter, the in-phase and quadrature components are selected, then the video signal is digitized in the ADC (the speed does not exceed tens of kHz) and is fed to further processing by the signal processor.

In the next generation, which appeared in the 2000s, the process of digitizing the signal began at the intermediate frequency. For example, a signal at an intermediate frequency of 40 MHz can be digitized by an ADC operating at 125 MHz. See a diagram of such a device in Fig. 4.57.

Here, the first stage of digital signal processing includes a DDC (digital down converter), which converts the signal to video frequency for further processing on the signal processor.

In this receiver architecture, compared with the first generation, the volume of conversions of the signal in digital form is larger.

In the last, third generation, the digitization of the signal came close to the antenna-feeder path (see Fig. 4.58).

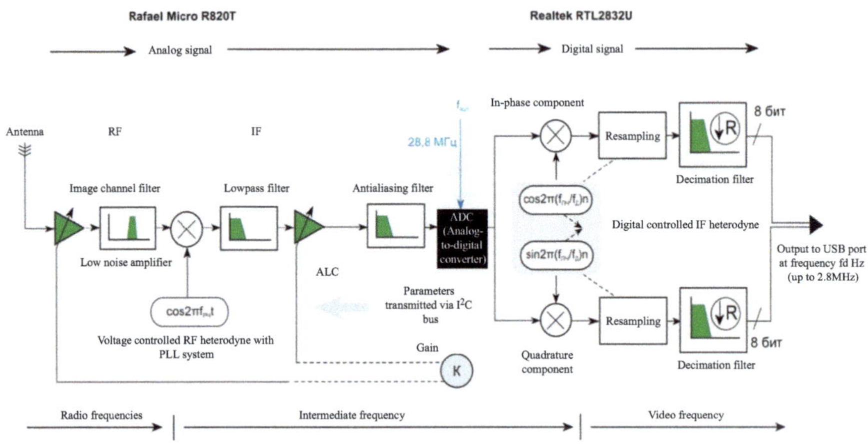

Fig. 4.59 Block diagram of an SDR receiver based on RTL2832U and R820T chips

As we can see, in this case there is no signal processing stage at an intermediate frequency, the drift occurs immediately—from radio frequency to video frequency.

It should be noted that such an option is difficult to implement in practice, since such high-speed ADCs are not very common due to their manufacturing complexity and high cost.

Consider the circuitry and operation of a specific SDR receiver based on Realtek RTL2832U and Rafael Micro R820T chips.

Initially, this device was the receiver of digital television format DVB-T. The RTL2832U chip was released by the manufacturer as a demodulator of COFDM signals used in digital television. But radio engineers discovered an undocumented mode of operation of the chip, in which the demodulation stage is turned off, and the chip turns, in fact, into an 8-bit ADC with quadrature output with a sampling frequency of 28.8 MHz. This, in fact, turned the DVB-T digital television receiver into a software-defined radio receiver.

Figure 4.59 shows a block diagram of the receiver under consideration.

The radio signal from the antenna-feeder path enters the R820T chip. Here it is transferred to the intermediate frequency with the help of a voltage controlled heterodyne, which is regulated by the RTL2832U chip through the I^2C bus. After passing the automatic gain control (AGC), the signal at an intermediate frequency should be converted to a video signal. The classical conversion method is to transmit a signal at an intermediate frequency through a low-pass filter, then digitize it in an ADC, then the digitized signal is transferred to the video frequency using a digital-controlled quadrature heterodyne. Thus, at the output of the receiver there are in-phase and quadrature components of the signal with a bit depth of 8 bits. The maximum possible signal bandwidth is limited by the characteristics of the R820T chip and is 2.8 MHz.

During an experiment, a classic antenna-feeder path was built, incorporating a TV channel of wave channel type, a GALAMP-102 TV amplifier, and an RG-58/U coaxial cable 30 m long (see Fig. 4.60) [64].

The transition from the classical analogue antenna-feeder path to the digital one should become a qualitative leap to increase the efficiency of receiver use. One of the most simple and affordable ways to implement the digital path is the use of active USB-8P8C converters (in Russia often referred to as RJ-45) and vice versa, and, accordingly, the use of twisted pair cable instead of the usual coaxial cable. One prominent advantage of this method is the immunity to analog interference due to the fact the signal is transmitted from the antenna to the receiver in digital form. It should be noted, however, that not all converters allow passage of a signal with 3.2 MHz band. The converter hardly provides transmission of a signal with a band of 1 MHz $\times$ 2 channels $\times$ 8 bits $=$ 16 Mbit/s, although the manufacturer of the equipment declares different values.

An even better path can be obtained by using USB interface-to-optical signal fiber converters. But this path variant will still be expensive to manufacture due to the high price of optical signal converters.

Another possible way to increase the reception efficiency is to build a coherent two-channel receiver based on two SDR receivers. In this case, for example, radio engineers simply solder the quartz from one receiver and then connect the freed contacts with the quartz of the second receiver by wire. This ensures coherence of reception. However, it should be noted that implementation of this variant would require, in addition to installing a second receiving antenna and organizing a second antenna-feeder path, the use of complex multichannel (two-channel) signal processing algorithms, which would entail development of its own software.

As a result, it can be said that a software-configured system is the most promising structure of a modern radio. It provides invariance to the type of communication equipment and types of protocols, implements complex signal processing algorithms for different data transmission channels. Such a system can be easily improved by updating the software, and the software can be easily transferred to a new, more modern hardware platform. Such unification and ease of transition to new equipment versions significantly reduces the operating budget of receiving/transmitting systems, expands the resource and capabilities of the already existing infrastructure. In the near future, universal software-configurable systems will define standards for cellular and wireless communications with the ability to work simultaneously in various communication networks using a single radio interface.

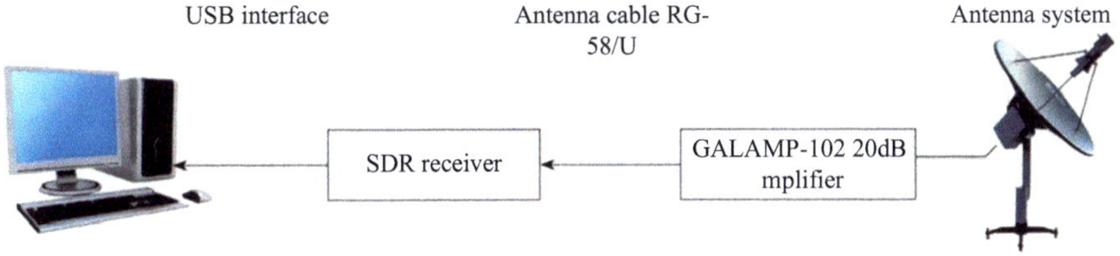

Fig. 4.60 Structure of a receive path using SDR

4.8 Cable Assemblies of the SHF Range

4.8.1 Purpose, Classification, and Features of the Use of Cable Assemblies

Cable assemblies are widely used in microwave devices for inter- and intra-block connections with fulfillment of modern electromagnetic compatibility requirements, when other methods of connection are technically and structurally impossible or impractical.

The term came to Russia from foreign technical literature, where it has long been used in the IEC (International Electrotechnical Commission) standard [68]. Most Russian experts in the field of electronics have a single concept of a cable assembly. However, since the term “cable assembly” is absent from the Russian ESKD classifier, the Russian design documentation usually uses the term “cable” permitted by the classifier (the classification characteristic code is 685671), which leads to a nonsense concept: cable made of cable. It is quite obvious that the need to introduce the term “cable assembly” into the ESKD classifier has long been overdue.

The main standards on this topic [69, 70] provide the following definitions of cable assemblies:

> Cable assembly is a combination of cable and connector (s) with additional protection and with or without marking, with established technical specifications.

The cable itself, used in radio frequency assemblies, is a rather complex composite structure containing internal and external metallic conductors, polymer or mineral insulation between them and a protective sheath. As an example, Fig. 4.61 shows the design of the PHOREEFLEX flexible RF cable from GORE (USA), used in measuring cable assemblies [70].

Flexible cable assembly is an assembly in which the cable may be subjected to multiple bends.

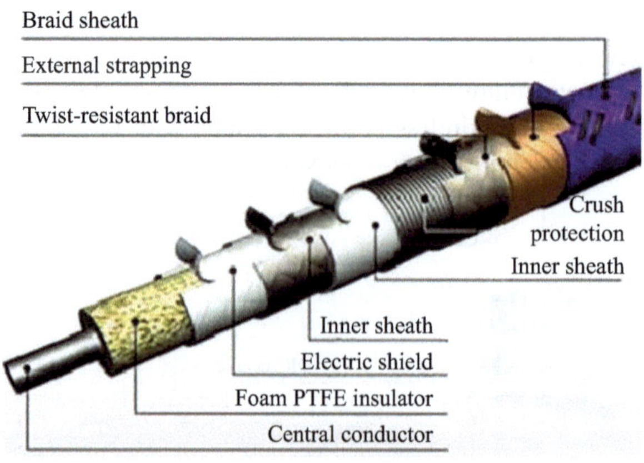

Fig. 4.61 The design of PHASEFLEX radio frequency cable

A semi-flexible cable assembly is an assembly that is not intended for multiple cable bends during operation, but for which bends or molding are permissible to facilitate installation. Exterior view of a Rosenberger semi-flexible cable assembly with a hand-molded cable and connectors is shown in Fig. 4.62a [71].

Semi-rigid cable assembly is an assembly not intended for bends or kinks after fabrication. Any bends or kinks during installation or operation may degrade its performance. Figure 4.62b shows a semi-rigid cable assembly with an SiO_2 dielectric (Meggitt Safety Systems, USA) [72], in which the distance between connectors can be changed due to elastic deformation of the cable.

Modern design and technological solutions of cables and connectors allow to create assemblies with lengths from several millimeters to several tens of meters with shielding efficiency up to 120 dB. Special attention is paid to minimizing changes in electrical and mechanical characteristics with external influencing factors, especially with repeated bends and temperature changes.

For modern aerospace engineering, a special class of cable assemblies has been developed with improved phase stability, with a range of operating frequencies up to 40 GHz and with the possibility of expanding it to even higher frequencies.

The radiation-resistant cable assemblies required by the nuclear power and space engineering enterprises deserve separate consideration.

The terminology, general requirements, and test methods of cable assemblies are set out in IEC Standard [68].

Manufacturers produce assemblies to customer specifications, including electrical parameters, resistance to mechanical and climatic influences, reliability during operation and after storage. When ordering a cable assembly, you should specify the cable brand, types of connectors, as well as the exact geometric length of the assembly. The definition of geometric length of a cable assembly in accordance with the IEC standard [69] is shown in Fig. 4.63.

The geometric length of a cable assembly is defined as the distance between reference planes of straight connectors and between the axes of corner connectors.

Depending on the design of insulation between the inner and outer conductors, coaxial cables are divided into two groups:

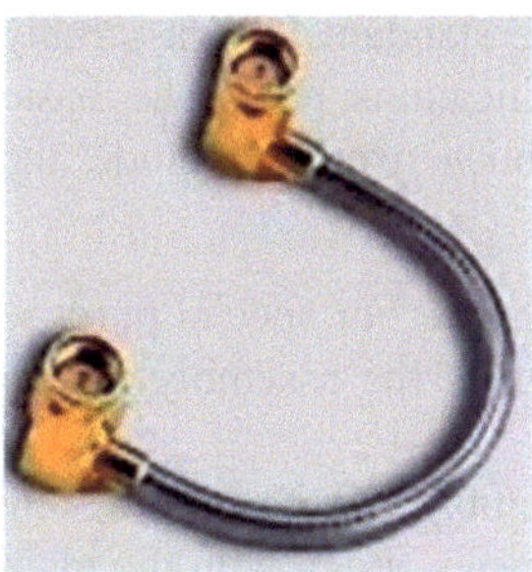

a)

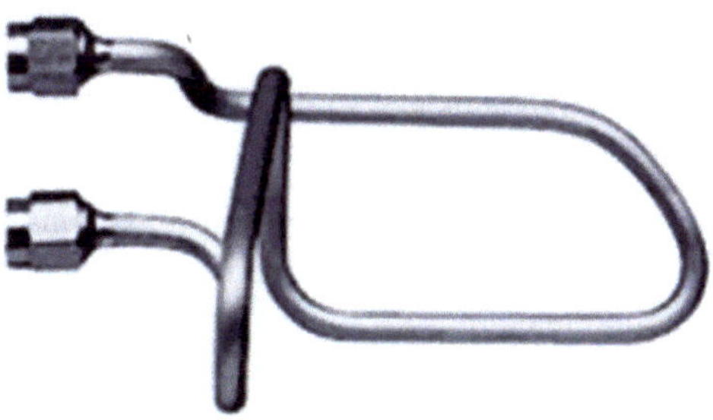

b)

Fig. 4.62 Appearance of cable assemblies: (**a**) semi-flexible cable assembly; (**b**) semi-rigid cable assembly

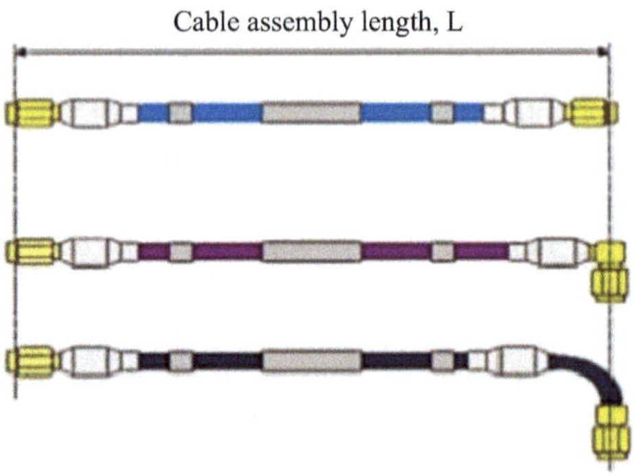

Fig. 4.63 Geometric length of various cable assemblies

1. Cables with solid insulation, in which the entire space between the inner and outer conductors is filled with insulating material. Such insulation may be uniform or consist of two or more layers with the same or different properties. Solid insulation can be obtained by extrusion of a molten dielectric, by plunger pressing a fine dielectric with its subsequent sintering using a patented technology, or by winding the inner conductor with thin strips of dielectric.
2. Cables with air-dielectric (semi-air) insulation, in which supporting insulation elements of various configurations at certain intervals or a cordell spiral are mounted on the inner conductor, and a tube of insulating material is placed on top of that. The most promising direction is filling the space between conductors with a porous dielectric obtained by physically foaming a dielectric during extrusion, plunger pressing an emulsion of a fine dielectric with sintering using a patented technology, using winding dielectric tapes of a porous structure.

4.8.2 Basic Specifications of Radio Frequency Cable Assemblies

The quality of coaxial cable assemblies is determined by a large number of parameters and test methods, which are governed by the IEC standard [68], assembly manufacturer specifications [71, 73], and general specifications [74]. The basic electrical parameters of radio frequency cable assemblies usually include the following parameters:

1. Nominal wave resistance and maximum deviations from it. For special purpose assemblies, uniform wave impedance along the length of the assembly is also important.
2. The range of operating frequencies. A HF cable assembly can be used at any frequency, starting from DC. However, at high frequencies, along with the main T (TEM)-wave, unwanted higher-order waves can be excited in a coaxial line. Therefore, the operating frequency range of assemblies is limited to the maximum operating frequency.

The theoretical maximum frequency (the critical frequency or cutoff frequency) is calculated using the formula [75]:

$$f_{cr} \cong 190.85/\sqrt{\varepsilon \times (D+d)}, \tag{4.2}$$

where f_{cr}—frequency, GHz; D and d are the inner diameter of the outer conductor and the outer diameter of the inner conductor of the coaxial line, mm; ε is the equivalent dielectric constant of the insulation material.

The maximum frequency of a cable assembly depends on the maximum frequencies of the cable and connectors and is limited to the lower limit of frequency of either of them. The limiting frequencies of cable connectors of different types, depending on the design and brand of the cable used, are given in [8].

3. The attenuation coefficient α at a given frequency at 20 °C and its change due to temperature. The attenuation coefficient of the assembly consists of the attenuation coefficients of the cable and each connector. These data are given in the specifications for cables and connectors. The temperature attenuation coefficient is calculated by the formula:

$$TCA = (\Delta\alpha/\alpha \times \Delta \mathrm{T}) \times 10^3 \left[\%/^{\circ}\mathrm{C}\right], \tag{4.3}$$

where $\Delta\alpha$ is the change in α when exposed to temperature, dB/m; ΔT is the temperature range, °C.

4. Voltage standing-wave ratio—VSWR. The VSWR of a cable assembly depends on the VSWR of the cable and connectors [75], the values of which are given to the specifications for these products.
5. Shielding attenuation. The shielding effectiveness of a cable assembly is determined by the design of the cable, but may also be limited by the design of the connector. To measure the attenuation of shielding, NPP "Spetskabel" uses a specialized measuring complex based on the triaxial line CoMeT (coupling measuring tube), jointly produced by Bedea and Rosenberger (Germany) Fig. 4.64 [71].

The lateral dimensions of the triax line and the software of this complex allow measurements at frequencies up to 3 GHz. The limitation of the upper limit of measurements is caused by the appearance of higher types of vibrations in the line.

6. Electrical length of cable assembly in electrical degrees. The radio frequency signal passing through the assembly can be represented as a rotating voltage vector, the angular velocity of which is 360 times the frequency:

$$V_{\text{ang}} = 360 \times f, ^{\circ}/c. \tag{4.4}$$

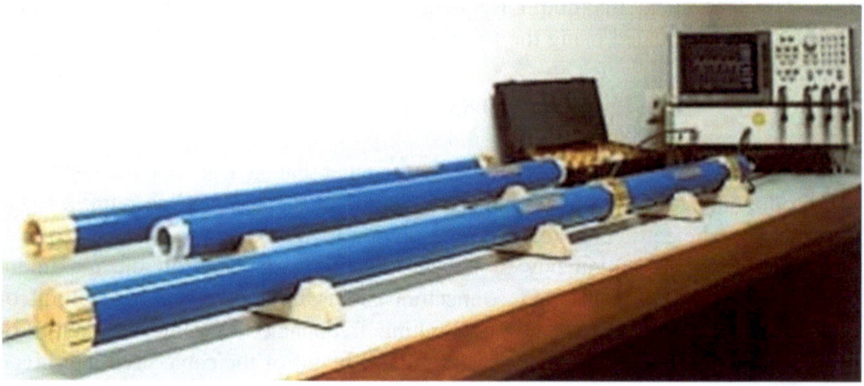

Fig. 4.64 Measuring complex for measuring attenuation of shielding of cables and cable assemblies

The signal propagates in the cable assembly during the delay time, which is calculated by the formula:

$$t_{\text{del}} = L \times \sqrt{\varepsilon/c} = L \times \xi/c, \tag{4.5}$$

where L is the geometric cable length, m; ε is the dielectric constant of the cable insulation; ζ is the coefficient of shortening the wavelength in the cable; $c \sim 3 \times 10 * 8$ is the speed of light in vacuum, m/s.

The product of the angular velocity of the voltage vector and the delay time corresponds to the electrical length of the cable assembly [75, 76]:

$$\psi = V_{\text{ang}} \times t_{\text{del}} = 360 \times f \times \left(L \times \sqrt{\varepsilon/c}\right).$$

For example, for cable assemblies with a length of 0.63 m with solid PTFE insulation (ε—2.04) and $t_{\text{del}} = 3$ ns, the electrical assembly length at a frequency $f = 9.5$ GHz is equal to:

$$\psi = 360 \times f \times t_{\text{del}} = 360 \times 9.5 \times 10^9 \times 3 \times 10^{-9} = 10260^{\circ}.$$

7. Signal propagation velocity in a cable assembly V_p.

The velocity of propagation V_p,% (VoP) shows how much the velocity of propagation of an electromagnetic wave in a cable assembly is based on the velocity of its propagation in a vacuum:

$$V_p = 1/\xi. \tag{4.6}$$

With an increase in V_p, a smaller amount of cable loss is achieved, since the value of ε approaches 1—the dielectric conductivity of air. In the best modern cord construction cables, the V_p value reaches 88%. Knowing the propagation velocity V_p, it is possible to calculate the value of the equivalent dielectric constant of the cable assembly ε and its electrical length λ.

As an example, consider a cable assembly with a geometric length $L = 1$ m, with an operating frequency $f = 18$ GHz and with a signal propagation velocity $V_p = 82\%$. The wavelength of the radio signal in the cable assembly is:

$$\lambda = c \times V_p/f = 3 \times 10^8 \times 0.82/18 \times 10^9 = 0.0137\text{м}.$$

The length of the cable assembly includes the wavelengths.

$$1/0.0137 \approx 73$$

Since a single wavelength corresponds to an electrical length of 360 electrical degrees, the electrical length of the cable assembly will be equal to (in °):

$$\psi = 73 \times 360 = 26280^{\circ}.$$

Assembly electrical length Ψ, expressed in electrical degrees, indicates how much the phase of the signal at the assembly output relative to the input signal has changed. The change in the F value depends on factors affecting the cable assembly, primarily on temperature and mechanical stress. For numerical evaluation of such changes, the concept of cable phase stability at a specific destabilizing effect is introduced. The temperature and phase characteristic is the most informative parameter in assessing the phase stability of a cable and, accordingly, of the cable assembly of this cable [76]. In Russian general technical conditions for radio frequency cables [74], the parameter "temperature coefficient of the phase" is introduced as a separate item.

8. Phase temperature factor (PTF) is the relative change in the electrical cable length λ with a temperature change of 1 °C. PTF is measured in the temperature ranges specified in the standards or technical specifications for cables of specific brands, and is calculated by the formula [74]:

$$PTF = \Delta\Psi \times \text{c}/360^{\circ} \times \xi \times L \times f \times \Delta T = \Delta\Psi \times 1 \times 10^6/\Psi \times \Delta T, \quad (4.7)$$

where $\Delta\psi$ is the relative change in the electrical cable length, °; $c \approx 3–10^8$ is the speed of light in vacuum, m/s; ζ is the wavelength shortening coefficient in the cable; L is the geometric cable length, m; f is the frequency, MHz; ΔT is the temperature range in which measurements are made, °C.

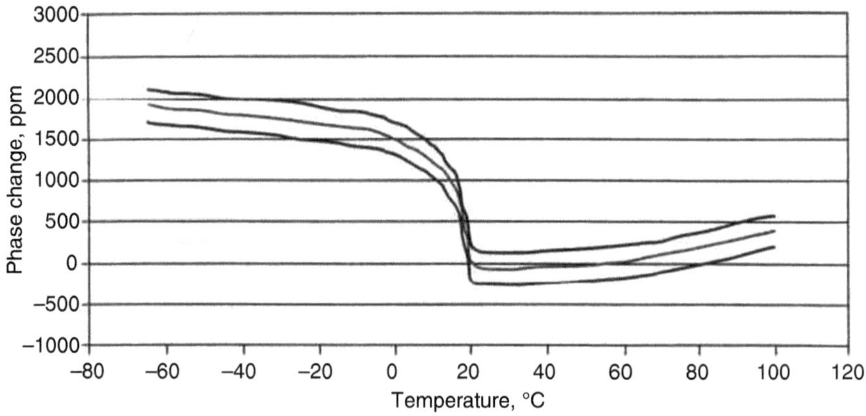

Fig. 4.65 PTF of a cable assembly with UFA210V cable (blue color indicates nominal phase change)

In foreign technical literature, PTF is expressed in relative dimensionless units ppm (parts per million). Figure 4.65 shows, as an example, the PTF characteristic of a cable assembly with a limiting frequency of 26.5 GHz manufactured by Micro-Coax, made of a UFA210B cable [73].

9. Operating temperature range. During operation, cable assemblies must be resistant to temperature changes. The maximum allowable temperature of the cable assembly is considered the maximum allowable temperature of its least heat-resistant element, which is established due to environmental heating and transmitted along the assembly by power. The operating temperature range of an assembly is always indicated in its specification.
10. The dependence of the phase change on bending of the cable. When bending, most European companies are guided by the recommendations of IEC standards (IEC) [68, 69]. The standard [68] recommends making 2 bends on the mandrel of a certain diameter of 180° clockwise and counterclockwise with straightening to the initial position after each bend. Other companies are guided by the recommendations of the MIL-C-17 standard. For example, Micro-Coax uses a single 360 cable bend (winding) around a cylindrical mandrel in accordance with this standard. The graph (Fig. 4.66) shows the limits of permissible phase change depending on the frequency for the UFA210B cable when it is wound around a mandrel with a diameter of 76.2 mm [6].

    Along with the above characteristic of the phase change from bending, most manufacturers of cable assemblies guarantee the preservation of electrical parameters of cable assemblies, in addition to phase changes, after other types of deformation as well. Thus, the following criteria are set in the specification for the Micro-Coax UFA210B cable [73]:

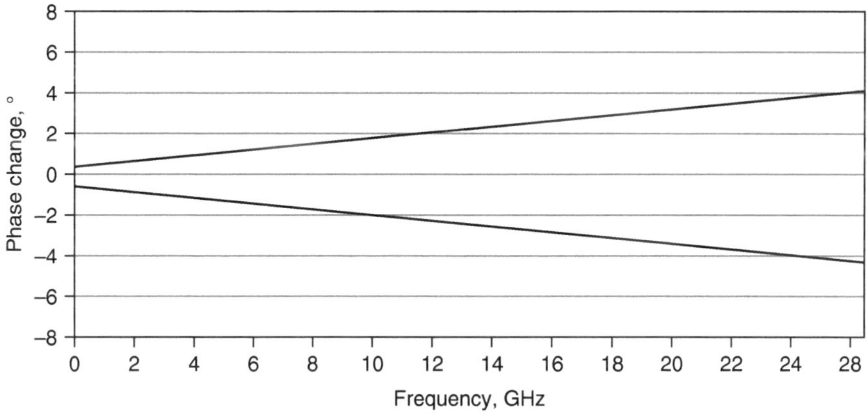

Fig. 4.66 Permissible limits of phase change when bending the UFA210B cable

- the minimum static bending radius of a cable with an outer diameter (5.33 ± 0.13) mm must be greater than 9.65 mm. After a single 360° winding of a cable around a mandrel with a diameter of 12.7 mm, the characteristic impedance of the cable should not change by more than 0.5 Ohm;
- the minimum dynamic bend radius of the cable is 50.8 mm. The cable must withstand 250,000 kinks with an angle of ±90° around a mandrel with a diameter of 101.6 mm (the so-called tick-tock-test). After this, the VSWR and attenuation values should not exceed the values guaranteed for this cable assembly;
- the cable must withstand 250,000 natural kinks (bending strength of the cable, "snake-test") with the same requirements for VSWR and attenuation.

11. The degree of cable assembly phase matching. This parameter is especially important for phase-identical cable assemblies. There is absolute and relative phase matching of cable assemblies [73]. Absolute matching entails matching of two or more assemblies to an absolute electrical length, including tolerance. Relative matching is the coincidence of two or more assemblies with respect to each other, and not with respect to the absolute electrical length. The minimum phase characteristic equivalence limit of above-mentioned cable assemblies made of UFA210B cable is ±0.5°/GHz [73]. The phase characteristic equivalence limit of other cable assembly manufacturers depends on the design of the cables used. The best result is ±0.4°/GHz.

In a number of special applications of cable assemblies, for example, for modules of phased antenna arrays and measuring systems, other additional parameters that characterize the degree of stability of the phase characteristics of cable assemblies are important.

12. Radiation resistance of cable assemblies.

    Radiation resistance is the ability of products and materials to maintain the original chemical composition, structure, and properties when exposed to ionizing radiation. It essentially depends on the type of radiation, the magnitude and power of the absorbed dose, the mode of exposure (continuous or pulsed, short-term or long-term), the size of the product and its specific surface, operating conditions (temperature, pressure, mechanical loads, magnetic and electric fields), and other factors. The quantitative characteristic of radiation resistance is the permissible absorbed radiation dose (Mrad, gray), at which the material becomes unsuitable for specific conditions of use or unacceptably changes certain characteristic parameter. For polymers, such parameters are most often the limits of strength and fluidity. For radio frequency cables, of equal importance is the change of dielectric properties of insulation when exposed to radiation.

It is necessary to draw the attention of developers using cable assemblies of various foreign companies to the differences in the methods of testing assemblies for the effect of destabilizing factors depending on the standard stated in the technical specifications for assembly.

RF microwave cable assemblies are manufactured by specialized companies from the USA, UK, France, Germany, Switzerland, South Korea, China, and Taiwan. The most promising technical solutions are offered by the following companies:

- Meggitt (www.stablecable.com), Times Microwave Systems, USA;
- AtlanTecRF (www.atlantechrf.com), UK;
- Radiall (www.radiall.com) and Axon (www.axon-cable.com), France;
- Rosenberger (www.rosenberger.de) and Teiegartner (www.telegaertner.com), Germany;
- Huber + Suhner (www.hubersuhner.com), Switzerland;
- Gigal.anc (www.gigalnne.com), South Korea.

The production of cable assemblies is a time-consuming process that imposes stringent requirements on the quality of cable and connectors and the assembly process itself; cable assemblies should be produced by qualified specialists in a specially equipped room. It is necessary to measure the electrical parameters of each assembly for compliance with technical requirements.

Often the question arises, whether to make the cable assembly at a specialized company or to make it yourself in order to save money. This is determined by the level of complexity of the equipment in which the assembly is used. In special-purpose equipment, especially that operating in the short-wavelength part of the centimeter and in the millimeter wavelength range, it is necessary to use assemblies that have passed the whole complex of tests. Before installation, many electrical parameters of assemblies should be measured: VSWR (S11 and S22), high-frequency losses (S12 or S21), permissible transmitted power, shield attenuation, capacitance, electrical resistance of conductors, insulation resistance, phase and phase temperature stability, intermodulation distortion. Assemblies must be tested for mechanical and climatic factors. In addition, it is necessary to check the permissible bending radius of the cable and the strength of its embedding into the connectors, the number of connections and disconnections of the assembly with mating connectors. For the manufacture of such assemblies, expensive installation equipment and tools are required, as well as highly qualified specialists.

In the case of the use of cable assemblies in less critical equipment, produced in small batches, independent manufacture of assemblies can be economically justified.

4.9 Basic Specifications and Design Solutions for Phase-Stable SHF Cables

This section is devoted to original design solutions for cables with high temperature and phase stability (http://www.spetskabel.ru/connect-special-1-18.htm) [63, 77–87].

Foreign companies are constantly working to improve single-complex design and manufacturing technology of phase-stable radio frequency cables, for which they widely use new materials and cable designs while simultaneously optimizing their manufacturing technology.

The main areas of work in the creation of phase-stable radio frequency cables are (http://www.spetskabel.ru/connect-special-1-18.htm) [63, 77–79]:

- Development of cable core design based on new, non-traditional insulation materials with minimal structural changes during temperature transitions. An ideal dielectric is air with a dielectric constant $\varepsilon = 1$; however, coaxial mounting of the inner conductor of the cable is required. The obvious solution is a semi-air insulation design of the core with the minimum equivalent values of ε and the dielectric loss tangent tg δ;
- Development of design and use of new conductor materials (primarily internal conductor) with minimal changes in linear dimensions with temperature.

4.9.1 Temp: Flex Radio Frequency Cables by Molex

To ensure strict requirements for phase stability of cables for aerospace engineering, as well as for precision measuring and medical equipment, Molex, USA (www.molex.com) has developed a series of flexible Temp-Flex RF cables and began their production starting 2013. Some data on Molex cable assemblies of 6 types are given in [63]; however, it does not reflect one of the main advantages of the Temp-Flex family cables—the absence of the undesirable "Teflon knee" effect in the temperature and phase characteristics.

In all Temp-Flex cables, high-temperature high-purity FEP (fluorinated ethylene propylene, Russian analogue—F-4 MB), and PFA (PFA-perfluoroalkoxy, Russian analogue—F-50) [78, 79], dielectrics are used, which have no "Teflon knee" in the temperature-phase characteristics.

The design of commercially available Temp-Flex Low Loss Microwave Coax family cables is shown in Fig. 4.67, and their main parameters are given in Table 4.4.

Temp-Flex Low Loss Microwave Coax cables of all models have an impedance of (50 ± 1) Ohms, a signal propagation velocity $V_p = 70\%$, a signal delay time $t_{\mathrm{del}} = 4.76$ ns/m, an electrical capacitance $C = 95$ pF/m, and an operating temperature range $\Delta T_{\mathrm{op}} = (-65$ to $150)$ °C.

In addition, miniature Temp-Flex Low Loss Microwave Coax 034SC-3201, 032SC-3301, and 020SC-3701 cables were developed. The core conductors of the 034SC-3201 and 020SC-3701 cables with a diameter of 0.20 and 0.11 mm are made of steel plated with copper, cladded with silver (SPCcS—silver plated copper clad steel), with a diameter of 0.18 mm, cores of 032SC-3301 cables are made of SPC. The cable insulator and sheath are made of FEP. The diameters of the insulator are, respectively, 0.66, 0.58, and 0.37 mm, shells—1.1, 1.0, and 0.71 mm. The 034SC-3201 and 020SC-3701 cables have a characteristic impedance of (50 ± 2) Ohms, 032SC-3301—(50 ± 1) Ohms.

Figure 4.68 shows the dependence of phase change during bending and due to temperature for cables 141SC-1901, 086SC-2401, and 047SC-2901.

A distinctive feature of the 047SC-2901 cable is a record-low bending phase change—Fig. 4.68c. Thinner cables appear to have similar characteristics of phase change during bending.

Comparative characteristics of the 141SC-1901 cable and similar cable MULTIFLEX_141 by HUBER + SUHNER (Switzerland) are given in Table 4.5.

All other parameters being equal, the main advantage of the 141SC-1901 cable is its almost linear temperature and phase characteristic (Fig. 4.68g). For the MULTIFLEX_141 cable, phase in the working temperature range changes more non-linearly, mainly due to the "Teflon knee" of the PTFE core insulator [80]. The insignificant differences in the attenuation coefficient, α, are apparently caused by the difference in the tan δ values of the solid core dielectrics.

To reduce α, Molex has developed a series of cables with an insulator (item 2 in Fig. 4.67) made of porous FEP. In terms of data transfer rate and other technological parameters, this series has intermediate characteristics between the Temp-Flex Low Loss and Temp-Flex Ultra Low Loss series.

Molex has carried out work to replace continuous FEP core insulator with PFA. Parameters of cables with insulator from solid PFA are given in Table 4.6.

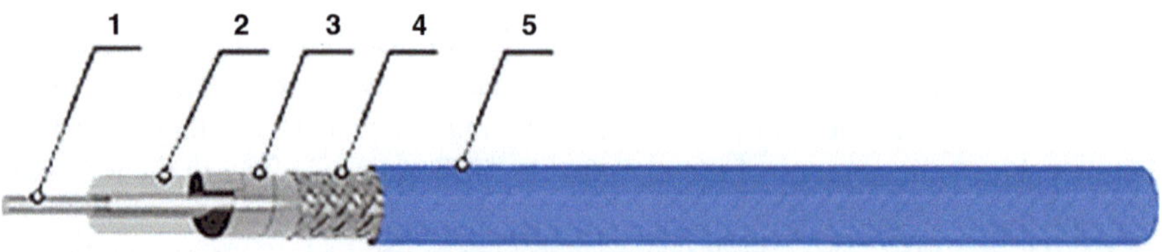

Fig. 4.67 Temp-Flex Low Loss Microwave Coax Cable Design: 1—inner conductor, SPC; 2—solid insulator of high-purity FEP; 3—internal tape shield, SPC; 4—external shield in the form of wire braid, SPC; 5—FEP sheath

Table 4.4 Temp-flex low loss microwave coax series cable parameters

		Diameter, mm; material							
No.	Cable model	Inner conductor	Insulator; FEP	Sheath; FEP	f_{max}, GHz	α, dB/m at $f = 18$ GHz	α, dB/m at $f = 40$ GHz	$\Delta\Psi$, ° at bending (f, GHz)	$\Delta\Psi$, ppm (ΔT, °C)
1	141SC-1901	0.91; SPC	2.97	3.99	35	2.13	–	Figure 4.68a	Figure 4.68d
2	086SC-2401	0.51; SPC	1.65	2.57	62	3.31	5.38	Figure 4.68b	Figure 4.68e
3	063SC-2701	0.36; SPC	1.17	1.90	–	–	–	–	–
4	047SC-2901	0.29; SPC	0.92	1.42	112	5.51	8.56	Figure 4.68c	Figure 4.68f

Note: *SPC* Silver plated copper, f_{max} theoretical maximum cable frequency, α attenuation coefficient, $\Delta\Psi$ phase change

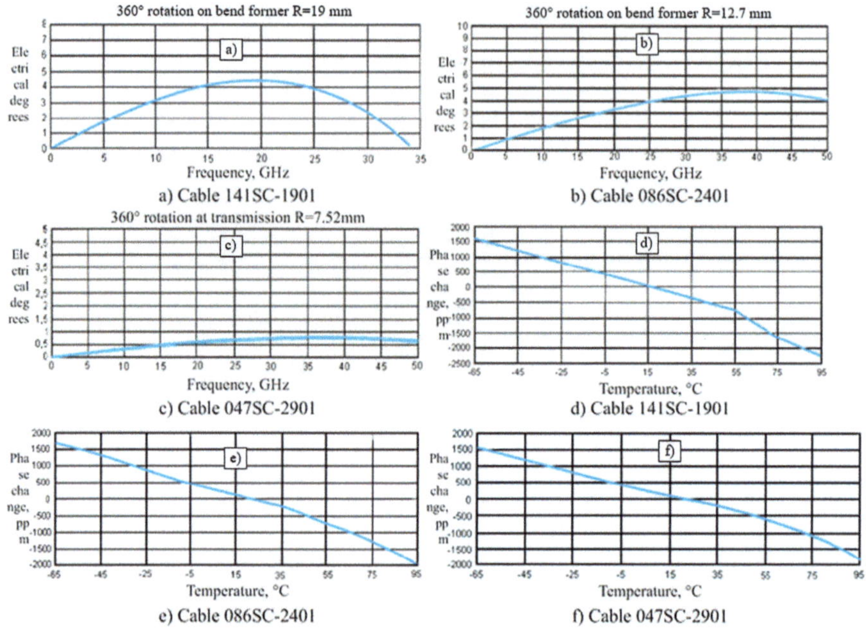

Fig. 4.68 The phase change dependence due to bending (**a–c**) and due to temperature (**d–f**) for cables 141SC-1901, 086SC-2401, 047SC-2901

Although the parameters of these cables given in Table 4.6 do not differ from the parameters of similar cables 086SC-2401 and 047SC-2901 (Table 4.4), the advantages of cables with an insulator PFA are obvious: higher permissible core heating temperature (250 °C) and dielectric strength (35–40) kV/mm [78, 79].

An original technical solution of Molex is the creation of a series of Temp-Flex Ultra Low Loss Microwave Coax cables, the design of which is shown in Fig. 4.69 (US patent 4.812.401 B2, 2004 "Ultra-small high speed coaxial cable with dual filament insulator").

Two strands of high-purity PFA twisted together are wound in a spiral around the silver-plated copper (SPC) inner conductor. By changing the density of spiral winding of the twisted PFA strands, it is possible to adjust the dielectric constant of the core insulation and, consequently, the transmission parameters of the cable.

A transparent tube of high-purity FEP is placed on top of the spiral. In the patent US 7.795.536 B2, 2010 "Ultra high-speed coaxial cable," a high-purity FEP dielectric Daikin Neoflon® NP-1101 with a dielectric loss tangent of less than 5–10^{-4} at a frequency of 2, 45 GHz is recommended. Compared with conventional FEP, this provides a significant reduction in cable losses. The tube is coiled with an internal shield of silver-plated copper tape, over which a braid with a density of more than 90% of silver-plated copper wire with a diameter of 0.08 mm is wrapped. The cable is protected by an ordinary purity FEP sheath.

Table 4.5 Specifications of 141SC-1901 and MULTIFLEX_141 cables

Cable brand, company	Diameter, mm; material			t_{del} ns/m	V_p, %	C, pF/m	f_{max}, GHz	α, dB/m at $f = 18$ GHz	α, dB/m at $f = 33$ GHz	ΔTop, °C	$\Delta\Psi$, ppm (ΔT, °C)
	Inner conductor	Insulator	Sheath; FEP								
141SC-1901,MOLEX	0.91; SPC	2.97; FEP	3.99	4.76	70	95	35	2.13	3.14	−65 to 150	3750 (−60 to 95)
MULTIFLEX_141, HUBER+SUHNER	0.92; SPC	2.93; PTFE	4.14	4.70	70.6	95	33	2.09	3.06	−65 to 165	4200 (−60 to 95)

Table 4.6 Parameters of Temp-Flex cables with a solid PFA insulator

No.	Cable model	Diameter, mm; material			t del, ns/m	V_p, %	C, pF/m	ΔTop, °C
		Inner conductor	Insulator, PFA	Sheath; FEP				
1	100067-1086	0.51, SPC	1.65	2.57	4.76	70	95	−65 to 150
2	100067-1047	0.29, SPC	0.92	1.42				

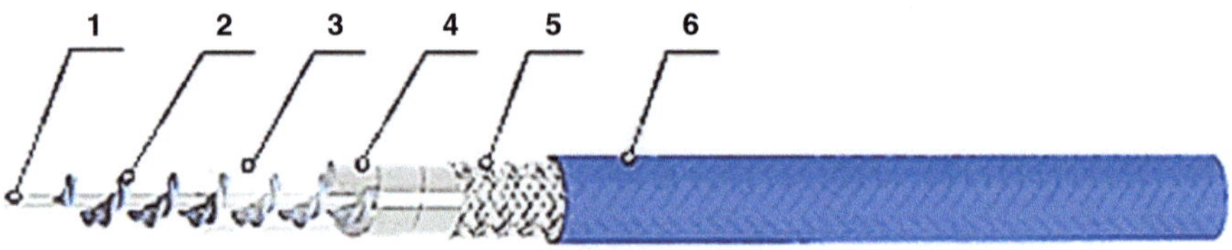

Fig. 4.69 Temp-Flex Ultra Low Loss Microwave Coax cable design: 1—internal conductor, SPC; 2—two twisted PFA threads; 3—FEP tube; 4—SPC tape inner shield; 5—external shield in the form of an SPC wire braid; 6—FEP sheath

The parameters of the Temp-Flex Ultra Low Loss Microwave Coax series cables are given in Table 4.7, and the dependencies of the phase change on bending and on temperature are shown in Fig. 4.70.

Along with cables with single-wire internal conductors, Molex also produces 50HCX24 Ultra Low Loss cables with a multi-wire internal conductor.

The use of FEP instead of PTFE in combination with the patented design of cordless-tubular insulation with a high air content made it possible to virtually eliminate the effect of the "Teflon knee." However, in the temperature and phase characteristics of the cables, a kink is observed at temperatures (35–55) °C, which is probably due to the peculiarities of the cord construction of the core.

Continuing to improve phase-stable cables, Molex developed cables similar in design to the Ultra Low Loss series cables, in which both elements of the insulator—the cord helix and the tube holding it—are made of PFA. The parameters of these cables are given in Table 4.8.

Comparative characteristics of cable 141-1701 and similar cable UFB142A by Micro-Coax (USA) with an insulator made of porous PTFE are given in Table 4.9.

The data in Table 4.9 show that the Ultra Low Loss type 141-1701 cable has a lower attenuation coefficient and a more monotonous temperature-phase response.

Molex also developed high-flexibility subminiature cables through using a multi-wire internal conductor. The inner conductor consists of 7 twisted copper wires coated with silver, the insulator and the sheath are made of FEP, the wire braid is made of copper wires coated with silver. The parameters of these cables are given in Table 4.10.

It should be noted that the use of Molex subminiature cables for interconnects in modern microwave equipment with a high density of layout, including connections with mobile units of equipment—antennas of transceiver–transmitting devices and medical equipment sensors, is promising.

Table 4.7 Parameters of Temp-Flex Ultra Low Loss Microwave Coax series cables

No.	Cable model	Diameter, mm; material			t_{del} ns/m	V_p, %	C, pF/m	f_{max}, GHz	α, dB /m at $f = 18$ GHz	α, dB /m at $f = 40$ GHz	ΔTop, °C	$\Delta\Psi$,° at bending (f, GHz)	$\Delta\Psi$, ppm (ΔT, °C)
		Internal conductor	Core; PFA + FEP	Sheath; FEP									
1	141-1701	1.15	2.97	4.01	3.82	87	76.4	41	1.41	2.46	−65 to 0.150	Figure 4.70a	Figure 4.70e
2	086-2201	0.64	1.66	2.57	3.85	86	77.5	72	2.68	4.25		Figure 4.70b	Figure 4.70f
3	063-2501	0.46	1.16	1.90	3.82	87	75.5	105	3.35	5.12		Figure 4.70c	Figure 4.70g
4	047-2801	0.32	0.82	1.42	3.92	87	76.4	143	4.69	7.22		Figure 4.70d	Figure 4.70h

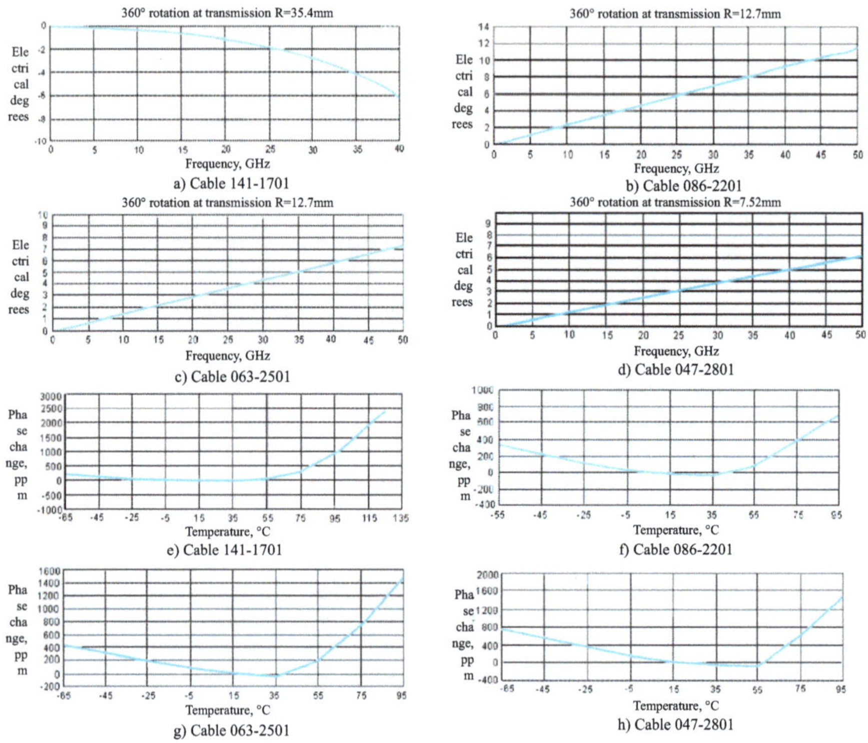

Fig. 4.70 Dependencies of phase change during bending (**a–d**) and due to temperature (**e–h**) of Temp-Flex Ultra Low Loss Microwave Coax series cables: (**a**, **e**) 141-1701; (**b**, **f**) 086-2201; (**c**, **g**) 063-2501; (**d**, **h**) 047-2801

Noteworthy is the statement from Molex that all Temp-Flex cables and cable assemblies using them can be made of nonmagnetic materials, which is important, for example, to reduce intermodulation distortion.

4.9.2 Radio Frequency Cables with Silicon Dioxide Insulation

The search for insulation material for radio frequency cables of special use led to an inorganic dielectric SiO_2 (silicon dioxide) of high purity. In 1957, a unique coaxial cable with an SiO_2 insulator was developed and patented in the USA, and in 1999, Meggitt Safety System USA [81] became the first manufacturer of serial coaxial SiO_2 cables and cable assemblies for most critical applications.

In addition, currently such cables and assemblies are manufactured by USA companies Times Microwave Systems [82] and Measure Tech [83], as well as the French company Thermocoax [84].

Table 4.8 Parameters of Temp-Flex cables with PFA insulation

No.	Cable model	Diameter, mm; material			t_{del} ns/m	V_p, %	C, pF/m	α, dB/m at $f = 18$ GHz	α, dB/m at $f = 40$ GHz
		Inner conductor; SPC	Insulator, PFA	Sheath; FEP					
1	100054-0008	1.15	2.97	4.0	3.82	87	76.4	1.48	2.43
2	100054-0007	0.32	0.82	1.42	3.82		75.5	4.69	7.22

Table 4.9 Parameters of 141-1701 and UFB142A cables

Cable brand, company	Diameter, mm; material			t_{del} ns/m	V_p, %	C, pF/m	f_{max}, GHz	α, dB/m at $f = 18$ GHz	α, dB/m at $f = 40$ GHz	ΔTop, °C	$\Delta\Psi$, ppm (ΔT, °C)
	Inner conductor	Core	Sheath, FEP								
141-1701, Molex	1.15, SPC	2.97, PFA + FEP	4.01	3.822	87	76.4	41	1.41	2.46	−65 to 150	800 (−60 to 95)
UFB142A, Micro-Coax	1.023, SPC	2.74, por. PTFE	3.607	4.08	83	80.4	40	1.74	2.64	−65 to 100	1000 (−60 to 95)

Table 4.10 Parameters of high-flexibility Temp-Flex cables

No.	Cable model	Diameter, mm: Inner conductor	Core	Wire braid; density	Sheath	Characteristic impedance, Ohm	*C*, pF/m	V_p, %
1	50CX-42	0.405	1.35	0.102; >95%	2.54	50 ± 2	98.4	70
2	50CX-41	0.255	0.84	0.079; >95%	1.8		98.4	
3	50CX-11	0.202	0.67	0.051; >90%	1.12		95.12	

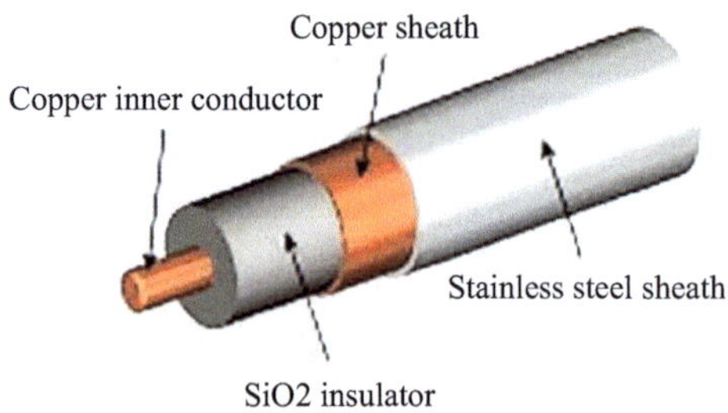

Fig. 4.71 Design of a cable with SiO_2 dielectric

Cables with an insulator made of ultra-pure SiO_2 are distinguished by extremely high values of operating temperature (from −270 to 1000 °C) and radiation resistance (more than 200 Mrad), and also have a close to linear temperature-phase characteristic. High reliability, electrical stability, and rigidity of the Meggitt Safety System's SiO_2 cable design provide an average time to failure of one million hours. This semi-rigid cable is 30% lighter than an equivalent cable with a PTFE insulator. Its bending radius is 1.5 times less than that of flexible cables of the same diameter.

SiO_2 coaxial cables are used in precision delay lines, in phased antenna arrays, in high-energy engineering and nuclear power engineering, in cryogenic installations, in measuring equipment, and also in space and military systems. The design of a cable with SiO_2 dielectric is shown in Fig. 4.71 [81].

Oxygen-free copper was used for the manufacture of the inner conductor of the cable, the outer conductor (sheath) is a stainless steel tube, clad inside with copper, the dielectric is ultra-pure silicon dioxide, SiO_2 (99.985%). Fine silica is mixed with a controlled amount of deionized water to form an extrusion paste mixture, which is applied to the inner conductor using plunger pressing. After air drying, the resulting core is inserted into a steel-copper sheath and heated in a vacuum furnace to remove residual contaminants. Then, the resulting core is compacted by compressing the outer sheath to a diameter that provides an impedance of 50 Ohms and the required dielectric strength. Directly after the manufacture of the cable of a required length, sealed connectors are installed on it using laser welding, resulting in an air-tight cable assembly. To identify the assembly, the stainless steel sheath is marked by ctching.

The effective dielectric constant of the formed semi-air dielectric with SiO_2 microparticles is 1.56–1.60. For comparison, the dielectric constant of an amorphous quartz glass is 3.75–3.80.

Cable assemblies from silicon dioxide cables have a number of special features in comparison with cable assemblies of other types:

Record wide operating temperature range. A cable with SiO_2 dielectric can operate in the temperature range (−270 to 1000) °C. However, in cable assemblies with this cable, the limiting heating temperature is limited by the permissible operating temperature of serial connectors (−273 to 200) °C. High-temperature connectors with a range of operating temperatures of (−270 to 600) °C have also been developed, they use an insulator made of ceramic composite glass in combination with cases made of Inconel nickel alloys 625, X750, or 718.

The need to use metal-glass sealed connectors for cable assemblies. The tightness of an assembly, determined by the flow rate of helium, is $1.3 \cdot 10^{-9}$ m^3 Pa/s. The connector is attached to the cable by laser welding—Fig. 4.72.

The most important advantage of a cable with an SiO_2 dielectric is a significantly higher temperature and phase stability compared to cables with a dielectric made of porous PTFE—Fig. 4.73 [81].

The disadvantage of cables with SiO_2 dielectric is a greater attenuation coefficient than that of cables with porous PTFE.

The characteristics of Meggitt Safety Systems cables with SiO_2 dielectric are given in Table 4.11 and Fig. 4.74. All cables have a characteristic impedance of 50 Ohms. The effective dielectric constant of the cable insulation is 1.56, the insulation resistance is $3.3 \cdot 10^{14}$ Ohm m at 20 °C and $3.3 \cdot 10^7$ Ohm m at 760 °C, the attenuation of shielding is at least 120 dB.

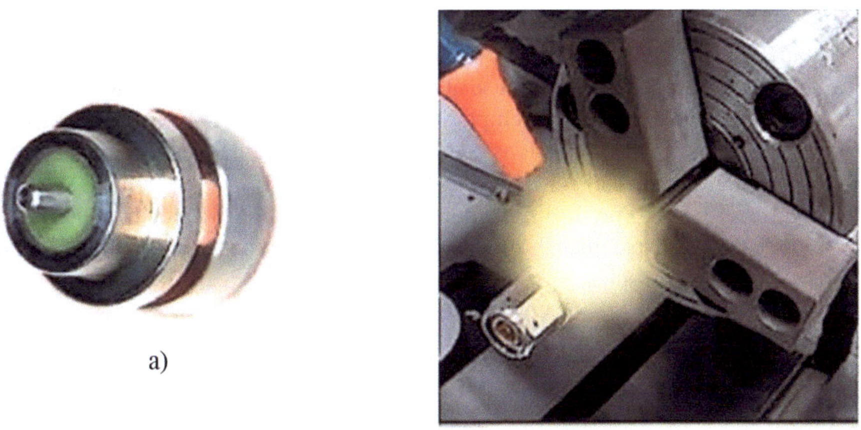

a) b)

Fig. 4.72 Sealed SMA connector (**a**), laser-welded connector installation on the cable (**b**)

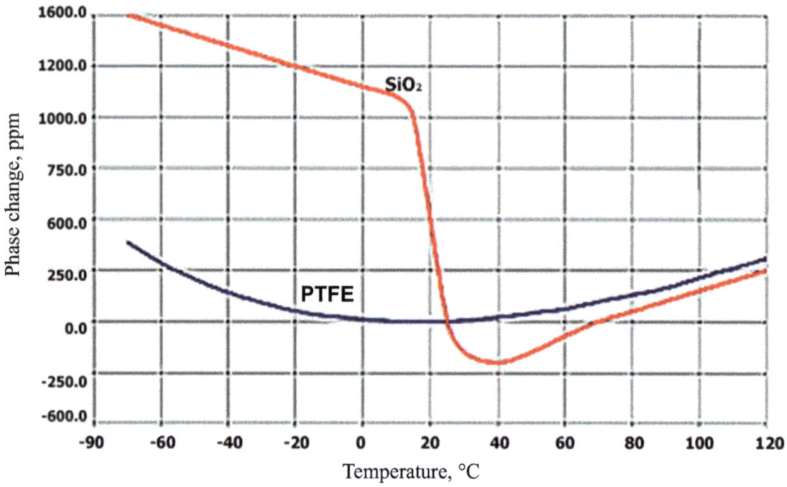

Fig. 4.73 Phase temperature dependence in cables with PTFE and SiO_2 dielectrics

Cables are designed for use in assemblies with SMP connectors, 2.4 mm, 3.5 mm, SMA, TNC, N, HN, SSMA, SC, and other types.

Times Microwave Systems manufactures four types of cables with SiO_2 dielectric with guaranteed attenuation of shielding (−110 to 120) dB. Cable assemblies with standard and high-temperature connectors are available. Characteristics of the cables are given in Table 4.12. The attenuation coefficient as a function of frequency is the same as in Fig. 4.74.

Thermocoax has developed a series of SiO_2 insulated signal transmission cables for operation at temperatures up to 1000 °C in aggressive environments [84]. The cables comply with MIL-T81492 standard. Specifications of the company's cables are given in Table 4.13.

Specifications of SiO_2 insulator cables of the leading companies: Meggitt Safety Systems, Times Microwave Systems, and Thermocoax are close. However, the first two companies are focused on the use of cables in APAAs and measuring equipment and therefore strive to ensure high phase stability of cables. Thermocoax develops cables for high-energy applications and therefore focuses on the ability of cables to operate at high temperatures.

Measure Tech [83] produces one brand of cable—SiO_2-0.142 (outer diameter of 3.6 mm), the main parameters of which coincide with the parameters of similar cables of other companies.

4.9.3 Phase Track Radio Frequency Cables

The company Times Microwave Systems (USA) has achieved an improvement in temperature and phase stability using original fluorocarbon porous dielectrics TF4

Table 4.11 Meggitt Safety Systems xCable specifications

No.	Cable model	Diameter, mm		Minimum bend radius, mm	t_{del} ns/m	V_p, %	C, pF/m	f_{max}, GHz	$\Delta\Psi$, ppm (T, °C)	α, dB/ft. (f, GHz)	Throughput power, kW	Breakdown voltage, kV	Weight, g/m
		Inner conductor	Sheath										
1	0.090	0.56	2.29	7.62	4.13	80	82	71.4	Figure 4.73	Figure 4.74	0.4	0.4	7.71
2	0.125	0.85	3.18	12.7				60.7			1.0	1.0	10.83
3	0.142	0.99	3.61	12.7				40.7			1.6	1.46	15.58
4	0.200	1.53	5.08	15.24				26.1			2.6	2.6	21.83
5	0.275	2.21	6.99	17.8				18.1			3.4	3.4	35.38
6	0.296	2.44	7.52	17.8				16.6			3.4	3.4	40.82
7	0.532	4.5	13.51	38.1				8.9			7.0	7.0	95.26

Note: The company has developed even more miniature cables of types 0.047 and 0.069 with an outer diameter of 1.19 and 1.75 mm, respectively

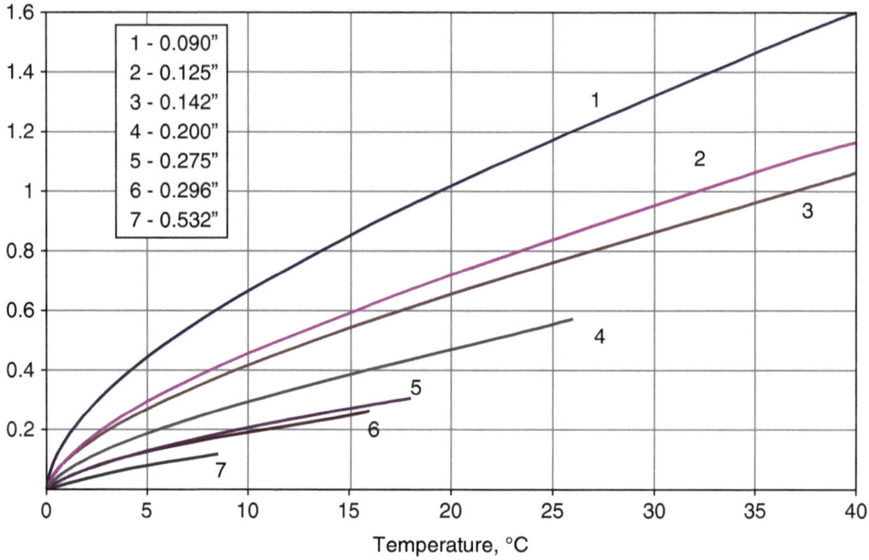

Fig. 4.74 Dependence of Meggitt Safety Systems cables attenuation coefficient on frequency at 20 °C (1 dB /ft. = 3.281 dB/m)

and TF5 of its own design in radio frequency cables. This made it possible to completely eliminate the effect of "Teflon knee" inherent in PTFE cables. In 2004, PT210 and PF402 cables were created, in which a TF4 dielectric was first used.

Based on the so-called TF4 technology, Times Microwave Systems has created several series of phase-stable RF cables.

Phase Track—flexible phase-stable cables. The design of the cables is shown in Fig. 4.75. The main parameters of the cables are given in Table 4.14 and in Fig. 4.74.

PhaseTrackSR—semi-rigid phase-stable cables (Figure 4.76). The design of the cables is shown in Fig. 4.77, their main parameters are given in Table 4.15 and in Fig. 4.78.

PhaseTrack PFlex—flexible, phase-stable cables. The design of the cables is shown in Fig. 4.79. The main parameters of the cables are given in Table 4.16 and in Fig. 4.80.

PhaseTrack LS is a flame-retardant, low-smoke (LS) flexible cable that uses an original porous dielectric TF5.

The design of the PhaseTrack LS cables is shown in Fig. 4.81. To reduce the weight, the inner conductor is made of copper-clad aluminum. The main parameters of the PhaseTrack LS cables are given in Table 4.17 and in Fig. 4.82.

PhaseTrack cables are far superior in temperature and phase stability to cables with porous PTFE.

Table 4.12 Times Microwave Systems cable specifications

No.	Cable model	Sheath diameter, mm	Minimum bend radius, mm	t_{del} ns/m	V_p, %	C, pF/m	f_{max}, GHz	$\Delta\Psi$, ppm (T, °C)	Breakdown voltage, KV
1	SiO_2–0.090	2.29	7.6	4.1	80	73.2	65	600 (−60 to 100 °C)	0.9
2	SiO_2–0.141	3.58	11.4				38		1.6
3	SiO_2–0.200	5.08	17.8				25		2.4
4	SiO_2–0.270	7.11	25.4				19		3.3

Table 4.13 Thermocoax cable specifications

Cable model	1CCAc10Si	1CCAc15Si	1CCAc20Si	1CCAc30Si	1CCAc36Si	1CCAc40Si
Construction materials and dimensions, mm						
Inner conductor	Copper					
Diameter	0.18	0.25	0.31	0.44	0.6	0.8
External conductor	Copper					
Sheath	Stainless steel 304 L or Inconel 600 (XH78T, XH60BT)					
Diameter	1	1.5	2	3	3.6	4
Electrical parameters						
Characteristic impedance, Ohm	50					
f_{max}, GHz	20					
Insulation resistance, Ohm·m, (at temperature)	$\geq 10^{13}$ (20 °C)				$\geq 10^{12}$ (20 °C)	
	$\geq 10^{7}$ (350 °C), $\geq 10^{3}$ (600 °C)					
C, pF/m at 20 °C	100				120	
α, dB/m at $f = 100$ MHz	0.08					
Breakdown voltage, KV	0.3	0.5	0.8	1	1.5	1.5
Weight, g/m	15	40	70	170	230	290
Length, m	10–80	10–60			20–60	Up to 50
Performance parameters						
Operating temperature range, °C	−273 to 1000					
Vibration	20 g, 2000 Hz					
Acceleration, m / s2	50 g					
Impact	7000 g, 10 kHz					

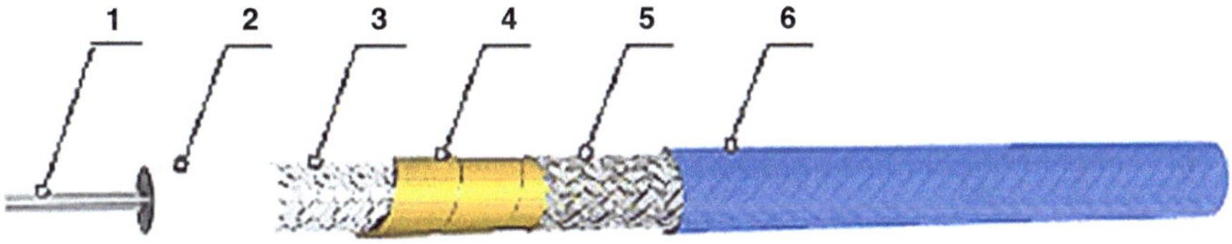

Fig. 4.75 Phase Track cable design: 1—copper or copper-clad steel silver-plated inner conductor; 2—porous dielectric TF4; 3—silver copper ribbon shield; 4—metallized polyamide film; 5—braid of silver-coated copper wires; 6—FEP sheath

4.9.4 Huber + Suhner SucoPearl Radio Frequency Cables

To reduce the "Teflon knee" effect, Huber + Suhner, Switzerland (www.hubersuhner.com), in collaboration with the European Space Agency (ESA), developed the original SucoPearl PTFE dielectric cable [85]. The designed cables have the following design features:

- The cable insulation is made in the form of a set of hollow PTFE "pearls" with a spherical surface, strung close to each other on the inner conductor, like pearls in a necklace;
- To fix the relative position of the "pearls," a tube made of high-purity FEP is used. This creates a core with a high air content, which, in combination with the use of a heat shrinkable tube made of solar cells, is designed to reduce the undesirable "Teflon knee" effect;
- To reduce the length change of the inner conductor due to temperature, it is made of Invar alloy (iron–nickel alloy, Russian analogue is 36H alloy) with a silver coating having a very low thermal expansion coefficient of -1.5 to 10^{-6} 1/°C in the temperature range (-80 to 100) °C;
- The outer conductor is made of a spiral winding with silver-plated copper tape, over which a braid of silver-plated aluminum wire is applied;
- The sheath is made of FEP.

The design of a cable with SucoPearl PTFE dielectric is shown in Fig. 4.83.

The PTFE pearl core design, although allows to reduce the effect of "Teflon knee," does not completely eliminate it. Theoretical analysis conducted by the developers showed that if the "pearls" were made from the PFA dielectric [78], which lacks the "Teflon knee" effect, the temperature-phase characteristic of the cable would become almost linear.

The parameters of the developed SP304_FEP and SP306_FEP cables with the SucoPearl PTFE dielectric are shown in Table 4.18.

For the sake of fairness, we note that the design of the SucoPearl cable is similar to that of the Russian-made cable described in [86, 87], where ceramic caps, partly interconnected, were used as insulating elements adjacent to each other. However, the use of modern polymeric materials in the design of SucoPearl can significantly improve cable parameters.

Table 4.14 The main parameters of Phase Track cables

No.	Cable model	Sheath diameter, mm; FEP	Minimum bend radius, mm	t_{del} ns/m	V_p, %	C, pF/m	f_{max}, GHz	$\Delta\Psi$, ppm (T, °C)	α, dB/m at $f = 18$ GHz	ΔT_{op}, °C
1	PT110	2.74	14	4.04	82.5	81	80	Figure 4.76	4.0	−55 to +150
2	PT150	3.68	19	4.04	82.5	81	52.4		2.3	
3	PT180	4.57	25.4	4.04	83.0	80.7	38.7		1.9	
4	PT210	5.33	28.6	4.04	83.5	80.0	29.0		1.6	
5	PT318	8.07	44.5	4.0	83.5	78.7	18.9		1.1	

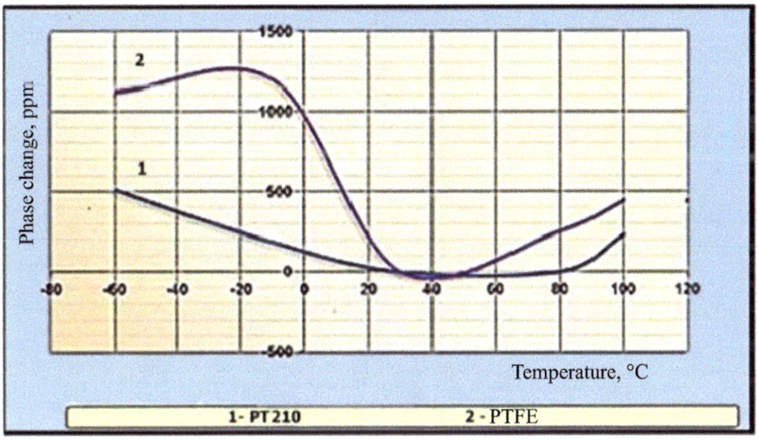

Fig. 4.76 Temperature and phase characteristics of PT210 cable

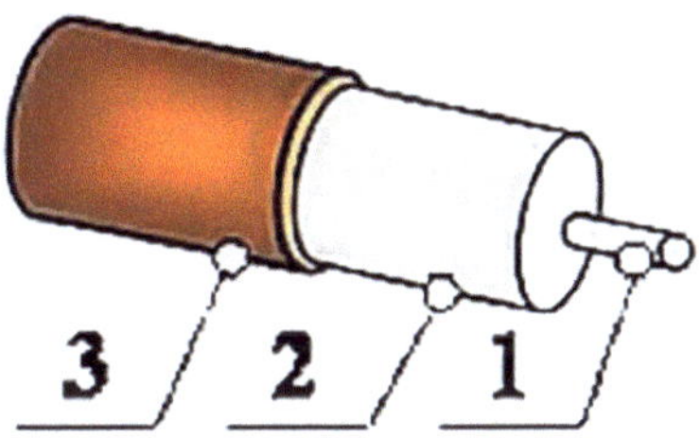

Fig. 4.77 PhaseTrackSR cable design: 1—copper or copper-clad steel silver-plated inner conductor; 2—porous dielectric TF4; 3—copper sheath tube

The technology of making cables is complex, which obviously will limit the construction length and lead to a rather high cost of the cable. The method has a fundamental disadvantage, due to the periodicity of the insulation structure. As a result, there is a probability of occurrence of reflection peaks of transmitted signals with wavelengths that are multiples of the linear dimensions of pearls. The described cables are not commercially available so far.

Summarizing this section, we can formulate the main conclusions as follows: In the construction of modern cables, new organic materials are used. Of all the cables with organic insulation, Molex cordless cables with no periodic insulation structure and Times Microwave Systems cables with TF4 and TF5 insulation are the most promising.

For high-temperature applications, phase-stable cables with SiO_2 dielectric are promising.

Table 4.15 Main parameters of PhaseTrackSR semi-rigid cables

No.	Cable model	Sheath diameter, mm; copper	Minimum bend radius, mm	t_{del} ns/m	V_p, %	C, pF/m	f_{max}, GHz	$\Delta\Psi$, ppm (T, °C)	α, dB /m at $f = 18$ GHz	ΔT_{op}, °C
1	PTSRB047	2.74	3.81	4.04	82.5	80.7	138.5	Figure 4.78	5.7	−55 to +125
2	PTSRB085	3.68	6.35				80.2		3.4	
3	PTSRB141	4.57	10.8				38.4		1.8	

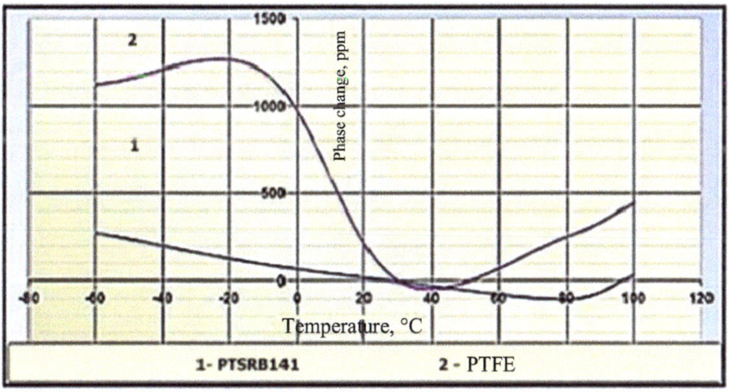

Fig. 4.78 Temperature and phase characteristics of PTSRB141 cable

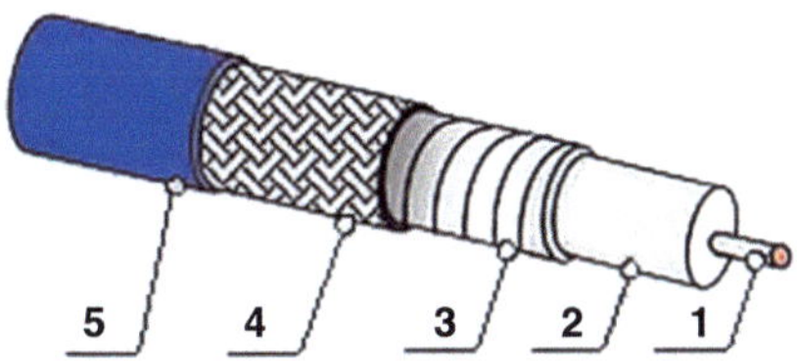

Fig. 4.79 PhaseTrack PFlex cable design: 1—copper or copper-clad steel silver-plated inner conductor; 2—porous dielectric TF4; 3—silvered copper ribbon shield; 4—braid of copper wires coated with silver; 5—FEP sheath

4.10 FOL in LAN and High-Speed Information Collection Systems: Selection of Physical Environment and Active Component Base

4.10.1 Main Trends in the Development of Fiber-Optic Technology

Consider some features of the use of fiber-optic technologies in areas related to the creation of high-speed local communication systems and information collection systems [88, 89].

Fiber-optic technologies are increasingly used in the fields of electronic technology, not related to the transmission of large amounts of information over considerable distances. First of all, there are various object systems of inter-interface telecommunication systems exchange, organization of process control, and many other applications in which the use of optical fiber provides significant advantages compared to traditional means of communication, both wired and wireless. Recently,

Table 4.16 The main parameters of Phase Track PFlex cables

No.	Cable model	Sheath diameter, mm	Minimum bend radius, mm	t_{del} ns/m	V_p, %	C, pF/m	f_{max}, GHz	$\Delta\Psi$, ppm (T, °C)	α, dB/m at $f = 18$ GHz	ΔT_{op}, °C
1	PF047	1.6	6.35	4.04	82.5	80	142.3	Figure 4.80	6.1	−55 to +125
2	PF405	2.4	12.7				79.9		3.6	
3	PF130	3.3	15.9				52.3		2.4	
4	PF402	4.1	19.1				38.7		1.9	

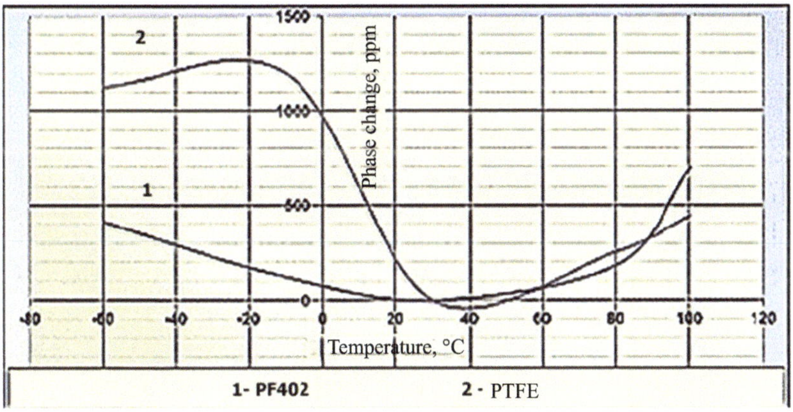

Fig. 4.80 Temperature and phase characteristics of PF402 cable

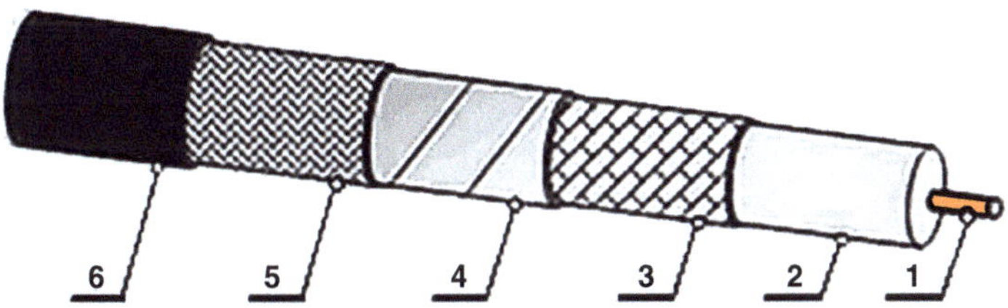

Fig. 4.81 PhaseTrack LS cable design: 1—copper-clad aluminum inner conductor; 2—porous dielectric TF5; 3—silver copper ribbon shield; 4—metallized composite tape; 5—braid of silver-coated copper wires; 6—non-flammable polyethylene sheath

fiber optics tools are increasingly used in systems for collecting information from remote sensors of physical values. In the future, the possibility is expected of creating information collection systems in which the measurement and transportation of the obtained data are carried out using only optical methods. This would significantly improve the safety of technological processes in the oil and gas industry, in the manufacture of explosive and chemically dangerous components. However, until recently traditional sensors of physical values remain the main measuring instruments, and fiber-optic technologies solve the problem of transporting the obtained data. The advantages of using communication lines using optical fibers follow from their main characteristics [88]:

- low intrinsic attenuation and lack of internal interference;
- large bandwidth;
- immunity to electromagnetic effects and absence of its own radiation;
- dielectric properties of fibers and their resistance to the effects of various corrosive media;
- relative difficulty of unauthorized access;
- relative simplicity of coordination with active components of optical cables (OC).

Table 4.17 The main parameters of Phase Track LS cables

No.	Cable model	Sheath diameter, mm	Minimum bend radius, mm	Weight, g/m	t_{del} ns/m	V_p, %	C, pF/m	f_{max}, GHz	$\Delta\Psi$, ppm (T, °C)	α, dB/m at $f = 10$ GHz	ΔT_{op}, °C
1	PTLS400	10.16	102	149	3.97	84	76.8	16.2	Figure 4.82	0.58	−40 to +85
2	PTLS600	15.24	152	238				10.0		0.37	

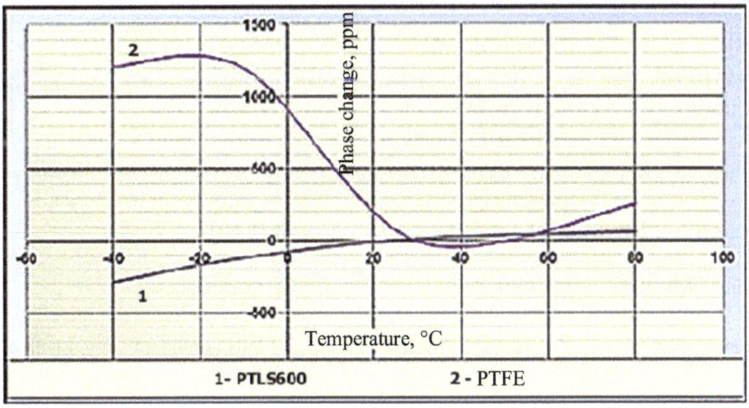

Fig. 4.82 Temperature and phase characteristics of PTLS600 cable

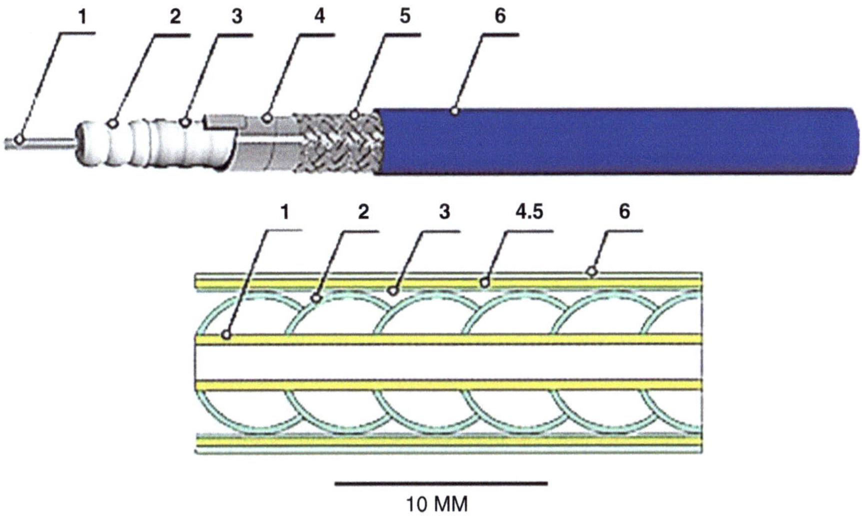

Fig. 4.83 Design of a cable with SucoPearl PTFE dielectric. 1—inner conductor, 2—PTFE pearls, 3—FEP tube, 4—inner ribbon conductor, 5—outer wire braid, 6—FEP sheath

Functionally, FOLs are an integral part of an information collection system, for the considered applications it consists of a passive part and active components. The passive part includes directly optical cables of a certain length with a different type of light-guiding core. This includes a whole arsenal of related components (connectors, splitters, etc.). The active part of a FOL are receiving and transmitting devices, produced as both single semiconductor LEDs, laser diodes, photodiodes, and as finished fiber-optic modules. The latter differ from each other using different wavelengths of the optical spectrum, output optical power, sensitivity, speed, format of the processed signals, principles of signal processing for solving problems of

Table 4.18 Specifications of SP304_FEP and SP306_FEP cables

Parameters	Parameter value	Note
Electrical parameters		
Upper frequency of use, GHz	SP306_FEP—19.5	
	SP304_FEP—28.3	
Maximum VSWR	1.15 (cable), 1.25 (cable assembly)	
Attenuation coefficient, dB/m	≤1.1, target—<0.65	At 18 GHz
Shielding attenuation, dB	<−90	Up to 18 GHz
Bend phase stability, °/GHz	<±0.4	2 winds around a mandrel with a radius of 85 mm
Maximum throughput, W	50	At a frequency of 18 GHz at a temperature of 40 °C
Maximum voltage, kV	2.5	Multipactor discharge
Insulation resistance, MOhm · m	>10	
Mechanical and operational parameters		
Weight, g/m	≤100	
Outer cable diameter, mm	SP306_FEP—7.55	
	SP304_FEP—5.38	
Minimum bend radius, mm	85	
Operating temperature range, °C	−65 to 185	

specific applications. It should be noted that to ensure communication over long distances, there are quite a few active components that integrate optical elements and integrated circuits in their composition that perform the functions of coordination and service. To solve specialized problems of collecting information from sensors of physical values using fiber technologies, as a rule, active components are required that can operate in modes different from the modes of operation of components manufactured to solve traditional communication tasks. These components can be used in the transmission of status signals and information in the form of individual batches and other unbalanced digital sequences; they can also be used to transmit analog signals. Currently, there are some changes in the use of FOLs in local information networks and information collection systems. Historically, the first industrially produced fibers were quartz and quartz-polymer fibers with a light guide core diameter of 200, 100, 62.5, and 50 microns. In these fibers, a multimode mode arises when optical radiation propagates at widely used wavelengths (0.63 μm, 0.85 μm, 1.3 μm, 1.55 μm) [89].

The appearance of fibers with such dimensions of the light guide was primarily associated with the problem of efficiently introducing optical radiation into the fibers in which single-mode operation is possible, as well as with technological problems in the production of fibers with a light guide diameter of 3 ÷ 10 microns. For this reason, single-mode fibers (SMF) were significantly more expensive than multimode fibers (MMF). Currently, the situation in this matter has changed dramatically. The problems of introducing optical radiation into single-mode fibers were solved with the advent of "end" laser structures, focons, and some other elements of non-linear optics. At the same time, the cost of the source material for the production of quartz multimode fiber cables was significantly higher than for the production of single-mode cables. For this reason, in local networks in which communication is implemented using optical fiber, it is increasingly common that optical cables based on single-mode fiber are given priority. When choosing a specific fiber type for use as a physical medium of information collection systems, the larger broadband of single-mode fibers and the absence of intermode combination noise are also taken into account, which in turn makes it possible to broadcast signals with a large dynamic range. FOLs based on single-mode optic fibers are increasingly used in the development of modern local area networks and information collection systems. Active optical components for local information networks using multimode optical cables were developed more than 20 years ago by a number of companies—Hewlett-Packard, Avago Technologies, etc. Today they are not used in development of advanced high-speed devices. In the products of the above companies, an LED with a sufficiently high pumping current (50 ÷ 100 mA) is used as a source of optical radiation to obtain optical power in a 10 ÷ 50 μW fiber. In turn, in order to provide an optical budget (the ratio of the optical transmitter output power to the receiver input threshold power) in the communication line of at least about 10 dB, such low output optical power values warranted the creation of a receiver with a threshold sensitivity of 0.1–0.5 μW. This situation is somewhat corrected by the emergence of new VCSEL laser structures characterized by low pump current (3 ÷ 15 mA) to obtain optical power in a 1 ÷ 2 μW multimode fiber. In this regard, it became possible to create optical radiation receivers with low sensitivity, but with high noise immunity and low consumption.

Thus, it can be assumed that the future of local information networks and information collection systems is bound with the use of single-mode fiber-optic lines [88].

Within this section, let us consider briefly two most popular and successfully tested fiber-optic modules for use both in systems that use single-mode fiber as a physical medium and in multimode applications. The need for the latter is due to their still wide spread, as well as the emergence of modern VCSEL laser structures with low pump currents (3 ÷ 15 mA). The optical power introduced into a multimode fiber, as already noted, is a few μW when they are used. The need to create such products is due to the fact that matching the electrical interface of a communication system with single fiber-optic components such as LEDs, laser diodes (LD), photodiodes, integrated receivers (photodiode + preamplifier), as a rule, was carried out directly by the developer in the process of creating a system. This required special knowledge from them, as well as availability of process and metrological tools. The experience of solving numerous problems of providing communication using FOLs inside limited objects shows that with a sufficiently large optical power input into the fiber from modern radiation sources, there is no need to create complex receiving devices with high sensitivity and dynamic range of input signals. This is due to the fact that the attenuation of optical radiation in modern fibers, depending on the type of the light guide and the wavelength used, is insignificant and ranges from ones to tenths of decibels per kilometer (dB/km). It was also found that for most existing object systems, an optical budget of about 20 dB is sufficient. When determining the requirements for the active FOL components developed for systems of this class, it is necessary to take into account that the information in them is usually transmitted in the form of batches, single signals, and even potential levels, that is, it has an unlimited format. Often in these systems there is an urgent need to minimize their own consumption, since often, especially in data collection systems, sensors of physical values are installed in hard-to-reach places and have autonomous or intermittent power supply. Often systems also lack the means of additional encoding/decoding of broadcast information in order to minimize costs, dimensions, and own consumption. Output and input electrical signals of the controllers in accordance with standard exchange protocols (UART, SPI, etc.) are directly converted into optical signals and vice versa.

Thus, the functionally complete active FOL components for use in high-speed local information networks and collection systems must have a set of specific properties, namely [89]:

- no restrictions on the format of the transmitted signals (single, permanent, batch signals, inverse, etc.);
- minimum consumption in power supply circuits while ensuring the necessary speed of information transfer;
- coordination, both in electrical and optical standard interfaces (standard logic levels, optical radiation wavelength, optical cable characteristics, connector types, dynamic range, etc.);
- sufficient stability of the main electro-optical parameters of the active components in the specified ranges of operation;
- minimum overall dimensions of component casings, corrosion resistance, high degree of electromagnetic protection, possibility of using various types of standard optical connectors.

4.10.2 Standard Specifications of the Basic Set for FOL

The standard kit for use in high-speed networks using optical cables made on single-mode fiber (SMF) includes the transmitting module OMTD-06 and the receiving module OMRD-04. The transmitter uses a high-performance laser module consisting of a laser diode (LD) with an optical radiation wavelength of 1.3 μm and an internal feedback photodiode (IFPD). Circuit solution when creating a transmitting fiber-optic module for the purpose of stabilizing optical parameters involves the use of an internal IFPD. The use of this element allows to provide high stability of the above parameters, both in the temperature range and in the time interval (during the service life). However, due to the significant variation in the parameters of individual laser modules (LD + IFPD) supplied by the manufacturer, their individual adjustment is required during the manufacturing process. Since the solutions used must also provide the ability to transfer input signals having an unlimited format, the circuit includes a device for compensating for the change in the duty cycle of the input signal. The working point of the laser diode is not chosen in the middle of the linear part of the watt-ampere curve, but at the characteristic point at which the transition from the "superluminescent" mode to the "laser" mode occurs. In the parameters of laser diodes, this is the so-called threshold current. The IFPD photo current and the current from the duty ratio compensation circuit are summed with different signs at the input of the photo current amplifier, which is included in the negative feedback loop between the photodiode and the laser diode. This solution provides the ability to process signals of any format while simultaneously stabilizing the position of the LD operating point on the watt-ampere curve. A simplified input signal duty cycle compensation circuit is shown in Fig. 4.84. In some modifications of the transmitting module for cases of critical decisions, the circuit contains indication and service devices. Although the reliability of semiconductor lasers manufactured now, characterized by the time between failures, has reached such a level that the need for these devices is questionable. A functional diagram of a fiber-optic transmitting module, which meets the above requirements, is shown in Fig. 4.85 [88].

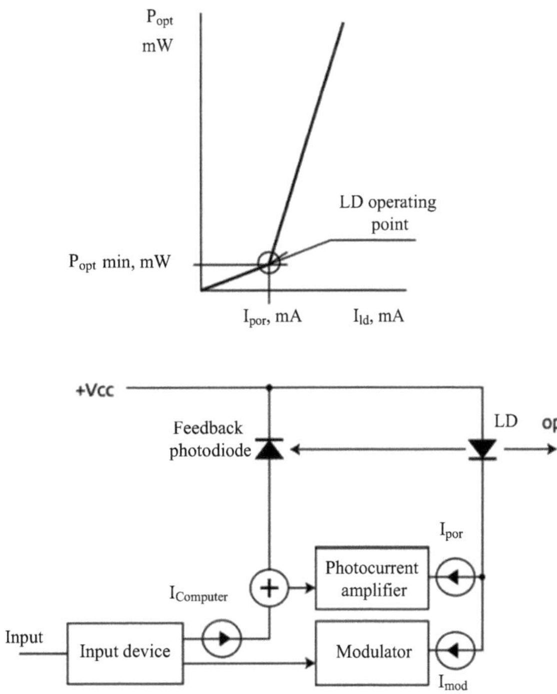

Fig. 4.84 The position of the LD operating point on the watt-ampere curve and a simplified input signal duty cycle compensation circuit

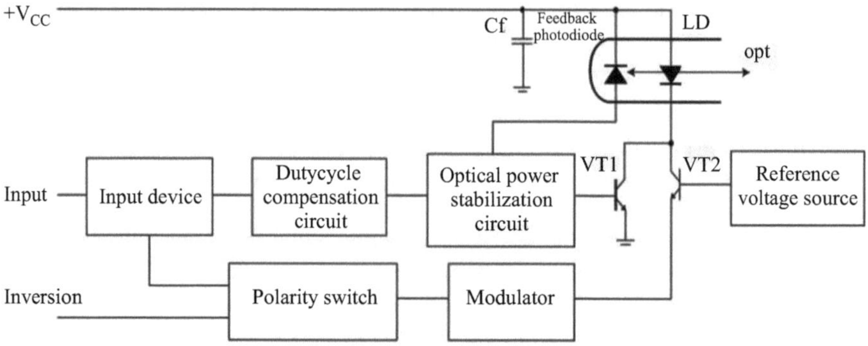

Fig. 4.85 Functional diagram of a transmitting module using IFPD to stabilize the output optical power

When developing the OMRD-06 photodetector circuit used for processing optical signals from FOLs, the fact that the receiver is the most vulnerable link in the system's transmit and receive path, often operating under external electromagnetic interference, was taken into account. This is primarily due to the need to receive rather weak signals, as well as the specifics of their subsequent processing with an arbitrary nature of the digital sequence. Operation with an unlimited data transfer format determines the requirement for the receiver to be in the maximum sensitivity mode, since all kinds of automatic gain control systems (AGC) have a certain reaction time and, therefore, do not allow to process signals having a constant component in their spectrum. Solving the problem of processing digital inputs with relatively high speed and sensitivity and an arbitrary format while obtaining satisfactory output parameters is problematic, and often not necessary. At the same time, special attention in the development of receiving optical modules for this class of tasks has to be paid to design characteristics, which is associated with their operation under climatic, mechanical, and electromagnetic effects. Currently, there are several ways to process arbitrary format signals in FOL receivers:

- fixing the level of determining the logical state;
- definition of the logical state based on the attack and the decay of the input pulse;
- using inverse coding of input signals.

For Implementation in the Receiving Module OMRD-06, a method for determining the logical state based on the attack and the decay of the input digital signal was chosen. This type of circuit design provides the minimum possible broadening of the output signals relative to the input ones, but this requires expansion of the analog band of the receiver and absence of significant dispersion in the FOL. For on-site systems, these requirements are fully met, since the length of fiber-optic lines is relatively small, and the operating frequency band can be increased by reducing sensitivity. The structure of this type of receiver includes a highly sensitive pin photodiode operating in the spectral region $1.1 \div 1.6$ μm, a current amplifier, and a highly sensitive Schmidt trigger. The signal from the current amplifier is differentiated, and short pulses control the operation of the trigger, which is based on a high-speed comparator, covered by positive feedback. A functional diagram of the OMRD-06 optical receiver is shown in Fig. 4.86.

Structurally, the above modules have a metal-glass housing made of a special alloy on a standard base and equipped with an optical connector to protect against external electromagnetic effects. The OMRD-06 receiving module is equipped with an FC type optical connector. The connectors of the OMRD-06-pg and OMTD-06 modules are made in the form of a fiber section terminated with an FC ("pigtail") fork. The appearance of the modules of this type is shown in Fig. 4.87.

A set of optical modules for use in local networks using optical cables with multimode fiber $\phi = 62.5/125$ μm consists of the transmitting module OMTD-04 and the receiving module OMRD-04. Structurally, the modules are housed in metal-glass cases on standard bases and are equipped with ST-type optical connectors. The appearance of the modules of this type is shown in Fig. 4.88 [88].

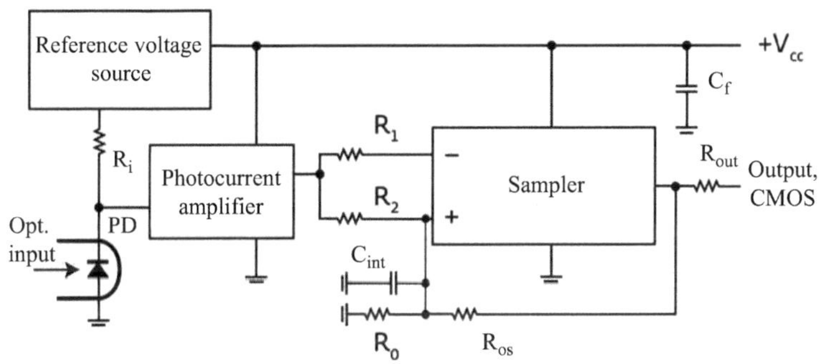

Fig. 4.86 A functional diagram of the OMRD-06 optical receiver [88]

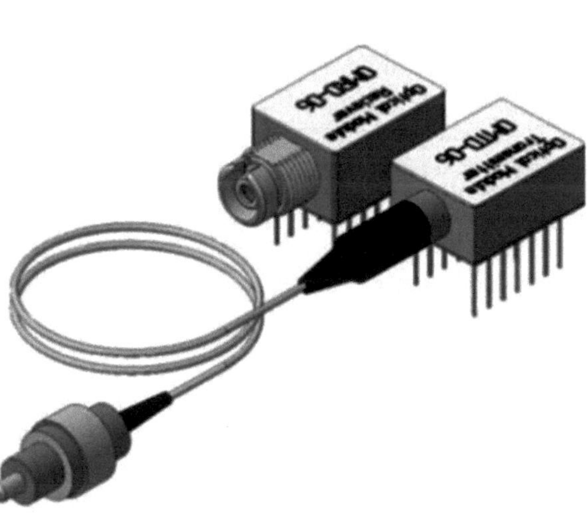

Fig. 4.87 The appearance of fiber-optic modules for use in systems using single-mode (SMF) FOLs

Fig. 4.88 The appearance of fiber-optic modules for use in systems using multimode (MMF) FOLs

The modules operate at a wavelength of 0.85 microns and in functional design (both circuit-wise and structurally) are no different from the OMTD-06/OMRD-04 kit. Thus, the main electro-optical parameters of OMTD-06 are [88]:

- optical power introduced into the fiber $\phi = 9/125$ microns,—0 dBm;
- attack/decay of optical radiation is 2.5 ns;
- input levels of electrical signals—TTL, CMOS;
- input signal duration—$\infty \div 10$ ns;
- supply voltage—5 V;
- current consumption—$\leq$40 mA.

For OMTD-04, these are, respectively,—ϕ = 62.5/125 μm—+3 dBm; $\leq$5 ns; TTL, CMOS; $\infty \div 10$ ns; 5 V; $\leq$15 mA. Both transmitting modules include a highly efficient VCSEL laser diode and a feedback photodiode.

The main electro-optical parameters of OMRD-06:

- threshold sensitivity is 17 dBM;
- minimum duration of processed pulses is 50 ns;
- dynamic range of input signals—20 dB;
- output levels of electrical signals—CMOS;
- supply voltage—5 V;
- current consumption—$\leq$10 mA.

For OMRD-04, these are, respectively, 13 dBm; 50 ns; 16 dB; CMOS; 5 V; $\leq$10 mA. In both these receiving modules, a highly sensitive silicon pin photodiode operating in the spectral region of $0.6 \div 0.9$ μm is used as a photodetector.

The growing demand for active components for local area networks and data acquisition systems using fiber optic in their composition is due to the advent of new optical elements and the tendency for wider use of single-mode fiber in them. In addition, the operating conditions of such systems are complicated by the development of new technological processes, and the range of tasks solved with their help is expanding. For the optimal solution of these tasks, the systems should use communication devices that have certain characteristics both at the receiving part and at the transmitting part. This applies primarily to the coordination of their dynamic range of optical signals, operating wavelength, fiber types, and optical connectors, as well as a single design. In order to obtain optimal characteristics of communication channels, we developed products for them in the form of kits. A kit consists of functionally complete receiving and transmitting modules that can be used independently. It should also be noted that in the presented developments, fundamental optical components are used, such as laser diodes, photodiodes, LEDs, and integrated circuits with the highest optical and electrical parameters achieved to date. For this reason, the presented products have advantages in terms of power consumption to optical output, ease of implementation of the receiving part, and often in terms of sensitivity and speed, over imported products previously developed and solving similar problems. The advantages of the developed products include their structural design, which allows standard connection, a high degree of electromagnetic protection and stability of performance characteristics.

References

1. Belous, A., Merdanov, M., & Shvedau, S. (2014). *Microwave electronics in radar systems and communications. Technical encyclopedia* (p. 688). Moscow: Technosphere.
2. Blackwood, M. (2014). Special requirements for powerful passive microwave components. *Electronic Components*, 1, 88–91.
3. Blackwood, M. *High power RF and microwave passive considerations and constraints*. www.highfrequencyeiectronics.com
4. Smolyaninov, I. (2014). Intellectual encoders with output switching signals. *Electronic Components*, 1, 80–83.
5. Novikov, S. (2014). State-of-the-art technologies of radio frequency components. *Electronic Components*, 7, 18–20.
6. Delisle, J.-J. (2014, May). GaN enables RF where LDMOS and GaAs can't. *Electronic Design Europe*.
7. Ward-Foxton, S. *The last piece of the puzzle*. electronicspecifier.com
8. Chernykh, M. (2014). The choice of design and technology parameters of high-power microwave MESFET-transistors based on silicon carbide. *Electronics: NTB*, 1, 160–164.
9. Lebedev, A., & Sbruev, S. (2004). SiC-electronics. Past, present, future. *Electronics: NTB*, 6.
10. Saddow, S. E., & Agarwal, A. (Eds.) (2004). *Advances in silicon carbide processing and applications* (218 p.). Artech House, Inc.
11. Gudkov, V., Mokeev, A., Zemlyakov, V., et al. (2007). A field-effect transistor with a Schottky gate based on 4H-SiC and an estimate of its microwave characteristics. *Electronic Equipment. 3*(491), (Series: Microwave Technology).
12. Blank, T., Goldberg, Yu., Posse, E., & Soldatenkov, F. (2010). The mechanism of current flow in an ohmic contact to n-4H-SiC. *Physics and Technology of Semiconductors*, *44*(4), 463–466.

13. Chernykh, M., Tsotsorin, A., & Kozhevnikov, V. (2015). *Modeling the impact of channel layer formation parameters on the electrical characteristics of a powerful microwave Schottky field-effect silicon carbide transistor* (Bulletin of Voronezh State University, No.2) (Series: Physics, Mathematics) (in Russian).
14. Chernykh, M., Tsotsorin, A., & Kozhevnikov, V. (2015). Simulating the influence of buffer layer parameters on the static characteristics of high-power microwave transistors with Schottky gate based on silicon carbide II Electronic equipment. *Editions 2–3*(236–237), 70–80. (Series 2. Semiconductor devices) (in Russian).
15. Mnatsakanov, T., Pomortseva, L., & Yurkov S. (2011). A semi-empirical model of carrier mobility in silicon carbide for analyzing its dependence on temperature and doping. *Physics and Technology of Semiconductors*. 35, Edition 4 (in Russian).
16. Gulyaev, Y. V., & Bagdasaryan, A. S. (2004). Acoustoelectronic hardware devices of communication systems. - Collection of works of the international scientific-technical conference. In *Information technologies and modeling of devices and technical processes to ensure quality and reliability* (No. 1, pp. 3–5). M.: MSUIECS. ISBN 5-8068-0344-9 (in Russian).
17. Bagdasaryan, A. S., Bagdasaryan, S. A., Dneprovski, V. G., & Karapetyan, G. Y. (2012). *To issue of surface acoustic wave piezoelectric devices development 180 c. Piezoelectric and related materials: Research and applications*. New York: Nova Science Publishers Inc.
18. Gulyaev, Y. V., & Bagdasaryan, A. S. (2003). SAW filters. Status and prospects of development. *Radio Engineering*, 8, 15 (in Russian).
19. Bagdasaryan, A. S., Gulyaev, Y. V., Nikitov, S. A., Bagdasaryan, S. A., Sinitsyna, T. V., Butenko, V. V., Mashinin, O. V., & Praporschikov, V. V. (2008). Narrow-band SAW filters in radio frequency identification systems (RFID). *Radio Engineering and Electronics*, *53*(7), 887–894 (in Russian).
20. Bagdasaryan, A., Bagdasaryan, S., Karapetyan, G., Mashinin, O., & Sinitsyna T. (2014). Impedance SAW filters for telecommunication systems. Russian priority. *Scientific and Technical Journal* "*Electronics*: *NTB*", 7, 48–60 (in Russian).
21. Bagdasaryan, A. S., & Sinitsyna, T. V. (2004). Selective acoustoelectronic devices based on unidirectional structures of surface acoustic waves. In *Academy of Engineering Sciences named after A.M. Prokhorov* (104 p.). M.: Publishing House. ISBN 5-7781-0033-7 (in Russian).
22. Kukk, K. I., Bagdasaryan, A. S., Zaitsev, V. E., Zlotnikova, E. A., Lokshin, M. G., Neretina, S. P., Sabitov, R. A., Samoilov, A. I., Sevastyanov, D. I., Sergeev, V. I., Taube, L. M., Khazarchiev, Y. D., Khlebnikov, V. I., & Shestakov, S. P. (2005). Development and implementation of scientific, technical and organizational solutions for the construction of a new generation state television network. *Science and Technology in Industry*, 2, 31–35 (in Russian).
23. Sinitsyna, T. V., Praporchtshikov, V. V., & Bagdasarian, A. S. (2004). Saw resonator filters for communications systems. In *Collection: 4th International Crimean Conference 2004: Microwave and Telecommunication Technology. - Conference Proceedings, CriMiCo'04 sponsors: Sevastopol National Technical University, Ukraine, FSUE, SCRRTI, Moscow, Russia, Interface-MFG Co.* Moscow, Russia, OJS SPE, Saturn, Kiev, Ukraine, NTUU KPI, SRI of Telecommunications, Kiev, Ukraine. Sevastopol.
24. Butenko, V. V., & Bagdasaryan, A. S. (2014). Radio frequency identification (RFID) tags on surface acoustic waves and systems based thereon. Patent Review. In *Proceedings of the Research Institute of Radio*, 3, 2–11 (in Russian).
25. Bessonov, L. A. (1974). *Linear electrical circuits* (523 p.). M.: Higher School (in Russian).
26. Bagdasaryan, A. S, Karapetyan, G. Y., & Kondratiev, S. N. Application for the invention No 5066042/22/039877 dd. 24/08/92. Favourable decision No.139 dd. 27/07/93. SAW device. (in Russian).
27. Bagdasaryan, A. S., Kondratiev, S. N., & Semenov, V. V. Application for the invention No 5066037/22/039878 dd. 24/08/92. Favourable decision No.140 dd. 27/07/93. SAW band-stop filter. (in Russian).

28. Bagdasaryan, A. S., & Karapetyan, G. Y. (1998). *Impedance SAW filters* (79 p.). M.: MPO. (in Russian).
29. Bagdasaryan, A. S., Karapetyan, G. Y., Mashinin, O. V., & Semenov, V. V. RF patent 2242838, dd. 20/12/04. SAW Filter. Published in bul. No.35 2004 (in Russian).
30. Bagdasaryan, A. S., Karapetyan, G. Y., Kondratiev, S. N., & Semenov V. V. (2002). RF patent 2195071, dd. 20/12/02. SAW band-stop filter. Published in bul. 35 (in Russian).
31. Cook, A. M., Farmer, J. O., & West, L. E. (1987). A new option in subscriber control. *Communications Engineering and Design*, 35–49.
32. Karapetyan, G. Y., Bagdasaryan, S. A., & Mashinin, O. V. (2003). Impedance SAW filters in master antenna television systems. In *Proceedings of the conference "Actual problems of electronic instrument manufacture"* (pp. 67–69), Saratov, (in Russian).
33. Mashinin, O., Bagdasaryan, A., Lvov, V., Praporschikov, V., Sinitsyna, T., & Bagdasaryan, S. (2008). Modular channel equalizers on SAW filters. *Electronics: Science, Technology, Business*, 2, 74–81 (in Russian).
34. Bagdasaryan, A. S., Sigov, A. S. RF Useful model patent No.120533 RUS, 20/09/2012. Head-end station. (in Russian).
35. Bagdasaryan, A. S., & Karapetyan, G. Y. (1998). Impedance SAW filters for satellite and high-definition television. *Telecommunications*, 6, 21–22 (in Russian).
36. Bagdasaryan, A. S., & Karapetyan, G. Y. (1998). Impedance SAW filters for satellite TV and HDTV. In *Proceedings of the 43rd scientific session devoted to the day of Radio* (p. 17) (in Russian).
37. Karapetyan, G. Y., & Bagdasaryan, S. A. (2003). Two-mesh impedance bridge SAW filters with increased reliability. In *Proceedings of the conference "Actual problems of electronic instrument manufacture"* (pp. 70–72). Saratov (in Russian).
38. Bagdasaryan, A. S., & Karapetyan, G. Y. (1999). Impedance SAW filters for cellular communication systems. In *Systems and means of communication, television and broadcasting* (1st ed., pp. 59–62) (in Russian).
39. Bagdasaryan, A. S., & Karapetyan, G. Y. (1998). Broadband impedance filter. In *Systems and means of communication, television and radio broadcasting* (1st ed., pp. 34–37) (in Russian).
40. Bagdasaryan, A. S., & Karapetyan, G. Y. (1999). The use of impedance SAW filters in broadband Fourier processors. In *Systems and means of communication, television and broadcasting* (1st ed., pp. 56–58) (in Russian).
41. Bagdasaryan, A. S., Mashinin, O. V., & Sinitsyna, T. V. (2004). Low loss SAW filters based on U-shaped coupler. *Telecommunications*, 2, 32–33 (in Russian).
42. Gulyaev, Y. V., Bagdasaryan, A. S., Sinitsyna, T. V., Mashinin, O. V., Praporschikov, V. V., Orlov, M. M., & Egorov, R. V. (2004). SAW filters in the input stages of transceiver devices. *Science and Technology in Industry*, 4 , 82–88 (in Russian).
43. Samoilovich, M. I., Belyanin, A. F., Zhitkovsky, V. D., & Bagdasaryan, A. S. (2004). Nanostructured carbon materials in thin-film technology. *Engineering Physics*, 1, 33 p. (in Russian).
44. Bagdasaryan, A., Bagdasaryan, S., Butenko, V., & Karapetyan, G. (2012). RFID tags based on SAW: Design and technology features. *Scientific and Technical Journal "Electronics: NTB"*, 7, 76–82 (in Russian).
45. Butenko, V. V., Bagdasaryan, A. S., Bagdasaryan, S. A., Karapetyan, G. Y., & Nikolaeva, S. O. (2013). Acousto-electronic ID tags in LTCC ceramics. *Proceedings of the Research Institute of Radio*, 1, 16–23 (in Russian).
46. Belyanin, A. F., Gulyaev, Y. V., Samoilovich, M. I., & Bagdasaryan, A. S. (2004). Equipment for the formation of diamond films and diamond-like materials for acoustoelectronics. *Science and Technology in Industry*, 4, 19 (in Russian).
47. Mayskaya, V. (2013). MEMS technology wins new frontiers. *Scientific and Technical Journal "Electronics: NTB"*, 1, 186–188 (in Russian).
48. *Tunable Impedance Matching Networks (TIM)*. www.wispry.com/products-TIM.php

49. *WiSpry envisions new types of tunable RF applications without frequency constraints for wireless handheld and infrastructure applications.* www.businesswire.com/news/home/20140227005288/en/WiSpry-Envisions-Types-Tunable-RF-Applications-Frequency
50. Moyer, B. *A new variable capacitor. Cavendish kinetics targets antenna frequency tuning.* www.eejournal.com/archives/articles/20130923-cavendish
51. Tornatta P. *A method to design an aperture-tuned antenna using a MEMS digital variable Capacitor.* www.microwavejournal.com/articles/2128 3-a-method-to-design-an-aperture-tuned-antenna-using-a-mems-digital-variable-capacitor
52. Zhang, S., Su, W., Zaghloul, M., & Thibeault, B. (2008). Wideband CMOS compatible capacitive MEMS switch for RF applications. *IEEE Microwave and Wireless Components Letters, 18*(9), 599–601.
53. Fouladi, S., & Mansour, R. R. (2010). Capacitive RF MEMS switches fabricated in standard 0.35μm CMOS Technology. *IEEE Transactions on Microwave Theory and Techniques, 58*(2), 478–484.
54. Fouladi, S., Domingue, F., Zahirovic, N., & Mansour, R. R. (2010). Distributed MEMS tunable impedance-matching network based on suspended slow-wave structure fabricated in a standard CMOS technology. *IEEE Transactions on Microwave Theory and Techniques, 58*(4), 1056–1064.
55. Kaynak, M., Ehwald, K. E., Drews, J. et al. (2009). BEOL embedded RF-MEMS switch for mm-wave applications. In *Proceedings of the IEEE International Electron Devices Meeting* (IEDM) (pp. 1–4).
56. Reinke, J., Fedder, G. K., & Mukherjee, T. (2011). CMOS-MEMS 3-bit digital capacitors with tuning ratios greater than 60:1. *IEEE Transactions on Microwave Theory and Techniques, 59*, 1238–1248.
57. Fouladi, S., & Mansour, R. R. (2009, October). Reconfigurable amplifier with tunable impedance matching networks based on CMOS-MEMS capacitors in 0.18μm CMOS technology. In *Proceedings of the 2nd Microsystems Nanoelectronics Research Conference* (pp. 33–34).
58. Fall, E. M., Domingue, F., Fouladi, S., & Mansour, R. R. (2011). Design of reconfigurable quad-band CMOS class AB power amplifier employing MEMS variable capacitors in 0.18μm technology. In *Fourth International Conference on Advances in Circuits, Electronics and Micro-electronics* (pp. 34–37).
59. Zahirovic, N., Mansour, R. R., & Yu, M. (2010). Piezoresistive position sensing for the detection of hysteresis and dielectric charging in CMOS-MEMS variable capacitors. *IEEE Transactions on Microwave Theory and Techniques, 58*, 3961–3970.
60. Belov, L., & Zhitnikova, M. (2004). RF MEMS components. *Scientific and Technical Journal "Electronics: NTB"*, 8, 18–25 (in Russian).
61. Chen, W., Fang, W., & Li, S. (2012). High-Q integrated CMOS- MEMS resonators with deep-submicrometer gaps and quasi-linear frequency tuning. *Microelectromechanical Systems, 21*(3), 688–701.
62. Richards, R., & De Los Santos, G. *MEMS devices for microwave applications: A new wave* (trans. by A. Fedyanovich). http://www.gaw.ru; http://catalog.gaw.ru/index.php?page=document&id=976
63. Prokimov, A., Lobanov, A., Dzhurinsky, K., & Kuznetsov, R. (2015). Phase stability of microwave cable assemblies with a PTFE dielectric. *Komponenty i Tehnologiya*, 6, 58–63 (in Russian).
64. Tumachek, A., & Kruglikov, D. (2014). Software-configurable radio is gaining momentum. *Modern Electronics*, 8, 66–71 (in Russian).
65. Bykhovskiy, M. A. (2014). The development of telecommunications. In *On the way to the information society: The development of satellite telecommunications systems* (p. 440). M. Hotline - Telecom (in Russian).
66. Qureshi, S. U. H. (1985). Adaptive equalization. In *Proceedings of the IEEE* (Vol. 73, No. 9, pp. 1349–1387), IEEE Press, New York.

67. Johnson, P. (1985, May). New research lab leads to unique radio receiver. *E-Systems Team, 5*, 6–7.
68. 1EC 60966-1 ed.2.0 (1999). *International standard. Radio frequency and coaxial cable assemblies - Part 1: Generic specification - General requirements and test methods.*
69. IEC 60966–2-1 ed.3.0 (2008). *International standard. Radio frequency and coaxial cable assemblies; Part 2–1: Sectional specification for flexible coaxial cable assemblies.*
70. *Gore cable assembly builder microwave assembly part number G5RO I RO 10 170*. www.goremicrowave.com
71. *Development and manufacture of coaxial connectors, adapters and cable assemblies*. www.spcable.ru, www.spetskabel.ru (in Russian).
72. Micro-coax specification cable assemblies, UFA210B low loss 26,5 GHz, 2004.
73. GOST National Standard 11324.0-78. Radio frequency cables. General technical conditions (in Russian).
74. Dzhurinsky, K. B. (2014). *Modern RF connectors and EMI filters* (in Russian).
75. MPD Microwave Product Digest. *Current innovations in phase stable coaxial cable design by Times Microwave System.* www.timesmicrowave.com.
76. Efimov, I. E., & Ostankovich, G. A. (1977). *Radio-frequency transmission lines. RF cables* (2nd ed.). M.: Communication (in Russian).
77. Prokimov, A., Dzhurinsky, K., & Kuznetsov, R. (2015). Microwave Cable Assemblies. Function, classification, application peculiarities. *Komponenty i Tehnologiya*, 5, 28–32 (in Russian).
78. Retnyuk, V. (2014). Solutions to cable assemblies of the microwave range. *Komponenty i Tehnologiya*, 7, 63–84 (in Russian).
79. *Teflon® PFA resin and film.* www2.dupont.com
80. *Fluoropolymer F-50.* www.plastpolymer.org (in Russian).
81. *Test+Measurement.* Edition 2013. www.hubersuhner.com
82. Meggitt Safety Systems. *SiO_2 cable systems*. www.stablecable.com
83. Times Microwave Systems. *SiO2 coaxial cable assemblies*. www.timesmicrowave.com
84. Measure Tech Inc. *Silicon dioxide cable assembly*. www.measure-tech.com
85. Thermocoax. *SiO_2 insulated signal transmission cables*. www.thermocoax.com
86. Karstensen, H., Koufogiannis, I., Sorolla, E., Kress, G., Mattes, M, Rupflin, M, Fuchs, J., & Wettstein, K. (2013, September 24–26). Phase stable RF cable for space applications. In *Space Passive Component Days, 1st International Symposium* (in Russian).
87. Efimov, I. E., & Ostankovich, G. A. (1977). *RF transmission lines. RF cables* (2nd ed., revised and updated). M.: Svyaz (in Russian).
88. Varava, N., Nikonorov, M., & Pronin, S. (2015). Active fiber-optic components for LAN and dedicated applications. Part 1. *First Mile*, 3, 84–91 (in Russian).
89. Varava, N., Nikonorov, M., & Pronin S. (2015). Active fiber-optic components for LAN and dedicated applications. Part 2. *First Mile*, 4, 56–61 (in Russian).

Chapter 5
Specialized Chips for Telecommunication Systems

This chapter describes in detail the main types of specialized microcircuits designed exclusively for use in high-speed telecommunication systems.

First of all, this refers to the chips of the SerDes, DCS-Control series (Sects. 5.1, 5.2, and 5.3). It is shown that the use of SerDes series microcircuits makes it easy to implement high-speed serial interfaces, which are characterized by ease of implementation, reasonable cost, and low power consumption. Using SerDes allows you to significantly simplify the wiring of a high-speed PCB device by reducing the number of communication lines between the chips on the board. Topics that are usually taught in various training courses using a variety of manuals are put together in this chapter to provide enough information for high-speed device designers.

Chip designers with experience in digital logic usually consider high-speed SerDes units solely as elements of digital data input/output. Here we will show that when using high data transfer rates, it is necessary to take into account the analog nature of the process in order to reduce the likelihood of errors and ensure high-quality transmission of digital signals.

Section 5.4 presents a specific example of using a system on chip (SoC) to create a high-speed bidirectional communication line with a data transmission frequency of at least 60 GHz.

In Sect. 5.5 the characteristics of a number of specialized microcircuits for building an RS-485 interface are considered and practical recommendations on their use as part of high-speed electronic devices are given.

In Sect. 5.6 the technical features and ways of using LVDS and DV1 high-voltage interfaces are considered. The largest section of the chapter in Sect. 5.7 considers the analysis of microwave chips designed for use as part of high-speed information processing systems. Here you can find not only reference data and recommendations on the use of the most popular microcircuits but also advanced technologies for their manufacture.

A. Belous, V. Saladukha, *High-Speed Digital System Design*,
https://doi.org/10.1007/978-3-030-25409-4_5

5.1 Principles of Construction and Main Specifications of Telecommunication Chips of the SerDes Series

5.1.1 Basic Provisions of the Basic Standard IEEE 802.3 for Ethernet Telecommunication Network

The rapid growth of information content requires development and implementation of high-performance solutions in the field of high-speed transmission of multimedia data. Below we consider the principles of construction and performance of telecommunication IS of the SerDes [1] series, designed to work in 40/100G networks and provide high rates of performance, reliability, and power consumption.

Global Internet traffic is increasing every year, and each year the amount of transmitted data almost doubles compared to the previous year. Ethernet has long been transformed from a data transmission technology into a global ideology of building telecommunication networks. The development of Ethernet occurs both in the direction of expanding applications and in the direction of increasing data transfer speed (throughput). IEEE 802.3ba (Media Access Control Parameters, Physical Layers, and Management Parameters for 40 and 100 Gb/s Operation) published in June 2010 is an important step in the development of this technology.

The main objective of the development was to extend the 802.3 protocol to ultra-high data transfer rates and at the same time ensure maximum interface compatibility with the 802.3 standard in order to preserve previous investments in network infrastructure [2]. The new standard was adopted by the IEEE Institute as an add-on to IEEE 802.3 Ethernet, it describes the physical and MAC levels and supports two data rates: 40 and 100 Gbit/s. The high bandwidth requirements of the data transmission medium greatly exceed the existing capabilities of Ethernet. In accordance with regulatory documents, the IEEE 802.3ba standard supports full-duplex mode and is focused on various physical level (PHY) types (media).

The main tasks that the IEEE 802.3b standard should solve were formulated as early as 2006 and with a number of changes remained in the final version [1]:

Preserve the format of the 802.3 frame used at the MAC layer.
Preserve the minimum and maximum frame size of 802.3 standards.

Ensure the reliability of data transmission at the MAC level: Bit error rate (BER) should not exceed 10 ~ 12.

Compatibility with optical transport networks (OTN)—a highly reliable medium for transmitting heterogeneous traffic, standardized in the recommendation ITU-T G.709.

Providing physical (PHY) level specifications for transmission in different environments: single-mode optical fiber (SMF), multimode optical fiber (MMF), copper twisted pair, and electrical backplane.

The standard is applicable for "short" connections with a length of one meter (connecting panels) and "long" connections of hundreds of meters (local area networks), as well as kilometers (access network) and tens of kilometers. At a speed of 100 Gbit/s, there are long-range (LR, up to 10 km) and extremely long-range (ER, up to 40 km) modes of operation. They differ only in line budget requirements: extremely long-range mode in fact implies the need to use optical amplifiers in fiber optics.

At the current level of development of semiconductor technology, the speed of about 100 Gbit/s can really be achieved only using the principal innovation of the standard—the transition from serial signal transmission to parallel one through several streams. At the same time, the system at the logical level (MAC level) uses a single data transfer rate (that is, all network processing devices are oriented at this speed), and the physical information flow is divided into a number of parallel "subflows," the data transfer rate which depends on the used transmission medium and communication range. Typically, the speed in each such physical channel is either 10 or 25 Gbit/s. So, for 100-Gbit/s interfaces, it is possible to transmit ten 10-Gbit/s streams at once, either using ten twisted copper pairs or ten MMFs in each direction. To operate using SMF, four parallel flows of 25 Gbit/s are usually used, separated by wavelength and transmitted in one optical fiber. For 40-Gbit/s interfaces, as a rule, only 10-Gbit/s streams are used that are transmitted over the backplane conductors, copper pairs, or optical fiber.

In accordance with this standard, data communication equipment should automatically perform a number of special procedures. This said, the division into different "levels" (physical, channel, etc.) and their "sublevels" is rather arbitrary. This functional division allows to create unified components, standardized not only by function but also by interface. To perform the parallel-serial/serial-parallel conversion, developers widely use various specialized SerDes chips, called serializers/deserializers [1, 3]. Semtech, the world leader in the design and manufacture of analog and digital–analog electronics, discussed below, has developed its series of telecommunication circuits consisting of components that fully meet the requirements of this standard and ensure reliable operation of telecommunication systems at data rates of 40 and 100 Gbps.

5.1.2 *Basic Specifications of SerDes Series Chips by Semtech*

The use of modern highly integrated serial-parallel and parallel-serial converters makes it easy to implement high-speed serial interfaces, which are characterized by ease of implementation, as well as reasonable cost and power consumption. The use of this approach greatly simplifies the layout of a printed circuit board of a high-speed electronic device by reducing the number of I/O lines and conductors between converters and microcontroller chips or programmable FPGA logic.

The Semtech SerDes series chips [1, 2] described below are characterized by a low level of output phase noise (jitter) not more than 4 ps, with an error rate of 10×15 error bits, they operate with a serial data stream at speeds up to 113 Gbit/s and are aimed for use in modern industrial telecommunication equipment, instrumentation systems, routers, etc. (Table 5.1). The main purpose of this series of chips

Table 5.1 Semtech's SerDes chips [1]

Characteristics		Data transmission rate, Gbit/s	Number of input data lines	Number of output data lines	Input interface	Output interface	Jitter, ps
Serializers	SMI10021	100–113	10	4	SFI-S	CML	3.7
	SMI10022	111.6–128.4	10, 20	4	SFI-S	CML	3.7
	SMI4028	45–49	16	2	SFI5	CML	3.7
	SMI4029	40–45	16	2	SFI5	CML	3.7
	SMI4024	39.8–44.6	16	1	SFI5	CML	4
	SMI4026	39.8–44.6	16	1	SFI5	CML	4
	SMI4027	39.8–44.6	16	1	SFI5	CML	4
Deseralizers	SMI10031	100–113	4	10	CML	SFI-S	3.7
	SMI4038	45–49	2	16	CML	SFI5	3.7
	SMI4039	40–45	2	16	CML	SFI5	3.7
	SMI4034	39.8–44.6	1	16	CML	SFI5	4
	SMI4036	39.8–44.6	1	16	CML	SFI5	4
	SMI4037	39.8–44.6	1	16	CML	SFI5	4

is the use of integrated circuits in the organization of extra-long communication channels with a capacity of 100 Gbit/s, for example, in underwater optical highways. Such components as SMI4024/SMI4034 and SMI4026/SMI4036 are mainly intended for use in SONET/OC-769 networks.

Functionally, SerDes series converters are organized as follows: A multiplexer with a built-in clock multiplier unit (CMU) and an integrated DQPSK (differential quadrature phase shift keying) encoder on the transmitter side, and a demultiplexer with a double clock and data recovery (CDR) circuit with an integrated DQPSK decoder on the receiver side [4]. Accordingly, microelectronic components of this series are always produced in pairs designed for the same bandwidth and number of input/output channels, for example, SMI10021/SMI10031 (Fig. 5.1) [1]. It should be noted that with the SMI10022 circuit, the width of the input data interface can be independently configured by the user (20 or 10 input lines).

It should be particularly noted that the applied Forward Error Correction (FEC) technology can significantly improve the quality of data transmission, since using FEC coding, you can get the required bit error rate (BER) with significantly lower signal-to-noise ratio. In particular, in such a system, a weak high-speed signal can be received at a higher channel noise level, which would be impossible to receive at all without the use of coding. Among other features, it is necessary to note the function of skew data adjustment, high-speed oscillator outputs with CML logical levels at full or half operating frequency, an integrated error correction circuit, and a test signal generator. This said, the test signal generator (PRBS) can be used at the request of the customer, for example, in the process of debugging or while checking the operability of the device. On the receiving side of such a designed system, among

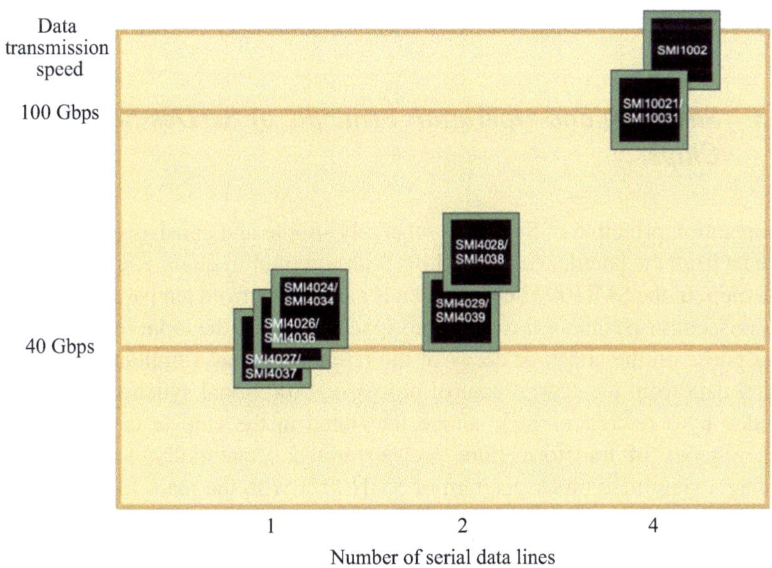

Fig. 5.1 Semtech's SerDes series IS

Fig. 5.2 Appearance of used cases

other things, it is possible to use a hardware indication of CDR circuit configuration completion.

All these chips are manufactured using BiCMOS technology based on silicon–germanium, produced in modern ceramic cases for surface mounting (BGA), and provide a high level of reliability and performance (Fig. 5.2) [1]. Small size and compliance with RoHS and WEEE standards help reduce energy consumption and parasitic losses.

A standard stabilized low-voltage source with output voltages of +1.2, +1.5, and +2.8 V can be used to power the chips. Most of the chips in this series consume no more than 2.9 W of power [5]. In order to save energy, it is possible to automatically switch to standby mode when there is no signal. In addition, the IS can be switched to the "sleep" mode and thus achieve an additional reduction in energy consumption. As for reliability indicators, for operating temperature range from 0 to +80 °C, the manufacturer guarantees the service life of SerDes of at least 20 years.

5.1.3 Structure and Operation Principle of SerDes Series Chips

The operation principle of SerDes is relatively simple and consists of the following. The data from the parallel bus, switching with a certain frequency, is fed to the input of the chip. In the SMI10021 chip, which is a converter from ten parallel data lines to four consecutive channels, these data are synchronized in the input registers based on the response on the attack or decay of the reference signal simultaneously with the parallel data from the central control processor. Additional synchronization is also provided by a reference clock source integrated in the chip (a crystal oscillator). Next, a series of transformations are performed sequentially, see Fig. 5.3 [1], showing a simplified block diagram of SMI10021 with the main functional blocks.

It is necessary to note that "long" sequences of zeros or ones can lead to the appearance in the spectrum of the output signal of spurious RF components, which can cause a receiver and transmitter to go out of sync. Therefore, the coding and formatting of the transmitted information avoids various spectral distortions and

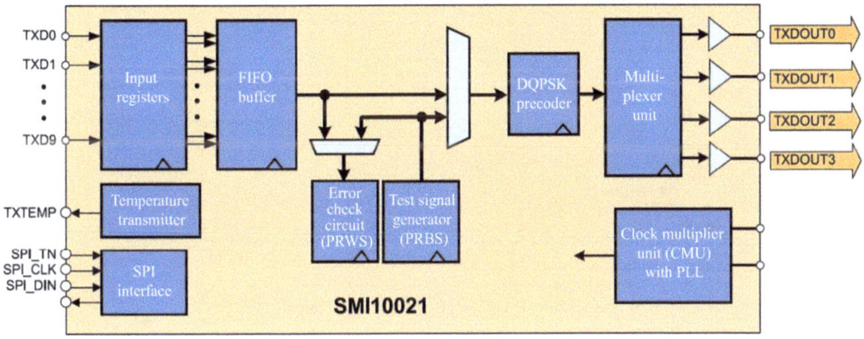

Fig. 5.3 A simplified block diagram of SMI10021

ensures a uniform distribution of the signal (zeros and ones) over the entire transmission band. The channel coding process includes not only the digital coding of the waveform, but also the addition of the necessary redundant bits to protect against noise and possible signal distortion.

Another well-known problem is related to the possibility of losing frame boundaries in the process of converting information flow into a sequential form. Since the serial transmission in this case is carried out simultaneously over several streams in order to restore the original stream to the receiver later, reliable time synchronization should be ensured between these parallel streams. Considering the speeds and distances foreseen by the above standard, physical mismatches are inevitable. To compensate for the time shifts (skews) arising in the transmission line, special leveling markers are used. By measuring the time shift of their arrival, the receiver is able to automatically restore the necessary synchronism of the streams and thereby ensure reliable equivalence of the transmitted and received information.

On the side of the receiving microcircuit—deserializer—the procedures of inverse transformations are performed [6]. Sequential data streams enter the input buffer of the differential signal, in which the incoming bits (in the form of voltage amplitude) are automatically converted into digital data. The main function of the receiver is to restore the clock synchronization of the incoming stream. A CDR unit is used to perform this task. The recovered signal is used to synchronize the incoming stream, converted into a parallel form. The stream of parallel bits is decoded into the original data sequence, which is fed to the parallel output registers, while the clock signal is also parallelized and transmitted along with the data.

It should be noted that the clock multiplier unit is an important component of any SerDes device, since it generates a high-speed clock signal to control the serial transmitter, as well as the receiving path of the device [6]. One of the main technical characteristics that determine the high quality of SerDes device operation is the stability of the clock frequency, expressed in terms of phase jitter. A significant amount of jitter in the reference signal usually leads to a lot of jitter in the stream of transmitted serial data. If the reference signal is unstable, the same characteristics appear in the data stream. Thus, the SerDes device's reference signal must meet very stringent requirements for jitter, clock pulse offset, and pulse attack/decay times.

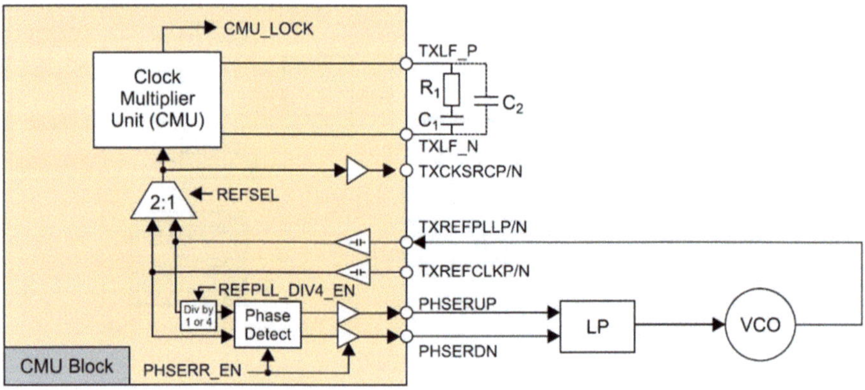

Fig. 5.4 Clock multiplier unit diagram

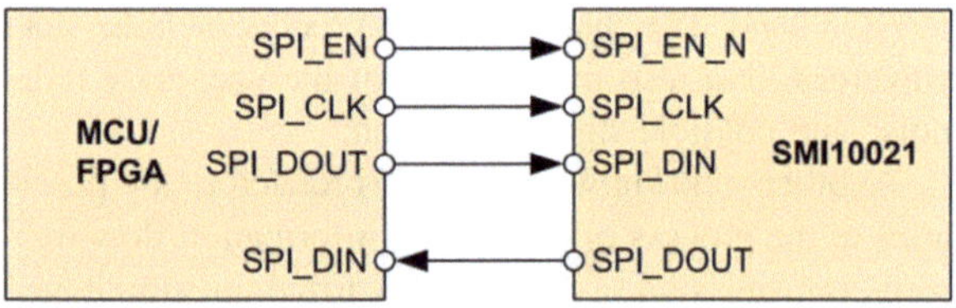

Fig. 5.5 Connection of SMI10021 to the controller

Highly stable clock sources can be connected by using two special pins of the CMU circuit, which are selected by the user: TXREFCLKP/N or TXREFPLLP/N. To obtain a reference signal with a minimum level of jitter, the PLL circuit is also used (Fig. 5.4) [1]. Phase jitter filtering is achieved using a special phase stabilization circuit consisting of an integrated phase detector with an additional frequency divider by 4, an external precision voltage-controlled oscillator (VCO), and a filter in the feedback circuit.

Separately, it is necessary to consider the features of using the SPI interface [1].

Virtually every functional unit can be configured according to system requirements. In the technical documentation for microcircuits, the list of control and signaling registers occupies more than one page, and the serial SPI interface is used for communication between any used external master device (microcontroller, FPGA, etc.) and serializer/deserializer IS (Fig. 5.5). Note that SerDes chips always function in Slave mode, where 16-bit format is used to transmit commands, the most significant bit always being sent first.

When issuing any command or in read mode, at least two 16-bit words are used. The first one contains the address of the registers (from the 0th to the 13th bit) and mode selection (0—read, 1—write), the second, respectively, the data or the command. When transferring more than one 16-bit word, the parcel address is automatically incremented.

As an additional option, it is possible to use the built-in sensor to measure the temperature of the crystal, whose readings can be read both in digital form using an SPI interface and in analog form using TXTEMP (Fig. 5.6).

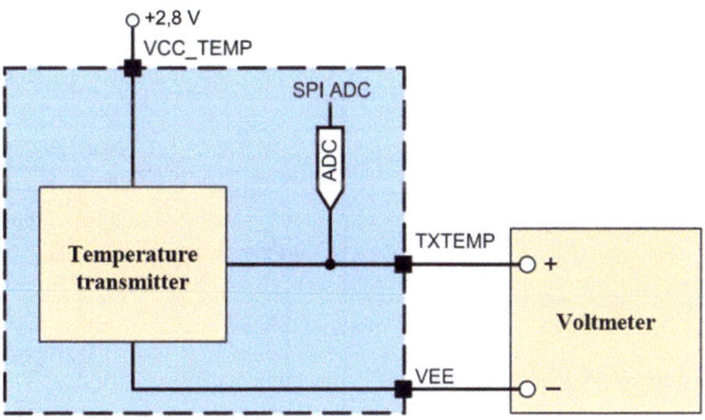

Fig. 5.6 Using the built-in temperature sensor

Multiple debug kits are usually offered for evaluating capabilities and developing various applications. Based on the requirements for the developed product, the user can choose the most suitable solution for them. Such debug kits provide everything developers need to get started. The standard kit includes an evaluation module and a CD with all the necessary documentation, including electrical schematics, printed circuit board topologies, and operation manuals. In addition, an easy-to-use evaluation board configuration utility with a graphical user interface (GUI) is always available, allowing to customize converter options.

5.2 Methods for Designing Pulse Voltage Converters with SEPIC Topology

5.2.1 Analysis and Calculation of Parameters of the Standard SEPIC Converter

Despite the fact that the standard SEPIC circuit consists of a small number of elements, its analytical description using formulas is rather difficult. Consider the calculation method proposed in [7, 8]. In this paper, in a relatively brief and in an accessible form, a detailed description of the operation of the circuit is given, as well as a calculation of a large number of parameters of the components used (compared, for example, with the sources) [8, 9].

To illustrate the presentation, we will use Fig. 5.7. The parasitic resistances of the elements L1, L2, S_w, and C_f are denoted on it, respectively, as R_{l1}, R_{l2}, R_{sw}, and R_{cp}.

Suppose that the values of the ripple currents and voltages are much less than the values of the constant component, and the voltage drops on the coils L1 and L2 in the steady state do not have a constant component (if we neglect the voltage drops on their parasitic resistances). Therefore, one of the plates of the capacitor C_p is under the potential V_{in} (through the coil L1), and the other is under the potential of the "ground" (through L2). Then the expression for the average voltage on the capacitor C_p will be:

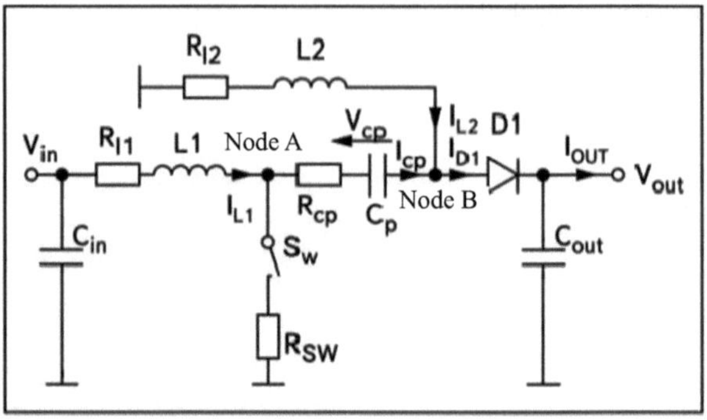

Fig. 5.7 SEPIC converter diagram with parasitic parameters

$$\left(V_{\mathrm{cp}}\right)\text{mean} = V_{\mathrm{in}}. \tag{5.1}$$

Assume that T is the period of one cycle of the converter operation. α is the part of the period T, during which the key S_w is closed, and $1 - \alpha$ is the remaining part of the period. Since the average value of the voltage on the coil L1 in the steady state is zero, the voltage on L1 during the time $To = \alpha T$, during which the key is closed, is fully compensated by the voltage during the time $T_{\mathrm{off}} = (1 - \alpha)\,T$, during which the key is open:

$$\begin{aligned} TV_{\mathrm{in}} &= (1 - a)T\left(V_{\mathrm{out}} + V_{\mathrm{d}} + V_{\mathrm{cp}} - V_{\mathrm{in}}\right) \\ &= (1 - a)T\,(V_{\mathrm{out}} + V_{\mathrm{d}}), \end{aligned}$$

where V_d is the forward voltage drop across the diode D1 for the sum of the currents through the inductances L1 and L2, and V_{cp} is equal to V_{in}:

$$(V_{\mathrm{out}} + V_d)/V_{\mathrm{in}} = a/(1 - a) = \mathrm{Ai}. \tag{5.2}$$

Ai is the amplification factor, where "*i*" means the ideal case when the parasitic resistances are zero. If we neglect in the first approximation the voltage drop across the diode V_d, we see that the ratio of V_{out} to V_{in} can be both greater and smaller than 1, depending on the value (and they are equal when $\alpha = 0.5$). This ratio demonstrates the difference of the SEPIC topology from the classical step-down topology. A more accurate expression for the real amplification factor Aa takes into account the parasitic resistances in the circuit [7]:

$$\begin{aligned} \mathrm{Aa} = &\left[V_{\mathrm{out}} + V_d + I_{\mathrm{out}}\left(\mathrm{Ai}\,R_{\mathrm{cp}} + R_{\mathrm{L2}}\right)\right] \\ &/\left[V_{\mathrm{in}} - \mathrm{Ai}(R_{\mathrm{L1}} + R_{\mathrm{SW}})I_{\mathrm{out}} - R_{\mathrm{SW}}I_{\mathrm{out}}\right]. \end{aligned} \tag{5.3}$$

This formula allows to calculate the minimum, typical, and maximum V_{in} amplification factors ($\mathrm{Aa}_{\mathrm{min}}$, $\mathrm{Aa}_{\mathrm{typ}}$, and $\mathrm{Aa}_{\mathrm{max}}$). The formula is recursive ("A" appears in both the left and right parts of the expression), but a few iterative calculations will asymptotically lead to the result. The expression does not take into account the losses in the switch S_w and the reverse current of the diode D1. These losses are usually negligible, especially if the S_w is a metal oxide semiconductor field-effect transistor (MOSFET) and the voltage range at its drain (V_{in} + V_{out} + V_d) is within 30 V (typical limit for modern MOSFET with low losses).

In some cases, you should also consider the reverse current of the diode D1 and the loss in the core from a large induction gradient. We can extrapolate the corresponding values of α from (5.2) [7]:

$$a_{\mathrm{XXX}} = \mathrm{Aa}_{\mathrm{XXX}}/(1 + \mathrm{Aa}_{\mathrm{XXX}}), \tag{5.4}$$

where xxx is the minimum, typical, or maximum level.

The constant component of the current through C_p is zero. This means that the output current flows only through L2:

$$I_{\mathrm{out}} = I_{\mathrm{L2}}. \tag{5.5}$$

There is no hard requirement for power dissipation of L2, since the average current through L2 is always equal to the output current and does not depend on changes in the input voltage V_{in}. To calculate the current through L1 (ILl), recall that a direct current cannot flow through C_p. Thus, the charge flowing at the time (interval αT) when the key is locked is fully balanced by the charge that flows at the time when the key is unlocked (interval $(1 - \alpha)$ T). When the key is locked, the potential of node A is fixed at 0 V. In accordance with expression (5.1), the potential of the node B is -V_{in}, which corresponds to the reverse bias of diode D1. The current through C_p is the current through L2. When the key is unlocked at time $(1 - \alpha)$ T, current L2 flows through D1, while current L1 flows through C_p: $\alpha T \times \mathrm{IL2} = (1 - \alpha)$ $T \times \mathrm{IL1}$. Knowing that $\mathrm{IL2} = I_{out}$, we put down:

$$I_{\mathrm{L1}} = \mathrm{Aa}_{\mathrm{XXX}} \times I_{\mathrm{out}}. \tag{5.6}$$

As input power is equal to output power divided by efficiency, IL1 depends only on V_{in}. For a given output power, IL1 increases if V_{in} decreases. Knowing that IL2 (and, therefore, I_{out}) flows through C_f during αT, we will select a C_p so that the voltage pulsations on it ΔV_{cp} are much less than V_{cp} ($\gamma = 1$–5%). The worst case will be for the minimum V_{in}.

$$C_p \geq I_{\mathrm{out}} \alpha_{\mathrm{min}} T/(\gamma \mathrm{Vin}_{\mathrm{min}}). \tag{5.7}$$

Use of a high-frequency controller in combination with the latest achievements in the field of creating multilayer ceramic capacitors MLC allows using a small non-polar capacitor as C_p. At the same time, it is necessary to make sure that it can dissipate the power of P_{cp} released on its parasitic internal resistance R_{cp}:

$$P_{\mathrm{cp}} = \mathrm{Aa}_{\mathrm{min}} R_{\mathrm{cp}} I_{\mathrm{out}}^{2}. \tag{5.8}$$

R_{sw}, usually consisting of a drain-source resistance of a MOSFET switch connected in series with a shunt to limit the maximum current, introduces the following losses:

$$P_{sw} = \mathrm{Aa}_{\min}(1 + \mathrm{Aa}_{\min})R_{sw} I_{out}^2. \tag{5.9}$$

Pr_{l1} and Pr_{l2} losses on the internal resistances of L1 and L2 are easily calculated:

$$Pr_{l1} = \mathrm{Aa}_{\min}{}^2 R_{l1} I_{out}^2, \tag{5.10}$$

$$Pr_{l2} = R_{l2} I_{out}^2. \tag{5.11}$$

When calculating the losses on the diode D1, we calculate V_d for the sum of the currents IL1 + IL2:

$$\mathrm{PDI} = V_d \times I_{out}. \tag{5.12}$$

L1 is chosen so that the proportion (β) of current pulsations in it (ΔIL1) is 20–50% of the average current IL1. The worst case for β is when V_{in} is maximal because DIL1 is maximal when IL1 is minimal. Assume ($\beta = 0.5$):

$$\mathrm{L1}_{\min} = 2T(1 - \alpha_{\max})Vin_{\max}/I_{out}. \tag{5.13}$$

Choose a coil with a standard inductance nominal closest to the calculated value, and make sure that its saturation current satisfies the following condition:

$$\begin{aligned} I_{\mathrm{L1sat}} &>> I_{\mathrm{L1}} + 0.5\,\Delta I_{\mathrm{L1}} \\ &= \mathrm{Aa}_{\min} I_{out} + 0.5\,T\alpha_{\min}\,Vin_{\min}/\mathrm{L1}. \end{aligned} \tag{5.14}$$

The calculations for L2 are similar to those for L1:

$$\mathrm{L2}_{\min} = 2T\alpha_{\max}\,Vin_{\max}/I_{out}, \tag{5.15}$$

$$\begin{aligned} I_{\mathrm{L2sat}} &>> I_{\mathrm{L2}} + 0.5\,\Delta I_{\mathrm{L2}} \\ &= I_{out} + 0.5\,T\alpha_{\max}\,Vin_{\max}/\mathrm{L2}. \end{aligned} \tag{5.16}$$

If L1 and L2 are wound on the same core, it is necessary to choose the largest of the two values of inductance. Two windings with the same number of turns and, therefore, with the same inductance can be wound on one core. If the inductances are different, the voltages on the windings will be different, and the difference will be closed through C_p. This useful property is used in power factor correctors to reduce the ripple of the mains current due to the energy in C_p, that is, when co-winding L1 is wound with greater scattering relative to the core than L2 (or if there is a ready choke with identical windings, then L1 is artificially increased) [10].

If the windings do not differ from each other, then the cost of the product can be reduced by winding them together in one operation.

The output capacitor C_{out} must smooth out large current pulses coming through the diode D1 during T_{off}. Therefore, C_{out} must be a high-quality component, as well as in the Flyback topology. Fortunately, modern ceramic capacitors have a very low equivalent series resistance (ESR). The minimum capacitance of C_{out} is determined from the allowable pulse value of output voltage ΔV_{out}:

$$C_{out} \geq \mathrm{Aa}_{min} I_{out} \alpha_{min} T/\Delta V_{out}. \tag{5.17}$$

In reality, an output capacitor of substantially larger capacitance may be required, especially if the load current is pulsating a lot. The input capacitor can be very small due to the filtering properties of the SEPIC topology. C_{in} can usually be ten times smaller than C_{out}:

$$C_{in} = C_{out}/10. \tag{5.18}$$

The overall efficiency η is calculated on the basis of V_{in} and Aa. The result, however, may turn out to be too optimistic, since it will not take into account losses in the key and in the core:

$$\eta = V_{out}/\mathrm{Aa}\, V_{in}. \tag{5.19}$$

In different sources, you can find different formulas for calculating the inductances of the SEPIC converter. For example, the document [8] contains formulas similar to (5.13) and (5.15), but the coefficient 2 in them is in the denominator, not in the numerator. This obviously gives four times smaller inductance ratings. However, computer simulation of the circuit operation shows that the inductances calculated according to document [8] are sufficient for the converter to operate in continuous current-mode. A third variant is given in [9], the results of calculations for which are closer to the results of [8]. This article provides a complete calculation method from [10], because of all the methods found by the author [1], it is the most detailed and allows you to calculate the largest number of parameters.

And finally, the key S_w and the diode D1 should be chosen so that their breakdown voltage is greater, respectively, than V_{ds} and V_r:

$$V_{ds} > 1.15\,(V_{out} + V_d + V_{in}), \tag{5.20}$$

$$V_r > 1.15\,(V_{out} + V_{in}). \tag{5.21}$$

As an example, we find the component ratings for the following low-power application: $Vin_{min} = 2.7$ V, $Vin_{typ} = 3.5$ V, and $Vin_{max} = 5$ V; $V_{out} = 3.8$ V, $I_{out} = 0.38$ A, $T = 2$ µs, and $V_d = 0.4$ V. Rounding the obtained values, we get: LI = L2 = 47 µH, RL1 = RL = 120 mOhm, Rep = 50 mOhm, and $R_{sw} = 170$ mOhm.

Using the expression (5.2), we first calculated the ideal amplification factor Ai for the minimum, typical, and maximum input voltages V_{in}, obtaining, respectively,

1.555, 1.2, and 0.84. Substituting these numbers into expression (5.3), we obtain more accurate coefficients Aa, equal to 1.735, 1.292, and 0.88. And the corresponding fill factors are obtained from the expression (5.4): 0.634, 0.563, and 0.468.

The current through the coil L2 (IL2) is equal to 0.38 A in accordance with the expression (5.5), and IL1 varies depending on V_{in}. Using the formula (5.6), we obtained the values of IL1 equal to 0.659, 0.491, and 0.334 A for the given values of the input voltage.

From the expression (5.7) we obtained for a given value of $y = 5\%$ the minimum capacitance C_p equal to 3.5 μF. The calculated voltage for C_p is obtained from (5.1). Modern multilayer ceramic capacitors provide Rep up to 50 mOhm, while the power loss calculated by the formula (5.8) is 12.5 mW.

The following parameters are calculated for the worst case, that is, for the minimum input voltage V_{in} [7]:

- A key with a resistance of 170 mOhm must dissipate 116.5 mW in accordance with the expression (5.9), which allows the use of a transistor in an SOT23 package or even in an even smaller SC70.
- Expressions (5.10) and (5.11) give a loss of 52.2 and 17.3 mW in inductors L1 and L2. In this case, we must make sure that the cross section of the wire winding L1 is larger than that of L2.
- Having calculated by the formula (5.12) the power dissipation of the diode D1, equal to 152 mW, we see that this diode is the main source of losses. Therefore, it is very important to choose a diode with a small voltage drop.
- For L1, the expression (5.13) gives a minimum inductance of 28 μH, which with a sufficient margin corresponds to the standard nominal value of 47 μH. For normal operation with this inductance value, expression (5.14) suggests a peak current of 0.69 A.
- According to expression (5.15), we have a minimum inductance of L2 equal to 24.6 μH. Here, the inductance with a nominal value of 47 μH can also be used. In accordance with formula (5.16), L2 must withstand a peak current of 0.43 A.
- For output voltage ripple of 38 mV, according to expression (5.17), the output capacitor capacitance must be at least 22 μF. The input capacitor C_{in}, according to (5.18), should have a capacity of 2.2 μF.
- Despite the significant values of parasitic parameters, expression (5.19) predicts a rather good efficiency of 81% for the worst case, when the input voltage is minimal. If we consider switching losses, the real efficiency will be somewhat smaller.

5.2.2 *Analysis and Calculation of Bipolar Voltage Source Based on the SEPIC Converter*

Based on the described above standard pulse voltage converter with the SEPIC topology, a bipolar symmetric voltage source can be designed by adding several passive components.

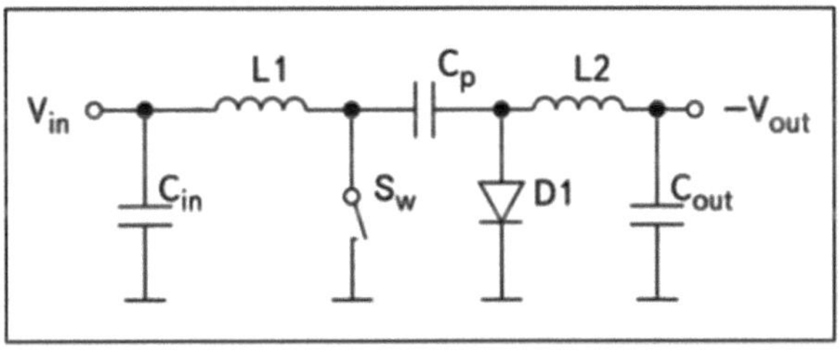

Fig. 5.8 Chuka converter circuit

To obtain a negative voltage from the positive, most commonly, the Chuka converter circuit is used (Fig. 5.8).

If we compare this circuit to the SEPIC converter circuit shown in Fig. 5.7, it can be seen that the left sides of these circuits are the same up to the isolation capacitor. The output voltage of the SEPIC converter can be calculated using the formula [7]:

$$V_{\text{out}} = V_{\text{in}} \times \alpha/(1 - \alpha). \tag{5.22}$$

The formula for the output voltage of the Chuka converter is

$$V_{\text{out}} = -V_{\text{in}} \times \alpha/(1 - \alpha). \tag{5.23}$$

Thus, by connecting the components D1, L2, and C_{out} from Fig. 5.8 to the interface L1, L2, S_w, and C_p of the SEPIC converter circuit (Fig. 5.7) through the separation capacitor C_p2, we get a symmetric bipolar voltage source (Fig. 5.9), which uses only one key and, therefore, only one controller or converter chip.

The elements of such a circuit are calculated in the same way as the SEPIC converter described above. The difference is that the ratings of inductances L2 and L3 must be doubled compared to the calculated ones, since the current flowing through L1 is distributed to two outputs.

Consider an example of a practical implementation of the bipolar voltage source discussed above. This example is part of a device that was developed in reality by the author of device [7], for the power supply of which a low-power, compact secondary source was required.

The requirements for the product being developed are:

- The device should work with a variety of standard DC voltages of the secondary power source, adopted in electronic equipment, including the input voltage range from +4.5 to +12.5 V. The power source should operate on standard voltages of +5, +9, and +12 V.
- The output of the source should be +11 and −11 V.
- Load current for both output voltages is 0.1 A.
- Operating temperature range—industrial, −40 to +85 °C.
- Device dimensions should be minimal; vertical dimensions should not exceed 3 mm.

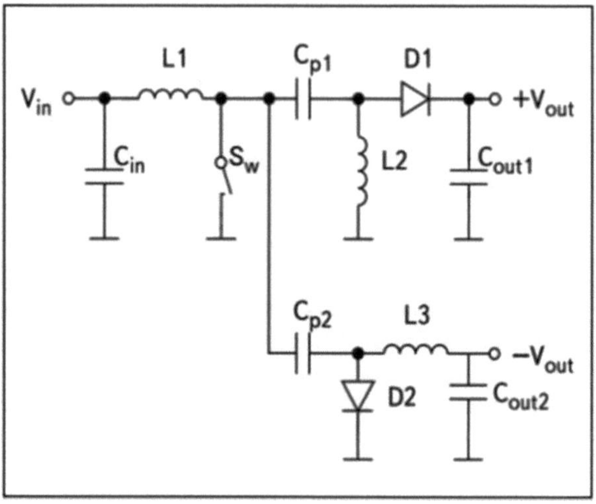

Fig. 5.9 Bipolar voltage source based on a SEPIC converter [10]

- The source must be able to turn off when the entire device is switched to a low consumption mode.
- The source is designed to power accurate analog circuits, so the level of interference should be minimized. In particular, the specifics of the operation of these circuits require external synchronization of the converter.

The design algorithm is as follows:

1. Select the converter chip.
2. Determine the parameters of other components of the circuit.
3. Simulate the converter to verify the correctness of the calculations.

5.2.3 Selection of the Converter Chip

Selection of the converter chip is determined by the set of requirements for the circuit, and first of all it is required to obtain the smallest possible dimensions. At the present level of miniaturization of semiconductor devices, the overall dimensions of the device will mostly be influenced by inductance dimensions. These, in turn, are inversely proportional to the frequency of the converter or the controller chip. Therefore, first of all, we need to learn the range of available transducers, determine the maximum frequencies at which they can operate, and choose the most suitable type. Other parameters of the converters, determined by the requirements of the device, should also be taken into account: input and output voltage ranges, operating temperature range, maximum permissible key current (for a converter with a built-in key), and accessibility. Also, special requirements should be considered: possibility of external synchronization and shutdown.

On the Internet, it is often possible to come across the statement that any chip designed for use in boost converters can be used to build a converter with the SEPIC topology. This is, however, not true. For example, the TPS61120 chip by Texas Instruments is definitely not suitable because of the characteristics of its launch algorithm: First, output key is opened (in a normal, non-synchronous circuit there is a diode), then you need to wait until the output voltage rises to the input. In the SEPIC topology, there is no direct path of direct current passing from input to output, consequently, the output capacitor will never be charged this way [10]. There is a specific reference to this on the company's website. The most reliable way is either to focus on chips for which the reference data specify the possibility of operation in SEPIC topology or to clarify this point in the manufacturer's support desk. Otherwise, it is necessary to make a study of the reference data.

The greatest number of types of chips for construction of switched-mode power supplies is made by such companies as Texas Instruments (www.ti.com), National Semiconductor (www.national.com), Maxim-Dallas (www.maxim-ic.com), Linear Technology (www.linear.com), and On Semiconductor (www.onsemi.com). The websites of these companies are provided with convenient search systems that enable selecting and sorting the chips by preset parameters.

After having considered the available assortment of chips, it can be concluded that the highest frequencies of operation of switching converters produced in 2015, are within the range of 500–1500 kHz, so it is necessary to focus on the converters that can operate at such frequencies.

A considerable number of chips will be rejected in practice due to strict requirements for the range of input voltages: input voltage shall not exceed 10 V for most of the high-frequency transducers.

In particular, study [7] demonstrates that many chips made by Linear Technology were rejected after check for availability. Despite the fact that products of this company have good parameters and they are quite widespread in Russia, their versions with the temperature range named Extended Commercial are available for sale. The operating temperatures from −40 to +85 °C are specified for Extended Commercial range but there is a clause that proper operation of the product is not guaranteed under negative temperatures. It is rather challenging to buy versions with the original industrial range of operating temperatures that have been produced in small amounts. Furthermore, the products of this company are generally sold at very high prices.

So, for example, based on search results, the developers [7] determined three appropriate chips: LM3488MM made by National Semiconductor and a pair of chips MAX668EUB and MAX669EUB made by Maxim-Dallas. These chips have much in common: they are produced in small-sized 10-output packages of the same size; they require one external transistor; many parameters are similar; prices are not significantly different. However, the maximum conversion frequency of LM3488MM is twice as high as that of MAX668EUB/MAX669EUB: 1 MHz vs. 500 kHz. That's why LM3488MM was chosen.

5.2.4 Determining Parameters of Other Components of the Circuit

Since the detailed calculation of SEPIC converter circuit parameters proposed in the work has already been described above, there is no need to repeat it in this section. We use the more brief method of calculation based on the technique specified in the reference data for LM3488MM chip [11].

We determine the conversion frequency. First of all, we determine the minimum fill factor relevant to the preset peak input voltage using the following formula:

$$D = (V_{\text{out}} - V_{\text{diode}})/\left(V_{\text{out}} + V_{\text{in}} - V_q + V_{\text{diode}}\right). \tag{5.24}$$

where D is a fill factor; V_{out} and V_{in} are output and input voltages of the converter, respectively; V_{diode} is a diode voltage drop; V_q is a voltage drop on the open switching transistor. Since our circuit is designed for relatively low current, the voltage drop on the power field-effect transistor may be neglected in this case. By plugging the peak input voltage of 12.5 V, output voltage of 11 V, and Schottky diode voltage drop of 0.4 V in the formula, we obtain the minimum fill factor of 0.48. Based on the reference date for LM3488MM chip, it is known that the minimum time of output transistor connection for it is 550 ns. Then, the minimum period of conversion is 550/0.48 = 1146 ns; therefore, the conversion frequency shall not exceed 873 kHz. The nearest frequency of external synchronization that can be reached in the device under development is 750 kHz. Therefore, further calculations will be done for conversion frequency of 750 kHz.

Now we define the minimum allowed values for coil inductances under which the circuit operates in the mode of continuous currents. The formula for L1 inductance is as follows [7]:

$$\text{L1} > \left(V_{in} - V_q\right) \times (1 - \text{D})/(2 \times I_{\text{out}} \times F_s). \tag{5.25}$$

where Fs is a conversion frequency. It is clear that the maximum inductance is relevant to the minimum fill factor. Disregarding the value of V_q, we plug the values of other parameters in the formula and obtain the minimum allowed inductance of 43 μHY.

Then, we calculate the output coil inductance according to the formula:

$$\text{L2} > \left(V_{\text{in}} - V_q\right) \times D/(2 \times I_{\text{out}} \times F_s). \tag{5.26}$$

The maximum inductance is obtained when the maximum fill factor is equal to 0.72. It is equal to 39 μHY. This value should be doubled for the bipolar source selected as an example; we get 78 μHY.

To choose the certain induction coils, it is necessary to determine the maximum current which flows through them in the circuit. It consists of the DC component and ripple amplifier. The DC component for the input coil will be equal to [7]:

$$\text{Iavg} = D \times I_{\text{out}}/(1 - D). \tag{5.27}$$

The DC component for the output coil is equal to output current. Double current ripple amplifier in the coil is equal to:

$$\Delta I_L = \left(V_{\text{in}} - V_q\right) \times D/L/\text{FS}. \tag{5.28}$$

It follows that the maximum current through the input coil will be equal to 0.563 A and to 0.148 A through the output coil.

Taking into account the range of operating temperatures, maximum allowed height and availability for developers, the induction coils of B82462G4 family made by EPCOS were chosen: B82462G4473M with the rated inductance of 47 μHY and active resistance up to 0.34 Ohm, as well as B82462G4104M with the rated inductance of 100 μHY and active resistance up to 0.58 Ohm.

Next, the field-effect transistor can be chosen. It shall withstand the maximum voltage:

$$V\text{sw}_{\max} = V_{\text{in}} + V_{\text{out}} + V_{\text{diode}}. \tag{5.29}$$

The maximum current through this transistor is equal to:

$$I\text{sw}_{\max} = I_{\text{L1}}\text{avg} + I_{\text{out}}(\Delta I_{\text{L1}} + \Delta I_{\text{L2}})/2. \tag{5.30}$$

It is 0.25 + 0.1 + 0.1/2 = 0.4 A for our circuit. In the examples of [12], it is recommended to use transistor IRF7807. However, this transistor is rather bigger in size (SOIC-8 package) and designed for current up to 6.6 A that is clearly excessive for this circuit. FDN5630 transistor made by Fairchild Semiconductor will be a good choice. It is produced in package similar to SOT-23; it is designed for current up to 1.7 A and maximum drain-to-source voltage of 60 V. In addition, its performance is higher than that of IRF7807; therefore, losses due to switching will be lower.

The diode used in the circuit shall withstand the same current as the switching transistor does, and its peak reverse voltage shall exceed the sum of input and output voltages. Besides, it shall be Schottky diode to reduce losses. Small MBR140SFT diode made by On Semiconductor having the maximum current of 1 A and reverse voltage up to 40 V was chosen from a wide range of diodes offered by different manufactures.

The coupling capacitor is calculated as follows:

$$C_p > \text{L1} \times {I_{\text{out}}}^2/\left(V_{\text{in}} - V_q\right)^2. \tag{5.31}$$

Now that might be the end of circuit calculation and we can start the circuit simulation. The calculation of circuit elements specific for LM3488MM chip won't be considered herein. To determine their parameters, it is recommended to apply approaches specified by the reference data, as well as Webench interactive design maker placed on the website of National Semiconductor company.

5.2.5 *Brief Description of the Modeling Process of the Converter SEPIC*

Modeling will be performed in Micro-Cap software. It is a widely known, accessible, simple-to-learn, and quickly running program.

First of all, we create a modeling framework (Fig. 5.10) [7].

In this framework, resistors Rl, R2, and R9 represent the resistance parasitics of coils LI, L2, and L3; resistors R4 and R7 represent the resistance parasitics of multilayer ceramic capacitors C2 and C4. Resistors R4 and R8 simulate the load. Resistor R5 is a current-measuring resistor for current-limit circuit LM3488MM. A record in the form of L(L2) near L3 means that L3 inductance is equal to L2 inductance.

As it is accepted in Micro-Cap software package, when setting up and displaying of various values, it is required to use letters indicating the unit fractions: *n* is nano, *μ* is micro, and *m* is milli. For example, a record of 10 μ near the capacitor means 10 μF.

Element V2 is an impulse generator (Pulse Source bean). It will simulate LM3488MM chip. Prior to modeling, it is necessary to preset its parameters (Fig. 5.11). Most parameters should be the same as they were defined by default. It is necessary to define only levels and timing parameters of impulses. The levels are preset in VZERO field (the lower level) and VONE field (the upper level). It is clear that the lower level should be equal to zero. The upper level should be equal to 4 V in order to monitor the reliable opening of FDN5630 switching transistor when supply voltage is 4.5 V. Fields P1-P4 set the timing parameters of impulses. Figure 5.11

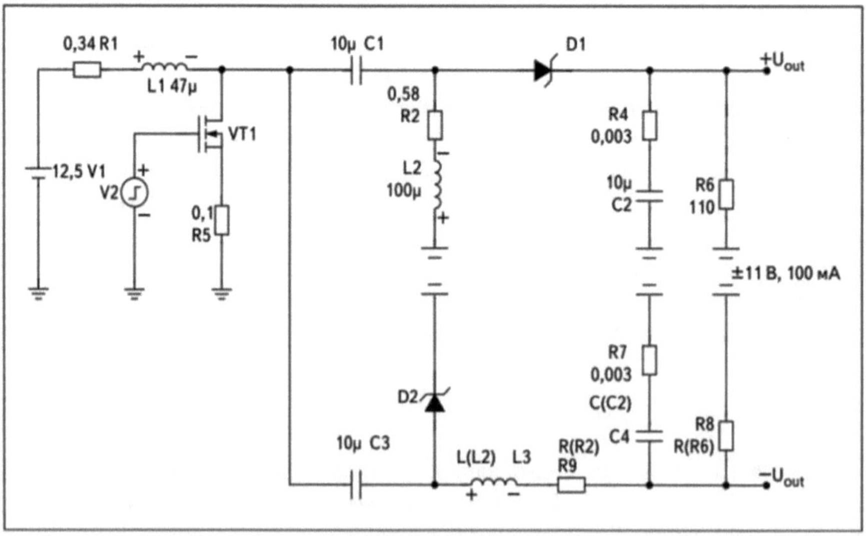

Fig. 5.10 Framework for modeling of bipolar converter

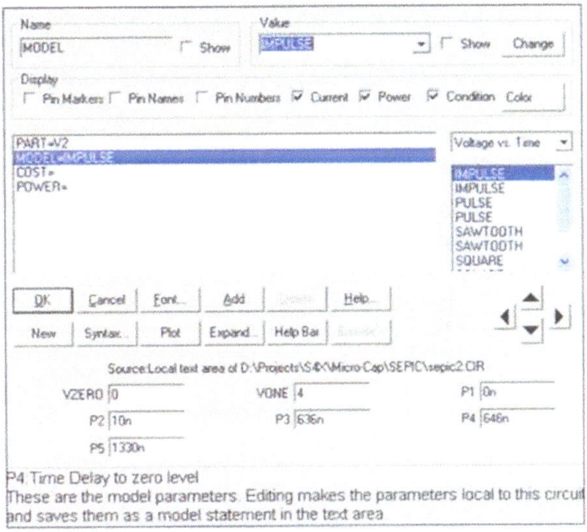

Fig. 5.11 Generator parameter setting

illustrates these parameters for the lowest fill factor. The purpose of any field is easily determined by aiming a mouse cursor: in this case a prompting message appears at the bottom of parameter setup window.

The next step is to draw up the task on modeling. Here it is reasonable to apply a transient-response analysis (Transient Analysis). The window to manage this type of analysis is called in menu option "Analysis–Transient" or using keys Alt+1. Then, the task drawn up will be displayed as it is illustrated in Fig. 5.12.

As a result of implementation of this task, we will get the time-based historical graphs of output voltages (the graph above, number 1 is in column P) and of currents through induction coils (the bottom graph). The voltage in negative output and current through its induction coil should be plotted as inverted ones so that all graphs are above the *X*-axis and look bigger. To start-up a task for implementation, click the Run button. The result of calculation of this example is illustrated in Fig. 5.13 [7].

It is clear that after circuit connection, in-rush current of 2 A or greater illustrated in Fig. 5.13, is smoothed in reality by the smooth start and current-limit circuits built into LM3488MM chip. The graph above shows that the calculated output voltages are close to the design value of 11 V.

Now it is necessary to enlarge the image along the axes of time, voltages, and currents applicable to the steady-state mode (Fig. 5.14).

The upper graph shows that the maximum double peak-to-peak output voltage ripple (in positive voltage output) doesn't exceed 7 mV. The bottom graph shows that the circuit operates in the mode of continuous currents since the current exceeds 36 mA through any coil at any time.

Furthermore, it is possible to experiment on the model as much as desired by changing its parameters and monitoring changes in the circuit operation. For example, we try to simulate the asymmetric loading. To do that, it is necessary to reduce

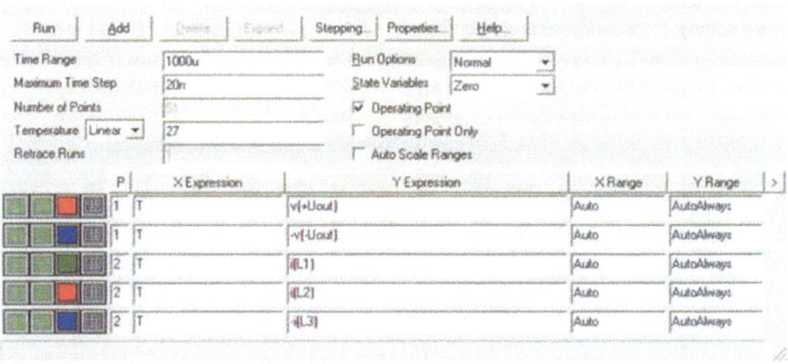

Fig. 5.12 Drawing up the task on modeling

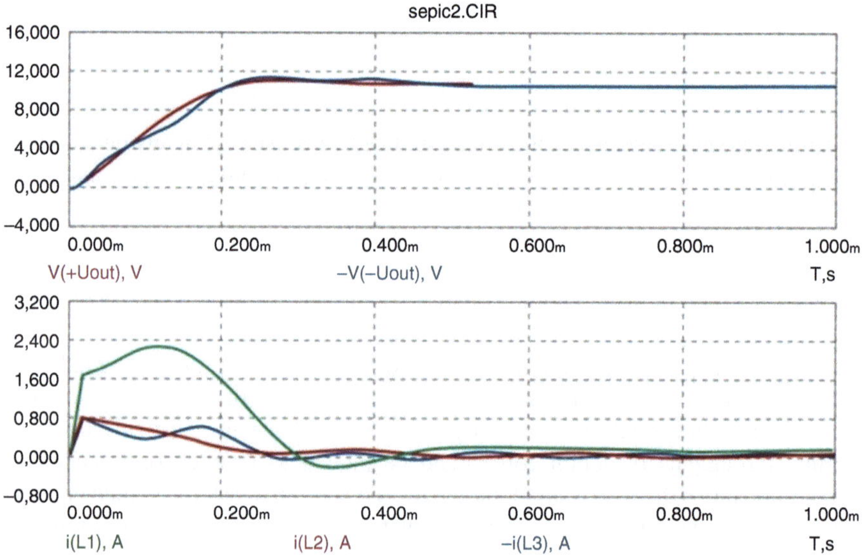

Fig. 5.13 The result of modeling of the transient process in the converter circuit

by half the upper load resistor R6 in order that current tapped from the high-side power doubles. Such situation occurs when power is supplied from the positive source of certain digital chips. The result is illustrated in Fig. 5.15.

Figure 5.15 shows that average difference between modules of positive and negative output voltages doesn't exceed 125 mV or about 1%. Such asymmetry is regarded as acceptable to the majority of practical applications.

Article [13] discusses the result of application of SEPIC topology for development of small-sized low-power secondary power supply source. The power source of high quality was received that meets all the above-mentioned requirements (Fig. 5.16).

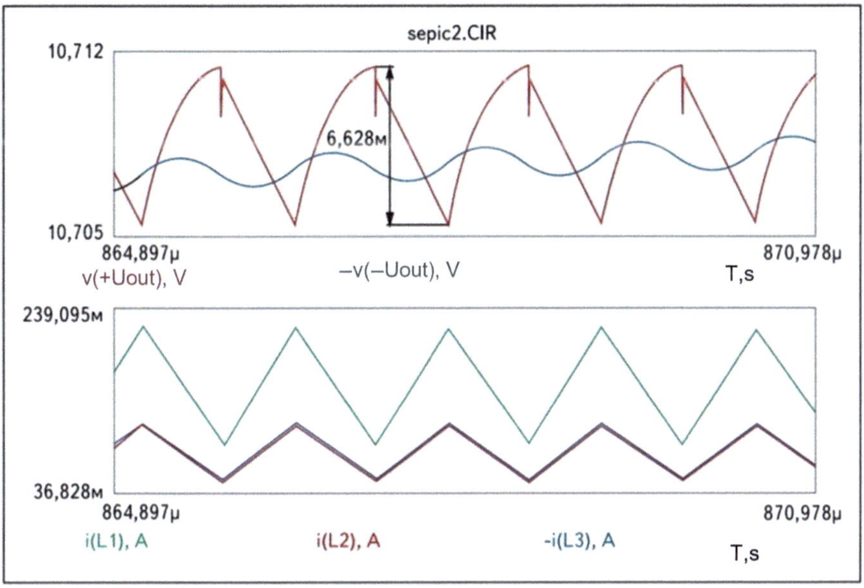

Fig. 5.14 The detailed survey of modeling outcomes

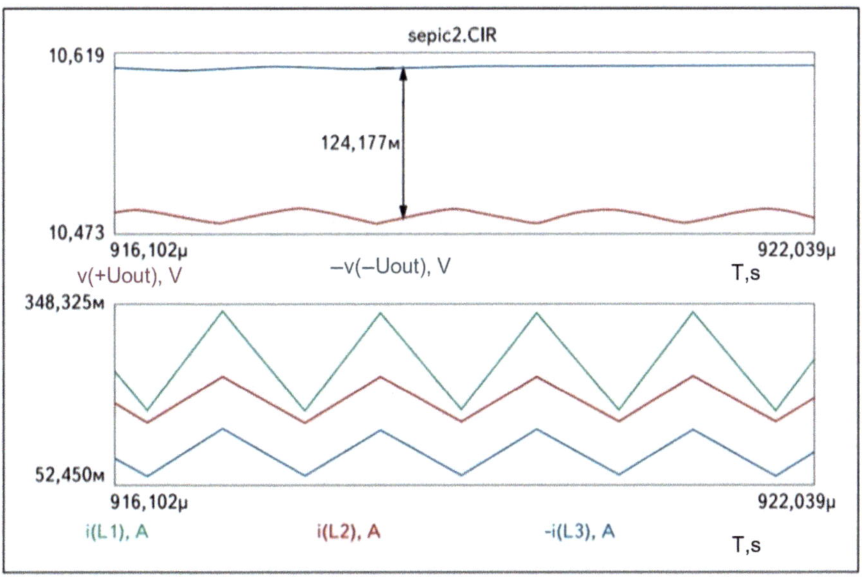

Fig. 5.15 Modeling of bipolar source with asymmetrical load

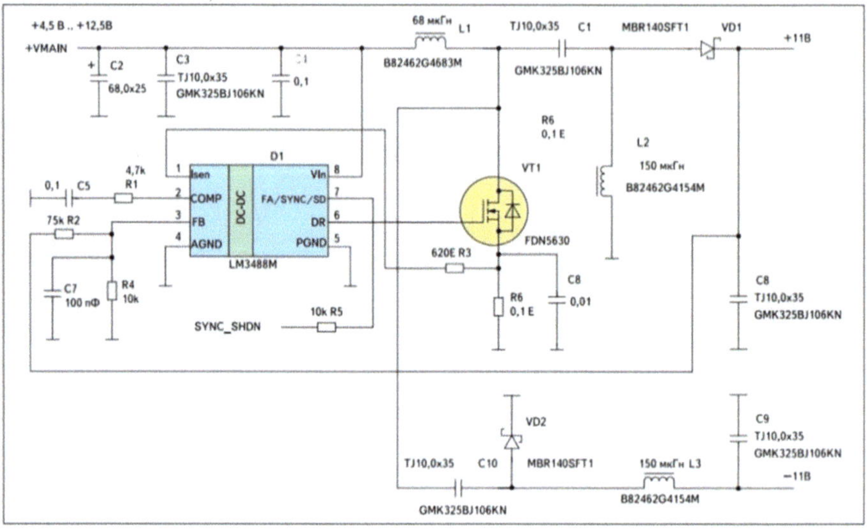

Fig. 5.16 The schematic circuit diagram of designed power supply.

5.3 Principles of Organization and Application of Chips of Converters with DCS-Control Topology

5.3.1 Basic Principles of Operation of the Chips of DCS-Control Converters

As seen from the above, the general technical capabilities of controllers and their performance in the widest possible range of loads are constantly increasing which makes these devices universal and suitable for a wide range of applications including for telecommunication system equipment that is highly-sensitive to noises and interferences. Such functioning can be ensured by transition of the controller to certain mode of regulating and managing in dependence to its loading as described in publication [14]. Texas Instruments Company developed a new topology under the trademark of DCS-Control (Direct Control with Seamless Transition into Power Save Mode) for widespread non-insulated synchronous DC/DC Step-Down Converters. Distinctive features of this topology (that is reflected in its name) involve regulation by method of direct management with seamless transition (without significant transient processes in the form of output voltage surges and drops) from the mode of high output currents to the energy-saving mode.

Management topology of DCS-Control type represents an optimal solution on constant-voltage regulation that combines the advantages of voltage-mode, current-mode with hysteresis management topologies [14] ensuring the best loading and transient characteristics. In this case the final solutions on such controllers show extremely low level of output voltage ripple even in the energy-saving mode as well as smaller generation of noises and electromagnetic interferences. Besides, by using an additional feedback loop for voltage-sensitive control, this topology ensures the high accuracy of DC/DC converter in the static mode. DCS-Control topology reduces the need for external filter components minimizing the space occupied by the solution on the circuit board and diminishing the value of final solution.

In general, DCS-Control topology represents a hysteresis topology [14] that includes a few more internal cascades providing the advantages of topologies with voltage- and current-sensitive control. The block diagram of controller for DC/DC step-down converter of DCS-Control topology is illustrated in Fig. 5.17. There are two inputs in the converter topology for implementation of DCS-Control: feedback input (FB terminal) and input for direct tracking of output voltage (VOS terminal). FB input operates and is used just as in the majority of DC/DC converters. It is connected either to the error amplifier with high input impedance or to the operating operational amplifier designed for error signal generation by comparing the voltage in FB terminal with the voltage of the internal reference V_{REF}.

Like in other types of DC/DC converters, the error amplifier ensures the fine control of output voltage. The voltage divider installed in FB input between converter output and common wire ("ground") programs the required output voltage. For implementation of controller of TPS6213x series with fixed output voltages, the FB input is connected via the voltage divider between VOS terminals to GND but FB terminal itself doesn't participate in output voltage control (it is connected to the grounded internal resistor and it can be designed for other purposes). Such solution reduces the quantity of external components by two items, as well as it lowers the FB input sensibility to inductive interferences. Besides, for ensuring the error amplifier stability, the appropriate compensation is cut in a circuit of its negative voltage feedback that increases the stability of controller operation.

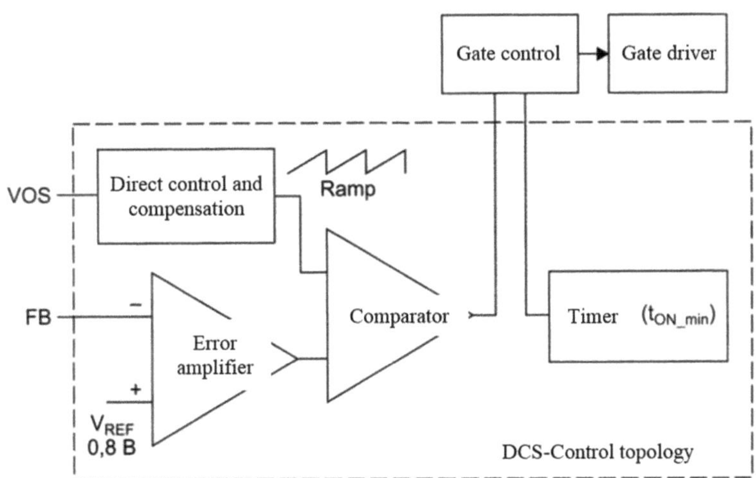

Fig. 5.17 The block diagram of controller for DC/DC step-down converter of DCS-Control topology

It is important to directly connect the VOS terminal to the output voltage converter at the point of connection of output capacitor using a communication line of minimum length. Similar to FB input, this input is performed with high resistance and it is closed in the control loop. However, as opposed to FB input, the VOS terminal is connected not to the error amplifier but to the circuit generating the line-to-line saw-tooth voltages that was developed by specialists of Texas Instruments Company. Then, the voltage is aligned with the mismatch signal from the error amplifier as it usually happens in the voltage- or current-sensitive control modes. This part of internal controller circuit from VOS input to comparator ensures a prompt response commonly found in DCS-Control topology. Changes in the output voltage in VOS input are directly transferred to the comparator and immediately influence the device operation. That's why the VOS input is rather sensitive to noises and interferences, so the circuit of its connection to the point of output voltage should be the shortest and direct; as well as the VOS input should be connected directly to the output capacitor. Besides, the appropriate compensation is provided here for ensuring the stability of this control loop.

Then, the comparator generates a signal which goes to the control circuit and report whether it should feed the next control pulse to the gate driver that controls the MOS transistor on the high-voltage side. To ensure a prompt response to the transient processes, the comparator is combined with timer circuit and controls the controller switching frequency.

Based on the V_{OUT}–V_{IN} ratio, the timer sets the minimum time (t_{ON_min}) by which the switch-on time from the comparator can be increased. The minimum switch-on time set by timer is specified in the technical description of the device, and the switch-on time is calculated in accordance with equation:

$$t_{ON} = V_{OUT}/V_{IN} \times 400\ \text{ns}.$$

In this example based on TPS62130 controller (an option with adjusted output), the minimum switch-on pulse length is 400 ns. Thus, the switching frequency is equal to 2.5 MHz. The adjusted switching frequency remains unchanged within the entire range of input and output voltages with V_{OUT}/V_{IN} factor what as we said before, and sets the minimum switch-on time (t_{ON_min}) based on the ideal operating cycle for step-down converters. Thus, the switch-on time equation can be recorded as t_{ON} = DxtPeriod that is the exact definition of switch-on time for any step-down converter.

The MOS transistor control on the low-voltage side is rather simple. After switching-off the MOS transistor on the high-voltage side, the MOS transistor on the low-voltage side is activated and current in the inductor is efficiently reduced. The MOS transistor on the low-voltage side is switched-off when current in the inductor drops to zero or when the comparator gives a command to switch-on the MOS transistor on the high-voltage side again. For avoiding of dangerous steady leakage currents arising due to mismatching of the MOS transistors on the high- and low-voltage sides, the controller control system forms a pause, some "dead time" during which the transistor connection control is blocked.

5.3.2 *Features of Operation of Converters in Energy-Saving Mode*

The key feature of DCS-Control topology is energy-saving mode. As a rule, most of the energy-saving modes are activated at low currents of load and enable to improve efficiency (coefficient of efficiency) of conversion by skipping of switch-on pulses and reduction of current consumption by device itself (static current). The pulse skipping takes the device to the discontinuous conduction mode (DCM) thus eliminating the negative current of the inductor (current flowing from the output to the input) that otherwise, is inevitable at light loads. Some switching cycles are skipped at low currents of loading that results in additional losses what negate the converter efficiency. Reduction in static current in its turn increases the coefficient of efficiency only under very low loads.

Energy-saving mode in DCS-Control topology is very simple. There is no change-over during switch from PWM-control mode into energy-saving mode in contrast to some other topologies where control is implemented via switching between control method for energy-saving mode and control method for PWM mode with standard operating load. Such switch between modes provides an opportunity for instantaneous power failures and results in random interferences arising as a reaction in the form of a transient process when switching between modes.

Energy-saving mode in DCS-Control topology is arranged simply enough: if comparator doesn't require a switching pulse, then such pulse isn't generated. So, if the output voltage exceeds its presets (it is identified by the error amplifier), then when the coil current drops to zero, the controller doesn't generate a new pulse. Instead of this, it reduces its current consumption down to the static current condition and goes into energy-saving mode. The controller is waiting until the error amplifier reports to the comparator that output voltage has been reduced down to the preset value (it is hysteresis mode, as we said earlier) and now it may be increased. Then, the controller generates a switching pulse and if the output voltage level has increased enough to appear within the preset range of adjustment, then this pulse remains with the minimum duration of switching-on. The minimum propagation delays of analysis and control circuits ensure the high efficiency and good stability (adjustment) of the output voltage in energy-saving mode.

Such single-shot switching pulses with minimum duration supply the smallest amount of energy to the converter output. Thus, they generate the minimum output voltage ripples. When increasing the current in light loading mode, single-shot pulses will be generated more often and located closer to each other, thus expanding the switch frequency that appears higher than audio frequency range in a shorter period of time than the period showed by other topologies. In energy-saving mode, such topologies use pulse trains or bursts that due to control loop delays (they are purposefully entered) [14], transfer much more energy during such surges than it is required for support of the converter output. Because of this, the output voltage requires more time to be reduced down to the preset value (the output capacitor should be run down, its discharge time is finite), the pulse trains are located rather far from each other, and they keep longer the effective recurrence rate that is within audio range. Single-pulse architecture of DCS-Control enables more efficient operation above the audio range under lower currents of loading than alternative topologies.

In case of increase in load current up to the level where there is no space for a pause between pulses, before the comparator issues a command to switch-on the MOS transistor on the high-voltage side, the inductor current doesn't go back to zero anymore.

This load current condition is at the edge of DCS mode; and in this case, the converter changes its energy-saving mode into standard PWM-control mode.

Completely predictable operation in energy-saving mode (a single pulse with the minimum pulse length with frequency regulation) and transition to standard PWM-control mode when zero current of the inductor is reached, makes the DCS-Control topology more flexible than other topologies. Said solution makes it possible to simplify the DC/DC converter configuration for the requirements of the end system. Take the following as an example: output voltage ripples in energy-saving mode for a system with input voltage of 12 V and with output voltage of 3.3 V. For this, we used an estimator made by Texas Instruments Company and based on TPS62130 controller [14] operating at a frequency of 2.5 MHz. The purpose is to demonstrate how to reduce the output voltage ripples due to increase in the external inductance and output capacity. Open-circuit operation, i.e., when the converter is not loaded, was used to demonstrate the worst level of output voltage ripple in energy-saving mode.

Figure 5.18a illustrates the initial level of output voltage ripples being equal to 26 mV (semiconductor)—0.8% of rated output voltage of 3.3 V. This level of ripples is defaulted. The increase in output capacity reduces output voltage ripples since the same amount of energy is transferred during each switching impulse. Unutilized energy accumulated in the capacitor has big output capacity (in this case, the converter stability isn't impaired) that results in lower switch frequency and lower rate of voltage ripples per unit time (Fig. 5.18b). From the other hand, the increase in inductance reduces the peak pulse current since the switch-on time remains

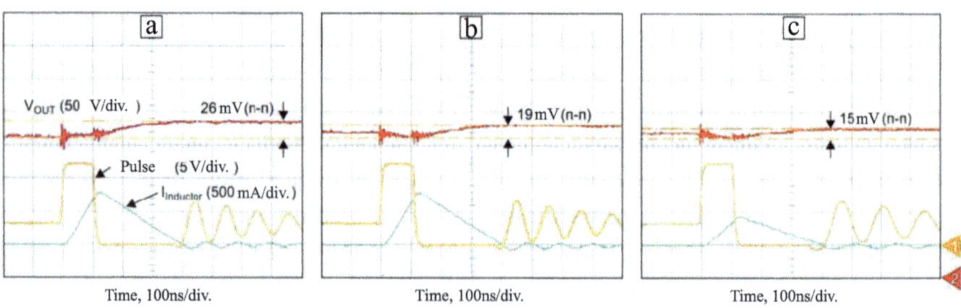

Fig. 5.18 The output voltage ripple through the example of DC/DC converter designed on the basis of TPS62130 controller: (**a**) circuit by default; (**b**) circuit with an auxiliary output condenser with a capacity of 22 μF; (**c**) circuit with an auxiliary output condenser with a capacity of 22 μF and increased inductance of the inductor up to 2.2 μHY

unchanged. The limits of inductance variations for the mode under consideration are described in publication [14].

Since the lower peak current accumulates the less energy in the inductor ($E = 1/2xLxp$) then the smaller amount of energy is supplied to the output that results in decrease of output voltage ripple level (Fig. 5.18c). Pay attention that switch-on pulse length is the same for all three outlined circuits since it is set by the controller circuit; it is fixed and it can't be changed by means of any external components. By changing the inductance, a research and development engineer sets the required minimum current of load under which the transition into energy-saving mode is carried out, i.e., it changes the DCS-control bound [14]. High inductance of the inductor reduces its current ripples. It means that the inductor current remains above zero up to the lower levels of the converter output current.

Such ability to adapt a point of DC/DC converter transition in energy-saving mode and to adjust the level of output voltage ripple to the level preset for specific needs allows to use this topology in various applications including in topologies that are extremely sensitive to noises and interferences in power supply circuits. Examples of application of DC/DC converters based on controllers using DCS-Control technology include wireless low-power transmitters of wireless systems and receivers of medical and industrial purpose, power supply systems for universal portable devices, as well as power supply systems for solid state drives (SSD). It is especially important that such controllers can be used in power supply circuits of medical equipment, since DC/DC converters based on DCS-Control topology generate lower values of electromagnetic interferences, the requirements for threshold values of which in medical equipment are constantly becoming more and more stringent [14].

Since the DCS-Control topology defines the current value of voltage output through VOS input connected directly to the converter output, it is well suited for quick response to the transient processes related to gradual changes of load current (so-called load drop/rise). As illustrated in Fig. 5.17, the output voltage is supplied directly to the comparator but it doesn't arrive there until turn-on time is adjusted for processing in the error amplifier with limited transmission bandwidth (bandwidth limitation in such control mode is inevitable) [14]. Being hysteresis by its nature, the DCS-control topology reacts instantly to fast transient processes related to rapid change in load current of DC/DC converter. This feature allows to increase an operating cycle of the controller up to 100%.

In such mode, the device keeps the MOSFET on the high-voltage side switched on for as long as necessary to recover preset level of output voltage. In other words, the requirement for turn-on time received from the comparator is completely satisfied. Figure 5.19 illustrates the response of the evaluation board of DC/DC converter based on TPS62130 controller to the transition process caused by 1 A load rise directly due to no-load conditions. This load rise is compensated by transition via the mode of 100% working cycle. The response time delay of 300 ns between the start of transient process and switching on the MOS transistor on the high-voltage side shows that the response to transient process is restricted almost completely due to problems of high signals (in particular, it is related to inductance of the inductor) but not due to low-signal problems (directly by the control topology itself). Thus, the

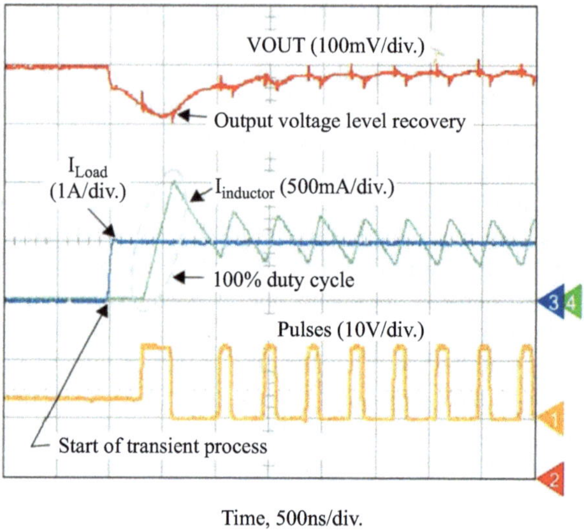

Fig. 5.19 An illustration of mode of 100% operating cycle during load rise using the example of DC/DC converter based on TPS62130 controller

DCS-control topology isn't the principal restriction of capabilities as regards a response to the transient process caused by the instantaneous change of loading. On the contrary, it enables to improve considerably a transient characteristic of DC/DC converter for the selected components of the postfilter. The response to load drop/rise is illustrated in Fig. 5.20 in a general view.

5.3.3 *Organization of Seamless Transition Between Controller Modes*

As noted above, DCS-Control topology contains one common control circuit which is used both for PWM-control, and for energy-saving mode. It enables to implement not only fast but also smooth seamless transition between two modes. Such approach when the circuit operating conditions are close to the limit between modes, ensures higher characteristics of the converter including its performance. This has been possible because there is no mode switch in the converter topology, so there are no "glitches" at its output during transition from one operation mode into another one.

Figure 5.21 illustrates behavior of the controller with DCS-control during a slow smooth transition from the mode of almost no-load condition to the mode of full-load condition. Figure 5.22 illustrates comparison of the controller behavior with the controller using other control topology. Here, load current (the lower route of green color) is linearly varying from 10 mA to 1 A and back up to 10 mA. Level of the transient processes or interferences arising when switching an operation mode of controllers, was monitored using the inductor current and output voltage.

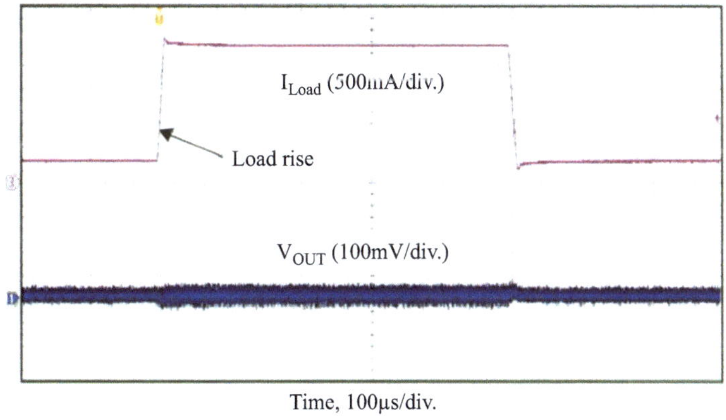

Fig. 5.20 The DCS-Control topology ensures a prompt end of transient processes related to a ramp change in load current increase without use of additional control and compensation circuits

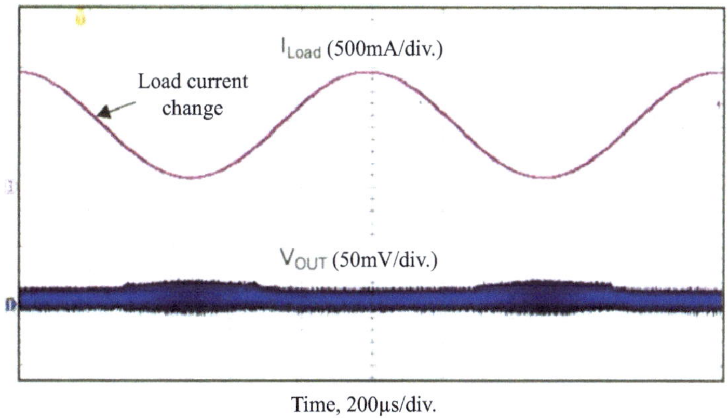

Fig. 5.21 The DCS-Control topology provides a smooth transition between PWM and energy-saving modes with very low-energy reaction of output voltage to such transition

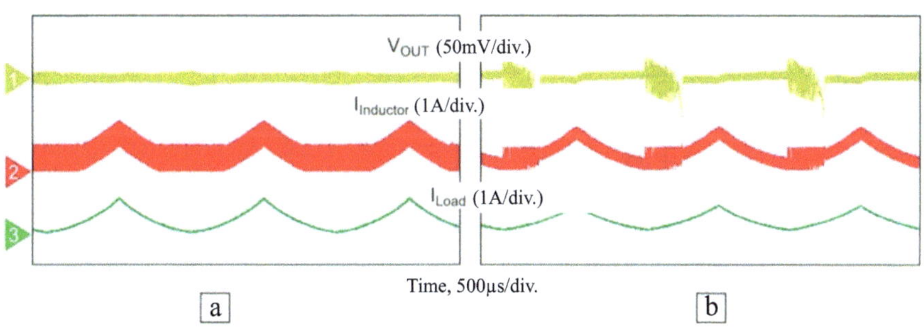

Fig. 5.22 Comparison of behavior of controllers of different types during their transition from PWM-control mode to energy-saving mode and back: (**a**) use of TPS62130 controller based on DCS-Control topology; (**b**) use of the controller based on alternative topology

Table 5.2 Controllers are made by Texas Instruments Company and based on DCS-control topology [14]

Controller	Range V_{IN}, V	I_{Load_max}, A	Range V_{out}, V	Characteristics	Design variants for field of application: Household appliances and wireless devices	Industrial/ telecommunication	Automobile, Q100	Package	Price
Of general purpose, small-sized, with high coefficient of efficiency									
TPS62230	2–6	0.5	Fixed output voltages	Required area on the board is 12 mm^2, height of 0.6 mm	•		•	1 × 1.5 × 0.6 mm SON	0.48
TPS62080	2.3–6	1.2	0.5–4	6.5 μA current consumption (sleep mode), Power Good flag		•	•	2 × 2 mm WSON	0.69
TPS62097	2.5–6	2	0.8—V_{IN}	iDCS-Control, Power Good flag, adjusted smooth start/tracking, selection of operating frequency		•		2 × 2 mm VOFN	0.81
TPS62085	2.5–6	3	0.8—V_{IN}	HotRod package; discharge of the output capacitor, Power Good flag		•	•	2 × 2 mm QFN	0.87
TPS62090	2.5–6	3	0.8—V_{IN}	Smooth start, tracking mode, CTPS62095 pin compatibility			•	3 × 3 mm QFN	0.87
TPS62095	2.5–5.5	4	0.8—V_{IN}	Smooth start, tracking mode, pin compatibility with TPS62090				3 × 3 mm VQFN	1.05

Extended range of input and output voltages									
TPS62177	4.7–28	0.5	1–6	Sleep mode with 5 μA current consumption				2 × 3 mm WSON	0.6
TPS62170	3–17	0.5	0.9–6	Power Good flag			•	2 × 2 mm WSON	0.53
TPS62160	3–17	1	0.9–6	Power Good flag			•	2 × 2 mm WSON, 3 × 3 mm VSSOP	0.75
TPS62150	3–17	1	0.9–6	Selection of operating frequency, smooth start/ tracking, Power Good flag			•	3 × 3 mm WQFN	0.79
TPS62140	3–17	2	0.9–6	Selection of operating frequency, smooth start/ tracking, Power Good flag				3 × 3 mm VQFN	0.88
TPS62130	3–17	3	0.9–6	Selection of operating frequency, smooth start/ tracking, Power Good flag			•	3 × 3 mm WQFN	0.96
With extended operational life of batteries, the lowest own consumption									
TPS62120	2–15	0.075	1.2–5.5	Coefficient of efficiency 96%; extended UVLO-hysteresis				2.9 × 1.63 mm SOT-23, 2 × 2 mm DFN	0.5
TPS62125	3–17	0.3	1.2–10	Adjustable threshold and enable input hysteresis (EN)				2 × 2 mm WSON	0.55
TPS62730	1.9–3.9	0.1	Fixed output voltage	Bypass, for radio frequency devices and devices with ultra-low power consumption	•			1 × 1.5 × 0.6 mm USON	0.52

(continued)

Table 5.2 (continued)

Controller	Range V_{IN}, V	I_{Load_max}, A	Range V_{out}, V	Characteristics	Design variants for field of application			Package	Price
					Household appliances and wireless devices	Industrial/ telecommunication	Automobile, Q100		
TPS62740	2.2–5.5	0.3	1.8–3.3	360 nA current consumption, 4 current discharges for adjustment of output voltage, load-interrupt switch	•			2 × 3 mm WSON	0.94
TPS62770	2.5–5.5	0.3	1–3	Back-to-back converters with load-interrupt switch in one device, 3 current discharges for adjustment of output voltage	•			1.58 × 1.58 mm WCSP	1.5
TPS62745	3.3–10	0.3	1.3–3.3	400 nA current consumption, 4 current discharges for adjustment of output voltage, input voltage switch				2 × 3 mm WSON	1.04
Comprehensible parts with integrated inductor									
TPS82740	2.2–5.5	0.2	1.8–3.3	360 nA current consumption, 3 current discharges for adjustment	•			2.3 × 2.9 × 1.3 mm MicroSiP	1.36

				of output voltage, load-interrupt switch					
TPS82084	2.5–6	2	0.8—V_{IN}	Hiccup mode, Power Good flag, discharge of the output capacitor				3 × 2.8 × 1.3 mm MicroSiP	1.65
TPS82085	2.5–6	3	0.8—V_{IN}	Hiccup mode, Power Good flag, discharge of the output capacitor				3 × 2.8 × 1.3 mm MicroSiP	2.15
TPS82130	3–17	3	0.9–5	Power Good flag, smooth start/tracking		•		3 × 2.8 × 1.5 mm MicroSiP	2.35

The bullet symbol is used to indicate the field of application for which the controller type in the first column is designed.

TPS62130 controller is based on DCS-control topology (Figs. 5.21 and 5.22) and characterized by much more smooth forms of output voltage and inductor current than that of the device applying an alternative control topology. Besides, TPS62130 controller shows the lower level of output voltage ripples within the full range of load currents. To be perfectly honest, ripples increase slightly when low currents of load but since the controller goes into energy-saving mode, this increase has considerably smaller impact than that of the devices based on other topologies. As can be noted from the example under consideration, the most important is that there is a short needle-form drop of output voltage (with a wide range of interferences) for the device with other topology.

As we can see, this drop occurs while current is increasing at a time when the device based on another topology different from DCS-Control topology goes out of energy-saving mode and goes into PWM-control mode. Naturally, the similar phenomenon is undesirable for sensitive loads or systems, and it is eliminated with the use of controller based on DCS-Control topology in the DC/DC converter.

The nomenclature of controllers with DCS-Control topology offered by Texas Instruments company (Table 5.2) can be found on the website www.ti.com.

5.3.4 Examples of Practical Application of DC/DC Converters with DCS-Control Topology

First, consider the example of use of typical DC/DC converters with DCS-Control topology of TPS6208x family [14]. Controllers are designed in HotRod QFN package that allows to reduce the area of final solution by 65 mm^2 in comparison with analogs. Figure 5.23 illustrates a typical circuit of a DC/DC step-down converter based on TPS62085 controller, with the recommended option of allocation of its components on the printed circuit board.

In general, the TPS62085 controller represents a high-frequency synchronous step-down converter optimized for small-sized solutions. Its own size of the controller is only 2 $\times$ 2 mm, and that is when load current is up to 3 A. The integrated compensation enables the controller to work steadily with output capacitors from 10 to 150 μF. As implied in DCS-Control topology, the controller operates in the PWM mode on average and high currents of load and automatically goes into energy-saving mode in case of low loads to maintaining a highly effective performance. The graph illustrated in Fig. 5.23d shows the load current-dependence of the converter coefficient of efficiency (Fig. 5.23a) for the solution based on the converter under consideration; its efficiency remains at very high level in the widest range of currents from 1 mA to 3 A. It is implemented without any external switching commands, external circuits and control circuits, as well as without unpredictable transient processes in case of change in loading in the full range of operating currents.

One more feature of this controller as well as some other controllers described in this article (Table 5.3), is availability of overcurrent protection called as hiccup

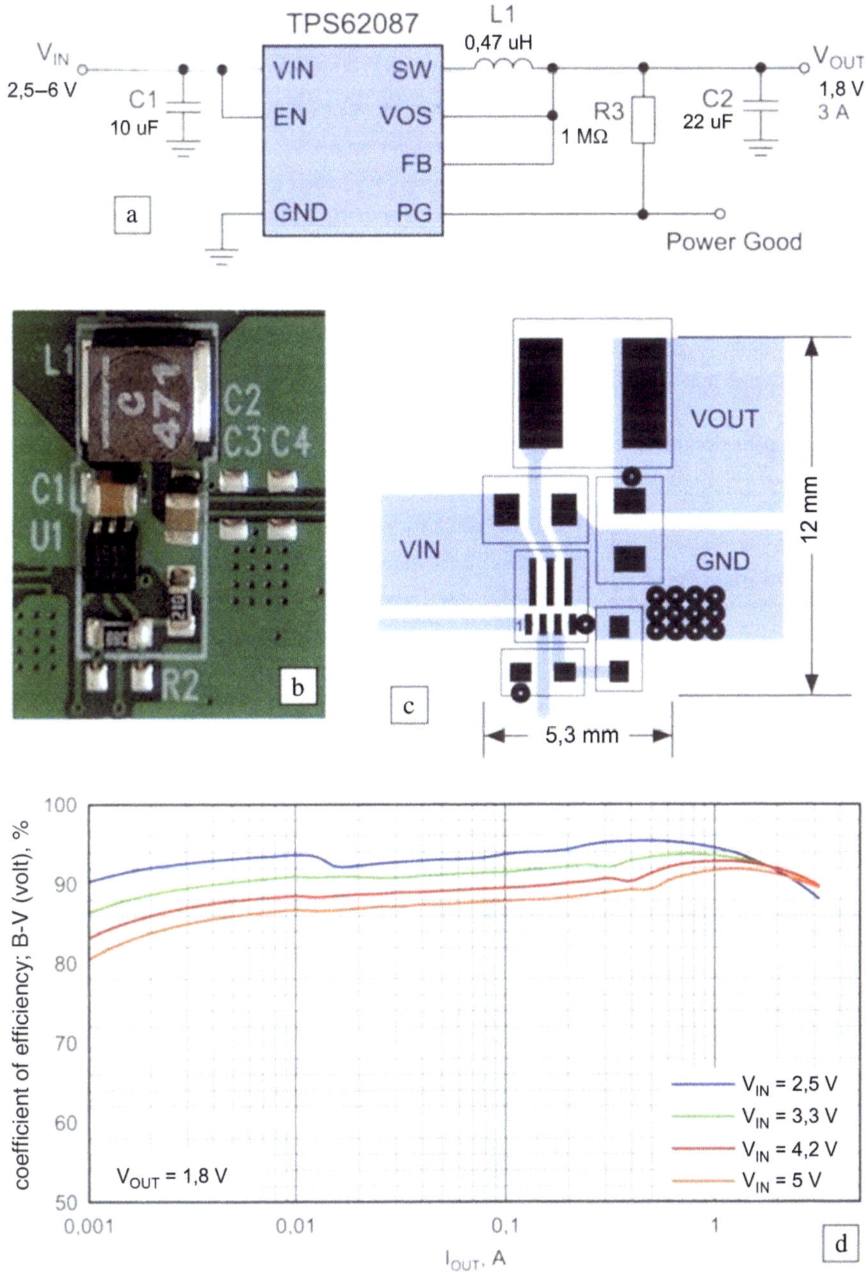

Fig. 5.23 Typical use of the TPS62085 controller designed in the HotRod QFN package [14]: (**a**) typical connection circuit; (**b**) part of the evaluation printed circuit board; (**c**) recommended printed circuit board layout; (**d**) typical coefficient of efficiency of the circuit (**a**)

Table 5.3 List of basic elements of the protection device made by MORNSUN company [15]

Name of the component	Value
B1	Input voltage surge protection, ratings of transceiver and capacitor depend on voltage value
Inductor L1[a]	10 μHY
Capacitors C2, C3[a]	1 μF
Capacitor C4	1 nF, 2 kV
Resistor R1	10 Mohm
Resistor R4, R5	10 Mohm, 2 W
Transils TVS5–TVS7	SMBJ15CA or equivalents with breakdown voltage of 16–20 V
Gas-discharge tube GDT2	B3D090L-C or equivalent with breakdown voltage of 90 V

[a]The standard electromagnetic filter can be used as B2

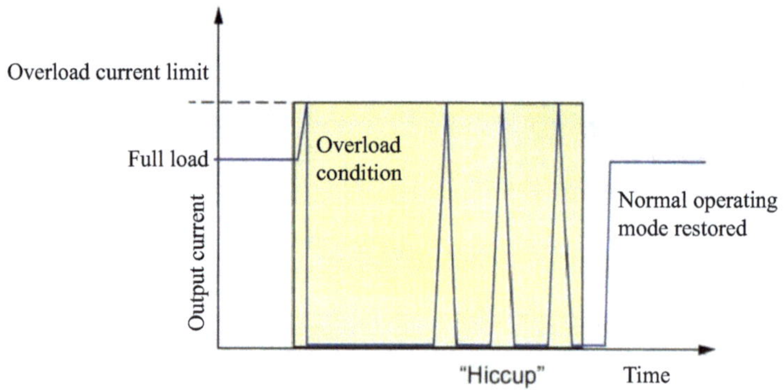

Fig. 5.24 Pulse mode of overcurrent protection is "hiccup"

which literally means "hiccups." The operation concept of similar protection is illustrated in Fig. 5.24 [14] in a general view.

Such protection protects the controller against high current of the inductor and short circuit in output; as well it limits the current consumption from the power supply source of the controller itself. Once the internal current limitation has actuated 32 times, the TPS62085 controller stops, turns off, and restarts the mode of smooth start-up. Typical turn-on delay is 66 μs that is enough for discharge of output capacity, then the controller starts up automatically. The device repeats this mode until the overcurrent condition isn't eliminated.

Advantages of DCS-Control topology are best shown by example of arrangement of power supply of the micropower microcontroller (Fig. 5.25). It is known that for optimization of balance between the power and productivity, the microcontrollers with low own energy consumption can be configured for operation in various modes, usually in sleep mode (SLEEP) with transition to operating mode. Thus, the current consumed by microcontrollers of systems of integrated electronics can vary from several micro-amperes (1.3 μA in sleep mode) to tens of milliamperes (30 mA when clock frequency of 80 MHz), depending on a few factors determined by the end application. These factors can provide different operation modes, clock frequencies, the number of peripheral devices which are clocked at any given time, control of GPIO input–output interfaces of general purpose and the number of GPIO that indirectly influences on general energy consumption. Thus, arrangement of power supply with the use of DC/DC converters designed in TPS62177 controllers with DCS-Control topology and having their own sleep mode (Fig. 5.25), is the ideal solution which is suitable for microcontrollers with various modes of energy consumption. Such converters meet the power supply requirements of embedded systems operating either from 2 to 6 lithium batteries, 3–6 alkaline batteries, USB port, or using higher voltage from AC voltage adapters.

The converter coefficient of efficiency at currents comprised between 23.9 and 0.71 mA is within 88–84%. When using an option of sleep mode for TPS62177 controller, the current consumed by the circuit from the battery is only 9.5 μA, and without this mode it increases up to 28.7 μ.

It is known that one of the modern areas for development of technologies is wide spread of LED light sources. However, the use of LEDs for these purposes imposes certain requirements to the control circuit—driver, i.e., it is necessary to provide the adjustable brightness and stability of preset current in a wide range of input voltage and temperatures. TPS62150 chip representing the controller of the step-down converter with input voltage of 17 V and adjustable smooth start, can serve a base for a compact, simple, and manageable circuit of the high-brightness LED drivers (Fig. 5.26). In case of change in temperature, the required current through LEDs remains at the constant level via its control using output voltage of the converter. Current and, therefore, light intensity of LED are adjusted in dependence of analog or PWM dimming. To reach the required stabilization and current control, it is necessary to use voltage drop across the known resistance Rcs. The Rcs resistor is connected between the feedback input (FB) and common return wire (GND). Here, it should be taken into account that due to rather high voltage in FB input, the power supplied from the Rcs resistor can reduce the solution coefficient of efficiency and shorten the battery service life if it is used as a source of supply voltage. Installation

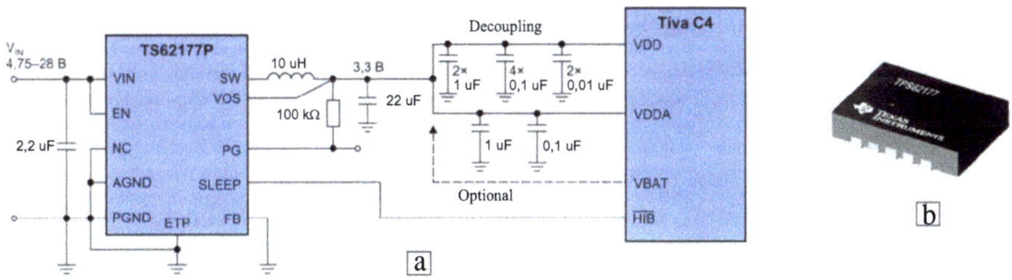

Fig. 5.25 (**a**) arrangement of power supply with the use of controller based on DCS-Control topology for the microcontroller with various modes of energy consumption in a wide range of power supply; (**b**) 2 × 3 WSON mm

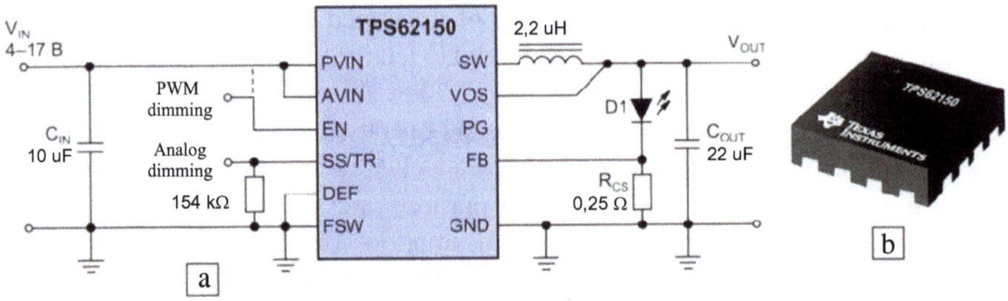

Fig. 5.26 (**a**) LED driver based on TPS62150 controller in dimming mode; (**b**) 3 × 3 mm VQFN [14]

of an additional resistor from SS/TR pin to GND pin is capable to reduce the voltage level required for functioning in FB input. LEDs are connected to FB contact directly from the converter output. Dimming mode in this application can be implemented by means of analogue control (via SS/TR input) or PWM-control (via enable input EN).

This circuit can be used as stroboscope or in alarm systems, as well as it can be applied to controllers TPS62130, TPS62140, and TPS62150.

In the modern power supply systems, preference is increasingly given to PoL technology (point-of-load, it's a system in which the power supply source being as close as possible to load) and to power distributed system when several buses of output voltage are formed. Here, there is a problem related to power supply sequence; for example, this situation is typical for power supply of motherboards of industrial computers [14] or for synchronous proportional power supply. The latter helps to avoid unwanted imbalance of power supply when switching the equipment on. Controllers of synchronous step-down DC/DC converter of TPS62097 family are ideal for this purpose. These controllers have:

- embedded flag Power Good for indication of the device readiness state;
- adjustable smooth start-up for slew rate control at start-up;
- Enable input for the device engage control.

Each of these functions is useful for tracking and setup of the switch-on sequence.

Issues of application in systems with distributed power supply of the Power Good function, smooth start, and voltage tracking are described in [14], and the circuit of the solution is illustrated in Fig. 5.27.

However, controllers can keep radiometrical, said to be proportional, start-up mode when all output voltages increase simultaneously but in proportion to each other, without imbalance. SS/TR pins should be interconnected for radiometrical start-up mode.

The question arises how to arrange the negative voltage from the positive one, for example for power supply of the operating amplifiers or for generation of bias voltage. Then, Fig. 5.28 illustrates the implementation of such solution using the TPS62120 controller with DCS-Control topology [14] in 8-pin package SOT-23 (2 × 2 mm). This controller represents a synchronous step-down DC/DC converter designed for low-powered solutions such as power supply of microprocessors with

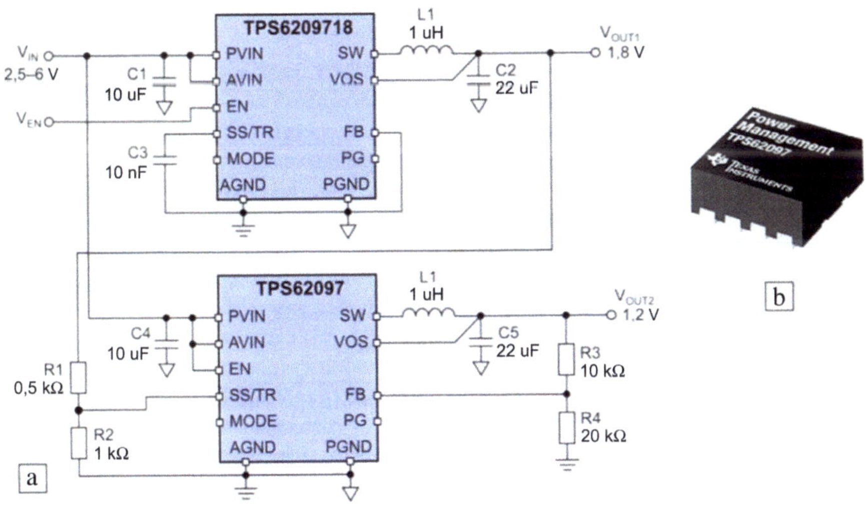

Fig. 5.27 (**a**) The circuit of the solution on arrangement of power supply system with sequential mode of output voltage switch-on; (**b**) 2 × 2 mm VQFN

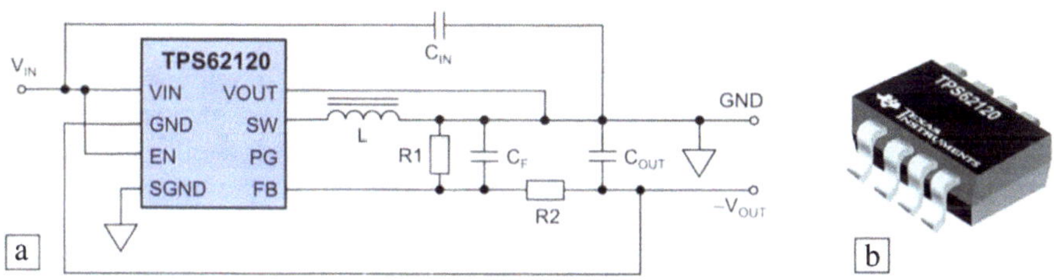

Fig. 5.28 (**a**) The circuit of negative output voltage shaping on the controller based on DCS-Control topology; (**b**) 2.9 × 1.63 mm SOT-23

ultra-low energy consumption for devices powered with free energy and for low-powered RF applications. The extended low range of interlock voltage enables longer operation of applications powered with free energy while using the energy accumulated in the buffer capacitor. Besides, the TPS62120 controller can be configured in inverting switch-on with the use of step-up and step-down topology where output voltage is inverted and becomes negative in relation to the common wire ("ground").

The developers of modern low-powered devices and widespread wireless sensors face with one more problem relating to prolongation of operation time of such devices without service staff involvement. One of such solutions with the use of controllers based on DCS-Control topology is illustrated in Fig. 5.29 [14]. TPS62745 controller was chosen for implementation of this solution. Its features are own current consumption being only 400 nA and output voltage graded 0.1 V is set using the binary code which is delivered to VSEL outputs, thus it

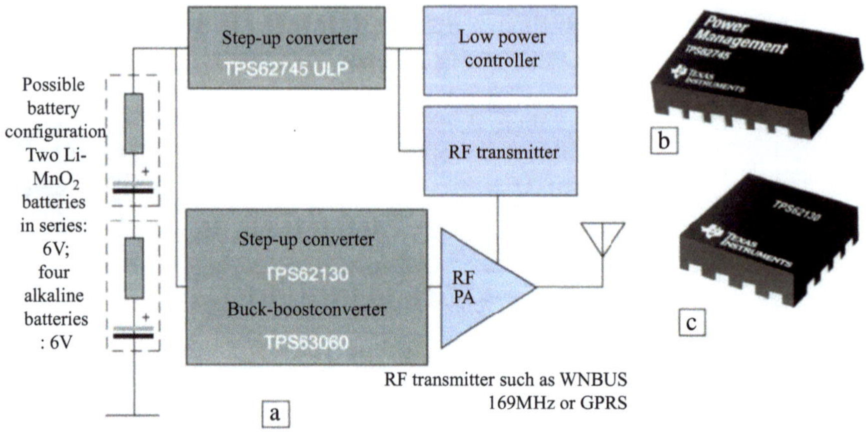

Fig. 5.29 (**a**) Implementation of the radio frequency device with the use of controllers based on DCS-Control topology; (**b**) 2 × 3 mm of WSON; (**c**) 3 × 3 mm of WQFN

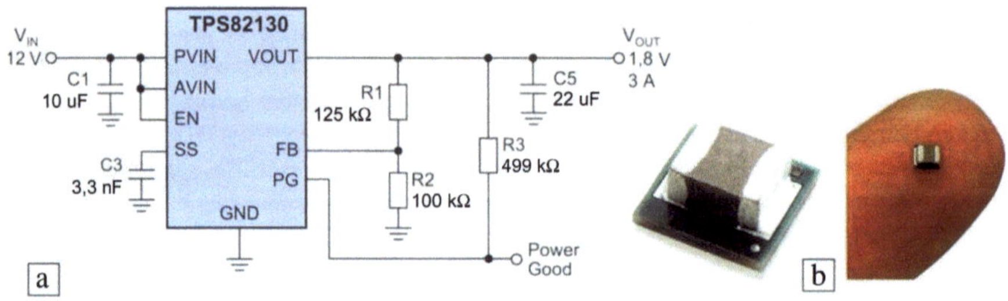

Fig. 5.30 (**a**) The example of switching on of MicroSiP-module; (**b**) 3 × 2.8 × 1.5 mm MicroSiP

supports functioning of dynamic voltage scaling. The advantage of this option is that after the battery voltage reaches the output voltage, the device can switch the level of its output voltage to a step-down and the system continues functioning without going into mode of high ripples that prevents radio frequency interferences. DC/DC converters based on TPS62745 controller are powered from two Li-Ion-batteries or lithium batteries such as Li-SOCl2, Li-S02, Li-MnO-, either from four or six alkaline batteries of 1.5 V. Besides, the output voltage range up to 10 V allows DC/DC converters based on TPS62745 controller to be powered from USB ports and thin-film solar cells.

As can be seen, the use of IBS-controller based on DCS-Control topology gives rise to no particular difficulties. The same applies to MicroSiP modules based on the same topology. But it is even much simpler for users. The fact is that such modules represent terminated DC/DC converters using the "System-in-package" technology with an embedded inductor. The example of a typical connection circuit of TPS82130 module is illustrated in Fig. 5.30 [14]. The undeniable advantage of such modules is their compactness. So, the described solution requires less space by

40 mm^2 on the printed circuit board that is by 40% less in comparison with the discrete design of converter based on the TPS62130 controller.

Finally, let's make a few comments on one more interesting representative that is TPS62097 controller. It is of some interest because developers of Texas Instruments company were not satisfied with what has already been achieved; they developed an extended version of the topology described here that is called as iDCS-Control and intended for industrial purposes. Its representative is TPS62097 controller that is distinguished by forced PWM-control. Description of this technology is not within the scope of this article. Information about this technology can be found in blog "What can iDCS-Control do for your industrial system" written by Chris Glaser on the website TI E2E Community www.e2e.ti.com.

In conclusion, additional functional capabilities of controllers based on DCS-Control topology are as follows:

- Adjusted smooth start/tracking.
- Inputs (discharges) for discrete choice of output voltage value (margining).
- Inputs for setup of operating switching frequency.
- Power Good flag.
- Embedded bypass or load tripping.
- Interface 12C for dynamic voltage scaling.
- Differential connection of remote sensor.
- Extended UVLO-hysteresis.
- Adjustable threshold of Enable option and its hysteresis.
- Overcurrent protection, including hiccup.
- Start-up on load with preliminary displacement.
- Forced discharge of the output capacitor.
- 100% operating cycle.

5.4 Use of System-on-Chip (SoC) for Creation of Bidirectional Communication Lines in the 60 GHz Millimeter Wave Band

5.4.1 Millimeter Wave Band Modem

More and more strict requirements to data transfer in cellular networks make operators look for ways to increase their capacity which is expected to increase by 5000 times by 2030 [16].

Such cells will be installed in buildings; in this case, the best solution for communication lines are optical channels. But such optical channels are unavailable or too expensive for outdoor cells, so it is required to use wireless repeaters.

It is known that the unlicensed band of 5 GHz is available and it can be used beyond the line-of-sight distance but its throughput is limited; as well, interferences from other users are almost guaranteed because of heavy traffic and wide antenna directional pattern.

Currently, the communication lines in the range of 60 GHz are considered as the main option to form data transfer channels between many thousands of cells which are located in outdoor applications and needed to keep the required network capacity. This frequency range is unlicensed too, but—unlike the frequencies lower than 6 GHz—it contains up to 9 GHz of available bandwidth throughput. Moreover, high frequency allows to use very narrow-beam antennas that are insensitive to interferences.

Articles [16, 17] describe the specific technical solution of this problem developed for bidirectional communication lines at the range of 60 GHz by the Xilinx and Hittite Microwave companies (now it is a branch of Analog Devices company). Xilinx company developed for this platform a digital modem Analog Device that is a transmitting and receiving part in the millimetric range.

Two basic blocks are required for creation of such communication line (Fig. 5.31) [16]. Each of them includes a transmitter with a modulator and related analog output circuits, as well as a receiver with a demodulator and analog ingress path.

The modem board integrates analog and digital parts: the DPLL module is intended for exact frequency synthesis; but all digital functions are performed by means of FPLD and SoC. The data-pump operating at one frequency maintains the modulation operation modes from QPSK to 256 QAM with traffic-carrying capacity up to 500 MHz when transmission rate is up to 3.5 Gbit\s. The modem provides full-duplex modes both with frequency division duplex (FDD) and time division duplex (TDD); it has a reduced level of phase noise and a protective coding of LDPC type that increases the communication line performance.

Let's consider the main components of this retranslator in more detail.

Technical solutions for a series of modems of millimetric range developed by Xilinx company allows to create flexible, cost-optimized, and customizable communication lines for wireless networks. This solution is based on complex programmable logic devices Xilinx Zynq-7000 SoC or Kintex-7 FPGA. Both platforms come as a part Xilinx FPLD family based on the advanced 28-nanometer technology.

This technical solution easily adapts to user's tasks. It has low power consumption and small sizes. It can be used both indoors and outdoors for high-speed microwave connections of "point-to-point" and "point-multiple-points" type. Like semiconductive devices produced by Xilinx company, R&D Roadmap of

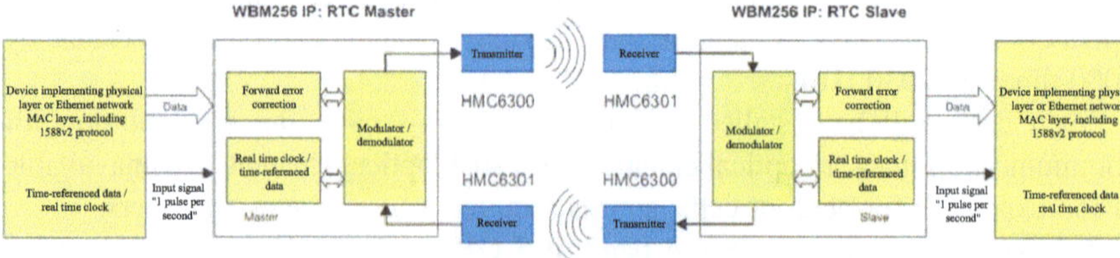

Fig. 5.31 Flow diagram of high-speed bidirectional communication line

millimeter-wave radio modems is very competitive; it provides operators with the possibility to deploy systems that are scalable and updatable in the field.

Flow diagram of digital modem on the platform Zynq SoC is illustrated. Along with programmable logic (PL), the scaling processor system (PS) of this platform includes the dual kernel ARM Cortex-A9 with integrated memory controllers and multi-standard I/O pins for peripherals.

Such system based on crystal (SoC) is an extremely flexible platform. In this case, it is used for various functions relating to data transfer and control, as well as it implements hardware acceleration. Figure 5.32 illustrates the final solution for the millimeter-wave modem that includes a physical layer (PHY), controller, system interfaces, and packet processing processor. Nevertheless, remaining within the provided architecture, it is possible to insert, remove, and change separate modules. For example, there is an opportunity to implement XPIC (Cross Polarization Interference Canceller), combiner, in order to use the modem in the cross polarization mode along with other modem. The solution is implemented on programmable logic (PL) where input–output cells are used for implementation of various interfaces, such as interfaces between the modem and packet handler, between the packet handler and memory; as well as they are used for operation with DAC/ADC.

It is worth noting, above all other important features of Xilinx modem, some special functions: transparent and error-free switch between conditions using Adaptive Coding and Modulation (ACM) algorithms for retention of channel functionality; adaptive digital predistortion (DPD) in a closed-loop system for improvement of radio frequency power amplifier efficiency and linearity; and the use of Reed-Solomon codes or low-density parity-check codes (LDPC) with forward error correction (FEC). The choice of a specific method is based on the requirements for the design. LDPC method is used by default for wireless transport network while Reed-Solomon coding is used for such systems as video processing. All these functions are implemented as IP-cores.

Implementation of LDPC algorithms is highly optimized and uses parallel computations which are included in FPLD for coding and decoding. As a result, significant increase in signal-to-noise ratio was reached (SNR). Decoding circuit can be optimized by size and power consumption using the appropriate quantity of parallel operating cores or the same quantity of iterations on one core. Thus, it is possible to find the optimal solutions based on bandwidth and speed limit of the specific channel.

The Xilinx modem is delivered along with the powerful graphic user interface (GUI) both for information display, and for debugging. It makes possible to control both high-level functions, for example to set up bandwidth and type of modulation, and low-level functions such as an access to separate hardware registers. The functional cores operate at a clock frequency of 440 MHz to reach a bandwidth of 3.5 Gbit/s for the solution illustrated in Fig. 5.31. Besides, gigabit transceivers (GTs) for the interface with DAC/ADC, data transfer through 10 Gb Ethernet or CPRI interface (Common Hublic Radio Interface) are used.

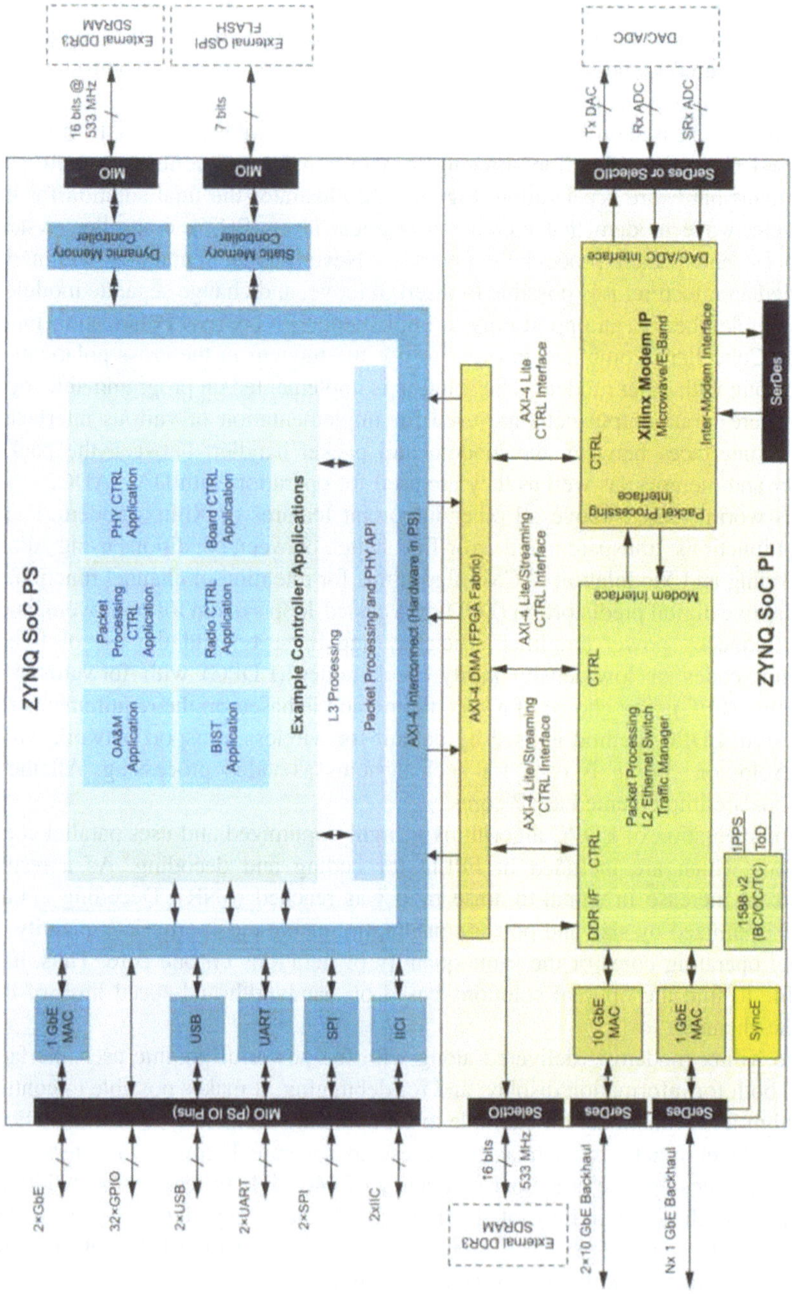

Fig. 5.32 Flow diagram of digital modem on the platform Zynq using programmable SoC [16]

5.4.2 *Basic Set of Transceiver Chips for the Millimeter Wave Band*

At the end of 2014, Analog Devices company released a basic set of silicon-germanium devices (SiGe) of the second generation for 60-GHz frequency [17]. This set (chipset) with considerably improved parameters is optimized for application in repeaters of small cells. The device of HMC6300-type is a terminal analog modulating step-up millimeter-wave converter. The improved frequency synthesizer covers the range from 57 to 66 GHz with a step of 250 MHz and low phase noise. In this case, modulation up to 64 QAM is available. Power output has increased approximately by 16 dBm (of linear power), and the built-in power detector controls the power supply so that it doesn't increase the rated values.

The transmitter supports analog or digital control of boost circuits for intermediate- or radio frequency paths. Sometimes, analog control of gain is required when using high-order modulators if the discrete changes in gain can be taken by mistake for amplitude modulation that leads to bit errors. The digital gain controller is supported by embedded interface SPI.

External phase-lock loop or voltage controlled oscillator with lower phase noise can be used instead of internal synthesizer when modulation of higher order is required for narrow channels. Figure 5.33 [17] illustrates the flow chart of such device—it is HMC6300.

The transmitter supports a bandwidth up to 1.8 GHz. Modulator option with minimum frequency manipulation (MSK) enables data transfer at rate up to 1.8 Gbit/s with small losses and without use of expensive DACs which consume a lot of power.

The receiver chip of HMC6301 type is complementary to this transmitter and it is optimized to meet the strict requirements for connection of small cells. The receiver sensitivity is significantly increased up to −20 dBm; nonlinear distortions of ingress

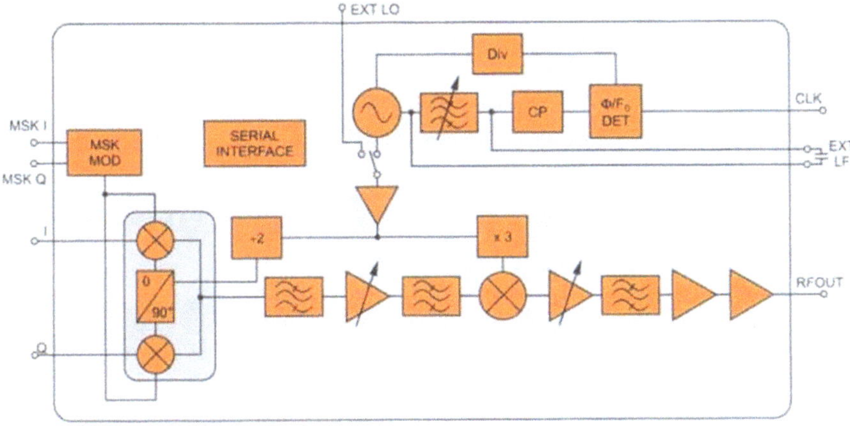

Fig. 5.33 Block diagram of HMC6300 chips

path is increased up to −9 dBm that enables operation in short lines with high level of a signal in input.

Other functions include low 6-dB noise level during maximum gain; adjustment of the lower and upper frequencies of modulating filters; new synthesizers that is the same as synthesizers in the transmitter which support modulation of 64 QAM type in the frequency band of 57–66 GHz, as well as analog or digital control of boost circuits for intermediate- or radio frequency paths.

Block diagram of HMC6301 receiver is illustrated in Fig. 5.34 [16]. It should be noted that the receiver is provided with AM detector for demodulation of amplitude-modulated waveforms such as binary amplitude-modulated waves (on/off keying; OOK). There is a frequency-sensitive detector for demodulation of simple PM and MSK-modulated waves. All this is an addition to the IQ-demodulator that is used for recovery of quadrature-modulated waveforms QPSK and QAM.

Both HMC6300 transmitter, and HMC6301 receiver are produced in BGA packages 4 × 6 mm in size. They are identified as HMC6300BG46 and HMC6301BG46.

Fig. 5.35 illustrates a flow diagram of one of options how to build a millimeter-wave radio modem. In addition to FPLD, modem software, and a set of millimeter-wave transceiver, the design of this modem contains some other components. They include a dual-channel 12-bit 1-Gsample/s ADC AD9234; four-channel 16-bit with a speed up to 2.8 Gsample/s DAC AD9144; a frequency synthesizer of HMC7044 type with an ultra-low jitter and support of serial interface JESD204B which is used for DAC/ADC control.

It should be noted that Xilinx and Analog Devices companies created a demo platform of the KC705 modem based on board FPLD where there is a mezzanine card of FMC standard containing DAC/ADC, frequency synthesizer, and two separate boards with receiving and transmitting modules (Fig. 5.36). The demo platform includes a laptop for modem control and display, as well as an adjustable

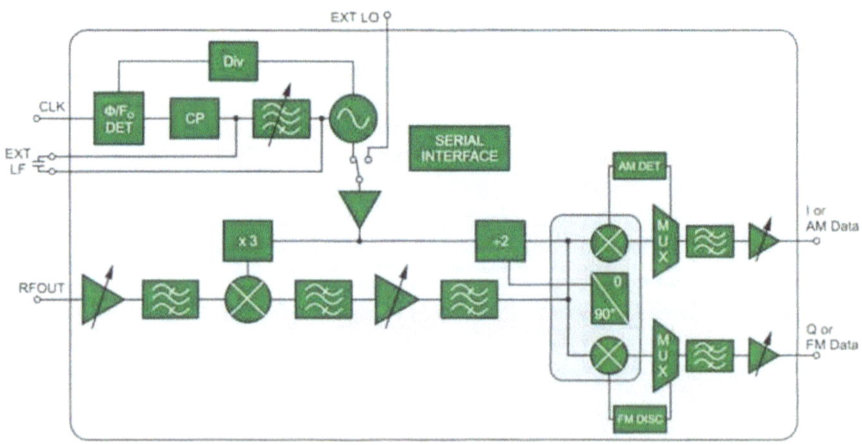

Fig. 5.34 Block diagram of HMC6301chip

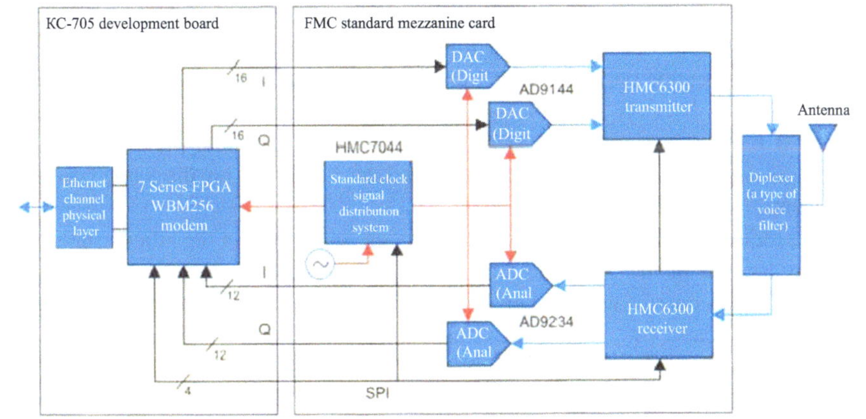

Fig. 5.35 Example of building of a radio modem using chips of NMS series

Fig. 5.36 Demo platform is based on KC705 Xilinx

HF-attenuator for setup of losses specific to the millimeter-wave line. Board KC705 made by Xilinx company contains a FPLD chip Kintex-7 XC7K325T-2FFG900C in which firmware of IP-cores of modem WBM256 is downloaded. Radio channel boards are connected to the FPLD board via intermediate mezzanine module of the FMC standard.

Millimeter-wave modules are installed on boards of modulators/demodulators. They have connectors of MMCX type for 60 GHz interfaces, as well as connectors of SMA type for connection of additional external oscillator.

The platform contains all hardware and software required for demonstration of duplex "point-to-point" connection with rate up to 1.1 Gbit/s in channels having bandwidth of 250 MHz in each direction with frequency division.

Indeed, FPLD is increasingly used in different wireless solutions for transit connections since the platforms based on them differ in modularity and adjustability, thereby reducing the total cost for OEM-manufacturers. Due to significant increase in efficiency of FPLD/SoC of the seventh series and application of high-performance high bandwidth IP-cores, many experts believe that this solution for a millimeter-wave radio modem will become the main one in repeaters for small cells. FPLD and

SoC made by Xilinx company are specialized for high-speed and energy efficient constructions; high-speed serial GT transceivers can be effectively used for wide-band communication and other switching functions. The above-considered technical solution Xilinx is easily scaled for support of several options of a product, from the simplest models having speed of a few hundreds of megabits per second to the solutions of the high end class having speed of 3.5 Gbit/s on the same hardware platform.

In a radio frequency path, the transceivers are integrated in the silicon integrated circuits located in the packages for surface mounting that allows to produce the printed circuit boards with rather low expenses. The chipset of millimeter-wave transceivers made by Analog Devices company meets the needs of small-cell repeaters and provides high rates of performance, power, size, flexibility, and functionality. Analog Devices company produces the data converters that are the best in the industry, and clock frequency control chips which are also used in the radio modem under consideration.

5.5 Specialized Chips for Construction of the RS-485 Bus

5.5.1 Functional Features of the Transceivers of the Interface RS-485

Given that the standard RS-485 interface is used in many applications such as industrial automatic equipment, electric drive, digital video recorders, as well as in standard buses—ModBus, ProfrBus, BACnet, CompoNet, a number of electronic companies produces customized components for implementation of this interface.

Figure 5.37 illustrates the functional capabilities of the main types of interfaces in “cable length”—“data transmission rate” coordinates. [15] As Fig. 5.37 shows, though the highest performance is provided by LVDS interfaces (TIA/EIA 644A), the differential interface RS-485/422 is very popular with developers that is caused by a successful combination of data transmission rate and cable length.

Figure 5.38 [15] shows the overview diagram of RS-485 interface. Considering cable length and its multiway design, it should be noted that there are strong common-mode voltage and electromagnetic interferences with high noise ratio in a long multiway cable, especially in industrial applications. These two features are capable not only to nullify all advantages of RS-485 but also to make it inoperative. However, a way of dealing with them is known—it is galvanic isolation of circuits (signal circuits, power supply circuits, and data links).

As typical examples of arrangement of RS-485 chips, the main technical solutions of MORNSUN company, one of leaders in this sphere of microelectronic components, are discussed below. As a rule, RS-485 line has high common-mode voltage and noises that might damage a transceiver chip, so it is necessary to take special precautionary measures for its protection. The standard circuit providing such protection is illustrated in Fig. 5.39; and basic elements of the circuit are listed

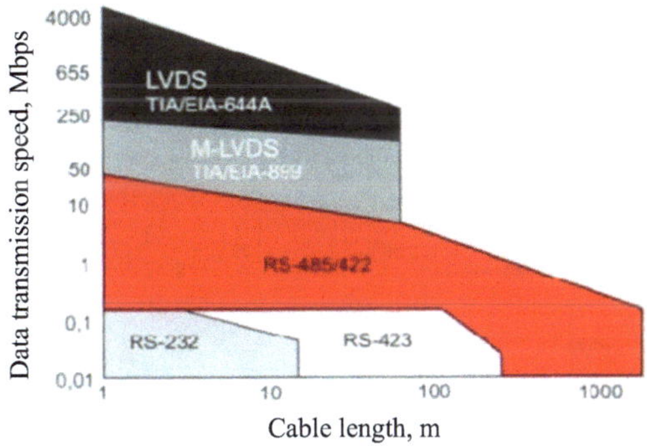

Fig. 5.37 Functional capabilities of interfaces

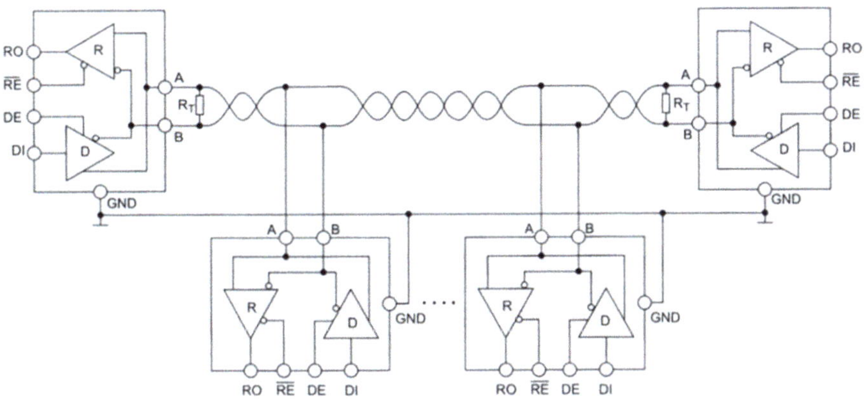

Fig. 5.38 Overview diagram of RS-485

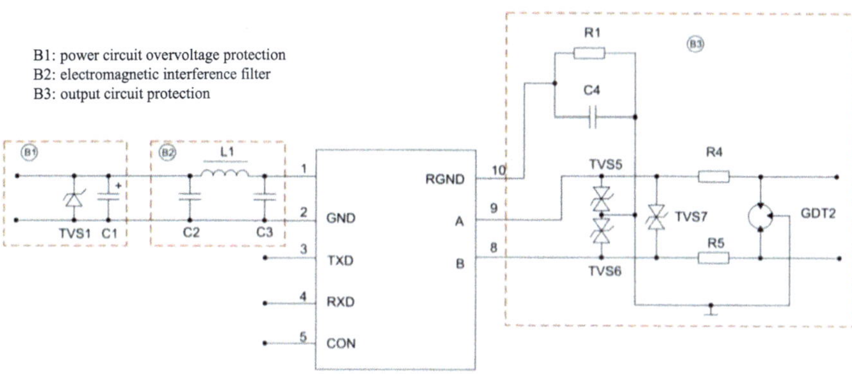

Fig. 5.39 The equivalent electric overvoltage protection circuit of transceiver RS-485

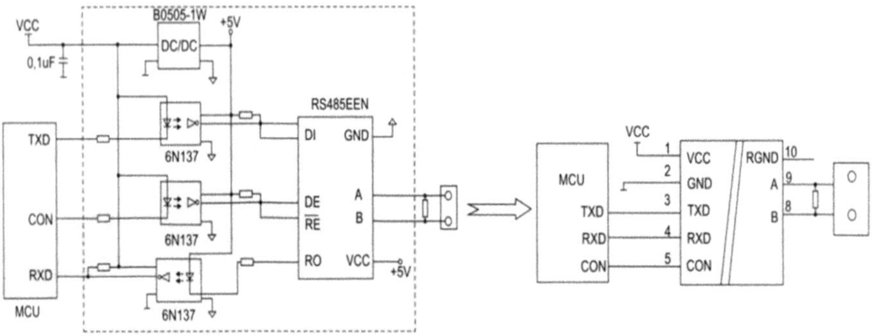

Fig. 5.40 The equivalent electric circuit of interface RS-485 when using normal transceivers (left) and the similar solution when using MORNSUN components (right)

Table 5.3. It should be noted that if the receiver/transmitter (transceiver) with two buses RS-485 is used, then output circuit protection system B3 is applied to each output channel.

The main advantage of transceivers made by MORNSUN company is galvanic isolation of input and output circuits, as well as power supply circuits that is provided due to a built-in insulating DC/DC converter. Thus, when using these transceivers, the number of the required external components is reduced, the topology of printed circuit board becomes much simpler and, therefore, the space occupied on the board decreases. Figure 5.40 [15] confirms the foregoing by illustrating the circuit solution with standard transceivers (left) and the similar solution with MORNSUN components (right).

All these chips meet the requirements of standards IEC/EN61000-4-2/6 on electromagnetic compatibility and passed tests according to standards CISPR22/EN55022 CLASS A. From the input side, chips are compatible to the serial interface UART for 3.3 or 5 V depending on the supply voltage used by the developer. Input-to-output signal propagation delay doesn't exceed 100 μs. Operating temperature range is −40–850 °C. The main technical characteristics of MORNSUN transceivers RS-485 are listed in Table 5.4. Transceivers differ with the truth diagram of control and signal transmission, with availability or absence of pull-up resistors, and etc. All these data are described in the manufacturing documentation.

5.5.2 *Practical Recommendations on the Use of Interface Transceivers*

Considering the limited scope of this book, we will consider only practical examples of the use of these transceivers in industrial automatic equipment and in railway transport [15] since the increased resistance to noises and high reliability are required to be ensured for these applications.

Table 5.4 Key parameters of transceivers RS-48S made by MORNSUN company [15]

Item	The quantity of channels	Maximum data transmission rate, kbit/s	The quantity of transceivers per one node	I/O insulation strength, V	Type of package	Size of package mm	Note
TD301D485[a], TD501D485[a]	1	9.6	32	2500	DIP10	19.9 × 16.9 × 7.05	
TD301D485H, TD501D485H		115.2					
TD301D485H-A, TD501D485H-A		115.2					Built-in ESD-protection
TD301D485H-E, TD501D485H-E		500					Built-in ESD-protection
TD301P485H-E, TD501P485H-E		500	256	4000	DIP24	31.8 × 20.3 × 9.5	
TD311D485H, TD511D485H		115.2		2500			
TD1211D485H[b], TD2411D485H[b]		115.2					Pins with isolated power supply cable of 5 V, not more than 0.4 W
TD31IP485H, TD51IP485H	2	115.2		input-output: 2500; output-input: 2500 V	DIP24	31.8 × 20.3 × 9.5	Built-in ESD-protection. Pins with isolated power supply cable of 5 V, not more than 0.15 W
TD312P485, TD512P485		9.6					Pins with isolated power supply cable of 5 V, not more than 0.15 W
TD312D485H, TD512D485		115.2					
TD312P485H, TD512P485H		115.2					

[a]The first digit in designation specifies the supply voltage: 3—supply voltage of 3.3 V; 5—supply voltage of 5 V
[b]The first two digits in designation specify the supply voltage: 12—supply voltage of 12 V; 24—supply voltage of 24 V

So, Fig. 5.41 shows the fragment of the structural circuit of standard control system in industrial automatic equipment. TD501D485H transceiver is used to connect the microcontroller of control system to the central processing computer. Due to galvanic power and data link isolation, the transceiver, figuratively speaking "cuts off" the data bus with the high common-mode voltage and noises from the microcontroller. In this case, the choice of transceiver with high data transmission rate (115.2 kbit/s) depends on high-volume information transfer. In case of low-volume information transfer, it is better to choose more economic transceiver TD501D485 with the rate of 9.6 kbit/s.

Two separate power modules are used in this circuit in addition to the transceiver. The first AC/DC module LH20-10B24 forms the standard 24-V bus from the mains voltage of 220 V, and the second DC/DC module URB2405ZP-6WR2 is connected to this bus and provides a stable power supply of 5 V.

The main features of these power modules are briefly listed. Digit 20 in designation LH20-10B24 means the power output of the converter expressed in watts. This series includes a few dozen modifications with a capacity of 4–25 W. Besides, converters of this series differ in the quantity of output channels and voltage output (in our case, it is one channel with voltage of 24 V).

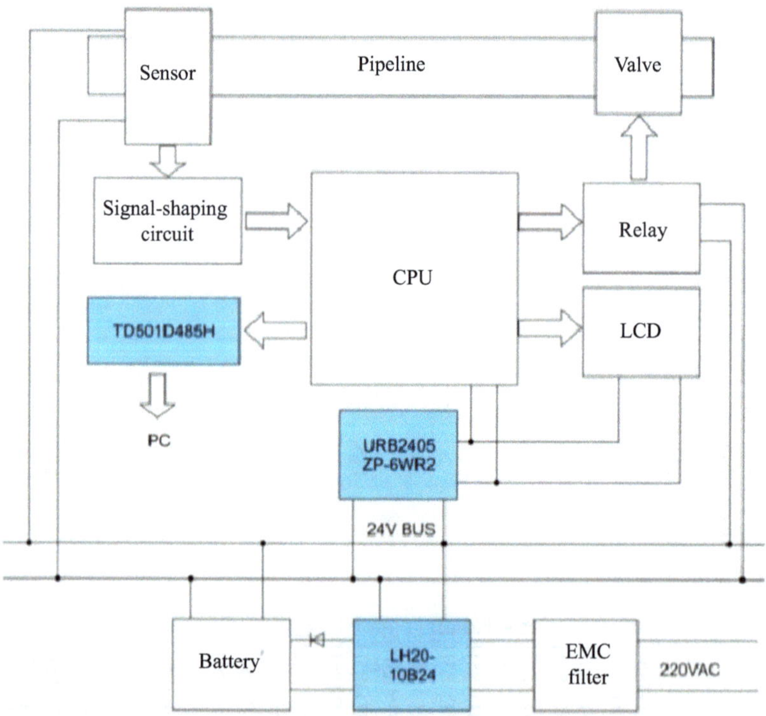

Fig. 5.41 The example of building of interface RS-48S in industrial automation control systems [15]

Depending on modification, output voltage of single-channel converters varies within 3.3–48 V. There can be both positive and negative voltage in output of two-channel converters. Bidirectional voltage within the range of ±5 to ±24 V is generated at outputs of the second and third channels in three-channel converters. I/O insulation strength is 3 kV.

DC/DC module URB2405ZP-6WR2 generates voltage of ±5 V at the maximum output current of 600 mA. This family of MORNSUN company includes 14 modifications which differ in operating range of input voltages, in quantity of channels and value of output voltage. I/O insulation strength is not less than 1.5 kV.

Figure 5.42 shows the example of building of the railway interface network. This circuit implements the known GSM-R standard. The platform based on the RS-485 interface ensures reliable communication at the motion speed up to 500 km/h. It is designed for communication of trains with control centers for traffic management. Prefix R in designation of the standard means Railway. As you can see in Figure, this system includes transceivers RS-485 (one of them, transceiver TD501D485H is considered in the previous example) and CAN (TD501D CAN).

TD1211D485H transceiver is intended to connect the host with the display terminal MMI (man–machine interface). Since the supply voltage of TD1211D485H is 12 ± 0.6 V, it can be connected directly to 12-V bus. It should be noted that there is a 24-V modification in this family, too. Data transmission rate is 115.2 Kbit/s. The feature of this TD1211D485H is an output of isolated supply of 5 V (4.75–5.8 V) with a capacity up to 0.4 W.

Standard bus CAN is used for communication with train radio station and data transmitter. The host is connected to the bus CAN via transceiver TD501DCAN. Similar to receivers/transmitters R5-485, transceivers are built-in power supply decouplings and data link decouplings. Insulation strength is 3 kV, and data

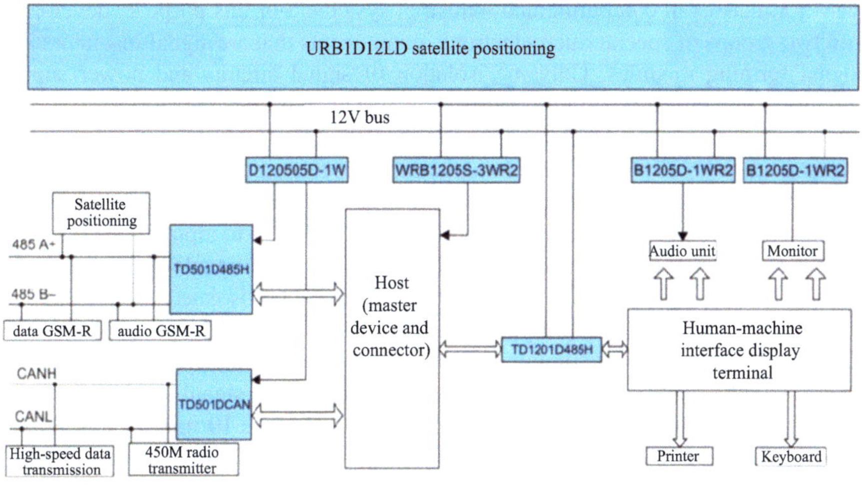

Fig. 5.42 Example of building of the RS-485 interface in railway management systems

transmission rate reaches 1 Mbit/s. The company's product range includes four transceivers of CAN-bus.

As shown in figure, the use of these transceivers allows to "cut" the network into separate galvanically uncoupled transmission lines and equipment. Certainly, such approach is justified for long distance communication lines and the environment with high noise contamination by electromagnetic interferences that always accompany the power supply equipment.

However, power supply decoupling is required for full protection of system against interferences and increased common-mode voltage. Generally, isolated power supply modules are used for this purpose. The first of these modules, a 15-W URB1D12LD-15W module, is connected to 110-V bus (40–160 V) that is a typical bus for railway transport and forms 12-V bus. The maximum output current of the module is 1250 mA. This family includes nine single-channel modules. They differ in power (15 and 20 W) and output voltage: 3.3–24 V. Insulation strength is 1.5 kV.

It is known that high-speed chips can influence on each other via the common power bus on which there are noises repeating the HF signal generated by the chip. In the above example, such mutual influence can appear when using transceivers TD501DCAN and TD1211D485H. That's why the power supply decoupling is required. The special double-channel module D120505D-1W is produced for such purposes. Its both outputs are galvanically isolated from the input and from each other. Insulation strength is 1 kV, minimum resistance is 1000 Mohm when test voltage is 500 V. This family includes 20 chips differing in power (5, 12, and 24 W), input voltage (5, 12, and 24 V), and output voltage (5, 9, 12, 15 V).

The display terminal is powered from two modules B1205S/D-1WR2. The company developed a series of 20 DC/DC converters with a capacity of 1 W for this family. They differ in input voltage (3.3; 5; 12; 15; 24 V) and output voltage (3.3; 5; 9; 12; 15; 24 V). The insulation strength is not less than 1.5 kV.

In addition to digital data link decoupling, a need for galvanic isolation of analog waveform arises in computing and industrial systems. For this purpose, developers use two groups of special microelectronic components that are signal insulators and signal forming modules. Galvanic isolation of signal circuits and power supply circuits is implemented in these components just as in transceivers.

The design of signal insulators provides for mounting on DIN rail. This product group consists of 12 families. Each family includes a few modifications. Depending on circuit-based design flows, either an analog waveform is supplied to input of insulators or thermocouple (thermal resistance) is connected to input. In the insulator output, there is a signal of current loop 0/4–20 mA or voltage signal of one of standard voltage band: 0/1–5; 0/2–10 V. Some insulators have 24-V isolated voltage pin.

Signal forming modules are produced in 12-, 16-, 18-, and 24-pin packages SIP DIP, SOIC. This group of components includes 19 families. Each family consists of many modifications. In the module output, there is a signal of current loop 0/4–20 mA or voltage signal of any voltage band: from ± 10 mV to ± 10 V and 0–15 V. In the output of such module depending on its modification, there is a signal of current loop 0/4–20 mA or voltage signal of one of standard voltage band: Some special modifications have a separate pin of isolated power supply voltage of

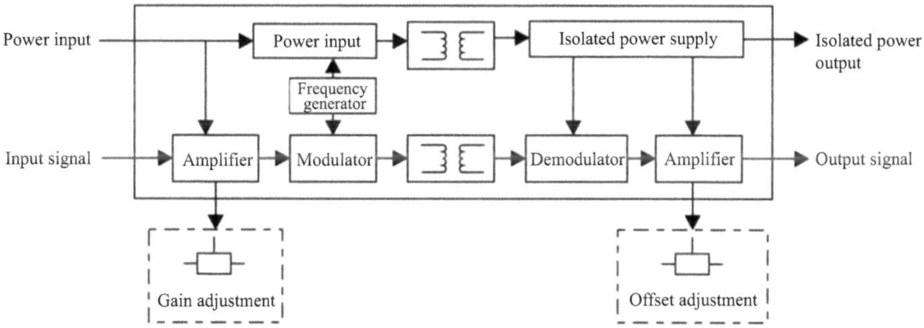

Fig. 5.43 Flow chart of signal forming module

3.3–24 V. Figure 5.43 [15] shows a general flow chart of signal forming module of this family.

Summing up the results of this section, it should be noted that the use of the above-mentioned recommendations and practical examples can help developers of systems with the RS-485 interface to avoid errors and reduce time spent on a system being designed.

5.6 Specialized Chips for High-Voltage LVDS and DVI Interfaces

5.6.1 Basic Technical Solutions for Building of a LVDS Interface

The main technical solutions used in chips intended for building of high-speed interfaces LVDS and DVI [18] are considered below.

LVDS and DVI are one of the most widespread video interfaces for displays and LC panels (LCD). The feature of these interfaces is the possibility to use them for serial data transmission that significantly reduces the required quantity of signal conductors. In addition, both interfaces use differential lines for signal transmission that significantly reduces the influence of external crosstalk and allows data transfer for long distances. Milandr company developed a series of chips with LVDS and DVI interfaces for various areas of application.

LVDS (Low-Voltage Differential Signaling) represents a standard developed under the aegis of the Telecommunications Industry Association (TIA) and Electronics Industries Association (EIA). It defines electrical specifications of low-voltage differential data transfer between integrated circuits where high-bit-rate binary signal exchange takes place with high bandwidth between data input devices and data processing terminal equipment.

The levels of voltage prescribed by this standard are defined in such a way that the maximum flexibility is ensured for implementation of low-power high-speed differential interfaces. In this case, output characteristics of transmitting part don't depend on power supply voltage that allows to build up a system using supply voltage of 5.0 V, 3.3 V, or 2.5 V with no loss in quality. In addition, the use of signals with small range of voltage allows to significantly reduce the power dissipation and mutual electromagnetic interference of signals while transferring data.

According to the standard, the interface consists of a generator connected to matched line at the end of which there is a terminator and receiver (see Fig. 5.44) [18]. The interface configuration is described in terms of the required voltages and currents at different values of loads. However, the standard doesn't provide clear description of transmitter and receiver functionality, so they can be either of inverting or non-inverting modification in dependence of their purpose, or they can have functions of serialization and deserialization to increase the data transmission rate. This standard doesn't define other interface functions such as signal quality, data transfer protocol, data transmission rate, signaling line pattern, and time response characteristics of signal propagation.

Against this background, LVDS interface was implemented in various chips, the functionality of which is rather diverse. For example, 5559ИН-19 represents a LVDS-transceiver in which the conversion of CMOS-signal into LVDS format and back takes place. One of new developments of the company implements a chip in which a LVDS signal transferred by four channels with the subsequent deserialization of input data stream is received. In addition, options when the LVDS interface is a part of the microcontroller as an additional means of commutation with peripherals were developed. The general concept applied in the design of this type of devices is illustrated in Fig. 5.45 [18].

The current generated in the transmitter flows through the terminator in forward or reverse direction depending on the control voltage on the keys. Thus, depending on the direction of current flowing, the differential voltage of different polarity relevant to logic "1" or "0" is generated in inputs of the receiver. In this case, for reliable signal reception, it is enough when peak-to-peak voltage in input of the receiver is about 350 mV that significantly reduces the energy consumption by the system. In this case, the standard clearly determines the range of input signal bias (VCM = 0.05 V ÷ 2.35 V), the minimum amplitude of differential signal at the level of 100 mV, and maximum amplitude of differential signal at the level of 600 mV. These conditions impose certain requirements to the input differential stress cell of

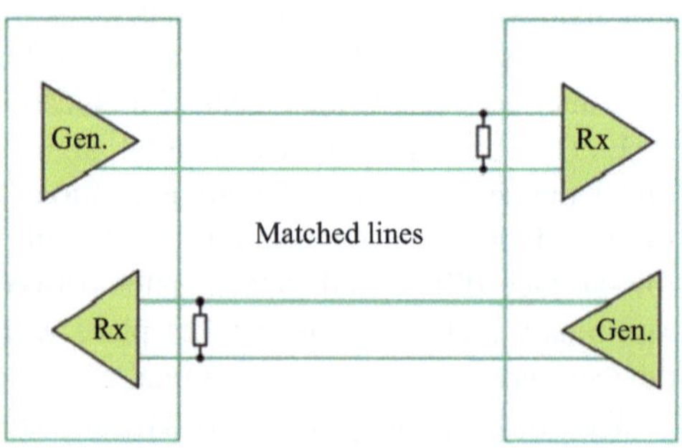

Fig. 5.44 General description of LVDS

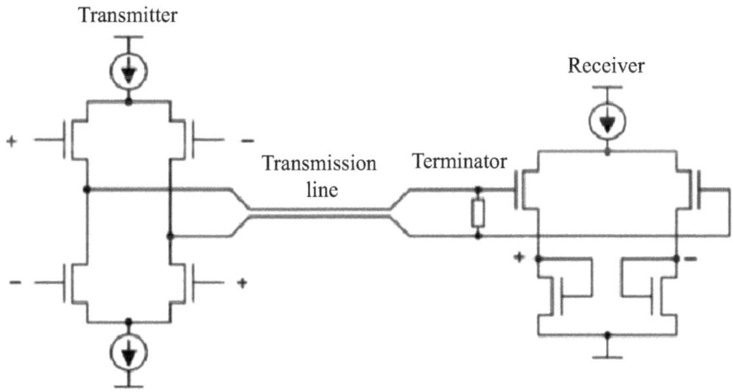

Fig. 5.45 General structure of LVDS-interface

the receiver. One of options to deal with this issue is to use two differential stress cells based on parallel-working transistors of different type. In this case it is possible to significantly increase the range of operating differential input signal bias.

One more parameter which is specified by the standard is voltage unbalance of zero offset in output of the transmitter during continuous switching from one to zero and back (1, 0, 1, 0, . . .). This requirement imposes considerable restrictions to the feedback system responsible for stabilization of the level of output signal bias of the transmitter. One of solutions for this problem is the use of circuit illustrated in Fig. 5.46 [18].

In its turn, optimization based on the key size–reference current ratio in the transmitter solves another problem defined in the standard, namely, a problem of the restrictions imposed on the rising and falling front of the output differential signal. In this case, special attention should be paid to the form of signals that control the keys and, if necessary, the special circuit should be used for generation of controlling signals.

In the design, special attention should be paid to other parameters of LVDS interface such as:

- input current in the receiver inputs;
- balance of input currents at different input voltages;
- current flowing in case of short circuit of the transmitter outputs;
- receiver sensitivity to offset of output line impedance.

5.6.2 *Basic Technical Solutions for Building of a DVI Interface*

Standard DVI ver. 1.0 provides data transfer in a wide range of image formats—from VGA (640 × 480) to QXGA (2048 × 1536) that is a reason for its popularity. The main idea of information transfer is explained in Fig. 5.47 [18].

Fig. 5.46 Solution for problem with stabilization of offset voltage of the transmitter

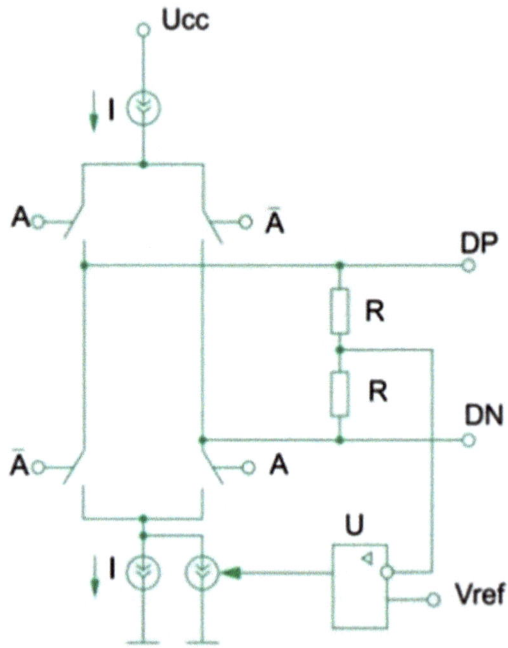

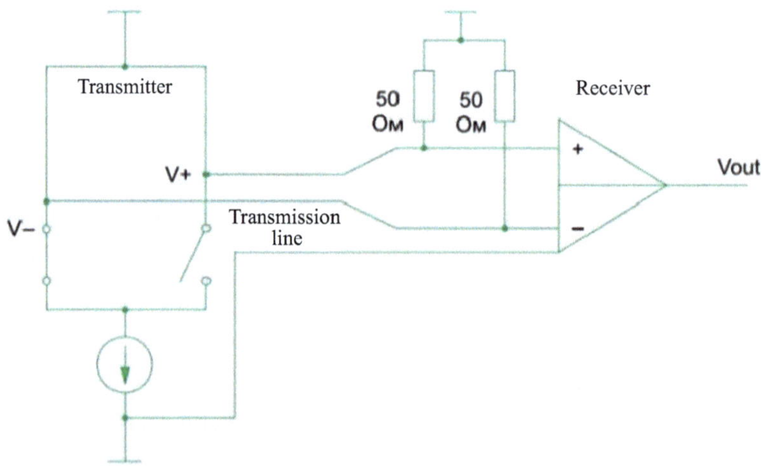

Fig. 5.47 Concepts of the DVI standard

Signal transmission is performed with the use of differential stress cell. The difference in voltage drops on load resistors of 50 Ohm is differential voltage that gives useful information about the signal. The protocol part of the DVI standard is described in Figs. 5.48, 5.49, and 5.50.

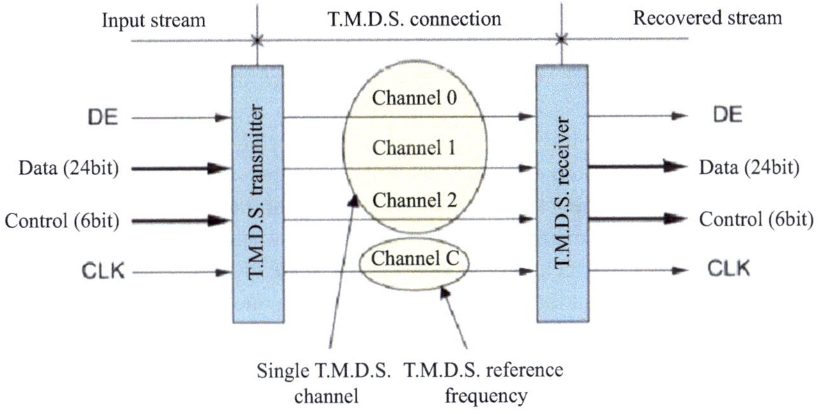

Fig. 5.48 Protocol DVI (general view)

Fig. 5.49 Data in input of the transmitter and in output of the receiver

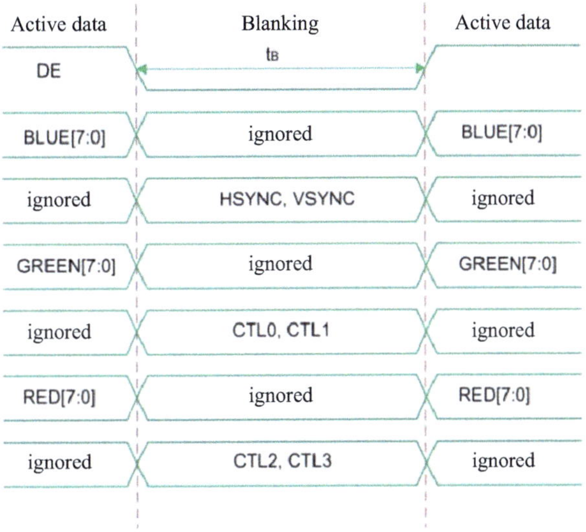

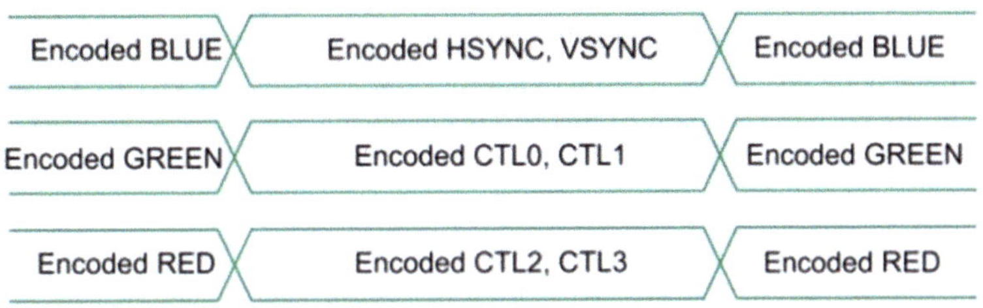

Fig. 5.50 Data in input of the receiver (T.M.D.S.)

Sync signal, video data (24 bits, red, green, and blue ingredients), true data tracking strobe, and service information are arrived in input of the transmitter. In the simplest case, they are signals of vertical and horizontal scanning VSYNC and HSYNC. The transmitter codes the input information according to the technology T.M.D.S. (transition minimized differential signaling). This technology of differential signal transmission with the minimum quantity of transitions allows to reduce the influence of electromagnetic interferences during signal transmission and to perform proper recovery of data even in the conditions of a low-quality transmission line.

The main idea of this technology comes down to the fact that 8-bit active data are encoded by 10 bits so that the quantity of transitions in one coded word can't exceed 5. Service information (HSYNC, VSYNC) is encoded by four possible 10-bit words: "1101010101," "0010101010," "1101010100," and "0010101011", the quantity of transitions is at least 7. These words are called as service digits. Coded active data and coded service information are transmitted through one channel T.M.D.S., i.e., through one differential stress cell. Video data are transmitted during DE = 1; service information is transmitted during DE = 0. 10 bits of a code are transmitted per one clock tick starting from the least significant bit. Now the task of the receiver is to recover information from this flow. Engineers of CJSC PKK Milandr developed a chip that performs data recovery of one channel from a flow. The purpose of this work was to develop a chip with such minimum functionality which would allow to check the circuit-based solutions for applicability in order to create a fully featured receiver of a DVI signal.

In case of the successful solution of this task, creation of a fully featured DVI receiver would come down to adding of two data recovery channels and the clock unit between them. The result of work performed by Milandr company is a chip designed according to the technology TSMC 180 nm, the structure of which is illustrated in Fig. 5.51 [18]. The circuit is designed in plastic 32-terminal package QFN.

The chip is powered with one voltage of 3.3 V. The chip contains a power control unit with external resistor having 1% precision of performance to generate the 1.8-V internal power supply voltage of the digital part and reference currents of analog units. There is possibility to set up the internal power supply voltage of 1.8 V externally via terminal UCC1p8.

The input signal arrives at the buffer converting a differential signal into CMOS-signal of 1.8 V. The difficulty of designing the input buffer resulted from that the input signal itself has wide variation of parameters. Two-cascade circuit was used for ensuring the proper operation of the buffer within the full range of rise/fall, amplitude, duty ratio of output signal, and minimum inserted jitter. The first cascade is a differential amplifier converting the differential voltage into currents. The second cascade is a current comparator operating in 1.8-V domain. Thus, the buffer circuit doesn't contain a voltage converter in its classical representation that made it possible to satisfy the technical requirements for frequency, duty ratio, and power consumed by the buffer.

Figure 5.52 shows the results of the input buffer modeling for bit sequence with different bit length. Generally, data and sync signal arriving from the buffer have

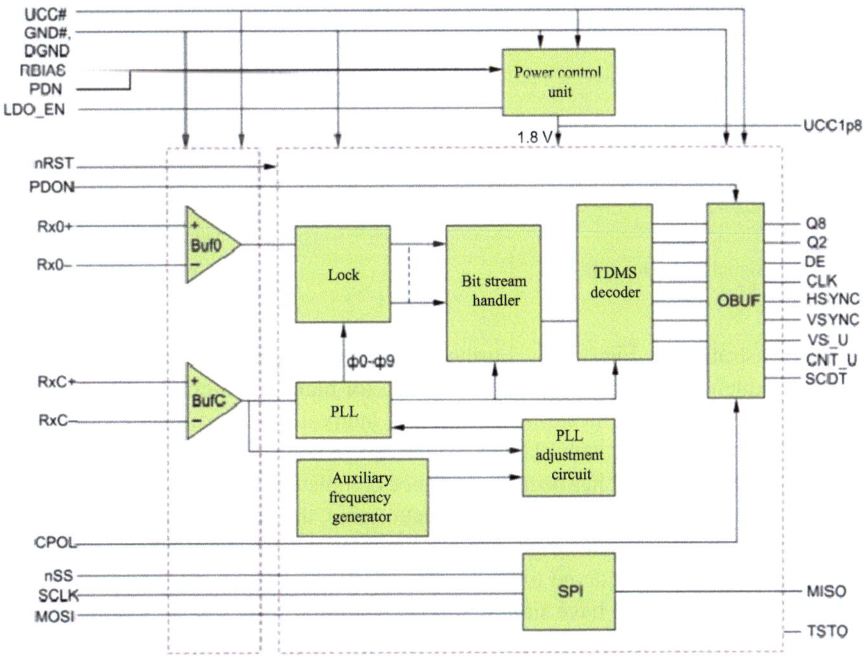

Fig. 5.51 The chip structure

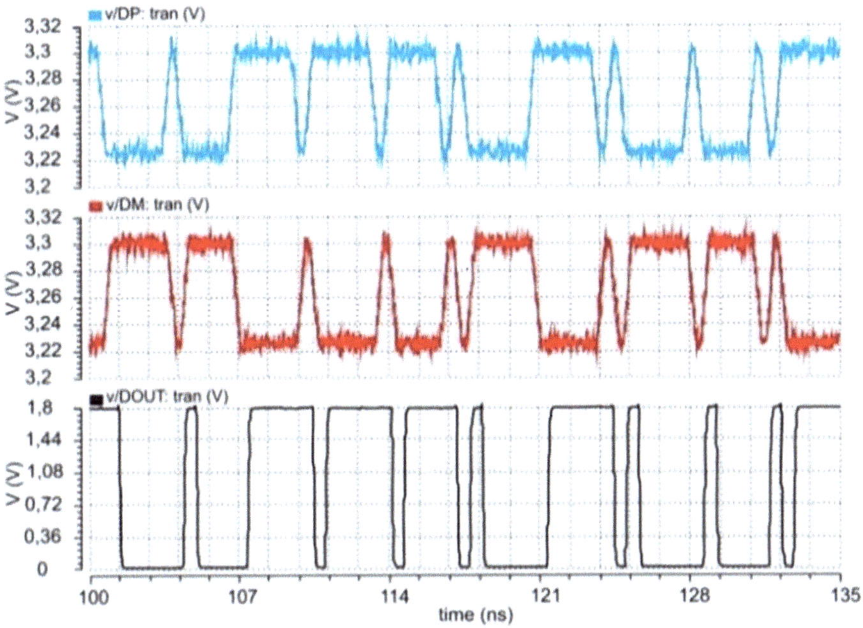

Fig. 5.52 The results of the input buffer modeling

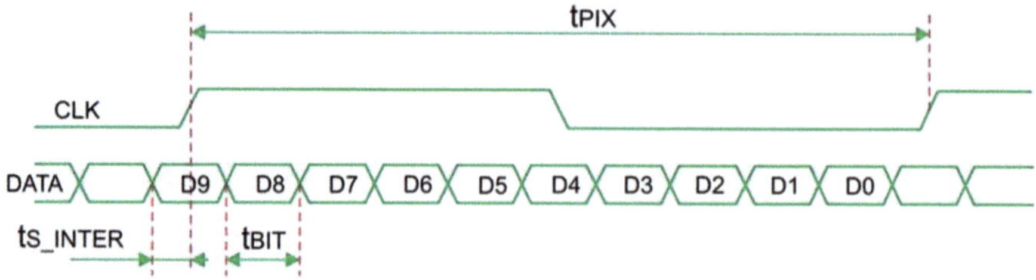

Fig. 5.53 Signals from input buffers

forms illustrated in Fig. 5.53. Further processing comes down to search for synchrosequence, and, more usually, to search for blanking interval t_B.

According to the specification, the blanking interval should consist of 128 service digits at least. As soon as unbroken sequence of 128 service digits is found, it means that the blanking interval has been found; once completed (the first non-service digit) active data will follow. As soon as active data arrive, the decoder should set a signal DE = 1 until the next service digit arrives. When DE = 1, the decoder gives decoded active data that are considered to be the true ones.

This algorithm would have simple implementation if there was a rigid bound of a sync signal rise/fall to a bit data stream. However, according to specification, the value of phase difference of sync signal and data signal can reach 0.6∗tPIX, i.e., there is no bound of a sync signal rise to some certain bit of data. That's why the following algorithm of search for blanking interval was chosen. The phase-lock loop unit having the frequency in its input from the sync signal output buffer, should generate ten signals of tenfold input frequency shifted relative to each other by $\pi/5$. Tenfold frequency was formed from the quantity of bits in one encoded word; phase difference $\pi/5$ was formed from the preset maximum input jitter of sync signal, data, and condition that at least one phase strobes exactly the truth data. Thus, each data bit will be sampled in sequence by ten rises of these signals. As a result, we will have 100 bits in output for the following analysis. Such sampling is carried out continuously, so the stranded data should be saved for further analysis. This process is performed in the "clamper" unit that consists of ten 10-bit shift registers and one 100-bit parallel-load shift register. Each of shift registers loads data by its phase edge $\varphi 0$–$\varphi 9$. After all 10 bits were sampled by ten phases, clkout signal edge, the value of shift registers is recorded in the parallel-load shift register. It is clear that there are rather stringent requirements for setup and hold time of sequential- and parallel-load shift registers performing the above functions (at high frequencies of operation) that's why the special dynamic flip-flops fulfilling these requirements were developed for generation of these registers.

The requirements imposed in this project for phase-locked loop unit are rather stringent, too. It is necessary to provide simultaneously the minimum jitter, stable loop line, octave bandwidth, and conversion factor stability of voltage-controlled generator within the full range of frequencies (21–165 MHz). The auxiliary unit—a PLL adjuster—using the auxiliary 1-MHz frequency generator generates a band

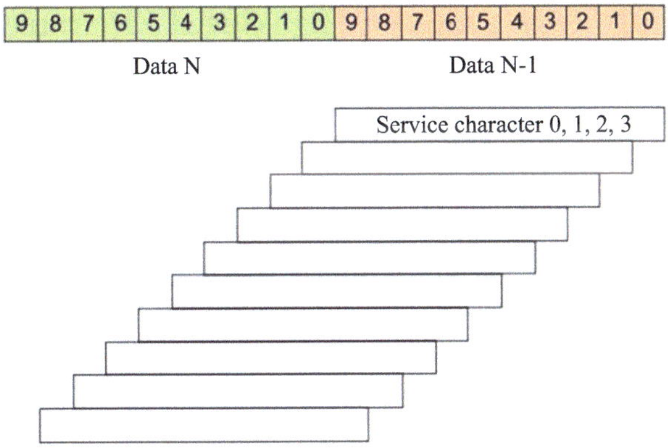

Fig. 5.54 Search for service digit

signal of high or low level up to 40 MHz and more depending on the range of input sync signal.

After the clamper has formed a 100-bit register of sampled data in output, the further processing takes place in the bit stream handler at the clkout frequency that is equal to the frequency of input sync signal. The intent of processing is explained in Fig. 5.54 [18].

Data N and Data N-1 in Fig. 5.54 are current and previous content of the output register-clamper for one phase. If that particular phase has successfully got on the true data but not in any transient state, and 10 bits sampled in sequence have given one of service digits, then the bit number starting this service digit is recorded; and this phase is marked as successful one. In this case, the counter goes into action that is incremented every clock period if the bit number and digit type match the ones which have been found in the previous clock period. If two or more numbers are found for one phase, then the error is recorded and the counter resets to zero. Once the counter reaches value 128, the flag is set that indicate the synchrosequence to be found. The same algorithm is applied to the rest of 9 phases. After two clock periods passed after setting the first flag, analysis of all flags from all phases is carried out. As a result, we have an interval from phase 3 to phase 7 where the service digit has been found. The number of operating phase is calculated as the midpoint of successful interval (in our case, it is 5), and then the data from the register relevant to this very phase are taken for decoder T.M.D.S. The data from decoder T.M.D.S. go to output buffers operating at frequencies up to 165 MHz with load up to 10 pF. All output buffers are switched off by external terminal PDON to decrease energy consumption.

To save the quantity of external terminals, the active data are represented by one bit on which all data bits are multiplexed in accordance with the value of register available via SPI interface. The chip health checking was performed on the debug board including FPLD for data generation and outcome analysis, chips of transmitters Texas Instrument TFP410, and chip of receiver Texas Instrument TFP401. The chip demonstrated the stable signal recovery within the full range of frequencies. Thus, the solutions tested in this chip can be used to build a chip for fully featured DVI receiver.

5.7 SHF Chips for High-Speed Electronic Devices

5.7.1 *Development of Gallium Nitride Technology and the Prospects for Its Application in SHF Electronics*

Gallium nitride (GaN) is long-interesting for researchers and developers of semiconductor devices. GaN heterostructures and its solid solutions have physical properties which provide the electronic devices with optical, power, and frequency characteristics allowing to apply them in various areas of semiconductor electronics.

Structures based on semiconductive nitrides (GaN, A1N) and some compounds of AlGaN and InGaN type have been considered as non-conventional materials for electronic and optoelectronic technologies for approximately 20 years [19–21]. It is possible to control the concentration of two-dimensional carrier gases in these materials. This possibility allows to optimize their properties for certain purposes [22]. Structures based on GaN can be used as active medium in laser diodes and LEDs in a short wavelength range. They help to obtain structures with Schottky barriers for devices of power electronics.

One of research areas of these materials is creation of LEDs of short wave (blue–green) area of visible spectrum and near ultraviolet range, as well as white light-emitting diodes based on the crystal-luminophor system (the semiconductor crystal covered with luminophor). Professors Isamu Akasaki and Hiroshi Amano from the Nagoya University (Japan) together with Shuji Nakamura from the University of California (USA) were awarded the 2014 Nobel Prize in Physics for the invention of efficient blue light-emitting diodes based on heterostructures of GaN and its solid solutions, as well as white light sources [23].

However, it should be noted that GaN is an advanced material not only for optoelectronics but in particular for LED structures. It serves as a basis for successful development of other areas of modern electronics, including creation of SHF-transistors.

Transistors based on GaN heterostructures hold promise for application in SHF-transmitters. The main advantage of GaN-based SHF-transistors is high specific capacity that allows to significantly simplify the topology of power amplifier integrated circuits, improve efficiency, decrease weight, and make dimensions better. Over the past few years, development of GaN-based technology has leaded to significant practical results and application of power SHF-transistors and monolithic integrated circuits in the industrial production [24].

Optimization of GaN-structure technology for SHF-transistors.

Improvement of characteristics of GaN-based structures used for creation of SHF-transistors has become possible due to development of new process technologies.

One of such technologies is a gate recessing by plasma-chemical etching that usually is combined with a slot etching process in dielectric [19, 22]. It results in improvement of many parameters. For example, increase in transistor steepness due to decrease of gate-to-channel distance, reduction of drain-source resistance caused by absence of depletion regions of the transistor gate-source and gate-drain, minimization or even elimination of transient processes when the transistor is switched on, due to mitigation the effects of traps in the gate-drain region since the surface where they are placed can be moved away at a safe distance. In this regard, technologies offer to grow a dielectric passivation layer immediately after growing of all layers of heterostructures [22].

Furthermore, in recent years there has been a noticeable development of researches aimed at search for new surface-passivating materials for transistor heterostructures based on GaN and its solid solutions. The use of new materials allows to increase more than twofold the pulse current of the transistor and its steepness, as well as significantly reduce switch-on time due to compensation of interface states [22, 25].

Researchers and developers of Cree, TriQuint, Northrop Grumman, and other companies have achieved high-frequency response characteristics of the transistor configurations that became the base for development and creation of efficient power amplifier integrated circuits operating in different ranges. These chips outperform by more than 10 times the integrated circuits based on gallium arsenide (GaAs) in mass-dimensional parameters [22]. The above-mentioned manufacturers have already started mass production of power amplifiers based on GaN heterostructures with frequencies up to 100 GHz; QuinStar Technology company in cooperation with HRL company is developing transmit-receive modules for radiolocators in 94-GHz range with power output more than 5 W [19].

Thus, important industrial and technological issues have been solved over the past few years. They kept down the GaN-based transistors and monolithic integrated circuits from reaching the commercial level [24]. Currently, GaN-based technology is actively implemented in the sphere of SHF devices. Rates of commercial exploitation of this new technology exceed considerably (by 2–3 times) the rates of development of technologies for creation of devices based on silicon and GaAs. In coming years, silicon- and GaAs-based devices can get into frequency range of 1–100 GHz. However, it should be noted that in comparison with GaAs, technology based on GaN has important advantages in characteristics and properties while technology based on GaAs has lower costs of production and higher level of utilization in mass production [19].

It should be expected that costs of production will decrease according to progress in introduction of GaN-based technology in mass production.

5.7.2 GaN-Based SHF Chips by RFHIC for Wireless Communication Systems

Optimal solutions in the area of GaN-based technology are offered by leading global producers of SHF devices such as RFHIC company [26].

RFHIC company (South Korea) is one of leaders at the markets of telecommunication and radar equipment. This supplier of GaN-based SHF chips has full production cycle that includes the product development and assembly during which the crystal installation on the base, cable splicing, packaging, and sealing operations are performed.

RFHIC company produces a wide range of GaN-based products for SHF-electronics ranging from cable TV amplifiers to power amplifiers for radiolocation. The basic specifications of products made by this company are discussed below.

First of all, it is worth noting the amplifiers made by RFHIC company for wireless communication systems [26] which are intended for application in modern networks using different data transmission format including LTE, CDMA, WCDMA, and WIMAX. They can be conditionally divided into groups by power and operating voltage range (Fig. 5.55).

The first group includes SHF-amplifiers with a rated power of 5–10 W and operating voltage of 28 V. These amplifiers are intended for operation at the frequency range of 2–3 GHz and they have a gain factor of 40 dB.

The following group includes amplifiers with a rated power of 28 W and operating voltage range of 48–50 V. These amplifiers are intended for operation at the frequency range of 0.8–3 GHz and they have a gain factor of 44.5 dB.

The third group includes amplifiers with a rated power of 56 W and operating voltage of 48–50 V. These amplifiers are intended for operation at the frequency range of 0.8–3 GHz and they have a gain factor of 47.5 dB.

The fourth group includes SHF-amplifiers with a rated power of 80 W and operating voltage range of 48–50 V. These amplifiers are intended for operation at the frequency range of 0.8–3 GHz and they have a gain factor of 55 dB.

In addition, amplifiers for wireless communication systems include hybrid amplifiers with a power of 1–5 W and operating voltage of 28 V, as well as amplifiers based on Doherty circuit with a power of 7 W and operating voltage of 31 V. The specified amplifiers are intended for operation at the frequency range of 0.8–3 GHz and with a gain factor of 27–38 dB; as well as at the frequency range of 1.5–3 GHz and with a gain factor of 14–16 dB, respectively.

The basic types and technical characteristics of SHF-amplifiers made by RFHIC company for wireless communication are listed in Tables 5.5, 5.6, and 5.7 [26].

Fig. 5.55 The appearance of SHF-amplifier made by RFHIC company for pulse-modulated radar

Table 5.5 The basic types and technical characteristics of GaN-based amplifiers made by RFHIC company for wireless communication

Type	Frequency, MHz	Gain, dBm	Output power, W	Coefficient of efficiency, %	Dimensions, mm
Amplifiers with a power of 5–10 W (28 V)					
RTP 21005-11	2110–2170	45	37	40	
RTP 21010-11	2110–2170	45	40	40	100 × 50 × 20
RTP 26010-N1	2570–2690	40	40	38	
Amplifiers with a power of 28 W (48–50 V)					
RTP 08028-20	869–894			42	
RTP 18028-20	1805–1880	44.5	44.5	42	125 × 90 × 20
RTP 21028-20	2110–2170			42	
RTP 26028-20	2496–2690			40	
Amplifiers with a power of 56 W (48–50 V)					
RTP 08056-20	869–894				
RTP 18056-20	1805–1880	47.5	47.5	42	150 × 90 × 20
RTP 21056-20	2110–2170				
RTP 26056-20	2620–2690				
Amplifiers with a power of 80 W (48–50 V)					
RTP 08080-20	869–894				
RTP 18080-20	1805–1880	55	49	42	170 × 100 × 20
RTP 21080-20	2110–2170				
RTP 26080-20	2620–2690				

Since these amplifiers are GaN-based, they can operate at high values of temperature and voltage.

5.7.2.1 GaN-Based SHF-Amplifiers

Consider below the solid-state power amplifiers designed according to GaN technology for production of GaN-HEMT-transistors that are used in SHF amplifiers for air-space and defense industry [27].

Table 5.6 Technical characteristics of hybrid GaN-based amplifiers with a power of 1–5 W (28 V) made by RFHIC company for wireless communication

Model	Frequency, MHz	Gain factor, dB	Output power, W	Coefficient of efficiency, %	Working voltage, V
HT0808-15A	869–894	37	33	26	28
HT1818-15A	1805–1880	33	33	25	28
HT1818-15M	1805–1880	34	33	27	5/28
HT2121-15A	2110–2170	33	33	24.5	28
HT2121-15M	2110–2170	32	33	26	5/28
HT2626-15A	2610–2690	30	34	25	28
HT2626-15M	2610–2690	30	33	25	5/28
HT2627-15A	2650–2750	27	33	31	28
HT0808-30A	869–894	38	37	28	28
HT0909-30A	925–960	36	37	27	28
HT1818-30A	1805–1880	38	37	26	28
HT1919-30A	1930–1995	38	37	26	28
HT2121-30A	2110–2170	36	37	26	28
HT2626-30A	2610–2690	32	37	28	28

Table 5.7 Technical characteristics of hybrid GaN-based Doherty amplifiers with a power of 7 W (31 V) made by RFHIC company for wireless communication

Model	Frequency, MHz	Bandwidth, MHz	Gain factor, dB
RTH18007-10	1805–1880	75	16
RTH20007-10	1930–1995	65	16
RTH21007-10	2110–2170	60	15
RTH26007-20	2620–2690	70	14

As shown above, the solid-state power amplifiers and wide-band amplifiers are important elements of equipment for space and radar systems for various purposes that's why particularly high requirements are imposed on them. Power SHF-amplifiers for radars are based on high-power transistors or modules are devices defining the critical system parameters such as radiated and consumed power, radio frequency band, size and weight, life duration, reliability, and cost.

Development of wide-band gap semiconductive materials (silicon carbide SiC and gallium nitride GaN) and devices based on them became a revolutionary tendency in the area of power SHF-elements [28, 29]. That's why the leading global producers of elements used in solid-state power amplifiers develop primarily the technologies based on gallium nitride GaN [30].

GaN devices based on GaN-on-SiC HEMT-technology are developed in the following areas:

- increase of maximum power, power density, and maximum voltage;
- increase of the upper range of super high frequencies (advanced developments are aimed at achievement of C, X, Ku wavelength ranges);
- cost reducing down to one dollar for one watt;
- improvement of reliability and radiation stability (see Table 5.8);
- reduction of the overall dimensions of products.

History of GaN technology development has been going on for 30 years. It should be noted that the first light-emitting diode GaN LED appeared in 1993, and the first sample of the GaN-transistor and samples of amplifiers based on it were produced in 1997. Military and state programs allowed GaN technology to be actively developed and implemented. First of all, it is a well-known American program WBGSTI, later these are European MARCOS, TIGER, KORRIGAN, and Japan NEDO [27].

The first serial GaN-transistor was produced in 2001 [28]. Further, all global leading electronic companies earlier related only to production of GaAs-components, started investing in new GaN technology particularly for production of amplifiers and MMICs. The first commercial series GaN-products appeared in 2006–2007: universal packaged power transistors in L and S ranges (2–4 GHz) with a power output from 5 to 50 W (later, from 120 to 180 W).

Commercial market entry pioneers are such companies as Eudyna (Sumitomo) [31], Nitronex [32], Cree [33], RFHIC [27], later Toshiba, RFMD, M/A-COM Tech., TriQuint [34], OKI [35], Microsemi [36], NXP, and etc.

It is commonly known that the width of the band gap of semiconductive material is in direct proportion to the permissible operating temperature and the shift of operating range of devices based on corresponding semiconductor materials to the short-wave region of the spectrum. For example, the maximum operating temperature for GaN-based devices reaches 350–400 °C; and the maximum operating temperature for C (diamond)-based devices reaches 500–600 °C and higher. The band gap width correlates well to the material melting temperature. These two values increase with growing of atomic-binding energy in the crystal lattice that's why the wide-band gap semiconductive materials are characterized by high melting temperatures that causes certain difficulties in creation of pure and complete-structured mono crystals for such semiconductive materials.

Of course, such physical parameter as current carrier mobility considerably defines the frequency responses of any semiconductor device. Therefore, SHF devices are produced with the use of semiconductor materials with high values of charge carrier mobility which are capable of operating under high temperatures and high level of radiation that's very important for space industry.

Table 5.8 Electronic properties of semiconductive materials Si, GaAs, InP, 3-4-6H-SiC, GaN, C (Diamond) [26]

Properties of material	Units of measurement	Si	GaAs (AlGaAs/ InGaAs)	InP (InAlAs/ InGaAs)	3C-SiC[a]	4H-SiC[a]	6H-SiC[a]	GaN (AlGaN/ GaN)	*C* (Diamond)
The width of energy band gap, Eg	electron volt when 300 K	1.12	1.42	1.34	2.4	3.26	3	3.39	5.47
Electron mobility, In	300 K, $cm^2/V^a s$	1500	8500	4600	1000	950	500	2000	2800
Hole mobility, Ip	300 K, $cm^2/V^a s$	600	400	150	40	120	80	200	2100
Drift velocity for electrons at saturation, vsat[a] 107	cm/s	1.0	2.1	2.3	2.5	2.0	2.0	2.7	1.5–2.0
Critical electric field, Ec	MV/cm	0.025	0.4	0.5	2.0	2.2	2.5	5.0	20.0
Thermal conductivity coefficient, K	$W/cm^a K$ when 300 K	1.5	0.55	0.7	3.0–4.0	3.0–4.0	3.0–4.0	1.3	24.0
Dielectric conductivity, ε	–	11.68	12.8	12.5	9.7	10	10	9.5	5.7
C FoM[b]	–	1	8.5	21	–	250	–	660	75000

[a]3C, 4H, 6H—supposed crystal structures of SiC materials

[b]C FoM (Combined Figure of Merit)—material quality factor in relation to silicon (Si) for power and frequency

Therefore, a wider band gap increases stability of operation of an SHF transistor (see Table 5.8) under high temperatures and high level of radiation, while a higher concentration of electrons increases current density in the transistor channel section, which results in a higher gain factor. The higher the maximum critical electric field of the used semiconductor material is, the higher the maximum voltage of the SHF-transistor drain (50–100 V) and its breakdown voltage (100–300 V) are, which improves reliability and service life of the product. SHF-transistors designed according to GaN technology have high specific capacity up to 10 W and the gate width exceeds by 1 mm; the gate width and specific capacity of SHF-transistors are designed according to GaAs-technology [27, 37]. Heatsink from active structure of crystal and growing of epitaxial GaN-structures remain challenging but addressable issues of GaN-technology [27].

It is well-known that earlier solid-state power amplifier was based on GaAs-technology for APAR—Active Phased Array Radars or AESA—Active electronically scanned array. GaAs-technology provides low density of power (0.5–1.5 W/mm, see Table 5.9) in comparison with high power density of GaN-HEMTs-technology (4.0–8.0 W/mm of SHF-transistor gate width) [9].

High power density of SHF-transistors based on GaN technology makes it possible to significantly reduce dimensions and weight of solid-state radar amplifier that is very important for aerospace applications where it is required to minimize the dimensions and weight of radars, including radars with AESA. For example, instead of five GaAs LDMOS-amplifiers, it is possible to use only one GaN-amplifier providing principally better technical characteristics of the product [28].

Physicists know that thermal conductivity of GaN-material is 8–10 times more than that of GaAs-material. It provides better and faster heat removal from MMIC chip and higher power density (see Table 5.8) [28].

It is always of relevance to meet challenge of production of space communication systems and radars based on new GaN-products for C, X, Ku frequency ranges.

Solid-state amplifiers and other products based on GaN technology offer a broad prospect for development and update of advanced devices operating in significant frequency ranges of 1–4 GHz, 2–6 GHz, 4–12 GHz, 6–18 GHz, and 2–20 GHz which are competitive for microwave tube devices (triodes, teratrons, magnetrons, klystron, and etc.) in output power, coefficient of efficiency, dimensions, reliability, and cost.

Table 5.9 Comparative analysis of basic parameters of GaAs and GaN materials

Material properties	Units of measurement	GaAs	GaN
Output power density, p(Pout)	W/mm	0.5–1.5	4.0–8.0
Working voltage, Vds	V/mm	5–20	28–48
Reverse voltage, Vbr	V/mm	20~40	>100
The maximum current density, p(Imax)	A/mm	~0.5	~1.0
Thermal conductivity coefficient, K	W/(m × K)	47	390(z)/490(SiC)

It is clear that GaN technology has an acknowledged place at the market of military and space systems in challenges of changing of GaAs MMIC-products in AESA transmit-receive modules of centimeter and millimeter wave bands [31].

In this regard, developers of SHF devices are interested in consideration of technical parameters of SHF-amplifier based on GaN technology and produced by Sumitomo company, one of acknowledged leaders in this sphere.

Sumitomo company was established in 1897 and initially catered to make copper products. In the early 1990s, Sumitomo company was among the first to master the GaN technology. Solid-state product department Sumitomo was renamed Eudina Devices Inc. in 2004, then Eudyna Devices Inc. was fully renamed Sumitomo Electric Device Innovations, Inc. in 2009. As of the date of publishing this book, Sumitomo Electric Device Innovations, Inc. (later Sumitomo) is one of leaders in the sphere of development, design, and serial production of SHF devices for space-communication systems, radioastronomy, radars of general and special purpose for various industries and specialized applications.

Developers of space system are recommended to learn and use information on the products made by this company (Tables 5.10 and 5.11).

The product range includes:

- GaN-HEMTs-transistors: for base stations, radars, and general application;
- monolithic integrated circuits MMICs: Low-noise circuits of Ku and V frequency ranges; high-power circuits of Ka frequency range; circuits of C and V frequency ranges;
- power GaN-amplifier on pallets and MMICs;
- converters of Ku and Ka frequency ranges;
- multiplexers of Ku and V frequency ranges;
- generators with wide dynamic range;
- GaAs FETs transistors: of high-power, power and low-power, crystal-based.

SHF-transistors are produced according to various technologies for space, aviation, ground, and above-water radars (AESA), wireless communications systems, and identification systems.

Sumitomo company produces the most efficient and promising SHF-transistors of L, S wavelength ranges according to technology based on GaN (GaN-HEMTs, GaN-on-Si, GaN-on-SiC). Figure 5.56 illustrates the main products developed by Sumitomo company according to GaN-HEMTs technology.

The use of terminal amplifying pallet-modules with an impedance of 50 Ohm remains of interest at this time. They are used in aviation, wireless communications systems, and radars of S and L frequency ranges (Table 5.10).

In many cases, two and more discrete components are installed on such pallet for reaching the amplification that can't be reached when using one component. The pallet is designed according to special technology [31] that allows to build compact devices. The small-sized pallets are ideal solutions for air-space applications.

As well, Sumitomo company actively develops products based on GaN-on-SiC technology (gallium nitride on silicon carbide) [38]. Products based on technology are applied in military industry.

Table 5.10 The main areas of application of Sumitomo products based on GaN technology

Products	Technology	Line	Areas of use/pulse duration	Wavelength range	Operating frequency range, GHz	Output power, W/dBm	Voltage (V)/impedance (Ohm)/ coefficient of efficiency (%)
Transistors	GaN-HEMTs	EGN13	Impulse radars/3 ms/10%, 1.5 ms/25%	VHF-L	1.2–1.4	170	50/50/
Transistors	GaN-HEMTs	EGN	Impulse radars/5 ms/10%, 750 μs/25%	VHF-S	2.7–3.5	120–600	50/50/
Transistors	GaN-HEMTs	SGN	Impulse radars/300 μs/ 10%	VHF-S	2.9–3.5	150–600	50/50/
Transistors	GaN-HEMTs	EGNB	Continuous wave radars (CW)	VHF-S	3.5	/40	50/50/
Transistors	GaN-HEMTs	EGN31	Continuous wave radars (CW)	VHF-S	3.1	/45.5	50/50/
High-power amplifiers on pallets	GaN-HEMTs	SMC	Impulse radars/300 μs/ 10%	VHF-S	2.9–3.5	150–600	50/50/50
High-power amplifiers on pallets	GaN-HEMTs	EMC	Impulse radars/5 ms/10%, 750 μs/25%	VHF-S	3.1–3.5	100	50/50/50
Amplifiers New	GaN-HEMTs	SGNC	Wireless communication and base station	UHF-S	0.9–2.6	300/66	50//
Amplifiers	GaN-HEMTs	EGNC-EGN3	Wireless communication and base station	UHF-S	0.9–3.5	70~270/ 47~53.5	50//
Amplifiers	GaN-HEMTs	EGNB	Of general purpose	UHF-S	0.9–3.5	10–90/41–53	–
Amplifiers	GaN-HEMTs	SGNE	Of general purpose	UHF-S	0.9–3.5	10–90/ 40.5–51	–

Table 5.11 The main areas of application of Sumitomo products including MMIC based on GaAs-technology

Products	Technology	Line	Areas of use/pulse duration	Wavelength range	Operating frequency range, GHz	Output power, W/dBm	Voltage (V)/impedance (Ohm)/coefficient of efficiency (%)
Amplifiers, converters	GaAs WLCSP[a] MMICs	SMM/ EMM	Communications system	C-E	12.7–30.0	/26–33	6
Amplifiers	GaAs MMICs	SMM/ EMM	VSAT and transceivers of communication systems	S-C-Ka	3.4–30.0	/30.0–33.5	6/14–29
Amplifiers	GaAs MMICs	EMM/ SMM/ FMM	VSAT and transceivers of communication systems	S-C-V	3.4–64.0	/26–34	3–7/17–29
Amplifiers	GaAs MMICs	EMM/ SMM/ FMM	VSAT and transceivers of communication systems	Ku-V	12.0–64.0	/7–20	/13.5–23
Converters	GaAs MMICs	SMM/ FMM	Satellite radio communication	Ku-Ka	12.0–32.0	/5.0	/10–12
High-power amplifiers	GaAs FETs	FLM/ELM	Radio communication	L-C	2.0–15.3	/39	10/9.5–11.5
High-power amplifiers	GaAs FETs	FLU/FLL/ FLC/FLX/ FLK	Mobile and cell communication, WCDMA, LTE, and WiMAX	L, S, C, X, Ku	2.0–15.0	40–80/	10/6.0–13.0
Low-power amplifiers	GaAs FETs	PSU/FSX	Middle-power amplifiers and generators	C	8.0–14.5	/15–24	8–10/10/18
Amplifiers, mixers, converters	GaAs HEMTs	FHC/FHX	DB S-converters, wireless and mobile communication, radioastronomy, and etc.	S-C	4.0–12.0	-	2.0/10.0–15.5

[a]WLCSP—Wafer Level Chip Scale Package

SMC2935L3012R* Pallet
GaN-HEMTs amplifier on pallet,
Pout. >300W in the 2.9-3.5GHz
frequency range (S-band),
Gain = 12.8dB,
Vds = 50V, Ids = 1.5A,
P/D = 300μs/10%,
impedance 50Ω,
Eff = 48%.

SMC2933L6012R Pallet
GaN-HEMTs amplifier on pallet,
Pout. > 600W in the 2.9-3.5GHz
frequency range (S-band),
Gain = 12.8dB,
Vds = 50V, Ids = 3.0A,
P/D = 300μs/10%,
impedance 50Ω,
Eff = 50%.

SGN2933-600D-R
GaN-HEMTs amplifier on pallet,
Pout. > 600W in the 2.9-3.3GHz
frequency range (L-band),
Gain = 12.8dB,
Vds = 50V, Ids = 3.0A,
P/D = 300μs/10%,
impedance 50Ω,
Eff = 50%.

Fig. 5.56 The main products developed by Sumitomo Electric Europe, Ltd according to GaN-technology [38]

As known, the following parameters are important for SHF-amplifiers:

- frequency range (MHz);
- supply voltage (V);
- output power (W);
- value of input/output impedance;
- gain factor (dB);
- coefficient of efficiency (%);
- fill factor (%) (see Table 5.10).

High-power SHF-transistors operate in wide range of frequencies and provide the maximum load power.

GaN-transistors popularity is due to their real advantages at the system level such as:

- simplicity and cheapness of circuit-based implementation of high-power amplifiers;
- simplicity in the broadbanding amplification;
- overlapping of a few sub-bands of space communications station by one high-power amplifier;
- reduction in energy consumption by the device (radar) and associated costs;
- reduction of complexity and cost of cooling system.

For example, the amplifier with a power of 10 W of L-range has coefficient of efficiency more than 70–80%; and amplifiers with a power of 100–600 W has coefficient of efficiency of 50%.

Developers of devices of space purpose have an obvious question—how to reach compromise in GaN- and GaAs-technologies, or how to plan the use of new technologies based on diamond for perspective space systems?

To answer this question and compare the possibilities of application of GaN- and GaAs-transistors and MMICs in the circuits of broadband power SHF-amplifiers, as well as the possibility to make optimal technical decision, it can be used a method of migration from one material SiC to the other one; and it can be performed a simple qualitative analysis of their specific parameters (relating to 1-mm width of the transistor gate). For this purpose, it is offered to apply well-known evaluations for amplifiers of A class with the maximum output power Pmax and optimum load resistance Ropt of the transistor [27].

For example, as known, developers of GaN-devices have a keen interest in wideband communication systems. First of all, this is the case of application of GaN-transistors in the base stations of standards W-CDMA, UMTS, and Wi-MAX. It is known that in the total volume of GaN-device production in 2014, 63% of GaN-devices were used in wideband commercial communication and TV systems vs. 26% of GaN-devices were used in space and military systems [39].

As of the date of publishing this book, centimeter and millimeter wave bands became the arena of serious struggle and compromise of two industrial technologies for production of high-power solid-state monolithic integrated circuits MMICs, discrete SHF devices, and amplifiers on pallets that are almost identical in frequency and amplifying properties. One of these technologies (GaN) has advantages in the maximum electrodynamic and the minimum weight and dimensional parameters and the minimum cost of power supply sources, and the other technology (GaAs) has the minimum costs of production and the maximum level of utilization in mass production [27, 40].

It should be recalled that it is a natural alternation of generations of technologies that ensures improvement of parameters of solid-state power amplifiers of space and commercial purpose. There's still a chance to implement many projects based on GaN technology. But a clock is ticking, and new technological solutions have appeared. These are technological solutions based on diamond (Diamond FETs) [40].

Indeed, the artificial synthesis of diamond monocrystals has been studied since the early 1990s. Thus, from 1990 onwards, Sumitomo Electric Industries, Ltd. applied in US Patent and Trademark Office and received a few key patents on development of field-effect transistors based on diamond.

Due to its unique properties, diamond is considered to be an ideal material for creation of new generation of nano-sized electronic components [40–42]. It was found that diamond gets sheet conductivity of hole-type after being processed in hydrogen plasma. Such properties of hydrogenized diamond made possible the creation of diamond-based field-effect transistors. For example, in 2009, the gate width of diamond-based transistors became equal to 50 nm that is by 1000 times less than the width of human hair and by 2 times less than the sizes of the previous world record-breaker produced by Japan company NTT [40, 42].

Conventional materials GaN and GaAs have its own strengths and weaknesses while the diamond is almost universal. Diamond FETs can be widely applied not only in medical tomographs of terahertz range, car's safety system, and etc. but also in devices of space purpose [30].

First of all, it should be noted that high radiation stability of diamond allows it to be a promising material for on-board UV-detectors intended for space research. As well, CVD-diamond can serve as a material for high energy particle detectors (alpha particles, gamma quantum, neutrons) applied in space vehicles for deep-space research.

Currently, no one semiconductive materials can be competitive to diamond in parameters (see Table 5.8 and Fig. 5.57) [30]. The original material for creation of such diamond-based transistors is produced by means of chemical hydrogen-vapor deposition. The technology of electron-beam lithography is used to form the structure of such transistors on the surface of diamond films. Due to their big sizes, diamond films are very advantageous for creation of position-sensitive and microstrip UV-radiation and X-ray detectors that are installed in the orbital interplanetary transport vehicles, as well as in the vehicles intended for transporting on the surface of planets, satellites, and asteroid.

5.7.3 SHF Chips by Hittite Microwave

5.7.3.1 SHF Chips of Attenuators

One of global leaders in the sphere of microwave electronics is American company Hittite Microwave which was established for development and production of analog and superhigh-speed monolithic microwave integrated circuits and modules of radio

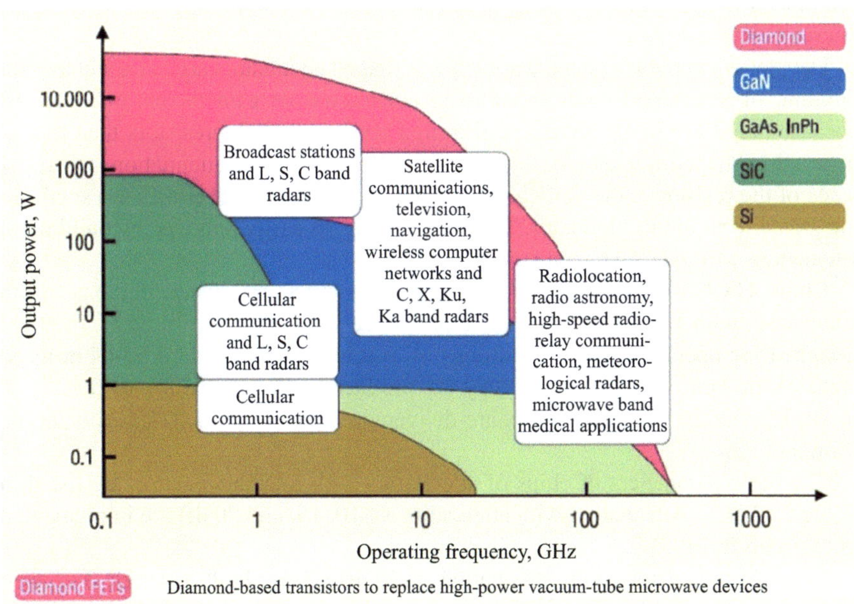

Fig. 5.57 Generalized diagram of dependence of output power and operating frequency of SHF-systems and devices on the type of semiconductive material applied.

frequency, microwave, and millimeter wave bands overlapping direct current frequencies up to 110 GHz for military and civilian usage [43].

Monolithic and hybrid film integrated chips are produced by this company using transistors of MESFET, PHEMT, MHEMT, and HBT-type based on GaAs, InGaP/GaAs, InP, SiGe.

Functional parts and integrated chips are in die form or mounted in mini surface-mount cases of more than 20 uniform types made with the use of special plastics and ceramics. Some products represent pressurized modules with coaxial SMA-connectors. The corporation adheres to the modern environmental requirements for materials and technologies in accordance with RoHS directive (Restriction of Hazardous Substances Directive) of the European Union and produces RoHS-versions of many its products.

A wide range of Hittite components of various purpose includes: *amplifiers, attenuators, mixers, amplitude and phase modulators, phase-shifters, frequency multipliers and frequency dividers, voltage-controlled oscillators, switches, sensors*, and other components intended for equipment of mobile, cell and satellite radio communication, fiber-optic communication lines, and radars.

Let's consider in more detail the SHF-chips of attenuators with fixed attenuation. Resistive attenuators with fixed attenuation as well as with attenuation controlled by analog and digital signals are widely applied in radio sets and measuring instruments. Generally, these are coaxial structures based on T- and inverted-U-shaped structure. Hittite Microwave is one of the few companies of the world that produces subminiature resistive SHF-attenuators represented by integrated chips with a frequency band from direct current up to 50 GHz. In this case, the coaxial structure was abandoned in favor of microstrip structure that is more miniature and suitable for microelectronic design.

Developers of radar systems know that T-shaped attenuators are more suitable for building of attenuators with small fixed attenuation of gain factor up to 10 dB because they have narrower range of design resistance of the resistors than that of inverted-U-shaped attenuators. However, in case of greater attenuation, the resistance of the resistor in the vertical part of T-shaped structure becomes too small for integrated form of the attenuator. This disadvantage is easily removed by building of attenuators with two T-shaped parts.

Chips HMC650/651/652/653/654/655/656/657/658 are linear ultrabandwidth attenuators with fixed attenuation that are based on T-shaped structure. They are designed for operation under conditions of matching with standard 50-Ohm resistance of the generator and load. They are produced with attenuation of 0, 2, 3, 4, 6, 10, 15, and 20 dB. Attenuators are delivered either in mini LP2-cases or in die form as a chip.

The basic parameters of chips of attenuators with fixed attenuation are listed in Table 5.12 [43]. Attenuators with attenuation of 10, 15, and 20 dB are implemented in two-step form.

Chips of attenuators with attenuation up to 10 dB are built as single T-shaped attenuators (Fig. 5.58a). Chips of attenuators with attenuation of 10, 15, and 20 dB are built as two-step T-shaped attenuators (Fig. 5.58b). It allows to limit the ratings of vertical resistors using permissible values.

Table 5.12 Parameters of chips of attenuators with fixed attenuation

Type of chip	Bandwidth, GHz	A of gain factor, dB	Gain factor, dB	Quantity of terminals	Dimensions, mm	Package type
HMC650	DC-50	±0.2	0.15	–	17 × 18	Chip
HMC651		±0.3		–	23 × 18	Chip
HMC652		±0.2	2	27	17 × 18	Chip
HMC652LP2E	DC-25	±0.5			–	LP2
HMC653	DC-50	±0.2	3	26	17 × 18	Chip
HMC653LP2E	DC-25	±0.5			–	LP2
HMC654	DC-50	±0.2	4	25	17 × 18	Chip
HMC654LP2E	DC-25	±0.5			–	LP2
HMC655	DC-50	±0.2	6	26	17 × 18	Chip
HMC655LP2	DC-25	±1.5		25	–	LP2
HMC656	DC-50	±0.1	10		17 × 18	Chip
HMC656LP2	DC-25	±1.5			–	LP2
HMC657	DC-50	±0.4	15		17 × 18	Chip
HMC657LP2	DC-25	±2			–	LP2
HMC658	DC-50	±0.5	20		23 × 18	Chip
HMC658LP2	DC-25	±2			–	LP2

Figure 5.59a [43] shows the dependence of gain factor on frequency (in the range of 0–50 GHz) for chips with small (2, 3, 4, and 6 dB) attenuation. Figure 5.59b shows the similar dependence diagrams for chips with attenuation of 10, 15, and 20 dB with two-step attenuators.

Developers need to know the dependence of reverse losses on frequency in order to evaluate exactly the attenuator behavior in the radar station under design. Such dependence diagrams are given in the manufacturer's descriptions of chips. So, dependence diagrams for one-step attenuators are shown in Fig. 5.60 [43].

Generally, chips of attenuators are designed for connection between two 50-Ohm transmission lines in the impedance matching mode. It is illustrated in Fig. 5.61. It is technologically implemented by means of gold wires.

Fig. 5.58 Typical functional diagram: (**a**) chips of single T-shaped attenuator; (**b**) chips of two-step T-shaped attenuators

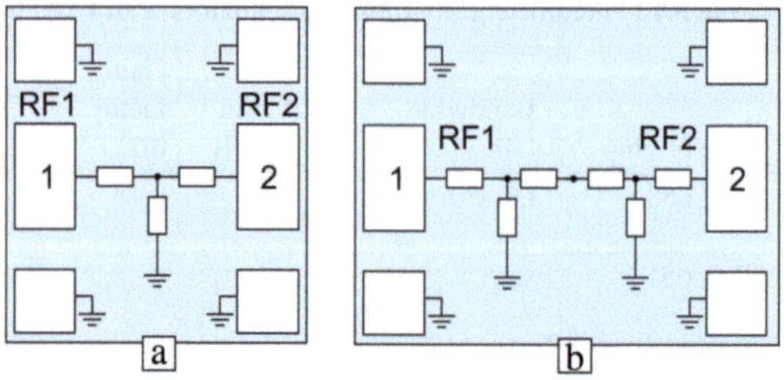

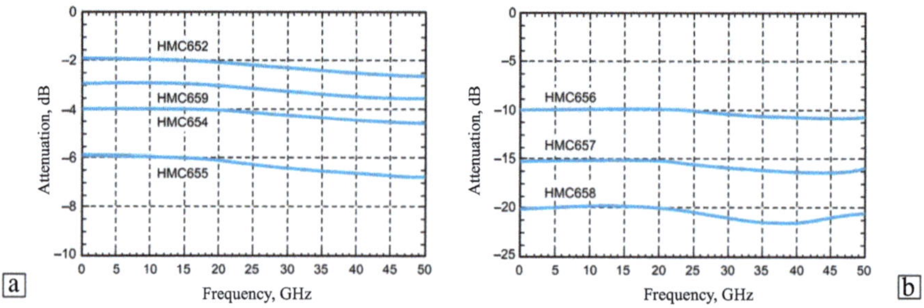

Fig. 5.59 Attenuation-frequency dependence: (**a**) one-step attenuators; (**b**) two-step attenuators

Fig. 5.60 Dependence of reverse losses of one-step attenuators on frequency

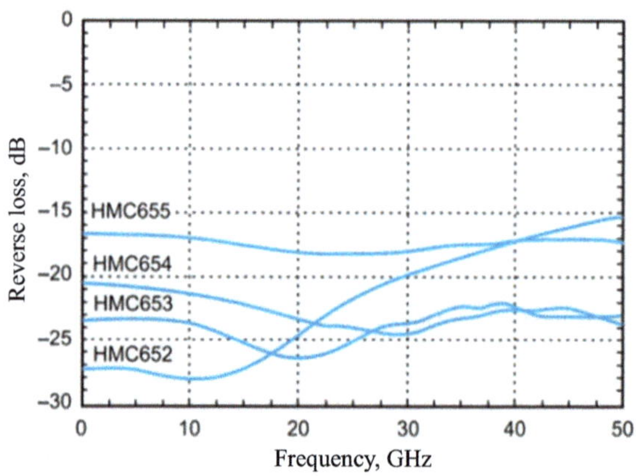

5.7.3.2 Standard Attenuator Chips with Analog Control

Hittite Microwave produces a series of attenuators with analog voltage control attenuation (Analog VVA). They are designed on inverted-U-shaped circuits with a variable horizontal resistor. Basic technical specifications of such attenuators are listed in Table 5.13 [43].

Typical package of an adjustable attenuator chip is miniature and designed for surface mounting. Such package provides small parasitic inductance of terminals. Functional diagram of chip HMC346 is illustrated in Fig. 5.62. The chip contains

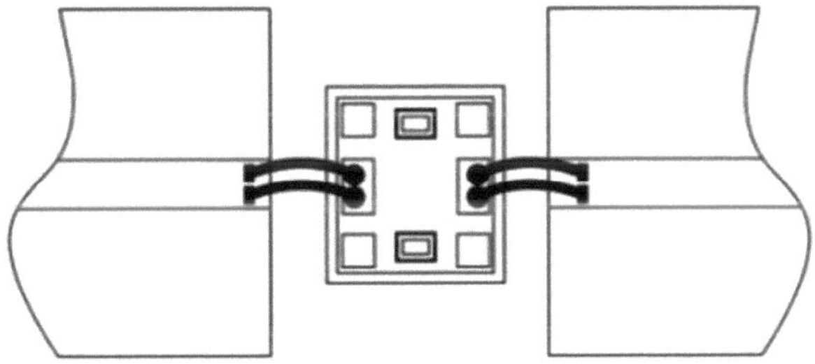

Fig. 5.61 The standard connection of attenuator chip between two 50-Ohm transmission lines

Table 5.13 Basic parameters of attenuators of Analog VVA type

Chip	Bandwidth, GHz	Losses, dB	Gain factor, dB	II3, dBm	Control voltage, V	Type of case
HMC-VVD102	17–27	1.5	0–22	17	−4 to +4	Chip
HMC-VVD104	70–86	2	0–14	–	−5 to +5	
HMC-VVD106	36–50	1.5	0–22	17	0/+4	
HMC346	DC-20	2.2	0–25	10	0. −3	
HMC346C8	DC-8	2	0–30			C8
HMC346G8						G8, H
HMC346LC3B	DC-18	1.5				LC3B
HMC346LP3	DC-14	2				LP3
HMC346MS8G	DC-8	1.5	0–32			MS8G
HMC473MS8	0.45–2.2	1.9	0–48	20	0. +3	MS8
HMC712	5–30	2.5	0–30	32	0. −3	Chip
HMC712LP3CE	5–26.5	3.5	0–28			LP3C
HMC812LC4	5–30	2	0–30	28		LC4
HMC973LP3E	0.5–6	2.5	0–26	35	0. +5	LP3
HMC985	0–50	3	0–35	33	0. −3	Chip

one controlled resistor and implements U-shaped circuit of the attenuator. It should be emphasized that it is the way how most of attenuators are implemented.

Figure 5.63 shows the typical attenuation-frequency dependence for chip HMC346. Generally, these diagrams take the form of horizontal lines but in case of greater attenuations, the characteristics are no longer uniform.

Figure 5.64 shows the general view of printed circuit board with mounted in its middle the chip HMC346 of attenuator with adjustable attenuation [43]. It should be noted that the upper part of the chip contains a microstrip line segment similar to the specified line segments connecting the attenuator chip. It is used to evaluate the line attenuation without attenuator chip that is practically used during calibration of attenuator chips. Connectors J1 and J2 are mini coaxial 50-Ohm connectors of PCB type.

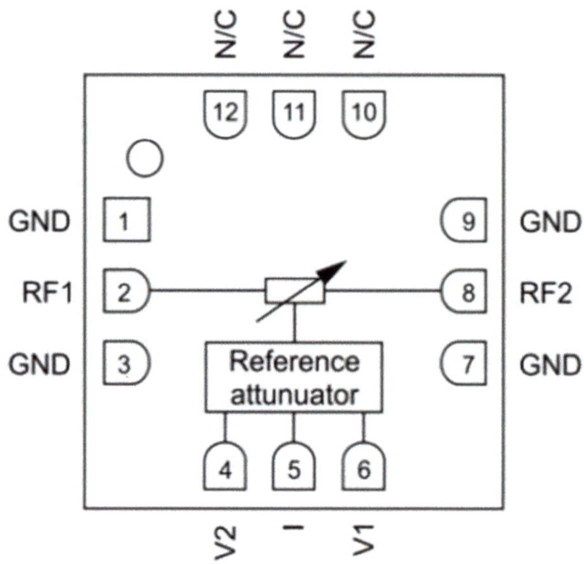

Fig. 5.62 Simplified functional diagram of chip HMC346

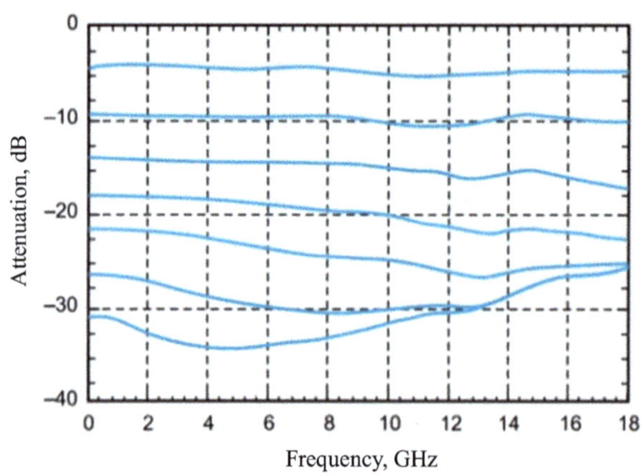

Fig. 5.63 Typical attenuation-frequency dependence for chip HMC346

5.7.3.3 Attenuator Chips with Digital Control

For modern radar stations, it is necessary to use SHF-chips of controlled attenuators with digital control that represent a set of attenuators with fixed attenuation switched-over by means of special digital control circuit. Generally, the bit capacity of a control digital signal is from 1 to 7 bits. It determines the range of attenuation variation. For example, 6-bit chip HMC424 has a DC frequency bandwidth up to 13 GHz, reverse losses of 4 dB, and attenuation adjustment from 0.5 to 31.5 dB. The chip is in die form, power is 5 V [43].

Chip HMC939 is distinguished by the highest operating frequencies from 0.1 to 50 GHz. It is a 5-bit digital attenuator based on GaAs-technology. The chip contains

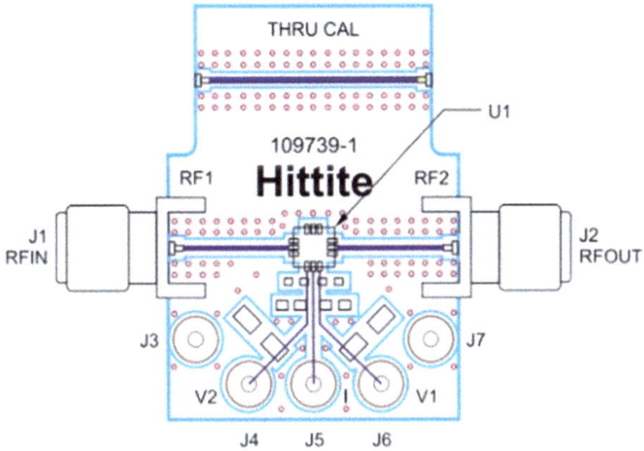

Fig. 5.64 General view of printed circuit board with mounted in its middle the chip of attenuator with adjustable attenuation

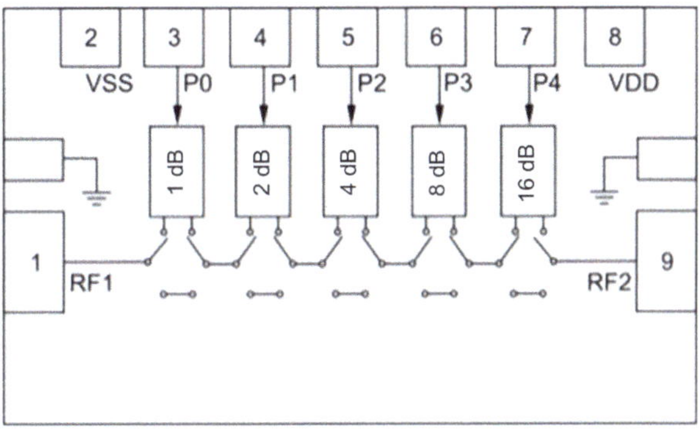

Fig. 5.65 Simplified functional diagram of chip HMC939

a set of attenuators with attenuation of 1, 2, 4, 8, 16 dB, the switching of which results in attenuation up to 31 dB with resolution of 1 dB. The chip crystal is 2.36 × 1.02 × 0.1 mm in size. The functional diagram of the chip is illustrated in Fig. 5.65 [43].

Typical attenuation-frequency dependence at difference levels of digital control signal is illustrated in Fig. 5.66.

1-bit chips (commutated attenuators) have a special place among digital attenuators. So, Fig. 5.67 shows the schematic circuit diagram of commutated attenuator with chip HMC800LP3E.

5.7.3.4 SHF and Ultra-Wideband Amplifiers with Fixed Gain

Here, first of all, it is necessary to consider chip HMC471MS8G/HMC471MS8GE. It is a typical two-channel wideband dual amplifier based on SHF heterotransition germanium silicon transistors. The chip is placed in mini package M28G with low-inductance beam leads (Fig. 5.68). Here, each channel has a bandwidth from 0 to 5 GHz when a gain factor of 20 dB and the maximum output signal of 20 dBm. Hittite company produces a wide range of such chips with different parameters.

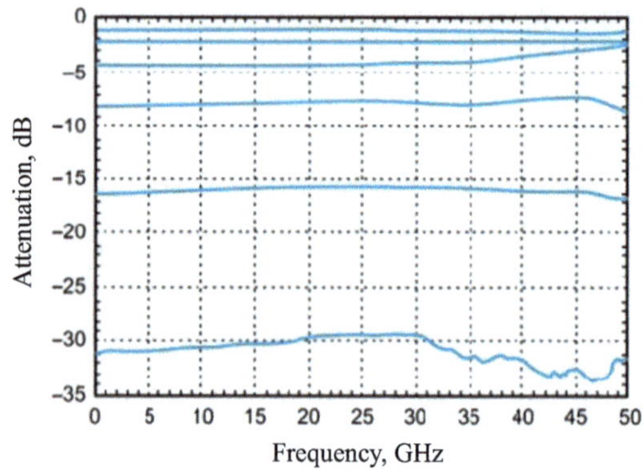

Fig. 5.66 Typical attenuation-frequency dependence at difference levels of digital control signal

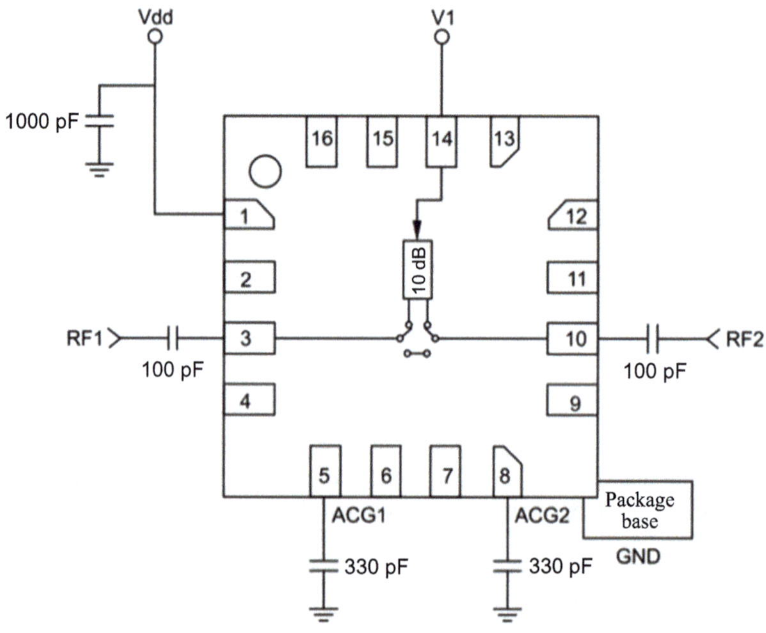

Fig. 5.67 Schematic connection circuit diagram of 1-bit digital attenuator—chip HMC800LP3E

Figure 5.68 shows that amplifiers of both channels are not differential and have one input and one output. Actually, each channel of this chip represents a compound transistor with open output. Such drawing a circuit provides the most flat amplitude–frequency response with zero low-frequency cutoff but there is stable offset voltage set by external circuits in input and output.

Typical dependence of this chip gain factor on the operating frequency for three values of temperature is illustrated in Fig. 5.69. It is not difficult to see that temperature stability of gain factor is rather higher that results from application of transistors on wide-band-gap semiconductive materials.

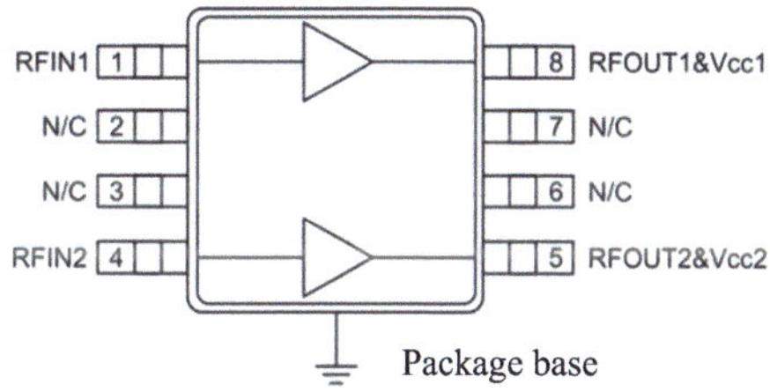

Fig. 5.68 The functional diagram of chip HMC471MS8G/HMC471MS8GE

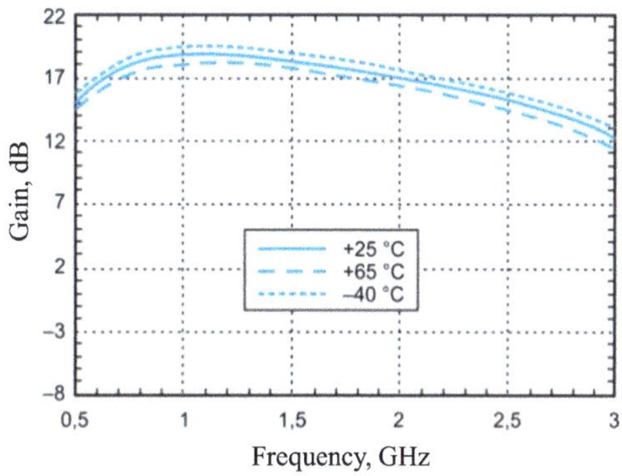

Fig. 5.69 Typical dependence of gain factor on the operating frequency and temperature of each channel of chip HMC471MS8G/HMC471MS8GE

Figure 5.70 shows the dependence of the maximum output power in the first harmonic of sinewave on frequency and three values of ambient temperature.

Two channels in this chip allow to form the wideband balanced power amplifiers with lower level of the third harmonic of output signal and increased power of output signal. Typical circuit of this amplifier is illustrated in Fig. 5.71.

Figure 5.72 shows the general view of printed circuit board of balanced amplifier based on chip HMC471MS8G/HMC471MS8GE that is built on packageless components with leads in the form of thin gold wires.

The maximum output power of such amplifier depends on the power supply voltage value. The firm offers a row of amplifiers with frequencies up to 86 GHz and with a power up to 2 W.

SHF-amplifier-driver HMC-AUH256 (Fig. 5.73) has frequency band from 17.5 to 41 GHz and it is based on gallium-arsenide (GaAs) field-effect transistors with Schottky barriers as a gate. This circuit has a gain factor of 21 dB, the maximum output power of +20 dBm, power supply voltage of +5 V, and consumption current up to 295 mA. Frequency-gain factor dependence is shown in Fig. 5.74.

Like many key producers, Hittite Microwave company produces a series of *wideband and ultraband small-signal amplifiers with standardized noise coefficient.* There is a common pattern: the wider bandwidth of amplified frequencies, the less gain reached. Devices with reactive circuits of impedance matching (total resistance) of input and output with their circuits included in amplifier have the maximum gain. Developers of transmit-receive modules for radar stations attach importance to the fact that ultra-wideband amplifiers are characterized with detailed organization of connection of the chip input and output to their terminals and HF-connectors (along

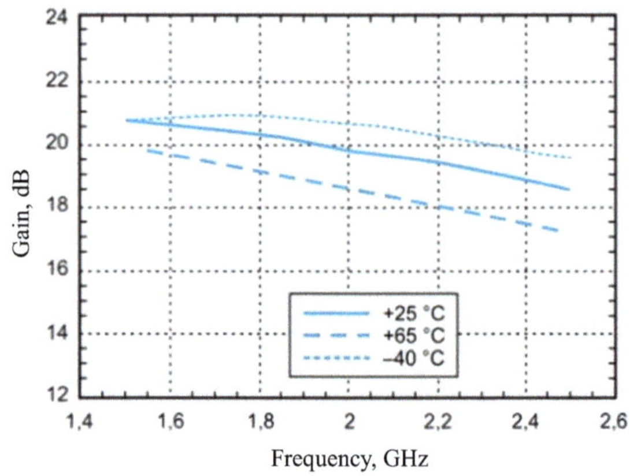

Fig. 5.70 Typical dependence of the maximum output power in the first harmonic of sinewave on frequency and temperature for chip HMC471MS8G/ HMC471MS8GE

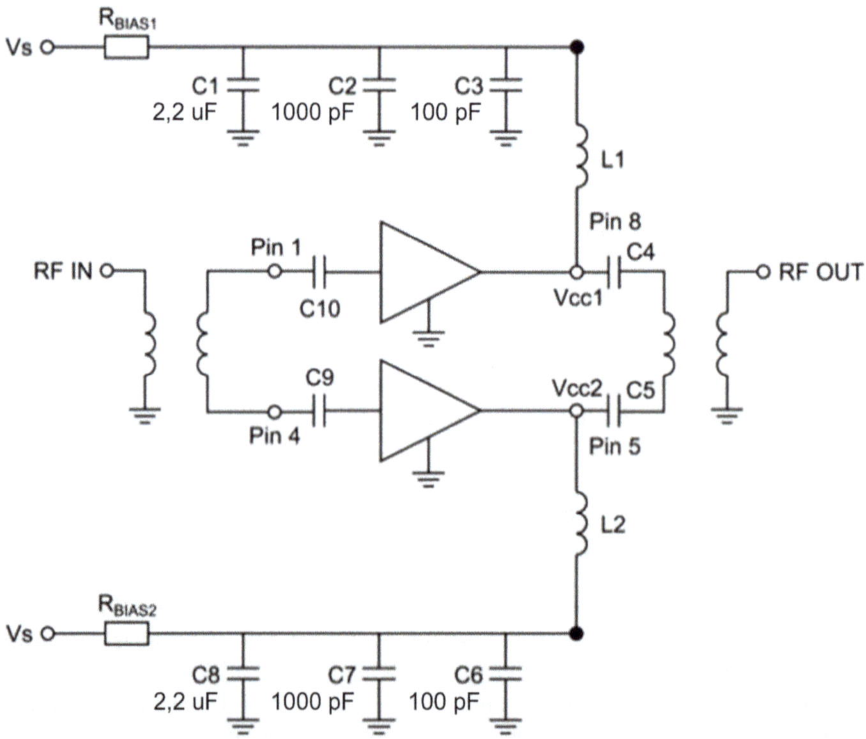

Fig. 5.71 Typical recommended circuit of balanced amplifier based on chip HMC471MS8G/ HMC471MS8GE

with band broadening). Generally, the system of s-parameters is used for description of such chips.

Figure 5.75 illustrates the functional diagram of SHF-chip HMC-AUH320 with bandwidth of amplified frequencies from 71 to 86 GHz when small-signal gain factor is 5 dB. The maximum output power of this circuit is 15 dBm.

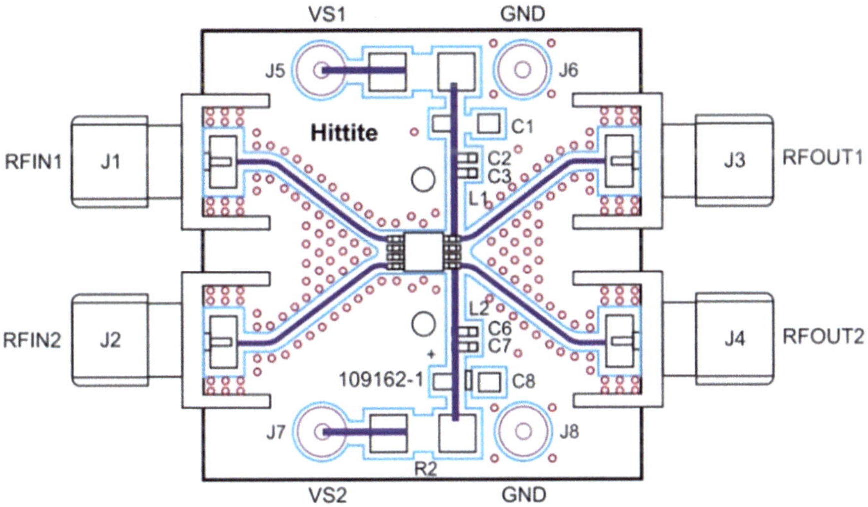

Fig. 5.72 The general view of printed circuit board of balanced amplifier based on chip HMC471MS8G/HMC471MS8GE

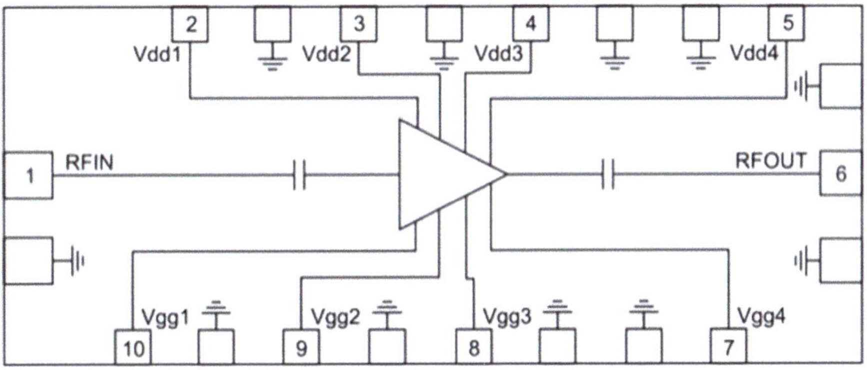

Fig. 5.73 Simplified structural diagram of SHF-amplifier-driver HMC-AUH256

Frequency-gain factor dependence of chip HMC-AUH320 is shown in Fig. 5.76. As we can see, nonuniformity of this dependence in this operating frequency band doesn't exceed 3 dB.

The developers have to optimize accurately the structure and break head over the choice of materials applied in the most of ultra-wideband amplifiers with bandwidth of a few units or dozens of GHz. Signal circuits are usually designed as matching and 50- or 75-Ohm circuits. In this case, all power supply circuits should have blocking circuits (for example, by means of capacitors with very low inductance of leads).

One more ultra-wideband chip HMC-AUH232 (Fig. 5.77) belonging to Wideband class has very wide bandwidth of amplified frequencies (from DC voltage up to 43 GHz) when small-signal gain factor is 12 dB.

Fig. 5.74 Typical gain factor-frequency dependence of amplifier-driver HMC-AUH256

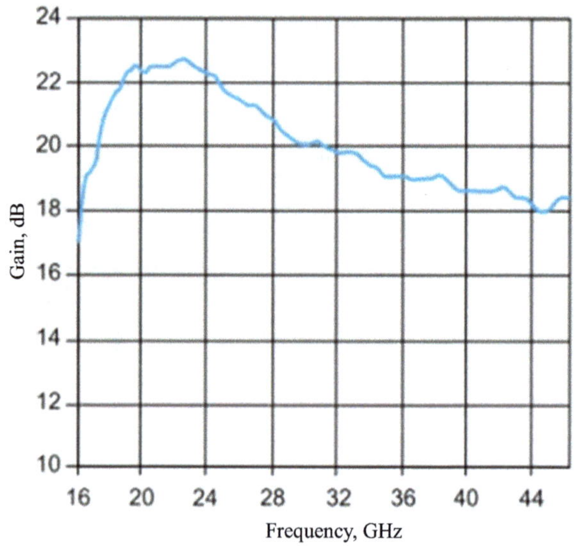

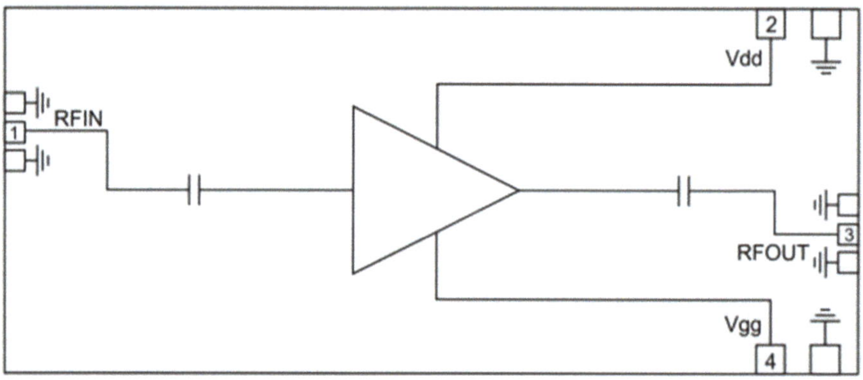

Fig. 5.75 Simplified functional diagram of SHF-chip HMC-AUH320

Typical amplitude-frequency response of chip AUH232 is illustrated in Fig. 5.78a; and dependence of noise factor on operating frequency is shown in Fig. 5.78b.

Analysis and measurement of characteristics of above-described SHF integrated chip are the practical issues for developers. These are the so-called eye circuits which are used for control of wideband amplifiers. Ultra-wideband digital storage oscilloscopes made by Agilent, Tektronix, or LeCroy are used for building of such circuits (www.hittite.com) [44].

Figure 5.79 shows a typical example of circuit of ultra-wideband amplifier based on chip HMC-AUH232 that is the most widely used by developers.

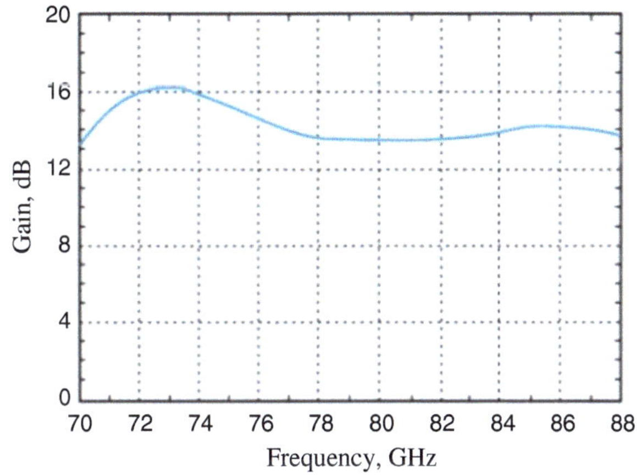

Fig. 5.76 Typical gain factor-frequency dependence of chip HMC-AUH320

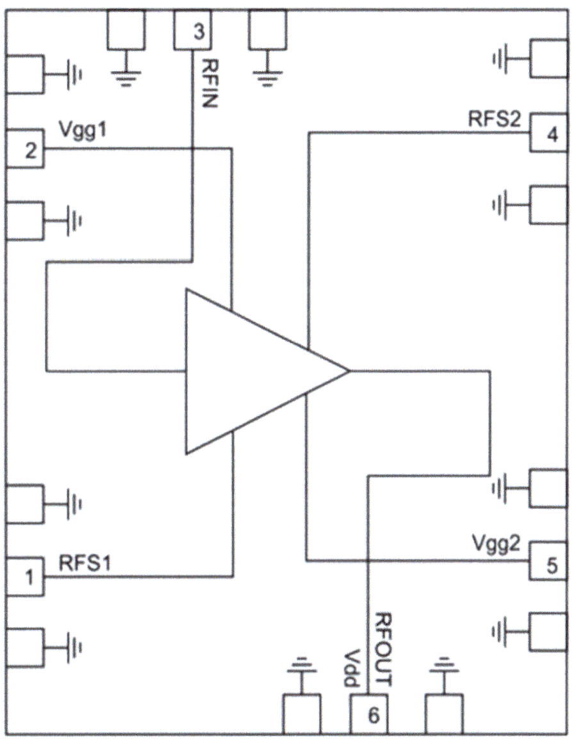

Fig. 5.77 Standard structure of ultra-wideband chip HMC-AUH23

Compared to standard chips of operating amplifiers, the wideband and ultra-wideband chips rarely contain differential stages since they have more complex circuit and narrower width of amplified frequencies. Figure 5.80 shows a typical example of the most widely used circuit of ultra-wideband amplifier based on chip with differential input. Frequency band of this circuit is from 0.04 to 1 GHz; small-

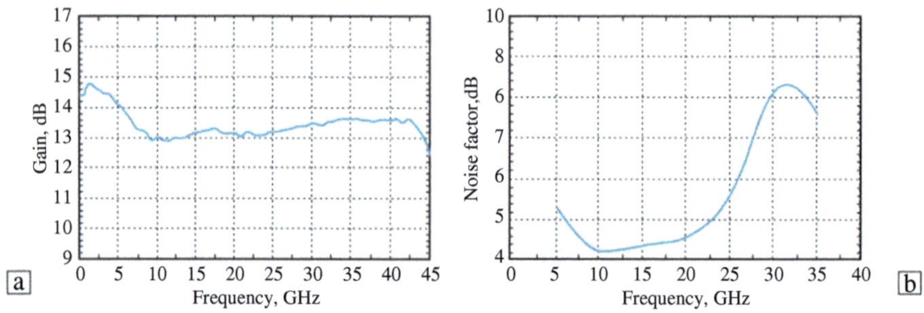

Fig. 5.78 Typical dependences for chip HMC-AUH232: (**a**) gain factor-frequency dependence; (**b**) noise factor-frequency dependence

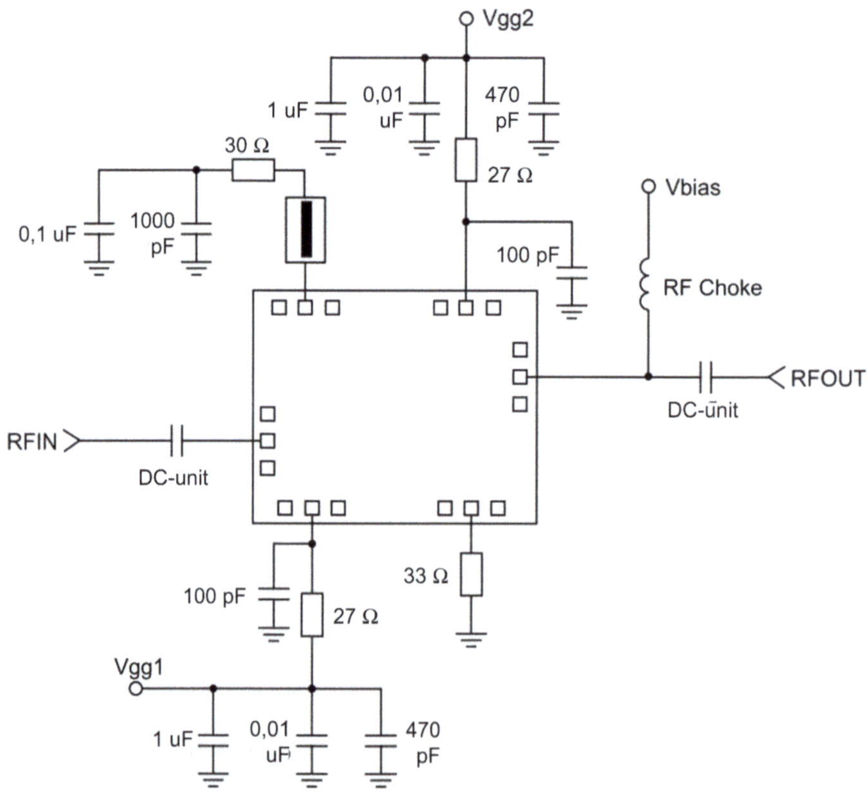

Fig. 5.79 Typical example of electric circuit of ultra-wideband amplifier based on chip HMC-AUH232

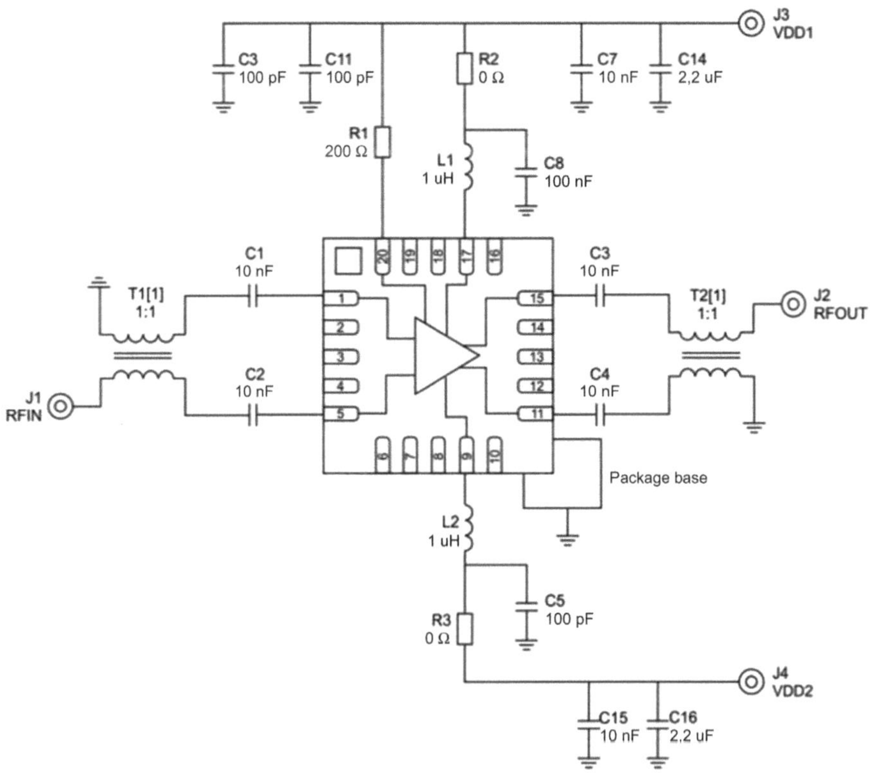

Fig. 5.80 Typical circuit of building of wideband amplifier based on chip HMC770LP4BE with differential input

signal gain factor is 15 dB. The chip has a low noise level: noise factor NF is only 2.75 dB.

Generally, the system of s-parameters is used for more complete assessment of behavioral features of such attenuator chips, as well as other SHF devices.

So, Fig. 5.81 shows the typical s-parameters-frequency dependence for differential amplifier chip HMC770LP4BE.

Developers of radar systems can find more details and s-parameters of chips made by Hittite Microwave corporation in Data Sheet documentation available on website of this corporation (usually in PDF-format).

It should be noted that Hittite Microwave corporation is one of global companies producing a wide range of SHF-amplifiers with a gain (transmission) factor controlled by electronic means. Such amplifiers are widely used in wireless systems (WiBro, WIMAX, 4G), mobile and cellular telephony, and testing and measuring instruments.

Amplifiers with adjusted gain belong to class of ICs Automatic Gain Control (for system with automatic gain control) and Variable Gain Amplifiers. The first class

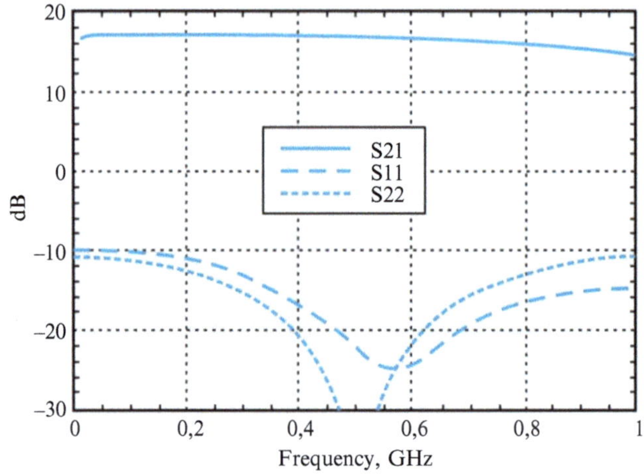

Fig. 5.81 Typical dependence of transmission factor on return losses (s-parameters) for chips of differential amplifier HMC770LP4BE

Table 5.14 The basic specifications of analogue control amplifiers

Chip	Bandwidth, GHz	Gain factor, dB	NF, dB	OIP3, dBm	P1, dBm	Power, V/mA
HMC287MS8	2.3–2.5	−9–21	2.5	7	3	3/9
HMC694	6–17	1–23	5	30	22	5/150
HMC694LP4	6–17	0–23	6	30	22	5/175
HMC972LP5E	0.5–6	−35–15	7.5	28	21	3/90

includes the popular model of amplifiers HMC992LP5E with relatively low frequency band: from 0.05 to 0.8 GHz. In its turn, amplifiers belonging to class of Variable Gain Amplifiers are divided into amplifiers controlled by analog signal and controlled by digital signal. The basic specifications of analogue control amplifiers are listed in Table 5.14.

Chip HMC287MS8 is of particular note: it has a frequency band of 0.2 GHz and a dynamic range of gain drift of 30 dB at amplified signal medium frequency of 2.4 GHz. This chip is designed in package MS8; and it has low power consumption: consumption current is only 9 mA at power supply voltage of 3 V (see Table 5.14).

Functional circuit HMC287MS8 (Fig. 5.82) is very simple and contains only one SHF-amplifier with an adjusted gain factor and built-in coupling capacitors.

Typical gain factor-operating frequency dependence at deferent control voltage for this chip is illustrated in Fig. 5.83a. It is qualitatively different for various control voltages: when VCTRL is close to 0, the gain factor slowly drops while the frequency is increasing; and when VCTRL is more than 1 V, the gain factor becomes greater while the frequency is increasing.

Typical gain factor-operating frequency dependence at deferent control voltage for chip HMC287MS8 is illustrated in Fig. 5.83b. When VCTRL is changing from

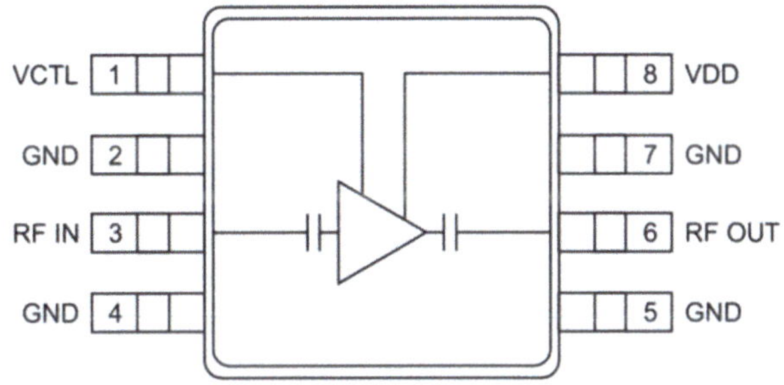

Fig. 5.82 Functional structure of chip HMC287MS8

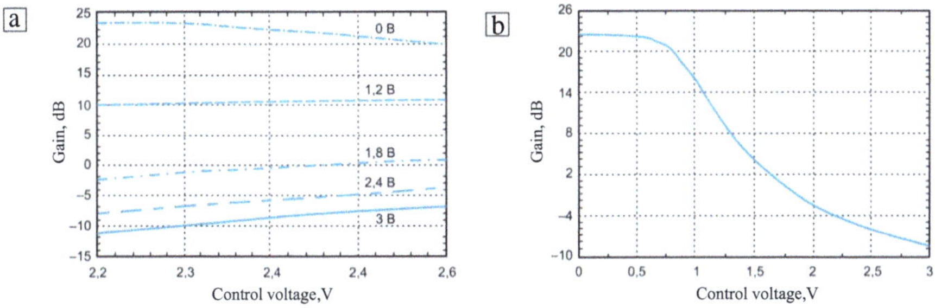

Fig. 5.83 The gain factor dependence for chip HMC287MS8: (**a**) dependence on frequency at deferent control voltage; (**b**) dependence on control voltage

0 to 1.2 V, the gain factor remains almost unchanged, and then it decreases according to nonlinear law.

Functional circuit of more wideband chip (bandwidth from 4 to 17 GHz) is illustrated in Fig. 5.84. According to specification, it is similar to the circuit as previously described for chip HMC287MS8. Bandwidth expansion up to 6–17 GHz results in significant increase of consumed power: consumption current is 175 mA at power supply voltage of 5 V. Parameters of chip HMC287MS8 are very similar to parameters of other chip HMC694.

Typical gain factor-frequency dependence at deferent control voltage for chip HMC694LP4 is shown in Fig. 5.85. The graphs have quite noticeable irregularity (especially at frequency lower than 7 GHz) resulted from weak resonance.

It may be noted that chip HMC972LP5E positions itself at the intersection between the chips described above in regards to bandwidth (from 0.5 to 6 GHz). Consequently, it is applicable to consumed power: power supply voltage is 5 V at consumption current of 90 mA.

Finally, it is worth noting one more type of SHF devices such as amplifiers with digital control of gain factor. This type of wideband SHF-amplifiers (chips with digital control of gain parameters) differs with circuit-based features, notably the control circuit is characterized by number of bits and control type (with parallel interface P and serial interface S). The rest of parameters of digitally controlled amplifiers are the same as parameters of analog controlled amplifiers. Table 5.15 specifies the parameters of four digitally controlled amplifiers that are most widely used:

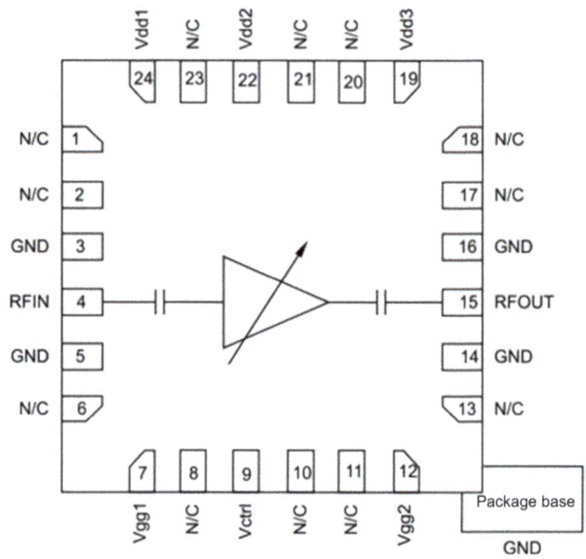

Fig. 5.84 Simplified functional diagram of SHF-chip HMC694LP4

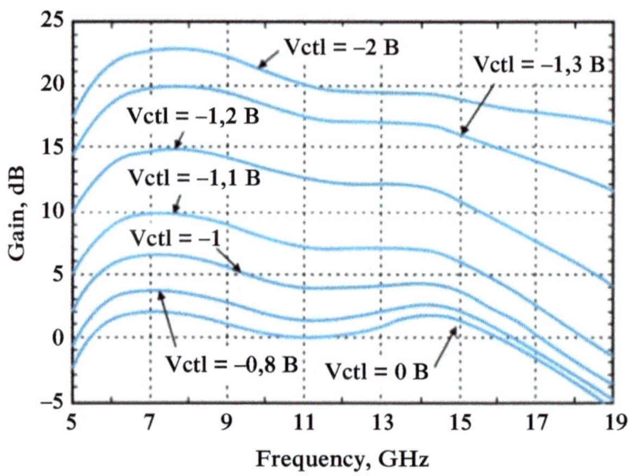

Fig. 5.85 Typical gain factor-operating frequency dependence at deferent control voltage for chip HMC694LP4

Table 5.15 The basic parameters of digitally controlled amplifiers

Chip	Bandwidth, GHz	Gain factor, dB	NF, dB	Interface, bit/type	OIP3, dBm	P1, dBm	Power supply, V/mA
HMC625LP5	DC-6	−13.5–18		6/S, P	33	19	5/88
HMC626LP	DC-1	8.5–40		6/P	36	20	5/176
HMC694LP4	0.07–4	−19.5–12		7/S, P	33	152	5/150
HMC960LP4E	DC-1	0–40		6.3/S, P	30	13	5/70

Figure 5.86 shows the functional diagram of chip that is a digitally controlled amplifier HMC625LP5. This ultra-wideband chip has a bandwidth of amplified DC frequencies up to 6 GHz. The gain factor-frequency dependence at deferent control voltage for chip HMC625LP5 is shown in Fig. 5.87.

5.7.3.5 Monolithic SHF Chips of Full Frequency Synthesizers

Monolithic Chips PLL of SHF Synthesizers For production of monolithic integrated chips of SHF synthesizers, the company uses the modern element base of heterotransition superhigh-speed transistors based on the wide-band gap semiconductive materials (www.hittite.com).

Frequency synthesizers of these devices are implemented on one-two monolithic chips of very small size with the minimum quantity of external packageless components that enables their usage not only in radar systems but also in modern devices of mobile communication.

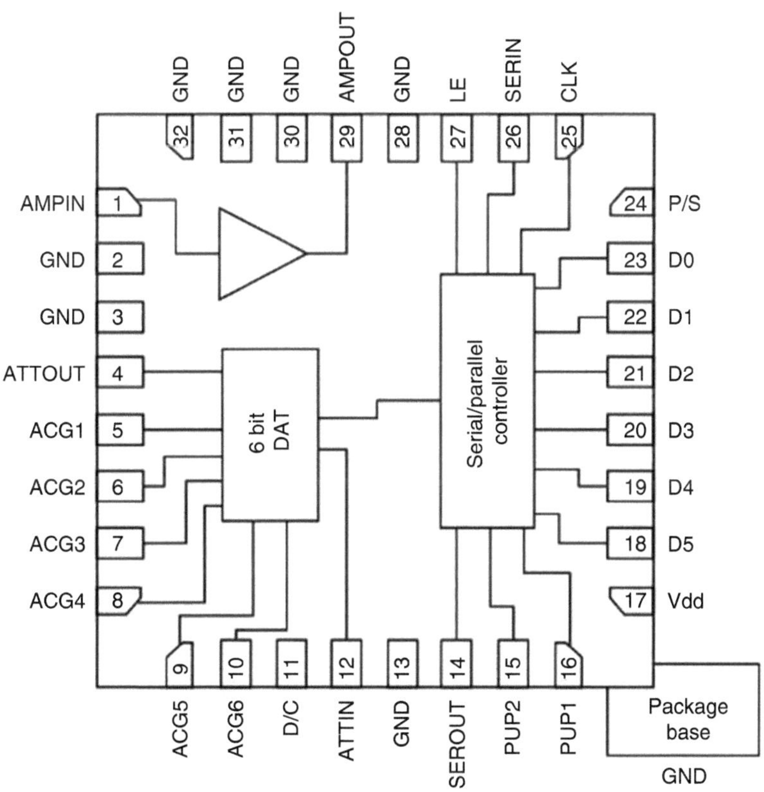

Fig. 5.86 The functional diagram of chip HMC625LP5

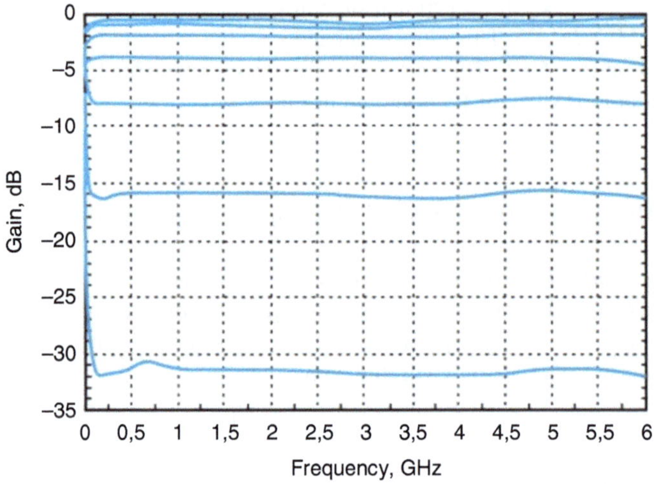

Fig. 5.87 Typical gain factor-operating frequency dependence at deferent control voltage for chip HMC625LP5

Table 5.16 The chips of microwave band synthesizers made by Hittite Microwave company

Type of chip	Frequency, GHz	Phase noise, dBc/Hz	Pout, dBm	Jitter, *f*sec/•
HMC764LP6CE	7.3–8.2	−101/−140	15	196/0.58
HMC765LP6CE	7.8–8.8	−101/−140	13	193/0.61
HMC783LP6CE	11.5–12.5	−99/−134	10	181/0.81
HMC807LP6CE	12.4–13.4	−98/−132	8	175/0.84

The company produces a series of chips of quite completed synthesizers in a very high-frequency range of frequencies which are one or another implementation of the main classical circuit of a synthesizer [45]. The ratio of frequencies (of output and frequency reference signals) can be either integer or fractionally rational values. Let's consider in more detail the monolithic chips that allow to design the quite completed frequency synthesizers in the widest frequency range using only 2–3 chips.

The basic characteristics of chips of microwave band synthesizers (Microwave) are listed in Table 5.16.

In addition to usual parameters, they are characterized by jitter (timing instability). Jitter is measured in femtoseconds and degree portions. Generally, its root-mean-square value (RMS) is given in fractionally rational mode of chip operation.

The typical example of this chip class is chip HMC807LP6CE (Fig. 5.88) with a frequency range 12.4–13.4 GHz. Compared to the other chips, it has the highest frequency, slightly worse phase noise, and the smallest power of output signal. Nevertheless, it is quite sufficient for most part of applications. This chip has the minimum jitter.

The phase noise depends on the synthesizer operation mode. It is lower when one frequency divider is used (the other frequency divider is not used and that's why it doesn't make noise, its dividing ratio is taken to be equal to 1), as well as the integer mode is implemented. Figure 5.89 shows the dependence of SSB phase noise on frequency offset in relation to the central frequency in single-side band measurement mode during the integer operation of the synthesizer and at different signal frequencies. Figure 5.90 shows the dependency at different ambient temperatures (and at a frequency of 12.9 GHz).

In case of operation with two frequency dividers in fractionally rational mode, the SSB phase noise increases since two frequency dividers make noise. Figure 5.91

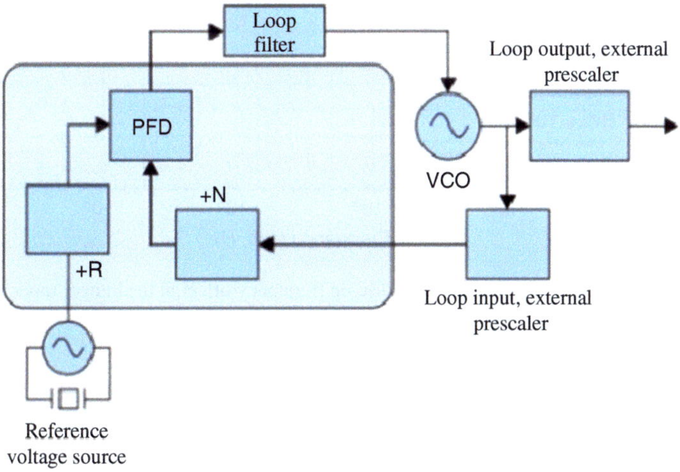

Fig. 5.88 Simplified functional diagram of SHF-chip HMC807LP6CE

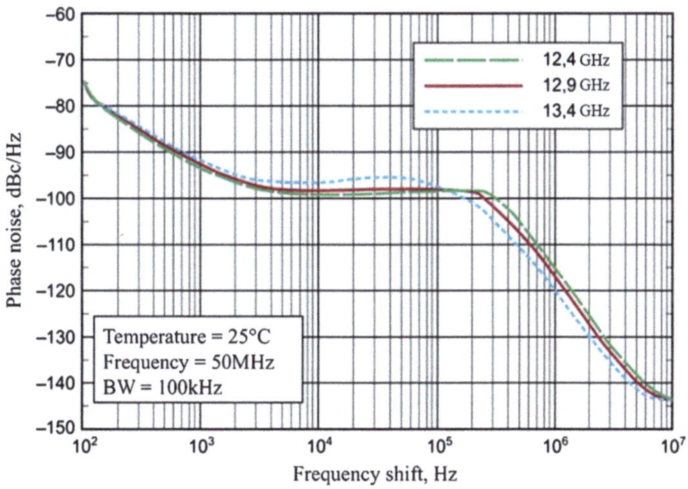

Fig. 5.89 The dependence of SSB phase noise on frequency offset in the integer mode at different signal frequencies

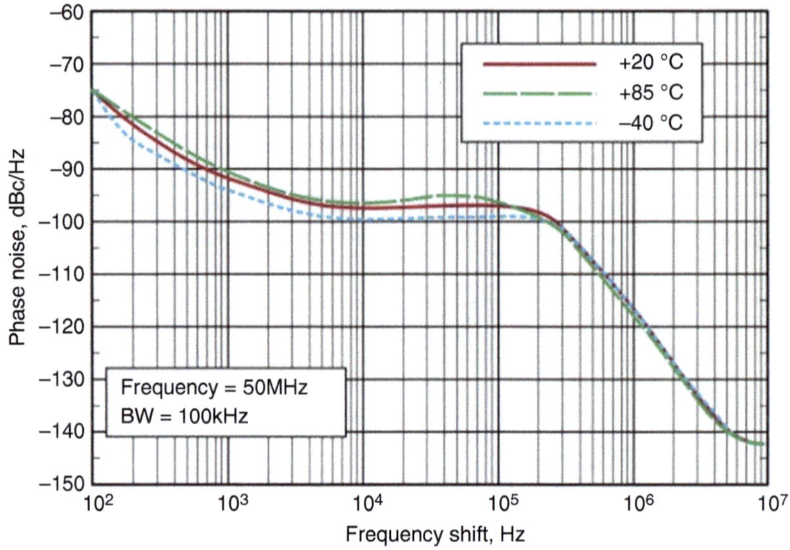

Fig. 5.90 The dependence of SSB phase noise on frequency offset in the integer mode at different ambient temperature

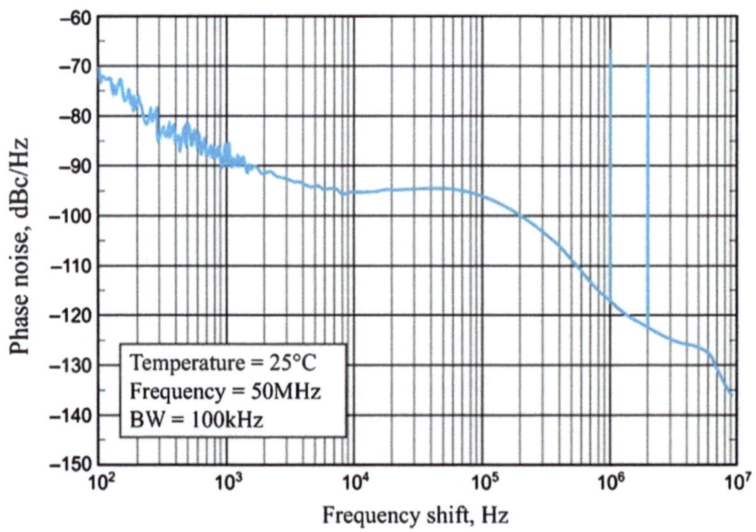

Fig. 5.91 Dependence of SSB phase noise on frequency offset when signal frequency is equal to 12.902 GHz in fractionally rational mode

shows the dependence of SSB phase noise on frequency offset when signal frequency is equal to 12.902 GHz in fractionally rational mode.

Of course, such chips are characterized by many other parameters relating to the built-in VCO. Their detailed description is given in documentation that is available on the website of Hittite Microwave company (www.hittite.com).

Three-Band Monolithic Chips of Synthesizers Hittite Microwave company produces a series of PLL synthesizers having a built-in VCO with three frequency bands (Tri-Band RF VCO). They are characterized by the fact that the first frequency divider is replaced by the circuit generating three frequencies: *f*0/2, *f*0, and 2*f*0. Thus, three variable-frequency ranges can be reached by changing the dividing ratio of the second frequency divider (and even by making it to be a frequency multiplier). Table 5.17 specifies the data of chips when using the medium frequency *f*0.

The complete data of these chips are available on the company's website.

Let's consider a chip HMC840LP6CE as an example of the most high-frequency three-band synthesizer. It has a built-in VCO and three frequency bands: 1310–1415, 2620–2830, and 5240–5660 MHz, as well as a built-in testing program that carries out tests and calibrations. The chip is designed in 20-pin package LP6CE, 6 × 6 mm in size, and 40 mm^2 in area.

Functional diagram of chip of three-band synthesizer HMC840LP6CE is shown in Fig. 5.92. Its composition and operation are described above. Figure 5.93 illustrates the level of phase noise of the chip. The level of phase noise in fractionally rational mode is a bit higher than in the integer mode. The reasons of this have already been explained.

Figure 5.94 shows the typical adjustment curves of synthesizer HMC840LP6CE. They cover all the frequency range of this synthesizer.

Wide-Band Monolithic Chips of Synthesizers As well, Hittite Microwave produces wide-band chips of Wideband RF VCO class having wider frequency range than that of chips previously described. One of such chips is HMC829LP6GE (Fig. 5.95) has three extended frequency ranges: 45–1050, 1400–2100, and 2800–4200 MHz. The low frequency limit of the first range is up to 45 MHz, the frequency ratio is equal to 2 at other ranges. Package of SMT type with 40 pins has

Table 5.17 Chips of wideband synthesizers Hittite Microwave (the data are specified for the medium frequency range)

Type of chip	Frequency, GHz	Phase noise, dBc/Hz	Pout, dBm	Jitter, *f*sec/•
HMC828LP6CE	1285–1415	−112/−143	10	180/0.09
HMC822LP6CE	1330–1650	−112/−142	6.5	180/0.11
HMC838LP6CE	1590–1890	−118/−143	7.5	180/0.12
HMC821LP6CE	1720–2080	−112/−141	6.5	180/0.13
HMC837LP6CE	2050–2300	−117/−141	10.5	180/0.15
HMC839LP6CE	2100–2410	−115/−140	7.5	180/0.16
HMC820LP6CE	2190–2550	−112/−141	6.5	180/0.17
HMC840LP6CE	2620–2830	−118/−139	9	180/0.1

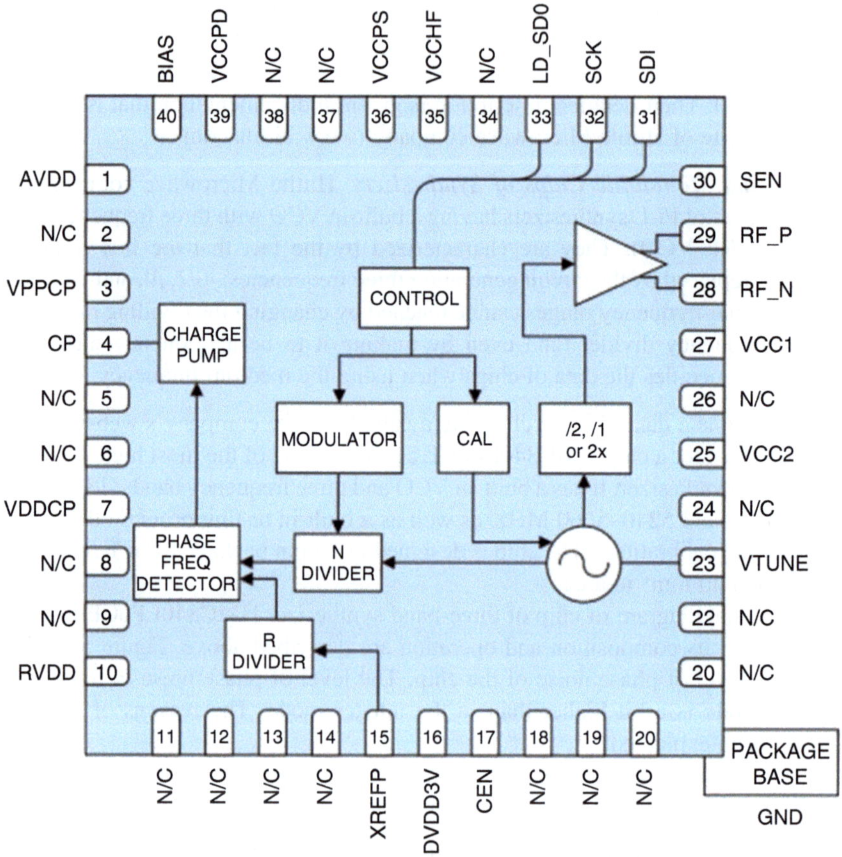

Fig. 5.92 Functional diagram of chip of three-band synthesizer HMC840LP6CE

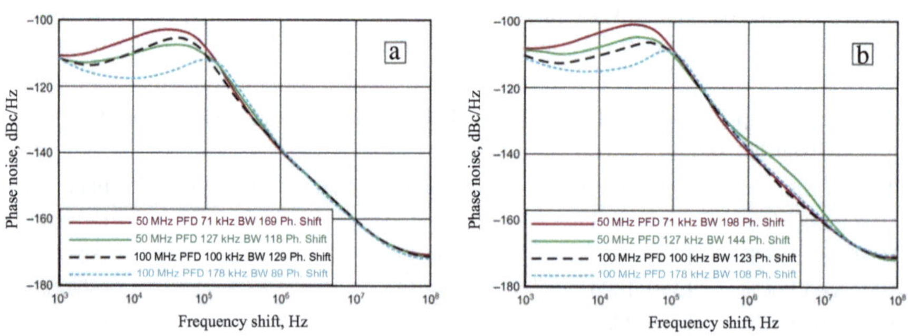

Fig. 5.93 The phase noise in the integer mode (**a**) and fractionally rational mode (**b**) at a frequency of 2.6 GHz for chip HMC840LP6CE

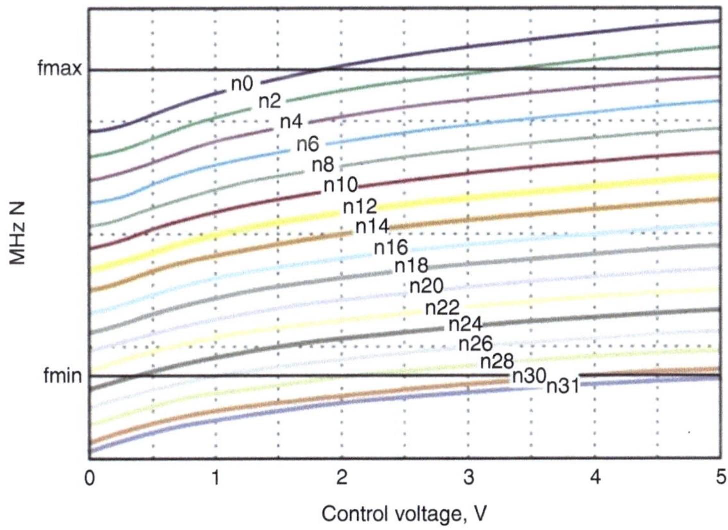

Fig. 5.94 The typical adjustment curves of chip HMC840LP6CE

its sizes are 6 × 6 mm. The chip is powered from the constant-voltage source of 3.3 and 5 V.

The chip phase-lock detector provides the capture area of 100 MHz. The chip provides the phase noise of −110 dBc/Hz when frequency offset is 100 kHz.

24-bit capacity provides frequency resolution up to 24 Hz. There is possibility to carry out digital test of chip operation and its calibration. Figures 5.96 and 5.97 show the phase noise of chip HMC829LP6GE in the integer mode and fractionally rational mode. In general, it's very low and ensures high-frequency stability of synthesizers based on this chip.

Figure 5.98 shows the dependency of output signal power on frequency. Power change within each frequency sub-band is small.

Figure 5.99 shows the assembly of chip HMC829LP6GE on printed circuit board. Packageless resistors and capacitors are used. Data Sheet for this chip provides a detailed specification of the circuit board elements.

As well, chip HMC830LP6GE is produced. It provides a frequency range 25 MHz to 3 GHz. The lower limit of frequency range is up to 25 MHz and it covers the short-wave region of output signal. It is provided the frequency resolution of 3 Hz. The functional diagram of this chip is similar to the diagram shown in Fig. 5.95. The specified frequency range is provided by the choice of the range of fundamental generator frequencies from 1500 to 3000 MHz.

Figure 5.100 shows the dependency of output power HMC830LP6GE on output frequency at three values of temperature. The feature of this chip is a jump-over of power value at a frequency of 1500 MHz.

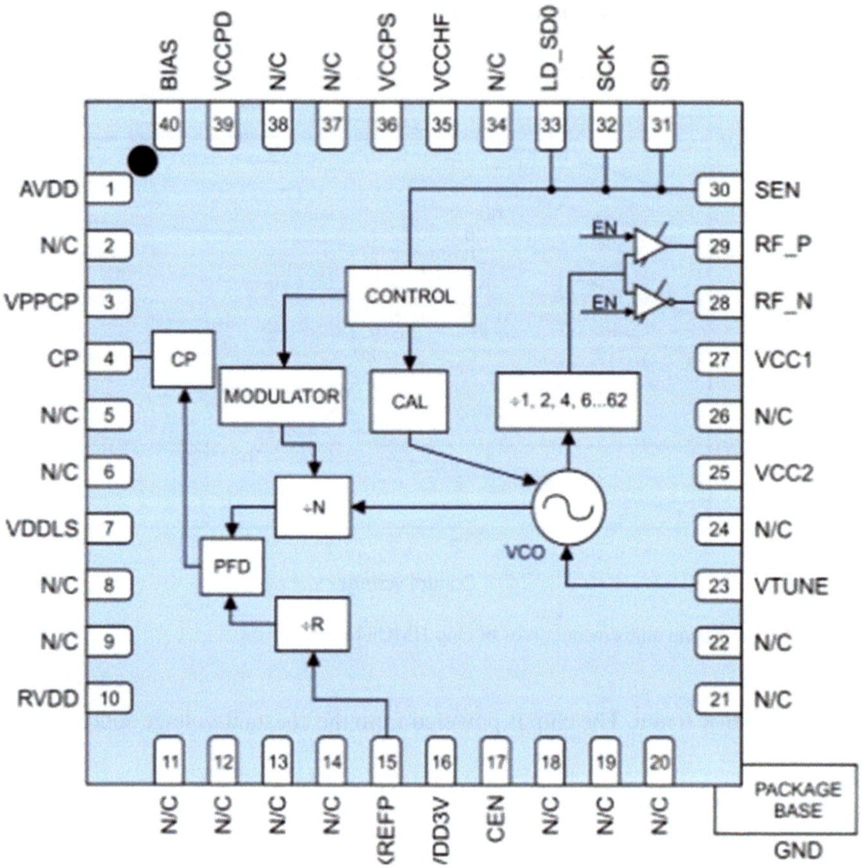

Fig. 5.95 The functional diagram of chip HMC829LP6GE

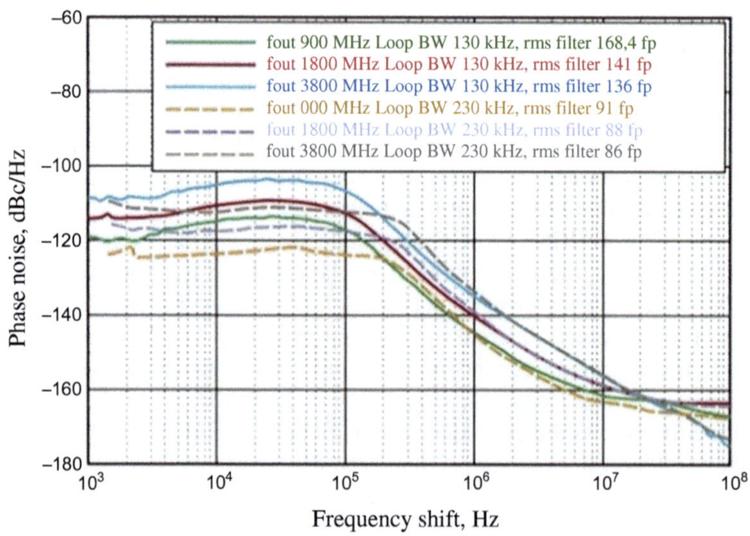

Fig. 5.96 The phase noise of chip HMC829LP6GE in the integer mode

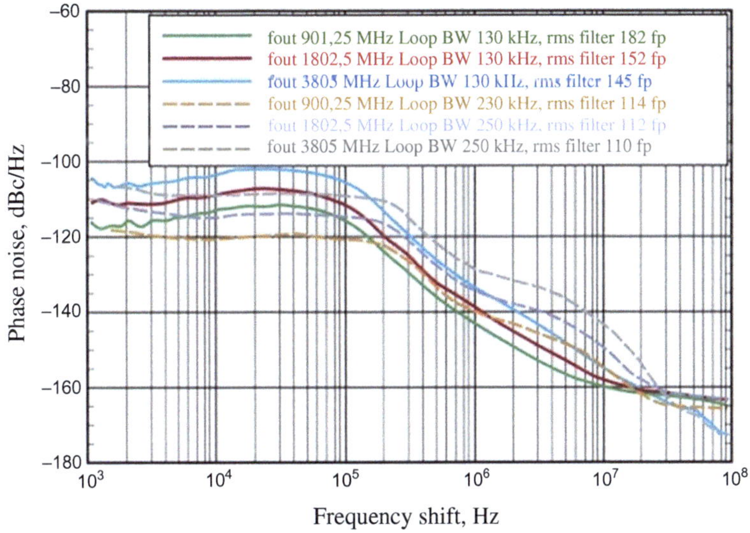

Fig. 5.97 The phase noise of chip HMC829LP6GE in the fractionally rational mode

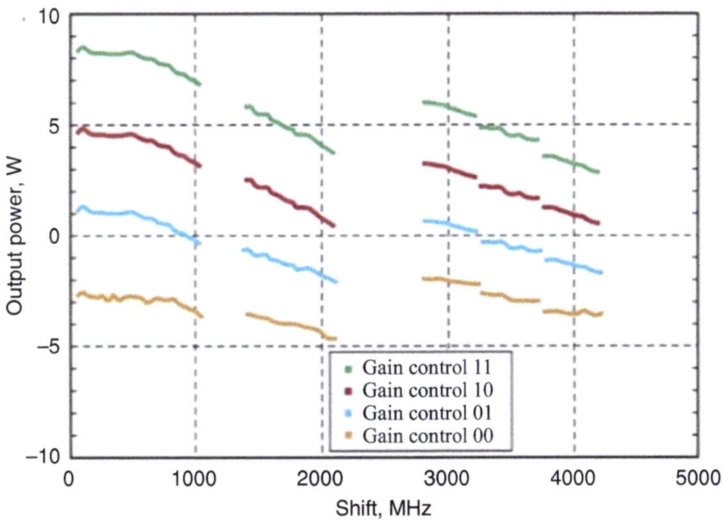

Fig. 5.98 Typical dependence of output signal power on frequency

The detail characteristics of this chip and the user guide can be found in www.hittite.com.

The cascade circuit can be used to ensure the quartz crystal control and to increase the reference frequency of synthesizers. Its example is shown in Fig. 5.101. The synthesizer in such circuit acts as a frequency multiplier.

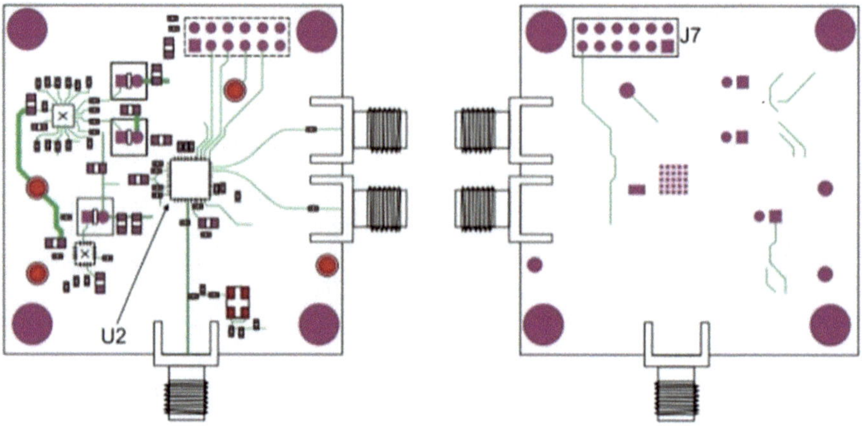

Fig. 5.99 Assembly of chip HMC829LP6GE on printed circuit board

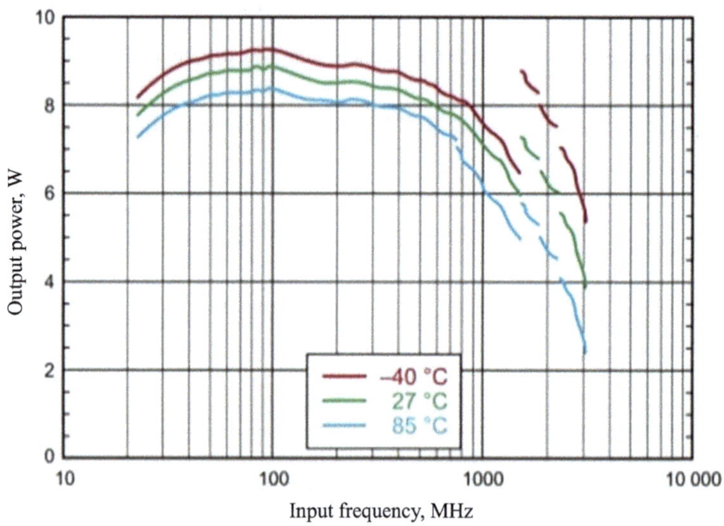

Fig. 5.100 Dependency of output power on output frequency for chip HMC830LP6GE at different values of temperature

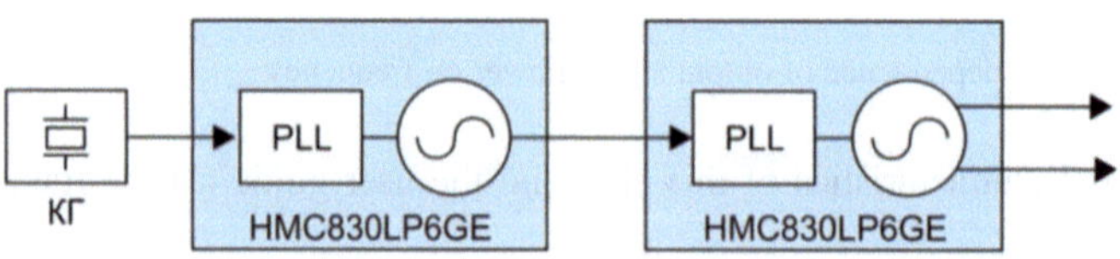

Fig. 5.101 Connection of quartz crystal-controlled generator to synthesizers

Chips HMC983 and HMC984 Are for Building of Ultra-Wideband Synthesizers Hittite Microwave produces a special-purpose pair of chips for phase-locked loop. These are chips HMC983 and HMC984 with a bandwidth from 0 (DC) to 7 GHz. These chips are characterized by their zero lower cutoff frequency, which allows them to create synthesizers with the lower cutoff frequency capturing the ranges of sonic and ultrasonic frequencies. Perhaps, the principal scope of application of such synthesizers is measuring instruments: Currently, many series sinusoidal wave generators and spectrum analyzers produced by global leading companies have the frequency range that is typical for these chips.

Chip HMC983LP5E is a wideband synthesizer with improved parameters. The chip has 32-pin package LP5, 5 × 5 mm in size.

Supply voltage -is 3.3 and 5 V; the values of the consumed current depend on a chip condition; they can be found in Data Sheet.

The chip contains the 20-bit frequency divider. The maximum reference frequency of the chip is 350 MHz, the maximum frequency of the phase detector PFD is150 MHz. The frequency resolution is 0.18 μHz when the reference signal frequency is 50 MHz. The chip contains 38-bit 100-Hz ΔΣ-modulator (DSM). The functional diagram of chip HMC983LP5E is shown in Fig. 5.102.

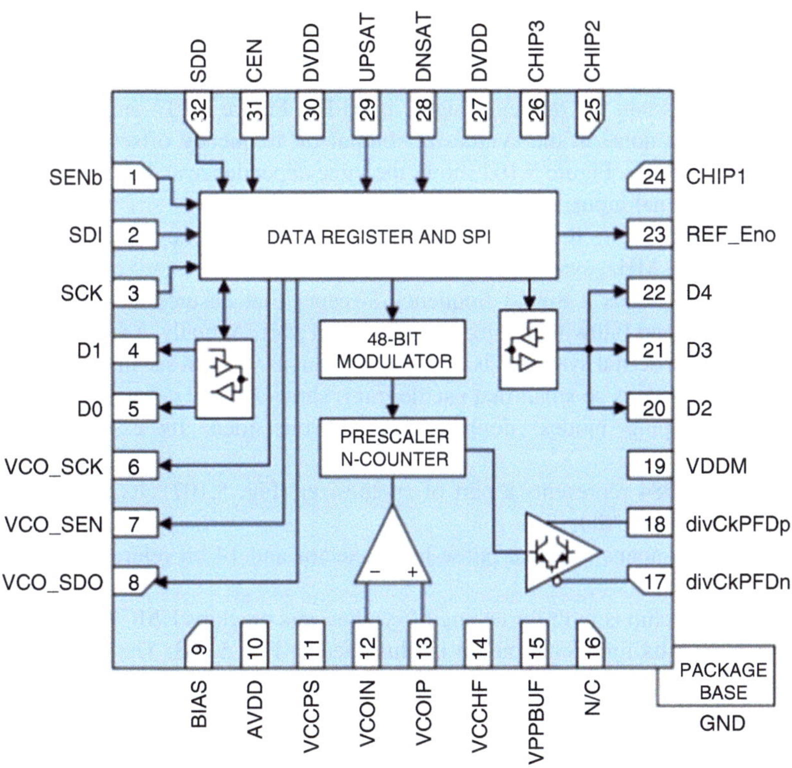

Fig. 5.102 The functional diagram of chip HMC983LP5E

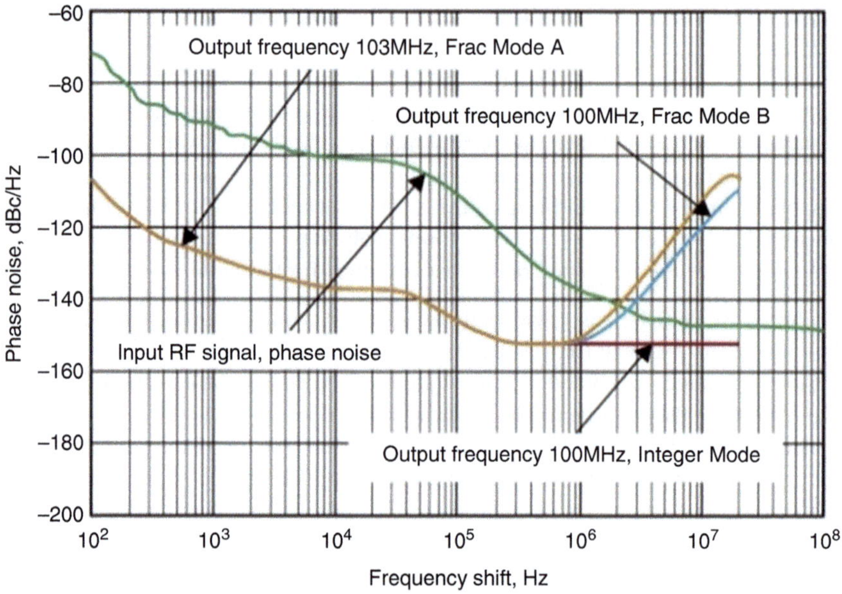

Fig. 5.103 The phase noise of chip HMC983LP5E when the reference signal is 100 MHz

This chip is characterized by low level of phase noise reaching the record value: −160 dBc/Hz when the reference signal is 00 Hz. Figure 5.103 shows the dependence of phase noise in the synthesizer output on frequency offset at the central frequency of 6 GHz. Figure 5.104 shows the same dependence at different frequency of reference signal input.

Figure 5.105 shows the oscillograph records of the frequency divider (reference frequency is 66 MHz, synthesis frequency is 6.5 GHz).

The chip contains a digital frequency sweeper that ensures the linear law of frequency rise and fall with automatic and manual start. Actually, a sweeping signal received during digital synthesis is multiple-step but as you can see in the Fig. 5.106, the step amplitude is so small that oscillograms show a signal as analog one. There are two sweeping modes: double-sided and one-sided. Its essence is clear (Fig. 5.106).

Chip HMC984 represents a part of synthesizer (Fig. 5.107). Its second part is based on chip HMC984.

This chip contains a digital phase-lock detector and 14-bit reference frequency divider.

The special chip is built for testing of synthesizers on chips HMC983, HMC984, and HMC507. Its functional circuit is illustrated in Fig. 5.108. The appearance of printed circuit board is shown in Fig. 5.109.

Data Sheet for this chip provides a detailed specification of elements of the printed circuit board.

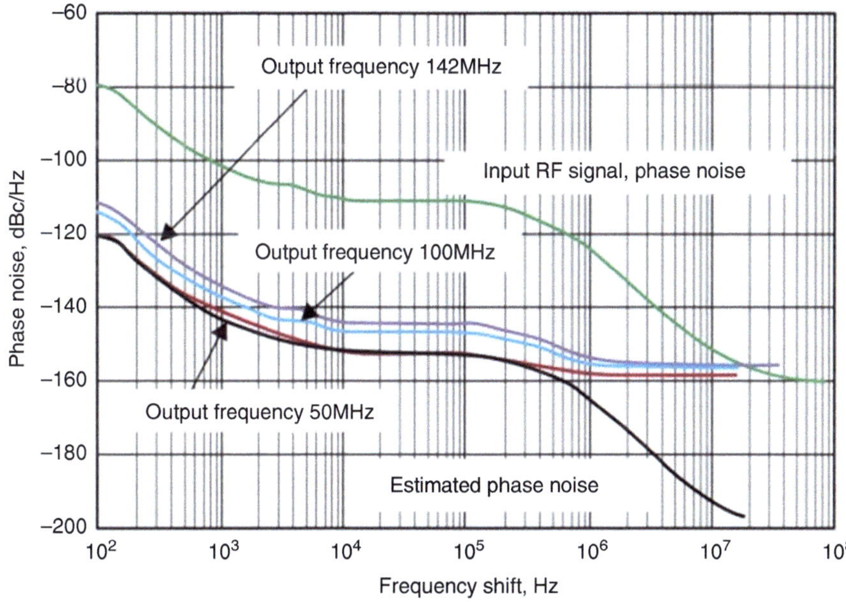

Fig. 5.104 The phase noise of chip HMC983LP5E when the reference signal is with a frequency of 50, 100, and 100 MHz

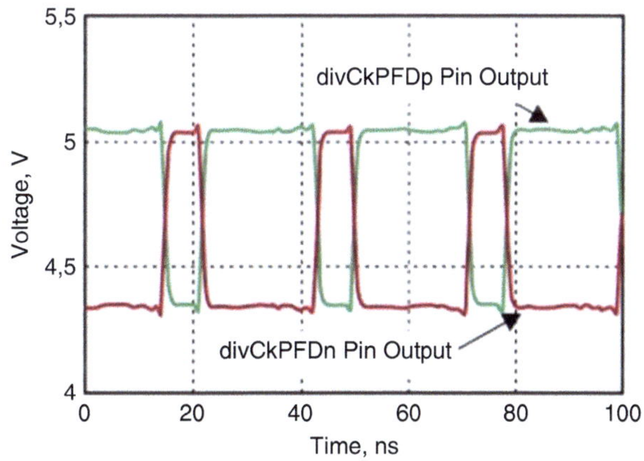

Fig. 5.105 Oscillograph records of the frequency divider of chip frequency divider

Figure 5.110 shows the functional circuit of wideband synthesizer based on chips HMC983 and HMC984.

As mentioned above, chip HMC507 can be used as a voltage controlled oscillator. Thus, the full synthesizer is implemented on three monolithic chips.

Board (Figs. 5.109 and 5.110) is used as a part of the computerized testing system shown in Fig. 5.111. The system allows to receive all characteristics of synthesizers that are described above, as well as many others. It is required for design of synthesizers of different types based on the described integrated chips.

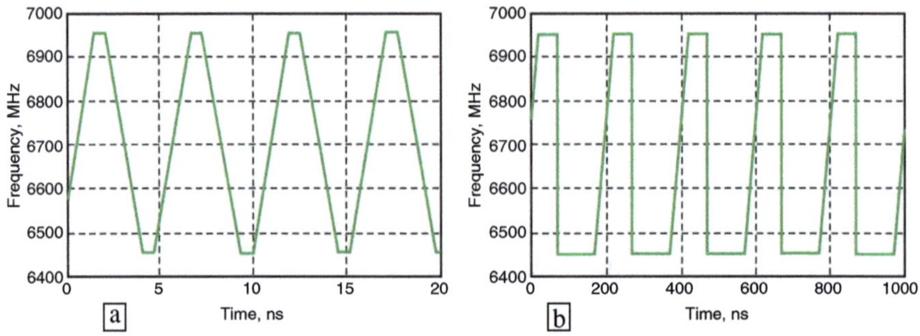

Fig. 5.106 Frequency-time dependency in two sweeping modes: (**a**) double-sided; (**b**) one-sided

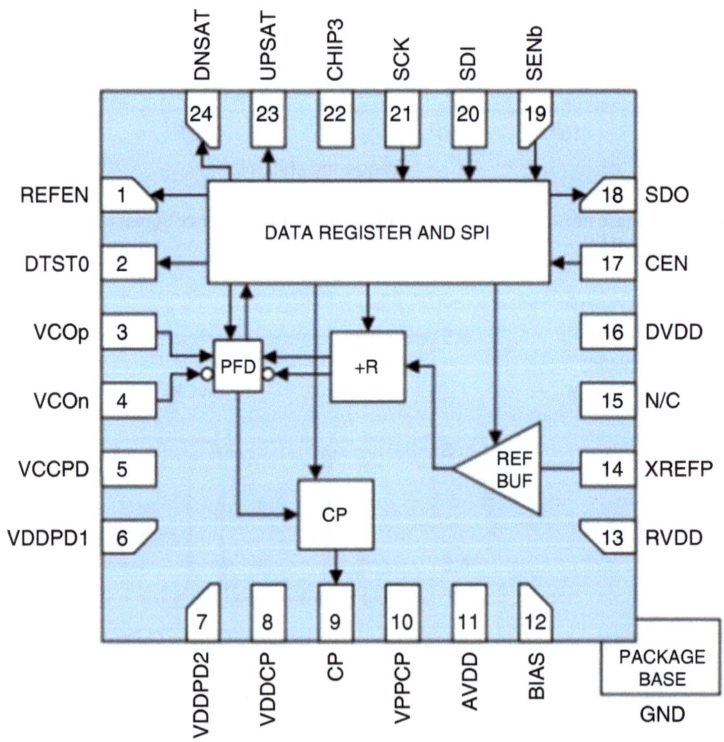

Fig. 5.107 The functional diagram of chip HMC984

Description of the system interface and user's manual are available on the company's website (www.hittite.com).

Monolithic Chips of Wide-Band Frequency Synthesizers HMC701/702/703 Hittite Microwave company produces a series of monolithic chips for wide-band frequency synthesizers. One chip of such synthesizer includes all nodes for external

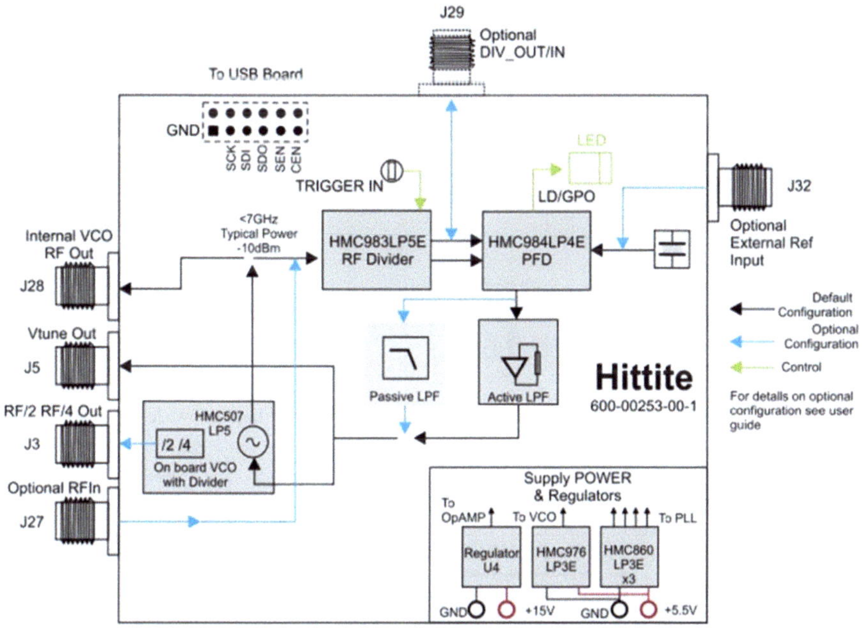

Fig. 5.108 Functional diagram of board circuit for testing of chips HMC983 and HMC984

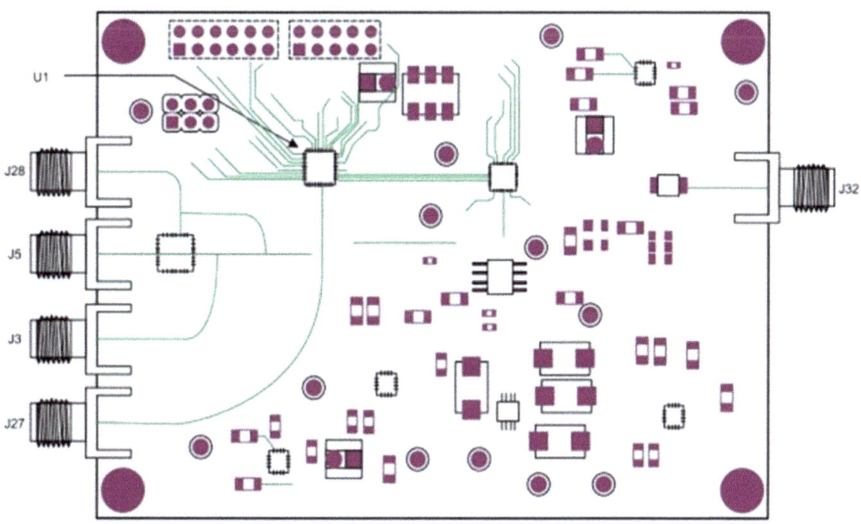

Fig. 5.109 Printed circuit board for testing of chips HMC983 and HMC984

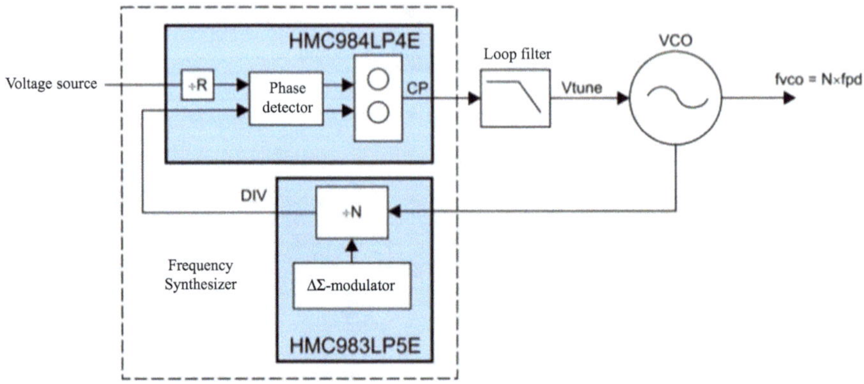

Fig. 5.110 The use of chips HMC983 and HMC984 for building of wide-band frequency synthesizers

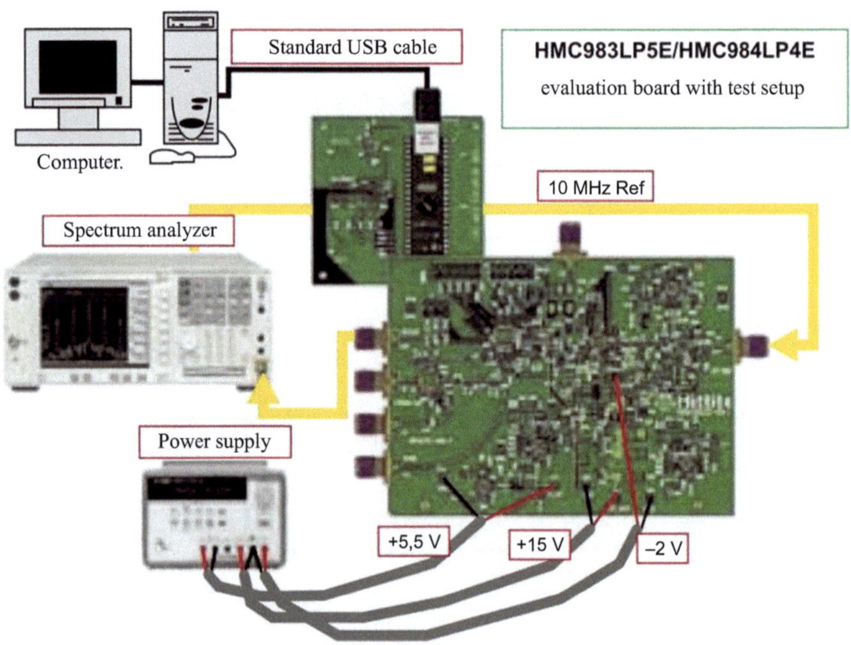

Fig. 5.111 Computerized testing system of HMC983 and HMC984

control VCO (two frequency dividers, phase-lock detector, frequency sweeper of program control).

The basic parameters of such chips are listed in Table 5.18.

This table specifies the frequency range, maximum frequency of the phase detector Fpfd max, phase noise, frequency resolution, supply voltage, and current.

Table 5.18 The basic parameters of monolithic chips for terminated wide-band frequency synthesizers

Type of chip	Frequency range	Fpfd max, MHz	Phase noise, dBc/Hz	Resolution, Hz	Power supply
HMC701LP6CE	10 kHz–8 GHz	75	−221/−227	3	+5 V when 37 mA, +3.3 V when 90 mA
HMC702LP6CE	10 kHz–4 GHz	100	−221/−227	6	+5 V when 37 mA, +3.3 V when 136 mA
HMC703LP4E	DC–8 GHz	100	−230/−233	3	+5 V when 6 mA, +3.3 V when 52 mA

Two of these chips have the lower limit of frequency range 10 kHz, so they cover even the part of audio frequency range. Such cutoff frequency relieves from the necessity to use coupling capacitors of high capacity in the synthesizer.

Functional diagram of chip HMC702LP6CE is illustrated in Fig. 5.112. The chip has the maximum frequency of the phase detector being equal to 75 MHz and frequency resolution of 3 Hz and 16-bit reference signal divider.

The phase noise is equal to −98/103 dBc/Hz (in the integer and fractionally rational modes) when frequency offset is 20 kHz. The mentioned phase noise (www.hittite.com) is specified in Table 5.18. Figure 5.113 shows the dependence of phase noise on frequency offset in relation to the central frequency for integer (at a frequency of 13 GHz) and fractionally rational (at a frequency of 12 GHz) mode of synthesizer operation.

Comparative analysis of phase noise for a few chips at different frequency of output signal is shown in Fig. 5.114.

Synthesizers with phase-locked loop are inertial devices. Figure 5.115 shows the cycle of time change of frequency for synthesizers with phase-locked loop of voltage-controlled generator. It's clear that transient period can reach tens of microseconds.

Chip HMC703LP4E is an ultra-wideband synthesizer with a frequency band from direct current up to 8 GHz. Its functional diagram is illustrated in Fig. 5.116. The chip is designed in mini package LP4E 2 × 2 mm in size with 24 pins. It includes frequency and phase modulators, control system with internal and external start-up, frequency sweeper, and phase-lock detector.

The level of chip phase noise is equal to −112 dBc/Hz in fractionally rational mode at a frequency of 8 GHz when frequency offset is 50 kHz. The specified phase noise is equal to −230 dBc/Hz in fractionally rational mode and to −233 dBc/Hz in integer mode.

The maximum frequency of the reference-frequency generator is 100 MHz; jitter is less than50 fc (RMS).

Typical functional circuit of wideband synthesizer based on chip HMC703LP4E with external VCO is illustrated in Fig. 5.117.

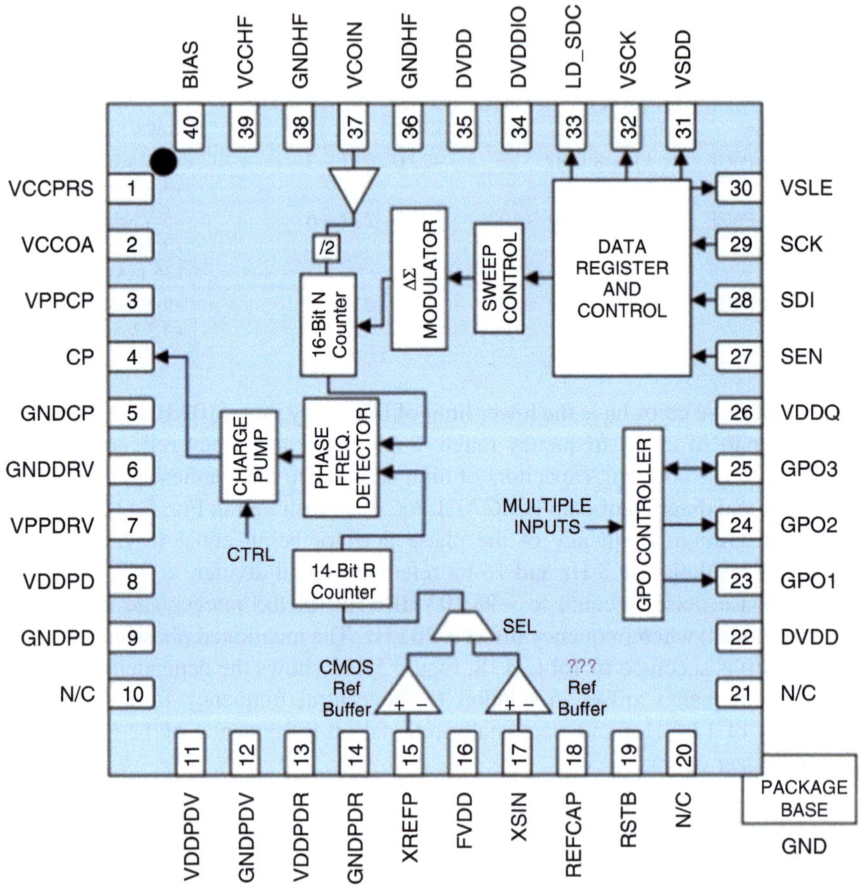

Fig. 5.112 The functional diagram of chip HMC702LP6CE

Figure 5.118 shows the functional diagram of board circuit for testing of synthesizer chips HMC703 with controlled generator circuit HMC508.

The board represents a completely built-up synthesizer on the specified chips.

Testing board topology is shown in Fig. 5.119. The complete specification of its components can be found in documentation available on the website (www.hittite.com).

The dynamic properties of the synthesizer can be optimized by improvement of the filter installed in output of phase detector. In particular, the active filter based on the integrated operational amplifier can be installed instead of usual filter. Functional circuit of such synthesizer is illustrated in Fig. 5.120.

Software for monolithic chips of synthesizers is for ensuring proper operation of frequency synthesizers based on monolithic chips, Hittite Microwave company produces software and firmware in addition to testing boards (for example, that are

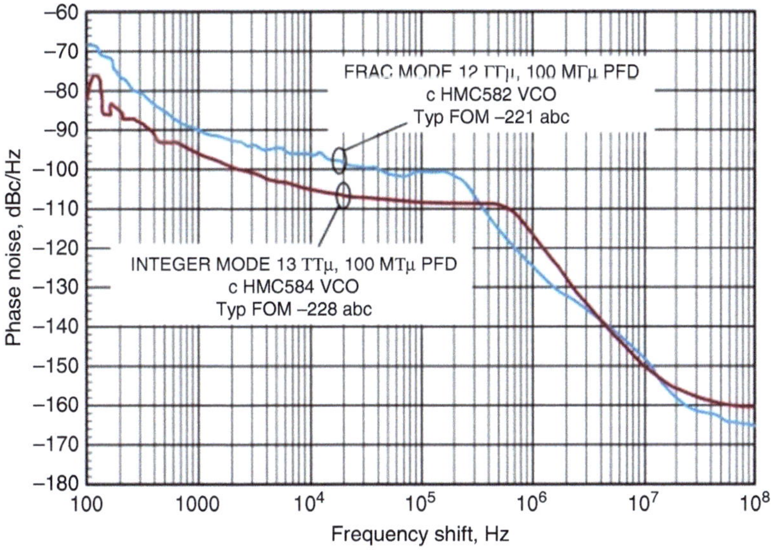

Fig. 5.113 The phase noise of chip HMC702

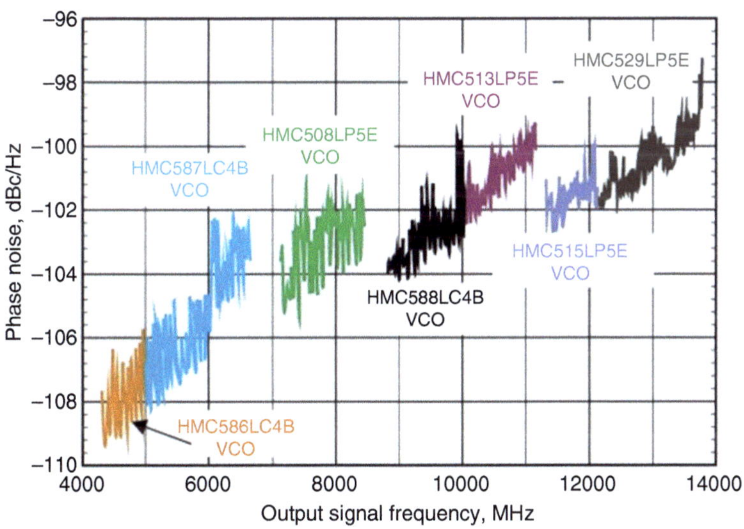

Fig. 5.114 Phase noise of different chips of synthesizers

shown in Figs. 5.118 and 5.119). Installation of type shown in Fig. 5.121 is used for firmware adding to storage of monolithic chips of frequency synthesizers.

Software for synthesizer control is implemented with graphic user interface (GUI). Its operation can be started with setting up the synthesizer type in window shown in Fig. 5.122.

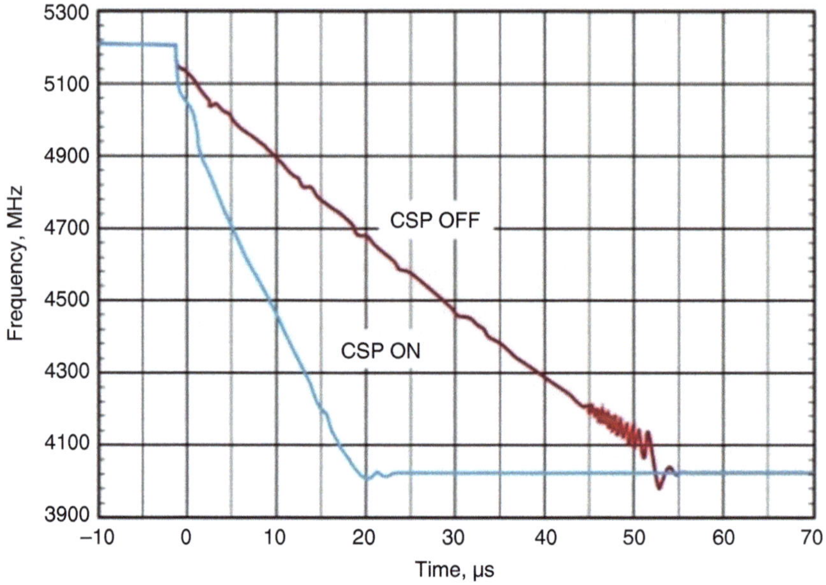

Fig. 5.115 Frequency change cycle of synthesizer based on chips HMC983 and HMC984 from 5200 to 3950 MHz

Window view for different synthesizers can have significant but not fundamental distinctions. Figure 5.123 shows a window of operation with synthesizers PLL.

GUI window of parameter setup and testing of various models of frequency synthesizers with enhanced capabilities is shown in Fig. 5.124. Detailed description of synthesizer programming and peculiarities of their testing are given in User Manual that is available on the website (www.hittite.com). It describes some features of one or another circuit of the synthesizer [46].

Thus, scope of application of the above-considered monolithic chips of frequency synthesizers and frequency-stable signal generators is very diverse. Scope of application includes microwave radio transmitters and radio receivers, base stations of mobile communication systems (GSM, PCS, DCS, CDMA, WCDMA), wireless communication lines (Wireless LANs, WIMAX), high-speed measuring and testing equipment (including in microelectronics and nanotechnologies), cable television, sensor systems, phase arrays AESA, SHF-jammers, FMCW radar systems, and etc. Small overall dimensions and weight of monolithic chips made by Hittite Microwave company, as well as their high (record) frequency rates allow to apply these chips in the modern communication systems, radio technical apparatus, and electronic and measuring equipment.

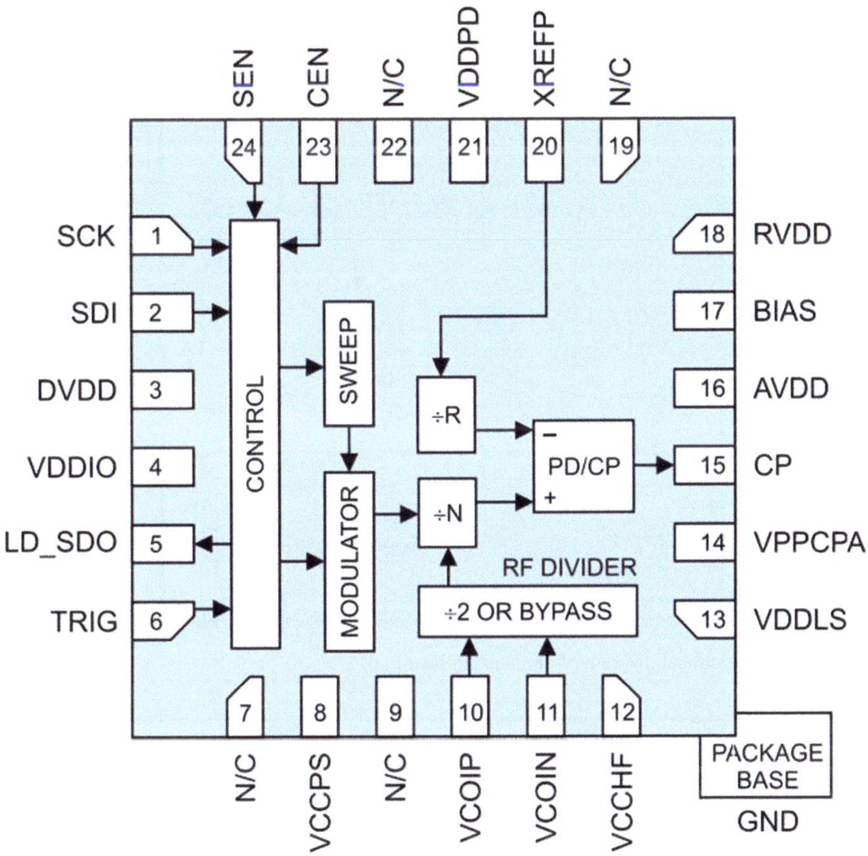

Fig. 5.116 The functional diagram of chip HMC703LP4E

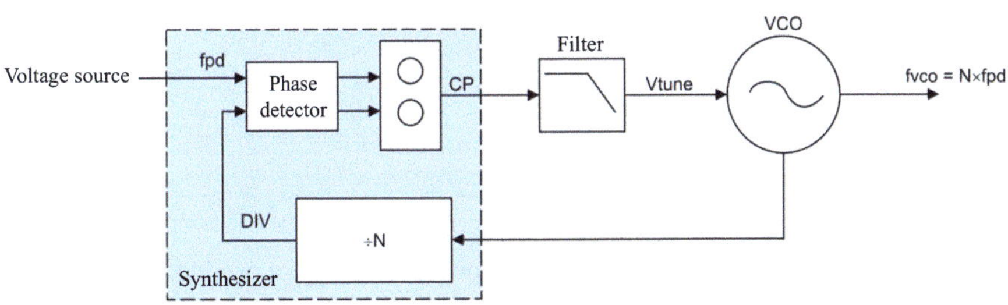

Fig. 5.117 Functional diagram of full synthesizer based on chip HMC703LP4E with external VCO

5.7.4 Powerful Gallium Nitride Transistors

Until quite recently, the theoretical advantages of powerful gallium nitride transistors over silicon transistors couldn't be implemented in practice since these transistors are normally open in routine operation. It became possible to produce transistors

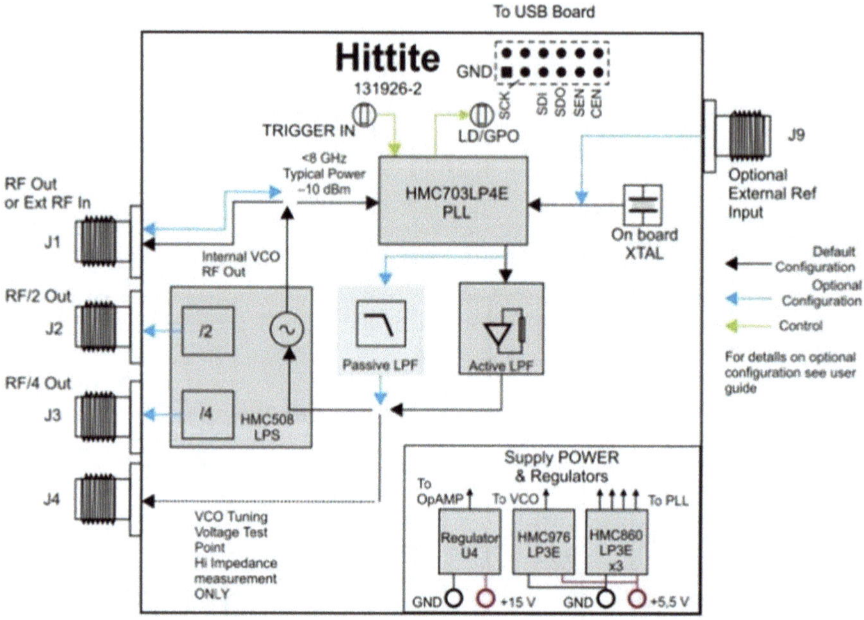

Fig. 5.118 Functional diagram of chip testing board

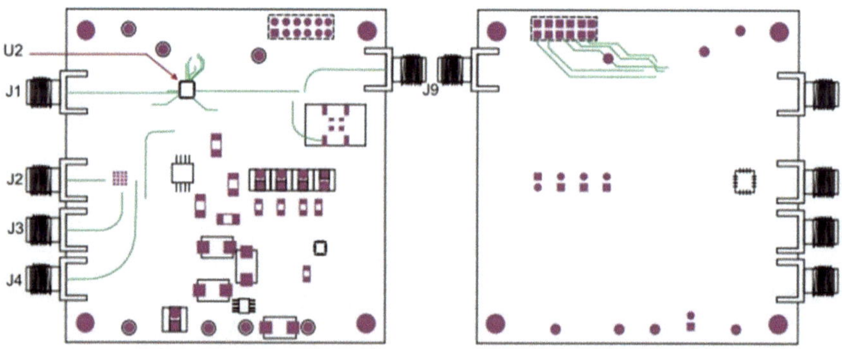

Fig. 5.119 Testing board topology

normally closed only due to recently developed technology of production of gallium nitride cell. The first practical applications for these transistors became DC/DC transformers for space and commercial applications, power amplifiers of D class, inverters, wireless power transmission systems, lidars, systems of HF-amplifier power supply adjustment (Envelope Tracking), system with increased temperature, and radiation stability.

As of the date of publishing this book, the vast majority of market belongs to silicon semiconductors. However, a sharp increase in the volume of power components based on gallium nitride (GaN) and silicon carbide (SiC) has begun over the past few years.

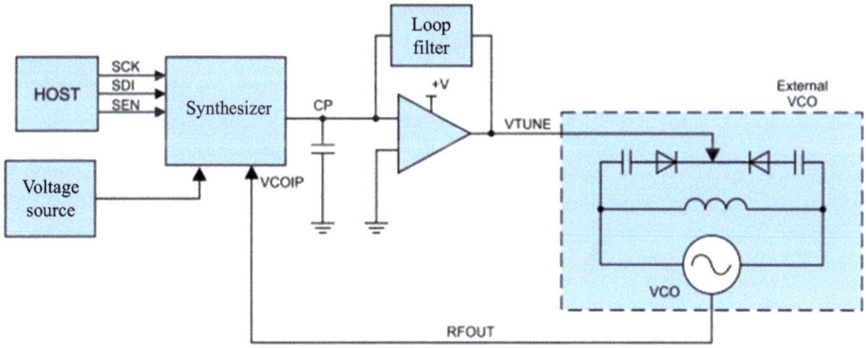

Fig. 5.120 Functional diagram of synthesizer with active filter in output of phase detector

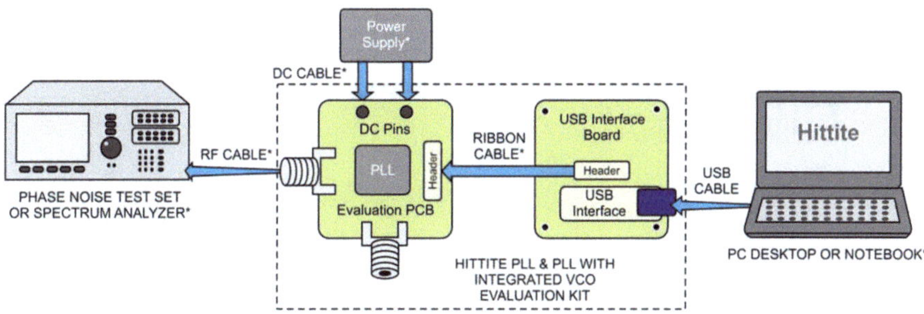

Fig. 5.121 Installation for firmware adding to storage of monolithic chips of frequency synthesizers

Fig. 5.122 Window of synthesizer type selection

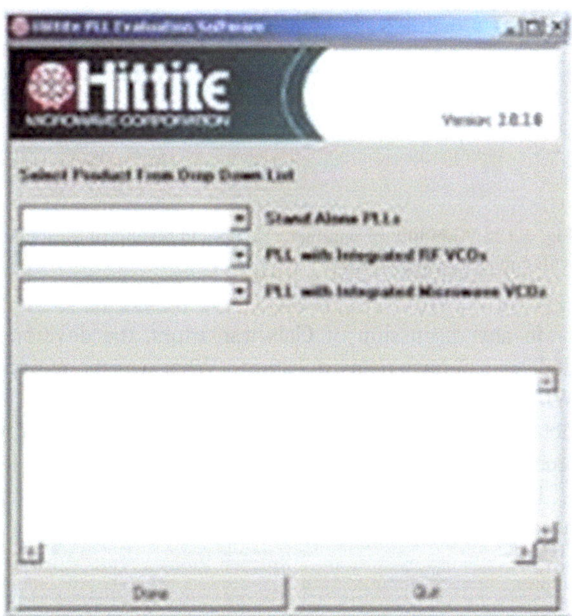

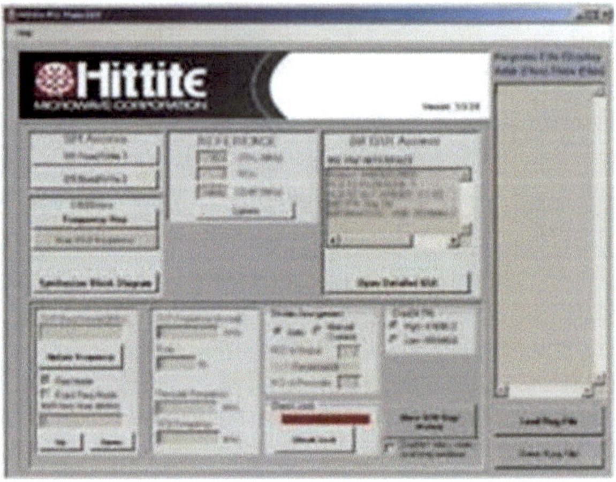

Fig. 5.123 GUI window for PLL synthesizer parameter control

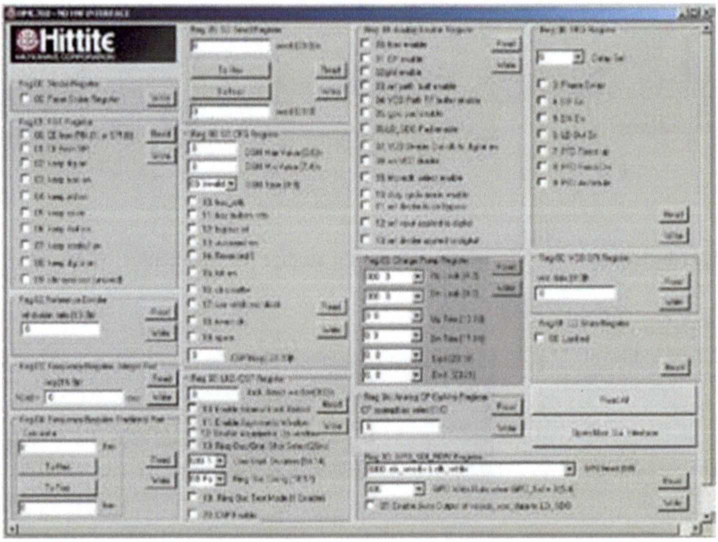

Fig. 5.124 Window of parameter setup and testing of various models of frequency synthesizers with enhanced capabilities

In any discussion of GaN-transistors, the developers of high-speed electronic devices can raise many questions. What was the reason for the choice of that GaN? Why has such rapid development begun just recently? What characteristics have the obtained production samples? What are the areas of their application? Does this technology have development prospects?

Let’s try to answer the listed questions using (http://epc-co.com/epc/Markets.aspx; http://epc-co.com/) [47–53].

Table 5.19 Characteristics of some semiconductive materials

Parameter	Material		
	GaN	Si	SiC
The band gap width, electron volt	3.4	1.12	3.2
Critical magnetic field strength, MV/cm	3.3	0.3	3.5
Charge-drift velocity, ×107 cm/s	2.5	1	2
Mobility, $cm^2/(V•s)$	990–2000	1500	650
Dielectric capacitivity	9.5	11.4	9.7

It must be said that victory march of power silicon semiconductors has lasted over 30 years since the power MOSFET appeared in the late 1970s. For a long time, silicon had no competitors since other known semiconductors (germanium and selenium) were considerably inferior to it practically by all important parameters. Later, semiconductor properties were discovered in new materials such as arsenide and gallium nitride, silicon carbide, and so forth.

Indeed, it should be noted that GaN is not so "new." Its special properties were first discovered by T. Mimura in 1975, and their detailed research was done by M. Cann in 1994.

Already at that time, the research results showed that GaN was much more advanced material than silicon (Table 5.19) (http://epc-co.com/epc/Markets.aspx) [48].

High critical field strength of GaN enables implementation of higher voltage devices. The wider band gap of semiconductive material provides the highest stability of its properties during temperature change or radiation exposure that is extremely important for space and military electronics, as well as for devices operating under harsh conditions.

High electron mobility and high drift velocity determine the much lower resistance in conducting state and high specific capacity as compared with Si.

The analysis of specific resistance-breakdown voltage dependence for various semiconductors is rather demonstrative (Fig. 5.125) [49]. This parameter is almost linear for each of them but GaN resistance is significantly lower at the same breakdown voltage.

As a result, GaN-transistors have specific capacity ten times higher that leads to the considerable reduction of overall dimensions in comparison with traditional silicon elements.

It is necessary to solve a few significant technological problems for organization of production.

To understand these reasons, we will consider the simplest traditional structure of GaN-transistor (Fig. 5.126a) [49]. Protective layer AlN is grown on the silicon wafer. GaN/AlGaN heterostructure is formed on it. Then, the protective dielectric layer and electrodes are formed.

As known, GaN and AlGaN have polar nature that's why spontaneous interfacial polarization with generation of surface charges occurs during their growth. In

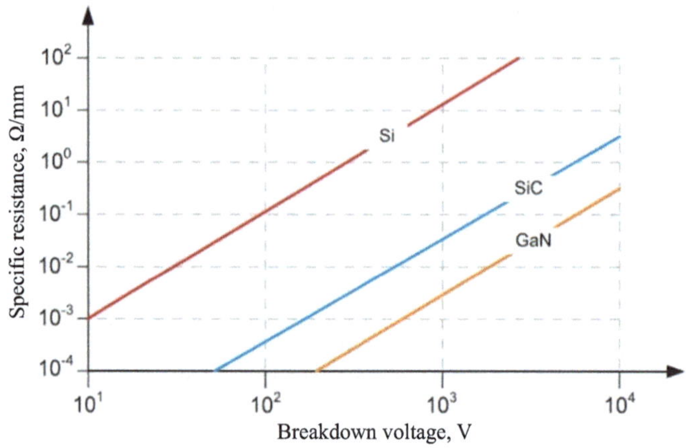

Fig. 5.125 Specific resistance-breakdown voltage dependence for various semiconductors

addition, GaN has pronounced piezoelectric properties. That's why it is additionally polarized under influence of deformation and mechanical stresses. Since GaN and AlGaN crystal lattices have mismatch, such voltages appear at their interface [49].

That's why poling processes cause the charge generation in the form of two-dimensional surface (so-called two-dimensional electron gas, 2DEG) [48].

When voltage is supplied to the drain and source electrodes of such base cell, the current starts flowing even if pinch-off voltage is equal to zero. That's why research and development engineers called this device "normally open."

To stop current flowing to the gate, negative voltage should be supplied in relation to the source.

It is clear that such transistor is not easy-to-use. First, to avoid electronic circuit zapping, it is necessary to reliably switch off the transistor before main power-on. Secondly, it is necessary to have an additional source of negative voltage.

Then the problems with a "simple" cell are far from over since the above-represented structure has a very simplified form. Actually it is much more difficult.

So, it is required to have a substrate for efficient heat extraction from GaN layer; and all standard materials (Si, SiC, sapphire) have lattice mismatch with GaN. That's why additional matching layers are inserted to reduce mechanical tensions of materials. Likewise, it is necessary to add additional layers between other dissimilar materials of the cell. As a result, the structure is complex [48].

These additional problems cause the need for carrying out additional researches for detection of optimum materials, optimum thickness of layers, and so on.

In this regard, it is appropriate to consider the main technological solutions developed by one of world leaders in this area—Efficient Power Conversion company (EPC).

Engineers of this company offered the improved structure of the transistor cell which was called eGaN® FET (Enhancement Mode) (Fig. 5.126b). Fundamental changes have affected the forming of gate and subgate area [49].

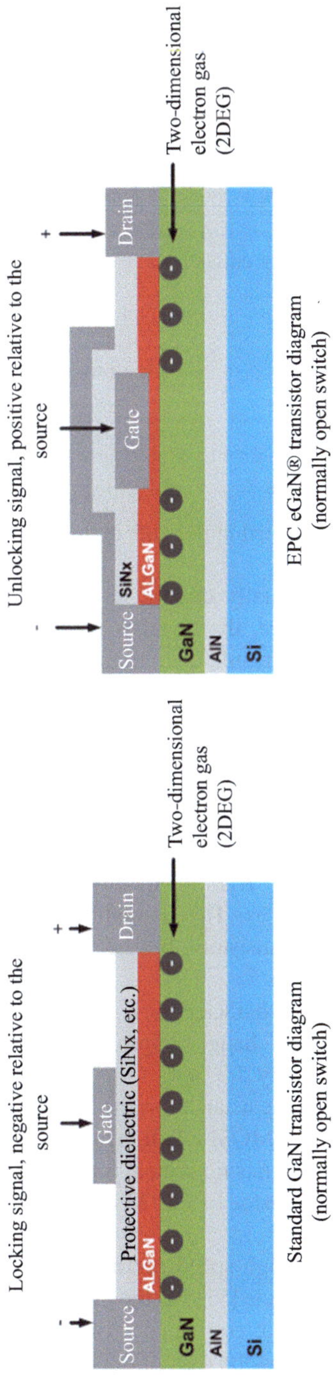

Fig. 5.126 Conventional cell of **(a)** GaN-transistor and **(b)** eGaN® FET produced by EPC

Table 5.20 Comparison of characteristics of power keys with voltage of 100 V

Parameter	Typical silicon MOSFET of 100 V	eGaN EPC2022 100 V
Drain-to-source on-state resistance (RDS(on)), mOhm	units-dozens	2.4
Resistance change RDSon during temperature change	2.2	1.4
Energy for switching-over	high	low
Bypass diode recovery time	long	absent
Threshold gate-source voltage, V	2–4	0.7–2.5
Change of threshold gate-source voltage during temperature change 125 °C/25 °C	0.66	1
Maximum voltage, Ugate, V	±20	−1.5
Operating temperature, °C	150	150
Input resistance of gate *Rg*, Ohm	a few	0.3
Input current Iзи	a few nA	1 mA

AlGaN depletion region is formed under gate in this cell that ensures forming of a gap in 2DEG electron gas region.

As a result, transistor is normally closed. In this case, to form a conductive channel between drain and source, the positive voltage should be applied to the gate in relation to the source. As a result, operation of eGaN is absolutely identical to the operation of usual N-MOSFET.

Of course, the described structure is most simplified, and the real cell has much more difficult form.

Table 5.20 gives the results of comparing the characteristics of these new transistors and their silicon precursors including pros and cons of representatives of each technology [50].

Here, open channel resistance RDSon, mOhm is one of the most important parameters determining keys losses. The standard value of resistance is close to that of serial samples of both transistors, but here it is necessary to make a few remarks.

First, GaN has more stable characteristics. For GaN, the RDS value in the temperature range of 25–125 °C changes approximately by 1.4 times. Resistance of MOSFET changes more than by 2.2 times.

Secondly, the dependence of resistance on the maximum operating voltage for GaN is much weaker than that of MOSFET. It is related to the fact that the increase in "Drain-Source" channel length has no critical impact on the resistance value. For example, resistance of 30-V transistor EPC2023 is 1.3 mOhm, and 200-V transistor EPC2034 is only 10 mOhm.

Capacity C, pF, defines the high-speed performance of the transistor. The flat eGaN FET structure offered in article [51] has the minimum values of CGD and CDS capacity. This allows the devices to commutate voltage of hundreds of volts with a gigahertz frequency while reducing the overall dimensions of voltage converters.

Threshold voltage Uзи thr. V, is only 0.7–2.5 V for GaN.

It is important to note that the maximum value of voltage on the GaN-transistor gate is only +6 V/−4 V.

Input resistance of the gate *Rg*, Ohm, determines the input capacitance priming speed. This resistance is rather low for GaN that results in high-speed performance and improvement of protection dv/dt. At the same time, the values of input currents increase and as a result, the control power grows.

The research and development engineers use a bypass diode in such systems. However, the typical circuit of GaN-transistor has no bypass diode in principle. However, there is reverse conduction mechanism that performs its function. An interesting fact is that there is no accumulation of minority carriers, unlike MOSFET during reverse direction of current. And it means that there are no time losses on the bypass diode resistance recovery.

Thus, eGaN FET transistor has excellent electrical characteristics but it is behind MOSFET in availability of control; therefore, this type of power keys will have advantage over MOSFET but not always. Nevertheless, it is possible to specify the perspective applications for eGaN FET: DC/DC transformers for various space and commercial applications, power amplifiers of D class, inverters, wireless power transmission systems, lidars, systems of HF-amplifier power supply adjustment (Envelope Tracking), and any systems with increased temperature and radiation stability.

To show advantages of use of eGaN FET, it is possible to specify the results of testing the DC/DC converters of 12 V/1.2 V with operating frequency of 1 MHz and 48 V/12 V with operating frequency of 300 kHz (Fig. 5.127). These graphs show that advantage of eGaN FET over MOSFET increases while raising the operating frequencies. The best results showed GaN-transistors, so-called transistors of the fourth generation: if the maximum coefficient of efficiency for the first converter is more than 91%, then that of the second converter is more than 98%. It is much higher than similar DC/DC values on MOSFET keys.

It should be noted that four generations of power components have been created since devices of this class appeared. Since 2015, series production of representatives of two generations Gen2 and Gen4 has been started.

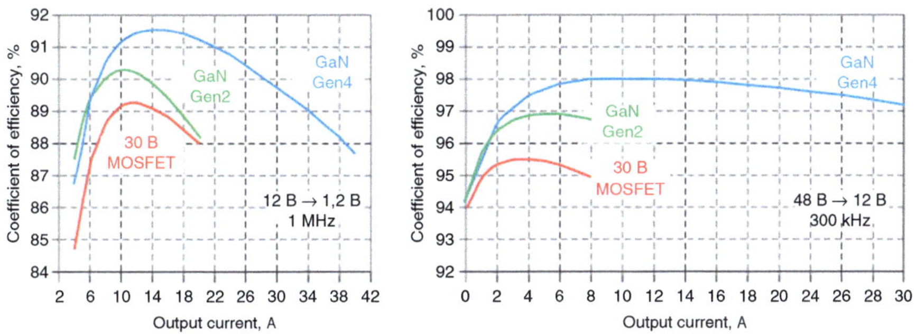

Fig. 5.127 The experimental results of comparison tests of DC/DC converters

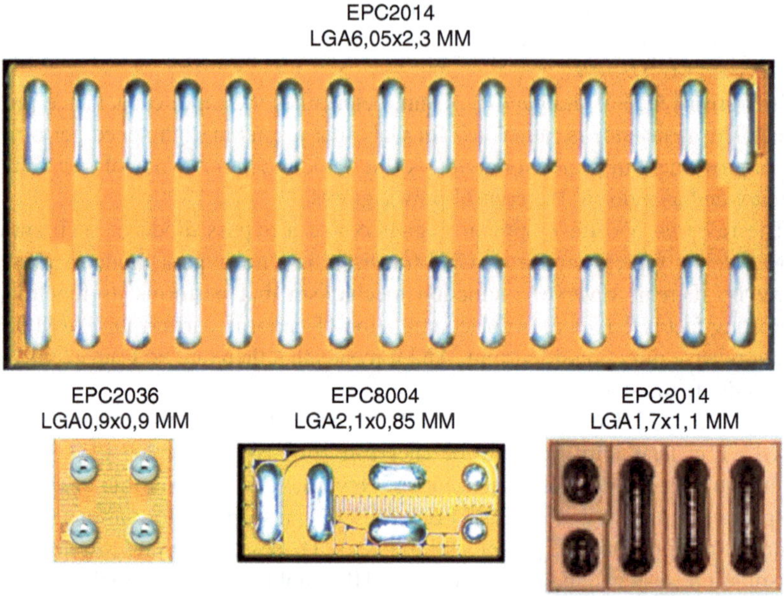

Fig. 5.128 Examples of packaged discrete eGaN® FET

It is necessary briefly to review a wide range of eGaN FET produced by EPC.

The EPC company's product range includes three main groups of power components:

- discrete power eGaN FET;
- eGaN FET for HF-applications;
- integrated circuit packaging eGaN FET.

All these elements are produced in LGA-packages (Fig. 5.128).

A large group of discrete eGaN® FET includes the representatives of two latest-generations of transistors with operating voltages of 30–450 V (Table 5.21).

All above-described power keys have low resistance. Record Resistance Drain-Source (RDS) is 1.3 mOhm for EPC2023—key of 30 V. In this case, resistance-operating voltage dependence inside this segment is not so strong as such dependence in MOSFET.

Generally, the maximum root-mean-square currents of this element group are tens of amperes, and pulse currents are tens and hundreds of amperes. Despite high power, all transistors are designed in mini packages, the biggest of which is LGA-package 6.1 × 2.3 mm in size, the most compact is LGA-package 0.9 × 0.9 mm in size (Fig. 5.128).

The group of products eGaN FET EPC800x is intended for operation in the subgigahertz range. It should be noted that it was in HF-applications where

Table 5.21 Discrete eGaN® FET of general purpose

Name	Max permissible drain-to-source voltage, V	Max permissible gate-source voltage, V	Max. RDS(on), mOhm when gate-source voltage = 5 V	Gate charge, nK	Gate-source charge, nK	Gate-drain charge, nK	Drain current, A	Impulse drain current, A	Tj max, °C	LGA-package, mm
EPC2023	30	6	1.3	20	5.8	1.9	60	590	150	6.1 × 2.3
EPC2024	40	6	1.5	19	6.4	2	60	550	150	6.1 × 2.3
EPC2030	40	6	2.4	18	5.2	3.4	31	495	150	4.6 × 2.6
EPC2015C	40	6	4	8.7	3	1.4	36	235	150	4.1 × 1.6
EPC2015	40	6	4	10.5	3	2.2	33	150	150	4.1 × 1.6
EPC2014C	40	6	16	2	0.7	0.3	10	60	150	1.7 × 1.1
EPC2014	40	6	16	2.5	0.67	0.48	10	40	150	1.7 × 1.1
EPC2020	60	6	2	16	5	2	60	470	150	6.1 × 2.3
EPC2031	60	6	2.6	17	5.2	3.2	31	450	150	4.6 × 2.6
EPC2035	60	6	45	0.88	0.25	0.16	1	24	150	0.9 × 0.9
EPC2021	80	6	2.5	15	3.8	2.1	60	420	150	6.1 × 2.3
EPC2029	80	6	3.2	13	4	2.5	31	360	150	4.6 × 2.6
EPC2022	100	6	3.2	13	3.7	2	60	360	150	6.1 × 2.3
EPC2032	100	6	4	14	4.2	3.1	31	340	150	4.6 × 2.6
EPC2001C	100	6	7	7.5	2.4	1.2	36	150	150	4.1 × 1.6
EPC2001	100	6	7	8	2.3	2.2	25	100	125	4.1 × 1.6
EPC2016C	100	6	16	3.4	1.1	0.55	18	75	150	2.1 × 1.6
EPC2016	100	6	16	3.8	0.99	0.7	11	50	125	2.1 × 1.6
EPC2007C	100	6	30	1.6	0.6	0.3	6	40	150	1.7 × 1.1
EPC2007	100	6	30	2.1	0.52	0.61	6	25	125	1.7 × 1.1
EPC2036	100	6	65	0.7	0.17	0.14	1	18	150	0.9 × 0.9
EPC2033	150	6	7	10	3.5	1.7	31	260	150	4.6 × 2.6
EPC2018	150	6	25	5	1.3	1.7	12	60	125	3.6 × 1.6

(continued)

Table 5.21 (continued)

Name	Max permissible drain-to-source voltage, V	Max permissible gate-source voltage, V	Max. RDS(on), mOhm when gate-source voltage = 5 V	Gate charge, nK	Gate-source charge, nK	Gate-drain charge, nK	Drain current, A	Impulse drain current, A	Tj max, °C	LGA-package, mm
EPC2034	200	6	10	8.5	2.6	1.4	31	140	150	4.6 × 2.6
EPC2010C	200	6	25	3.7	1.3	0.7	22	90	150	3.6 × 1.6
EPC2010	200	6	25	5	1.3	1.7	12	60	125	3.6 × 1.6
EPC2019	200	6	50	1.8	0.6	0.35	8.5	42	150	2.7 × 0.95
EPC2012C	200	6	100	1	0.3	0.2	5	22	150	1.7 × 0.9
EPC2012	200	6	100	1.5	0.33	0.57	3	15	125	1.7 × 0.9
EPC2025	300	6	150	1.85	0.61	0.3	4	20	150	1.95 × 1.95
EPC2027	450	6	400	1.7	0.6	0.25	4	12	150	1.95 × 1.95

GaN-transistors occupied a dominant position since they ensure the operation in systems with hard modes of power key switching and operating frequencies from tens to hundreds of MHz.

Characteristics of EPC800x family with working voltage of 40–100 V are listed in Table 5.22. All representatives of this group are packed in the same way—LGA-package 2.1 × 0.85 mm (Fig. 5.128).

The main applications for products of EPC800x series became systems of HF-amplifier power supply adjustment (Envelope Tracking) and wireless power transmission systems, as well as devices of space purpose.

One more category of products is a group of integrated assemblies eGaN FET that includes six components with an operating voltage of 30–100 V (Table 5.23). All these assemblies represent classical half-bridge circuits (Fig. 5.129).

It should be noted that integrated eGaN FET-assemblies can have symmetric or asymmetric structure (Fig. 5.130) [52] that is important to users.

Sizes of transistor crystals of both arms are equal in symmetric configuration. Consequently, resistance of upper and lower keys is equal. Such circuit is used for amplifiers of D class and motor drives where both transistors have equal loading range.

Size of crystal of upper transistor is approximately four times less than that of lower transistor in asymmetric configuration. Their resistances are different, too. Such asymmetric configuration is used in DC/DC converters operating at small pulse length and big differences between input and output voltage. In such cases the lower key is much more loaded.

The use of integrated assemblies instead of discrete components allows the hardware engineers to make more compact solutions with the minimum values of parasitic inductances that provide additional advantages at higher frequencies (Fig. 5.131) [52].

Of course, special drivers should be used to control such eGaN FET having a few features.

These drives eGaN FET generate the relevant control currents and voltages and have some additional functions. It is applicable to the controllers of use.

First, they should have the elevated operating frequencies. Secondly, they should differ with the minimum own losses. Thirdly, transistors should be switched on/off with safe levels of dv/dt and di/dt.

Control chips recommended by EPC company are listed in Table 5.24. Here, the most appropriate choice is the use of drivers LM5113 and LM5114 produced by Texas Instruments company.

Of course, creation of power converters and HF-devices requires great skill and special care [53]. That makes sense to use ready solutions, debugging, and development tools produced by EPC to minimize errors at the first development stages.

So, EPC company provides developers with all necessary informational support: documents for components, models (PSPICE, TSPICE, LTSPICE), mounting seat libraries for Altium Designer, heat models. However, a great number of peculiarities

Table 5.22 eGaN® FET for HF-applications

Item	Max permissible drain-to-source voltage, V	Max permissible gate-source voltage, V	Max. RDS(on), mOhm when gate-source voltage = 5 V	Gate charge, nK	Gate-source charge, nK	Gate-drain charge, nK	Drain current, A	Impulse drain current, A	Tj max, °C	LGA-package, mm
EPC8004	40	6	110	0.37	0.12	0.047	2.7	7.5	150	2.1 × 0.85
EPC8007	40	6	160	0.302	0.097	0.025	2.7	6	150	2.1 × 0.85
EPC8008	40	6	325	0.177	0.067	0.012	2.7	2.9	150	2.1 × 0.85
EPC8009	65	6	130	0.37	0.12	0.055	2.7	7.5	150	2.1 × 0.85
EPC8005	65	6	275	0.218	0.077	0.018	2.7	3.8	150	2.1 × 0.85
EPC8002	65	6	530	0.141	0.059	0.0094	2*	2	150	2.1 × 0.85
EPC8010	100	6	160	0.36	0.13	0.06	2.7	7.5	150	2.1 × 0.85
EPC8003	100	6	300	0.315	0.11	0.034	2.7	5	125	2.1 × 0.85

Table 5.23 Integrated eGaN® FET of general purpose

Name	Type	Max permissible drain-to-source voltage, V	Max permissible gate-source voltage, V	Max. RDS(on), mOhm when gate-source voltage = 5 V	Gate charge, nK	Gate-source charge, nK	Gate-drain charge, nK	Drain current, A	Impulse drain current, A	Tj max, °C	LGA-package, mm
EPC2100	Asymmetric	30	6	8; 2	3.5; 15	1.4; 4.6	0.57; 2.6	9.5; 38	100; 400	150	6.1 × 2.3
EPC2101	Asymmetric	60	6	11.5; 2.7	2.7; 12	1; 3.7	0.50; 2.5	9.5; 38	80; 350	150	6.1 × 2.3
EPC2102	Symmetric	60	6	4.4	6.8	2.3	1.4	23	215	150	6.1 × 2.3
EPC2105	Asymmetric	80	6	14.5; 3.5	2.5; 10	1; 3.2	0.50; 2	9.5; 38	75; 320	150	6.1 × 2.3
EPC2103	Symmetric	80	6	5.5	6.5	2	1.3	23	195	150	6.1 × 2.3
EPC2104	Symmetric	100	6	6.3	7	2	1.2	23	165	150	6.1 × 2.3

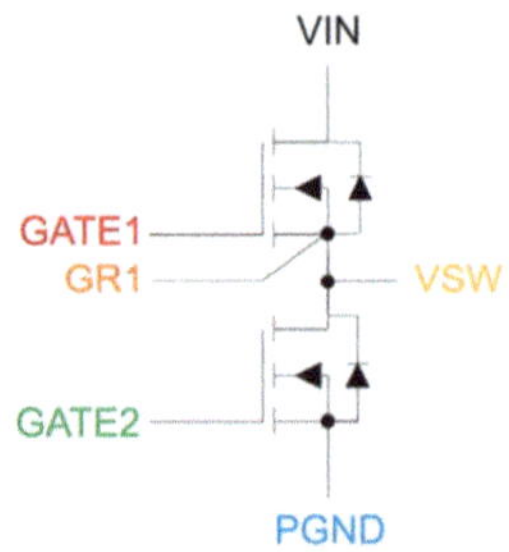

Fig. 5.129 Electric circuit analogy of eGaN FET-assemblies

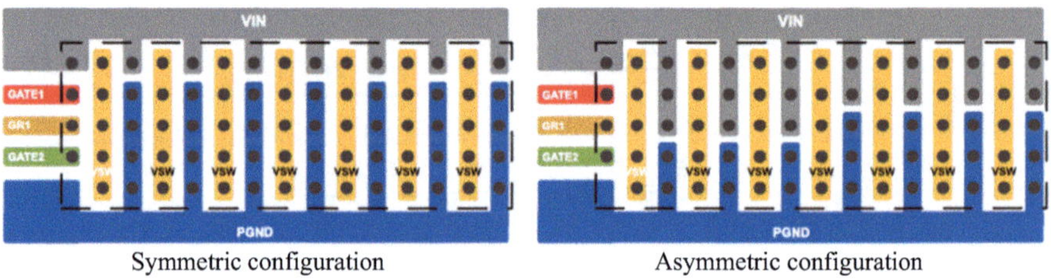

Fig. 5.130 Typical configurations of eGaN FET-assemblies produced by EPC

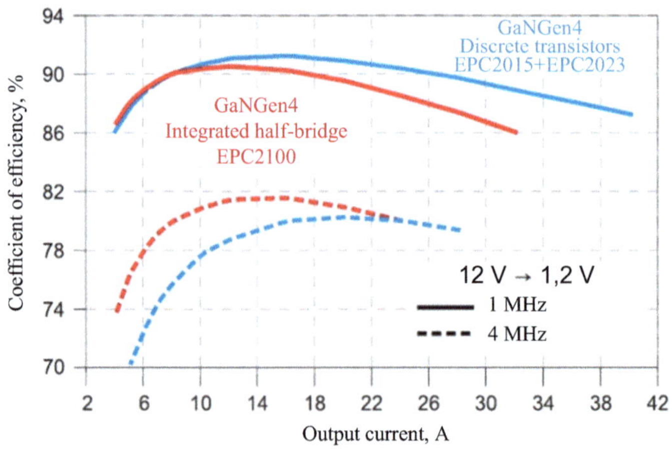

Fig. 5.131 Advantages of eGaN® FET-assemblies at high frequencies

can cause defects and ever errors in the design. To avoid this and quickly master new devices, the ready solutions should be used such as:

- Debug boards;
- Demo sets;
- Terminal modules DrGaNPLUS.

Table 5.24 Recommended drivers and controllers eGaN® FET

Item	Functionality	Manufacturer	Description
LM5113	Driver	Texas Instruments	5-A, 100-V half-bridge driver eGaN FETs
LM5114	Driver	Texas Instruments	7.6 A single-channel low side driver
UCC27611	Driver	Texas Instruments	4A/6A high-speed 5-V single-channel driver
ADP1851	Controller	Analog Devices	Step-down controller 2.75 V/20 V
ISL6420	Controller	Intersil	Synchronous step-down controller 4.5 V/16 V
LM27403	Controller	Texas Instruments	Synchronous step-down controller 3 V/20 V
LTC3833	Controller	Linear Technologies	Step-down controller 4.5 V/38 V
LTC3891	Controller	Linear Technologies	Synchronous low-energy step-down controller 60 V
MAX15026B	Controller	Maxim	Synchronous step-down controller 4.5 V/28 V
MCP19118/19	Controller	Microchip	Step-down PWM controller 4.5 V/40 V
SC419	Controller	Semtech	Step-down controller with integrated diode 3 V/28 V
TP253219A	Controller	Texas Instruments	Synchronous step-down controller 4.5 V/25 V
TPS40490	Controller	Texas Instruments	Step-down PWM controller 4.5 V/60 V
UCC24610	Controller	Texas Instruments	Synchronous step-down controller for secondary coil

Debug boards eGaN FET (Tables 5.25 and 5.26) represent ready printed circuit board with half-bridge circuit, driver, additionally required passive components, and logic. There is a debug board nearly for each eGaN FET. It applies to discrete transistors and integrated assemblies (Table 5.25), as well as to eGaN FET for HF-applications (Table 5.26).

For your information, EPC company produces a series of demo sets: step-down converters, wireless power transmission sets, amplifiers of D class (Table 5.27).

Particularly worth mentioning are modules DrGaNPLUS that represent ultra-small boards intended for direct insertion in ready devices. Their sizes are only 11 × 1 mm (Fig. 5.132). The operating voltage is 30 V (EPC9201) or 80 V (EPC92013) (Table 5.28).

Table 5.25 Basic list of eGaN® FET debug boards of general purpose

Item	Description	Max permissible drain-to-source voltage, V	Max drain current, A	Type of transistor
EPC9036	Debug board based on integrated half-bridge	30	25	EPC2100
EPC9031	Half-bridge circuit with driver	30	40	EPC2023
EPC9018	Half-bridge circuit with driver for applications with the minimum fill factor	30	35	EPC2015/ EPC2023
EPC9016	Half-bridge circuit with driver for applications with the minimum fill factor	40	25	EPC2015
EPC9032	Half-bridge circuit with driver	40	35	EPC2024
EPC9005C	Half-bridge circuit with driver	40	7	EPC2014C
EPC9005	Half-bridge circuit with driver	40	7	EPC2014
EPC9001	Half-bridge circuit with driver	40	15	EPC2015
EPC9037	Debug board based on integrated half-bridge	60	22	EPC2101
EPC9038	Debug board based on integrated half-bridge	60	20	EPC2102
EPC9033	Half-bridge circuit with driver	60	30	EPC2020
EPC9049	Half-bridge circuit with driver	60	4	EPC2035
EPC9046	Half-bridge circuit with driver	80	22	EPC2029
EPC9034	Half-bridge circuit with driver	80	27	EPC2021
EPC9041	Debug board based on integrated half-bridge	80	20	EPC2105
EPC9039	Debug board based on integrated half-bridge	80	17	EPC2103
EPC9019	Half-bridge circuit with driver for applications with the minimum fill factor	80	20	EPC2001/ EPC2021
EPC9040	Debug board based on integrated half-bridge	100	15	EPC2104
EPC9035	Half-bridge circuit with driver	100	25	EPC2022
EPC9006	Half-bridge circuit with driver	100	5	EPC2007
EPC9010C	Half-bridge circuit with driver	100	7	EPC2016C
EPC9010	Half-bridge circuit with driver	100	7	EPC2016
EPC9050	Half-bridge circuit with driver	100	2.5	EPC2036
EPC9002	Half-bridge circuit with driver	100	10	EPC2001
EPC9047	Half-bridge circuit with driver	150	12	EPC2033
EPC9014	Half-bridge circuit with driver	200	4	EPC2019
EPC9017	Half-bridge circuit with driver for applications with the minimum fill factor	100	20	EPC2001
EPC9013		100	35	EPC2001

(continued)

Table 5.25 (continued)

Item	Description	Max permissible drain-to-source voltage, V	Max drain current, A	Type of transistor
	Debug board of four parallel half-bridge circuits			
EPC9004C	Half-bridge circuit with driver	200	3	EPC2012C
EPC9004	Half-bridge circuit with driver	200	3	EPC2012
EPC9014	Half-bridge circuit with driver	200	4	EPC2019
EPC9003C	Half-bridge circuit with driver	200	5	EPC2010C
EPC9003	Half-bridge circuit with driver	200	5	EPC2010
EPC9042	Half-bridge circuit with driver	300	3	EPC2025
EPC9044	Half-bridge circuit with driver	400	1.5	EPC2027

Table 5.26 List of debug boards eGaN® FET for HF-applications

Item	Description	Max permissible drain-to-source voltage, V	Max drain current, A	Type of transistor
EPC9024	Half-bridge circuit with driver	40	4.4	EPC8004
EPC9027	Half-bridge circuit with driver	40	3.5	EPC8007
EPC9028	Half-bridge circuit with driver	40	2.2	EPC8008
EPC9022	Half-bridge circuit with driver	65	1.6	EPC8002
EPC9025	Half-bridge circuit with driver	65	2.2	EPC8005
EPC9029	Half-bridge circuit with driver	65	3.5	EPC8009
EPC9023	Half-bridge circuit with driver	100	2.2	EPC8003
EPC9030	Half-bridge circuit with driver	100	3.2	EPC8010

5.7.5 *SHF Devices Based on Heterogeneous Integration Technology*

5.7.5.1 COSMOS Project

Compared to digital chips that historically developed only as silicon-based technology, since the very beginning of their development, the SHF devices were based on compound semiconductor technology: first, on gallium arsenide (GaAs), then on gallium nitride (GaN), silicon carbide (SiC), and etc.

Table 5.27 Main demo sets eGaN® FET

Item	Description	Uin, V	Uout, V	Max drain current, A	Type of transistor
EPC9101	Step-down converter 19 V/1.2 V, 1 MHz	8–19	1.2	18	EPC2015/ EPC2014
EPC9102	Step-down converter 48 V/12 V	36–60	12	17	EPC2001
EPC9105	Step-down converter 48 V/2 V, 1.2 MHz	36–60	12	30	EPC2001/ EPC2015
EPC9106	150-W/8-Ohm amplifier of D class	–	–	–	EPC2016
EPC9107	Step-down converter 28 V/3.3 V	9–28	3.3	15	EPC2015
EPC9111	Demo set of wireless energy transmission conforming to requirements A4WP	8–32	Uin	10	EPC2014
EPC9112	Demo set of wireless energy transmission conforming to requirements A4WP	8–32	Uin	6	EPC2007
EPC9115	Step-down converter 48 V/12 V	48–60	12	42	EPC2020/ EPC2021
EPC9118	Step-down converter 48 V/5 V, 400 kHz	30–60	5	20	EPC2001/ EPC2021
EPC9506	Demo board of wireless amplifier of D class	8–32	Uin	10	EPC2014
EPC9507	Demo board of wireless amplifier of D class	8–32	Uin	6	EPC2007
EPC9508	Demo board of wireless amplifier of D class	7–36	Uin	3	EPC8009/ EPC2007

Fig. 5.132 The appearance of ready modules DrGaNPLUS

Table 5.28 Basic specifications of modules DrGaNPLUS

Item	Description	Max permissible drain-to-source voltage, V	Max drain current, A	Type of transistor
EPC9201	Half-bridge circuit on discrete transistors	30	20	EPC2015/ EPC2023
EPC9203	Half-bridge circuit on discrete transistors	80	40	EPC2021

Generally, these technologies developed thanks to financing of the US military agency called DARPA by different programs known in public sources as MIMIC [54], WBGS-RF and devoted mainly to development of GaAs- and GaN-based SHF devices of military purpose.

New technology called as heterogeneous integration has been developed within these programs since 2014 [54]. It is based on integration in one semiconductor crystal of different technologies both of silicon and compound semiconductors. The final purpose of a section of this program is to create such integrated technology that allows to optimize the operation mode of each device based on compound semiconductors in such a way that they would be able to perform necessary specialized operations (functions) by means of certain "digital tuning." If this purpose is reached, this technology will have a significant impact on development of modern SHF-electronics.

As shown in work [54], SHF IC technologies based on GaAs and GaN materials obtained as a result of implementation of MIMIC and WBGS-RF programs laid the foundation for building of the entirely new electronic warfare systems including radar systems with AESA [55–59]. One of the strategic task of DARPA program is to drastically improve the parameters of typical SHF IC and modules, as well as further enhance the efficiency of modern radioelectronic armament and military equipment control systems using heterogeneous integration technology.

Each type of compound semiconductors has its own features used when designing the devices that operate in SHF-range. So, InP makes it possible to obtain transistors with the maximum operating frequency more than 1 THz; GaN ensures high breakdown voltages and high output power. All compound semiconductors are still behind silicon in integration level (quantity of transistors in the circuit). So, being the most promising material in this regard, InP is still behind silicon at least in five orders, not to mention the GaN that is characterized by small-scale integration. At the same time, according to Moore's law, silicon CMOS technology is continuously growing the integration scale and high-speed performance of integrated circuit with high-yield that is still unachievable when using compound semiconductors. In addition, silicon integrated circuit allows to place the built-in compensation and parameter optimization circuits and devices operating simultaneously with high frequency and digital signals on one crystal. The perspective of further development of the integrated SHF-electronics is associated with combination of technologies of different types that are based on compound semiconductors and silicon, i.e., their heterogeneous integration. A number of special projects [60–63] were implemented

Table 5.29 The basic specifications of SHF integrated circuits implemented in project COSMOS [54]

Basic specifications	Type of chip		
	DA (Differential amplifier)	DAC (Digital-to-analog converter)	ADC (Analog-to-digital converter)
Percentage yield of heterogeneous junctions, X (it is determined on a separate test plate with a great number of series-connected heterojunctions)	>99	>99.9	>99.9
Chip reliability, % (it is determined as a part of remained yield IC after 100 thermal cycles in the interval −55 to 85 °C and hold time at each extreme temperature not less than 10 min)	>50	>95	>95
The minimum length of heterojunction connecting CS- and SI-transistors, μm	<5	<5	<5
The minimum distance between power buses of CS- and SI-transistors, μm	<25	<5	<5
Percent conductivity of SHF-transistors measured before and after heterointegration process, %	>80	>90	>90
Percent conductivity of Si CMOS transistors measured before and after heterointegration process, %	>80	>90	>90
SHF IC yield ratio, %	>25	>50	>50

within the basic program DARPA in order to develop the technology of heterogeneous integration of SHF IC; the main results received during the program implementation are considered below.

COSMOS Project (Compound Semiconductor Materials on Silicon) was implemented between 2007 and 2013 [60]. Generally, the issues of integration of crystals of compound semiconductors and silicon are solved by assembly of a few types of IC in special multi-crystal modules (micro-assemblies) although the module parameters decline inevitably. It results from action of different effects of mismatching due to long length (areas) of conductors connecting the heterogeneous crystals and devices. Therefore, in COSMOS subprogram, the integration process was raised to the level of transistors based at least on one type of compound semiconductor which are produced along with plate-based transistors according to the upgraded CMOS process. In this case, length of connections between different types of transistors and distance between conductors suitable to various transistors didn't exceed a few microns (Table 5.29). In COSMOS sub-program, InP was taken as compound semiconductor material; and standard chips such as differential amplifier, digital-to-analog converter, and analog-to-digital converter were selected as pilot prototypes of devices.

As known from literature sources [54], three processes of heterointegration were studied within COSMOS project.

Actually, it was due to the choice of three companies, each of which offered its technologies. These three companies are Northvop Grumman, Raytheon and HRL Laboratories.

That's why these technological solutions are as follows:

- heteroepitaxial growth of InP HBT-transistors coplanarly with silicon CMOS-structure on multilayer silicon platform (Raytheon company);
- micrometric assembly when ready chiplets with InP-transistors are mounted on fully treated silicon wafer (group of companies headed by Northrop Grumman);
- coating of only epitaxial InP-heterostructures by printing technique on treated silicon wafer with subsequent separate processes of production of InP-transistors and IC (HRL Laboratories company).

The compulsory requirement for all project participants is to demonstrate certain integrated circuit of increasing degree of complexity that were produced and operates at the stage of project handover. So, theoretical and experimental research of all methods of heterogeneous integration was done at the first stage (2007–2009); and at that time the first IC of differential amplifier with record parameters was created. This chip contained ten heterogeneous junctions, five InP HBT-transistors and four CMOS transistors. The basic requirement of the second stage (2009–2011) is to increase the percentage yield. As a result of the second stage, the circuit of 13-bit DAC was produced. It included more than 500 heterojunctions, among them 400 InP HBT-transistors and 3200 CMOS transistors. At the third stage of program (2011–2013) using the example of ADC, it was demonstrated that the developed process of integration can be used for production of analog-digital LSI. Experimental circuit of ADC included about 1000 InP HBT-transistors, 18000 CMOS transistors, and more than 1800 heterogeneous junctions connecting 1-pR-chip with master silicon crystal.

Combination of submicron technology of silicon IC and high-speed In-P-transistors made it possible to apply a few methods of calibration and self-healing which were impossible when using only one InP-technology. In its turn, InP HB-T-transistors ensured high-speed performance and higher breakdown voltages, as well as good matching of transistors in comparison with classical CMOS technology. That's why being developed during implementation of this project, the ultra-wideband ADC designed on 130-nm CMOS process turned out to be comparable in effective number of bits to ADC based on pure silicon 32-nm CMOS process. In this case, signal-to-noise-and-distortion ratio exceeded 30 dB in the frequency ranges 2.75–8.75 and 14.25–20.25 GHz that corresponds to the best values obtained at that moment for silicon-based chips.

When implementing COSMOS project, each participant used his own design and technological solutions. In particular, Raytheon company developed an integration process for GaN HEMT and Si CMOS transistors based on modified master plate "silicon-on-insulator" which represents base substrate Si (111) and instrument layer Si (100) (Fig. 5.133). The epitaxial layer of GaN HEMT-transistors was grown on the surface of base substrate Si (111) through the window formed in the silicon dioxide layers. This solution served as the basis for creation of circuit of power

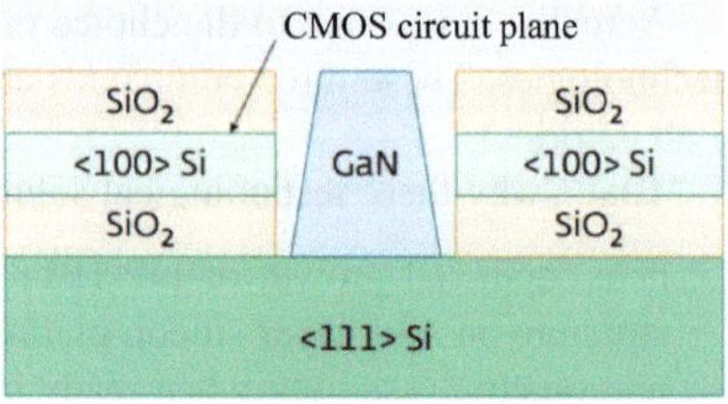

Fig. 5.133 Sketch of the design of multilayer silicon plate for monolithic integration of GaN HEMT and CMOS IC

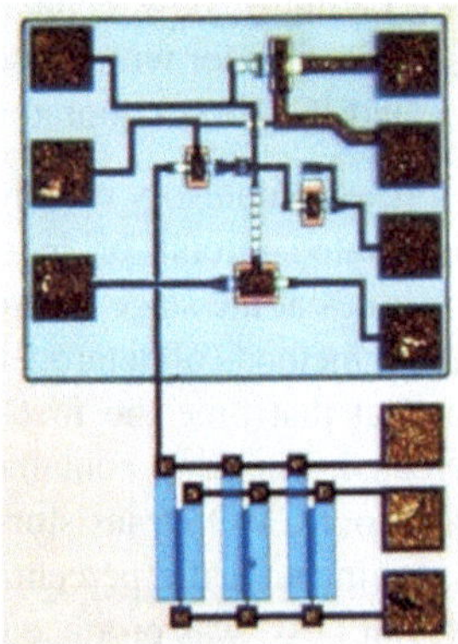

Fig. 5.134 Sketch of the topology of power GaN HEMT-amplifier integrated with CMOS IC

GaN-amplifier in GHz range that included GaN HEMT-transistors which were monolithic-integrated with silicon circuit for transistor offset control (Fig. 5.134). Output power and coefficient of efficiency of amplifier based on silicon wafer turned out to be close to the parameters of identical amplifier based on silicon carbide wafer (SiC).

The multiproject wafer method was applied in COSMOS project. This method provides that a few different IC structures are implemented in one common plate that allows to largely reduce the expenses for production of masking means for experimental prototypes of IC. In particular, one of project participant, the laboratory of US Air forces, created a four-stage low-noise amplifier of millimeter wave band with the noise ratio not exceeding 7.2 dB in the frequency range 75–100 GHz and with a gain factor more than 20 dB. In this case, the amplifier dissipation power was only 19.0 mW.

5.7.5.2 NEXT Project

When building the complex IC, capabilities of GaN as one of basic materials used for heterogeneous integration are rather limited [62] in 2009–2012. DARPA funded the development of another new GaN technology on the project Nitride Electronic NeXt Generation (NEXT). Its purpose was the improvement of integration level and high-speed performance of IC based on GaN HEMT-transistors due to design and circuit-

Table 5.30 The fundamental results of implementation of NEXT project on improvement of parameters of CaN HEMT-transistors

Device parameter	Parameter value
Cutoff frequency in depleted (D) mode, GHz	500
Maximum frequency of oscillation in D-mode, GHz	550
Johnson's factor in D-modes, THz × V	5
Cutoff frequency in enhancement (E) mode, GHz	400
Maximum frequency of oscillation in E-mode, GHz	450
Johnson's factor in E-modes, THz × V	5
Transistor yield ratio, %	95
Root-mean-square deviation of threshold voltage, mV	30
Root-mean-square deviation of cutoff frequency, GHz	30
Root-mean-square deviation of the maximum frequency of oscillation, GHz	5
Hours in operation, h	>1000

based solutions, and particularly, the reduction of transistor dimensions; as well as the use of new epitaxial structures. As a result of this project implementation, the transistor cutoff frequencies corresponding to gain factor equal to 1 (unity-gain cutoff frequency, fT) and maximum generation frequencies (f_{max}) were increased in chip samples. As known, in case of simple physical reduction in sizes, the maximum permissible breakdown voltage automatically declines. So-called Johnson's factor is used for evaluation of optimum ratio of these parameters. It is a simple product of cutoff frequency and breakdown voltage. Johnson's factor was successfully increased up to 5 THz during NEXT project (Table 5.30).

Of course, GaN HEMT-transistor operation modes had been modified during research of possibilities to improve parameters of GaN. As known, GaN HEMT-transistors operate in D-mode but for operation with digital signals it is required to use transistors in E-mode since the simultaneous use of transistors of both types significantly simplifies the building of SHF IC and enables creation of different directly coupled logic circuits with the possibility to use more than thousands of such basic elements on the plate.

Thus, the results of COSMOS project confirmed the possibility to significantly improve IC parameters using heterogeneous integration that was experimentally proved by the example of only one type of compound semiconductors (InP).

5.7.5.3 DAHI Project

Since 2013, the effective development of COSMOS program by American agency DARPA [62–65] has resulted in implementation of the next project Diverse Accessible Heterogeneous Integration Program (DAHI). Its main purpose was the development of a technique to promote the use of heterogeneous integration methods in the entire wide range of SHF devices including GaN, MEMS-devices, thermal

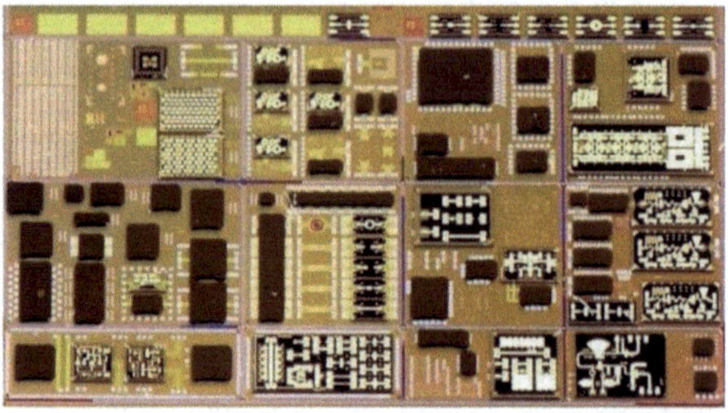

Fig. 5.135 The fragment of multiproject that is a plate with a few IC of different design

control models, and etc. The ultimate purpose of DAHI project is to organize the industrial-scale production of SHF IC of military and commercial purpose.

As a result of analysis of work being done in DAHI project, one method of assembly of integrated device made of separate heterogenous fully processed chips was chosen by experts among all the heterointegration methods being studied within COSMOS program. Actually, this method requires no significant changes in the basic silicon technology, as well as in technology of integrated devices based on compound semiconductors. All other methods under consideration are related to essential modifications of technology before or during the process of heterointegration require significant time and financial costs. Just to be fair, some countries make good use of these technologies especially in the sphere of military and space applications but there is little information about this in public sources.

In particular, in one of other works performed within DAHI project, two more technologies of compound semiconductors were implemented on one silicon plate (Fig.5.135). Being a part of millimeter-wave generator-amplifier chain, the generator circuit was designed on 0.25-μm InP HBT-transistors; and the amplifier circuit was designed on 0.2-μm GaN HEMT-transistors with a gain factor of GaN-amplifier being equal to 15 dB.

Since these results that were achieved in the process of developing the heterogeneous integration technology, create the basis for significant improvement of parameters of electronic warfare systems, the results of further research in this field are not published in open sources. In any case, the further development of this technology will depend on the development of appropriate IC design techniques. Continuous toughening of requirements for modern SHF systems of military and commercial purpose certainly suggests the further complication of structure of application-specific integrated circuits. With relatively low consumption that is typical of military applications, the design cost of such application-specific integrated SHF circuits can make an lunacceptably high percentage of the total cost of system (device) development. In this case, developers have to make tough choices between the required IC parameters, design time, and the total cost of this project. Of course, experience shows that reduction in cost and design time of radioelectronic devices can be achieved due to the use of field programmable gate arrays. However, practice in the design of such reprogrammable IC is still underdeveloped. For getting the

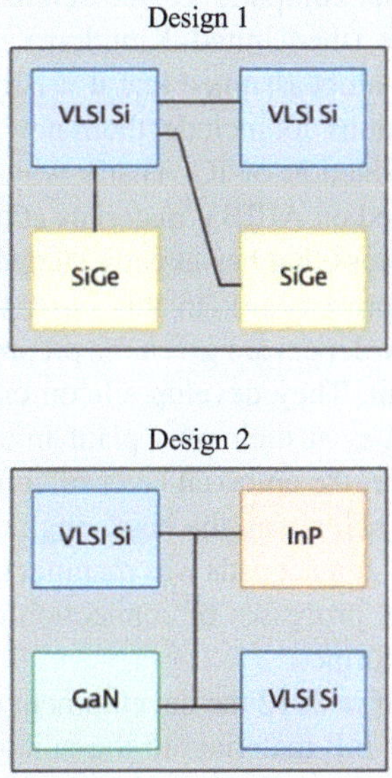

Fig. 5.136 Variants of IP-block assembly as an interposer on the common standard substrate

heterogeneous integration methods into actual practical use, it is required to develop an entirely new approach ensuring the economic efficiency of the development. One of ways is to create the standardized reusable IP design blocks based on semiconductor crystals made with the use of various compound semiconductors. Such blocks that are interconnected by means of assembly on the common standard substrate as an interposer, could be multifunctional and meeting the requirements for different military and space systems depending on the tasks to be addressed (Fig. 5.136).

Implementation of this approach will require to create the unified system for managing the processes of IP-block creation using different complex semiconductor materials. In addition, the essential condition is to develop a design flow including simulation and verification that can be easily integrated with the known commercial means of design. Of course, for wide commercial exploitation of this heterogeneous integration it is necessary to provide a mechanism for process flow exchange between different companies by means of standard interfaces that is rather difficult for organization.

5.7.5.4 SMART-LEES Project

As known, not only the military are interested in the possibility to create new IC based on AmBv materials and silicon. In early 2012, governments of Singapore and the USA founded a special group that worked in SMART-LEES Project on the similar task in the commercial field. The main organizations are such highly respected program members as MTI (USA), National University of Singapore (Singapore), and Nanyang Technological University (Singapore). Partners for silicon production are such renowned companies as Global Foundries (Singapore),

Japan company Tower Semiconductor-Panasonic, and others: Aixtron (Germany), IQE (the United Kingdom), EVG (Austria), and Samco (Japan). The program initiators assumed that it was rather difficult in technological and economic considerations to include their new types of devices into the basic process of silicon production of IC. That's why it was made a decision to develop IC with devices based on AIIIBV materials at first separately, and later to integrate them into CMOS process that has already started in the processing line (so-called, a modular method of processing). In this case, the following order of cooperation was established. Developers are given the preprepared silicon plates 200 mm in diameter at the silicon plant. They develop silicon circuits using the standard CMOS-circuit design techniques at their own plant intended for production of plates 200 mm in diameter. Here, the epitaxial layer of compound semiconductors was applied to the plates and in such a way the basic engineering substrate with all the required dielectric layers and contact pads was manufactured. Then they returned to the silicon plant where the final processes of connection of silicon and AmBv-devices inside the crystal were performed.

If in 2012 the development team started its work from application of GaN, GaAs, and InP materials to the silicon substrate, then since 2014 InGaAs and GaN have been used in production; and they serve as the base for transistors InGaAs HEMT, GaN HEMT, and light-emitting devices GaN LED. Such companies as IQE (the United Kingdom), Lumileds (the Netherlands), and Analog Devices (the USA) joined the project in 2015. Due to serious financial support of this program, the first test chips have already been produced in 2016. Developers consider that one more important purpose of this program is to create a set of means for end-to-end design of IC on AmBv and CMOS devices, as well as commercial market launch of new types of IC.

Currently, it is possible to highlight two main technological directions of heterogeneous integration: the method of micrometric assembly of ready CS-devices based on silicon substrate (DARPA, the USA) and the method of production of CS-devices directly on Si-substrate with subsequent final processing in the standard silicon production. Although both technologies are technologies of dual-use application, each of them has its own primary objectives: these are military and space applications in the USA; but these are space applications in Singapore.

As well, the European countries do research of technologies of compound semiconductors and silicon. Thus, in the United Kingdom, the area for development of compound semiconductors is considered to be one of the most important areas determining the perspective industrial and economic leadership of the country [66, 67]. So, in 2015 the scientific and industrial cluster was established at the premises of the Institute for Compound Semiconductors at the Cardiff University and IQE company, the largest manufacturer of semiconductor plates. The scientific research in this cluster is combined with production [66, 67]. It is planned to develop the commercial SHF- and power devices using the epitaxial wafers GaN-on-silicon (GaN/Si) 150 mm in diameter, and develop the photonic devices based on GaAs- and InP-plates of the same diameter which are perspective for military applications. The large semiconductor clusters which exist in Europe, apply silicon technologies;

but the newly established cluster will be the first one specializing in compound semiconductors. It is hoped that its establishment will initiate a process of return movement of high-tech manufacturing from competitors from East Asia to Europe.

5.8 Methodological Features of Design of Radio Frequency Chips

Simulation is the key element of design process of radio frequency chips for high-speed electronic devices. Here, two primary approaches are applied: simulation of radio frequency VLSI as a part of radioelectronic system under design; and simulation of analog/radio frequency VLSI as a part of existing radioelectronic system.

The end-to-end simulation of element base, modules, and systems is a means of presentation of development results which is generally accepted in the global electronics industry; and it ensures vertical integration and compatibility at different levels. Here, the prime tool is a model (of component, device, system). For fast, exact, and unambiguous usage of the results of design of lower level, each developed component, block, or system should be described by a model which can be used when designing a system of higher level. The model involves the description of functions, parameters, and behavior of the described element (component, system) when exposed to signals, temperatures, radiation, and other factors. Spice and VHDL descriptions serve as an example of commonly used formats of models compatible to many CAD-tools.

Vertical integration and high efficiency of use of ECB development results can be ensured if such model formats are selected that:

- are widespread and supported by many existing CAD-tools;
- allow to simulate devices (components, systems) of different degree of complexity;
- allow all types of the description—text, schematic, physical, and etc.;
- can be used for analog, digital, and mixed chips.

Such approach to the vertically integrated design based on models is not only standard now, but also the only one possible in the modern conditions of the electronic equipment market.

In case when the system is at the stage of mathematical simulation, the model of the existing large-scale integrated circuit can be built in the system model that reflects the principle of vertical integration. Algorithmic (VHDL, Verilog A) or in-circuit approaches, can be used to describe model along with complex simulation system Agilent ADS or others.

When the high-performance electronic system under design is at the stage of physical breadboarding, the LSI model is usually described in the ADS system (structural, schematic, physical, VHDL methods, and their combinations can be used for the model description). If this LSI is made by the third-party developer or

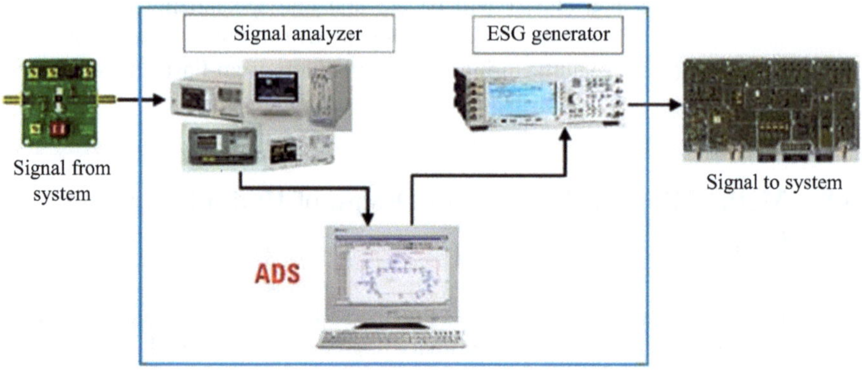

Fig. 5.137 Example of connection of ADS model to the physical model of radioelectronic system

vendor, then its description can be provided by the developer. Signals of the existing physical electronic system (a system prototype) which should arrive at LSI inputs are supplied to the analyzer, and then the digital data stream from its output is entered into the LSI computer model. The digital data stream from the output of computer model arrives at the generator; electrical signals are generated at the generator output and then they arrive at the physical system again. Thus, the LSI model is connected to the circuit breaking where the physical LSI should be installed. It allows to perform debugging of the internal structure of LSI as well as to debug the system at the physical level with no need for available LSI produced (Fig. 5.137) [26].

Such approach can be used when the electronic system is at the stage of mathematical simulation and it proves useful in solving two problems such as check of manufactured electronic unit being a part of virtual system, or simulation of the system, a part of units for which already exists.

The device that already physically exists is selected in the system simulated in ADS. The model of a signal in input of this device is downloaded via interface into the generator. The electric signal is supplied from the generator output to the physical device; and then this signal arrives from the device output to the analyzer (Fig. 5.138). The digital data stream being signal mathematical description is downloaded in the model from generator output via the interface. Thus, the physical device (model) is connected to the model of radioelectronic system.

The important stage of the design process is generation of test signal sequences (test vectors) with variable parameters.

This approach should be used when a developer has no standard solution for generation of a test signal with the specified characteristics. The example of this approach is an addition of narrow-band interference with the specified power and spectrum to satellite navigation signal (Fig. 5.139).

Automatic formation of component models based on instrument measurements should be used for fast and exact description of behavior of components and devices that have already been produced (Figs. 5.140 and 5.141).

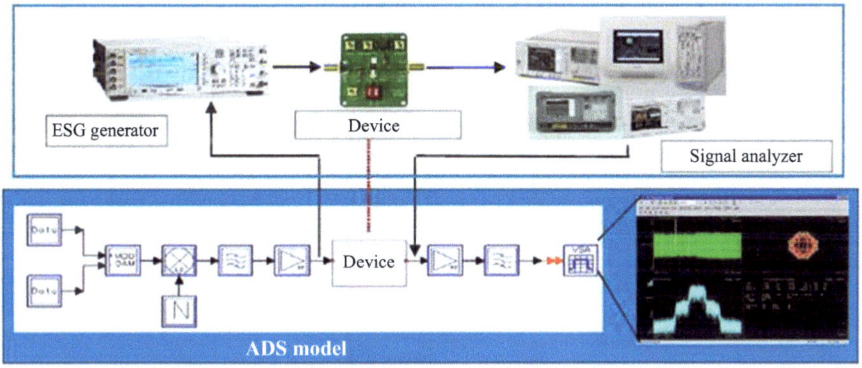

Fig. 5.138 Diagram of the physical device connection to the model of radioelectronic system

Fig. 5.139 The structure of a typical test system with random tuning of test signal

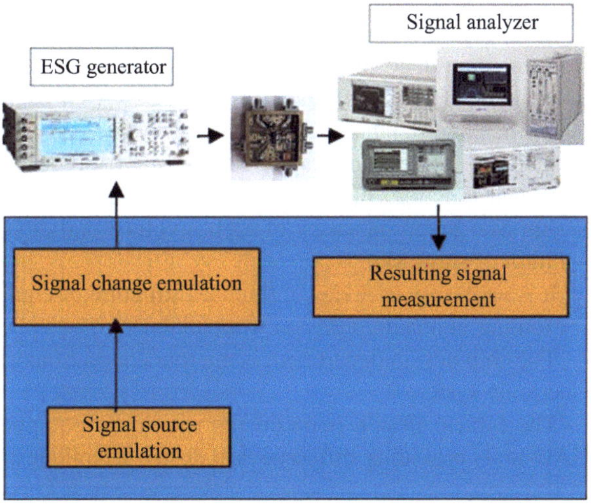

Fig. 5.140 The principle of automatic formation of model of radioelectronic components of high-speed electronic devices

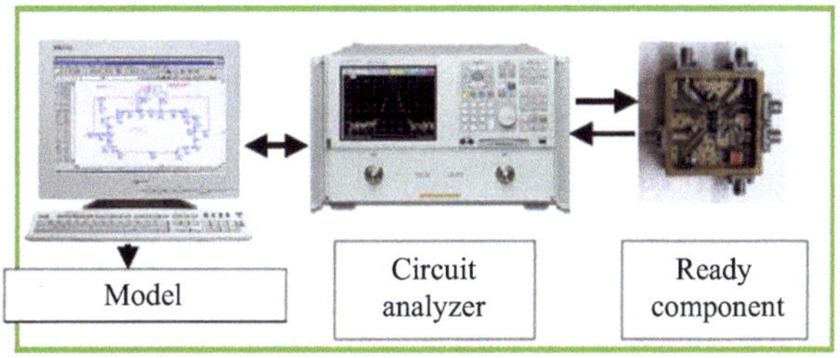

Fig. 5.141 The principle of automatic formation of radioelectronic device model

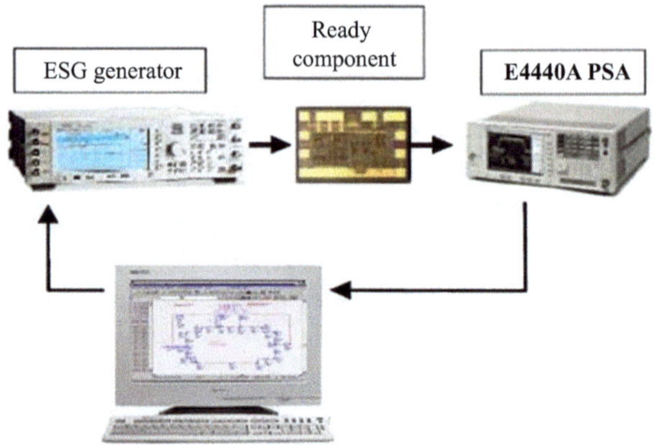

From a practical point of view, the internal information control and exchange systems should be used for efficient monitoring of the project and planning of occupation of separate developer teams, as well as for effective coordination of departments of standard design center. Most objectives of such system are solved by specialized software package such as Microsoft Project Manager. The additional important functions of this project are prediction of design time and calculation of design cost based on salary of each involved participants and on time required for solution of sub-tasks.

It is necessary to review a standard structure of equipment and software required for tooling of standard design workstation

The chip design department is the foundation of any design center for high-speed electronic system

The typical design departments should be equipped with the world's leading CAD-tools providing different chip design techniques:

- semicustom analog-digital integrated circuit based on master matrix crystal;
- semicustom digital integrated circuit based on master matrix crystal;
- digital custom VLSI (ASIC) based on design kit library and IP-blocks;
- analog full custom chips (ASIC) based on design kit library and IP-blocks;
- digital-analog full custom chips (ASIC);
- digital custom VLSI based on FPGA;
- custom HF- and SHF-chips.

Figure 5.142 [26] shows the structure of base content of such typical hardware of High-Speed Chip & Device Design Center.

All computing resources and peripherals of the center should be connected in local area network. The network includes a network server, print server, compute server, application and database server, workstations, personal computers, admin terminals, network devices, peripheral devices for obtaining hard copies of documents, and archiving of information. For cost optimization, some servers can be

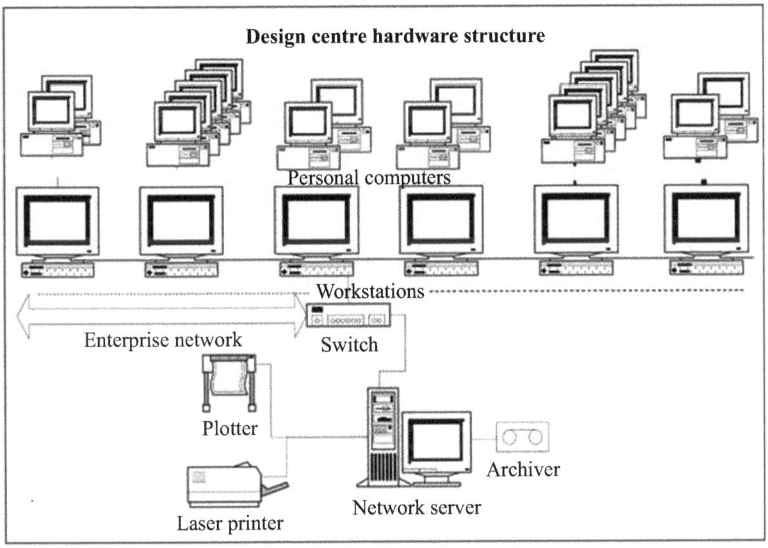

Fig. 5.142 The hardware structure of a typical chip design center

physically localized in one server without compromising the total operating performance. The server features depend on information flows in network and resources required for implementation of mission critical applications such as crystal simulation at the transistor level. Resources for such application essentially depend on software used, quantity of transistors, and design techniques; and they can be rather different in some cases.

When drawing up the detail specifications, specialists optimize not only server features but also the quantity of workstations, personal computers, and terminals. PC workstations can be used either as remote terminals workstations under the operating system HP-UX, or as an independent workstation under the operating system MS Windows or Linux. Increasing of the functional capabilities of CP hardware is possible due to increase in the number of workstations and PC connected to them, as well as due to additional connection and/or replacement of processor boards and other options of workstations.

Installable software (floating licences of software units) is located on the server and downloaded to the specific workplace on developer's command. In addition, these programs can be downloaded to computers with different operating systems.

Instrumental-process simulation software is installed to support the process of development and adoption of new process technologies of production of IC crystals. In this case, it will be necessary to deal with the challenges of adaptation and/or adjustment of element base SPICE-models during modification of process technology.

As a rule, for organization of effective work of design departments, the executive staff of design center install special software for design process management, statistics collection, and quality control. It allows to solve the following task:

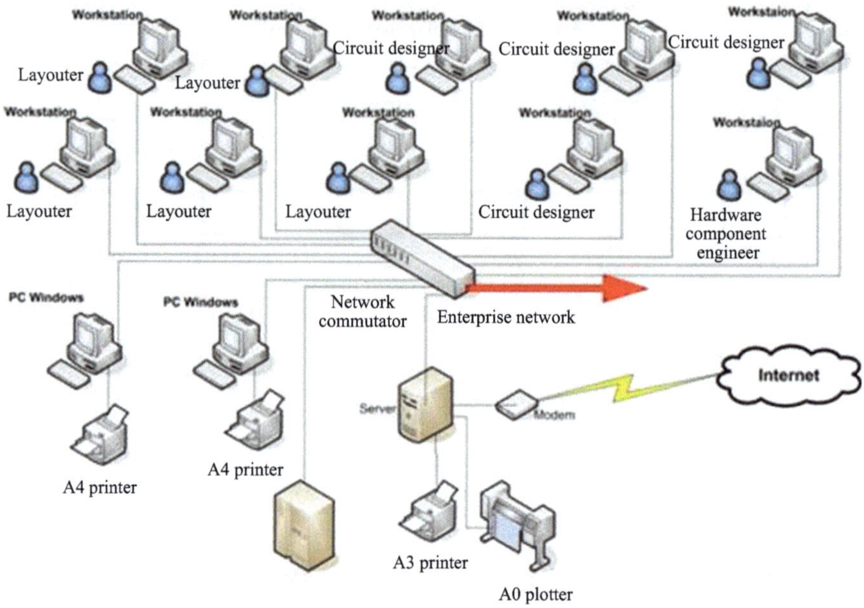

Fig. 5.143 Typical structure of computing resource distribution within the chip design department

- collection of the data on deadlines for completion of separate design stages, resource usage, and amount of work completed;
- design flow tracking with the appropriate recording of objective signs of completion of each procedure;
- quick training of new developers in specific design flow and reporting requirements, as well as transfer of best practices of lead specialists to young specialists.

Distribution of computing resources within a typical department is shown in Fig.5.143 [26].

Measurement and test department of standard design center is equipped with appropriate inspection, measuring and testing equipment having the required software, measuring and testing design hardware, metrology, and validation means.

First of all, these are measuring complexes made by Teradyne, Agilent, Sentry, and others. As well, the department should be equipped with modern probers for measurement of HF and SHF chips made by Cascade Microtech company with full tooling package, as well as with instrumentation kits produced by such companies as Agilent Technologies, Hewlett-Packard, Fluke, DMT, Hart Scientific, Tektronix, Keithley, QualMark, and National Instruments. A part of this equipment will be used in analytical laboratories of the design departments.

It should be noted that the top priority in the instrumental infrastructure of the Measurement and test department is a complex measurement system of IC functional control made by above-said companies. It is offered to use a Flexible Measurement System (a measurement system of resource type) that includes standard measuring

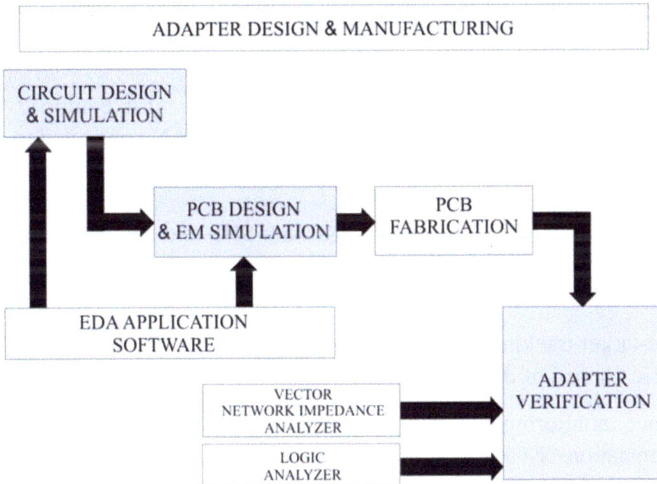

Fig. 5.144 The flow diagram of organization of adapter production and debugging

tools with switching elements. And it is controlled by dedicated software via standardized interface.

Software shell of flexible systems provides fast transfer of measuring programs to the high-performance industrial testers. A wide choice of hardware components (voltage/current sources, generators, measuring tools, electronic loads, switching apparatus, and etc.) controlled by single software shell enables implementation of functional tests of any devices in the wide range of supply voltages and load currents.

Further, it's worth noting the significance of such measurement element as adapter. The key component of qualitative measurements is a test adapter implementing an interface "measured device—measuring system." Figure 5.144 shows the generalized process of production of such adapter and specifies the required standard software and hardware, as well as the process equipment for adapter production.

Two analytical laboratories of typical design departments are usually equipped with the most advanced tools for measurement and analysis of all possible parameters of chips in plates and devices. Such tools include high-precision testers of transistor current voltage characteristics, LCR testers including for HF and SHF ranges, multibeam HF- and SHF-oscilloscopes and spectrum analyzers, transistor characteristic monitoring devices, high-stable voltage sources and wide-band generators, and etc. Two laboratories should be equipped with probe fine manipulators for analyzing of crystals on the plate.

The laboratories should be equipped with such optical equipment as ^{1000}-power microscopes capable of being connected to the computer, taking photos of image and its digitizing.

As well, special equipment is required for research of SHF chips.

Quality control department is equipped with computer equipment and special programs for collection and processing of statistics.

Documentation development and archiving department is equipped with computer equipment and safe information storage devices LAN-connected to the server.

Figure 5.145 shows the main fields of application of SHF chips developed by the center.

The main military technology fields in which SHF and RF chips are used today:

Missile guidance and launch control systems, anti-missile systems.

Radars:

- multiple-target tracking systems built on AESA;
- electronic control of data arrays (AESA).

Aerospace monitoring and control systems.

Communications system:

- satellite communications;
- software defined radio;
- voice recognition and recording systems.

Electronic security systems and electronic countermeasure means:

- high-speed data transmission;
- image output for high-precision displays of different class (military imaging);
- munitions and targeting;
- hyper-spectral imaging.

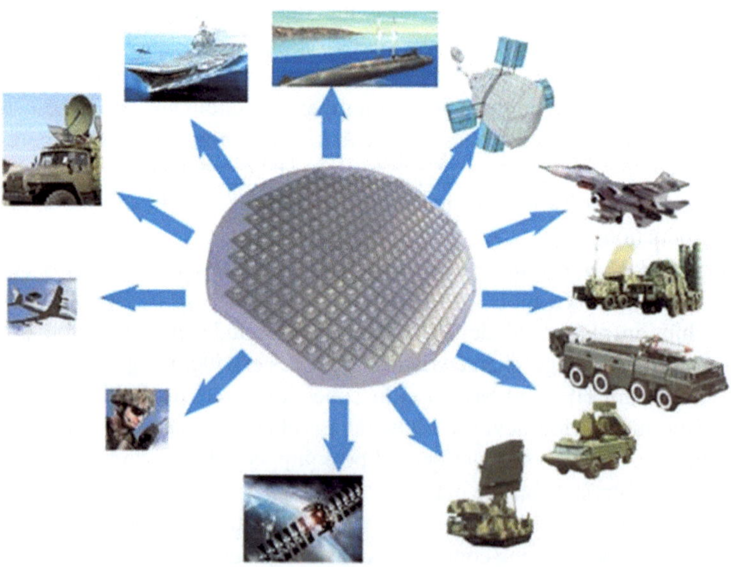

Fig. 5.145 Fields of application of radio frequency chips

Avionics, ground infrastructure, and battle-field:

- individual navigation aids (GLONASS);
- portable enciphered communication systems
- motor system monitoring;
- visualization.

The latest technologies 90, 65, 45 nm in size and smaller based on the most advanced materials of GaN, SiGe, GaAs, Si-on-isolator types, and others should be used to implement chips for the above-listed applications.

References

1. Verulevsky, K. (2013). Semtech Serializers/Deserializers for high-bandwidth data networks. *Components & Technologies*, 3, 75 (in Russian).
2. Goloshchapov, S., & Shakhnovich, I. (2011). 100 GB Ethernet: Basic principles. *The First Mile*, 3 (in Russian).
3. Smirnov, M. (2010). Modern methods of data transmission. *Electronic Components*, 11 (in Russian).
4. SMI4028: 45 to 49 Gbps 16:2 multiplexer/CMU with DQPSK precoder. product brief. www.semtech.com
5. SMI10031:100-113 Gbps 4:10 DQPSK CDR/demultiplexer. Datasheet, 28.10.2011, Rev. 1.8. www.semtech.com
6. Jitter Clean-up PLL. Application note. Rev. 0.2, 17.12.2010.
7. Ioffe, D. (2009). Designing pulse voltage converters based on the SEPIC topology. *Components and Technologies*, 9, 12 (in Russian).
8. *Auxiliary negative* voltage output with SEPIC. *Topic on the Internet forum*, www.electronix.ru/forum/index.php?showtopic=19981&st=0&p=147092&#entry147092 (in Russian).
9. *SEPIC equations and component ratings* (Application Note 1051). Maxim-Dallas. www.maxim-ic.com.
10. Betten J., & Kollman, R. Guess what: Underutilized SEPIC outperforms the flyback topology. *Planet Analog*. www.planetanalog.com/showArticle.jhtml?articleID=165600702
11. High efficiency low-side N-channel controller for switching regulators. LM3488. *National Semiconductor*. www.national.com
12. SMI10021:100-113 Gbps 10:4 Mux/CMU with dual-DQPSK precoder. datasheet, 2012, Rev. 1.0. www.semtech.com
13. Palczynski, J. Versatile low power SEPIC converter accepts wide input voltage range. *Texas Instruments*. www.ti.com
14. Rentiuk, V. (2016). Texas Instruments DC/DC converters based on the DCS-Control™ topology. *Components & Technologies*, 8, 26–34 (in Russian).
15. Golyshev, V. (2016). MORNSUN's components *for* building RS-485 bus. *Electronic Components*, 11, 22–25 (in Russian).
16. John, K., Robbie, S., & Manish, S. (2016). 60 GHz millimeter-wave repeater. *Components & Technologies*, 4, 65–68 (in Russian).
17. *Evolutionary and disruptive visions towards ultra high capacity networks*. IWPC, April 2014.
18. Demidov, D., Krasnobrov, D., & Odnolko, A. (2015). Solutions for LVDS and DVI high-speed interfaces. *Electronic Components*, 3, 48–52 (in Russian).
19. Turkin, A. N. (2011). Gallium nitride (*GaN*) as one of the most promising materials in modern optoelectronics. *Components & Technologies*, 5, 6–10 (in Russian).

20. Balakirev, A., & Turkin A. (2015). Development of gallium nitride technology and the prospects for its application in SHF electronics. *Modern Electronics*, 4, 28–32 (in Russian).
21. Yunovich, A. E. (1996). LEDs based on hetero-structures of GaN and its solid solutions. *The Lighting Journal*. 5/6, 2–7 (in Russian).
22. Zolina, K. G., Kudryashov, V. E., Turkin, A. N., & Yunovich, A. E. (1997). The luminescence spectra of blue and green LEDs based on InGaN/AlGaN/GaN quantum-well multilayer heterostructures. *Semiconductors/Physics of the Solid State Journal (FTP)*, *31*(9), 1055–1061 (in Russian).
23. Fedorov, Y. (2011). Wide-gap heterostructures (Al, Ga, In) N and devices based on them for the millimeter wavelengths. *Electronics: NTB*, 2, 92–107 (in Russian).
24. Kishchinsky, A. A. (2009). Solid-state microwave amplifiers on gallium nitride - the status and development prospects. In *Proceedings of the XIX Crimea conference "Microwave and Telecommunication Technologies"*. Sevastopol: Weber (in Russian).
25. Turkin, A. N., & Yunovich, A. E. (2015). 2014 Nobel Prize winners: In physics - I. Akasaki, H. Amano, S. Nakamura. *Nature*, 1, 75–81 (in Russian).
26. Belous, A. I., Merdanov, M. K., & Shvedov, S. V. (2016). *Microwave electronics in radar systems and communications*. Moscow: Technosphera (in Russian).
27. Kishchinsky, A. A. (2009). Solid-state microwave amplifiers on gallium nitride - the status and development prospects. In *Proceedings of the 19th Crimea conference "Microwave and Telecommunication Technologies"*. Sevastopol: Weber (in Russian).
28. GaN Technology for Radars. In *CS MANTECH conference*, April 23rd–26th, 2012, Boston, MA.
29. Pharmicone, G. et al. (2007). High-power microwave LDMOS-transistor technology for L-band radar transmitters and aeronautical applications. *Components & Technologies*, 10 (in Russian).
30. *SELEX GaAs-GaN enabling technologies for microwave*. Sapienza Universita di Roma. 4–5 October, 2012.
31. *Wireless device products 2012–2013*. Sumitomo Electric Europe, Ltd.
32. NITRONEX Corp. (2008). *GaN Essentials™ AN-012 thermal.*
33. *Rad effects in emerging GaN FETs*. NASA. 11 July 2012.
34. Reese, E., Allen, D., Lee, C., & Nguyen, T. (2010). *Triquint_Wideband power amplifier MMICs utilizing GaN on SiC*. 978-1-4244-6057-1/10/$26.00 ©2010 IEEE.
35. Hoshi, S., Okita, H., Morino, Y., & Itoh, M. (2007). Gallium nitride high electron mobility transistor (GaN-HEMT) technology for high gain and highly efficient power amplifiers. *Oki Technical Review*.211 *74*(3).
36. *Gallium nitride (GaN) versus silicon carbide (SiC)*. Microsemi PPG. April 2010.
37. Curtis, D. (2007). Designing multi-stage class C amplifiers for pulsed radar applications. *Modern Electronics*, 1 (in Russian).
38. *SUMITOMO Research & Development*. 2012.
39. Kishchinsky, A. A. (2010). Broadband transistor amplifiers of microwave power - succession of generations. *Electronics: Science, Technology, Business*, 2 (in Russian).
40. Kasu, M. (2010). Diamond field-effect transistors as micro wave power amplifiers. *NIT Technical Review*, *8*, 1–5.
41. Model of a diamond microwave transistor. *Technology and Design in Communications Electronics Equipment*, 6, 2011 (in Russian).
42. Nippon Telegraph and Telephone Corporation. (2003). Diamond semiconductors operate at highest frequency ever-A step closer to diamond devices for communication satellites, broadcasting stations, and radars. *R&D Information*, *1*(7).
43. Dyakonov, V., & Hittite Microwave Corp. (2011). Monolithic Microwave Integrated Circuits (MMICs) for attenuators and amplifiers. *Components & Technologies*, 10, 14–22 (in Russian).
44. Afonskij, V. P., & Dyakonov, V. P. (2011). *Electronic measurements in nanotechnology and microelectronics*. M.: DMK-Press (in Russian).
45. Dyakonov, V. P. (2011). Picosecond digital monolithic microcircuits of Hittite Microwave Corporation. *Components & Technologies*, 9 (in Russian).

46. Mikhalev, P. (2006). Analog Devices' PLL-circuits and synthesizers on their basis. *Components & Technologies*, 4 (in Russian).
47. Vance, A. (2015). *The semiconductor revolutionary*. Bloomberg Business.
48. Fedorov, Y. (2011). Wide-gap heterostructures (Al, Ga, In) N and devices based on them for the millimeter wavelength range. *Electronics NTB*, 2 (in Russian).
49. Lidow, A., & Strydom, J. (2012). *WP001 gallium nitride (GaN) technology overview*. EPC.
50. Lidow, A. (2011). *AN001 is it the end of the road for silicon in power conversion?* EPC.
51. Lidow, A., & Reusch, D. (2014). *AN017 fourth generation eGaN® FETs widen the performance gap with the aging MOSFET?* EPC.
52. Lidow, A., Reusch, D., & Strydom, J. (2015). *AN018 GaN integration for higher DC-DC efficiency and power density*. EPC.
53. Mehta, N. (2014). *Design considerations for LM5113 advanced GaN FET driver during high-frequency operation* (Application Report SNVA723). Texas Instruments.
54. Vikulov, I. (2016). Heterogeneous integration as a new stage in the development of integrated microwave electronics. *Electronics: NTB Scientific and Technical Journal*, 1, 104–112 (in Russian).
55. Vikulov, I., & Kichaeva, N. (2007). GaAs Technology of Monolithic Microwave Integrated Circuits in foreign military equipment. *Electronics: NTB*, 2, 56–61 (in Russian).
56. Vikulov, I., & Kichaeva, N. (2007). GaN-technology: A new stage in the development of microwave circuits. *Electronics: NTB*, 4, 80–85 (in Russian).
57. Vikulov, I. (2009). WBGS-RF program. Phase II: Results and intentions. *Electronics: NTB*, 8, 62–65 (in Russian).
58. Vikulov, I. (2012). Monolithic microwave integrated circuits: AESA technique *Electronics: NTB*, 7, 60–73 (in Russian).
59. Vikulov, I. (2015). Microwave electronics today: Directions and challenges. *Electronics: NTB*, 3, 64–72 (in Russian).
60. Rosker, M. J. (2008). The DARPA compound semiconductors on silicon (COSMOS) program. In *CS MANTECH conference*.
61. Rosker, M. J., et al. (2010). DARPA's GaN technology trust. In *IEEE microwave symposium technical digest* (p. 1214).
62. Green, D. S., et al. (2014). The DARPA diverse accessible heterogeneous integration (DAHI) program: Status and future directions. In *CS MANTECH conference*.
63. Green, D. S., et al. (2015). Compound semiconductor technology for modern RF modules: Status and future directions. In *CS MANTECH conference*.
64. Green, D. S., et al. (2015). Heterogeneous integration for revolutionary microwave circuits at DARPA. *Microwave Journal*, *58*(6), 22.
65. *Delivering the future*. http://www.compoundsemiconductor.net/article/98066-delivering-the-future.html
66. *CSC formally launched as first compound semiconductor cluster*. http://www.semiconductor-today.com/news_items/2015/nov/csc_201115.shtml
67. *IQE: Semiconductor cluster ambition*. http://www.compoundsemiconductor.net/article/98247-iqe-semiconductor-cluster-ambition.html

Chapter 6
Methods and Means of Ensuring Interference Resistance of High-Speed Electronic Devices

One of the most important requirements for electronic devices and systems is to ensure their high reliability. In a broad sense, the term "reliability" describes the ability of a device to perform all the required functions for a long time without failures and without errors.

For up-to-date high-speed electronic devices, to meet this requirement is a very complex problem. Indeed, short fronts of the processed and transmitted signals lead to the emergence of many undesirable effects, including distortion of the forms of these signals and the generation of various interferences, which ultimately can lead to various kinds of errors and failures. In order to avoid these undesirable phenomena, a designer at the very early design stage should apply special technical solutions, which exclude the occurrence of such failures and errors. But for this you need to very well understand the physical essence of these phenomena.

Thus, in Sects. 6.1 and 6.2, the character of electromagnetic interference is analyzed in detail, and an explanation of electromagnetic compatibility of electronic devices is given along with the means of ensuring it.

Section 6.3 is dedicated to methods of ensuring electromagnetic compatibility of microprocessor counter units of high-speed electronic devices.

In Sect. 6.4, methods of reducing the level of intermodulation distortion in high-speed electronic devices are examined.

Section 6.5 is dedicated to protective SHF devices for telecommunication systems.

Section 6.6 contains an overview of special components for filtering and suppressing electromagnetic interference in telecommunication systems.

In separate Sect. 6.7, for the first time in the literature, by theoretical and practical examples we have demonstrated the need for taking into account the reliability of mechanical components when calculating the reliability of the entire system, in addition to the reliability of electronic equipment.

A. Belous, V. Saladukha, *High-Speed Digital System Design*,
https://doi.org/10.1007/978-3-030-25409-4_6

6.1 Electromagnetic Compatibility of Electronic Devices: Terms, Definitions, Classification

6.1.1 The Nature of Electromagnetic Interference

In general, the term "electro-magnetic compatibility (EMC)" refers to the ability of equipment to work correctly in its electromagnetic environment without affecting this environment, to which other electric devices also belong, in an unacceptable manner.

Currently, strong emphasis is being put on the issues of ensuring electromagnetic compatibility (EMC) of high-speed electronic devices and modules with their individual assemblies and components. There is an increase in requirements for further improvement of EMC characteristics due to constant widening of scope of electronic device application. System solutions based on microelectronics and semiconductor electronics are used in all industries, household, and transport. Currently, the evaluation of products regarding to EMC is needed even more than at earlier stages of electronics development. The basic concepts of electromagnetic compatibility consider the effects of both radiated and conductive interferences (the so-called crosstalks) propagated through conductors (for example, crosstalk on power supply circuits), as well as the sensitivity of electrical equipment to the effects of interference (immunity). In this case, the electromagnetic compatibility characteristics can be determined in the frequency range 0–400 GHz. Interrelation of basic electromagnetic compatibility concepts is given in Fig. 6.1.

Electromagnetic interferences arise from natural phenomena or as a result of technical processes. Examples of natural interferences include atmospheric disturbances (electromagnetic impulses are generated during lightning strike) or electrostatic discharges (ESD). The latter are particularly important for semiconductor electronics. Both in industrial equipment and in electronic modules, switching processes in electrical circuits associated with a very rapid change in currents and voltages are the main source of noise.

And this, in turn, leads to the occurrence of electromagnetic noise, either periodic or random. The effects of these interferences can be of conductive nature (as current or voltage crosstalks in conductors) or of radiative nature (under influence of alternating magnetic field).

Type of conducted interference is called as symmetric or differential mode interference when the current induced in conductors has a sign, i.e., it flows with the same amplitude both in forward and backward directions. If the interference current is grounded or it flows in one direction, then such interference is called the asymmetric or common-mode interference.

Electromagnetic coupling between interference source and receiver can result from:

- galvanic coupling (the most commonly encountered case) that creates differential mode interferences;

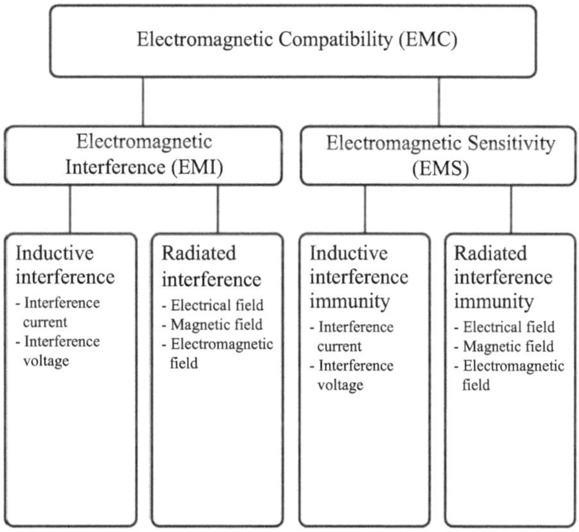

Fig. 6.1 Various aspects of electromagnetic compatibility [1]

- capacitance coupling induced by alternating electric field affecting stray capacitances;
- inductive coupling caused by presence of conductor through which the current is flowing, in the alternating electric field;
- electromagnetic coupling that can be of conductive nature (occurs as crosstalk between conductors inside cable bundles or between conducting tracks of the printed circuit board); or it is propagated by radiation (if the gap width between the interference source and receiver exceeds 0.1 of radiation wavelength λ).

6.1.2 Interference Signal Representation Forms

Interferences that are of periodic or random nature in certain time interval can be mathematically represented as overlapping of sinusoidal and cosinusoidal signals of different frequency and amplitude. Typical signals, noise and their spectral representations are shown in Fig. 6.2 [1].

Semiconductor switch contacts, logical integrated circuits, microcontrollers are the sources of broadband interferences caused by internal processes in these devices and by operation of clock-pulse generators. Signals of the specified interferences are of periodic nature and they can be decomposed in frequency spectrum using Fourier transformation. Approximating expressions that can be used when calculating the amplitudes and break frequencies of trapezoidal waveforms are given below in this section. The example of waveform is given in Fig. 6.3; its relevant approximating spectral envelope is shown in Fig. 6.4. Graph of spectral dependence of signal amplitude on frequency parallels the frequency axis from the base frequency of a signal to the first break frequency f_{g1}. The amplitude decreases with a slope of

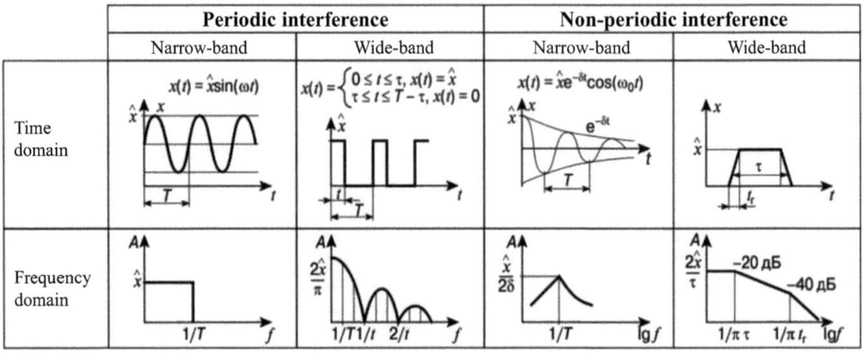

Fig. 6.2 Illustrations of different interference signals in the frequency and time domains

Fig. 6.3 Signal switch-over process represented as a trapezoidal function

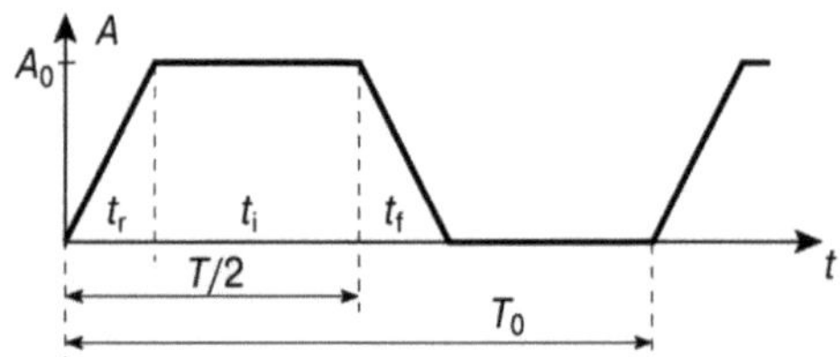

Fig. 6.4 Spectral characteristic envelope shape

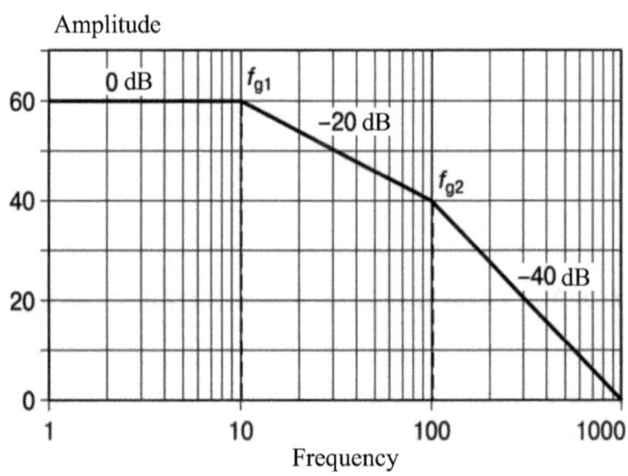

20 dB/decade in the spectral region between the first and second (f_{g2}) break frequencies, and with a slope of 40 dB/decade in the last region.

The following notations are used in the formulae below:

- A_0—original signal amplitude;
- A_n—n-th harmonic amplitude;
- $TO\_$—signal fundamental frequency period;
- t_i—pulse length;
- t_S—switch-over time ($t_r = t_f$);
- n—harmonic number of the fundamental frequency;
- n_{g1}—harmonic number of the first break frequency;

- n_{g2}—harmonic number of the second break frequency;
- f_0—fundamental frequency;
- f_{g1}—the first break frequency;
- f_{g2}—the second break frequency.

The signal amplitude doesn't depend on the frequency [1] in the first spectral region within the frequency range f_0–f_{g1}:

$$A_n \approx \frac{2 \cdot A_0 \cdot t_i}{T_0}. \tag{6.1}$$

In the second spectral region within the frequency range f_{g1}–f_{g2}, the signal amplitude falls with a slope of 20 dB/decade:

$$f_{g1} = \frac{1}{\pi \cdot t_i}, \tag{6.2}$$

$$A_n \approx A_0 \cdot \frac{2}{\pi} \cdot \frac{1}{n}. \tag{6.3}$$

In the third spectral region (frequency is higher than f_{g2}), the signal amplitude falls with a slope of 40 dB/decade:

$$f_{g2} = \frac{1}{\pi \cdot t_s}, \tag{6.4}$$

$$A_n \approx A_0 \cdot \frac{2}{\pi^2} \cdot \frac{1}{n^2} \cdot \frac{T_0}{t_s}. \tag{6.5}$$

6.1.3 Measurement of Electromagnetic Radiation

Measurements of electromagnetic radiation are performed in the frequency domain using a test pattern receiver or spectrum analyzer. These devices allow for evaluation of various interference characteristics by measuring their peak, quasi-peak, average, or effective values. In this case, measurements are performed in relation to the defined level providing for appropriate indication when measuring physical values from μV to V. The device readings correspond to logarithmic ratio of signal amplitude to reference value (this reference value is commonly referred in communication engineering to as effective power or interference power). The power level is determined as common logarithm of the measured signal power to actual power ratio multiplied by ten. It is measured in decibels [1]:

$$p = 10 \cdot \log_{10} \frac{P_X}{P_0}, \ \text{dB}. \tag{6.6}$$

In communication engineering, it is accepted that the reference (actual) power P_0 is 1 mW. To emphasize this, letter "m" is added to the power level UOM (dBm, i.e., decibels are measured relative to 1 mW). The level of voltage or field strength received during measurement of electromagnetic radiation has dimension dBμV or dBμV/m, respectively, and the formula for it can be determined from the expression for power level. The reference value of 1 μV corresponds to the value of 0 dBμV. Therefore, the measured power level value of 0 dBm corresponds to the voltage level of 107 dBμV for systems with 50 Ohm characteristic impedance.

Voltage level formula [1]:

$$p = 20 \cdot \log \frac{U_X}{U_0}, \mathrm{dB\mu V}. \tag{6.7}$$

The sensitivity of test receivers of electromagnetic interferences depends on the operating frequency which is determined according to the requirements of CISPR16 standard. Frequency is set using different bandpass filters, depending on the required range of operating frequencies as is specified in Table 6.1.

If the reception channel of the receiver "captures" only one harmonic of interference signal, then the measured value doesn't depend on the transmission bandwidth (B_W) and the indicator characteristics. Such interference is called narrowband. On the contrary, the broadband interference occurs when the signal fundamental frequency is low; therefore, there are several signal harmonics at once in the frequency band of the test receiver, and the measured value depends on the receiver transmission bandwidth (B_W) and quality of signal harmonic analysis. In this situation, the "integrated" procedure for signal measurement is inapplicable since it doesn't reflect its actual amplitude–frequency characteristic (AFC). Such interference occurs when measuring the broadband signals at the boundaries between operating frequency ranges of the test receiver (0.15 and 30 MHz) where step changes of their amplitude happen.

6.1.4 EMC Terms, Definitions, and Classification

Electromagnetic compatibility of radioelectronic systems (RES) is the ability of these devices to operate simultaneously under actual operating conditions, when exposed to spontaneous electromagnetic interference (SEI), without generating any intolerable electromagnetic interference (IEI) for other devices.

Table 6.1 Selecting transition bands for measuring bandpass filters according to the operating frequency range

Bandwidth (B_W)	200 Hz	9 kHz	120 kHz
Frequency range	0.01–0.15 MHz	0.15–30 MHz	30–1000 MHz

The scope of EMC issues under consideration includes almost all the radio frequency resource (RFR) that is divided into the following bands according to the standards of International Radio Consultative Committee:

- 9–535 kHz;
- 535–1605 kHz;
- 1605–4000 kHz;
- 4–29.7 MHz;
- 29.7–100 MHz;
- 100–470 MHz;
- 470–2450 MHz;
- 2450–10,500 MHz;
- 10,500–40,000 MHz;
- 40,000–275,000 MHz;

Technical characteristics of any RES can be divided into two large groups:

- *Functional characteristics* such as radio-transmitted power and receiver sensitivity;
- *The characteristics influencing on the EMC level.* For example, the extraneous emission power and the receiver susceptiveness on the side channel.

Specialists distinguish the following EMC levels:

- Intra-hardware EMC (inside the unit, node, etc.);
- Intrasystem EMC (inside the radioelectronic complex);
- Intersystem EMC (between different systems and complexes).

Practically, RES developers often use the term "electromagnetic environment (EME)," meaning determination of real conditions under which the specific product operates or will operate with or without the desired signal in input during SEI action through this input or apart from it. According to the three above-mentioned EMC levels, three types of EME are determined, respectively.

Methods and ways of ensuring EMC are divided into the following groups:

- *Organizational* measures generally belong to the level of intersystem EMC and include the rational choice of radio channel frequencies for different radio services, as well as determination of location for devices, etc.
- *Technical* measures are divided into design-engineering and circuit-based ones. Design-engineering measures are implemented mainly at intra-system and intra-hardware levels of EMC provision. EMC provision requirements should be included in the design specification and in the subsequent production specifications during production and operation.

For a deeper understanding of the essence of the material presented in this chapter, it is necessary to define a few terms commonly used in this field.

Electromagnetic interference is an electromagnetic, electric, or magnetic process generated by any source in space or by conducting medium and influencing on operations with a desired signal in radioelectronic equipment.

Electromagnetic interference (EMI) generated in free electromagnetic field is called *radiated interference*; EMI generated in conducting medium is called *conductive* (or inductive) interference.

Spontaneous electromagnetic interferences (SEIs) are caused by random processes occurring in the interference sources; that's why they are of probabilistic nature and described by statistical tools.

Receptor is any engineering device reacting to an electromagnetic desired signal or to IEI.

Electromagnetic environment (*EME*) is the totality of electromagnetic, electric, magnetic fields, currents, interference voltages and signals in the defined region of space which influences or can influence on the receptor operation.

EME is quantitatively defined by EMC characteristics of the receptor.

Radio frequency resource is the totality of radio electromagnetic fields that is possible to use for data communication or energy transmit and receive.

Receptor sensibility is a metric for reaction to the external IEI with or without desired signal.

Sensibility threshold is the maximum allowed SEI level when the receptor works correctly.

Receptor immunity is a feature to resist to external and internal EMI due to the choice of the desired signal structure and receptor design concept.

Interference immunity of the receptor is a feature to resist to external and internal EMI due to circuit and design means which are not impairing the selected structure of the desired signal and receptor design concept.

Shield is a structural element of radioelectronic equipment in the form of metal grounded cover with high electrical or magnetic conductivity. It serves for attenuation of EMI in certain region of space within the wide range of frequencies.

Shielding is a way of EMI attenuation using metal cover (shield) with high electrical or magnetic conductivity.

Filtration is attenuation of voltage and interference current by using an electric circuit (filter) damping in the specified ranges and in the specified frequency bands.

Grounding is an electric circuit that possesses the property of preserving the minimum (ground) potential.

There are two levels of a hierarchy in the EMI classification: EMI classification by type and by kind. EMI classification by type is presented in Fig. 6.5: radiostation, industrial, and natural interferences. In its turn, each of this types has its own "subclasses," first of all "radiated" and "conductive," etc.

The following terms are accepted in communication engineering for the most commonly occurring EMI:

So, *radiostation* interferences are generated by antenna of radio transmitting center.

Industrial interferences are generated by electrotechnical, electronic, and radioelectronic household appliances, industrial, medical, and scientific devices and equipment, etc.

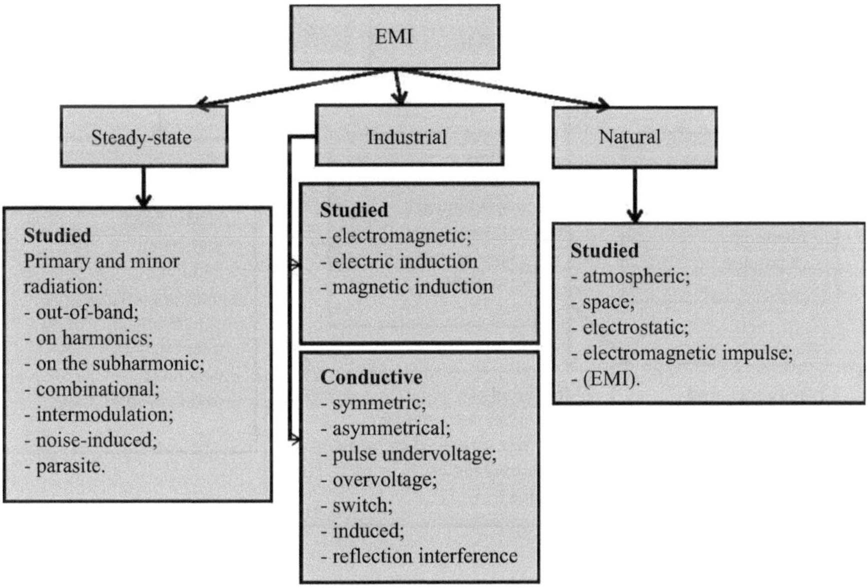

Fig. 6.5 EMI classification by type

In view of the diversity of interference sources, the industrial interferences are the most common type; it covers a wide range of frequencies (from tens of Hertz to a few gigahertz).

Natural interferences are caused by natural physical processes such as electromagnetic radiation (space and atmospheric noise, relict radiation, solar radio emission, atmospherics, electrostatic fields of different atmospheric formations, aircrafts, etc.).

Contact interferences are caused by re-emission from current-carrying mechanical contacts with non-linear current conductivity when these contacts are exposed to radiation from rather high-power radio transmitter.

These interferences are typical for moving objects (ships, tanks, airplanes, cars, etc.), and the level of such interferences grows with speeding-up of the object.

EMI classification by type [1] presented in Fig. 6.5 is considered conventional.

Second level classification of EMI—by their kind—is presented in Fig. 6.6.

So, interferences differ in frequency and spectrum (H4, B4, CB4, sinusoidal, impulse, modulated, noise, impulse-noise interferences), in time (constant, continuous, short, regular interferences), in effect level (permissible, intolerable, crosstalk, blocking interferences), in interference relation to the receptor (narrowband, wideband, additive, multiplicative, external, internal, coherent, non-coherent interferences), radiated (from one source, a few sources, contact, intermodulation interferences).

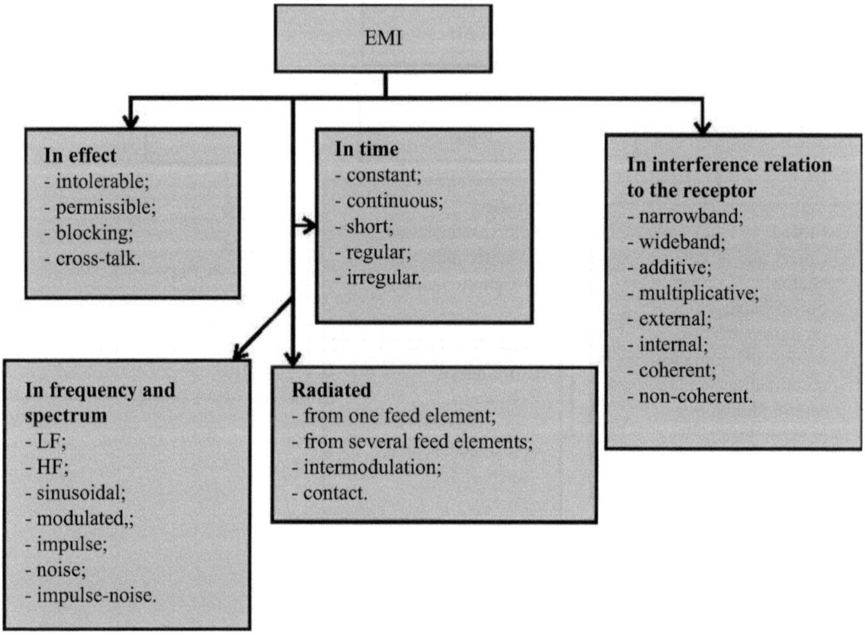

Fig. 6.6 EMI classification by kind

Basic methods (means) of achieving EMC are presented in Fig. 6.7. These methods include shielding, filtering of interferences, electrical grounding, choice of special wired-logic connections and relevant element-component base, lightning protection (high-power electromagnetic pulse). Methods and means of RES protection against SHF-emissions and EMI are detailed below.

6.1.5 Regulations and Standards for Electromagnetic Compatibility

There are many regulations and requirements for electromagnetic compatibility of equipment [2]. They are divided into standards regulating the measuring equipment specifications, test system parameters, and measurement procedure for interferences of different nature. Defining a procedure to test the electric devices for electromagnetic compatibility, these regulations prescribe the criteria which can provide the grounds for conclusion that the tested devices meet EMC requirements.

Work on standardization of requirements for electromagnetic compatibility is carried out at the international, European, and national levels. ISO (International Organization for Standardization), IEC (International Electrotechnical Commission), and CISPR (International Special Committee on Radio Interference) share the bulk of the burden at the global level. At the European level, this work is

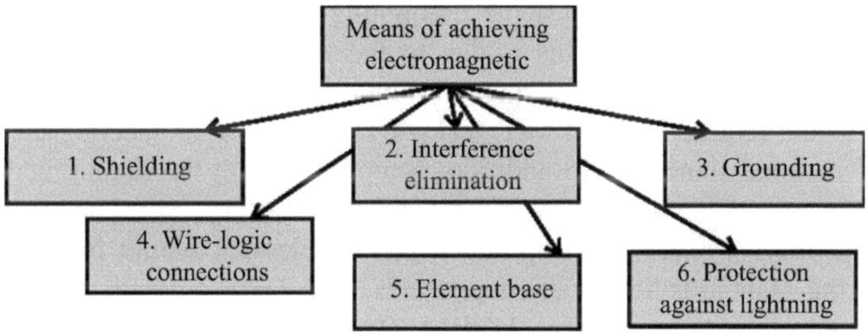

Fig. 6.7 Methods of achieving electromagnetic compatibility

performed by CEN (European Committee for Standardization) and CENELEC (European Committee for Electrical Standardization) and ETSI (European Telecommunication Standards Institute).

Semiconductor integrated chips are relatively new object of EMC-standardization that requires adoption of special regulations applicable only to these devices. Electromagnetic compatibility requirements for them are almost identical to those for other devices and components. However, individual IC components are rarely used in one field only. The IEC developed two groups of regulations, which standardize the procedures for measuring electromagnetic emissions (IEC 61967) and achieving IC electromagnetic immunity (IEC 62132).

The peculiarity of EMC is an unintentional nature of generated and analyzed interferences. In addition, characteristics of interferences and effects are prescribed for strictly specified conditions of operation of vehicles. For these reasons domestic and international practice in this field is characterized by a high level of regulation of interference measurement techniques and immunity testing. The introduction of Technical Regulations of the Customs Union TP TC 020/2011 "Electromagnetic Compatibility of Technical Equipment" eloquently emphasizes the importance of such regulation, as well as of achieving EMC in a modern environment saturated with electric, electronic devices, and radio aids. The Regulations apply to civilian products. Naturally, the described interference and immunity criteria of the Customs Union, as well as the assessment methods included in the Regulations are similar for products of defense and may other industries, such as aviation and automotive industries and communications industry. It is obvious that having own specific additional EMC requirements, the military standards will be stricter than regulations applied to the equipment of non-military nature (Tables 6.2, 6.3, and 6.4).

Trend analysis of changes in national defense industry standards shows that they are mainly based on similar foreign standards and often function as their analogues. Such situation has led to the fact that almost in every field, the current national standard is the second to latest version of a similar foreign standard. That's why for assessment of current requirements for national defense products, it is necessary to consider and rely on the relevant foreign standards.

Here, the governing standard is MIL STD 461 and MIL STD 462 partially. Despite the fact that standard MIL STD 461 was adopted in the interests of the US Department of Defense, it is applied in different spheres, especially in the area of development of management systems, computing systems, components of aviation and automobile equipment. This is due to the fact that the requirements of the specified document are tougher than that of civil standards; the compliance of vehicles to the requirements of MIL STD 461 is regarded as meeting the requirements of national standards. In addition, compliance with the requirements of the standard ensures the use of vehicles in the defense industry, increasing competitive ability and market share of the product. Nevertheless, it would be beneficial to consider the standards for aviation products (DO-160), as well as the relevant automotive standards specified in UNECE Regulation No. 10. This is due to the fact that dual-use devices often become subject matter of these standards, making the requirements specified therein important for defense industry products in general.

Trend analysis of changes in basic requirements specified in these standards is given in tables. The analysis revealed the considerable similarity of physical nature of tests by showing the possibility to bring together the test methods specified in different regulatory documents. To make the point stronger, MIL STD 461 (version C) will be introduced any time soon. It is assumed that the new version of the standard will include new tests, in particular CS117, RS106, RS108, which are very similar to the tests specified in DO-160. Thus, CS117 test will be similar to the test described in Sect. 22 of DO-160 with some changes:

- it is assumed that immunity testing will be performed only with regard to repeated impact and repeated flash;
- interference cable entry should be the only test method;
- limit the application of this test to airborne electrical equipment.

The RS106 and RS108 tests will be similar to the RS105 tests and Section 23 of RTCA/DO-160, respectively.

The main trends in requirements for EMC of defense products are as follows:

- frequency extension for evaluation of emission up to 18 GHz and stability up to 40 GHz;
- increase the level of complexity of immunity testing: immunity to radiated interference—up to 200 V/m for narrowband interference and up to 7 W/m for impulse interference and for some vehicles.

No significant changes in other tests.

It can be said that three specified standards fully cover the entire list of regulations and methods for estimation of EMC values; and vehicle compliance to the regulations specified in these standards makes it fully satisfying the EMC requirements. The standards offer the alternative methods of estimation that is convenient for manufacturers and crash testers of vehicles.

The majority of EMC requirements are easily fulfilled, provided that the necessary equipment is available. Many products manufactured by different foreign companies (such as SOLAR, AMETEK, TESEQ, Keysight, and others) are currently available in the market. They include universal automatic test oscillators, impulse-noise generators, various interference input devices, measuring antennas, as well as measuring receivers, measuring sensors, or current collectors. Special

Table 6.2 Trends of changes in the requirements of MIL STD 461 standards (designations in frequency ranges: U—voltage, I—current, E—electric field strength, H—magnetic field strength; frequency without dimension, expressed in Hz) [1]

		MIL STD, version/year					
		461 A	461 V	461C	461D	461E	461F
Test	Test description	1968	1980	1986	1993	1999	2007
CE101	Power supply lines, conductive interferences	I:0.03–20 kHz	CE101 I:30 Hz–15 kHz		I:30–10 kHz		
CE102	Signal cables, conductive interferences	I:30–20 kHz			Power supply lines: U: 10 kHz–10 MHz		
CE103	Power supply lines, conductive interferences	I:20 kHz–50 MHz	CE103 I:15 kHz–50 MHz				
CE104	Signal cables, conductive interferences	I:20 kHz–50 MHz					
CE105	Inverse filter method	I:30–50 MHz					
CE106	Antenna inputs	10 kHz–12.4 GHz	10 kHz–26 GHz		10 kHz–40 GHz		
CE107	Power supply lines, pulses, time domain		CE07				
CS101	Power supply lines, conductive LF-interferences	30–50 kHz				U:30–150 kHz	
CS102	Power supply lines	50 kHz–400 MHz					
CS103	Antenna ports, intermodulation	30–10 GHz	15 kHz–10 GHz				
CS104	Antenna ports, undesired signals (2)	30–10 GHz	30–20 GHz				
CS105	Antenna ports, cross-modulation	30–10 GHz	30–20 GHz				
CS106	Impulse interferences, power supply lines	100/10	200/10; 0.5				400/5
CS107	Switching interferences	CS107					
CS108	Antenna ports, undesired signals (1)	30–10 GHz					
CS109	Structural currents		60–100 GHz				

(continued)

Table 6.2 (continued)

		MIL STD, version/year					
		461 A	461 V	461C	461D	461E	461F
Test	Test description	1968	1980	1986	1993	1999	2007
CS110	Damped sinusoids, contacts 10 kHz–100 MHz			CS10			
CS111	Damped oscillating interferences, cables			CS11			
CS114	Conductive high-frequency interferences, current				I:10 kHz–400 MHz	I:10 kHz–200 MHz	
CS115	Conductive pulse interferences, front				2/30 ns, 5 A		
CS116	Damped oscillating interferences, cables				10 kHz–100 MHz (6 frequencies)		
RE101	Magnetic field, emission	30 Hz–30 kHz	30–50 kHz		H:30–100 kHz		
RE102	Electric field, emission	14 kHz–10 GHz			E:10 kHz–18 GHz (per 1 m)		
RE103	Harmonic and parasite signals, antennas	10 kHz–40 GHz					
RE104	Magnetic field, emission	20–50 kHz					
RE105	Automobiles, motors, emission	E:150 kHz–1 GHz					
RE106	Overload of power supply lines	E:14 kHz–1 GHz					
RS101	Magnetic field, stability	30 Hz–30 kHz	30–50 kHz		30–100 kHz		
RS102	Electric/magnetic-field crosstalk	Power frequency magnetic field and impulse		Magnetic field/electric field			
RS103	Electric field, stability	14 kHz–10 GHz		14 kHz–10 GHz (40 GHz)	10 kHz–40 GHz		2 MHz–40 GHz
RS104	Harmonic magnetic field crosstalk	14 kHz–30 MHz					
RS105	Pulsing electromagnetic field			RS05	RS105, 50 kV/m, front 75 ns		

Table 6.3 Principal differences in the latest version of DQ-160D document (by chapters DO 160D) [1]

Chapter number	Test	Purpose and special aspects	Basic amendments to the version
15	Magnetic field	Assessment of influence of the test object on readings of compass or on readings of magnetic-field sensor	Detailed figure illustrates the test set and test methods
16	Input power supply	Immunity to power supply change under normal and emergency condition of the aircraft, as well as distortions and effects on the aircraft power supply system	Requirement for testing the resistance to instant power interruptions and characteristics of interruptions Emergency test of resistance to impulse interference for each AC phase
17	Voltage impulses	Impulse voltage–resistance in AC/DC supply lines, analogue CS 106 MIL STD 46 1F	At least 50 impulses of positive and negative polarity are required
20	Radio frequency interferences (conductive and radiated interferences)	Resistance to conductive and radiated interferences with amplitude and impulse modulation. Tests are, respectively, similar to method MIL STD 46F CS114 for frequencies of 10 kHz–400 MHz using a current probe, and to method MIL STD 46 1F RS103 for frequencies of 100 MHz-18 GHz using semi-anechoic or reverberation chamber	The wording and terms were revised to match the content of section with the requirements of other standards Requirement regarding to the exposure of all front surfaces to radiation Reference to equality of the distance to antennas during calibration and tests Method in the reverberation chamber was changed from continuous rotation mode to discrete rotation of the mixer; new relevant techniques were adopted
21	Emission of radio frequency interferences (conductive and radiated interferences)	Emission of conductive interferences in the connecting cables and power supply lines and of radiated interferences Conductive interferences are measured within the range of 150 kHz–152 MHz using a current	New limit of interferences for high-frequency and navigation receivers was added The frequency was changed from 1 GHz to 960 MHz at which the transmission band ranges from 100 kHz to 1 MHz

(continued)

Table 6.3 (continued)

Chapter number	Test	Purpose and special aspects	Basic amendments to the version
		collector; radiated interferences are measured within the range of 100 MHz-6 GHz in echo-free or reverberation chamber	The option to use a 10-kHz transmission band at frequency points higher than 960 MHz was deleted; and it was noted that the pre-amplifier could be used
22	Impulse interferences induced by lightning	Resistance to impulses induced by lightning in the contact connectors, connecting cables, and supply lines The contact entry (for damage stability) or cable entry (for disruption stability) The section considers the following possible impulses: simple impact, multiple impact, and multiple flash	The method of resistor to determine the source impedance for contact entry was excluded Impulse form 6 for cable entry was added only for multiple flash tests
23	Direct lightning effects	Resistance of the external electrical and electronic equipment to the direct lightning effects. Since these tests result usually in breakage, then the object isn't powered. High voltage is required up to levels of thousands of kilovolts and/or hundreds of amperes	There are no changes as compared to the previous version
24	Electrostatic discharge	Equipment resistance to air electrostatic discharge. This test is intended to determine whether this object can function during or after air electrostatic discharge. Action of 10 impulses of positive and negative polarity of 15 kV	Amendments include the explanation of possibility to use test points, in particular, confirmation that connector contacts are not checked

attention is paid to tests for resistance to high-power interferences which require to install high-power amplifiers and anechoic shielded chambers in conventional configuration that makes them extremely expensive. At the same time, reverberation chambers and field generators (for instance, GTEM cells) are widely used abroad to

Table 6.4 Difference in EMC requirements specified in national and foreign standard for vehicles [1]

Tests	Requirements for parts	
	GOST R 41.10-99 (Rules No. 10, 1997)	UNECE Rules No. 10, 2012
Emission of interferences from vehicle and electronic component assemblies	30–1000 MHz	30–1000 MHz
Vehicle resistance to the radiated interferences	30–1000 MHz frequency range Level of 24 V/m in 90% frequency points, including not less than 20 V/m within the full range of frequencies	20–2000 MHz frequency range Level of 30 V/m in 90% frequency points, including not less than 25 V/m within the full range of frequencies
Resistance of electronic component assemblies to the impulse interferences in supply lines	Absent, the relevant tests are specified in GOST	Impulses 1, 2a/2b, 3a/3b, 4 (third level of rigidity) according to ISO 7637-2 (2004)
Emission of impulse interferences to the power supply lines	–	Section based on ISO 7637-2 (2004) is included
Electrostatic discharge	–	Reference to ISO 10605 is included
Tests specific for hybrid cars	–	Tests for resistance to impulse microsecond, nanosecond interferences, distortions of supply line, etc.

abandon the use of anechoic chambers and high-power amplifiers. So, primary qualification of reverberation chamber TESEQ showed that only 1 W of input power is required for field of 200 V/m in the frequency range up to 18 GHz. Thus, there is an opportunity for significant reduction in cost of estimation of resistance to the radiated interferences.

Another trend in designing advanced hardware and software with EMC in mind is a natural desire of manufacturers and test engineers to develop complex systems, providing for a possibility to perform tests according to several standards simultaneously and in one device. Since when testing of EMC characteristics, the check for resistance is more expensive, this is an area where complexation of test is considered.

Tables 6.5 and 6.6 contain the results of comparing test parameters subdivided into three groups—low-frequency RF conducted interference immunity, high-frequency RF conducted interference immunity, and radiated RF interference immunity. Closeness of frequency ranges and test toughness is obvious. Noticeable similarity of test characteristics is used to form the complex test systems which provide for requirements to perform checks according to several standards simultaneously. The tables specify an approximate combination of suggested assemblies using the example of equipment of famous and popular manufacturers such as TESEQ, Keysight, and others.

Table 6.5 RF conducted interference immunity testing [1]

Documents	Effects (level, range, signal)	Equipment
Low-frequency RF conducted interference immunity		
GOST R 51317.4.16	Up to 15 V, in the range of 15–150 kHz narrowband	NSG 4060 (TESEO) is a dedicated test system consisting of a generator (15–150 kHz), a power amplifier (up to 200 W), and monitoring hardware Coupling–decoupling networks at option (TESEQ) Additional modules (TESEQ) for high-current supply lines
MIL STD 461F, CS101	Up to 6 V, in the range of 30 Hz–150 kHz or from the second harmonic of power supply frequency	
DO-160, section 18	Up to 4 V, in the range from 700 Hz to 32 kHz for AC power and in the range from 10 Hz to 200 kHz for DC power	
UNECE Rules No. 10	Absent	
High-frequency RF conducted interference immunity		
GOST R 51317.4.6	Up to 10 V, in the range from 150 kHz to 80 MHz, with amplitude modulation and increasing peak values	NSG 4070, TESEO is a dedicated test system consisting of a generator (9 kHz—1 GHz), monitoring hardware; three-channel power meter with an optionally built-in power amplifier (up to 75 W) within the full range or in the range from 150 kHz to 230 MHz Power amplifier (TESEQ), 100 W Coupling–decoupling networks (TESEQ) at option depending on type of lines (with line interruption) Current probe (TESEQ) for input of interferences of symmetric or asymmetric currents (without line interruption) Matching attenuator (TESEQ) Accessories and software
MIL STD 461F, CS114	Up to 109 dBμA (300 mA), in the range from 10 kHz to 200 MHz, meander and impulse modulation with maintained peak values	
DO-160, section 20 (clause 1)	Up to 300 mA, in the range from 10 kHz to 200 MHz, meander and impulse modulation with maintained peak values	
UNECE Rules No. 10, ISO 11452-4	Up to 60 (100) mA, in the frequency range of 1–400 MHz, AM with preserving maxima	

Figures 6.8, 6.9, and 6.10 [1] show technical configurations of the unified test units suggested by manufacturers.

Configurations of a test unit based on three amplifiers and two radiating antennas are presented in Fig. 6.10. It is clear that the quantity of amplifiers can be increased depending on the required frequency range of tests and characteristics of antennas. As well, instead of antennas, field generator as GTEM cells or reverberation chambers can be connected. As the result, the configuration universality is observed. Each of illustrated test units is a test system, so it is subjected to validation as test equipment. Validation processes including calibration techniques are described in detail in the relevant standards; and they don't raise any technical problems during their implementation.

Table 6.6 High-frequency radiated interference immunity testing (device housing) [1]

Documents	Effects	Equipment
GOST R 51317.4.3	Up to 10 V/m, in the frequency range of 0.08–6 GHz, AM with increasing maxima	ITS 6006 (TESEO) is a dedicated test system that consists of a generator 80–6000 MHz, monitoring unit, three-channel power meter, commutator, and switching device Power amplifiers up to 6 GHz (TESEQ with power output up to 100 W) Radiating antennas (0.08–6 GHz), GTEM cells (TESEQ, 10 kHz–20 GHz), reverberation chambers (TESEQ, ranging 0.5–18 GHz) Accessories (commutators, couplers, power meters at option) Software applications (general and dedicated to certain requirements and configurations)
UNECE Rules No. 10	Up to 30 V/m (75 V/m for device), in the frequency range of 0.02–2 GHz (1 kHz, 80%), with preserving maxima	
DO-160, section 20 (clause 2)	Up to 200 (300) V/m, in the range of 100 MHz–18 GHz, meander and impulse modulation with preserving maxima	Generator (for example, Keysight N5183A with options up to 20 Hz and modulation) Power amplifiers up to 3 GHz (TESEO with output power up to 1 kW), up to 18 GHz (IPI, with output power up to 100 W in the frequency range up to 18 (40) GHz). Radiating antennas (0.03–40 GHz), GTEM cells (TESEQ, 10 kHz–20 GHz), reverberation chambers (TESEQ, ranging 0.5–18 GHz). Accessories (commutators, couplers, power meters at option) Software applications (general and dedicated to certain requirements and configurations)
MIL STD 461F, RS103	Up to 200 V/m, in the range of 2 MHz–18 (40) GHz, meander and impulse modulation with preserving maxima	

Trend analysis of requirements for EMC shows their similarity in terms of observing regulations and test procedures. So, the complex unified systems for EMC tests can be created. The research of available technical means shows the practical opportunity to implement the complex test systems. This factor is useful to have in mind when organizing and making up a set of test laboratories intended for testing the defense products and technical means of general purpose. As a result, laboratories with a fairly wide scope of accreditation are established.

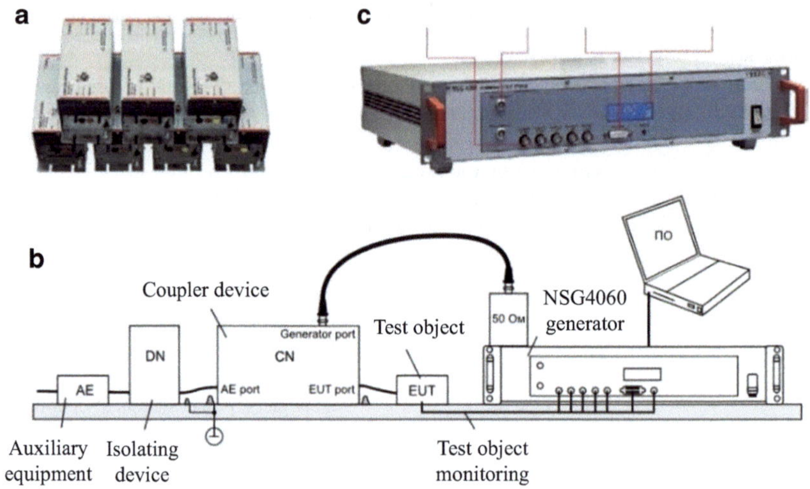

Fig. 6.8 Technical means and configurations for checking low-frequency RF conducted interference immunity: (**a**) generator NSG 4060; (**b**) coupling–decoupling networks; (**c**) test configuration

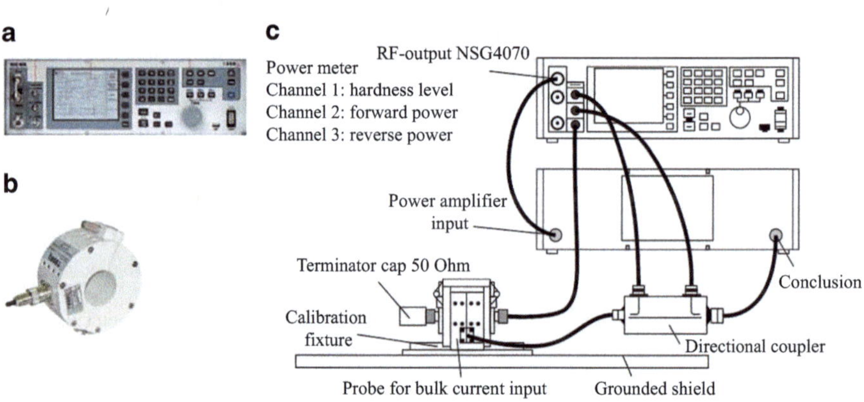

Fig. 6.9 Technical means and configurations for checking high-frequency RF conducted interference immunity: (**a**) generator NSG 4070; (**b**) probe; (**c**) assembly configuration

6.2 Assessment of the Tolerance of the Semiconductor Element Base of High-Speed Electronic Devices to the Effects of Electromagnetic Interference

The issues of protecting communications electronics equipment against the impact of HF and SHF electromagnetic interference (EMI) are especially relevant due to rapid expansion of fields and areas of application of radioelectronic devices with semiconductor devices (SD) and microelectronic integrated circuit (MIC) as their

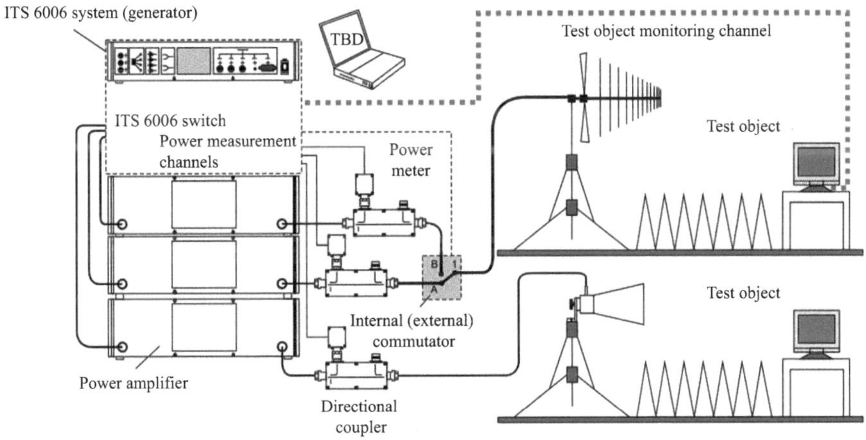

Fig. 6.10 Technical means and configurations for checking radiated RF interference immunity (using radiating antennas)

main components. Key factors attracting particular attention to this issue are the following [3–13]:

- constant increase in power levels of modern radio transmitters up to hundreds of kilowatts in the continuous-wave mode, and up to tens of megawatts in the pulsed mode. In this regard, many radioelectronic devices have to operate in electromagnetic fields, whose power flow density reaches a few tens of kW/m^2 in decimetric and centimetric wavelength ranges (Fig. 6.11);
- increase in the integration level of modern integrated chips. In this case, there is a reduction in levels of voltages and currents consumed. The desired signal energy of microelectronic devices is relatively similar, and sometimes it is even lower than energy of EMI. Furthermore, due to improvement of high-speed performance of modern digital systems and a decrease in operating signal duration down to units and even fractions of nanoseconds, the computing devices themselves can generate HF interferences, the level of which is sufficient for failures to occur.
- constant complication of radioelectronic systems (RES). The task that often faces the developers is to saturate a small-sized unit with radioelectronic equipment of various functional purpose. Inevitable close spacing leads to crosstalk and aggravates a problem of electromagnetic compatibility (EMC);
- enhancement of capabilities of electronic warfare devices. The presence of superpower SHF-signal sources based on gyrotrons, klystrons, magnetrons, lasers, beam-plasma oscillators enables remote target impact on radio equipment in order to put it out of order by damaging its sensitive element base.

Traditional methods of improvement of radioactive equipment immunity are conditionally divided into design-technological, circuit-based, and systematic ones. It's a common practice to start considering carefully the issues on equipment

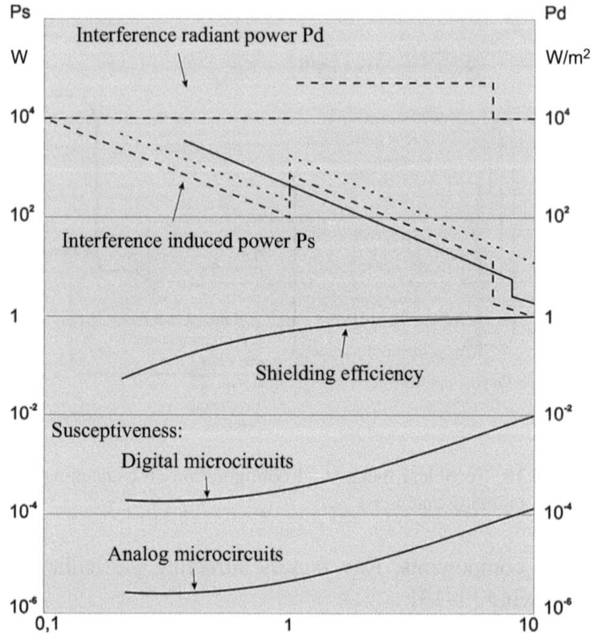

Fig. 6.11 Electromagnetic environment within the operating area of communications electronics equipment: summary of analysis results

protection against interference attack only at the stage of engineering study of the product. That's why the simplest engineering methods are the most commonly used. Here, various shields are most frequently used for protection against interferences. In theory, by the use of ideal continuous shields, it is possible to reach the shielding efficiency of 100 dB and higher. However, in practice, such attenuation is difficult to be obtained, as it was shown above. The presence of various irregularities (holes and cracks for ventilation and indication, leads, joints, covers, etc.) creates additional channels for penetration of electromagnetic fields in the shielded area. The more effective shields are those of special design with holes for leads, ventilation, and indication as below-cutoff waveguide; however in this case, the cost, dimensions, and weight of the device increase significantly. The known circuit-based methods involve the use of various additional components such as filters, chokes, transformers, excess-voltage suppressors, optoelectronic elements to reduce the levels of interferences.

Effective means for immunity improvement are system methods related to the use of various error control codes for correction of system and group errors. However, when interference has an impact not only on the transmission line but also on all controlling electronic devices, processing units, and data contained in them, they prove ineffective.

Therefore, additional means of improving radioactive equipment immunity include identification and consideration of susceptibility level (sensitivity to the EMI impact) of the semiconductor element base, in particular, its discrete semiconductors (SD) and microelectronic integrated circuits (IC).

Systemic approach to the communications electronics equipment design presupposes the need to address the issues of its operational reliability at the initial stage of design, i.e., when selecting an element base and designing PCBs. Since sooner or later induced interference signals get into the circuits of radioelectronic devices, affecting their elements, the immunity of communications systems, radio command systems, and data processing systems largely depends on susceptibility of the element base, in particular, its discrete semiconductors (SD) and microelectronic integrated circuits (IC). The results of studies performed by national and foreign scientists show that in some cases, the choice of an element base which is less susceptible to interference makes it possible to reduce the material and financial expenses on EMC by tens and hundreds of times at the stage of equipment operation (Fig. 6.12) [2, 3].

Susceptibility of discrete semiconductors and microelectronic integrated circuits to interference was studied as early as in the 1970s, due to problems encountered during operation of special equipment. The results of the first studies of RF interference impact on the characteristics and parameters of discrete semiconductors and microelectronic integrated circuits are detailed in [4, 5]. Nowadays, these studies were given a new impetus [6], due to further development of special electronic-countermeasures receivers, as well as due to requirements for safe operation of industrial equipment, transport, communication systems, and household appliances.

When choosing a method to study the susceptibility of semiconductor devices and integrated chips to the EMI impact, one should be guided by the following basic criteria: receiving the maximum information on change of discrete semiconductors and microelectronic integrated circuits characteristics and parameters under the influence of interferences; the accuracy of determination of the interference signal

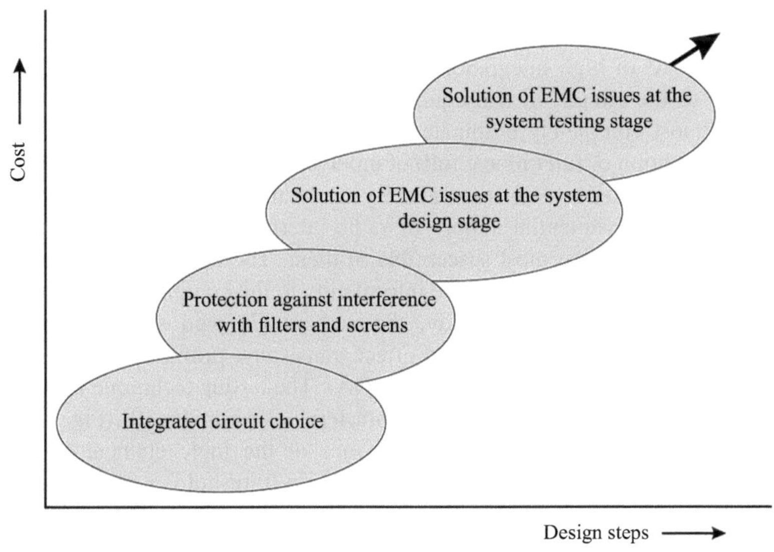

Fig. 6.12 Dependence of EMC-related costs on the design stage

levels and values of monitored parameters; minimization of costs on research practice; making the experimental conditions similar to the actual interference environment.

When evaluating the susceptibility of radioelectronic systems and units, the radiation method is used for EMI impact simulation. An interfering signal is supplied to the antenna that transmits it in the direction of the item under study. Generally the latter is located at a distance of 1 m from the feeding antenna. The same method was used during immunity testing of discrete semiconductors and microelectronic integrated circuits [4, 7, 8]. The evaluation criterion was taken as the level of exposure which leads to the irreversible failure, i.e., breakdown of the element. The studies demonstrated [7, 8] that most of the time degradation changes occurred in the parameters of microelectronic integrated circuits either having longer leads or oriented perpendicular to the electromagnetic field vector. Degradation changes accumulated when exposed to series of SHF-impulses. Most often, the failure reason was breakdown of Schottky input diodes and p–n junctions of transistors of Schottky transistor–transistor logic elements. However, when using such method, it is very difficult to exactly determine the interfering signal energy absorbed by transistor, IC, or p–n junction; when calculating the levels of EMI crosstalk at terminals of SD and IC and their external circuits, many assumptions have to be made due to uncertain positioning of conductors of various designs of test specimens with reference to the direction of the interference field vector. Furthermore, such tests are expensive. Therefore, in this case, the radiation method which is effective for testing of separate nodes and units should be replaced with a conductive method of EMI supply to the terminals of SD and IC. By supplying the radio-interference through separating capacitance directly to the terminal of element under test, the interfering signal energy absorbed can be quite exactly determined by known methods. However, when supplying EMI to one of the terminals, it isn't possible to prevent its impact on others. It is difficult to take into account the degree of relation between them, especially for IC of high integration level. It depends on chip package design, its connection to the circuit. In consequence, there is uncertainty of the experiment. But it is almost completely eliminated if the worst-case scenario is simulated to which the condition of full (almost full) of input EMI power absorption in the circuit of the terminal under test corresponds. By analyzing the changes in SD and IC parameters during sequential supply of radio-interferences to all terminals, it is possible to determine the most susceptible of them. The testing technique is much cheaper, the test results provide more information to the equipment developers.

At the initial stage of the study, the authors [3] paid special attention to the susceptibility of bipolar and field-effect transistors, primitive logic gate (LE) Not-AND, elementary operating amplifier (OA). The testing technique involved the analysis of changes in voltage–current characteristics of diodes [9] (Fig. 6.13) and transistors [10], voltage-transfer characteristics of the logic element (Fig. 6.14) and the operating amplifier. The sensibility criteria (threshold) were taken as EMI level leading to the circuit failure or to certain change of static parameters SD and IC.

During tests of integrated circuits, it was found that it was more appropriate to evaluate their sensibility in the dynamic mode—when supplying test operating signals to the input [11]. During studies conducted at the Belarusian State University

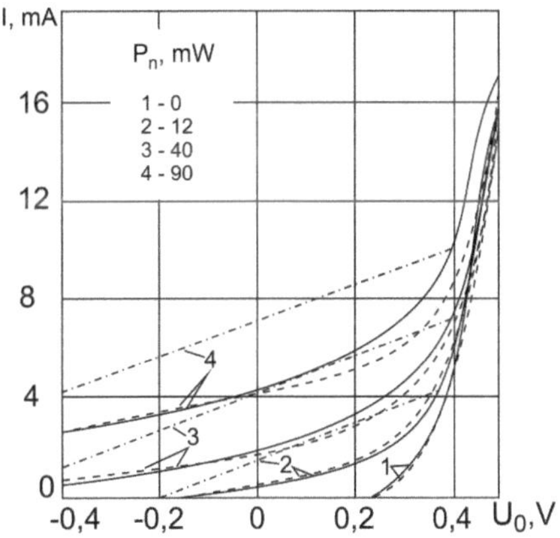

Fig. 6.13 Experimental and rated changes in characteristics of diode Д9А when exposed to EMI at a frequency of 400 MHz

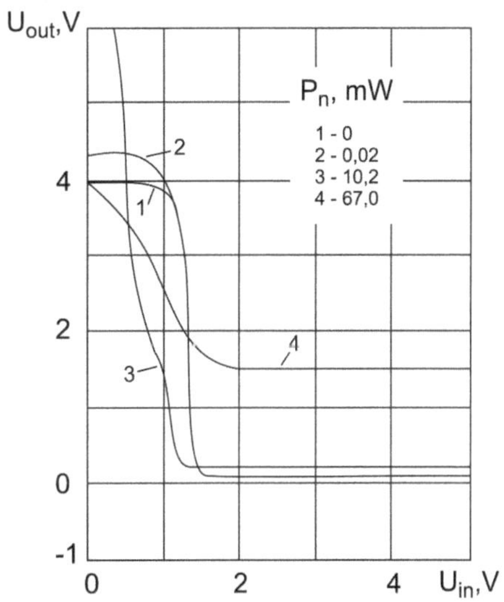

Fig. 6.14 Change in the transfer characteristic of IC К155ЛА3 when its output is exposed to EMI at a frequency of 95 MHz

of Informatics and Radioelectronics (BSUIR Minsk), several prototypes of monitoring and metering instruments (protected by copyright certificates and a patent) [12] were designed and manufactured to determine susceptibility thresholds in both simple and complex microelectronic integrated circuits. The integrated circuit under study is placed in the measurement chamber; impact of interference on other circuits of the device is prevented by means of decoupling filters. Change in parameters of the integrated circuit under study is monitored by comparison of their parameters with parameters of the identical chip of reference. The sensibility analyzers can work both in manual and automatic mode.

Depending on the frequency range, a high-frequency voltmeter, an oscilloscope, or an SHF absorbed power meter may be used to measure the level of EMI impact. The display processing units are designed in such a way that enable to study the sensibility using a "certain failure" criteria related to change in the output state of

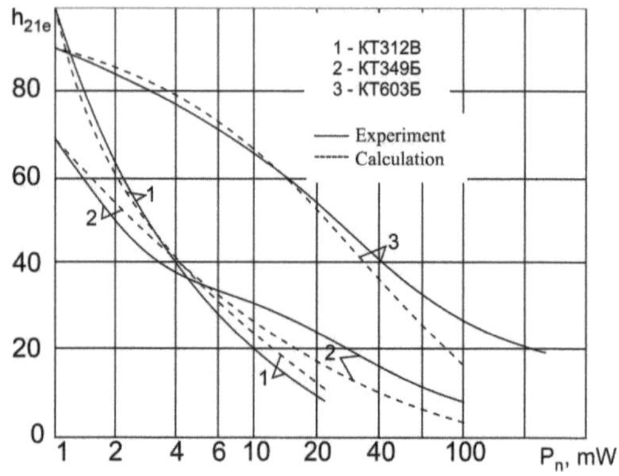

Fig. 6.15 Experimental and rated dependencies of a change in the transmission factor of a 450 MHz interfering signal using bipolar transistor current flow from the absorbed power level with impact on the base circuit

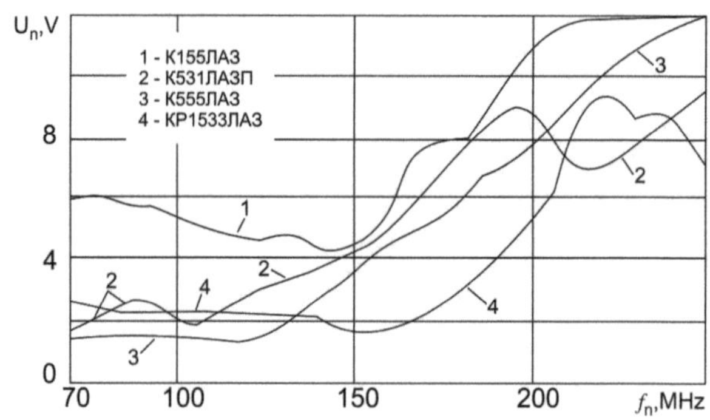

Fig. 6.16 Dependence of the susceptibility threshold of TTL and Schottky transistor–transistor logic elements on the jamming frequency affecting the input

the integrated circuit, as well as a "possible failure" criteria that is a deviation of one or a few parameters from the permissible limits. It is possible to carry out monitoring simultaneously by a few outputs of the integrated chip under study.

Figures 6.15 and 6.16 show the experimental and rated dependencies of some transistors and digital chips with regard to their susceptibility to interferences. The test results were used in special handbooks for developers of radio equipment operating in the high interference signaling environment.

Studies of various types of diodes, transistors, and microelectronic integrated circuits [9–11, 13] have shown:

- susceptibility of SD and IC to interference decreases with increasing frequency of interfering signals. Sharp increases or decreases of the sensibility threshold can be observed at some frequencies. This is due to influence of parasitic inductance and capacitance of packages, and distinctive features of IC connection;
- with increasing level of integration of IC, i.e., with a decrease in the size of components and an increase in the density of their arrangement, the susceptibility to interference is higher. So, during tests of bipolar transistors, it was found that devices with large emitter perimeter, i.e., having interdigitation of the emitter-to-base junction, are less sensitive;

- sensitive circuits of operating amplifiers, low-power bipolar transistors, high-speed digital microelectronic integrated circuits are very susceptible to radio interference. Field-effect transistors, CMOS IC, chips with increased load-carrying capacity have better interference immunity;
- in microcircuits of various series, the dynamics of changes in parameters with an increase of interfering signal is not uniform. So, in the metric wave band of EMI, most often LE AND-NOT of 155 (TTL), 531 and 555 (TTLS) series have unacceptable changes in dynamic parameters such as turn-off delay; when exposed to interferences, elements of 1533 (TTLS) series have the most critical changes in static parameters: logic-0 level and logic-1 level.

The developed technique of SD and IC testing can serve as a basis for experiments conducted to evaluate the sensibility of individual units and the entire devices [13]. Test results classification allows using them not only as handbooks for developers of modern radio equipment but also as tutorials for training specialists in EMC.

6.3 Ensuring Electromagnetic Compatibility of Microprocessor Control Units of High-Speed Electronic Devices

6.3.1 Design Features of Printed Circuit Boards Optimized in Terms of Electromagnetic Compatibility

Thus, HF electromagnetic interference should be suppressed in its source (at the level of a semiconductor element); otherwise, it will be necessary to take measures on their attenuation either at the circuit level or in the terminal device affected by such interference. Only it will cost much more. Although the IC design with due regard to the known concept of on-chip interference suppression requires some spending, this solution is much more efficient than any emergency measures on protection of external devices against interference. EMC parameter optimization for systems based on microcontrollers should be performed at the chip level, as well as at the level of printed circuit board.

Before proceeding with the EMC parameter optimization for electronic systems, a number of tests should be performed at the level of certain IC, separate modules and systems of the devices under development, taking into account the following factors:

- level of electromagnetic interferences generated by the integrated circuits;
- interference immunity at the component level (IC level);
- level of interferences generated by separate systems and devices;
- interference immunity at the system level;
- level of interferences generated during operation of electronics;
- interference immunity of all equipment in whole.

The modern microcontroller architectures apply a synchronous timing. It means that the active pulse edge of system clock frequency with respect to which the microcontroller logic circuits are switched-over should arrive simultaneously (as far as it is possible) at all points of the printed circuit board. As a result, there are very short and steep current impulses with high rate of di/dt rise reaching 100 mA/ns in the system. High currents of overcharge are required for generation of clock frequency impulses rectangular in shape (in practice, these impulses have trapezoidal shape with maximum steep rising and falling edges). The specified requirement is applied to all the transistors which must be switched over simultaneously. Due to flow of high switching currents, high-frequency electromagnetic radiation is generated; and the higher the radiation is, the more the energy is consumed by the system from the external power supply. It should be recalled that the high-frequency component of this energy dissipates in the environment; in this case, conductive paths and other elements of printed circuit board act as transmitting antennas. That's why it is very important that the source of high charging current that can ensure the required switching rate of clock frequency impulses of the microcontroller is located as close as possible to the radiation source, i.e., to the microcontroller. Then, energy radiated by structural elements of the printed circuit board will be lower than in the situation when the "switching" energy source is a constant-voltage regulator located at a certain distance from the microcontroller. In the latter case, HF-interference propagates on the following path: IC internal power bus—IC package—power bus on the printed circuit board (Fig. 6.17).

For this reason, a large number of buffer capacitors ensuring the current required for switching are located on the printed circuit board between the power and ground buses in close proximity to the microcontroller. The capacitors are charged in the period of time between two active edges of clock pulses. Unfortunately, the use of discrete capacitors on the printed circuit board fails in that they have non-zero contact inductance (ESL is equivalent series inductance) and contact resistance (ESR is equivalent series resistance). Their values (of about 2 nH and 30 mOhm) are rather high to make capacitor lose its capacitance properties at the highest frequencies. On the contrary, the contact inductance of built-in capacitor located in close proximity to the internal IC power bus is negligible. Therefore, the more quantity of such internal capacitors can be integrated directly in the chip, the higher effect of HF radiation suppression is.

The level of electromagnetic radiated interferences can be reduced by using of the appropriate circuit-based and engineering solutions when developing the printed circuit board. They include: the use of inductive filters (for example, chokes and ferrite beads) operating as low-pass filters; building of the printed circuit board with an additional grounded layer without through-holes in it; wiring of connections for HF signals as strip lines; the use of additional capacitors and electromagnetic shields. However, all these additional elements increase cost of electronic devices that is critically important in the mass production.

If for economy reasons, it is decided to use the printed circuit board only with two layers, then one of these layers should be dedicated to grounding (GND and V_{SS}) as much as practical. If it is not reasonably practical, positive supply voltage buses (V_{DD}) and V_{SS} should be placed as close to each other as possible or one above the

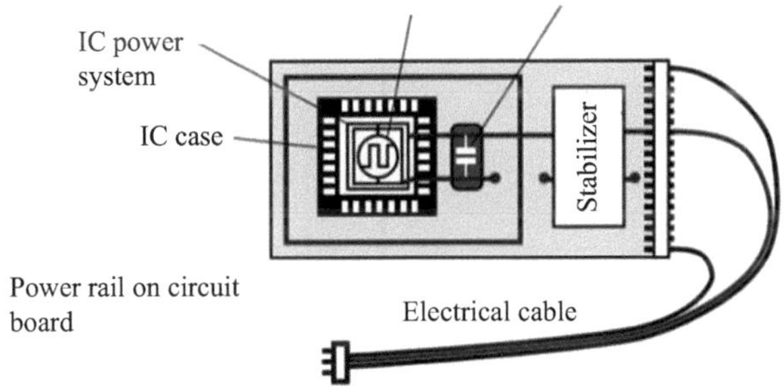

Fig. 6.17 Interference propagation paths in clocked systems

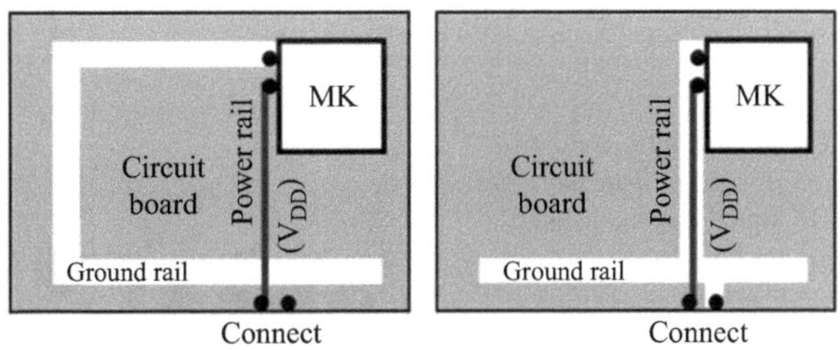

Fig. 6.18 Power wiring on double-layer printed circuit board

other in order to minimize the return current flow path and, consequently, reduce common-mode interference (Fig. 6.18).

When designing multilayer printed circuit boards, two separate layers should be foreseen exclusively for supply voltage buses (V_{DD} and V_{SS}). Any other supply voltage in the circuit should be connected to "islands" located just below those modules that use these supply voltages. It is desirable to ensure the maximum parasitic capacitance between supply layers in order to minimize the level of electromagnetic radiated high-frequency interferences. It can be achieved by placing the PCB internal layers just one above the other (Fig. 6.19).

For example, the second layer can be dedicated to signal ground in the 6-layer printed circuit board that ensures the preset value of impedance with respect to high-frequency data signal. Here, similar to wiring of power buses, special attention should be paid to that the return current flow paths are as short as possible. It is particularly important to avoid any electrical contact of the ground layer with the adjacent layers through thru-vias (Fig. 6.20).

The highest priority should be given to location of a few IC on the printed circuit board. First of all, it is necessary to lay the power buses from the constant-voltage regulator. Voltage should be supplied separately to high-speed logic circuits, "low" logic circuits, and sensitive analog cascades in order to minimize the cross-effect of

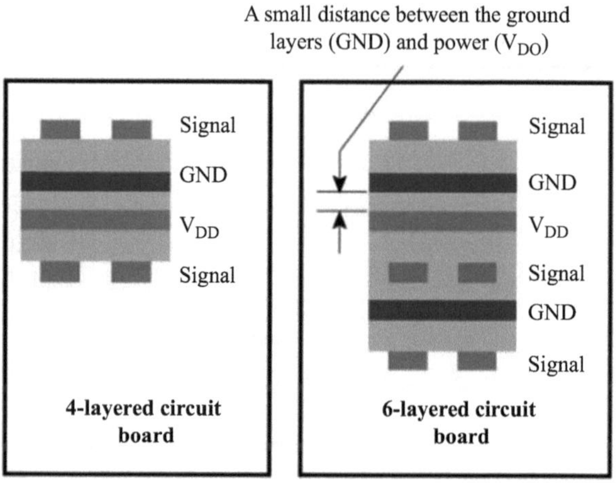

Fig. 6.19 Distribution of layers in a multilayer printed circuit board

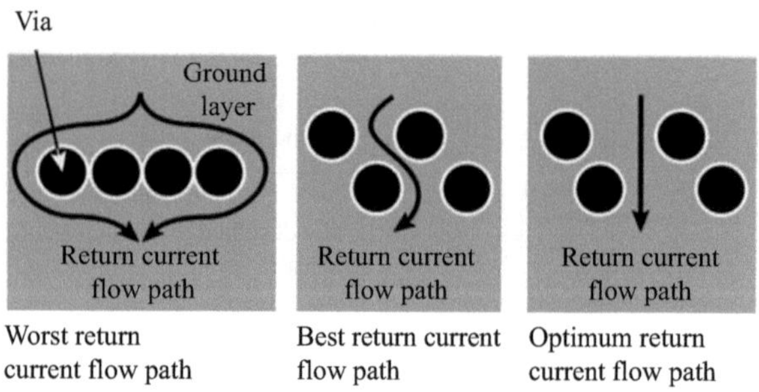

Fig. 6.20 Optimization of return current flow paths for HF signals on the printed circuit board

these circuits. The relevant power lines (paths) should be connected according to the "star" schema at one point in the constant-voltage regulator (Fig. 6.21).

When using buffer capacitors, they should be located on the printed circuit board in such a way that switching current is supplied to IC on the shortest path. In this case, current flows from the power source and through the capacitor to IC (using V_{DD} or V_{SS} buses, depending on the IC connection diagram). It is illustrated in Fig. 6.22, which gives examples of correct and wrong connection of buffer capacitors to the power terminals of the microcontroller. Two upper schemes in Fig. 6.22 illustrate a situation when components are located on one side of the board; and two lower schemes—a situation when the microcontroller and the capacitor are located on the front and back sides of a 4-layer printed circuit board (PCB).

When connecting to the ceramic or quartz crystal oscillator that is built in the microcontroller, special attention should be paid to creating a conductive area ("island") connected to the microcontroller V_{SS} output (or to its ground) just under the main terminal of the oscillator on the printed circuit board. It is the only way to avoid the return current flow on the surface of ground layer of the printed circuit board.

In addition to the buffer capacitor concept, in terms of reduction of the HF radiation level and ensuring the desired signal integrity, "canalization" is of paramount importance during transmission of HF desired signals using data buses.

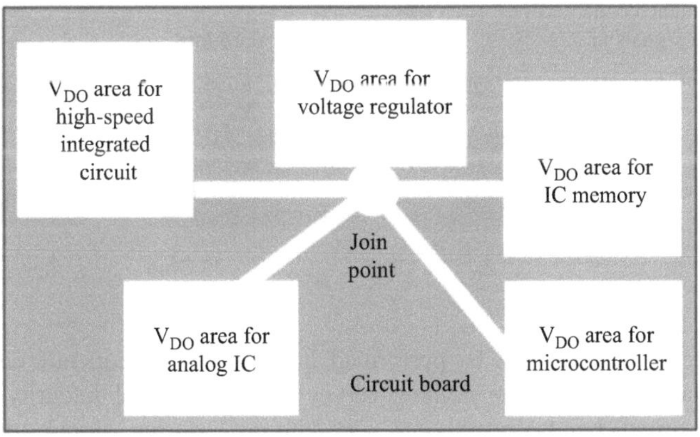

Fig. 6.21 Wiring IC power buses of different types according to the "star" schema

Fig. 6.22 Wrong (**a**) and correct (**b**) placement of a buffer capacitor

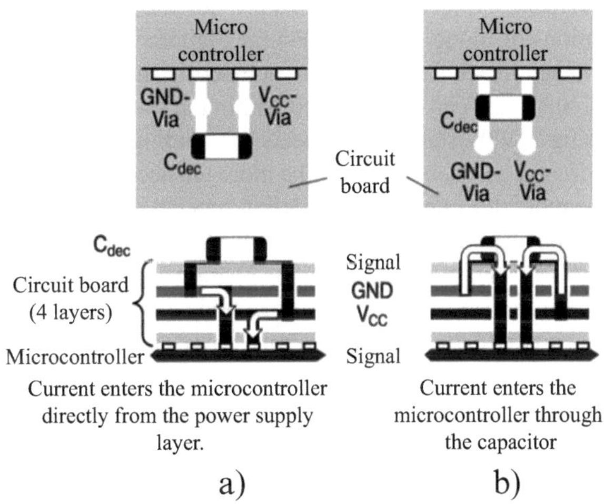

Radiation or HF interference entry in the power supply circuit can be efficiently avoided by shielding of signal tracks. In this case, a microstrip line (when using single-layer shielding) or full strip line (when using double-layer shielding) is formed. In both cases, strip lines can be designed with the preset impedance in order to match them with output impedance of the driver of signals. Examples of such lines with different impedances are shown in Fig. 6.23.

The line and driver impedance matching is aimed at improving signal integrity during its transmission over the line while reducing the level of radiated interferences. On the contrary, mismatch between impedances leads to the signal reflection from line to driver which causes the signal rise-time distortions (signal bursts) or even generation of parasitic HF oscillations ("ringing effect") at the moment of their switching over. This effect is considered in more detail in Sect. 6.3.3.

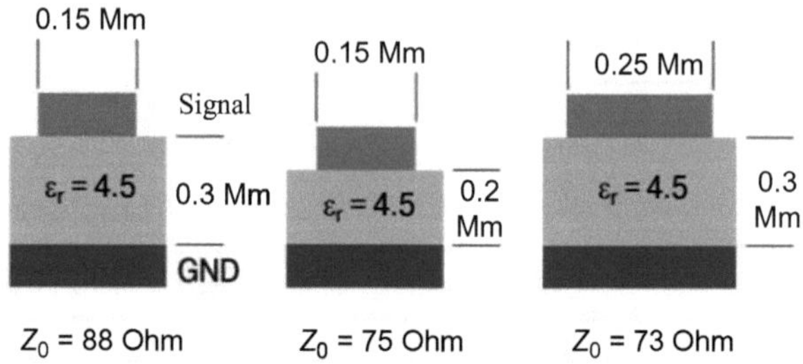

Fig. 6.23 Characteristic impedance of signals in microstrip lines

Computer simulation can be performed in the time or frequency domain. The picture obtained in the time domain serves to assess signal integrity, i.e., edge steepness, signal bursts/undershoots, or "ringing" during signal transmission. The results of computer modeling help identify "oversize" drivers, extremely long connecting wires, and impedance mismatching. To ensure technical support of their products, the IC manufacturers provide the so-called computer IBIS-model (specifications of input/output buffer) for digital switching circuits that describe switching characteristics for all drivers included in this IC. Structural elements of the printed circuit board can be described within the so-called transmission line model (TLM) or using 2D/3D standard elements (depending on the PCB type) in order to simulate different geometric forms. The more unsuccessful design of the printed circuit board, the more thoroughly the modeling process of this PCB and its surrounding elements should be performed, the more data volume is processed and the more computational resources are required. Simulated parameters are recorded as complex RLCG arrays, which are solved using electromagnetic field solver programs. Using these programs, it is possible to make calculations at the conductor level, as well as at the field level. Examples of such analysis in the time and frequency domain are given in Fig. 6.24.

6.3.2 *Features of Measurement of the Level of Interference Emitted by Microcontrollers*

In assessing the level of interference radiated by microcontrollers, the measurement technique conforming to standard IEC 61967 is applied. This standard describes 5 different methods of measurement of conductive (IEC 61967-4, the former standard VDE 767.13) and radiated (IEC 61967-2, the former standard SAE 1752/3) interferences. Besides, the national standards include additionally various measuring techniques which were offered by national omissions on EMC issues and currently co-exist with standard IEC 61967. Detailed description of interference measuring techniques standardized in accordance with IEC 61967 is given above.

Two of these measuring techniques can be implemented using a test board (Fig. 6.25). Structurally, a universal test board for measurements by TEM-chamber method is 4-inch square (10.14 cm). The microcontroller IC under test is located on

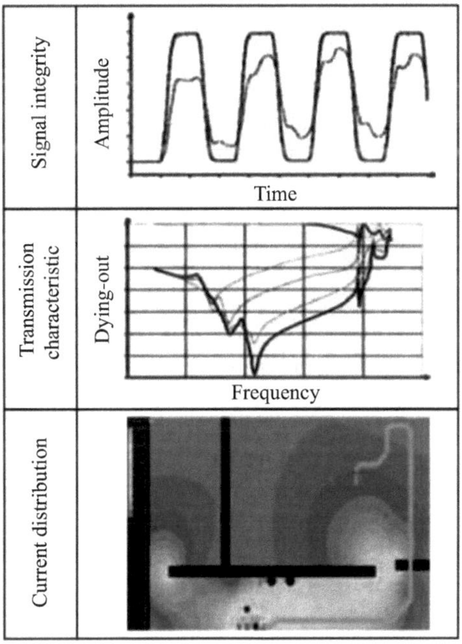

Fig. 6.24 Computer analysis of signal integrity in the time domain (upper chart), transmission factor (S21) in the frequency domain (chart in the middle), and local current distribution on the printed circuit board (lower chart)

the shielded back side of the printed circuit board, and peripheral items (memory, clock-pulse generator, external input/output ports) on the face side. The same board together with a near-field scanner can be used when measuring the electromagnetic field distribution along the surface of IC under test. Such test is useful to determine the microcontroller terminals which are critical in terms of radiated interferences.

In a typical spectrum, the interferences radiated by the microcontroller stand out sharply by harmonic amplitude of the system clock frequency. When comparing the levels of radiated interferences obtained at different stages of IC design, the signal spectrum envelope is used. A typical spectrum of such interferences radiated by the microcontroller operating at a clock frequency of 40 MHz is shown in Fig. 6.26.

The microcontroller consists of many functional modules (Fig. 6.27). First of all, there is a central processing unit (CPU) responsible for calculations and data transfer process control.

Peripheral modules or simpler "peripherals" of the microcontroller include memory modules (random-access memory, programmable read-only memory, electrically alterable read-only memory, flash memory), timers (for example, timers of capture/comparison), series and parallel data transfer interface (ASC, SSC, SPI, parallel port), special modems (CAN, J1850), interruption controllers, and external memory interface (EBU is an external bus unit). For ensuring the maximum computational power of the microcontroller, data buses should have the maximum transmission capacity and operate at the highest possible clock frequency. The high clock frequency of data transmission means that the microcontroller has little time for data processing, that's why the duration of latent phase of data transmission cycle

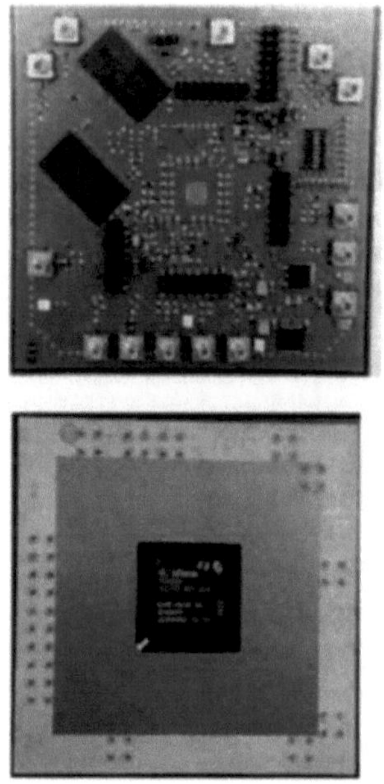

Fig. 6.25 Test board for measurement of the level of interference radiated by the microcontroller. At the top—upper side of the board; at the bottom—lower side of the board with IC under test

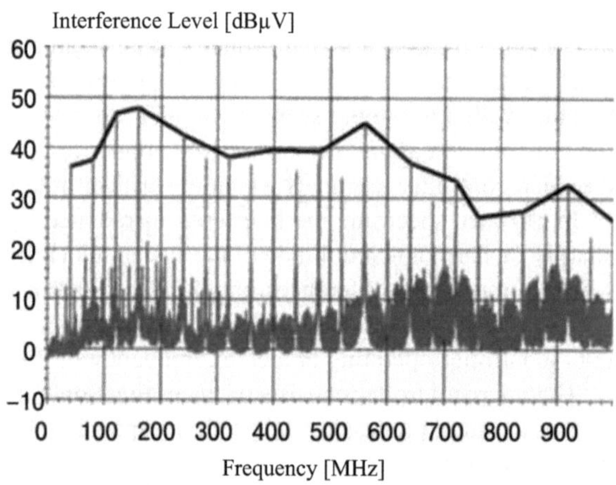

Fig. 6.26 Spectrum of interferences radiated by the microcontroller which was measured by TEM-chamber method in accordance with IEC 61967-2. Microcontroller clock frequency is 40 MHz

should be set as much as possible. Consequently, it is necessary to provide the extremely steep edge of clock frequency signals; moreover, these signals should arrive synchronously within the whole system.

Since during switching process all microcontroller transistors consume current flowing in the inner power bus, it is the main HF-interference propagation path (it is counted that up to 60% of HF-interferences penetrate the surrounding systems using the power supply circuit of the microcontroller). Around another 30% of interferences propagate through input/output ports; and only the rest of 10% are emitted directly in the environment. Those high-frequency interferences which

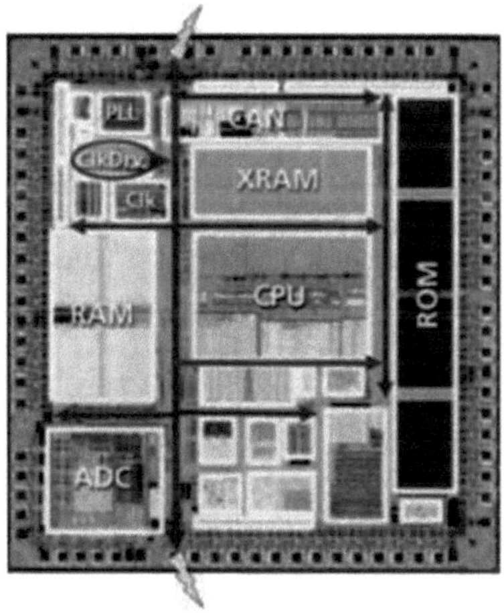

Fig. 6.27 Layout of functional modules in the microcontroller housing using a centralized clock synchronization system

aren't suppressed by circuit-based techniques at the chip level (for example, due to the use of decoupling capacitors integrated in the crystal) penetrate the systems surrounding the microcontroller through the contact pads of the power buses. The external systems exposed to impact of interferences include a printed circuit board where the microcontroller is located. This HF-interference can be partially attenuated by discrete decoupling capacitors which should be located as close to the power terminals of the microcontroller as possible. These capacitors can efficiently suppress interferences at rather low frequencies up to 10 MHz, despite their non-optimal characteristics of contact inductance (ESL—equivalent series inductance of about 2 nH) and contact resistance (ESR—equivalent series resistance of about 30 Mohm). Then, high-frequency interferences aren't attenuated. In this case, it is necessary to ensure the suppression of electromagnetic HF interferences inside the chip by minimizing their sources and propagation paths.

The clock pulse synchronizing system is the main source of narrowband high-frequency electromagnetic interferences. They overlap with secondary clock pulses and aperiodic signals "responsible" for producing broadband interference in the spectrum (Fig. 6.28). Modern complex and high-performance microcontroller system use impulse synchronization for control of all logic circuits, i.e., clock impulse signals should simultaneously arrive at all points of the system. Current impulses of very large amplitude (about a few amperes) are generated during simultaneous switching of hundreds of thousands of transistors in the power supply system of the microcontroller. Ideally, the bus has low impedance for avoiding of large voltage drops on it, that's why the major contribution to the HF-interference generation is made by the current flowing in the power supply circuit. Depending on organization of synchronization of internal logic circuits, current impulses emerge one or two times over a period of clock frequency. They are characterized by a very steep edge and are gradually attenuated as the signal passes through different logic circuit cascades (Fig. 6.28).

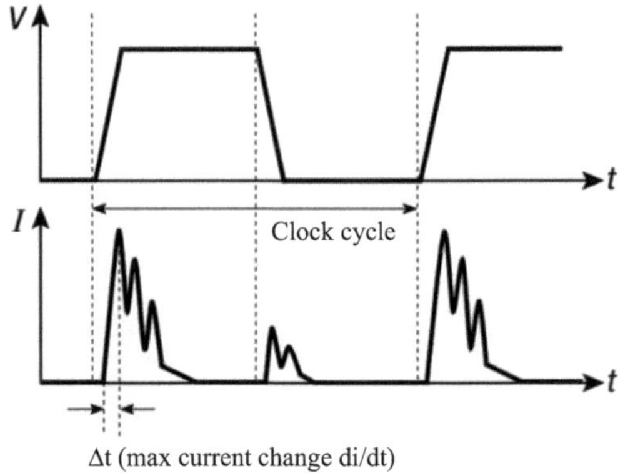

Fig. 6.28 Current and voltage waveform for a clock signal

Switching currents of logic circuits which are located on the semiconductor crystal of the microcontroller are supplied from the external power source in normal operating mode. However, simultaneously switching noises (SSN) are generated due to high steepness of current pulses (high di/dt value) and the presence of parasitic resistances and stray inductances in the feed circuit. High-frequency component of these noises penetrates the near electronic systems or it is radiated in the environment. That's why the power supply system of microcontrollers includes decoupling (buffer) capacitors which are located as close to the current consumers, i.e., integrated circuits, as possible. Ideally, it means that each microcontroller is supplemented by capacitor located directly on its power terminals. It is practical to use these capacitors from two points of view:

- First, they serve as local current sources at the moment of switching of clock frequency impulses;
- Second, they limit on some extent the high-frequency interference propagation from the microcontroller to the constant-voltage regulator or vehicle accumulators.

6.3.2.1 Circuit-Based Methods of Preventing Interference Radiation from Microcontrollers

The closer to the EMI sources the elements intended for interference suppression are located, the higher their efficiency. EMC characteristics of electronic devices can be improved by the following measures: limit of pulse steepness during switch of transistors and logic circuits (elements), as well as location of additional charge sources (capacitors) in close proximity to the switched logic circuits. It is allowed to reduce the steepness of clock pulse edges only if it doesn't contradict to requirements for the microcontroller performance. Simultaneous working transistors and logic elements should not generate impulses with extremely steep edges. In the world today, circuit synthesis methods have become widespread in which the appropriate circuit components are chosen with the use of software tools.

Most manufacturers of microcontrollers integrate capacitors in chips which are intended for current impulse limiting during switch. In combination with its series low-resistance series, such capacitor forms a low-pass RC-filter. Thus, the high-frequency component of switching current is attenuated inside the chip and it isn't allowed to pass through to the outside. The use of series inductance would be more efficient (due to lower series resistance), but an integrated inductance of about 100 nH requires a very large surface area of the semiconductor chip. Systematic studies dedicated to the operation of microcontrollers have demonstrated that their design with due regard to the implementation of similar RC structures can significantly reduce the level of radiated interference. Figure 6.29a–c is a graphic representation of the results of two measurements of the level of electromagnetic interference radiated by a 16-bit microcontroller.

In the first case, no any measures aimed at improvement of EMC characteristics are provided in the microcontroller; in the second case, the microcontroller is designed using the concept of built-in decoupling capacitors. It is clear that the interference level in a critical frequency range of 100 MHz (used for FM broadcasting) can be reduced by 10 dB.

Interferences appearing during signal switching in output of input/output port driver of the microcontroller contribute to the total level of interferences generated by the microcontroller. Signal lines of a printed circuit board connected to port

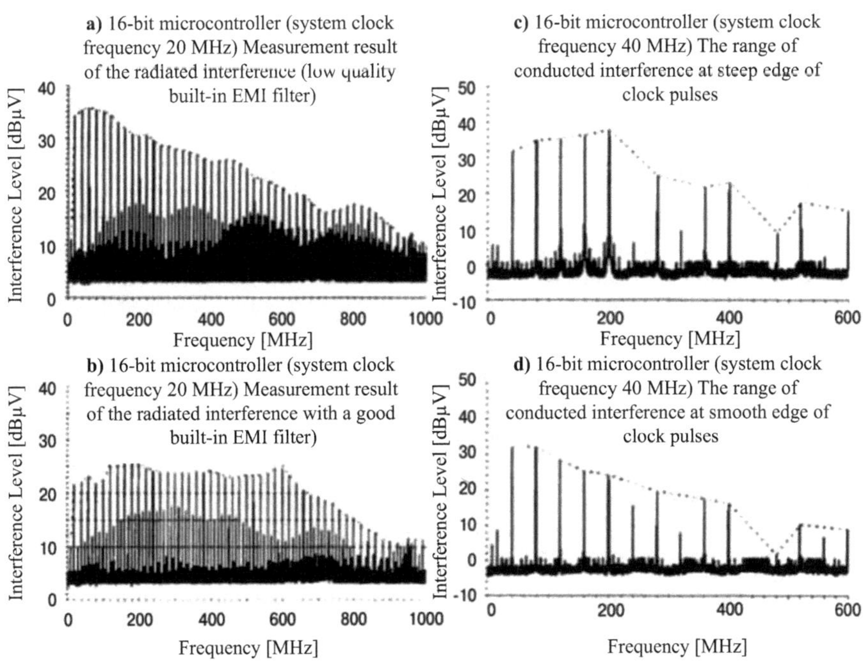

Fig. 6.29 Reduction of the level of electromagnetic interference radiated by microcontroller using on-chip methods. Left—the spectrum of radiated interferences without (**a**) and with (**b**) decoupling capacitor. Right—the spectrum of conductive interferences in output of input/output port driver during steep (**c**) and smooth (**d**) edges of clock signals

terminals are very effective antennas up to a few centimeters long (as are the power buses considered above). Some circuit-based measures should be taken to minimize the interferences generated by output stages (drivers) of input/output ports. The most common methods are the reduction of driver output power or leading edge smoothing (control of signal rise rate). Since the parameters of driver output transistors are usually chosen with something to spare (with a view to the worst operating conditions), as a rule, it is possible to use impulses with less steep edges. It is particularly applicable to the cases when the driver operates at low clock frequency or with low capacitive load. Figure 6.29c, d shows in a spectral view the results of measurements of conductive interference generated by a 16-bit microcontroller in output of input/output port driver. In the first case, edges of clock impulse signals have more steepness, and in the second case, their steepness is reduced. The use of such method provides for interference reduction in FM range (100 MHz) approximately by 6 dB.

Another circuit-based method for improvement of EMC characteristics of the microcontroller systems is the use of generators of frequency-modulated clock signals collectively known as a "generator with distributed spectrum." Even with very limited level of modulation (0.5–0.1%), it becomes possible to considerably reduce the amplitude of narrowband interferences which are harmonics of clock frequency. At the same time, the interference level is slightly increased in the sideband. Although such method does not reduce the total radiated energy, its distribution over the spectrum is more uniform, making it possible to limit the maximum amplitude of electromagnetic interference.

The level of interferences generated by the microcontroller can be reduced by improvement of its packaging and terminal layout diagram. Leadframe-based IC packages are currently widely used (Fig. 6.30a). In such packages, the signal lines and power buses connecting the IC semiconductor crystal to its terminals are laid as "spidery legs." Packages of DIL, PLCC, M-QFP, and T-QFP types are designed according to this technology. The more distance is between the crystal semiconductor substrate and external terminal of the package, the higher contact inductance of this connection is. It consists of leadframe inductance and inductance of wire connection which ensures the electric contact between the crystal and leadframe. When designing the microcontroller terminal layout diagram, special attention should be paid to wiring of power terminals that should be coupled and located as close to the center of the relevant package side as possible. Such location can minimize the inductance of internal connections and create the optimal conditions for blocking of interferences.

While the signal line wiring layout in leadframe packages is preset and not subject to change, the use of packages performed using BGA technology (ball grid arrays, see Fig. 6.30b) may optimize critical signal connections (from the point of view of ensuring EMC) and supply lines by developing the corresponding templates. Here, terminals are performed on the lower part of the package as small balls which are welded to the contact pads of the printed circuit board by pinpoint soldering method. The electric contact between the IC semiconductor crystal and the relevant paths of the support plate is performed either using wire connections or directly through bumps in the variant of "face-down" crystal.

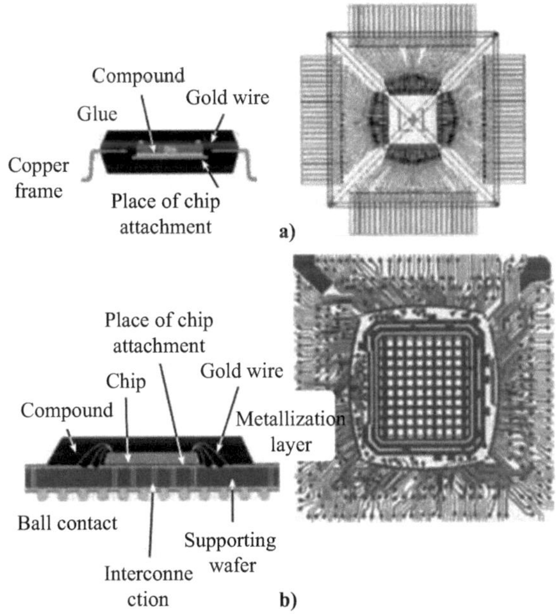

Fig. 6.30 Sectional plan view of leadframe packages (**a**) and BGA packages (**b**)

6.3.2.2 Computer Simulation of EMC Characteristics

As compared to design of printed circuit boards, where computer simulation of EMC parameters has been widely used for some time, the technique of modeling and software-based simulation of EMC characteristics of microcomputers is still at the initial stage of development. The main reason of this is the great complexity of microcontroller. The software-based simulation of its operation at the transistor level requires extremely large data arrays and takes much time. The more practical solution is the software-based generation of equivalent current sources which simulate the microcontroller "behavior" in whole or its separate modules when switching the currents.

Various standard program models have been developed for carrying out the system-oriented computer simulations of EMC parameters of the microcontroller. For example, IBIS-model describes only characteristics of input/output ports. The so-called integrated circuits electromagnetic emission model (ICEM) determines the equivalent current sources connected to the IC power terminals. Although the interference level at these terminals of the microcontroller is essential and it needs to be primarily evaluated, the simulation model should contain the data on package type and PCB layout in close proximity to the IC (for example, taking into account the so-called blocking concept). At the chip level, this model should describe equivalent current sources corresponding to the microcontroller functional modules and include a precision RLC model of the voltage power source (Fig. 6.31).

The quality of a computer model depends directly on its ability to correctly calculate dynamic switching currents and accurately extract parasitic RLCs. If the model is used only for qualitative assessment of EMC characteristics, for example, during modernization of the device, then its quantitative accuracy may be compromised to some extent.

Up until now, the issue on the microcontroller interference immunity has not gained much attention since the power supply voltages of 3.3–5 V exceed by many times the level of external electromagnetic interference; this issue becomes relevant when the supply voltage is reduced down to 1.5 V. Assessment of the microcontroller interference immunity by using the appropriate measurement techniques (for example, error ratio measurement during signal transmission) is extremely expensive since prior identification is required for those circuit elements which perhaps improperly operate under the influence of interferences, and develop a specialized (custom) program to control the process of measurements. Since methods for testing the printed circuit boards were developed, some efforts have been made to standardize the IC immunity measuring techniques at the international level. They refer mainly to radiated HF fields and induced HF currents.

Currently, work is under way on implementing the units for testing IC immunity to the action of impulse interferences. The microcontroller immunity to the action of electrostatic discharge (ESD) is important. In this regard, it is worth noting that none of the microcontroller terminals (not including the power terminals) is connected directly to connectors which are intended for external connections, that's why the requirements for protection of the microcontroller against external electrostatic impulses are not so strict as for other IC or the whole system.

The greatest impact on the microcontroller immunity is due to interferences occurring in the power supply circuit during flow of high switching impulse currents in it (the so-called SSN-noises). These interferences are caused by processes inside the microcontroller and by influences of external devices, especially by high-power semiconductor elements; and they can significantly impair the signal-to-noise ratio.

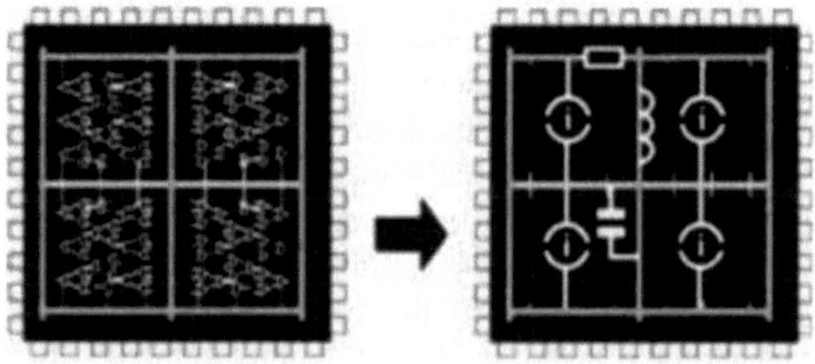

Fig. 6.31 Modeling complex IC using equivalent current sources and RLC extraction

6.3.3 *Ensuring Electromagnetic Compatibility in Wired Lines of Telecommunication Networks*

6.3.3.1 General Requirements for Electromagnetic Compatibility of Telecommunication Systems

The term "wired communication systems" refers to a wide range of electronic devices and information technologies. It includes elements of telecommunication networks, switching centers, and LAN routers, as well as subscriber devices for private use, for example, telephones, personal computers, and network termination devices.

However, as described above, the requirements for EMC in the area of communication systems refer to the entire telecommunication systems, rather than their individual components. Therefore, electromagnetic compatibility should be mainly tested at the level of external interfaces of telecommunication systems:

- signal lines (check of HF voltage values and current levels);
- supply lines (check of HF voltage values);
- packages (check of electric and magnetic field strength).

Data transmission rates in the telecommunication systems are becoming higher. Therefore, there is an increase in the width of analog signal transmission band dedicated to provision of various services. Thus, for example, now ADSL and VDSL services use the same frequency range as wired radio. The latter should be protected by the appropriate EMC regulations. Frequency ranges used by various wired communication services are shown in Fig. 6.32.

It is rather difficult to meet the EMC requirements when operating in this frequency range. One does not just add a filter to interface since in this case, the desired signal transmitted over a communication channel can be impaired too. Figure 6.33 shows a typical signal spectrum on ADSL line. Common-mode voltage was measured by EMI receiver using standard coupling circuit for telecommunication lines.

ADSL signal spectrum ranges from 30 kHz to 1.1 MHz. Two graphs are shown in Fig. 6.33. The first graph shows measured peak values (Pk), the second graph shows average values (Avg) of a signal. It conforms to a standard defining the different limit values for each type of measurements.

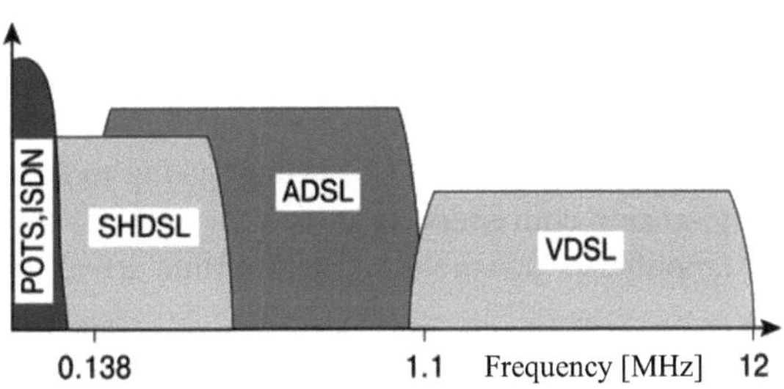

Fig. 6.32 Frequency ranges allocated for various broadband services in wired communication systems

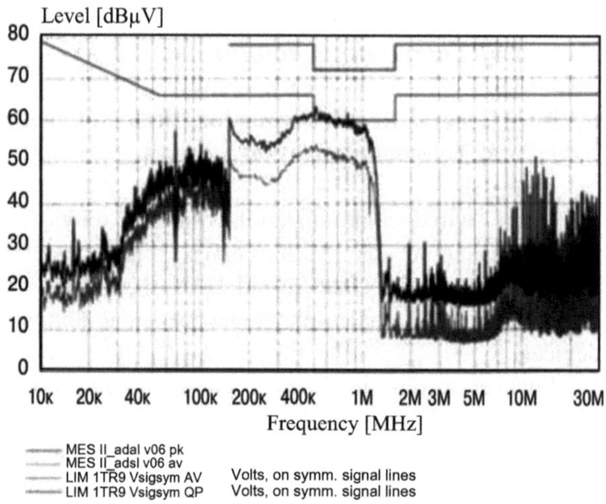

Fig. 6.33 Signal spectrum on ADSL line

But since that DSL protocol is symmetrical with respect to ground, no signal should be generated, when measurements are taken in common mode. The reason for its presence in the graph is some asymmetry of the used test circuit. There is a narrowband interference in the signal spectrum. Its source is a voltage converter on the system board or clock frequency impulses and their harmonics. These interferences from the system board must never be allowed to penetrate ADSL line.

It is clear that the presence of such "mixture" of the desired signal and interference raises some difficulties when choosing the best strategy to reach the required level of EMC. Sooner or later, it becomes clear that the best way to solve this issue is a circuit-based improvement of the system itself, including at the PCB level and at the component level.

A few engineering solutions to ensure proper signal passage inside the system are considered below.

6.3.3.2 Systems, Components, and Basic Concepts

As discussed in Chap. 1, telecommunication devices generally consist of the following main components:

- one or several power sources;
- digital front-ends (DFE), such as digital signal processors (DSP) and microcontrollers;
- analog front-ends (AFE), such as analog-to-digital converters (ADC) and digital-to-analog converters (DAC);
- amplifiers—transmitter amplifier (line driver) or receiver amplifier.

Typical block diagram of a primitive wired communication system is shown in Fig. 6.34.

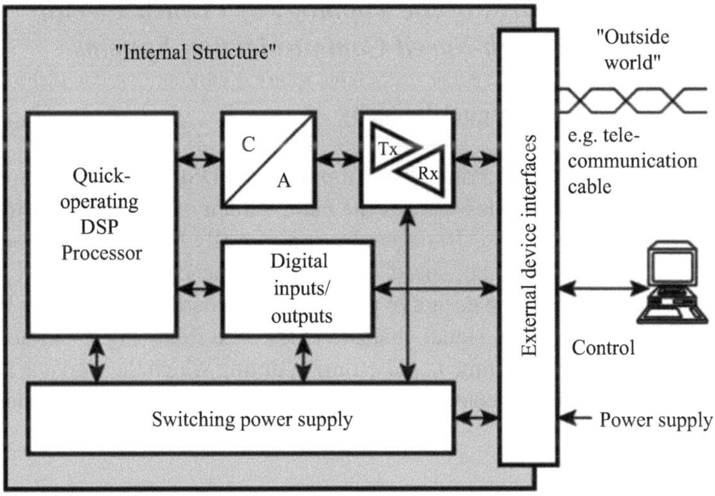

Fig. 6.34 Typical block diagram of a wired communication system

Each communication system should meet the certain EMC requirements; moreover, these requirements should be specified at the system design stage. Here, the key points are as follows:

- determination of EMC characteristics for new products;
- system design: breaking down the system into small functional units (modules) if EMC calculation at the system level is difficult to be performed;
- modular design: prior to start designing the system in whole, EMC characteristics of separate modules should be determined using the computer simulation, software emulation, or by testing the ready prototypes;
- check for compliance of EMC characteristics of products under development to the established regulations for all possible scopes of application;
- consideration of EMC parameters in quality assessment (QA).

EMC characteristics are influenced by any changes in module, such as replacement of components or connecting cables. EMC characteristics can be changed even as a result of error correction in the software:

- distribution of products (developed system) only at that market for which they were intended;
- receiving of reports on faults related to the EMC issues; organization of feedback with users that contributes to the process of improvement of products both ready and under development.

6.3.4 Design Features of the Topology of Printed Circuit Boards for High-Speed Communication Systems

6.3.4.1 The Concept of Signal Integrity

As can be seen in Fig. 6.34, a digital signal processor (DSP) with a clock frequency cycle duration of 10 ns and less carries the main burden of signal processing in the system. But can we take these 10 ns as criterion of really high speed of operation?

The meaning of term "high-speed" depends on the context where it is used. However, since it is about the design of printed circuit boards, this term can be rather clearly defined using formal signal characteristics. The following notations will be used below: rise time t_r, fall time t_f, pulse time t_p during which the signal is in HIGH logical state. Frequency spectrum of such signal can be easily calculated using Fourier transformation:

$$\lambda = \frac{c \cdot \pi \cdot t_{fr}}{\sqrt{\varepsilon_r}}, \tag{6.8}$$

$$f_{\text{highspeed}} = \frac{1}{\pi \cdot t_{fr}}, \tag{6.9}$$

where $t_{fr} = \min(t_r, t_f)$, c—speed of light in vacuum, ε—specific inductive capacitance of the printed circuit board.

If length of the PCB conducting track l_{track} is consistent with signal wavelength λ (for instance, it exceeds the value $\lambda/20$), we can assume that the signal is high-speed. In this regard, it is worth noting that spectrum of digital signals with high steepness of edges includes components of very high frequency even at low frequency of a clock signal. It means that the impulse signal characterized by a rise time of 500 ps should be considered as high-speed if the length of relevant conducting track of the printed circuit board exceeds 11.75 mm. On one side, that's good: If we design a track shorter than 11.75 mm, then the signal passing through it will no longer be a high-speed signal. But, unfortunately, there is a flip side of the coin when designing the printed circuit board: it isn't always possible to "keep" the desired length of all tracks. In this case, it is especially important to take into account the requirements for EMC and ensure signal integrity.

What exactly is "signal integrity"? On the website of Signal Consulting Inc. (http://signalintegrity.com), this term is defined as follows: "... it is a research area occupying the intermediate position between digital design and analog circuit building theory. Here, parasitic damped oscillations ("ringing"), crosstalk noises, ground instability and power source interferences are considered. Here it is studied how to build actually high-speed digital equipment that really works. Here, practical real-life solutions are developed for designing the high-speed systems." Signal integrity refers to ensuring "... the maximum performance and minimum cost of interconnect processes used in high-speed digital devices."

Thus, the term “signal integrity” refers to a small but important part of EMC, particularly to the internal structure of the designed system (Fig. 6.34). In a sense, this concept is more abstract than the laws of electrodynamics. With increasing high-speed performance of IC, engineers have to pay more attention to the issues of electromagnetic compatibility [14–16]. However, developers of digital devices aren’t usually well-grounded in analog or high-frequency technologies; as a rule, they don’t want bother with studying the basic laws of electrodynamics.

To ensure signal integrity, special attention should be paid to the following factors:

- Crosstalk;
- Parasitic damped oscillations (“ringing”);
- Ground instability;
- Power source interferences.

This section describes in more detail how to ensure good signal integrity, which, in turn, provides a good basis for EMC.

6.3.4.2 Crosstalk

Crosstalk is an electromagnetic interconnection of signal tracks located inside the semiconductor circuits, on the printed circuit board or in the multicore telephone cables.

In electrotechnics, crosstalk is considered as an undesirable, but inevitable phenomenon. However, it can be kept within certain limits to prevent its impact on the system performance.

Conducting tracks of the printed circuit board or telephone lines are the so-called shared circuit elements; and their characteristics can’t be determined by formulae intended for description of elements with lumped parameters (of resistors, capacitors, impedance coils, and transformers). However, in electrical engineering, a method is used whereby circuits of elements with lumped parameters are used to represent distributed elements. This approach is acceptable, because a conducting $_{\text{track}}$ of length l and a chain of n conductors each of length $dz = l_{\text{track}}/n$ may always be regarded as identical to each other. If the length Δz of conductor strand is rather small, then as mentioned above, signal that passes through it can’t be considered as high-speed one. The conductor strand (dz) can be represented as its equivalent elements with lumped parameters (resistors, capacitors, chokes, and transformers).

A model of this kind is shown in Fig. 6.35, illustrating two loss free conductors. The c_1–l_1 chain is used as a model of the first loss free conductor, and the c_2–l_2 chain—as a model of the second one. Elements of capacitive coupling (c_m) and inductive coupling (l_m) are used to represent the cross coupling between conductors.

Solution of simultaneous partial differential equations for this model describes a wave which is propagated in forward and backward directions:

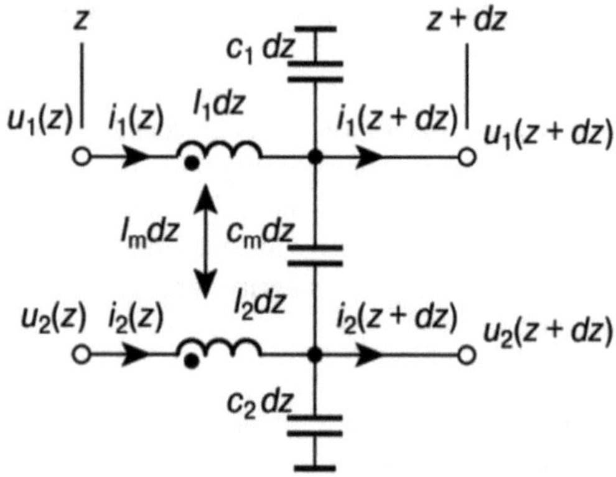

Fig. 6.35 A simple crosstalk model based on two loss free conductors (conductor length—*dz*)

$$u_i(z, t) = A_i u_i(z + v_i t) + B_i u_i(z - v)_i t) \text{ for } i = 1.2. \tag{6.10}$$

This solution can be obtained by linear algebra methods using a diagonalizing matrix. Types of crosstalk are shown separately in schemes below in order to simplify their explanations.

6.3.4.3 Inductive Crosstalk

Figure 6.36, explaining the effect of forward crosstalk and backward crosstalk, shows a conductor conventionally "broken" into n unit cells. Each of them can be represented as a transformer and a delay element $\tau = dz/v$. This discrete model really simplifies the description of crosstalk, which can be either forward or backward.

The forward crosstalk pulse amplitude increases proportionally to the Z-parameter, and these pulses are inverted with respect to the "aggressor" conductor. Let us assume that measurements in the "victim" conductor are performed at point $z = 1 \times dz$. The result of measurements corresponding to time point $t = 1 \times \tau$ will be a short negative pulse.

However, measurements at point $z = 2 \times dz$ will show a double amplitude pulse corresponding to time point $t = 2 \times \tau$. This doubling is related to pileup of the pulse converted in unit cell no. 2 and the pulse propagated through the "victim" conductor from point $z = 1 \times dz$ to point $z = n \times dz$.

Backward crosstalk pulse length (width) increases with a decreasing Z-parameter, but the pulse amplitude remains unchanged; and this pulse is not inverted with respect to the "aggressor" conductor. Let us assume again that signal measurements are performed at different points along the conductor. Measurements at point $z = (n - 1) \times dz$ show a single short positive pulse generated at time point $t = (n - 1) \times \tau$ and propagated backwards through the conductor towards point $z = 0$. Two pulses will be marked at point $z = (n - 2) \times dz$: the first pulse—at time point $t = (n - 2) \times \tau$; the second one—at time point $t = n \times \tau$. This additional pulse is a backward-moving pulse, occurring at point $z = (n - 1) \times dz$ and "arriving" at point $z = (n - 2) \times dz$ with a delay τ. N pulses will be marked at point $z = 0$. Lag time of

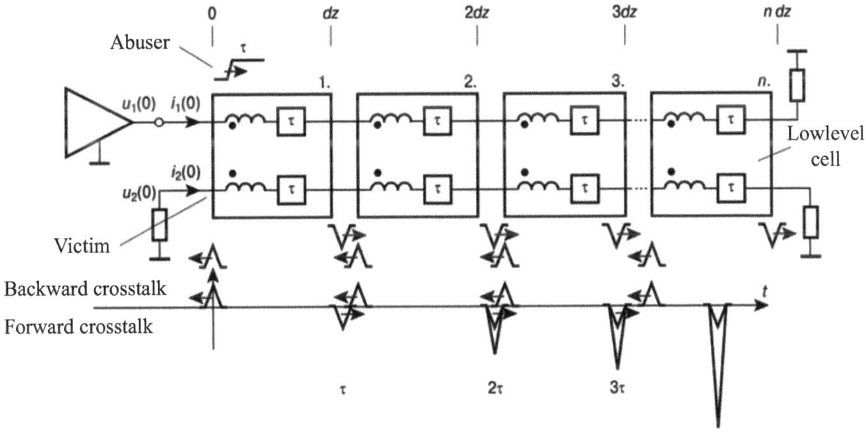

Fig. 6.36 Inductive crosstalk

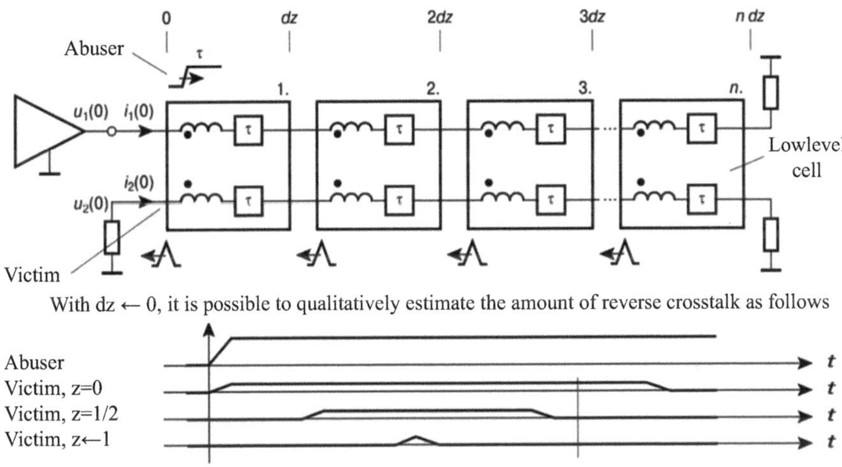

Fig. 6.37 Backward inductive crosstalk

each of these pulses will be $2 \times \tau$ due to delays in both the "aggressor" conductor and the "victim" conductor. When $dz \rightarrow 0$, impulses become inseparable and combine in one long impulse with a small amplitude.

Figure 6.37 shows qualitative dependencies of crosstalk values at different points of the conductor (Z-parameter) at time point t.

6.3.4.4 Capacitive Crosstalk

The model in Fig. 6.38 is very similar to the above-considered model for inductive crosstalk. In this case, however, an inductive transformer was replaced by a capacitor transformer (i.e., capacity coupling between conductors is implemented).

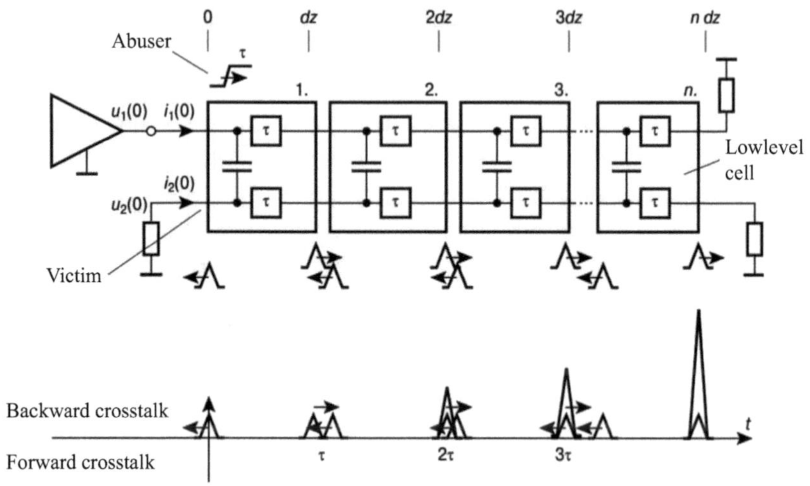

Fig. 6.38 Capacitive crosstalk

The forward crosstalk pulse amplitude increases proportionally to the Z-parameter, and these pulses are not inverted with respect to the "aggressor" conductor. Let us assume that measurements in the "victim" conductor are performed at point $z = 1 \times dz$. The result of measurements corresponding to time point $t = 1 \times \tau$ will be a short negative pulse. However, measurements at point $z = 2 \times dz$ will show a double amplitude pulse corresponding to time point $t = 2 \times \tau$. This doubling is related to pileup of the pulse converted using a capacitor transformer in unit cell no. 2 and the pulse propagated through "victim" conductor from point $z = 1 \times dz$ to point $z = n \times dz$.

Backward crosstalk pulse length (width) increases with a decreasing Z-parameter, but the pulse amplitude remains unchanged (this pulse is not inverted with respect to the "aggressor" conductor). Let us assume again that signal measurements are performed at different points along the conductor. Measurement at point $z = (n - 1) \times dz$ shows a single short positive pulse generated at time point $t = (n - 1) \times \tau$ and propagated backwards through the conductor towards point $z = 0$. Two pulses will be marked at point $z = (n - 2) \times dz$: the first pulse—at time point $t = (n - 2) \times \tau$; the second one—at time point $t = n \times \tau$. This additional pulse is a backward-moving pulse, occurring at point $z = (n - 1) \times dz$ and "arriving" at point $z = (n - 2) \times dz$ with a delay τ. N pulses will be marked at point $z = 0$. Lag time of each of these pulses will be $2 \times \tau$ due to delays in both the "aggressor" conductor and the "victim" conductor. When $dz \rightarrow 0$, impulses become inseparable and combine in one long impulse with a small amplitude.

6.3.4.5 Combination of Inductive and Capacitive Crosstalk

If inductive and capacitive crosstalk are well-balanced as is the case when signal conductors on the printed circuit board are shielded by power or ground layers (i.e., they are performed as "full" strip line), then the forward crosstalk value is close to zero. If conducting tracks of the external signal layers of the printed circuit board are shielded by the ground layer only from one lower side (i.e., they are microstrip lines), the capacitive coupling is weaker than inductive one; and there is a forward crosstalk inverted with regard to the "aggressor" conductor in the "victim" conductor. The reason of capacitive coupling attenuation lies in the fact that most electric-field lines are completed through air but not through dielectric material of which the printed circuit board is made.

Inductive backward crosstalk is always non-inverted with regard to the "aggressor" conductor and its value depends on the geometrical pattern according to which the signal conductors of the printed circuit board are performed as microstrip (asymmetric) lines or strip line. Figure 6.39 shows the difference between these two options.

On the basis of all the above-stated, it is possible to define the following rules by which it should be guided to reduce the crosstalk level.

- Limit the steepness of signal impulse edges (i.e., rise/fall time should be increased; and *dv*/*dt* value should be decreased).
- Properly perform the wiring layout of the printed circuit board (special attention should be paid to the geometrical pattern of the conducting tracks):
  - Use the tracks of the minimum possible length;
  - Provide as much track spacing as possible;
  - Do not place transmission lines, which are sensitive to signal crosstalk, in parallel to tracks, where signals are characterized by high steepness of edges;
  - Differential signal transmission lines should be placed in parallel to each other with as small a gap as possible; it should be done in order that each conductor of the differential line is influenced by the same interference that ensures their pulse-to-pulse cancelation in the differential signal.
- The ground layers should be added in such a way as to avoid of forming the ground islands. Use the ground islands only when you are sure that tracks dedicated to high-speed signal transmission don't cross the ground islands.

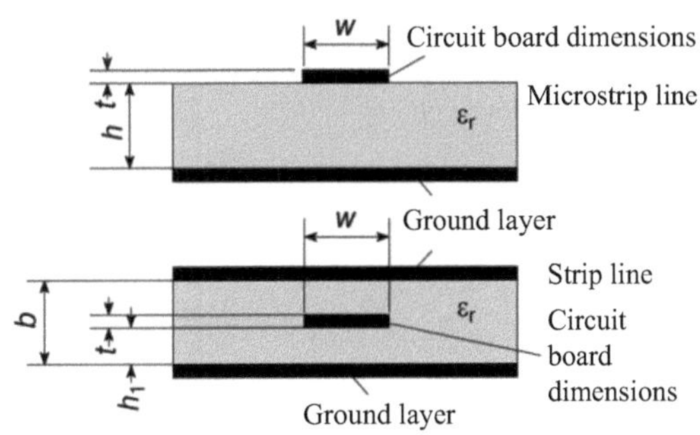

Fig. 6.39 Comparison of designs of microstrip line and "full" strip line

- The ground island located under the conducting track on which the high-speed signal is transmitted doesn't allow the return current to flow on the shortest way, i.e., value l_m will be increased anyway.
- The external signal layers are worse than internal ones in terms of interference immunity:
  - strip lines are more efficient than microstrip lines;
  - c_m and l_m values are balanced when strip lines are used; therefore, forward crosstalk is insignificant;
  - when using microstrip lines, electric-field lines pass mainly through the air but not through dielectric material (unlike the strip line), that's why the capacitive coupling is weaker than inductive one.

The crosstalk level increases when signal is reflected at the leads of conductor. The ways to reduce these reflections are considered below.

6.3.4.6 Relaxation Oscillation (Ringing)

The so-called ringing, shown in Fig. 6.40, represents parasitic oscillations that often occur when transmitting digital high-speed signals.

The reason of these oscillations is the impedance mismatching of signal source and receiver. Due to this, a part of signal energy (or even all this energy) is reflected back to the line. The ringing amplitude is determined by reflection coefficient, and the frequency of these parasitic oscillations depends on the conductor length l_{track} (Fig. 6.41).

Here, the following expressions are used. The frequency of parasitic oscillations is calculated by formula:

$$f_{\text{ring}} = \frac{c}{2 \cdot \sqrt{\varepsilon_r} \cdot l_{\text{track}}}, \tag{6.11}$$

coefficient of reflection from the signal source side

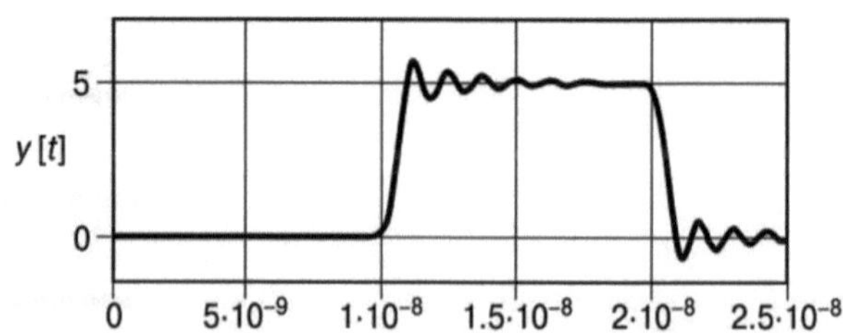

Fig. 6.40 Example of relaxation oscillation (ringing) during transmission of digital high-speed signals

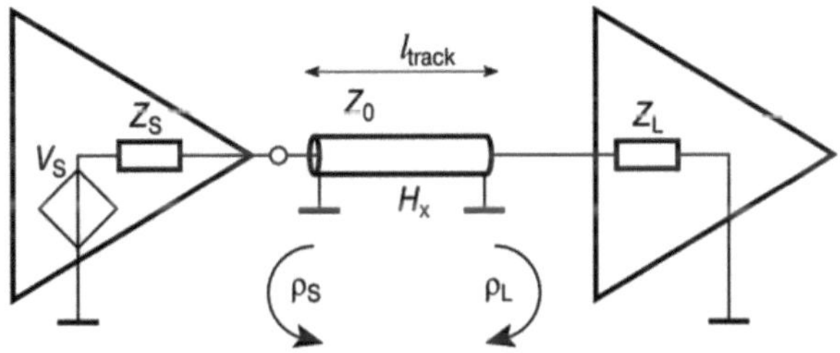

Fig. 6.41 Signal reflection model

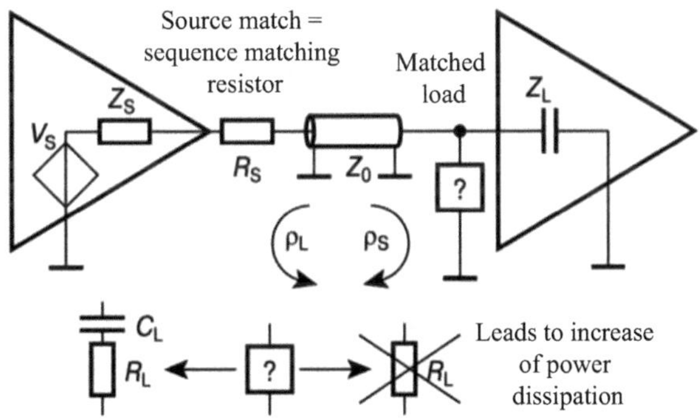

Fig. 6.42 Recommended connection diagram of a termination resistor during transmission of digital signals

$$\rho_s = \frac{Z_s - Z_0}{Z_s + Z_0}, \tag{6.12}$$

coefficient of reflection from the signal receiver side

$$\rho_L = \frac{Z_L - Z_0}{Z_L + Z_0}. \tag{6.13}$$

Since the width of digital high-speed signal transmission band should be rather wide, it is not possible to decrease the reflection coefficient down to zero. However, there are some other ways to reduce parasitic oscillations down to the level that meets the system requirements (Fig. 6.42).

It is recommended to calculate the values of relevant resistors and capacitors using the following formulae:

$$R_S = \mid Z_0 \mid - \mid Z_0 \mid \;\; (\text{usually } |Z_0| > |Z_0|); \tag{6.14}$$

$$R_L = \mid Z_0 \mid ; \tag{6.15}$$

$$\mid \rho C_L \mid >> \mid Z_0{}^{-1} \mid, \text{where } \rho = j \times 2p \times f_{fr}; \tag{6.16}$$

$$C_L \approx 10.\frac{1}{2\pi \cdot f_{fr}|Z_0|} = 5.\frac{t_{fr}}{|Z_0|}, \tag{6.17}$$

where $t_{fr} = \min\,(t_f, t_r)$, i.e., corresponds to the value of critical rise time.

Taking into account these boundary conditions, the “ringing” level can be reduced using the following rules:

- Find the critical (i.e., the most high-speed) signal in the system.
- The length of the PCB conducting tracks intended for high-speed signal transmission should be minimum (shorter than $\lambda/20$ for cutoff frequency corresponding to the steepest pulse edge).
- The conducting track impedances should be matched.
- Use the continuous layers of ground (but not separate ground “islands”).
- Place the series termination resistor as close to the signal source as possible (i.e., to the line driver).
- Place the load termination resistor as close to the signal receiver (to the termination) as possible.
- Matching load is intended for AC matching of impedances, i.e., DC current shall not flow through it (to prevent increased power loss during signal transmission).

Crosstalk and parasitic oscillations (“ringing”) have a direct impact on the transmitted signals. However, when designing systems, it is necessary to take into account other aspects of EMC, for instance, the ones related to arranging appropriate ground planes or selecting a proper system power supply.

6.3.4.7 Ground Instability

Ground instability is undesired voltage drop of a signal in the conductors connected to the ground. This phenomenon can appear at the level of the printed circuit board, as well as inside the chip (Fig. 6.43). As a rule, di/dt ratio is very high in digital high-speed system that is related to very small values dt (time of signal switching from HIGH level to LOW level and conversely). Here, the most critical factor is the parasitic inductance value of the conductor connecting the chip with ground. Due to high speed of signal switching, the inductive component of the complex impedance of the conductor exceeds its resistive component significantly. It means that the current amplitude in the conductor isn’t very large.

To reduce the effect of ground instability, the following rules should be observed:

- Decrease the steepness of signal edges (i.e., increase the impulse rise/fall time).
- Use the connecting wires of low-inductance types as grounding wires. The entire conducting layer (or layers) of the printed circuit board should be used as ground but not separate regions or areas on any layer (since the return current tends to flow on the path with the lowest impedance).
- Use a great number of additional decoupling capacitors; each of them should be located in close proximity to the relevant power or ground terminals of IC.
- Apply the method of parallel connection of several capacitors with different capacity ratings (for instance, 10 pF, 1 nF, and 100 nF) to ensure “overlapping” of the entire range of switching frequencies which can be used in the system. Pay attention that in real conditions, the capacitor with the highest rated capacitance

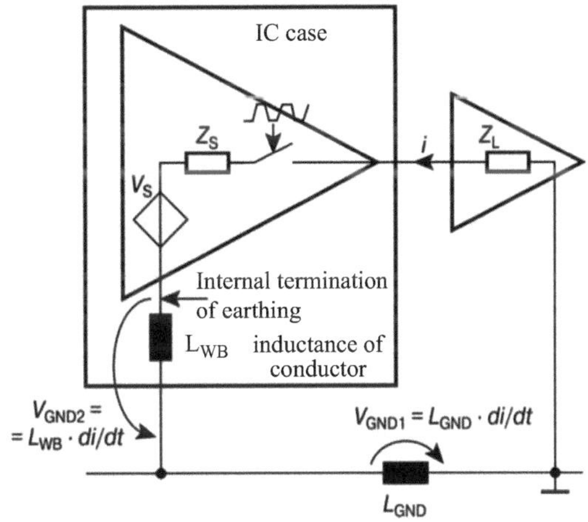

Fig. 6.43 Ground instability effect production chart

value has the worst capacitance parameters at the high frequencies and even it can behave similar to inductor.

- IC developers are recommended to install read buses inside them for internal use; since the current flowing through these lines is very small, additional buses will not increase the ground instability effect. Many internal ground terminals should be located along the perimeter of the IC semiconductor crystal.

6.3.4.8 Power Source Interference

System power source interference may occur for several reasons: due to current switching processes in its different components and as a result of penetration of interferences from DC/DC converter operating in the key mode in the power supply circuit. During operation of DC-to-DC converters, pulses of greater amplitude and higher rise rate (*di*/*dt*) are generated. The main problem is that interferences which are propagated along the power supply circuit impair the efficiency of operation of analog circuits (for example, ADC, DAC, or operating amplifiers) since the value of power supply rejection ratio (PSRR) of these circuits can't be indefinitely great. Therefore, interferences penetrating the power source circuit cause the emergence of interferences in IC outputs; attenuation degree of these interferences depends on PSRR value.

Interferences of similar type can be reduced by using the following methods:

- Use the conducting tracks with low impedance for power buses (for example, conducting tracks can be widened or a ground layer can be used that's even better). It can minimize voltage drop in the power bus.
- Use ferrite beads or resistors for reduction of interferences in the power buses. Make sure that their parameters (I_{max}, f_{20dB}, R_{DC}, etc.) correspond to the specified operating range. Of course, it is necessary to use local energy storage units

(capacitors) which should be located as close to the components consuming high current as possible.
- Decoupling capacitors should be located in close proximity to the power or ground terminals of IC.
  - It is necessary to take into account the fact that real capacitors contain parasitic elements such as equivalent series inductance (ESL) and equivalent series resistance (ESR). These elements have strong impact on the operational efficiency of the decoupling capacitor.
- Carefully select the parameters of the power source, i.e., such operating frequency of DC/DC converter, which makes it possible to:
  - meet the IC performance and noise level requirements (it should be noted that PSRR value is reduced with an increase in operating frequency),
  - select an operating frequency of the DC/DC converter outside the frequency range used by the data transmission system,
  - if possible, use the linear power supply.

Many recommendations on optimal design of printed circuit boards in terms of ensuring the EMC and signal integrity are applicable for IC design at the crystal level. They concern not only "solid-state circuit engineering" but also take into account the application aspects (for example, how to dedicate the IC terminals correctly, choose the timing parameters, etc.). Development of chips meeting the requirements for functionality and EMC is not the sufficient condition for development of the entire systems which satisfy these requirements. The solution of this task supposes close cooperation among IC developers, system developers, and users. Taking into account all of the above, the following recommendations on system development can be offered:

- The conducting tracks on the printed circuit board should be done as short as possible (first of all, it concerns the high-speed signals).
- Use separate layers of multilayer printed circuit board for wiring the power and ground buses; insulation regions on these layers should be avoided.
- The adjacent layers should be used for wiring the power and ground buses since together they form a shared circuit capacitor.
- Decoupling capacitors should be located in close proximity to the power or ground terminals of IC.
- Check of the reference data of decoupling capacitors. Their parasitic parameters (ESR, ESL, resonance frequency) should be taken into account when designing the high-speed systems.
- To minimize the signal reflection, the corresponding impedances should be matched.
- Use the properly selected resistors for load matching of the high-speed signal lines (of the PCB conducting tracks).

- Make sure that path of return current flowing is minimal. The level of return HF current is much lower than the current level in the signal line (the loop with the lowest impedance has the smallest inductance at high frequencies).
- The loops, where pulse currents flow with high rise rate (di/dt), should be as short as possible. In such a way, it is possible to reduce the crosstalk level, as well as electromagnetic emissions and other interferences related to EMC.
- When designing the printed circuit board, the tracks of high-speed digital signals should be located away from the tracks on which signals of sensitive analog devices pass.
- Parallel location of the printed circuit board tracks can have both positive and negative effect:
  - avoidance of parallel location of tracks if it is necessary to have the minimum level of crosstalk;
  - otherwise, for differential signal transmission, it is necessary to use parallel tracks which are located as close to each other as possible.
- Increase the spacing between those tracks where cross impact of signals is undesirable.
- Wiring of trucks for the most high-speed signals should be performed on internal layers of multilayer printed circuit board.
- Work on system design should be performed in close cooperation among IC design engineers, system developers, and developers of printed circuit boards.

6.4 Methods of Reducing the Level of Intermodulation Distortion in High-Speed Electronic Devices

6.4.1 Mechanisms of the Occurrence of Passive Intermodulation Events

Intermodulation distortion level (IDL) of SHF broadband signals is a critical parameter affecting the characteristics of communication systems, as well as their compliance with EMC standards [17]. A specific type of IDL is passive intermodulation (PIM) occurring in nodes and structures traditionally considered to be linear ones: in substrates of printed circuit boards, microstrip lines, radio frequency cables and connectors, in antenna structures and items surrounding the antenna. Increased signal power, frequency-division multiplex, increased space density of mobile base station equipment arrangement, and other factors [18] impose high requirements to the level of PIM. PIM causes, measurement techniques, and reduction methods are considered in detail in the study [19].

The intermodulation phenomena occur in electric circuits during non-linear conversion of the amount of fluctuations of a few frequencies or one signal with continuous frequency spectrum. In addition to distortions within the range of input signal frequencies, the out-of-band interfering intermodulation components with frequencies that were absent in input occur in the output circuit of such node. PIM products can get in the transmission band of the receiver of this communication system or in the frequency bands of other communication systems, thus violating the EMC requirements. For users of telecommunication systems, it means that the signal-to-noise ratio decreases, while data reception quality deteriorates [20].

If two sinewave oscillations with frequencies of f_1 and f_2 occur at the input of the electronic assembly, then signal components with combination frequencies of $f_K = \pm nf_1 \pm mf_2$ appear at the output, where n and m are the positive whole numbers 0, 1, 2, etc. If one or both of the transmitted signal components (Tx) have continuous spectrum with emission frequency band (Δ), then receiver interferences (Rx) are structured as a third-order intermodulation band (IM3) based on combinations of the form $2f_1$-f_2 and $2f_2$-f_1, a fifth-order intermodulation band (IM5) based on combinations of the form $3f_1$-$2f_2$ and $3f_2$-$2f_1$, as well as intermodulation bands of higher odd order.

The levels of components with frequencies higher or lower than transmission bands (IM^+ and IM^-, respectively, as shown in Fig. 6.44) are identical, provided that non-linearity is inertia-free; they may differ, if a memory effect is caused by amplitude-phase conversion in the inertial non-linear element. The ratio between the levels of intermodulation products of the third, fifth, and higher orders varies as the input signal power P_{inp} increases at frequencies f_1 and f_2: IM3 power increases at the rate of 3 dB/dB; IM5 power increases at the rate of 5 dB/dB, etc.

As a result of severe lack of spectrum, IMI components may get in separate bands of a low-noise receiver of the same or similar frequency standard and impair its operation [21]. For instance, receive/transmit bands (890–915)/(925–960) MHz are assigned to the communication network of GSM-900 standard. For transmission frequency values of 930 and 955 MHz, the third-order intermodulation component with a frequency of 2 × 930–955 = 905 MHz penetrates the band for reception of this standard. For standard 3G-UMTS with frequencies of (1920–1980)/(2110–2170) MHz, the seventh-order intermodulation component with a frequency of $4f_1$-$3f_2$ penetrates the band for its reception.

The situation is more difficult, if antennas of different standards are installed on the same cell tower [22]. For example, the second-order intermodulation products

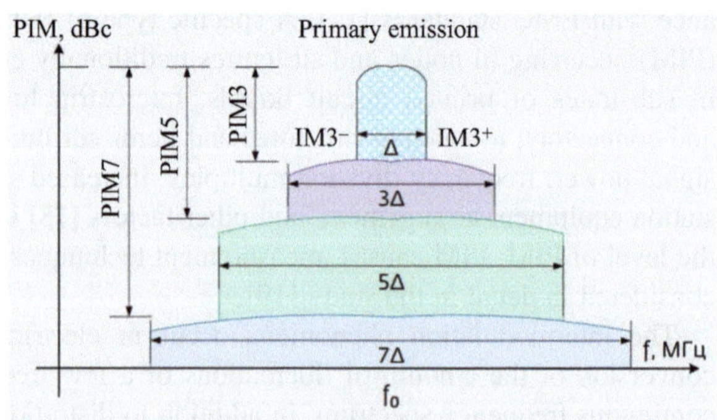

Fig. 6.44 Intermodulation products in output of non-linear node next to the operating frequency band. f_0—signal central frequency

of GSM-900 system penetrate the band for GSM-1800 reception; the fourth-order intermodulation products of GSM-900 system and the third-order intermodulation products of GSM-1800 system interfere the reception of signals of UMTS standard. With the development of 4G broadband systems with frequency bands ranging from (832–839.5)/(791–798.5) MHz to (2530–2540)/(2650–2660) MHz, as well as digital television (DTV) in Russia, the risk of undesirable effect of interfering intermodulation products has increased. Widespread use [23] of co-siting solutions (shared use of equipment by different operators and for different communication standards) results in the pollution of frequency bands of cellular base stations caused by radiation due to passive intermodulation.

6.4.2 Methods of Reducting the Level of Passive Intermodulation in Radio Frequency Connectors

In coaxial connectors [24], non-linear phenomena causing PIM can appear due to insufficiently tight contact in the current-carrying "metal-to-metal" connections, as well as due to non-linear processes in the connector housing or in the surface-covering material.

Cavities, irregularities, microcracks, and dirt accumulation can appear on the metal surface at the contact area along the center conductor or on the internal surface of the coaxial line due to coaxial connector design features or manufacturing technology, insufficient force at the contact area, deficiencies in cable-end sealing. In this case, the conductive path becomes longer, the potential voltage barrier appears; microscopic sparking can occur with generation of intermodulation products. Surface metallic oxides act as metal-oxide detectors with non-linear characteristic ("diode effect"). The metal surface accumulates moisture, dust, organic impurities, and residues of various metal salts after operations of chemical etching and coating deposition. They also have non-linear current–voltage characteristic (VAC) causing PIM. The sealing contact is required to avoid corrosion from exposure to ambient oxygen since the oxide layer of even small thickness has non-linear voltage–current characteristic.

The resistive and magnetic-resistive effects in dissimilar conductors inside connector cause a local resistance change at the contact area. Steel (including stainless steel), iron, nickel have ferromagnetic properties with possible effects of hysteresis causing the non-linear phenomena and PIM. Therefore, high-quality connectors should be made from expensive but nonmagnetic materials such as brass, beryllium, and phosphorous bronze. If the corrosion-resistant coating is applied to the material having magnetic properties (for example, as a result of silver-plating or gold plating), it is necessary to take into account the skin layer depth for HF currents passing through contacts. In case of insufficient thickness or unevenness in coating, the currents penetrating the basic material cause PIM. If there are strict requirements for PIM, it is inadmissible to use electrodeposited nickel as a corrosion-resistant coating of the connector. The connector housing made of aluminum alloys reduces the product weight but the oxide film on the surface of this material can generate an inadmissible level of PIM.

"White bronze" (nonmagnetic corrosion-resistant wear-resistant alloy of copper, tin, and zinc) and silver are widely used for coating the connectors optimized by PIM level. IM3 level in high-quality connectors with white bronze coated housing and silver-coated central contact shall not exceed 155 dBc during dual-frequency test with a signal power of 2 × 43 dB/mW. IM3 level measurements for typical coaxial connectors using a test signal with a power of 2 × 43 dB/mW at frequencies of 463 and 468 have shown that the level of IM3 self-interference for SMA connectors was 108 dBc, for BNC connectors—118 dBc, for "standard" connectors—123 dBc, for "N silver-coated" connectors—163 dBc, for DIN 7–16 connectors—at least 175 dBc. It was found that direct connectors have lower level of PIM than right-angle connectors.

It is difficult to detect the low-quality preparation and cable sealing in the coaxial connector when inspecting with the naked eye. The connector quality by PIM parameter decreases as quantity of connect–disconnect cycles growths due to wear and tear, microcracking, and force weakening in connections "metal–metal."

6.4.3 PIM in the Printed Circuit Board Material

The substrate material and manufacturing technique of the printed circuit board are essential to reduce the level of PIM. Leading manufacturers of clad dielectric materials (laminated plastics) for SHF board substrates optimize the properties of their products by reducing dielectric loss and diversely matching COTE of the insulating material with COTE of metal coatings [25, 26]. For example, special-purpose PIM-laminated material of DiClad 880-PIM type manufactured by Arlon-MED Company on the basis of multilayer glass-fabric-reinforced foiled PTFE has a dielectric permittivity of $\varepsilon_f = 2.20$ at a frequency of 10 GHz, a thermal factor of $TK_\varepsilon = -160$ ppm/°C, and low attenuation of $tg\delta = 0.0009$. Laminated material has high-thermal conductivity $\lambda = 0.261$ W/mK within the temperature range from −40 to 150 °C. Products made from this material are tested by a dual-frequency signal with a power of 2 × 20 dB/mW in the frequency range of 1.9 GHz. The test results show that IM3 relative level does not exceed 155 dBc.

6.4.4 PIM in Band, Coaxial and Waveguide Transmission Lines

The PIM level in transmission lines is increased due to cold welding, twisting, tie wrap, impurities on the dielectric plate, cracks in the semi-rigid cable sheath, ambient temperature fluctuations, heating of conductors from flowing direct and high-frequency current.

In the microstrip transmission lines [27], the PIM level may increase due to the following reasons: the use of PCB dielectric of insufficiently high quality; the use of coating interlayers made of nickel having ferromagnetic properties; occurrence of temperature distribution gradients in the board material caused by heating of printed conductor.

It has been established that in models of coaxial cables that are not aimed at reducing the PIM level, with a dual-frequency test signal within the range of 1.8 GHz, a power of 40 dBm (10 W), IM3 level is −78 dBm. If a certified cable and DIN 7–16 connectors are used, the level of such interference is reduced by 30 dB and does not exceed −125 dBm.

The fluctuations in ambient temperature caused by differences in thermal-expansion coefficients for substrate, coatings, and metallization lead to a decrease in tightness of joint and to formation of local zones with oxides having non-linear voltage–current characteristic. The printed conductor heats up from the current flowing through the transmission line [28], which causes a different thermal expansion of the substrate and coating insulating material.

The specific PIM source—multipactor effect—is typical for passive transmission lines included in satellite repeaters operating in vacuum, for presence of electrojet current causing the secondary electron emission.

6.4.5 PIM in Directional Couplers, Frequency Duplexers, and Transformers

In directional couplers, valves, and circulators containing ferrite materials [29], non-linear effects are produced by hysteretic nature of their magnetic properties. With the increasing power of an input signal, the magnetic-flux density increases that leads to the saturation mode as a result of which the PIM level becomes higher.

In addition to non-linearity caused by ferrite metal, the increased level of PIM in the transformers with ferrite cores is due to magnetic reluctance of the core. The manganese-zinc cores of low resistance are used for typical passive components. They are usually coated with insulating material to prevent short circuit. Winding wires are insulated to prevent short circuit between winding turns. It has been found that during standard assembly operations, the insulation of wires and the core may be damaged, leading to local short circuits and, as a result, the PIM level increases [30]. The use of special materials and insulation quality control in the structures optimized by PIM level reduce the probability of such effects.

The architecture, materials, and manufacturing technique of the frequency-splitting filters–diplexers play an important role in reducing the impact of PIM on operation of data transmission system. To ensure the squareness of amplitude–frequency characteristics in the transition zone between *Tx* and *Rx* bands, use high-order resonant circuits with fine tuning. The use of tuning elements in the form of steel screws with insufficiently dense silver coating can significantly increase the PIM level due to metal oxides in the threaded joints, and lead to the dynamic changes in the PIM level under influence of mechanical vibration.

6.4.6 External PIM Sources

The PIM sources external in relation to the transmitting device and the antenna path [31] are structural components of the antenna assembly, cell towers, items, and facilities surrounding the antenna. They include steel bolts, couplings, contact joints having non-linear voltage–current characteristics and being capable of re-emitting the power of incident SHF-signal in different directions at different combination frequencies. As a result, the signal in artificial antenna can have the permissible level of PIM, but in the far-field region, the PIM level increases up to impermissible values. This phenomenon known as "rusted bolt effect" is typical for antennas and structures which are inadequately protected against impact of aggressive environment. During operation of 3C/4G network base stations in common band, the PIM from external sources leads to loss of operational efficiency and reduction of coverage area (Fig. 6.45) [31].

Modern antennas are complex electrical and mechanical structures which support a few bands and polarizations, as well as they have a remote-controlled directional diagram. Consequently, there is a strong possibility that mechanical failures occur during signal transmission. The antenna is usually checked for PIM level prior to installation. For exact measurement of antenna characteristics under operating conditions, its location needs to be taken into account since there is a potential for PIM occurrence from external sources. Ideally, the antenna should be located on the metal structure not less than 0.5 m in height above ground, and its main beam should be pointed to the open space. The place around the antenna should be free of any metal structures (fences, elevators, and towers) which can cause PIM. The airports, sports

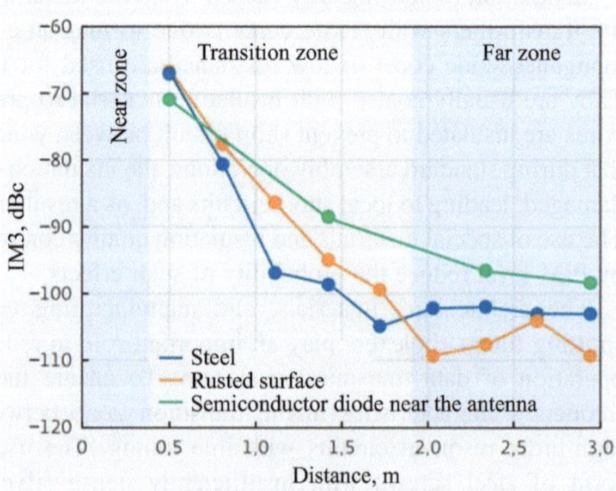

Fig. 6.45 PIM level change depending on the distance between the reflecting object and the antenna (anechoic chamber, test 2 × 20 W)

facilities, and long tunnels can use the distributed antenna systems where signals from base stations of all operators are combined and use the common antenna and distributing system [32]. In such cases, PIM prevention is a critical task which cannot be solved even with a filtration of high quality.

For search and minimization of interference from external PIM sources, beam azimuth direction and pattern elevation angle are used; it is necessary to remove non-linear units reradiating IM components, for example, metal beams prone to oxidation.

The frequency duplexers (splitters of incident and reflected waves of close frequency bands) with increased squareness of amplitude–frequency characteristics in the boundaries of frequency bands as well as structures containing materials and components that may lead to non-linear conversions of transmitted signals are used to decrease the PIM level. Measures are taken to reduce the peak-factor of the sum signal power of different frequency bands in common parts of antenna path; and the type of polarization of transmitting and receiving antennas is changed. Recommendations have been developed for location of antennas of different standards and frequency bands in one tower in order to minimize their interrelation and impact, as well as for design and location of indoor antennas taken into account the design features of building and equipment.

6.4.7 PIM Level Evaluation Methods

In literature and in practice, great attention is paid to PIM measurements and means of their reduction for improvement of energy performance of power amplifiers [30]. Recommendations of International Electrotechnical Commission (IEC) 62,037 (Committee 46, Working Group 6) [33] determine the procedure for measuring the PIN level by testing the circuit or its component using a signal in the form of a sum of two sinewave oscillations of offset frequencies f_1 and f_2 having the same power of 2 × 10 W, 2 × 20 W, or 2 × 30 W, as well as by measuring the spectral component level at one of the IM3 frequencies—$2f_1$-f_2 or $2f_2$-f_1.

The PIM level can be measured by two methods: "at reflection" (Fig. 6.46a) and "at transmission" (Fig. 6.46b) The cable assemblies, filters, mixers, and antennas can be used as units under test. The spectrum analyzer acts as a receiver.

When measuring low-level PIM, care should be taken to properly weaken the coupling between the test signal synthesizers, for instance, with a bridge adder; to ensure the low level of own PIM components occurring in cables, connectors, frequency filters, in high-power matched load of the measuring circuit (not more than 170 dBm for a test signal of 2 × 43 dBm).

To characterize the PIM level of a measured component, it is recommended in [33] to use absolute power of a signal at IM3 frequency in the logarithmic scale [dB/m] in relation to the power of 1 mW or its power value in relation to the power of each of the two test signal sources in the scale [dBc] with obligatory reference to

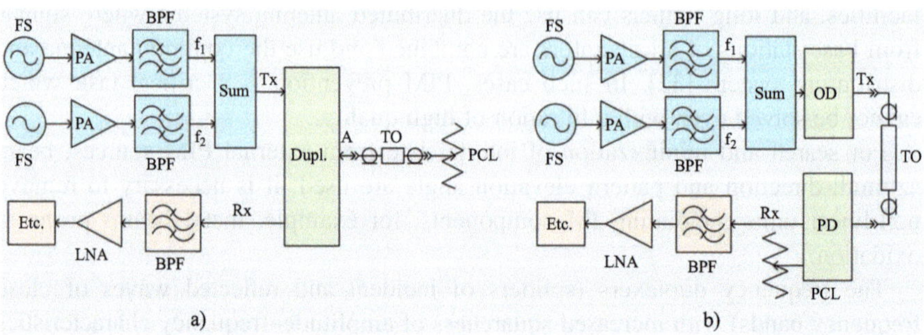

Fig. 6.46 PIM level measuring diagram “at reflection” (**a**) and “at transmission” (**b**). *FS* frequency synthesizer, *PA* power amplifiers, *BPF* band-pass filters, *Summ* bridge signal summator (combiner hybride, dyslexer), *Dupl* duplexer of the frequency distribution of the input signal Tx in the direction of the test object (TO) and the signal reflected from it in the direction of the receiver (Pru), *LNA* low noise amplifier, *PCL* powerful consistent load

Fig. 6.47 Combination interference total power depending on the power of a double-frequency test signal

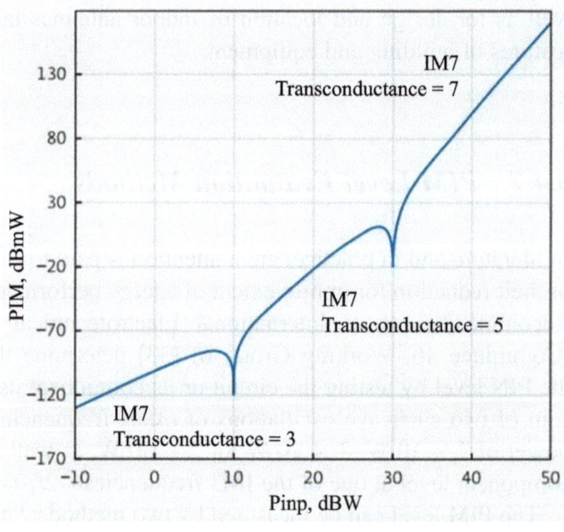

the test signal power. Therefore, evaluations of PIM −100 dBm and −153 dBc are of equal value when the test signal is 2 × 43 dBm.

Since the PIM level can change within very wide ranges, to detect serious failures in the radio frequency path, test signals of lower power are used (for example, 2 × 2 W). The test signal power is increased up to 2 × 4 W and more for fine measurements and detection of additional sources.

In case of increased operating power of radio transmitting equipment, the greatest interferences for the receive path can be generated by intermodulation components of higher odd order such as IM5, IM7, etc. Their power increases much faster with the increasing test signal power (Fig. 6.47).

In systems with many carrier frequencies, it is necessary to control the level of operating signal power peak factor for adjusting the test source power in accordance with it. When using the transmission technology with extended spectrum, the PIM-interference is wideband too that should be taken into account when adjusting the receiver.

The method "at reflection" has higher sensitivity; it is used for items with one and two ports. According to [33], the residual level of own intermodulation products and other elements of measuring circuit should be 10 dB lower than the level of products of the tested device. This level must not exceed −170 dBc. It is offered to carry out a dynamic test (Fig. 6.48) to check the mechanical integrity and structural stability. For this purpose, the tapping of components and junctions, filters, back-boards, antennas, cable bending is performed. Tapping is used to identify defects of contact connections, which can increase the PIM level when exposed to heat, cold, wind, vibration, fluctuating loads.

The primary test of the PIM level is performed by method "at transmission," then method "at reflection" is used; in this case, the nearest point with significant level of PIM is determined. Each time when the source is found, the search precision is increased in terms of the value of the PIM source. Tests will be repeated until all considerable dynamic effects of PIM are eliminated (Fig. 6.49).

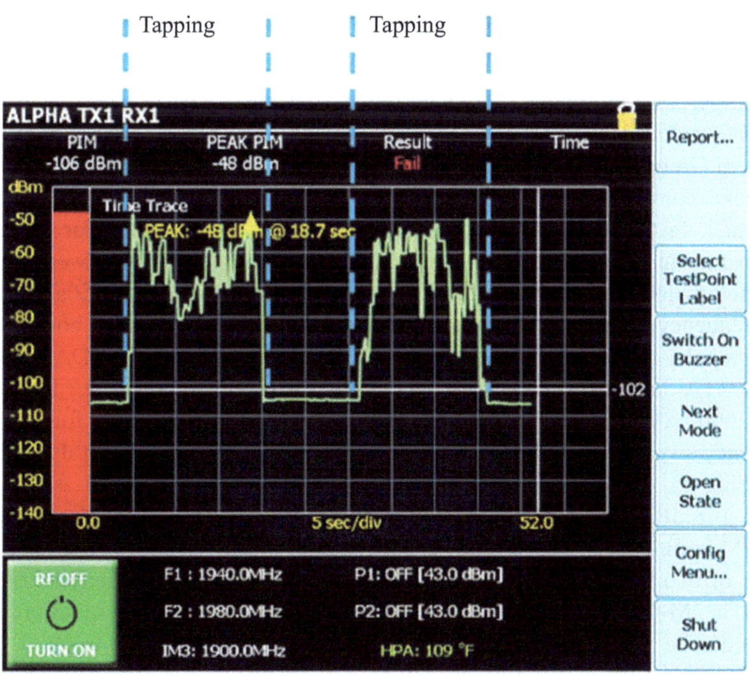

Fig. 6.48 Dynamic PIM test during component tapping with test signal

a)

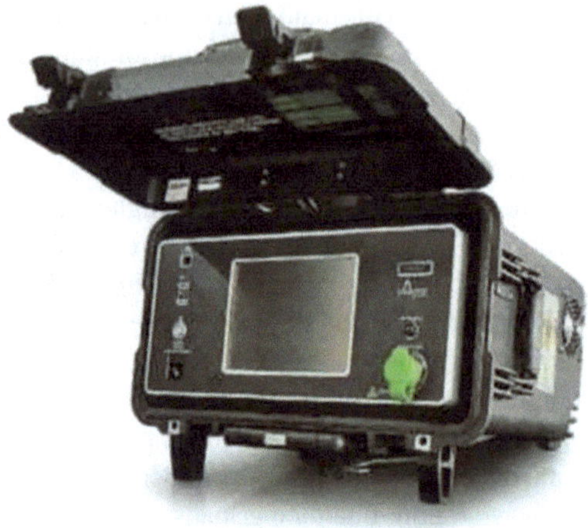

b)

Fig. 6.49 PIM level meters by Kaelus: (**a**) battery-powered tester of iTA series; (**b**) portable analyzer of iOA series

6.4.8 Specialized Equipment for PIM Level Testing

Special-purpose equipment to test the circuits, devices, and systems for the PIM level is produced by such companies as Kaelus, Agilent, Rosenberger, Wireless Telecom Group, AWT Global, Anritsu, etc.

Kaelus company (department of Smiths Interconnect company) is a successor of Summitek Instruments, a well-known manufacturer of equipment for testing the PIM-parameters of the network base stations. The portable battery-powered passive intermodulation analyzer of iTA series made by this company (Fig. 6.50a) combines two frequency ranges. It is designed for operation at the top or at the bottom of the antenna pole, on the roof, in the building with phased-array multiple DAS-systems. Series iTA includes model for standards LTE-800, GSM-900, DCS-1800, LTE-2600, AWS, etc. equipped with built-in modes of monitoring for spectrum, frequency sweeping, and of tracking the PIM changes by time during dynamic test according to standard IEC 62037. Their self-generated PIM level is not more than -168 dBc during test 2×20 W. The receiver sensitivity enables measuring the PIM components with the power level from -50 to -128 dB/mW. The input power of two test sources is set within the range 20–43 dB/mW in increments of 1 dB. The IM-components of the third-, fifth-, optional seventh-, and ninth-orders are checked. The weight of tester iTA is 22 kg; the shock resistance is 40 G. Being powered by integrated batteries, the device operates up to 2 h at a temperature -10 to 45°C. Range-to-fault and distance-to-PIM modules promptly determine distances to the irregularities in the path and to a few PIM sources, respectively.

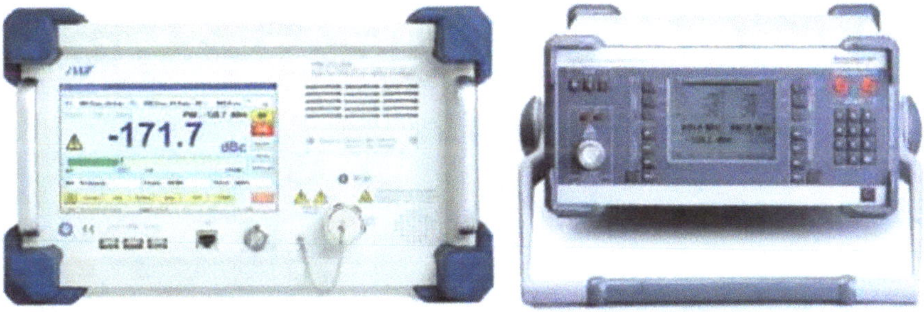

a) b)

Fig. 6.50 Laboratory PIM-analyzers: (**a**) of S1L series made by AWT Global Company; (**b**) of Portable Type series made by Rosenberger Company

The analyzers of iQA series (Fig. 6.50a) enable monitoring of intermodulation components of odd order from IM3 to IM11. They have the test signal power from 33 to 43 dB/mW. They are powered by the mains and operate within the ranges *Tx*/*Rx* from the lowest band of standard 700 NHz—(728–746)/(698–716) NHz to the frequencies of standard UMTS—(2110–2170)/(1920–2080) MHz.

Laboratory analyzer of S1L series made by AWT Global Company (Fig. 6.50a) is intended for diagnostics of PIM-phenomena at frequencies from 730 to 2690 MHz. Sensitivity to PIM components in the analyzer S1L is -172 dBc during test 2×43 dB/mW; the power of test signal sources is set within 15–46 dB/mW. The device enables measuring of intermodulation components of odd order from IM3 to IM25; it has 96 dB; it records the results; it is equipped with distance-to-PIM and distance-to-fault measurement modes. The device is equipped with cables, matched loads, connectors, and adapters with low PIM level, test distorted signal source PIMGEN with the PIM level up to -80 dB/mW.

The PIM level analyzers are produced by Rosenberger company in portable (Fig. 6.50b) and rack-mounted design. They are intended for operation with standards AMPS, EGSM, DCS, PCS, UMTS, WiMAX in bands (3500–3600)/(3400–3500) MHz. They check the IM components of the third to ninth order using double-frequency test signal with a power of 37–46 dB/mW. Their self-generated PIM level is not more than -168 dBc.

Tester MW82119A of PIM Master series made by Anritsu company [34] (Fig. 6.51) was developed for calibration of PIM level in the frequency band of 900 MHz in networks GSM, UMTS-900, and LTE. It is characterized by high-precision measurements and equipped with distance-to-PIM mode to determine the distance to the non-linearity source and PIM vs. time mode for dynamic testing. The GPS-receiver is optionally available. It allows to provide the recorded information with exact coordinates of measuring point. In addition to standard power 2×20 W, this device has an option to test using a signal with an increased power of 2×40 W for more precise determination of problems which may go undetected when tested by a signal of standard power.

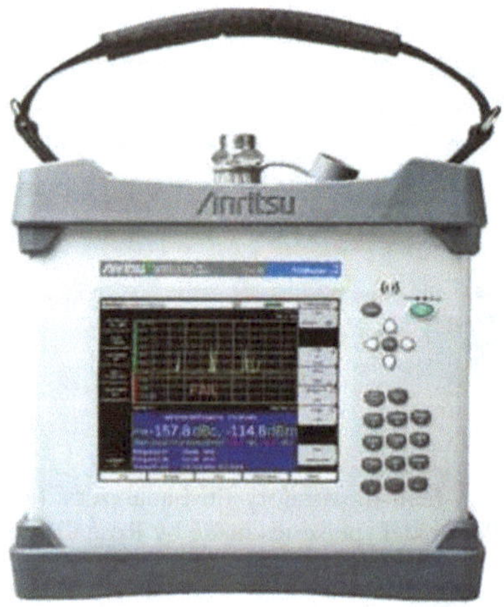

Fig. 6.51 PIM level analyzer MW82119A of PIM Master series made by Anritsu company

It should be noted that modern telecommunication systems depend more than ever before on the characteristics of radio frequency infrastructure, the implementation of which is directly related to the design of components and quality of their couplings. PIM-testing has become an integral part of creation of new telecommunication systems and technologies. The archive of test results can be used for network monitoring in case of degradation of communication quality and call blocking.

Problems arising from the increased level of PIM and violation of system internal requirements for EMC should be approached in their entirety: measures should be taken to optimize the linearity of materials and components within the intervals of changes in the environmental parameters; frequency bands, architecture, and manufacturing technique of frequency-splitting circuits, as well as design and layout of antennas should be selected taking into account the requirements for PIM.

6.5 Protective SHF Devices for Telecommunication Systems

6.5.1 Classification and Application Features of Protective SHF Devices for Communications Electronics Equipment

Currently, SHF protective devices (PD) are widely used both in military and in civil radar location, communication, and navigation complexes. They are designed to protect ingress paths of radar station receivers against high levels of SHF-power of their own and "foreign" transmitters.

Previously, the protective devices were used only in the ground, mobile, fixed, marine, and airborne radar sets. Most often, the locator had one common antenna for transmission of a large signal and reception of the same target-reflected signal.

That's why one of the master nodes of antenna-feeder devices (AFD) is an antenna switch with protective device.

In communication and navigational systems, a probing or transmitted signal arrives at the retranslator on the satellite which as a rule forms a backward signal at other frequency. The issue of protection of the receiver against its own transmitter is solved by the presence of diplexer in AFD. That's why the protective device isn't usually installed in the synchronous satellite communications and navigational systems. However, the protective device is required for communication and radar location systems when intended interferences are generated. Marine radio communication and location systems are the most vulnerable because the breaking signal sources can approach the ship at rather short distance. It is possible to imagine such situation in the air when using the modern means of electronic countermeasure.

The protective devices which protect the receiver input against its own transmitter can be controlled by supplying to them a control signal synchronized with a transmitter signal. PDs intended for protection of receivers against "foreign" signals are commonly referred to as limiters; limiters operating without power voltage supply are commonly referred to as stand-alone limiters [35–37]. Implementation of stand-alone protective devices is more difficult than controlled protective devices because they have to operate under action of SHF-signal, for which the thinner structure (for example, solid-state structure) is required. When possible, the stand-alone protective devices are used instead of controlled ones because in this case, the protection reliability increases. The stand-alone protective devices will be considered in this section.

Developers of transmit/receive equipment have been dealing with the issues of developing and improving PDs from the very beginning [37]. Over the decades, as improvements are introduced into receivers and high-capacity power supply sources, the requirements for protective devices become more and more difficult to implement.

Due to emergence of modern technology of semiconductor transistors used in the receiver stage, it became possible to reduce distances between their electrodes that is required for implementation of low gain-to-noise-temperature. However, this reduces the permissible level of SHF-power supplied to the receiver input stage; so currently, it is equal to the very few mW. At the same time, the presence of superpower pulse generators, which can be used to destroy signal sources, leads to the need to design protective devices, which are capable of resisting the action of SHF pulse power reaching units of kW and SHF continuous power reaching hundreds of W.

The mentioned requirements become difficult to implement especially in the upper part of the frequency range where communication and radar location systems operate; currently, it is cm- and mm-range but in perspective, it will be submillimeter-wave range. In this case, it is necessary to ensure low inserted losses of the protective device; otherwise, it is not practical to reduce the receiver noise due to the use of very vulnerable transistor in its input. Typical required loss of a protective device does not exceed 0.5 + 2.5 dB depending on the value of average operating frequency.

The protective device parameter set wouldn't be complete without consideration of spike leakage energy limitation and time of protective device recovery once the SHF-power action on it is over. Spike leakage energy can deactivate the receiver input transistor because of breakdown of its interelectrode spacings. The permissible values of spike leakage energy are within 10–20 nJ. Recovery time for locating the dead zone is tens and hundreds of nanoseconds (when the recovery time is 1 μ, the distance to dead zone is 150 m).

For specific applications, there is a variety of additional requirements complicating the implementation of protective device: capability to withstand severe mechanical and climatic effects, small dimensions, etc.

Despite the significance of protective device for operation of complex hardware, this class of SHF devices has not been given sufficient attention throughout its historical development. It was reflected in explanation of physics of processes which occur in semiconductor protective devices. There are well-grounded monographs on transistor amplifiers [38, 39], and only recently, several studies of protective devices of polemical nature have been published [40]. Protective devices have been traditionally analyzed based on the equivalent-circuit concept [41]. Other examples could be mentioned to explain the lack of consideration for protective devices. Semiconductor transistors intended for input amplifiers and power amplifiers are implemented on multilayer structures based on different materials (silicon, gallium arsenide, indium phosphide, etc.). Precision process equipment was specially developed for production of semiconductor substrates and application of electrodes. So it should seem that this equipment could become the basis for improvement of semiconductor diodes used in the protective devices but this has not yet happened. The industry continues manufacturing the diodes developed 10 and more years ago.

Ashkenazi et al. [36] attempt to trace the major milestones of PD development, find causes of their underdevelopment, as well as discuss the prospects for their improvement.

Among many SHF protective devices, it is possible to highlight three of the most widely used types: gas discharge, semiconductor protective devices, their combinations and vacuum protective devices. They are the focus of further consideration in this section.

6.5.2 Gas Discharge Protective Devices

Gas discharge protective devices were developed first and still occupy their rightful place among stand-alone protective devices designed to withstand high power levels [35, 36]. There are coaxial and waveguide protective devices. The most common type of waveguide gas discharge protective devices is shown in Fig. 6.52. In one of the first variants, a containment area is formed in the waveguide duct (Fig. 6.52a). In general, it contains a resonator with a gas-discharge gap or a resonator system with several gas-discharge gaps. The pressurized volume is filled with gas under certain pressure. At the low level of power, gas isn't ionized; and the total SHF-power

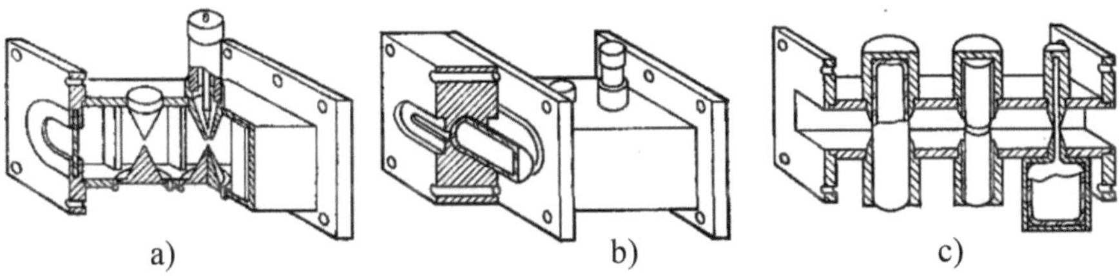

Fig. 6.52 Gas discharge protective devices: (**a**)—with pressure-tight resonators; (**b**, **c**)—with inserted resonators

passes through the discharger; at the high level of power, discharge occurs due to gas ionization in the discharge gap; current flowing in it makes the discharge gap shorter that leads to off-resonance cavity, as well as the forward SHF-power bounces off it. Gas discharge protective devices shown in Fig. 6.50a are perfectly suitable for use in the millimeter wave region.

The gas dischargers have gone through many phases of improvement. In particular, there has been an important shift towards the use of inserted capillary dischargers. In dischargers of this type, only the discharge gap area is filled with gas, not their entire volume (Fig. 6.52b, c). It makes possible to significantly increase in high-speed performance of gas discharge protective devices reaching the fractions of μs. The inserted discharger contains a quartz envelope filled with chemically active chlorine [42]. The inserted dischargers are used in dm- and cm-ranges of wavelength.

A major leap forward was made with the introduction of radioactive elements (for example, tritiated titanium film) in the area filled with gas to ensure spontaneous ignition and gas discharge [43]. Previously, for ignition, it was necessary to supply high external voltage (up to hundreds of volts) through special electrode that was rather inconvenient, particularly for protection of disconnected equipment.

The gas dischargers are produced in many options and can differ in allowable input power and in leakage power. Depending on this, there are dischargers of preliminary and final protection of the receiver.

The main disadvantage of gas discharge protective devices is their slow speed of response. Both gas discharge and semiconductor protective devices are characterized by discrepancy between the response time and forward SHF-power. The recovery time is a limitation parameter determined by plasma-discharge time since the end of the transmitter impulse. Another disadvantage is high power consumption due to insufficiently high conductivity of gas discharge. Currently, the pace of improvement of gas discharge protective devices has decreased significantly; the great leap in their development could hardly be expected. The primary focus is on other types of protective devices.

With the advent of semiconductor protective devices it became possible to add stages of semiconductor protective devices to stages of gas discharge protective devices. In these additional stages, the spike leakage energy can be reduced due to their increased high-speed performance. Now, such combined protective devices are the main type of protective devices.

6.5.3 Semiconductor Protective Devices

Semiconductor protective devices have the greatest potential for improvement, thanks to their high reliability (at least, it is no lower than reliability of quiet transistors located behind them). First of all it concerns the creation of rational electrodynamic systems in which the properties of semiconductor structure are used to the fullest extent possible. The waveguide protective devices are commonly used because low-noise amplifier in waveguide design works better than coaxial (or stripline) low-noise amplifier, thanks to filtering properties of waveguide duct itself. Besides, being waveguide structure, the waveguide duct itself is less sensitive to impact of high SHF-power and it has lower losses than coaxial and stripline amplifiers.

For building the semiconductor protective devices, the waveguide has one more advantage: it easily houses an assembly of many semiconductor structures that enables a proportional reduction in electrical load on each structure and increase in permissible input SHF-power.

The following milestones can be distinguished in the history of development of semiconductor protective devices. One of them is "make-up" of switching diodes. One path houses a semiconductor limiting VD1 pin-diode with rather thick base, which functions as the main switching link, and a low-power "feeding" VD2 p–n diode, which is loosely connected to the path and electrically connected to the current circuit of the pin-diode [44] (Fig. 6.53). The current circuit includes resistors R_1 R_2 and chokes $L_{\text{др}}$ which ratings are selected depending on duration of SHF-power pulses $P_{\text{пад}}$. It ensures a few advantages in comparison with usual version of semiconductor protective devices. First, the use of rather "thick" pin-structures ensures an increase of protective device power; second, these pin-structures can be used at higher frequency in the stand-alone operation mode. With the advent of "makeup" in the semiconductor protective devices, it became possible to create combined gas discharge-semiconductor protective devices in which the specific weight of semiconductor stages was more significant.

Another way to improve the protective devices is search for such solutions which can boost the permissible input SHF-power without sacrificing other important parameters of the protective devices (first of all, such parameters as high-speed performance and inserted losses in working band). Over the past three decades, a variety of original solutions have appeared. These solutions provide for parallel, serial, and parallel-serial SHF-current connection of a few diode structures in cross section and longitudinal section of the waveguide.

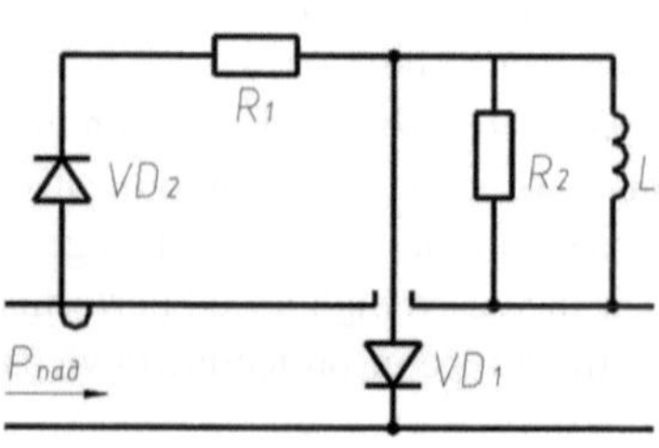

Fig. 6.53 Infeed limiter analog circuit

Possible solutions are schematically shown in Fig. 6.54. The diodes and their connecting conductors are located in cross section of the rectangular waveguide.

In case of series connection of diodes (Fig. 6.54) [45, 46], it is possible to increase the diode capacitance in proportion to the quantity of diodes in the series SHF circuit without changing the diode structure. In this case, the total impedance of diode assembly is equal to the impedance of one diode; and permissible power can be increased in proportion to the quantity of diodes. However, this advantage of series circuit couldn't be implemented because it was difficult to provide appropriate heat removal from all diodes if there are more than two diodes.

Parallel connection of diode structures (Fig. 6.55a, b) [47, 48] seems impractical, especially while attempting to reach the millimeter wave region, because of the impedance reduction in the assembly of diodes. Indeed, to keep the impedance being equal to the impedance of a single-diode protective device, it is required to reduce the capacity of each diode in proportion to the quantity of diodes. In this case, SHF-current flowing through each diode is reduced by the same number of times. As a result, the permissible power of the protective device with parallel-connected diodes is similar to that of a single-diode protective device.

For this reason, solutions schematically shown in Figs. 6.54 and 6.55 can be considered a prototype of efficient SHF resonant circuits with parallel-series connection of diode structures, as shown in Fig. 6.56a–d [49–51]. Series-parallel switching enables distribution of SHF-power absorbed in protective mode between many diode structures while keeping unchanged the impedance properties of a single-diode protective device. Options [49–51] known as "resonant arrays" (Fig. 6.56b, c) have worked well as cascades right behind gas discharge protective devices. Proper heat removal from diode structures installed on the side walls which are exposed to the waveguide, and equal current distribution between them ensure the increase in permissible forward power to the first approximation in proportion to the quantity of diodes.

Competitive PD solution based on pin-diodes is shown in Fig. 6.56d [51]. A series-parallel connection of pin-diodes is also applicable. They are distributed throughout the entire cross-sectional area, which makes it possible to use more

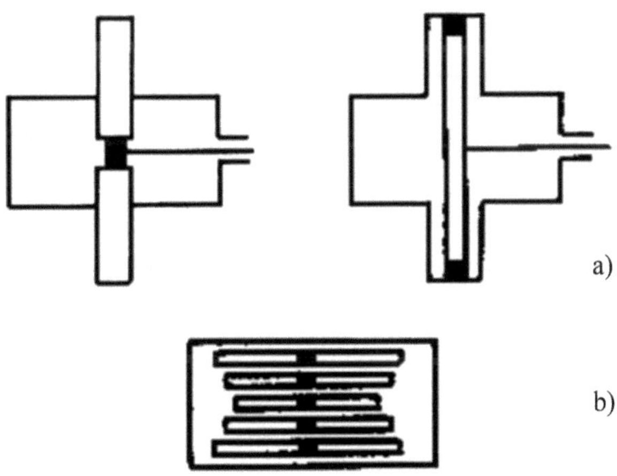

Fig. 6.54 Series connection of diodes in limiters: (**a**) diodes with rods; (**b**)—nipin – diodes

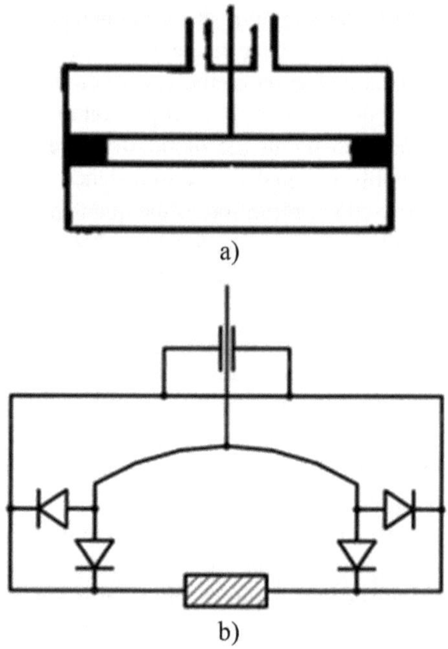

Fig. 6.55 Parallel connection of diodes in limiters

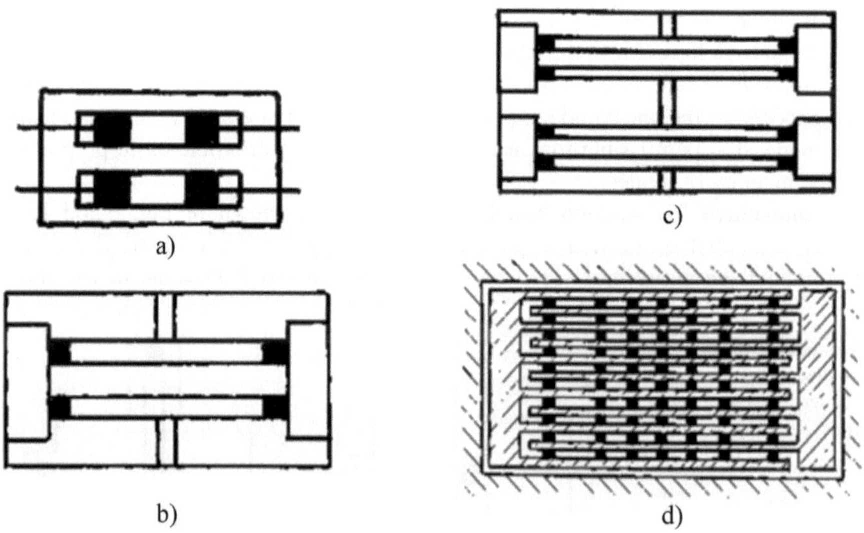

Fig. 6.56 Parallel-series connection of diodes in limiters: (**a**)—nipin - diodes; (**b**, **c**)—resonant arrays; (**d**)—waveguide window

diodes as compared to the design in Fig.6.56b, c, but heat removal from each pin-structure deteriorates. As a result, advantage of such design in forward power is minor.

The search for successful solutions for semiconductor protective devices is still under way, although there is a sense of some saturation in this direction.

In a wide range of resonant arrays, some of which are shown in Fig. 6.56 [49, 50], one solution [52, 53] concerning a "semi-array" design (Fig. 6.57) deserves special attention. Although, in this case, the purpose was to connect in-series "makeup" and switching diodes but not identical diode structures, it should be noted that this solution can get a definite development. Both in a resonant array and in a semi-array, an SHF circuit with diodes connected in series has excellent heat removal characteristics (this includes all diode structures in it). It can be useful to build the protective devices with many in-series structures.

Possible increase in the quantity of diode pin-structures between which SHF-power is evenly distributed in case of proper building, created an illusion about removal of gas discharge stages in the protective devices and about an increase of much-needed high-speed performance.

However, along with indisputable advantages of many diode protective devices, there are some disadvantages which haven't been overcome by now, and coping with which is actually problematic.

First, it is difficult to place a great number of diodes with equally good heat removal in the waveguide. Because of this, some groups of diodes will be in the worst conditions, so the efficiency of an increase in their quantity decreases.

Second, in any sequence of diode switching, the failure of one diode leads to uneven distribution of the SHF-power absorbed by other diodes, thereby beginning a snowballing process of destruction of the entire protective device.

Third, non-identity of diode parameters is the reason of failure of one of the many diodes, especially if they aren't produced in unified process cycle similar to large matrix crystal.

The possibilities to overcome the specified disadvantages of many diode versions of protective devices can be found in the following directions. It should be noted the possibility to increase the permissible input power of single lumped or quasi-lumped element of the protective device. In this case it can include many diode structures and even a few types of interconnected diode structures. Monolithic form of such assembly ensures its dimensions much less than operating wavelength. At

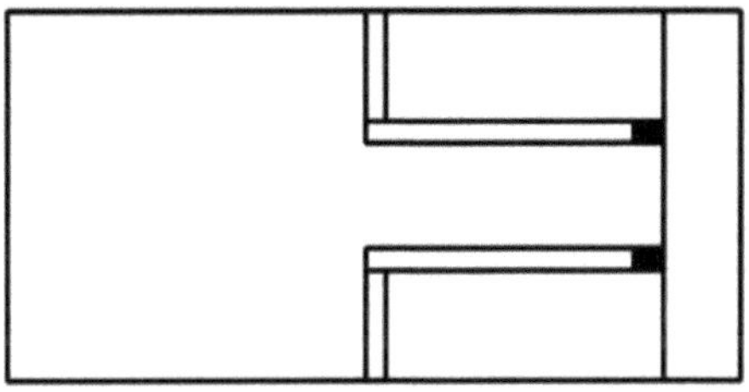

Fig. 6.57 Resonance principle of building of "semi-array"

the same time, the identity of all single elements included in such monolithic device is ensured. The similar conditions of heat removal from each diode are automatically provided.

The matching circuits of lumped design can be used in these monolithic devices, if needed. Similar solutions are known in the semiconductor electronics. The most impressive results are related to the use of the integrated diamond heat removal in generators [54], as well as the recently developed transistor amplifiers with a continuous power of tens and hundreds of W.

For example, taking into account that at least 50% of delivered power dissipate on the transistor structure itself in the transistor amplifier, and not more than 20–25% of delivered power dissipate in the diode protective devices, it should be expected that the permissible input power of single-element (monolithic) protective devices can exceed the power of transistor power amplifiers by two times at least.

It would seem promising to combine functions of "feeding" and switching diodes in one element [53–55]. During researches which were significant in their duration and scope, the favorable ratio between capacities of main and makeup diodes was found; and this principle was implemented in practice using low-frequency models.

Development of semiconductor protective devices isn't satisfied by creation of high-power and superpower stages which could replace "slow" gas discharge protective devices. The frequency barrier of semiconductor protective devices, especially stand-alone devices is one of the most important problem that hasn't been completely solved yet. Actually, as operating frequency increases, it is necessary to reduce thickness of the base of the conventional pin-diode and simultaneously to reduce its capacity, as a result of which the protective device can't perform its protective functions because of too low value of permissible operating power. The specified combination of functions of makeup pin-diode and switching pin-diode in one diode structure partially solves the task to reach short wave band. Even greater opportunities in this direction are offered by creation of diodes with compound section of lifetime-killer doping. In [54], five-layer structures are suggested. The principle of their operation is shown in [37]. Figure 6.58 schematically shows the five-layer diode structure.

Compared to a conventional pin-diode designed as layers 1, 4, 5, layers 2, 3 were introduced in the five-layer structure which perform function of hole injector, and they are located between base 4 and the first doped layer; the lifetime-killer doping of these areas is performed in steps. ($N_4 \ll N3$; $N_2 < N_3$, where N is a doping concentration). Thanks to the presence of layers 2, 3, it is possible to limit the time of transit through injecting area, enable passage of the minority carrier, and maintain the constant capacitance value at which such voltage can be generated that opens the diode structure under the impact of input SHF-signal.

The study by Ropiy et al. [37] contains important explanations of the principles of behavior of various types of diodes operating in SHF range. Moreover, an equivalent circuit of diode arrangement is not limited to capacitance and resistance connected in parallel. Instead, a series connection of two circuits of this type is provided (see Fig. 6.59). In addition, this work describes a mechanism of changing of all elements of the equivalent circuit under action of applied SHF-field, also it shows

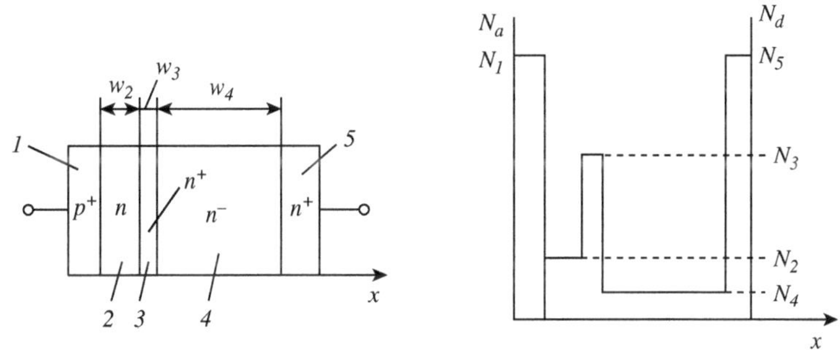

Fig. 6.58 Concentration profile of carriers in a five-layer limiting system

Fig. 6.59 Analog circuit of a clamp diode in SHF range

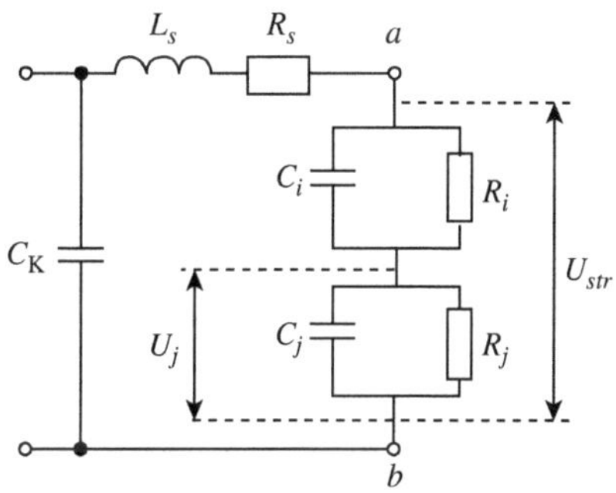

the dependence of their value on the selected diode structure in the total frequency range. This model can explain a variety of interesting processes in diodes used in the protective devices, in particular, "hysteresis" in the amplitude characteristics. It is hoped that the suggested representation of diode operation at SHF provides new opportunities for synthesis of promising diode structures not only for protective devices but also for other types of SHF devices.

Attempts to improve parameters of semiconductor limiters using available technological capabilities stimulate the use of FET gate-to-drain junctions instead of pn- and pin- structures [56, 57]. These gate-to-drain junctions are based on GaAs and they have higher speed of operation than commercial silicon diodes.

In a way, it confirms the lack of attention of developers to protective devices. Since the choice of pin- and pn-diodes is very limited, they tend to use more diverse and technologically advanced transistors. Moreover, part of these devices is not used that's unreasonable.

Currently, there are prerequisites to consider the behavior of complex structural synthesis of semiconductor diodes to be progressive for limiters.

Semiconductor protective devices have wide prospects for further development. The key to this is a wide range of ideas for their improvement, appearing recently. These ideas include the use of "trigger" properties of two-diode semiconductor chains [58–60]. The development of the principle of contactless protective devices made in the form of waveguide inserts of dielectric wafers on which diodes are placed that are not conductively connected to the walls of the waveguide is of interest [61]. In this case, the insertion loss of the protective devices decreases, which is important for the formation of the last relatively low-power cascades most sensitive to this parameter.

The implementation of promising ideas at the level of R&D and subsequent serial production, apparently, will be determined by the peculiarities of the political and economic situation in the country, affecting the relevance of complexes containing protective devices.

6.5.4 Vacuum Protective Devices

The idea of the emergence of vacuum protective devices was dictated by the need to create ultra-high-speed protective devices capable of responding to superpulse impulses with nanosecond pulse fronts. The first vacuum protective devices appeared in the 1960s [59] and were called "protective devices based on secondary electron resonance" (SER). The principle of operation of these devices lies in the fact that in an SHF-resonator, pumped off to extreme vacuum and containing in the microwave field a source of electrons with secondary emission, an electron beam arises under the influence of an external microwave field. This beam increases if the secondary emission coefficient is greater than 1 and the condition of equality of the time of flight of electrons between the electrodes of the resonator for the half-period of the microwave field oscillations is satisfied. In this case, the resonator is detuned, and the main part of the microwave power is reflected from the resonator [2]. Recent attempts to improve the SER protective devices relate to the 1970s. With their indisputable advantage—high speed (less than 15 not in 3-cm wavelength range), they have disadvantages that were too difficult to overcome—narrowband, high consumption of microwave power, heating of resonators, and difficulty in maintaining secondary emission at high density of microwave current.

It is possible that SER protective devices will develop a "second wind" in the future with the development of field emission technology and technology for building wideband resonators, but at present they have been superseded by relatively new vacuum protective devices, known as cyclotron protective devices [19, 62, 63].

The principle of operation of the cyclotron protective device is illustrated in Fig. 6.60 [19]. The tape electron flow formed by the electron gun passes through the input resonator, the separation section, the output resonator and settles in the collector. An extensive capacitive gap is made in each of the resonators. When a longitudinal magnetic field B_0 and a small microwave signal (or absence thereof) are applied the conditions for the passage of a tape electron beam and the transmission of the microwave signal from the input resonator to the output one are created. Starting from the threshold microwave signal and further as it increases, the electron beam is deflected and settles in the capacitive gap of the input resonator. In this case, the input resonator is not consistent and most of the input microwave power (up to 90%) is reflected from it.

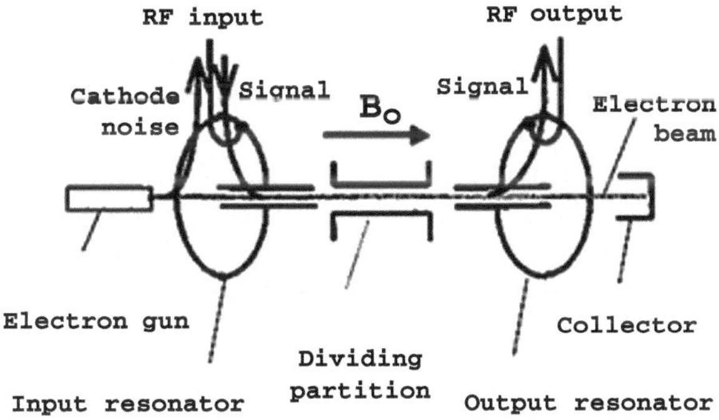

Fig. 6.60 Equivalent circuit of cyclotron protective device

Cyclotron protective devices are produced by the industry complete with solid-state low-noise amplifiers. The protective device speed does not exceed 20 ns. In the 3–4 cm wavelength range, the maximum permissible power per pulse is at least 10 kW with an average power of 300 W.

It may be desired that this relatively young type of vacuum protective devices continues to improve at a faster rate. It is necessary to overcome the traditional disadvantages for resonant vacuum devices that use a magnetic field—narrowband and high weight. The hope of success in this direction is instilled by a good theoretical analysis of the passage of an electron beam through a cyclotron protective device with a non-uniform magnetic field applied [64].

Thus, microwave protective devices, being an important node of many location and communication complexes, have a history rich in inventions and stages of technological improvement. A great contribution to this fruitful process was made by our scientists and developers: I.V. Lebedev, S.I. Rebrov, D.Ya. Ashkenazi, A.M. Starik, K.K. Shutov, O.S. Orlov L.M. Yakuben, and others.

6.6 Special Components for Filtering and Suppressing Electromagnetic Interference in Telecommunications Systems

6.6.1 Ceramic Condensers

One of the main technical solutions to the problem of ensuring electromagnetic compatibility (EMC) of high-speed electronic devices of modern information and switching systems is the suppression of electromagnetic interference.

For this purpose, various low-pass filters, filter connectors and assemblies, filters for surface mounting, etc., which are widely used in communication systems and telecommunications, in aerospace equipment, in industrial electronics, medical equipment, are used.

A brief look at the main technical characteristics of these components on the example of the nomenclature of products of the company Spectrum Control already mentioned above, which is one of the major global manufacturers of this type of components [65].

The main types of Spectrum Control products are shown in Fig. 6.61.

Spectrum Control uses two types of capacitors in its production: tubular and disk multilayered. They are the main elements of noise suppressing low-pass filters and are intended for shunting electromagnetic interference to the ground. The main parameters of the capacitors are presented in Table 6.7.

Spectrum Control produces 55 sizes of tubular capacitors: 31—C-type, including multilayer, 24 Pi-type, as well as 35 sizes of multilayer disk capacitors. The advantage of these capacitors in comparison with standard capacitors with wire leads is a small amount of their own inductance.

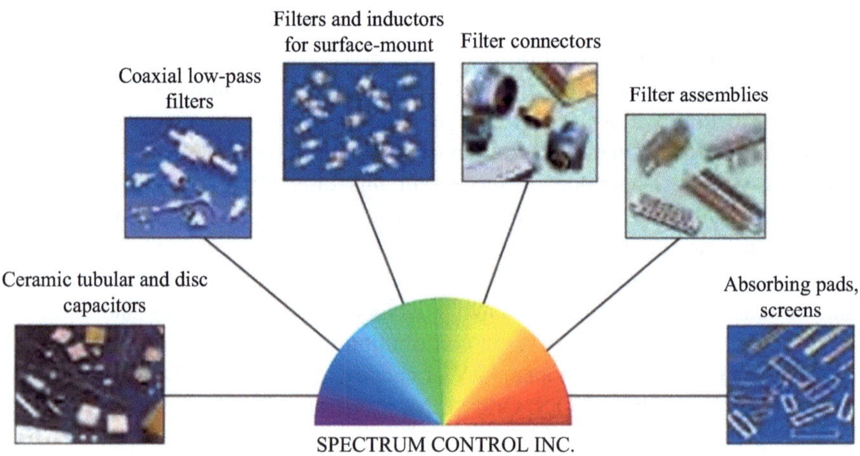

Fig. 6.61 Spectrum Control products

Table 6.7 The main parameters of ceramic capacitors

Type of capacitor	Dimensions, mm			Electrical capacity	Working voltage, V
	Outside diameter	Inner diameter	Length (thickness)		
Tubular C-type	1.6; 1.83; 2.06; 2.11; 2.41; 2.67; 2.79	0.94; 1.14; 1.19; 1.75	4.39; 8.38; 10.1; 10.8; 11.43	From 270 pF up to 0.82 μF	50, 100, 200
Tubular pi-type	1.55; 1.6; 1.85; 1.91; 2.49; 3.18; 3.86	0.51; 0.66; 0.81; 0.86; 1.14; 1.73	7.62; 7.75; 1.57; 9.4; 10.8; 11.3	From 4700 up to 30,000 μF	
Multilayer disc C-type	2.54; 3.35; 4.95; 6.3; 8.58; 11.04; 15.04	1.01; 1.4; 1.57; 1.93; 2.41	1.0; 1.5; 1.78; 1.9; 2.79; 3.05; 3.81	From 50 pF up to 3.8 mcF	50, 100, 150, 200, 300, 400

The simplest C-type capacitor is a tube with metallized wafers of one of the ceramic materials (NPO, X7R, BX, Z5U, etc.) with different dielectric constants. Coating of capacitor wafers is made of wear-resistant silver nickel barrier layer. Pi-type tubular capacitors are a combination of two capacitors connected in parallel.

Disk multilayer capacitors have a higher level of electrical capacitance (up to 10 μF) compared with single-layer capacitors. They are used in the construction of noise suppressing filters with an operating voltage of up to 400 V and with a small transition area of the frequency dependence of the insertion loss. It should be noted that the creation of multilayer ceramics made it possible to develop not only capacitors, but also inductors, chokes, ceramics with a low sintering temperature, microstrip microwave outputs, and other components with a high level of parameters.

The operating range of capacitors of all types is from −55 to +125°C.

6.6.2 *Coaxial Low-Pass Filters*

The design, principle of operation, electrical circuits of coaxial low-pass filters are considered in [66]. Spectrum Control produces several hundred sizes of pass-through noise suppression filters with C-, L-C-, Pi-, and T-electric circuits (Fig. 6.62).

By design, these filters can be attributed to one of the four typical groups:

1. Miniature, sealed with metal-glass joints, nonthreaded, soldered into the device housings filters (miniature solder-in filters);
2. Sealed with compound, nonthreaded, soldered into the device housings (solder-in filters);
3. Sealed with compound, threaded, including for chains with high voltages and currents (resin sealed bolt-in filters);
4. Sealed with metal-glass joints, threaded (bolt-in style hermetic filters or glass seal filters).

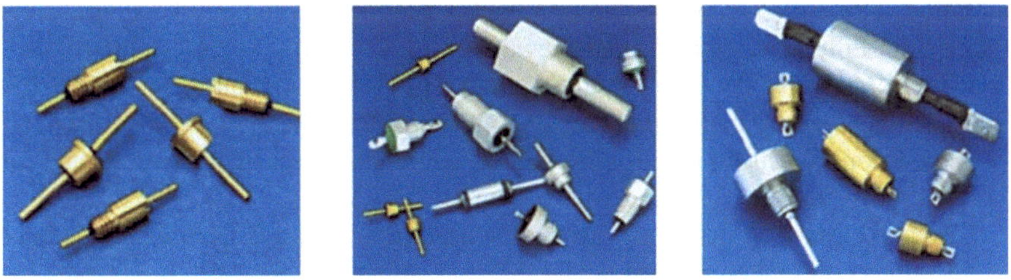

Fig. 6.62 Spectrum Control EMC filters

Filters of all types are available for both civil and military use (in accordance with the standards MIL-F-15733 and MIL-F-28861). The main parameters of the filters of all groups are given in Table 6.8.

The filters of the first group are a capacitor placed in a metal sleeve with a flange, sealed at one of the ends with the metal-glass joint. The opposite end of the sleeve is sealed with epoxy compound. The designation of filters in which the joint is located on the side of the flange is supplemented with the letter “R” at the end (Reverse).

To indicate the capacity of filters the color marking is applied. The red color of the compound corresponds to the capacity of 2700 pF, white 1200 pF, brown 500 pF, etc. The metal surfaces of the filters are gold plated. Filters of this group have increased reliability and withstand heating to a temperature of 260 °C when soldered into the housing of products.

A new series of SCI-9925 C-type filters, mounted in the housing of products by pressing, has also been developed. The flange sleeve of such filters is knurled, and the metal-glass junction is made on the side opposite to the flange. Filters are pressed into holes with a diameter of 3.5–3.8 mm in the housing of the product using a special tool SCI-9925-200.

C-type filters are used in high impedance circuits, unless a very large amount of electrical capacity is required.

The second large group of filters are compound-sealed nonthreaded filters of various designs. They have found application in devices with a dense arrangement of components, in which the possibility of using a tool for mounting filters is limited.

The third, largest group includes threaded filters, sealed with compound. It includes threaded filters on the 4-40 UNC-2A, 6-32 UNC-2A, 6-40 UNF-2A, 8-32 UNC-2A, 10-32 UNF-2A, 12-28 UNF-2A, 12-32 UNEF-2A, 5/16-24 UNEF-2A housings, etc., as well as filters for high currents and voltages. Spectrum Control manufactures threaded filters with C-, L-C-, and Pi-electric circuits. The threaded filters are either screwed in and then soldered into the housing of the product or passed into a smooth hole in the wall of the housing and secured on the reverse side with a nut and a lock washer.

To date, a huge range of filters of this group is developed and produced. The new series 54-874 of miniature filters with 2-56 UNC-2A thread without hex head attracts attention. These filters are screwed into the housing of the product like a corkscrew (Spec Spin Filters) using a special screwdriver 54-874-020.

Filters are convenient for products that require simultaneous installation and quick replacement of many filters located close to each other.

In addition, Spectrum Control produces a numerous series of filters designed for high currents (up to 100 A) and voltages (up to 2.5 kV).

Threaded filters of this group are widely used in microwave technique, telecommunications equipment, control devices.

The last, fourth, group includes threaded filters, sealed with metal-glass junction. They are designed for large currents and voltages, have high electrical capacity, and provide for the suppression of electromagnetic interference at frequencies ranging from units of kilohertz.

Table 6.8 Main parameters of filters [6]

Series numbers of filters	Number of sizes/ constructive options	Electrical circuit	Electrical capacity	Working voltage, V	Operational current, A	Characteristics of the structure
I. Miniature sealed						
SCI-9900, SCI-9910, SCI-9920	16/2	C	10–30,000 pF	50; 100; 200	5	Sleeve diameter is 3.25 mm. Center conductor diameter is 0.76 mm. The length of the sleeve is 2.79 mm
SCI-9925 (new)	7/1					The same, knurling length is 1.02 mm
II. Compound sealed nonthreaded filters						
54803004X5R 54-785, 54-786	12/5	C	100 pF 0.3 μF	10; 25	50; 100; 250;750	Sleeve diameter is 3.18; 3.96; 4.83; 5.59 mm. Center conductor diameter is 0.81; 1.02; 1.12; 1.59 mm
51-703, 51-704, 51-713, 51-723, 51-750	29/8	P_i	1300 pF 0.022 μF	10; 15; 25	50; 100; 250;500	Sleeve diameter is 3.56; 3.97; 4.95; 5.94 mm. Center conductor diameter is 0.81; 1.02 mm
III. Compound sealed threaded filters						
SCI-9110, SCI-9112, SCI-9120, SCI-9122, etc.	31/4	C	10 pF–0.056 μF	50; 10	3; 100; 200	Thread 4-40 UNC-2A. Thread length is 3.56; 5.59 mm. Central conduct diameter is 0.41; 0.46; 0.51; 0.64; 0.81 mm
SCI-3102, SCI-3112, SCI-3122, etc.	16/1	L-C, P_i	1500 pF 0.075 μF		3; 5; 10	Thread 4-40 UNC-2A. Thread length is 5.59 mm. Center conductor diameter is 0.41; 0.46; 0.64 mm
54-874 (new)	18/2	C	10–10,000 pF	50	5	Thread 2-56 UNC-2A. The length of the nonthreaded part of the sleeve is 2.67 mm, the length of the sleeve is 5.63 mm. Center conductor diameter is 0.61 mm

(continued)

Table 6.8 (continued)

Series numbers of filters	Number of sizes/constructive options	Electrical circuit	Electrical capacity	Working voltage, V	Operational current, A	Characteristics of the structure
51-726, 54-779, etc.	14/3	C, L-C, P_i	10 pF–0.05 μF	50;100;200	3;10	Thread 6-32 UNC-2A, 6-40 UNC-2A. Thread length is 5.59; 5.72 mm. Center conductor diameter is 0.46; 0.81 mm
SCI-9210, SCI-9220, etc.	17/4	C	80 pF 0.21 μF	50;70;100;150;200	10	Thread 6-32 UNC-2A. Thread length is 5.59; 6.77 mm. Center conductor diameter is 0.81; 1.02; 1.3 mm
51-712, 1250, etc.	22/6	L-C, P_i	65 pF–0.031 μF	50;70;100;125;200	10;15;20	Thread 8-32 UNC-2A. Thread length is 4.15; 6.17; 6.77; 8.0 mm. Center conductor diameter is 0.81; 1.02 mm
54-761, 54-786, etc.	6/2	C, P_i	0.01–0.08 μF	50;100;200	10	Thread 8-32 UNC-2A. Thread length is 4.19; 6.17; 6.77; 8.0 mm. Center conductor diameter is 0.81; 1.02 mm
9910-381, 54804002X5R	8/4	C	100 pF 1 μF	35;50;100;250	10;15	Thread 10-32 UNC-2A. Thread length is 5.94; 8.33 mm. Center conductor diameter is 0.81; 1.02 mm
51-714, 51-719, etc.	33/6	P_i	1300 pF 0.15 μF	50;70;100;200;300;350	10	Thread 12-28 UNC-2A, 12-32 UNC-2A. Thread length is 5.94; 10.31 mm. Center conductor diameter is 0.81; 1.02; 1.3 mm
SCI-9550, 51-702, etc.	12/3	C, P_i	1000 pF 0.1 μF	100;200;400;500;700	25	Thread 5/6-24 UNC-2A; 5/6-32 UNC-2A; 5/6-32 UNEF-2A. Center conductor diameter is 1.45 mm
54-844, 54-763, 54-853, etc.	9/5	C	1000 pF 0.22 μF	60;200;600;750;1250	25;50;100	Filters for high voltages and currents. Thread 7/16-28 UNEF-2A; 5/8-24 UNEF-2A; 5/16-24 UNEF-2A, etc. Center conductor diameter is 1.3; 1.45; 1.58 mm

1280-060	1/1	P_i	1500 pF	2500	25	Thread 3/8-32 UNEF-2A. Thread length is 9.53 mm
IV. Sealed threaded filters						
51-359	96/2	L-C	0.15–1.4 μF	80;100;150;200;250;300	0.06;0.15; 0.45;0.5; 1;2;5;10;15;25	Thread on the filter housing 1/4-28 UNEF-2A

Filters are produced in versions with C-, L-C-, Pi-, and T-electric circuit. Their nomenclature is extremely diverse with over 250 items. Typical filter housing diameters are 9.78 mm (0.375′∅ group) and 17.45 mm (0.690′∅ group). As an example, Table 6.10 shows the parameters of the L-C-filters of the most numerous 0.375′∅ group. The magnitude of the attenuation of these filters at frequencies above 1 GHz exceeds 70 dB.

Filters of this group are used in power supplies, ignition systems, equipment for the coal and oil industry, in medical technology.

6.6.3 Surface Mount Filters

The accelerated development of mobile, wireless and telecommunications systems, computers, peripherals, aerospace systems, and other areas of technology has led to microminiaturization, high-density mounting of microelectronic microwave devices, and the introduction of high-performance surface-mount technology. Such mounting involves group methods of automated assembly and soldering of device components.

But for this reason it was necessary to create the appropriate radio components, including noise suppression filters in a chip-performance.

Spectrum Control has developed and produces ceramic three-lead chips filters of C, L-C, and Pi types (Fig. 6.63).

They have the shape of parallelepipeds of length L, width W, height T with a width of leads at the ends of LE and with a width of central lead LM. C-type chips are available in the following groups: SF 0805, SF 1205, and SF1806 and L-C-type of the group LC 0805 and LC1206.

The group number is formed by a combination of the length and width of the chips, measured in inches. The sizes and nominal capacity of filters are given in Table 6.9.

Spectrum Control produces 8 chips of the SF 0805 group, 9 SF 1205 and 10 SF 1806. You can select chips with a nominal capacity from 22 to 220,000 pF.

The filters of the first two groups are designed for a working voltage of 50 V and a current of 0.3, 0.4, and 1.0 A, the third group, respectively, is designed for 100 V and 0.3 A. Filters operate in a temperature range from −55 to 125 °C. The level of attenuation of all filters increases with increasing of their capacity. At a frequency of 1 GHz for filters with a capacity of more than 220 pF, the value of the insertion attenuation is approximately 40 dB.

L-C-type filters for surface mounting, depending on the values of capacity and inductance, have a cutoff frequency (the frequency at which the insertion loss is 3 dB) from 10 to 220 MHz with a tolerance of 20%. These filters are rated for 25 V operating voltage and 10 mA operating current.

All chips are mounted on the board by soldering at a temperature of 220–240 °C. After soldering, ultrasonic cleaning of flux residues is necessary: oscillation frequency is not more than 28 kHz, time is not more than 5 min.

Fig. 6.63 Three-lead ceramic filters

Table 6.9 Parameters of chip filters of C and LLCC types

Group of chips	Dimensions, mm					Nominal capacity, pF
	L	*W*	*T*	LE	LM	
C-type						
SF 0805	2 ± 0.2	1.25 ± 0.2	0.8 ± 0.2	0.3 ± 0.2	0.6 ± 0.2	22–22,000
SF 1205	3.2 ± 0.2		0.7 ± 0.2	0.4 ± 0.3	1.1 ± 0.3	22–47,000
SF 1806	4.5 ± 0.3	1.6 ± 0.3	1.0 ± 0.3	0.5 ± 0.3	1.4 ± 0.3	22–22,000
L-C-type						
LC 0805	2 ± 0.2	1.25 ± 0.2	0.8 ± 0.2	0.3 ± 0.2	0.6 ± 0.2	–
LC 1206	3.2 ± 0.2	1.6 ± 0.3	1.0 ± 0.3	0.4 ± 0.3	1.1 ± 0.3	–

In addition, Spectrum Control produces two series of high-power surface-mounted filters: PSM and SSM, designed for operating currents of 10 and 20 A. Filters are used in power amplifiers, power supplies, motor control devices. The appearance of filters is shown in Fig. 6.64.

The PSM (power surface mount) series filters can have a C or Pi-electric circuit.

The rated operating current of the capacitive filters is 20 A, 10 A for Pi-filters, the nominal DC voltage is, respectively, 200 and 100 V. The capacity of PSM filters can be selected in the following row: 68, 100, 130, 470, 820, 1000, 1500, 2500, 4000 for the Pi circuit and 10,000 pF in addition to the indicated capacities for circuit C.

The SSM (square surface mount) series filters have only a Pi-electric circuit, they are designed for an operating current of 10 A and a DC voltage of 100 V. Their electrical capacity can be 100, 500, 1500, 2000, and 4000 pF.

Chip filters of all types are supplied usually packed in plastic tape for automated installation. Finally, Spectrum Control continues to produce high-frequency three-lead ceramic filters for mounting on PCB-based board. Filters provide an insertion attenuation of 20 dB at frequencies from 100 MHz to 1 GHz and have a cutoff frequency of 8 MHz. The minimum electrical capacity of the filters is 800 pF, the maximum operating current is 10 A, and the voltage is 50 V.

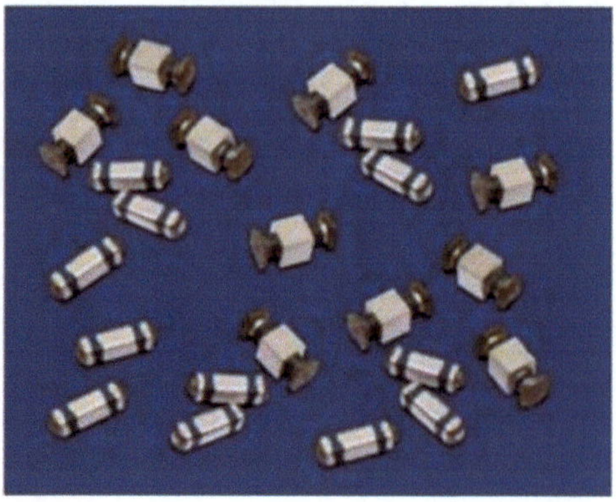

Fig. 6.64 PSM and SSM Series Filters

In addition to the above components, other products should be noted—filter connectors and multi-lead assemblies, the range of which is huge, a large group of filter pads of different profiles, sections, and protective shields.

6.7 Methods for Assessment of the Resource of High-Speed Electronic Considering the Reliability of Mechanical Components

As shown above, high-speed electronic devices have an extremely wide range of applications: from simple household appliances to complicated devices of military and space technology, since the reliability of all such devices is extremely important, you need to be able to most accurately assess both the probability of uptime, operating time, and resource. In particular, this is due to the fact that the orbital components of modern telecommunication systems, including the entire space technology, are by definition non-recoverable and after the onset of a limiting state (i.e., exhaustion of the resource) the probability of failure starts to increase very quickly.

Resource is the total operating time of the object from the start of its operation or its resumption after repair to the transition to the limiting state, the gamma percentage resource is the total time during which the object does not reach its limiting state with probability γ expressed as a percentage, and the average resource is the expectancy of a resource. The limiting state is the state of the object in which its further operation is unacceptable or impractical or the restoration of its operational state is impossible or impractical [32, 66–68].

Another important point in determining the value of the resource of electronic devices, including microwave devices, is the consideration of mechanical components. It is worth noting that failures can be interconnected: malfunctions of mechanical parts lead to the failure of the electronic component, and vice versa. Structural damage, deformations can lead to signal distortion, and electrical disturbances due to the failure of the ECB can lead to unsoldering of wires from the contact pads, temperature increase, etc. Thus, neglecting the failures of mechanical components, the developer involuntarily overestimates the resulting values of the reliability and durability indicators of the designed high-speed electronic devices.

Resource assessment methods are clearly regulated by OST [32], resource values for limiting and lite operating conditions are given in the technical conditions for domestic development. However, they do not describe the transition from these values to those specified for the product, i.e., the developers are forced to evaluate the product resource at the lower boundary, since space technology operates in the conditions different from the limiting and lite conditions. Also, the time schedule of the product is usually not taken into account when the product undergoes various checks on the ground before being directly launched into space, is transported to the launch site, stored, etc. Detailed coverage of these issues is presented in one of the first works of the team of authors of the Moscow Institute of Electronics and Mathematics of the National Research University Higher School of Economics (Moscow). It is shown here how to estimate the resource of a product functioning in certain conditions, and how to take into account the uneven use of the resource from the moment of commissioning.

Active provisions of this paper are presented in the following as close to the original text as possible. So, in order to take into account the uneven consumption of a resource, it is necessary to normalize values for different groups of equipment with respect to one, arbitrarily chosen (for definiteness, let us call it "resource under normal conditions"—$T_{r.n.c.}$). As a result, we obtain the proportionality coefficient for the use of the resource $K(T_r)$: for $T_{r.n.c.}$, it will be equal to one, for lite operating conditions, the coefficient will be strictly greater than one, for more severe conditions, the coefficient will be strictly less than one. Indeed, if we turn to the technical conditions, then for lite conditions, the resource is greater than for the limiting ones, and hence the result of normalization with respect to $T_{r.n.c.}$.

$$K(T_r) = \begin{cases} > 1, \text{at}\, T_r > T_{r.n.c.} \\ 1, \text{at}\, T_r = T_{r.n.c.} \\ < 1, \text{at}\, T_r < T_{r.n.c.} \end{cases} . \tag{6.18}$$

Then, using this coefficient, the consumed resource is calculated during the operation of the product under the conditions of each group of equipment, i.e., for "normal conditions," the amount of consumed resource is equal to the time of operation under these conditions, and, for example, for more severe conditions, more resource is used up for the same time. For periods of operation in the conditions of each group of equipment, the difference between the actual value of the resource and the normal one is determined and subtracted from the $T_{r.n.c}$. Then the true value of the resource T_r of the product in the given conditions of groups of equipment can be determined from the following expression:

$$T_r = T_{r.n.c.} - \sum_{i=1}^{n} t^{(i)} \cdot \left(\frac{1}{K_i(T_r)} - 1 \right), \tag{6.19}$$

where $T_{r.n.c.}$ is the value of the resource for the selected "normal" operating conditions; n is the number of groups of equipment operating conditions (excluding "normal"); $t^{(i)}$ is the operation time in the conditions of the i-th group of equipment; $K_i(T_r)$ is the proportionality factor of resource consumption for the conditions of the i-th group of equipment.

The method of calculating the resource for certain conditions of groups of equipment is given in [67]. In domestic technical conditions, resource values for limiting and lite operating conditions are usually given, while the article discusses how to switch from the values specified in the technical conditions to real ones:

$$T_r = \frac{T_r(\mathrm{LS})}{U_f \cdot L_f}, \tag{6.20}$$

where T_r (LS) is the resource value specified in the technical conditions for the limiting conditions of operation; U_F—utilization factor; L_F—load factor (critical parameter).

In equality (6.20) U_F is determined according to formula (6.21), and L_F is determined according to formula (6.22):

$$U_f = \frac{t_{\Sigma \mathrm{dc}}}{t_{\Sigma \mathrm{dc}} + t_{\Sigma \mathrm{st}}}, \tag{6.21}$$

where $t_{\Sigma \mathrm{dc}}$, $t_{\Sigma \mathrm{st}}$—the total time spent in the duty cycle and standby for the period of operation:

$$L_f = \frac{\Psi_{\mathrm{dc}}}{\Psi_{\mathrm{nom}}}, \tag{6.22}$$

where Ψ_{dc} is the critical parameter value in the duty cycle; Ψ_{nom} is the nominal value of the critical parameter according to the technical conditions.

The authors consider a typical product of onboard electronic equipment for space technology—a microwave switch as an example. During operation, the switch is exposed to various external influencing factors (EIF). First, the product is stored indoors (in the warehouse) for one year (8760 h), which corresponds to group 1.1 [69], and is further transported by any type of transport (group 1.3 for automotive, 3.1 for aviation, 1.9 for railway) to the launch site of the carrier rocket for 5 days (120 h), after which the rocket is launched and placed into orbit (group 8.1) for 2 days (48 h) and, finally, performs its main functions (reception and transmission in the microwave range) until the end of the active shelf life (ASL) in the conditions of group 5.3 (Fig. 6.65). According to the technical conditions and article [37], the product resource for each of the equipment groups will be different. Taking into account the above information about storage, transportation, and direct operation of the product, we conclude that the resource is spent unevenly.

After analyzing the technical documentation data on the switch, a failure tree was built (Fig. 6.66) [32], where the following events are marked with numbers: *1*—the product is inoperable; *2*—product failure; *3*—no signal processing; *4*—violation of fastening and sealing; *5*—failure of the electronic part; *6*—no signal at the input of the circuit; *7*—no signal processing in the boards; *8*—no output signal; *9*—failure of input cascades; *10*—no signal at input 1; *11*—no signal at input 2; *12*—failure of the elements of input no. 1; *13*—failure of W1; *14*—failure of XW1; *15*—burnout of the valve W1; *16*—burnout of the connector XW1; *17*—failure of the elements of input no. 2; *18*—failure of W2; *19*—failure of XW2; *20*—burnout of valve W2; *21*—burnout of connector XW2; *22*—board failure; *23*—lack of power to the diode of the XN1 board; *24*—lack of power to the diode of the XN2 board; *25*—failure of one of the diodes VD1-VD8; *26*—failure of one of the diodes VD1-VD8; *27*—output cascade failure; *28*—no signal on bridges XN8, XN9; *29*—failure of the outputs of the product; *30*—failure of the bridges XN8, XN9; *31*—no signal of bridge XN8; *32*—no signal of bridge XN9; *33*—damage to the conductive paths of the bridge XN8; *34*—damage to the conductive paths of the bridge XN9; *36*—no signal in the output channels; *37*—failure of output connectors XW3-XW5; *38*—failure of the circulator W3; *39*—no signal on connectors XW3-XW5; *40*—burnout of the circulator W3; *41*—failure of the connector XW3; *42*—failure of the connector XW4; *43*—failure of the connector XW5; *44*—burnout of the connector XW3; *45*—burnout of the connector XW4; *46*—burnout of the connector XW5; *47*—destruction of welded, brazed, and glued joints and depressurization of plugs and gaskets; *48*—failure of screws fastening valves W1, W2; *49*—solder connection failure of jumpers on the XN1, XN2 boards; *50*—failure of the welded connection of jumpers of XN5, XN6 boards; *51*—stub failure; *52*—failure of adhesive bonding of boards XN4, XN6; *53*—cover screws failure; *54*—cover gasket failure; *55*—loosening of the fastening of valves W1, W2; *56*—destruction of the solder

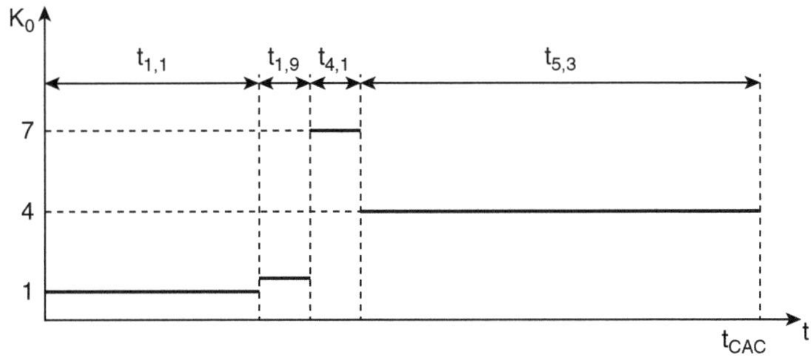

Fig. 6.65 The time schedule of the operation of the electronic switch

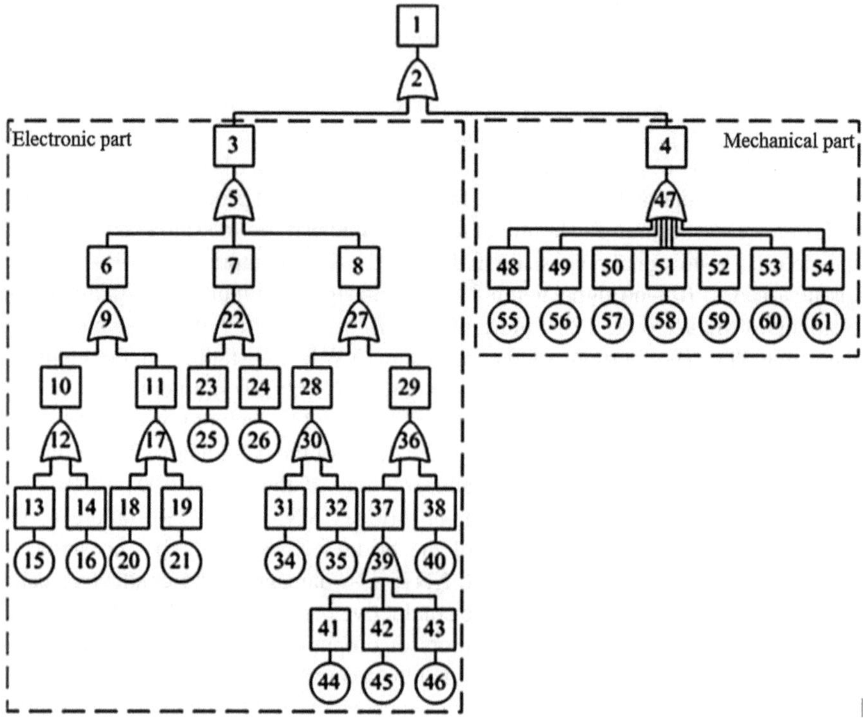

Fig. 6.66 Electronic switch failure tree

connection of the jumpers of the XN1, XN2 boards; *57*—destruction/failure of the welded joint of jumpers of the XN5, XN6 boards; *58*—deformation of the plug; *59*—peeling of the adhesive film in the connection boards XN4, XN6; *60*—loosening of the cover; *61*—deformation of the gasket cover. As you can see, the failure of the product is equally dependent on the failure of electronic and mechanical parts.

The results of the calculation of reliability indicators carried out in accordance with the constructed tree are given in Table 6.10, and Fig. 6.67 shows the contribution to the overall reliability of the electronic and mechanical components of the switch. The calculation of the electronic part was carried out on the basis of the data given in [69], the mechanical part—on the basis of [70].

As can be seen from the figure, the reliability of the "mechanical" part is comparable to the "electronic" part and, neglecting it, we unwittingly "improve" the reliability by 2 times. In other words, it is quite obvious that mechanical components should be taken into account when calculating the reliability and durability of microwave devices. The obtained values of reliability indicators, and the resource is calculated taking into account non-stationary conditions of operation using the expression (6.19).

Table 6.10 The results of the calculation of the reliability of the electronic switch

Name of the component	Operational failure rate, 1/h	The failure rate in standby mode, 1/h	Gamma percentage resource, h
Electronic part	7.27e-07	5.56e-08	23,400
Mechanical part	9.92e-07	7.26e-08	26,200
Switch	1.71806079201876E-6	1.281957368E-7	23,400

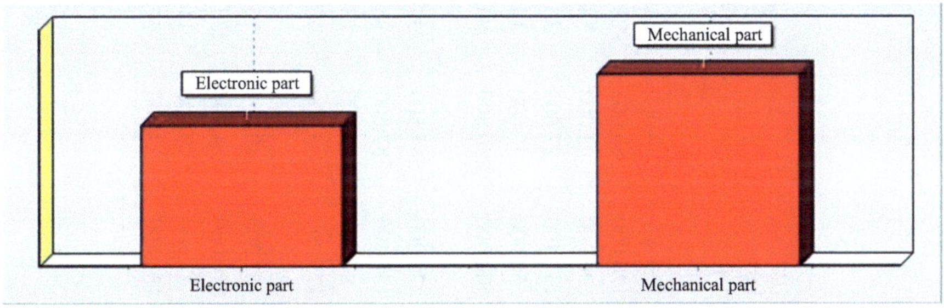

Fig. 6.67 Contribution to electronic and mechanical switch reliability

When calculating the final resource from Table 6.11 using the expression (6.19), the conditions for group 5.3 were taken as "normal" conditions and, as can be seen, the total value is higher than for group 5.3 on average by 4000 h (a little less than 6 months).

The calculation carried out by the authors confirmed the need to take into account the functioning in the conditions of various groups of equipment when assessing the resource. In connection with the tightening of requirements for equipment, it is not always possible to afford the luxury of estimating the lower boundary (the worst case), since such requirements imply accurate calculations. The proposed method of calculating the resource takes a step towards increasing the accuracy of the resource calculations. Also an important fact is a visual demonstration of the influence of mechanical components of microwave devices on reliability (Fig. 6.64). It is noteworthy that taking into account the mechanical components worsens the resulting values of reliability indicators to the same extent as electronic ones. Thus, all the proposed measures for more accurate calculations confirm the need to develop methods for accurate calculations with stricter requirements for microwave devices.

Summarizing this section, we can draw the following conclusions:

- when assessing reliability, in particular durability, it is necessary to take into account the mechanical components of electronic devices;
- the numerical value of the resource depends on the specific group of equipment, i.e., from the set of EIF, and in assessing the durability indicators it is necessary to take into account the nonstationarity of the operating conditions of the electronic device.

Table 6.11 Microwave switch reliability indicators

Type of device	Operational failure rate	Failure rate in standby mode	Resource				
			1.1	1.9	4.1	5.3	Total
Electronic part							
ФПВН2-88	7.30e-08	8.76e-10	120,000	80,000	21,800	60,000	64,300
ФЦП2-13	1.90e-07	8.93e-10	120,000	80,000	21,800	60,000	64,300
СРГ-50 (ВР0.364.026ТУ)	5.49e-09	7.69e-09	200,000	33,000	53,300	100,000	104,000
EEE-wave soldering	6.90e-10	6.90e-12	131,400	87,600	35,000	65,700	70,000
2A517A-2H, Б-2H	4.50e-08	1.68e-09	50,000	47,600	10,000	17,800	23,400
Printed circuitry	3.40e-11	3.40e-13	500,000	230,800	95,200	178,600	184,000
Mechanical part							
Valve mounting screw	1.00e-09	1.00e-11	876,000	584,500	233,600	438,000	442,400
Jumpers soldering	1.20e-09	1.20e-11	131,400	87,600	35,000	65,700	70,000
Welding wires to contact pads	2.40e-10	2.40e-12	87,600	58,400	23,300	43,800	48,100
Fluoroplastic connector plug Ф-4 (or Ф5)	3.60e-07	3.60e-08	175,200	116,800	46,700	87,600	91,970
Glued board connection	2.40e-08	2.40e-09	61,300	40,800	16,300	30,600	34,900
Cover mounting screw	6.00e-09	6.00e-11	876,000	584,500	233,600	438,000	442,400
Cover gasket	2.88e-07	2.88e-08	43,800	29,200	11,600	21,900	26,200
Cover soldering and sealing	2.40e-07	2.40e-09	131,400	87,600	35,000	65,700	70,000

References

1. Smirnov, A., & Basalaev, K. (2015). The development of requirements to the characteristics of electromagnetic compatibility of general industrial and defense products. *Components & Technologies*, 2, 98–101 (in Russian).
2. Belous, A. I., Merdanov, M. K., & Shvedov, S. V. (2016). Microwave electronics in radar systems and communications. In *Technical encyclopedia in 2 books. Book 1*. Moscow: TECHNOSPHERA (in Russian).
3. Titovich, N. A., & Polzunov, V. V. (2015). Investigation of semiconductor devices susceptibility to the effects of electromagnetic interference. *BSUIR Reports*, *2*(88), 114–118 (in Russian).
4. Brigidan, A. M., Titovich, N. A., & Kirshchov, V. M., et al. (1992). *Electronic equipment II*. Ed. No.1 (148), 3–13 (in Russian).
5. Mardiguian, M. (1988). A handbook series on electromagnetic interference and compatibility. In *Electromagnetic control in components and devices* (Vol. 5). Virginia.
6. Avdeev, V. B., Avdeeva, D. V., & Berdyshev, A. V. (2011). Information countering the threats of terrorism: Technological Institute of the Federal State Educational Establishment of Higher Professional Education "Southern Federal University", 16, 11–16 (in Russian).
7. Titovich, N. A. (1995). *Thesis report of science and technology conference "Modern problems of radio engineering, electronics and communications"* (pp.189–191). Minsk (in Russian).
8. Titovich, N. A., Polzunov, V. V., & Murashkina Z. N. II Mater. (2011, September 27–29) *XV International Science and Technology Conference (ISTEC) "Modern means of communication"*. Minsk (in Russian).
9. Titovich, N. A., & Brigidin, A. (1992). Radio Engineering and Electronics *21*, 157–160 (in Russian).
10. Titovich, N. A. (1993). *II International symposium on EMC: Book of scientific reports* (pp. 171–175). St. Petersburg (in Russian).
11. Titovich N.A. (1991). *Radio Engineering and Electronics II*. 20, 84–87 (in Russian).
12. Titovich, N. A., Buevich, E. A. *Device for controlling the parameters of digital integrated circuits when exposed to electromagnetic interference*. The Patent of the Russian Federation 2069865 (in Russian)
13. Titovich, N. A., & Polzunov, V. V. (2009, March 19). *Mater international science and technology conference, dedicated to the 45th anniversary of BSUIR (MRTI)* (pp.73–74). Minsk (in Russian).
14. Korotkov, S. (2016). Electromagnetic interference and signal integrity on printed circuit boards. *Electronic & Components*, 5, 14–16 (in Russian).
15. Zakharov, D. (2016). Signal integrity in high-speed communication lines. *Electronic & Components*, 8, 60–62 (in Russian).
16. Peresadin, A. (2016). Ensuring the integrity of signals at all stages of design. *Electronic & Components*, 8, 26–31 (in Russian).
17. Belov, L., Kochemasov, V., & Stroganova, E. (2015). Passive interpolation in microwave circuits: Appearance mechanisms, measurement methods and ways to reduce. *Electronics: NTB*. 3(00143), 80–91 (in Russian).
18. Hartman, R. (2011, May 11) Passive intermodulation (PIM) testing moves to the Base Station. *Microwave Journal*.
19. Budzinsky, Yu. A., Bykovsky, S. V., & Vildanov, S. A., et al. (2001, September 14) Completed amplifiers with cyclotron protection for radar receivers. In *The book of 11th International Conference "Microwave Appliances and Telecommunication Technologies"*, Sevastopol, pp. 190–191 (in Russian).
20. Cannon, N. (2011, October 7). PIM testing crowing in importance as 4C rolls out in Europe. *Microwave Journal*.
21. Strigin, I. (2013). Passive intermodulation interference in mobile networks. *Technologies and Communications*, 2, pp. 39–40 (in Russian).

22. Stroganova, E. P. (2008). Smart antennas for 3G networks. *Technologies and Communications*, 6, 42–45 (in Russian).
23. Stroganova, E. P. (2008). Cositing: Pros and cons. *Technologies and Communications*, 4, 68–70 (in Russian).
24. Dzhurinsky, K. (2006). *Miniature coaxial radio components for microwave microelectronics* (2nd ed.). M .: Technosphera (in Russian).
25. Jargon, J. A., & DeGroot, D. C. (1999, October). *Comparison of passive intermodulation measurements for the U.S. wireless industries* (NIST Technical Note 1515).
26. Bogdanov, Y., Kochemasov, V., & Khasyanova, E. (2013). Foiled dielectrics - how to choose the best option for printed circuit boards of high-frequency/microwave ranges. *Printed Circuit Wiring*, Part 1, 2(0043), 156–168; Part 2, 3(0044), 142–147 (in Russian).
27. Wilkerson, J. R., Lam, P. C., & Card, K. C., & Steer M. B. (2011). Distributed passive intermodulation distortion on transmission lines. *IEEE Transactions MTT*, *59*(5), 1190–1205.
28. Rocas, E., Collado, C., Orloff, N. D., Mateu, J., Padilla, A., O'Callaghan, J. M., & Booth J. C. (2011). Passive intermodulation due to self-heating in printed transmission lines. *IEEE Transactions MTT*, *59*(2), 311–322.
29. Collins, S., & Flynn, K. (1999, November 1). Intermodulation characteristics of ferrite-based directional couplers. *Microwave Journal.*
30. Belov, L. A., Kondratov, A. S., Rozhkov, V. M., & Romashchenko, K. V. (2012). Improving the linearity and energy efficiency of wideband power amplifiers. *Telecommunications*, 5, 23–25 (in Russian).
31. Bell, T. (2013–2019). *Mitigating external sources of Passive Intermodulation* (11410-0756, Rev. A). Anritsu Company.
32. OST 4G0.012.013-84. Electronic equipment. Definition of durability indicators (in Russian).
33. The IEC standards for measuring PIM. IEC 62037: Passive RF and microwave devices, Intermodulation level measurement, parts 1–8, IEC 62037–2 Ed. 1.0 b: 2013 RF connectors, connector-cable assemblies and cables. *Intermodulation Level Measurement* (in Russian).
34. Bell, T. (2013–2019). *Mitigating external sources of Passive Intermodulation* (No. 11410-0756, Rev. A.). Anritsu Company.
35. Alybin, V. G. (2002, September 9–13). The problems of creating microwave radiation protection shields for radar and communications. In *12th International Crimean Conference "Microwave Appliances and Telecommunication Technologies". Conference proceedings.* Sevastopol: Weber (in Russian).
36. Ashkenazi, D. Y., Belyaev, V. P., Brodulenko, G. I., Lagovier, B. B., Lebedev, I. V., Mandelstam M. Y., Peretz, R. I., Rebrov, S. I., & Starik A. M. (1976). *Resonant gaps of antenna switching units* (274 p.). I. V. Lebedev (Eds). M: Soviet Radio (in Russian)
37. Ropiy, A. I., Starik, A. M., & Shutov, K. K. (1993). *Ultrahigh-frequency shields* (128 p.). M: Radio and Communication (in Russian).
38. Sze, S. (1984). *Physics of semiconductor devices in 2 books* (translation from English under the editorship of Suris R. A.) (456 pp.). M: Mir (in Russian).
39. Shur, M. (1991). *Modern devices based on gallium arsenide* (632 p.) (translation from English under the editorship of Levinshteyn M. E. and Chelnokova V. E.). M: Mir (in Russian).
40. Lebedev, I. V. (1995). New views on the impedance characteristics of semiconductor microwave devices. *MPEI Reporter*, 2, 19–29 (in Russian).
41. Lebedev, I. V., & Alybin, V. G. (1978). Resonant grid application for the creation of microwave devices. News of the USSR Universities. *Radioelectronics*, *XXI*(10), 24–31 (in Russian).
42. Soushworth, G. C. (1950). *Wave guide system*. U.S. Patent No. 2514678 (in Russian).
43. Goldie, H. (1972). Radioactive (tritium) ignitor for plasma limiters. *IEEE Transactions*, *ED-19*(8), 917–928.
44. Gawronsri, M., & Goldie, J. (1977). 20WMIC L-band receiver protector Microwave J., 5, 43–46.
45. Butson, P. S., & Thompson, G. T. (1964). Useful microwave hybrid circuit. *Proceedings of the IEEE*, *111*(7), 1281–1282.

46. Liberman, L. S., Sestroretsky, B. V., Shpirt, V. A., & Yakuben, L. M. (1972). Semiconductor diodes for controlling microwave power. *Radio Engineering*, *27*(5), 9–24 (in Russian).
47. Kirillov A. V., & Karchevsky E. V. (1982). Authorship certificate (A. C.) No. 1 042110 (USSR) cl. H01R1/15 (in Russian).
48. Ramay, R. L. (1972). *Active irises and windows*. US patent 3649935 (in Russian).
49. Lebedev, I. V., & Alybin, V. G. (1977). A.C. 566297 (USSR). *Power adder circuit for adding powers of microwave devices*. BI number 27 (in Russian).
50. Lebedev, I. V., & Alybin, V. G. (1978). Resonant grid and its application to create microwave devices. *News of USSR Higher Educational Establishments - Radio Electronics*, 21(10), 24–31 (in Russian).
51. Bakeman, P. E., & Armstrong, A. L. (1976). Fast high power octave band width X-band waveguide microwave switch. In *IEEE/MTT-S international microwave symposium* (pp. 154–156). Cherry Hill: Dig. of Techn. Papers-1976.
52. Lebedev, I. V., Ugnichev, L. V., & Skorobogatov, D. V. et al. (1992). A. C. 1737571 (USSR) MKI H01P1/15. Microwave power limiter. Discoveries. Inventions, No. 20 (in Russian).
53. Prokhorov, R. A., Lebedev, I. V., & Skorobogatov, D. V. (1993). Patent 1827041 (USSR), MKI H01R1/15, H01R1/22. Microwave power limiter. Discoveries. Inventions, 25 (in Russian).
54. Rucken, C. T., Wamoss, J. W., Hill, C. N., & Cox, N. W. (1977). Series-connected GaAsSi IMPATT-diodes chips. *Electronics Letters*, *13*, 331–332.
55. Lebedev, I. V., Shnitnikov, A. S., Ugnichev, D. V., & Rebrov, A. S. (1989). A. C. 1483518 (USSR), MKI H01P1/15 Limiter. Inventions. Discoveries. 20 (in Russian).
56. Lazunin, Yu. A., & Fefelov, A. G. (1990). Resonant microwave power limiter on a field-effect transistor. *Journal of the Russian Universities. Radioelectronics*. *33*(5), 83–84 (in Russian).
57. Krutov, A. V., Mitlin, V. A., & Rebrov, A. S. (2002, September 9–13). Low power protection device. In *The book of 12th international Crimean conference "Microwave Appliances and Telecommunication Technologies"*. Sevastopol: Weber (in Russian).
58. Lebedev, I. V., & Shnitnikov, A. S. (1993). Patent of the Russian Federation 2003 208 C1, MKI H0129/06. Semiconductor limiting diode. Inventions. Discoveries, 41–42 (in Russian).
59. Lebedev, I. V., Shnitnikov, A. S., & Dyakov, I. V. (1997). Trigger-type diode control microwave devices. *Radio Engineering*, 1 (in Russian).
60. Lebedev, I. V. (1997). Trigger properties of two-diode chains at high and ultrahigh frequencies. *Radio Engineering*, 9, 54–60 (in Russian).
61. Lebedev, I. V., Shnitnikov, A. S., Drozdovsky, N. V., & Drozdovskaya, L. M. (1995). Patent application number 94011538 (RF), MKI H01R1/15 Microwave limiter. Decision to grant a patent (in Russian).
62. Forrez, M., & Milazzo, C. (1962). Duplexing and switching with multi- pactor discharge. *Proceeding of IRE*, 4, 442–451.
63. Budzinsky, Yu. A., Bykovsky, S. V., & Kantyuk, S. P., & Mastryukov, M. A. (1999). Electronic microwave devices on a fast cyclotron wave of electron flow. *Radio Engineering*, 4, 32 (in Russian).
64. Golenitsky, I. I., Dukhanin, N. D., Budzinsky, Yu. A., & Bykovsky, S. V. (2001, September10–14). Simulation of a three-dimensional electron-optical system cyclotron protective device. In *The book of 11th international conference "Microwave Appliances and Telecommunication Technologies"*, Sevastopol, pp. 216–217 (in Russian).
65. Karapuzov, M. A. (2014). The influence of external disturbing factors on the durability of microwave devices. In M. A. Karapuzov, S. N. Polessky, & V. V. Zhadnov (Eds.), *Reliability and quality of complex systems* (No. 2(6), pp. 14–20) (in Russian).
66. Dzhurinsky, K., & Filippov, A. (2007). Products of Spectrum Control company for the suppression of electromagnetic interference. *Components and Technologies*, 7, 100–103 (in Russian).
67. Dzhurinsky, K. (2006). *Miniature coaxial radio components for microwave microelectronics*. Moscow: Technosphera (in Russian).

68. GOST 27.003-89. Reliability in technology. Composition and general rules for setting requirements for reliability (in Russian).
69. *Reliability of electrical radio products: Handbook.* http://www.twirpx.com/file/1062157/ (in Russian).
70. NSWC-11. (2011). *Handbook of reliability prediction procedure for mechanical equipment.*

Chapter 7
Basics of Designing Structures of Printed Circuit Boards of High-Speed Electronic Devices

This chapter is devoted to the concerns and peculiarities of printed circuit boards' structural designing for high-speed electronic devices. Designers of modern high-speed electronic devices, looking through the table of contents of the book, will almost certainly miss this chapter—well, what else is interesting and useful they can learn about printed circuit boards? Usually they choose numerous options of this class of products currently offered by the market and use them only as a kind of physical medium for the structural design of their device (apparatus, system) being designed.

The designers of high-speed electronic devices usually address the issues of analyzing the structure of the board or the technology used for its manufacture only when in practice they encounter the aforementioned "black magic effects": when everything is designed and simulated as if "right," but under tests the device behaves "wrong." The matter is that manufacturers of numerous types of printed circuit boards are not obliged, unless specifically requested, to inform their customers on the numerous design and technological features of their products, although the reasons for the occurrence of these "mystical" effects often lie in these "differences" and "features" rather than anywhere else.

In this chapter, a number of design and technological features of modern printed circuit boards (Sects. 7.1, 7.2, 7.3, and 7.4), which are directly related to the reliability of the high-speed devices in which they will be used, are considered in detail based on the analysis and generalization of the positive and negative experience that designers already have in their possession.

In addition, various methods for ensuring the integrity of signals on high-speed transmission lines (Sects. 7.5) are considered; methods and formulas for calculating optimal structures and layouts of conductor (Sects. 7.6, 7.7, 7.8, and 7.9) are given, including empirical rules (Sects. 7.10) for designing layouts for the boards with high-speed signals.

A. Belous, V. Saladukha, *High-Speed Digital System Design*,
https://doi.org/10.1007/978-3-030-25409-4_7

7.1 Printed Circuit Board Design and Manufacturing Technology

7.1.1 History of Printed Circuit Boards

All modern electronic devices are made using printed circuit boards. Based on the technology of manufacturing printed circuit boards, standard prefabricated assemblies of microelectronic products are made—hybrid modules containing components of various functional purposes and degree of integration [2]. Multilayer printed circuit boards and electronic components of a high degree of integration reduce the weight and dimensions of electronic products and computing devices.

A printed circuit board (PCB) is a plate made of electrical insulating material (paxolin, fabric-base laminate, fiberglass plastic and similar dielectrics) with thin metallized electrically conductive strips (printed conductors) applied on its surface and pads for connecting add-on radio elements, including modules and integrated circuits.

The main structural element of the printed circuit board—a rigid or flexible insulating base with conductors on its surface. Additional elements are used to install add-on components and connect them to conductors: pads, plated vias and component holes, connector lamellae, special surface areas for heat removal, shields and current-carrying surfaces, etc.

Transition to the use of printed circuit boards marked a qualitative leap in communications electronics equipment design. A printed circuit board functions both as a carrier of radio elements and their electrical connector. The latter function can only be performed, if a sufficient level of insulation resistance is provided between conductors and other conductive elements of the printed circuit board. Consequently, the PCB substrate shall perform the function of an insulator.

A brief history of the creation and development of printed circuit boards is presented in the literature (http://www.belplata.by/pechatnye_platy/pechatnye_platy/istoria).

At the beginning of the twentieth century, a German engineer and telephone industry pioneer Albert Parker Hanson created a device that is considered a prototype of all types of printed circuit boards known today. The birth year of printed circuit boards is considered to be 1902, when this inventor filed an application with the patent office of his native country.

Hanson's printed circuit board was represented by a simple pattern formed on a bronze/copper foil by cutting or stamping. The resulting conductive layer was glued to a dielectric, which was a type of paraffin-impregnated paper. Even then, taking care of placing more conductors, Hanson stuck the foil on both sides, creating a two-sided printed circuit board. The inventor also used PCB through-holes. In his works Hanson describes the creation of conductors by electroplating or using conductive ink, which is a powdered metal mixed with an adhesive carrier.

Initially, only additive process was used to manufacture printed circuit boards, i.e., drawing elements were glued to/sprayed on a dielectric.

Thomas Edison had similar ideas. There is a letter to a colleague, where Edison describes three ways of forming a conductive pattern on paper.

1. A pattern is formed using adhesive polymers by applying graphite or bronze powders on their non-hardened surface.
2. A pattern is formed on the dielectric itself. Lapis (silver nitrate) is used to form a pattern. Then silver is simply recovered from silver salt.
3. The conductor is gold foil with a pattern formed on it.

Even though Edison never used the term "printed circuit board," almost all of the above-mentioned ideas have been applied in today's engineering procedures. Based on the first of them, today's thin-film technologies have been formed. The second method is widely used for coating by reduction of metal salts.

In 1913, Arthur Berry patented a print-and-etch method of manufacturing printed circuit boards. The developer suggested covering the metal base with a layer of resist material and removing the unprotected parts from the surface by etching. In 1922, Ellis Bassit from the USA invented and patented the first method based on using photosensitive materials in the manufacture of printed circuit boards.

In 1918, a Swiss inventor Max Schoop introduced the technology of flame-spraying metal onto a board, but back then it was unpopular due to the cost of production and uneven metal deposition.

An American developer Charles Ducas patented the original technology of electroplating conductors, which can be summed up as follows: channels were drawn and subsequently filled with metallized conductive pastes in a soft dielectric (for instance, wax) by electrochemical action.

The same patent included an etching technology, i.e., electrolytic deposition of metal (silver, gold, or copper) through a contact mask on a low-temperature alloy plate. The plate with the deposited pattern was heated up, and all parts of the alloy uncovered with silver were removed. Charles Ducas also placed conductors on both sides of the insulative substrate.

Ducas further designed multilayer printed circuit boards and proposed several interesting solutions for vias.

A French inventor Caesar Parolini went back to using the additive process as a conductive layer manufacturing method. In 1926, he formed a pattern on a dielectric using an adhesive material with copper powder sprayed on it and polymerized it at high temperature. It was Parolini who started using jumper wires in printed circuit boards. They were installed before the material polymerization.

In 1933, the works of Erwin Franz were published. They include virtually all methods of manufacturing flexible printed circuit board, which are used today. In particular, he managed to form a conductive pattern on a cellophane film, using a liquid polymer with graphite filling.

Engineer Paul Eisler was among the first to introduce printed circuit boards in specific radioelectronic devices in the UK. During the Second World War, he was looking for process engineering solutions to start the mass production of printed circuit boards, primarily using printing technologies. After the war, in 1948, Eisler founded his own company for the manufacture of printed circuit boards—Technograph Printed Circuits.

In the 1920s to 1930s, many other patents were issued on the design and methods of manufacture of printed circuit boards. The first PCB manufacturing techniques remained largely additive (development of Thomas Edison's ideas). But the present form of the printed circuit board appeared due to the subsequent use of manufacturing methods borrowed from the printing industry. Printed circuit board is derived from the term "printing plate" (meaning a form, a matrix). However, the true founder of the printed circuit board is the Austrian engineer Paul Eisler, who first came to realize that printing (subtractive) technologies can be used for mass production of printed circuit boards. In subtractive technologies, an image is formed by removing unnecessary fragments. Paul Eisler developed a method of galvanic deposition of copper foil and its etching with ferric chloride. Mass production technology of printed circuit boards was becoming more popular during the Second World War. Since the mid-1950s, printed circuit boards have become an essential element of radio equipment actively used not only in military but also in consumer electronics.

7.1.2 *Designs of Printed Circuit Boards*

7.1.2.1 Multilayer Printed Circuit Boards

The predecessor of printed circuit boards—a simple insulated wire—had a major drawback. To prevent physical impact and vibration, additional elements were required for its fixation inside the communications electronics equipment. Various carriers of radio elements and structural elements were used for intermediate connections, as well as for fixing the wires. It was the so-called three-dimensional wiring.

New generation printed circuit boards are flawless in this regard—their conductors are rigidly fixed on the surface, their position is stable, allowing for mutual coupling calculations. Initially printed circuit boards were single-sided or double-sided depending on the arrangement of conductive tracks.

A single-sided printed circuit board is an insulated plate with printed conductors located on one side only. In double-sided printed circuit boards, printed conductors were placed on the originally empty side of the plate. A number of options were proposed to form a conductive path from one side of the board to the other with plated vias being the most common of them. Fragments of single-sided and double-sided printed circuit boards of the simplest design are shown in Fig. 7.1.

The use of double-sided PCB instead of single-sided ones was the first step in the transition from plane to volume. Putting aside the substrate of a double-sided printed circuit board, one gets three-dimensional conductors. By the way, this step was made at once. In his patent application Albert Hanson indicated the possibility of placing conductors on both sides of the substrate and connecting them through thru-holes.

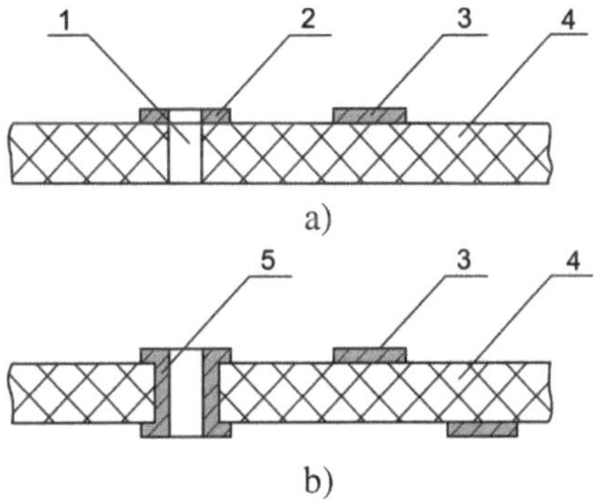

Fig. 7.1 PCB design fragments: (**a**) single-sided and (**b**) double-sided: 1—component hole; 2—pad; 3—conductor; 4—insulative substrate; 5—plated via

Further development of electronics and microelectronics resulted in the introduction of multi-terminal components (some chips have more than 200 terminals). New electronic components appeared. In its turn, the use of digital microcircuits and their high-speed performance led to the development of new requirements for their shielding and power distribution to components. For this purpose, special shielding conductive layers were included in multilayer PCBs in digital devices (for instance, computers). All this led to growth and sophistication of conductive paths, resulting in multiple layers. Some modern printed circuit boards have much more than ten. In a sense, a multilayer printed circuit board has become three-dimensional.

Let us consider a typical design of a multilayer PCB.

In the first variant—and it is the most common one—the inner layers of the board are made from double-sided copper-laminated fiberglass plastic, which is called the "core." The outer layers are made from copper foil and pressed with the inner layers using a binding substance—a resinous product called "pre-preg" (abbr. from "pre-impregnated"). These are composite raw materials. This ready-to-process pre-impregnated product is partially cured. It contains a binder and reinforcing woven or non-woven materials. They are obtained by impregnating the reinforcing fiber base material with evenly distributed polymeric binders. Impregnation is carried out so as to maximize the physico-chemical properties of the reinforcing material. Pre-preg technology is aimed at obtaining monolithic products of complex shape with minimal tooling. Pre-pregs are made in the form of a sheet covered with polyethylene film on both sides and rolled.

After pressing at high temperature, a multilayer PCB "pie" is formed, where holes are drilled and coated with metal. The second variant is less common: the outer layers are formed from "cores" held together with pre-preg. This is a simplified description forming the basis for many other designs. However, the basic principle is that pre-preg is always used as a binder between the layers. Obviously, a situation with two adjacent "cores" without a pre-preg layer is hardly possible. However, a "foil–pre-preg–foil–pre-preg" structure is often used in boards with complex combinations of blind and buried vias.

Multilayer printed circuit boards now account for two-thirds of the world production of printed circuit boards in terms of price, although in quantitative terms they are not as widespread as single-sided and double-sided PCBs.

A design fragment of a modern multilayer printed circuit board is shown schematically (in a simplified form) in Fig. 7.2. Conductors in such printed circuit boards are placed not only on the surface, but also in the bulk of the substrate. At the same time, the number of layers and the arrangement of conductors have been preserved (due to the use of planar printing technologies). Layering is inevitably present in the names of printed circuit boards and their elements—single-sided, double-sided, multilayered, etc. In essence, layering reflects the structural element and the PCB manufacturing techniques corresponding to this structural element.

In reality, the design of multilayer printed circuit boards differs from the one shown in Fig. 7.2 [2].

In their structure, multilayer PCBs are much more complex than double-sided boards, and therefore their manufacturing process is much more complicated. And their structural arrangement differs significantly from the one shown in Fig. 7.2. They include additional shielding layers (ground and power), as well as several signal layers.

In reality, they look more like the ones shown in Fig. 7.3 [2].

Vias and microvias are used to interconnect the layers of a multilayer PCB (see Fig. 7.3a). Vias can be performed in the form of through-holes connecting the outer layers between themselves, as well as with the inner layers. Blind and buried vias are used, too.

A blind via is a plated through-hole, visible only from the top or bottom sides of the board.

Buried vias are used to interconnect the inner layers of the board. Their use simplifies the wiring of boards, for instance, a 12-layer PCB can be substituted by an

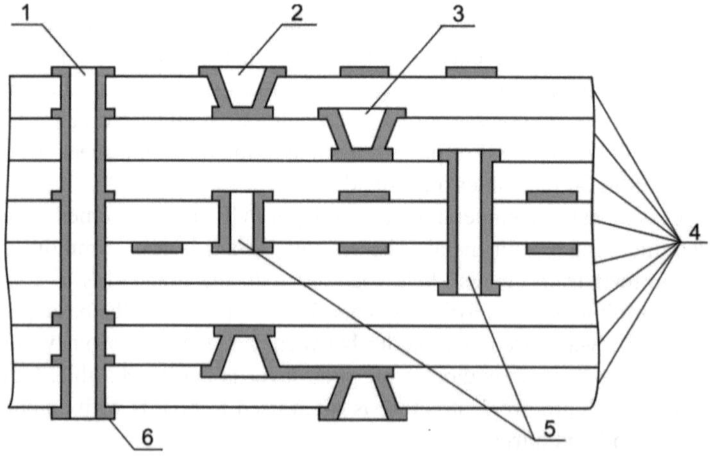

Fig. 7.2 Multilayer printed circuit board design fragment: 1—plated through-hole; 2—blind microvia; 3—buried microvia; 4—layers; 5—buried interlayer holes; 6—pads

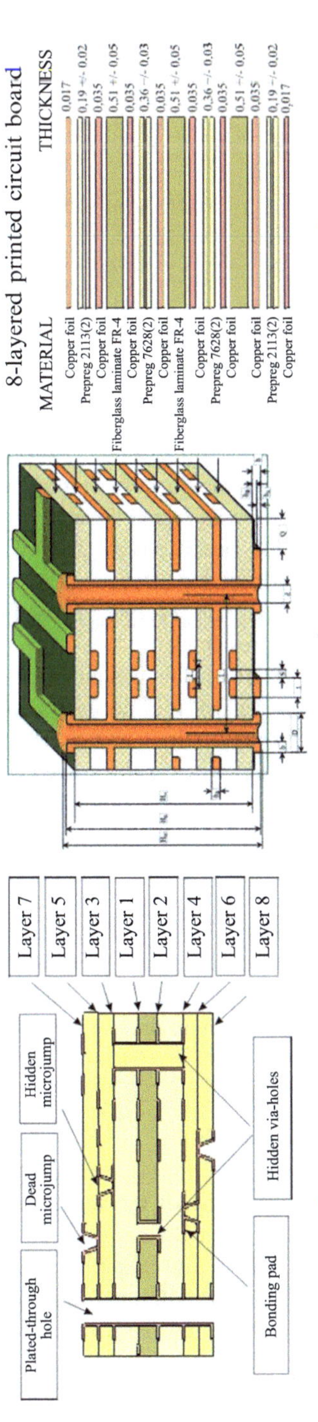

Fig. 7.3 Multilayer PCB typical design: (**a**) a simplified multilayer PCB design sketch; (**b**) a simplified multilayer PCB isometric image; (**c**) a multilayer PCB design example: vertical cross section

Fig. 7.4 Flexible PCB appearance

equivalent 8-layer cross-plugging. For the purpose of surface mounting, microvias have been developed to connect pads with signal layers.

To manufacture multilayer printed circuit boards, several foil-laminated insulating materials are combined using adhesive pads—pre-pregs.

Pre-preg is marked in white in Fig. 7.3b. Pre-preg is used to adhere the layers of the multilayer PCB during thermal pressing.

Total thickness of multilayer PCBs is increasing disproportionately with the number of its signal layers. In this regard, it is necessary to take into account the large ratio of board thickness to through-hole diameter, which is a fixed parameter for the process of through-hole plating. Nevertheless, even considering the difficulties of small diameter through-hole plating, manufacturers of multilayer PCBs prefer high-density installation, providing more relatively cheap layers, as compared to fewer but more expensive high-density layers.

Figure 7.3c shows the approximate arrangement of layers of a multilayer printed circuit board, as well as their thickness.

The PCB technology is decidedly and constantly evolving and, as is clear from the above-mentioned commonly known material, the familiar term "printed circuit board" no longer reflects its internal structure and manufacturing technology.

7.1.2.2 Flexible Printed Circuit Boards

Rigid printed circuit boards represent one of the most popular products used in electronics, but there are also flexible printed circuit boards (Fig. 7.4), whose scope of application is increasingly expanding. An example is the so-called flexible flat cable. Such printed circuit boards perform a limited range of functions (excluding the function of a substrate for radio elements). They serve only to combine conventional printed circuit boards while replacing the traditional cables. Flexible printed circuit boards acquire elasticity due to their highly elastic polymer "substrate."

Modern flexible printed circuit boards have two degrees of freedom, they can even form a Moebius strip.

Some traditional rigid printed circuit boards have one or even two degrees of freedom, but this freedom is very limited, considering the hard (glassy) polymer matrix of their substrate. The degrees of freedom are achieved by reducing the substrate thickness. One of the advantages of 3D printed circuit boards made from thin insulating materials provides for a possibility for their "rounding." Thus, it becomes possible to match their shape with that of the objects (for example, nanosatellites), where they can be placed. The result is additional internal volume of products.

Their major disadvantage consists in the following: flexibility of such printed circuit boards decreases with each layer; bending of such printed circuit boards with non-flexible components leads to high mechanical loads at the points of their connection with a flexible printed circuit board; thus, their reliability deteriorates.

"Ancient" printed circuit boards, consisting of rigid elements, which are folded like an accordion, fall in between rigid and flexible printed circuit boards. Modern flex-rigid printed boards have a different design. It concerns mainly multilayer printed circuit boards. They combine rigid and flexible layers. A printed circuit board with flexible layers placed outside hard ones consists of both rigid and flexible fragments. Another option—a combination of two rigid fragments with a flexible one.

Classification of printed circuit board designs based on the "conductive pattern layering" comprises most of the designs, but not all of them. For instance, not only printing equipment, but also weaving machinery was necessary to manufacture fabric PCBs and flexible flat cables. Such "printed circuit boards" have three degrees of freedom. Just like regular fabric, they can have the most bizarre outlines and shapes.

7.1.2.3 Printed Circuit Boards on Bases with High Thermal Conductivity

Recently, there has been an increase in the heat release of electronic devices, which is associated with a big leap in their functionality, as well as with the use of high-performance computer systems, the need for high-power switching devices, and the use of electronic components with increased heat release (http://www.belplata.by/pechatnye_platy/pechatnye_platy/istoria) [3–8].

The latter is most clearly manifested in LED lighting, where the interest in creating light sources based on high-power ultra-bright LEDs has increased. The luminous efficiency of semiconductor LEDs has already reached 100 lm/W. Such ultra-bright LEDs replace conventional incandescent lamps and are applicable in almost all areas of lighting technology: street lamps, automotive lighting, emergency lighting, advertising signs, LED panels, LED indicators, LED scrolling signs, traffic lights, etc. These LEDs have become indispensable in decorative lighting and in light-dynamic systems due to their monochrome color and switching time. Their use

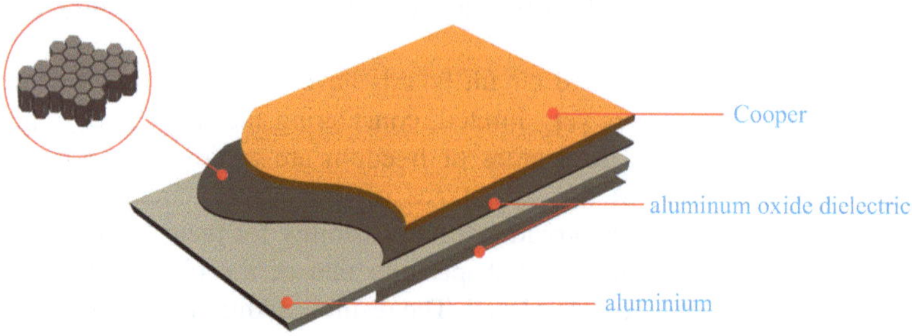

Fig. 7.5 PCB detail with high thermal conductivity

is advantageous, where it is necessary to save electricity; where frequent maintenance is expensive; and where electrical safety requirements are high.

Studies of LEDs have shown that approximately 65–85% of the electrical energy is converted to heat. However, provided that the thermal conditions recommended by the LED manufacturer are observed, LED service life can be up to 10 years. However, if the appropriate thermal conditions are not met (usually in case of operation at a transition temperature of more than 120–125 °C), LED service life can be ten times shorter! In case of gross non-observance of the recommended thermal conditions, for instance, if an emitter type LED has been on without a radiator for more than 5–7 s, the LED may fail during the first power-up. Moreover, an increase in the transition temperature leads to a decrease in the light intensity and an operating wavelength shift. Therefore, it is very important to correctly determine the thermal conditions and dissipate the LED-generated heat as much as possible.

Major manufacturers of high-power LEDs, such as Cree, Osram, Nichia, Luxeon, Seoul Semiconductor, Edison Opto, etc., have long been making LED modules and clusters on printed circuit boards with a metal substrate (according to the international classification, IMPCB—insulated metal printed circuit board or AL PCB—aluminum printed circuit boards) to enable quick start-up and extend LED applications (Fig. 7.5).

These AL PCBs have a small and fixed thermal resistance. When installed on a radiator, they provide heat dissipation from the p–n junction of the LED and ensure its operation during the entire service life.

Copper, aluminum, and various types of ceramics are used as printed circuit board base materials with high thermal conductivity.

7.1.3 Basic Technological Features of the Production of Printed Circuit Boards

During the entire history of printed circuit boards, developers and manufacturers have been trying to ensure their quality and reliability.

Printed circuit boards manufactured by through-hole printing, despite their wide application, have a major drawback associated with the peculiarities of the junction point of studs in vias and conductive layers (pads). A path connecting the stud with the conductive layer goes along the end of the pad. The length of the path is determined by the copper foil thickness and is usually 35 μm or less. Galvanic metallization of the via walls is preceded by a stage of chemical metallization. As opposed to galvanic copper, chemical copper is loose. Therefore, the connection of the stud with the end surface of the pad occurs through an intermediate sublayer of chemical copper, which is structurally weaker. Coefficient of thermal expansion of fiberglass plastic is much higher than that of copper. When reinforced in epoxy resin, this difference increases dramatically at the glass transition temperature. The PCB connection is subject to extremely high mechanical loads and often fails for a variety of reasons, including technical effect or thermal shock. As a result, the corresponding electrical circuit is broken and the operation of the entire device is impaired.

To increase the reliability of vias in multilayer PCBs, technologists generally treat the insulating material with a caustic liquid, i.e., partially remove it in vias before plating. In this case, studs are connected with pads not only at the end, but also partially—along the outer annular areas of these pads (Fig. 7.6).

A higher reliability of plated vias of multilayer PCBs can be achieved, provided that layer-by-layer electroplating is used to manufacture multilayer PCBs (Fig. 7.7). In this method, the paths between the conductive elements of the printed layers are formed by copper electroplating in the insulating layer holes. In this case, as opposed to the method of through-hole plating, vias are filled with copper entirely. There is a much larger connection area between the conductive layers, and the geometry is different. Such connections are not easy to break. Nevertheless, this technology is also far from perfect, since it still has the "galvanic copper–chemical copper–galvanic copper" structure.

Printed circuit boards made by via plating shall withstand multiple resoldering (at least 4 times; multilayer PCBs—at least three times). 3D PCBs may be resolded up to 50 times. According to experienced developers, plated vias in 3D printed circuit boards increase their reliability, instead of decreasing it. Within the 3D PCB manufacturing technology, both the conductive layers and the studs connecting them constitute a single (simultaneous) technological cycle. Therefore, the "galvanic copper–chemical copper–galvanic copper" structure is absent.

Technologists assume full responsibility for the formation of a barrier layer at the junction of pad ends and plated eyelets. Nevertheless, they managed to solve this problem. Revolutionary changes in the PCB manufacturing technology have led to the introduction of direct metallization techniques, which are limited to the preliminary surface activation, excluding the stage of chemical plating. Moreover, direct metallization processes are implemented in such a way that a conductive film appears only where necessary—on the surface of the insulating material. As a result, a barrier layer in the plated vias of the printed circuit boards manufactured using direct metallization techniques is absent as such.

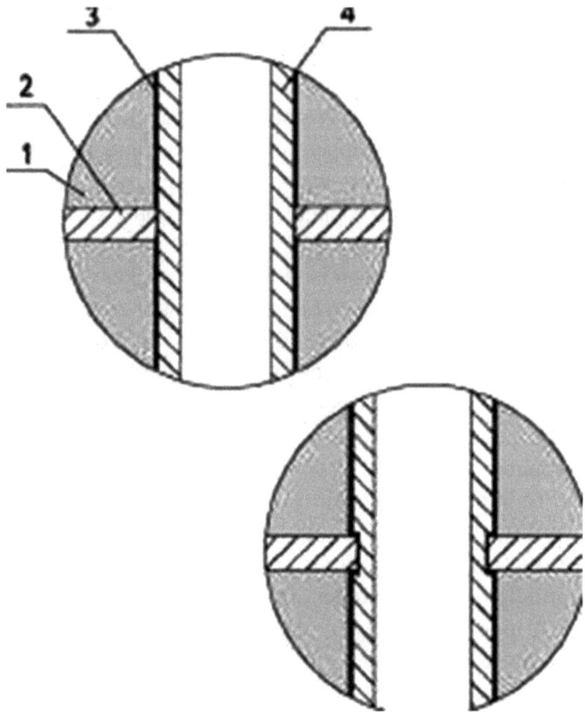

Fig. 7.6 Vias in multilayer printed circuit boards: (**a**) without additionally treated insulating material; (**b**) with additionally treated insulating material: 1—insulating material; 2—inner layer pad; 3—chemical copper; 4—galvanic copper

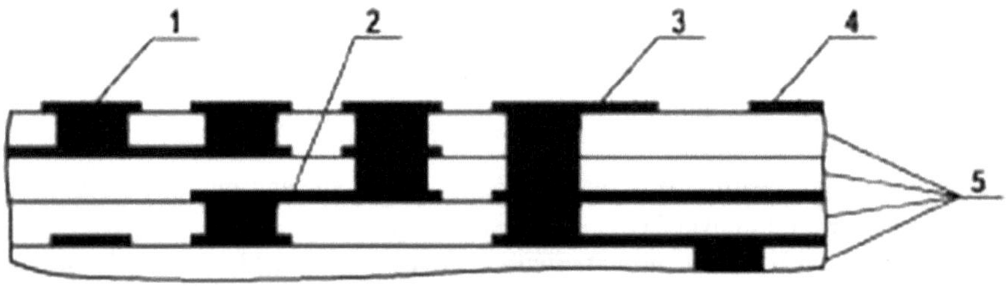

Fig. 7.7 Fragment of a multilayer printed circuit board manufactured by layer-by-layer electroplating: 1—via; 2—inner layer conductor; 3—mounting pad; 4—outer layer conductor; 5—dielectric layers

Another problem concerns via plating. Plated vias may become a weak spot of printed circuit boards. Via wall coating thickness should always be uniform throughout their height; otherwise, reliability problems may occur. Physical and chemical electroplating processes prevent this.

The ideal and real profiles of plated via coating are shown in Fig. 7.8. The hole depth coating may be less thick than that on the surface for various reasons: irregular current density, cathodic polarization, insufficient electrolyte exchange rate, etc. In modern printed circuit boards, the diameter of plated vias passed the point of 100 μm long time ago. In some cases, the height-to-diameter ratio of the hole reaches 20:1.

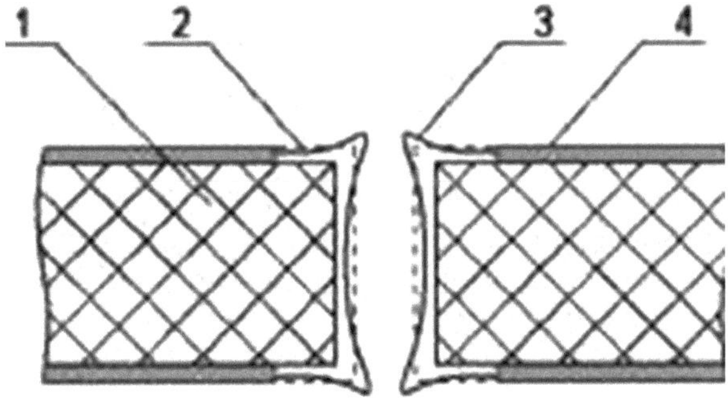

Fig. 7.8 PCB plated via cross section: 1—insulating material; 2—ideal profile of via plating; 3—real profile of via plating; 4—resist

The existing physical methods (use of ultrasound, increased intensity of fluid exchange in PCB holes, etc.) are practically at the end of resources. Even the electrolyte viscosity is beginning to play a significant role.

Traditionally, this problem has been solved through the use of electrolytes with levelers, which are adsorbed in the areas of higher current density. The sorption of such levelers is proportionate to the current density. The levelers create a barrier layer, preventing excessive electrodeposition on the sharp edges and in adjacent areas (closer to the PCB surface).

Theoretically, another solution to this problem was known long time ago. However, it was brought to life only recently—after the industrial production of high-power switched mode power supplies. This method is based on the use of pulse (reverse) power supply mode of electroplating baths. Most of the time direct current is supplied. Simultaneously, deposition of coating takes place. Reverse current is supplied once in a while. At the same time, the deposed coating dissolves. In this case, uneven current density (it is higher at acute angles) is only beneficial. For this reason, the coating dissolution occurs primarily and to a greater extent at the board surface. This technical solution uses a whole bunch of methods for sorting out technical differences: use a partially redundant action, apply the transition from a continuous process to a pulse one, do the opposite, etc. A certain combination of the duration of direct and reflected pulses makes it possible to obtain a coating thickness in the depth of the hole greater than that on the board surface. This technology has proven to be indispensable for filling blind vias with metal, while the density of the PCB interconnects increases about twice.

Other problems that technologists face are solved in the process of development, while consumers do not even think about them. We obtain multilayer printed circuit boards for our needs and use them.

7.1.4 Problems of Microminiaturization of the Elements of Printed Circuit Boards

Initially, when radioelectronic devices were developed, the same components were used on printed circuit boards that were used for three-dimensional wiring of communications electronics equipment, but with some improvement of outputs and their reduction in size. However, the most common components were installed on printed circuit boards without any alterations.

The advent of printed circuit boards and their miniaturized components resulted in a decrease in the levels of operating voltages and currents consumed by them.

With the advent of miniature semiconductor amplifying devices—transistors—printed circuit boards were widely used in household appliances, then in the industrial sector. With the advent of electronic circuit fragments combined on a single chip—functional modules and microchips—their design provided for the installation of exclusively non-printed circuit boards.

With the continuous reduction of active and passive components in size, the concept of "microminiaturization" appeared.

In electronic components, this was reflected in the emergence of LSI and VLSI circuits containing millions of transistors. Their introduction made it necessary to increase the number of field ties (see the standard graphics processor pad in Fig. 7.9a), which, in its turn, resulted in the appearance of more complex interconnections of the current-flow lines of its board (see Fig. 7.9b).

Such a board of the graphic processor/the CPU is in fact a small multilayer printed circuit board, which includes the processor chip itself, the interconnections of the chip outputs with the contact region, and some add-on components (generally, filter capacitors of the power distribution system).

Virtually every modern central processing unit (CPU) manufactured by Intel or AMD is also some kind of a multilayer printed circuit board (Fig. 7.9c).

Microminiaturization and high concentration of elements on a PCB surface are general trends in the development of printed circuit boards and (micro-)electronic equipment at large. In this case, by "elements" one should understand both our own components and components of printed circuit boards (conductors, substrates, vias,

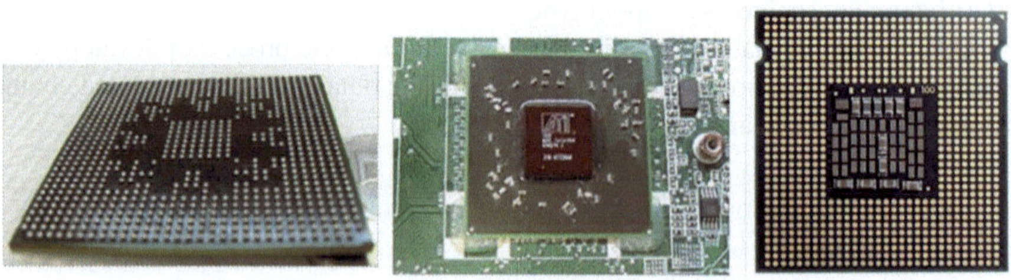

a b c

Fig. 7.9 Graphic processor contact surface—(**a**); interconnections of the current-flow lines of the board—(**b**); standard CPU contact surface—(**c**) etc.), as well as the elements of a supersystem (printed board assembly)—radio elements. By the speed of microminiaturization, the latter are ahead of PCBs.

The original methods of manufacturing printed circuit boards were based on attaching copper foil conductors to the surface of the insulative substrate.

It was assumed that the width of conductors and the gaps between them were measured in millimeters. In this design, this technology seemed quite efficient. The subsequent miniaturization of electronic equipment required the creation of new PCB manufacturing methods. The most important versions (subtractive, additive, semi-additive, combined) are still used today. These technologies allowed for the implementation of printed circuit boards with the size of elements measured in tenths of a millimeter.

The introduction of printed circuit boards with a resolution level of approximately 0.1 mm (100 μm) was a milestone event. On the one hand, it marked a step "down" to another level. On the other hand—a kind of a quantum leap. Why? The insulative substrate of the majority of modern printed circuit boards is fiberglass plastic—laminated plastic with a fiberglass reinforced polymer matrix. Reduction of the gaps between the conductors of the printed circuit board led to the fact that they became commensurable with the thickness of the glass threads and their knots in the woven fiberglass. Conductors "closed" with such knots are not at all an unlikely scenario. As a result, the formation of specific capillaries in fiberglass plastic "closing" these conductors is just as real. In high humidity conditions, these capillaries ultimately lead to a deterioration of insulation between the PCB conductors. More precisely, it happens even under normal humidity conditions. Condensation in the capillary structures of fiberglass plastic is observed under normal conditions. Humidity always reduces the level of insulation resistance.

Since such printed circuit boards have become a common phenomenon in modern communications electronics equipment, it may be deduced that the developers of basic materials for printed circuit boards managed to solve this problem using traditional methods. Samsung specialists have mastered the new technology of PCB manufacturing with a width of conductors and gaps between them of 8–10 μm. And this is not the thickness of the glass thread, but that of fiber!

Ensuring isolation in the ultra-small gaps between the conductors of the existing PCBs and especially in the printed circuit boards of the future is a great challenge. Both traditional and non-traditional methods to address it (if addressed at all) will be developed over time.

There were difficulties in obtaining ultra-small (ultra-narrow) conductors in printed circuit boards. For many reasons, subtractive methods of manufacturing printed circuit boards have become widespread. In subtractive methods, an electrical circuit pattern is formed by removing unnecessary foil fragments. During the Second World War, Paul Eisler developed the technology of etching copper foil with ferric chloride. This plain technology is still used by radio amateurs. Industrial technology did not advance much neither. Except that the composition of etching solutions changed, and the process was automated with the introduction of new elements.

The principal disadvantage of an absolute majority of all etching technologies is that etching goes not only where it should (towards the surface of the insulating

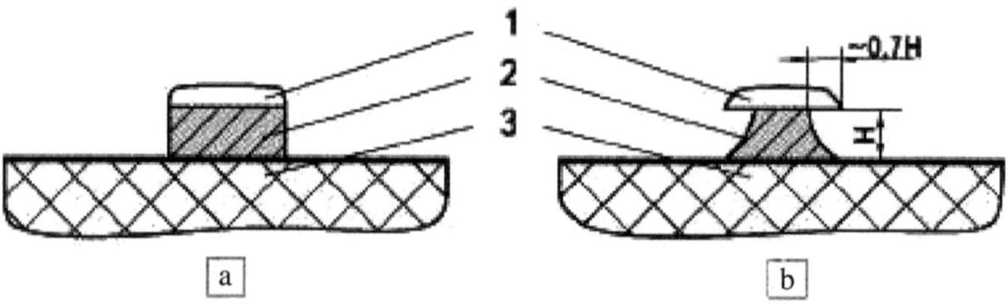

Fig. 7.10 Copper foil etching profiles: (**a**)—ideal profile; (**b**)—real profile; 1—protective layer; 2—conductor; 3—insulating material

material), but also in the unwanted transverse direction. Lateral undercutting of conductors is commensurate with the copper foil thickness (about 70%). Generally, a mushroom-like profile is obtained instead of the ideal conductor profile (Fig. 7.10). If conductors are wide (in the simplest printed circuit boards, the width is measured in millimeters), the lateral undercutting simply goes unnoticed. However, if the width of the conductors is commensurate with their height or even less than it (current situation), then the "lateral character" of such technologies casts doubt on their practicability.

In fact, the area of laterally undercut printed conductors can be reduced to some extent. This is achieved by increasing the etching rate using flow coating (etching agent flows in the desired direction—perpendicular to the sheet plane), as well as in other ways. However, when the width of the conductor approaches its height, the effectiveness of such improvements is clearly insufficient.

But the achievements of photolithography, chemistry, and technology can now solve all these problems. Originally, these solutions were developed in microelectronics technology [1].

7.1.5 Printed Circuit Boards for Components with High Heat Dissipation

Designing printed circuit boards for components with a high heat dissipation has its own features [4–6]. These features are most clearly expressed in the boards for highly luminous LEDs.

Unlike conventional incandescent lamps, LEDs do not radiate heat into the surrounding space, but direct it from the p–n junction to the heat sink in the LED housing (generally either an LED output or a special metal plate). Therefore, heat removal process is more complex and specific (see Fig. 7.11) [6]. The heat removal path has a variety of thermal resistances: "p–n junction–housing heat sink," "housing heat sink–printed circuit board," "printed circuit board–radiator," "radiator–environment." As a result, the use of high-power LEDs will most probably result in an

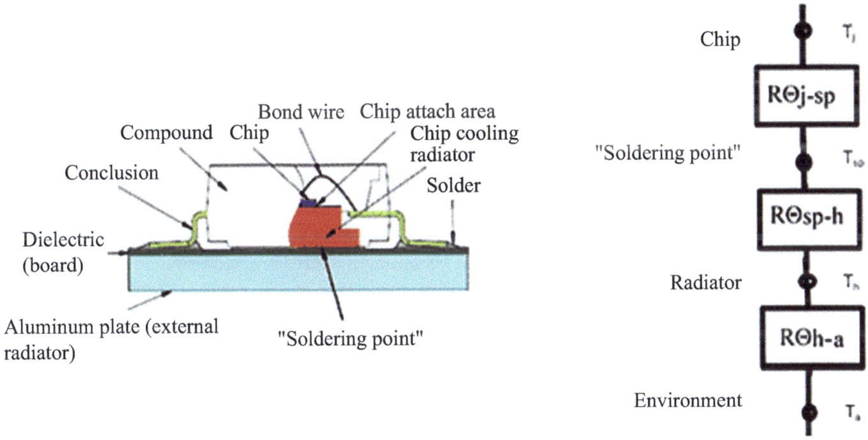

Fig. 7.11 LED circuit board design sketch

excessive increase in the junction temperature, while service life, reliability, and light characteristics directly depend on this parameter.

When dissipating high power, it may sometimes be necessary to use special IMPCBs with improved thermal conductivity and a thermal resistance ranging from 0.45 to 1.5 K/W.

Let us consider a printed circuit board used for high-power LED clusters—an IMPCB. In such a board, a metal substrate is used as a base with one or several conductive layers of copper foil glued on it using woven-glass-fabric pre-preg. Such boards are used in products with increased local thermal loads or with thermal loads relating to the entire area. The simplest and therefore the most cost-efficient type of PCB with surface-mounted components is a single-layer AL PCB. This board is, in essence, a single-layer printed circuit board glued onto an aluminum substrate. The generated heat easily passes through the insulating material and then quickly dissipates through aluminum—a firmly glued radiator serves as a PCB component (Fig. 7.12).

A standard copper foil with a thickness of 35–350 μm is used for the production of printed circuit boards.

Insulating material—pre-preg—a 50–150 μm fiberglass impregnated with epoxy resins.

Both a standard epoxy fiberglass (FR-4) and a special heat-conducting material (T-preg) may serve as a pre-preg, which has the best thermally conductive and insulating properties. T-preg has a peculiar chemically resistant structure. It is characterized by a high thermal conductivity. It is 75–200 μm thick and made from a special insulating material—a mixture of a polymer with special ceramics. The polymer is selected based on its insulating properties, whereas the ceramic filler is designed to improve thermal conductivity, so the material has excellent insulating properties and an extremely low thermal resistance. For instance, this layer is made

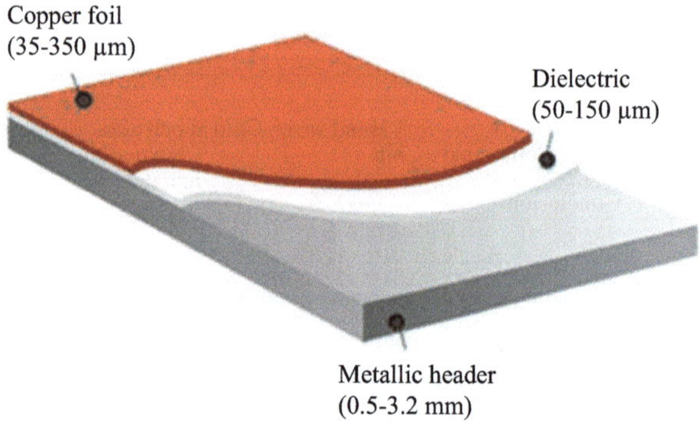

Fig. 7.12 Materials used in IMPCBs

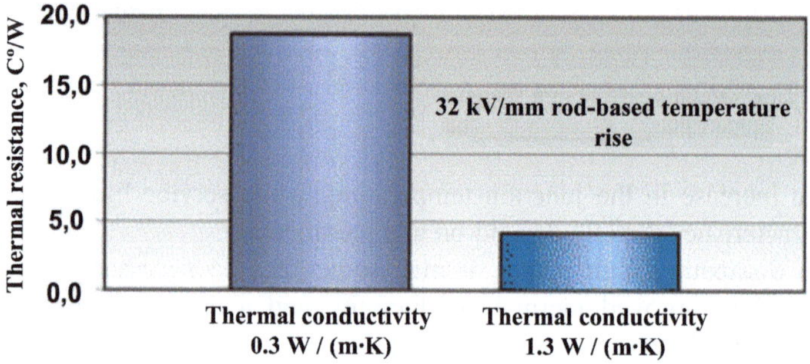

Fig. 7.13 Total thermal resistance: Fr4 versus MP

from Bond Ply Thermal Clad IMS (Bergquist) material with the volume resistivity of at least 10 14 Ω • cm. With the insulating material thickness of 75 μm, the allowable service voltage between its layers is not lower than 5.5 kV AC, thermal conductivity—not lower than 1.3 W/m K, which is enough for most applications (Fig. 7.13).

In IMPCBs, the insulating layer is indispensable, since it serves to connect the copper layer to the bottom metal (aluminum or copper) substrate, which serves simultaneously as a radiator for the entire printed circuit board and as a heat conductor from the top layer to the bottom one (Table 7.1).

Metal substrate thickness—0.5–3.2 mm. Metal substrate is generally produced from aluminum and copper. Their performance comparison is given in Table 7.2.

In the design of IMPCBs, coefficient of thermal expansion (CTE) of the substrate material plays an important role. The use of materials with a high CTE at high temperatures leads to the internal mechanical stresses in the structure. Therefore, for high-temperature applications, where this parameter is critical, materials with a low carbon steel substrate (1 and 1 mm thick) and a low CTE are used.

Although copper has higher heat conductivity, aluminum remains the most commonly used material in IMPCBs, since it is cheaper and, most importantly, light (Table 7.3).

Table 7.1 Main comparative characteristics of insulating materials

Parameter	FR-4	T-preg
Thermal conductivity, W/mK	0.3	1–5
Dielectric strength, kV/mm	20	24–31

Table 7.2 Performance comparison of metal substrate materials

Parameter	Aluminum	Cooper
Thermal expansion coefficient, ppm/°C	24	18
Thermal conductivity, W/mK	173	260
Density, g/cm^3	2.7	8.9

Table 7.3 Thermal conductivity of the used aluminum substrates

Material	Thermal conductivity, W/mK
Aluminum 1100 (AD analogue)	222
Aluminum 5052 (AlMg2.5 analogue)	138
Aluminum 6061 (AD33 analogue)	167

To date, several large manufacturers of thermally conductive electrically insulating materials produce basic materials for the manufacture of IMPCBs: Bergquist (USA), Laird (Thermagon) (USA), Denka (Japan), Totking (China), Ruikai (China).

A wide list of supplied materials with different performance characteristics is able to satisfy the most demanding developers and technologists of communications electronics equipment. It is economically beneficial both at the production stage and while in use. The materials meet both the commercial and military standards and can be used in almost any field: from home appliances to military equipment.

For instance, T111 (Totking)—a widely used material from a heat-conducting polymer based on ceramics with an aluminum substrate—has the following characteristics (Table 7.4).

Most of the IMPCB manufacturing processes, such as etching, masking, hot air solder leveling (HASL), marking, are similar to the processes of manufacturing traditional FR-4 boards and differ only in the modes of mechanical circuit processing and drilling.

IMPCBs can be used in any product with special requirements for heat dissipation and dimensions. The use of such boards considerably simplifies the design of communications electronics equipment, especially those with high heat dissipation, since heat removal does not depend much on the arrangement of elements and the board area around them—heat dissipates through the substrate. In some cases, provided that heat dissipation is moderate, there is no need for additional heat sinks—radiators, bars, etc. As a result, there is a higher degree of integration of board elements, while PCBs are becoming smaller. Some of these boards are shown in Fig. 7.14.

Table 7.4 Characteristics of the T111 material (Totking)

Parameter	Value
Aluminum base thickness, mm	1.5
Insulating material thickness, μm	100
Copper foil thickness, μm	35
Insulating material thermal conductivity, W/mK	2.2
Insulating material thermal resistance, °C/W	0.7
Aluminum substrate thermal conductivity (Aluminum 5052, or AlMg2.5), W/mK	138
Breakdown voltage, KV	3
Glass transition temperature (Tg) °C	130
Volume resistivity, Ω • cm	10^8
Surface resistance, Ω	10^6
Maximum operating voltage (CTI), V	600

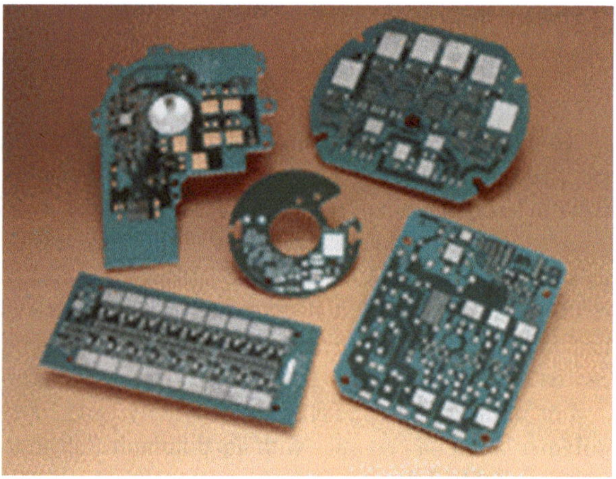

Fig. 7.14 Examples of printed circuit boards with additional heat sinks

IMPCBs have many advantages over conventional PCBs:

- dissipate heat without the use of additional radiators, special thermal conductive pastes.
- reduce/eliminate the need for forced air cooling fans.
- add mechanical stiffness to the product.
- increase the degree of integration of high-power equipment elements operating with high currents and voltages at high operating temperature.
- reduce the effect of heat stress of all components, thereby increasing the product's durability and the service life of its elements.
- the cooling properties of such boards enable heat removal, which favorably affects the unit cost of products.
- significantly expand the device memory due to some circuit configuration.
- IMPCBs have excellent electromagnetic compatibility and shielding characteristics.
- The use of such boards improves the reliability of devices and their MTBF.

7.2 Design Features of the Printed Circuit Boards for High-Speed Transmission Lines

7.2.1 *Classification of the Main Types of Transmission Lines of High-Speed Electronic Devices*

The generally accepted classification of signal transmission lines of high-speed electronic devices was discussed in detail above in Chap. 2. Below we consider only the main types of high-speed SHF signal transmission lines on PCBs, as well as some practical issues related to determining the technical specifications of conductors in similar transmission lines on PCBs. At the time of publication of this book, the most detailed discussion of the existing problems and possible solutions was provided in the work of an experienced design engineer S. Vladimirov, whose main ideas will be described below, as close as possible to the original text [9].

As shown in Chap. 2, all transmission lines have their peculiar topological structure and specifications, which makes it difficult to unify their manufacturing processes. As is well-known, these lines transmit electromagnetic energy to separate components of telecommunications systems. In general, transmission lines (microstrip structures in the form of flat antennas for wireless mobile devices or large waveguides of high-power radars) merely serve to transmit high-frequency signals with or without insulative substrates.

Thus, all transmission lines based on standard PCB materials and substrates for semiconductors have a characteristic impedance, which, as a rule, is 50 Ohm. The characteristics of the transmission lines are described using a standard model with lumped parameters—resistance (R), capacitance (C), inductance (L), and slope (G).

In single-sided (SS) PCBs, only one layer of foil is applied on one side of the insulating material;

- double-sided (DS) PCBs have two layers of foil;
- in multilayer (ML) PCBs, foil is applied on both sides of the board and forms the inner layers of the insulating material.

Such multilayer printed circuit boards are produced by gluing together several single-sided or multi-sided boards.

Depending on the properties of the PCB substrate material, the boards are traditionally divided into three groups:

- rigid printed circuit boards,
- flexible printed circuit boards,
- thermally conductive printed circuit boards.

Thus, a flexible printed circuit board generally uses one or more layers of insulting material forming at least one electrically conductive circuit (electronic circuit). It is designed to connect individual electronic components or assemblies into a single operating device.

Flexible printed circuit boards are bendable, allowing for their installation in hard-to-reach places and their use as "flexible connectors."

The substrate materials are extremely important for determining the characteristic impedance of an individual line. Thus, certain parameters of these materials (permittivity, dielectric loss, and even the surface roughness of copper conductors) have a direct influence on the performance of electronic circuits made from these materials.

Here, first and foremost, we should consider the main design features of microstrip lines as the most promising signal transmission lines in high-speed electronic systems.

Microstrip line—one of the most common types of transmission lines (of the three mentioned above), due to its high efficiency and relatively simple manufacturing method. As shown in Fig. 7.15, such a signal transmission line generally consists of an upper conductive layer and a lower ground plane with an insulating layer between them.

The use of such microstrip lines will increase, considering the universal trend of growth of wireless applications in number and volume. As for the design of these types of transmission lines, it is the rectangular microstrip antennas that are used in various wireless devices.

As a rule, the performance of such microstrip antennas considerably improves, if thick substrates with low permittivity are used, since these materials ensure the required directional pattern, high efficiency, and a large bandwidth. The use of thinner substrates with a higher permittivity generally results in bandwidth narrowing and lower efficiency. However, the use of materials with a sufficiently high permittivity may significantly reduce the size of the circuit for a given characteristic impedance.

Strip transmission lines are also widely used in the so-called fast applications. As compared to microstrip lines, where electromagnetic waves go through the insulating material and partially propagate in the surrounding space, a strip line conductor is insulated from all sides (see Fig. 7.16).

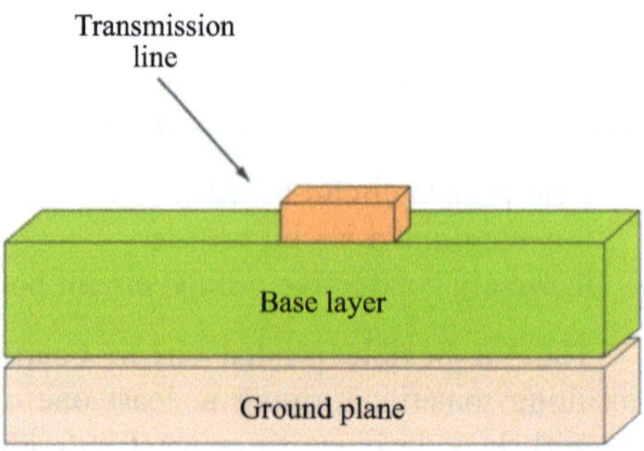

Fig. 7.15 Standard microstrip line design

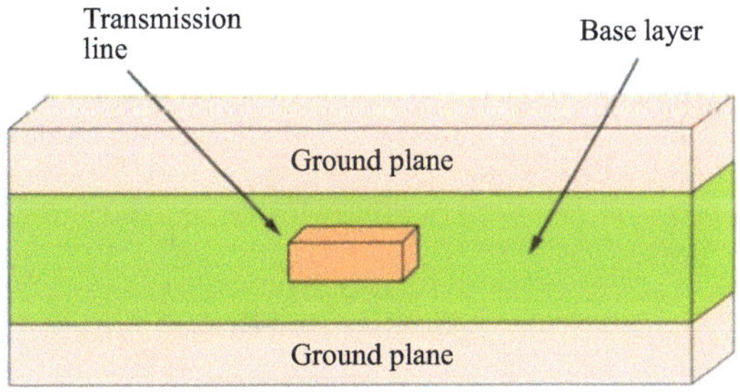

Fig. 7.16 "Strip line" design sketch

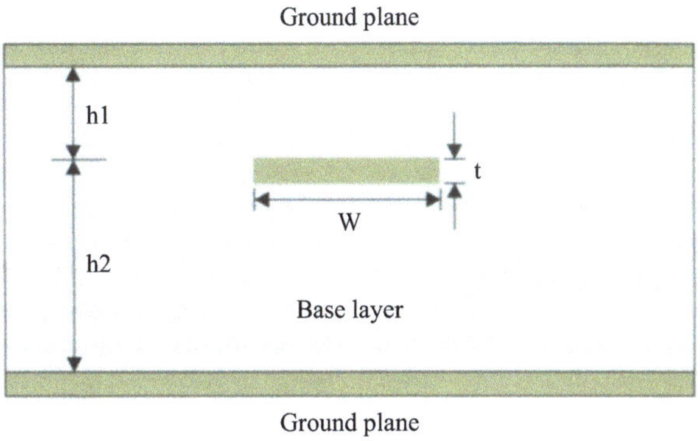

Fig. 7.17 Offset strip line design sketch

One of the most common types of standard strip lines are the so-called offset lines, whose conductor is closer to one of the two ground planes (Fig. 7.17). Another type is the so-called suspended strip line, whose conductive layer covering the insulating layer is in the upper air layer, while its ground planes surround the upper and lower air layers (Fig. 7.18). In this case, an air gap also functions as an insulating material. At the moment, lines of this type are actually the most common ones, but the manufacture and assembly of circuits using strip lines of this type, as is well-known, is difficult and more expensive than the production of other types of strip lines. However, due to this design, a larger bandwidth is provided, the directional pattern is improved, and the circuit losses are rare.

Another strip line is widely used in the industry: generally, it is formed by two conductors of different widths with one placed on the surface of the other. They are surrounded by insulating layers, which, in turn, are surrounded by ground planes (Fig. 7.19). Strip lines of this type may be considered as two microstrip circuits with unequally thick conductors.

Surely, it is technologically quite difficult to manufacture all of the above-mentioned types of strip lines. Since their components are located between ground planes or other structures, for instance, between plated through-holes, it is necessary to ensure reliable electrical connection with the individual elements and microcircuits surface-mounted on microstrip lines.

Fig. 7.18 Suspended strip line design sketch

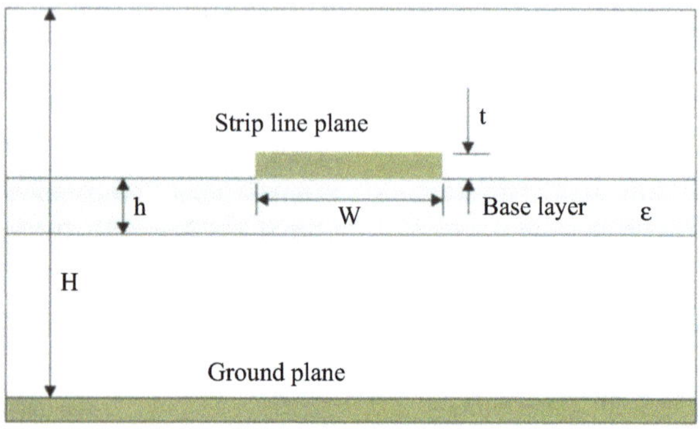

Fig. 7.19 Double conductor strip line design sketch

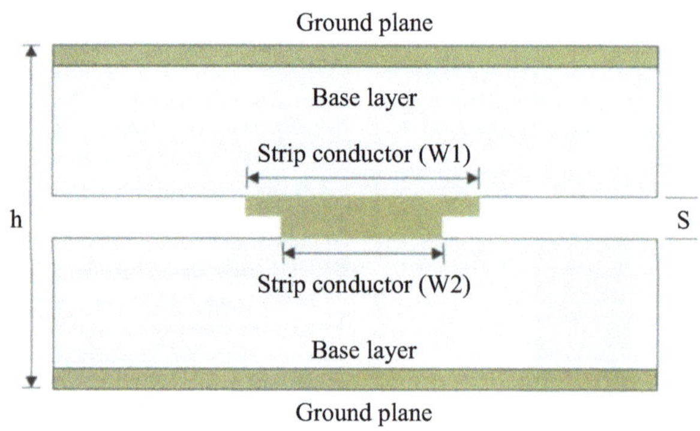

In a real-case scenario, other types of transmission lines are used, too. These types of transmission lines that have traditionally been manufactured for HF/SHF applications, for instance, include *coplanar waveguides, slot lines, fin-lines, and waveguides* integrated into the substrate. In a coplanar waveguide, the conductor and the ground planes on either side of it are usually located on top of the insulating material (Fig. 7.20). Theoretically, electromagnetic field in this case propagates through the conductor and the insulating material with minimal losses. It should be noted that a coplanar waveguide can easily be manufactured as a standard grounded structure with an additional ground plane located on the underside of the substrate.

Mobile coplanar waveguides are widely used in high-speed electric installations, because, according to experts, they provide for the installation of add-on devices "above" the circuit, just as microstrip lines. Coplanar waveguide circuits can operate at the millimeter-wave frequency while ensuring minimal losses. Several such "coplanar" waveguide circuits can be easily connected to the ground plane.

Coplanar waveguides generate heat. The solution of the heat dissipation problem usually depends on the PCB material thickness and performance. Effective permittivity of a standard coplanar waveguide substrate is usually calculated as an average

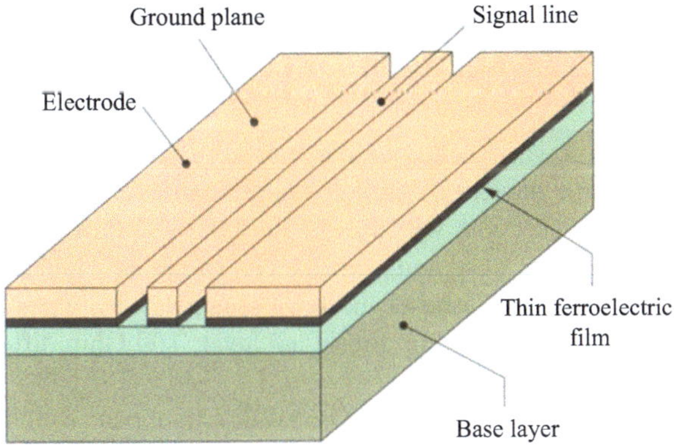

Fig. 7.20 A standard coplanar waveguide is designed as a conductor with grounding planes on both sides, which are placed on top of the insulating material

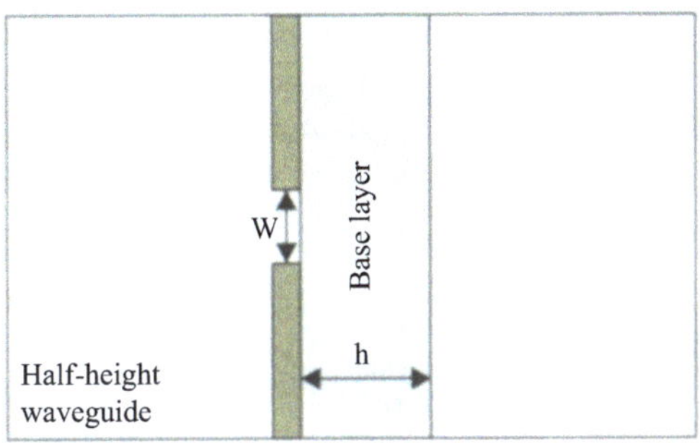

Fig. 7.21 Standard fin-line design

value of the insulating material and air, because, just as in a microstrip line, one half of the lines of the effective electric field is always on the available surface, while the other half—directly in the insulating material. There is usually a "slot" located on the surface of the insulative substrate in the slot line between two conductors. As in the case of microstrip lines and coplanar waveguides, various active and passive components can be mounted on the open transmission lines.

Another type of such lines is a fin-line (Fig. 7.21), generally consisting of two conductors and ground planes surrounded by insulating material. There is a slot space between the conductors. These lines, generating a circular electromagnetic field, are often used in systems with circulators and insulators. However, communication lines of this type are rather complicated and expensive to manufacture. Nevertheless, serial rectangular-shaped waveguides integrated into the substrate are often used in telecommunications systems operating at a frequency above 100 GHz.

7.2.2 Constructive Features of Designing High-Speed Signal Transmission Lines

As is known, 50 Ohm conductors are used to transmit digital signals in high-speed electronic devices of printed circuit boards (PCBs).

In this section, we consider how the width of conductors on a multilayer board affects their cost and impedance, which is vital for the development of TCS [9]. Figure 7.22 shows the trace channels of a telecommunications system of the same width located on the signal layer: 100 Ohm differential pair, 50 and 60 Ohm unbalanced lines.

Here, the so-called 100 Ohm differential pair, which is usually determined before single-wire lines, is located in the main wiring channel (between through-holes) and should not contain any inhomogeneities, which may interfere with the transmission of digital signals at high speed. After calculating the width of the conductors of the 100 Ohm differential pair and the distance between them, usually the system designer determines the width of the conductors of 50 and 60 Ohm single-wire lines of the same layer. It should be noted that if only the width of single-wire lines is changed, the impedance of conductors will be different.

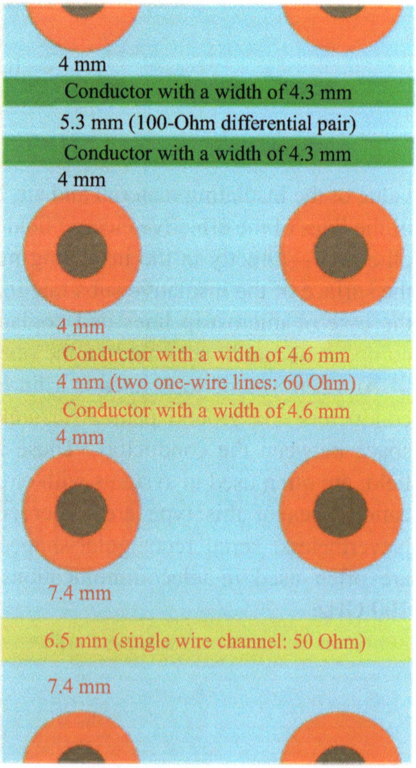

Fig. 7.22 Visual representation of the wiring channels of the same width on the signal layer of the board: 100 Ohm differential pair, 50 and 60 Ohm single-wire lines

Typical case: the following results are obtained as a result of wiring [9]:

- top view: one 100 Ohm differential pair with a 4 mm wide conductor; gap: 5.5 mm;
- in the center: two 60 Ohm conductors with a width of 4 mm; gap: 4 mm;
- bottom view: one 50 Ohm conductor with a width of 6.5 mm; gap: 7.4 mm.

In the description of methodological aspects of developer's work and possible challenges [9], the importance of the use of reference designs and IS manufacturer documentation containing special recommendations is emphasized.

In other words, the principal methods for designing high-speed digital circuits outlined in these documents have been repeatedly used before for manufacturing end products. For instance, the decision to use a memory interface without matched load should be made by the developer based on the memory range specified in the original circuit. However, in order to save money, the developer can choose completely different devices (with different performance of the data input/output buffer) from those that are usually presented by the customer in the reference design.

The so-called reference designs constitute an integral part of the modern process of designing printed circuit boards. However, it is necessary to understand well the principles and limitations of the methods outlined above, which is essential for the consumer to make a final optimal decision—both in terms of the cost of the device and its consumer properties.

7.2.3 *Features of Calculation of Characteristics of Conductors and Transmission Lines on Printed Circuit Boards of SHF Devices*

In determining the technical characteristics of different layers of conductors for the transmission of digital signals on a printed circuit board, the following main factors should be considered [9, 10]:

- pulse rise (t_R) /fall time of a digital signal source and the ability to control them;
- output impedance (Z_0), parameters of current drivers, and other buffers;
- PCB signal transmission time (t_{PCB});
- internal matched load of the signal source and the receiver;
- external matched load of the signal source and the receiver.

Figure 7.23 shows the main types of PCB conductors, providing quick connection between the digital signal source and the receiver. Table 7.5 lists the types of PCB conductors that may be used for this connection.

The so-called electrically long conductors ($t_{PCB} \geq 0.2_{tR}$) serve as a PCB transmission line base (PCB TL). Here, electrically short conductors ($t_{PCB} < 0.2_{tR}$) work as an LC circuit with lumped parameters. The parameters of such conductors with controlled impedance are usually specified in the original PCB design file. Thus,

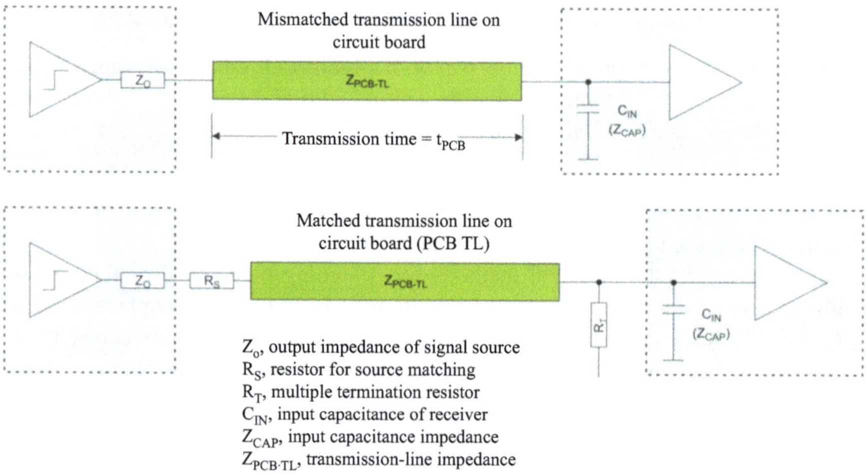

Fig. 7.23 The analog circuit of the main types of conductors on printed circuit boards, providing a connection between the digital signal source and the receiver

Table 7.5 The main types of conductors used are typical printed circuit boards of high-speed electronic devices [9]

Defining the categories of transmission line conductors	Typical signal transmission/rise time (t_{PCB}/t_R)	Matched load		Electric model type
		Signal source	Receiver	
Mismatched transmission line	≥0.2	No	No	Distributed model (Z_c, reflection, transmission time)
Transmission line matched at the far end (in parallel)		No	Yes	
Transmission line matched to the source (in series)		Yes	No	
Matched transmission line		Yes	Yes	
Mismatched transmission line conductor	<0.2	No	No	Model with lumped parameters (RLC, ringing effect, attenuation of a signal)
Conductor of the transmission line matched at the far end		No	Yes	
Conductor of the transmission line matched to the source		Yes	No	
Conductor of the transmission line matched at both ends		Yes	Yes	

Table 7.6 Typical transmission line impedance values for load matching [9]

Transmission line	ZPCB-TL conductor impedance	Coefficient of reflection at signal source	Coefficient of reflection § receiver
Mismatched	See note	$(Z_0 - Z_{PCB\text{-}TL})/(Z_0 + Z_{PCB\text{-}TL})$	≈1
Matched at the far end	R_T	$(Z_0 - Z_{PCB\text{-}TL})/(Z_0 + Z_{PCB\text{-}TL})$	0
Matched to the source	$Z_0 + R_S$	0	≈1
With matched impedance for a single-ended channel	$Z_0 + R_S = R_T$	0	0
With matched impedance for a symmetric channel	$Z_0 = R_T$	0	0

impedance of the main conductors shall always remain within the allowed range. Surely, in accordance with the international code, the manufacturer of printed circuit boards can measure the actual impedance of the conductors at all time and submit a report at the customer's request.

It should be noted that the PCB signal transmission time (t_{PCB}) is the single most important parameter related to its particular topology. Table 7.6 shows how to calculate the transmission line impedance to match the load (Fig. 7.23).

As is known from [10, 11], the PCB transmission lines with matched impedance reliably ensure the integrity of the transmitted signal, mainly due to the fact that reflections between the source and the receiver are excluded. However, the matching resistors installed on printed circuit boards of these high-speed devices increase the cost and power consumption. In many ways, the cost of power consumption depends on the load, generally used in multichannel lines, and its matching to the source—on the use of unidirectional signals, for instance, clock signals. It should be noted that the following facts were obtained based on the analysis of the current state: if the ranges of values of logic voltage and timing in the used electronic systems are broad enough, then many digital interfaces use mismatched transmission lines.

The "good" or "bad" behavior of printed conductors on such boards depends not only on their linear length, but also on their physical and energy characteristics, including the input/output of the digital signal source and the receiver, as well as the nature of the used load. The width of the printed conductor determines its impedance and bandwidth capability. The simplest rule for high-speed PCB developers can be formulated as follows: it is required to use more PCB layers to design a TCS channel with low bandwidth capability. Developers of such systems should always have a choice between several PCB transmission line impedance values, so that they do not always use lines with 50 Ohm conductors.

7.3 Design and Technological Features of the Design of Printed Circuit Boards for High-Speed Electronic Devices

7.3.1 Methodological Aspects of the Design of Printed Circuit Boards of High-Speed Devices with Consideration of the Insertion Loss

As is known from the theory of transmitting HF signals [10], each of the sinusoidal harmonic components of the transmitted signal's digital pulse is usually just a HF signal of a certain frequency. For instance, a pulse transmitted at a speed of 10 Gb/s consists of the sum of various HF signals with frequencies of 5 GHz (fundamental component), 15 GHz (third harmonic), 25 GHz (fifth harmonic), and 35 GHz (seventh harmonic).

A simplified example (Fig. 7.24) [12] below shows why developers sometimes have certain difficulties in developing digital HF applications. One of the most significant of them is the so-called insertion loss—if a seventh harmonic loss of the transmitted HF signal is bigger than the third harmonic loss, the digital pulse will be distorted and the result will be different.

A typical example given in Fig. 7.24 refers to the practical case, where signals are transmitted at a speed of 10 Gb/s. Here, for the 28 Gb/s signal, the corresponding

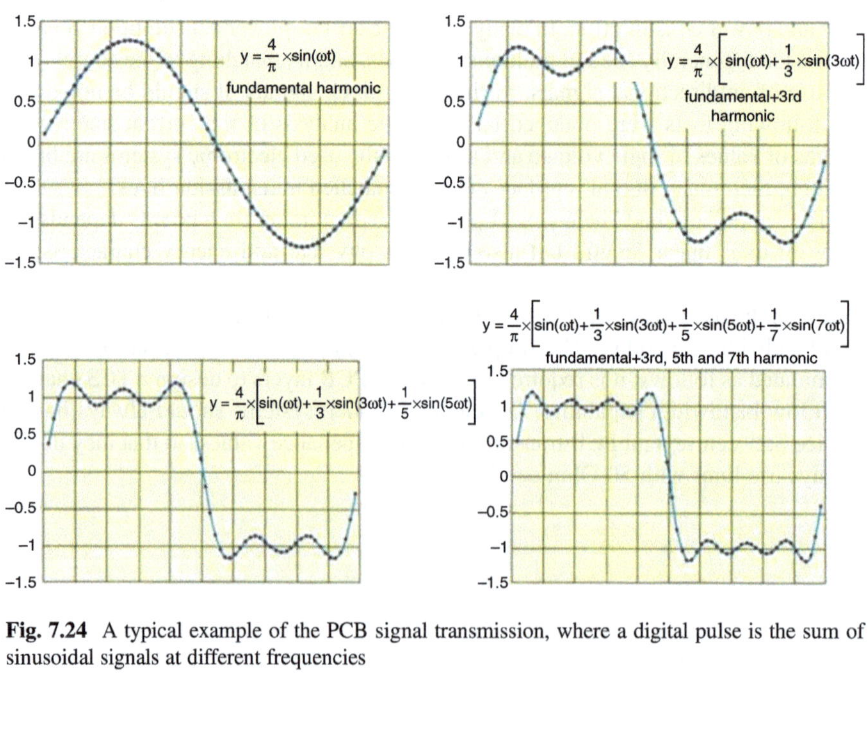

Fig. 7.24 A typical example of the PCB signal transmission, where a digital pulse is the sum of sinusoidal signals at different frequencies

harmonics are 14, 42, 70, and 98 GHz. As is known from generally accepted sources, as the frequency of data transmission increases, these losses tend to increase, too. For instance, the difference between losses at 14 and 98 GHz may turn out to be quite significant. However, inexperienced designers of SHF systems do not always treat them as the impedance value of printed circuit boards. Generally, a time domain reflectometer is used to estimate the actual impedance value of the designed printed circuit board. However, not all manufacturers have the necessary equipment to test the insertion loss of printed circuit boards, to determine their basic performance. As a rule, a standard circuit analyzer is used, although experienced developers recommend using special time domain reflectometers.

A number of important facts related to the organization of the production of final products should also be noted [12].

Thus, materials suitable for the use in digital HF devices that are sensitive to insertion loss are generally determined by the dielectric loss tangent. Another key factor the losses depend on is roughness of the copper surface, which was previously ignored. So far, digital HF applications use materials characterized by an average level of these losses. These materials have a dielectric loss tangent of less than 0.01. Since these materials have very large losses at a speed of around 10 Gb/s, the task was to find equivalent substitutes for them.

For materials with insignificant losses, which are typically used in applications with a transmission speed of about 10 Gb/s, the dielectric loss tangent is less than 0.005. If, as predicted, next generation high-speed digital devices will operate in the range of 25–28 Gb/s, other materials with insignificant losses and a loss tangent of less than 0.003 will be required. For 56 Gb/s lines, this figure will be even less—0.0015.

Below, we will consider in more detail the influence of the PCB rough surface on the transmission quality of HF signals. This mechanism of influence was ignored by the developers before.

Indeed, at frequencies below 1 GHz, the influence of the roughness of the PCB copper surface on dielectric losses was negligibly small. However, as shown in [13], at higher frequencies these losses increase noticeably due to the surface effect. As the effective length of any conductor with an uneven surface increases, since the current flows along its contour, the effective resistance of copper (the main material used) increases at high frequencies due to the greater distance that the current has to overcome. In this case, the total energy losses include the so-called conductivity losses and dielectric losses. An important point should be made here. It may be new for developers of boards for electronic devices. It may puzzle them.

For instance, in case of a widely used copper foil with a mirror finish on both sides, the adhesion between the foil and the resin deteriorates. This significantly increases the likelihood of foil separation under thermal stress arising in the process of manufacturing printed circuit boards and their subsequent assembly. To prevent this, a slightly non-uniform oxide coating is usually applied on the inner layers of the foil. As a result, the adhesion of resin in pre-preg slightly improves. During treatment, it begins to flow at increased temperature and pressure.

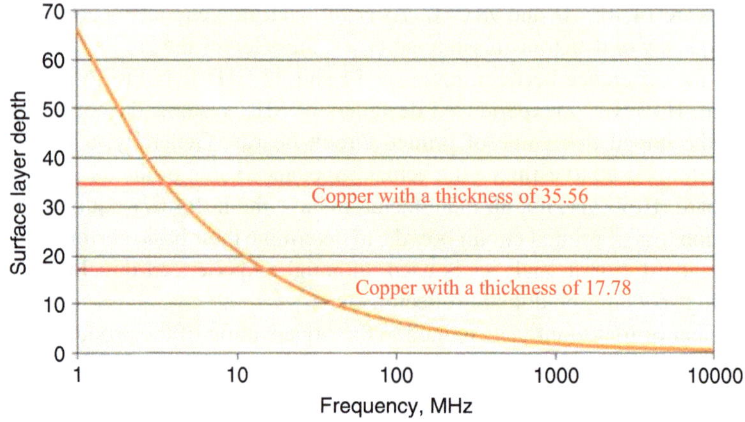

Fig. 7.25 Dependence of the depth of the surface layer on frequency

Due to the surface effect, the alternating current is distributed in such a way that its density increases as it approaches the conductor's surface. The depth of the surface layer is determined as follows:

$$\delta = \sqrt{\frac{2}{2\pi f \mu \sigma}}, \tag{7.1}$$

where δ is the depth of the surface layer, μm; f—frequency, MHz; μ—magnetic permeability (4π–10–7 H/m); σ—specific conductivity of copper (5.6–107 cm/m).

Thus, the depth of the surface layer is approximately equal to:

$$\delta = 66\sqrt{\frac{1}{f}}. \tag{7.2}$$

As it can be seen from this equation, the depth of the surface layer decreases at higher frequencies (Fig. 7.25). If at low frequencies (1 MHz) the depth of this layer is usually 66 μm, then at 10 GHz it is equal to 0.66 μm, i.e., one order less.

As can be seen from Fig. 7.26, at a higher frequency of the signal (10 MHz), this signal passes through conductors with a copper coating thickness of 17.78 μm. It does not occupy the entire cross section, since a noticeable surface effect is already observed at this frequency.

Due to some specific ferromagnetic properties of nickel contained in ENIG coating, conductor losses increase at high transmission frequencies. It would seem that since nickel is deposited on the surface of the strip line, it should have only a slight impact on the properties of conductors. However, at a frequency of about 2.7 GHz, insertion losses increase significantly due to this nickel layer. It is for this reason that all HF boards without exception shall be treated under the so-called solder mask over bare copper procedure (SMOBC).

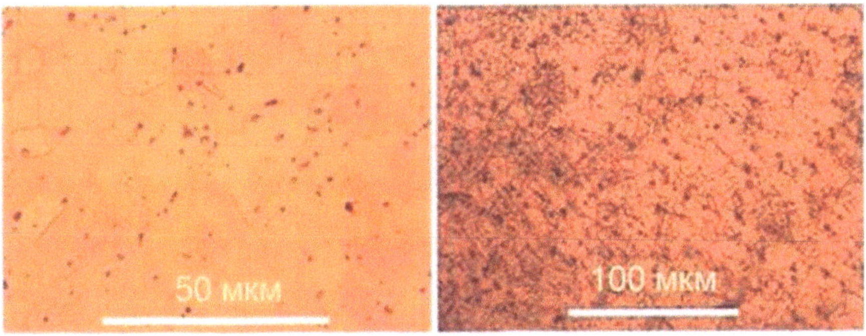

Fig. 7.26 Rolled annealed copper (left) and etched copper (right) under the electron microscope [14]

In 2014–2016, rolled annealed (RA), electrodeposited (ED) copper, or reverse-treated foil (RTF) were used in the majority of copper-coated substrates. Such rolled annealed copper with a smooth surface (Fig. 7.26) is uniform in thickness. It should be noted that electrodeposited copper has the roughest surface (Fig. 7.26), but depending on the application and frequency requirements, foil of this type may turn out to be the most suitable material. Reverse-treated foil has a smoother surface, and its cost is about the same as that of electrodeposited copper. However, since RTF has a drawback—a rather low tensile strength—it is subject to flaking when the board is in use. Thus, to select a foil material, the developer should find a balance between performance, price, and reliability of the board.

In fact, electrodeposited copper is a standard material used to manufacture various laminates. Usually in the USSR, this foil was made by deposition of copper from a salt solution on a drum cathode at a certain constant voltage. In this case, the granular (dendritic) structure is usually formed on the side of the foil directly facing the bath with a solution.

A very smooth surface is formed on the side that is in close contact with the polished surface of the drum. Figure 7.27 shows a cross section of such a standard copper foil layer. There is a visible roughness on its upper and lower surfaces.

Rolled copper is known to be manufactured using traditional metalworking techniques: a copper bar passes through a series of metal rollers, until a foil of a given thickness is obtained. The surface of the rolled copper is smoother and becomes even more flexible if annealed. Due to its greater smoothness, the adhesion of the foil to the laminate depends only on the quality of treatment and the adhesive properties of the used resin.

The granulated structure of the rolled annealed copper differs from that of the electrodeposited copper and has a different etching rate. As a rule, rolled annealed copper, which is most often used to manufacture "flexible" boards, is attached to the polyamide film using acrylic resin.

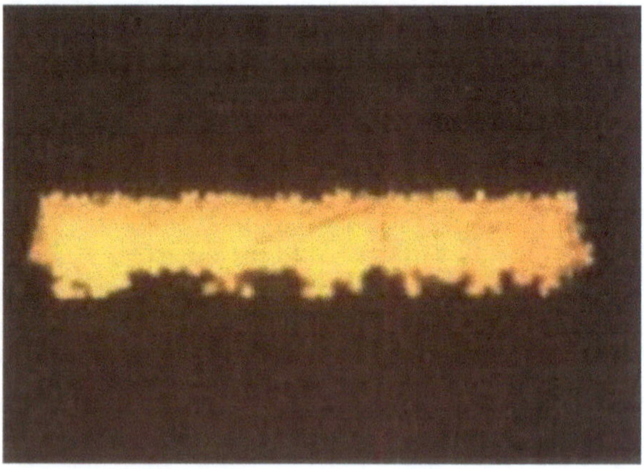

Fig. 7.27 Copper foil layer cross-section design sketch

The smooth surface of ED copper on the reverse-treated foil is subjected to further treatment. The thickness and roughness of treated layers is insignificant, increasing the adhesion of the foil base to the insulating material.

At first glance, laminated plastic with copper foil has a smooth surface, but all materials have an uneven "studded" surface under the microscope. If an average slope of such a "stud" on the conductive surface of the board is 60° and, consequently, its shape is close to an equilateral triangle, then the sheet resistance increases by about two times as compared to a perfectly smooth surface. For other sizes, resistance is increased to a lesser extent. However, a more accurate mathematical characteristic of the surface roughness is the root mean square surface roughness height (h_{RMS}), since the exact profile irregularities are very difficult to establish. Assuming that the irregularities have equilateral edges, then the RMS surface roughness is approximately 0.29 of the average roughness height (from base to top).

The frequency ω, which is closely linked to the rough surface effect, is calculated using the following formula [12]:

$$\omega = \frac{2}{\mu \sigma h_{RMS}^2} \tag{7.3}$$

where μ is the magnetic permeability ($4\pi \cdot ^{10-7}$H/m); σ—the electrical conductivity of copper ($5.6 \cdot 10^{-7}$ cm/m), and h_{RMS}—the root mean square surface roughness height of electrodeposited copper.

Substituting the values in this formula, we get $\omega = 1.28$ GHz for electroplated copper. The smoother the surface of copper, the higher the frequency producing the effect in question.

Additionally, it should be noted that effective surface permittivity and, consequently, capacitance increase along with the increasing surface area of spuds. In its turn, characteristic impedance slightly decreases. The capacity of microstrip conductors increases by approximately 5%, and in case of strip line conductors with a rougher surface—by 10%. As a result, impedance decreases by 2% and 5%, respectively.

Despite the fact that generally manufacturers do not specify surface roughness in technical details, we know from experience that typical values range from 3 μm for outer surface to 6 μm for inner surface. For instance, permittivity (D_k) of Panasonic's base material Megtron6 is 3.4, and loss tangent (D_f) is 0.004 at 18 GHz. There are different modifications of this material taking into account a profile: ED, VLP, SVLP, and HVLP (very/super/hyper low profile). In this case, the average size of the foil irregularities is 7–8, 3–4, and 1.5–2 μm, respectively.

The well-known company Isola, as a rule, uses VLP copper coating for boards transmitting high-frequency signals. The size of irregularities in RTF copper coating is 4–8 μm. This is in reference to FR408HR film-clad dielectric for multilayer printed circuit boards and IS415 laminate. VLP2 material type (Isola labeling) is used in state-of-the-art products that require minimal insertion losses. The size of irregularities in this case is 2 μm.

Thus, HVLP materials have the smoothest surface. Since the average roughness height of "studs" is 1.5 μm, undesirable effects do not appear up to 20 GHz. Therefore, when selecting materials for a particular application, the size of the copper surface irregularities at frequencies above 1 GHz should be considered.

Summarizing the above, here is a list of facts formulated by an experienced developer for practical use [12]:

- At frequencies below 1 GHz, copper surface roughness has almost no effect on dielectric losses, but at frequencies above 1 GHz these losses increase noticeably due to the surface effect.
- If smooth surfaces are used, the probability of foil separation significantly increases under thermal stress arising in the process of manufacturing printed circuit boards and their subsequent assembly.
- Due to the surface effect, the alternating current is distributed in such a way that its density increases as it approaches the conductor's surface.
- The intensity of the surface effect decreases at higher frequencies.
- The ferromagnetic properties of nickel contained in ENIG coating increase the insertion loss at high frequencies.
- At a frequency of about 2.7 GHz, the insertion loss increases due to nickel. It is for this reason that all HF boards shall be treated under the so-called solder mask over bare copper procedure.
- In most copper-coated substrates, rolled annealed, electrodeposited copper or reverse-treated foil were used.
- For electrodeposited copper, the surface roughness effect begins to manifest itself at a frequency of 1.28 GHz.
- Typical values of irregularities range from 3 μm for outer surface to 6 μm for inner surface.
- HVLP materials have the smoothest surface. Since the average roughness height of irregularities is 1.5 μm, undesirable effects do not appear up to 20 GHz.

7.3.2 *Finishing Coating for HF Boards*

Finishing coating is a metallic or organic coating that protects copper pads from corrosion, ensuring "solderability" of the metal layer after storing printed circuit boards in different environmental conditions for some time.

Finishing coatings vary in type, cost, availability, shelf life, reliability, and applicability during assembly. Electroless nickel immersion gold (ENIG) coatings are traditionally considered to provide the smoothest surface and lead-free soldering for SMT boards. Unfortunately, conductivity of nickel is three times worse than that of copper. Additionally, due to the ferromagnetic properties of nickel, the insertion loss increases in the high-frequency range [12].

PCB manufacturers solve this problem using a nickel–gold alloy (NiAu). Since conductivity of gold is slightly lower than that of copper, and it does not possess ferromagnetic properties, gold has a relatively smaller impact on losses in conductors at high frequencies.

Characteristics of microstrip (outer) layers of multilayer printed circuit boards depend significantly on variations of parameters such as width and thickness of conductors. Parameter spread is due to the imperfection of the technological process of electrodeposited coating of through-holes. Since the minimum thickness of copper cylindrical holes, as a rule, should be 25.4 μm, an additional layer of copper is deposited on their surface to obtain this parameter. Unfortunately, part of it inevitably gets on the surface of conductors. Due to the fact that their thickness and width vary, one should not put conductors with strictly specified impedance on microstrip layers.

Additionally, it is necessary to prevent excess copper from getting on the signal layers of the board. This will lead to a significant change in the impedance of conductors.

There are several types of protective coatings:

- hot air solder leveling (HASL);
- organic solderability preservative (OSP), for instance, Entec 106 or Shikoku;
- nickel coating obtained by the method of chemical reduction (ENIG);
- electrodeposited gold over nickel (NiAu);
- immersion silver (IAg);
- immersion tin (ISn).

All conductors are coated with NiAu before applying solder mask, since the NiAu layer, as a rule, performs the functions of a resist during outer layer etching. Thus, a layer of nickel covers the conductors of outer layers throughout the length. However, as already mentioned above, the use of nickel is undesirable at high frequencies.

Alternatively, ENIG coating may be applied on all conductors, but most of the time it is used selectively, since it is applied after solder mask. In the second case, nickel is not deposited on the conductors. However, the pads are protected by coating since they are exposed to environmental factors even before assembly. This is the aforementioned solder mask over bare copper procedure. To limit the

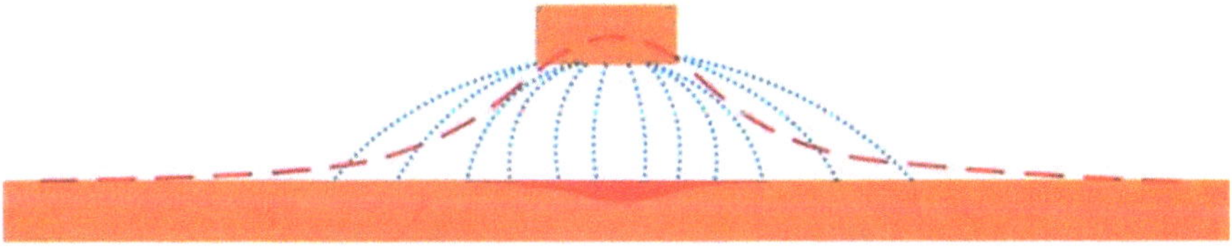

Fig. 7.28 Electric fields of the microstrip line and the nature of reverse current distribution

metal plating surface and reduce the consumption of expensive gold, a liquid photoimageable solder mask (LPISM) is applied before the ENIG process.

In a microstrip configuration (Fig. 7.28), electric fields arise between the conductors and the ground plane. At frequencies above 1 GHz, the maximum current density is dash-lined on the edges and on the coating surface as a result of the surface effect. Thus, it is necessary to take into account any structures whose edges are in contact with each other on the outer layers. This contact is more pronounced, for instance, if all copper elements are already ENIG coated before applying solder mask. However, this effect can be minimized as long as the pads (and not the conductors) are finished.

The thickness of the protective ENIG coating for printed circuit boards is determined under the IPC-4552 standard. The protective gold layer is usually thin and amounts to 0.075–0.125 μm. On the other hand, the thickness of the nickel layer, preventing the interdiffusion of copper and gold, as a rule, is 3–6 μm, depending on the characteristics of the technical process. To apply a thicker layer of gold, a longer dwell time or a higher solution temperature is required.

Finishing coatings can affect the total loss in the data transmission line due to the lower conductivity of the metallization layer covering the surface of conductors. Silver coating in this regard is an exception due to the fact that its conductivity is close to that of copper. However, this coating has not become as widespread as, for instance, ENIG coating or immersion tin.

The miniaturization of electronic devices decreases the size of chips, pads, as well as the gaps between them. As a result, the risk of a short circuit increases. In this case, Pd/Au or ENIG coatings are much more suitable for forming a metal layer on elements with a gap of 15 μm between them. Due to the absence of nickel, these coatings allow for a greater reduction in losses in the transmission lines. ENIG coating is also successfully used in microconnections between the chip and the PCS base, the lead frame, etc.

As a rule, to form a solder joint between the solder and the copper layer, solder paste is distributed over the PCB surface. Alternatively, wave soldering is performed. The only exception to this rule is ENIG coating, when the solder dissolves in a thin layer of gold to form a compound with a lower layer of nickel.

The principle of the immersion process is shown schematically in Fig. 7.29 [12]. Here, the substitution reaction takes place in the metal layer located on top of the copper layer of the printed circuit board. The copper atoms donate electrons, thereby reducing the concentration of positively charged metal ions present in the

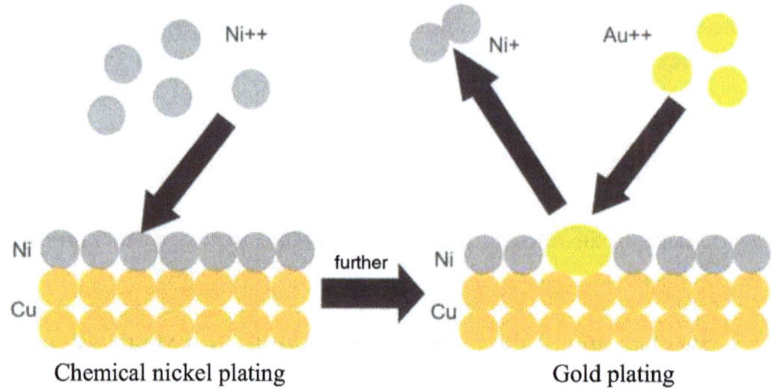

Fig. 7.29 Schematic representation of the principle of immersion gold plating

solution. As the immersion layer thickens, the deposition rate automatically decreases. Thus, it is a self-limited process.

If ENIG coating is used, the following aspects should be considered [12–14]:

1. *Do not apply ENIG coating on BGA pads, which are defined by solder mask.* This prevents: (a) the appearance of brittle intermetallic compounds formed during nickel soldering; (b) the risk of appearance of "black pads" (due to the unbalanced chemical composition of ENIG coating), especially in small contact BGA elements.
2. *The problem of blocking through-holes on one side.* As with most finishing coatings, one should avoid partial blocking of through-holes on one side. Additionally, these through-holes should not be placed near SMD pads to avoid electrolyte ingress, resulting in the compound contamination and reduced solderability.
3. *Connection elements in solder mask between SMD pads.* Like immersion tin coating, ENIG has a profound impact on solder mask. To reduce this effect, in case of SMT components, connection elements should be as large as possible and closely spaced between outputs.
4. *Consideration of the influence of high frequencies on losses in conductors.* To design HF circuits, the choice of a metallic coating with low losses is just as important as the choice of insulating material. At a frequency of about 2.7 GHz, insertion loss increases due to resonance phenomena caused by the ferromagnetic properties of nickel in ENIG coating. Thus, not all conductors on the HF board outer layer should be ENIG coated. In fact, the third or fifth harmonics may be in the high-frequency range, while the main signal frequency is three or five times less, respectively. Consequently, all boards for HF applications must undergo the solder mask over bare copper procedure.

Summarizing the above information on the problems of finishing coatings in modern HF boards, a number of practical recommendations should be made [12, 14]:

- A nickel–gold alloy (NiAu) is used to neutralize the effect of increasing insertion loss (due to nickel).

- Microstrip (outer) layers of multilayer printed circuit boards suffer from variations in the width and thickness of conductors, which means from the spread of impedance values.
- Make sure that excess copper does not get on the signal layers of the board.
- ENIG coating ensures the smoothest surface and lead-free soldering of SMT boards.
- ENIG coating may be applied on all conductors, but it should be used selectively, since it is applied after solder mask.
- Pd/Au or ENIG coatings are more suitable for forming a metal layer on elements with a gap of 15 μm between them.
- If ENIG coating is used, the following aspects should be considered [12–14]: BGA pads defined by solder mask; through-hole blocking on one side; the size of connection elements in solder mask; and the effect of high frequencies on losses in conductors.
- To design HF circuits, the choice of a metallic coating with low losses is just as important as the choice of insulating material.
- At a frequency of about 2.7 GHz, insertion loss increases due to resonance phenomena caused by the ferromagnetic properties of nickel in ENIG coating. ENIG coating should not be applied on all microstrip conductors operating at high frequencies. All boards for HF applications must undergo the solder mask over bare copper procedure.

7.4 Main Trends in the Development of Printed Circuit Board Design for High-Speed Electronic Devices

7.4.1 Main Trends in the Development of Innovative Technologies of Printed Circuit Boards

Over the past decades, there has been a steady trend towards the ever-increasing complexity of PCB design and manufacturing technology, since new design and process engineering solutions immediately pass from the stage of prototype development to mass production.

If in microelectronic technology we observe a generally accepted trend for the evolutionary transition from the classical Mur's law to the new law—more than Mur—then there are similar trends in the technology of printed circuit boards, traditionally designed for compact placement of up-to-date microelectronic components. Following Mur's law, we must admit that printed circuit boards in modern telecommunication devices have actually ceased to perform the single function of switching the signals between the outputs of microelectronic components. In recent past, a standard printed circuit board was designed as a flat board of a homogeneous insulating material with numerous dielectric and conductive layers and vias between them. The so-called 3D structure of modern boards has various design differences:

Electronic device type	Mobile phone	Smartphone	System-on-board	All boards
PCB dimensions	125x55 mm	85x20 mm	30x30 mm	20x20 mm
PCB thickness	1 mm	0,25 mm	0,13 mm	0,06 mm
Ratio of geometrical dimensions of the board	100/100 μm	40/40 μm	30/30 μm	20/20 μm

Fig. 7.30 Evolution of design of boards for telecommunications systems [15]

grooves, cavities, multi-level internal installation of uncased components, and different types of insulating materials. According to the leading manufacturers of printed circuit boards, in 7–8 years after this book's publication, advanced developments and technologies will reach the point of a 100% internal installation of electronic modules, meaning, among other things, that soldering will not be used.

Figure 7.30 shows the main trends in the development of design and technological features of printed circuit boards in high-speed electronic devices over a relatively short period (from 2005 to 2017 and until 2020 according to the forecast) in a simplified form [15]. These trends include mobile phones, which revolutionized the industry in 2000–2005, smartphones, which replaced them (but not for all consumers), Systems-on-Board, which entered the world market, and many other applications, such as nanosatellites.

So, if in a standard mobile phone manufactured in the early 2000s, all of its electronic components were placed on 125 × 5 mm boards, then in just 10 years the entire "electronic brain" of its successor—a smartphone—was placed on a 85 × 20 mm board. Thus, with a significant increase in the functional capacity and the number of transistors on a single chip, the area of arrangement of this entire "microelectronic base" on the board was reduced by as much as four times (from 6875 mm^2 to 1700 mm^2). And in two more years, the industrial production of the "system-on-board" modification made it possible to unite all the previous design and schematic microelectronic solutions in a single system on a 30 × 30 mm chip, reduce the effective area by almost two times (1.89). By 2020, according to experts, it will be possible to place various microelectronic devices for mobile and other applications with different levels of functional complexity on a standard (up-to-date) board with an area of only 2020 mm^2 [15]!

No less important is the dynamics of changes in the PCB thickness—from 1 mm in 2005 to the expected 0.06 mm in 2020 [15]—this is a very important "technical breakthrough" that should be taken into account when designing high-speed systems.

7.4.2 Certain Technological Features of the Implementation of Embedded Components of Printed Circuit Boards

It is commonly known that embedding active and passive components inside a printed circuit board allows the developer of modern telecommunications systems to immediately and quickly solve many problems associated with miniaturization and reliability of devices, ensuring high-speed signal transmission and heat removal.

In particular, there are two types of passive components built into the printed circuit board (resistors, capacitors):

- components formed in the inner layers from special foiled insulating materials or using pastes;
- housing components installed on the inner layer of printed circuit boards.

If the technologies of formed embedded components have been created and applied for several decades, the evolutionary technologies for embedding housing components have been developed and implemented over the past decade. The leading companies for the production of hardware components, manufacturing equipment, electronic devices have successfully implemented a number of innovative programs, resulting in the emergence of technologies of saleable built-in housing components that have found a use in mass production. In general, all known patented technologies and options for the placement of passive and active components can be classified based on:

- orientation of the component leads in relation to the pads on the layer—"face up" or "face down";
- connection of the pads to the component leads by applying solder paste (soldering) or through plated thru-holes;
- installation in the cavity or on the layer surface.

The main technological characteristics of the embedded component technology implemented in printed circuit boards are presented in Table 7.7. Here are the names of the twelve world leading companies in this area, name/designation of the used manufacturing process (EOMIM, E-B2IT, aEASI, etc.), indication of the type of installation: "face up" or "face down," and indication of the type of installation procedure (soldering or mounting using copper-plated vertical channels (through-holes)), as well as information on the possibility of installing microcomponents into the cavity (yes/no).

When this book is published, the designs using the traditional "face down" installation and soldering are considered to be the simplest and the least time-consuming to understand and implement at the existing place of production. But these designs have limitations when it comes to using uncased components and chips. For such embedded components, it is preferable to connect to component leads using the innovative technology of galvanic metallization in blind vias.

Table 7.7 Embedded component technology implementation features

Company	Process	Installation face up/down	Soldering or metal, through-holes (Cu)	Installation in the cavity
Taiyo Yuden	EOMIN	Up	Cu	Yes
DNP	E-B2IT	Down	Soldering	No
ASE	aEASI	Up	Cu	Yes
Panasonic	SIMPACT	Down	Soldering	No
Imbera	IMB	Down	Cu	No
LG	Innotek	Down	Soldering	No
Samsung	SFC	Up	Cu	Yes
At&S	ECP	Down	Cu	No
Fraunhofer	CIP	Up	Cu	No
Intel	BBUL	Down	Cu	Yes
Verdant	OCCAM	Down	Cu	No
Hofmann	AML	Down	Soldering	No

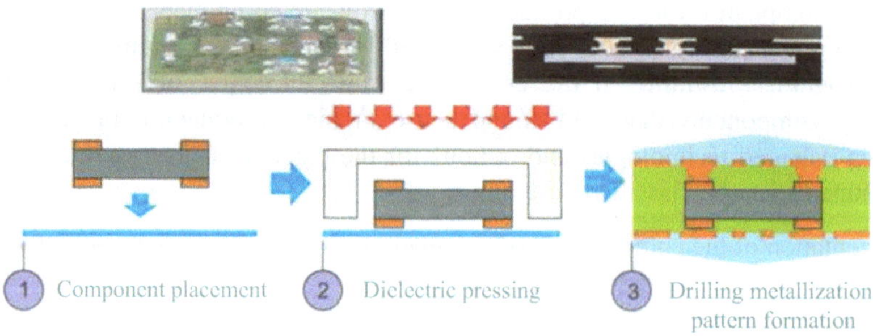

Fig. 7.31 Embedded Component Packaging (ECP) technology flowchart

Let us consider the example of the so-called embedded component packaging (ECP) technology used by the well-known Austrian company AT&S (Fig. 7.31). In 2010–2015, it released more than 100 million electronic modules with unpackaged chips for various purposes (the main of them are listed in Table 7.8).

It should be noted that the ECP technology is most likely to find application not only in high-speed data transmission channels, but also in power electronics. For instance, in 2015, the European Union's EMPOWER project was implemented by several research centers with the participation of AT&S, whose goal was to develop an element base for electronic devices, as well as a manufacturing technology with integrated power electronic components and a power consumption of more than 2015 W.

Table 7.8 Main design features of electronic micromodules with AT&S integrated components

Identification	Area, mm^2	Size reduction, %	Sketch of structure	The main advantage of using embedded components
Voltage converter	7	40		The smallest 600 mA DC/DC converter on the market
Charge control	20	40		Multilayer silicone micro-assembly to control the charge of lithium-ion batteries
Media module	20	30		Integrated module—discrete passive microassembly on eWLP
Silicon microphone	5	>50		High performance MEMS with very small dimensions
Mobile TV	20	50		Mobile TV tuner in one module
Identification	60	Under development		Integrated Biometric Sensor
Position sensor	60	50	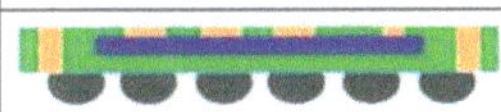	Highly precise Hall sensor for micro joystick
Wireless module	20	40		Small-sized multilayer micro-assembly

7.4.3 Some New Applications of Copper in Interconnects of Printed Circuit Boards of High-Speed Electronic Devices

As is known, copper has been within view of developers of communication lines for microelectronic devices due to a number of its obvious advantages.

Here, let us consider another interesting development in this area, which was proposed by the well-known company IntrinsiQ (USA). The powder from copper nanoparticles produced by the company (particle size up to 10 nm) in printed circuit boards was used to solve two problems at once: deposition of an ultrathin conductive copper layer on the insulating base and filling the PCB blind holes and through-holes. In fact, the properties of the same material in a monolithic crystalline state and in the form of a powder of nanoparticles can vary significantly. For example, the melting point of crystalline copper is +1200 °C. Copper powder in the form of nanoparticles can be sintered at temperatures below +200 °C. Thus, IntrinsiQ has developed a special installation for deposition and sintering of copper nanopowder on the surface of flexible polyimide. As a result, the copper layer is less than 1 μm thick, although due to its porous structure, the conductivity of the obtained copper

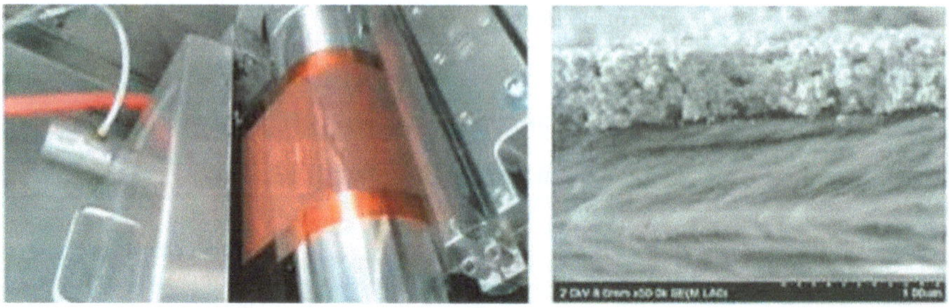

Fig. 7.32 The structure of a nanoparticles layer deposited on polyimide, sintered at a temperature of 200°C

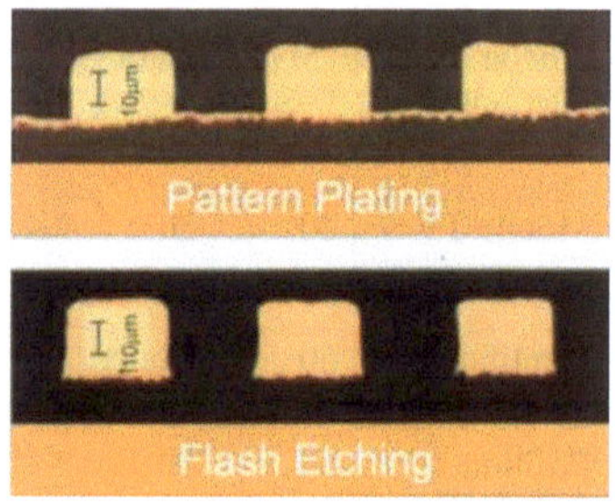

Fig. 7.33 Semi-additive process for the formation of thin conductors using an ultrathin layer of copper from sintered nanoparticles

coating was only 25% of the conductivity of electrodeposited copper as shown in Fig. 7.32.

Here is a list of the main advantages of such coating:

- a simple application and sintering technology;
- a homogeneous boundary between copper and polyimide preserves the integrity of the signal for high-speed applications;
- a possibility of double-sided application;
- low manufacturing cost;
- semi-additive process for the formation of thin conductors not less than 10 μm.

As shown in Fig. 7.33, the technology of "directional etching" of copper (referred to as "flash etching" in literature) makes it possible to implement the technology of forming thin conductors using an ultrathin layer of copper consisting of a set of "fused" nanoparticles of this material, which is quite a complex scientific and technical problem in itself. As can be seen from Fig. 7.34, scientists and technologists managed to successfully resolve this problem, resulting in a possibility to manufacture printed circuit boards with a similar technical solution.

In fact, copper nanoparticles can be effectively used to fill the PCB blind holes, as an alternative to the previously widely used pastes or copper electrodeposition. Filling with copper nanopowder has the advantages of both methods, since, as with the use of such a paste, copper particles can be used to fill large diameter blind holes and through-holes. In this case, the hole is filled entirely with copper

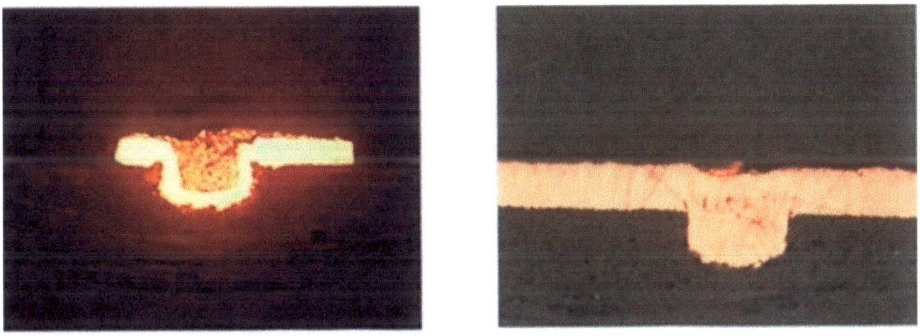

a) b)

Fig. 7.34 Blind hole filled with copper nanopowder cross-section design sketch: initial addition of copper to fill the entire hole (**a**) and the structure of filling after a series of thermal shock tests (**b**)

without binder, as in the case of copper electrodeposition. Reliability of the holes filled with copper nanopowder was confirmed by a thermal shock test: 5 cycles, $T = +288$ °C, $t = 10$ s (Fig. 7.34).

7.4.4 *Methods of Designing of Special Substrates for Microassemblies of Electronic Devices*

Developers of boards for high-speed electronic devices should keep in mind that according to experts' forecasts, by 2020, the following miniaturization of topological features is predicted in substrates for microassemblies: hole diameter up to 20 μm, conductor width up to 3 μm, deflection not more than 30 μm. Since it is impossible to fabricate substrates with such topological features using screen printing, and LTCC ceramics nowadays fails to ensure both the necessary dimensional stability and the absence of substrate deformation, a material such as glass, which has been successfully used for many years to produce masks for photoexposure of various micromodules, perfectly matches these requirements. Indeed, coefficient of thermal expansion of glass does not exceed that of the silicon chip. Therefore, Georgia Tech—Research Package Center (USA) and Fraunhofer IZM (Germany) independently developed industrial technologies to produce such glass substrates for microassemblies. The techniques outlined below should help technologists solve their particular problems, namely ensure the formation of plated holes in glass and work with large-format blanks in order to put them into mass production in future. Thus, Fig. 7.35 shows in a simplified form the PCB substrate design evolution in next generation high-speed electronic devices, which is very important for developers and manufacturers of printed circuit boards. As you can see, serious changes are coming: instead of cores from organic materials, well-known to the PCB developers and widely used in most electronic devices, cores from glass will be used soon. Naturally, this advanced technology has its drawbacks, but experts know

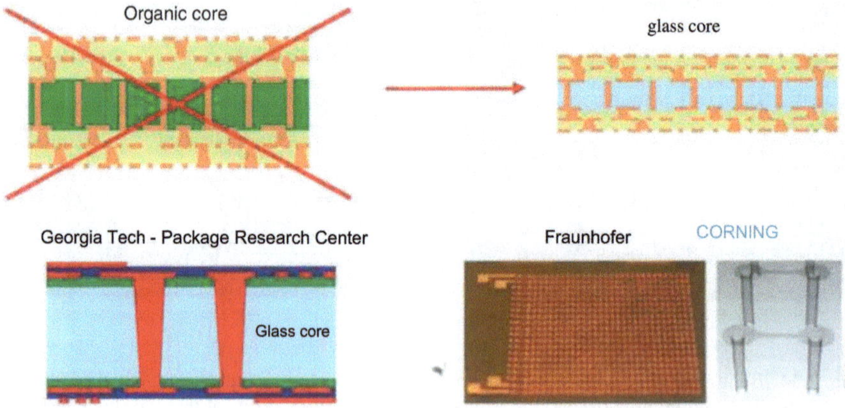

Fig. 7.35 Substrate design evolution—glass substrates replace ceramic and resin ones [15]

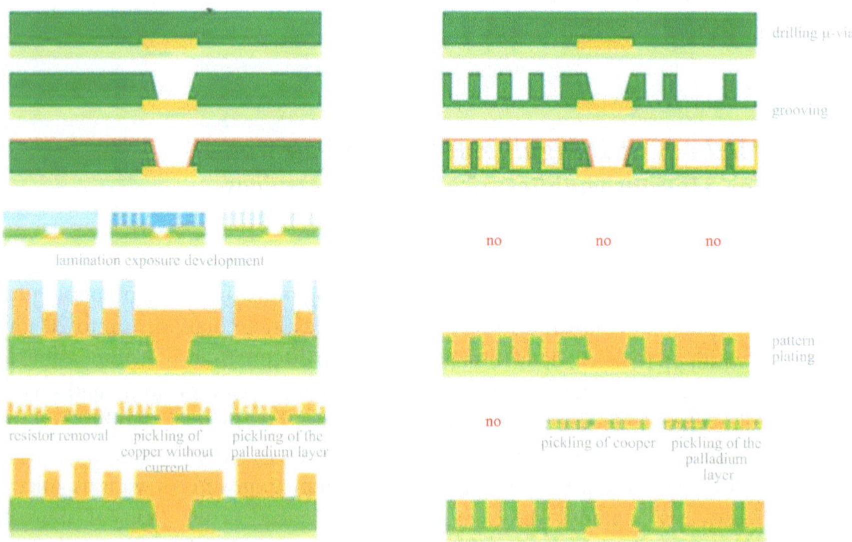

Fig. 7.36 Process flow for the formation of conductors in the entire glass substrate [15]

the solutions and ways to eliminate them. For example, to eliminate the undesirable unstable thickness of the conductive pattern, accompanying the positive galvanic technology of its formation on the surface (pattern plating) during operation [15], a panel plating solution has been presented. A similar technology is already used in the manufacture of printed circuit boards and substrates for galvanic filling of blind holes (Fig. 7.36). After panel plating, "extra" copper is simply etched off the surface, ensuring perfect planarity. Additionally, due to the exclusion of photoresist and operations associated with its treatment, it is possible to shorten the entire production cycle.

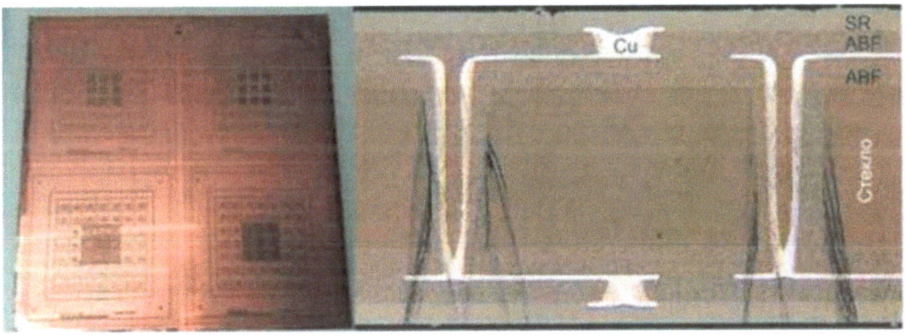

Fig. 7.37 A 500 × 500 mm blank of the 4-layer substrate with a glass core (transverse microsection)

Qualcomm, Corning, and Unimicron have obtained a pre-production prototype of a 500 × 500 mm blank with a glass core and the following structure and specifications (Fig. 7.37):

- 4 conductive layers;
- glass core, external dielectric layers of epoxy resin;
- a 100 μm thick 4-layer blank;
- via-in-via holes; distance between the walls of the holes in the glass and the plated walls in the resin is 200 μm;
- conductor/gap width—8 μm.

7.4.5 Features of Placement of Optic Networks on Boards

It is commonly known that the performance of processors is constantly increasing. Thus, the capacity of electrical transmission channels in modern electronic systems is gradually approaching its physical limit. At first, it concerned only electrical cables between computers and workstations (see Chap. 1), which was replaced by fiber-optic communication. Today, the same capacity is achieved by electrical circuits in printed circuit boards starting from the point of connection of the optical cable to the processor. The maximum capacity of modern electrical circuits is 40 Gb/s. A further increase in the capacity of digital circuits is possible only if special optical circuits are used inside electronic device boards (Fig. 7.38).

It should be noted that such board designs and technologies for their production have long been researched and developed by a number of leading global companies [16], namely:

- under the TERRABUS project (IBM in association with EMCORE, HP Labs, Avago, and DARPA).

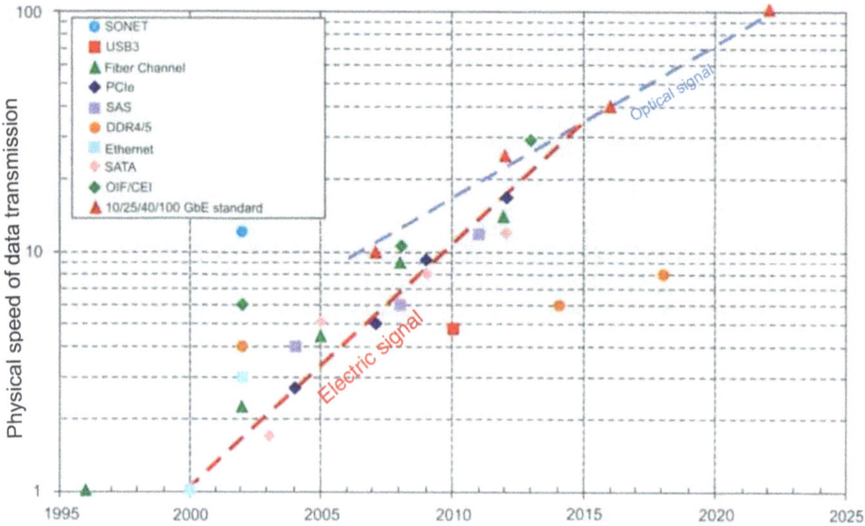

Fig. 7.38 Performance of electrical and optical circuits: results of the comparative analysis [15]

- during the implementation of the European project—EOCB-EU. University of Berlin, University of Paderborn, University of Dortmund, Helsinki University, Fraunhofer Institute, ILFA, IBM-Zurich, Siemens, Bosch, Andes, Mikropacks, OECA, Aspocomp, vario-optics; coordinated by: BMBF (German Dept. of Education and Research).
- during the implementation of the project of HDP international industrial group (40 companies and corporations).

Thus, IBM is implementing the TERRABUS project in four stages. Three of them have already been completed (Fig. 7.39). In 2015, the technology for forming optical conductors running from the optical connector to the processor terminals was developed. By 2020, it is planned to introduce an optical signal on the contact pads of the chip.

According to the published reports [16], two processors installed on special optical modules in test boards are connected by a bus of polymeric optical conductors passing inside the PCB (Fig. 7.40).

In this case, optical modules convert optical signals into electrical signals and back through a special system of microlenses contacting the ends of optical conductors extending to the board surface. 24-channel optical modules with a capacity of 20 Gb/s per channel have been developed. Total capacity of such modules is now almost 480 Gb/s. The main direction for the introduction of optical-electrical boards are supercomputers and server platforms.

And finally, it should be noted that the well-known project of HDP international industrial group is implemented in three areas:

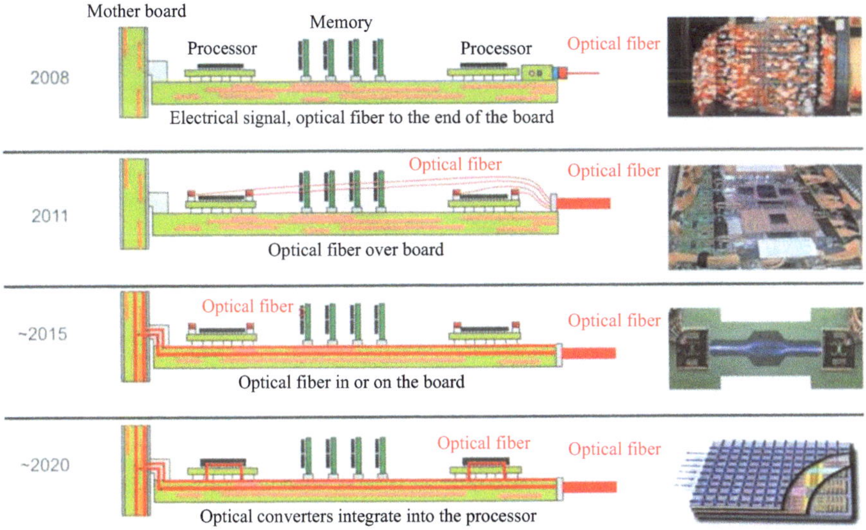

Fig. 7.39 Main stages of implementation of the TERRABUS project by IBM

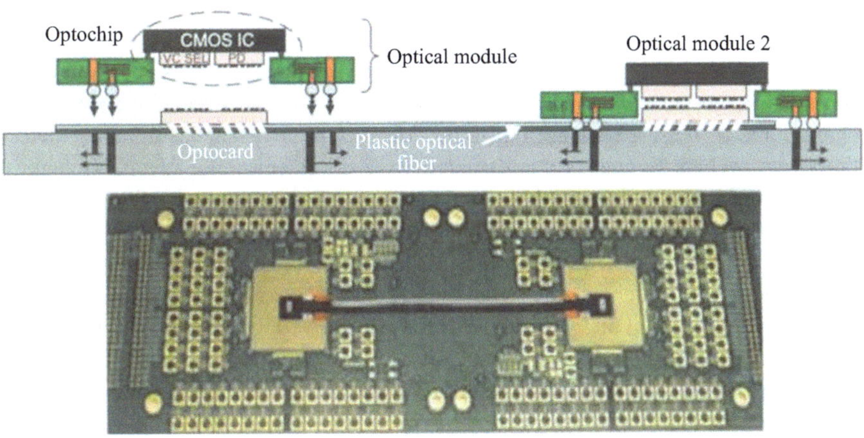

Fig. 7.40 Structure of the test board with optical modules and polymer optical conductors in it

- flexible optical cable technology development (Sumitomo, Seagate, Optical Inter Links);
- embedded glass waveguide technology development (Fraunhofer IZM, Seagate);
- polymer optical waveguides in organic substrates technology development (TTM, FCI, DOW Corning). Figure 7.41 shows the main directions in the development of this technology.

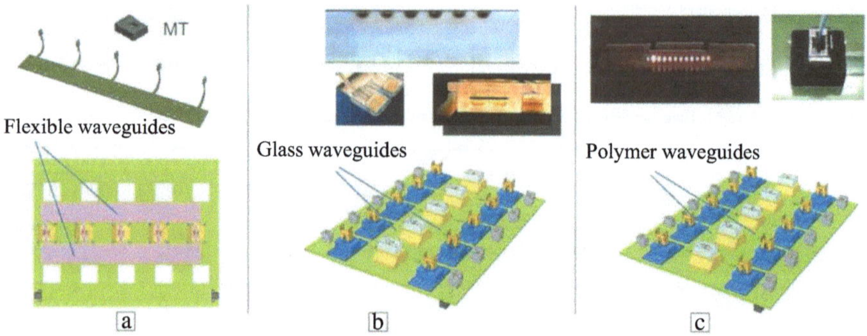

Fig. 7.41 Directions in the development of embedded optical waveguides technology by HDP industrial group: (**a**) flexible; (**b**) glass; (**c**) polymeric

7.5 Methods for Ensuring Integrity of Signals in High-Speed Communication Lines

7.5.1 What Is Signal Integrity?

Signal integrity is a set of quality characteristics for determining the quality of an electrical signal. Any digital signal is essentially analog, i.e., it is represented by special voltage current waveforms. Naturally, the form of the analog signal can change due to noise exposure, non-linear distortion, crosstalk disturbance, reflections, etc. When signals are transmitted over short (meaning electrical length) distances and at a low rate, there is no impact on the accuracy of the received information. With an increase in the transmission rate or in the length of the transmission path (i.e., with an increase in the electrical length), these effects may distort the signal to such an extent that the transmitted information will be erroneous [1]. The work of signal integrity engineer may be divided into two stages. The first is signal integrity analysis, i.e., identification of the very effects leading to signal distortion. The second stage consists in the prevention of these distortions.

To describe the history of signal integrity as a branch of radio engineering, you may use the chronology [17] proposed by Douglas Brooks:

First stage. The integrity of signals is unheard of, so its impact on the transmission is not considered.

Second stage. There are problems with stray inductances.

Third stage. There are problems caused by the fact that the resistance of the transmission line can no longer be considered a constant, and it is necessary to take into account its dependence on frequency.

Fourth stage. Numerous problems have arisen due to the fact that critical lengths (transmission lines) have become too short, and it is extremely difficult to find technical solutions to eliminate them, even though these problems are known.

Let us consider the main problems preventing the integrity of signals [10, 17].

System Bandwidth The branch of radio engineering dedicated to the integrity of signals appeared relatively recently, especially in Russian science and engineering. Consequently, signal integrity engineers most often come from the SHF field. As a result, they often fail to grasp the digital signal bandwidth concept. Any digital signal is characterized by its data rate. Let us take the example of an arbitrary pseudo-random digital signal transmitted at a speed of 1000 Mbps. Does this mean that the spectrum of the signal also lies in the frequency range up to 1 GHz? Spectral analysis using the fast Fourier transform (FFT) tool may provide an answer. As it turns out, the signal spectrum is much wider than its clock frequency. The upper limit corresponds to the frequency of the "break," which is determined by the steepness of the signal edge. This frequency is calculated as follows [1]:

$$F_{\text{knee}} = \frac{1}{2T_t} \tag{7.4}$$

where T_t is the pulse rise/fall time.

Reflection is the phenomenon of the return of a part of the wave energy back to the transmission line with an unmatched load. A load is considered to be matched if its impedance is equal to the complex characteristic impedance of the line. The greater the difference between these two impedances—the greater the reflection. The reflection phenomenon becomes noticeable with the electrical length increase, i.e., when the wavefront time is comparable to the line length. The two extreme cases of unmatched lines are a short circuited line ($Z_0 - 0$) and an open-circuit line ($Z_0 \to \infty$).

Wave impedance is the resistance that an electromagnetic wave encounters when propagating along a uniform line without reflections. It is defined as a voltage-to-current ratio of the incident wave:

$$Z_0 = \frac{U_{\Pi}}{I_{\Pi}}. \tag{7.5}$$

For printed conductors, wave impedance of the line depends on its width and the distance to the backup layer (earth or power). Wave impedance of printed conductors usually ranges from 50 to 75 Ohm (there may be deviations, both upwards and downwards, but they are associated with a highly specialized task).

Losses in Transmission Lines. Ohmic Losses Ohmic losses in transmission lines are caused by the imperfection of the used conductors with finite conductivity (electrical conductivity of copper is 5.7E+7 S/m). Since conductors have a well-defined thickness in the manufacture of printed circuit boards (18 μm for signal conductors and 36 μm for ground and power layers), ohmic resistance is easily calculated knowing the resistivity per square. For copper conductors with a thickness of 18 μm, this value is 0.25 mΩ/square. For instance, a 150 μm wide and 15 mm long conductor will have a resistance of 25 mΩ. A 75 μm wide and 7.5 mm long conductor (since both conductors have an "area" of 100 squares) will have the same resistance.

Surface Effect For each electrical parameter, consider the frequency range where it applies. The same is true about series resistance. Direct current and low-frequency current are evenly distributed in a parallel section, i.e., current density remains the same both in the center of the conductor and at the surface. At high frequencies, current density increases at the surface of the conductor and decreases to almost zero in the center. It has been established that current density exponentially decreases along the radius from the surface of the conductor to its center as a result of the surface effect. Naturally, with such a distribution, effective resistance of the conductor will increase at high frequencies. Thickness of the skin layer, where current will flow, depends on the frequency [1]:

$$s = \left(\frac{2\rho}{\omega\mu}\right)^{\frac{1}{2}}, \tag{7.6}$$

where ρ is conductor resistance, ω—frequency (in radians per second), μ—absolute permeability of the conductor.

Dielectric Loss A lossless wave propagates only in vacuum. There are losses in any real dielectric. The magnitude of these losses depends on the type of dielectric, frequency and is determined by the dielectric loss coefficient. Naturally, the higher the dielectric loss, the more expressed is the signal attenuation during propagation. In the material specifications, loss tangent is generally indicated instead of loss factor. For convenience, let us consider an ideal capacitor. In such a capacitor, the current vector is 90° greater than the voltage vector. If a loss occurs in a conductor, an angle shifts by δ, called loss angle. The tangent of this angle is recorded in the material specification. Sometimes the inverse value is used—quality factor:

$$Q = \frac{1}{\mathrm{tg}\,(\delta)} \tag{7.7}$$

The standard loss tangent for the FR-4 material is 0.025.

Crosstalk Disturbance Crosstalk disturbance is the appearance of a signal in the conductor exposed to adjacent conductors. It is impossible to avoid this phenomenon in modern digital devices, but it can be minimized. Speaking of crosstalk disturbance, two terms are introduced—aggressor and victim. Aggressor is a conductor experiencing an impact; victim is a conductor providing a response. In a real-case scenario, when different buffers switch constantly, every conductor is both an aggressor and a victim. To study the crosstalk disturbance, low-level or high-level voltage is supplied to the victim, and a square signal—to the aggressor. Alternatively, the latter can receive a pseudo-random sequence with specified levels of logical “0” and “1,” rate of pulse rise/fall, and a given clock frequency.

Additionally, a number of other challenging issues should be taken into account: power supply noise, simultaneous switching noise, jitter, etc.

Problems of ensuring the integrity of signals were studied in a fundamental work [1], which was repeatedly re-edited over the past 10 years. This book is well-known to Russian specialists—it came out in Russia and in Ukraine in 2006 under the intriguing name "High Speed Digital Design: A Handbook of Black Magic" by Howard Johnson and Martin Graham. The second volume by the same authors was published, too. In fact, the issue of signal integrity has become more complicated since the first book came out.

Thus, over the past 15 years, data transfer rate increased from 7 to 28 Gb/s. Modern PCIe buses already have more than 700 tracks. Each of them operates at a speed of at least 8 Gb/s, and the capacity of current systems is much higher than 10 year ago. Therefore, there is a need to develop new measures to ensure the integrity of signals in modern high-speed communication lines.

7.5.2 Signal Integrity

All signals transmitted through PCB conductors or using a coaxial cable are influenced to some extent. As the frequency spectrum of a signal expands, the impact becomes more and more significant. For instance, the capacity of a printed circuit board is generally very small (as a rule, several pF) and has practically no effect on low-frequency signals. But as frequency grows, this capacity causes sneak circuits to appear between two signal lines, as well as between signal lines and the ground. Another kind of impact is a change in impedance, which results in reflections affecting the amplitude of the transmitted signal.

A change in the signal-to-noise ratio (SNR) affects the channel capacity. Equation (7.8) determines this index according to the Shannon–Hartley theorem. As the SNR system value decreases, this ability deteriorates. Thus, all factors contributing to signal degradation (assuming that the channel noise level is constant) affect the ability of the channel to accurately transmit data [17].

$$C = \mathrm{BW}.\log\left[1 + \frac{S}{N}\right], \tag{7.8}$$

where C is the channel capacity, bps; BW—channel bandwidth, Hz; S—average signal power; N—average noise power.

Transmission losses depend on a combination of numerous factors (see Fig. 7.42) [1, 17]. These include linear losses (blue line), impedance breaks in connectors or vias, which result in peaks and troughs on the insertion loss line (red curve), as well as crosstalk disturbance between signals. All these factors cause insertion loss, which is characterized by scattering parameters 527 or S12. The most noticeable loss is limitation of the bandwidth of the PCB conductors or cables.

Transmission bandwidth depends on the dielectric loss, whose magnitude is directly proportional to the signal frequency, as well as on the surface effect, limiting the depth of current penetration and leading to a decrease in the conductor's effective

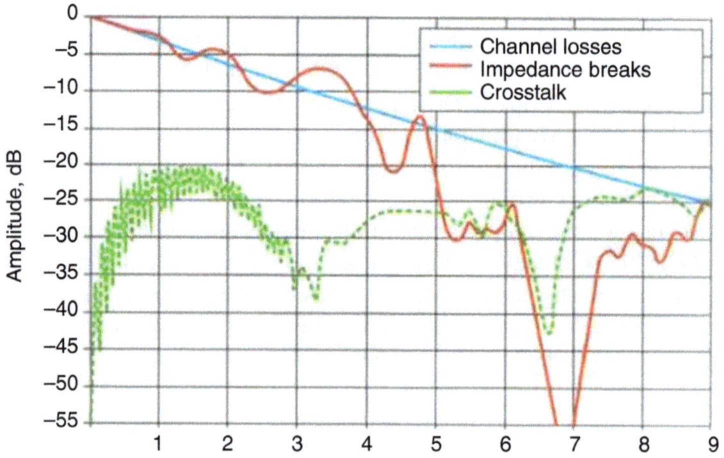

Fig. 7.42 Impact of various factors on the data transmission line bandwidth

cross section. Penetration depth is inversely proportional to the square root of the current frequency. Decreasing the conductor's effective cross section leads to an increase in its resistance, which increases the loss. It should also be noted that dielectric loss in the insulating material varies with frequency.

There are several ways to solve the problem with insertion loss [1]. The first and easiest of them is to choose a material with low loss. For instance, materials for multilayer printed circuit boards NY9000 or MEGTRON7 have excellent performance. These materials are based on polytetrafluoroethylene (PTFE). They have very small loss, but high cost and limited availability. Choosing their low-cost analogs, it is necessary to fully understand the features of the designed board.

Active compensation on the receiver side or preemphasis of the transmitted signal may be used to neutralize insertion loss as an alternative to costly insulating materials. In fact, in both cases, the amplitude of HF signal components increases. They are then absorbed by the data transmission line, which makes the amplitude–frequency response in a given bandwidth flat. The use of devices, such as the DS80PCI402 repeater manufactured by TI, which is designed to compensate for signals in PCIe channels (gen 3), is often determined by the requirements for the physical layer or protocol.

Channel losses may also be influenced by jitter [2] or signal desynchronization between the transmitter and the receiver, when front crossing time does not coincide with the sampling instant. In an ideal data transmission line, there are no impedance breaks resulting in reflections. In fact, even in case of a small defect in the printed circuit board or the transmission line passing through an insulating boundary, impedance discontinuity causes reflections. Errors are summarized along the signal path up to connectors, which are known to cause impedance mismatches due to defects in the mechanical structure or poor coupling with transmission lines.

Jitters may be deterministic and random. Deterministic jitters are subdivided into periodic data-dependent jitters and limited uncorrelated jitters. Occurrence of periodic jitter is due to pulsed power sources and sync signals penetrating into the valid signal. Data-dependent jitter is caused by sources of intersymbol interference (151) or duty-cycle distortion (DCD), which affect the ability of the receiver to correctly detect the bit state. The limited uncorrelated jitter depends mainly on the relative proximity of the channels. If conductors are compactly packed between channels, crosstalk disturbance occurs. All types of jitter are corrected either by the receiver's compensation unit or using preemphasis in the signal generator circuit. In some cases, both methods are used.

Sources of random jitter, representing a Gaussian unlimited component, are, for instance, thermal noise, caused by random movement of charge carriers in conductors. Random jitter becomes noticeable as the data transfer rate increases, when the so-called unit interval (UI) decreases, reducing the time for the receiver to recognize the bit. Compensation methods do not provide for a possibility to get rid of this type of jitter. If the requirements are sufficiently strict, 10 Gb/s Ethernet lines use clock signal, including special microcircuits (retimers), for instance, TI's DS100RT410, to resynchronize data. As a result, all jitter is eliminated, but there is a slight delay associated with the use of these devices. The use of microcircuits that restore signals is not provided by SATA protocols (serial advanced technology attachment—serial data exchange interface with data storage devices) or SCSI (small computer systems interface). In this case, you have to monitor the operation of drivers and receivers, making the necessary adjustments.

7.5.3 Protocol Selection as a Means of Increasing the Signal Integrity

With the advent of standardized physical layers and protocols, transmission protocols (standards) were developed, partially ensuring the interoperability of different systems. Most relevant standards were developed to regulate the electrical parameters of the physical layer. Some standards are based on CML (current-mode logic) or LVDS (low voltage differential signaling) specifications, which define the drivers, receivers, and type of termination used. Other parts of these standards define the channel or transport layer with the appropriate protocols for correcting (training) channels or for synchronizing data fields. This has an impact on the choice of components used to ensure signal integrity. Some standards provide for the use of physical connectors with a small form factor, such as SFP and QSFP, determining the dimensions of interconnect cables.

The use of advanced materials for the manufacture of boards and active signal repeaters improves the performance of most high-speed channels. However, some protocols allow for effective channel training due to the variability of data transmission lines. For instance, according to PCIe protocol, adapter boards are unaware of

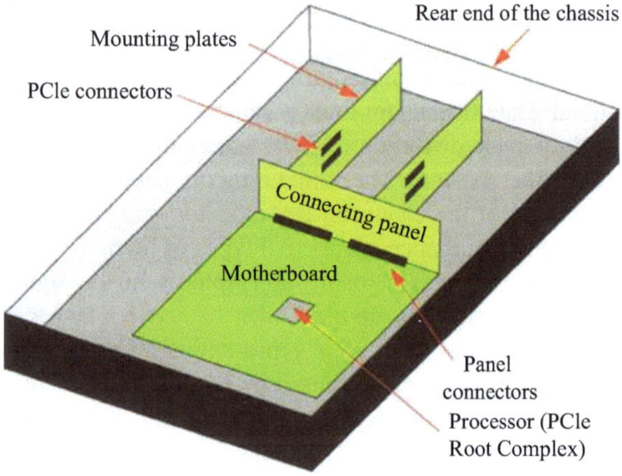

Fig. 7.43 Server chassis with a connecting panel and mounting plates

the channel characteristics until the adapter is switched on. In a server, communication adapter boards are generally installed at the rear of the chassis, along the riser card, which is connected to the panel (see Fig. 7.43). Adapter board or end node (EN) and the processor part contain a root complex (RC) device. After the initialization of tracks, the EN is connected to the RC device using serial low-frequency switches over a high-speed channel. This signaling is called out-of-band, because the connection is established at a slower rate than normally.

Active devices that are designed to improve the integrity of signals should not prevent their propagation; otherwise, the connection will be interrupted. For example, TI's DS80PCI402 repeater allows out-of-band signals to pass through, ensuring normal interaction between RC and EN devices. Additionally, losses at both ends of the transmission line are reduced.

Instead of full restoration of signals between RC and EN devices, the transmission line is divided into two independent PCIe channels [17]. This method presupposes a rather long delay (above 30UI) in each direction, which leads to deterioration of communication parameters during short transactions, for instance, when reading a single memory cell. This line is separated using special devices. However, they are characterized by high power consumption, since they have to restore the signal of RC and EN devices in all PCIe tracks in both directions.

Some simpler protocols cannot dynamically control the channels. These protocols are used in point-to-point applications, regulating, among other things, data transmission between routers of 10 Gb/s Ethernet lines and QSFP connectors with a small form factor. The length of such lines is known and invariable. Signal transmission problems may arise due to the strict jitter requirements of the connected optical modules. If these requirements are not met, a buffered repeater may be used to ensure line losses before the signal enters the connector.

If, however, the repeater fails to cope with the task due to excessive jitter, a signal regenerator is installed in front of the connector on the edge of the chassis. In this

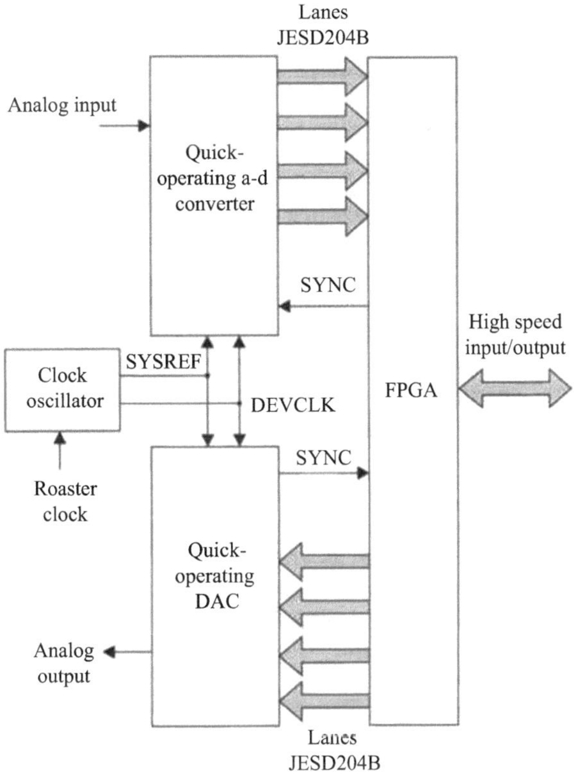

Fig. 7.44 JESD204B system with high-speed tracks and synchronization

case, this device is used in the absence of out-of-band signals. The use of clock regenerator has a double advantage: first, it recreates the output data without jitter; second, many regenerators are equipped with a signal integrity control function. This function is used in modern routers to assess the state of the channel, which eliminates the need for checking the bit error rate (BER), requiring offline operation.

Many protocols, including SATA, SAS, Fiber Channel, Ethernet, and PCIe, are used in data storage and communication systems. Other interface standards [17] are used, too. Among them—CPR (Common Public Radio Interface) or JESD204B, defining data transmission between functional units. For example, the JESD204B standard simplifies the interaction between high-speed ADC and DAC. This standard defines the physical layer (CML) and several protocol layers (for instance, channel and transport ones).

Each track operates in "one direction" using a transmitter and a receiver. This means that there are several tracks on the ADC side (between this converter and the signal data processor) and another set of tracks that are used to transfer data to the DAC, for instance, from a signal data processor, as well as sets of clock and control lines (see Fig. 7.44). The maximum speed provided by the data lines is 12.5 Gb/s. In future, this figure is expected to grow to 16–25 Gb/s.

In HF systems with direct sampling, it is advisable to install data converters on the edge of the printed circuit board, while a PLD or a processor—closer to the digital elements. The JESD204B standard does not specify the requirements for the frequency correction of drivers or the receiver's compensation unit, but their performance may be indicated for specific implementations. If signal integrity is compromised due to excessively long conductors or an impedance rupture, printed circuit boards made from advanced materials are used.

If active devices are used, please consider the following two problems. First, if an application requires a phase coherence of all data converters, it is necessary to establish deterministic latency (or calibrate the system) to make sure that the information from coherent devices is received. Any delay in this path results in a delay in data processing Second, the advantage of the JESD204B standard is that since it does not require out-of-band data transmission, in the event of excessive jitter, clock regenerators are installed in the system.

Summarizing the above, it should be noted that with an increase in the speed of data transmission, the need to ensure the integrity of signals increases. Many problems of this kind are solved by choosing printed circuit boards manufactured from advanced materials, and topology, which makes it possible to avoid impedance breaks. If the cost of such solutions is high or path losses are large, and the necessary materials are not available, active solutions with repeaters or regenerators are used to restore signal integrity.

7.6 Electromagnetic Interference and Methods for Ensuring Integrity of Signals on Printed Circuit Boards of High-Speed Electronic Devices

To ensure signal integrity in printed circuit boards of high-speed electronic devices, it is necessary to take into account the individual specifications of printed circuit boards, the connections between them, while the transmitted signal should have the relevant performance characteristics, including time ranges, power supply voltage, and operating conditions [1, 18]. These requirements became decisive, when data transfer rates in serial I/O systems increased to tens of GHz. According to the previous section, the focus is on signal reflections, crosstalk disturbance, ground plane potential jumps, and power supply decoupling. The board topology and temporal relationships should be carefully considered.

In this section, electromagnetic compatibility (EMC) is considered on the scale of the entire system designed, taking into account the requirements not only for printed circuit boards, but also for enclosures, cables, and supply equipment. The main issues are noise level, noise suppression, and mutual compatibility of digital and analog circuits, drive control modules, relays, etc. It is necessary to thoroughly work out the board topology, provide for grounding and shielding, filtering and protection from transient phenomena. The apparent complexity of the problem lies in the fact that permissible noise levels are measured in millivolts and milliamperes, and protection circuits work with low voltages and currents [18].

Interaction of different circuits is carried out at the level of the printed circuit board and local interconnects, while the levels of the main signals under consideration are generally very different. To ensure signal integrity, the noise levels should be significantly lower than the signal levels, i.e., measured in millivolts in case of digital circuits. However, to ensure effective protection against electromagnetic noise, their values should be measured in microvolts and microamperes, i.e., as a rule, be three orders of magnitude lower than the acceptable self-noise level. Protection circuits must resist external noise in the kilovolt-ampere ranges, which are several orders of magnitude higher than the values set for logical units and analog devices.

The developer shall be able to resolve the obvious contradiction—the values of the parameters that are quite acceptable and ensure signal integrity can be significantly higher than the permissible noise levels and significantly lower than the values of external noise of the protection circuits. Figuratively speaking, one should address the twin challenge—to protect against electromagnetic noise, it is necessary to eliminate stray crosstalk, while to ensure signal integrity, the losses during its propagation should be reduced.

Let us consider the design features of printed circuit boards, taking into account such factors as impedance, wiring, noise, and signal circuit losses [18].

Impedance of grounding conductors is the decisive factor for almost all problems related to signal integrity and electromagnetic noise. In both cases, the impedance value should be low WRG. This is easily achieved using a continuous ground plane, but it is very difficult to implement, if conductors are used in a double-layer PCB.

Impedance of grounding conductors is an important factor affecting not only signal integrity, but also electromagnetic noise, especially at high frequencies. The ground plane acts as a return signal circuit, provided that the ground under the signal path is continuous. However, despite the continuous return circuit, a sufficiently large potential fall takes place on the ground plane, resulting in common-mode voltage. Its impact on signal integrity is marginal, but it is the main cause of electromagnetic noise in the signal layer, ground layer, and power ground.

In the cases considered, in-phase currents are completely foreign. They do not affect the signal quality, but they are difficult to eliminate as sources of electromagnetic noise. Differential currents pass through the signal path and affect its integrity more than electromagnetic noise. In fact [2], inductive impedance of the signal or return circuits is proportional to their area, as is the antenna's coefficient of efficiency. However, the areas of the signal and ground circuits of a multilayer board are small; as a result, the return circuit of the ground plane is continuous and generally does not cause problems with electromagnetic noise.

The thickness of the copper layer is irrelevant, since at high frequencies the surface effect prevails. Thus, currents flow near the surface of conductors.

The main problem associated with ground impedance is its discontinuity, when the signal passes through the reverse signal path, which has a major impact on the characteristic impedance control. Therefore, to ensure signal integrity in the gigahertz range, the characteristic impedance should be carefully monitored. Its impact on the electromagnetic noise control is less significant. To effectively protect against electromagnetic noise, it is usually sufficient to minimize the surges and AC failures, especially at the PCB output. To preserve signal integrity, it is most difficult to ensure the continuity of the return circuit. Let us consider some specific cases in more detail [18]:

1. Ideally, a continuous copper plane should be placed directly below the signal path. Due to this arrangement, impedance control comes down to ensuring matching, usually at the load end. In applications operating at lower frequencies, where electromagnetic noise control is mainly required, it is generally sufficient to match the source with the transmission line to prevent the generated noise currents from propagation outside the chip. Matching the signal source reduces the rate of pulse rise/decay, which may be unacceptable for high-speed applications.
2. The continuity is most affected when the plane of reference changes, the signal passes from the ground plane to the power plane (see Fig. 7.45). Naturally, when a return circuit is created, it is virtually impossible to use vias connecting the ground plane to the power plane. The only option is to install decoupling capacitors along the perimeter of the board to provide a low-impedance reverse HF path at its edges. Unfortunately, this solution is acceptable only for relatively low-frequency signal circuits.
3. Continuity is unlikely to be broken when the signal passes from one ground plane to another. In this case, the return circuit between the planes should be continuous with controlled impedance. Typically, this is achieved by providing vias between ground planes along the perimeter of the signal hole, as well as by choosing the appropriate constraint zone using the keep out layer, the size of the mounting pads, through-holes, and the length of conductors to match the impedances.
4. Developers of printed circuit boards may face certain challenges, when the signal passes from one side of the ground plane to the other one (see Fig. 7.46). Since this process is not followed by the plane change and the impedance discontinuity is minimal, grounding holes are unnecessary. To ensure signal integrity at

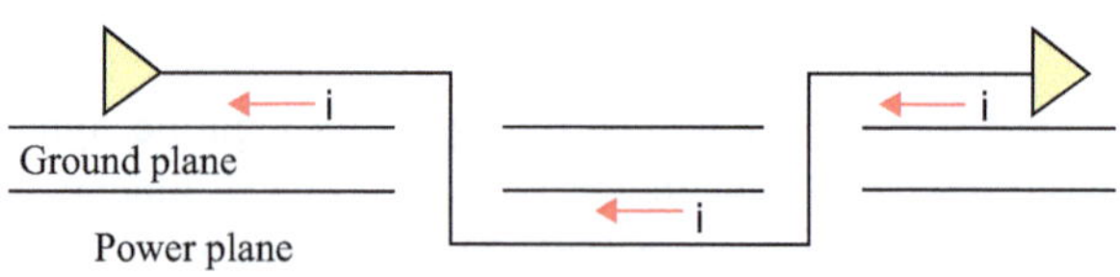

Fig. 7.45 Discontinuity of the reverse current path when changing planes of reference

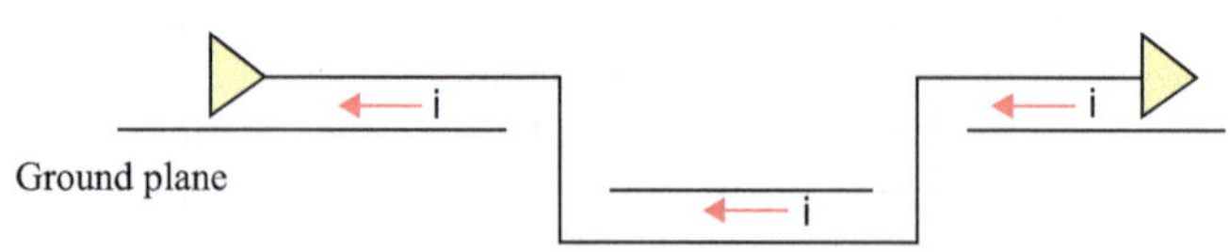

Fig. 7.46 Impedance discontinuity is minimized if the current continues to flow along the same plane of reference

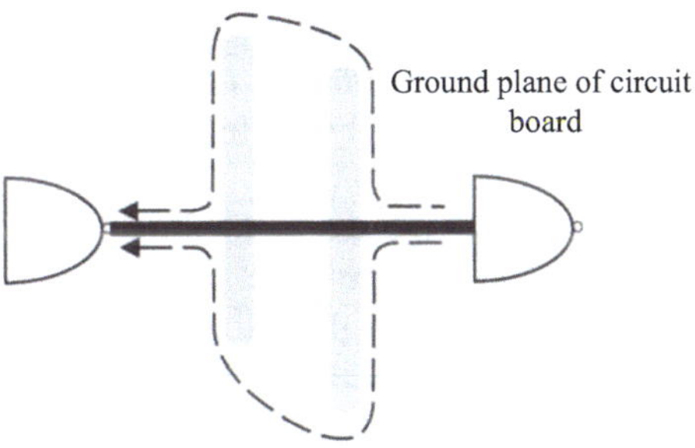

Fig. 7.47 In addition to the problem of the return signal circuit discontinuity, there are gaps in the ground plane

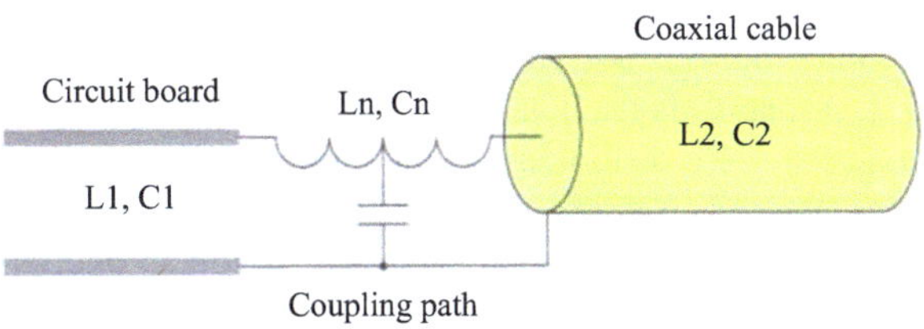

Fig. 7.48 Coaxial cable and PCB impedance matching

maximum speed, one should minimize the impedance discontinuity by regulating the size of through-holes—their diameter and depth.

5. Discontinuity in the return current signal path is due to the breaks in the ground plane (see Fig. 7.47). The current in this case flows around the gap, which increases the value of the characteristic impedance of the path. The gap works as a notch antenna, if there is another conductor by the edge of the divided plane or next to the contour of the connector.
6. Since the signal path may have inevitable discontinuities, it is necessary to control the impedance at the corresponding circuit edges. First of all, breaks occur at the junction of the printed circuit board and the connector (see Fig. 7.48) and are generally noticeable if the cable impedance is not matched with the PCB impedance. In this case, the impedance matching circuit should be installed near the edge, which is achieved by controlling the copper layer parameters. For large gaps, the inductance increases. A thicker copper layer provides for a higher capacitance at the junction of the board and the connector.

Speaking of the PCB layout design, it should be noted that determining critical conductors is mandatory to establish control over electromagnetic noise and signal integrity [18]. In both cases, most of the problems arise because developers pay attention to a very limited number of conductors due to lack of time or insufficient space on the board. The conductors that have the greatest impact on signal integrity and on electromagnetic noise are usually different.

Signal integrity is determined by several conductors with relatively high frequencies of the transmitted signal. First of all, we are talking about the conductors transmitting high-speed serial data. To design such boards, one should focus on the signal and reverse paths, as well as to the adjacent metal elements.

To protect against electromagnetic noise, one should primarily pay attention to the lines at the input/output of the printed circuit board. The main sources of noise of this kind are lines with high-speed clock signals and data, as well as spurious coupling with relatively low-speed signal lines, power lines, and ground lines [18]. The main receivers of radio noise are analog lines at the input. Digital lines are subject to impulse noise.

Having identified these lines, one can proceed with the installation of chips on the board, thus simplifying the wiring layout. The simpler the circuit consisting of critical conductors, the easier it is to ensure signal integrity and establish control over electromagnetic noise.

Let us say a few words about signal decoupling. As a rule, the problem of signal disintegration is solved using the correctly calculated tolerance on supply voltage [1]. It manifests itself at the level of electromagnetic compatibility in the sense that external noise distorts the voltage signal of the power source or that of the built-in regulators. The noise from the clock signal, appearing on the voltage planes and propagating through the power cable, results in interference, even if its amplitude is within the microvolt range. However, this problem does not affect signal integrity until the noise amplitude is within the millivolt range [18]. Thus, the requirements for decoupling due to electromagnetic noise are much tougher than those aimed at ensuring signal integrity.

Chip manufacturers recommend using decoupling capacitors to reduce Vcc. This means that the range of frequency set points during decoupling does not exceed the clock frequency, while the noise at its harmonics typically exceeds its value by ten times or more.

Thus, the requirements for decoupling due to noise are much stricter than those aimed at ensuring signal integrity. This requires a smaller inductance, rather than a large output capacitance. Modern computers operate at frequencies, resulting in HF harmonics, which are inevitably higher than the resonant frequency of a series oscillatory circuit with a typical decoupling capacitor. An increase in the inductance at the outputs of all decoupling capacitors by 1–2 nH will result in a much greater impedance than is required for effective filtration. If impedance exceeds 1 Ohm, a more effective filtration should be provided. Alternatively, decoupling capacitors should be installed parallel to each other. At frequencies above 200 MHz, the interlayer capacitance of a multilayer PCB becomes a crucial factor.

The developer shall be able to deal with such factors as crosstalk disturbance [2]. Crosstalk disturbance, resulting from unintentional contact with the adjacent metal elements, disintegrates the signal leading to electromagnetic noise. Typically, this noise affects the adjacent signal, power, or ground paths.

Crosstalk disturbance refers to field communication between adjacent lines. This problem occurs not only when cables are used, but also at the level of printed circuit boards between conductors. Any connection between high-frequency (in gigahertz) lines degrades the adjacent conductor's signal quality. Crosstalk disturbance makes a significant contribution to electromagnetic noise of the I/O lines, when crosstalk has an impact on clock signals or susceptible lines on the board. Generally this problem is solved by dividing the indicated lines. In the resulting gap, lines less susceptible to crosstalk may be placed. In any case, the larger the gap, the better, since crosstalk decreases with increasing square root of the distance between the conductors.

In addition to losses due to crosstalk disturbance, other losses should be taken into account. The main ones are dielectric losses and losses resulting from series resistance of interconnects.

Thus, the signal circuit losses appear due to the series resistance of the conductor and the dielectric loss in insulating materials. Most often, these losses are insignificant at the PCB level, unless a HR signal circuit is involved. These losses increase remarkably, if cables are used. First of all, losses of this kind deteriorate signal integrity, leading to its loss. When it comes to electromagnetic noise, this issue is unimportant. However, it is obvious that if the signal is attenuated, data errors occur with less noise.

Loss of stability resulting from crosstalk disturbance becomes noticeable for differential signals, in case of an increase in the serial data transmission speed. As noted above, instability is the result of different signal propagation time from the source to the receiver over communication lines of different lengths. This problem is much more reflected in signal integrity than in electromagnetic noise control.

Strong connection between elements of adjacent boards results in electromagnetic noise. Typically, clock signal noise of the microprocessor chip operating at high frequency through capacitive coupling affects the adjacent printed circuit board and propagates in the external environment. A similar situation is observed, if the internal cable is located next to the same chip. To neutralize the electromagnetic noise, the chip is shielded. This problem rarely affects signal integrity.

7.7 Methods for Ensuring Integrity of the Signal During All Stages of Designing of Printed Circuit Board

Typically, the issue of signal integrity (SI) is addressed later on while designing a high-speed electronic device (system), when the PCB topology development stage is over. For this purpose, a complex and time-consuming calculation of 30 parameters and modeling of high-speed data transmission lines are usually performed. Here, we consider the basics of this approach [18, 19]. The issues of ensuring the SI are dealt with at the very beginning of design.

Unfortunately, little attention is usually paid to the issue of signal transmission integrity while developing the high-speed electronic device's topology. As a result, many problems of this kind are often identified during modeling. To solve them, numerous successive approximations are required, which delay the run-out production and increase expenditures. Therefore, the built-in control is widely investigated as a unit that automatically controls the presence/absence of standard errors in the project, leading to the signal disintegration. Standard errors include: absence of a shield, impedance discontinuity in the reverse path, presence of closed loop circuits on the voltage plane/ground plane, unmatched impedances, etc. Even after this test, circuit modeling is required for the final test. However, fewer problems solved at this late stage of development make it possible to bring products to market faster while reducing the engineering costs.

It should be noted that there are certain difficulties in ensuring the SI in modern high-speed electronic devices [19].

Increased data transfer rate, addition of new I/O ports, and complexity of devices are the main factors leading to signal integrity and electromagnetic noise (EMN). For instance, if a volumetric packaging technology is implemented for microcircuits, vias make wire connections unnecessary and reduce the length of interconnects in vertical configurations of the chip. As a result, signals are transmitted at a higher rate, power consumption is reduced, but there is harmful radiation. The requirements for ensuring electromagnetic compatibility (EMC) in high-speed systems are tightened. The current trend to combine multiple sources of radiation in a single electronic system creates additional difficulties.

Almost all PCB design tools are equipped with built-in design rule checkers (DRC). However, these tools are used to analyze manufacturability, rather than signal integrity. Design is usually followed by modeling, when it is possible to simulate system performance by solving Maxwell's equations for electromagnetic interactions. For this reason, all too often SI problems are identified rather late during design, when introduction of changes is very costly. As a rule, the cost of changes increases by at least one order of magnitude during phased implementation of the project.

All most advanced PCB computer-aided design systems, for instance, CR-8000 by Zuken, are already equipped with means to verify signal integrity and analyze the topology for possible problems related to SI and EMC. At the same time, characteristic impedance, delay time per unit length, mutual inductances and capacitances between lines are calculated. The SI control device is activated using the system menu command; an external circuit is not required (a part of the circuit may be selected). Lists of components, exchange files, and translators are unnecessary. Users are able to configure every rule by setting certain circuit parameters. This option is very important, because every time printed circuit boards are designed, a balance should be found between performance, cost, and project deadlines.

The option to configure rules makes it possible to manage the process taking into account the specifics of a particular project. The user sets the rules according to the application needs at his or her own discretion. At the same time, it is possible to analyze the project using the "what if" technique, completely changing the parameters and adjusting the project rules. Additionally, identifying problems at the earliest stages of design minimizes further project changes and reduces expenses.

Let us consider typical design errors that are most often encountered in practice.

Shielding Sensitive circuits, including wiring diagrams for transmitting high-frequency and clock signals, need shielding for protection against HF fields. For instance, all electronic products must comply with the CE certificate to enter the markets of the European Union. Thus, they are tested in the field with a strength of 1–10 V/m and a frequency range of 8 MHz to 0.6 GHz. In such segments of the

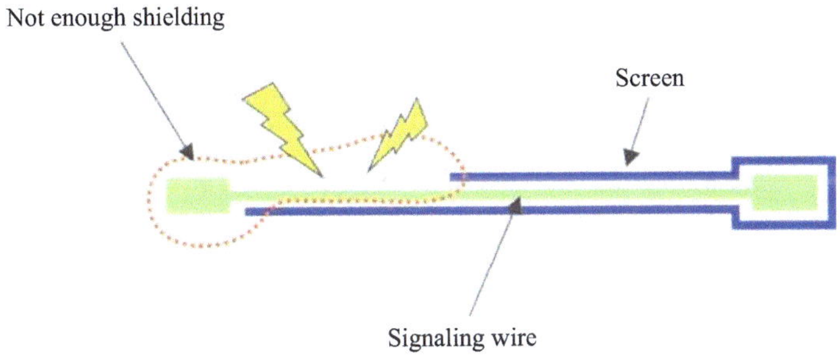

Fig. 7.49 Conductor shielding test

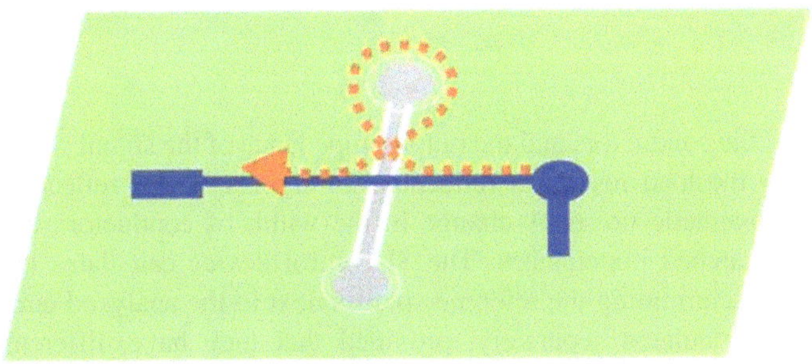

Fig. 7.50 Invalid reverse path occurrence

market as automotive electronics and aerospace engineering, products must pass even more stringent tests. The SI control device checks for the availability of shields near the corresponding through-holes through which the signals pass (see Fig. 7.49) [19]. If the shielding level is below the one set by the user, it is recommended to take measures and protect the relevant circuits—for instance, install shields and connect them to ground or to power.

No breaks in the reverse path. Since the signal current always flows along the path with the lowest impedance, it can flow in an unintended direction of an undesirable return circuit (see Fig. 7.50). If it has a form of a loop, signal transmission is delayed or radiation creates noise in the circuits of the device. In such cases it is recommended to minimize the area formed by reverse current in the circuits of clock signals, HF signals, as well as in feed circuits. The SI control device determines the required signal path, the wire connection to the terminal load, and the return circuit that is closest to the signal path between the ground terminals of the signal generator and the receiver. Then the SI control device proceeds with the search for an invalid reverse path and determines error weight according to the formula: error rate = 1—reverse path length/error path length.

Impedance Mismatch As the signal frequency increases, transmission lines are subject to excessive noise. In particular, the discrepancy between the characteristic

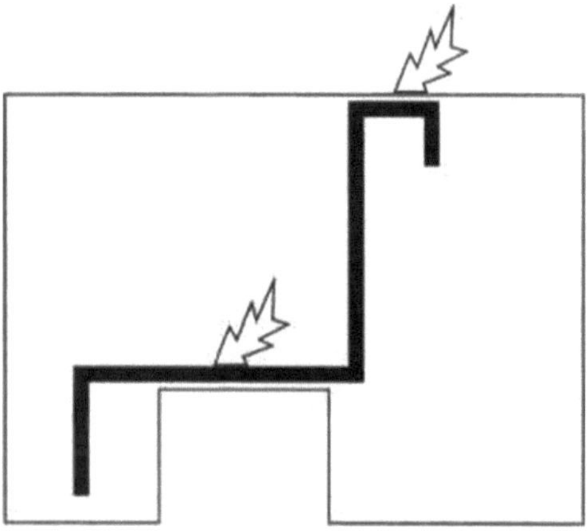

Fig. 7.51 Example—the conductor runs along the edge of the board

impedance of the conductors and the output impedance of the signal source/the input impedance of the load results in reflections of signals. These reflections, in turn, cause electromagnetic noise. A change in the width of conductors is one of the causes of unmatched impedances. The SI control device calculates characteristic impedance by determining the reference planes next to the analyzed circuit. Wiring diagrams are calculated separately, provided that they have different reference planes.

High-speed signal wires, passing very close to the edge of the reference plane, may become sources of interference (see Fig. 7.51). The EMC control device determines the frequency, leading edge pulse time and voltage, which exceed the values set by the user, thus identifying the risks of electromagnetic noise affecting HF signals.

The Density of Placing Through-Holes If through-holes are scarce or unevenly distributed on the surface of power or ground planes, any part of these planes functions as an antenna and generates undesirable noise. An insufficient number of through-holes in the reference plane circuit may result in the formation of large return circuits and increase switching noise. The control device determines how evenly the outputs are distributed across the PCB plane (see Fig. 7.52) [19], highlighting the circles with a radius set by the user.

In Fig. 7.52, unshaded surface areas are considered problematic. As soon as the ratio of total to shaded surface area reaches a given level, the control device notifies the user of the fact. This tool identifies many other errors in the layout design. These include: resonance of conductors, electromagnetic incompatibility of the PCB layers, insulated copper areas, overlapping power planes, high impedance of these planes, etc.

After checking the compliance with the selected rules, the control device represents data in the form of a histogram, which determines operational performance

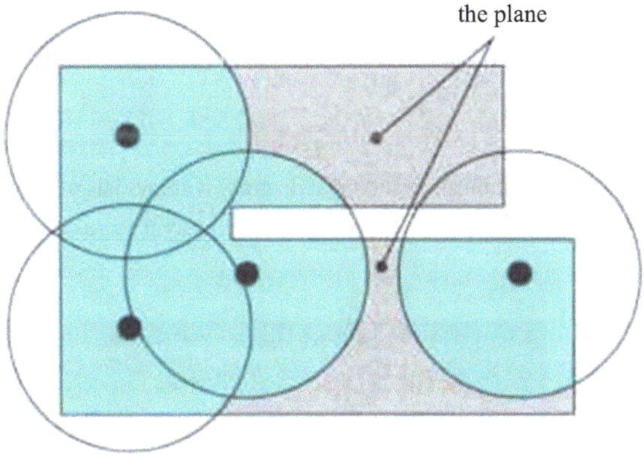

Fig. 7.52 The principle of placing through-holes in a plane

under the project and the scale of the problem. This tool provides complete control using any combination of rules and project elements.

Let us now consider the recommended design methodology [19] in more detail. Traditionally, independent groups of developers were engaged in the design of printed circuit boards. For instance, one of them specialized in circuit implementation, the second one worked on PCB logic, and the third one tested the project using analytical tools. This approach has a number of drawbacks. As a result, the obtained boards did not function due to a number of challenging problems related to the integrity of signal circuits and feed circuits, which were undetected by analysis tools. However, even after advanced analysis tools were used, there was a threat of falling behind the schedule, since a lot of time was spent on coordination of issues by different groups of developers.

Modern PCB design methodology is more focused on teamwork. Thus, all groups of developers ensure the integrity of signal circuits and feed circuits based on the existing constraints and identify existing problems early on during design.

Here, a preliminary analysis also constitutes an important part of the project. Companies seeking to go beyond the traditional constraints use tools providing a certain degree of integration. These tools use a constraint system integrated into SI testing tools at all design stages. To determine the physical separation of the system into parts, choose components, IP solutions, and requirements for feed circuits, a preliminary layout analysis is performed. It aims to assess the dimensions of the product and the corresponding constraints according to the power supply mode. As part of the development process, this analysis saves time and prevents any signal integrity problems.

The preliminary analysis of the project plan is mainly associated with a layout analysis, which makes it possible to compare different installation options (see Fig. 7.53) [19]. Developers have an opportunity to launch a project from the first stage or to extract design information on the circuits. In any case, one gets an idea of

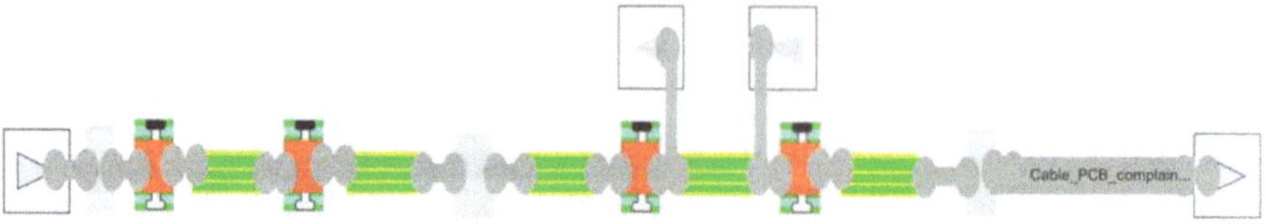

Fig. 7.53 Preliminary topological diagram with models of W-elements and vias, in which connections are established between printed circuit boards

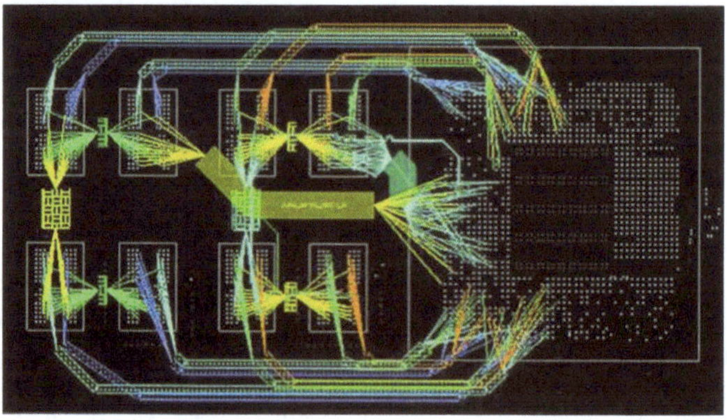

Fig. 7.54 Circuits are presented as bundles illustrating traceability management

the main project interfaces. Models of active devices should be as accurate as possible, allowing for interconnects to be evaluated, parameterized, and tested.

Layout data calculated based on the schematic circuit diagram and the project requirements provide all developers with access to the same information, which is necessary to generate a flowchart used to create a layout or define constraints. Additionally, engineers have an opportunity to prevent the reflection of signals and solve topological problems along with the developers who ensure signal integrity. The obtained topology may be used for a design flow, as well as to determine the corresponding constraints (see Fig. 7.54) [19].

The preliminary analysis of the project plan often overlooks the possibility of virtual prototyping. A virtual model may be used by the engineer, who is responsible for ensuring signal integrity, to create a prototype in the absence of a netlist or a project database (see Fig. 7.55) [19]. Ideally, the PCB layout requires available project and library data, including the number and order of layers, geometric descriptions of all outputs and dimensions of the board. Virtual layouts make it possible to find out more about the component arrangement and wiring in order to perform subsequent simulations and define constraints, which would be impossible to achieve with a finished layout. Naturally, a virtual layout does not give a comprehensive picture of the project, but it really helps to analyze the interfaces.

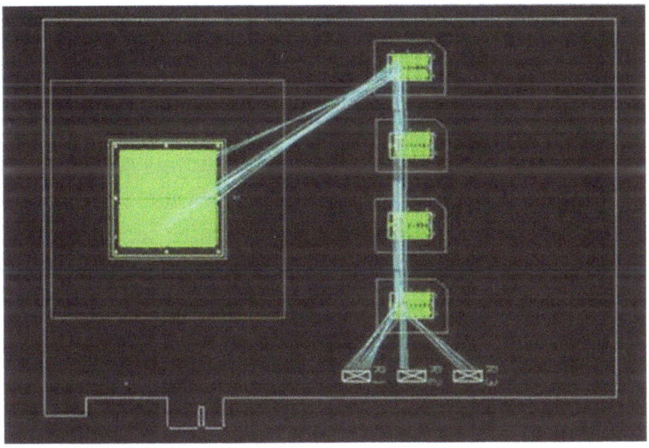

Fig. 7.55 DDR3 interface virtual prototype and development of a process flow for an address bus

Both topology and virtual layout should be supported by structured device model management. If an analytical model is repeatedly managed, time is saved for preparatory work. The preliminary analysis of the project plan reveals the need for new models, which can be tested and sent to the central library. It is even better to include analysis models in the central library management system. Knowing the exact data describing a particular model, one can track a SI model in the library data during design: from the schematic circuit diagram to the finished layout.

The stage of preliminary analysis is the design starting point. It is followed by a special floor plan, determining physical location of the elements and the connections between them. When a netlist is drafted, arrangement options, form factor, requirements for mechanical structure or thermal characteristics are analyzed. Once the netlist has been prepared and uploaded to the corresponding application, the work on the floor plan goes on. It is at this design stage (before wiring) that signal integrity should be analyzed, followed by workload reduction.

To ensure signal integrity, it is necessary to check the circuit for the presence of parameters beyond the set limit values, identify reflections and crosstalk disturbance, set the delay value, find the best layout for a particular application, as well as determine the need for terminal load.

The only above-mentioned parameter that requires tracking is crosstalk disturbance. At this stage, there is a perfect opportunity to model the entire project and, in particular, the interfaces that have not been modeled with finished layout in order to identify SI problems. Data assignments of the device model are included in the netlist and sent to the design database. Simulation with non-wired circuits is not time-consuming, since interconnects are represented in a simple model based on Manhattan distance and impedance. In some cases, Manhattan distance is replaced with design data for a more accurate analysis of some circuit connections, which can improve the accuracy of the SI analysis prior to wiring.

Analysis at the stage of drawing up a floor plan may require time-consuming changes or editing after wiring. Thus, the constraint system and SI analysis, which use the same design data, simplify the entire process implementation.

It is necessary to analyze the problem of protection against noise on its own. At this design stage, various rules and restrictions are usually verified. This verification actually leads to a revision of the previous rules, developed and implemented at the previous stages of automatic or manual wiring using a design rule checker (DRC). This design methodology using constraints was developed about 20 years ago.

During this time, data transfer rates have grown from tens of Mbit/s to several Gb/s for parallel interfaces. Meanwhile, serial interfaces with even higher speeds have appeared. As a result, by now means of checking design standards have become widespread along with electrical rule checks (ERC) and simulation. Thus, checks have become large-scale as compared to the traditional physical verification of the rules for the same data used in the SI analysis.

At this stage, developers have a perfect opportunity to identify many potential signal integrity problems prior to the project post-analysis. It is possible to determine the impedance, as well as the presence of crosstalk and take appropriate action while skipping the complex preparation.

The issue of the effect of signal and feed circuits on the SI is essential for modern projects. As power distribution networks evolve, it is necessary to take into account the interaction of feed circuits, signals and through-holes.

Many constraints appear early on during design—before planes and through-holes. As these elements are added, verification of physical rules should not be based on the assumption of perfect performance of planes. Analyzing the impact of noise from feed circuits on signals, impedance, interactions between conductors, crosstalk disturbance, delays, phase shift, and similar parameters should be considered.

Verification facilities serve to analyze the interaction between signals and feed circuits, while a linear effect manifests itself with regard to numerous signal circuits. The results of signal quality test, including noise coupling, are issued without delay, making the time-consuming non-linear simulation unnecessary. These results reveal problems that the general examination of design norms after wiring has failed to detect, based on the existing assumption of ideal reference planes. For instance, noise pickup from plane through-holes may result in crosstalk disturbance, which increases as the conductors are separated. Not only does it seem unreasonable, it also contradicts the generally accepted rule, based on the assumption of perfect reference planes. This problem can be identified only if noise from the plane and through-holes is taken into account.

At this stage, it seems unnecessary to use all types of verification facilities to check the device models available in the project plan. However, it is more convenient if the same tools and solvers are used for checking the rules as for analyzing the SI.

At the final design stage, having taken measures for protection against noise to eliminate significant signal reflections or crosstalk disturbance, one should proceed to the final stage of analysis. As a rule, during this analysis, a thorough simulation is performed in order to identify compatibility problems in high-speed interfaces. In this case, DDR memory interfaces and serial data lines are mainly considered.

Based on the fact that the analysis started at the preliminary stage, when a layout was created, detailed data on interconnects should be obtained, replacing

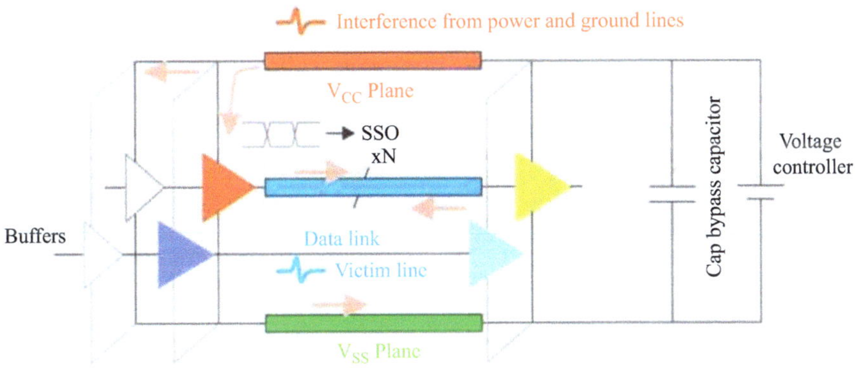

Fig. 7.56 SSN simulation

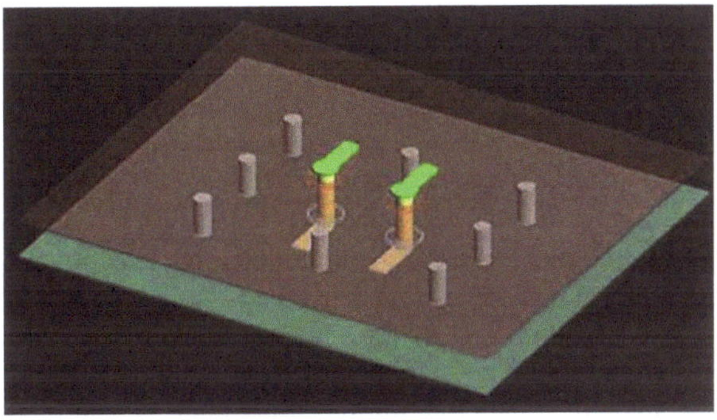

Fig. 7.57 Three-dimensional array of thru-vias

"hypothesized" W-elements and models of through-holes with semi-valuable S-parameters, double-checking the results of simulation and compatibility using limiting design parameters. To design modern DDR interfaces, one should obtain data on the power distribution network (PDN) along with the signal data in order to take into account real effects, such as simultaneous switching noise (SSN) (see Fig. 7.56) [19].

The most efficient way to do this is to use a "hybrid" electromagnetic solver, dividing copper structures into through-holes, data lines, and other elements, calculating them, and presenting the overall result as a S-parameter model. Real power line effects considered as DDR simulations prevent potential problems.

Data for high-speed serial interfaces should be calculated with pinpoint accuracy. Uniform conductors of printed circuit boards are easily modeled using a 2D field solver. As a rule, 3D simulation data are usually required for simulating through-hole arrays, if modeling of interlayer vias is accurate (see Fig. 7.57).

Since these interlayer vias often cause significant impedance discontinuities in the signal path conductors, vias play a major role in obtaining an accurate channel model. Full-wave three-dimensional solutions require a lot of computational resources, for instance, several hours for calculating a number of coupled differential pairs.

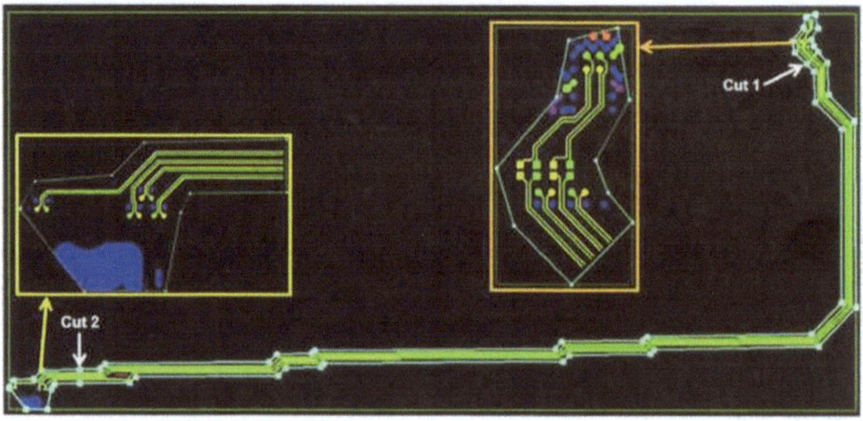

Fig. 7.58 "Cutting and stitching" method

Three-dimensional mesh spacing of long uniform conductors of a printed circuit board is very time-consuming. It is unnecessary if large two-dimensional structures are modeled. In more complicated cases, the "cutting and stitching" method is used (see Fig. 7.58) [19]. A full-wave representation, obtained using a combination of several field solvers, can be automated by selecting a serial data line in the circuit, splitting it into zones and assigning them to the solver for calculation. Upon its completion, the data on these zones are combined in a single S-parametric representation. After obtaining accurate S-parameters, pretesting can be performed, allowing for quick data updating with detailed interconnect models used in the analysis.

The device shall be modeled with due care. In multi-gigabit interfaces, compensation is usually used to ensure the required signal quality. In case of data transfer through the interface at a speed higher than 10 Gb/s, a compensation is applied, whose characteristics vary over time. Traditional time-based simulation of circuits fails to provide such data traffic that would allow setting the compensation, generate enough samples for visual representation and provide for bit error ratio (BER) extrapolation. Interfaces of these types require channel modeling with pulse responses used to simulate high-capacity lines.

In these types of device models, executable software modules are used to simulate different types of adaptive compensation, for instance, continuous time linear equalization (CTLE), decision feedback equalization (DFE), and feed forward equalization (FFE).

IBIS-AMI behavioral models are much more developed and suitable for description of SerDes devices, than the ones used a few years ago. However, they are still inadequate for description of a particular device, despite the fact that general compensation parameters are known, including the number of equalizer taps with FFE or amplitude–frequency response of CTLE filter. In these cases, modern tools are designed to quickly synthesize the IBIS-AMI model for simulation and model an interface at the system level (see Fig. 7.59) [19].

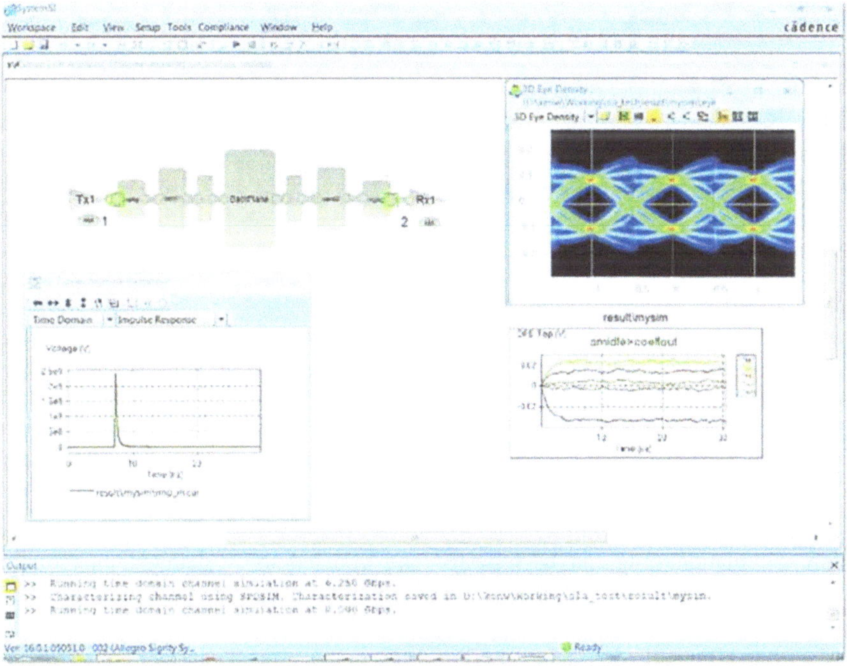

Fig. 7.59 Serial data line analysis results

Interconnect models and devices are intended for high-speed interface compatibility tests. DDR interfaces are evolving. Starting with DDR4, it has become possible to model memory interfaces using serial data lines and analytical methods. Since memory controllers are currently equipped with FFE equalizers and CTLE filters, which have been used for many years in serial data lines, developers start using IBIS-AMI models in these controllers. According to the requirements specified in new standards, such as DDR4, there is a shift from the traditional definition of compatibility according to data setup time and data hold time to shadow mask compatibility according to voltage and time synchronization, as well as to the specific BER requirements.

Since testing all memory interfaces for signal quality and mask compatibility (see Fig. 7.60) [19] may be rather time-consuming, there is a need to automate the subsequent signal processing and generate detailed reports on the work performed.

In addition to mask compatibility test, other unique requirements for the interface are generally imposed, including requirements for frequency-domain interconnect performance.

Summing up the above, it should be noted that by checking signal integrity at all stages of the PCB design, starting from the very beginning, the need for time-consuming step-by-step calculations for the model, including in subsequent iterations, is minimal.

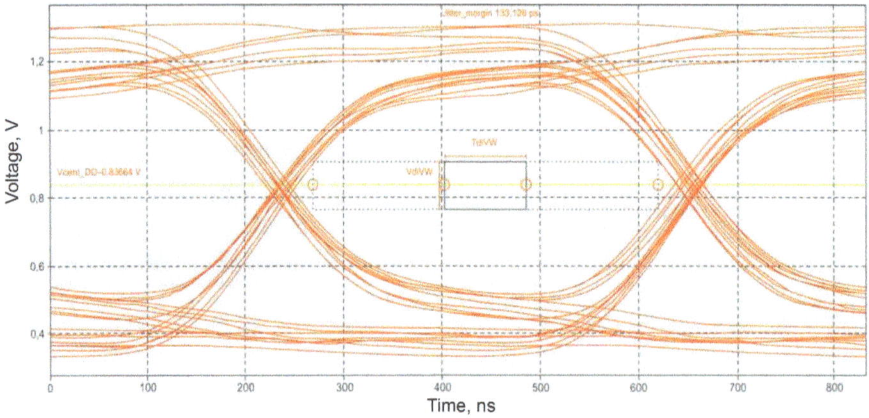

Fig. 7.60 Mask compatibility of the 00Я4 interface signals

The full verification process starts with the preliminary analysis of the project plan. Simulation tests can be repeated and updated using detailed interconnect models obtained from the design circuit. To calculate the DDR interface bus power parameters, electromagnetic solvers are required, while full-wave methods are used to efficiently determine interlayer vias in high-speed interfaces of serial data lines. Automated analysis of interface compatibility is required at the final design stage, confirming the developed device production possibility.

7.8 Method for Calculation of Zigzag Delay Lines for High-Speed Printed Circuit Boards

7.8.1 Calculation of Crosstalk in Connected Transmission Lines

Printed circuit boards of modern high-speed digital devices use special delay lines to synchronize signals [2]. The most popular solution for such delay lines is the cyclic inclusion of a zigzag section called a loop (a meander) in the communication line, ensuring that the necessary time delay of signals is effortlessly achieved [20].

Normally, when this loop is designed, standard solutions are used to increase noise immunity, for instance, the distance between the loop sections is 3–4 times greater than the signal line width. However, this results in inefficient use of the PCB area. In modern systems, considering an increase in the circuit density and miniaturization of printed circuit boards, the solution of problems associated with the reduction of the delay line area is very important.

The number of bends of zigzag lines and their size affect the delay and quality of the transmitted signal. Analysis and calculation of the delay line can be performed

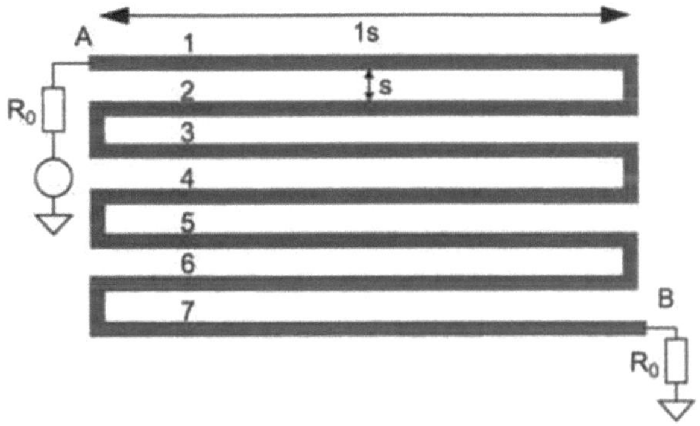

Fig. 7.61 Example of a delay line (loop) consisting of 7 sections: Is—section length; *s*—distance between sections

using EMF modeling systems: quasistatic or full-wave 3D modeling. The latest systems are the most accurate and complete, although they require large computational resources, a good understanding of the results (for the circuits under study) and are quite expensive.

To use modeling systems correctly, one should know the geometric parameters of the investigated delay line as input data. If no samples are used, preliminary engineering design of the line is necessary.

A simple engineering method for calculating a delay line for printed circuit boards in the form of a zigzag line (loop, meander) is proposed in [20].

Analysis and calculations are performed based on the following conditions and requirements for the delay line:

- input signal—voltage drop (step, trapezium) with rise time (T_t);
- minimum loop area for a given board stack;
- required delay time and margin.

When a signal passes through zigzag delay lines (Fig. 7.61), crosstalk between the line sections (1–7) occurs.

Let us consider a loop bend as two coupled lines perfectly matched at their ends (Fig. 7.62), when the segment's characteristic impedance is equal to the load at both ends, i.e., $Z_o = R_o$.

If the active line (aggressor) is excited by the input signal V_{in}, induced noise occurs in the connected passive line (victim), due to the capacitive and inductive coupling of the lines (Fig. 7.63).

Crosstalk is present at the near end of the passive line (V_b) and at the far end of the passive line (V_f).

Mathematical expressions for calculating near-end crosstalk coefficient (K_b) have been obtained by Feller [20]:

$$K_b = (1/4tpd)(L_m/Z_o + C_m/Z_o) \tag{7.9}$$

or by substituting expressions as follows: $Z_o = \sqrt{L_o/C_o}$ and tpd $= \sqrt{L_o \times C_o}$.

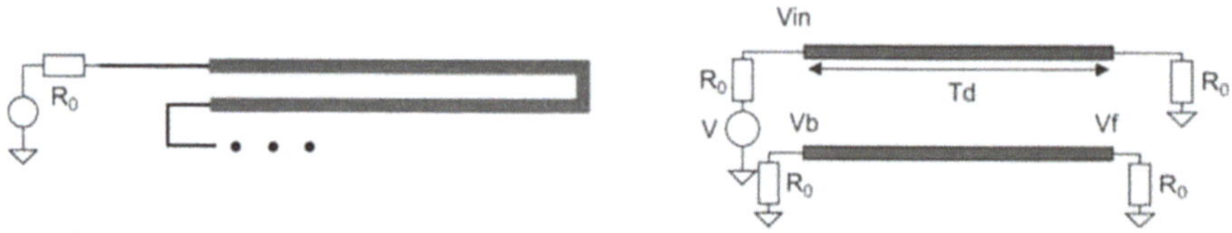

Fig. 7.62 Equivalent electrical circuit of a single loop/meander bend

Fig. 7.63 Noise waveforms in coupled lines: T_d—signal propagation time through the line; T_t—signal pulse rise time

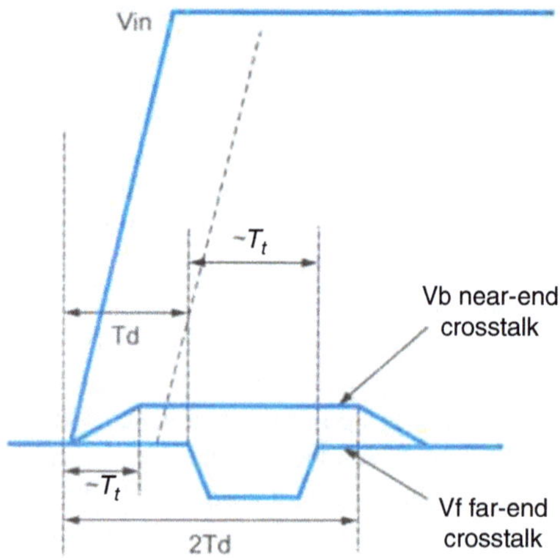

$$K_b = 1/4(L_m/L_o + C_m/C_o) = 1/4(K_1 + K_c), \tag{7.10}$$

where tpd is constant propagation delay; Coт—per-unit-length mutual capacitance; C_o—self-capacitance per unit length; L_m—per-unit-length mutual inductance; L_o—self-inductance per unit length; Z_o—characteristic impedance of the line; K_1—inductive coupling coefficient; K_c—capacitive coupling coefficient.

Near-end crosstalk coefficient (K_b) determines the ratio of the near-end crosstalk amplitude to the active line signal magnitude: $K_b = V_{bl}V_{in}$; and the crosstalk form depends on the incoming signal pulse front.

Let us consider the input signal noise V_{in} (t)—voltage drop (steps) in Fig. 7.64 [20].

If we introduce the function [20]:

$$f_{in}(t) = V_k/T_r(t), \tag{7.11}$$

where V_k—maximum input signal.

Then input signal voltage (V_{in}) will be of the form:

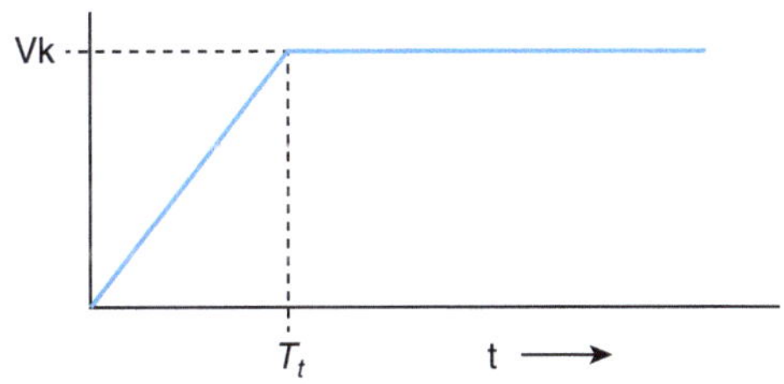

Fig. 7.64 Input signal waveform, where T_t—leading edge rise time

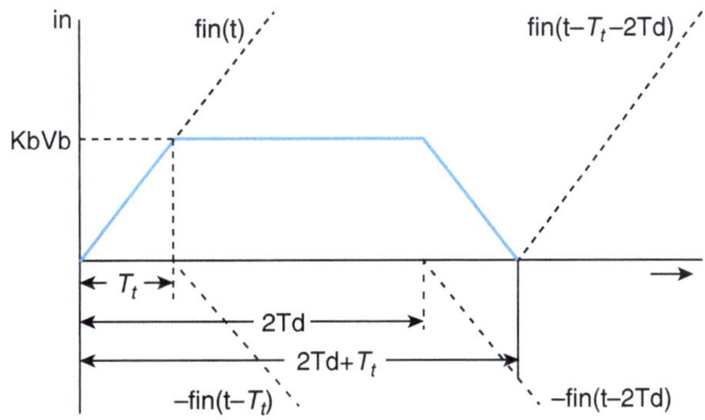

Fig. 7.65 Near-end crosstalk waveform, long line case

$$V_{\text{in}}(t) = f_{\text{in}}(t) - f_{\text{in}}(t - T_t). \tag{7.12}$$

Near-end crosstalk voltage (V_b) for a staircase waveform (Fig. 7.65) can be written as [19]:

$$V_b(t) = K_b[V_{\text{in}}(t) - V_{\text{in}}(t - 2T_d)], \tag{7.13}$$

where T_d—signal propagation time in a line of length *l*, defined by the expression:

$$T_d = tpd \times l,$$

where *tpd*—delay time depending on the relative dielectric constant (*εr*) of the dielectric in the board stack.

Substituting (7.12) into (7.13), we obtain an expression for the near-end crosstalk for a given input signal:

$$V_b(t) = K_b[f_{\text{in}}(t) - f_{\text{in}}(t - T_t) - f_{in}(t - 2T_d) + f_{\text{in}}(t - T_t - 2T_d). \tag{7.14}$$

If the length of the interconnected area is greater than the critical length (long line case), near-end crosstalk in the victim line reaches its maximum (Fig. 7.65).

In this case, maximum near-end crosstalk does not depend on the interaction length and is equal to:

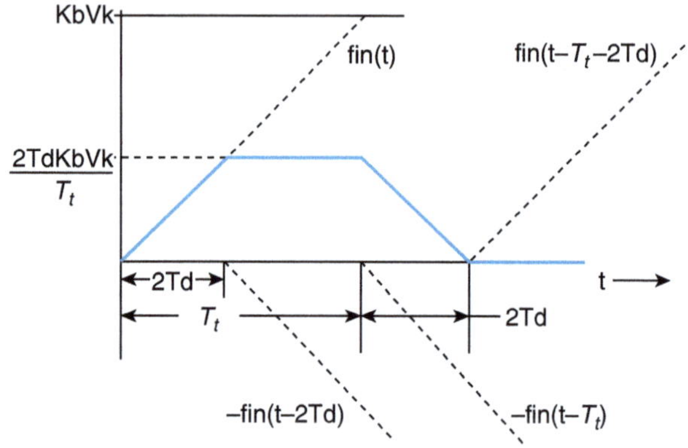

Fig. 7.66 Near-end crosstalk voltage waveform, short line case

$$V_{b\max} = K_b \times V_{\text{in}}.$$

The criterion for calculating the critical length of the line is the ratio:

$$l_{\text{cr}} = T_t/2T_d.$$

For a short line, the near-end crosstalk voltage expression remains the same (7.14), but the maximum amplitude depends on the rise time of the leading/trailing edge of the input signal and the interaction distance between the lines (Fig. 7.66):

$$V_{b\max} = V_k \times K_b \times (2T_d/T_t). \quad (7.15)$$

If the signal edge does not change, noise amplitude of the short line is directly proportional to its length.

7.8.2 Mechanism of Interference Formation in the Loop

A noise generation diagram of a three-section loop is shown in Figs. 7.67 and 7.68 [20].

When a staircase waveform appears at the input of the first section, the signal edge induces leading near-end crosstalk at the left end of the second section (Fig. 7.67). This noise precursor will appear at the output of the loop T_d (period of time) earlier than the output signal. It has a duration of $2T_d$. When a signal appears at the right end of the second section, leading N2-2 noise is induced and mixed with N1 noise and N2-1 echo noise. N2-1 noise appears at the output along with the signal and has a duration of $2T_d$. When a signal appears on the third section, N3 echo noise is induced to be mixed with N2-1 noise.

Similarly, far-end crosstalk is induced (Fig. 7.68).

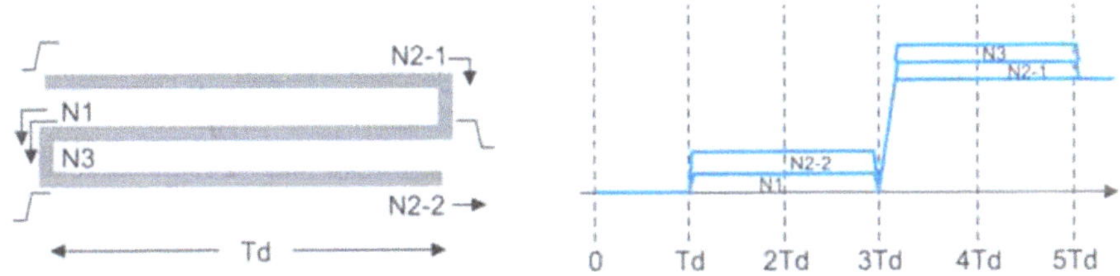

Fig. 7.67 Loop near-end crosstalk generation

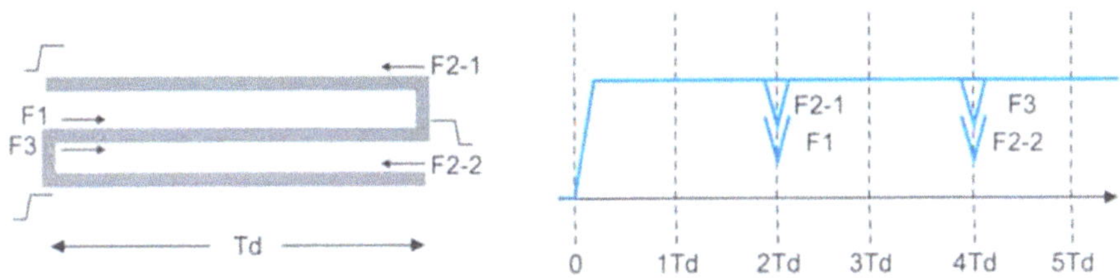

Fig. 7.68 Loop far-end crosstalk generation

Near-end crosstalk propagates along the loop to the output. Far-end crosstalk propagates to the input source.

If loop matching at the input/output is considered ideal (i.e., there are no reflections from the input source), far-end crosstalk may be ignored in loop calculations—it is absorbed by matching impedance at the input (point A in Fig. 7.61).

Thus, in a three-section loop, the output signal will contain the leading total noise precursor with an amplitude of V_b x 2 and a sum of signal and echo noise of V_b y 2.

Near-end crosstalk occurs not only from the adjacent loop sections, but also from further sections. It is a staircase waveform preceding the main signal represented by leading noise (Fig. 7.69) and noise following the signal edge [20].

The graphic form of the loop output signal (Fig. 7.61), taking into account the near-end noise precursors, is shown in Fig. 7.69. Echo noise has the form of "steps" as well, but in reverse order. It is mixed with the output signal as soon as it reaches point B (Fig. 7.61).

For a line containing $2N + 1$ ($2N$) sections, the number of steps is equal to the number of loop bends (N); duration of each step is $2T_d$ (Fig. 7.69).

When it comes to noise, the level of steps is determined as follows [19]:

$$V_b i = 2i \times K_b(p) \times V_{\text{in}}; \tag{7.16}$$

$$p = (2(N - i) + 1) \text{ for } i = 1 \ldots N, \tag{7.17}$$

where $K_b(p)$—reverse crosstalk coefficient between the loop sections located at a distance of pxs, where s—distance between adjacent sections.

Near-end crosstalk form, level, and duration affect the quality of the loop output signal.

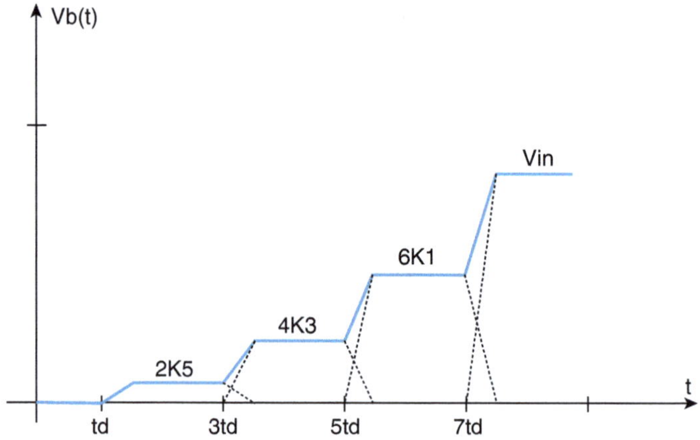

Fig. 7.69 The form of noise precursors at point B for a step input signal

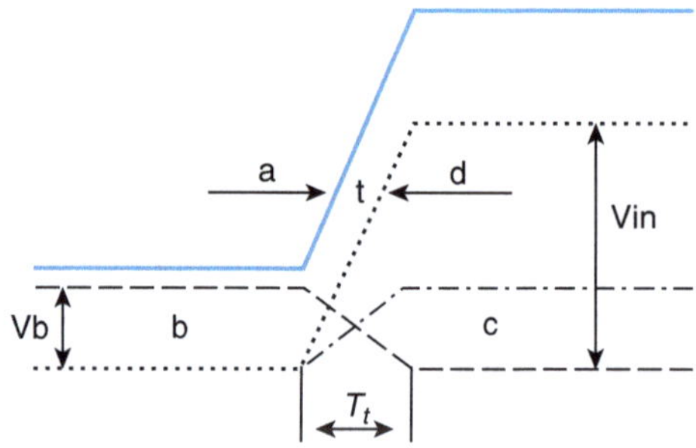

Fig. 7.70 Lead time of the output signal due to noise chart

7.8.3 *Equations of the Loop Output Signal*

In case a signal propagates along the loop, it appears at the output before the expected arrival time [20]. This is due to that fact that the noise and the input signal are summed up.

In Fig. 7.70, the sum of the edges of noise precursor (b), echo noise (c), and input signal (d) results in an output signal (a), which appears at the output ahead of time (t).

According to the study of delay lines, if the loop sections are electric long lines, the output signal edge will be severely damaged [21]. A delay line with small sections "improves" the output signal edge—it becomes steeper, compensating for the "collapse" of the signal edge due to resistive and dielectric losses in the communication line. For these reasons, the section should be less than 1cg long (critical) [22].

Let us consider in more detail how "short" noise is combined with the main signal at the output of the loop. Figure 7.71 shows a stack of strip and microstrip printed interconnections. Here, h—insulating material thickness, w—width of the printed conductor, s—distance between conductors.

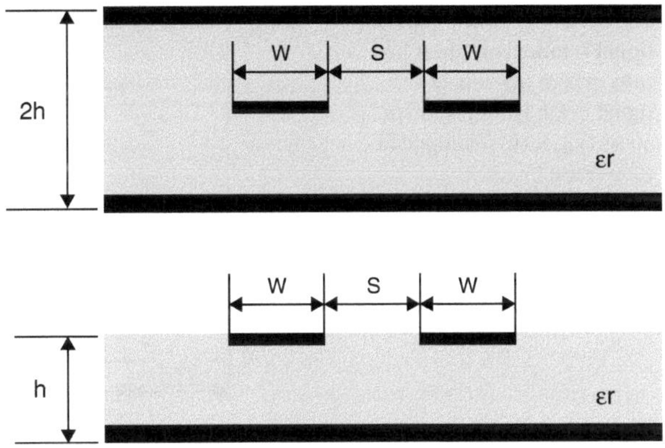

Fig. 7.71 Stack of strip and microstrip lines

After selecting the PCB stack, only the distance between the sections can be varied. The real ratio of the amplitudes of the noise "steps" differs from the one shown in Fig. 7.69. The maximum energy (95%) is concentrated in the "steps" with coefficient K1; inclusion of the "steps" with coefficient K3 makes it possible to take into account virtually 100% of the noise in zigzag delay lines [20].

This noise energy distribution has two main reasons:

- noise between sections is inversely proportional to the square of the distance between them [23];
- thc summed noise in each "step" decreases along with its coupling coefficient (i.e., a smaller "*r*" value).

For this reason, noise from non-adjacent sections should not be taken into account in the obtained loop output signal equation. Such noise may be relevant in the event of closely coupled interface of segments, for instance, if the distance between sections is equal to or less than the width of the communication line.

Figure 7.72 shows the near-end crosstalk steps overlapping with the input signal, when sections of the loop are short lines:

$$2\,t_d < \mathrm{T_t}.$$

Point 0 on the time axis corresponds to the moment of occurrence of the total adjacent noise precursor—V_{Ipr} (Fig. 7.72a, b) at the output of the loop. The main signal, V_{in} (Fig. 7.72a), and the integrated echo noise, V_{po} (Fig. 7.72c), appear in $2T_d$. Duration of each noise is equal to TV, since the segments are short lines.

The resulting signal will be equal to:

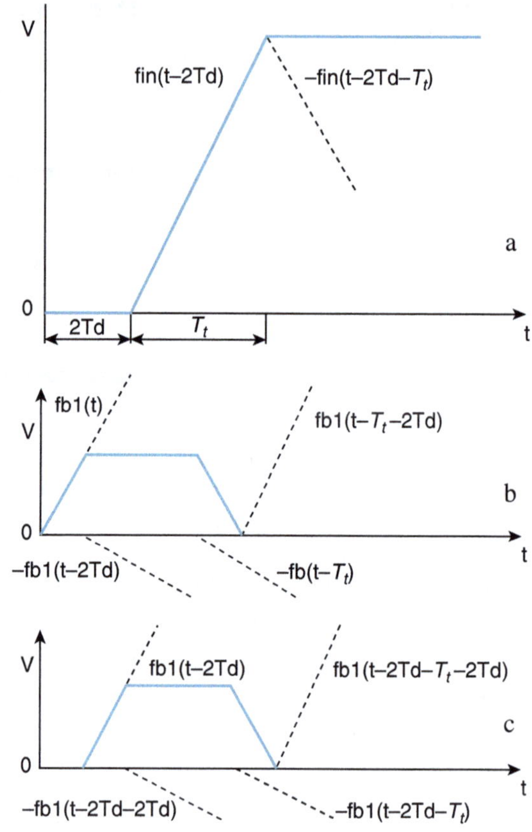

Fig. 7.72 Noise and input signal summed up, short line case, where (**a**)—main signal (V_{in}); (**b**)—precursor noise (V_{Ipr}); (**c**)—integrated echo noise (V_{po})

$$V_{\text{out}} = \text{V}1_{\text{pr}} + V1_{\text{po}} + V_{\text{in}},$$

where

$$\begin{aligned} V_{\text{in}} &= f_{\text{in}}(t - 2T_d) - f_{\text{in}}(t - 2T_d - T_t); \\ V1_{pr} &= fb1(t) - fb1(t - 2T_d) \\ &\quad - fb1(t - T_t) + fb1(t - 2T_d - T_t); \\ V1_{\text{po}} &= fb1(t - 2T_{\text{d}}) - fb1(t - 2T_{\text{d}} - 2T_{\text{d}}) \\ &\quad - fb1(t - 2T_d - T_t) + fb1(t - 2T_d - T_t - 2T_d). \end{aligned} \tag{7.18}$$

These equations can be supplemented with a noise "step" for K_{b3}, if the reference point is shifted to the left along the time axis by $2T_d$ more:

$$V_{\text{out}} = V3_{\text{pr}} + V3_{\text{po}} + V1_{\text{pr}} + V1_{\text{po}} + V_{\text{in}}. \tag{7.19}$$

Keeping track of the remaining steps does not affect the result due to very small coupling factors.

When the input signal waveform is a voltage step or a trapezium (leading edge), then fin, fb\ can be replaced by the corresponding slope coefficient of the line: $f_{\text{in}} = K_{\text{in}} = V\text{in}_{\text{maxl}}T_t$—for the input signal; fb\ = ΛT.

Let us determine total noise coefficient of the near (adjacent) "step"—$K1$.

If a loop contains N bends, then the near step is step no. N. For this step, $p = 1$ (from Eq. 7.17), reverse crosstalk coefficient is K_{b1}. Total noise coefficient forming the near step (from Eq. 7.16) is 2N × K_{bl}. For each term of Eq. (7.14), near-end crosstalk coefficient is equal to $K_b \times f_{\text{in}}$. After the corresponding substitutions, we obtain an expression for $K1$:

$$K1 = 2N \times K_{b1} \times K_{\text{in}}.$$

For step no. $N - 1$, we obtain coefficient $K3$, which is equal to:

$$K3 = 2(N - 1) \times K_{b3} \times K_{\text{in}}.$$

Substituting the K_{in}, $K1$ coefficients in the expression (7.18), we obtain the following equations for a step-type input signal (leading edge of the trapezium):

$$V_{\text{in}} = K_{\text{in}}(t - 2T_d) - K_{\text{in}}(t - 2Td - T_t); \tag{7.20}$$

$$\begin{aligned} V1_{\text{pr}} &= K1(t) - K1(t - 2T_d) \\ &-K1(t - T_t) + K1(t - 2T_d - T_t); \end{aligned} \tag{7.21}$$

$$\begin{aligned} V1_{\text{po}} &= K1(t - 2T_d) - K1(t - 2 \times 2T_d) \\ &-K1(t - 2T_d - T_t) + K1(t - 2 \times 2T_d - T_t). \end{aligned} \tag{7.22}$$

The terms of the equations with the $(t–x)$ factor are zero, if $t < x$.

Similarly, we can obtain equations for the expression (7.19) with the K1 and K3 coefficients.

7.8.4 *Method for Determining the Distance S and Coefficient of Communication Between Neighboring Sections*

To design delay lines, when the board stack is defined and only the distance between the line sections can be changed, to select s, either template solutions (for instance, s = 4x or s = 5 xh) or background experience are used, including original projects or experience of other companies.

Consider the possibility of calculating the distance between sections for a given stack, input signal edge and permissible distortion of the output signal. To calculate s, one should know the required coupling coefficient—K_{b1}.

As shown above, the loop sections should be designed as short lines. In this case, the following dependencies can be noted.

Let the required delay A f correspond to a straight line of length L. We will use this length as a reference value. Dividing the reference length (L) by the critical section length (1 cg), we obtain the exact number of the loop sections corresponding to the critical section length—N_{scr}.

$$N_{scr} = L/l_{cr}.$$

The critical number of bends is equal to the incomplete quotient:

$$N_{cr} = N_{scr}/2.$$

We do not take into account the orthogonal elements of the loop, considering that they are much shorter than the section. Maximum noise with coupling coefficient K_{b1} (maximum "step") occurs from adjacent sections, the other noise (K_{b3}, K_{b5}, etc.) is much smaller and further away from the signal edge. Total noise from the sound with coefficient K_{b1} determines an excessive amplitude of the output signal in relation to the input signal—V_b (Fig. 7.70).

For N_{cr} bends of the loop, the level of total adjacent noise from sound will be:

$$V_b = V_{b1} = V_{in} \times (2 \times N_{cr}) \times K_{b1}.$$

If the input signal edge remains unchanged, then the noise level between sections (4) decreases proportionally to the section length, but the number of the loop sections corresponding to the "reference" length L increases by the same amount. Reducing the section length by x times, the total noise level will be:

$$V_{b1} = V_{in} \times (2 \times N_{cr} \times x) \times (K_{b1}/x).$$

Thus, excessive amplitude coefficient of the output signal (K_b) due to noise in the loop will be equal to:

$$K_b = 2 \times N_{cr} \times K_{b1}. \tag{7.23}$$

In order to simplify further analysis, we will consider lines without losses.

Studies conducted in [24] have demonstrated that there is a difference of about 2% as a result of modeling delay lines with losses as compared to delay lines without losses.

Integrated echo noise from the near end, coming along with the signal edge, causes a signal amplitude spike (V_b), which may affect the operation of the load logic circuits. The strength of the spike should be limited by noise immunity of the load circuits, for instance, 15% of the input signal amplitude.

This tolerance determines the K_b value for the designed loop:

$$K_b = V_b/V_{\text{in}}.$$

Knowing K_b and the number of critical sections, we calculate the coupling coefficient value between adjacent sections for a given delay time:

$$K_{b1} = K_b/N_{\text{cr}}. \tag{7.24}$$

The corresponding distance 5 between the loop sections can be calculated from K_{b1} and a given stack of the board.

7.8.5 Analytical Expressions for Determination of the Distance Between the Loop Sections

The following expressions, specified in standard 1PC 2251, are required for the developer to calculate distance 5 between the loop sections [22]. Study [25] analyzes formulas for calculating coupling coefficients based on this standard and the results of noise modeling in large commercial simulators. The formulas are 10% accurate. The same accuracy was obtained by the authors. These formulas were compared with the results of HyperLynx LineSim (Mentor Graphics).

For a strip line, the distance between the sections is calculated as follows:

$$s = -([\ln(K_C/0.55) \times h + B1 \times w]/A1).$$

For a microstrip line, the value s is found by numerical minimization of the function $F(s)$:

$$K_{b1} = 1/4(K_L + K_C);$$
$$F(s) = K_L + K_C - 4K_{b1};$$

где

$$K_L = 0.55e[-(A2(s/h) + B2(w/h))];$$
$$K_C = 0.55e[-(A1(s/h) + B1(w/h))].$$

Distances 5 can be calculated using one of the mathematical packages (R, Matlab, Mathcad, etc.) or by writing the corresponding simple program.

Figure 7.73 shows the equation solution in the Mathcad package for microstrip lines with a stack:

$$E_r = 3.9;\ h = 10\,\text{mil};$$
$$w = 8\,\text{mil};\ K_{b1} = (K_0) = 0.05.$$

The calculated distance between sections: $s = 14.77$ mil.

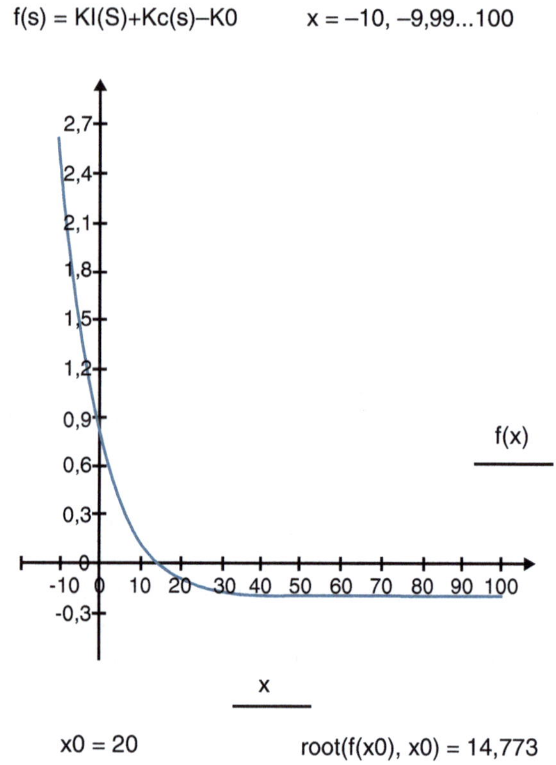

Fig. 7.73 Example of calculating s for a microstrip line in Mathcad

The calculated distance between sections predetermines the minimum area for the designed loop.

7.8.6 *Practical Recommendations for Selection of the Section Length*

In the literature, different techniques are recommended for selecting the loop section length. In most works, it is recommended to choose the section length on the assumption that:

$$I_s < 15\text{h and } I_s < 1/3 \text{ GHz [26].}$$

However, for a loop with short sections, maximum length should be selected based on the actual distortion of the input signal during its transmission in the loop. Therefore, the section length shall satisfy the following basic conditions:

- $I_s < I_{cr}$ (this contributes to the improvement of the signal edge);
- the section length shall correspond to the input signal bandwidth.

Since the transmission line reduces the signal bandwidth by delaying the signal edge in order to reduce the losses, it is necessary to check the loop bend for the so-called stop band criterion, and the stop band frequency shall be not less than the maximum frequency in the signal spectrum [27].

Since the maximum frequency in general is determined by the bandwidth of the input signal:

$$BW_{\text{in}}(\text{GHz}) = 0.35/\text{T}_{\text{t}}(\text{ns}).$$

then the stop band frequency of the analyzed loop depends on the bend length and is equal to [23]:

$$BW_s = 3 \times 10^8/\left((2ls + s)/\sqrt{\varepsilon r}\right),$$

where Is and s are measured in meters, and frequency—in hertz.

Developers dealing with these issues in practice may refer to work [27], where it is recommended to use stop band frequency allowing for changes, for instance:

$$BW_s = 2 \times BW_{\text{in}}.$$

Ensuring the ratios will eventually simplify the selection of the maximum length of the loop section.

The minimum length of the loop section should be set, too. If a segment is too short, it increases the total noise, making currents flow outside the loop and significantly accelerating the signal appearance.

Data of the program calculating the loop output signal based on the nearest total noise precursor, the input signal, and the integrated echo noise overlapped are presented in Fig. 7.74.

If a section is too short, at some point the sum of signal and noise will be greater than the planned $V_{\text{in}} + V_b$ (Fig. 7.74b, d).

When the section length is equal to or less than half of l_{cr}, the sum of precursor noise and echo noise doubles in amplitude and remains at this level with a further decrease in the section length. As a result, the sum of signal edge and doubled noise level at some point will exceed $V_{\text{in}} + V_b$. Let us determine the minimum allowable k_{min} section length providing for noise to exceed the planned level.

For the input signal described by the expression (7.12), the sum of "precursors" and "echoes" is a kind of trapezium, whose upper base crosses the inclined line constituting the signal edge (Fig. 7.74c, d). The maximum sum of the signal edge and the upper base of the trapezium in this case will be at the start point of the noise precursor decrease (right side of the trapezium). This point corresponds to the moment of time $t = T_t - 2T_d$. For this moment, let us compose an equation, where the sum of input signal and doubled noise exceeds the level of $V_{\text{in}} + V_b$:

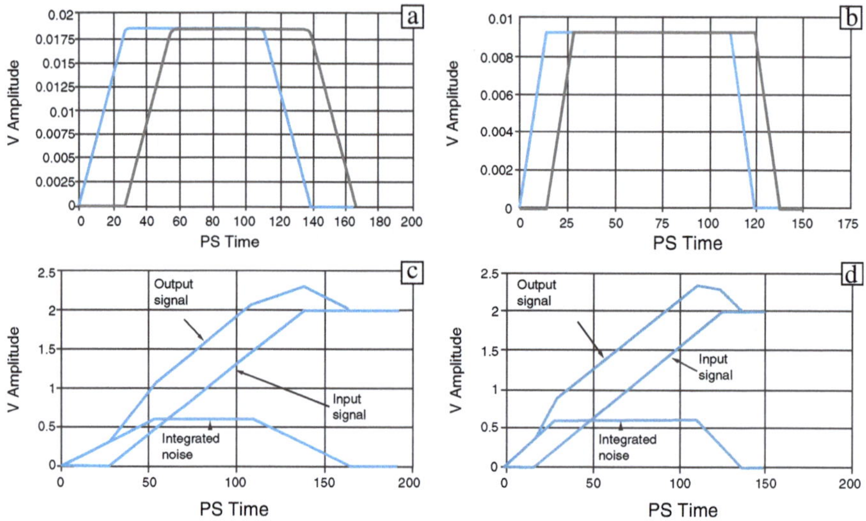

Fig. 7.74 Output signal waveform taking into account the noise from adjacent sections: (**a**, **b**); near-end crosstalk and far-end crosstalk of the first bend; (**c**, **d**) signal, sum of signal and noise, total noise at the output of the loop

$$(V_{\text{in}}/T_t) \times t + 2V_b = V_{\text{in}} + V_b;$$
$$t = ((V_{\text{in}} - V_b)/V_{\text{in}}) \times T_t,$$
$$t = T_t - T_t \times K_b \text{ or } T_r - t = T_r \times K_b$$
$$\text{or } 2T_d \min = K_b \times T_t = 2(tpd \times ls \min).$$

Knowing 2*Td*mia, we calculate the minimum section length (Is min):

$$2(tpd \times ls \min) = K_b \times T_t,$$
$$ls \min = (K_b \times T_t)/2tpd.$$

When the section is shorter than *fc*min, the total noise increases significantly, and the currents may flow outside the loop segments [28]. Thus, the length of the loop section (*ls*) shall be within the range:

$$I_{\text{cr}} > ls > (K_b \times T_t)/2tpd. \tag{7.25}$$

Individual section length is selected based on the size of the platform designed for the loop.

7.8.7 Method for Calculation of the Magnitude of the Lead Time of the Output Signal Due to Interference

Typically, such a signal lead time is determined at level 0.5 of the input signal.

Analytically, the expression for calculating the lead time of the output signal can be obtained from Eqs. (7.20)–(7.22) and Fig. 7.72. Thus, level 0.5 of the input signal for reference point 0 in Fig. 7.72 generally corresponds to the time:

$$t05 = 2T_d + 0.5T_t. \tag{7.26}$$

Then total output signal will have an amplitude equal to 0.5 of the input signal after some time t. Let us determine this time from the above equation for the output signal.

Since the terms of equations with the (t–x) factor are usually zero and the time $t < x$, only the terms that are in the time range (7.26) can be included in the equation to calculate the output signal.

Then it is easy to obtain the following expression, which is convenient for practical calculations:

$$\begin{aligned} V_{\text{out}} = {} & K1(t) + K_{\text{in}}(t - 2T_d) - K1(\text{t} - 2T_d) + K1(\text{t} - 2T_d) - K1(\text{t} - T_t) \\ & - K1(\text{t} - 2 \times 2T_d) \end{aligned}$$

or

$$\begin{aligned} 0.5 = {} & K1(t) + K_{\text{in}}(t - 2T_d) - K1(t - T_t) \\ & -K1(t - 2 \times 2T_d). \end{aligned} \tag{7.27}$$

Coefficient K1 is calculated from (7.23) and (7.24):

$$\begin{aligned} K1 &= 2 \times N \times K_{\text{in}} \times K_{b1} \\ &= 2 \times N_{\text{cr}} \times K_{\text{in}} \times (K_b/N_{\text{cr}}). \end{aligned}$$

Since the term of sum Kl (t – T_r) can be left in Eq. (7.27), for the standard case T_r < (2TU + 0.5 Gg), the second term will be:

$$K1(t - 2 \times 2T_d),$$

if

$$2 \times 2T_d < (2T_d + 0.5T_t)$$

Then from the resulting equation one can easily determine the time t, and then the lead time Δfc:

$$\Delta t = (2T_d + 0.5T_t) - t.$$

To determine the lead time at any level, it is better to construct a summary curve, for instance, using a simple program (Fig. 7.74).

7.8.8 Method for Final Calculation of the Delay Line

After determining the distance between sections (*s*), the section length (*Is*), and the number of sections (*Ns*), total length of the loop (*Ls*) should be recalculated. Thus, for Ns sections of the board, it will be equal to:

$$Ls = Ns \times (ls + s).$$

Then the delay factor ΔtLs of the reference line of length *Ls* will be determined from a simple expression:

$$\Delta tLs = Ls \times tpd.$$

And the delay for one section of the loop will be equal to:

$$\Delta ts = \Delta tLs/Ns.$$

Generally, an important delta delay of the reference line *I* and *Ls* of the loop Δt will be:

$$\Delta t1 = \Delta t - \Delta tLs.$$

Obviously, to compensate for the AL, a certain number of sections in the loop shall be added/removed to obtain n_{add}:

$$n_{\text{add}} = \Delta t1/\Delta ts.$$

It should be mentioned that in practice a change in the number of these sections shall correspond to the permissible time error for all signals.

In fact, in the latest TCS developments, a loop usually does not go along the entire delay line. Generally, only a few small loops are designed along the intended delay line. This design is widely used due to lack of place, as well as to reduce the radiation effect of the loop on neighboring transmission lines.

Surely, the best and obvious solution is to break the loop with N bends into small loops with 3–5 bends and bending angles under 45° [23].

In study [29], the results of such engineering design of board loops in ultrafast electronic devices are compared with the modeling results obtained by other authors.

Here, study [30] was mostly used to compare the results. It represents a thorough investigation of loop-type delay lines. These studies were conducted using commercial 3D simulators of electromagnetic fields (3D FEM EM simulation tool). In this work, the dependencies of the loop lead time on such parameters as the input signal, the distance between sections, the length of sections, the number of bends were also described. In fact, this indicator was chosen by the authors of this study, since it was defined by all the parameters of the delay line and the input signal. The results are presented in Table 7.9.

In the table, the results of the study [31] are specified in columns "Study ps," the results of calculations by the procedure described in the article—in columns "Res. ps" [20].

More accurate results of the loop layout calculation by this procedure can be found on the website (www.supercompri.ru).

Summarizing the above, we may conclude that the analysis of delay zigzags confirmed the possibility of a methodological approach to organizing engineering design aimed at determining the geometric parameters of any such delay line based on given parameters, namely board stack, input signal edge, and required delay.

Because of high-density installation and limited board area, the best loop layout of a standard configuration (meander, accordion) is usually achieved, when the section is shorter than the critical length for a given input signal edge.

At the same time, according to the permissible distortion of the signal, such important parameters as coupling coefficient and distance between the sections can be calculated, which usually makes it possible to determine the minimum area of the designed loop. It is recommended to use the above calculations of the range of variations of the loop section length, as well as the above formulas for calculating the numerical value of the loop lead time.

7.9 Practical Recommendations for Eliminating Undesirable Effects Caused by Hanging Vias

As developers of high-speed electronic devices know, the so-called via stubs may significantly distort the signals passing through the interconnects and reduce the bandwidth of all types of transmitted signals. All via stubs can be divided into two types (see Fig. 7.75) [32], which are designated as A and B. In via stubs of the type A, a signal comes from the top layer of the board to the inner one, and a stub is a segment between the junction point of the inner layer with the through-hole and the PCB lower level. Alternatively (type B via stub), a part of the through-hole serves to transmit a signal between the two inner layers of the board. The first stub starts at the junction point of the first inner layer and ends at the top layer of the board; the second stub starts at the junction point of the second inner layer and ends at the bottom level of the board.

Table 7.9 Results of experimental studies of the loop [30]

Initial stack data: strip line, $er = 4$, $w = 4$, $h = 1/2(h1 + h2) = 1/2(8.6 + 3) = 5.8$
Loop configuration: number of segments Ns = 16, number of angles = 32[a]
Input signal is a step with $Tt = 100$ ps

Segment length, Is/Loop length, L	$s = w = 4$ $K_b = 0.101$			$s = 2w = 8$ $K_b = 0.0435$			$s = 3w = 12$ $K_b = 0.018$			$s = 4w$ $K_b = 0.008$		
Parameters	No.	Study ps	Res. ps	No.	Study ps	Res. ps	No.	Study ps	Res. ps	No.	Study ps	Res. ps
Is = 1.27 mm $L = 20.32$ mm	1	40	41	4	23	23.3	7	12	9.6	10	8	4.5
1 = 2.54 mm $L = 40.54$ mm	2[b]	62	51 + 14 = 51.6	5	33	34.3	8	17	18.6	11	9	8.58
1 = 5.08 mm $L = 81.08$ mm	3[b]	105	72 + 26 = 98	6	44	48	9	22	23	12	12	13

[a]Each segment of the loop has 2 corners
[b]In these cases, K_{b3} noise was also taken into account due to the strong interconnection

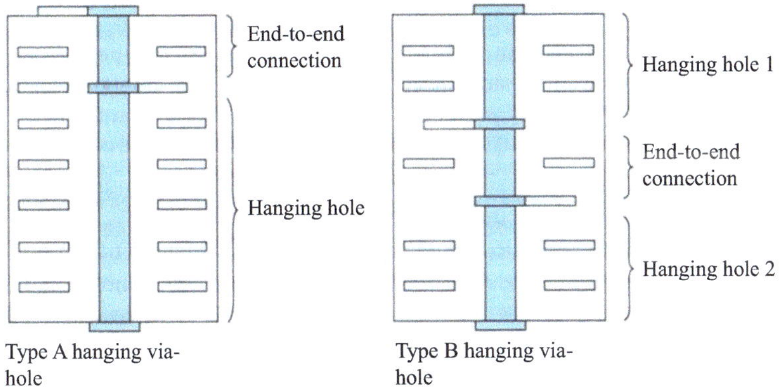

Fig. 7.75 Classification of via stubs [32]

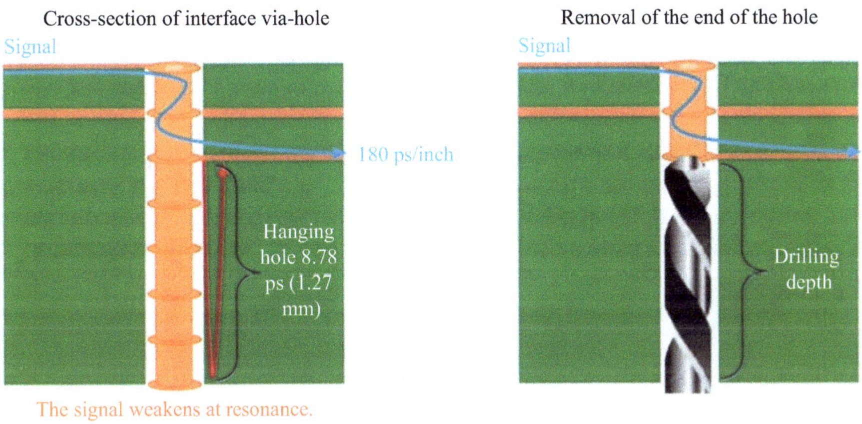

Fig. 7.76 Explanation of the method for removing via stubs: initial situation (**a**) and removal process (**b**)

Resonant frequency of these via stubs, essentially working as antennas built into the transmission line, is determined by a quarter of the wavelength of the transmitted signal. At this frequency, the transmitted signal is attenuated by 3 dB. When low-frequency signals are transmitted, via stubs do not cause serious problems, since the frequency of the signals is lower than the resonant frequency of these unused via stubs.

However, at higher frequencies (above 1 GHz), this problem is becoming very noticeable, since the frequency of the transmitted signals in high-speed electronic devices is approaching the resonant frequency of the via stubs. Odd harmonics of the fundamental frequency of the signal can be noticeably attenuated, too.

As a rule, in practice this problem has a technological solution. Back-drilling technique is used to remove a part (or the entire plated end hole) (see Fig. 7.76) [32].

If the leading edge of the transmitted pulse has a duration comparable to the time it takes for the signal to pass through a thru-via, then it works as an additional shunt capacitance. In general, the entire via contributes to the path capacitance, and the inductance is determined only by the part through which the current flows. If a via stub is long enough, the capacitance becomes significant; therefore, it cannot be ignored. Thus, to avoid undesirable effects, it is strongly recommended that the signal would pass all the way through the plated thru-holes.

If the length of a via stub is equal to a quarter of the wavelength of the signal, this signal passes through the conductor to the end of the via stub. Then it is reflected from the extreme point of the open circuit and returns to the conductor, having completed a shared path of half the wavelength. As a result of the phase shift of the signal by 180°, resonance occurs at the end of the via stub.

This means that the amplitude of the reflected signal with a shifted phase reaches its maximum, when the signal itself has a minimum amplitude, and vice versa.

Obviously, in the well-known Nyquist bandwidth of discrete signal, defined as half of the sampling rate, the spectral density of the signal must be much higher than outside it. Additionally, the power spectral density of the signal's harmonic components up to the fifth harmonic may be high at frequencies above the Nyquist frequency.

Resonant frequency of the unused portion of the via is inversely proportional to the square root of the permittivity of the material on the walls of this via, while the signal wavelength in it is four times the length of the via according to the following expressions [32]:

$$t_{\text{via}} = l\frac{\sqrt{\varepsilon_r}}{c},\ f_{\text{res}} = \frac{1}{4lt_{\text{via}}}, \tag{7.28}$$

where t_{VIA}—time it takes for the signal to travel through the via stub (about 180 ps for fiberglass); 1—via stub length; ε_r—material permittivity; c—speed of light; f_{RES}—resonant frequency of the via stub.

If resonant frequency is approximately equal to the Nyquist frequency, one or several frequency components of the signal transmitted through this via stub are greatly weakened, leading to impedance drop. This phenomenon is especially noticeable at frequencies above 1 GHz, while at frequencies of about 4–5 GHz, this problem cannot be ignored, since the bandwidth is reduced due to signal attenuation.

Specific comparative results are presented in Fig. 7.77: impedance of a standard plated through-hole and a hole without end portion measured with a reflectometer [32]. In case a high-speed backboard is used, the observed effects become very noticeable at a Nyquist frequency of about 5 GHz and a substrate thickness of 6.35 mm.

As already mentioned above, through-holes create inductive and capacitive inhomogeneities, which degrade the signal performance as it passes through a thru-via. One way to neutralize them is to design a via divided into small capacitive and inductive elements, interacting with planes and with each other.

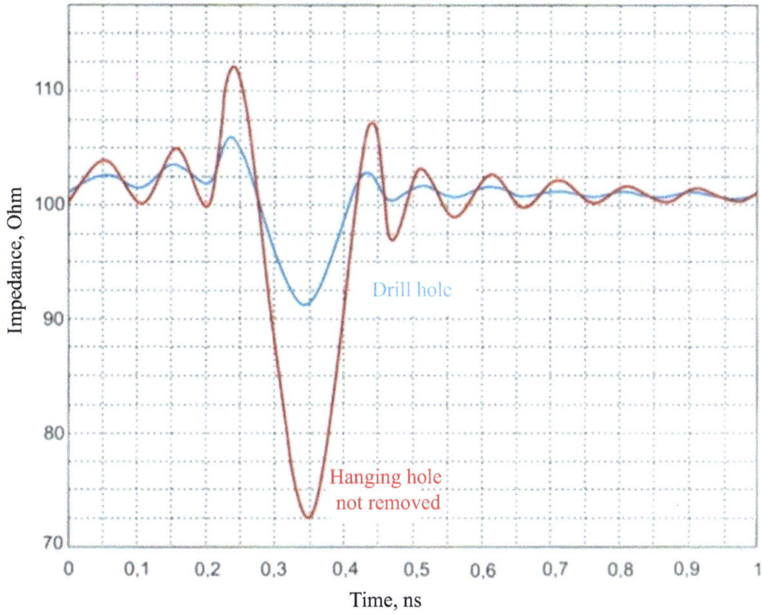

Fig. 7.77 Differential impedance (100 Ohm) of a standard thru-via (curve 1) vs. impedance of a back drilled via (curve 2) [32]

Using this method, it is difficult to achieve an accurate result, since the fringing fields are large. In addition, it is extremely difficult to coordinate discrete elements taking into account these overlapping fringing fields. Figure 7.78 demonstrates simplified π-models with LC components (non-functional pads are absent) with lumped parameters. The results of analysis of the latter show the influence of the parasitic capacitance and inductance of the via.

Although this model (measurement results (b)) [32] is applicable only in specific cases, namely if the time delay while the signal is passing through a thru-via is less than 1/10 of the leading edge of the pulse, it still gives a better understanding of the contribution of parasitic components. The attenuated signal with a phase shift (curve 1) shows the degrading quality of the signal passing through the via stub, while at the drilled hole (curve 2) the signal is not distorted and the frequency band is wider.

There are several other sources of heterogeneity in the channel to be considered by developers of high-speed devices. Thru-vias intended for the signal to pass through are among them. As a result of jitter and a decrease in the eye diagram opening, the receiver may incorrectly interpret some part of the received data. The length of the through-hole is the main factor affecting inductance of the hole, depending on the circuit complexity, the number of layers and, consequently, the total thickness of the printed circuit board. It is clear that the length of the plated through-hole is almost the same as the total thickness of the board. As a rule, in high-speed data transmission systems, board thickness ranges from 1.0 to 1.8 mm. In more complex cases, when special backboards designed for operation in harsh

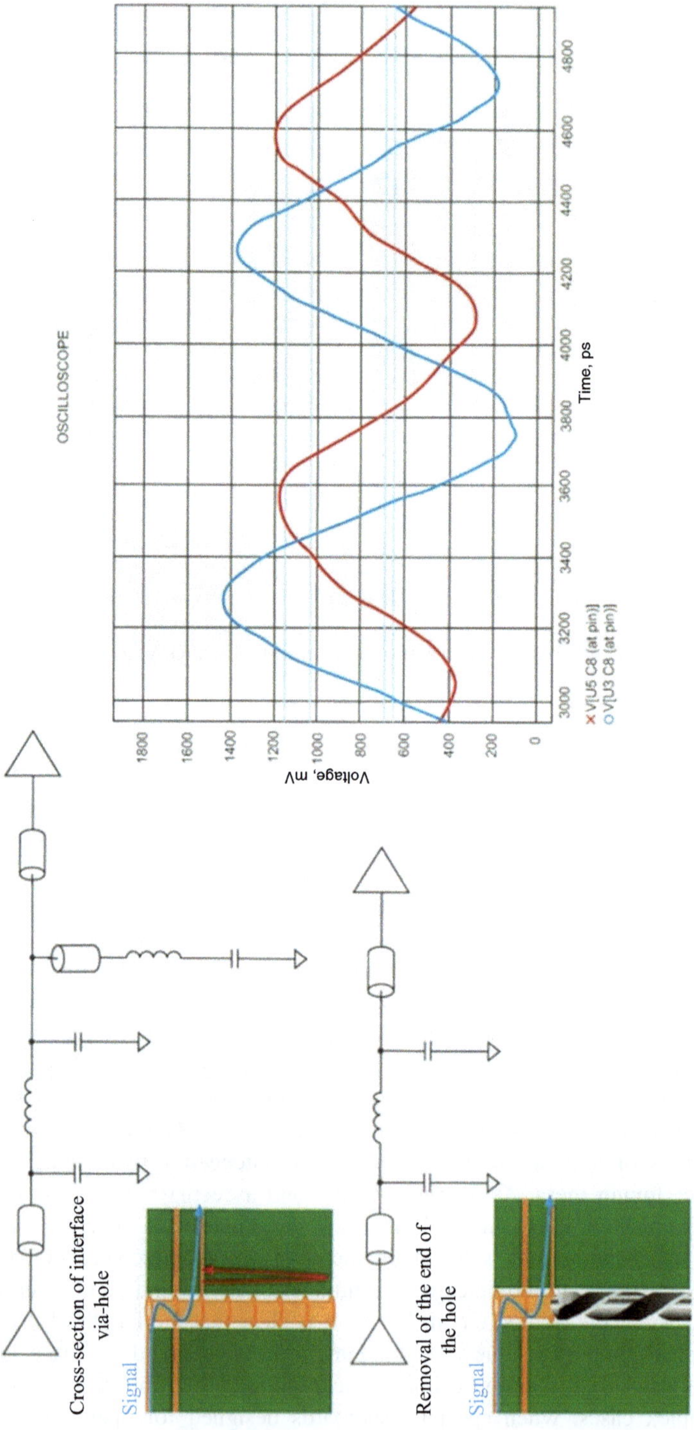

Fig. 7.78 Simplified π-models with LC components with lumped parameters: with a via stub (**a**) without a via stub (**b**) [32]

environments are used, printed circuit boards are thicker due to a larger number of layers. To weaken the effect of via stubs on signals, it is necessary to:

- minimize the inductance of vias by eliminating via stubs or reducing their number; shorten thru-vias;
- minimize the capacitance of vias by reducing the size of the pads; remove non-functional pads; increase the size of antipads—unoccupied space between the plated component holes and the adjacent conducting plane.

Among the main design and technological methods to solve these problems are the following [32]:

- removal of the unused portion of the via;
- use of blind vias;
- removal of non-functional pads;
- increase of the gap between the plated hole and the conductive plane;
- via stub impedance matching;
- use of plating with lower permittivity;
- via plating using a material with high resistive losses.

Let us briefly consider each of the above-mentioned methods.

The unused portion of the plated via is removed after the board has been manufactured. The diameter of the back drilled via shall be greater than that of the original thru-via. In fact, this method is often used instead of manufacturing blind vias in order to remove unused portions of vias in very thick backboards for high-speed systems. Since in modern printed circuit boards the standard diameter of back drilled vias in the signal layer is 0.20 mm, there is always a small unused part of the thru-via. The substrates of high-speed SerDes interconnects are generally thicker, due to the system architecture and the requirements for connecting boards using tightly fixed connectors.

Back drilling of via stubs in boards with a thick substrate shortens these via stubs at a signal transmission speed above 3 Gb/s (1.5 GHz). However, at a speed above 10 Gb/s (5 GHz), drilling is not enough to reduce jitter and bit error rate (BER). Figure 7.79 [32] shows how sufficiently long unused vias affect the voltage of a high-speed differential pair. On the left, there is a diagram of this pair, obtained by simulation using a pseudo-random bit stream (PRBS) in transmission lines with losses. Note that the diagram has a relatively high eye opening ratio. On the right, a simulated case is presented: parasitic elements affect the signal passing through the via; in addition, the signal is affected by the resonance state of the via. High-frequency harmonics decrease, leading edge pulse time decreases, signal is distorted, bandwidth is narrowed, eye is closed.

Thru-vias can create inductive and capacitive inhomogeneities. These parasitic elements degrade the signal quality as it passes through the via. To design high-frequency backboards with thick substrates, these problems shall be addressed first.

Naturally, special high-precision equipment is required for back drilling of via stubs. This process is costly and time-consuming due to regular breakdowns of drills and yield reduction, but usually all costs are offset by increased reliability of high-speed devices.

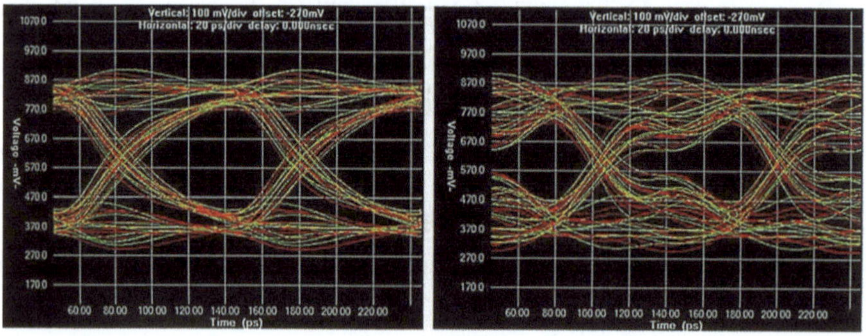

Fig. 7.79 Simulation of a type of high-speed differential signal transmitted along a resistive line without resonance (on the left) and with resonance (on the right). The charts were obtained as a result of simulation in HyperLynx [32]

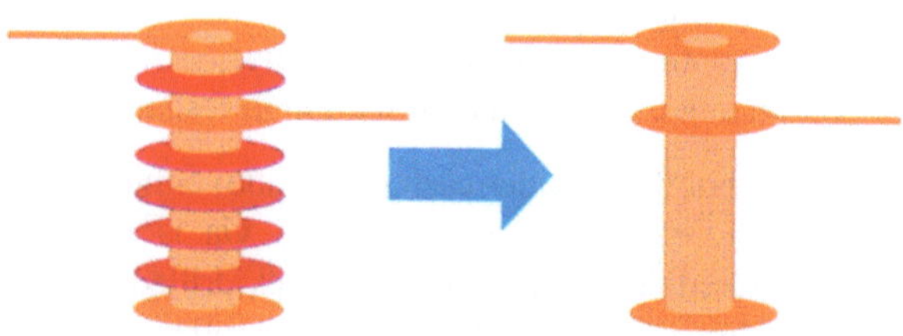

Fig. 7.80 Removal of non-functional pads

Blind vias connect outer microstrip layers with one of the inner strip lines. A central ground plane may pass between the signal layers, providing a return circuit with very low inductance. The diameter of these laser-drilled vias is usually 0.08–0.10 mm. Blind vias are characterized by lumped capacitance and very low inductance. Since blind vias have a very small diameter, their capacitance is lower than that of standard plated thru-vias.

On the other hand, blind vias are only used to establish connection between the inner layers of the strip line. The unused portions of these vias are very short, and the capacitance is higher than that of standard plated thru-vias with a typical diameter of 0.2 mm. The unused portions of blind vias may be too short to drill them; moreover, these vias may not be suitable for high-frequency applications.

Non-functional pads are located on the inner layers of the board, which are not connected to the signal layer/plane at the appropriate level (see Fig. 7.80) [32]. The issue of the impact of these pads on the reliable operation of printed circuit boards, especially on the fatigue of plated through-holes with a high height-to-diameter ratio, remains unresolved. Generally, these pads are removed to simplify the manufacturing technology and to increase the yield by controlling thermal capacity, which affects the board's expansion in the vertical direction. Additionally, one should exclude the possibility of short circuits on the board as a consequence of filling the gap between conductors with pre-preg in the absence of pads.

To remove antipads, one should remember that, according to IPC standards, an antipad should be 0.5 mm larger than the diameter of the thru-via pad. However, this requirement is rarely fulfilled in the event of very dense arrangement of BGA components on a multilayer printed circuit board. For a thru-via with a diameter of 0.20 mm, a standard pad diameter is 0.40 mm, while a standard antipad diameter is 0.66 mm. Thus, the size of the gap around the contact pad is 0.23 mm, which corresponds to a relatively large via capacitance. Most of it is concentrated near the upper plane. With an increase in the size of antipads via capacitance decreases and the ground plane resembles Swiss cheese. As a result, DC voltage drop increases, while the magnitude of the instantaneous current flowing in simultaneously switching devices is reduced, which may lead to the power supply integrity degradation.

There is an alternative solution, but it does not seem practical. The gap could be increased at the expense of the conducting plane next to via stubs in order to reduce their capacitance. However, it would have to be performed manually using the PCB design tools. As a result, wide holes would appear near the end parts of these vias, which would partly eliminate the effect of using a single continuous ground plane. Thus, since increasing the size of antipads is not a viable option, elongated antipads could be used as an intermediary solution. In this case, parasitic capacitance decreases markedly.

At the same time, reverse current starts to flow directly between thru-vias, reducing the circuit inductance and preserving the ground plane continuity. Most of the PCB design tools provide for an elongated shape of antipads on the inner layers.

In case of differential pairs, elongated antipads shall be placed around both of them (see Fig. 7.81) [32]. Grounded vias shall be placed on opposite sides of elongated antipads to reduce the inductance of the reverse path. Due to the elongated shape of these pads with terminals, the continuity of the reverse current path along the ground plane is maintained.

Note that via stubs work as mismatched transmission lines. If a matching element is placed at the lower end of this via, the signal reflection will be minimal. As matching elements, one or several resistors, capacitors, and inductors are used. They

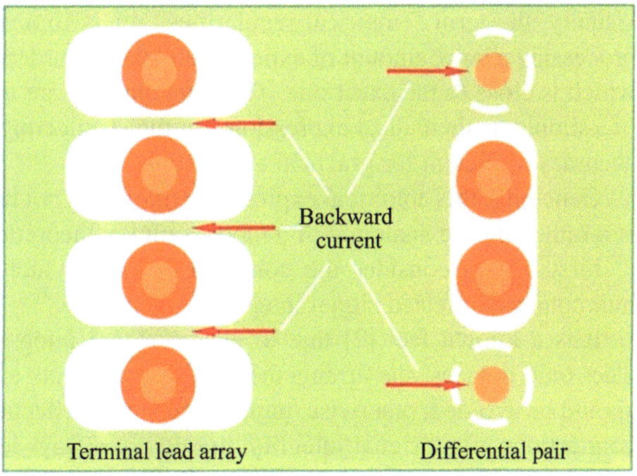

Fig. 7.81 Elongated antipads reduce the capacitance of thru-vias

are installed between the via stub and the ground plane. The most appropriate solution is found by simulation. Matching elements can be formed on a printed circuit board or mounted on its surface. For instance, a resistor with a thru-via characteristic impedance of 50 Ohm can be installed between a conductor, a single-ended signal, and a ground/power plane. For a differential pair, a 100 Ohm matching resistor is used.

In [33], it is suggested to use materials with a lower permittivity around thru-vias. For this purpose, non-plated vias are installed near the interconnects. In these vias, permittivity of air is 1, while permittivity of fiberglass plastic is 4. As a result, resonant frequency of via stubs increases. By increasing this frequency so that it does not coincide with the frequency of the transmitted signal, the cause for its attenuation can be eliminated. However, this may result in a ground plane similar to Swiss cheese, which is undesirable in high-speed systems.

Finally, a method for reducing the Q-factor of via studs was described in [34]. Resonance may be reduced by coating their walls with a more resistant material. For instance, you can use tin instead of copper. Coating the walls of the vias with tin contributes to a better neutralization of the resonance effect inside them.

Thus, via stubs distort high-frequency signals and reduce the bandwidth. Since the capacitance of thru-vias changes in proportion to their total size and the diameter of antipads, small vias with large antipads should be used, where practical. Elongated antipads are designed to reduce the capacitance and maintain the continuity of the ground plane. To eliminate undesirable effects from via stubs, it is recommended to use blind vias and remove via stubs on the backboards. Moreover, elongated antipads should be used as the most cost-effective solution for high-speed applications.

7.10 Basic Rules of Thumb for Designing Topology of Boards with High-Speed Signals

Usually, the term "empirical regularities" refers to a certain dependence based on processing a large amount of experimental data in order to get an approximate result, which is close to the exact one. These regularities are usually expressed in the form of a simple mathematical expression (formula) reflecting the observed results with an accuracy sufficient for practical application.

Generally, this empirical expression (rule) does not have a strong conclusion, or it is a rather simple analogue of a more complex theoretical ratio.

First, let us consider the conditions, under which the so-called angle plated interconnects affected signal integrity.

It is a known fact [2] that to ensure signal integrity on boards of high-speed electronic devices, the arrangement of interconnects at a right angle to each other should be avoided; otherwise, unintended signal reflections will occur. Indeed, angle connections affect signal integrity, but do not always lead to its degradation.

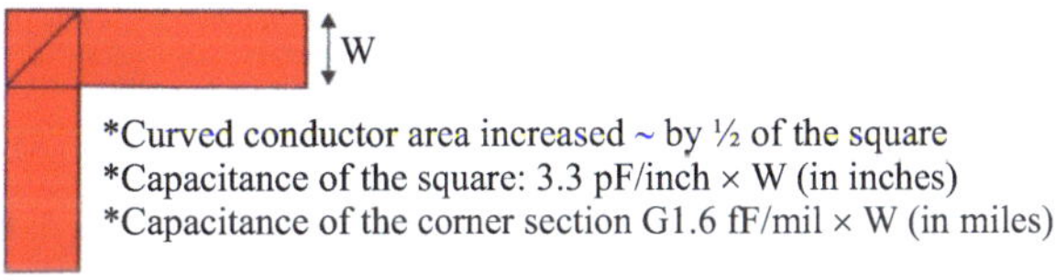

Fig. 7.82 Determining additional capacitance of a curved conductor

Let us consider the cases, where curved interconnects are critical and the cases, where they are permissible. Sometimes manufacturers of boards in instruction manuals for developers recommend placing conductors at an angle of 45° instead of 90°, primarily based on the need to ensure reliable operation of the device, rather than signal integrity. This is explained by the PCB manufacturing technology. If interconnects are placed at an angle of 90°, sometimes there are acid residues from etching next to the conductors, which are hard to remove. They destroy the conductors during long-term storage of the board in a humid environment. Manufacturers are better off avoiding such residues.

Let us consider the simplest case of a rectangular conductor, bent at an angle of 90°. Since its width is larger at the point of bending, impedance of such a line decreases, provided that the thickness of insulating material remains unchanged. The bend here functions as an additional discrete capacitance, which is smaller compared to a conductor of uniform thickness. Figure 7.82 [35] shows how this capacitance is calculated.

For simplicity, we assume that the conductor's angle section area is approximately half as large as the square area (see Fig. 7.82), as compared to the conductor, whose width remains unchanged at the point of bending. What is the capacitance of this section? Since the answer to this question (or any other question concerning signal integrity) depends on many factors, one should be able to quickly assess capacitance.

In 50 Ohm signal transmission lines, capacitance of conductors is about 3.3 pF/inch (about 1.3 pF/cm). This value does not depend on the width or any other parameter of the 50 Ohm line on the printed circuit board with a fiberglass substrate.

The side of the square is equal to the width of the line. To increase the width, the insulating material thickness should be increased accordingly for the impedance to remain unchanged, i.e., equal to 50 Ohm. Capacitance per unit length shall remain constant. Additional capacitance is approximately 1.6 (fF/mil) × W (mil).

For experimental verification of this simple empirical rule, the authors of [35] produced a 50 Ohm line with four bends on a two-layer plate (see Fig. 7.83). The thickness of the board was 0.8 mm, and that of the conductor—1.5 mm. The estimated capacitance of the angle section in each bend was about 120 fF. There has been a failure in the reflected signal of the reflectometer used to measure this capacitance.

Each reflected signal failure corresponds to one angle section of the conductor. The leading edge pulse time is 40 ps, and the scale is 5%/div. As a result of measurements, it was established that the capacitance of the angle section was 96 fF, which is quite consistent with the above simple ratios.

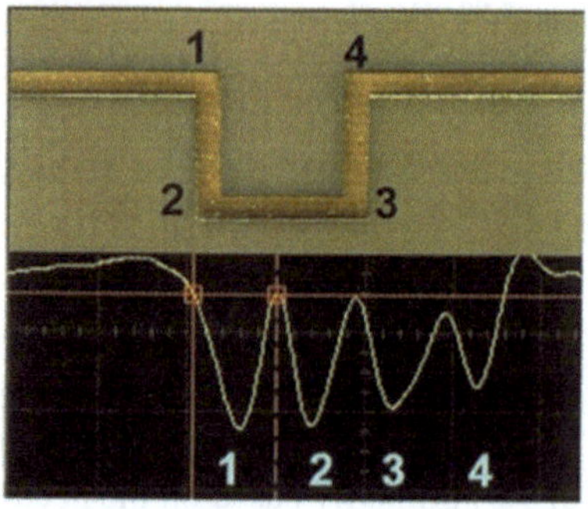

Fig. 7.83 Tested board and reflected signal types

Once the conductor's angular capacitance is determined, the issue of preserving signal integrity while using curved conductors can be addressed. Hence, the first empirical rule can be formulated as follows: "If the capacitance expressed in femtofarads is more than 10 times longer than the signal's leading edge pulse time (LEPT) expressed in picoseconds, then the capacitance shall be reduced." To do this, use a simple ratio [36]:

$$1.6\ (\mathrm{fF/mil}) \times \mathrm{W}\ (\mathrm{mil}) > 10 \times \mathrm{LEPT}\ (\mathrm{ps}). \tag{7.29}$$

For instance, if LEPT of the signal is 10 ps, signal integrity is preserved until the 50 Ohm line width exceeds 50 mil (about 1.3 mm).

The second empirical rule concerns the effect of adjacent conductors on the line impedance. The electrical specifications of interconnects are determined by the boundary conditions of conductors and the geometry of dielectrics. Maxwell's equations are a basis for calculating these specifications.

Let us consider an example of a microstrip conductor with metallic ground planes on both sides or another signal line in order to determine the impact of adjacent metallic surfaces on the signal line impedance. It is believed that the signal line impedance can be significantly reduced by placing a ground plane next to it.

Indeed, in most cases, these decisions are generally justified, although there are exceptions one needs to be aware of. Since the effect of interaction between the signal line and its adjacent conductors in the presence of a reverse path is much less pronounced compared to the fringing fields, it is necessary to use a field solver to study these complex phenomena.

According to the case presented in Fig. 7.84, we will assume that the characteristic impedance is approximately 50 Ohm, since the adjacent metal conductors are placed at a distance that is 20 times greater than the width.

In addition, we will proceed from the fact that $H1 = 3$ mil (0.075 mm); substrate permittivity: 4.3; $W1 = W2 = 5$ mil (0.13 mm); $T1 = 1.2$ mil (0.003 mm); $O1 = 100$ mil (2.54 mm) [35].

The resulting impedance is 50.11 Ohm. Under the same conditions, the impedance of a single insulated microstrip line (calculated using Polar Instruments SI9000 field solver) is 50.06 Ohm. Thus, the results coincide with an accuracy of 0.1%.

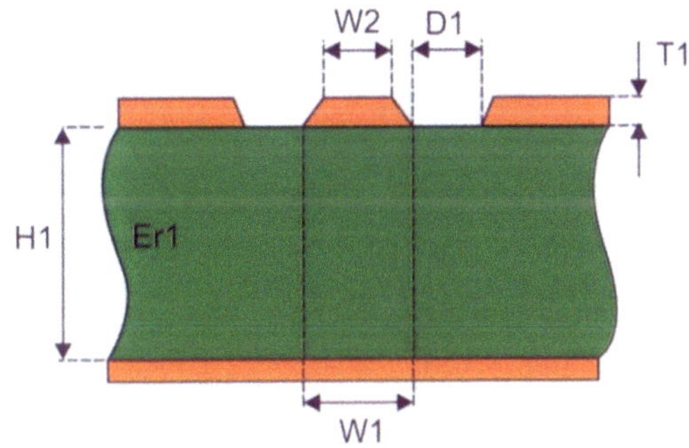

Fig. 7.84 The conductors on both sides of the signal line on the upper layer are connected with the return current path at the bottom

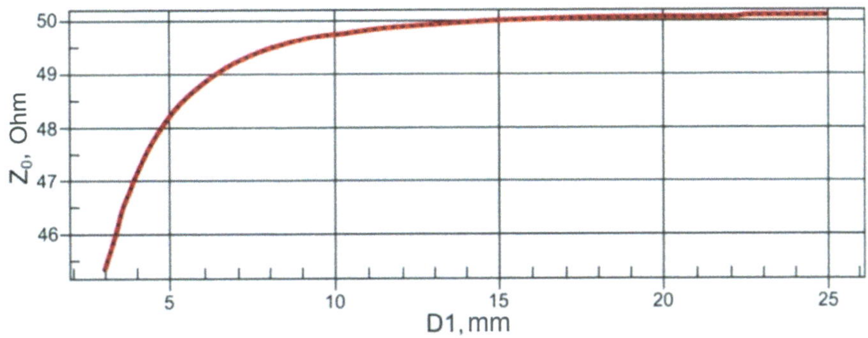

Fig. 7.85 Nature of change in the microstrip impedance with increasing distance to the adjacent conductor

When the adjacent conductor approaches the signal line, one can expect that, due to the return current path of this conductor, the line impedance will decrease. This is what happens in practice, and the degree of this impact is shown in Fig. 7.85.

Even if the gap D1 between the line and the adjacent conductors is two times its width, the line impedance decreases by only 0.27 Ohm or 0.5% in relation to the asymptotic value. With a further decrease in the distance, the proximity effect, which has the same nature as the well-known surface effect, is manifested. When the gap is equal to the width of the 50 Ohm line, its impedance changes by no more than 2 Ohm or 4%. If there is a conductor on one side of the signal line only, the impedance is reduced by less than 2%.

If the gap between the signal line and the adjacent metal conductor is three times the line's width (15 mil or 0.38 mm), there is practically no interaction between them. Thus, we can formulate the following empirical rule: "If the gap between the signal line and the adjacent metal conductor exceeds its thickness, the line impedance changes by no more than 2%."

However, it should be remembered that the impact of the adjacent conductor on the characteristic impedance of the line is just one of the many other factors: when passing the adjacent conductor on the board surface, one should also take into account the resonant interaction of crosstalk and the so-called ringing effect.

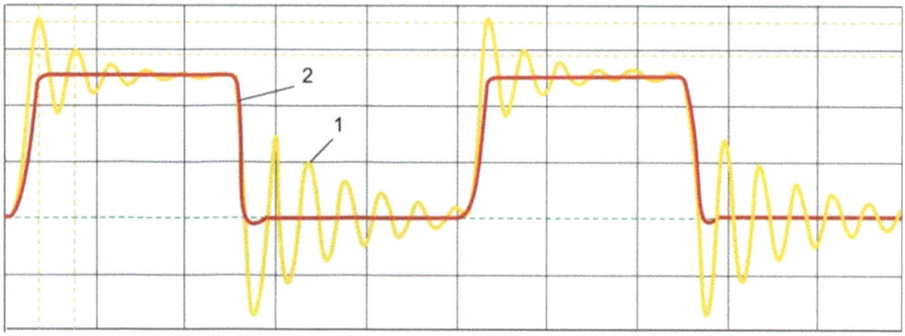

Fig. 7.86 Signal waveforms in mismatched (curve 1) and matched (curve 2) receivers

On the receiving side, the ringing effect appears as a result of signal integrity degradation due to reflections in the mismatched transmission line. Figure 7.86 shows signal waveforms in the line in two cases: curve 1—a signal in the mismatched receiver, and curve 2—a signal in the matched line of the same receiver.

This figure illustrates the two main features of the output signals. When an n-channel field-effect transistor is turned on at the output of the signal buffer with the signal changing from a high logic state to a low one, the transistor's turn-on time is generally faster and the impedance is lower as compared to the p-channel transistor, whose signal state changes from low to high. Due to both of these features, the ringing amplitude increases at the trailing edge of the signal propagating along a mismatched line.

The main cause of ringing consists in reflections occurring in the receiver due to mismatched line impedances and RX high input impedance, as well as due to mismatched line impedances when the reflected signal passes to the transmitter with low output impedance.

The receiver's reflection coefficient is always close to 1, but it is negative in the transmitter. Its signal path is as follows: the signal reaches the receiver and is reflected. Meanwhile, the voltage on the receiving side is high. The reflected signal returns to the transmitter, whose impedance changes from high to low, and is reflected again, while the voltage of the reflected signal is negative.

This reflected signal with negative voltage enters the receiver again and re-reflects, and since it is negative, the voltage of the signal in the receiver portion decreases. The voltage of the signal reflected from the receiver is still negative. Then the signal returns to the transmitter, reverses the polarity when reflected and returns to the receiver.

Thus, a positive voltage signal needs some delay time (B3) to return from the receiver to the transmitter, and the same amount of time—to be reflected to the receiver. Then just as much time is needed to return to the transmitter as a positive voltage signal.

Thus, the total time between two consecutive peaks due to reflections in a mismatched line is 4B3. The signal propagates on a printed circuit board with a woven-glass reinforced substrate at a speed of about 15 cm/ns. This means that the time (ns) between two consecutive peaks (oscillation period) is [35]:

$$4B3 = 4 \cdot \frac{\text{length}}{15\ (\text{cm/ns})} = \frac{2}{3}\text{of length (ns)},$$
$$\text{length (ns)} = 1.5 \cdot \text{period (ns)}$$

For instance, if the measured ringing period is 1 ns, the length of the mismatched line is about 3.8 cm. Naturally, to observe this ringing, the leading edge pulse time shall be within the ringing period.

Next, let us consider another parameter, namely the aspect ratio for a 50 Ohm microstrip line. As a rule, the majority of asymmetric interconnects with controlled impedance are designed for lines with a characteristic impedance of 50 Ohm, resulting from the minimum attenuation in the coaxial cable. Using 50 Ohm lines in printed circuit boards is a reasonable solution. Balance is achieved between manufacturability, power dissipation, control over crosstalk, and compatibility with signal generators. Surely, the characteristic impedance of the transmission line is completely independent of its length and is determined only by the aspect ratio and the dielectric constant (D_k). Control over the line impedance is exercised by setting linear dimensions of the cross section of components, as well as by selecting the appropriate materials. For this purpose, developers use 2D field solvers to successfully design transmission lines of finite thickness.

The results obtained using such a 2D field solver can be formulated as an empirical rule allowing for a quick verification of any other calculations with the line width and the insulating material thickness being first-order terms. The conductor thickness is a second-order term.

As the conductor width increases, its characteristic impedance should decrease. However, the characteristic impedance should increase proportionally to the increasing thickness of the insulating material. If you do both, keeping the ratio of these parameters constant, the impedance will remain more or less the same.

Consequently, all 50 Ohm microstrip lines, including those with a fiberglass substrate, should have the same aspect ratio, which is normally 2:1. Thus, another empirical rule goes as follows: "For a 50 Ohm microstrip line with a fiberglass substrate, the ratio of the line's width to the thickness of the insulating material is 2:1."

The validity of this rule can easily be verified using a standard Polar Instruments SI9000 field solver. It is used to select the assembly parameters. The other parameters are calculated so that the line impedance is 50 Ohm. An example of the main geometrical dimensions of the microstrip line is given in Fig. 7.87.

In this example, $D_k = 4.0$, and conductor thickness is 0.6 mil (0.015 mm). Insulating material thickness variation range: 2–20 mil (0.051–0.508 mm). Thickness of the line was calculated for a single-ended configuration with an impedance of 2 Ohm. After exporting the parameter values as an Excel spreadsheet, an aspect ratio was obtained for the indicated insulating material thickness and a characteristic impedance of 50 Ohm (see Fig. 7.88) [35].

Fig. 7.87 Typical microstrip cross section

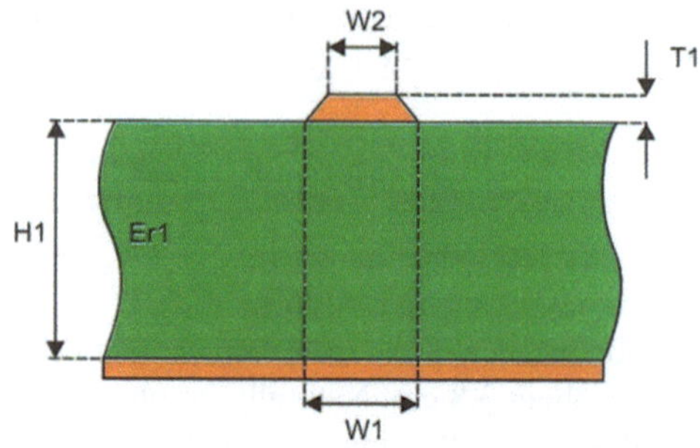

Fig. 7.88 Calculated values were obtained using Polar Instruments SI9000 field solver for a 50 Ohm microstrip line

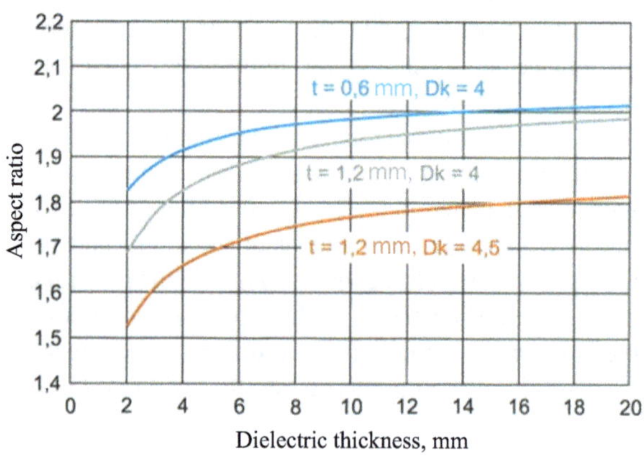

Figure 7.88 shows that at $D_k = 4.0$, with conductor thickness of 0.015 mm, the aspect ratio for a 50 Ohm line is almost 2:1. If D_k was larger or the conductor was thicker, this ratio would change accordingly.

Let us consider the justification of a similar empirical rule regarding the aspect ratio of a 50 Ohm strip line. As noted above, the aspect ratio of all 50 Ohm microstrip transmission lines implemented on printed circuit boards with a fiberglass substrate is the same. Consequently, if the bandwidth is doubled, the insulating material thickness shall be doubled as well in order to maintain the characteristic impedance of 50 Ohm.

This empirical rule has two important consequences. First, in a 50 Ohm microstrip line with a fiberglass substrate, the ratio of the conductor width to the insulating material thickness is 2:1. Thus, developers can quickly assess the conditions for arranging conductors on the surface.

The second consequence concerns crosstalk: its impact is also determined by the width of the line, i.e., the effect on the conductors on the surface is easy to assess.

Crosstalk is determined by the fringing fields, which, in their turn, depend on the insulating material thickness. When it increases, the fringing fields propagate further, making crosstalk grow, while the gap between conductors remains the same.

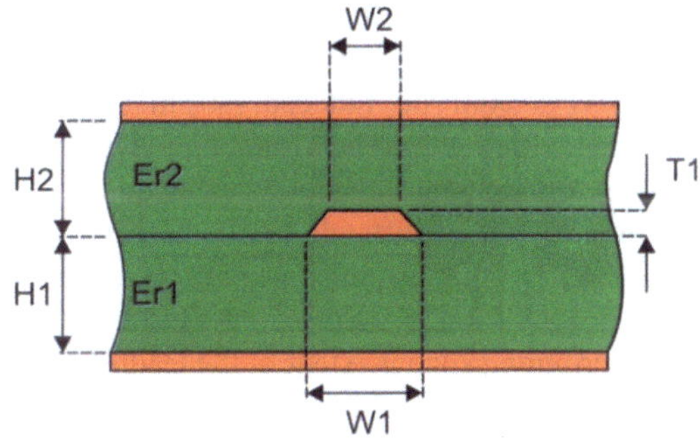

Fig. 7.89 Balanced strip line cross-section design sketch (for 50 Ohm conductors)

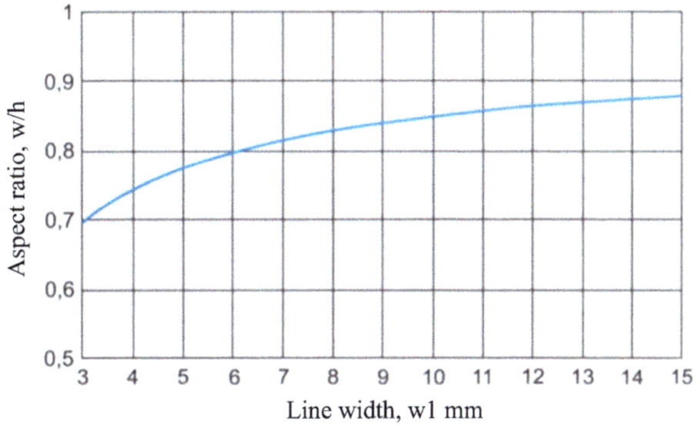

Fig. 7.90 Dependence of the aspect ratio on the width for a balanced 50 Ohm strip line [35]

As the line width changes proportionally to the insulating material thickness, crosstalk changes proportionally to the line width. It is a known fact [1] that in order to prevent crosstalk at the near end of a single-wire 50 Ohm bus from exceeding 5%, the worst case gap between the edges must be twice the width of the line.

Similarly to a microstrip line, the conductor width to the insulating material thickness aspect ratio in a 50 Ohm balanced strip line with a fiberglass substrate is calculated using a 2D field solver.

For instance, we assume that coefficient $D_k = 4.2$ and conductor thickness is 0.6 mil (0.015 mm). The same insulating material is used in the main layer, in the pre-preg, and in the balanced line H1 = H2 (see Fig. 7.89).

The calculation data are presented as a dependency diagram with the aspect ratio depending on the insulating material thickness (see Fig. 7.90).

Based on the obtained results, we can formulate another empirical rule: "If a balanced line conductor has a width of about 5 mil. (0.13 mm), the line aspect ratio is 0.8. Even if the width increases to 15 mils (0.38 mm), the increase in this indicator will not exceed 10%."

Thus, for a balanced 50 Ohm line with a fiberglass substrate, the insulating material thickness for a 5 mil wide conductor on its top or bottom surface is 6 mil (0.15 mm).

Knowing the aspect ratio in these specific initial conditions, one can assess the rule effectiveness in other conditions. When $D_k < 4.2$, the insulating material thickness should be reduced, so that the characteristic impedance remained unchanged (50 Ohm). The insulating material thickness should be increased proportionally to the increasing conductor width.

Naturally, it is difficult to estimate the exact change in the dielectric layer thickness, if other parameters have been changed, based on this empirical rule only. In such cases, a field solver is used.

Determining spatial extent of the leading edge is another issue that developers of high-speed board should address.

According to rough estimates, a signal propagates along the transmission line with a fiberglass substrate ($D_k = 4$) at a speed of about 15 cm/ns [37].

The distance that the signal covers passing through the interconnect with a finite leading edge pulse time (LEPT) is Δx. In this area, voltage between the signal and the reverse paths changes. As a rule, to determine the boundaries of this area, points of 10% and 90% are selected (see Fig. 7.91).

The speed of the signal leaving the transmitter is 15 cm/ns. Meanwhile, the rising edge covers the following distance:

$$\Delta \mathrm{x} = 15\ (\mathrm{cm/ns}) \cdot \mathrm{LEPT}\ (\mathrm{ns}).$$

For instance, in a standard CMOS transmitter used by developers, the leading edge pulse time is 1 ns. It follows that its spatial extent is 15 cm. In DDR3 storage devices, LEPT = 0.3 ns. During this time, the leading edge covers a distance of 4.5 cm. When a signal is transmitted at a speed of 28 Gbit/s, LEPT = 20 ps, and the required distance is 3 mm.

As the leading edge of the signal is passing through the interconnect with a uniform structure, LOS is inevitable. In this case, another rule is applied: "If there is a heterogeneity on the signal path, whose linear dimension does not exceed 1/3Δx, its presence does not affect the signal. If it is greater than the specified value, the signal is most likely to be distorted."

Using simple calculations, we can easily prove that the length of the inhomogeneity should not exceed 1/10 of the wavelength of the highest frequency component

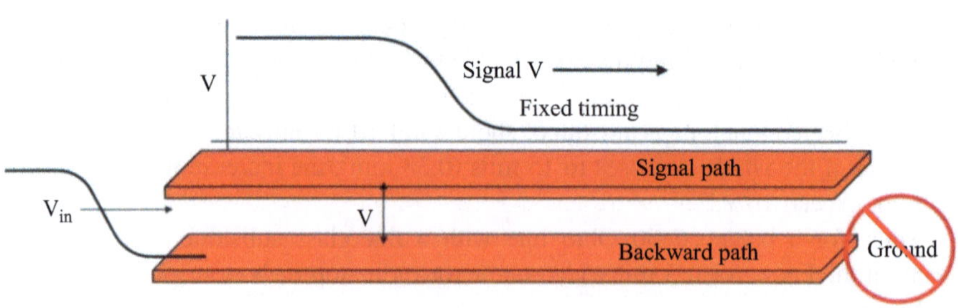

Fig. 7.91 Rising edge of the signal propagating through the interconnect

of the signal. For instance, if a signal is transmitted at a speed of 28 Gb/s, the inhomogeneity should not be longer than 1/3·3 = 1 (mm). This is one of the main reasons why it is much more difficult to develop a 28 Gb/s system than, for instance, a 8 Gb/s system.

And last but not least is the resonant frequency of the cavity. It is a known fact that signal paths are resonators with low Q-quality to prevent ringing. However, in cavities formed, for example, by the power and ground planes, Q-quality is very high. This means that even small levels of crosstalk from signal paths may cause resonance and noise inside a cavity. Knowing the resonant frequency of the cavity, it is possible to prevent potential problems in advance.

If there are large impedance breaks at the ends of the interconnect, the signal is reflected there, creating noise. At some frequencies, each reflected signal is in phase with a new incoming wave. As a result, several signals are generated. These frequencies are resonant.

Resonance effect analysis is best performed in the frequency domain. If a sine signal is sent to one edge of a cavity, it proceeds to the next edge and is reflected from it back to the source. When a signal is reflected from an open far end, its phase does not change. The signal reaches its source and is reflected again without a phase change. Total signal transmission time is defined as follows [35]:

$$t_{\mathrm{TOTAL}} = \frac{2L}{v} = \frac{2L}{c}\sqrt{D_k},$$

where c—speed of light in a vacuum (v—speed of light in a cavity); L—distance that the signal covers while passing to one edge; D_k—permittivity of the cavity filling material.

Phase change of the signal, which has passed into the cavity during this time, is determined as follows [35]:

$$\Delta\Theta = f_{\mathrm{RES}} \cdot t_{\mathrm{TOTAL}} = f_{\mathrm{RES}} \cdot \frac{2L}{c}\sqrt{D_k}.$$

The wave reflected from the far end of the interconnect is added to the wave entering the cavity when their phases coincide. The resulting wave amplitude increases. Further amplitude increase is limited by the energy entering the cavity, as well as by signal loss in the cavity.

Amplitudes of the waves are added up provided that their phases differ by a whole number of cycles. This means that half of the signal period is spent on its passage in one direction, i.e., cavity length is half the wavelength. For resonance to occur, the cumulative phase change must be divisible by a whole number n of periods [35]:

$$n = \Delta\Theta = f_{\mathrm{RES}} \cdot \frac{2L}{c}\sqrt{D_k}.$$

For fiberglass cavities, resonant frequencies are defined as follows [35]:

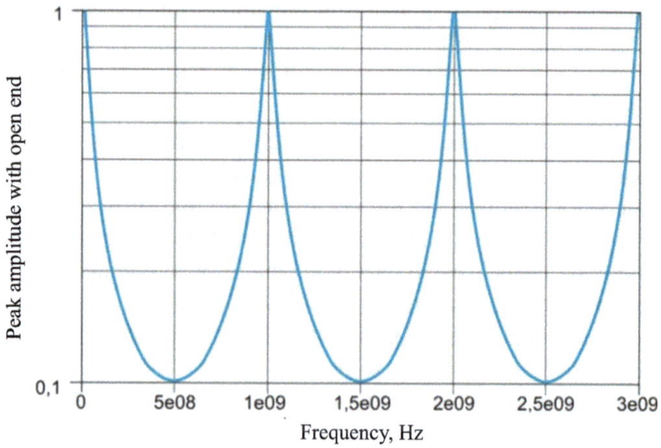

Fig. 7.92 Amplitude change at the far end of the cavity depending on the input signal frequency

$$f_{\mathrm{RES}}(\mathrm{GHz}) = \frac{c}{\sqrt{D_k}} \cdot \frac{n}{2L} = 6\ (\mathrm{inch/ns})\frac{n}{2L} = n\frac{7.5}{L(\mathrm{cm})}.$$

Resonant frequency of a 7.5 cm long cavity is about 1 GHz. Figure 7.92 shows the amplitude obtained using a field solver at the far end of the cavity at varying frequency of the input signal. Note that these amplitudes are repeated every 1 GHz.

Resonant frequency of a multilayer integrated circuit housing with a cavity length of 35 mm is equal to [37]:

$$f_{\mathrm{RES}} = n\frac{7.5}{3.5\,(\mathrm{cm})} = 2.1\,(\mathrm{GHz}).$$

If this IS has a clock frequency of about 2 GHz, noticeable noise occurs in the cavity between the power and ground planes. Its occurrence is one of the reasons why reverse paths are chosen as VSS layers. In this case, several through-holes leading to the reverse path between the VSS planes can be placed next to the signal through-holes to minimize the possibility of resonance in the cavity.

In case of developing a specific board design, it is recommended to determine resonant frequencies of all cavities based on their maximum linear dimensions. If the clock frequency or harmonic components coincide with the resonant frequencies of the cavities, a connection may be established between any signals passing through these cavities. This situation must be avoided.

References

1. Johnson, H., & Graham, M. *High speed digital design: A handbook of black magic.*
2. Belous, A. I., & Emelyanov, V. A. (2013). Fundamentals of IC microassembly (p. 316). Moscow: DMK Press (in Russian).
3. *7 Key considerations for effective chip-package thermal co-design a high-level 'how to' guide mentor graphics, whitepaper*. January, 2014.
4. State of automotive the art automotive thermal design by DENSO. *Engineering Edge*, 2(1), Mentor. Graphics Corp. 2013.
5. *10 Tips for streamlining PCB thermal design. A high-level "how to" guide Mentor Graphics white paper*. January, 2014.
6. Kondratyev, V. (2016). Calculation of the thermal regime of the components. *Electronic Components*, 2, 17–24 (in Russian).
7. Bodnar, D. (2014). Metallic and composite heat-conducting materials for high-power semiconductor packages. *Components and Technologies*, 12, 155–160 (in Russian).
8. Eblen, M. (2006). *Transient thermal modeling techniques for WBG device packaging*. Duluth, GA: Kyocera.
9. Vladimirov, S. (2016). Features of the design of the transmission line on printed circuit boards. *Electronic Components*, 2, 30–33 (in Russian).
10. Belous, A. I., Merdanov, M. K., & Shvedov, S. V. (2016). *Microwave electronics in radar systems and communications. Technical encyclopedia in 2 books*. Moscow: Technosphere (in Russian).
11. Zhang, P. (Senior Design Engineer) & Morant, K. (Lead Design Engineer) (2016, January 12). *Nuvation Engineering*.
12. Vladimirov, S. (2015). Features of the topology of printed circuit boards for RF applications. *Electronic Components*, 3, 20–24 (in Russian).
13. Coondor, J., & Rogers Corporation. *Understanding PCBS for high-frequency applications*. Printed Circuit Design Fabrication.
14. Warner, J. (2012, March 9) How surface roughness impacts high-performance PCBS. *Microwave Journal*.
15. Savenko, S. (2016). Innovative trends in the technology of printed circuit boards. *Technologies in the Electronics Industry*, 7, 22–27 (in Russian).
16. *Proceedings of the EPIC conferences*, 2014–2016 (in Russian).
17. Zakharov, D. (2016). The integrity of signals in high-speed communication lines. *Electronic Components*, 8, 60–63 (in Russian).
18. Korotkov, S. (2016). Electromagnetic interference and signal integrity on printed circuit boards. *Electronic Components*, 5, 14–16 (in Russian).
19. Peresadin, A. (2016). Ensuring the integrity of signals at all stages of design. *Electronic Components*, 8, 26–31 (in Russian).
20. Gakharia, V., & Gakharia L. (2016). Analysis and calculation of delay lines in printed circuit boards of digital devices. *Technologies in the Electronics Industry*, 5, 6–11 (in Russian).
21. Brooks, D. (2004, January) Crosstalk effects in serpentine traces. Adjusting signal timing (part 2). *UltraCAD Design*.
22. Guo, W.-D., et al. (2004). Comparison between flat spiral and serpentine differential delay lines on TDR and TDT. In *IEEE 13th topical meeting, electrical performance of electronic packaging* (pp. 147–150).
23. IPC-2251 A standard developed by IPC-2003.
24. Feller, A., Kaupp, H. R., & Digiacomo J. J. (1965). Crosstalk and reflections in highspeed digital systems. In *Proceeding fail joint computer conference*.
25. Wu, R. B., & Chao, F. L. (1995). Laddering wave in serpentine delay line. *IEEE Transactions Components, Packaging and Manufacturing Technology B*, *18*(4), 644–650.

26. Kerinami, M. H., & Ramahi, O. M. (2003). Effects of segment length and number of turns on designing a precise meander delay line. In *IEEE international symposium on electro-magnetic compatibility* (Vol. 2).
27. Bogatin, E. (2003). *Signal and power integrity- simplified.* Prentice Hall.
28. Rubin, B. J. (2000). Study of meander line delay in circuit boards. *IEEE Transactions on Microwave Theory and Techniques.* 48(9).
29. Soh, W.-S., See, K.-Y., Chang, R. W.-Y., Oswal, M., & Wang L.-B. (2009, December 7–10) Comprehensive analysis of serpentine line design. In *Microwave conference.* Asia Pacific: APMC.
30. Lee, H., & Kim, J. (2002). Unit cell approach to full- wave analysis of meander delay line using FDTD periodic structure modeling method. *IEEE Transactions on Advanced Packaging*, *25*(2).
31. Shin, J., & Michalka, T. (2012). Comprehensive, scalable design guidance for serpentine time delay variation in digital system. In *Electronic Components and Technology Conference (ECTC).*
32. Krasnov, S. (2016). How to rneutralize undesirable effects from hanging vias. *Electronic Components*, 12, 17–21 (in Russian).
33. Olney, B. *How to Handle Danglers.* iconnect007.uberflip.com
34. Mutnury, B. M. *Circuit board having holes to increase resonant frequency of via stubs.* US 8542494 B2. IBM Corporation. https://patents.google.com/patent/US8542494B2
35. Dorokhov, V. (2016). Empirical rules for designing the topology of boards with high-speed signals. *Electronic Components*, 12, 22–26 (in Russian).
36. Eric, B. *Bogatin's rules of Thumb.* www.edn.com
37. Medvedev, A. M. (2005). *Printed circuit boards. Constructions and materials.* Moscow: Technosphere (in Russian).

Chapter 8
Packaging Technologies for Electronic Devices

8.1 Trends in the Development of Chip Packaging Technology

Modern production is characterized by trends of miniaturization and increasing the density of functional elements, the so-called splicing of traditional electronics and microelectronics.

In early 2000s, there was a clear separation of the concepts of "electronic components" (microchips, discrete active and passive components) and "printed circuit," "electronic assembly," or "product." As of the time of publication of this book, there is blurring of the boundaries between the actual production of electronics (development, assembly and testing of printed circuits, and electronic units) and microelectronics (production of electronic components).

This is well illustrated by the space industry, where the processes of evolution of onboard electronic systems of spacecraft are associated with an increase in specific heat liberation. And if earlier the heat was successfully dissipated from relatively large surfaces of microcircuit packages, today, it is necessary to introduce special elements of a conductive heat sink in the design of modern packages.

The development of microelectronic components is going in the direction of increasing the degree of their integration, productivity, and functionality, this process is characterized by an increase in the density of active elements on the chip by about 75% per year, which makes it necessary to increase the number of their leads on the package by about 40% per year. Firstly, this is due to the ever-growing demand for new microcircuits packaging methods, and secondly, to an increase in the interconnect density on a printed circuit board, not only for high-speed devices, but also in numerous other technical areas.

It should also be noted that physical dimensions (volumes) of electronic devices annually decrease by 10–20%, with the density of interconnections constantly increasing and the elements of printed mounting becoming smaller.

A. Belous, V. Saladukha, *High-Speed Digital System Design*,
https://doi.org/10.1007/978-3-030-25409-4_8

In creation of the first microcircuits, measured chips were placed inside ceramic or plastic packages. Until about 1980, all the microcircuit packages had leads designed for their placement and fastening in the metallized holes (PTH, Plated Through-hole—the metallized hole) of the mounting substrate and the subsequent soldering. At the same time, the standard pitch of mounting holes on printed circuit boards was formed: 2.54 mm (1/10 in.).

Later, with an increase in the degree of integration, the package leads began to be made in the form most suitable for surface-mounting technology (SMT—surface mounting technology), which has become the most popular technology for mounting electronic components. Passive or low output components for surface mounting essentially increase layout density by simply reducing the size. Modern passive components (resistors and condensers) are usually designed as microchips. However, the process of increasing density of the microcircuit layout cannot keep following such a simple way forever; there shall always be certain limited geometrical space for placing multiple leads on the component package and mounting area of the board.

In practice, there are two main ways of placing the leads of the microcircuit packages:

- peripheral lead distribution (assignment) when they are located at the edges (at the periphery) of the chip or the microcircuit package;
- matrix arrangement of leads on a flat surface of the package.

Most types of modern microcircuits have peripheral lead distribution (assignment). The peripheral lead pitch is limited to 0.3 mm (almost 0.4 mm), which allows placing up to 500 leads on microcircuits with large-sized packages. However, when a pitch of leads is less than 0.5 mm, the mounting process becomes much more complicated. The configuration of peripheral leads in this case can take all kinds of forms: the so-called gull wing, J-shaped (with soldering under the package), I-shaped (for butt soldering), lead-free with soldering along the package face.

Components with a matrix arrangement of leads have an even greater variety [1, 2]:

- CSP (Chip-scale Packages—package that is commensurate with the size of the chip);
- PBGA (Plastic Ball Grid Array—plastic packages with ball matrix leads);
- CBGA (Plastic Ball Grid Array—ceramic packages with ball matrix leads);
- PPGA (Plastic Pin Grid Array—plastic packages with matrix pads);
- CCGA (Ceramic Column Grid Array—ceramic packages with column matrix leads).

It is known that microcircuits with the number of leads from 150 to 200 are usually performed in packages with peripheral leads. For microcircuits with more than 200 leads, a matrix system of leads is preferable, since it makes it possible to place a large number of leads in a limited area.

Constructions of leads in the matrix system have a smaller variety [1, 2]:

- column leads from solder reinforced with thin-wire "tangled turnings" for microcircuit packages and multi-chip MCM modules (multi-chip module), often called pad grids, land grids;
- matrix of ball leads (BGA—Ball Grid Array) with a pitch of 1.50 mm (0.060 in.), 1.27 mm (0.050 in.), 1 mm (0.040 in.), 0.8 mm (0.03 in.), and less.

The matrix system of leads provides a number of advantages:

- minimal area of the substrate mounting field;
- better conditions for ensuring the functional performance of electronic modules due to reduction of the parasitic effects occurring at higher frequencies;
- availability of additional areas for placing elements of the heat sink on the substrate.

Figure 8.1 shows the history of development of microcircuit technology in graphical form.

It should be noted that the technology of installing components with matrix outputs is associated with additional investments to ensure precise positioning of outputs on the board and testing of soldered connections that are not available for visualization. The limited length of the pins does not provide damping necessary to compensate for the effect of mismatch of thermal expansions of materials in the system of interconnects (Fig. 8.2) and does not provide the resistance to mechanical effects associated with the deformation of substrates, which is necessary for special conditions (space-related application). Thus, the list of materials for use in this interconnect system is very limited. The current trend towards a further reduction in lead pitch (<1.0 mm) further exacerbates this technical problem.

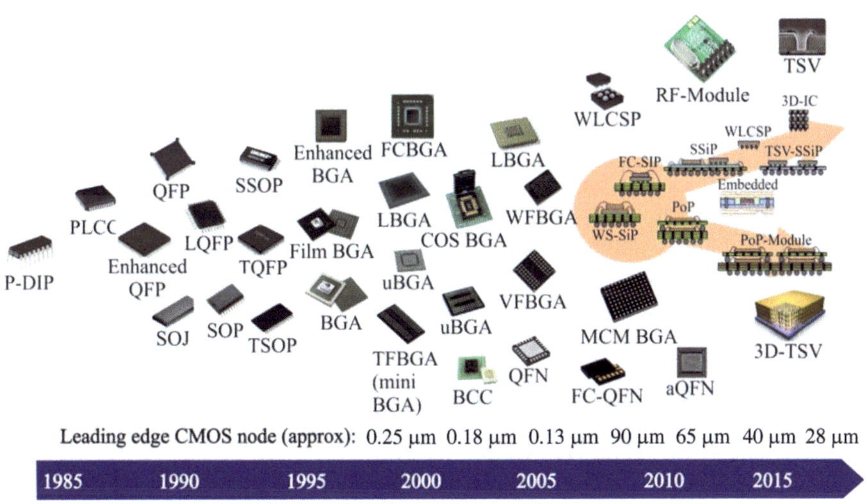

Fig. 8.1 History of development of microcircuit assembly [1]

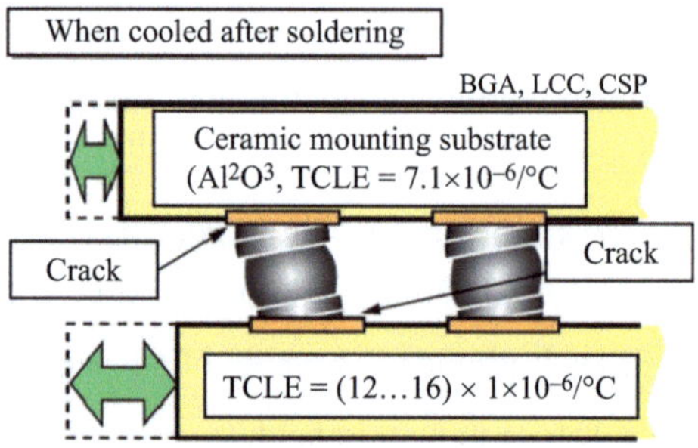

Fig. 8.2 Explanation of the problem of mismatch between thermal expansions of materials in the system of interconnections of BGA components

Therefore, microcircuit manufacturers often use other materials: damping polyimide films (Tape BGA—TBGA), ceramic substrates (CBGA—Ceramic BGA or PGA—Pin Grid Array). The use of ceramic substrate makes it possible to achieve high resistance to thermal cycles in CBGA packages not mounted on boards.

Let us take a brief look at the features of the BGA technology.

8.2 Main Trends in the Development of Packaging Technology for High-Speed Electronic Devices

Let us take a brief look at the summary list of the main trends in the development of packaging technology for microelectronic products, presented in graphical form [1, 3, 4] in Fig. 8.3. 3D-technology of assembly is the most promising option for implementation of high-speed, energy-efficient and miniature electronic devices.

However, it takes long-term efforts, from development of the relevant system architecture, design, technology, testing, and packaging to bespoke applications of using 3D IC and receiving positive outcome.

Therefore, developers of electronic devices often use other solutions as well:

- Fan packaging at wafer level (FOWLP);
- Embedded Die Technology (EDT);
- Fan-In technology;
- Fan-Out technology.

Figure 8.4 shows the main trends of innovative development of another rapidly developing direction of high-speed microelectronic device assembly technologies—Embedded Packaging Technologies, which include some already mentioned assembly directions (Fan-In/Fan-Out, Chip-First/Chip-Last, Face-Up/Face-Down).

Embedded Packaging Technologies are developing in two directions: Embedding Die (Die Embedding in Organic Laminate) and Fan-Out (Die Embedding in Epoxy Molding Compound).

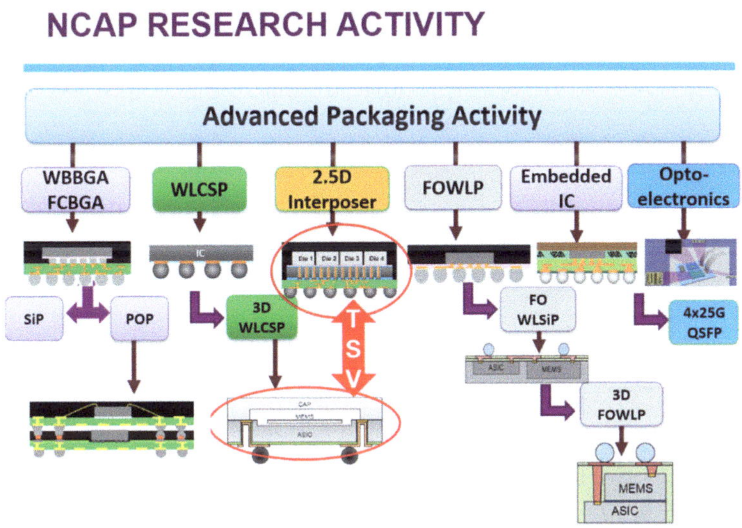

Fig. 8.3 The main trends in the development of mounting of microelectronic products

Further, there are various types of technologies for each of these directions, which differ in terms of complexity, cost, thermal resistance, duration of the technological process, testing features, list of used materials, and so on.

In order to demonstrate the possibilities of modern assembly of microcircuits, Fig. 8.5 shows the typical structure of an interposer; Fig. 8.6 contains the list of materials used for its production—glass, silicon, and organic materials.

There is a wide range of bars, from large-sized ball matrices (BGA) (connections to the substrate) to very small micro-bars used in memory modules packaging.

According to the experts, glass has the best material properties that are most favorable for miniaturization, and ensures high resistance to mechanical impacts.

Developers of high-speed electronic devices need to know a number of topological features of the 3D assembly, the ignorance of which can entail very unpleasant consequences. This concerns the reliability of the microelectronic module and, ultimately, can lead to the failures that are sometimes referred to as "black magic effects" by experts. So, we all know that adhesive mounting a chip is widely used in space instrument making in the manufacture of 3D products. However, it is necessary to know that during adhesion of the components of microelectronic elements in the module, it is necessary to avoid the occurrence of stresses after curing of the so-called adhesive line. It is known that when such mechanical stresses appear during the operation of a spacecraft in orbit or interplanetary flight due to climatic and mechanical loads, such an adhesive joint can degrade—crack, delaminate, and ultimately, can break due to the difference in temperature coefficients of materials, chips, and substrate.

The classic process of adhesion of two different materials occurs due to the coupling of the adhesive film to the surfaces of these materials. As a result of complex chemical reactions occurring in the adhesive film, this film eventually

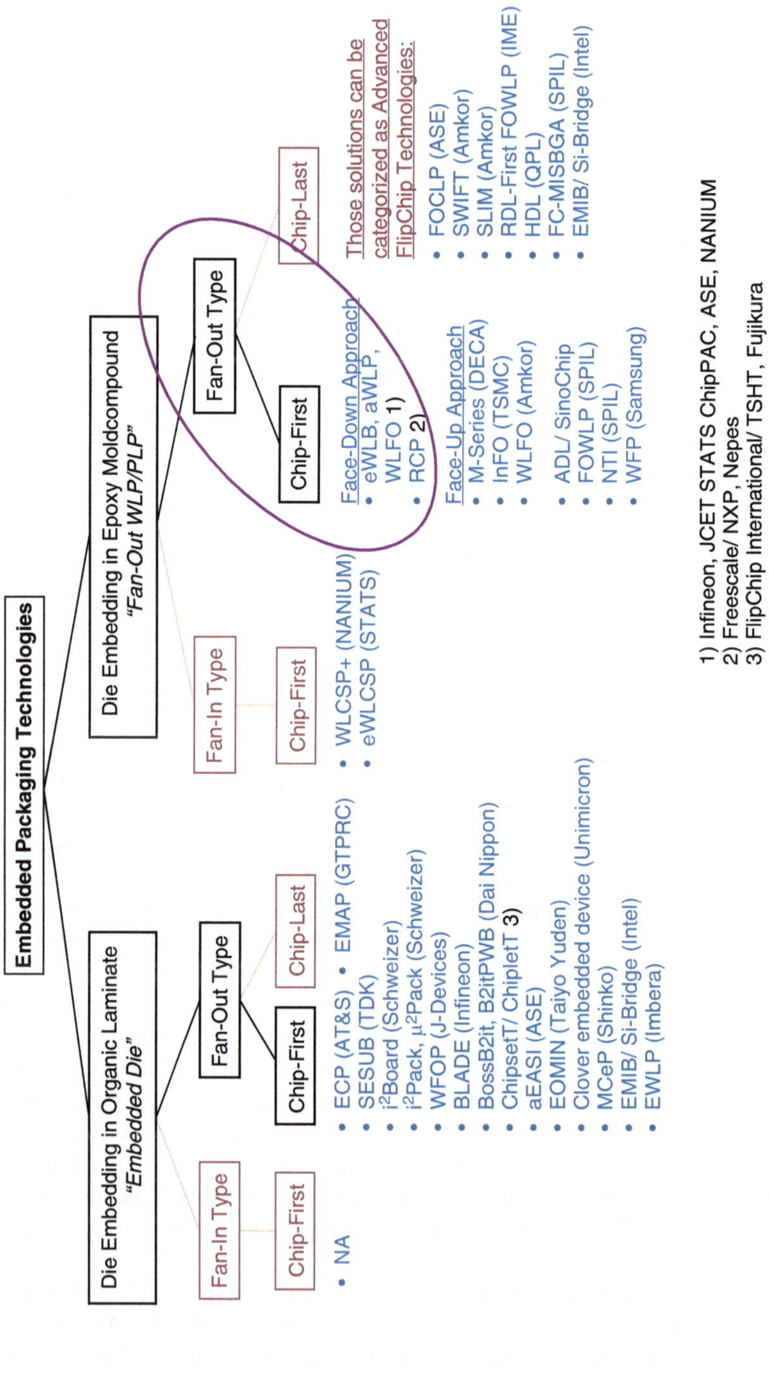

Fig. 8.4 Directions of development of embedded packaging technologies

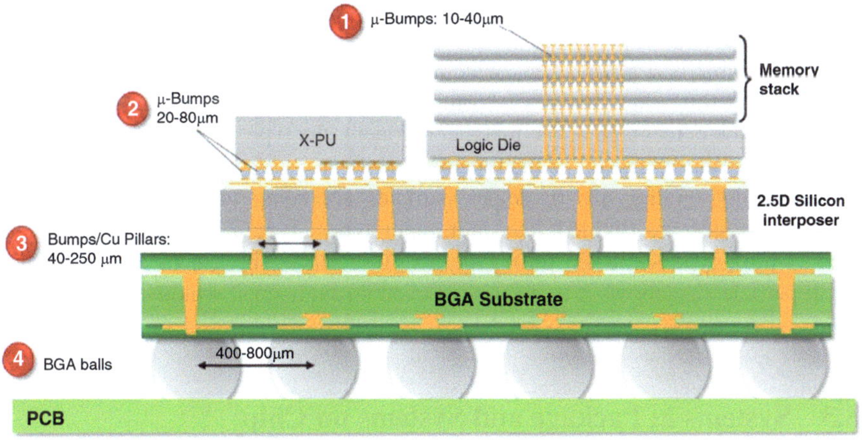

Fig. 8.5 Typical design of a 2.5D interposer

Desired properties	Glass (Borosilicate)	Silicon	Organic
TTV <5 pm			
Buckling <2 pm/20 mm			
Insulation resistance High			
Optical transparency Optical I/O			
Surface roughness <5 nm			
TCE 3.2-8.0 ppm/C			
Hermetic connectors for military specifications.			

Fig. 8.6 Main materials used for interposer production

turns into a solid substance that firmly connects the materials being bonded. It was found that the coupling of the adhesive film to the surface arises not only due to chemical reactions, but also as a result of the appearance of intermolecular and electrostatic forces between the adhesive film and the surfaces of the materials being bonded. Unfortunately, it has not yet been possible to create the corresponding physical and mathematical models of such effects.

One of the causes of failures due to the degradation of the electrical parameters of 3D microcircuits are residual thermomechanical stresses in chips due to the differences in temperature coefficients of linear expansion (TCLE) in the chip-package system. In order to reduce the magnitude of the residual stresses in the chip and the temperature of its installation in some products, the use of thermosetting (epoxy, phenolic, and rubber) and thermoplastic (polyamide) adhesives is recommended.

In microelectronic modules of onboard electronic devices of spacecrafts, epoxy adhesives with a silver filler are widely used, which makes these materials conductive to ensure low resistance between the chip and the substrate.

Developers of high-speed microelectronic devices also often use connections made by means of solder pastes.

Solder paste is a pasty mass, which includes powdered solder with spherical particles and flux—ligaments. The composition and properties of the solder paste depend on the percentage of the metal component, the type of alloy, and the size of the solder particles, as well as the type of flux.

Depending on the size of solder particles according to IHC/EIAJ-STD-005 standard, pastes are divided into four main types: type 2 (75–45 μm); type 3 (45–25 μm); type 4 (38–20 μm); type 5 (25–15 μm). The use of a paste with a small particle size allows it to be applied even through small windows in a stencil.

8.3 Systems in Package and Systems on Chip

Let us consider the main trends in the field of Systems in Package and Systems on Chip as well as features of the most frequently used WLP and 3D-integration technologies.

As evidenced by [1], the following obvious trends in the development of packaging technologies can be identified:

- decrease of the minimum pitch of leads;
- increase in the number of leads in a package;
- integration of several microelectronic components into one package;
- the transition from the location of the leads *on the perimeter* of the package to the location of the leads *under* the package (Figs. 8.7 and 8.8).

It should be noted that the decrease in the lead pitch of the PBGA and QFP components actually reached its limits (0.65 and 0.3 mm, respectively). The minimum lead pitch of the FBGA components will continue to decline; according

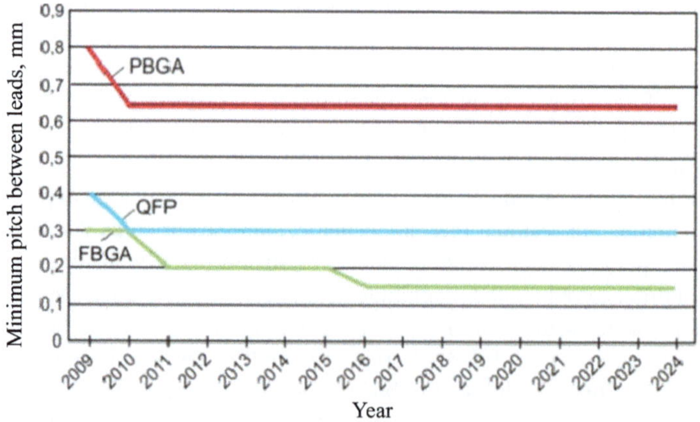

Fig. 8.7 Trends of change in the minimum pitch between leads for leads of types QFP, FBGA, and PBGA [1]

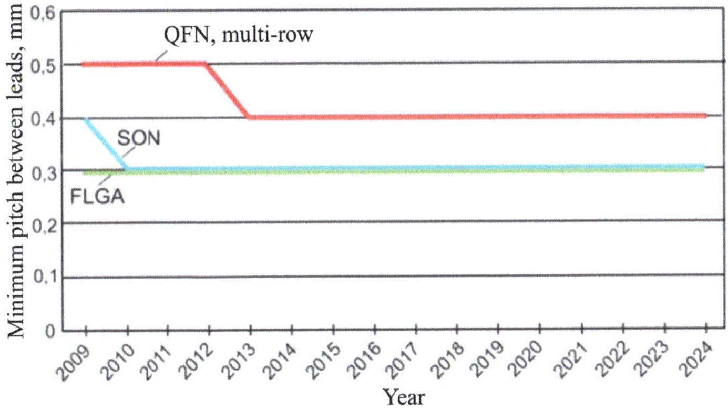

Fig. 8.8 Trends of change in the minimum pitch between leads of components with contact pads for packages of types SON, QFN, and FLGA

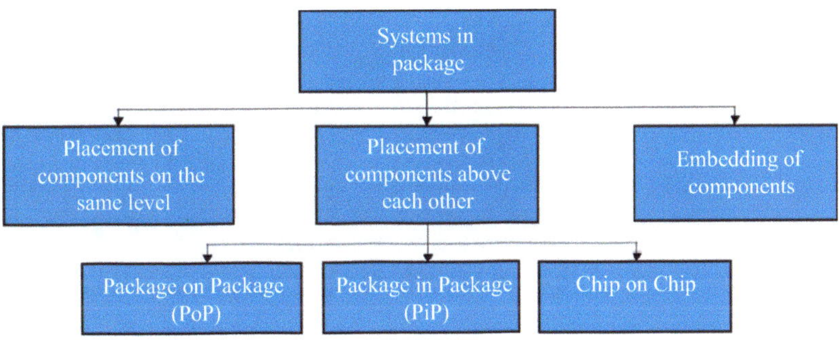

Fig. 8.9 Classification of systems in package

to forecasts of authoritative experts, after three years it will be halved and reach a value of about 0.15 mm.[1].

At the time of publication of this book, we are observing rapid development of such direction of mounting as System in Package (SiP). SiP is a combination of several active electronic components of different functionality, assembled into a single module that provides the implementation of various functions usually performed by the system or subsystem. System in Package may include passive components, MEMS, optical components, and other packages and devices. Combining these components in one package has significant advantages: the design becomes smaller, lighter, safer, and cheaper. Classification of such systems in package is shown in Fig. 8.9; schematic images of variants of design solutions—in Figs. 8.10, 8.11, and 8.12.

Figures 8.13 and 8.14 show the graphs of changes in the maximum number of chips in package and maximum number of leads for systems in package designed for application in various industries [1].

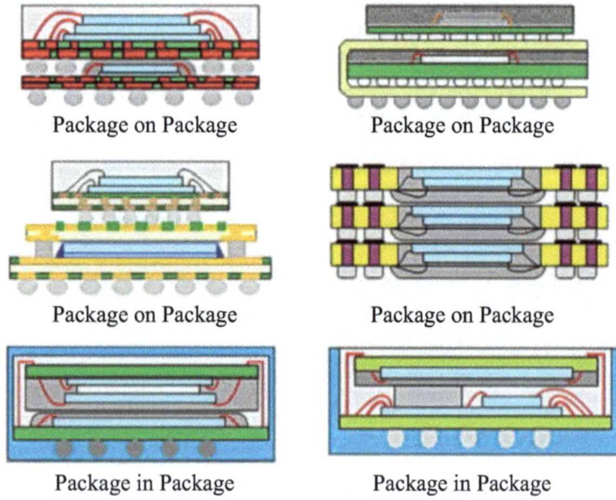

Fig. 8.10 Schematic representation of the main options for the design of high-speed microelectronic systems in package

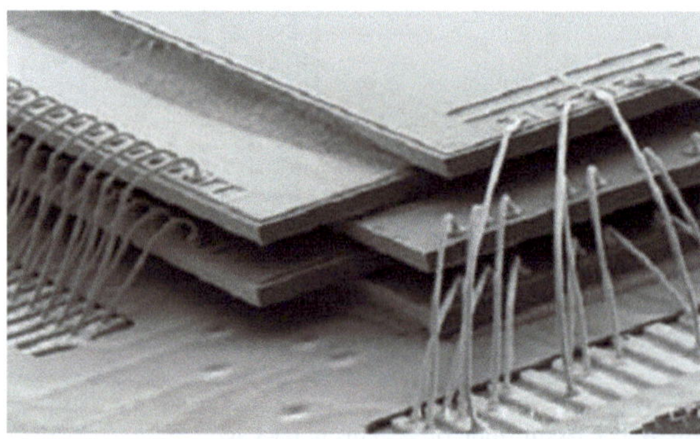

Fig. 8.11 Example of lead splicing in a system in package [1]

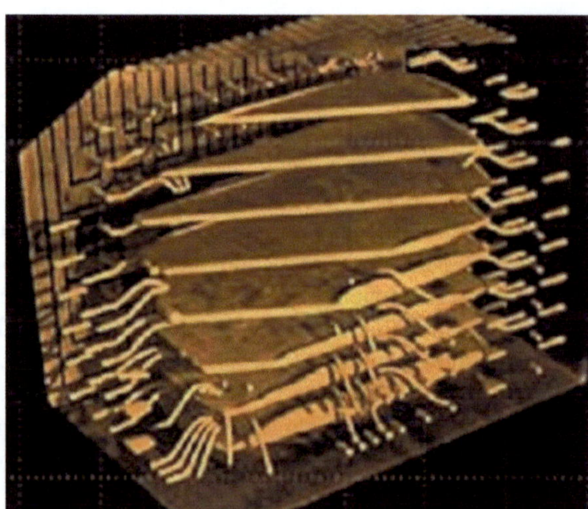

Fig. 8.12 Image of a system in package (image received by means of X-ray tomography using the company's equipment Phoenix|x-ray) [1]

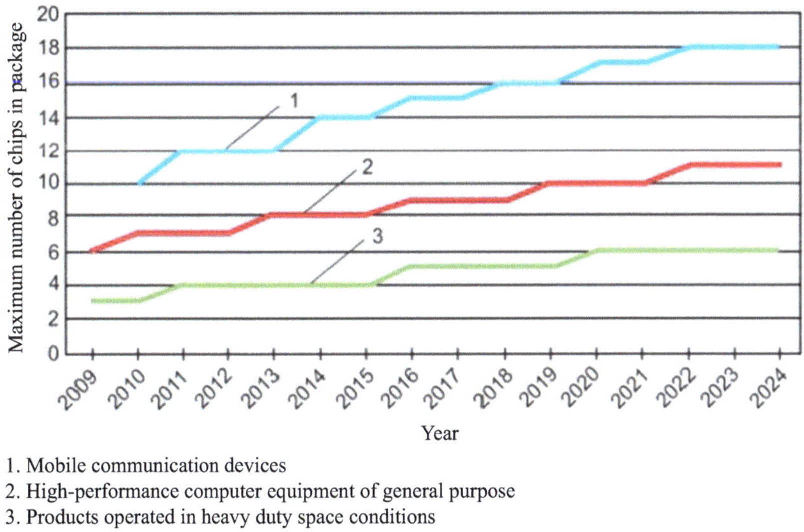

Fig. 8.13 Trends of changes in the number of chips in SiP for various applications

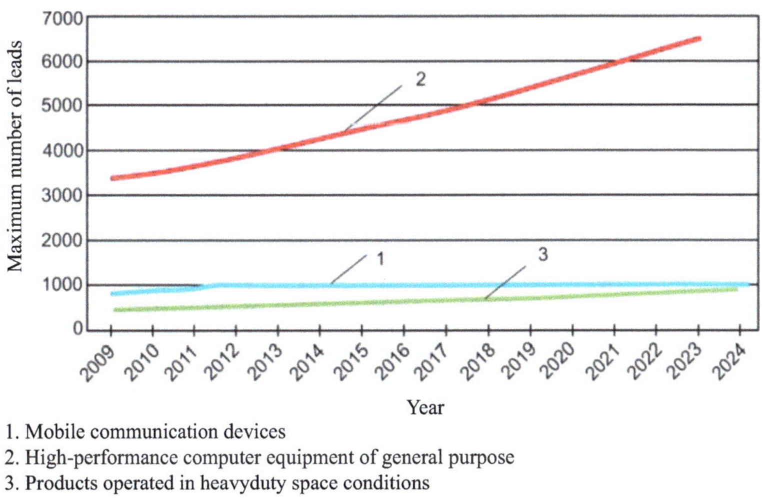

Fig. 8.14 Generalized graph of change in the number of SiP leads for various applications

Systems on Chip (SoC), all elements of which are contained in a single semiconductor chip, are also widely used for implementation of modern high-speed electronic devices.

By comparing systems on chip with systems in package, it is possible to identify the main advantages and disadvantages listed in Table 8.1.

Table 8.1 Comparative analysis of advantages and disadvantages of SiP and SoC

	System in package	System on a chip
Advantages	• Time before commercial distribution—from 3 to 6 months • Ability to embed active and passive components • The ability to replace individual components • Reuse of individual items	• Higher miniaturization • Higher interconnect density • Higher reliability (with the exception of very large chips) • Higher yield of valid products if the technology is developed
Disadvantages	• More complicated build • High power density when the chips are located above each other	• Time before commercial distribution—from 6 to 24 months • Difficulty in making changes • Product capabilities are limited by selected technology • Limited yield of valid products for complex large chips

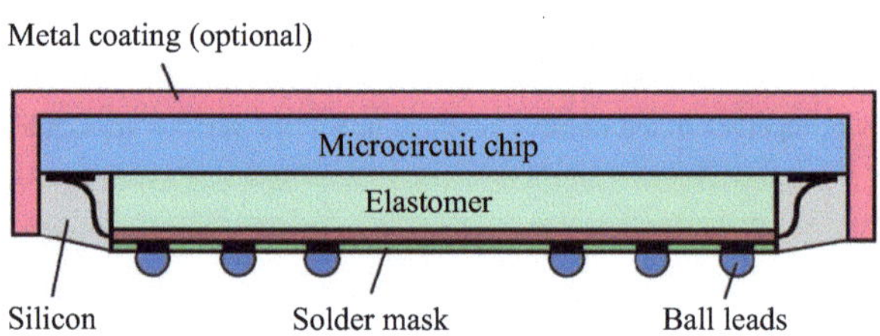

Fig. 8.15 Design of μBGAmicropack (CSP)

8.4 Technology of Installation of Chips on the Board (Chip-on-Board (COB) Technology of Mounting)

Further development of the technology of microcircuit assembly made it possible to place chip inside a package commensurate with the size of the chip itself. In modern terminology, they have several names: Chip Scale Package—CSP, and Chip Size Package—CSP, as well as μBGA, and finally, Slightly Larger than an IC Carrier—SLICC ("slightly larger than a microcircuit chip"). Figure 8.15 proves that this description is actually true. The size of a regular μBGA micropack normally does not exceed 15–20% of the chip size.

Another variation of package design called μBGA (Fig. 8.15) is a micropack containing the chip of the microcircuit. In order to prevent thermal expansion of interconnections, the chip is covered with a neutral, heat-resistant elastomer (e.g., silicone). It contains a polyimide carrier of interconnections of TAB type, the leads of which are connected to peripheral leads of the microcircuit chip by means of micro-welding. A matrix of NiAu coated pads is placed on the carrier field. In this form, microcircuits are tested and then delivered to the consumer. By customer's request, ball leads can be soldered on the contact pads. Modern technologies make it possible to use balls with a diameter of 85 μm; pitch of leads in the matrix can be selected from the following options: 1.27 mm, 1.0 mm, 0.8 mm, 0.65 mm, 0.5 mm. It should be noted that the size of the micropack is usually only 0.3 mm larger than the size of the chip.

The SLICC package mentioned above uses a glass-epoxy base as a mounting substrate (Fig. 8.16). Here, connection of the chip with the substrate is carried out through the balls of the leads placed directly on the surface of the microcircuit chip. Modern printers allow precise positioning and rationing of doses of solder paste to semiconductor plates ("wafers"), on which solder balls are formed after thermal reflow. They are usually used for mounting the chip on a glass-epoxy substrate. In this case, the mounting substrate in SLICC packages, in fact, serves to increase the lead pitch from the microcircuit to the board. The disadvantage is that in this constructive solution different thermal expansion coefficients are not compensated by anything: the system, which is heterogeneous in terms of thermal expansion rations, is sealed by filling the space between the chip and substrate with a compound.

Trends of further reduction of the weight and dimensions of modern electronic equipment designs have led to the development of direct chip mounting technology (DCA—Direct Chip Attach): This is a "chip on board"—COB (Chip-on-Board) or multichip modules (MCM—Multi-Chip Module).

When installing the chips on the mounting substrate, it is not always possible for the manufacturer to pre-verify proper workability of the chips before mounting them on the substrate. Therefore, several solutions have been developed to solve this problem: these include using a Known Good Die (KGD), as well as using micropacks, the size of which only slightly exceeds the size of the chip. The use of micropacks makes it possible to test a microcircuit before installing it on a mounting surface. A typical example of micropack is the CSP-package (CSP—Chip-Scale Packaging).

Since for such CSP packages the lead pitch of the matrix sometimes is 0.5 mm (0.020 in.) or less, the use of special technologies for the production of printed circuit boards is required to ensure optimal wiring of signal circuits in very narrow spaces between the elements of the mounting field.

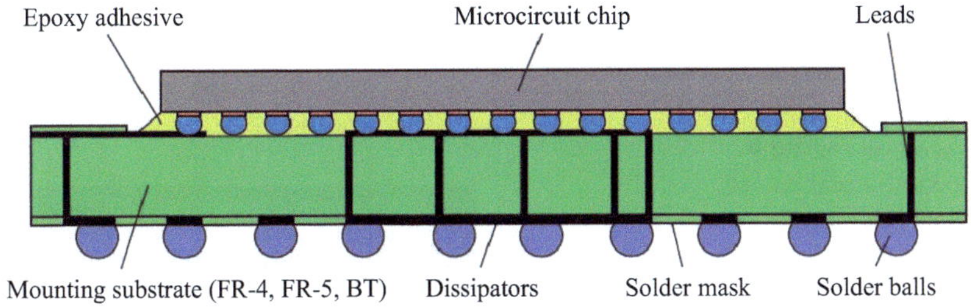

Fig. 8.16 Design of SLICC package

8.5 Wafer Level Packaging (WLP) Technology

The use of the WLP technology for production of high-speed electronic devices suggests that all operations of chip packaging process are performed before separation of the wafer. During creation of the first versions of the WLP technology, the following condition had to be met: all leads needed to be located within the limits of the chip. The package of the component actually corresponded to the size of the chip; such components were referred to as WLCSP. However, a significant limitation of such version of the WLP technology was the number of leads that could be placed under the chip. WLP technology can be applied when the requirements for further reducing the size of components, increasing the operating frequency, and reducing the cost cannot be met by traditional packaging technologies: by unwinding with a wire or mounting a chip using flip-chip technology. Thus, components have been designed (Fig. 8.17) that did not correspond to this classic definition of WLP. Here the semiconductor wafer is divided into chips before packaging, after which the chips are placed in a polymer matrix in such a way that each chip is surrounded by a polymer around the perimeter. Then the polymer matrix with the installed chips is subjected to the operations of the classical WLP technology. The key advantage of such WLP components manufactured using polymer matrices is the placement of a larger number of leads on the component (Figs. 8.18 and 8.19).

Today, there are the following main technologies of production of 3D-integrated structures:

- chip on chip. individual chips are combined and connected with each other;
- chip on wafer: wafers with chips of the same level are separated, and the chips are combined and connected to a wafer of another level, after which the wafer is separated;
- wafer on wafer: wafers are combined and connected to each other, and then separated.

Figure 8.20 shows the main areas of application and trends for 3D assembly of chips.

Another promising area is printed circuit boards with embedded components.

Embedding active and passive components in printed circuit boards of microelectronic devices made it possible to implement new interconnection technologies, without using splicing, providing improved thermal and electrical characteristics.

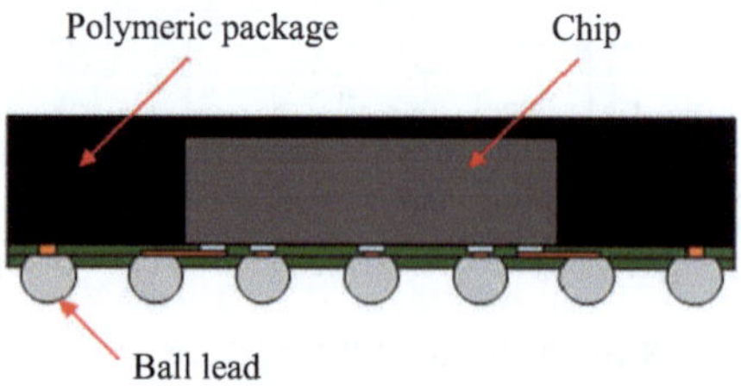

Fig. 8.17 Draft of the design of Fan-Out WLP components manufactured using polymeric matrices

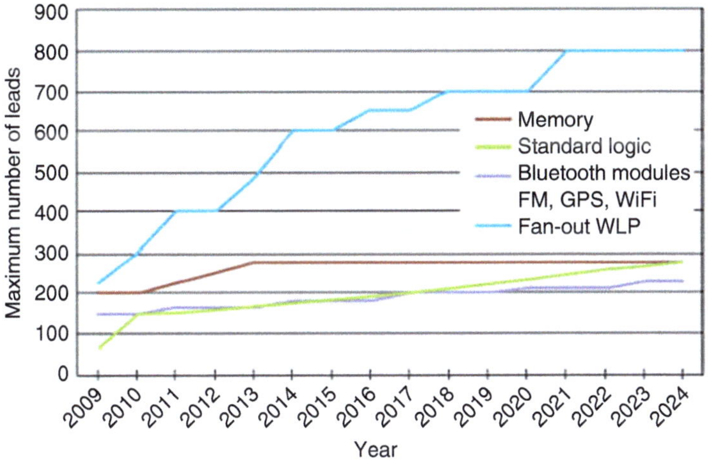

Fig. 8.18 Trend of change in the maximum number of leads of components implemented using the WLP technology [1]

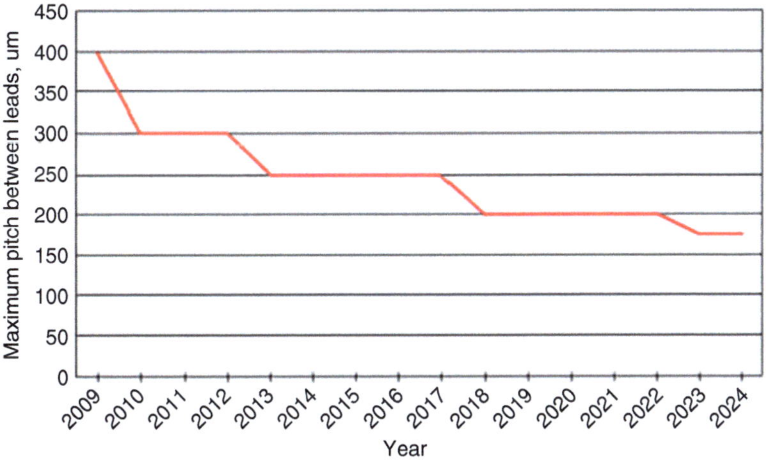

Fig. 8.19 Trend of reducing the minimum lead pitch of the components when using WLP technology (through the example of memory microcircuits) [1]

For embedding passive and active components in a printed circuit board, one of the two processes is used, the sequence of operations of which is shown in Fig. 8.21. When the component is embedded with the pads upwards, a good heat sink from the component is provided. Unfortunately, however, the capabilities of this technology have problems with embedding components of different heights. Embedding a component with pads down has the following advantages: high accuracy of combining the component with the base and the possibility of embedding several components of different heights (about 100–350 μm).

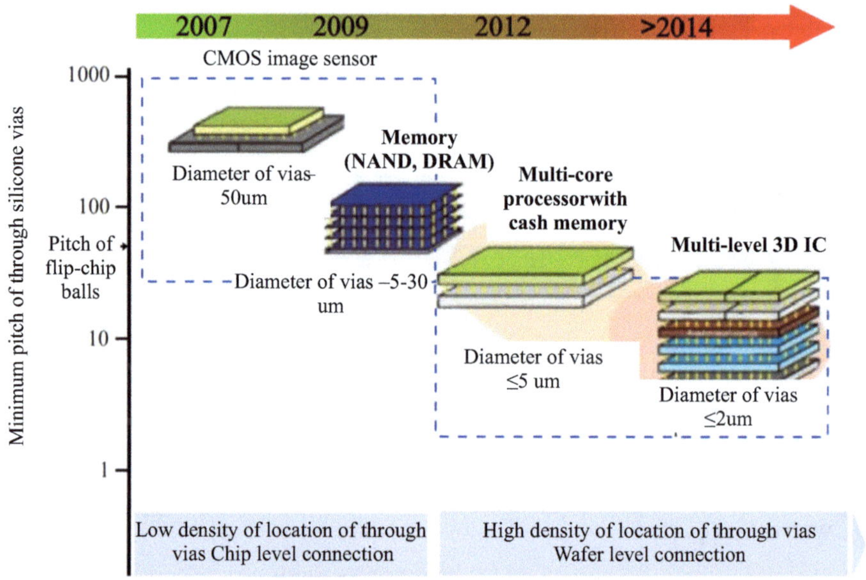

Fig. 8.20 Main applications of 3D-assembly

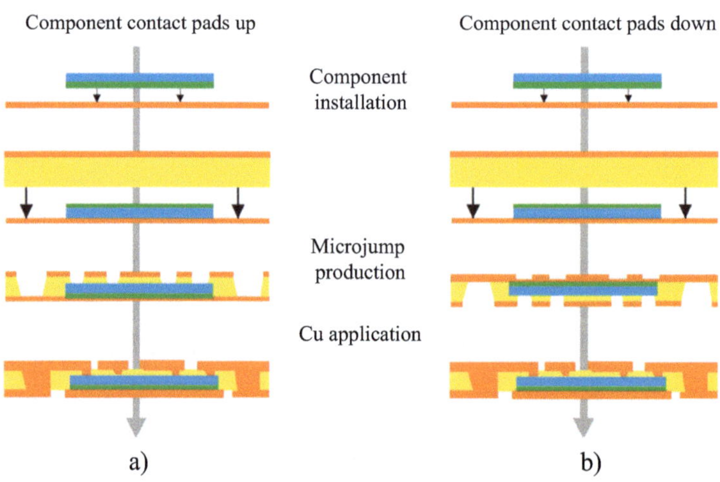

Fig. 8.21 The main options for the sequence of operations when embedding components into a printed circuit board: (**a**) contact pads of the component up, (**b**) contact pads of the component down [1]

The same layers with embedded components can be used as the inner layers of multilayer printed circuit boards of high-speed electronic systems, on both sides of which various additional components of surface mounting can be installed (Fig. 8.22).

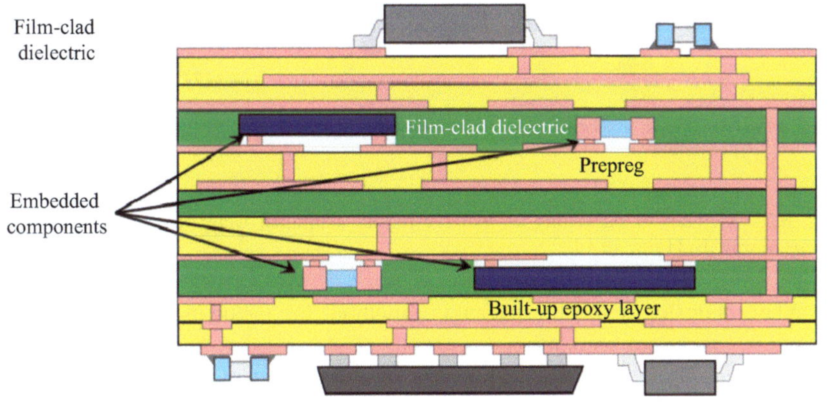

Fig. 8.22 Schematic representation of the printed circuit board design with embedded components for high-speed microelectronic systems [1]

8.6 Features of Packaging of SHF Chips on Wafers

As evidenced by the previous sections in this chapter, packaging of electronic devices (modules) is one of the most labor-consuming and expensive technological operations. However, in practice, not all components of the module require packaging, but only its most sensitive microelectronic components, such as, for example, microwave monolithic circuits (MMIC). Thus, the use of the MMIC packaging technology directly on the wafer (Wafer-Level Packaging—WLP) significantly reduces the requirements for hermeticity of the device as a whole and in some cases eliminates the cost of packaging at higher levels of equipment. WLP technology is a group one and is fully compatible with standard processes for the production of microwave circuits based on semiconductor compounds of the A_3B_5 group. In addition to packaging, it allows for three-dimensional integration of microwave MMIC made on different substrates, which provides significant weight savings and size reduction in the assembly of multifunctional input modules. Of the first companies to use WLP technology for space applications, Northrop Grumman Space Technology (NGST) should be mentioned [5–7] (www.3d-mid.de, www.harting-mitronics.ch).

Packaging on the wafer according to the WLP method can be performed in two ways (Fig. 8.23):

- using the cavity alreadye existing in the substrate or the cavity formed in the additional film, which is covered with a wafer-cover made of a material similar to the substrate (Fig. 8.23a);
- the creation of a microcapsule around the circuit, the open edge of which is closed by the subsequent application of a film (Fig. 8.23b).

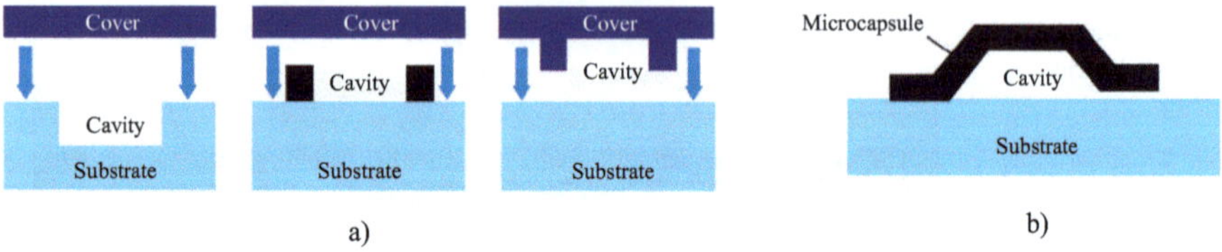

Fig. 8.23 Methods of packaging on wafer using an additional wafer (cover) (**a**) and using a microcapsule (**b**) [1]

When choosing a specific packaging method, the following significant factors should be taken into account.

Compatibility with MMIC technology. In the process of packaging, you should consider the temperature at which the connection of the wafers occurs, the profile of temperature change, and the duration of thermal exposure. Any of these factors may affect the performance and reliability of the temperature-sensitive components of MMIC. Many MMIC containing molecular beam epitaxy (MBE) structures on semiconductor compounds of the A^3B^5 group are very sensitive to temperature. Too high temperature or a longer process time can cause mutual diffusion of thin MBE layers and deterioration of the reliability of the MMIC. It is also necessary to take into account the mechanical stresses that occur during heating between the attached wafer and any contacts of heterogeneous (dissimilar) materials, such as semiconductor-metal or dielectric-metal. To eliminate such stresses, the temperature of the packaging process should be as low as possible.

Hermeticity Ensuring hermeticity in the process of MMIC packaging ensures their stable operation when exposed to humidity and undesirable organic substances. Humidity inside the package can cause moisture condensation in the active area of the MMIC, which leads to corrosion of the structure and degradation of the circuit parameters.

Certain requirements are imposed on the materials used. Materials such as polymers and adhesives can break the seal due to gas exhalation.

In some cases, to improve the reliability of an instrument intended for space applications, it is necessary to create a special controlled environment in the instrument package. The best medium for microwave MMIC is oxygen-free (nitrogen). And special MEMS-based devices, such as resonators and pressure sensors, necessarily require a vacuum or a certain level of pressure inside the package.

Reliability of Packages Connections made in the packaging process must remain chemically and physically stable throughout the life of the instrument. The release of gases and the effects of organic adhesives can lead to a violation of the physical integrity of the package and a violation of the parameters of the MMIC. One of the important factors affecting the reliability of the package is the mismatch of the mechanical properties of the materials of the package and the instrument, which can lead to strong residual stresses, and subsequently to deformation and destruction of the package. In order to avoid these phenomena, it is necessary to carefully design the package and carry out the packaging process at a low temperature.

The process of packaging a MMIC on a wafer, first developed by Northrop Grumman Space Technology, is a group process performed after the completion of standard technological operations for manufacturing a MMIC. During the processing of the front side of the GaAs wafer, the main elements of the monolithic circuit are formed. At the final stage of the process, end-to-end contacts to the circuit are created on the back side of the wafer. The process of packaging on the wafer does not affect the topology of the circuit or the structure of the back side of the wafer and thus does not change the parameters of the microwave MMIC.

WLP process combines the advantages of technological processes of low-temperature brazing and soldering with a thermodynamically stable alloy (eutectic). It also allows to enclose a set of several wafers and is compatible with standard assembly processes. The temperature in the packaging process usually does not exceed 180 °C.

The basic packaging process involves combining two wafers, one of which serves as the base, and the other as a cover (Fig. 8.24). First, microcircuits are manufactured separately on the front surfaces of both wafers using standard group processes. Then, for the subsequent alignment of the wafers, metal connecting rings are applied to each of them. When the wafers are compressed in a special device during the low-temperature WLP process, the rings melt and closed cavities form between the wafers, which are sealing the microcircuits formed on their front sides. After that, the reverse sides of the wafers are processed, on which end-to-end contacts to the circuits are created: DC, RF, and for grounding (www.harting-mitronics.ch) [7].

At the end of the operation, the package of wafers with enclosed microcircuits (Fig. 8.25a) is divided into separate chips (Fig. 8.25b). A separate enclosed cavity with a microcircuit and end-to-end contacts, as well as microphotographs, is shown in Fig. 8.25c, d, respectively [8].

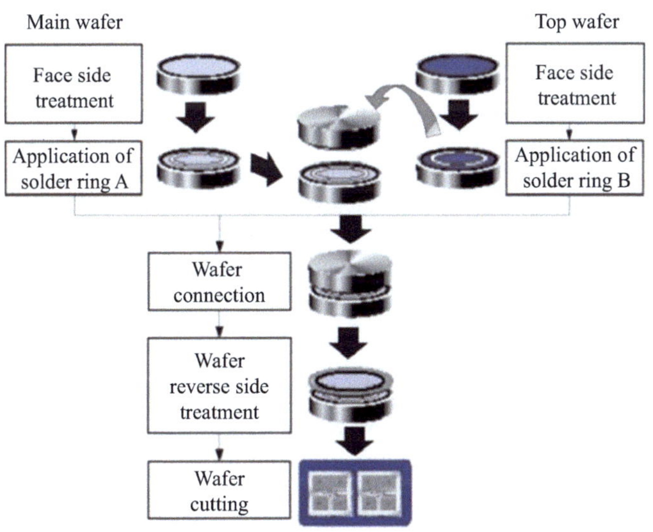

Fig. 8.24 The sequence of operations of the packaging process for the microwave MMIC on the wafer

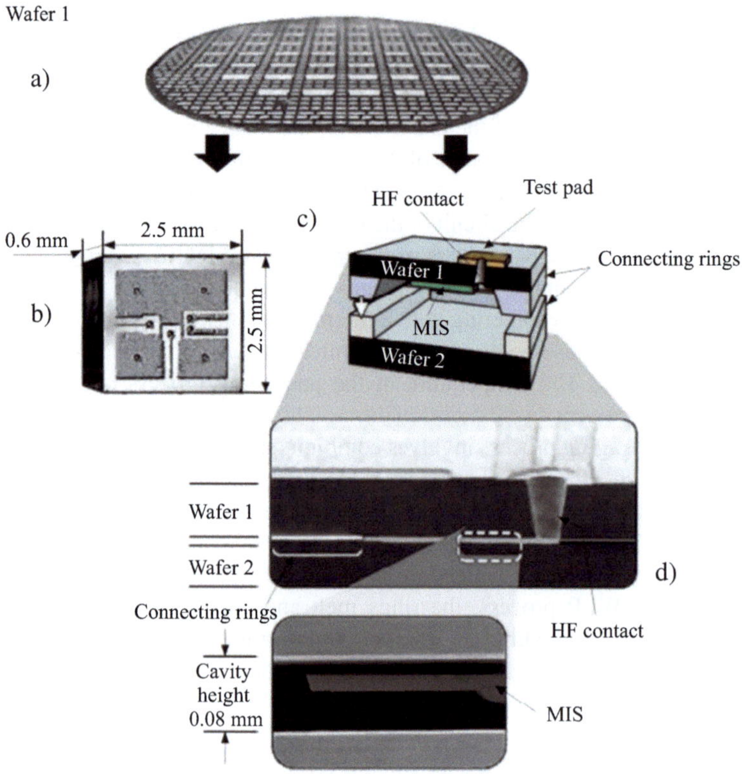

Fig. 8.25 The final part of the process of packaging the microwave IC: (**a**) connected wafers with enclosed MMIC, (**b**) enclosed chips after cutting of wafers, (**c**) cross section of an enclosed cavity, (**d**) its microphotograph [1]

One of the undoubted advantages of the packaging method considered here is the possibility of three-dimensional integration of various circuits or functional units made on different semiconductor materials. So, in a single package, low-noise amplifiers optimized by noise factor and, as a rule, manufactured on InP, and power amplifiers optimized for efficiency and power, usually manufactured on GaAs or GaN, can be combined. Integration of microwave MMIC with silicon digital circuits is also possible.

The basis of the technology of such a multi-layer assembly (Wafer-Scale Assembly—WSA) is the creation of vertical communication elements between circuits, or so-called intracavitary connections (Intra-Cavity Interconnection—ICIC). ICICs are formed between the two inner surfaces of the wafers inside the cavity formed during the packaging process on the wafer (Fig. 8.26). ICIC elements are created simultaneously with connecting rings, and therefore no additional manufacturing operations are required to obtain them. The ICIC elements, which provide the connection of DC circuits, are simple metal bridges, which can also be used to create insulating walls between the individual sensitive microwave components of the circuits.

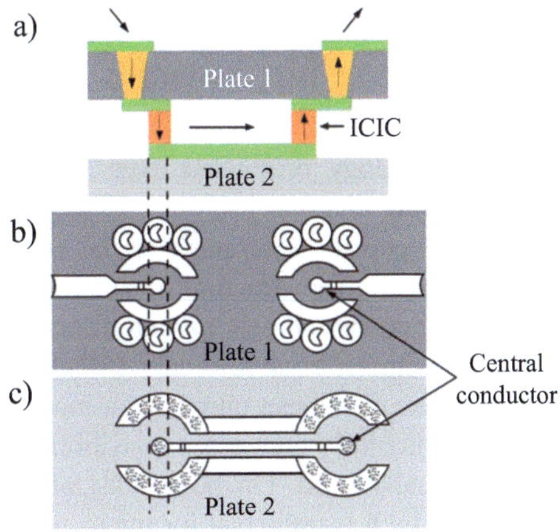

Fig. 8.26 Intra-Cavity Interconnection (ICIC): (**a**) transition design, (**b**) inner side of wafer 1, (**c**) inner side of wafer 2

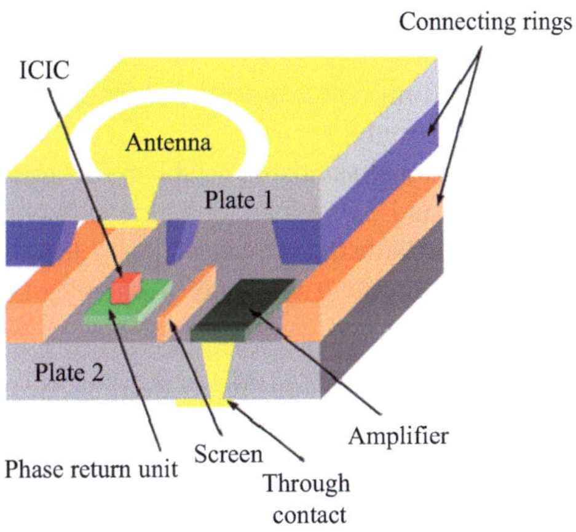

Fig. 8.27 An example of the design of the WLP module integrated with the antenna [1]

ICIC technology allows to significantly reduce the number of wired connections in circuits, to reduce losses and spurious effects, especially at high frequencies.

This WLP technology served as a basis, for example, for various versions of low-noise amplifiers for frequency ranges from 300 MHz to 50 GHz and power amplifiers for Ku and W ranges.

Figure 8.27 [9] shows an example of the construction of such an input Q-band enclosed module containing an antenna, an amplifier, and a 3-bit phase shifter.

Placement of active elements in close proximity to the emitter significantly reduces the mass and volume of the microwave device. It should be noted that the WLP process of packaging and assembly of microcircuits require from the developer mandatory knowledge and use of three-dimensional modeling, taking into account the influence of such factors as the very small height of cavities formed during packaging and the proximity of brazed connecting rings to active elements of microcircuits.

8.7 BGA Technology of Chip Assembly a Ball Grid Array (BGA) Chip Assembly Technology

BGA technology originated over 30 years ago in the IBM Company and was suggested for internal use only. Later, Motorola purchased the license from IBM and developed its technology to the version called OMPAC (Overmolded Plastic Array Carrier). The structure of such component is shown in Fig. 8.28.

After that, the chip is mounted on a mounting substrate made of a composite material with an organic binder (FR-4, FR-5, BT), having a high glass transition temperature. For these purposes, a material based on BT (Bismaleimide Triazine) with a particularly high glass transition temperature and excellent electrical properties is usually used. On both sides of the substrate, a NiAu finishing coating is applied, which is universal for micro-welding (unwinding of wires from the microcircuit to the substrate) and soldering ball leads.

After that, the chip is bonded to the substrate with a special heat-conducting adhesive based on epoxy resin. The substrate used has elements of a conductive heat sink from one side to the other. Part of the ball leads is designed for thermal connections only.

International standards establish the next pitch of the grid matrix leads: 1.5 mm, 1.27 mm, and 1.0 mm and a ball lead diameter of 0.6 mm. Theoretically, design of 50 × 50 mm substrates makes it possible to install 2401 leads with a pitch of 1.0 mm.

The microcircuit chips are normally mounted on a substrate using one of four methods:

1. Thermocompression micro-welding (wire-bonding) is the oldest, most flexible and widely used method (Fig. 8.28).
2. Attaching chips to tape terminal leads (Fig. 8.29), or TAB (Tape-automated Bonding). This method is used for automatic mounting of chips with a small lead pitch on an intermediate carrier. Possible advantage of the method is that it provides the ability to pre-test the chips before its final installation on the mounting substrate.
3. Attaching a flip chip through ball leads (Fig. 8.30). The advantage of this method lies in improved electric characteristics of the system of metallized interconnections.

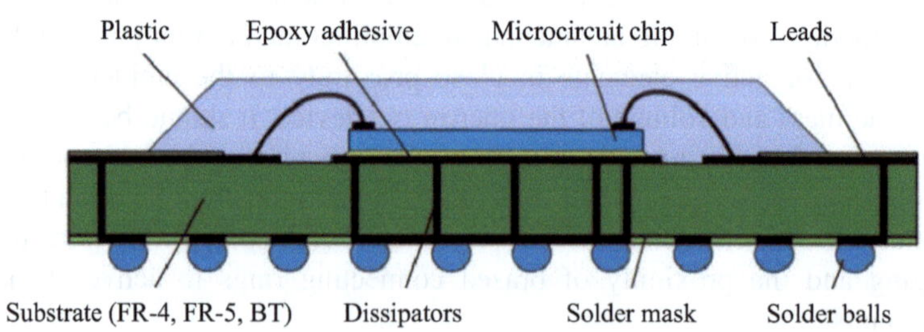

Fig. 8.28 Design of the OMPAC component by Motorola [1]

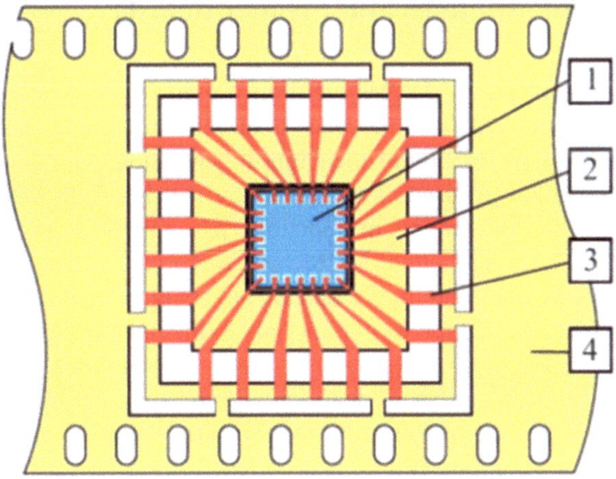

Fig. 8.29 Tape media of microcircuit chip: 1— microcircuit chip; 2—tape frame; 3—contact pads for mounting media on a mounting substrate; 4—transport tape to automate the installation of the chip on the mounting substrate

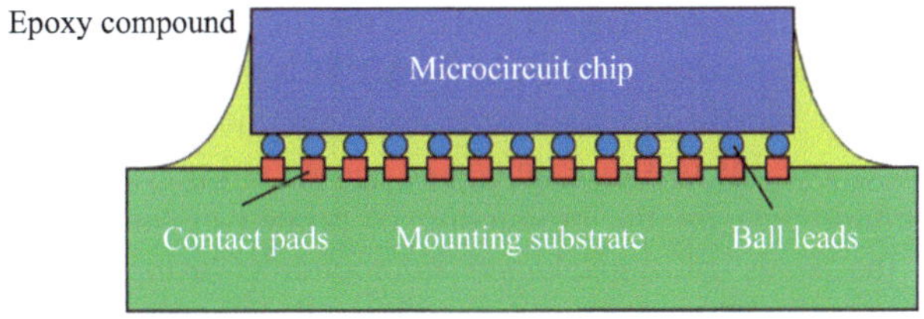

Fig. 8.30 Mounting the microcircuit on a substrate using the inverted chip method (flip chip)

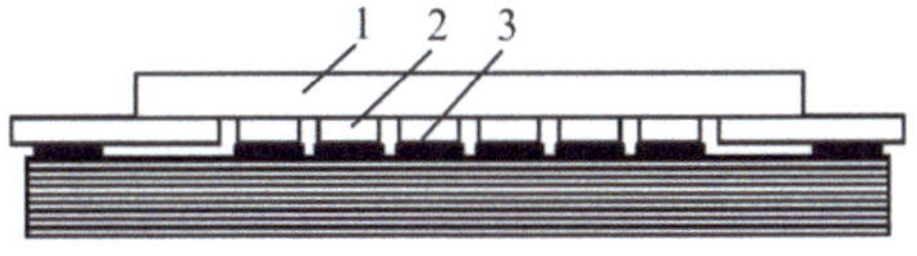

Fig. 8.31 Mounting of the microcircuit using beam leads: 1—microcircuit chip; 2—beam lead; 3—contact pad of the mounting substrate

4. Attaching a chip by means of the so-called beam leads (Fig. 8.31). In this method, technologies of thermocompression and ultrasonic micro-welding of beam leads to peripheral contact pads on a chip and then beam leads to a substrate are used.

8.8 Multi-Chip Modules and Printed Boards

Earlier, multi-chip modules (MCM) were only used in the field of military and spacecraft equipment. IBM was the first to succeed in commercial use of MCM while organizing the production of their fourth-generation computer series. Multi-layer ceramic boards were used as the mounting and heat-removing base, due to which the degree of integration of functions per unit of board area increased significantly. Studies in the field of MCM were performed at that time in Germany (the University of Rostock and the Technical University in Berlin), where the use of MCM technology increased the layout not only in the module plane, but also along its vertical (Fig. 8.32).

In the 1990s, the use of MCM was actually the only commercially available solution for increasing integration and a corresponding increase in the functionality of military and space equipment.

The main problem in creating MCM lies in the difficulty associated with organization of microcircuit chip testing. If the probability of getting into the composition of a three-chip module is usually equal to 95% of suitable chips, the output of suitable MCM is no more than 85%. Then comes the question of identifying suitable and unsuitable components of the module. The profitability of MCM production is directly related to the percentage of suitable microcircuits in the output.

In any case, the most expensive part of the MCM is the base (substrate). Table 8.2 shows the characteristics of the main versions of the mounting bases of MCM [1]. Only for the three options of substrate design shown in Fig. 8.33.

Modern technologies for the production of printed circuit boards are capable of mounting the leads of unpackaged microcircuits using the technology of thermocompression welding or using tape media (TAB-technology).

When using packages with a small matrix lead pitch, the situation is complicated by the fact that the signal connections from the internal leads of the matrix must be

Fig. 8.32 Photographs of experimental samples of MCM elements made at the Berlin Technical University

Table 8.2 Comparative characteristics of the main types of multi-chip modules

Parameter/technology	MCM/L	MCM/C	MCM/D
Type of substrate	Figure 8.10a	Figure 8.10b	Figure 8.10c
Material	Glass epoxy composite	Aluminum, aluminum nitride	Polyimide on ceramics, silicone, aluminum
Metallization technology	Foil dielectrics	Thick film substrates	Thin film substrates
Metalization material	Cu	W, Mo	Cu
Conductor width, μm	75–750	90–125	15
Gap width, μm	100–1000	125–375	25
Minimum diameter of blind holes, mm	0.3	0.13	0.025
Minimum pitch of blind holes, mm	0.6	0.225	0.050
Layer thickness, mm	0.36	0.15	0.006

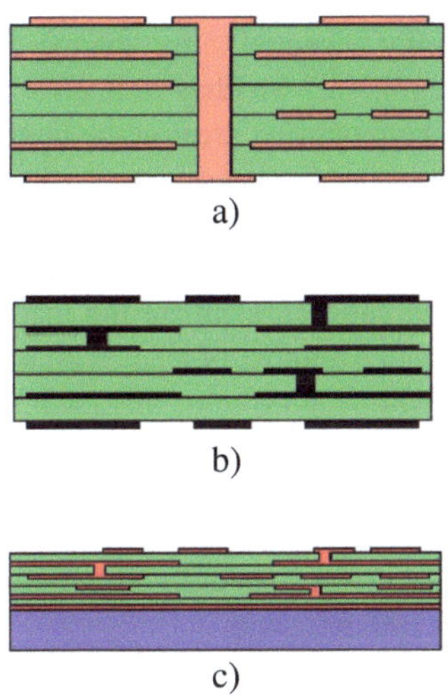

Fig. 8.33 Drafts of designs of the main substrate variants (for Table 8.2)

drawn out between the contact pads of the matrix. Usually, in this case it is possible to place a maximum of two conductors between contact pads. Therefore, in most cases, the conductors from the internal leads of the matrix are held on the inner layers of multilayer printed circuit boards.

For the installation of such components, special layers with "blind" metallized holes are added to the multilayer printed circuit board (PCB) design, on which the wiring of circuits from under the micropacks or from under the unpackaged chips of the microcircuits is realized.

Such thin additional specialized layers are pressed onto the multilayer PCB, after which blind (non-through) metallized holes are made in them. Therefore, this method is called SLC (Surface Laminar Circuit). It can be seen that the method of metallization of through holes and the method of layer-by-layer building are actually connected in it.

A few words shall be said about the interrelation of the number of leads and the degree of integration of space-related application of integrated microcircuits. It is clear that, when mounting the chips on the substrate of the package and the package on the mounting substrate (board), the areas used for this will inevitably increase. This is due to the need to allocate a certain physical space to accommodate the leads. As is well known, the number of leads is subject to the general trend of their increase with increasing integration of microcircuits according to a known law [1]:

$$n = qNR, \tag{8.1}$$

where n is the number of leads, q is the coefficient of connectedness of microelements in the structure of the microcircuit, N is the degree of integration of the microcircuit, R is the rent indicator.

In contrast to the generally accepted term "degree of integration," we call this effect the "degree of disintegration," which is estimated by the ratio of the density of microelements referred to the mounting area on the board, with their density on the chip. For example, if a processor chip has a size of 10 × 10 mm, and the mounting field of its package on the board occupies an area of 5000 mm, this design of the interconnect system is characterized by the degree of disintegration with a ratio of 50. And by what value then can you estimate the matrix of 800 leads? Since the peripheral arrangement of such a number of leads with a pitch of 0.4 mm is occupied by a minimum mounting field of 8000 mm, this means that the degree of disintegration of such a design is 100.

It should be noted that this "degree of disintegration" grows as the hierarchical level of the structure increases: chip–microcircuit–printed circuit–module–unit. For example, disintegration ratio in the onboard computer unit of a modern spacecraft can reach and even exceed 100 thousand.

Table 8.3 shows the data of the author's assessment of the degree of disintegration using different methods of packaging and installing the chips of the processor microcircuits on the mounting base (board) [1].

Table 8.3 Assessment of the degree of disintegration of the microelectronic module depending on the level during the transition to the next level of interconnects

Installation option	Area of the mounting field on the substrate, mm	Degree of disintegration
Chip	10 × 10	1
Flip∗chip	12.5 × 12.5	1.5
Wire∗bond	14 × 14	2
Flip∗TAB	24 × 24	6
TAB	42 × 42	18
Flat∗pack	72 × 72	52
Package with peripheral leads	105 × 105	110

It is obvious that the degree of integration and disintegration should correspond to the current technical level of production development. The use of marginal possibilities of production will cause the cost of the product to become unreasonably high due to the large volume of waste for scrap. Moreover, the reliability of such structures will also not be guaranteed.

As shown in Chap. 2, development of dense assembling of elements in production is ensured by development of the corresponding technologies of production of printed circuit boards with high density of interconnections.

8.9 Assembly of 3D Products Using Flip-Chip Technology

The basis of the inverted chip method is the so-called bar-shaped leads located on the metallized contact pads of IC chips. Usually, when assembling using the flip-chip technology, three options are used to form the bar leads on the chip:

1. Placement of finished solder bars.
2. Manufacturing of solder bar leads.
3. Receiving bar leads by melting the wire.

Figure 8.34 depicts the simplest structure of such a 3D microelectronic module consisting of two chips. The bottom chip is produced using the "flip-chip" technology, and the top one using the usual splicing of wire leads.

To improve the reliability of such unpackaged IC, it is usually recommended to form solder balls not on a chip, but on a board. The sequence of technological operations is as follows: applying metallization to the board and welding acquiring of contact pads; forming a mask from a photoresist layer; deposition of a barrier metal layer; applying solder paste by screen printing; removing the photoresist, and melting the paste to form solder balls of the same height.

Contact pads, on which the bar leads made of the solder are applied, are usually obtained using only two photolithography operations. First, a film of metal nonwettable by solder (Cr, Ti, Zr, etc.) is deposited on the Si wafer with contacts in the form of aluminum metallization, then a photoresistive mask is applied, in the windows of which a film of metal wetted by solder (Cu, Co, Ni, etc.) is deposited

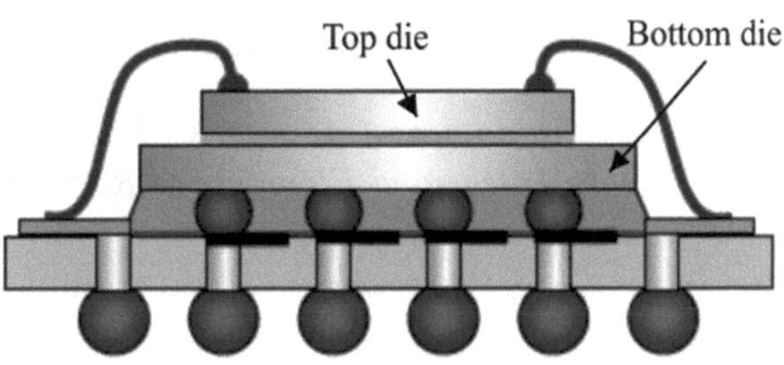

Fig. 8.34 Standard design of 3D products manufactured using flip-chip technology

with a sublayer from a combination of wetted and nonwettable metals. After removal of the first photoresist mask, a mask with large-sized windows is formed from a thick negative photoresist. Solder is deposited galvanically or by sputtering into these windows. After removing the second photomask, etching with metallization nonwettable by solder is carried out using the precipitated solder as a mask. The Si wafer is then heated to the melting point of the solder, which, due to capillary forces, takes the form of balls on the wetted part of the contact pad.

For mounting chips using an inverted chip, a metallization system is usually used, consisting of a layer of composition $WSi_{0.4}$ $N_{0.2}$ with a layer of nickel wetted with solder. Bar leads are formed by successive deposition of layers of gold and tin in the ratio of 80:20 (wt.,%) for the formation of solder eutectic composition with a thickness of 10 μm. The melting of solder is carried out by pulsed heating to a temperature of 330 °C at a rate of 4 °C/s [1].

The AuSn and Au_5Sn phases formed at the initial stage of heating are characterized by mutual diffusion in the solid state. Further heating leads to their melting with the formation of a homogeneous eutectic melt, which forms bar-shaped leads of homogeneous composition when cooled.

In order to ensure the protection of such a module from various electromagnetic fields, interference and reduce parasitic effects, a special design of bar-shaped leads is usually used. Here, the leads consist of a central ledge and a rim of the same height (15–30 μm) coaxially located with it. In this case, the central ledge should be connected to the signal conductor, passing in the middle of the multilayer switching structure, and the rim—with the upper grounded metallization layer. The leads are usually made of indium solder, galvanically deposited on thin-film metallization from a Nb layer, protected by a Pd barrier layer, and additionally coated with an Au layer. The surface of the chips used in the manufacture of multichip modules with operating frequencies of 250 MHz is usually protected by a thin-film dielectric SiO_2.

Sometimes, in order to improve the quality of mounting of devices with bar-shaped leads, a special polymer lacquer is applied to the planar surface of the semiconductor wafers, the thickness of which usually ranges from a minimum thickness of 8 μm to a thickness of 2/3 of the lead height.

8.10 Special Features of Designing Electronic Devices for Spacecraft Applications

8.10.1 Technologies of Assembly of Microcircuits with Integrated Elements of Radiation Protection

Even at the dawn of the space age, it was found that various electronic devices break down in space much earlier and more often than their counterparts in ordinary earth conditions. This is due to the fact that on the orbit these devices are under intense influence of various types of ionizing radiation, the main sources of which are galactic cosmic rays, the solar wind, and the Van Allen radiation belts [10].

Therefore, the task of ensuring high levels of operational reliability of instruments and equipment under the conditions of radiation exposure (electrons, protons, heavy charged particles, X-rays, and gamma rays) is very topical. The radiation resistance of the equipment determines the duration of its active use and trouble-free operation. This is especially true for electronic systems that are part of the onboard equipment of spacecraft. Especially, it is important to consider when creating new generations of spacecraft (SC), since the increase in the level of their operational reliability in conditions of a long stay in extraterrestrial conditions is perhaps the most significant factor.

It is known that solar flares (proton fluxes) and powerful magnetic storms (electron fluxes) are the main hazards for onboard electronics of spacecrafts. The main range of the energy spectrum of protons is in the range from 1 to 10 MeV, electrons from 0.1 to 1.4 MeV. The ratio between the penetrating abilities of protons with energies $E_p = 10 \div 50$ MeV and electrons with energies $E = 1 \div 5$ MeV for $Z = 3 \div 80$ can be described by the expression [11]:

$$E_p = 15 \times E_e^{0.75} \tag{8.2}$$

For conception of irreversible changes in semiconductor instruments, the absorbed doses of radiation must be 105 $\div$ 106 rad and more.

In order to increase radiation resistance of electronic equipment, the following main methods are used [12]:

- *technological* (use of specialized technological processes and materials in the manufacture of integrated circuits (IC) and other components of the onboard equipment);
- *constructive* (specialized IC packages, local protection methods);
- *sheet-oriented* (methods of majorization, etc.);

Structural protection is not an alternative to the listed methods, but serves only as an additional method of reducing the radiation load on all objects aboard spacecraft (SC). However, when using standard materials of structural protection (aluminum and its alloys), a sufficiently high level of radiation exposure is maintained, therefore local protection is recommended for the most vulnerable elements.

Local protection is a special additional screen designed to protect critical components and equipment. The use of such a screen does not entail a serious increase in the mass or dimensions of spacecraft units; it is one of the most effective and economical approaches to ensuring the radiation protection of an IC. A promising direction for the development of local protection methods for ICs is the creation of new types of packages with integrated radiation protection screens (RPS) integrated into them.

Local screening of IC was proposed in 1979; the first studies confirming its effectiveness were conducted in the 80s [1].

The effect of radiation on elements and apparatus of a spacecraft is determined by the type of radiation, energy, intensity, and penetrating power of primary and secondary particles in the volume of a substance. RPS are effective for protecting electronic equipment from exposure to the soft component of cosmic radiation (electrons, protons), which is unable to penetrate deep into matter.

As for ionizing radiation with a higher penetrating power (X-rays, gamma rays, muons, heavy charged particles (HCP), etc.), the protection of electronic systems through screens is less effective. It should be noted that when using one method it is impossible to provide a sufficient level of protection of spacecraft equipment from the entire spectrum of space effects. At the same time, the RPS protect the spacecraft elements from the cumulative (dose) component of the radiation load, while high-energy particles, including HCP, creating single radiation effects, have a probabilistic nature of the impact and require other ways to increase radiation resistance.

One of the design solutions for creating radiation-resistant electronic devices is the creation of special packages with integrated radiation protection screens. Integrated radiation protection (IRP) allows to significantly increase the marginal dose of ionizing radiation of chips by shielding.

The effectiveness of such protection of the chip from the effects of radiation depends on the parameters of the orbit on which it will be used and may differ for hundreds of times. The GEO electron trap zone is 36,000 km high, the MEO proton trap zone is 2000 $\div$ 10,000 km high, the LEO proton trap zone—500 $\div$ 800 km high [11].

So, in a satellite orbit, the GEO package with an IRP allows a 600-fold increase in the accumulated dose, providing the components with the ability to withstand radiation of more than 600 Crad.

When using an electronic device in MEO orbit, where the predominant particles are protons, packages with IRP can enhance the protection of the chip from hard cosmic radiation up to four times.

The use of such packages also provides protection against X-rays in the range of 1.3 $\div$ 1.5 times [13].

The use of combinations of technical solutions using radiation-protecting packages allows the developers of integrated circuits to be more likely to achieve the desired results.

As screens, in order to increase their protective properties, materials created in the form of spatial structures consisting of separate layers of two (or more) homogeneous metals with different atomic number Z are used.

The combination of substances with small and large Z in one heterogeneous structure creates conditions for multiple reflections of particles and γ rays from interfaces between different substances layers and the return of photons to layers filled with high-consistency substance without going beyond the outer limits of the structure. Due to its specific design, such a structure is a kind of trap for photons.

Since 2014, production of two types of packages equipped with radiation protection screen has been organized in Belarus: 100-lead CQFP and 144-lead CQFP.

The package design is shown in Figs. 8.35 and 8.36, the main technical characteristics are given in Table 8.4 [11].

The package is sealed by soldering the upper protective screen with solder based on gold alloy at a temperature of 280–320 °C.

The principal difference in the design of these packages is the way they are sealed. The design of the 100-lead package provides sealing by the method of seam-roller welding, and the 144-lead package design provides for sealing by soldering with an upper protective screen using solder based on gold alloy at a temperature of 280–320 °C. In both versions of the packages, the lower protective screen acts as a mounting platform for the installation of the chip, which makes it possible to significantly reduce the thermal resistance of the package.

A number of semiconductor production companies in the production of radiation-resistant elemental base for military and space technology and weapon systems are already actively using packages with integrated protective screens of their own original design: Maxwell (USA) (Fig. 8.37a), Actel (USA) (Fig. 8.37b), 3D Plus (France) (Fig. 8.37c).

It should be noted that the packages with integrated protective screens manufactured by Actel, 3D Plus, and Maxwell were intended for sealing the internal volume only by the soldering method. Therefore, the use of local protection methods, in particular, the creation of special packages with integrated radiation-protective screens, is one of the promising areas of development of radiation-resistant electronic equipment. Such technologies allow the use of cheap chips of commercial and industrial grade microcircuits instead of expensive radiation-resistant ones, which gives developers the opportunity to reduce the cost of onboard equipment and expand the range of components used.

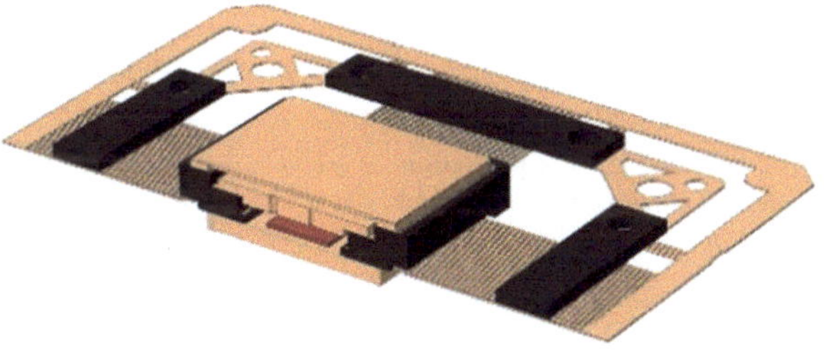

Fig. 8.35 Specialized 100-lead planar metal-ceramic package with integrated radiation protection screens

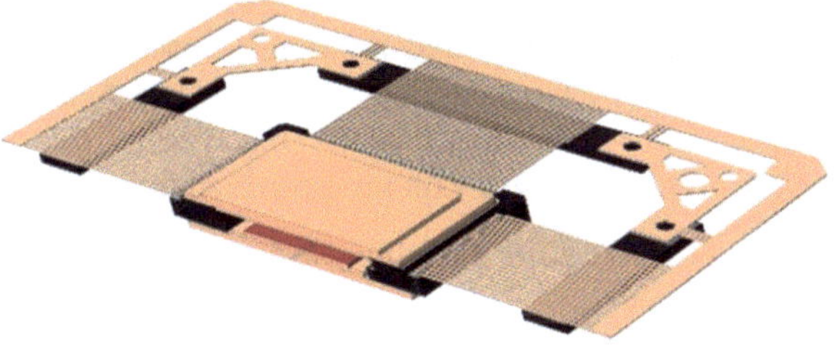

Fig. 8.36 Specialized 144-lead planar metal-ceramic package with integrated radiation protection screens

Table 8.4 Main technical characteristics of packages with radiation protection

Parameter name	100-lead planar MK package	144-lead planar MK package
Number of leads	100	144
Lead pitch, mm	0.5	
Dimensions of package body, mm	≤18.65 × 18.65 × 5.06	≤24.20 × 24.20 × 4.35
Dimensions of mounting pad, mm	≥7.5 × 7.5	≥12.5 × 12.5
Depth of mounting well, mm	0.5 ± 0.05	
Method of sealing	Seam-roller welding	Soldering
Lead location	Evenly on the four sides of the package	
Design features	The rim is electrically connected to lead No. 100 The lower protective screen (MP) and the upper protective screen are electrically connected to lead No.1	The lower protective screen (MP) is electrically connected to lead No.1 The upper protective screen (cover) is electrically connected to lead No. 144
Coating of metallized surfaces and metal parts of the base	N231.1,8 (nickel 2 μm, gold 1.8 μm)	
Cover coating	Chemical N3	–
Insulation resistance, Ohm	≥109	
Insulation strength (max. test voltage), V	≥200	
Resistance of conductive elements, Ohm	≤1.0	
Capacity of conductors (between the MP and each lead and the rim and each lead), pF	≤2.0	
Capacity between adjacent leads, pF	≤2.0	
The provided level of protection against radiation (max. absorbed dose) in space conditions, crad	≥100	

Specialists also designed new packages for IC with integrated radiation-protective screens, made on the basis of the composite material W-Cu.

Two variants of the metal-ceramic package were investigated, which differ in the methods of sealing the "under-package" space (Fig. 8.38). In the first case, sealing is carried out by the method of soldering (Fig. 8.38a), and in the second case—by the method of seam-roller welding (Fig. 8.38b).

The use of packages with integrated radiation protection allows to:

- provide enhanced radiation resistance of IC, electronic components and equipment;
- use electronic components of commercial and industrial classes for space applications;

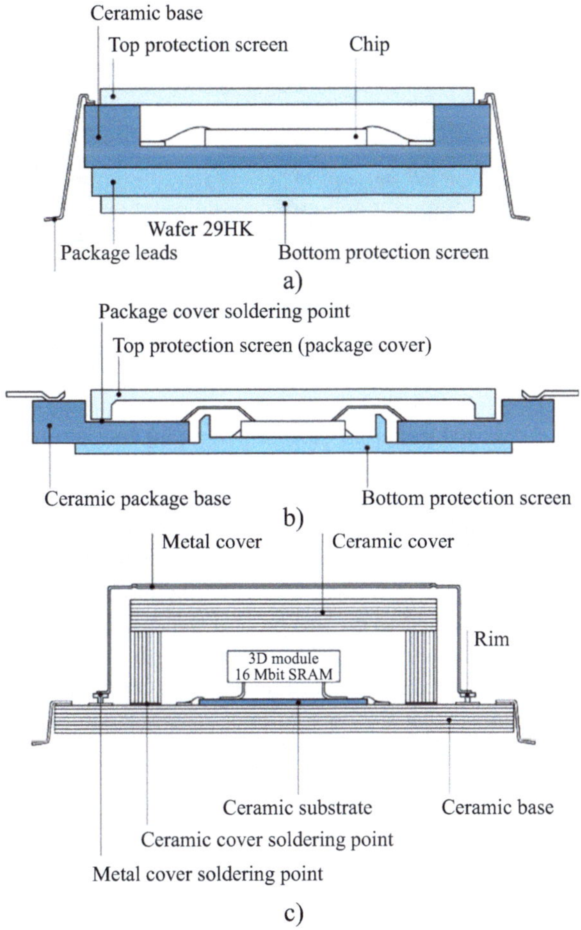

Fig. 8.37 IC package designs with protective screens integrated into them: (**a**) package design with REE f. Maxwell, (**b**) package design with REE f. Actel, (**c**) package design with REE f. 3D Plus

- expand the range of IC used and thereby reduce the cost of equipment in the manufacture of space equipment;
- ensure the reduction of weight and dimensional parameters compared with standard methods of structural protection.

As recent studies have shown, multilayer structures and composite materials have the greatest protective properties against radiation exposure, which allow reducing the dose loads on spacecraft equipment elements several times while reducing weight and size characteristics (www.3d-plus.com/radiation-assurance.php) [14]. Such RPS include WALOPACK materials, which are structures with alternating layers of Al_2O_3 ceramics and tungsten powder. In studies [15–17], it was shown that for the geostationary orbit (GSO) the three-layer protection composition is most effective, in which a material with a high atomic number Z is located between layers of a material with a low Z, in particular, composition Al-W-Al.

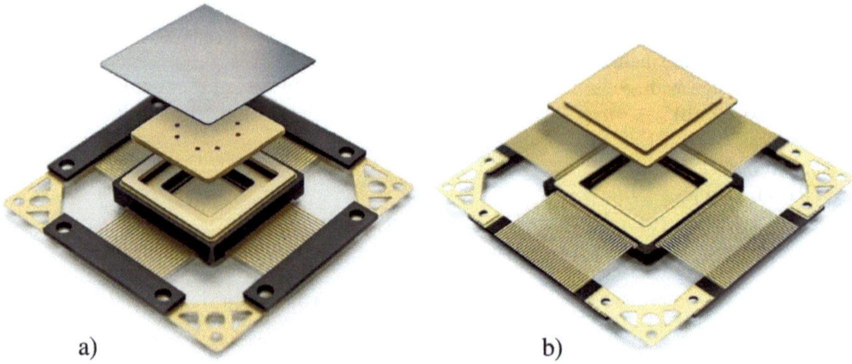

Fig. 8.38 Variants of metal-ceramic package with integrated RPS based on composite material W-Cu: (**a**) 100-lead planar package with sealing by soldering, (**b**) 144-lead planar package with sealing by seam-roller welding

The configuration and size of the RPS should correspond to the standard IC packages. From this point of view, the technology of applying RAD-COAT protective coatings is universal, which, unlike the RAD-PAK and WALOPACK technologies, is not tied to the design of a standard IC package. RAD-COAT coatings are made on the basis of a plastic matrix with a filler in the form of tungsten powder. However, taking into account the density of these coatings (13 g/cm^3), it can be assumed that they are inferior in efficiency to protection like RAD-PAK and WALOPACK.

8.10.2 Technologies of TSV Assembly of Chips for Aerospace Applications

TSV technology (through-silicon via) is used more and more in the production of high-speed electronic 3D-devices; in this technology, the operation of splicing chips in the package is excluded due to the formation of through silicon interconnects (vias) in silicon wafers. This provides for the maximum possible level of IC integration. TSV technology includes the processes of formation of joints, deposition, filling, removal of metal from the surface, wafer thinning, joining/stacking, inspection, and testing.

The TSV technology also makes it possible to significantly increase the amount of information that can be recorded in the memory microcircuit of the electronic device. Since each chip in the memory module has the same topology, it is possible to assemble 3D modules consisting of absolutely identical chips. This allows to repeatedly increase the amount of memory for the microcircuit compared to single-chip microassemblies.

The typical design of the final product manufactured using TSV technology is shown in Fig. 8.39.

In addition to the process of making through vias in silicon wafers, additional operations of wafer thinning, deep plasma etching, and viral filling operations are used.

Depending on the sequence of technological operations used, two main approaches in the production of 3D space and military information systems are distinguished:

- Via First (making vias before forming structures);
- Via Last (making vias after the formation of structures).

When the vias are filled with conductive material (for example, copper) before the formation of structures (Via First), the process of filling the holes is not limited by the maximum allowable working temperature limit for CMOS (450 °C). However, filling of the holes is an extremely difficult task from the point of view of technology.

Anatoly Belous, one of the authors of this book, in cooperation with Vladimir Dubin, member of the Minsk Radiotechnical Institute, suggested the idea of creating a new technology similar to TSV. But the problem was not only the development of the technology for forming these "through holes," but also the choice of material and the technology of filling the through holes with this material. As a result, Vladimir Dubin created such a technology at AMD, where he worked for many years as one of the leading managers of the company eventually organizing mass production of microcircuits with copper metallization of interconnections, It was confirmed by a number of US patents obtained by him and AMD employees. It should be noted that one of the first practical results of this collaboration was the creation of a manufacturing concept, one of the first designs of interposers, the manufacturing technology of which was based on the ideology of the TSV-assembly.

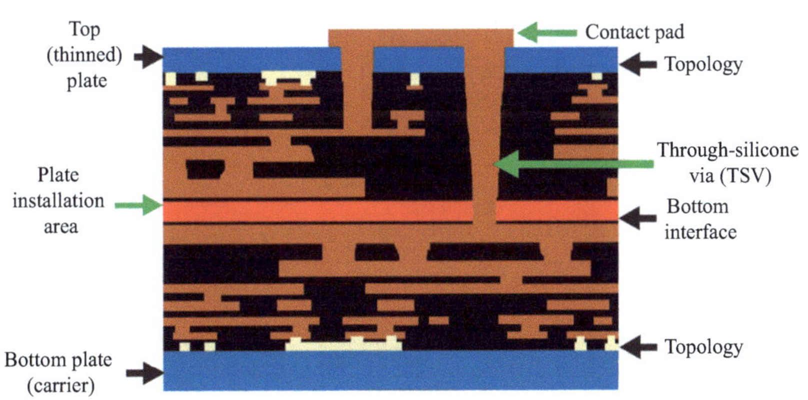

Fig. 8.39 Typical design of a microelectronic product made using TSV technology [1]

It should be noted that the beginning of the production technology of 3D-IC was laid in PoP-type products (Package on Package). Here, the reduction of the dimensions of microelectronic devices was achieved by mounting the packages in three-dimensional (3D) assemblies using interconnections and switching layers placed directly on the packages. This technological method, which is a logical development of the previously existing surface-mounting technology, was only used as intermediate, since developers were not able to deal with multiple challenging limiting factors at once: optimal heat distribution (which was especially important for space and military applications), low reliability of connections available at the time, and the number of elements in the microassembly. However, this method allowed the developers of electronic devices to significantly reduce the time of transmission of the signal from one microcircuit to another.

Figure 8.40 shows a graphical interpretation of the evolution of various types of design of microelectronic modules of the 3D type, which shows all the diversity of technologies and design solutions used by developers.

Practically in parallel, with this idea of stacking in a three-dimensional assembly of enclosed elements, many developers, mainly the ones related to military and space themes, had the idea of performing the same operation for chips and then packaging into a single package. Such an assembly, later called "stacked die" and allowed to significantly reduce the amount of space "stuffing" by mounting the chips "on top of each other" and their subsequent interconnection by a known method of unwinding the wire leads. Thus, a natural logical continuation of this technique was the technology of performing TSV (Through Silicon Vias—vias in silicon), which made it possible to completely eliminate the unwinding operation from the classical technological chain, ensuring the highest possible level of integration of the IC.

Thanks to TSV technology, it has also become possible to drastically increase the capacity of memory microcircuits for onboard computers of spacecraft. Due to the fact that each chip in the memory module has the same topology, it has become possible to assemble 3D modules consisting of many identical chips, but allowing a multiple increase in the memory size compared to single-chip microassemblies.

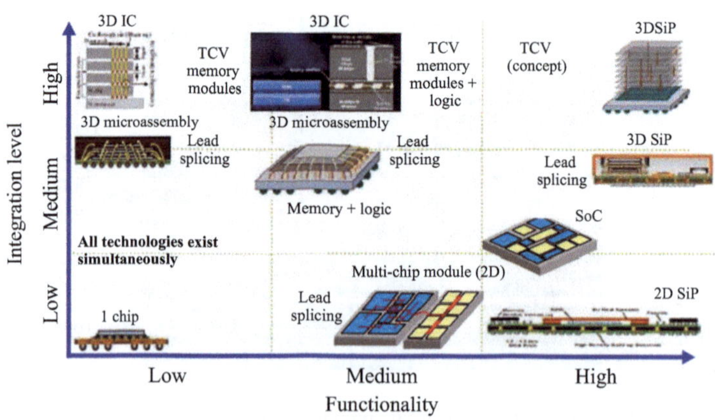

Fig. 8.40 Evolution of types of SD microassemblies [1]

The use of TSV technology in the assembly of such hybrid modules provided an additional impetus of SiP-type devices (System in Package), a logical continuation of which was the so-called interposer—a single element assembled at a chip level without the use of intermediate operations of splicing wire leads for switching elements of the microassembly. The concept of an interposer for aerospace applications deserves separate consideration.

To complete this section on the technical features of 3D technology, refer to Fig. 8.41, where the basic phases of the implementation of a typical 3D-IC production process are presented in a simplified form. First, the silicon wafer passes the

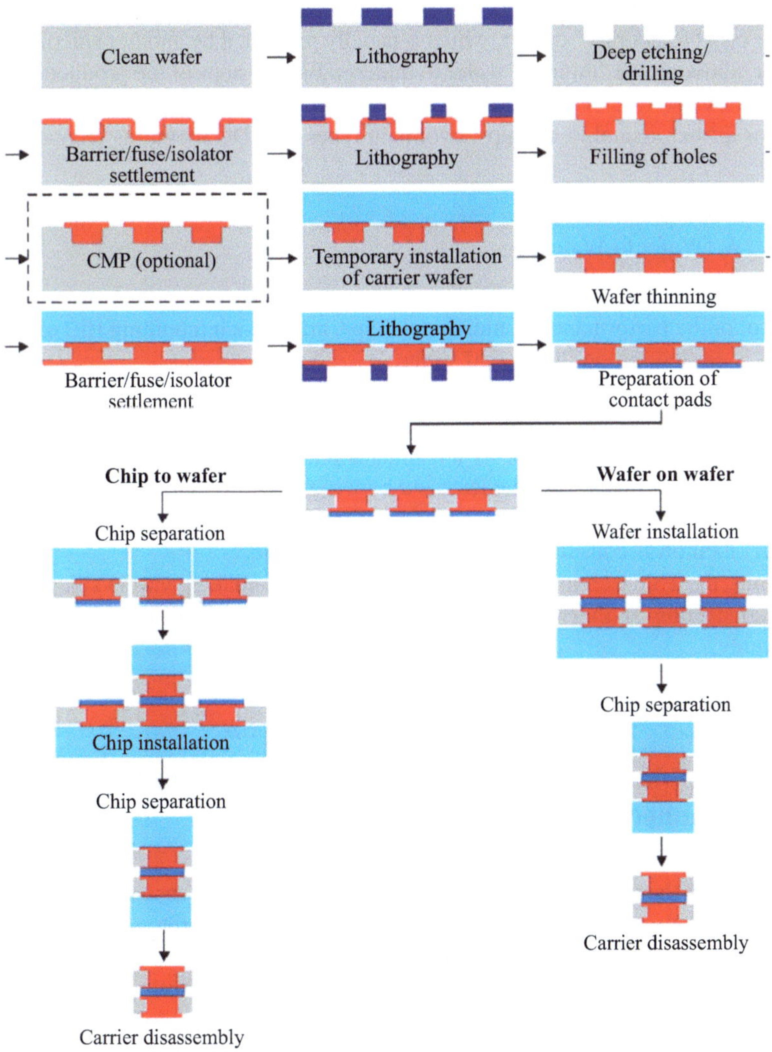

Fig. 8.41 Typical technological process of production of 3D IC [1]

required number of lithography cycles, then, according to the topology developed in advance, the vias are etched. They are filled with a conductive metal (most often copper, as recommended by V. Dubin), either by galvanic technique of the filling of conductive holes with the use of accelerating or retarding additives, or by vacuum deposition. After filling the holes, the operation of chemical-mechanical polishing should be performed.

The operation of reducing the thickness of the silicon wafer is designed to open the "blind" vias, which ultimately allows to increase the performance of the micro-electronic device and improve the heat removal from the chip (for example, SOI—Silicon on Insulator—used by the leading developers of microprocessors to improve performance). The thinning operation is preceded by the operation of mounting the wafer on a "temporary" carrier, which is usually presented by another silicon or glass wafer, allowing the "thinned" wafer to undergo further steps of the production cycle. On the reverse side of the thinned wafer the contact pads for installation are put.

For further assembly of chips, two main types of 3 DC mounting technology are used:

- C2W (Chip to Wafer);
- W2W (Wafer to Wafer).

In the first case, the thinned wafer is subjected to mechanical or laser cutting and control of the performance of individual chips, after which mounting (fit) of suitable chips is carried out on the second wafer.

In the second case, the wafers are mounted directly to one another, and only after that they come in for cutting.

Each of these options has its advantages and disadvantages. For example, the "chip-to-wafer" method is characterized by a higher yield ratio, but with a decrease in the chip size and an increase in the required mounting accuracy, the "wafer-to-wafer" method is considered to be more cost-effective.

Thus, the TSV technology (Through Silicon Vias—making vias in silicon) allows interconnecting at the level of wafers or chips mounted in a three-dimensional assembly.

8.11 Problems of Ensuring Reliability of Soldered Connections

In the end of the twentieth century, the most common way to ensure the reliability of a developed electronic product was experimental testing of prototypes—an iterative process of making samples, testing them in the conditions of maximum severity and eliminating the identified design weaknesses. This process lasts until the stable state of the device is reached, or (in the worst case scenario) until depletion of funds and/or time, as shown in Fig. 8.39.

However, this method has many drawbacks—time consuming, high cost, and often the inability to use/fix devices after finding the error/failure. The method does not always allow to identify design errors, besides the conditions of generation of tests in many cases do not meet the actual conditions of operation. In early twenty-first century, an alternative (or rather an addition) to the above method appeared—the so-called design for reliability (Fig. 8.42).

This method is based on a complex of precise reliability prediction methods, first of all— for solder joints (SJ). Such methods based on the provisions of physical theory of reliability make it possible to significantly reduce the time of production cycle of the product and its final cost, as well as to increase the reliability of the results predicted for operation of the product in real conditions.

It should be noted that predicting reliability of soldered connections is a complex task that requires application of knowledge from various fields of physics, which shall be properly understood and taken into account during generation of adequate equations of dependence of soldering properties on various operating conditions, varied geometry of connection of the mechanical system including chip, component, and printed circuit board; these equations shall subsequently be used to establish parameters of the technological process during production. Most fatigue failures can be attributed to thermomechanical stresses mainly due to the difference in coefficients of thermal expansion (CTE) of the component, solder, and substrate/printed circuit board (PCB) [18]. However, the cause of failure (destruction of SJ) may also consist in the impact of other factors that influence both individually and jointly [19]. The main types of applied loads and the processes arising from them in SJ are shown in Fig. 8.43.

During operation of the electronic device, these impact fields are temporarily and spatially dependent; besides, some of them are interdependent. For example, Joule heating is determined by the current density, which can vary due to electromigration, and the state of mechanical stress which arise due to the difference in CTE, stresses developed due to the migration of atoms/ions due to the action of "electron wind" forces, and external loads operating on this site of the designed electronic device. Of all the above, the process of formation of conditions reducing reliability of SJ is most influenced by the structure of strength and rigidity of the design, thermomechanical fatigue, electromigration, and thermomigration. If lead-free SJs are used, this is supplemented by the effect of growth of whiskers and, as a result, by compressing

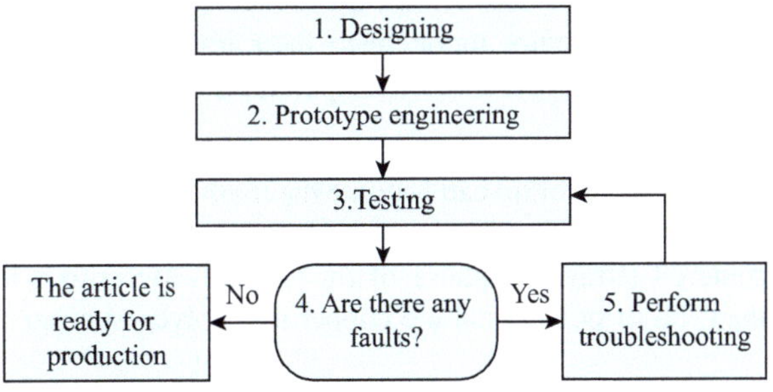

Fig. 8.42 Standard method of bringing the developed device reliability to the required level

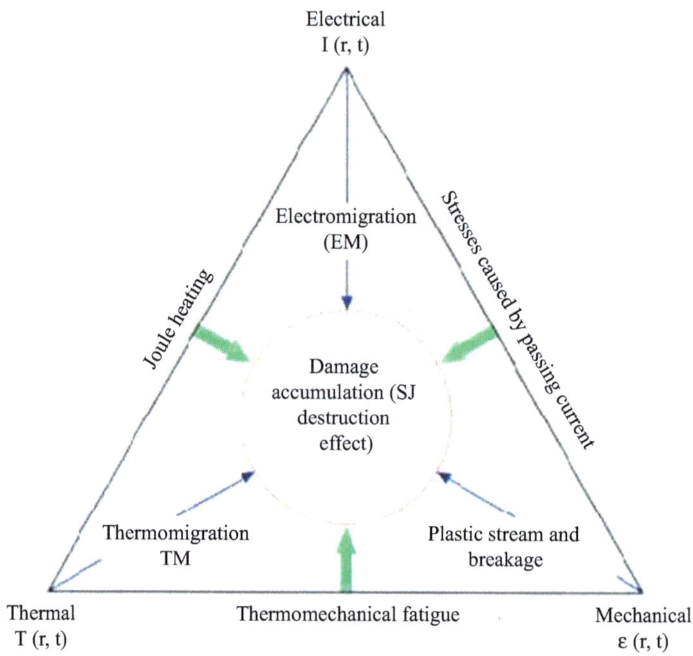

Fig. 8.43 Pyramid of fields of loads is an illustration of the complex process of accumulation on damages in full connections due to the effect of various fields of loads occurring during operation

internal stresses. Of the above five processes, electromigration, thermomigration, and growth of "whiskers" are the main problems of SJ reliability due to the increasing degree of miniaturization of electronic devices. Further development of this trend will cause the combined effect of fields of loads to contribute more and more to the accumulated damage of SJ.

Let us consider the essence of these five problems in more detail and state the main ways to solve them [1, 2, 20, 21].

Thermomechanical Fatigue (TMF) Thermal effects are among the most powerful factors capable of causing significant damage to SJ, affecting its reliability. Accumulation of damage caused by TMF in SJ occurs due to a strongly expressed inhomogeneity of the stress distribution caused by the difference in thermal expansion coefficients between the different elements forming the SJ. Here, all parameters of the temperature mode have their impact: operating temperature, heating and cooling rates, difference of temperatures in the adjacent areas of the structure, etc.

There are more complicated aspects to this problem. For example, it is known that the anisotropy of tin can become the main reason contributing to the accumulation of damage, since the difference in the CTE between the a- and c-directions of space-centered tetragonal phase of tin is nearly two times larger than the difference in the CTE of polycrystalline copper and polycrystalline tin.

The manifestation of damage due to TMF during tests usually occurs after several hundred temperature cycles, while the residual mechanical and electrical characteristics deteriorate significantly from the earliest stages of TMF. The prevailing damage regimes due to TMF are the sliding of grain boundaries and decohesion destruction caused by microstructural enlargement. These events take place in the entire volume of SJ; however, the greatest amount of damage is accumulated in the areas adjacent to the substrate and in intermetallic layers, which ultimately causes the formation of localized deformations. During the later stages of TMF, when residual properties tend to normalize, surface damage progresses in individually distributed (micro) fractures and causes sometimes fatal destruction.

Currently, there is a special mathematical device for predicting reliability of SJ, which is precisely focused on accounting for the influence of the thermomechanical fatigue. The main problem, as it turned out [19], was the enlargement of intermetallic compounds, affecting the reliability of SJ.

Electromigration (EM) Recently, electromigration has become an important issue. Previously, the values of current density in electronic interconnects did not reach the values causing effects in SJ, which need to be considered from the point of view of reliability. The ongoing miniaturization of electronic components together with an increase in the operating temperatures of electronic equipment has made EM one of the potential problems of SJ reliability. Negative phenomena associated with EM are primarily due to the materials used. SJ is a multiphase alloy, whose properties, changing under the influence of EM, will significantly depend on the presence of atomic particles, mutual diffusion in the solid state of substances in the composition of SJ, morphological features of microstructural components and phase stability.

The currently developed computational methods for studying EM do not allow to accurately predict the time of the appearance of a failure due to the impact of this phenomenon. However, there are tools that help to model it and even suggest ways of improving the device design if negative EM consequences exceed certain critical values.

Mechanical Deformations (MD) As part of an electronic module, SJ can be subjected to deformations caused by mechanical loads on the equipment, exposure to external temperature, heating as a result of the passage of electric current, and the influence of other environmental factors acting during the manufacture and operation of the equipment. It should be understood that reliability studies carried out on assembled devices and even more so on modules consisting of many electronic devices cannot adequately assess the problems of mechanical states arising at the level of materials, design parameters of the current-conducting board structure, geometric and physical–chemical characteristics of SJ. On the other hand, studies that are carried out on samples of SJ without any restrictions, taking into account the configuration of SJ, do not provide a complete picture of the prerequisites for possible failure of SJ for mechanical reasons.

So, to study the mechanical factor of damage development in SJ components in BGA-, CSP- and other packages, tests are carried out in which a shock load is applied to each SJ ballpoint,—Charpy impact test (ASTM E23, ISO 148-1, EN 10045-1 standards). However, in real-life conditions of using electronic equipment, solder joints, as a rule, are influenced by a shock load applied at any other points of the module constructive—not those that are loaded during the specified standard tests. Consequently, such tests cannot provide the necessary detailed information on stress states, types of faults, etc., which is crucial in structural design or robust design [22].

Another obvious reason why standard tests cannot answer all the questions about the durability and reliability of SJ is the absence of tests that correspond to specific situations of actual operation in the regulatory documentation. For example, in the automotive and aerospace industry, there are several scenarios where an accidental collision can lead to repeated shock loads that are not simulated in any way in the tests regulated by current standards.

When studying the reliability of SJ as a function of its mechanical state, one has to take into account a large number of various reasons for possible failure due to the nuances of the applied manufacturing technology. A good example of this kind is the dependence of the reliability of the soldering of the BGA components on the way in which the pads were obtained on the printed circuit board.

The fact is that there are two main types of contact pads where the BCA packages are installed: pads limited by metal (metal defined—MD), and pads limited by solder mask (solder mask defined—SMD). There are gaps in the solder mask around the MD pads, and when using SMD technology, the mask partially comes to the edges of the contact pad. The diameter of the SMD pads in equal other conditions is larger than the diameter of the MD pads, that is, they are better adhered to the board surface and, according to this particular indicator, are preferable in terms of SJ reliability. However, a different kind of problem is associated with them: the material of the solder mask, expanding from heating during the soldering process, deforms the ball leads of the BCA component, distorting the geometry of the solder joint, and creating a stress concentrator in the ball, which can cause cracks on the mounting surface (Fig. 8.44). This effect, which is almost microscopic in size, is nevertheless so serious that certain standards expressly prohibit "the use of components of the BGA type, in which the contact pads are limited by a solder mask," when assembling printed circuits of class C (electronic equipment for important purposes) [1, 2, 23].

Speaking about BGA-type components, one can also note the important role that the size of ball leads plays. Depending on the size of the balls, the reliability of such SJ in different conditions can vary greatly. For example, in [24], the effect of stencil thickness and shape of a stencil on the mechanical reliability of BGA SJ are shown.

Thermomigration (TM) Thermomigration is an effect similar to the well-known thermodynamic effect of Sore in liquids [25], which consists in the appearance in the mixture of a gradient of concentration of solid components due to the temperature difference between its different parts. As a result of thermomigration in SJ, due to the effect of high temperatures, microstructural enlargement of phases can occur.

During the accelerated electromigration tests, Huang et al. [26] found that SJ, which at the moment are not under current, but are adjacent to those through which

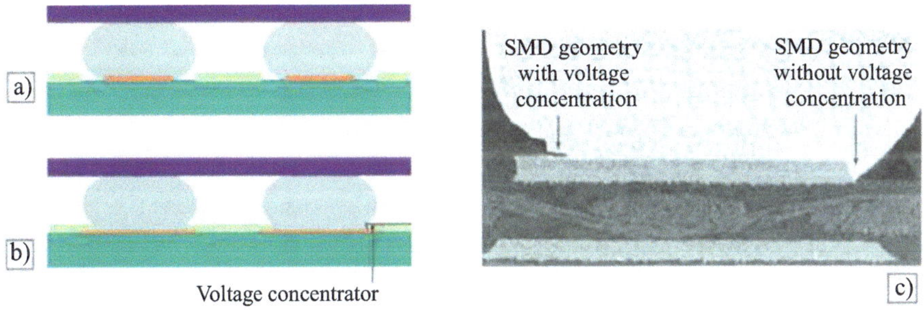

Fig. 8.44 Two types of pads for mounting BGA components: (**a**) MD pads—solder can flow freely around the edge of the pad; (**b**) SMD pads—the solder mask, expanding when heated, advances deep into the pads and deforms the ball output of the component; a voltage concentrator is created, the contact area of the ball output is reduced; (**c**) comparison of the geometry of the solder joint using the MD and SMD technologies

current was started, may also be subjected to thermal migration. They calculated that in order to observe the thermomigration of Sn and Pb, a temperature gradient of 1000 °C/cm is required. Chen et al. [27] found that a temperature gradient of 400 °C/cm is necessary for thermomigration of Cu. Moreover, such thermomigration is observed when SJ is in the solid state; in the molten state, diffusion will go many times faster, respectively, a much smaller temperature gradient will be needed.

It should be noted that the redistribution of particles in SnPb follows the rule: the flux of Pb particles tends to a colder zone, and Sn to a hotter zone (Fig. 8.45). Since in the experiment shown in Fig. 8.45 [28], the Joule heating source was between leads 1 and 4, the coldest zone of lead 1 was on the left, and lead 4 was on the right; accordingly, the lead-rich phase migrated in these directions.

EM and TM often act simultaneously. On the PCB, in the immediate vicinity of this SJ, a heated element can be located, creating a temperature gradient in this SJ, and thus activating a thermomigration process in it. In the SJ itself, the current flowing through it causes electromigration. In this case, the separation and enlargement of the phases occur under the combined influence of TM and EM.

For the interaction of EM and TM, the initial presence of a temperature gradient caused by sources external to the SJ is not necessary. Sometimes the geometry of SJ can be unfavorable from the point of view of electromigration, so that the latter leads to a local increase in Joule heating in such a SJ [29]. The difference may be very small, but with a high degree of miniaturization, it can cause significant temperature gradients that will provoke additional microstructural changes and damage. This problem shall be considered by developers as a component of design for reliability.

Growth of "Whiskers" It has long been known that tin is prone to the formation of filamentous growths called "whiskers" (Fig. 8.46). The ongoing miniaturization of electronic components led to a decrease in the distance between conducting paths to the values commensurate with the dimensions of whiskers; now, the space between them can be 100 nm or less. If the "whiskers" on each of these tracks grow to 50 nm, a short circuit or electrical breakdown of air between adjacent "whiskers"

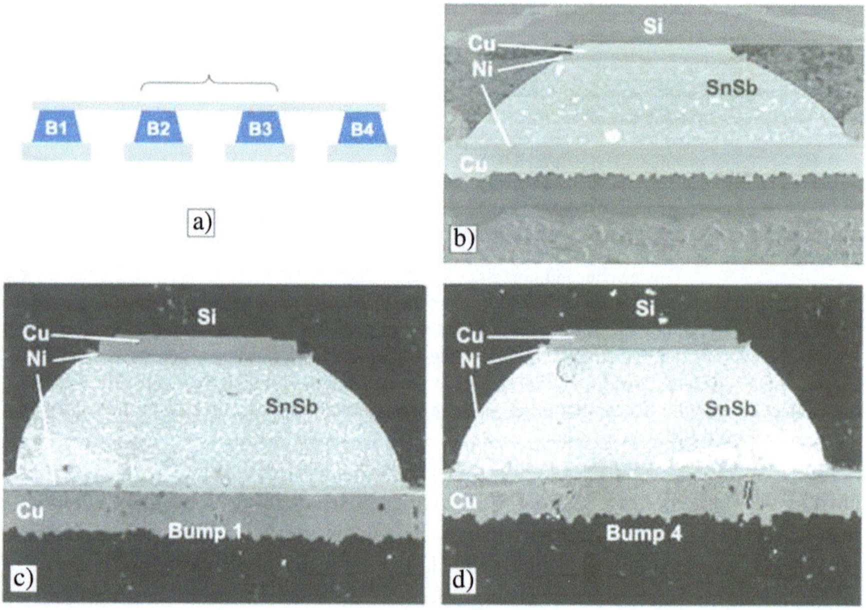

Fig. 8.45 Thermomigration caused by the heat of Joule heating: (**a**) scheme of conditions for conducting accelerated tests for thermomigration: current is passed through leads 2, 3, thermal migration to SJs of leads 1 and 4 is analyzed (Bump 1, Bump 4); (**b**) prior to the start of thermomigration; (**c**, **d**) thermomigration in two leads after 6 h at room temperature and a current of 1.88 A

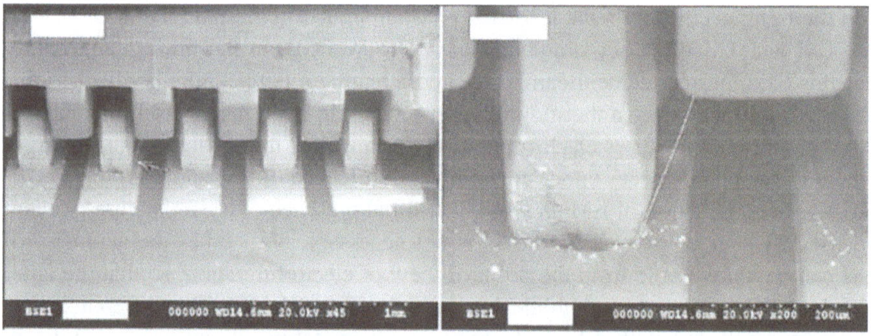

Fig. 8.46 "Whisker" formed on the tin coating of the leadless connector (ZIFF) (photo: nepp.nasa.gov)

may occur. As can be seen from the Table 8.5, failures due to the formation of "whiskers" occur in different areas of application of electronics, from spacecraft to nuclear power plants. These examples show the dangers inherent in the formation of "whiskers" [31].

Several models have been suggested to describe the process of growth of "whiskers." It is assumed that the "whiskers" grow due to the action of compressive

Table 8.5 Most frequent cases of failures due to tin "whiskers" [1, 30]

Year	Object	"Whiskers" growth place
2000	Galaxy VII spacecraft (total loss)	Relay
2000	Military missile program	Coating (leads) of the chip
2000	Power management modules in industrial equipment	Connectors
2000	Solidaridad 1 spacecraft	Relay
2001	Nuclear power plant	Relay
2002	Aircraft GPS transmitter	RF module cover
2002	Nuclear power plant	Potentiometer-type rheostat
2003	Telecommunication device	RF module cover
2005	Optus B1 spacecraft	Relay

stress. In tin-based SJs, compressive stresses may arise as a result of the formation of an intermetallic Cu-Sn layer at Sn grain boundaries—such intermetallic compounds are formed as a result of the diffusion of copper from the material of the contact pad. For long-term growth of the "whiskers," the stresses must be present on an ongoing basis, that is, for a long time without relaxation. Consequently, any external application of stresses, or the resulting stresses due to volume changes due to the formation of intermetallic compounds, or those stresses that are formed as a result of electromigration, contribute to the growth of "whiskers."

Unfortunately, as of the time of publication of this book, there is no uniform reliable solution to the problem of formation of "whiskers." A significant part of the difficulties arises because of the uncertainty where and when they will be formed, respectively, it is difficult to assess the effectiveness of the decisions made to prevent the growth of "whiskers." Currently, in testing the growth of "whiskers," the IEC 60068-2-82 standard is applied, which allows to only roughly estimate the time until the "whiskers" reach a critical length (over 50 nm).

The Problem of Polymorphism Tin has polymorphism—it is the ability of some substances, depending on external conditions (for example, temperature) to form various crystalline structures (modifications). As a result of polymorphic transformation, the atoms of a substance are rearranged from a crystal lattice of one type into a crystal lattice of another type. Polymorphic modifications are denoted by Greek letters α, β (γ and α correspond to the lowest temperature modification).

The β-Sn modification is a tetragonal space-centered structure (lattice) that stably exists at temperatures above 13 °C. At temperatures below 13 °C, a cubic space-centered structure is stable—α-Sn modification. Thus, tin present in SJ usually has a space-centered tetragonal structure. Transformation from one structure to another occurs very slowly, the process parameters strongly depend on the purity of the metal. If a SJ undergoes extremely low temperatures for a considerable time (for example, a cold backup chain in the spacecraft equipment), then there is a transformation from the β- to α-phase, which can cause a significant increase in the volume of SJ (up to 26–27%) [12]. Tin in the α-phase, called "gray tin," is an extremely

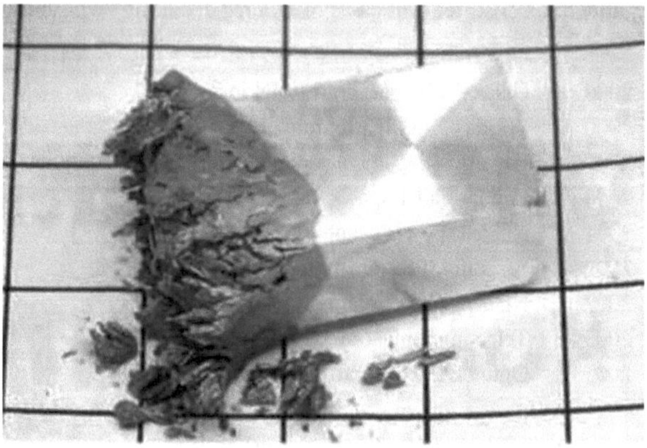

Fig. 8.47 The left part of this tin bar went into the α-phase, while the right remains in the β-phase [30]

fragile material, and such an increase in its share in the SJ volume causes extensive cracking and peeling (Fig. 8.47).

This process can be a serious problem of reliability for the aerospace industry and equipment for any purpose, operated in extremely cold regions, for example, in the polar regions. However, a solution was found: to prevent transformation, it is necessary to introduce into the tin no more than 2% of impurities such as elements, for example, of bismuth, antimony, or germanium.

Let us draw up brief conclusion to this section: it is necessary to remember the following:

– reliability of equipment is built in during the design stage; however methods of research and testing to determine and confirm the reliability of SJs, used today and recorded in the relevant regulatory and technical documents, do not yet allow us to create a complete and reliable picture of the behavior of SJs in all possible operating conditions of the equipment containing them.

References

1. Belous, A., Saladukha, V., & Shvedau, S. (2017) *Space microelectronics Volume 2: Integrated circuit design for space applications* (p. 720). London: Artech House. ISBN: 9781630812591.
2. Belous, A. I., Emelyanov, V. A., & Turtsevich, A. S. (2013). *Fundamentals of IC microassembly technology* (p. 316). Minsk: CMD Press (in Russian).
3. Ingram, B. (2008). WaferCap-packaged amplifiers drive innovation into RF designs. *Microwave Journal*, 51, 114, 116, 118.
4. *RF newsletter, millimeter wave special*, November 2008. www.bfioptilas.com/wafercap
5. Eloy, J. (2010). Look for MEMS sector to double revenues, triple units shipped by 2015. *MEMS Trends*, 3.
6. Brizoux, M., Grivon, A., Maia Filho, W. C., Monier-Vinard, E., Stahr, J., & Morianz, M. (2010). Industrial PCB development using embedded passive and active discrete chips focused on process and DfR.

7. Vikulov, I. (2009). Packaging of microwave ICs on wafer: Technology, advantages, results. *Electronics: NTV*, 3, 74–79 (in Russian). http://www.electronics.ru/files/article_pdf/0/article_203_853.pdf
8. Chang-Chien, P., et al. (2007, May 14–17). MMIC packaging and heterogeneous integration using wafer-scale assembly. In *CS MANTECH conference* (pp. 143–146), Austin, TX.
9. Bonnet, B., et al. (2008, October). 3D packaging technology for integrated antenna front-ends. In *Proceedings of the 38th European microwave conference* (pp. 1569–1572), Amsterdam, The Netherlands.
10. Raikunov, G. G. (Ed.) (2013). *Ionizing radiation of outer space and its effect on the onboard equipment of spacecraft* (p. 256). Minsk: Fizmatlit (in Russian).
11. Vasilenkov, N., Maksimov, A., Grabchikov, S., & Lastovsky, S. (2015). Specialized radiation protection packages for microelectronics products. *Electronics: NTB*, 4, 50–56 (in Russian).
12. Belous, A. I., Merdanov, M. K., & Shvedov S. V. (2016). *Microwave electronics in radar systems and communications: Technical encyclopedia: 2 books* (p.1415). Minsk: Technosphere (in Russian).
13. Bogatyrev, Yu., Vasilenkov, N., Grabchikov, S., et al. (2014). Screens for local radiation protection of microelectronic products. *Problems of Atomic Science and Technology*, *4*, 53–56 (Physics of radiation effect on radio-electronic equipment) (in Russian).
14. Gulbin, V. (2010). Development of composite materials modified by nanopowders for radiation protection in atomic energy. In *IX All-Russian conference physical and chemical ultradispersed nanosystems*, Izhevsk (in Russian).
15. Millward, D., & Strobel, D. (1990, May). The effectiveness of RAD-PAK™ ICs for space radiation hardering. In *Proceedings of 40-th ECTEC conference*, Las Vegas.
16. Fan, W., Drumm, C., & Boeske, S. (1996). Shielding considerations for satellite microelectronics. *IEEE Transactions on Nuclear Science*, *43*(6), 2790–2796.
17. Mangeret, R., Carriere, T., & Beacour, J. (1996). Effects of material and/or structure on shielding of electronic devices. *IEEE Transactions on Nuclear Science 43*(6), 2665–2670.
18. Lau, J. H. (1991). *Solder joint reliability theory and applications*. New York, NY: Van Nostrand Reinhold.
19. Subramanian, K. N. (2012). *Lead-free solders: Materials reliability for electronics edited.* Chichester: Wiley.
20. Nisan, A. (2011). Eight trends that will change electronics. *Technologies in the Electronic Industry*, 2 (in Russian).
21. *International technology roadmap for semiconductors*. Assembly and Packaging. 2016.
22. Tu, K. N. (2007). *Solder joint technology: Materials, properties and reliability*. New York, NY: Springer.
23. GOST °P°56427-2015. Soldering of electronic modules of radio-electronic equipment. Automated mixed and surface mounting using lead-free and traditional technologies. Technical requirements for the implementation of technological operations (in Russian).
24. Liang, Y. H., Mao, H., Yan, Y. G., & Lee J. K. Study on solder joint reliability of fine pitch CSP. In *IPC APEX EXPO conference proceedings*.
25. Soret, C. (1879). *Archives des Sciences Physiques et Naturelles*. Geneve 3.
26. Huang, A. T., Gusak, A. M., Tu, K. N., & Lai, Y. S. (2006). Thermomigration in SnPb composite flip chip solder joints. *Applied Physics Letters*, *88*, 141911–141913.
27. Chen, C., Hsiao, H.-Y., Chang, Y.-W., Ouyang, F., & Tu, K. N. (2012). Thermomigration in solder joints. *Materials Science and Engineering: R: Reports*, *73*(9–10), 85–100.
28. Dandu, P., Fan, X. J., Liu, Y., & Diao, C. (2010). Finite element modeling on electromigration of solder joints in wafer level packages. *Microelectronics Reliability*, *50*(4), 547–555.
29. Leidecker, H. (2006). *NASA Goddard Jay Brusse*. QSS Group, Inc.
30. Tikhomirov, K. (2016). Actual problems of reliability of solder joints of surface mounting. *Technologies in the Electronic Industry*, 8, 43–47 (in Russian).
31. Andrae, A. S. G. (2010). *Global life cycle impact assessments of material shifts*. London: Springer.

Chapter 9
Protection of High-Speed Electronic Devices from Electromagnetic Interference

As is known, electromagnetic radiation can degrade the characteristics of high-speed electronic devices, regardless of what specifically their sources are and what devices they are generated by—by one of the modules of our device or by the other adjacent devices. The main methods and means of protection against various types of electromagnetic radiation were discussed in detail in Chap. 6 above, but here we will consider only the main structural methods of protection against interference.

One of the basic reasons for the unstable operation of high-speed electronic devices is electromagnetic interference (EMI), i.e. the mutual increase in the resulting amplitude of two or more coherent waves when they are superimposed on each other.

Section 9.1 reviews the basic and most effective structural and technological methods of protection against electromagnetic interference. Section 9.2 gives practical recommendations for choosing reliable filters to protect microwave devices from electromagnetic impulses of natural and artificial (anthropogenic) origin. Section 9.3 focuses on some aspects of designing protective devices for radio electronic devices with variable-frequency drives which are rarely discussed in scientific-technical publications, but important for practical use. Section 9.4 considers practical recommendations for solving the problem of protection against conducted (induced) interference. In the last Sect. 9.5, the peculiarities of protection of the spacecraft onboard equipment from the effects of quasistationary magnetic fields are described.

A. Belous, V. Saladukha, *High-Speed Digital System Design*,
https://doi.org/10.1007/978-3-030-25409-4_9

9.1 Constructive and Technological Methods of Protecting Electronic Devices from Exposure to Electromagnetic Radiation

As shown in [1], electromagnetic radiation can degrade the characteristics of electronic equipment, regardless of what is its source and what device produces them: one of the units of radio and electronic equipment or other systems located nearby.

The main methods of protection against EMR are described in detail in Chapter 13 of this paper, but here we will consider the main constructive methods of protection against electromagnetic interference [2].

Electromagnetic interference (EMI) is one of the main reasons for the unstable operation of many systems. Therefore, it is necessary to provide the necessary measures to weaken it. In particular, it is necessary to prevent leakage currents that cause electromagnetic interference to other parts of the circuit.

There are two main approaches to the reduction of electromagnetic radiation, protection from it. The first implies protection at the level of the PCB circuitry techniques. The second involves the use of a protective screen.

At the circuit level, any conductor on a printed circuit board or active element can work as a local antenna, radiating electromagnetic waves. Signal interconnections also create unwanted electromagnetic radiation. At the system level, it is more difficult to reduce EMI, since there may be several sources of interference, for example, the local oscillator in the receiver, and they may be in close proximity to the elements that are sensitive to EMI (mixers, ADC, etc.). When all these elements are protected by a metal case, the EMI acting on them is significantly weakened.

Any circuit can be subject to EMI; therefore, a large number of protective materials, such as seals and protective screens (BLS—board level shielding) installed on the element box are used. Electromagnetic protection is created by any obstacle located between the source of the electromagnetic wave and the sensitive element and attenuating this radiation.

The degree of attenuation of electromagnetic radiation depends on the electrical and physical properties of the barrier: dielectric constant, conductivity, thickness, frequency of electromagnetic radiation, as well as the distance between the radiation source and the protective barrier. In addition to protective equipment, it is important to ensure uniformity of electrical contacts between individual parts of the system, for example, a microwave element and a wall, to minimize reflections of electromagnetic energy, and to prevent leakage through connections or sealing in parts with water cooling.

The edge connections on the waveguides should have a uniform electrical contact. Figure 9.1 [2] shows the solder plug-in ring, where the end of the outer conductor has one broken pin. It can be seen that the rest of the ring has no damage.

Below we consider the protective devices used both at the board level and at the system level. The most common protective device is usually made in the form of a screen with five edges (see Fig. 9.2) [2]. The shape, size, and height of the screen

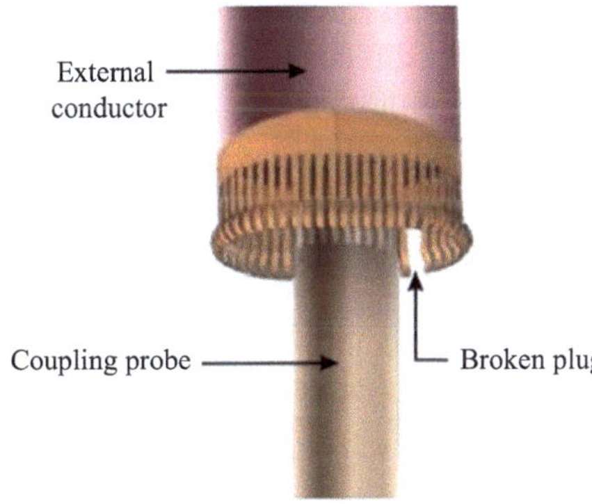

Fig. 9.1 Ring with pins on the end of the outer conductor

Fig. 9.2 Types of protective screens

may be different. Such a screen is placed above the component of the printed circuit board, which must be protected from EMI. It weakens EMI to an acceptable level. To ensure high protection performance when designing a protective screen, consider the factors discussed below.

Effects of the Near Field The so-called near field effects are associated with the relative position of the screen body and the protected hole and take into account the effects of autoresonance.

When a protective screen is installed inside the so-called close interaction field, the system protection level changes depending on the frequency of electromagnetic radiation used, the field configuration, the position of the interferer, and the value of the distributed inductance and capacitance. Thus, the developer, instead of just the radiation source, should consider the entire system of interactions, including the interaction between the radiation source and the protective screen, the effects of interaction between the individual elements, the end sealing of the protective screen, the grounding method. You should also take into account the currents penetrating the screen, its folds and edges. The optimal shape of the screen and the size of the holes in it are very difficult to calculate, and approximations are used when evaluating the effectiveness of protection.

The Location of the Screen and Holes The effectiveness of the protective screen is largely dependent on the specific topology of the PCB, including the topology of the ground plane on the board. The number of communication lines and the distance between them, vias, slots from the protected area to other components of the board, the size and number of holes, thermal effects also have a significant impact on the effectiveness of protection.

It should be noted that capacitive and inductive couplings have a greater influence than the size of the holes in the screen.

Resonance Phenomena Another problem associated with working at high frequencies is the resonance effect caused by the autoresonance of various structures. So it is established that a 2×1 inch screen resonates in a first-order mode of about 12 GHz. Even weak interactions at such high frequencies can lead to strong oscillations that can affect any point on the screen.

Thermal Effects The higher the frequency of the microwave component, the more energy it emits. Heat management is very important. As a rule, it is carried out with the help of heat sinks and heat-conducting gaskets and is not already subject to consideration by the developer of radio and electronic equipment.

Protective Seals In general, most of the leaks through the protective screen or the case occur through the joints (seams, junctions, folds, etc.). To reduce this effect, special seals are used. The effectiveness of the protection of a compound depends on the material, pressure in the contact, and the surface area of the contact. Seals provide conductive contact along the mating surfaces. To attenuate electromagnetic interference, it is necessary to ensure complete contact between the mating metal surfaces. If it is missing, a leakage current gap arises. A seal should also be installed in such places.

Conductive surfaces must be free of any insulating contaminants and coatings. If the mating surfaces are close, at a distance of less than 25 mm from each other, you can do without a seal, but it is still recommended to install it to reduce the number of connectors and compensate for mechanical vibration or heterogeneity of the connection.

In turn, the seals can be divided into two large groups according to the principle of operation: acting on the basis of compression and on the basis of friction (see Fig. 9.3) [2]. When the seal is installed under a flat covering panel and compressed, the protection of the joint is ensured by pressure. The second type of use implies protection against sliding off the contacting surfaces. Special edges, borders, gaps, etc. are made. Protective seals are made of an alloy of beryllium and copper (BeCu) and a metallic material.

Developers of high-speed electronic devices are recommended to use high reliability seals made of copper and beryllium alloys, which provide a high level

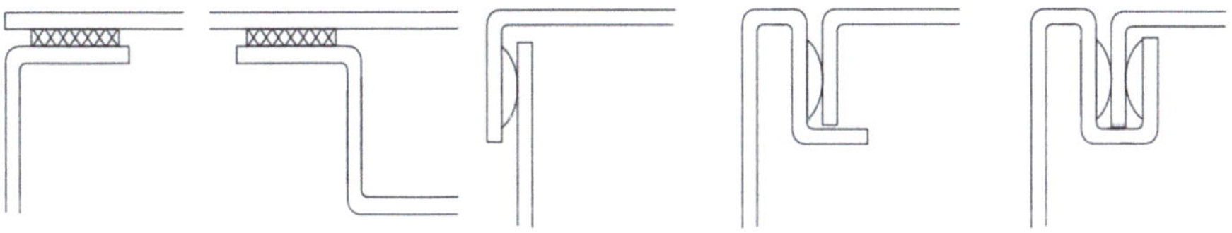

Fig. 9.3 The principle of using seals: to ensure contact and to prevent slipping

Fig. 9.4 Beryllium and copper seal

of EMI suppression in a wide frequency range (see Fig. 9.4) [2]. They are used in both variants, both as a compressive material and against slipping. Solid pins have a large cross-sectional area and, consequently, greater conductivity. The shape of the pin provides a large contact area when connected to the ground plan. The parasitic inductive connections are weak. When connecting the pin rubs on the surface, which helps to remove the oxide from the contact surface. At the same time, the design is resistant to compression and is not subject to breakage due to excessive force. A potential source of problems, depending on the frequency range, is the distance between the pins. At sufficiently high frequencies, radio frequency currents pass through the gaps between the pins.

Metallized seals are also a very effective option.

These seals are made of conductive material deposited on a foam base (see Fig. 9.5) [2]. Their electrical conductivity can be quite low, and then they provide a high level of EMI suppression. It depends on the configuration of the seal and the number of conductive elements, as well as the force of compression. The seal can have a different shape: rectangular, square, D-shaped, bell-shaped, knife, etc. Seals of different shapes provide different levels of compression.

As a recommendation for choosing specific types of amplifiers, it should be noted that the seals provide effective protection against EMI regardless of type. The choice of seal, as a rule, is made on the basis of mechanical characteristics and cost. In order to ensure effectiveness of the seal, it is necessary to take into account

Fig. 9.5 Foam based seal with metallization

Table 9.1 Electrode potential of some materials (as the exposure to corrosion decreases, starting with the most susceptible)

Material	Electrode potential, V
Magnesium	+2.37
Beryllium	+1.85
Aluminum	+1.66
Zinc	+0.76
Chromium	+0.74
Steel	+0.44
Cadmium	+0.40
Nickel	+0.25
Tin	+0.14
Lead	+0.13
Cooper	−0.34
Silver	−0.80

such parameters as its height, impedance, material compatibility (for soldering), compressive strength, compressibility, compression shrinkage, external conditions, etc. Impedance control requires high conductivity and low inductance. Beryllium-copper alloy seals are characterized with the highest conductivity. Corrosion can reduce the effectiveness of protection because it causes an insulating effect.

There are two types of corrosion: galvanic, due to the presence of contact between two dissimilar metals in the presence of moisture, and electrolytic, due to the current flowing between the metals in the presence of an electrolyte. For commercial systems, materials with an electrode potential of 0.5–0.6 V are allowed. Electrode potentials of some materials are given in Table 9.1 [2].

The height and diameter of the seal should be sufficient to compensate for the heterogeneity of the connection of the mating surfaces by the application of force. Compression force is the force that must be applied to obtain maximum protection efficiency. The greater the pressure or force of compression, the lower the impedance. It is recommended to apply a minimum pressure sufficient to seal off the oxide or corrosion from the contact surface to the seals, thus preventing slipping. In this case, the surface resistance will be small, and the protection will be as effective as possible. If the pressure is not sufficient, the effect will be reversed.

9.2 Information on Filters for EMI Protection Required by the Developer

9.2.1 Practical Recommendations on Selecting Reliable Filters for Protection of SHF Devices from Electromagnetic Pulse

Protection of modern radio equipment from the effects of electromagnetic radiation of various nature has always been relevant, and recently has become one of the main problems of developers of complex high-performance electronic systems, both commercial and military. For example, protection from electromagnetic impulses of a high-altitude nuclear explosion, as shown in [1], from the sphere of potential hazards goes into the category of important, unfortunately—extremely urgent problems of the modern stage of development of electronic equipment in general, and in particular, microwave systems and devices being the "heart" of modern telecommunication systems, electronic warfare systems, military equipment control systems and weapon systems, in particular, modern high-precision weapon control systems.

This problem will be especially acute for developers after the exotic for the average man, but already quite real for the developers and the first few "users" of the arsenal of modern microwave weapons, both space and ground. A number of objective reasons contributing to an increase in investor interest in the microwave systems being developed are listed in Chapter 9 of [1].

It is known that the main classical methods of protection from EMI exposure on modern electronic equipment are only two main areas—the electromagnetic *shielding* of the equipment itself and all possible interfaces (cables) connected to it, as well as the *suppression* of this very EMI pulse by connecting a number of filters in the right places specially designed for this purpose and described in detail in the special scientific and technical literature.

To perform its function, these filters should be switched on directly between the protected equipment, external devices, and systems.

Figure 9.6 [3] shows the equivalent circuit drawing of such a filter, where L1, L2, L3 are the designations of these channels of communication of equipment with external devices.

As shown in [1], there are a lot of such filters on the market today, into the development and manufacture of which dozens of "large" and hundreds of "small" companies are engaged.

The list of foreign manufacturers and developers of these filters includes such well-known companies as ETS-Linden, MPE, European EMC Product LTA, AP

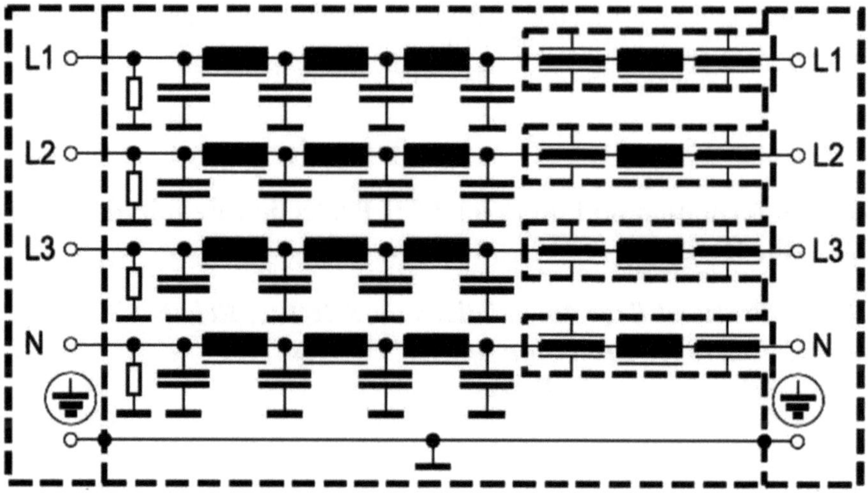

Fig. 9.6 Equivalent circuit drawing of the EMI NE filter, consisting of a set of DC links

Tochnobogies, Meteolabor-EMP, LCR Electronics, Fi-Coie, EMI Solutions PVT. Ltd., Astrodyne TDI Corp., RFI Corp, as well as many other companies.

These companies have been working in the market for many years, and the sales volumes of their products increase annually, which indicates their obvious demand, i.e. their products are actively bought and put into various electronic equipment, the customers of which, among other requirements, indicated to the developers the need to ensure resistance to EMI.

And here, very often the developer has a question about the actual effectiveness of all these products [3].

It is clear that in the last 10–20 years, no one (neither the USA, nor the USSR (Russia), nor China) has carried out any real tests to evaluate the effectiveness of these devices—this is prohibited by numerous international agreements and protocols, numerous expert services keep a close eye on this (except special services) that deal with these issues professionally).

And here for developers (and most importantly—for customers reading such books) a number of questions arise, some of which the author tried to answer [3].

The first question is: do these all numerous filters really protect the electronic device from an electromagnetic pulse?

The authors of this book, understanding the technical nuances of the problem, familiarized themselves with the paper [3], where the Russian specialist, Candidate of Technical Sciences V.I. Gurevich studied this topic quite deeply, his main analytical conclusions and practical recommendations coincided with the position of the authors of this book, since the links on the analyzed literature of this author coincided with the available information and position of the authors.

Therefore, below, as close as possible to the text of his paper [3], the main technical problems and ways of solving this issue are described.

So, the key question is: do all these numerous filters of reputable companies really protect electronic components from electromagnetic impulses?

And here a number of unpleasant questions for these manufacturers arise, which the developer and the customer of the microwave system should know about.

The first obvious question can be formulated as follows: all of the above companies advertise their filters as highly effective means of protection against EMI, citing full compliance of these filters with the well-known military standard MIL-STD-188-125. However, most of them indicate the parameters of the test pulses that were used in the process of filters testing, which are significantly different from those specified in the standard, and, usually, in the "worse" direction.

What deserves a special notice is if this standard provides for testing pulses with 20/500 ns of a well-defined amplitude for a standard load of 60 Ohms, then it follows from technical documentation that these filters were tested by manufacturers with current pulses from 8/20 μs to a load of only 1 Ohm!

Consequently, the characteristics of the test pulse and the characteristics of the load are completely different and in no way correspond to the standard MIL-STD-188-125.

Of course, specialists (whose number is unfortunately low) know the simple answer to this question—the 8/20 μs pulse (which has nothing to do with the EMI factors) is a standard pulse reproduced by numerous types of test equipment for assessing resistance to electrical discharge of a lightning!

Of course, such tests are by an order of magnitude cheaper, because testing with current pulses of 20/500 ns at 60 Ohms load requires special developed and expensive test equipment, and such tests themselves are very expensive and, as a rule, filter manufacturers do not have means for it [3, 4].

Developers of microwave systems should know one more thing which is very important to ensure the reliability of programmable systems.

This point is that MIL-STD-188-125 provides for testing the object with current pulses in two modes: when current flows between all common inputs and ground (common mode) and between each input separately and ground (wire-to-ground mode). However, in a high-altitude nuclear explosion, a high-voltage impulse can be applied not only between the inputs of the equipment and the ground (this mode in other standards is designated as "common mode"), but also between different inputs of the equipment isolated from the ground ("differential mode"). Standard MIL-STD-188-125 does not provide for such tests, they are considered in other standards. In this regard, some companies that advertise their filters as EMI NE filters do not at all provide them with elements of limiting impulse voltages, referring to the same MIL-STD-188-125, but at the same time they claim that since their filters were tested by a current pulse with an amplitude of several thousand amperes and recognized as compliant with the MIL-STD-188-125 standard, this means that they provide complete protection from EMI NE. Indeed, in this standard, pulse voltage suppressors are presented as completely independent elements unrelated to filters (Fig. 9.7).

With this approach, the filters are really not required to protect against over-voltages at the inputs. But is it possible with such an approach to assert that these

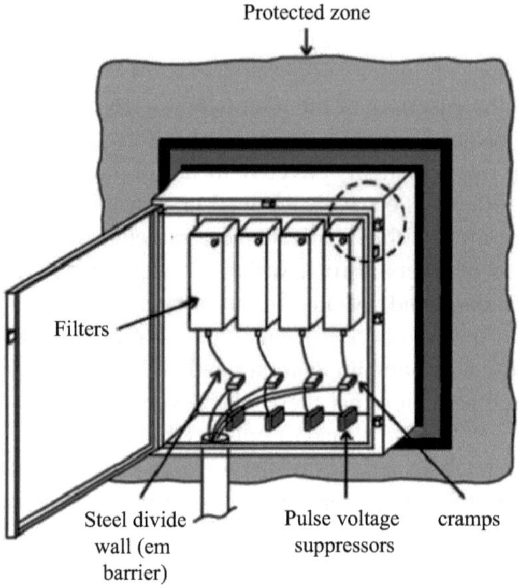

Fig. 9.7 The device of the input box for connecting an external cable to an object protected from EMI NE (version MIL-STD-188-125)

same filters are a complete protection against EMI NE? Obviously, understanding the problem, many manufacturers still supply their filters with surge protection elements installed at the inputs. In their opinion, now such filters can rightfully be called filters for protection against EMI.

However, upon careful consideration of the protective elements used in these filters, serious doubts arise as to their effectiveness. The most common and cheap pulse voltage limiters used in filters are gas dischargers and zinc oxide varistors (Fig. 9.8). As you know, they are relatively "slow" elements, coping well with the suppression of standard 8/20 μs pulses, but fail to work when exposed to a short high-voltage pulse E1 EMI NE with parameters 2/25 not [5] (or 5/50 not according to [6]). A consumer, having read that the filter he chooses is designed to protect against EMI NE and passed tests with current pulses of 8/20 μs in full compliance with the military standard MIL-STD-188-125, is unlikely to look for this standard and check whether this is the very pulse.

However, it's too early to make claims to manufacturers. In [3], it is explicitly stated that insufficiently fast varistors installed in filters and even slower gas dischargers are not designed to protect against EMI NE, but are elements of protection against lightning discharges and switching overvoltages. Like this: in filters for the protection against EMI NE voltage limiters are applied, designed to protect against lightning, but not from EMI NE! However, some types of MPE filters with protective varistors are named in advertisements as filters specifically designed to protect against the E1 component. However, a careful analysis of the parameters

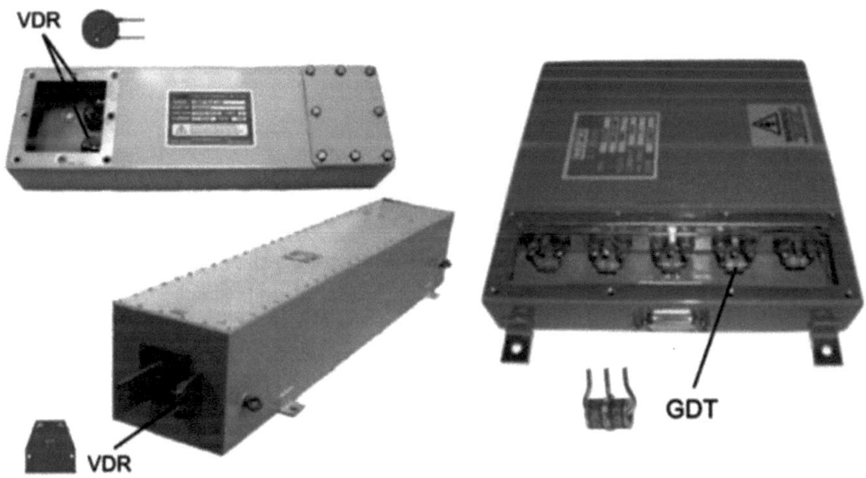

Fig. 9.8 MPE filters with input protection elements, which use varistors (VDR) and gas discharge protectors (GDT)

of these filters showed that they do not differ from the rest of the filters of this company, except for the name in the title, that is, in fact, they are protected from lightning discharges, and not from the E1 EMI NE component. Otherwise, it will be necessary to recognize that the parameters of the lightning discharge are no different from the parameters of the E1 component, which in fact is completely untrue.

Despite the fact that the varistor itself is not a very fast element, in conjunction with the L-C filter elements, its effectiveness becomes sufficient to protect against the E1 component. A number of publications on this topic [7, 8] state that joining even short external conductors with very low inductance to a protective element reduces its speed. It turns out that the reaction time of a protective element on a voltage pulse applied to it is very much dependent on the design of the case of this element and on the form (length) of its leads. Moreover, in [9] it is stated that it is the design and the length of the external leads that determines the reaction time of the protective element. Knowing this, manufacturers of protective elements often indicate in promotional materials the reaction time not of a completely assembled element in a case with leads, but only of the material used to manufacture this protective element [8]. However, manufacturers are working to improve the design of the leads of the protective elements and are making significant progress [9].

From the above it becomes clear that objective data on the speed of a type of protective element can only be obtained by conducting our own independent tests of finished products offered on the market, although some indirect data on the results of such tests [9] allow a preliminary comparative assessment. For example, according to [9], the dynamic resistance and reaction time of a protective element based on an avalanche diode (TVS diode) are almost ten times less than that of a varistor.

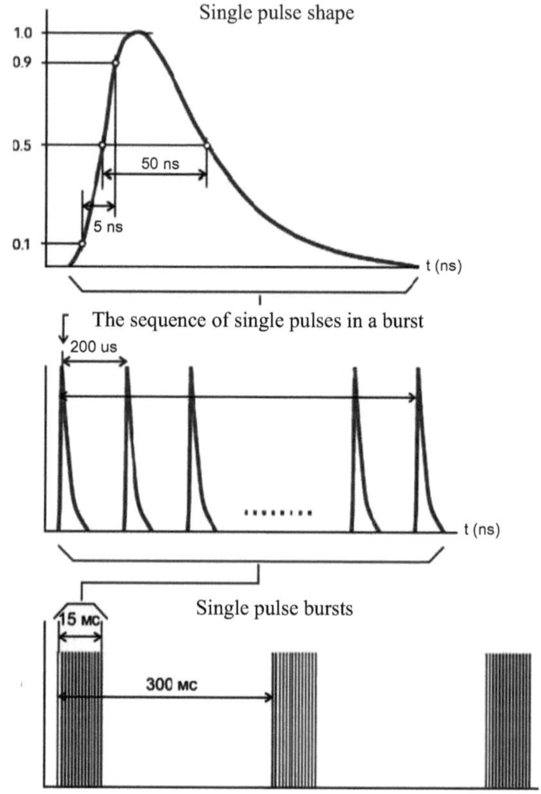

Fig. 9.9 Fast pulse (IEC 61000-4-4)

Another problem associated with the use of pulse voltage protection elements is the connection circuit of these elements, adopted in most filters (Fig. 9.6), in which each such element is connected between the input and the grounded filter case. With this inclusion, it turns out that between the filter inputs two series-connected elements are included, causing double residual voltage, which can be dangerous for the protected electronic circuit.

Technical requirements for equipment resistance to such voltages and its test methods are described in IEC 61000-4-4 [6] and IEC 61000-4-25 [10] standards. The test pulse of such a voltage means the so-called electrical fast transient (EFT)—a fast pulse whose parameters and test methods are described in the IEC 61000-4-4 standard (Fig. 9.9). The method of selecting the parameters of test pulses based on these standards for a specific example—a microprocessor relay protection device (MRPD) is described in [11], for which the amplitude of the pulse voltage EFT was 8 kV.

Filters with pulse voltage protection elements designed to protect against EMI NE should be subjected to such tests in addition to the tests prescribed by MIL-STD-188-125, moreover, with the application of test voltages both between inputs and ground, and between individual inputs.

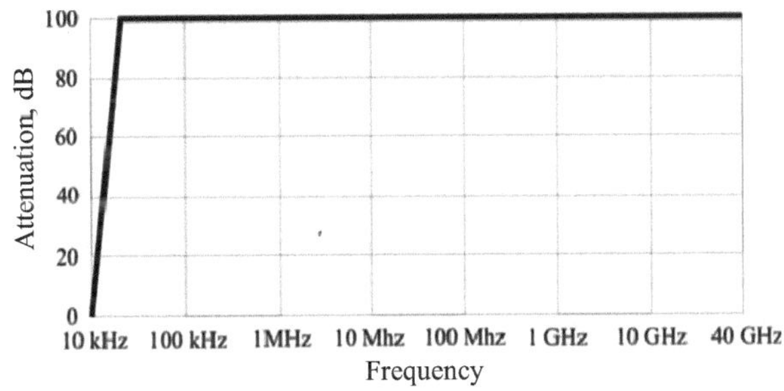

Fig. 9.10 Typical amplitude–frequency characteristic of high-quality filters EMI NE

Table 9.2 Frequency characteristics of filters from leading manufacturers

Note	Attenuation DB	Frequency range		Manufacturer of filters EMI NE
		Min, kHz	Max, GHz	
–	100	14	1	LCR electronics
For a standard filter	80	10	1	MPE
For a modified filter	100	14	18	MPE
–	100	14	1	Fi-coil
–	100	14	10	Captor Corp.
–	100	14	40	ETS-Lindgren
For PLP-type filter	80	200	1	Meteolabor
For USP-type filter	–	–	– <	Meteolabor

For developers of microwave antennas, such antennas with amplitude–frequency characteristic of filters are of interest. A typical characteristic of a high-quality filter designed to protect against EMI NE is shown in Fig. 9.10. How are the parameters of real filters associated with this typical characteristic?

Some companies (such as, for example, Meteolabor AG) do not generally provide in the technical description of some of their filters data on the frequency range and the attenuation introduced. Others (for example, MPE) give in their documentation a typical characteristic that filters should have according to the standard MIL-STD-188-125 (frequency range 14 kHz–40 GHz, attenuation in the whole range of 100 dB) and immediately give real characteristics of manufactured filters with parameters: 10 kHz-1 GHz with attenuation of 80 dB—for the filter of general quality and the same frequency range, but with attenuation of 100 dB—for the filter of improved quality. And where did the frequency range from 1 GHz to 40 GHz go? Employees of Astrodyne (LCR Electronics, Inc.) were smarter. They wrote in the technical documentation for their filters that they provide attenuation of 100 dB in the frequency range 14 kHz–10 GHz in accordance with the standard MIL-STD-220 [12], but if their filters are well shielded and isolated (read: installed in the "Faraday cage"), then their frequency range can be extended to the required 40 GHz, although it is obvious that for the filter to "correctly" protect against EMI NE, it must first be protected from this same EMI NE!

All the comic situation is reflected in the Table 9.2, in which the frequency characteristics of filters from leading manufacturers are given for comparison: all filters have the same purpose, all comply with the requirements of the standard MIL-STD-188-125, and at the same time they all have significantly different parameters. How can this be?

Another feature of the adopted circuit for including separate internal filter elements between each input and ground is not only double residual voltage on the pulse voltage protective elements, as mentioned above, but also double capacity and double inductance connected between the inputs compared to the inductance and capacity between each input and ground.

It follows that the frequency characteristics of the filters for a pulse applied between inputs will not be the same as for a pulse applied between input and ground. The question arises: "How would these characteristics be acceptable for protection against EMI NE?"

Which consumer will "dig" so deeply? Why he should not believe the statements of the manufacturer about the high efficiency of his products? And even if he does not believe, he still cannot, in most cases, independently verify the real effectiveness of the "operation" of such a filter. It is clear what will the effectiveness of the protection of its critical equipment by the filter be, if in a critical situation it turns out to be unable to protect against EMI NE.

Thus, today the situation is such that each manufacturer decides whether or not to include pulse voltage limiters in the filters; use or not to use elements that are cheap, but not suitable for its parameters; test filters with a standard "thunderstorm" pulse, or pulse with parameters specified in the military standard, mention or not to mention the frequency range at all, use the frequency range specified in MIL-STD-188-125 or MIL-STD-220 [12], and maybe indicate its own frequency range and at the same time provide a link to the well-known military standard.

The current state is a consequence of the absence of a special standard stipulating the requirements for the design and parameters of EMI NE filters, their test methods, and performance criteria.

9.3 Design Features of Protective Devices for Radioelectronic Equipment with Frequency-Adjustable Drives

In modern high-power commercial telecommunication systems, especially in dual-use and special-purpose PLC, as well as in applications related to the control of command complexes, industrial buildings, and facilities, so-called frequency controlled drives are widely used, which are usually powered from three-phase electrical networks.

A typical system of such an electric drive usually consists of three main components: motor (main component), frequency converter, and input suppression filter.

An indispensable condition for the reliable functioning of such systems is the need to include between the converter and the motor a special shielded cable up to

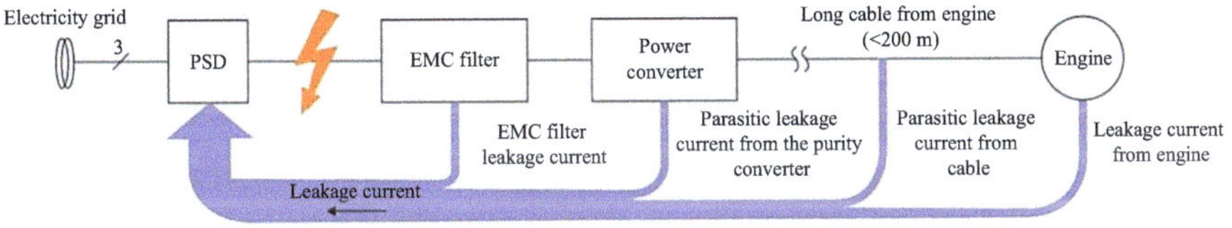

Fig. 9.11 The mechanism of the false positives of the protective shutdown device [13]

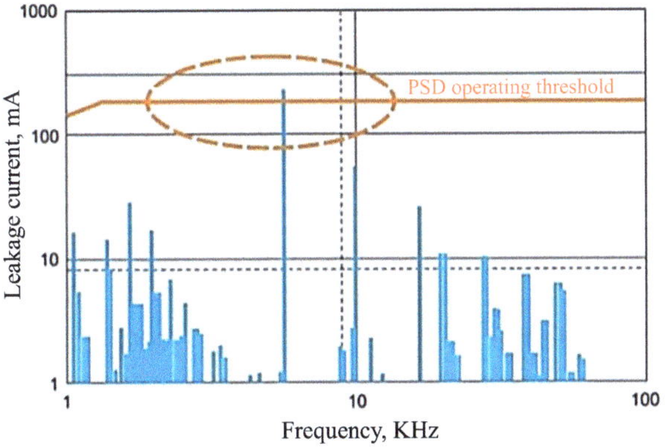

Fig. 9.12 Amplitude–frequency characteristic of earth leakage currents [13]

200 m long, although designers of control systems often "save" this, sharply reducing the reliability characteristics of such control systems.

To ensure the reliability of operation, all drive systems must be connected to the power grid not "directly," but only through protective shutdown devices (PSD) [13].

One of the serious problems of frequency controlled drives is the generation of leakage currents to earth during operation due to parasitic capacitive coupling. These currents depend not only on the system topology, but also on the maximum slew rate of the output voltage during switching, frequency, and amplitude. Under adverse conditions, the total leakage current to earth may exceed the threshold of the PSD (see Fig. 9.11).

Under adverse operating conditions, the total leakage current to earth may exceed the response threshold of the protective shutdown device. This often happens if the motor cable is long enough.

In standard PSDs, for example, the response threshold is 30 mA for currents in the frequency range up to 100 Hz, and at frequencies above 1000 Hz, this parameter reaches 300 mA. When using cables of particularly large length for connecting a motor, the response threshold exceeds 300 mA (see Fig. 9.12). As a result of the appearance of large leakage currents on the ground, unauthorized system shutdowns occur, which leads to costly production downtime at industrial enterprises.

When using long cables, large leakage currents occur in the frequency range above 1000 Hz, which can lead to false triggering of the PSD. There are two ways to prevent excessive leakage currents, thus avoiding the false triggering of the PSD.

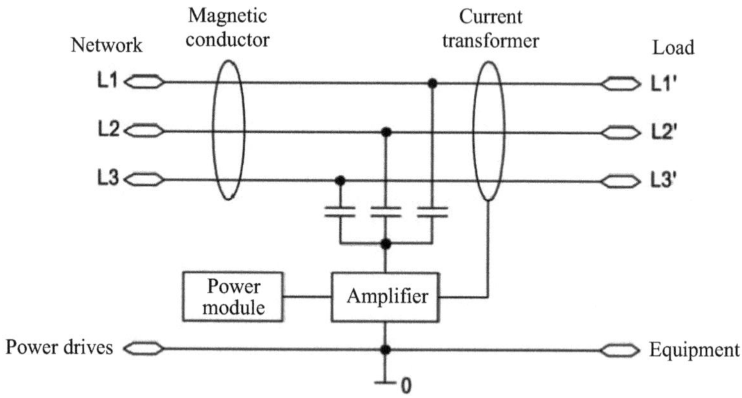

Fig. 9.13 Wiring schematic diagram of the LeaXield module

1. Interference filtering. The use of large filters allows to reduce high-frequency earth leakage currents.

For example, more effective interference-suppression filters at the input are used by experienced developers of microwave systems in combination with additional filters at the output, although at the expense of relatively high costs. The disadvantage of this approach is the space occupied, which is required for installing additional filters.

2. Using a PSD with a higher threshold. This approach is fraught with significant risks, since large earth leakage currents can lead to exceeding the maximum allowable touch voltage of 50 V AC. This may damage system components. Thus, both methods cannot be considered satisfactory; therefore, a number of experts [13] suggest using the LeaXield module for these purposes.

To specialists designed to eliminate earth leakage currents is installed between the PSD and the input interference-suppression filter.

The phases of the currents coming from the amplifier are shifted by 180° from the phase of the initial earth leakage current. Since the power module of the amplifier operates directly from three-phase current, there is no need for an additional power source.

The principle of operation of the LeaXield module is shown in Fig. 9.13. The current transformer is installed on the load side, i.e. where there are connections to the filter. The transformer detects three-phase currents and determines the earth leakage currents from the corresponding difference between the phases. This information comes to the amplifier. From the amplifier, currents of equal amplitude flow through the circuit of capacitors; the currents have three phases, which are shifted by 180° from the phases of the initial earth leakage currents. In such a way the original earth leakage currents are removed (see Fig. 9.13). The LeaXield module compensates for currents up to 1000 mA, which ensures reliable system operation.

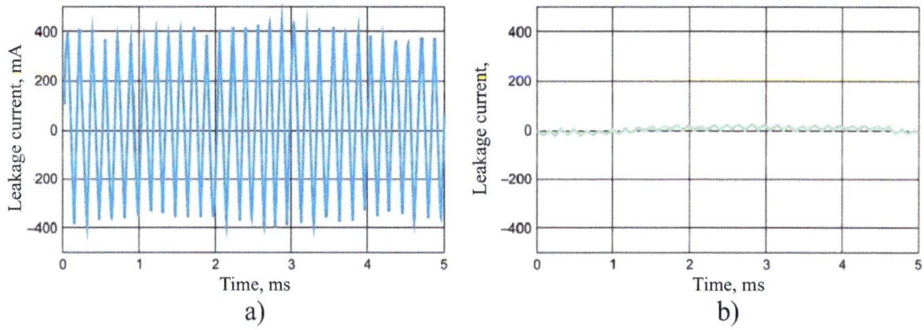

Fig. 9.14 Current leakage shape: without filter (**a**), with filter (**b**)

Fig. 9.15 View of the LeaXield module

Figure 9.14 [13] demonstrates an important advantage of the LeaXield module: high earth leakage currents (Fig. 9.14a) are almost completely eliminated with the help of this module (Fig. 9.14b), which prevents the false triggering of the PSD.

Common-mode current attenuation with the LeaXield module is another advantage provided by this device. Interference on wires, which, as a rule, are normalized in the range of 150 kHz–30 MHz, can be significantly reduced in the initial part of this range—150–500 kHz.

Tests in typical conditions showed that the LeaXield module allows to increase the EMC of the system based on the converter from the C2 level to the C1 level according to the EN 61800-3 standard. The LeaXield module (see Fig. 9.15) works particularly effectively at frequencies below 150 kHz. At 4 kHz, attenuation reaches 30 dB, at 10 kHz–40 dB, and at 150 kHz–15 dB. In particular, this advantage is effectively realized in new plants, where the high attenuation coefficient of LeaXield allows the use of more economical interference-suppression filters.

The main parameters of the EPCOS LeaXield module [13]

- voltage between cables: up to 520 V AC;
- load current (max.): 150 A;
- frequency: 50/60 Hz;
- compensation of earth leakage currents (max.): 1000 mA;

– common-mode attenuation (max.): 40 dB (10 kHz);
– dimensions: 258 × 80 × 100 mm.

Due to its small size (258 × 80 × 100 mm), the LeaXield module is also suitable for replacement—it is easily installed in existing systems as an interference-suppression filter.

9.4 Practical Recommendations for Developers of Radioelectronic Equipment on Solution of Problems of Protection from Conductive Interference

As a rule, radio and electronic equipment designed by developers should perform all the functions specified in the customer's terms of reference, regardless of factors such as radio interference, network effects, overvoltages, voltage fluctuations in the electrical network, electromagnetic effects, parasitic connections, background effects of the industrial frequency of 50 Hz, grounding effects, etc. To understand the problem, it is necessary to provide a list of a number of regulatory documents governing the techniques, methods, and ways of solving these problems. For example, the VDE 0870 standard (Society of German Electrical Engineers) defines EMC as "the ability of an electrical device to function satisfactorily in its electromagnetic environment without affecting this environment, to which other devices also belong, in an unacceptable manner"; The Russian GOST R 50397-92 defines electromagnetic compatibility as "the ability of a technical means to function with a given quality in a given electromagnetic environment and not to create unacceptable electromagnetic interference to other technical means" [14].

The basic concepts of electromagnetic compatibility consider the effects of both radiated and conducted interferences (so-called crosstalks) propagating through conductors (for example, crosstalk on power supply circuits), as well as the sensitivity of electrical equipment to the effects of interference (immunity).

The most typical examples of manifestations of the EMC problem can be such phenomena as:

- failures of NPP control and management systems;
- failures of control and management systems in production, including chemical;
- failures of onboard systems of spacecraft, airplanes, and airfield control systems of airlines;
- failures of medical equipment for diagnostics and life support;
- direct impact of electromagnetic radiation from various kinds of electronic equipment on human health, especially high-frequency (cell phones, computers, radio stations, microwave ovens, RF installations, high-voltage transmission lines, etc.).

Of all this variety of interference in this section, we consider only conducted interference [14].

Conductive interference is interference that propagates in a conductive medium (through wires, conducting surfaces, that is, interference transmitted by the contact method). High-frequency conducted interference can theoretically be attributed either to continuous oscillations or to aperiodic and oscillatory transients.

Depending on the mechanism of occurrence and the nature of the distribution, conducted interference is divided into the following types:

- symmetrical;
- asymmetrical.

Symmetrical interference occurs when a voltage is applied between the phase (line) and neutral wires. In other words, this is an interference that propagates similarly to the path of alternating current in a network. In DC circuits, a symmetrical interference voltage is applied between the positive and negative wires.

Asymmetric interference is the interference between the conductors and the case or the ground bus through a parasitic impedance (parasitic capacity) between these objects. In DC circuits, the voltage of asymmetrical interference is applied between the conductors (positive, negative) and the case.

The effect of such industrial electromagnetic interference is distributed over almost the entire frequency range from 0.011 to 1000 MHz, which in the work of ground, airborne, and mobile electrical equipment can cause various violations that can lead to their failure, disasters, accidents, and even catastrophic consequences.

We have not thought about the EMC problem for a long time, until massive failures in banking systems were detected when exposed to interference. This led to the emergence of the so-called Directive 336EC 89, which obliged all countries of the European Community to introduce common standards for electromagnetic compatibility and to develop an appropriate certification system. As a result, since 1996, the sale of technical equipment in Europe without this certificate of conformity to standards for electromagnetic compatibility is not allowed in Europe.

In Russia, until the beginning of 2001, only electrical and electronic equipment included in the relevant register was subject to mandatory EMC certification. Since Russia is now “close” to Europe and has introduced its system of standards and certification, virtually all electrical products fall under the mandatory EMC certification. The basic Russian standards for immunity to interference generally comply with IEC 1000-4. Because the standards for permissible interference levels are based on CISPR standards.

Consider further the electromagnetic compatibility of semiconductor converters of electrical energy (transistors), which are part of the power supply of various devices and make the main contribution to the component of the interference emitted by the device [14].

The mutual influence and connection of the power and information parts of a standard semiconductor converter of electrical energy requires special attention to the issues of their electromagnetic compatibility and protection against interference caused by surges and voltage and currents spikes when switching power switches.

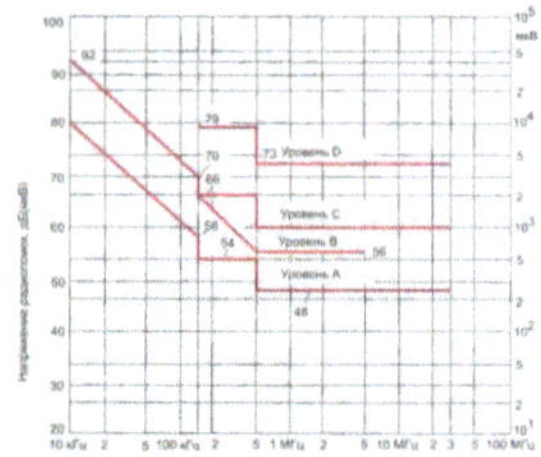

Fig. 9.16 The permissible voltage values of industrial interference according to Table 9.3

It is known that the frequency spectrum and the level of energy released during the switching of power switches in a high-power transducer are very large, and the generated interferences can either spread through the connecting wires or can be radiated into space. In this case, the propagation path of interference and crosstalks is difficult to identify. These may include, for example, mains power transformers for electrical substations, power transformers for the converter control system, pulse transformers for key management circuits, transformers and resistors for feedback circuits, parasitic mounting capacities between circuits belonging to different parts of the converter, etc. [15].

There are a number of relevant GOST standards for electromagnetic compatibility, but GOST R 51527-99 is best suited for a power source with the characteristics of a current source. "Electromagnetic compatibility of technical equipment. Stabilized DC power supplies. Conductive electromagnetic interference" [16] (Fig. 9.16).

There is an established classification of information technology equipment (ITE), divided into two categories: class A ITE and class B ITE.

Class A ITE refers to industrial devices, which in turn are used in factories and industrial areas. Class B ITE is intended primarily for use in a domestic environment and includes:

- equipment without a fixed place of use, for example, portable equipment powered by built-in batteries;
- communication terminal equipment powered by the communication network;
- personal computers and auxiliary equipment connected to them.

From the classification of the ITE it is clear that all power sources being developed are classified as class A ITE (Table 9.3). In this case, the characteristics of the corresponding radio interference meter should meet the standards given in Table 9.4 [14].

In Fig. 9.17 the structural diagram of measuring the voltage values of industrial radio interference with the device SMV 11 [14] is presented.

Measurement of interference was carried out on the entire frequency range of 0.011–0.15 MHz on QP I and from 0.15 to 30 MHz on QP II. The interference voltage is calculated using the following formula:

$$U = U_p + U_{\text{pr}} + 20\,\text{db}, \tag{9.1}$$

Table 9.3 Voltage standards of IRP on the network terminals of the ICU according to GOST 051527-99

Frequency	Radio interference voltage, dB (μV)			
	A	*B*	*C*	*D*
10 kHz	80	–	92	–
20 kH–20 kHz	74	–	86	–
50 kHz	66	–	78	–
150 kH–150 kHz	58	66	70	79
150–500 kHz	54	66–56	66	79
0.5–5 MHz	48	56	60	73
5–30 MHz	48	60	60	73

Notes: 1. At the cutoff frequency, the norm is a smaller value. 2. Norms do not apply to devices whose design does not have interference-suppression filters. The absence of built-in interference-suppression filters should be indicated in the technical documentation for the device

Table 9.4 Characteristics of the radio interference meter with a quasi-peak detector in accordance with GOST R 51527-99

Characteristics	Frequency band	
	10–150 kHz (QP I)	0.15–30 MHz (QP II)
Bandwidth at 6 dB, kHz	0.2	9
Detector charge time constant, ms	45	1
Detector discharge time constant, ms	500	160
Ballistic time constant of indicator device, ms	160	160

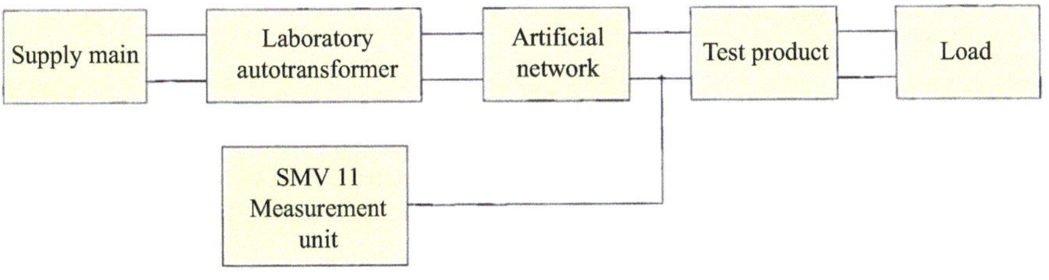

Fig. 9.17 Structural diagram of the measurement of the voltage of conductive interference with the device SMV11

where U_r is the level of attenuation of the voltage of the radio interference introduced by the switch of the measurement limit (6) and displayed on the indicator of the measurement limit (5); U_{pr}—indications of the indicator device (8); 20 dB is the attenuation introduced by the network equivalent.

A series of experiments has been carried out that fully reflects the effect of conducted electromagnetic interference of the device being developed on the supply network. All experiments were carried out in accordance with GOST.

In [17], an experiment was performed demonstrating the effect of power consumption on the conducted interference emitted by the device. To do this, with the help of a laboratory autotransformer, the output voltage was set at the 220 V network equivalent to which the tested power source was connected. Then, using a SMV 11 selective microvoltmeter, we measured the conducted interference emitted by the device into the AC mains on the phase conductor under various loads (100 and 250 W). The experimental diagrams are shown in Figs. 9.18, 9.19, 9.20, and 9.21.

When analyzing the results of this experiment, we can conclude that in almost the entire frequency range, the level of interference emitted by the device significantly exceeds the allowable limit set by GOST. The change in load had almost no effect on the level of radiated interference.

In [18], an experiment was implemented that reflects the effect of changes in the supply voltage on the level of interference emitted into the network. To do this, a load of 250 W was connected to the tested power source and measurements of conducted interference were emitted by the device into an AC network at various voltage levels of the supply mains (160 and 220 V). The diagrams of these experiments are presented in Figs. 9.22 and 9.23.

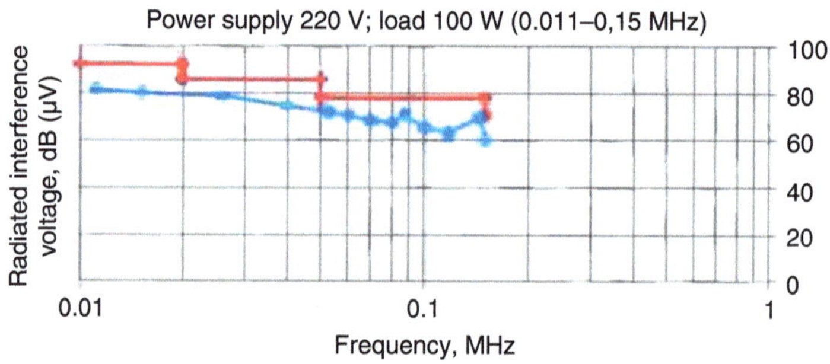

Fig. 9.18 The dependence of the voltage of the emitted interference on the frequency at a supply voltage of 220 V and a load of 100 W in the frequency range from 0.011 to 0.15 MHz

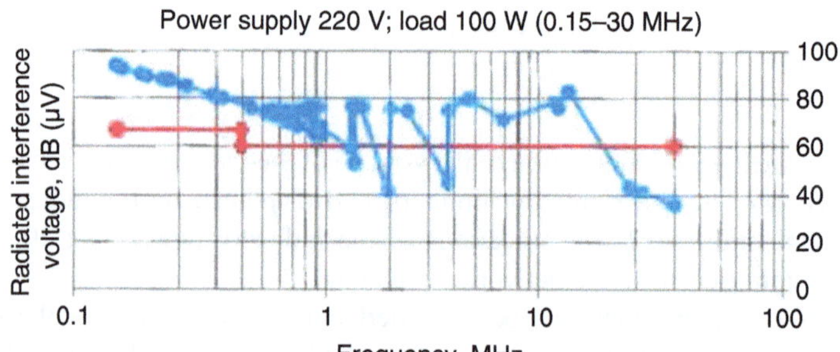

Fig. 9.19 The dependence of the voltage of the emitted interference on the frequency at a supply voltage of 220 V and a load of 100 W in the frequency range from 0.15 to 30 MHz

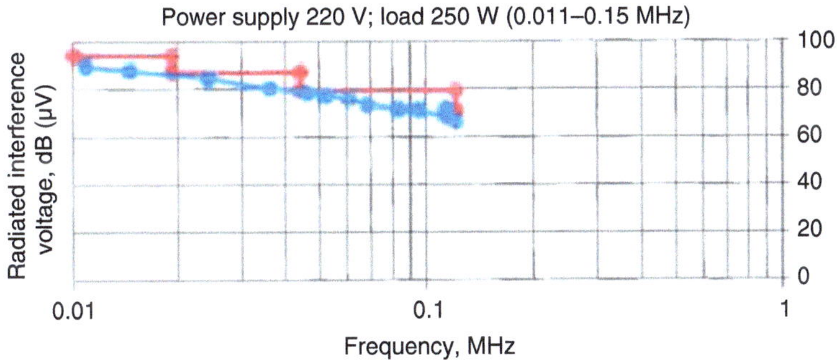

Fig. 9.20 The dependence of the voltage of the emitted interference on the frequency at a supply voltage of 220 V and a load of 250 W in the frequency range from 0.011 to 0.15 MHz

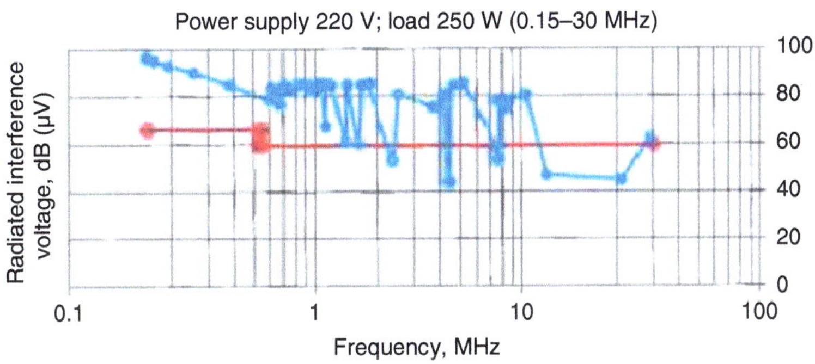

Fig. 9.21 The dependence of the voltage of the emitted interference on the frequency at a supply voltage of 220 V and a load of 250 W in the frequency range from 0.15 to 30 MHz

From the analysis of Figs. 9.20, 9.21, 9.22, and 9.23, it can be seen that the level of interference emitted by the device linearly depends on the voltage level of the supply network. Thus, it can be concluded that the level of interference emitted by the device into the network is influenced precisely by the magnitude of the supply voltage, and not by the level of current consumed by the device.

The above graphs indicate interference, exceeding the permissible values set by GOST.

To ensure that the level of conducted interference is in accordance with GOST R 51527-99 in the frequency range from 0.15 to 30 MHz, special measures must be taken, namely introduction of an interference protection filter into the circuit.

Making sure with a simple example that a filter is necessary, let us analyze a number of components that EPCOS offers on the market, these are B84773A (standard) and B84773M (medical) [14].

Both variants of the circuit configuration of the standard and medical filters are presented in Fig. 9.24. Table 9.5 shows the parameters of the components used in the filter.

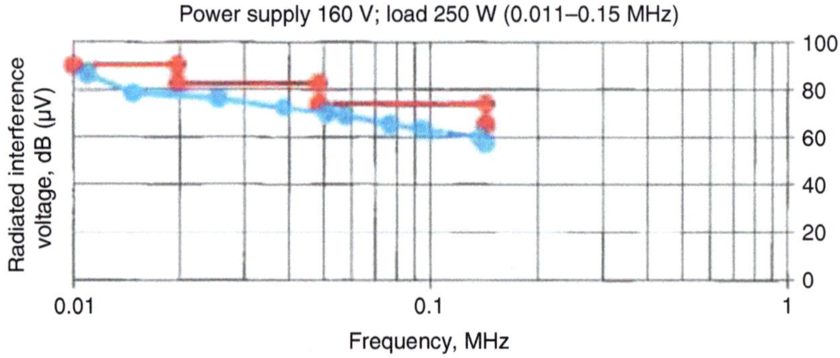

Fig. 9.22 The dependence of the voltage of the emitted interference on the frequency at a supply voltage of 160 V and a load of 250 W in the frequency range from 0.011 to 0.15 MHz

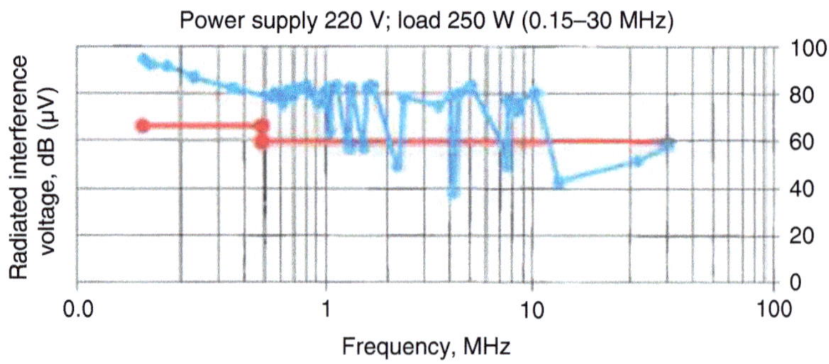

Fig. 9.23 The dependence of the voltage of the emitted interference on the frequency at a supply voltage of 220 V and a load of 250 W in the frequency range from 0.15 to 30 MHz

The optimal choice of the parameters of the components C1, L, C2, NW, R provides a degree of RF interference suppression of the order of 20–80 dB in the frequency range 0.15–30 MHz. Capacitor *C*1, connected between the phase conductor and the neutral, filters symmetrical interference, while capacitors *C*2 and *C*3 filter interferences of an asymmetrical type. Experimental studies have shown that up to hundreds of kilohertz symmetric interference prevails, and asymmetrical—above 1 MHz [19].

Using the calculation given in [14, 18], it is possible to estimate the value of the magnitude of the ground current I_3 for a standard filter, namely for the item B84773A:

$$\begin{aligned} I_3 &= C2(C3) \times 4\pi \times f_c \times U_{c\,\max} \\ &= 2200 \times 10^{-12} \times 4 \times 3.14 \times 50 \times 364 = 0.364 \ \text{mA}. \end{aligned} \tag{9.2}$$

Thus, the value of the ground current does not exceed 0.5–2 mA and is safe for humans. Focusing on the 1 MHz frequency, it is possible to calculate the minimum

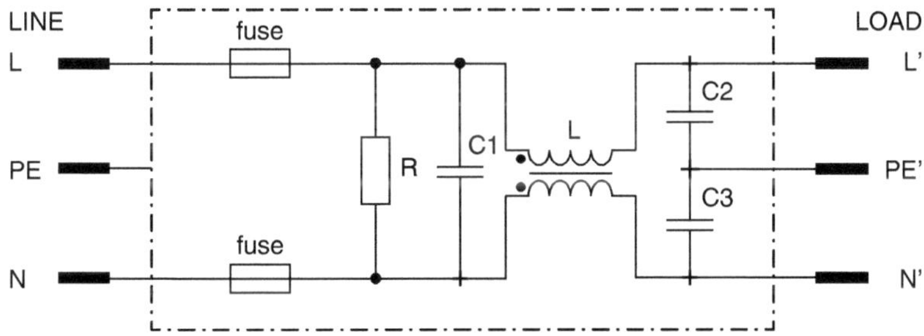

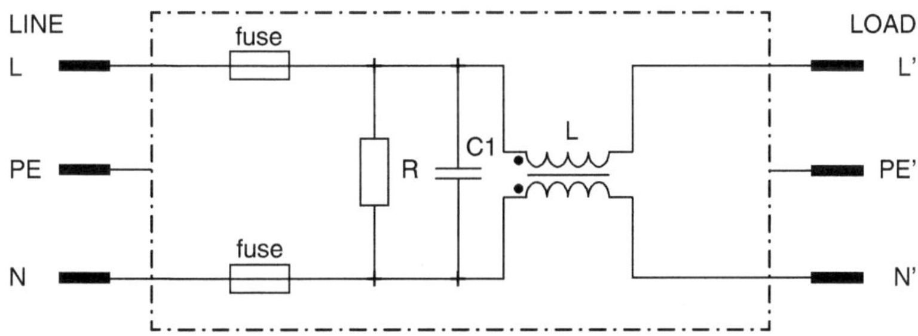

Fig. 9.24 Circuit configuration: (**a**) standard filter; (**b**) medical filter

inductance value of the inductor L based on the common-mode interference suppression coefficient Kpip taken from the graph for a specific filter B84773A0002A000 to 2 A (Fig. 9.25), recalculated using the formula $A = 20\lg C_{\text{PNP}}$ from dB.

Using the formula below, we define L and compare this value with the table value for EPCOS filters:

$$L_{\min} > \frac{K_{\text{пнп}}}{(2\pi f)^2 \times 2C_2} = \frac{200}{\left(2 \times 3.14 \times 10^6\right)^2 \times 2 \times 2200 \times 10^{-12}} = 1.15\,\text{мГн} \tag{9.3}$$

The tabular value is 2.7 mH for one coil, that is, it is 2.35 times the minimum value, which fully meets the requirements.

Table 9.5 Parameters of components used in network filters B84773* EPCOS

L_gA	C_RX2, μF	C_RY2, pF	L_R, mHz	I_{LK}, mA	R_{res}, MOM	Weight, g	Code of nomenclature	Compliance with standards		
								IEC 60939	UL 1283	CSA C22.2 No.8
VR = 250 V AC/DC										
1	1 × 0.1	2 × 2200	2 × 5.4	0.173	1	55	B84773A0001AOOO	+	+	+
1		–	2 × 5.4	0			B84773M0001AOOO			
2		2 × 2200	2 × 27	0.173			884773A0002A000			
2		–	2 × 2.7	0			B84773M0002A000			
4		2 × 2200	2 × 1.1	0.173			B84 773A0004AOOO			
4		–	2 × 1.1	0			B84773M0004AOOO			
6		2 × 2200	2 × 0.3	0.173			B84773A0006AOOO			
6		–	2 × 0.3	0			B84773M0006AOOO			
10		2 × 2200	2 × 0.2	0.173		75	B84773A0010A000			
10		–	2 × 0.2	0			B84773M0010A000			

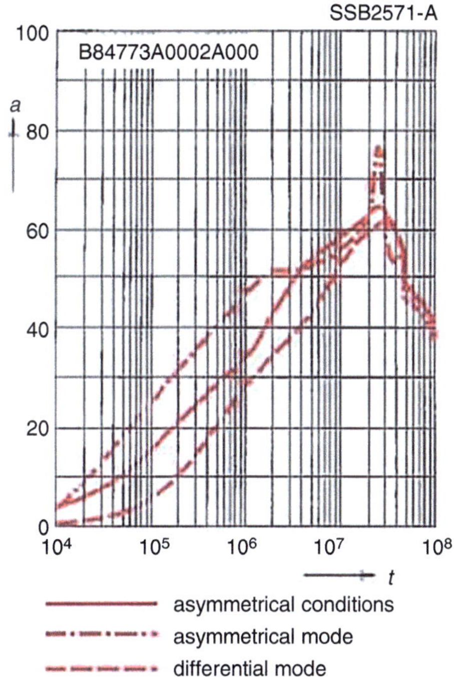

Fig. 9.25 Graph of frequency dependence for filter B84773A0002A000

The calculation of the suppressor capacitors for symmetrical interference $C2$ and C will be performed at a frequency $f_1 = 0.15$ MHz. According to the graph shown in Fig. 9.25, the level of suppression of symmetrical interference is 8 dB, then the C_{pnp} is equal to 2.5. Consequently,

$$C_{x\,\min} > \frac{K_{\text{пнп}}}{(2\pi f_1)^2 \times 2L_{\min}} = \frac{2.5}{(2 \times 3.14 \times 0.15 \times 10^6)^2 \times 2 \times 1.15 \times 10^{-3}} = 1.23\,\text{нФ}. \tag{9.4}$$

The tabular value is $C_{x\ \min} = 100$ nF, the stock is more than sufficient.

The simplest analysis performed on the attenuation curves, which are presented in the documentation for a specific group of filters, confirms the high quality of the EPCOS components.

9.5 Application of Multilayered Film Electromagnetic Screens for Protection of Onboard Equipment of Spacecrafts from the Effects of Quasistationary Magnetic Fields

When working in near-Earth space, scientific equipment is subjected to intense magnetic and electromagnetic fields. This causes errors in the work of fast front-end electronics and introduces errors in the measurement results. As noted above, for example, photomultiplier tubes (PMT) operating in scintillation detectors are sensitive to changes in the magnetic field, since it changes the direction of movement of electrons, i.e. creates defocusing, thereby reducing both the amplitude of the PMT pulses and the amplitude and temporal resolution of the spectrometer, respectively.

Roscosmos specialists were tasked with experimentally confirming the possibility of suppressing quasistationary magnetic fields with multilayer film screens (MFS), determining screening coefficients for prototypes of MFS, experimentally confirming the possibility of using MFS to protect PMT from weak quasistationary magnetic fields.

Before considering technical solutions, it is necessary to remind readers a number of features of the magnetic field in the vicinity of the earth. So, to estimate the magnetic field, under the action of which the cosmophysical scientific equipment will be in the case of its planned placement onboard the International Space Station (ISS), the data obtained in the experiments SPRUT-6 and Lazio-Sirad-Egle [20] were used.

The measurement of the magnetic field on board the Mir orbital station was carried out by a ferroprobe magnetometer, which was located there as part of the SPRUT experimental equipment [20]. The measurement data of the magnetic field in the framework of the SPRUT-6 experiment during the magnetic substorm from 17.00 UT on July 30, 1999 to 04.05 UT on July 31, 1999 is presented in Fig. 9.26. During the measurement, the station performed seven orbits around the

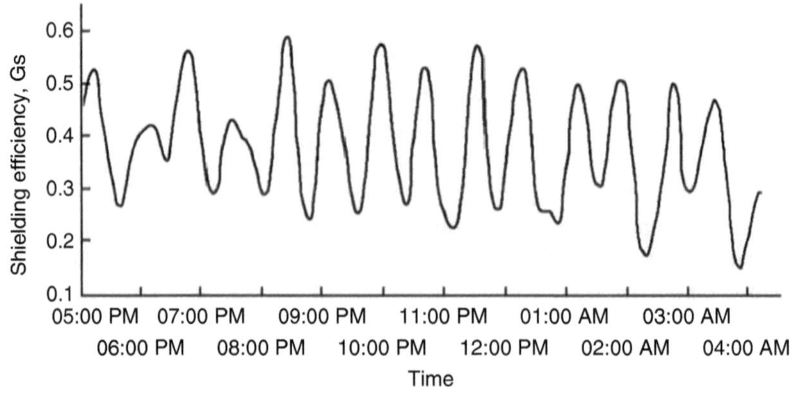

Fig. 9.26 Changes in the magnetic field aboard the Mir orbital station

Earth, each lasting 89 min. As can be seen from the figure, the maximum amplitude of the quasistationary magnetic field at the Mir orbital station reached 0.6 G, and the magnetic field fluctuations—0.4 G. Thus, for the ISS, taking into account the equivalence of its orbit, the quasistationary magnetic field can also reach 0.6 G (47.75 A/m), and its oscillations—0.4 G (31.8 A/m).

The EGLE wideband magnetometer, developed at the University of Roma-3 in Rome, Italy [21] and installed onboard the ISS as part of the experimental equipment Lazio-Egle.

On April 25, 2005 it was demonstrated that the maximum electromagnetic field strength reaches about 1 μG for the range of 20 kHz–40 kHz and about 100 μG for the range of 0–5000 Hz (the obtained data were filtered to exclude known signals generated by the ISS itself).

Before proceeding further, it is necessary to remind the reader some general questions of the theory of shielding. In part, these questions were raised above, but in order to facilitate the study of the problem by specialists in space physics spectrometers, the main points should be repeated here.

Shielding is a constructive means of suppressing electromagnetic fields (EMF) within a certain space in order to create normal conditions for the life of biological objects and the functioning of measuring equipment for various purposes, to ensure the requirements of electromagnetic compatibility. To quantify the effect of shielding two parameters are used—the shielding efficiency and the shielding factor. The shielding efficiency (*E*) is estimated from the ratio of the strengths of the electric or magnetic fields in the protected area of space in the absence of a screen (*E*0, *H*0) and in the presence of it (*E*, *H*) [20]:

$$E = E0/E = H0/H. \tag{9.5}$$

The shielding factor (*K*) is the reciprocal of *E* and is determined by the expression:

$$K = E/E0 = H/H0. \tag{9.6}$$

For shielding magnetic static and quasi-static magnetic fields (frequency from 0 Hz to 1000 Hz), ferromagnetic materials with high magnetic permeability are used.

The principle of shielding is based on the closure of power lines through a medium with high magnetic permeability. The shielding efficiency, defined as the ratio between the external and internal magnetostatic fields, for an infinitely long cylinder in a uniform transverse magnetic field can be approximately estimated from the formula:

$$E = 1 + \mu d/D, \tag{9.7}$$

where μ is the relative magnetic permeability of the magnetic screen, *d* is the thickness of the screen, *D* is the diameter of the cylindrical screen.

In the frequency range under consideration, the screen is usually bulky and expensive, besides the shielding efficiency of such a screen is not high. Achieving a satisfactory level of shielding efficiency is only possible with special materials with high μ values.

The following materials are mainly used for shielding low-frequency magnetic fields in scientific experiments: mu-metal or nickel–iron system alloys containing 80%; 36%; or 50% nickel; pure steel.

The so-called mu-metal is used in most scientific applications. This material has a high magnetic permeability in a weak magnetic field (some samples have $\mu \sim 500{,}000$ with an external magnetic field of 0.01 Oe), but when the external field is strengthened, it quickly becomes saturated and ineffective. The advantages of this material are excellent shielding of weak magnetic fields and the possibility of reducing the weight of magnetic protection due to the use of a smaller amount of material.

Pure steel is used with a strong external field, as it is highly resistant to saturation. Having a low magnetic permeability (maximum $\mu \sim 100$) steel, however, is usually used together with mu-metal. Pure steel reduces a strong external magnetic field, and mu-metal shields a residual weak magnetic field. Alloys with a content of 36% and 50% nickel are a compromise between mu-metal and steel.

Traditional technologies for manufacturing screens from massive sheet, tape, and foil materials based on iron and its alloys allow in many cases to solve the problem of protecting onboard equipment elements from the effects of external electromagnetic fields in the frequency range $f > 10$ kHz. In the field of low-frequency, quasi-static, and permanent magnetic fields, single-layer metal screens are not effective. Other materials science and technology solutions are required. In addition, the manufacture of screens according to traditional technology is an expensive, material-intensive and technologically complex path that increases the weight and cost of devices several times and does not always provide the required protection against permanent magnetic, low-frequency magnetic and electromagnetic fields (EMF).

To ensure high shielding efficiency in a wide frequency range, it is desirable to use multilayer metal screens consisting of layers with high magnetic permeability and layers with high electrical conductivity. This is due to the fact that in multilayer screens consisting of metals with different wave resistance, due to the multiple reflections of electromagnetic waves (EMW) between the layers, the screening efficiency increases. The reflection effect is observed at the interfaces between media with different wave resistances.

As it was already partially shown above, according to the available experimentally obtained results on multilayer screens, we can conclude:

- multilayer screens should contain layers of materials with high magnetic permeability and layers of materials with high electrical conductivity;
- in a two-layer screen, the EMI shielding is most effective in the low-frequency area with a ratio of copper and steel thicknesses dm/dm + dst = 0.5–0.6;
- in the low-frequency area (from 0.5 to 10 kHz), the optimum thickness ratio of the layers dm/dst = 1, because the copper layer switches to the electromagnetic mode

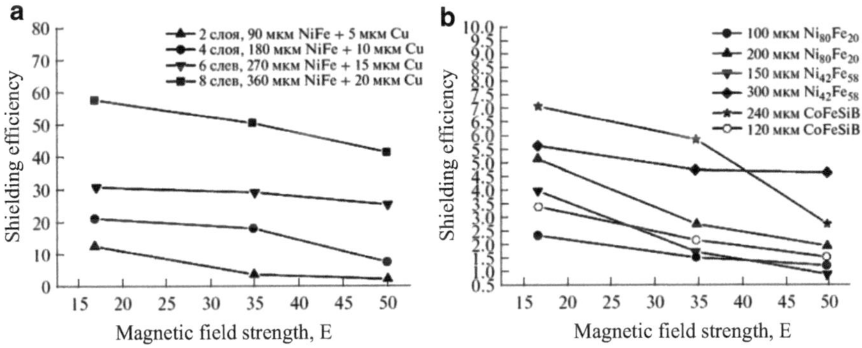

Fig. 9.27 Dependencies of shielding efficiency on the magnitude of the external magnetic field (the total thickness of the magnetic and nonmagnetic layers are indicated): (**a**) for cylindrical NiFe/Cu MFS with different thickness and number of layers; (**b**) for cylindrical single-layer screens made of Ni80Fe20, Ni42Fe58 alloys, and CoFeSiB foil amorphous alloy

of operation, and the steel layer continues to operate mainly in the magnetostatic mode;

- in the frequency range from 10 to 1000 kHz, copper and steel layers operate in electromagnetic mode. With increasing frequency, the optimal thickness dm decreases, and dst increases due to the greater effect of absorption;
- in the frequency range above 1000 kHz, the advantages of a multilayered screen are reduced, since homogeneous metal layers due to the high effect of EMW absorption provide high shielding efficiency.

Next, we briefly consider the main features of the construction of the MFS.

The Academy of Sciences of Belarus proposed MFS, which are manufactured by forming multilayer electrolytically deposited coatings on a metal casing of a shielded product. From Fig. 9.27 [20], it can be seen that the shielding efficiency of the constant magnetic field of an MFS is higher than that of single-layer screens of serially produced metallurgical alloys.

Structurally, the MFS is a multilayer film structure consisting of alternating layers with high magnetic permeability—magnetic layers and layers with high electrical conductivity (σ) or low resistivity (ρ)—nonmagnetic layers. The thickness of the nonmagnetic layer is selected in the range 1–25 μm, the thickness of the magnetic layer is selected in the range 6–250 μm, the total thickness of the magnetic layers is 300–500 μm, the total number of magnetic and nonmagnetic layers is in the range 3 < Nmagn. + Nnonmagn. < 99, with Nmagn. = Nnonmagn + 1. The nonmagnetic layer is a material based on copper or noble metals (Ag, Au). The material of the magnetic layers—alloys of the system iron–nickel–cobalt with a nanocrystalline or amorphous structure. In nanocrystalline materials with crystallite sizes of 10–100 nm, which is much smaller than the size of domains, a state with high magnetic permeability and low coercive strength is realized, i.e. magnetically mild condition. In amorphous alloys—due to the absence of crystallographic

Table 9.6 Samples used in experiments of PMT cases

No.	Name	Structure	μ	σ, ×107 cm/m	ρ, mcOhm cm	Weight, kg
1	Duralumin case with a prototype MFS No.1	4 layers of permalloy 55–60 μm each	12,000	1.54	0.65	0.05
		4 layers of copper with a thickness of 10 μm each	1	5.97	0.0175	
		Aluminum alloy AMG6 with a thickness of 1 mm	1	3.3	0.027	
2	Duralumin case with a prototype MFS No.2	10 layers of permalloy 25 μm each	12,000	1.54	0.65	0.05
		10 layers of copper with a thickness of 5 μm each	1	5.97	0.0175	
		Aluminum alloy AMG6 with a thickness of 1 mm	1	3.3	0.027	
3	Steel case	Steel-45 with a thickness of 1 mm	200	1	0.1	0.08
4	Duralumin case	Aluminum alloy AMG6 with a thickness of 1 mm	1	3.3	0.027	0.03

anisotropy, grain boundaries, and nonmagnetic inclusions—the magnetically mild state is also realized.

The technology of electrolytic deposition allows you to synthesize MFS with the required magnetic and electrical characteristics, the number and thickness of individual layers, provides high processability of the process, the possibility of forming MFS on cases and parts of complex geometric shape, on specified local areas of the part, rigid fixation of the screen structure that, in principle, solved with the help of assembly technology from sheet materials. When this occurs, the cost of manufacturing the MFS is reduced, as well as minimizing the overall and weight characteristics of the screen, with high protective properties.

For an experimental study of the shielding properties of the MFS and to confirm the possibility of using them to protect the PMT against a quasistationary magnetic field, measurements were made in [21] with prototypes of metal cases for the PMT is shown in Table 9.6.

So, under laboratory conditions, to study the shielding properties of a MFS in the Earth's magnetic field, measurements were made for the components and the modulus of the magnetic field inside the case with the sample MFS No. 2 (Fig. 9.28) and near its outer surface (Fig. 9.29). The measurements were carried out using a MAG3 digital three-component magnetometer with a measurement range of ±300 A/m and a resolution of 0.05 A/m. The screen was placed on a

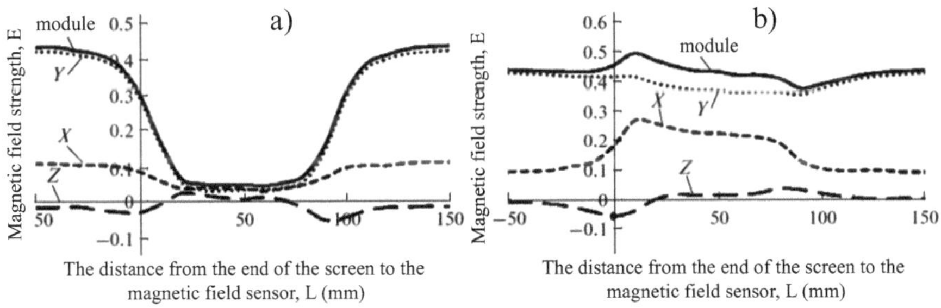

Fig. 9.28 Magnetic field measurement results: (**a**) inside the screen; (**b**) from the outside of the screen

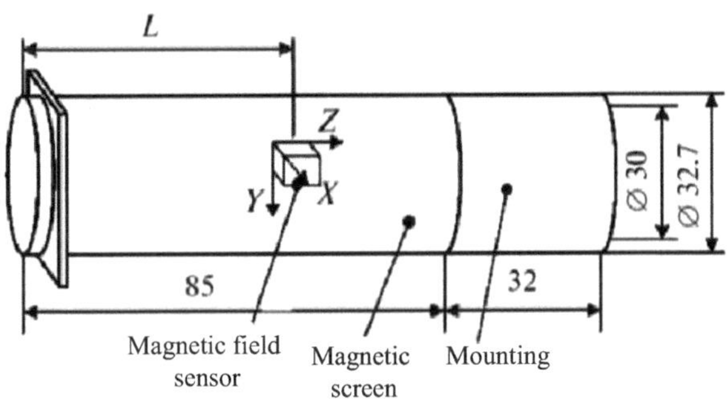

Fig. 9.29 The layout of the screen relative to the sensor of the magnetic field

horizontal surface. Its axis was directed perpendicular to the vector of the Earth's magnetic field (Fig. 9.29).

Analysis of the experimental results showed a high attenuation efficiency of the constant component of the Earth's magnetic field. The average attenuation coefficient of the magnetic field inside the test screen was $C_{av} > 7$.

To study the shielding properties of the MFS in a constant magnetic field, measurements of the components and the modulus of the magnetic field were carried out inside the MFS No.1 and MFS No.2 samples depending on the distance to the permanent magnet (Fig. 9.30). The measurements were carried out using a MAG3 digital three-component magnetometer. The magnet was located along a straight line connecting its center with the center of the sensor. The sensor was located in the center of the screen ($L = 42.5$ mm). The field of the magnet was directed along the strand X-axis of the sensor. The magnetic field strength of a permanent magnet, measured by a magnetometer, is shown in Fig. 9.31. The obtained shielding coefficients for the samples MFS No.1 and MFS No.2 are presented in Fig. 9.32 [20].

To assess the effect of the MFS on the operation of the FEU-85 in a magnetic field, an analysis of the spectra of LED radiation was carried out. A permanent magnet with an adjustable position relative to the center of the photomultiplier was

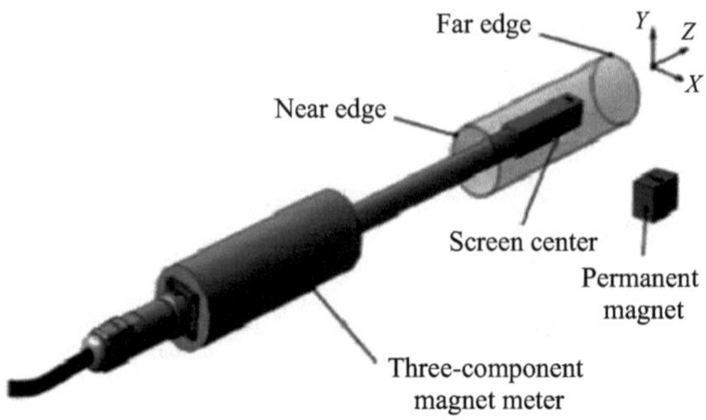

Fig. 9.30 The location of the magnet relative to the magnetic screen

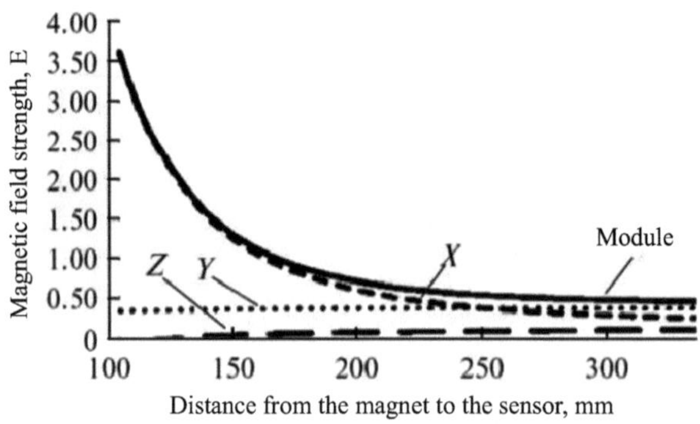

Fig. 9.31 Magnetic field of the magnet depending on the distance to the magnetometer sensor

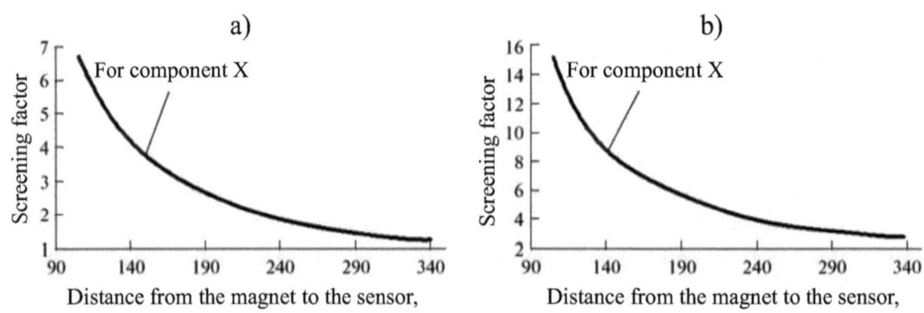

Fig. 9.32 Dependencies of shielding coefficients on the distance to the magnet: (**a**) for sample MFS No.1; (**b**) for sample MFS No.2

used as the source of the magnetic field. The magnet was located along a straight line connecting its center with the center of the magnetic screen, similar to the configuration shown in Fig. 9.30 [20]. Illumination of the photocathode of the PMT by photons from the LED was carried out in a pulsed mode with a frequency of 100 Hz.

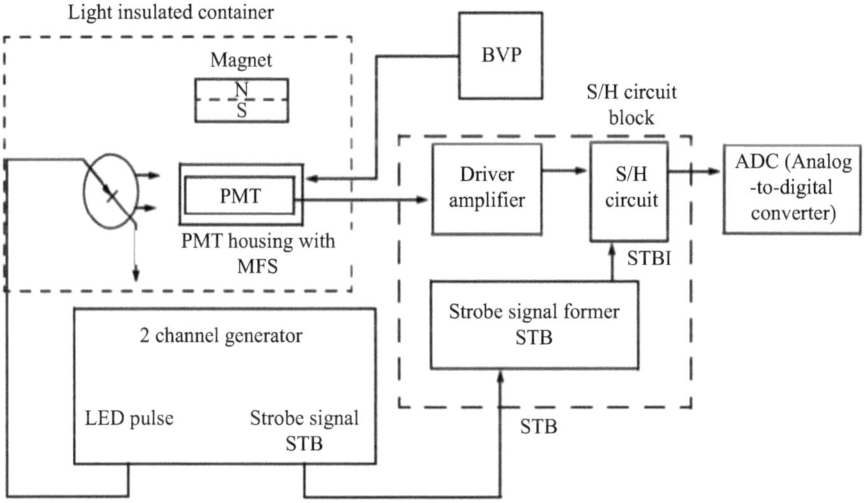

Fig. 9.33 Functional circuit of experimental installation

The AL-102B diode with a radiation maximum in the red area of the spectrum, switched on in the "direct" mode, was used as the LED. Before the measurements, a preliminary natural heating of the installation for temperature stabilization of the amplitude spectra obtained from the PMT was carried out (the relative change in the position of the maximum of the spectrum was no more than 0.1% for 10 min). Functional circuit of experimental installation is illustrated in Fig. 9.33.

The studies were conducted with all the PMT cases from the Table 9.6. The pulse operating mode of the installation was provided by a 2-channel Tektronix AGF 3252 generator. The signals from the PMT were received at the input of the sampling and storage device (SSD). The amplitudes of the signals from the SSD were digitized using the Advantech PCI—1714U analog-to-digital conversion (ADC) board installed in the PCI slot of the computer. As a result of processing the data obtained from the ADC, the amplitudes were determined corresponding to the maxima of the amplitude spectra of the LED radiation measured during operation of the PMT in various magnetic fields, which were created by a permanent magnet as it moved relative to the screen center. The error in determining these amplitudes, taking into account the operation of all components of the installation and software, is no more than 0.1%.

Figure 9.34 shows the dependence of the relative amplitude change corresponding to the maximum of the amplitude spectrum on the field strength when the PMT operates in various external quasistationary magnetic fields. The amplitude of the signal A (B0) was taken to be the amplitude of such a signal, which corresponded to the maximum of the amplitude spectrum measured during the operation of the PMT in the Earth's magnetic field only. As can be seen from the above dependences, the relative change in the signal in a magnetic field of 0.6 Oe for

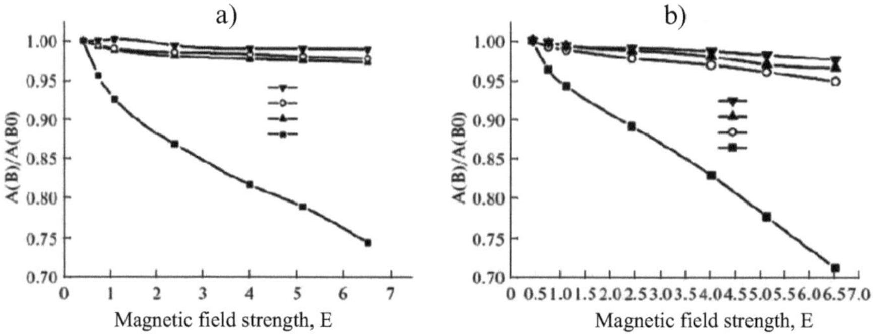

Fig. 9.34 The relative change of the signal depending on the external magnetic field for different types of cases: (**a**) for the vertical orientation of the dynodic PMT system; (**b**) for the horizontal orientation of the PMT system

a steel case is not more than 0.5%, for sample 2 it is 1%, for sample 1 it is 1.1%, and for ordinary aluminum case—about 7%. For scientific equipment representing scintillation spectrometers whose amplitude resolution can reach, for example, 7–9% (for spectrometers made on the basis of NaI (Tl), CsI (Tl) scintillators) or 3–4% (for spectrometers based on BrilLanCe scintillators), as well as for scientific equipment that requires regular calibration of the analog and time paths of its electronic support during operation, changing the amplitude of signals from a photomultiplier by 7% is a significant factor degrading both the amplitude resolution of the device and calibration results, which seriously affects the accuracy of measurements. The obtained results confirm the possibility of using the MFS to suppress the influence of a quasistationary magnetic field on the operation of a PMT with the aim of ensuring precision spectrometric measurements with an accuracy of about 1–2%.

As can be seen from the results of the experiments carried out and described above, the MFS are very promising for their use in the development of small-sized onboard scintillation (for example, cosmophysical) spectrometers as screens to protect the PMT from weak quasistationary magnetic fields (for example, the Earth's magnetic field and magnetic fields created by the equipment included in the spacecraft infrastructure). Their use also allows to provide a sufficient level of protection of the PMT from the effects of magnetic fields in near-Earth space caused by magnetic storms and other geophysical phenomena. Screens made by new technologies are compact, durable, and light, which is an important advantage for use in modern cosmophysical scientific equipment. It is also important that the cost of such screens is noticeably less than the average cost of the PMT itself (for example, the cost of imported screens for the R5946 PMT produced by Hamamatsu is equal to the value of the PMT).

References

1. Belous, A. I., Merdanov, M. K., & Shvedov, S. V. (2016). *Microwave electronics in radar systems and communications*. Technosphere (in Russian).
2. Nakauchi, E. (2012). Devices for protection against electromagnetic radiation. *Electronic Components*, 1, 6–8 (in Russian).
3. Gurevich, V. I. (2015). The lack of a standard for filters for protection against electromagnetic pulse makes it difficult to create an effective protection. *Electronics Info*, 6, 35–38 (in Russian).
4. Gurevich, V. I. (2014). *Vulnerabilities of microprocessor protection relays: Problems and solutions* (p. 256). Minsk: Infra-Engineering (in Russian).
5. MIL-STD-188-125-1 High-altitude electromagnetic pulse (HEMP) protection for ground-based C4I facilities performing critical, Time-Urgent Missions; Part 1: Fixed Facilities.
6. Nalhorczyk, A. J. (2008, November 26–27). HEMP filter design to meet MIL- STD-188-125 PCI test requirements. In *IEEE. 10-th international conference electromagnetic interference & compatibility*, pp 205–209.
7. IEC 61000-4-4 Electromagnetic compatibility (EMC)—Part 4-4: Testing and measurement techniques—Electrical fast transient/burst immunity test, 2012.
8. MPE. (2012, December). *HEMP filter maintenance and monitoring* (Rev. 1.). Application Notes Cat. 1. MPE Ltd.
9. Schneider Electric. (2011). *Surge protective device response time*. Application Note 9910-0003A. Schneider Electric.
10. *Surface mount power TVS diodes deliver optimal protection for power supply*. Application Note, Bourns, Inc. 7/14.e/ESD1435.
11. Goldman, S. J. (2010, June 1). *Selecting protection devices: TVS diodes vs. metal-oxide varistors*. Power Electronics.
12. Department of Defense. (1959). *MIL-STD-220B method of insertion loss measurement*.
13. Jehle, C. (2012). EPCOS. How to eliminate leakage currents on the ground and significantly increase EMC. *Electronic Components*, 1, 68–69 (in Russian).
14. Goreev, A. (2016). Protection against conducted interference. *Power Electronics*, 1, 7–12 (in Russian).
15. Glazenko, T. A., & Senkov, V. I. (1984). Prediction of high-frequency noise generated by transistor pulse-width converters. *Electricity*, 12 (in Russian).
16. *GOST P 51527-99 "Electromagnetic compatibility of technical equipment. Stabilized DC power supplies. Conductive electromagnetic interference*. Standards and test methods—State Standard of the Russian Federation (in Russian).
17. Veksler, G. S., Nepochetov, V. S., et al. (1990). *Suppression of electromagnetic interference in power supply circuits*. Kiev (in Russian).
18. Lantsov, V., & Eranosyan, S. (2007). Electromagnetic compatibility of pulsed power sources, Part 2. In *Power electronics*, 1 (in Russian).
19. Nasyrov, I. I. (2011). *Specialized current source for coagulation of aqueous solutions*. Thesis, Tomsk (in Russian).
20. Batishchev, A. G., Galper, A. M., Dmitrenko, V. V., Zamesov, A. Yu., Naumov, P. Yu., Grabchikov, S. S., Sosnovskaya, L. B., Cheledyuk, A. V., Sharapa, T. E., Grishin, S. A., & Petyuk, A. L. (2012). The use of multilayer film electromagnetic screens in spaceborne cosmophysical spectrometers. *Nuclear Physics and Engineering*, *3*(5), 1–8 (in Russian).
21. Sgrigna, V., Altamura, F., et al. (2005). *First data from the EGLE experiment onboard the ISS*. Rome: Università Roma Tre.

Chapter 10
Power Supply Systems of High-Speed Electronic Devices

This chapter is devoted to the design of one of the most important systems—the power supply system for high-speed electronic devices.

In Sect. 10.1, the core requirements to these systems and concerns of their design are considered. Section 10.2 focuses on composition, technical parameters, and features of application-specific integrated circuits (ASIC)—controllers and power management drivers, voltage pulsed regulators, supervisor ICs, control signals sequencers in power supply circuits, load switching ASICs, and "hot" plugging ICs for electronic equipment.

In Sect. 10.3, technical characteristics and application features of step-up and step-down DC/DC converters are considered separately and in detail.

Section 10.4 examines various effective methods for suppressing noise in the power supply circuits of high-speed devices, as well as the peculiarities of grounding systems connection for electronic equipment of electrical power generating facilities incorporated in telecommunication systems (Sect. 10.5).

10.1 Design Features of Power Supply Systems for High-Speed Electronic Devices

When creating any telecommunication system, various power distribution systems are widely used, the main purpose of which is to provide uninterrupted power supply to various equipment (servers, automatic telephone exchanges, etc.), which usually have different supply voltage levels. The most widely used is the so-called *power supply system with an intermediate bus* (Intermediate Bus Architecture—IBA), in which up to three stages of voltage reduction can be used [1].

Each of these levels uses its own circuit design and structural solutions for voltage level conversion devices.

A. Belous, V. Saladukha, *High-Speed Digital System Design*,
https://doi.org/10.1007/978-3-030-25409-4_10

At the highest (third) level of the hierarchy, the input converter (rectifier) converts the AC mains voltage into a stabilized DC voltage of 48 V.

The task of voltage converters of the second hierarchy level is to convert (lower) the voltage from 48 V to 24, 12, 8, 5, or 3 V depending on the specific types of microcircuits used on the boards of this level.

In this case, losses are significantly reduced due to the high coefficient of efficiency of the pulsed down stabilizer at the point of application of the load and reduction of power loss ($P = I^2R$) on the power bus conductors, as well as a decrease in the voltage drop across the power bus conductors.

And finally, at the first (lower) level of the three-step power supply system, the voltage of the buses on each of the boards is stabilized with the help of pulsed step-down voltage converters, which are considered directly at each system load.

Depending on the application characteristics of the converters used at the top of the power system hierarchy, they are called either a voltage stabilization module (VRM) or a voltage converter at the point of load (Point of Load—POL).

Let us consider the features of such the most widely used POL-converters in the organization of monitoring and control procedures in relation to synchronization and the sequence of connecting the voltage to the microcircuits of the board [1].

The more complex the microcircuit used, the more important it is for the system developer to properly organize both the power synchronization system and the power supply sequence.

Modern microcontrollers require the use of POL-controlled voltage converters to upgrade power systems.

The sequence of switching on the supply voltage requires a certain sequence of connection of the supply voltage, since when switched on they pass through a duty (sleep) mode and operating mode. Switching the power on and off switches between these two main modes. In addition to ensuring the correct sequence of voltage connections, control of the rate of rise of the output voltage is very important. Sometimes it is even more important than ensuring the sequence of connection of voltages. The sequence of switching on the supply voltage determines how these several voltages applied to the chip interact with each other and in what ways it is possible to eliminate the occurrence of dangerous transients.

Figure 10.1 [1] shows the signals of the supply voltage of the processor core and the supply voltage of the inputs-outputs when the Xilinx FPGA starts up (turned on) uncontrolled using a POL voltage converter with two outputs. Yellow beam—input voltage 5 V of POL voltage converter; the blue and violet beams represent 3.3 V input-output device power and 1.8 V core power supply voltage, and the green beam is 3.3 V power supply output current.

In the programmed FPGA data on the status of keys and triggers are stored in the RAM of the configuration cells. When the FPGA is enabled, before the configuration RAM cells are programmed, the keys and triggers are in a random state. After the configuration RAM cells have been programmed, the keys and triggers are programmed to the desired state. This causes a short-term current surge of more than 6 A with a duration of approximately 1.5 ms at the output of 3.3 V, after which the current drops sharply to almost zero.

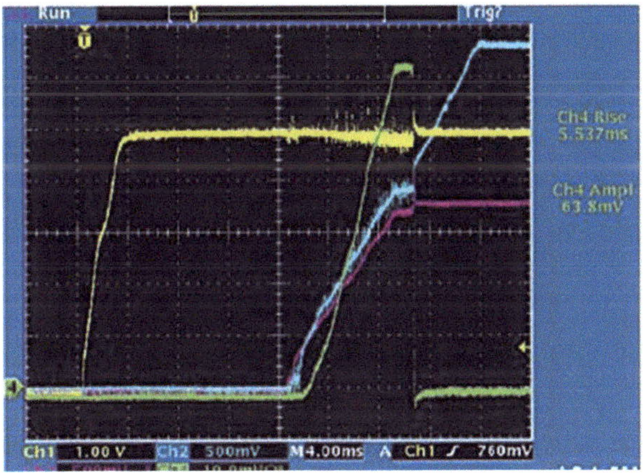

Fig. 10.1 Transient processes

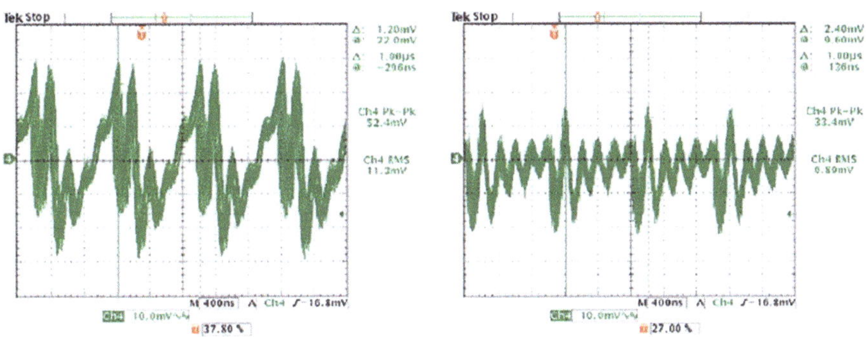

Fig. 10.2 Interference on the intermediate power bus

During this transient process, the voltage of the inputs and outputs and the voltage at the outputs of the triggers do not correspond to the voltages of the FPGA operating mode. In addition, due to the high current at the output of 1.8 V during the transient process, the voltage of +5 V at the POL input of the voltage converter drops briefly.

A few words should be said about process synchronization [1].

If the frequencies and phases of several POL voltage converters are different, they can interfere with each other and generate electromagnetic interference on the intermediate bus. Synchronization is simply necessary when multiphase POL voltage converters are used (Fig. 10.2). In the figure, the left oscillogram shows the input interference in a system with three POL voltage converters without synchronization; the right oscillogram shows the reduction of interference in the phase shift of the switching of the second and third POL voltage converters by 123.75° and 247.5° relative to the first.

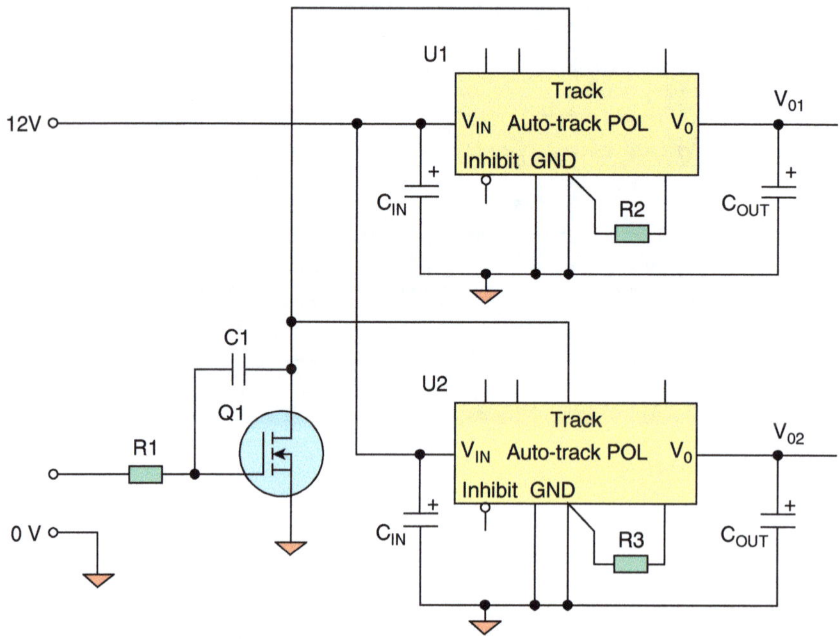

Fig. 10.3 The principle of simple correction of the output voltage

There is always a special technical monitoring service on powerful stationary telecommunication systems.

Monitoring provides two benefits: fault detection and failure analysis.

In the data registration and processing center, you need to know, firstly, that the board has failed and, secondly, where to send the equipment to replace it. However, it is always better to have more detailed information. It is better to immediately replace the module, the temperature, or output voltage of which is steadily rising, than to wait until it fails completely and leads to failure of other circuit elements. For failure analysis, it is preferable to use power supply performance monitoring for a long period of time.

In order to control POL converters that are used in the development of distributed power systems, one can use simple voltage correction circuits (Fig. 10.3) [1] or more advanced circuits designed for this purpose, which use I^2C remote control.

A simple correction is that the auto-tuning of the POL voltage converter, adopted by the Point of Load Alliance (POLA), which includes the manufacturing companies Astec, Artesyn, Ericsson Power and Texas Instruments, is made via the TRACK lead. The output voltage of the POL voltage converter varies with the voltage on this lead. The TRACK leads of several POL voltage converters are connected together and controlled by a common voltage through one MOSFET (Q1 in Fig. 10.3) [1].

The switching sequence and the slew rate of the output voltage of the POL voltage converters are usually controlled using specialized analog controllers.

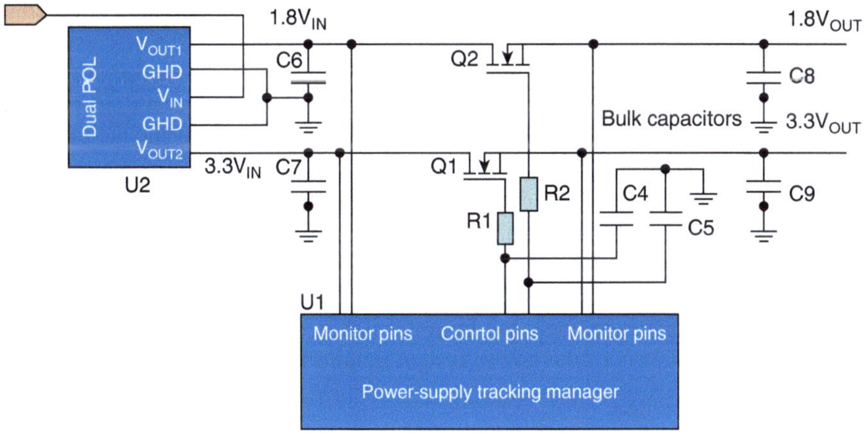

Fig. 10.4 Typical power management circuit with an analog microcontroller

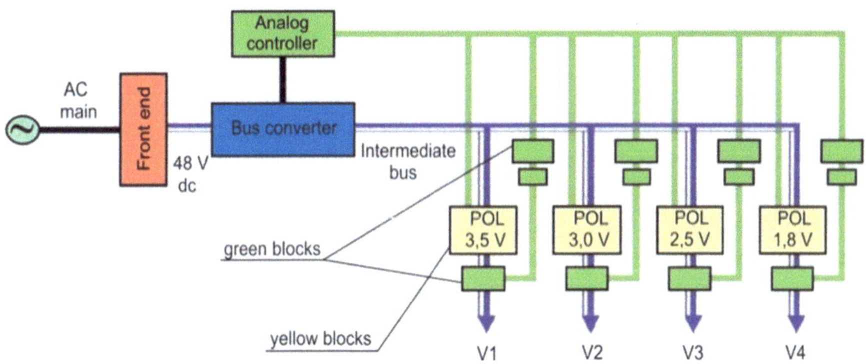

Fig. 10.5 POL analog control circuit

Analog power controllers (Fig. 10.4), offered by many manufacturers, allow to control the switching sequence and adjust the slew rate of the output voltage due to additional MOSFET transistors and time-sensitive resistors and capacitors.

Typical analog controllers can simultaneously monitor up to four POL voltage converters. They can provide minimal programmability and control through a JTAG interface or an I^2C bus. Vendors offer PC-based graphical interfaces that simplify parameter management.

The power management IC can control the voltage at the output of the converter and at the load (in this case, a POL voltage converter with two outputs used to power the FPGA) and control the order of switching them on and off through the series-connected MOSFET transistors. Turn-on delay time is set by simple RC-chains.

Analog power management (Fig. 10.5) [1] is effective up to a certain level of system complexity. With analog power management, an increase in the number of POL voltage converters (yellow blocks) leads to a dramatic increase in discrete elements (green blocks). The system, which has eight output voltages to be controlled, uses more than 200 additional passive components and discrete semiconductor devices, occupying approximately 64.5 cm^2 area of the printed circuit board. The development of such a distributed power board can take months. In systems corresponding to the high end level (see Table 10.1), it is more advantageous to use digital power management technology.

This technology allows building an intelligent control circuit of POL voltage converters, making them individually addressable (via a contact connection) and connecting to the controller via an interface via a digital bus (Fig. 10.6) [1]. The controller, in turn, communicates with the monitoring system via the I^2C bus. In this case, 32 POL voltage converters can be controlled in real time.

The main parameters of the power system monitoring:

- Output voltage range.
- Tripping thresholds for increasing and decreasing voltage and current, as well as increasing the temperature.
- On-off sequence and slew rate of output voltage.
- Frequency conversion.
- High and low limits for the “Power Good” signal.

For systems with many POL voltage converters on a board, the intelligent (digital) POL control method allows to reduce the number of elements used for power management and the size of the printed circuit board, and also provides more control and monitoring capabilities (Fig. 10.6).

Like analog controllers, digital power controllers have GUI-driven development and debugging tools that are more configurable than analog controllers.

Hazardous transient processes that may occur in the microcircuit, before modeling can be represented using models provided by the graphical user interface.

Let us note the main advantages of using the Z-One system [1]:

- The main advantage is a 90% reduction in the number of components, conductors of the printed circuit board, and development time.
- Reduces the number of components in the system, reduces the cost, reliability, and power density.
- Designing and modeling using a graphical interface greatly simplifies the design of a power system.
- Leading manufacturers in the field of power management and conversion provide the means to program the output voltages, the switching sequence, and the ability to adjust the characteristics of the output voltage and the protection response limits during operation.

Z-Series digital power controller:

- programs Z-POL converters and controls them;
- controls the intermediate power bus, processes the signals, and provides protection;

Table 10.1 Characteristics of power systems [1]

Application	Quantity of outputs	Output currents	Necessity of power management	Implementation	Price	Priorities
High end: • processing and transfer of data • test equipment • high end servers	More than 6	From 10 up to 100 A	High Systems are very sensitive to power conditions	The complex structure of modular converters, sometimes discrete	More than $10,000	Performance, digital control
Average level: • servers • complex equipment • telecommunications	From 4 up to 6	From 5 up to 10 A	Average or high: Depends on the application	DBA or IBA Modular and discrete converters are used	From $1000 up to $10,000	Cost
Low level: • medicine • high end computers • high quality household equipment	Not more than 3	Less than 5 A	Low or not needed at all Single microcircuit systems	Discrete	Less than $1000	Price Analog control

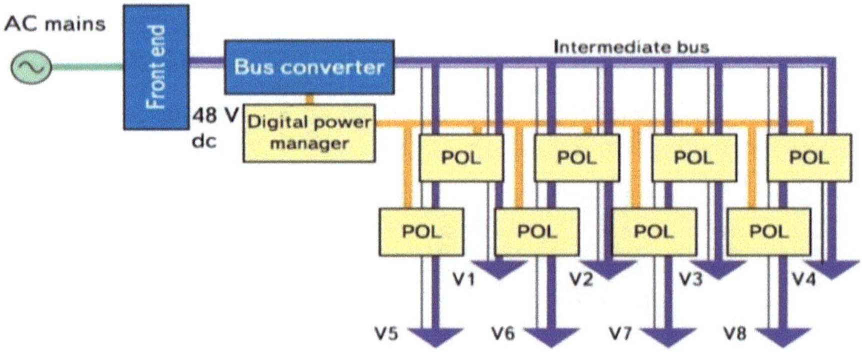

Fig. 10.6 POL (Z-One) digital control circuit

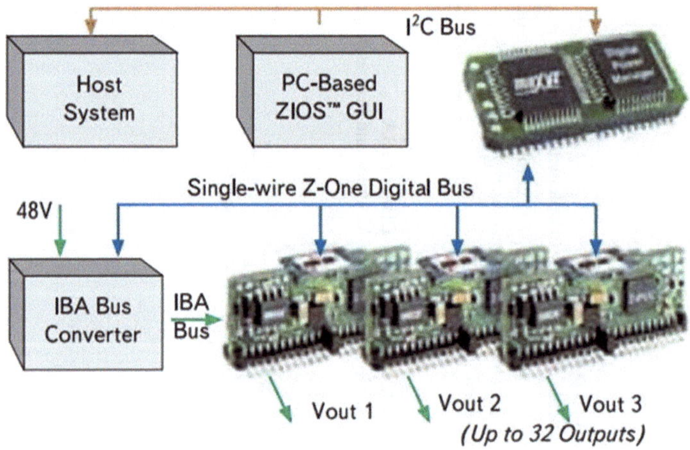

Fig. 10.7 POL implementation with Z-One bus

- facilitates telemetry of Z-POL performance for remote monitoring and troubleshooting.

Z-POL converters provide:

- programmable output voltage from 0.5 up to 5.5 V DC at input voltage from 3.0 up to 13.2 V DC;
- current density up to 39 A per inch^2;
- real-time voltage, current and temperature monitoring.

It can be seen from the above that the use of a modern digital control system for distributed power supply of complex telecommunication communication systems and industrial automation systems can significantly reduce the cost of creating power supply systems, reduce operating costs, and simultaneously improve the quality of power supply of such systems (Fig. 10.7).

10.2 Specialized Chips for Power Supply Systems of Electronic Devices

10.2.1 Generalized Structure of the Power Supply Circuit of Telecommunication Devices

Taking into account the limited volume of the book, in this section we consider only the main types of specialized microcircuits designed to control the power supply of modern information and communication devices by the example of the nomenclature series of the world leader in this field—Analog Devices [2]. We will give a brief overview of the IC for power management produced by this company, consider the features of their structural and circuit design, and give specific examples of applications.

To understand the complexity of the design problems of power supply systems of modern high-speed electronic devices, simply look at Fig. 10.8 [2]. Presented here is a generalized system diagram of the power supply system of a modern telecommunications system. Toned color here highlights the places occupied by the basic elements of this system. These include AC/DC and DC/DC converters, controllers and drivers, pulse controllers, pulse voltage regulators, linear voltage regulators, supervisors and various drivers of control sequences, power management IC, load switches of the "hot plug" microcircuit, and much more.

It is clear that the design of such systems can only be dealt with by well-trained specialists, for which, in fact, this book is intended.

Let us consider in more detail the typical representatives of this family of microcircuits [2].

10.2.2 Controllers and Drivers in Power Supply Control Circuits

Analog Devices, a pioneer in the use of digital power management, is also the largest supplier of AC/DC and DC/DC converters of power supply controllers and power supplies to the world market that meet all developer requirements and are easy to program and use. All controllers are supported by a graphical user interface that allows to quickly optimize the coefficient of efficiency and other parameters of power supply circuits and thereby reduce the time of transferring the product to market (Fig. 10.9).

So, the ADP1046 microcircuit is a digital controller with seven PWM outputs, which is relatively simply programmed via the I^2C interface using a graphical user interface. The controller is designed to control different power sources in different modes of operation, provides the ability to adapt the power source to changes in input voltage, which allows reducing transient processes in voltage stabilization circuits with feedback, reducing output interference, reducing losses by "dead" time

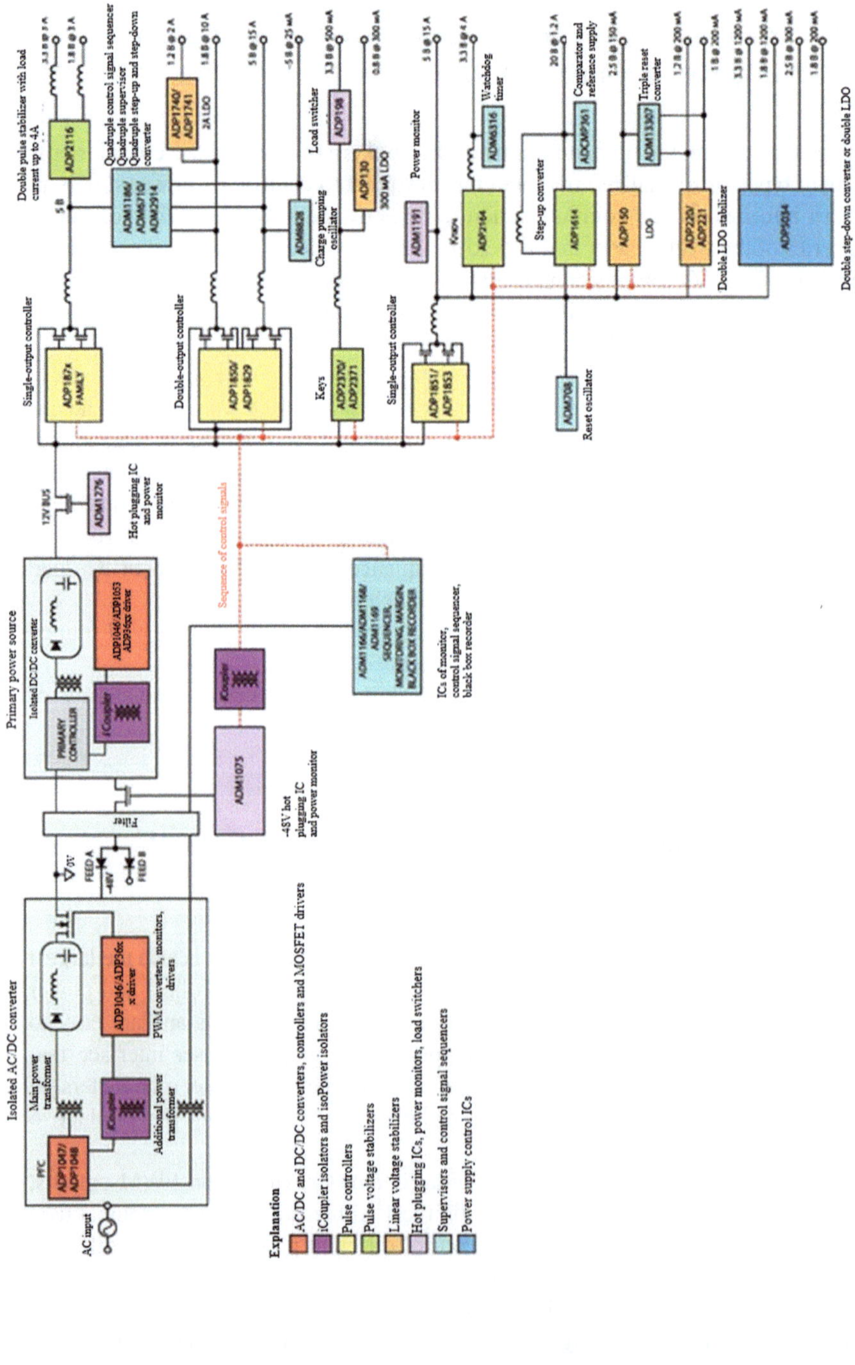

Fig. 10.8 Generalizing system diagram of the power supply system of an electronic device [2]

Fig. 10.9 Evaluation board ADP1046-100 W with a bridge synchronous rectifier [2]

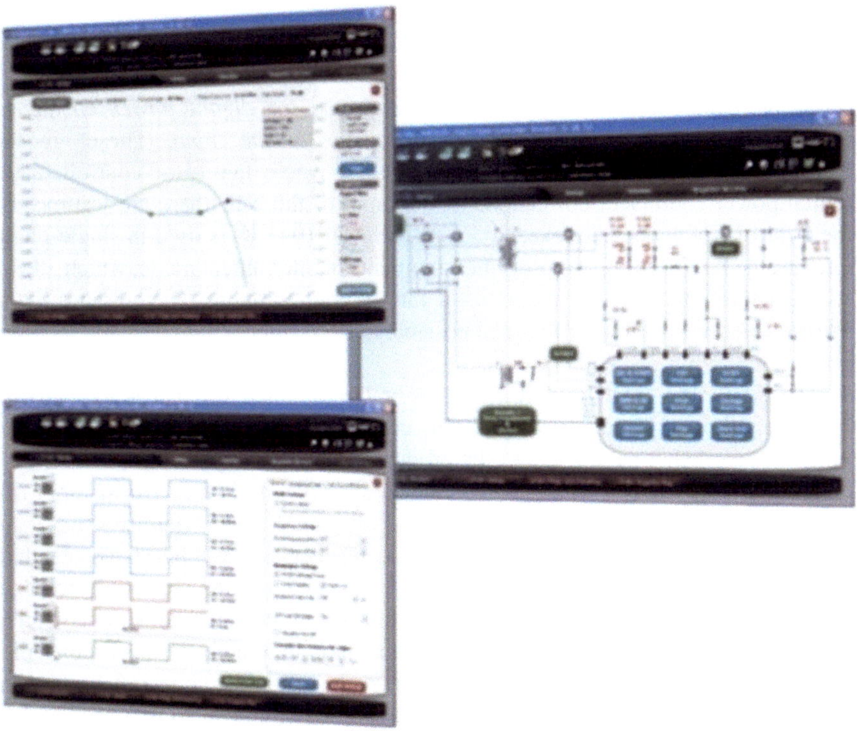

Fig. 10.10 An example of a graphical user interface for programming the ADP1046 controller [2]

when changing load current. The controller can be used in parallel or redundant power sources for separate control of external FET transistors through the OR switch (Fig. 10.10).

The next microcircuit is the ADP1053, a dual-channel DC/DC converter with eight PWM outputs, which is designed for use in various power sources. The ADP1053 has a PMBus interface and provides synchronization of multiple power sources (Fig. 10.11).

Fig. 10.11 Evaluation board of AC/DC converters with power corrector. AC input voltage 85–265 V, DC output voltage 385 V, output power up to 600 Wt. The board is made on the basis of the IC ADP1048 [2]

Of particular *note is a series of microcircuits for digital power coefficient correction (cos φ) with a precision wattmeter at the input.* These are the MMCADP1047 and ADP1048 microcircuits—a whole family of controllers for digital power coefficient correction. At the input of these IC there are wattmeters, the error of which does not exceed $\pm 1\%$. The ADP1047 IC is used in single-phase circuits, the ADP1048 IC supports measurements in intermittent mode, which allows its use in multiphase electrical circuits of various configurations. The presence of the PMBus interface in these IC makes it possible to adjust their parameters and measure power, input voltages, and currents.

The ADP1047, ADP1048 ICs are easily programmed using a graphical user interface, allowing developers of AC/DC converters to optimize the coefficient of efficiency and other parameters of the products being developed.

High-speed drivers for controlling powerful MOSFET transistors are represented by the ADP3654/ADP3634/ADP3624/ADP3630 family—these are dual high-speed drivers used with power controllers. These ICs have standard sizes and additionally contain key elements. In addition, they are highly reliable. The drivers of this family have two levels of protection against overheating (provide the formation of an alarm and the transition to a mode with low consumption). These drivers also have standard overvoltage and voltage fall protection, protecting the powerful FET transistor from overloading. The ADP3650 is a dual driver of MOSFET transistor for use in synchronous down converters.

Analog Devices manufactures a variety of *pulse converter controllers that provide high coefficient of efficiency and ease of use*—single and dual controllers with external MOSFET transistors to provide high coefficient of efficiency of power converters with a wide input voltage range and a minimum output voltage of up to 0.6 V. Analog Devices provides developers with automated tools for designing the power management circuits of ADIsimPower, which in a few steps allow to quickly develop a schematic diagram of the device and create a specification of components. All this allows to refer this company to the most reliable suppliers of IC for power management and tools for their design.

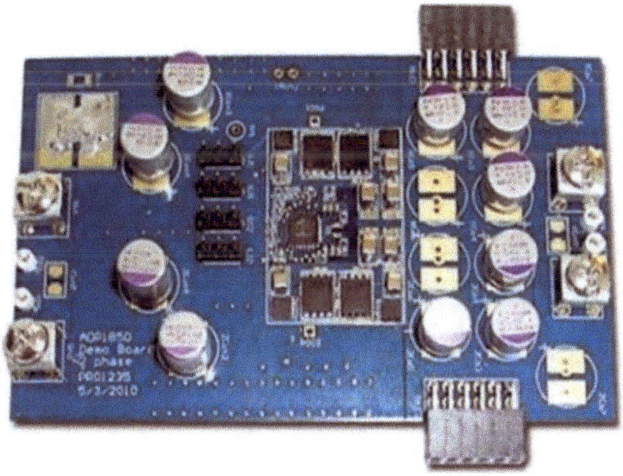

Fig. 10.12 Demonstration board with a pulse converter controller: input voltage range from 0.9 to 12 V, load current 50 A [2]

Analog Devices pulse converter controllers are designed for a wide range of load currents. ICs of the ADP187x family have a large input voltage range and are distinguished by a short response time. Single-channel controllers ADP1851 and ADP1853 have a large input voltage range and operate with a constant switching frequency. Both ICs can be configured for output current or voltage control. Synchronization of controllers can be provided by an external generator, the controllers support fast power on mode. The ADP1853 IC has a clock pulse output (CLKOUT).

The ADP1850 dual pulse converter controller (can work in intermittent mode) is a synchronous current controller and is designed for use in multi-level power sources. The ADP1829 dual pulse converter controller is a synchronous voltage controller and has the same basic functions as the ADP1853 controller (Figs. 10.12, 10.13, 10.14, and 10.15).

10.2.3 Pulse Voltage Stabilizers

Analog Devices supplies the electronic components with a large series of switching regulators with an integrated FET transistor for the global market. These stabilizers provide a decrease or increase in the voltage supplied to their input, while the output voltage can be fixed or adjustable. Pulse stabilizers are highly integrated ICs, and when they are used they practically do not require the use of external components. The availability of the ADIsimPower automated design system greatly facilitates the development of power supplies based on pulse stabilizers from Analog Devices.

Pulse stabilizers ADP2384/ADP2386:

- input voltage up to 20 V, load current up to 4/6 A;
- the best coefficient of efficiency among the closest analogues (up to 95%);
- the presence of a load current limiting system allows the use of small chokes;
- availability of special lead to control the switch-on time;

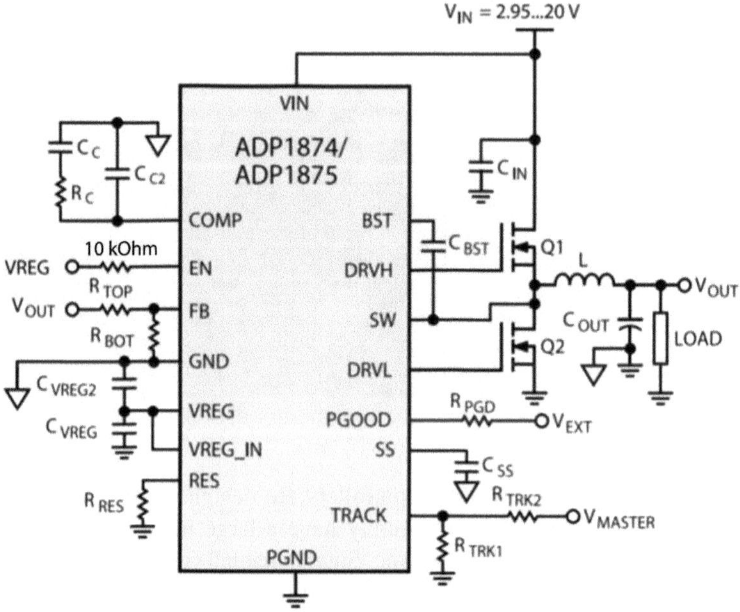

Fig. 10.13 Circuit of turning on the IC DP1874/ADP1875—single-phase controllers of pulse converters with one output and constant switching frequency [2]

- the ability to adjust the switching frequency and its synchronization in a multi-level power mode reduces the level of radiated electromagnetic interference;
- optimal input voltage of 12 or 5 V (Fig. 10.16).

Pulse stabilizers ADP2370/ADP2371:

- input voltage up to 15 V, load current up to 800 mA;
- ultra low current consumption: not more than 14 μA;
- three external components are used, the occupied area of the IC is about 53 mm^2;
- minimum discharge time of the output capacity (for ADP2371);
- high coefficient of efficiency even at partial load;
- synchronization of the switching frequency is possible;
- main application: battery packs with multiple batteries (Fig. 10.17).

Dual stabilizers ADP2323/ADP2325:

- input voltage up to 20 V, load current up to 3/5 A;
- for ADP2323, the load current for each of the two outputs is up to 3 A; when connected to one output, up to 6 A,
- for ADP2325, the load current for each of the two outputs is up to 5 A; when connected to one output, up to 10 A,
- programmable switching frequency,
- high coefficient of efficiency even at partial load;

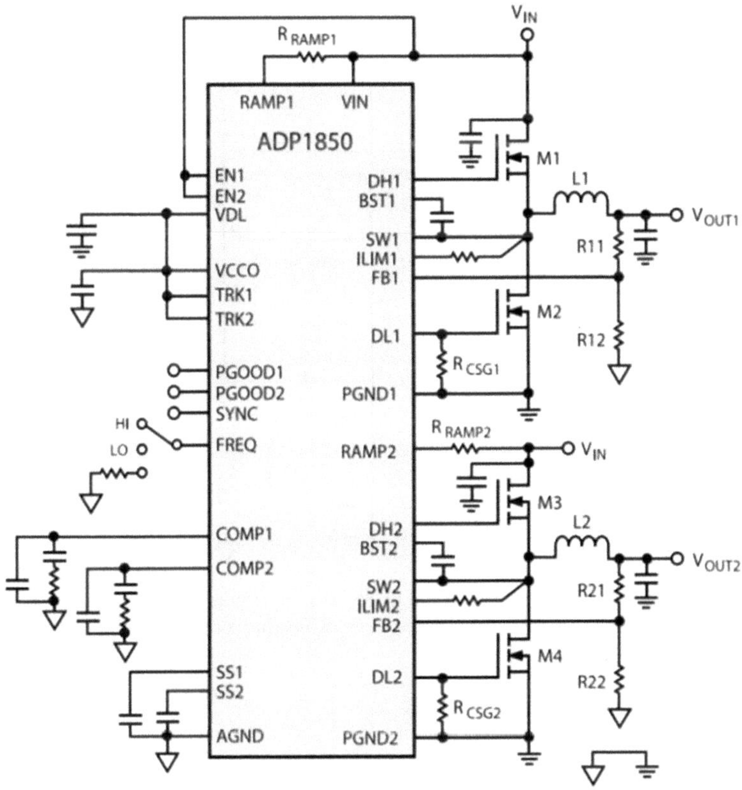

Fig. 10.14 Circuit of switching on the controller of pulse converters ADP1850, operating in single-phase and intermittent modes and having a low consumption mode [2]

- optimal input voltage of 12 or 5 V in multi-level mode (Fig. 10.18).

Boost highly integrated pulse stabilizer ADP1614:

- precise adjustment of current limit up to 4 A;
- used to design miniature power sources;
- switching frequency from 650 kHz up to 1.3 MHz, depending on the modification;
- adjustable output voltage up to 20 V;
- adjustable "soft" start (Fig. 10.19).

Pulse stabilizers ADP2441/ADP2442:

- input voltage from 4.5 up to 36 V, load current up to 1 A;
- wide range of input voltages;
- switching time up to 50 ns;
- the best coefficient of efficiency among the closest analogues: more than 95%;

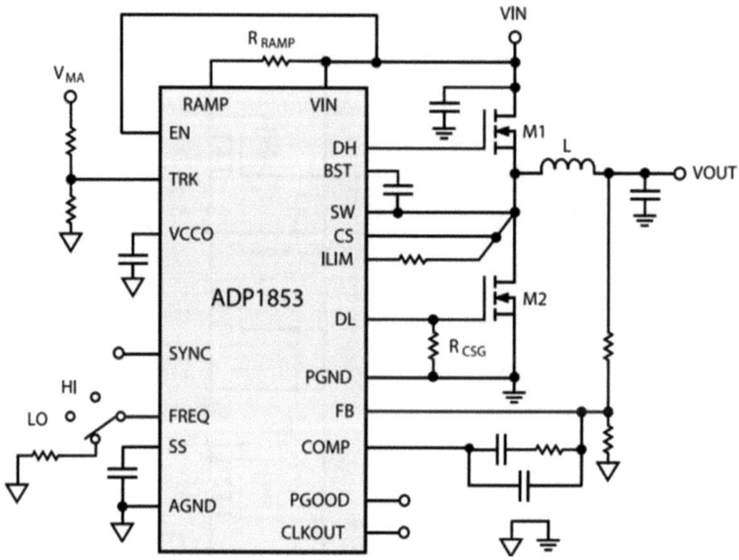

Fig. 10.15 Circuit of switching on the IC ADP1853—a single-phase controller of pulse converters with one output and a constant switching frequency, which has a low consumption mode [2]

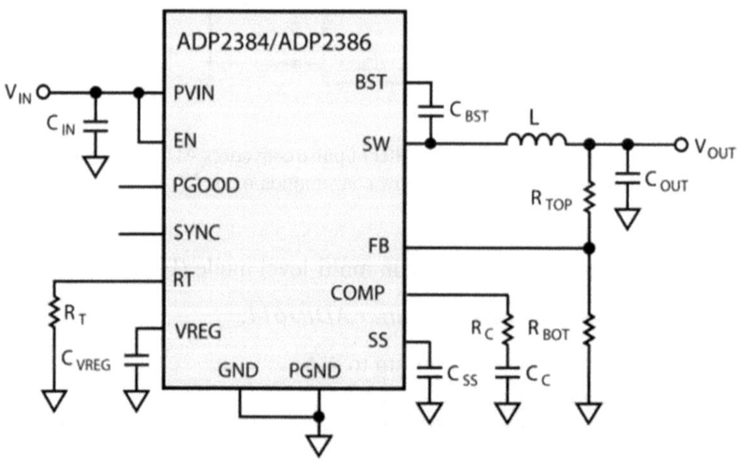

Fig. 10.16 Pulse stabilizers ADP2384/ADP2386: input voltage up to 20 V, load current up to 4/6 A

- for ADP2441 adjustable "soft" start, tracking input voltage;
- for ADP2442 high coefficient of efficiency even at partial load, the ability to synchronize the switching frequency;
- designed for a nominal value of input voltage of 24 or 12 V (Figs. 10.20 and 10.21; Tables 10.2 and 10.3).

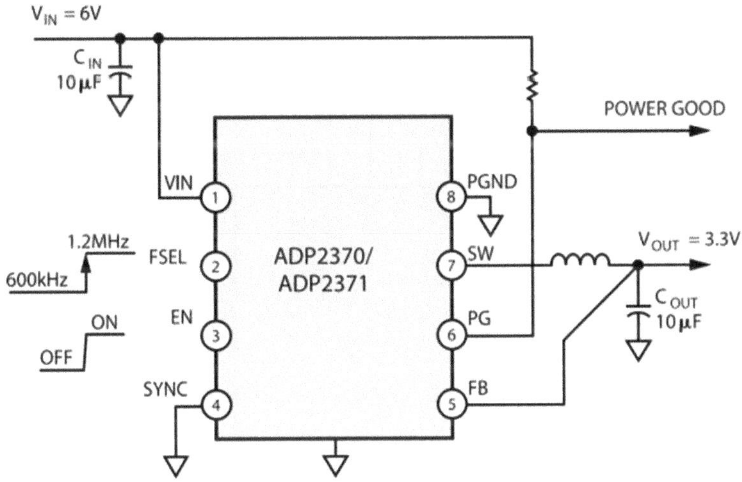

Fig. 10.17 Pulse stabilizers ADP2370/ADP2371: input voltage up to 15 V, load current up to 800 mA

10.2.4 LDO-Stabilizers for Voltage

Modern high-quality processors and systems based on FPGA-type have many levels of supply voltage: for the core, analog and digital input/output nodes, memory, etc. Typical processor systems have one primary pulse stabilizer and several LDO voltage regulators. This solution requires the use of a large area of the PCB to accommodate the power supply IC. Pulse and linear LDO stabilizers implemented in the same IC save the size of a printed circuit board while ensuring the required coefficient of efficiency and reliability of power sources used simultaneously for microprocessors, FPGA processors, and precision analog nodes [2].

ADP5034: dual step-down voltage stabilizer containing two pulse and two linear stabilizers

- dual pulse step-down stabilizer with a load current of 1.2 A and an input voltage range of 2.3–5.5 V;
- dual LDO-stabilizer with a load current of 300 mA and an input voltage range of 1.7–5.5 V;
- stabilizers with adjustable and fixed voltage levels are supplied;
- there is a lead for autonomous inclusion of each stabilizer;
- 24-LFDCSP type of case with dimensions of 4 × 4 mm or 28-TSSOP with dimensions of 9.8 × 6.4 mm (Fig. 10.22).

ADP5041: An IC step-down pulse stabilizer containing two LDO stabilizers, a watchdog timer and a supervisor

- step-down pulse stabilizer with a load current of 1.2 A and an input voltage range of 2.3–5.5 V;

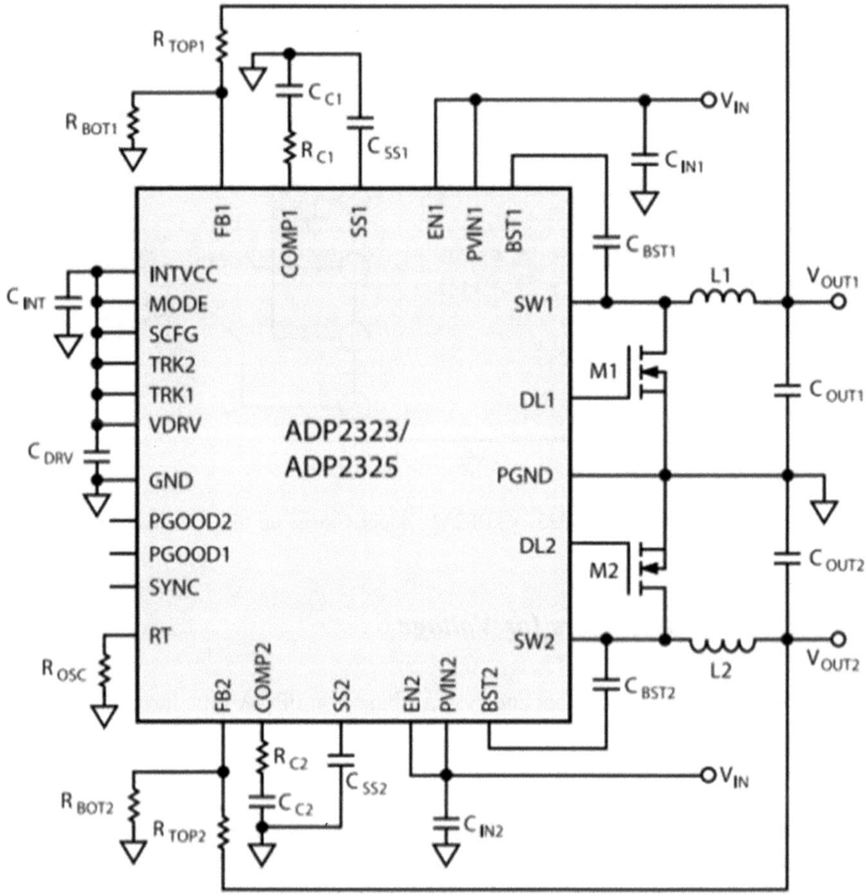

Fig. 10.18 Dual stabilizers ADP2323/ADP2325: input voltage up to 20 V, load current up to 3/5 A

- dual LDO-stabilizer with a load current of 300 mA and an input voltage range of 1.7–5.5 V;
- error of the supervisor ±1.5%;
- the presence of a watchdog timer and the ability to restart manually;
- case type 20-LFDCSP with dimensions of 4 × 4 mm (Fig. 10.23).

Linear LDO stabilizers are designed to create power sources for both analog and digital devices. Distinctive features of these stabilizers are high accuracy, low voltage drop across the regulating transistor, and as a result, high coefficient of efficiency. In addition, these stabilizers are available in miniature cases, have protection against overheating, output overload, protection against incorrect polarity of the connection, and can also go into an energy-saving mode. The stabilizers have a fixed or adjustable output voltage with a load current of up to 2 A and a voltage drop across the control transistor of no more than 80 mV.

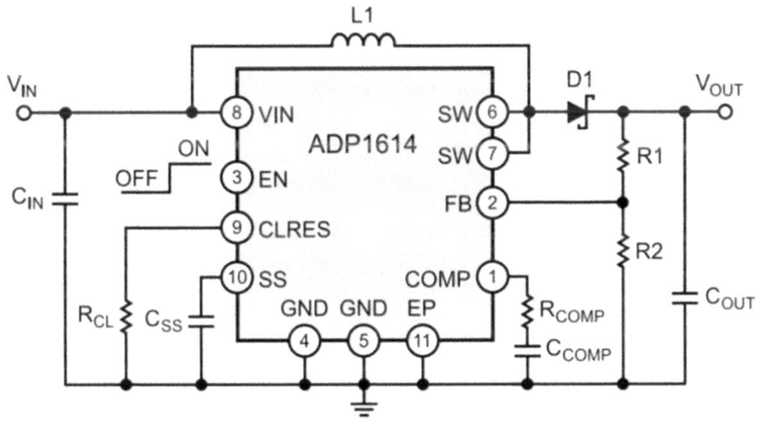

Fig. 10.19 Boost highly integrated pulse stabilizer ADP1614

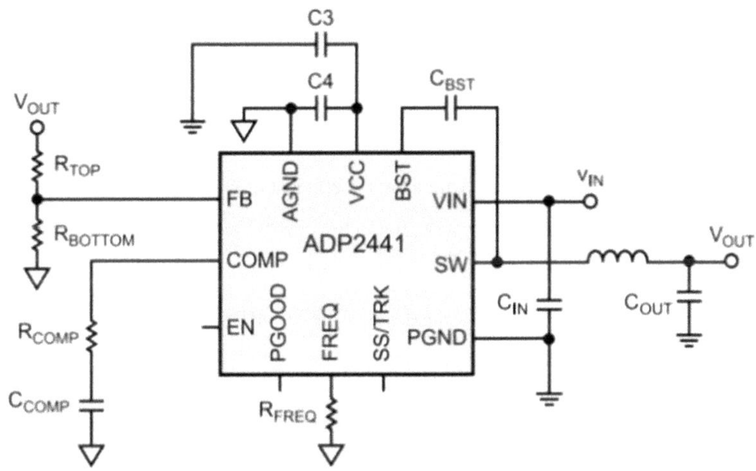

Fig. 10.20 Pulse stabilizers ADP2441/ADP2442: input voltage from 4.5 up to 36 V, load current up to 1 A

ADP7102/ADP7104 are LDO voltage stabilizers with a low noise level at the output and a load current of 300/500 mA:

- volume units noise level of 15 μV without an additional output capacitor;
- attenuation coefficient of input voltage instability is 80 dB at a frequency of 10 kHz with an output voltage of 5 V;
- input voltage range from 3 up to 20 V;
- voltage drop on the regulating transistor of 20 mV at a load current of 300 mA;
- IC with a fixed output voltage of 1.5, 1.8, 2.5, 3, 3.3, 5, 9 V;
- IC with adjustable output voltage from 1.22 V up to U;
- protection against incorrect polarity of the connection;

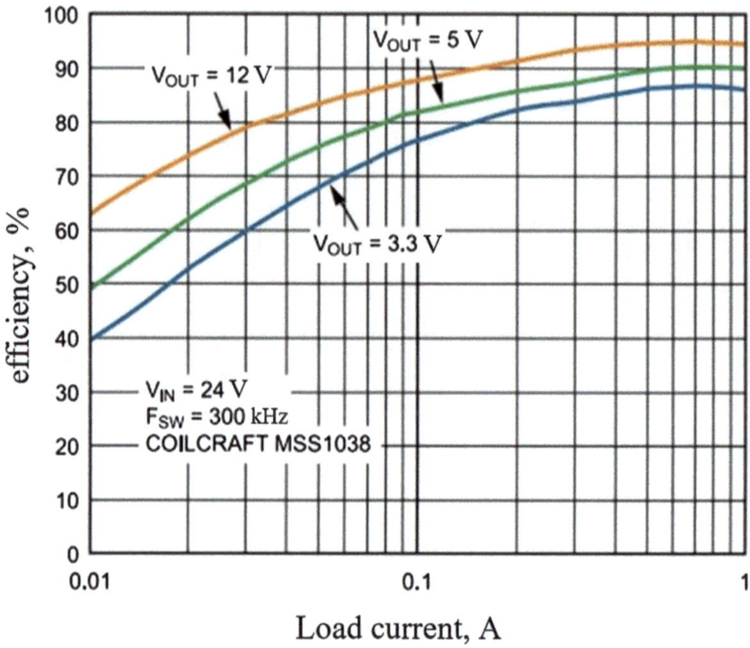

Fig. 10.21 Pulse stabilizer ADP2441 with a high coefficient of efficiency

Table 10.2 Parameters of boost pulse stabilizers

Type IC	U_{in}, V	U_{out}, V	Load current, max., A	Switching frequency, MHz	Type of case
ADP111	2.0–30	3.3, 5, 12	1.5	0.07	8-DIR 8-SOIC
ADP1612	1.8–5.5	U_{in}–20, adjustable	1.4	0.65–1.3	8-MSOP
ADP1613	2.5–5.5	U_{in}–20, adjustable	2	0.65–1.3	8-MSOP
ADP1614	2.5–5.5	U_{in}–20, adjustable	4	0.65–1.3	10-LFCSP

[a]FOB USA price in a batch of 1000 pcs

Table 10.3 Parameters of step-down/step-up pulse stabilizers

Type IC	U_{in}, V	U_{out}, V	I_{load}., max., A	I_{cons}., typical, μA	Switching frequency, MHz	Type of case
ADP2503	2.3–5.5	2.8, 3.3, 3.5, 4.2, 4.5, 5.0, adjustable (2.6–5.5)	0.6	38	2.5	10-LFCSP
ADP2504	2.3–5.5	2.8, 3.3, 3.5, 4.2, 4.5, 5.0, adjustable (2.6–5.5)	1	38	2.5	10-LFCSP

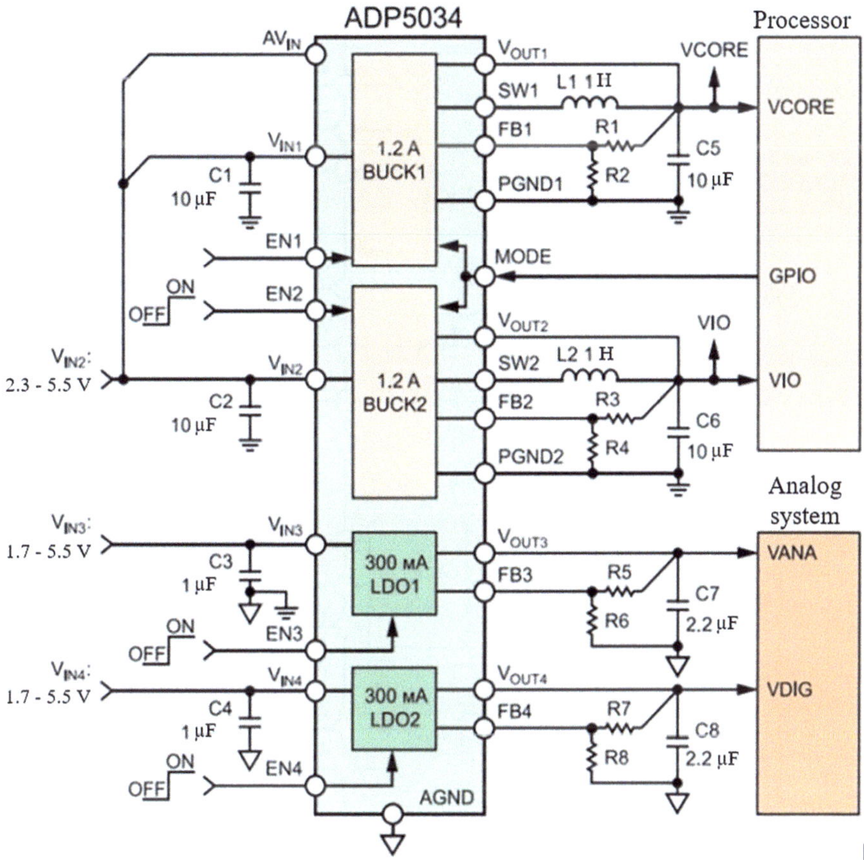

Fig. 10.22 ADP5034: dual step-down voltage stabilizer containing two pulse and two linear stabilizers [2]

- consumption current in power saving mode is 40 μA;
- error of the output voltage of the regulator ADP7102 setting is ±0.8%;
- current limiting at the level of permissible, protection against overheating;
- user-programmable level of control of switching on the stabilizer;
- indicator of normal operation (Figs. 10.24 and 10.25).

10.2.5 Supervisor Chips

Supervisors with an error of setting voltage threshold of no more than ±0.8% prevent failures and loss of information in high-quality data processing systems.

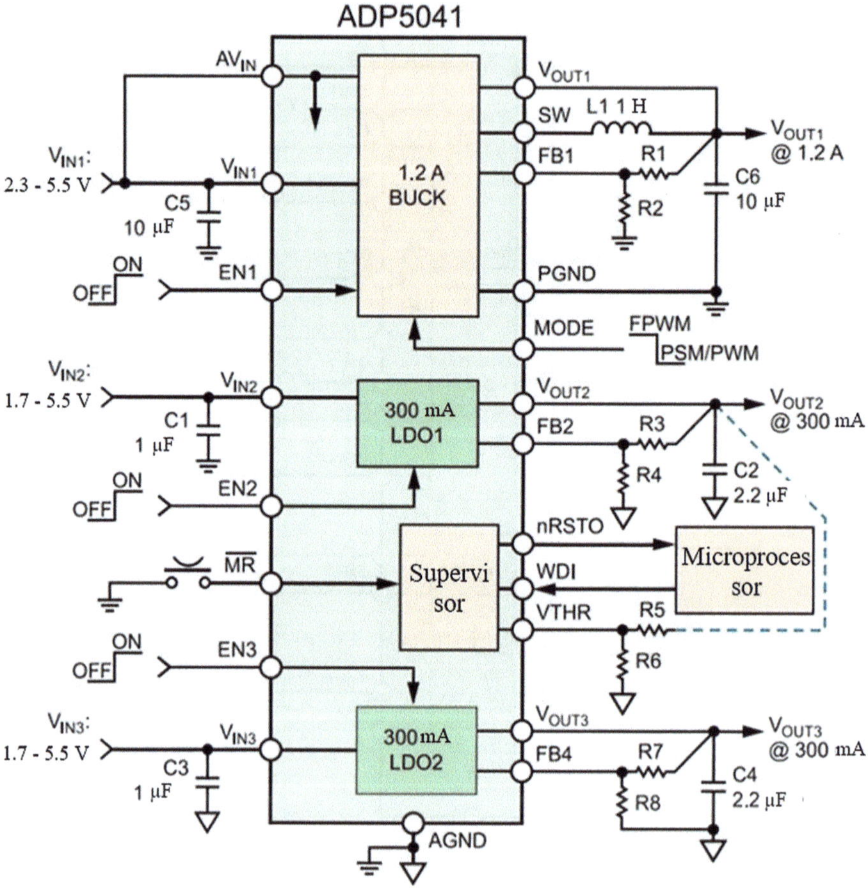

Fig. 10.23 ADP5041: An IC step-down pulse stabilizer containing two LDO stabilizers, a watchdog timer and a supervisor [2]

Supervisor for multi-level power sources (from 2 up to 12 levels) provides high accuracy of threshold setting (Fig. 10.26) [2].

High accuracy of voltage threshold setting is especially important at low supply voltage levels, for example, of the processor core. The accuracy of the level setting, which is supported by Analog Devices supervisors, is twice the accuracy guaranteed by similar ICs from other manufacturers, which ensures higher reliability of data processing systems based on these ICs (Figs. 10.27, 10.28, 10.29, and 10.30).

The energy efficient supervisor family, the ADM626, ADM6328, and ADM6348, is the IC of supervisors for monitoring the power supply voltage of portable microprocessor devices. The use of these supervisors increases battery life, because they consume no more than 500 nA and at the same time increase the reliability of the power supply (Figs. 10.31, 10.32, 10.33, and 10.34; Table 10.4).

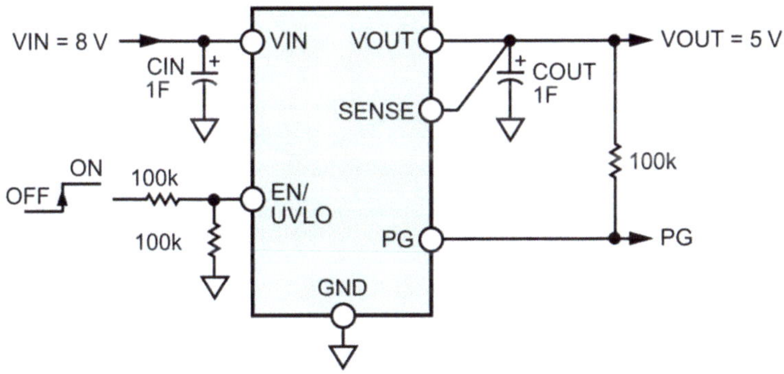

Fig. 10.24 ADP7102/ADP7104 are LDO voltage stabilizers with a low noise level at the output and a load current of 300/500 mA

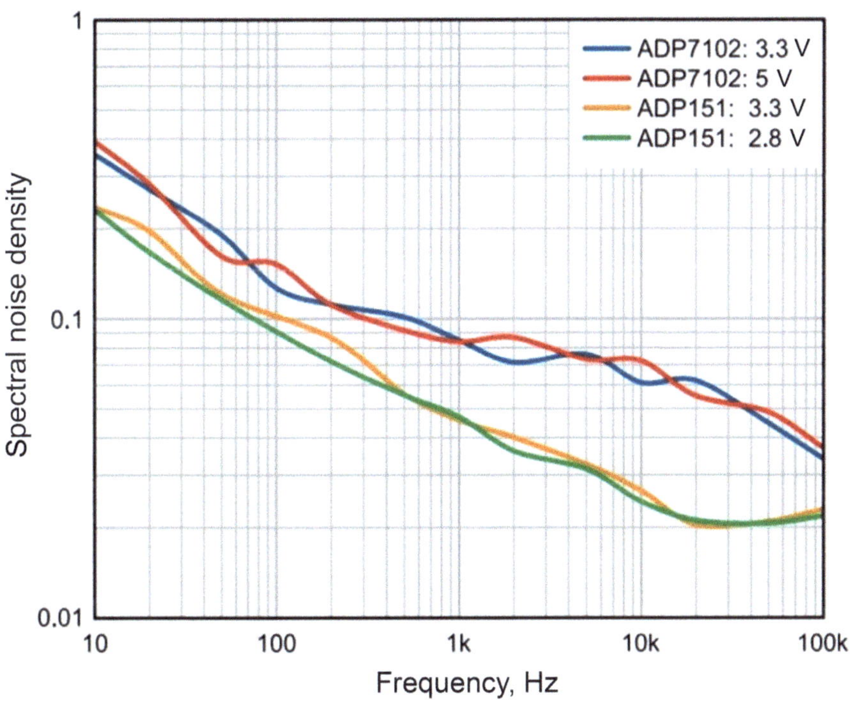

Fig. 10.25 The dependence of the spectral density of noise stabilizers ADP7102 and ADP151 from frequency [2]

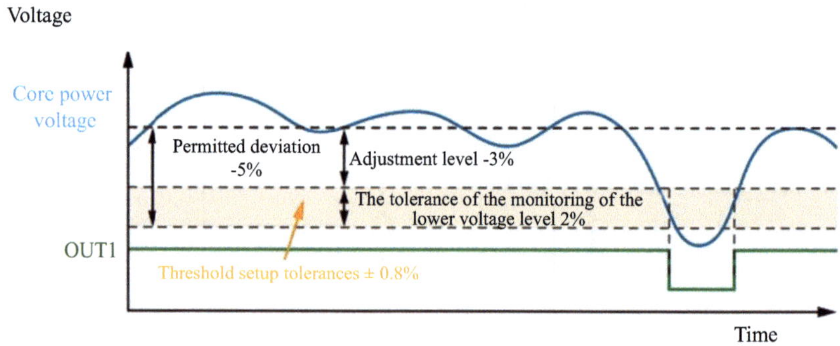

Fig. 10.26 Threshold setting error

Fig. 10.27 ADM6305: dual supervisor in 5-SOT23 case

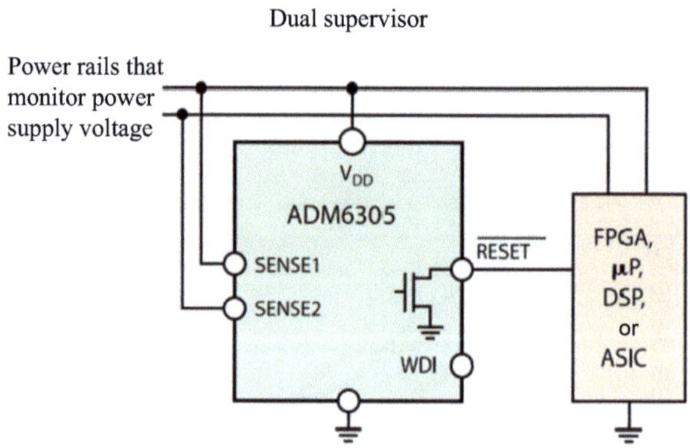

Fig. 10.28 ADM13307: triplex supervisor with an error of ±0.8%, made in the case of 8-SOIC

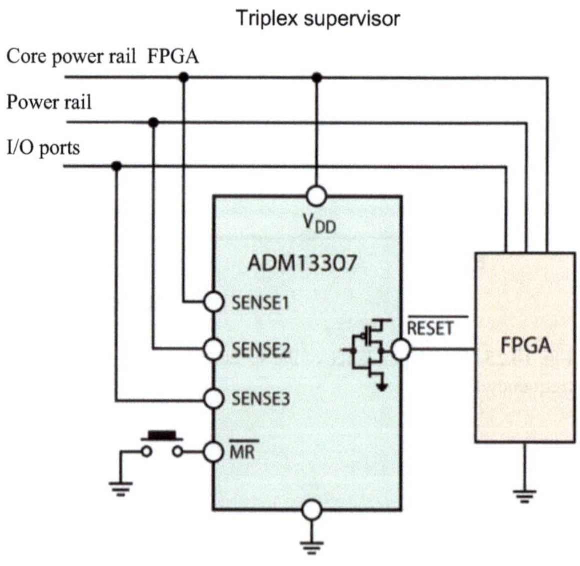

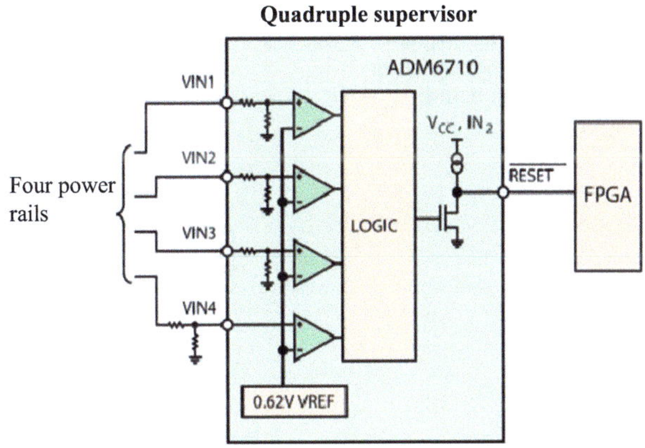

Fig. 10.29 ADM6710: triplex or quadruple voltage supervisor with an accuracy of ±1.5%, made in the 6-SOT23 case

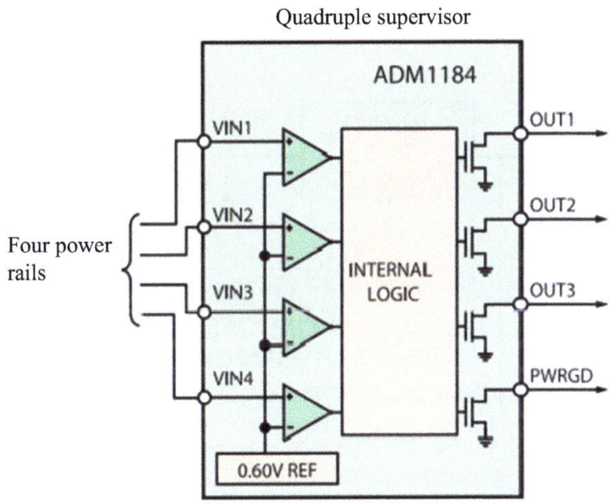

Fig. 10.30 ADM1184: voltage monitor with an accuracy of ±0.8%, made in a 10-MSOP case

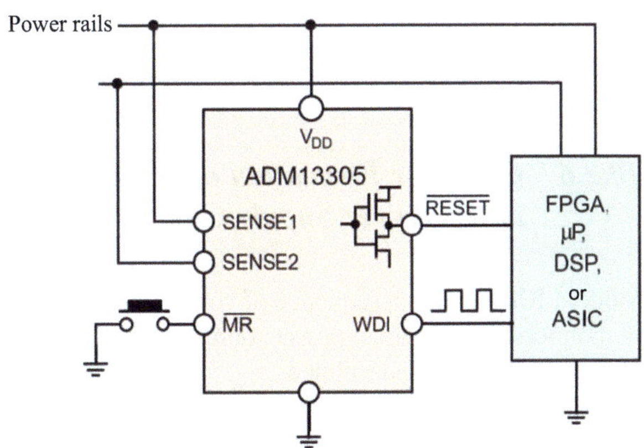

Fig. 10.31 ADM13305 supervisor switching circuit

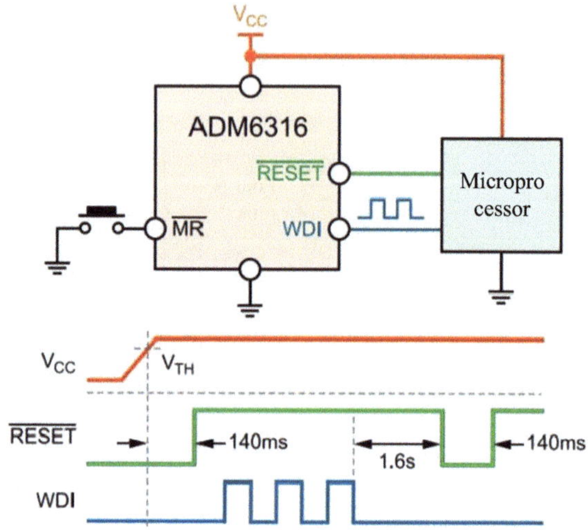

Fig. 10.32 Switching circuit and timing diagram of the work of the supervisor ADM6316 with built-in watching timer

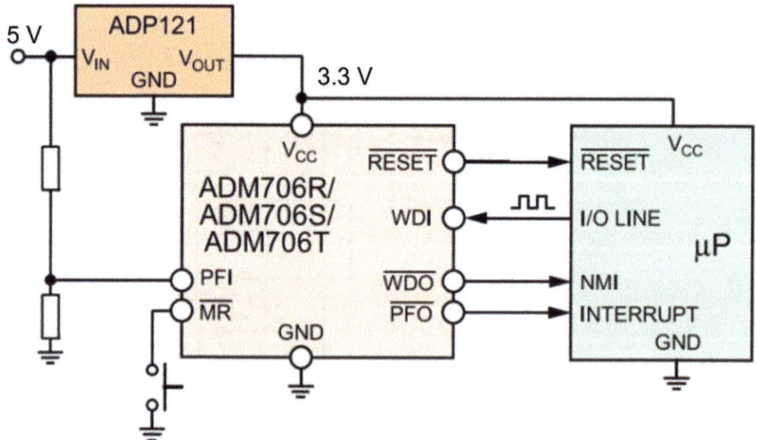

Fig. 10.33 ADM706R/ADM7065/ADM706T supervisor switching circuit

10.2.6 Chips for Formation of Sequences of Control Signals in the Power Supply Circuits of Electronic Devices

Modern ICs forming a sequence of control signals in power supply circuits and ICs of monitors of multi-level power sources (from 2 to 12) should have a high accuracy of setting the threshold voltage.

There are many similar IC of forming a sequence of control signals in the power supply circuits and the IC of monitors for software-controlled power supplies that significantly simplify power management, and can provide simultaneous control of multiple power sources (Fig. 10.35).

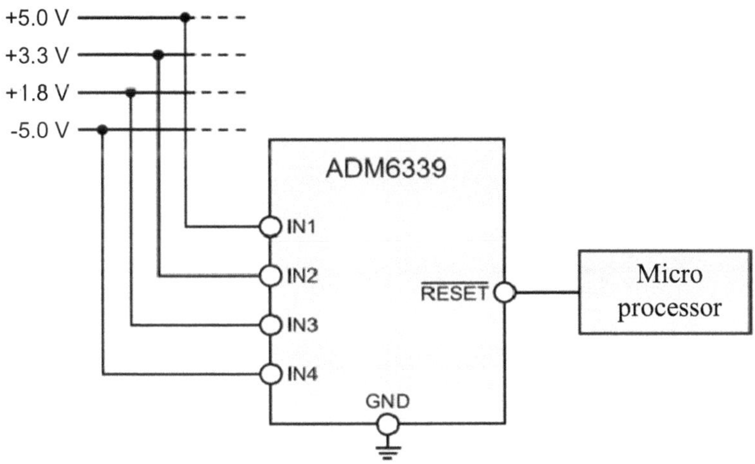

Fig. 10.34 ADM6339 supervisor switching circuit

A brief look at the main, the most widely studied microcircuits and their typical parameters [3].

First of all, we should mention the microcircuit ADM1085—this is the IC of forming a sequence of control signals in the power supply circuits, which is characterized by the following features:

- adjustment of the time interval by changing the capacity of the capacitor;
- the possibility of cascading multiple ICs;
- IC ADM6819 and ADM6820 have an output to control the FET transistor;
- allowable output voltage is not less than 22 V;
- main case type SC70 (Fig. 10.36).

The IC of forming a sequence of control signals in power supply circuits for multi-level power sources ADM1184/ADM1185 has the following parameters [3]:

- the error of setting the threshold voltage $\pm 0.8\%$;
- the ability to adjust the on/off time of the power source;
- the possibility of cascading multiple ICs to control the inclusion from 2 to 12 power sources;
- the possibility of adjustment of the time interval by changing the capacity of the capacitor;
- the minimum output voltage does not exceed 1 V;
- case type MSOP or QSOP (Fig. 10.37).

ADM1166 is another widely used IC of forming a sequence of control signals in power supply circuits, which has high characteristics:

- provides programming of the required sequence of control signals;
- the error of setting the threshold voltage $\pm 1\%$;
- use of feedback to control the sequence of control signals;

Table 10.4 Parameters of multi-level supervisors [2]

Type IC	The number of controlled levels of voltage	Threshold level, V	Min. Return time to the initial state, ms	Initial state		Possibility of manual installation	Typical delay time of the watching timer, ms	Type of case
				low level	high level			
ADM13305	2	0.6 (reg), 1.68, 2.25, 2.93, 4.55	140	Two-cycle		+	1600	8-SOIC_N
ADM6305	2	0.4 (reg), 1.23 (reg.)	1, 20, 140, 1120	Open Channel	–	–	–	5-SOT23
ADM6306	2	0.4 (reg), 1.23 (reg), 2.5–5.0 (26 types)		Open Channel	–	–	+	5-SOT23
ADM13307	3	0.6 (reg), 1.25 (reg) 1.68, 2.25, 2.93, 4.55	140	Two-cycle		+	–	8-SOIC_N
ADM6710	3 or 4	0.62 (reg), 1.58, 1.67, 2.19, 2.32, 2.63, 2.78, 2.93, 3.08, 4.38, 4.63	140	Open Channel	–	–	–	6-SOT23
ADM1184	4	0.6 (reg)	100	Open Channel	–	–	–	10-MSOP
ADM6339	4	−0.5 (reg), 0.62 (reg), 1.23 (reg), −4.38, −4.63, 1.58, 2.19, 2.63, 2.78, 2.93, 3.08, 4.38, 4.63	140	Open Channel	–	–	–	6-SOT23
ADM2914	4	0.5 (reg)	Regulated	Open Channel	–	–	–	16-QSOP
ADM12914	4	0.62 (reg)		Open Channel	–	–	–	16-QSOP
ADM8710	4	0.62 (reg), 1.73, 3.07		Open Channel	–	–	–	6-SOT23

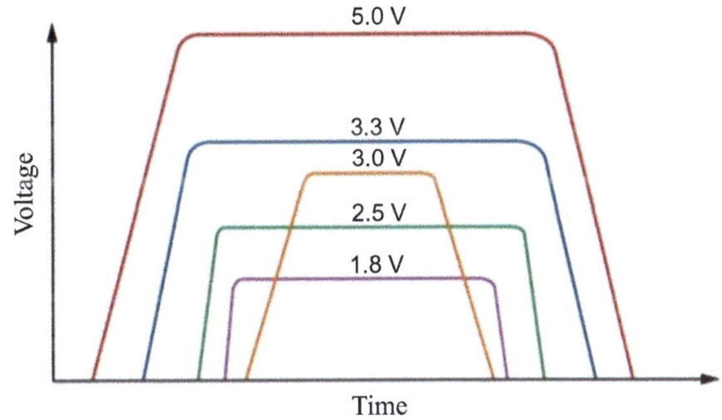

Fig. 10.35 The timing diagram of the operation of the IC of the formation of a sequence of control signals in the power supply when managing multiple power sources

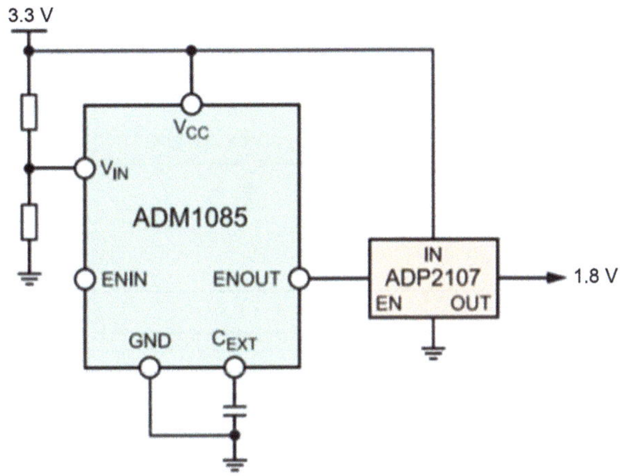

Fig. 10.36 The structure of the microcircuit ADM1085

- the ability to adjust the on/off time of the power source;
- non-volatile fixation of failures;
- the presence of the output to control the FET transistor;
- case type LFCSP or TQFP (Figs. 10.38, 10.39, and 10.40; Tables 10.5 and 10.6).

10.2.7 Specialized Chips for Provision of the Hot Mode for Connecting Electronic Equipment

As is known, ICs for hot plugging as part of a precision ADC (Figs. 10.41, 10.42, 10.43, and 10.44) [3] provide savings of electricity consumed by telecommunication systems. Amplifiers and data converters with hot plugging controllers provide monitoring of power consumption with an accuracy of up to ±2%, which reduces power consumption in data acquisition and processing systems.

- embedded power consumption monitor;
- embedded failures comparator in the power supply circuit;

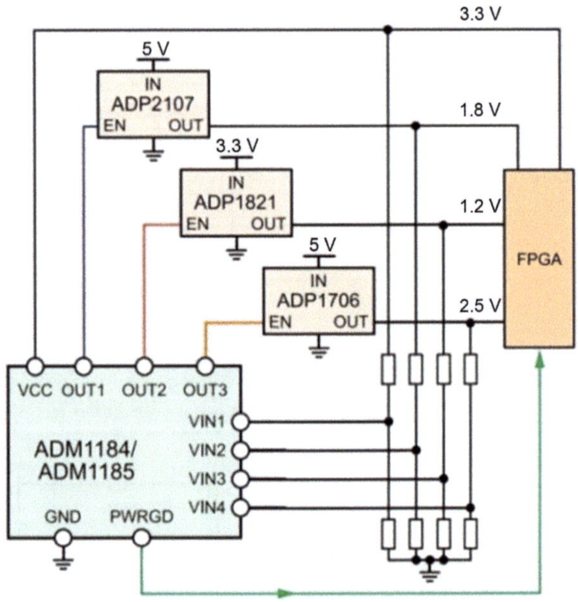

Fig. 10.37 The structure of the microcircuit ADM1184/ADM1185

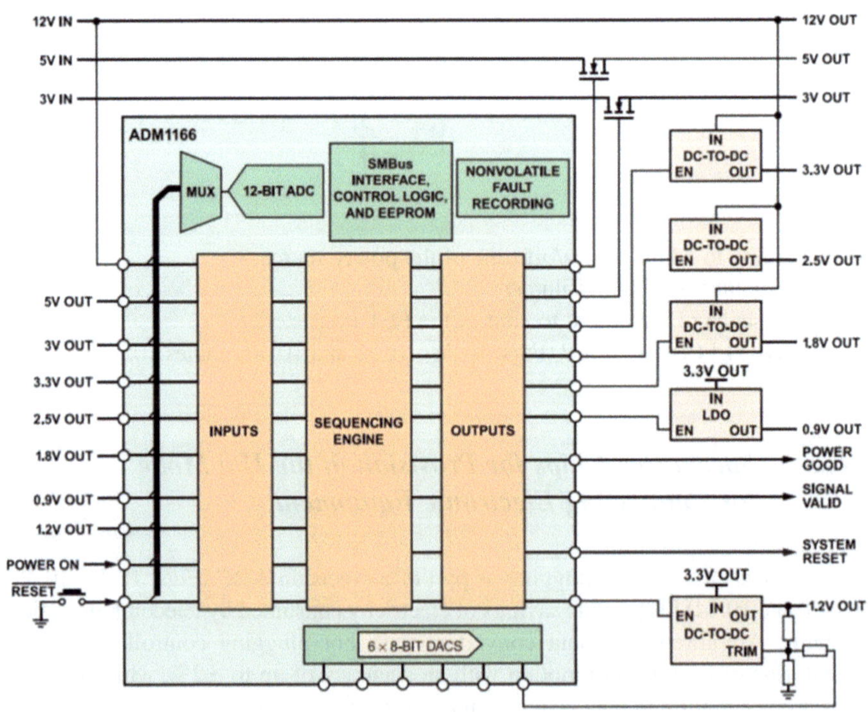

Fig. 10.38 The structure of the microcircuit ADM1166

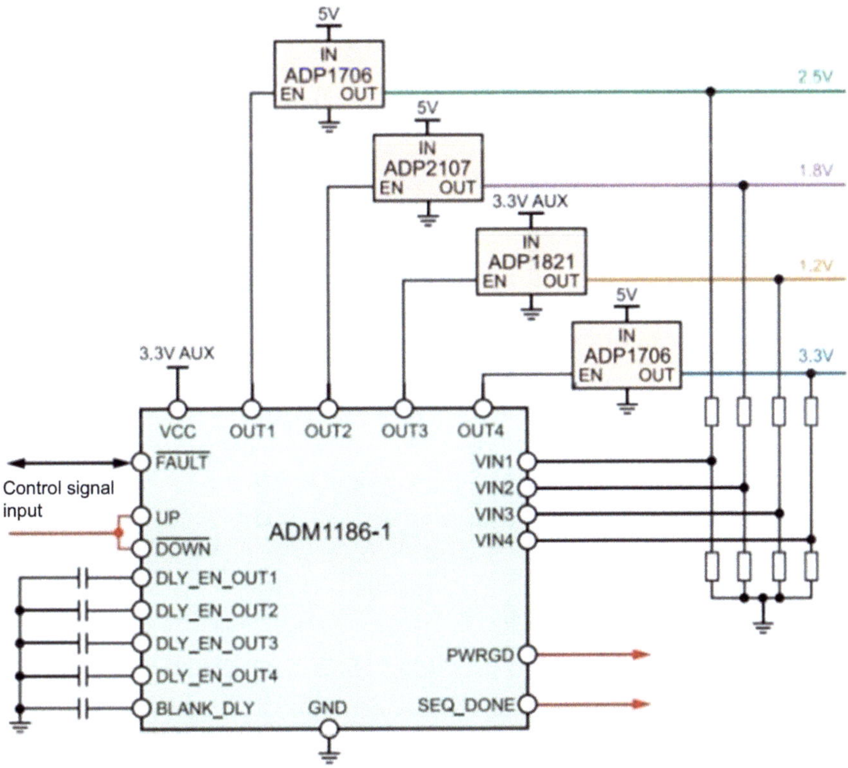

Fig. 10.39 Typical wiring scheme of the ADM1186-1 IC with providing switching on/off of power supply [3]

Fig. 10.40 Typical timing diagram of the IC ADM1186-1 [3]

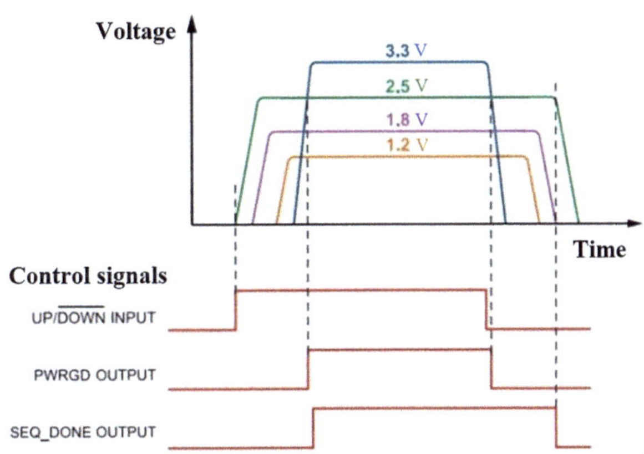

Table 10.5 Parameters of analog microcircuits for forming a sequence of control signals in power supply circuits [3]

Type IC	The number of controlled power sources	Cascading capability	Voltage monitoring error, %	Number of outputs	Type of output	Increase/decrease of control signal frequency	Type of case
AMD1085	1	+	7	1	Permissive	+/–	6-SC70
AMD1086	1	+	8	1		+/–	6-SC70
AMD1087	1	+	8	1		+/–	6-SC70
AMD1184	4	+	0.8	4		+/–	10-MSOP
AMD1185	4	+	0.8	4		+/–	10-MSOP
AMD1186-1	4	+	0.8	4		+/+	20-QSOP
AMD1186-2	4	–	0.8	4		+/+	16-QSOP
AMD6819	2	+	2.6	1	Driver	+/–	6-SOT23
AMD6820	2	+	2.6	1		+/–	6-SOT23

Table 10.6 Parameters of digital microcircuits for forming a sequence of control signals in power supply circuits [3]

Type IC	The number of controlled power sources	Cascading capability	Voltage monitoring error, %	Number of outputs	Embedded 12-bit ADC for voltage reading	The number of embedded 12-bit ADC/DAC	Type of case
AMD1060	7	+	2.5	9	–	–	6-SC70
AMD1062	10	+	1	10	+	1/6	40-LFSCP, 48-TQFP
AMD1063	10	+	1	10	+	–	
AMD1064	10	+	1	10	+	–	
AMD1065	10	+	1	10	–	–	
AMD1066	12	+	1	10	+	1/6	
AMD1067	10	+	1	10	–	–/6	
AMD1068	8	+	1	8	–	–	32-LQFP
AMD1069	8	+	1	8	+	1/4	32-LQFP, 40-LFSCP
AMD1166	12	+	1	10	+	1/6	40-LFSCP, 48-TQFP
AMD1168	8	+	1	8	–	–	32-LQFP
AMD1169	8	+	1	8	+	1/6	32-LQFP, 40-LFSCP

Fig. 10.41 The principle of the organization of hot plugging with monitoring of power consumption

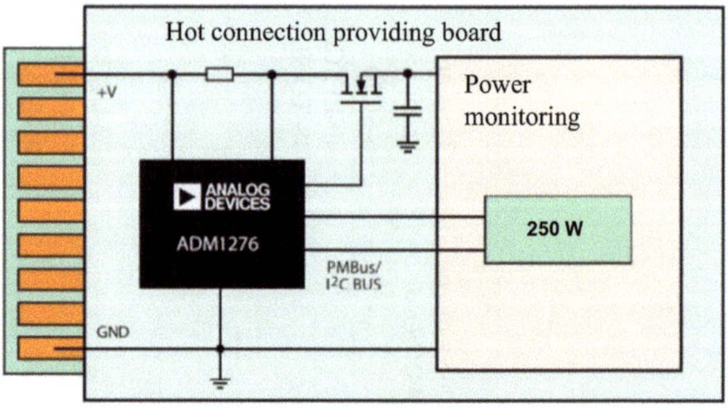

Fig. 10.42 Typical scheme of switching on ADM1172 hot plugging controller

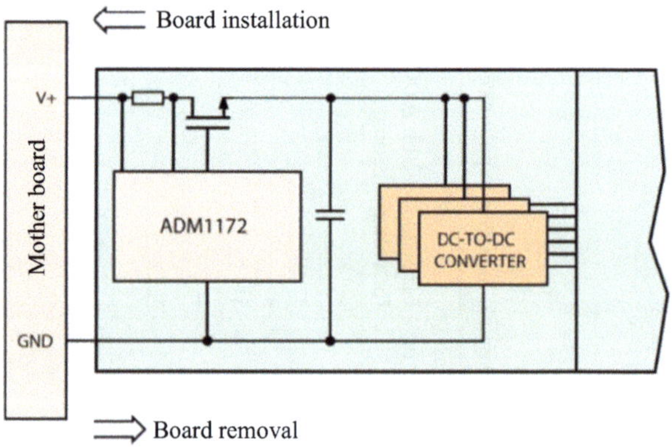

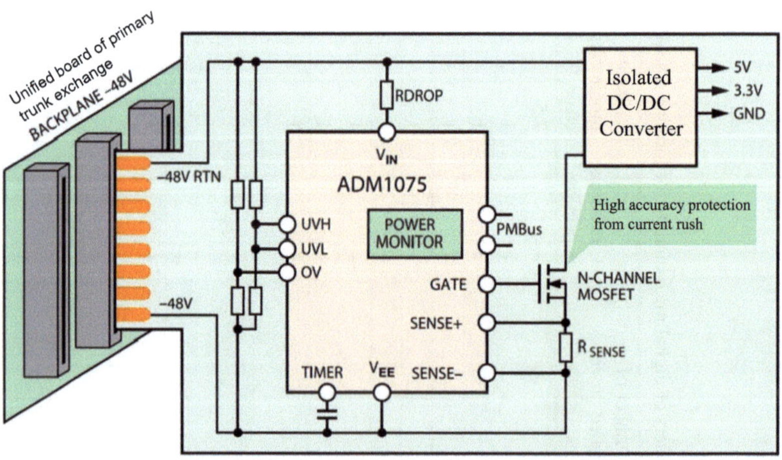

Fig. 10.43 Typical scheme of switching on ADM1075 hot plugging controller

Fig. 10.44 Typical scheme of switching on ADM1192 power consumption monitor

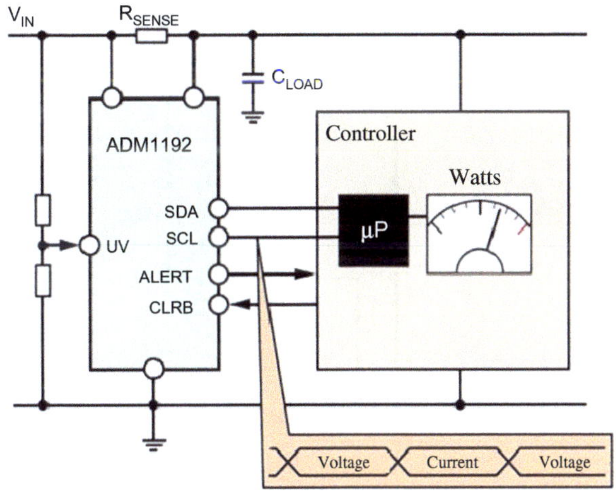

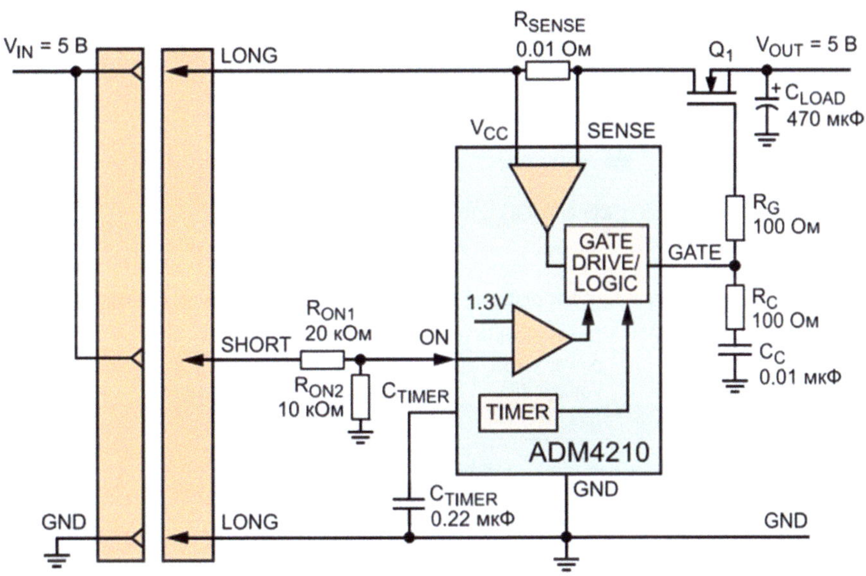

Fig. 10.45 Typical wiring scheme of a hot plugging controller for low positive supply voltages [3]

- case type TSOT.
- formation of a feedback signal to protect the equipment from inrush current;
- high-accurate measurement (with an error of up to 1%) of voltage and current;
- wide input voltage range due to the use of a by-pass jumper with a voltage drop of 25/50 mV.
- I^2C interface;
- the presence of the output of the interrupt signal of the processor (Figs. 10.45 and 10.46; Tables 10.7, 10.8, 10.9, and 10.10).

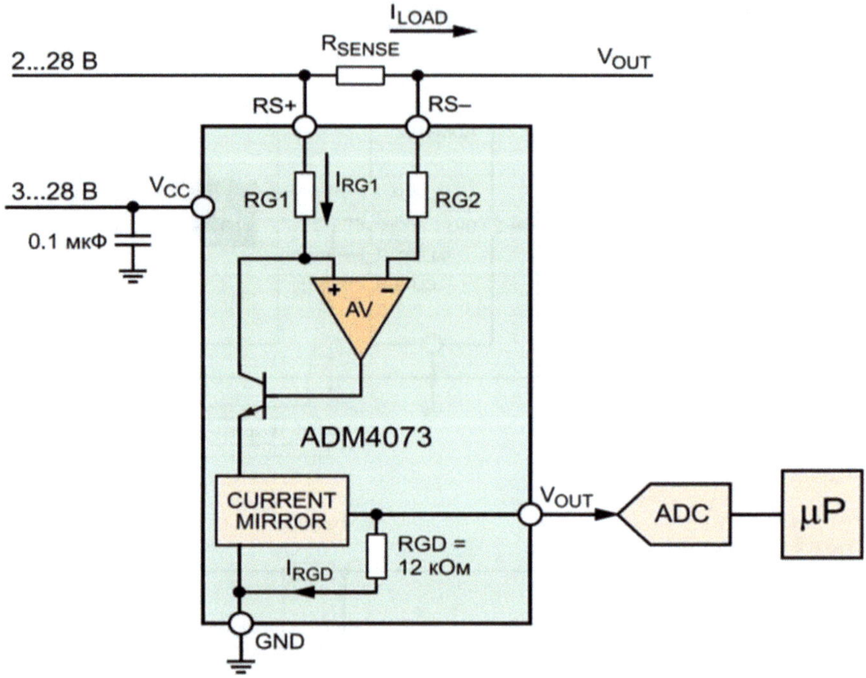

Fig. 10.46 Typical wiring scheme of a current reader amplifier [3]

Table 10.7 Key parameters of hot plugging controllers for low-range positive voltages [3]

Type IC	Tracking voltage range, V	Determination of power surge/ voltage dips	Interface type/ mapping	Additional inputs	Additional outputs	Type of case
ADM1170	1.6–16.5	–/+	–	–	–	6-SC70
ADM1171	2.7–16.5	–/+	–	–	CS_OUT	6-SC70
ADM1172	2.7–16.5	–/+	–	PFI	PFO	6-SC70
ADM1175	3.15–16.5	+/+	I^2C/4	CONV	–	10-MSOP
ADM1176	3.15–16.5	–/+	I^2C/16	–	–	10-MSOP
ADM1177	3.15–16.5	–/+	I^2C/4	–	–	20-QSOP
ADM1178	3.15–16.5	–/+	I^2C/4	–	ALERT	16-QSOP
ADM1275	2–20	UV, OV leads	PMBus/4	Enable CONV	GPO1, GP02, PWRGD	6-SOT23
ADM1276	2–20		PMBus/4	Enable	GPO1, PWRGD	6-SOT23
ADM4210	2.7–16.5	–/+	–	–	–	

Table 10.8 The main parameters of the hot plugging controllers for voltage—48 V [3]

Type IC	Tracking voltage range, V	Determination of power surge/voltage dips	Reading of voltage and current values	Additional features	Type of case
ADM1070	Regulated by by-pass jumper	+/+	–	–	6-SOT23
ADM1073		+/+	–	Soft start, lead for monitoring the voltage on the channel of the transistor, additional I/O-leads	14-TSSOP
ADM1075		+/–	Via PMBus interface	Soft start, feedback signal about the amount of power consumption, instantaneous transition to the shutdown mode, adjustment of the range of the load current limit	28-LFCSP, 28-TSSOP

10.2.8 Specialized Chips of Load Switches

Microcircuits of load switches are usually high-speed electronic devices themselves characterized by miniature size, low leakage currents and are designed for use in multi-level power circuits of modern telecommunications systems. They usually contain FET transistors, level shifters, and drivers. Additional features include: discharge of load capacity, restart, reverse current and overload protection, load current limiting, and overheating protection. These ICs are used in measuring devices and industrial systems to control the operation of power circuits in battery charge mode, in wireless measuring devices to optimize the working cycle, and in optical modules to protect against surges. Table 10.11 presents the main parameters of these ICs.

It is necessary to give a number of comments to this figure, which will be useful to the developers of power supply chains in their practical activities:

- resistance of the closed CMOS-key is 50 mOhm at $U_{in} = 3.3$ V and $I_{load} = 200$ mA;
- input range from 1.65 V ($U_{in.min} = 1.65$ V) to 6.5 V;
- current consumption of 2.5 μA at $U_{in} = 2.8$ V and $I_{load} = 0$ mA;
- blocking the reverse current of 5 μA at $U_{out} = 6.2$ V;
- current consumption in quiescent mode 100 nA;
- a miniature WLCSP package or an 8-LFCSP package with the dimensions 2 × 2 mm;
- programmable start time: 30, 200, 450, 1000 μS (Fig. 10.47).

Literally a few words should be said about the features of the use of iCoupler technology in the power supply circuits of telecommunication systems.

Table 10.9 Parameters of digital monitors of power consumption [3]

Type IC	Supply voltage range, V	Monitoring error current ±%	Pulse interference filter	CONV lead	CLR lead	Alarm	Interface type/ mapping	Package type
AMD1191	3.15–26	10	–	+	+	ALERT	I^2C/4	10-MSOP
AMD1192	3.15–26	3	+	–	–	ALERT	I^2C/16	10-MSOP

Table 10.10 Parameters of current reader amplifiers (from the high voltage side) [3]

Type IC	Common-mode range, V	Ky	Max. by-pass jumper voltage, mV	Supply voltage range, V	Typical error, %	Package type
ADM4073T	2–28	20	150	3–28	±1	6-SOT23
ADM4073F	2–28	50	150	3–28	±1	6-SOT23
ADM4073H	2–28	100	150	3–28	±1	6-SOT23

As is well known to specialists, the isolation of power supply circuits simultaneously with the isolation of signal circuits has until recently been provided with complex circuit solutions. Analog Devices isoPower technology has made it easier to solve this problem. Thanks to iCoupler technology film transformers are decoupled from both signal and power circuits, and this requires one IC, made in a miniature SMT package and providing insulation strength up to 5 kV.

So, on Figs. 10.48 and 10.49 [3] a number of such functional circuits are presented:

At the end of this section, a few words should be said about the features of the choice of microcircuits intended for organizing the power supply circuit of devices of the FPGA type.

The choice of the optimal voltage regulators for devices made on the FPGA type depends on the magnitude of the input voltages, maximum load current, coefficient of efficiency, and other parameters. Linear LDO voltage stabilizers provide a high coefficient of efficiency and lower interference level as compared with pulse voltage stabilizers (Tables 10.12 and 10.13) [3]. They have miniature sizes and are distinguished by low cost. Such stabilizers lower the voltage, and at high load currents they require the use of an external MOSFET transistor.

ADIsimPower's power supply design tools can be easily found on the Internet at: www.analog.com/power. They include a four-level design process using a simple graphical user interface. By applying these tools, the developer can optimize the cost, size, and other parameters of the designed power source.

Design tools include [3]:

- ADIsimPower, which include power supply design tools, including DC/DC converters and supervisors.

Databases in the Excel environment include:

- parameters of linear voltage regulators,
- power supply design tools based on IC of increasing stabilizers of the AD161x family,
- power supply design tools based on IC step-down stabilizers ADP1864,
- power supply design tools based on an IC of step-up/step-down pulse stabilizers ADP2503/ADP2504,

Table 10.11 Parameters of the switch load IC [3]

Type IC	Input voltage range, V	Resistance of the closed key, MOhm, with:			Load current, A	Current consumption, μA	Start-up time, μS	U_{EN}, *min*, V at $U_{in} = 1.8$ V	Package type (overall dimensions, mm)	Additional functions[a]
		$U_{in} = 1.8$ V	$U_{in} = 3.8$ V	$U_{in} = 5.5$ V						
ADP190	1.1–3.6	105	80		0.5	2	5	1.0	4-WLCSP (0.8 × 0.8)	
ADP191	1.1–3.6	105	80		0.5	2	80	1.0		QOD
ADP194	1.1–3.6	80	55		0.5	2	7	1.2		RCB RCB
ADP195	1.1–3.6	65	50 70		0.5 1.1	1 1	5 5	1.2	4-WLCSP (1 × 1) 6-LFCSP (2 × 2)	RCB RCB
ADP196	1.8–5.5	12 20	12 20	12 20	3.0 3.0	18 18	1000 1000	1.2	6-WLCSP (1 × 1.5) 6-LFCSP (2 × 2)	CL CL
ADP197	1.8–5.5	12	12	12	3.0	18	1000	1.2	6-WLCSP (1 × 1.5)	TDS
ADP198	1.65–6.5	110 110	–	40 40	1 1	2.5 2.5	30 1000	1.2	4-WLCSP (1 × 1) 4-WLCSP (1 × 1)	RCB RCB
	1.65–6.5	120		85	1	2.5	30, 200, 450, 1100	1.2	8-LFCSP (2 × 2)	RCB
ADP199	0.9–3.6	40	40		0.5	6	20	1.2	4-WLCSP (0.8 × 0.8)	

[a] *QOD* discharge of load capacity, *RCB* protection against reverse currents, *CL* current limitation, *TSD* protection against overheating

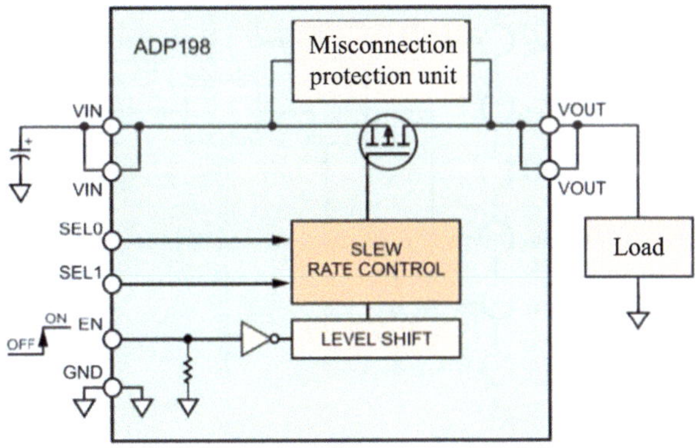

Fig. 10.47 Wiring scheme of the IC of the ADP198 load switch, providing protection against erroneous changes in polarity [3]

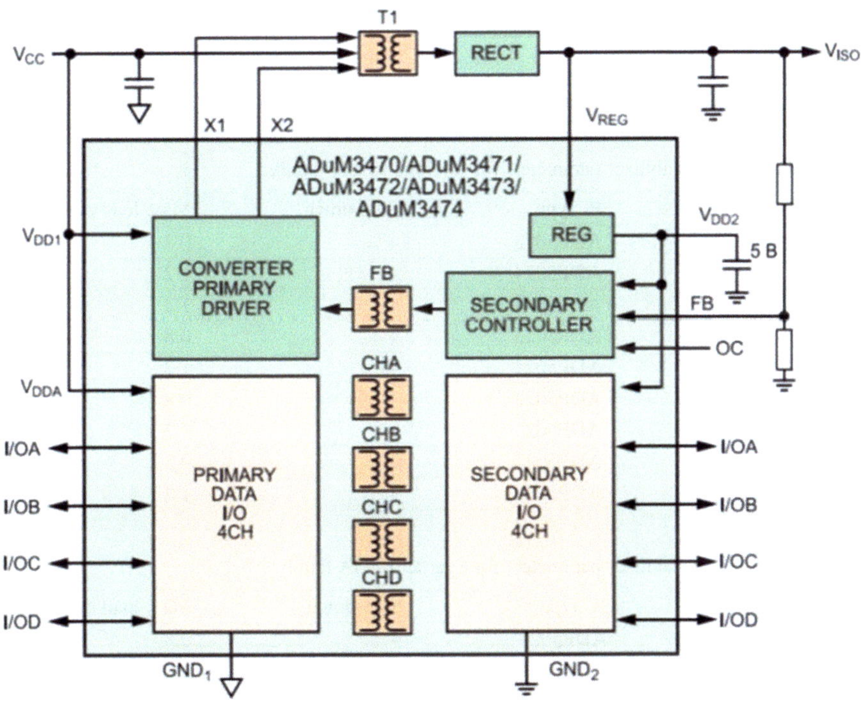

Fig. 10.48 Chain diagram of the IC of the isolator of signal and power supply circuits of the ADuM347x family with an external transformer and output power up to 2 W

- power supply design tools based on IC of pulse stabilizers of the ADP210x family,
- design tools for step-down DC/DC converters ADP2108,
- design tools for supervisors and other devices.

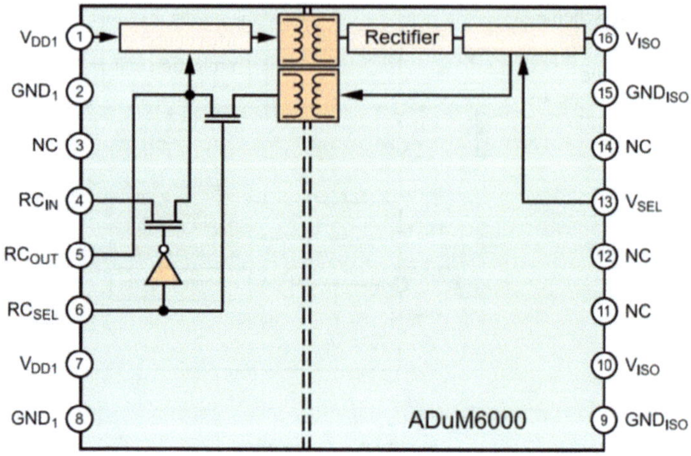

Fig. 10.49 Chain diagram of the IC of the supply circuits insulator ADuM6000 up to 0.5 W with built-in film transformer

Table 10.12 LDO stabilizer parameters for Cyclone FPGA family

FPGA type	IC type	Step-down	Max. load current, A
Cyclone III	ADP5023	+	0.8
	ADP5024	+	1.2
	ADP5024	+	1.2
Cyclone III LS	ADP5023	+	0.8
	ADP5024	+	1.2
Cyclone IV/E	ADP5023	+	0.8
	ADP5024	+	1.2
Cyclone IV/GX	ADP5037	+	0.8
	ADP5034	+	1.2

Table 10.13 LDO stabilizer parameters for Spartan FPGA family

FPGA type	IC type	Step-down	Max. load current, A
Spartan-3AN	ADP5023	+	0.8
	ADP5037	+	0.8
Spartan-3A/3A	ADP5024	+	1.2
	ADP5034	+	1.2
Spartan-3E	ADP5024	+	1.2
	ADP5034	+	1.2
Spartan-3	ADP5023	+	0.8
	ADP5024	+	1.2
	ADP5037	+	1.2

10.3 Step-Up and Step-Down DC–DC Converters for High-Speed Battery-Powered Electronics

Let's consider the main features of the construction of universal DC–DC converters that can work as a step-up or step-down device in the structure of high-speed battery-powered electronic devices [4].

The constant struggle to increase the coefficient of efficiency of DC/DC converters and more fully utilize the charge of the batteries causes the need to search for new circuit solutions to accomplish these tasks.

Since modern portable electronic equipment uses high-speed microprocessors with high power consumption and low core power supply, as well as a number of other devices whose supply voltage is 3.3–5 V, it becomes necessary to build both step-down and step-up voltage converters. Such converters are needed to power high-power LEDs operating in devices with low supply voltage and in many other cases.

Traditionally, step-up/step-down converters (buck–boost converters) are based on the SEPIC (Single-Ended Primary-Inductance Converter), Flyback, and ZETA [5] (http://en.wikipedia.org/wiki/ DC_to_DC_converter) topologies. As with conventional step-down or step-up converters, there is one active and one passive key element in these topologies. The ability to both reduce and increase the output voltage is achieved by using an additional inductor in converters like SEPIC and ZETA (also called Inverse SEPIC) or a transformer in Flyback topology (Fig. 10.50) [4].

The inverting converter [4], implemented according to the simplest scheme (Fig. 10.51a), or using the Cuk topology (Fig. 10.51b), which is actually a combination of the SEPIC and ZETA topologies, can be used as a step-down/step-up converter. Converters with a Cuk topology can be either unisolated or isolated due to the use of a transformer.

One of the world leaders in the SnC market segment, Analog Devices manufactures a large number of DC/DC converters [5] (http://en.wikipedia.org/wiki/ DC_to_ DC_converter; http://en.wikipedia.org/wiki/Buck-boost_converter), which allow providing a wide range of input and output voltages and currents.

As an example, consider an IC of a universal asynchronous converter supporting several topologies of this company ADP1111, the main characteristics of which are given below [5]:

- work in the mode of step-down, step-up, or inverting converter;
- input voltage range 2.30 V (2.12.6 V for a step-up converter);
- output voltage range 1.25.45 V;
- frequency of transformation is 72 kHz;
- consumption current in the Switch-Mode Power Supply (SMPS) mode does not exceed 500 μA;

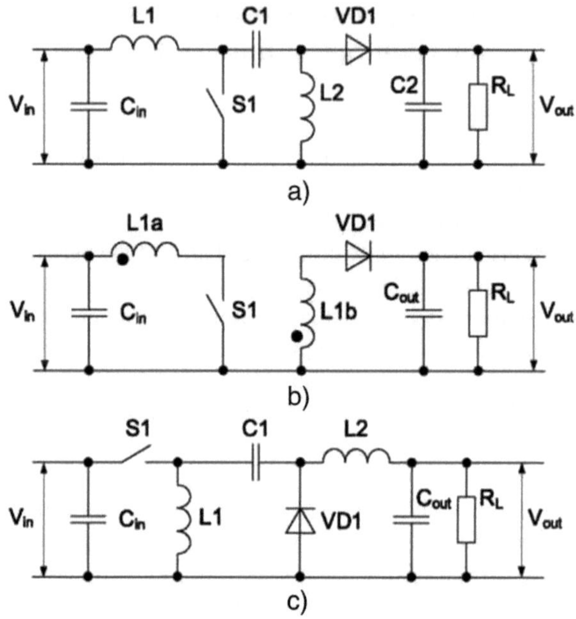

Fig. 10.50 Asynchronous DC/DC converters topologies: SEPIC (**a**), Flyback (**b**), and ZETA (**c**)

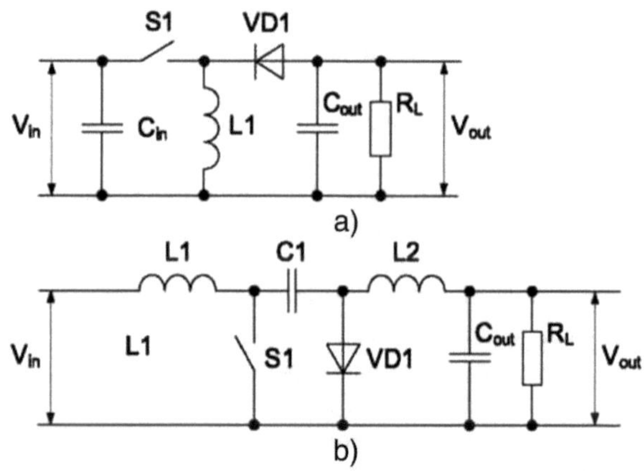

Fig. 10.51 Asynchronous inverting DC/DC converters topologies: the simplest (**a**) and unisolated converter (**b**)

- detector of low battery power;
- user adjustable output current;
- embedded switch that provides the output current of the converter up to 400 mA;
- fixed or adjustable output voltage;
- DIP-8 or SO-8 package.

There are three modifications of the microcircuit with a fixed output voltage: ADP1111-3,3 (+3.3 V), ADP1111-5 (+5 V), and ADP1111-12 (+12 V).

The functional diagram of ADP1111 is shown in Fig. 10.52. A typical wiring diagram in the step-up converter mode is shown in Fig. 10.53 The voltage from the

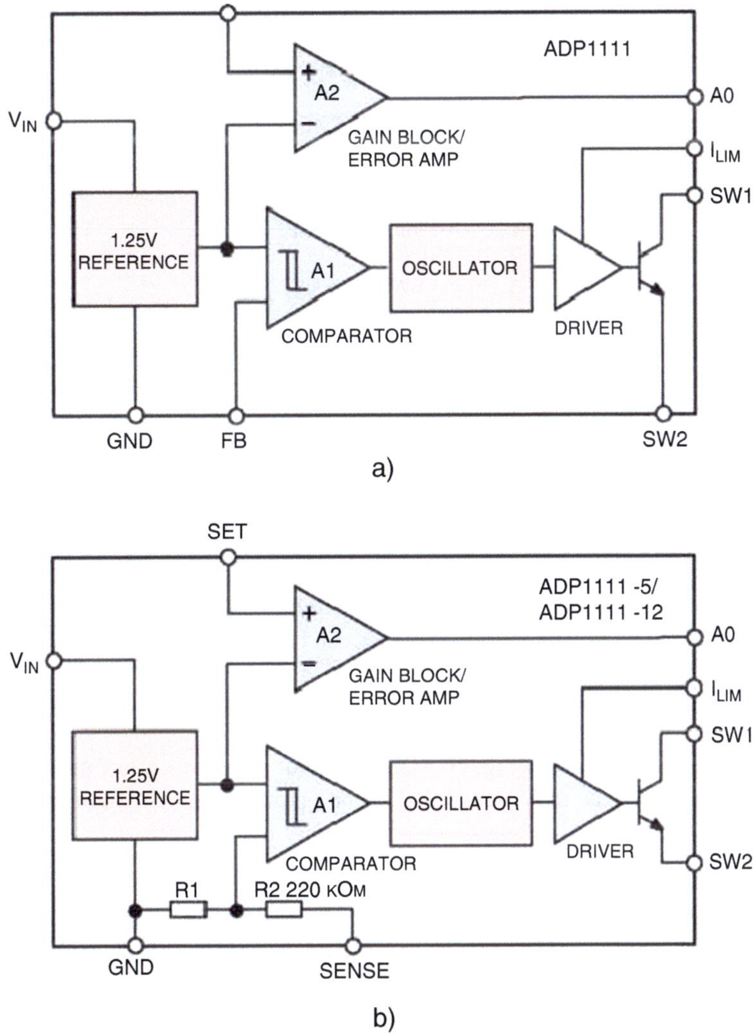

Fig. 10.52 Chain diagrams of step-up/step-down converters ADP1111 (**a**) and ADP1111-5/ ADP1111-12 (**b**) [4]

output of the converter is supplied to the SENSE lead and through the voltage divider to the inverting input of comparator A1 (Fig. 10.52), which compares it with a reference voltage of 1.25 V. The output voltage of the comparator controls the operation of the PWM modulator (OSCILLATOR in Fig. 10.52), the output of which is connected to the base of the transistor switch through a matching circuit (DRIVER). When the output voltage changes, the pulses duty cycle at the base of the transistor changes and the voltage stabilizes. To change the output voltage between the converter output and the SENSE input, you can turn on an additional resistor. In

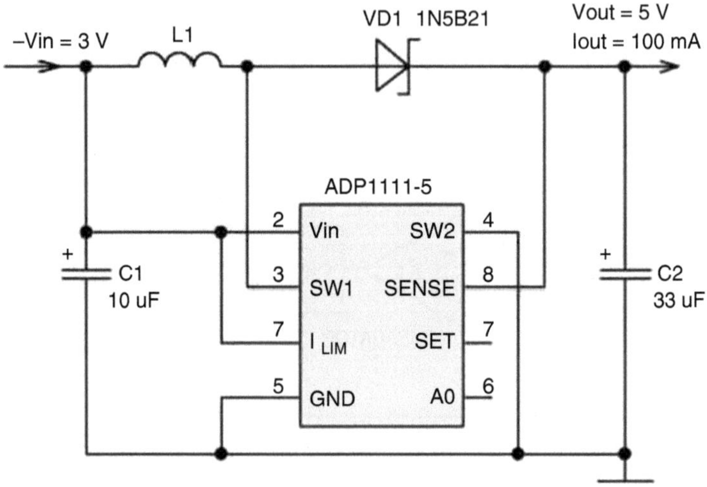

Fig. 10.53 Typical wiring scheme of ADP1111-5 as a step-up converter [5]

other versions of the ADP1111, the output voltage is regulated by an external divider. The output voltage of the converter can be calculated by the formula [4]:

$$V_{\text{out}} = 1.25\ (1 + R2)/R1, \text{V} \tag{10.1}$$

For a step-up converter (Fig. 10.54), the output voltage is calculated using the same formula, and the maximum value of the output current is determined by the relation:

$$I_{\text{max}} = (V_{\text{in}} - V_{\text{ce}} - V\text{out})\ 7\ \mu\text{s}/\text{L1}$$

where V_{in} and V_{out} are the input and output voltages of the converter, V_{ce} is the saturation voltage of the collector-emitter junction of the key transistor (not more than 0.5 V), L1 is the inductance of the coil, 7 μs is the turn-on time of the transistor.

When using the built-in transistor as a switch, the output current of the converter does not exceed 400 mA. To increase the output current, it is recommended to connect an external power transistor, as shown in Fig. 10.55. The voltage drop at the junction of the collector-emitter transistor VT1 is approximately 1 V. In (http://en.wikipedia.org/wiki/ DC_to_DC_converter), the parameters of the elements of the circuits of step-down and step-up converters for various values of currents and input/output voltages are given.

IC ADP1111 can also be used to form a voltage, the polarity of which is inverse to the polarity of the input voltage. Figure 10.56 shows the formation of a negative voltage at the output of the positive input (Fig. 10.56a) and positive from the negative input (Fig. 10.56b). The value of the output voltage of the converter with a negative output voltage is calculated by the same formula as for the previous circuits, and for the converter in Fig. 10.56b—according to the formula [4]:

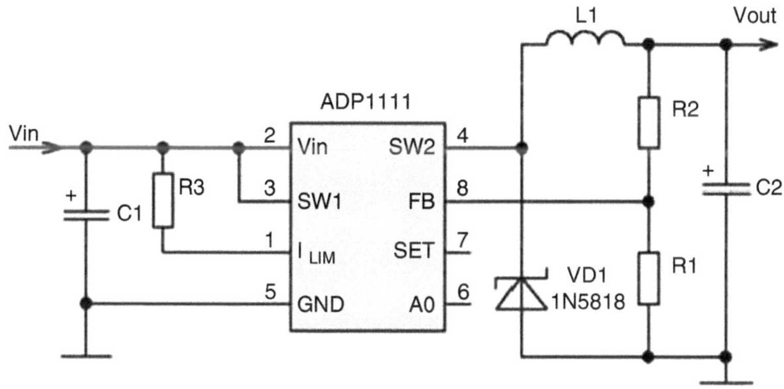

Fig. 10.54 Typical ADP1111 switch-on circuit as a step-down converter

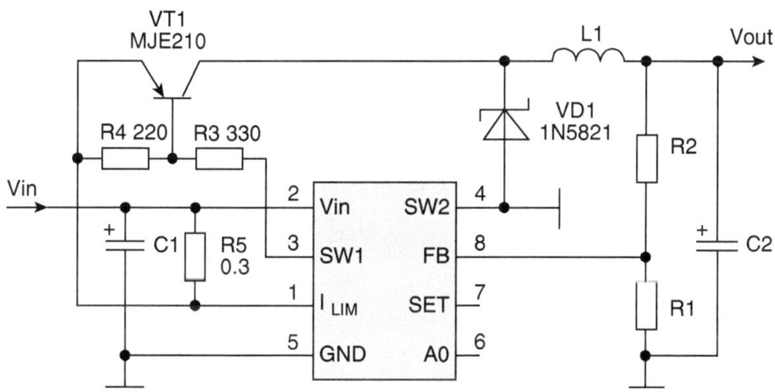

Fig. 10.55 Wiring scheme of ADP1111 with an external power transistor in the mode of step-down converter

$$V_{out} = 1.25\ (R2/R1)\ \mathrm{V} \tag{10.2}$$

The schemes below illustrate only some of the capabilities of the ADP1111 IC. In (http://en.wikipedia.org/wiki/Buck-boost_converter; http://www.analog.com/en/products/power-management/switching-power-converters/switching-regulators.html; http://www.analog.com/media/en/technical-documentation/data-sheets/ADP1111.pdf), examples of using ICs for constructing a current source, a battery discharge detector, voltage controlled polarity converters, etc., as well as calculated ratios are given.

As is known, the coefficient of efficiency of asynchronous DC/DC converters is less than coefficient of efficiency of synchronous (http://en.wikipedia.org/wiki/Buck-boost_converter; http://www.analog.com/en/products/power-management/switching-power-converters/switching-regulators.html; http://www.analog.com/media/en/technical-documentation/data-sheets/ADP1111.pdf; http://www.analog.com/media/en/technical-documentation/application-notes/AN-1075.pdf). When

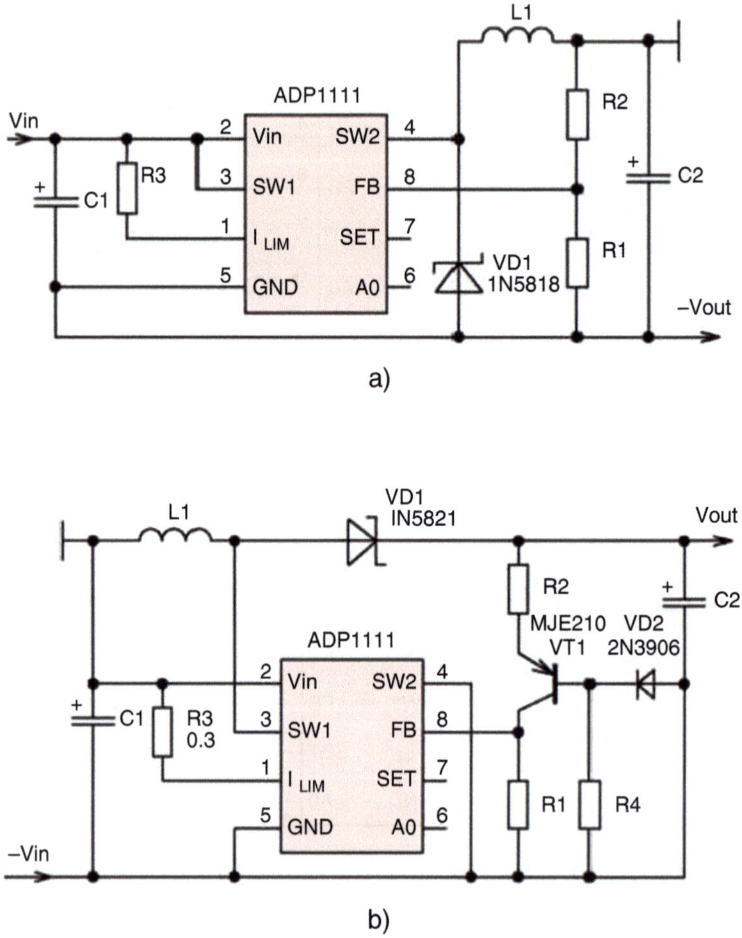

Fig. 10.56 Wiring schemes of ADP1111 for converting output voltage polarity from positive to negative (**a**) and inverse conversion (**b**)

using the ADP1111 IC, referring to a class of asynchronous converters, it is possible to ensure small dimensions of the device, but the coefficient of efficiency will be far from the maximum possible for DC/DC converters.

To provide a large amount of output current and increase the coefficient of efficiency of the converter, synchronous converters based on the DC/DC converter controller and external power transistors are used. An example of such controllers is the IC ADP1870/ADP1871 (http://en.wikipedia.org/wiki/Buck-boost_converter; http://www.analog.com/en/products/power-management/switching-power-converters/switching-regulators.html; http://www.analog.com/media/en/technical-documentation/data-sheets/ADP1111.pdf; http://www.analog.com/media/en/technical-documentation/application-notes/AN-1075.pdf) [6].

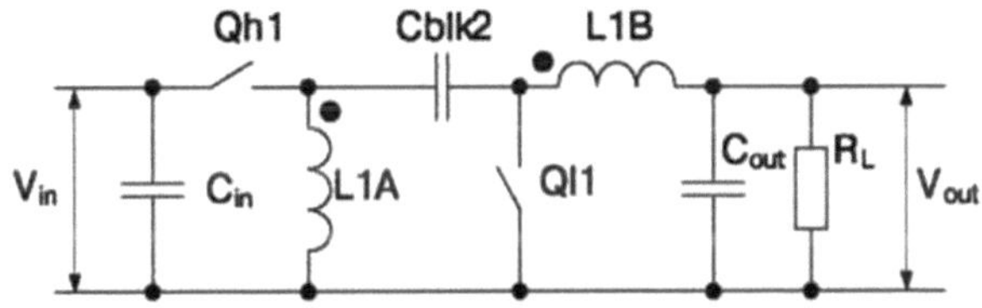

Fig. 10.57 Synchronous DC/DC converter built according to ZETA topology [4]

In (http://en.wikipedia.org/wiki/Buck-boost_converter; http://www.analog.com/en/products/power-management/switching-power-converters/switching-regulators.html; http://www.analog.com/media/en/technical-documentation/data-sheets/ADP1111.pdf; http://www.analog.com/media/en/technical-documentation/application-notes/AN-1075.pdf) there is information that the best result by the price-quality criterion can be obtained using converters built on the ZETA topology (Inverse CEPIK). It is for this topology that the ADP1870/ADP1871 and ADP1872 ICs are implemented. Figure 10.57 (the original designations (http://en.wikipedia.org/wiki/Buck-boost_converter) are preserved in the diagram) shows the topology of the synchronous converter with the ZETA topology (in Fig. 10.50, the topology of the asynchronous converter is shown). The principle of operation of such a converter is considered in detail in http://www.analog.com/media/en/technical-documentation/application-notes/AN-1075.pdf. The switches Qh1 and Ql1 work in antiphase. When the Qh1 switch is closed, the current flows in two ways (Fig. 10.58a). From the input capacitor Cin through Qh1, capacitor Cb1k2, coil L1B, and the resistance of the load on the common wire. The second way is from the input capacitor through the coil L1A to the common wire. In this cycle, energy is stored in the capacitor Cblk2. In the second cycle, the switch Qh1 is opened, and Ql1 is closed and two current paths are formed (Fig. 10.58b). The capacitor Cblk2 is discharged through the coil L1A and the closed switch Q11. The energy given off by the capacitor Cb1k2 is transmitted through an inductive coupling between the coils from L1A to L1B and given to the load.

The ratio of the output voltage to the input is determined by the ratio [4]:

$$V_{\text{out}}/V_{\text{in}} = D/(1-D)$$

where D is the duty factor, numerically equal to the ratio of the time of the closure of the switch Qh1 to the repetition period of the pulses controlling the switch Qh1. If $D > 0.5$, the converter is step-up (boost), and if $D < 0.5$ it is step-down (buck).

The capacitor Cb1k2 performs another function. It blocks the passage of direct current from the input source to the load, protecting it from a short circuit in it.

Analysis of the circuit shows that the current through the output capacitor flows continuously, and this leads to the fact that the ripple voltage at the load will be less than in the converter, in which the current flows through the output capacitor only part of the period. This allows to reduce the capacity of the output capacitor of the converter and thereby reduce its dimensions.

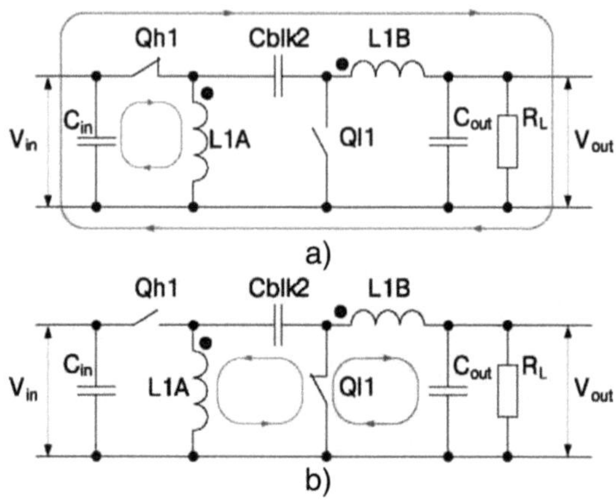

Fig. 10.58 The contours of the current flow formed by the closure of the switch Qh1 (**a**) and Ql1 (**b**) [4]

The chain diagram of the ADP1870/71 IC [6] is shown in Fig. 10.59. Initially, the controller was designed to build step-down synchronous DC/DC converters. However, in http://www.analog.com/media/en/technical-documentation/application-notes/AN-1075.pdf, a variant was proposed for using the ADP1870 to build a universal step-down/step-up converter using the ZETA topology, as shown in Fig. 10.60. A detailed description of the operation of such a converter and timing diagrams of signals at various points in the circuit are also given in http://www.analog.com/media/en/technical-documentation/application-notes/AN-1075.pdf.

The repetition period of the control pulses of the power transistors Qh1 and Ql1 is determined by the formula:

$$1/f_{sw} = a\left(\frac{V_{\text{out}}}{V_{\text{in}}} + 1\right), \tag{10.3}$$

where $a = 3.33{\cdot}10^{-6}$ s for the modification ADP187xARMZ-0.3-R7 (for ADP187xARMZ-0.6-R7 $a = 1.66{\cdot}10^{-6}$ s, for ADP187xARMZ-1.0-R7 $a = 1{\cdot}10^{-6}$ s).

The output voltage of the converter is determined by the formula:

$$V_{\text{out}} \geq 10\left[\frac{I_{\text{out}}D}{f_{\text{sw}}C_{blk2}} + aC_{blk2}\left(\frac{V_{\text{in}}D}{2L_{1B}f_{\text{sw}}} + I_{\text{out}}\right)\right] \tag{10.4}$$

In http://www.analog.com/media/en/technical-documentation/application-notes/AN-1075.pdf, detailed recommendations were given on the calculation of pulsations at the output of the converter, depending on the parameters of the elements used and the load current, as well as on the choice of power transistors and other circuit elements.

Figure 10.61 shows the graphs of the efficiency of the ADP1870 based inverters on the load current for various values of input and output voltages.

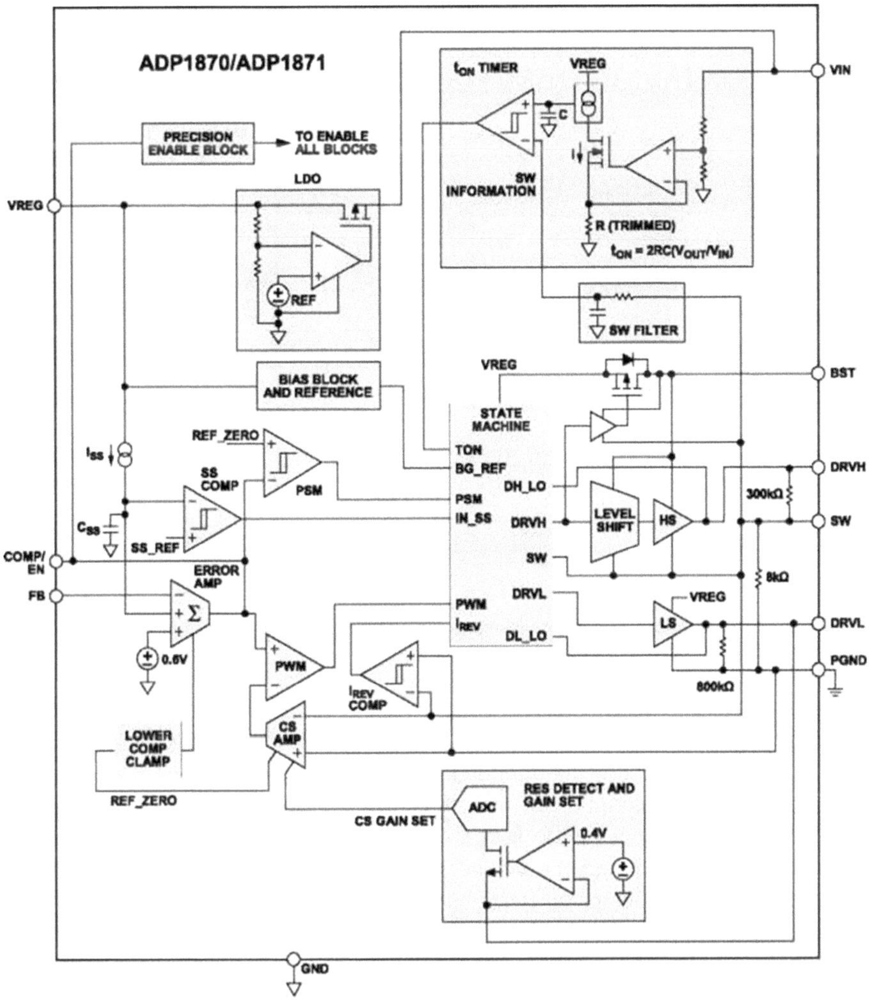

Fig. 10.59 Chain diagram of the controller ADP1870 [4]

Analysis of the graphs shown in Fig. 10.61 allows us to conclude that the maximum coefficient of efficiency of the converter reaches when the input and output voltages are equal, which agrees well with the results presented in [4, 5]. For illustration, in [5], typical dependencies of the efficiency of synchronous step-up and step-down converters on the input voltage for an output voltage of 3.3 V are shown (Fig. 10.62). Analyzing these dependencies, it is also easy to see that the highest values of the coefficient of efficiency are achieved when the input and output voltages are close in magnitude or equal. Under these conditions, the amplitudes of the current pulses in the inductance and the switches are also minimal [5] (http://www.analog.com/media/en/technical-documentation/application-notes/AN-

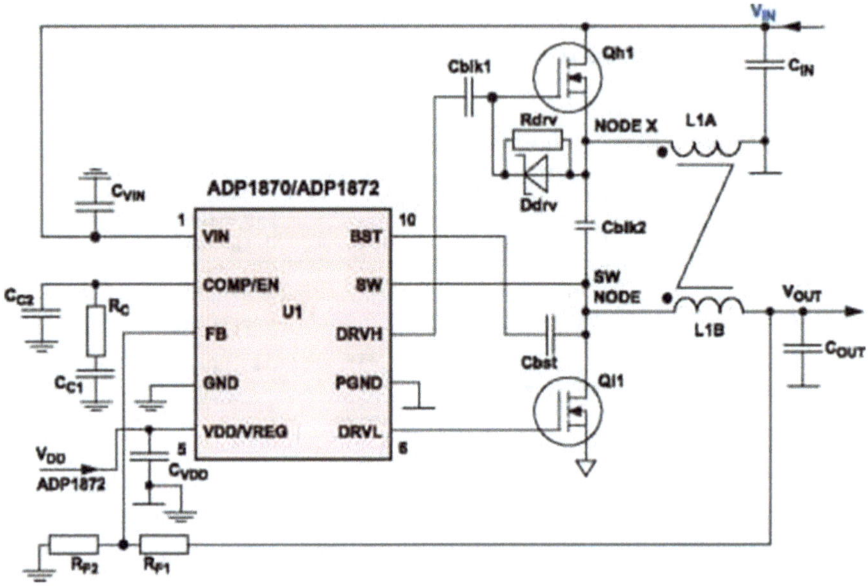

Fig. 10.60 Layout diagram of step-up/step-down synchronous converter implemented in ZETA topology based on ADP187x controller [4]

1075.pdf). This allows in this case to reduce the inductance of the converter coil and the maximum current flowing through it.

In [6], a brief description of the ADIsimPower programs for calculating and modeling power supplies built on ICs manufactured by Analog Devices is given. The examples of the design of a DC/DC step-down converter are considered, when it is optimized for various parameters—efficiency, size, cost, or the number of components used. All programs are free and available on the Analog Devices website [7].

The use of ADIsimPower™ programs allows you to significantly reduce the time required to develop power supplies, and to conduct a comparative analysis of options for constructing a circuit while optimizing it for various criteria.

10.4 Circuit-Based Methods for Suppressing Interference in the Power Circuits of High-Speed Electronic Devices

10.4.1 Analysis of the Typical Wiring Diagram of the Power Supply Circuit of SHF Device Board

As developers of radioelectronic equipment are all aware, noise generated in the electric circuits impairs the reliability parameters of any electronic device, and especially high-speed ones. In order to properly design a highly reliable and high-

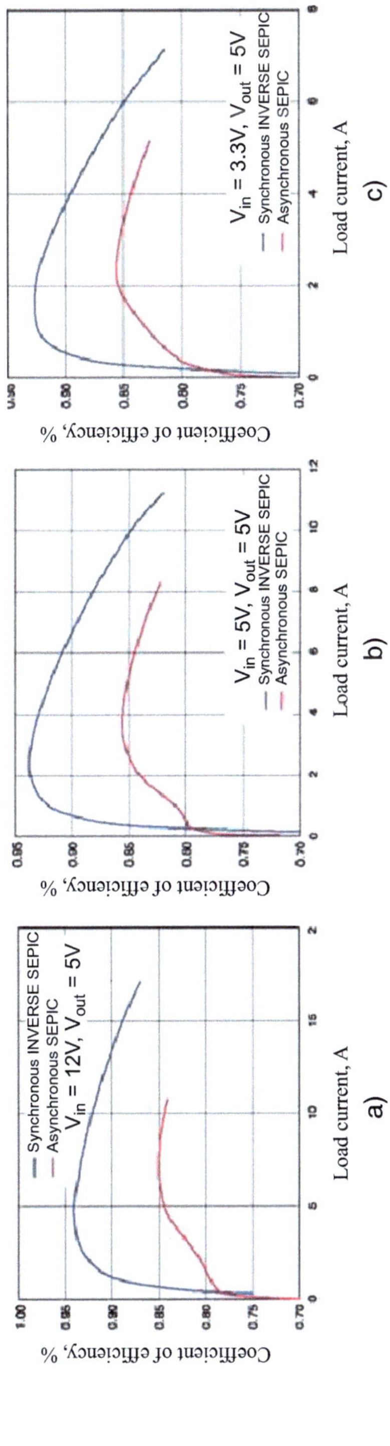

Fig. 10.61 Graphs of the coefficient of efficiency of the DC/DC converter based on the ADP187x controller on the load current at $V_{in} = 12$ V, $V_{out} = 5$ V (**a**); $V_{in} = 5$ V, $V_{out} = 5$ V (**b**); $V_{in} = 3.3$ V, $V_{out} = 5$ V (**c**)

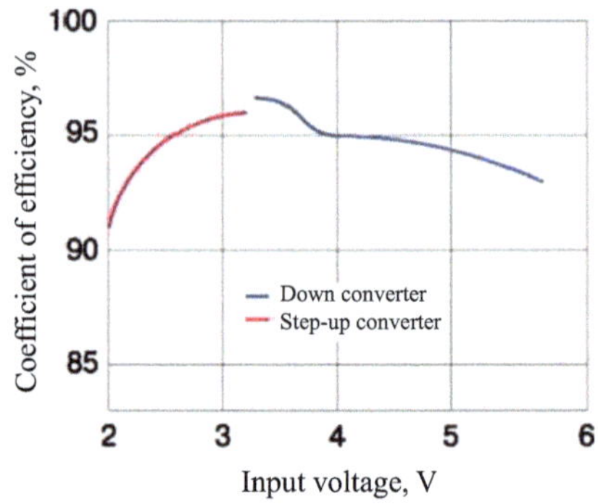

Fig. 10.62 Dependencies of the coefficient of efficiency of a synchronous step-down/step-up DC/DC converter on the input voltage with a fixed output [4]

speed electronic system, it is necessary to understand very well how the so-called noise forms and spread in the wiring diagram of the power supply circuits. Noise is minimal (and it is impossible to get rid of it completely) if the signal source, wiring and load are electrically matched with each other [8, 9].

The output resistance of signal generators, oscilloscopes, spectrum analyzers, and vector network analyzers is often equal to 50 Ohm. This equipment is connected to other units by means of 50-Ohm coaxial cables and 50-Ohm connectors.

Other standard levels of impedance are used in practice. For example, the value of 100 Ohm is often used in the wiring on the digital circuit boards; and the value of 75 Ohm is often used in closed-circuit television systems. It is strictly required to use only matched wiring layouts in all this equipment that improves the performance capabilities of applications and minimizes the noise. The same requirement is applicable to the design of common power distribution network of SHF-devices.

Let's perform a brief analysis of typical wiring diagram of the microwave system [8, 9].

Power distribution network (PDN) includes a voltage regulator module (VRM), interconnected layers of printed circuit board, connecting wires and loads to which power is supplied. Figure 10.63 shows a typical dependence of impedance of the well-balanced power distribution network on frequency. The impedance of this PDN is 3.3 Ohm. Here, the output impedance of the voltage regulator is equal to 3.3 Ohm and the load impedance (ZPCB) is 3.3 Ohm that ensures perfectly even characteristic of power distribution network (PDN).

If connections between the voltage regulator module and power distribution network are mismatched in any direction, they become sensitive to noise that impairs the system operating parameters. Figure 10.64 [9] shows the impedance-frequency characteristic of power distribution network in three cases when the voltage regulator impedance is more, less than, or equal to the impedance of the printed circuit board and load (Fig. 10.65).

If the voltage regulator impedance is too high, the resultant impedance of power distribution network is mainly capacitive in effect at frequencies lower than resonance frequency of the printed circuit board, and it is reduced at a rate of 20 dB/ decade. If the voltage regulator impedance is too low, the resultant impedance of

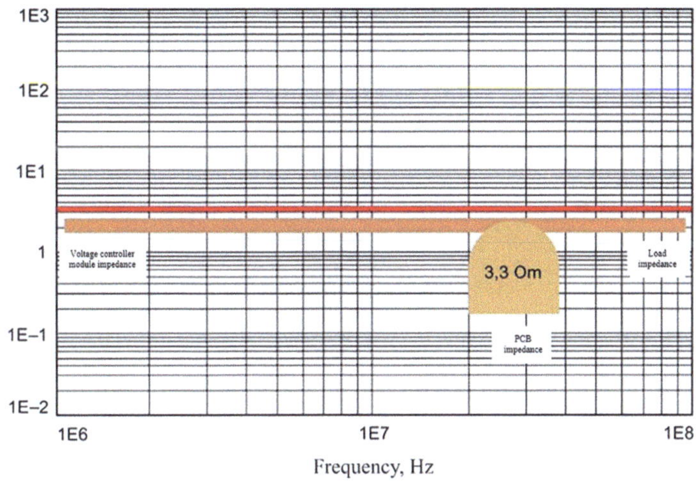

Fig. 10.63 In a well-balanced wiring diagram, the impedances of the voltage regulator, printed circuit board, and load are matched over the entire frequency range below the specified

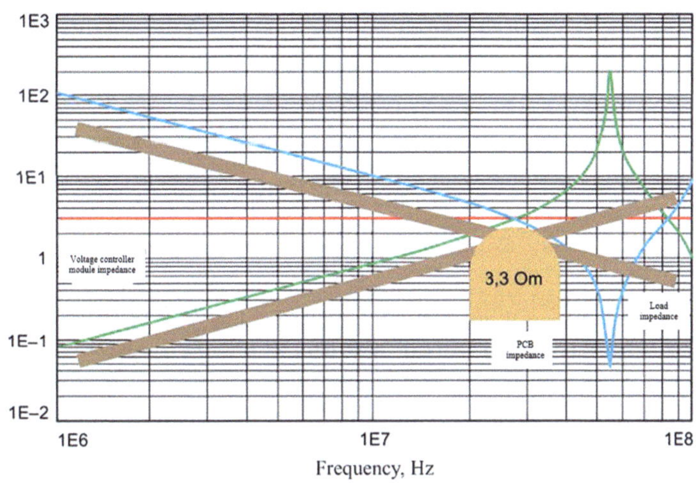

Fig. 10.64 Frequency response of the impedance of a typical microwave device [9]

power distribution network becomes inductive in effect at frequencies lower than resonance frequency of the printed circuit board, and it is increased at a rate of 20 dB/decade.

As shown in Fig. 10.66 [9], addition of a decoupling capacitor with high Q-factor allows to eliminate the first maximum; but at low and high frequencies, the peak values of impedance remain and exceed by far the impedance of the printed circuit board.

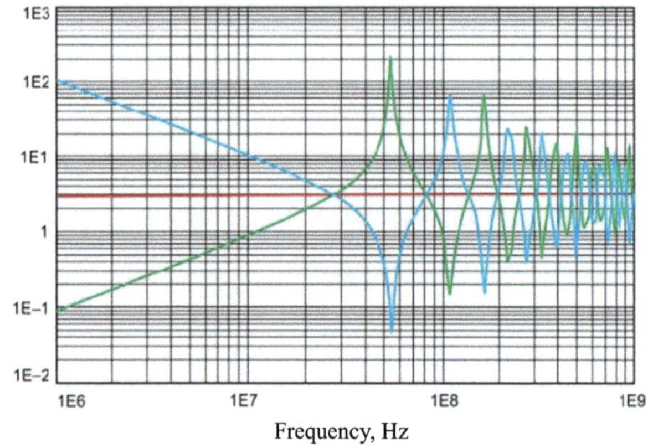

Fig. 10.65 If the voltage regulator impedance is higher or lower than impedance of the printed circuit board (PCB), the peak values of the voltage regulator impedance are much higher than PCB impedance

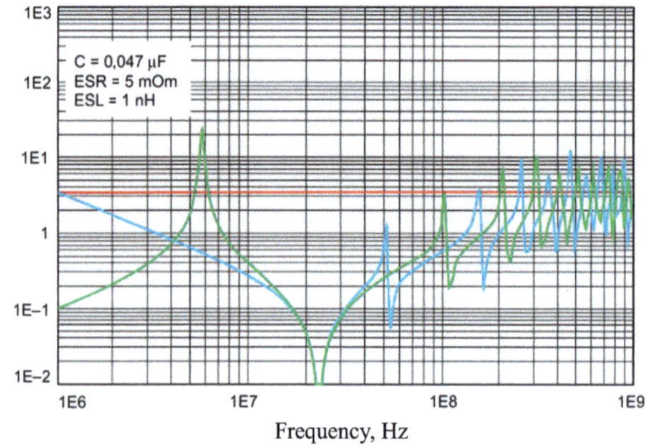

Fig. 10.66 Impedance-frequency characteristic of the board with decoupling capacitor

As the signal passes through the transmission line, the PCB impedance has a sharp peak near the load at a frequency which is twice as much as a resonance frequency of this board. In case of resultant inductive impedance of the power bus, the amplitude of this peak has a positive value; and in case of capacitive impedance, the amplitude has a negative value.

Figure 10.67 [9] shows that even if close to the load, there is a decoupling capacitor having ESR equal to PCB impedance and ESL lower than PCB impedance, the characteristic can't be considered optimal because of the presence of negative peaks in the range of high frequencies.

Figure 10.67. shows the frequency characteristics for three cases under consideration with different impedances of the voltage regulator. These curves are characteristics of transmission lines with cyclic-repeated positive and negative peaks. At higher frequencies, the characteristics of the board with resultant capacitive impedance have a negative peak followed by a positive peak, then these peaks cyclically revolve. At higher frequencies, the characteristics of the board with inductive impedance have a positive peak followed by a negative peak; by alternating, these peaks form similar cyclic-repeated sequence. In both cases, the maximum peaks of

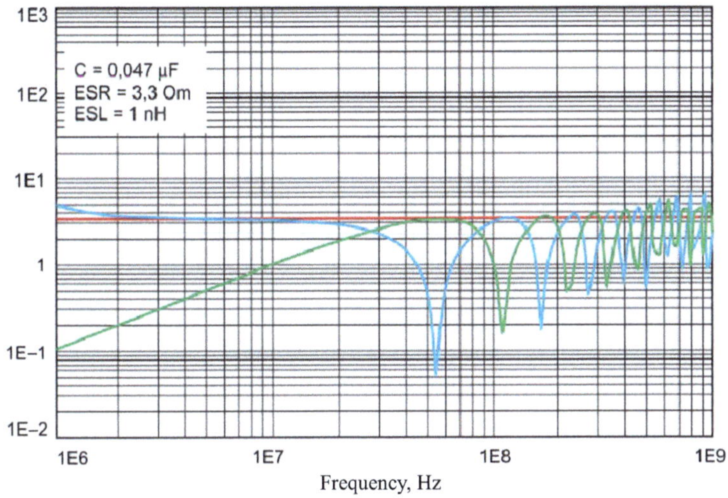

Fig. 10.67 The presence of negative peaks in the range of high operating frequencies

voltage regulator exceed by far the impedances of the load and printed circuit board; and the negative peaks are much lower than impedances of the load and printed circuit board.

Generally, a decoupling capacitor is added to the load in order to smooth the characteristics of resultant impedance of power distribution network and improve the system operating parameters. Figure 10.68 shows the results of use of the ceramic capacitor with a capacity of 0.47 μF.

10.4.2 Basic Circuit-Based Methods for Interference Suppression in Power Supply Circuits

Addition of the ceramic capacitor often reduces LF-impedance down to the value of PCB impedance; however, resulting in this case, the peak inductive impedance at lower frequency with smaller amplitude can exceed by far the PCB impedance. At higher frequencies, the output impedance of the regulator is increased again and it exceeds the PCB impedance.

In order that the peak impedance remains below the PCB impedance, the equivalent series resistance (ESR) of the ceramic decoupling capacitor should be equal to the PCB impedance within the full range of frequencies higher than the first resonance frequency; and the equivalent series inductance (ESL) of this capacitor should be equal to the PCB impedance at the highest frequency of the range under consideration. In this case, the capacitor ESR is equal to 3.3 Ohm, and its ESL should be less than 1 nH (Fig. 10.67).

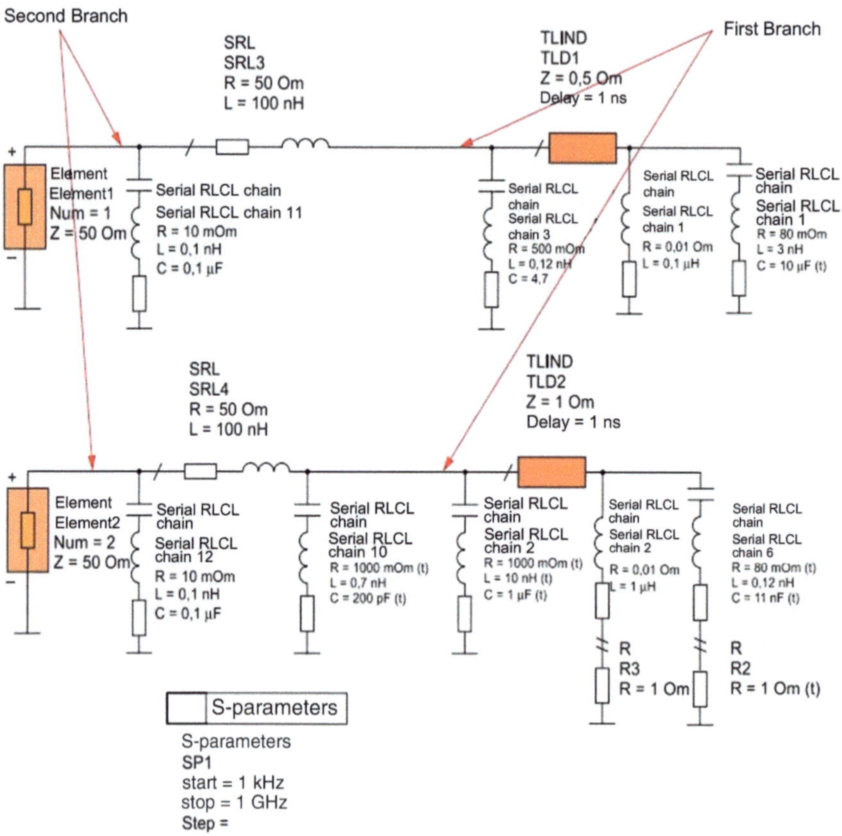

Fig. 10.68 Equivalent electrical power distribution network with a series-connected 1-Ohm circuit [9]

Negative peak values being much lower than impedance of the printed circuit board continue to impair the system performance. The impedance close to the load is an internal impedance of the power source for the next circuit. This impedance should be equal to the resultant impedance of the board in order to minimize the peaks in characteristics of the next circuit (Fig. 10.68).

As shown in Fig. 10.69, the tantalum capacitor with low ESR is installed in the top power distribution network close to the voltage regulator. The bottom power distribution network provides the maximum flat frequency characteristic.

Here we will list some core capabilities to use these principles in practice [9] in order to reduce the cost of project under development while improving the system performance.

Of course, the voltage regulator with lower impedance and better regulating characteristic costs more. If impedance of this device is lower than characteristic

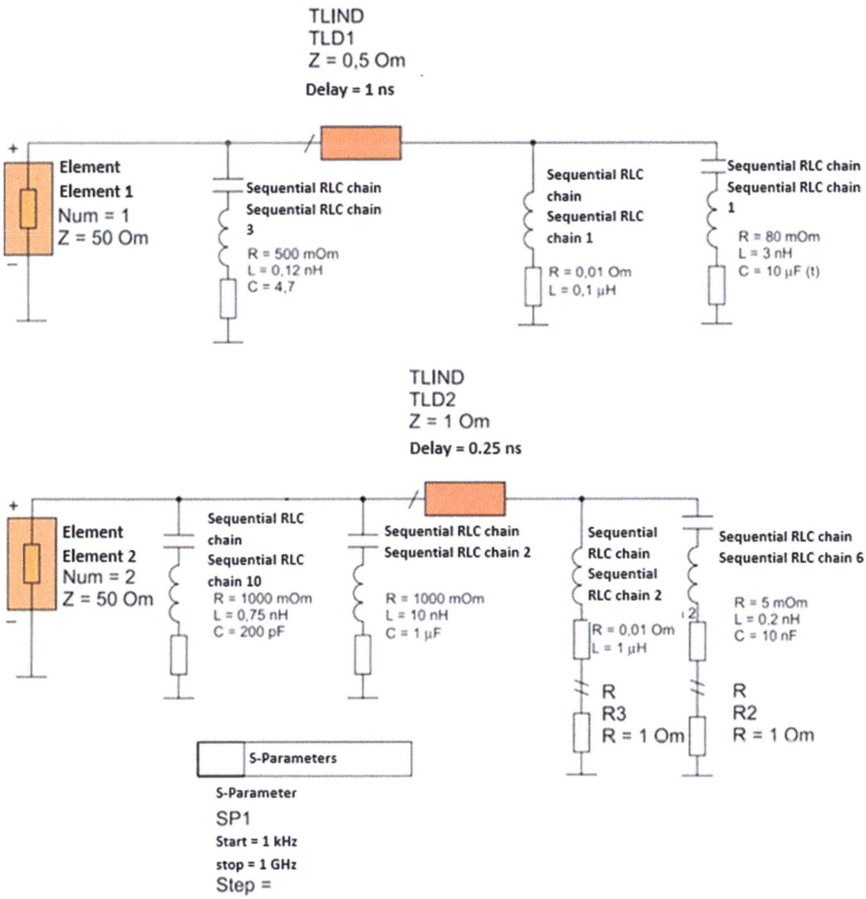

Fig. 10.69 Equivalent electrical power distribution network with included tantalum capacitor [9]

impedance of the printed circuit board, then under load the peaks can appear on the frequency characteristic of the impedance.

PDN on the printed circuit board should not be too complicated. To reduce its impedance, it is necessary to expand the board area that leads to unreasonable increase in the number of levels and, consequently, to the rise in the product cost.

When using the tantalum capacitors with low ESR, there is an increase in the system production cost. To provide impedance matching on the power distribution, a capacitor with a higher ESR value is required, which allows relatively inexpensive tantalum or aluminum capacitors to be used. Often this is also the case when parallel decoupling capacitors are installed in parallel with large-capacity capacitors. To obtain equality with impedance of power distribution it is required to use high capacity capacitors with a higher ESR value. In some cases an external resistor is

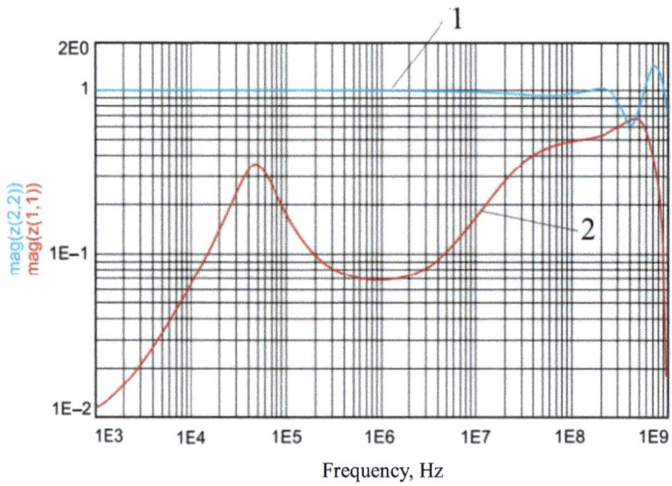

Fig. 10.70 Frequency response of a board with a tantalum capacitor with a low ESR [9]

installed in series with the capacitor for this purpose. This solution is the least expensive.

As seen in Fig. 10.70 [9], the characteristic with maximum linearity in almost the entire frequency range is shown by curve 1. Curve 2 illustrates the characteristics of the power distribution network which includes a tantalum capacitor with low ESR.

In cases where the regulator is linear, it does not need capacitors—they need to be installed only at the load.

The output impedance of all voltage regulators allows you to determine the stability—a critical parameter of any application. In addition, it is used to establish peak impedance values caused by a small margin of stability. Of course, it can be increased by changing the value of the compensating resistance or capacitor, it will be much cheaper than using smoothing filters.

For cascaded circuits, impedances should be equal. For example, if a circuit with a low impedance is followed by a circuit with a large impedance, the power distribution circuit can become inductive, which is difficult to correct, then, as with equal impedances, balance is maintained.

The noise is minimal in the system with matched impedances of the voltage regulator, printed circuit board, and load. Paradoxically, the series-connected resistor between the voltage regulator module and the printed circuit board reduces the range of voltage fluctuations near the load.

As shown in Fig. 10.71, more expensive PDN doesn't provide the required damping that results in the presence of a sharp peak on the impedance-frequency characteristics which impairs the system performance and requires an additional damping circuit [8].

According to Fig. 10.71, the power distribution network (PDN) operating in the frequency range of 0–1 GHz with the maximum impedance of 2 Ohm, includes a linear voltage regulator with inductance of 1 μH and series resistance of 10 mOhm.

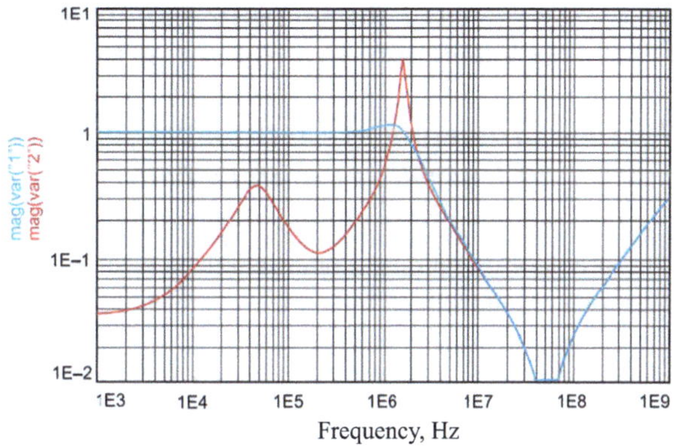

Fig. 10.71 Example of the frequency response of the power supply distribution [8]

These typical values for linear voltage regulator operating at current of 100 mA can be easily measured. When designing this PDN, two approaches are used (Fig. 10.69) [9]. The first approach provides the maximum linearity; the second approach is a typical solution in which along with a regulator, a capacitor with low ESR is installed that increases the dimensions of the printed circuit board.

Figure 10.70 [9] presents the results of simulating impedance from the load. The comparative cost of a solution that uses a larger printed circuit board and a capacitor with a small ESR is slightly lower, but it provides a lower impedance (albeit with a non-linear characteristic).

It should be noted that a more expensive solution may cause problems to other individual power supply circuits connected to this node due to insufficient damping below 1 kHz and in the range of 100 kHz–10 MHz. An additional 1-Ohm circuit consisting of inductance per 100 nH and a capacitor per 100 nF is connected to this node (Fig. 10.68). Figure 10.70 shows the results of modeling the characteristics of the impedance from the second circuit. Further improvement in performance is provided by using a 1-Ohm resistor, which is connected in series to 0.1 μF capacitors in the second circuit, if the impedance of the load connected to this node of the power distribution circuit does not reach 1 Ohm [8].

The exact matching of impedances of the voltage regulator module, printed circuit board levels, and load circuits with each other, as well as their conformity to the specified value enable to create the straight-line characteristic and minimize the area of the wire-circuit board. This optimization also reduces the use of expensive low ESR capacitors and low impedance voltage regulators.

10.5 Features of the Organization of Grounding Systems for Electronic Equipment of Power Facilities

10.5.1 Grounding of Electronic Equipment: Analysis of Unevident Contradictions

The problem of rational organization of electrical grounding circuits of modern electronic equipment in industrial, commercial, military, space, powerful telecommunication devices, and other electric power facilities, in radar stations with active phased arrays, etc. is usually considered only as a necessary and sufficient condition for ensuring their operability under the conditions of exposure to various kinds of electromagnetic interference—from lightning generated to so-called—deliberately destructive, deliberately generated in order to bring an object out of order (see Chapter 9 "Microwave weapons of space and ground application" of the book [8]).

In accordance with international regulatory documents, only two "classical" types of grounding are distinguished [10]: protective, performed in order to ensure electrical safety, and functional, performed only to ensure the efficiency of the electrical installation). Modern electronic control devices, protection devices, and special automatics used in electric power industry and systems for ensuring their uninterrupted power supply have their own internal electronic circuits, completely isolated from all external circuits. For example, microprocessor-based relay protection devices (MRPDs) contain built-in current and voltage transformers in input analog circuits, which isolate internal analog-to-digital converters from any external current and voltage circuits. The logic inputs of the MRPD are usually made on optocouplers that provide a high level of isolation of the inputs from the outputs of the electronic system. The output circuits of the standard MRPD are made on the basis of electromechanical relays, also providing isolation of the internal electronic circuit from external circuits. Own autonomous internal switching power supply of MRPD contains a special transformer that provides galvanic isolation of all internal circuit elements from external power circuits. As a rule, the DC power supply itself is in most cases completely isolated from the "ground."

Thus, on the one hand, modern complex electronic devices and systems, especially high-speed electronic devices, do not seem to require the obligatory presence of functional grounding to ensure their effective operation. On the other hand, the grounding of electronic equipment of such high-performance systems is often considered by the developers of these systems only as one of the measures to protect it from electromagnetic interference, that is, as an indispensable condition for its normal functioning in actual operating conditions when exposed to this very electromagnetic interference. Even in a number of quite professional foreign publications on this issue, an equal sign between electromagnetic compatibility (EMC) and "grounding" for automation and control systems is actually put [11].

In this connection, the problems of the "correct" design of electrical circuits for grounding electronic equipment require more detailed consideration.

As discussed in detail in [8, 10], five basic types of electromagnetic interference are usually distinguished:

- conductive;
- inductive;
- capacitive;
- electrostatic;
- combined.

Conductive interference is distributed in the presence of direct electrical contact between the electrical circuits and is divided into two types:

- interference of the “wire-ground” type, the voltage of which is applied between each of the conductors and the “ground,” which are also called asymmetrical, common mode, of a general form;
- interference of the “wire-to-wire” type, the voltage of which is applied between individual electrical circuits or between elements of the same electrical circuit, which are also called symmetric, antiphase, and differential.

Common-mode interference, unlike differential, as a rule, does not lead to failures in the process of electronic equipment operation, but can completely disable such equipment due to electrical breakdown of internal insulation (or pn-n junctions) in microcircuits and microprocessors.

On the other hand, if the electronic equipment is completely isolated from the “ground,” then the pulse interference and various electrical overvoltages relative to the ground (including common-mode asymmetrical interference of a general form) cannot affect this equipment “just as high voltage relative to the ground does not interfere with birds sitting quietly on high-voltage wires” [10].

Differential interference does not have any relation to the presence or absence of electrical ground circuits.

As for inductive interference propagating through electromagnetic fields, it is known that effective protection against interference of this type is carried out in practice by placing “sensitive” electronic equipment inside closed metal shells (the classic “Faraday cage”), whose role is carried out by the metallic cases of electronic devices. As is known, the grounding of the “Faraday cage” does not affect in any way the effectiveness of its inductive interference.

With capacitive interference, the situation is somewhat more complicated [8], since interference of this type is propagated through capacitive coupling between passing wires, between closely spaced metal cases, as well as between the listed elements and the “ground.” Grounding the shields of wires and metal cases (however, as well as reducing the capacity) allows you to eliminate common-mode capacitive interference and has virtually no effect on differential-type capacitive interference.

Static interference can occur as a result of the accumulation of static charge on a case insulated from the “ground” and resulting periodic breakdowns to the “ground.” Case grounding prevents static charge accumulation. However, to eliminate electrostatic (and even partially capacitive) common-mode interference, low-impedance grounding is not required at all; it is enough to connect the case to the grounding system through a high-resistance resistor, with a resistance of about 1 GOhm.

It is also quite obvious that the earthing of the MRPD case is in no way capable of affecting interference of all types that penetrates the internal electronic components through cables and wires connected to the MRPD. However, the internal "ground" of an electronic circuit in many types of MRPD, which serves as a "zero (reference) potential" circuit, is usually structurally connected to a metal case, and through it to an external grounding system. Sometimes, for especially sensitive electronics, a separate ground loop is used, which is connected at one point to a common ground loop (Signal Reference Subsystem), which does not fundamentally change the essence of such grounding. It is assumed that numerous electronic devices having electrical and informational connections between themselves will have a common zero (reference) potential, which prevents failures in the operation of highly sensitive electronics due to electromagnetic interference, creating an additional potential difference between the zero potential circuits, if they can't be grounded.

This practice, according to [12], "provides the reference potential for relay protection and other equipment and helps protect equipment and personnel from powerful high-frequency impacts." This is a generally accepted approach and common practice for providing EMC of electronic equipment. However, as noted in [13], "a lot of problems with electromagnetic interference arise because developers consider grounding ideal and do not pay enough attention to the actual characteristics of the grounding systems they use."

Below, using materials from [10], we will try as briefly as possible to consider the problems that actually exist here, so that the developers themselves can see the unreality of such an "ideal" grounding and still pay more attention to the actual characteristics of these systems.

10.5.2 Problems of Conventional Grounding Systems

As it is known to all students of professional colleges and institutes, the electrical potential of the grounding system of any electronic device with respect to the "ground" can increase significantly (by orders of magnitude) at the moments of lightning discharges. However, for some reason it is believed that if the total potential of the grounding system is maintained between all electronic devices, that is, there is no potential difference between the circuits of the "zero potential" of various devices, then this is a stepwise short-term increase in the total potential and its difference from zero, what is happening simultaneously in all devices is not capable of causing significant disturbances (malfunctions) in the operation of these electronic devices. On this assumption, the whole modern theory of grounding is actually based, prescribing to keep the resistance of the elements of the grounding system to a minimum, to apply the so-called equipotential surfaces and similar measures aimed at preventing the possibility of potential differences between the circuits of the "zero potential" of various devices removed from each other and therefore grounded at different points, but having electrical and informational connections between them.

At the same time, the developers of modern high-speed electronic devices for some reason do not publicly consider the question seemingly obvious even for a student of what happens in a single electronic device of an electronic system (telecommunication or radar) with such a short but significant increase in amplitude potential of the circuit "zero potential".

After all, it is obvious that any modern electronic circuit contains many non-linear elements and elements with their own equivalent capacity and inductance connected to the so-called zero potential circuit. Therefore, it is obvious that with a pulse increase in the potential of this circuit (the same lightning discharge), the voltages at different points of such an electronic circuit will not increase simultaneously (completely synchronously).

Students of culinary college can easily imagine this situation as some mechanical plate with a lot of blocks (weights) of different mass attached to it with a system of mechanical springs of different stiffness. Naturally, when the plate moves up smoothly, potential energy of all elements installed on it also increases gradually (simultaneously). However, if the plate moves rapidly, position and potential energy of each of these elements will change non-simultaneously (at different speed), and if they were interconnected mechanically, mechanical connections may be broken. This means that even equipotential surface and zero difference between chains of zero potential of different electronic devices of bigger electronic system do not guarantee absence of failures in the operation of highly sensitive electronic mechanisms under the influence of such factors. In real conditions of the operation of electronic mechanisms located on the objects with larger surface it is very difficult and sometimes totally impossible to preserve this ideal zero difference of potentials between the chains of zero potential.

Our foreign competitors know that. Thus, for example, under the date provided in situation [12–16], when several electric devices and separate electronic devices are located quite far from each other and earthed at their locations, it is inevitable to have high difference of potentials between earthing points which at lightning strikes may be up to 10 kW and more. High voltage between remote points of earthing appears as a result of voltage drop in earthing system elements when lightning current or short circuit current goes through them. Generally speaking, "the larger is the area of the object under protection, the higher is the potential for possible problems" (Whitaker [17]).

Even in the basic standard well known to all earthing specialists for earthing systems of electronic mechanisms [18] the following is noted: "In case of lightning or in emergency in energy system the difference of potentials between electronic equipment and earthed objects may be so important that it will become a threat to electric safety or damage electronic mechanisms."

From the above, it is evident that any so-called low-Ohm earthing of packages of the majority of sensitive control, automation and protection equipment only increases the risk of failure during operation of this equipment. Paradox?! And it's the result of lightning current discharge through earthing system under usual lightning discharges! But as lightning discharges are not the only source of powerful electromagnetic influences on electronic devices.

Lately, developers of strategically important electronic systems pay a lot of attention to the very relevant problem of the protection of their electrical power systems (electric feed systems) from so-called intended remote destructive impacts (ref. to Chapter 9 of technical encyclopedia "Wave electronics in radiolocation and communication systems" [8]), of which the most powerful one is electromagnetic impulse of high-altitude nuclear explosion. The nature of impact of the nuclear explosion of electromagnetic impulse on classic earthing system is different from the impact of lightning discharge "by spectral power density, pulse leading edge, current level, frequency of impulses, and zone of coverage" [19].

Firstly, lightning discharge is just a spot influence on the earthing system, while the element of E1 of the nuclear explosion of electromagnetic impulse creates by the ground surface an impulse of distributed electric field with the intensity up to 50 kW/m. This field has vertical and horizontal components and covers large area (hundreds and thousands of square kilometers).

More complex picture is created in cases when such impact is applied to electronic board systems of Earth remote probing spaceships and other space objects in outer space conditions—but the results of such research, as you can understand, are not published in open sources and will hardly be published in the near future.

Secondly, parameters of the component of E1 of the nuclear explosion of the electromagnetic impulse (2/50) are not really different from lightning impulse parameters (1.2/50 mcs).

Thirdly, if during a lightning strike into lightning discharger and earthing system, there is a task to reduce the resistance of earthing system elements to minimum to reduce drop of voltage on current-conducting elements when lightning current comes through them, in case of nuclear explosion of electromagnetic impulse the high-voltage impulse is applied simultaneously to all elements of the earthing system which becomes sort of big equivalent antenna absorbing energy from larger area. The lower is the resistance of the elements of this antenna and the bigger are its dimensions, the higher is its absorbing capacity and the bigger part of the energy gets directly to the mechanisms connected to this "antenna." And we are talking about impulse voltage with the amplitude of hundreds of kilowatts!

Note that considerable differences between lightning and nuclear explosion of electromagnetic impulse are often mentioned in a range of other sources. For example, in [20] it is explicitly stated that the term "grounding" is usually interpreted as part of a circuit having a relatively low impedance respective to the local surface of the earth. To protect against NE EMI, a conventional grounding system that meets this definition can obviously be much worse than even the absence of grounding." In [21], it is noted that "the technical means used to protect electronic equipment from lightning pulses with flat front are not effective enough for protecting against NE EMI pulses with a steep front." In [22] it is stated that "ideally, grounding should maintain the uniform potential of all system components. In practice, however, due to possible inductance loops, capacitive coupling, resistance of grounding connection elements, interference on the ground, as on the antenna, and other phenomena, high potentials may appear in the grounding system. Therefore, the correct choice of the concept of designing a grounding system is very important in solving the complex problem of protection from NE EMI."

Unfortunately, statement of the fact did not go further in this document either, and no special grounding systems capable of resolving the contradiction between the need for grounding and the danger posed by the grounding system when exposed to NE EMI, are proposed in open sources.

Even more categorical is the statement of authoritative professionals [23] who explicitly indicate the negative impact of grounding on the operation of electronic equipment: "If you do not take special measures, both types of grounding - both internal and external - can become the main sources of electrical 'interference' caused by the impact of EMI pulses. External grounding in general cannot serve as a 'tap' for EMI, since the latter is widely distributed in space and penetrates deep into the ground. At the same time, the absence of a closed circuit for external grounding does not exclude the possibility of interference from EMI." A conclusion similar in its essence is also given in [19]: "NE EMI may induce very large currents and voltages into the 'antenna' formed by the grounding system. These currents and voltages of the grounding system present a very serious risk of damage to electronic components ..."

The inconsistency of the situation in this matter is also reflected in the fundamental work [24] (more than a thousand pages!). The title of section 4.1.1 speaks for itself: "Grounding may not be the solution; rather, it may be part of the problem." And on page 935 of this document there are two completely contradictory statements:

> The purpose of grounding is to redirect the currents induced by NE EMI to the ground.

"The primary effect of NE EMI is to induce high voltages or currents in all extended systems and conductors such as power lines, cables, antennae, and grounding systems." According to the first statement, the grounding system is the receiving capacity absorbing the energy of NE EMI, and according to the second the grounding system is a source of high voltage, supplying NE EMI energy to the earthed equipment.

10.5.3 Topical Issues of the Use of Protective Filters for Grounding of Electronic Devices

Another very difficult problem for developers of high-speed electronic devices related to grounding is filters designed to protect sensitive inputs and power supply circuits of electronic equipment (Fig. 10.72) [10]. As this figure shows, in-phase asymmetrical filters which developers sometimes call single line are simpler than symmetrical differential (sometimes called two line) and contain fewer elements, smaller in size, and cheaper. For this reason, most manufacturers prefer to produce in-phase asymmetric filters for protection against NE EMI. Other types of filters are

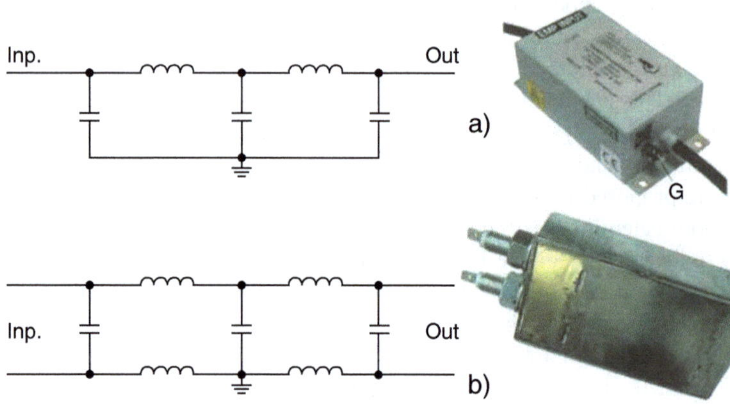

Fig. 10.72 Simplified electrical circuits and the appearance of LC filters designed for protection against NE EMI: (**a**) asymmetrical common-mode filters with common wire; (**b**) symmetrical differential mode filters

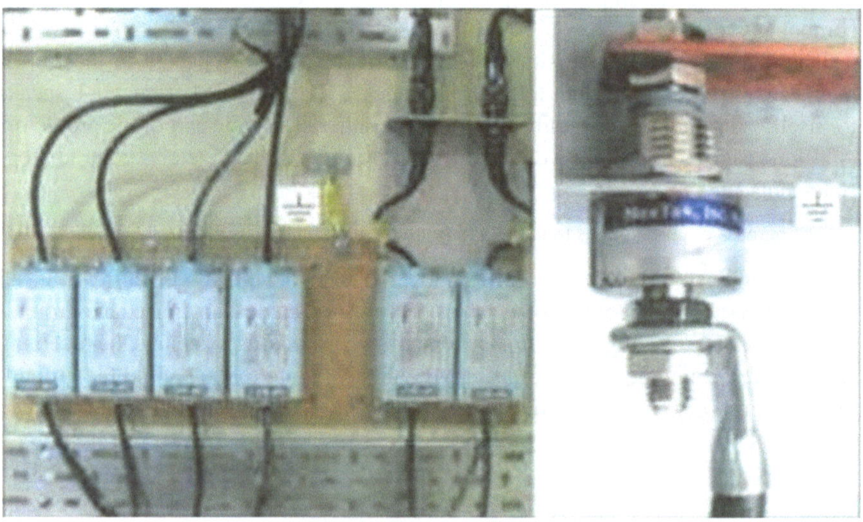

Fig. 10.73 NE EMI filters with a housing acting as a common grounding electrode

not even considered in the document [25] devoted to the analysis of different types of filters designed to protect equipment from NE EMI.

As follows from Fig. 10.72 [10], filters of this type attenuate the signal arriving at their input by introducing losses by series-connected inductances and bypassing the pulse signal to the ground by capacitors. The "ground" connected via the terminal located on the housing is a prerequisite for the filter to function. Since this terminal is usually not isolated from the housing, often the housing itself is a common grounding electrode (Fig. 10.73) [10].

Thus, these filters are designed to divert excess energy that falls on their inputs under extreme conditions into the area of zero potential, which is automatically understood as a classic grounding system, but which, as we have shown, it is not at all, One can only guess how such a filter will behave in case of the simultaneous high potential both at its input and at the common grounding electrode (instead of zero potential). In any case, it will apparently not work correctly in this mode. According to [23], such filters do not always have acceptable characteristics and are not recommended for protection against NE EMI. Perhaps it is for this reason that some filter manufacturers simply do not indicate their type in their technical specifications, and some, even if they do, create such confusion in the accompanying technical documentation that it is very difficult for a non-specialist in the field of filters to sort it out. For example, the specifications of a number of well-known companies (for example, MPE) provide descriptions of "two-phase" for alternating current and "two-line" for direct current filters intended for use in networks isolated from the "ground," but at the same time only the main technical characteristics of asymmetric filters are provided.

10.5.4 Floating-Type Grounding

If such grounding creates so many problems, maybe it is better not to ground the electronic equipment at all? After all, it is not by chance that in the report of the international authoritative specialized working group C4.206 SIGRE which specializes in the protection of electronic equipment of power systems from intentional electromagnetic impacts [26], the grounding of electronic equipment is not considered at all?!

This variant of grounding (or rather, "non-grounding") is also considered in some technical literature open for public where it is called "floating ground"—for example, in basic American military standard for grounding systems MIL-HDBK-419A [27], from where it moved to other military standards and instructions, for example to [22]. Documents [22, 27] state the following about this grounding option: "The effectiveness of floating ground depends on the degree of its real isolation from the electrically conductive parts of the equipment located nearby — that is, to be effective, the floating ground must really 'float in the air'. At large sites it is often difficult to create such completely isolated system, but even if such a system has been created, it is very difficult to maintain it. In addition, the floating ground has other limitations. For example, static charge which often accumulates in isolated electronic circuits can pose a serious threat during discharge, especially if such circuits are located near high-voltage power lines. Therefore, most types of modern electronic equipment powered by external power sources have zero reference potential in the form of grounding.

Another problem of 'floating ground' is the danger of high potential transfer to electronic circuits in case of damage in power circuits, as well as under the influence of lightning discharge. Without being connected together, parts of such a design of

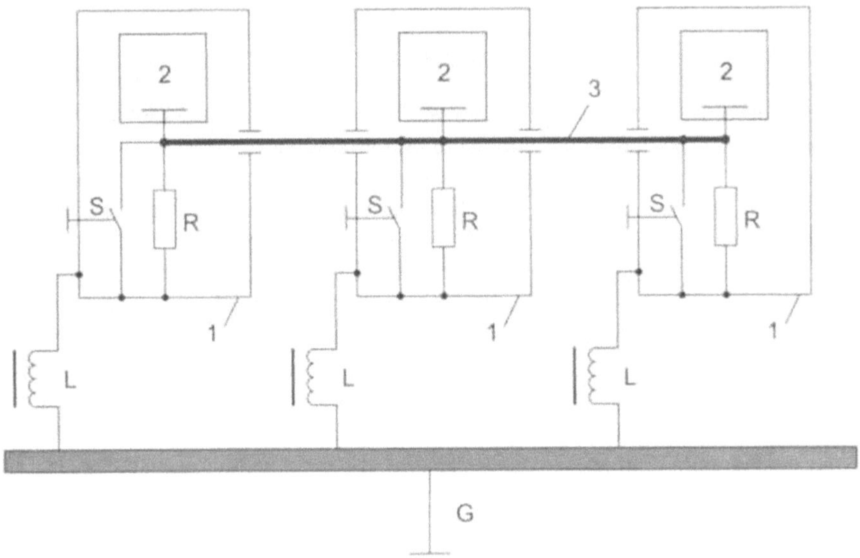

Fig. 10.74 Proposed grounding system is "special floating ground": 1—metal cabinets with electronic equipment; 2—electronic devices; 3—copper bus forming an insulated local area of total potential (local "ground"); R—high-resistance high-voltage resistors (1 GOhm); S—a switch with an electric blocking of a cabinet door; L—throttle; G—power object grounding system

the ground circuit can acquire a very high potential difference, followed by subsequent breakdown and even an electric arc."

In contrast to these statements, the specialists of Associated Power Technologies (APT), one of the leading manufacturers of powerful power supplies, believe [28] that: "Grounding systems are the source of a whole range of problems... The dark side of grounding systems is that they can disrupt functioning of an electronic system... Small ground loops inject noise into electronic systems and cause interruptions or temporary interruptions in data communication channels, such as RS-232 or GPIB interfaces. Large ground loops can damage electronic equipment and can even pose a threat to personnel in regard of safety when large currents flow through them." On the basis of the analysis carried out in [28], these specialists come to the conclusion "floating earth" is preferable for electronic systems, emphasizing that this approach is the standard for products manufactured by this company (Fig. 10.74).

So what is the way out of this highly controversial situation?

A partial solution to this problem can be the use of a special (improved) "floating land" (Fig. 10.75) [10], that is, a local area of total potential (so-called local ground) formed, for example, by a special copper bus (3) located on special insulators in a metal cabinet (1) with installed electronic equipment (2) (Fig. 10.74) [10]. Insulated (with a minimum capacity from the cabinet) with the help of insulators (4), separate housings of each electronic device (2) must be connected to this copper bus (3). All such copper buses of different cabinets located at a small distance from each other in the same room should be connected together by one powerful external copper bus located coaxially in a shielded casing (5).

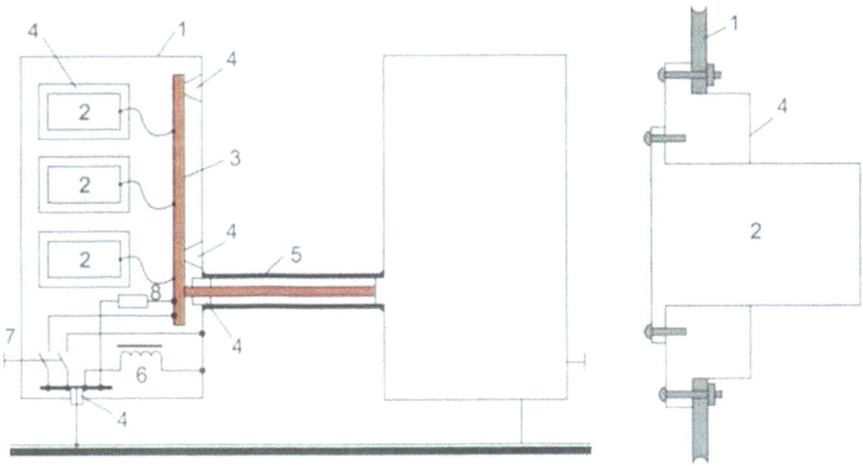

Fig. 10.75 Example of implementation of the "special floating earth" grounding system: 1—metal cabinet; 2—an electronic device, for example, an MRPD; 3—copper bus forming a local area of zero potential; 4—insulators; 5—screen casing; 6—throttle; 7—switch; 8—high-voltage high-MOhm resistor

Fig. 10.76 Appearance of standard high-voltage high-MOhm resistors

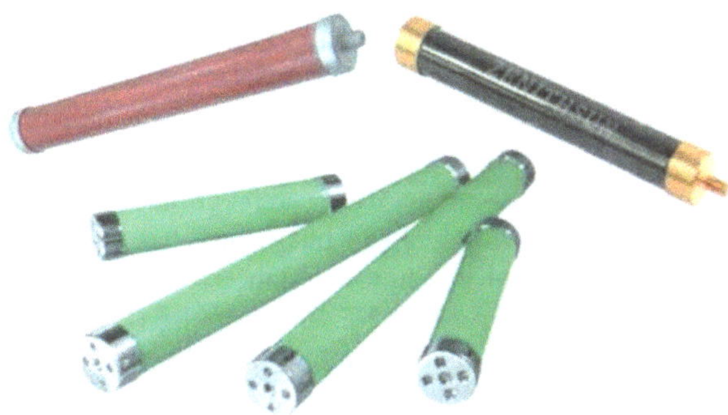

The local area of zero potential must be connected to an external grounding system only through a high-voltage high-MOhm resistor (Fig. 10.76) with a resistance of at least 1 MOhm which prevents the accumulation of static charge on the electronic devices.

The transfer of information between electronic devices located at a distance from each other in different rooms should be carried out only by optical fiber, and the transfer of discrete commands and on-off signals between such remote objects should be carried out by special insulated interfaces (high-speed reed switches with high-voltage insulation of the input from the exit), Fig. 10.77 [28, 29].

Fig. 10.77 High-voltage insulation interfaces for operating voltages between input and output 10, 20, 50, and 70 kV [28]

10.5.5 Optimal Solution for the Problem of Organization of Protective Grounding of Electronic Devices

As shown above, most modern electronic devices do not require functional grounding, so the above solution should not significantly affect the performance of electronic equipment. The problem of protective grounding, which in this concept ensures safety of personnel in case of internal damages to the equipment or cables in a cabinet, accompanied by damages to insulation and the escape of dangerous potential, can be solved by several measures.

First, the metal cabinet itself can be grounded through a powerful high-frequency throttle wound with a copper bus, which has very low resistance for direct current and 50-Hz current, but high resistance for a lightning impulse and especially high resistance for a very short NE EMI impulse (Fig. 10.78) [10].

Secondly, the cabinet must be equipped with a three-phase switch with powerful contacts and a handle from the cabinet (Fig. 10.79) [10], which closes for the duration of work performed by personnel with the equipment inside the cabinet.

One contact here is used to connect an insulated internal copper bus to an external grounding system, the second is to directly connect the cabinet to the external grounding system, the third is to unlock the cabinet door, which allows the door to be opened only after the switch is closed. It is also possible to use a two-pole switch with an additional block-contact for these purposes (Fig. 10.80).

In addition, if necessary, the developer of a highly protected electronic device can use portable grounding temporarily installed on each insulated internal copper bus during the works inside the cabinet; however, this measure is not effective enough and even redundant. The technical solutions considered above may free the floating grounding from its inherent flaws, which opens up the possibility of its widespread use in the power industry in order to increase the resistance of electronic equipment to powerful electromagnetic interference of natural and artificial origin.

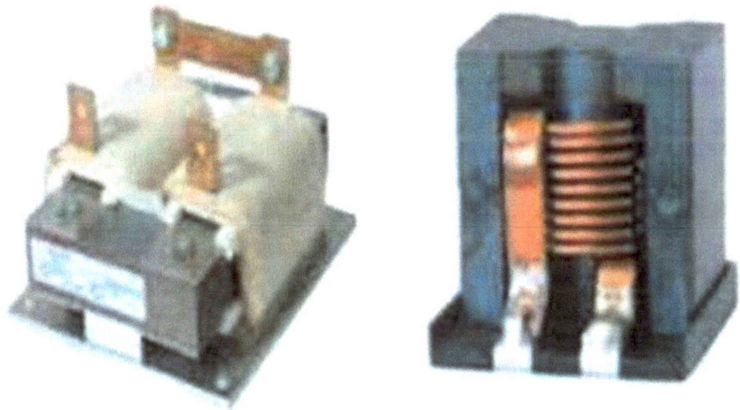

Fig. 10.78 Powerful high-frequency protective grounding circuit throttles

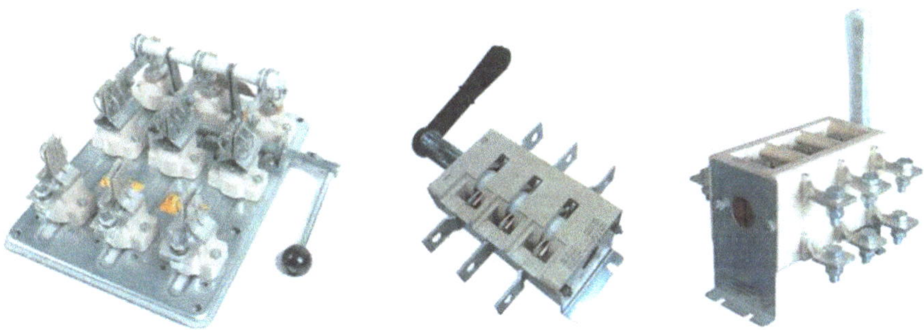

Fig. 10.79 Powerful three-pole switches with visible break of chains

Fig. 10.80 Two-pole switch with additional block-contact

References

1. Tuite, D. (2006). Use of POL-controlled voltage converters for the modernization of power supply system, *Power Electronics*, 4, 54–57.
2. Newsletter of ANALOG DEVICES company. *Electronic Components and Systems*, 6, 13–25. 2013. http://www.ekis.kiev.ua/UserFiles/Image/pdfArticles/6_2013/Bul_AD_Power_Management_part1_EKIS_6_2013.pdf
3. Newsletter of ANALOG DEVICES company. *Electronic Components and Systems*, 7, 4–16. 2013. http://www.ekis.kiev.ua/UserFiles/Image/pdfArticles/7_2013/Bul_AD_Power_Management_part2_EKIS_7_2013.pdf
4. Makarenko, V. (2011). Step-up and step-down DC/DC converters ANALOG DEVICES for portable equipment. *EKiS*, 2, 36–41.
5. Nueheusler, J. *Optimization of step-up/step-down converters in portable equipment.*
6. Makarenko, V. (2010). Designing DC/DC converters using ADIsim-power ™ software. *EKiS - Kiev: VD MAIS*, 8, 42–47.
7. http://www.analog.com/media/en/technical-documentation/data-sheets/ADP1870_1871.pdf
8. Belous, A. I., Merdanov, M. K., & Shvedov, S. V. Microwave electronics in radar systems and communications. In *Technical encyclopedia* (2 Vols., p. 1415) (in Russian).
9. Sadler, S. (2012). How to improve system parameters and reduce its cost. *Electronic Components*, 1, 18–21 (in Russian).
10. Gurevich, V. (2017). Problems of grounding electronic equipment of power facilities. *Components and Technologies*, 4, 106–111.
11. Barreto, R. M. (2013, March 19). EMC = Grounding on automation and control systems applications to eliminate electromagnetic interference in industrial plants. *Interference Technology*, 14–19.
12. Armstrong, K. (2013, July 1). Planes need to be grounded to something? *Interference Technology*.
13. Duff, W. G. (2011, November 6). Designing electronic systems for EMC: Grounding for the control of EMI. *Interference Technology*.
14. Kruse, V. J., Nickel, D. L., Bonk, J. J., & Taylor, E.R. (1991). *Impact of nominal nuclear electromagnetic pulse on electric power systems* (Report ORNL/Sub/83-43374/2, 94 p.). Oak Ridge National Laboratory.
15. IEC 60364-7-707: 1984. *Electrical installations of buildings - Requirements for special installations or locations - Earthing requirements for the installation of data processing equipment.*
16. Kuznetsov, M. B., & Matveyev, M. V. (2007). Protection against secondary lightning and ensuring equipment MP EMC at oil and gas facilities. *Energy Expert*, 2, 61–65.
17. Whitaker, J. C. (2001). *Electronic systems maintenance handbook* (2nd ed., 624 p.). Boca Raton/New York/London: CRC Press (Taylor & Francis Group).
18. IEEE Std. 1100–2005. (2005). *IEEE recommended practice for powering and grounding electronic equipment* (589 p.).
19. Pamphlet No. 1110-3-2 (1990). *Electromagnetic pulse (EMP) and tempest protection for facilities* (467 p.). Engineering and Design. Department of the Army, U.S. Army Corps of Engineers.
20. *The effects of nuclear weapons* (660 p.). U. S. Department of Defense and Energy Research & Development Administration, 1977.
21. *AD-A009 228 electromagnetic-pulse handbook for electric power systems* (341 p.). Stanford Research Institute, 1975.
22. Department of the Army. (2002). *TM 5-690 grounding and bounding in command, control, communications, computer, intelligence, surveillance and reconnaissance (C4ISR) facilities.* Washington, DC: Headquarters Department of the Army.
23. Ricketts, L. W., Bridges, J. E., & Miletta, J. (1976). *EMP radiation and protective techniques* (380 p.). New York/London/Sydney/Toronto: Wiley.

24. Joffe, E. B., & Lock, K. S. (2010). *Grounds for grounding: A circuit-to-system handbook* (1064 p.). IEEE Press, Wiley.
25. Report No. HDL-TM-81-9. (1981). *Application of filters for high-altitude electromagnetic pulse protection* (53 p.). U. S. Army Electronics Research and Development Command, Harry Diamond Laboratories.
26. Radasky, W. A., Bäckström, M., De Souza Bronzeado, H., Carpenter, D., He, J., Hoad, R., et al. (2014). *Protection of high voltage power network control electronics against intentional electromagnetic interferences (IEMI)* (Report WG C4.206 CIGRE, 27 p.)
27. *MIL-HDBK-419A grounding, bonding, and shielding for electronic equipment and facilities* (404 p.). U. S. Department of Defense, 1987.
28. *To float or not to float? Analysis of a floating vs. grounded output*. Associated Power Technologies Inc. www.aptsources.com/resources/pdf/Floating%20Output.pdf
29. Gurevich, V. (2003). *Protection devices and systems for high-voltage applications* (292 p.). New York/Basel: Marcel Dekker.

Chapter 11
Measurement and Analysis of SHF Devices

This chapter is devoted to the problems of measurement and analysis of modern microwave devices. As you know, even for complex digital microchips, the complexity and cost of solving reliable testing problems are already comparable with, and sometimes even exceed, the cost of their design and manufacture. For microwave devices, due to the specific aspects of the measured parameters and the methods used in this process, this task is much more complicated.

Appropriate hardware and software are required to study the characteristics and parameters of microwave devices in the process of their creation and verification of compliance of such devices with the specification requirements for the manufacturing release.

Therefore, the whole complex set of problems is considered in this case—from theoretical aspects to specific methods of using software and hardware, features of their practical application, necessary for a reliable assessment of the characteristics of microwave devices.

To that effect, the main features of measurement and analysis of circuits of the centimeter, millimeter, and submillimeter wavelength range are considered in the Sect. 11.1.

In Sect. 11.2, the tasks, methods, and tools for instrumental analysis of microwave circuits operating in the millimeter wavelength range have been explored in more detail.

Sections 11.3, 11.4, and 11.5 present the results of a comparative analysis of all known methods and means of measuring parameters of microwave circuits, development trends and areas of application of vector analyzers, methods and means of their methodological support (Sect. 11.6), features of non-linear microwave circuits measuring (Sect. 11.7).

Section 11.8 should be highlighted as it is dedicated to the hardware for measuring the parameters of microcircuits directly on semiconductor wafers. The specific character of this operation is that measurements on a semiconductor wafer in the millimeter wavelength range are fundamentally different from similar procedures performed for standard digital microcircuits. In particular, so-called coplanar testers

A. Belous, V. Saladukha, *High-Speed Digital System Design*,
https://doi.org/10.1007/978-3-030-25409-4_11

are required for this purpose (special probe devices). Only with their help the reliable measurements can be made on the wafers of such radio frequency microcircuits as amplifiers, switches, mixers, attenuators, phase shifters, etc.

In Sect. 11.9, the classification and principles of design of the main types of vector analyzers for circuits of a millimeter wave band, and in Sections 11.10, 11.11, and 11.12 the features of the most widely used circuits are reviewed in detail.

11.1 Features of Measurement and Analysis of Circuits in the Millimeter Wave Band

A special place in the field of radioelectronics is occupied by the creation and use of radioelectronic microwave systems and facilities, including centimeter, millimeter, and in recent years, submillimeter wavelengths. When creating such systems a variety of microwave devices and their connections are used. The study of the characteristics and parameters of microwave devices during their creation and verification of the compliance of such devices with the specification requirements during production, as well as many other tasks and studies require appropriate tools for instrumental analysis of microwave devices and their connections [1–23].

Both abroad and in Russia, numerous groups of specialists are working on solutions of these problems. This chapter is based on materials kindly provided to the expert authors of the Belarusian scientific school—A.V. Gusinsky, G.A. Sharov, and A.M. Kostrinsky.

The variety of types of transmission lines and devices used in the microwave ranges results in a variety of parameters and characteristics that describe their properties and require experimental definition. This, in turn, makes it necessary to solve a variety of measuring problems that are solved using appropriate measuring tools. The range of existing and newly created measuring instruments is very large, since it must ensure measurements for all systems of interest for developers, as well as parameters and characteristics of microwave devices of interest for other consumers in different frequency ranges and for different types of transmission lines used.

Modern methods of analysis and calculation of microwave devices and their connections are largely based on their "chain" representation, when a microwave device seems to be an equivalent multi-terminal described by a certain system of parameters [3, 9, 10, 14]. Connecting microwave devices to the desired microwave circuit is, respectively, considered as a connection of such equivalent multi-terminals. When studying such devices and connections it is important to know the extent of efficiency and amount of loss during transmission of microwave signals carrying useful information through them (how coordinated are the devices, what are the losses, etc.).

Therefore, along with other measurement tasks (measurement of power, frequency, spectrum, etc.), the task of measuring the parameters of microwave circuits in various frequency ranges was initially and continues to be one of the main and most important measurement tasks.

Among the most effective measuring tools designed to analyze the parameters of microwave devices (microwave circuits) are scalar and vector analyzers of microwave circuits which are modern high-performance information-measuring systems that allow to perform necessary measurements of the parameters of devices with guaranteed precision in wide frequency ranges with appropriate processing, presentation, and storage of the measured information about the parameters and characteristics of the tested devices [11–13, 17, 22]. Progress in the field of radioelectronics is largely determined by the development of new higher frequency ranges. In this regard, researchers and developers of radioelectronic equipment are increasingly interested in the range of submillimeter waves, which are intermediate between millimeter and optical waves.

Of particular note is the significant progress achieved recently in the industrial development of the millimeter wave range, including its shortwave part (with a wavelength of 3 and 2 mm), which has stimulated the accelerated creation of various measuring tools in this wavelength range, including those for circuit analysis.

Features of measurement and analysis of the circuit parameters in the millimeter wavelength range and, as a consequence, the tasks of developing and using appropriate scalar and vector analyzers of circuits (VAC) require consideration of factors specific to this wavelength range and having a significant impact on the technical and metrological characteristics of such VACs.

Firstly, such factors include large energy losses in transmission lines. As the results of calculations show, for the millimeter wavelength range power losses in single-mode metal waveguides reach 2 dB/m at a wavelength of $\lambda = 4$ mm, and at $\lambda = 1$ mm they rise up to 12 dB/m. Obviously, to compensate for these losses, a source of test signals with high output power level is required. Industrially produced swept-signal sources (SSSs) usually have an average output power level of 5–10 mW. With an average attenuation value in the microwave path (15–25) dB and the required attenuation measurement limit of 60–80 dB, this level of the SSS output power is not enough. To ensure the required measurement limit, it is necessary to reduce the attenuation introduced by the microwave path of the circuit analyzer which can be done by better matching the components of the path and/or its substantial structural simplification. Structural simplification of the microwave path in the millimeter range is also important because of the high cost of the nodes and elements included in it. All this imposes certain conditions and restrictions on the choice of an appropriate measurement method which should ensure the structural simplicity of VAC with the possibility of obtaining high metrological characteristics.

Secondly, with decreasing wavelength, phase errors in measurements are significantly increased, due to the non-repeatability of the set frequency value during measurement and calibration, as well as from measurement to measurement (both when sweeping the frequency and when measuring at the "point") and because of frequency instability. In general, it is more convenient to call them "error due to the undetermined frequency value." This error can have two components: first, related to the uncertain electrical length of the measurement object, and second, depending on the frequency characteristic of the measured parameter.

Quantitative estimate of the absolute error during phase shift measurement shows that with a relative error in setting the frequency $\delta f = \pm 0.002$ characteristic for most SSSs, with a waveguide length of 10 mm, at 100 GHz this error reaches $\Delta up = \pm 3.0°$. As the waveguide length increases, the error increases linearly.

The third problem arising in the millimeter wavelength range is related to the accuracy of manufacturing the channels of the waveguide elements of the measuring path and calibration waveguide standards (calibration measures) during their production. The studies conducted show that actual dimensions of waveguide channels, as a rule, exceed nominal values, and this "oversize" tends to increase with increasing frequency, especially noticeable at frequencies above 90 GHz. This can lead to serious errors in measurement results if the nominal calculated cutoff frequency of the waveguide is entered into the memory of the circuit analyzer as a constant and is not actually measured, as it may differ from the nominal one by several GHz.

Speaking about the accuracy of manufacturing waveguide channels in the shortwave part of the millimeter range ($\lambda \leq 2$ mm), it is necessary to note the importance of observing the shape of the waveguide aperture, in particular, in flanges, since there are deformations, corner rounding, other contour distortions, etc. which adversely affect measurement results.

Fourthly, a very serious problem, the solution of which largely depends on the accuracy of the measurement results on millimeter waves, is the problem of waveguide flange connections and the problem of tolerances on main mounting dimensions of flanges. These tolerances should be about $\pm$ 10 μm at a wavelength of $\lambda = 4$ mm and units of μm at $\lambda = 1$ mm. It can be shown that for a waveguide section 2.4 × 1.2 mm with a tolerance along the waveguide axis about ± 2–10^{-2} mm phase measurement error is about $\pm$ 10°. For the same error, the disconnection of flanges should be no more than ± 5–10^{-2} mm. The gap in flanges, in addition, leads to lack of heterogeneity and changes the conditions of approval. Therefore, to reduce this error, it is necessary to minimize the number of flanged connections of the measuring path, which is also associated with a constructive simplification of the microwave path, and, in addition, if possible, use of precision flanges with tighter tolerances on main mounting dimensions. First of all, this applies to the flanges of the waveguide elements of the measuring path and to the flanges of the calibration standards.

The problem of tolerances is closely related to the problem of ensuring the immobility and rigidity of the measuring path when connecting measurement objects and calibration standards (calibration measures). Even a small mechanical deformation of the path (about dozen microns) can lead to significant measurement errors. The solution to this problem is connected to constructive and technological tasks, but it also suggests that it is desirable to exclude any mechanical switches in the channels of the measuring microwave path which reduce the speed of the meter and cannot be taken into account by the calibration algorithm.

Other problems include the difficulties of metrological certification of a VAC of the millimeter range associated with the need to create and certify relevant model measures and measuring tools. Much more stringent requirements to the geometry and tolerances than to usual elements of the microwave path should be placed on the accuracy of manufacturing channels of waveguide model measures. Thus, the "oversize" waveguide component used as a calibration measure can cause a noticeable change in the dispersion taken into account by circuit analyzer, and result in an incorrect waveguide loss model. The shift between the connecting planes of the path and the calibration measure due to tolerances on the connecting dimensions of flanges during the calibration process can lead to parasitic pulsations in the measurement data that can significantly affect, for example, the level of reflection loss measurements, etc.

Summarizing the above, we can conclude that to successfully solve the problem of constructing a VAC in the millimeter wavelength range, it is necessary to apply such measurement methods that solve basic measurement tasks making it possible to simplify the microwave path as much as possible, minimize the number of flange connections, use precision components of the measuring path and calibration tools, use precision flanges, and also eliminate, if possible, the use of microwave switches.

11.2 Instrumental Analysis of SHF Circuits in the Millimeter Wave Range (Tasks, Methods, Tools)

11.2.1 Measured Circuit Parameters of SHF Devices

Mathematical tool for theoretical analysis of microwave circuits and their compounds are matrix and topological methods, and various systems of parameters (*Z*, *Y*, *A*, *G*, *H*, *T*, *R*, *S*) with clear connection between them [2, 14, 17, 19] can be used for the matrix description of a microwave device. The existence of such a connection makes the choice of one or another system of parameters for describing a microwave device unimportant. However, not all of these parameters have found wide application in practice. This is due to a number of circumstances and, first of all, to the possibility of experimentally determining and physically interpreting the parameters used. In the theoretical analysis and calculations of various microwave devices and compounds, the most preferable was the use of a system of *S*-parameters, which are elements of the scattering matrix describing the device or connection being analyzed. In some cases, for example, when analyzing cascade connections of microwave devices, it is convenient to use *T* (or *R*) parameters which are easily expressed in terms of *S*-parameters. Practice has shown that during experimental studies and measurements of the parameters of microwave devices it is most preferable to use a system of *S*-parameters and additional derivatives from them. When analyzing microwave devices using topological methods (the method of oriented graphs), *S*-parameters are also mainly used to describe the graphs.

Recall that S-parameters connect the a ith waves of microwave devices incident on i-inputs and the b ith waves reflected (scattered) from them, where $i = 1, 2, \ldots,$ n (n is the number of pole pairs of the equivalent microwave device of $2n$-terminal).

The connection of the incident and reflected (converging and diverging) waves in the general case can be expressed as

$$b_i(\omega) = \sum_{k=1}^{n} S_{ik}(\omega)a_k(\omega), \tag{11.1}$$

where $k = 1, 2, \ldots, n$,or in matrix form

$$[b] = [S][a]. \tag{11.2}$$

In the general case, S-parameters are complex quantities depending on frequency.

Parameters S_{ii} determine the reflection coefficients at the corresponding i inputs, and parameters S_{ij} $(i \neq j)$ define the coefficients of transmission between the i-th and j-th arms of the equivalent multi-terminal.

So, the main "chain" parameters of microwave devices to be measured, in general, are:

- reflection coefficient modules $|S_{ii}|$;
- phases of reflection coefficients arg S_{ii};
- modules of coefficients of transfer between the arms (inputs and outputs) of the device $|S_{ik}|$;
- phases of transmission coefficients arg S_{ik};

as well as parameters derived from them, such as

- standing wave voltage coefficient of i-th arm

$$k_i = \frac{1 + |S_{ii}|}{1 - |S_{ii}|}; \tag{11.3}$$

- insertion loss between arms i-k:

$$A_{ik} = 20lg|S_{ki}|; \tag{11.4}$$

- total input resistance Z_{ii} and conductivity Y_{ii} from the i-th arm

$$Z_{ii} = \frac{1}{Y_{ii}} = Z_{0i}\frac{1 + S_{ii}}{1 - S_{ii}}; \tag{11.5}$$

- delay time, or group delay time (GDT) when the signal passes between the arms i–k:

$$\tau_{ik} = \frac{d}{d\omega}(argS_{ki}); \tag{11.6}$$

– nonreciprocal phase shift between arms *i-k*:

$$\alpha_{ik} = argS_{ki} - argS_{ik} \tag{11.7}$$

and some other parameters (non-uniformity of GDT, etc.).

It should be noted that in the general case, the scattering parameters S_{ik} are functions not only of frequency ω but also of time t counted from the moment the incident waves α_i (ω) arrive at the input poles of the multi-terminal, and the power levels of the incident waves

$$P_i = \frac{|\alpha_i|^2}{Z_{0i}}. \tag{11.8}$$

Therefore, a rigorous formulation of the problem of determining the scattering parameters $S_{ik}(\omega)$ in (11.1) requires the following conditions:

$$\begin{aligned} &|\alpha_i(\omega, t)| = \text{const}, \\ &t = \tau_{ik}(\omega) \end{aligned} \tag{11.9}$$

11.2.2 Measurement Goals of the Analysis of SHF Circuits

The diversity of microwave products and devices used in modern electronic systems, tools, and equipment designed to work in different conditions on different types of transmission lines and in different frequency ranges, covering wavelengths from decimeter to fractions of a millimeter, predetermines the variety of measurement tasks it is necessary to solve in the process of their creation, research, and in determining the compliance of their parameters and characteristics with the requirements of specifications in the process of industrial production and operation.

The main measurement tasks for analyzing microwave devices that can be solved using various instrumental analysis tools include the following.

Analysis of Linear Passive Devices.
The extensive class of devices most widely used in microwave technology, first of all, include passive linear microwave devices, i.e. devices that do not contain amplifiers or microwave power sources, the external characteristics of which do not depend on the level of the incident (passing) microwave power (this level, of course, must remain within certain limits, for example, not exceed the electrical strength limits of the incoming elements). Therefore, among the most common measurement

tasks associated with the analysis of microwave circuits are the traditional *tasks of measuring S-parameters (and parameters derived from them) of passive linear microwave devices in the corresponding operating frequency ranges.*

Analysis of Non-Linear Devices. The next very significant group of microwave devices consists of various non-linear microwave devices.

When measuring the characteristics of non-linear microwave devices, a certain power level of incident waves $\alpha_k(\omega)$ determining the mode of operation of the device under investigation must be provided, which complicates the measurement task.

Taking into account the specifics of real non-linear microwave devices and the capabilities of measuring tools, the following *main measuring tasks* are distinguished *when analyzing non-linear circuits*:

– measurement of frequency characteristics of *S*- parameters $S_{ik}(\omega)$ *of* low-power non-linear devices in linear mode, i.e. at power levels at the input of the device

$$P_{\text{вх}} \leq \left(10^{-5} - 10^{-7}\right) \text{Вт}; \tag{11.10}$$

– measurement of $S_{ik}(\omega)$ *of* semiconductor devices at medium power levels at their inputs

$$P_{\text{вх}} \approx \left(10^{-3} - 10^{-1}\right) \text{Вт}; \tag{11.11}$$

– measurement of $S_{ik}(\omega)$ *of* powerful microwave devices at

$$P_{\text{вх}} \geq 1 \text{ Вт}; \tag{11.12}$$

– measurement of amplitude characteristics (determining the level of compression)

$$S_{ik}(\omega, \alpha_k) = S(\omega_0, P_{\text{вх}}). \tag{11.13}$$

Analysis of Frequency-Selective and Ultra-Wideband Devices.
Specific requirements may arise when measuring and analyzing frequency-selective and ultra-wideband microwave devices.

Currently used and newly created microwave devices and systems often include multiple elements and units distinguished either by frequency ultra-narrowband or by frequency ultra-wideband, due to which it is necessary to take into account the specifics of measurement tasks when studying their characteristics and relevant capabilities of the instruments used.

Usually, ultra-narrowband devices include those with

$$\frac{\Delta f}{f_H} \leq 10^{-3} \tag{11.14}$$

ultra-wideband devices include those with

$$\frac{\Delta f}{f_H} \geq k, \tag{11.15}$$

where Δf is the operating frequency band of the device; f_n—initial frequency of the operating band of the device; k—coefficient of the overlap of the operating frequency band, which takes on values depending on the measurement frequency area (e.g., in the centimeter range for $f_n \geq 1$ GHz typically $k \approx 10$, and in the shortwave part of the millimeter wave band for $fn \geq 60$ GHz this value will be $k \approx 1.0 \div 2.5$.

Analysis in Time Domain

Analysis of microwave devices in the time domain makes it possible to obtain additional information about their internal structure. For transition to the time domain, the time-frequency transform (inverse Fourier transform) of the measured frequency response of the device is used. The tasks of analysis in the time domain also include:

- determination of the time response of a non-linear device to a pulse signal of a given shape and amplitude, i.e. the problem of determining $b_{ik}(t)$ with

$$\alpha_k(\omega, t) = \sum_{m=0}^{\infty} \alpha_m \cos\left(\frac{2\pi}{T} m + \varphi_m\right), \tag{11.16}$$

- definition of components $S_{ik}(\omega, t)$:

$$S_{ik}(\omega, t) = \sum_{l=0}^{\infty} S_l(\omega) e^{-j\omega\tau l}, \tag{11.17}$$

where $S_l(\omega)$ are complex amplitudes and τ_l are delay times of the individual spectral components $S_{ik}(\omega)$.

Comparative Analysis of the Parameters and Characteristics of Microwave Devices

When creating some systems and products, for example, phased systems, it is necessary to be able to compare a range of devices with each other (or with a standard). In this regard, there is a problem of measuring non-identity of two devices, which often involves the measurement of non-identities of frequency dependencies of attenuations $\Delta A(\omega)$ and phases $\Delta \varphi(\omega)$ defined as

$$\Delta A(\omega) = 20lg\left|\frac{S_1(\omega)}{S_2(\omega)}\right|, \tag{11.18}$$

$$\Delta\phi = argS_1(\omega) - argS_2(\omega), \tag{11.19}$$

where $S_1(\omega)$ and $S_2(\omega)$ are complex transfer coefficients of the compared devices.

Measurement of the Characteristics of Phased Antenna Arrays, Microwave Microassemblies, Integrated Circuits, etc.

The measurement tasks listed above do not exhaust the whole variety of microwave device analysis. In recent years, the field of millimeter waves covers the development of such specific systems as *phased antenna arrays (PAA)* are developed, their characteristics measurements having certain specific features. As it is known, FAAs can contain from several dozens to several thousand receiving channels, each of which consists of an antenna, an input waveguide, a phase shifter, a low-noise preamplifier, automatically controlled selectors, frequency converters, SAW filters, etc. In the process of developing, manufacturing, and operating the FAAs, it is necessary to solve many of the above measuring tasks, including measurements of:

- *S*-parameters of passive (filters, phase shifters, phase splitters), active low-power (input amplifiers) and ultra-narrowband (SAW filters) devices;
- *S*-parameters of waveguide paths and frequency converters;
- $S_{ik}(\omega)$ | and $\tau_{ik}(\omega)$ of through paths from antenna inputs to the outputs of secondary transducers;
- non-identity of amplitude and phase-frequency characteristics of individual elements and channels in general.

The total number of measurements for one FAA can reach 10^{4}–10^{5} and more. The emergence of the most complex radio-engineering systems such as FAAs required the creation of multifunctional, high-performance measuring instruments for the parameters of microwave circuits that ensure the precision and necessary accuracy, particularly in the millimeter wavelength ranges.

In recent years, advances in semiconductor technology and microstrip transmission lines have led to the creation of microwave semiconductor microassemblies and integrated circuits operating at ever higher frequencies. For the analysis and research of their characteristics, *S*-parameter meters with appropriate positioning systems for microassemblies and adapters (adapter probes) are widely used to connect the meters to microassembly tips. To ensure their reliability, such measurements require not only the development and application of special technological equipment and devices, but also the development of appropriate error models, calibration methods, and tools, and the availability of appropriate mathematical methods and computational capabilities for correct processing of measurement results.

11.2.3 Certain General Requirements for Devices for Measurement of the Parameters of Millimeter Wave Band Circuits

Along with the solution of measuring tasks in the necessary dynamic and frequency ranges of measurement, a number of other general requirements are imposed on microwave circuit analyzers, including VACs.

The most important of them in the vector analysis of microwave circuits in the millimeter wavelength range is to ensure the required accuracy of measurement of the parameters of the tested devices. The practice of microwave measurements, confirming the constant growth of requirements for the parameters of microwave devices, makes it very important to *increase measurement accuracy*, which is solved both by improving the parameters of the elements of the measuring circuit, setting more rigid tolerances on the geometric dimensions of the waveguide devices of the measuring circuit, and by developing improved error models, improved calibration methods and tools, applying digital methods for processing, storing, and presenting measurement information.

Since analyzers of microwave circuits are measuring instruments (measuring instruments or information-measuring systems), they are subject to metrological requirements of the main types of measurements with appropriate model tools, standards, and certification methods, as well as verification means and methods of verification during operation.

An important requirement is the multifunctionality of analyzers, the possibility of solving many of the measurement tasks listed above. If it is necessary to conduct multiple, single-type measurements with high performance (as is the case with FAAs or microwave microassemblies), it is advisable to create specialized AISs based on multifunctional meters. Improving measurement performance is directly related to meter automation.

Huge achievements in the field of creating high-performance, high-memory, and low-cost computing equipment, primarily personal computers (PCs), as well as the development of various packages of relevant software that allow the processing of measurement information and its presentation in almost real time, make it possible to *use* (embed) these *computing facilities (PCs)* directly into the measuring complexes, significantly increasing the efficiency of gauges operation and improving the convenience of working with them. The use of new PCs in new microwave circuit analyzers is, in essence, also a common requirement.

A wide range of existing and potential consumers of gauges of millimeter waves chains parameters, the ever-increasing variety of measurement objects and the growth of measurement volumes also make urgent the task of creating sufficiently *massive* and *inexpensive* vector analyzers of circuit parameters that perform basic types of measurement tasks with sufficiently high accuracy and performance. This class of gauges includes the so-called homodyne vector analyzers of circuits, which are the main focus of this paper.

Before the advent of industrial development of such meters for the purpose of measuring single-mode waveguides of various "chain" parameters (impedances and VSWR, insertion attenuation, phase characteristics of reflection and transmission, etc.), a large number of methods and devices were developed, each with their benefits and applications. Let's make a brief retrospective analysis of the main methods and means of instrumental analysis of the parameters of microwave circuits, focusing primarily on the millimeter wave band.

11.3 Retrospective Analysis of Methods and Means of Measurement of SHF Circuits

11.3.1 Probe Methods

11.3.1.1 Methods of Measurement Based on Measuring Waveguide Line

The Essence of the Method With all the variety of methods existing for measuring the main parameters of microwave paths, the measuring waveguide line designed to measure at fixed frequency points the working range of the waveguide of the corresponding VSWR section, full impedances, reflection and transmission coefficients, and other characteristics of the tested devices, due to its universality, reliability, and simplicity was one of the first major devices that have already been used in the development of the millimeter wave range and its shortwave part (2–3 mm). The operating frequency range of waveguide measuring lines extends to 180–220 GHz.

The method of measuring the parameters of circuits using a measuring line is one of the *probe* methods, which is based on the fact that in a narrow slot cut in the outer conductor (in the upper wall) parallel to the axis of the waveguide and weakly disturbing wave types propagating in the waveguide, for energy extraction of a probe (Fig. 11.1), by changing its position along the slot, it is possible to obtain data characterizing the change in the field strength, i.e. remove the picture of the field distribution along the waveguide. Measuring microwave loads (of devices tested) with a measuring line in the end means comparing impedance of load Z_L with line characteristic impedance Z_0 which is a standard one. This comparison is based on the research of standing wave field distribution inside the transmission path of the measuring line.

Output voltage of probe $|b_s|$ applied to detector is proportional to the field voltage. Usually transverse component of electric field E_τ is registered.

At the same time,

$$|b_S| = K_1|E_\tau| = K_2|a+b| = K_2|a||1+\Gamma|, \tag{11.20}$$

where K_1 and K_2 are constants; α and b are amplitudes of oppositely travelling waves.

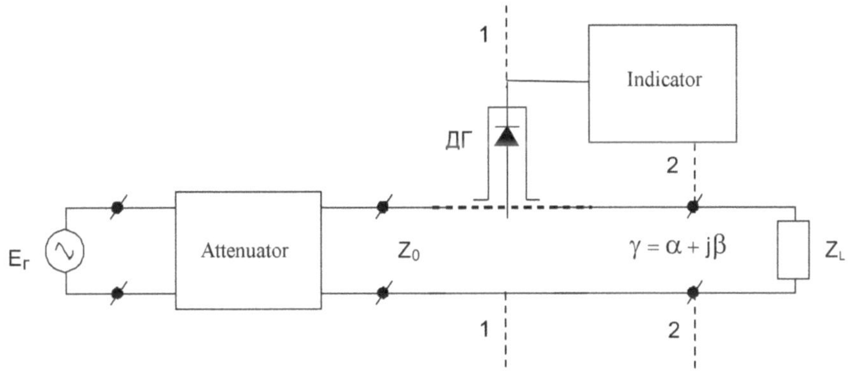

Fig. 11.1 Standard scheme of measuring line energization

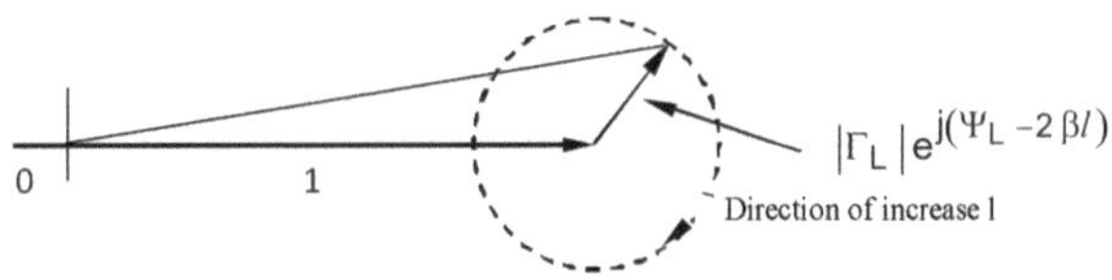

Fig. 11.2 Vector visualization of signal at probe's output

If a probe is in section 1, which changes its position, and load with deflection rate Γ_L is connected in section 2 (at distance *l* from section 1) which is fixed, at the probe's output:

$$|b_S| = K_2|\alpha_1||1 + \Gamma_1| = K_2|a_2|e^{\alpha l}\left|1 + |\Gamma_l|e^{j(\psi_L - 2\gamma l)}\right|, \tag{11.21}$$

where $\gamma = \alpha + j\beta$ is propagation constant in the line.

In *ideal* measuring line without losses.

$\left(\gamma = j\beta = j\frac{2\pi}{\lambda}, \quad \alpha = 0\right)$, we have the following

$$|b_S| = K_2|\alpha|\left|1 + |\Gamma_L|e^{j(\psi_L - 2\beta l)}\right|. \tag{11.22}$$

In expression (11.21) term$\left|1 + |\Gamma_L|e^{j(\psi_L - 2\gamma l)}\right|$ is a module of sum of a unit vector and vector $\Gamma_L\left|e^{j(\psi_L - 2\gamma l)}\right|$.

When position of the probe is changed, smaller vector in Fig. 11.2 rotates. At the same time, at some points it is in the phase or reverse phase with fixed vector of unit length.

Consequently, the ratio of signal maximum at the probe's output relative to its minimum is

$$\frac{|b_S|_{\max}}{|b_S|_{\min}} = \frac{1 + |\Gamma_L|}{1 - |\Gamma_L|} = K, \tag{11.23}$$

where *K* is VSWR.

Vector diagram shows that when the probe moves because of load, the first minimum of output probe signal only takes place when $l = l_m$, where l_m satisfies the following ratio:

$$\psi_L - 2\beta l_m = \pm\pi. \tag{11.24}$$

From there, phase angle ψ_L of the load deflection ratio is equal to

$$\psi_L = 2\beta l_m \pm \pi. \tag{11.25}$$

Full resistance Z_p in any point of the line is connected to full resistance of load Z_L with the following ratio:

$$Z_p = Z_0 \frac{Z_L + Z_0 th\gamma l}{Z_0 + th\gamma l}, \tag{11.26}$$

If there are losses in the line ($\alpha = 0$, $\gamma = j\beta$), the ratio looks as follows:

$$Z_p = Z_0 \frac{Z_L + jZ_0 tg\beta l}{Z_0 + jZ_L tg\beta l}, \tag{11.27}$$

Full resistance of the load connected to the transmission line end can be calculated from known VSWR of the line and position of voltage minimum related to the load. The expression provided above makes it possible to determine full resistance of the load Z_L via VSWR and length in the following way:

$$Z_L = Z_0 \frac{1 - jKtg\beta l}{K - jtg\beta l}, \tag{11.28}$$

It is easy to calculate active and reactive components of the load resistance:

$$R_L = Z_0 \frac{K(1 + tg\beta l)}{K^2 + jtg^2\beta l}, \tag{11.29}$$

$$X_L = Z_0 \frac{(1 - K^2) tg\beta l}{K^2 + \mathrm{tg}^2\beta l}, \tag{11.30}$$

Module of full resistance will be equal to

$$|Z_L| = Z_0 \sqrt{\frac{1 + K^2 + \mathrm{tg}^2\beta l}{K^2 + \mathrm{tg}^2\beta l}}, \tag{11.31}$$

In practice, full resistance based on VSWR and displacement of standing wave minimum I_m are more often determine with Smith plot.

Measuring line can be used to measure attenuation in transmitting lines and to measure complex ratios of two-port transmission without losses and other characteristics.

11.3.1.2 Analysis of the Measuring Line Errors

Real measuring line, unlike ideal one, is much more complicated, and its rigorous calculation is a complicated task. Measuring full resistances leads to errors. Common sources of errors during the measurement of standing wave amplitudes are proper deflections (proper VSWR, residual VSWR) in the measuring line, unbalanced connection of the probe with the field in the line and its shunting effect. During the measurement of standing wave unbalanced connection practically does not have any importance, but the error is added due to the error of calculation of probe position along the line. As lines are developed, different measures are taken to reduce component errors occurring due to the error sources indicated, but it is not possible to eliminate them completely.

Figure 11.3 shows simplified direct graph for the measuring line.

In the figure the followings symbols are used:

- $\Gamma_g, \Gamma_p, \Gamma_L$ are ratios of deflection of the generator, probe, and unknown researched load, respectively;
- Γ_S are deflections from slot and residual deflections of the measuring line;
- C is probe connection ratio;
- T_i and T_0 are ratios of slot section transmission before and after the probe;
- E_g is a source;
- D is a detector.

Based on the graph of Fig. 11.3 and using the nonintersecting loop rule, we determine the ratio of transmission from generator E_g to generator D:

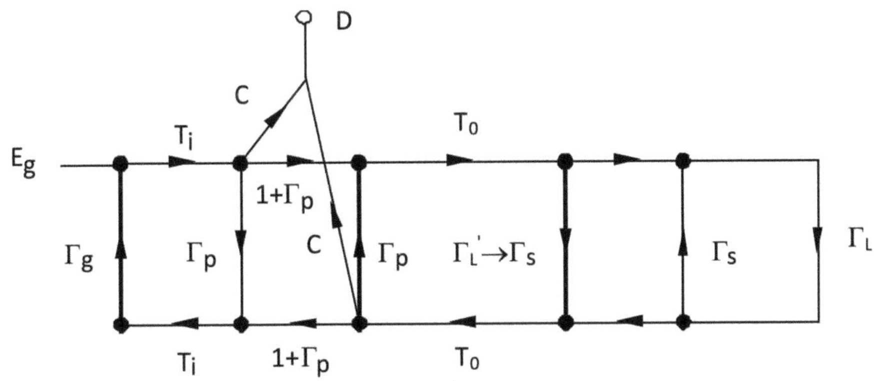

Fig. 11.3 Simplified direct graph used to analyze errors of the measuring line

$$T_{ED} = \frac{CT_i\left[1 + T_0^2 \Gamma_L{}'\right]}{1 - T_i^2 T_0^2 \Gamma_L{}' \Gamma_g (1 + 2\Gamma_p) - T_i^2 \Gamma_p \Gamma_g - T_0^2 \Gamma_L{}' \Gamma_p}, \tag{11.32}$$

where

$$\Gamma_L' = \frac{\Gamma_S + \Gamma_L}{1 - \Gamma_S \Gamma_L} \cong \Gamma_S + \Gamma_L + \Gamma_S \Gamma_L^2 \;(\text{for} \Gamma_S 1). \tag{11.33}$$

If via Q you determine expression

$$Q = \frac{1}{1 - T_i{}^2 T_0{}^2 \Gamma_L' \Gamma_g (1 + 2\Gamma_p)}, \tag{11.34}$$

then, as

$$Q\Gamma_p = 1 \tag{11.35}$$

we have the following

$$T_{ED} \cong CT_i Q \left[1 + T_0^2 \Gamma_L'\right] \left[1 + T_i^2 Q \Gamma_p \Gamma_g + T_0^2 Q \Gamma_p \Gamma_L'\right]. \tag{11.36}$$

Inserting value Γ_L' and excluding second-order terms, we have the following:

$$\begin{aligned} T_{ED} \cong CT_i Q\big[1 - T_i^2 Q \Gamma_p \Gamma_g + T_0^2 \Gamma_S + \Gamma_L T_0^2 \{ 1 + Q\Gamma_p (1 + T_i^2 \Gamma_g)\} \\ + \Gamma_L^2 T_0^2 (\Gamma_S + Q\Gamma_p T_0^2).\big]. \end{aligned} \tag{11.37}$$

Then we make the measurement relocating the probe in a way so that

$$|T_{i1}|^2 = -|T_{i2}|^2 = -|T_{i1}(1 + \Delta)|^2, \tag{11.38}$$

$$|T_{01}|^2 = -|T_{02}|^2 = -|T_{01}(1 - \Delta)|^2, \tag{11.39}$$

$$C_1 = C_2 + s \tag{11.40}$$

where Δ is losses at the line segment where the probe is located; s is the change of probe connection ratio due to the slope and unbalanced frequency feature.

Excluding second-order terms, we have the following:

$$\begin{aligned} \Gamma_m &= \frac{1}{2}(T_{ED1} - T_{ED2}) \\ &\cong 1 + (\Delta + s + Q\Gamma_p \Gamma_g + \Gamma_s) + \Gamma_L(1 - 2\Delta + Q\Gamma_p) + \Gamma_L^2(\Gamma_s) \end{aligned} \tag{11.41}$$

Thus, error of measurement Γ_L is equal to

$$\pm\Gamma_L \cong \left(\Delta + s + Q\Gamma_p\Gamma_g + \Gamma_s\right) + \Gamma_L\left(Q\Gamma_p - 2\Delta\right) + \Gamma_L^2(\Gamma_s). \tag{11.42}$$

First term in round brackets of expression (11.42) is equivalent to the term determining direction in reflectometry schemes. Second term in round brackets is the error of calibration preconditioned by deflections from quadratic characteristic of the detector, and error of indicator calibration. Third term determines component of the error due to deflections from the probe.

11.3.1.3 Measuring Line with Frequency Sweeping

Major disadvantages of measuring lines should be inability to make simultaneous measurements in frequency band—only in frequency range points, lack of automation, and respectively, high labor intensity and time input into manual measurements, as well as absence of full picture of measurement results. Labor intensive measurements by points are not only ineffective, but significantly restrict the amount of information that the user receives; resonances between measurement points may be missed, etc.

To ensure measurements in frequency band the method of VSWR measurement in centimeter wave range may be used which is based on the use of measuring line with frequency sweeping. The scheme of measurement by this method is provided in Fig. 11.4.

Signal source (swept-frequency source) is connected to the measuring line via some attachment which is in fact a short segment of the measuring line with the fixed probe. Output signal of the attachment probe is kept permanent with the scheme of automatic signal level control. This scheme effectively excludes the impact of frequency changes on the connection to the measuring probe. Each frequency of measurement corresponds to the specific point in horizontal axis of the oscillograph cathode-ray tube. When the carriage of the measuring line moves

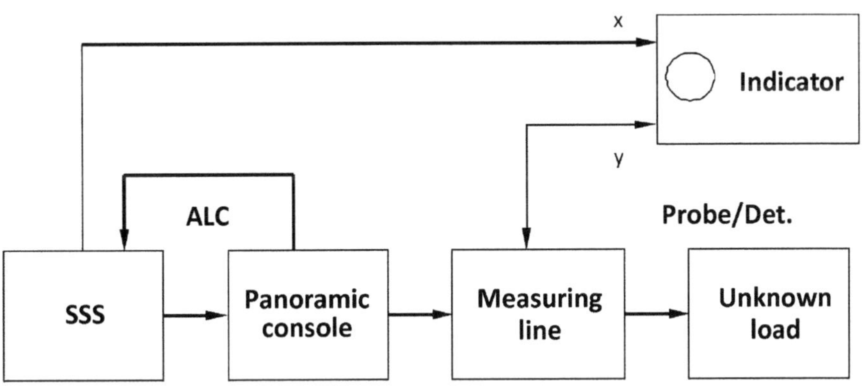

Fig. 11.4 Scheme of VSWR measurement by the method of measurement line with frequency sweeping

within half-wave at one measuring frequency, maximum and minimum values of the signal are recorded at the output of detector head which are shown as vertical line extremes on cathode-ray tube screen. Relation of these maximum and minimum values is VSWR. If high signal source frequency sweeping speed is chosen and the carriage is at the same time moved back and forth at least within half-wave, it will result in oscillograph record which can be used as a basis to record maximums and minimums at each frequency of the band and determine corresponding values of VSWR at each frequency of measurement in the required band. This measurement method has not become popular because of the complexity of the scheme, restricted information (VSWR only), and mainly due to newer ideas or more progressive technical solutions that appeared.

11.3.1.4 Multi-Probe Measuring Lines. Four Probe Method

To record standing wave field distribution in waveguide line, several fixed probes may be used instead of a moving one. The structure of measuring line was a basis for automated measuring instruments of full resistances ensuring full measurements in relatively wide frequency band (10–20%).

Multi-probe line is a system of three, four, or five fixed probes put along waveguide line at the distance approximately equal to $\lambda_{g0}/8$, where λ_{g0} is average length of wave in the waveguide of the measuring line. The method of four probes was used in centimeter wave range more often (Fig. 11.5).

As the distance between probes is fixed, electric phase shifts between them will be equal to$\frac{\pi}{4}$ only at average frequency λ_{g0}, while at other frequencies they will be equal to$\delta\frac{\pi}{4}$, which results in measurement error. Coefficient δ will be determined by formula

$$\delta = \frac{\lambda_{g0}}{\lambda_g}, \tag{11.43}$$

where λ_{go} is an average length of wave in the waveguide; λ_{go} is the current length of wave in the waveguide.

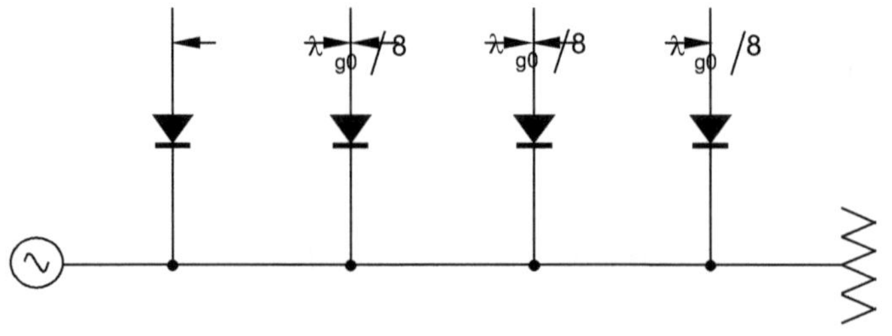

Fig. 11.5 Scheme of four-probe measuring line

In general case, when in the waveguide there are incident and reflected voltage waves in probes, they will be determined as follows:

$$\left.\begin{aligned}
E_1 &= |\mathrm{E}_\Pi| \cos\left(\omega t - \frac{3\pi}{8}\delta\right) + |E_0| \cos\left(\omega t + \frac{3\pi}{8}\delta + \varphi\right), \\
E_2 &= |E_\Pi| \cos\left(\omega t - \frac{\pi}{8}\delta\right) + |E_0| \cos\left(\omega t + \frac{\pi}{8}\delta + \varphi\right), \\
E_3 &= |\mathrm{E}_\Pi| \cos\left(\omega t + \frac{\pi}{8}\delta\right) + |E_0| \cos\left(\omega t - \frac{\pi}{8}\delta + \varphi\right), \\
E_4 &= |\mathrm{E}_\Pi| \cos\left(\omega t + \frac{3\pi}{8}\delta\right) + |E_0| \cos\left(\omega t - \frac{3\pi}{8}\delta + \varphi\right).
\end{aligned}\right\} \tag{11.44}$$

Since

$$|E_0| = |\rho|\ |E_\Pi| \tag{11.45}$$

using quadratic detectors and filtering constant component of signals recorded by detectors, the following voltages on detector loads are obtained:

$$\left.\begin{aligned}
U_{D1} &= k\frac{|\mathrm{E}_\Pi|^2}{2}\left[1 + |\rho|^2 + 2|\rho|\ \cos\left(\frac{3\pi}{4}\delta + \varphi\right)\right], \\
U_{D2} &= k\frac{|E_\Pi|^2}{2}\left[1 + |\rho|^2 + 2\ |\rho|\ \cos\left(\frac{\pi}{4}\delta + \varphi\right)\right], \\
U_{D3} &= k\frac{|\mathrm{E}_\Pi|^2}{2}\left[1 + |\rho|^2 + 2|\rho|\ \cos\left(\frac{\pi}{4}\delta - \varphi\right)\right], \\
U_{D4} &= k\frac{|\mathrm{E}_\Pi|^2}{2}\left[1 + |\rho|^2 + 2|\rho| \cos\left(\frac{3\pi}{4}\delta - \varphi\right)\right].
\end{aligned}\right\} \tag{11.46}$$

Loads from detector heads after gain and pairwise deduction are applied to cathode-ray tube plates (Fig. 11.6).

At the same time,

$$U_X = U_{D1} - U_{D3} = -k\ |E_\Pi|^2\ |\rho|\ \sin\frac{\pi\delta}{2} \sin\left(\varphi + \frac{\pi\delta}{4}\right), \tag{11.47}$$

$$U_Y = U_{D2} - U_{D4} = k\ |E_\Pi|^2\ |\rho|\ \sin\frac{\pi\delta}{2} \sin\left(\varphi - \frac{\pi\delta}{4}\right), \tag{11.48}$$

where k is proportionality coefficient.

If constant level of wave power is ensured at signal source output ($|E_\Pi|^2 = \text{const}$), position of the beam on tube screen will be determined by the coordinates

$$X = k\ |\rho|\ \sin\frac{\pi\delta}{2} \sin\left(\varphi + \frac{\pi\delta}{4}\right), \tag{11.49}$$

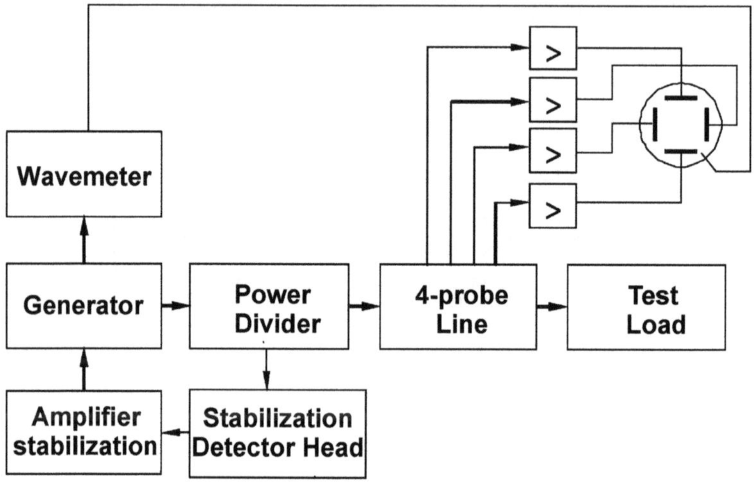

Fig. 11.6 Scheme of an automated installation using the four-probe method

$$Y = k\,|\rho|\,\sin\frac{\pi\delta}{2}\sin\left(\varphi - \frac{\pi\delta}{4}\right), \tag{11.50}$$

Deflection of the point on tube screen is proportional to the module of the deflection coefficient, and the angle between the vertical line and the vector between this point and tube center is proportional to the deflection coefficient phase

$$|\Gamma| = k\,|\rho|\,\sin\frac{\pi\delta}{2}\sqrt{\left(1 - \cos 2\varphi\,\cos\frac{\pi\delta}{2}\right)}, \tag{11.51}$$

$$\varphi = arg\Gamma = -\left(\varphi + \frac{\pi}{4}\right) = \mathrm{arctg}\frac{\sin\left(\varphi + \frac{\pi\delta}{4}\right)}{\sin\left(\varphi - \frac{\pi\delta}{4}\right)}, \tag{11.52}$$

The error of measurement of module and phase of the reflection coefficient depends on the parameter δ . With the frequency moving away from the average value, the error of measurement of module and phase of the reflection coefficient increases. Method error will be

- by module

$$\frac{\Delta\Gamma}{\Gamma} = 1\text{-}\sin\frac{\pi\delta}{2}\sqrt{\left(1 - \cos 2\varphi\,\cos\frac{\pi\delta}{2}\right)}, \tag{11.53}$$

- by phase

$$\Delta\varphi = \operatorname{arctg}\left[\frac{\sin\left(\varphi + \frac{\delta\pi}{4}\right)}{\sin\left(\varphi - \frac{\delta\pi}{4}\right)}\right] + \varphi + \frac{\pi}{4}. \tag{11.54}$$

In the ten percent frequency band for a waveguide the maximum modulus error of the method is 7% and by phase it is 4.6°.

The disadvantages of the method are: inconstancy of the position of the reference plane of the phase of the reflection coefficient relative to the output flange when operating in the frequency range; the presence of four detectors. In the millimeter wavelength range, where errors are increasing due to the non-ideal manufacturing and the requirements for dimensional tolerances are significantly tightened, the method has practically not been used.

In addition to probe sensors, other waveguide sensors were used to measure impedances in the waveguide paths of the centimeter range, for example, with inductive-capacitive connections (coupling with loops), but in the millimeter wavelength range, their use becomes problematic. Other methods are used in this range.

11.3.1.5 Parameters of Serial Measuring Lines of Millimeter Waves

Table 11.1 shows the main parameters of the waveguide measuring lines of the millimeter wave range, developed by enterprises of the USSR and foreign firms.

The following abbreviations of the parameters and names of the companies are used in the table:

- K_r—residual line VSWR;
- K_s—own line KSVN;
- δ—inconsistency of the connection of the probe with the line field;
- Y—relative shunt conductance;
- FMI—Flann Microwave (England);
- HIT—Hitachi (Japan);
- SPC—SPC Electronics (Japan);
- NK—Nihon Koshuha (Japan).

11.4 Reflectometric Methods for Measuring Parameters of SHF Circuits. Reflectometers

In the 60s, the reflectometric method for the analysis of microwave circuits, first proposed by R. Julian, became widespread. Its essence is to obtain information about the S-parameters of the measured devices using reflectometers on directional couplers oriented to the waves incident and reflected from the measured device (and/or transmitted through it). The magnitude of the ratio of the amplitudes of

Table 11.1 Measuring waveguide lines of millimeter waves

Line type	Country, company	Frequency range, GHz	Path (type, section), mm^2	Parameters			
				K_r	K_s	δ, %	Y
P1-13A	USSR	17.44–25.86	11 × 5.5	1.005	1.02	1.4	
P1-12A	USSR	37.5	7.2 × 3.4	1.007	1.02	1.4	
P1-30	USSR	17.44–25.86	11 × 5.5	1.008	1.03	2.0	0.048
P1-31	USSR	25.86–37.5	7.2 × 3.4	1.012	1.03	2.0	0.048
P1-39	USSR	37.5–53.57	5.2 × 2.6	1.013	1.04	2.0	
P1-40	USSR	53.57–78.33	3.6 × 1.8	1.021	1.04	2.0	
P1-41	USSR	78.33–118.1	2.4 × 1.2	1.03	1.05	3.0	
P1-42	USSR	118.1–178.4	1.6 × 0.8	1.03	1.05	3.0	
20,030	FMI	17.6–26.7	WR-42 (10.67 × 4.32)	1.01			
22,030	FMI	26,4-40,1	WR-28 (7.11 × 3.56)	1.01			
23,030	FMI	33.0–50.1	WR-22 (5.69 × 2.84)	1.02			
24,030	FMI	39.3–59.7	WR-19 (4.78 × 2.39)	1.02			
25,030	FMI	49.9–75.8	WR-15 (3.76 × 1.88)	1.03			
26,030	FMI	60.5–92.0	WR-12 (3.10 × 1.55)	1.03			
27,030	FMI	73.8–112.0	WR-10 (2.54 × 1.27)	1.04			
K1816	HIT	18.0–26.5		1.01			
A1816	HIT	22.0–33.0		1.01			
R1817	HIT	26.5–40.0		1.01			
Q1817	HIT	33.0–50.0		1.015			
F1817	HIT	40.0–60.0		1.015			
M1817	HIT	50.0–75.0		1.015			
E1817	HIT	60.0–90.0		1.015			
W1817	HIT	75.0–110.0		1.02			
V1818	HIT	90.0–140.0		1.03			
T1818	HIT	110.0–170.0		1.03			
3S502	SPC	18.0–26.5	WRI-220 (10.67 × 4.32)	1.015			
3S553	SPC	26.5–40.0	WRI-320 (7.11 × 3.56)	1.015			
3S581	SPC	33.0–50.0	WRI-400 (5.69 × 2.84)	1.015			
3S602	SPC	40.0–60.0	WRI-500 (4.78 × 2.39)	1.015			
3S631	SPC	50.0–75.0	WRI-620 (3.76 × 1.88)	1.015			

(continued)

Table 11.1 (continued)

Line type	Country, company	Frequency range, GHz	Path (type, section), mm^2	Parameters			
				K_r	K_s	δ, %	Y
3S652	SPC	60.0–90.0	WRI-740 (3.10 × 1.55)	1.015			
3S681	SPC	75.0–110.0	WRI-900 (2.54 × 1.27)	1.015			
3S721	SPC	90.0–140.0	WRI-1200 (2.03–1.02)	1.03			
3S731	SPC	110.0–170.0	WRI-1400 (10.65 × 0.83)	1.04			
3S740	SPC	140.0–220.0	WRI-1800 (1.30 × 0.65)	1.04			
WSD-245	NK	18.0–26.5	WRI-220 (10.67 × 4.32)	1.01			
WSD-345	NK	26.5–40.0	WRI-320 (7.11 × 3.56)	1.01			
WSD-505	NK	40.0–60.0	WRI-500 (4.78 × 2.39)	1.02			
WSD-605	NK	50.0–75.0	WRI-620 (3.76 × 1.88)	1.02			
WSD-755	NK	60.0–90.0	WRI-740 (3.10 × 1.55)	1.02			
WSD-955	NK	75.0–110.0	WRI-900 (2.54 × 1.27)	1.03			
WSD-1105	NK	100.0–140.0	WRI-1200 (2.03 × 1.02)	1.03			

the emitted waves (reflected and incident or transmitted and incident) determines, respectively, reflection (VSWR) and transmission (attenuation) coefficients.

The initial interest in reflectometers on directional couplers was caused by the search for alternatives to the measuring lines of solutions that would eliminate the problems associated with the gap and the load of the waveguide line with probes. In Fig. 11.7, there is a diagram of a typical reflectometer that allows measuring the VSWR of the device under study in the frequency band. It consists of two interconnected directional couplers, one of which is focused on the incident wave coming from the signal source, and the other on the wave reflected from the device under study. Detector devices are turned on at the outputs of secondary channels of the directional couplers.

First studies of such a reflectometric scheme, along with its advantages, also showed disadvantages, even when measured at fixed frequencies associated with limited directivity and residual reflection of directional couplers. Therefore, further there was a search for methods to combat the final direction, in particular, by adjusting the reflectometer, and ways to reduce reflections in the reflectometric system.

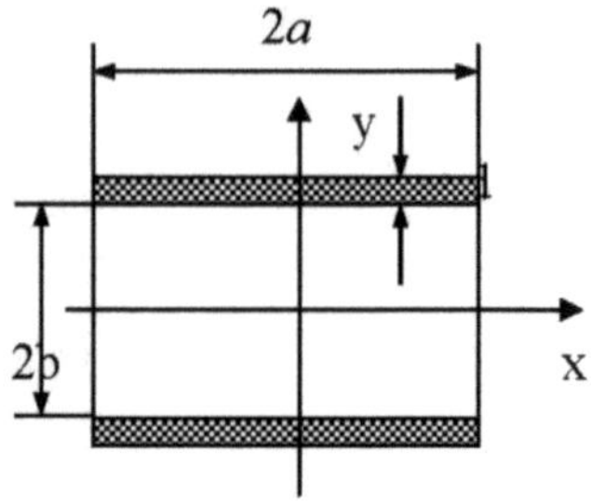

Fig. 11.7 Cross section of an oversize rectangular MDWG

11.5 Trends of Further Development and New Fields of Application of Vector Circuit Analyzers

11.5.1 Main Trends in the Development of VNAs

As one of the most effective types of measuring instruments in the centimeter and millimeter wavelengths, vector-based circuit analyzers are widely used in practice of measuring the parameters of microwave devices and systems of various purposes and applications both at research and development and in industrial production of various microwave equipment. The main capabilities of VACs are associated with measuring the *S*-parameters of microwave devices and presenting the results obtained in the frequency domain on the display.

The effectiveness of modern VACs is predetermined by high performance and measurement accuracy, convenience of presenting the information received in the formats you need, the ability to work in automated measurement systems and networks, and the wide functionality of the VAC that allows you to conduct research and measurements in the centimeter and millimeter wavelengths when solving a wide variety of technical tasks.

Many leading companies in the field of microwave measuring equipment have joined the creation of VAC, thanks to which in recent years significant progress has been made in creating universal, automated VACs with high metrological characteristics and wide functionality,

Modern development of the VAC is characterized by the following main trends and directions:

- advancement up the frequency range to the submillimeter-wave area (up to 1000 GHz) and the use of new measuring media (types of transmission lines) in the millimeter wavelength range;
- improvement of measurement precision and further improvement of methods and means of their metrological support;
- expansion of functionality, the use of AC for measuring parameters of active and non-linear microwave devices;

- creation of new automated technological measuring systems based on VAC, designed to measure the parameters of microwave microassemblies, substrates, technological inserts, etc. in millimeter wavelength range;
- further computerization, improvement of the software itself, and the development of various modern software products for specialized measurement tasks solved with the help of the VAC.

Let us briefly review these main areas of modern development of the VAC.

11.5.2 New Frequency Ranges and Measurement Media

Really mastered frequency bands of millimeter VAC reach 110–170 GHz, i.e. cover the shortwave part of the millimeter range—2- and 3-mm sections of it. As the main type of transmission lines of the VAC millimeter-wave measuring path, rectangular section waveguides of the type WR-15, WR-10, WR-08, WR-06 operating in single-mode are used. The use of rectangular single-mode waveguides as the main type of VAC transmission line in higher frequency ranges, for example, up to 325 GHz (WR-05, WR-04, WR-03), is associated with such significant design and technological difficulties that it becomes almost impossible. When substantially smaller cross-sectional dimensions of a single-mode rectangular waveguide, such as 1.1×0.55 mm, 0.86×0.43 mm, 0.71×0.355 mm, or less, it becomes extremely difficult to ensure correct shape of the cross section and tolerances on it, the negative impact on device parameters of tolerances on the connecting dimensions of flange joints is enhanced, the creation of functional units necessary for operation in such waveguide paths is difficult (if not impossible) due to the fact that the dimensions of active and passive elements used, required for placement in the waveguide (generator and detector-mixing diodes, thermoelements, absorbers, etc.) are large in size compared to the waveguide cross section, etc. In wavelengths shorter than 3–2 mm, the use of single-mode rectangular waveguides as transmission lines in many cases becomes unjustified due to a sharp increase (according to the law λ $-3/2$, where λ is the wavelength) losses in them. Thus, at a frequency of 150 GHz ($\lambda = 2$ mm) in the copper loss waveguide section 1.6×0.8 mm account for more than 6 dB/m at a frequency of 200 GHz (δ waveguide sections $1.1 = 1.5$ mm) δ 0.55 mm loss will be more than 10 dB/m and at a frequency of 300 GHz ($\delta = 1.0$ mm) in the waveguide section 0.7×0.35 mm they will make more than 20 dB/m. At the same time, in such waveguides the maximum permissible transmission power is sharply reduced. In addition, as noted above, due to the reduction in geometric dimensions and, accordingly, stricter requirements for manufacturing accuracy, the complexity of the design and technological implementation of devices and the cost of their manufacture increase enormously.

At the same time, industrial development of more high-frequency bands is getting on the agenda. The development of radioelectronics is constantly moving along the path of mastering new, not yet used frequency bands, since each new segment may be of scientific or practical interest in a particular area of knowledge. At present, the situation has arisen when between the well-developed range of centimeter and millimeter waves on the one hand and the optical range intensively mastered in the past two decades (infrared technology, fiber-optic communication, etc.), on the other hand, there remains a relatively poorly developed submillimeter range. The good prospects for using the shortwave part of the millimeter and submillimeter wavelengths in radiometry, biology and medicine, inter-satellite communications and navigation, for scientific research, and in many other industries make the task of mastering these ranges very promising and relevant.

The impossibility of creating single-mode rectangular waveguides with low losses led to the use of quasi-optical technology in the submillimeter wavelength range. Various types of transmission lines of the quasi-optic tract were studied, such as:

- oversize multimode metal rectangular waveguide with the main wave type H10;
- groove waveguide;
- quasi-optical beam guide;
- metal-dielectric waveguide (MDWG) of circular cross section;
- quasi-optical transmission line of the "channel in dielectric" type;
- metal-dielectric waveguide (MDWG) of rectangular (square) section, etc.

The use of oversize rectangular waveguides, providing a significant reduction in losses and an increase in the transmission power, is made difficult by the excitation of higher wave types in them, i.e. multimode transmission mode, in which, besides the possible initial existence of interfering types of waves, the basic working type of the wave is easily transformed into higher ones.

Among other types of transmission lines of a quasi-optical type (a flute waveguide, a quasi-optical beam guide, circular cross section MDWG, and others) are very promising in the range of short millimeter and submillimeter waves ($\lambda \leq 3$ mm) turned out to be the use of oversize rectangular (square) metal-dielectric waveguides (MDWG) with two impedance and two metal walls, which allow transmitting the operating type of wave with significantly lower losses ($\leq$ 0.03 dB at 150 GHz) and significantly large power compared to single-mode rectangular waveguides and at the same time provide a mode of high losses for 'non-working' types of waves. The principal point for the application of such waveguides in the equipment was the possibility of creating on their basis a complete set of functional elements, helped by the stability of wave polarization and the simplicity of the transition to other types of transmission lines, in particular to a single-mode rectangular waveguide.

So, the conducted studies have shown that the most preferred for use in equipment and various systems in the shortwave part of the millimeter and submillimeter range (at frequencies above 150 GHz) is an oversized metal-dielectric waveguide (MDWG) of square section.

MDWG (Fig. 11.7) represents a waveguide with layers of dielectrics on opposite walls forming impedance walls.

The dimensions of the waveguide are chosen much larger than the wavelength λ in free space: $2a \gg \lambda$, $2b \gg \lambda$, where $2a$ and $2b$ are the width and height of the internal channel.

Impedance walls in MDWG are usually formed by applying a thin dielectric layer (for example, a type of fluoroplast-4) on a metal surface. With the thickness of the dielectric layers close to $\frac{\lambda}{\sqrt{4\varepsilon' - 1}}$, it is possible to realize very small damping of the working mode (LM11). In this case, fields in the dielectric layers are very small, while attenuation for modes of orthogonal polarization (LE) increases significantly, since the conditions in the dielectric are close to resonance conditions, and the fields there are relatively large.

Practical applications were found, in particular, transmission lines on MDWG with geometrical dimensions (10 × 10) mm, for which all the main set of devices necessary for building measuring paths and circuits (modal filters, directional couplers, dividers, modulators, attenuators, detector and multiplying devices, etc.).

AB Millimetre (France) has created the measuring system MVNA-8-350, which allows measuring the transmission parameters (S21 and S12) of devices and materials in a wide dynamic range in the shortwave part of the millimeter and submillimeter wavelengths (at frequencies up to 1 THz). The system is used to study the properties of dielectrics and superconductors, to determine the parameters of resonators, filters, antennas, quasi-optical systems, etc. In the frequency bands above 110 GHz, the WR-6 rectangular waveguide (single mode only at 2 mm wavelength range) is used as the main type of transmission line of the measuring path with subsequent transitions depending on the object being measured to horn antennas (pyramidal or conical) after appropriate calibration of such a measuring (with horns) tract, they are applied to the object being characterized, for example, to the sample of the dielectric material under investigation. The use of such measuring paths of the VAC allows to obtain certain quantitative characteristics of the parameters of the objects under study, but requires further studies to assess the errors (uncertainties) of measurements, including due to the multimode of the path used.

In the centimeter wave bands, some types of flexible coaxial cables with a characteristic impedance of 50 Ohm are usually used as the main type of the transmission line of the measuring path of the VAC. As is known, the main type of wave in coaxial is the TEM-type wave. The working frequency range of the coaxial transmission line is determined from the direct current to the cutoff frequency FCP, over which other types of TEM-wave begin to propagate. At frequencies above FCR, other modes propagating in the coaxial dominate, and its characteristic impedance begins to depend on frequency. Based on this condition (non-proliferation of other types of waves), the cutoff frequency for a coaxial cable is determined by

$$F_{CP} = \frac{7.51}{\sqrt{\varepsilon_r}} \cdot \frac{1}{D + d}, \tag{11.55}$$

where D and d are the diameters of the outer and inner conductor, respectively; ε_r is the relative dielectric constant of the dielectric material used in the coaxial.

Wave resistance of a coaxial is known to be expressed by the formula

$$Z_0 = \frac{60}{\sqrt{\varepsilon_r}} \cdot \ln \frac{D}{d}, \tag{11.56}$$

Until recently, VAC used coaxial cables in the frequency range up to 50–65 GHz with coaxial connectors of types 2.4 mm and V (1.85 × 0.85 mm) (for example, VAC 8510 from HP).

Creating coaxials for higher frequencies is associated with significant structural and technological difficulties, determined by the need to reduce the diameters of the outer and inner conductors (D and d) while respecting their geometry, ensuring strict alignment of the inner conductor relative to the outer, tightening tolerances on the geometric dimensions used dielectric. At the same time, it is necessary to solve the problem of creating appropriate types of well-coordinated coaxial connectors. Despite the difficulties, the work carried out in this direction has now allowed to create coaxial cables and connectors operating in the frequency range up to 110 GHz (1 mm coaxial). Such coaxials and connectors are used in measuring systems 8510XF by Agilent Technologies. 1 mm coaxial connectors are rigid, reusable precision connectors that provide repeatability (reproducibility) of connections.

The advantage of using coaxial measuring paths is their broadband—in one sweep, the frequency across the entire oscillator tuning range (for example, from 18 to 110 GHz) is passed through, while in the case of a waveguide path, this range must be overlapped by five types of waveguide sections. However, a very serious problem when using coaxial transmission lines in the millimeter range are the design and technological difficulties of creating the necessary functional devices of the measuring path (taps, filters, bridge circuits, detector-mixing devices, loads, etc.) on these coaxials. Therefore, one of the promising directions of development of the VAC of the millimeter wave range in terms of the measuring paths is the creation of a complete coaxial measuring medium of the millimeter waves.

11.6 Improvement of Precision Characteristics of VNAs and Further Improvement of the Methods and Means of Their Metrological Support

Vector network analyzers are, first of all, measuring instruments, the main purpose of which is to measure complex S-parameters (and parameters derived from them—VSWR, group delays, etc.) of the tested devices. Therefore, to them, as to measuring tools, there are increasing requirements for the reliability of the measurements provided. The accuracy of the VAC measurements of the millimeter wavelength range, assessed depending on the approach used, the values of the errors, or uncertainties in measuring the relevant parameters, is determined by a number of factors, the most important of which are:

- deviations from the ideal (nominal in geometry and size) measuring path of the WAC, due to its real geometry and actual dimensions;
- real geometry and connecting dimensions of connecting interfaces (waveguide flanges, coaxial connectors) of the measuring path;
- completeness of coverage in the used error models of factors affecting the reliability of the measurement result (for example, non-directionality of couplers, source and load mismatch, isolation, non-quadratic detectors);
- methods and means of calibration of the measuring circuit of the VAC, i.e. metrological support of the VAC;
- methods for generating and measuring quadrature signals by the VAC measurement circuit;
- used algorithms for processing measurement information, accounting for noise characteristics, etc.

When using the millimeter-wave single-mode rectangular waveguides as the transmission paths of the VAC measuring paths, more stringent requirements are imposed on ensuring the correct geometric shape of the section of the path and on the tolerances on its dimensions defining the "degree of reference" of the measurement path. Meeting these requirements, depending on the waveguide channeling technology used, is important from the point of view of ensuring the nominal electrical parameters of the measuring path, namely its characteristic impedance, to which the measured parameters of the tested devices are normalized, and the critical frequency of the waveguide (cutoff frequency H10) waves in the waveguide and, accordingly, the accuracy of the VAC calculations provided by the algorithm based on the measurement results. The constancy of the geometric tolerances of the cross section of the waveguide channel with a significant decrease in the dimensions themselves, such as maintaining tolerances during the transition from the waveguides WR-10 to WR-5, WR-4, WR-3, leads to a significant increase in VAC measurement errors with increasing operating frequencies.

The most important elements of the VAC measuring waveguide paths that significantly affect the repeatability and accuracy of measurements are flange connections, which ensure that test devices and devices for calibration and verification are included in the path or connected to it. With an increase in operating frequencies, in particular with advance into the shortwave part of the millimeter range, in order to ensure acceptable measurement accuracy, requirements for tolerances on the dimensions and crosssectional shape of the waveguide channel of the measuring path, as well as requirements for design, connection dimensions, precision of channel building and the quality of processing of working (mating) surfaces of the flanges toughen significantly. The connecting flanges of the measuring paths of operating VACs constantly experience mechanical impacts when connecting and disconnecting the path to the devices under test, calibration, and/or verification measures, therefore the connecting surfaces of the flanges must be resistant to such effects, ensuring both reliable and perfect (without distortions, gaps and gaps) contact.

Despite the presence of various flange designs used by firms and variations of their types with standardized connecting dimensions, work continues on improving the designs and specifying requirements for the geometry and dimensions of flanges.

In recent years, in the measuring paths of the millimeter-wave devices, the designs of precision and "anti-transverse" flanges are increasingly used, which are nominally compatible with the corresponding standard flanges, but are manufactured with more stringent tolerance requirements and have additional, precisely positioned, leveling pins, and holes.

To ensure uniform compression of the flanges during contact (to avoid distortions and deformations), it is necessary to control the installation torque when tightening the screws (bolts) of the flanges. There are certain recommendations for the installation torque values for the different types of screws used with flanges. A number of companies produce a special tool that allows tightening the screws with a set torque value, which is especially important when working with waveguides of the smallest sections (WR-6, WR-5, WR-4, WR-3).

An important role in ensuring the accuracy of measurements is the calibration of the VAC and the sets of reference calibration measures used. Depending on the measurement problem to be solved, various calibration methods and corresponding calibration devices are used. The docking interfaces (coaxial connectors and waveguide flanges) and waveguide channels of calibration devices are subject to the most stringent requirements on the geometry and tolerances on the dimensions of the channels and the connecting dimensions. Along with the calibration measures of the manufacturers of the VAC, some companies produce special calibration kits for the VAC. For example, the company Maury Microwave Corporation (USA) produces a large range of such kits for coaxial and waveguide VAC, among them are calibration kits for waveguide VAC of millimeter waves from WR-28 (26.5–40.0 Hz) according to WR-10 (75.0–110.0 Hz), designed to work with a VAC type Agilent 8510A/B/C, Anritsu 360, Anritsu 37,000. The kits use the SSLLT (Short-Short-Load-Load-Thru) Calibration Technique and include precision loads (sliding and fixed), a precision transmission line segment, a quarter-wave insert and a fixed short circuit, as well as flanges with calibrated pins and other accessories. All component kit flanges have precision calibrated holes and calibrated pins, which ensure very high repeatability of the connections. Nevertheless, work continues in terms of improving calibration techniques, and in terms of creating new precision calibration standards (devices), due both to the requirements for improving calibration performance in the measurement environments already used and frequency ranges, and with the development of new measurement media (coplanar, microstrip, quasi-optical, etc.) and higher frequency ranges (above 110 Hz).

A prerequisite for the functioning of the VAC as a measuring tool is compliance with the requirements of uniformity of measurement, legalized by applicable international and national standards. Unity of measurements in the case of VAC is usually ensured by the user physically transmitting microwave devices used to calibrate the VAC to the appropriate metrological services for their certification. After receiving properly calibrated calibration devices, the user then calibrates his VAC with their help, making the certified devices ideal. There is a certain frequency of certification of calibration devices. The procedure for transferring calibration devices for certification involves a number of inconveniences for the user.

Modern VACs can be integrated into various automated systems, and the development of the Internet allows the implementation of VAC calibration schemes over the Internet, eliminating the need for periodic physical transfer of calibration devices from the user to their supervisor and back. Further development of such Internet calibration systems is promising, since it not only reduces the amount of work performed by the user, but also guarantees the use of the latest techniques and makes it possible to reduce the hierarchy of measurement chain to the only link related to national or international standards, ultimately reliability of measurement results.

11.7 Extension of Functional Possibilities and Application of Circuit Analyzers for Measurement of Parameters of Active and Non-linear SHF Devices

11.7.1 Frequency-Time Transformations

Along with measurements of *S*-parameters and parameters of microwave devices derived from them as functions of frequency, modern VACs allow to obtain additional information about the properties and characteristics of the devices under study using appropriate mathematical methods, in particular, the inverse Fourier transform transforming information from the frequency domain in temporary. To ensure these transformations in the VAC, time-stimulus signals are used stepwise or pulsed, while the device under test for the correct application of the inverse Fourier transform should be represented by a linear multi-pole. The presence of special conversion modes, such as diaphragming, use of weighing windows (cropping), and gating, allows, for example, to view in detail the measured reflection characteristic of the object under investigation, identifying specific reflection sites in the device with the corresponding reflected signals. Application and improvement in the VAC of the millimeter wave band of the methods of frequency-time transformations in order to increase the resolution of characteristics and range resolution, increase the time range of measurements, reduce the minimum noise level, improve identification capabilities in characterizing heterogeneities, etc., is a promising direction for further improvement of VAC [8, 11, 17, 20, 23, 24].

11.7.2 Modulation Vector Analysis of Circuits

Traditional VACs use sinusoidal signals as data carriers when measuring the scattering parameters of the tested devices. More recently, a new principle of measuring *S*-parameters was proposed, based on the use of modulated signals as stimuli. A device based on this principle is the so-called modulated vector network analyzer (MVAC).

MVACs act in the same way as usual VAC in the sense that in the error model of the direct direction a stimulus is applied to the same point a 1 M. However, in this case, the stimulus is not a sinusoid of one frequency, but a modulated signal with a continuous distribution of energy near the carrier frequency. In order to create such a modulated signal, the sinusoidal VAC source is replaced by a modulated source. In the case of MVACs, the most commonly used source is a synthesized RF signal generator with a $\frac{I}{Q}$ modulator, controlled by two high-speed random signal generators; however, any modulated or unmodulated RF signals may be used.

From the measurement point of view, narrow-band VAC receivers are being replaced by broadband receivers. Each receiver in MVAC is a high-speed variable sampler with 15 Hz bandwidth. While a normal VAC should move from frequency to frequency, performing measurements on each of them, the MVAC can cover up to 15 MHz of frequency filling simultaneously.

Comparison of algorithms for measuring and calculating sinusoidal and modulated parameters is shown in Fig. 11.8.

From the comparison it is clear that the modulated S-parameters are calculated very similarly to how they are calculated in the case of sinusoidal S-parameters. The main difference between MVACs is the ability to calculate S-parameters at many frequencies, based on a one-time acquisition of information instead of repeatedly performing step-by-step (stepwise) frequency measurements.

The ability to measure S-parameters with modulated stimuli can be very useful, in particular, for devices that behave differently when using sinusoidal and modulated stimuli. For example, good possible objects for the use of MVABs are amplifiers operating near their compression point, in particular, because the gain of S21 and the input matching of S11 may depend on the type of excitation. A typical problem is setting the input matching and amplifying the amplifier using a sinusoidal VNA, and then using such a device to amplify the modulated signal. At the same time, careful adjustment of the input matching, performed using a sinusoid, is unjustified due to the fact that the matching will be different if the modulated signal is then used. In contrast, the MVACs allow you to adjust the input matching and amplification of the amplifier, using the signals of interest that are close to the actual operating signals.

For devices with signal characteristics varying in power or in time, the MVAC provides a good way to determine S-parameters using signals of interest.

Devices that are not profitable to measure at MVAC are passive devices or linear devices. Thus, cables, connectors, filters, and attenuators do not give a difference when measuring their S-parameters using sinusoidal or modulated signals. This fact is somewhat comforting, confirming in fact that the modulated S-parameters are the same as the sinusoidal S-parameters in cases where the excitation sources are irrelevant.

The ability to measure S-parameters with modulated stimuli can be very useful in analyzing the behavior of active devices, especially those for which high linearity is a desirable characteristic of behavior. The MVAC provides accurate measurements of the characteristics of devices under conditions of actually used signals, that

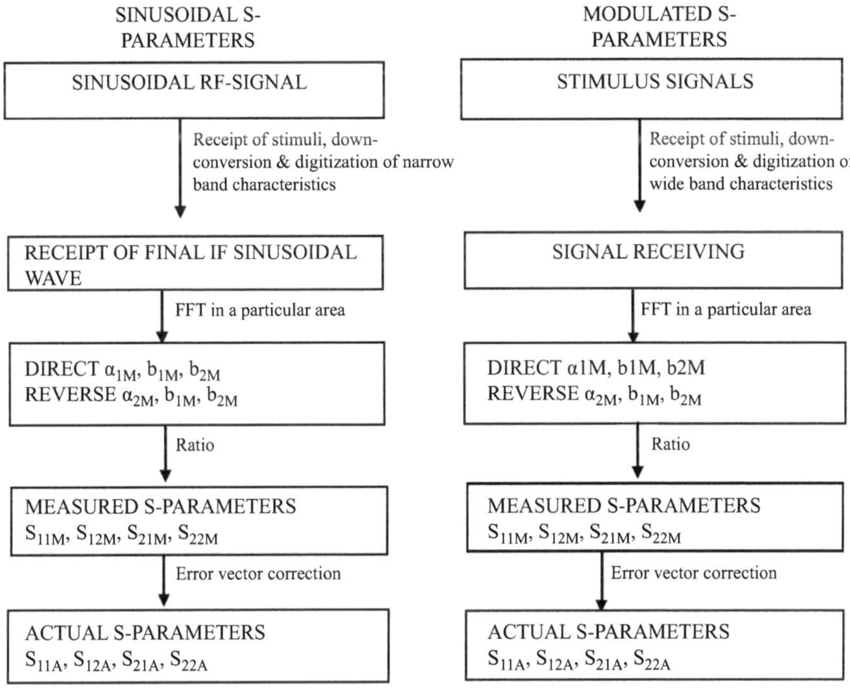

Fig. 11.8 The process of measuring and calculating modulated and sinusoidal parameters

is, test stimuli are consistent with the intended use of devices. The process of calculating the modulated *S*-parameters is similar to the usual method, except for the receiver bandwidth for data collection.

11.7.3 Features of Measurement of Parameters of Non-linear Microwave Circuits

Among the numerous microwave devices used by modern electronics, including those in the millimeter wavelength range, are devices containing semiconductor, ferrite, and integrated equipment that are non-linear as objects of measurement. At the same time, most methods for analyzing microwave circuits, including measurements using the VAC, are focused on measuring the *S*-parameters of linear circuits, i.e. circuits whose parameters are independent of the amplitudes of the measuring signals. The relevance, importance, and availability of analysis of the parameters of non-linear microwave circuits predetermined the interest in this problem and the numerous attempts to solve it with various assumptions. A serious attempt to formulate approaches to solving this problem was made in the doctoral thesis of I. I. Chuprov. It was shown that solving the problem of multifunctional analysis of microwave circuits requires, first of all, clarifying the concept of S-parameters of non-linear microwave circuits, considering the nature of their dependence on amplitudes and spectral composition of measuring signals and determining the possibility of specifying specific non-linear circuits through their S-parameters. In the case of determining the possibility of such a task, the generalized task of instrumental analysis of non-linear circuits is formulated as the task of measuring one or another set of relevant S-parameters describing the functional purpose of the circuit under given boundary conditions at its inputs. The implementation of direct methods for measuring S-parameters in many cases involves the need to overcome many technical difficulties, therefore various indirect methods are developed for determining S-parameters of non-linear circuits from measurements with a small measuring signal.

The solution methods used substantially depend on the characteristics of the non-linear microwave circuit under study (its broadband or narrowband, its inertia, etc.) and parameters to be measured. Setting the functional purpose of specific circuits through approximate models, determined by a small number of corresponding S-parameters, greatly simplifies the solution of the problem of measuring non-linear circuits. Nevertheless, the choice of method and means of measuring certain parameters of non-linear circuits require rigorous justification, taking into account the specific features of the circuit being measured.

The development and theoretical substantiation of appropriate methods for measuring the S-parameters of non-linear circuits and the creation of specialized meters, including for the millimeter wavelength range, are a very promising direction for solving the problem of instrumental analysis of microwave circuits. The solution of these problems, as well as the problem of metrological assurance of measurements of non-linear microwave circuits, requires further serious research.

11.8 Automated Hardware for Measuring the Parameters of Integrated Circuits on Semiconductor Wafers in the Millimeter Wave Range

11.8.1 Testing of SHF Integral Circuits on Semiconductor Wafers

The development of microelectronics technology led to the creation of various integrated circuits, including the microwave range, on semiconductor wafers, which made it possible to ensure their mass production. At the same time, the task of testing these integrated circuits with the characterization of their parameters or selection by the criterion of "fit/bad" was an actual task.

Semiconductor plates for testing RF products are an almost ideal measuring medium. Conventional RF circuits that are tested on semiconductor wafers are: amplifiers, switches, mixers, attenuators, phase shifters, various structures with connections, etc.

To ensure the vector or scalar measurements performed during testing of ICs, there are appropriate test interface technologies. The selection of the probe interface of the probe installation for checking the structures on a semiconductor wafer is based on the measurements performed and the required measurement environment determined by the specific product under test [24–29].

Until 1985, the standard approach for characterizing at microwave frequencies and assessing the quality of semiconductor wafers with crystals (integrated circuits, devices) before being delivered to users consisted in cutting crystals, choosing several elements (devices), usually one in each quadrant, mounting them in fixture and then check in fixture with registration of S-parameters or power levels. Often these elements were high-power transistors or very common RF/HF products that were used as samples. As test samples (coupons) for gallium arsenide monolithic microwave integrated circuits (MMIC), transistors or MMICs were used in the same way. Usually, the elements were mounted in metal-ceramic cases with leads, with epoxy or eutectic fastening, gold wires were attached to them by wedge microwelding for RF connection and constant displacement. The assembly was then manually placed in the test fixture and fixed. The armature was connected using an RF coaxial cable used to connect the 50 Ohm medium at its end to test equipment, usually a vector network analyzer or scalar power meter. This technique has many shortcomings, including due to the multiplicity of sources of uncertainties during calibration (testing):

- uneven placement of the element in the housing (fixture) and the profile of the loop of communication wires, manually executed by the operator the difference in the lengths of the wires of the conclusions and, therefore, the various approvals for the device under test;
- inaccuracies in the manufacture of hull models and, accordingly, their variability from case to case;
- RF and earth contacts, which, due to the difference in physical pressure of the fastening clamp applied to the body of the case and the leads, lead to non-constant (different) values of inductance and resistance of the leads and to potential fluctuations, mainly at microwave frequencies;
- An empirical model of recovery (extraction) of reinforcement for connectors and transmission lines used at RF ports;
- Calibration of test equipment at docking interfaces between RF cables and test fittings, and not in the measuring planes of the element or assembly.

Most of these technical uncertainties arise due to the fact that the calibration plane shifts from the true plane of the product, and intermediate compounds are not well characterized or not reproduced.

The main disadvantages of the tests in the fixture are:

- the impossibility of checking the entire product sent, except for a representative sample obtained by the method of destruction of the main sample. This can lead to the rejection of a large number of defective modules and products after assembly, causing large losses for the user, especially for MMIC, where the losses on release can be significant;

- costs for reinforcement tests; the need to sacrifice the elements and assemblies used for verification;
- long cycle time; as a rule, it takes a day or two to pack items into assemblies;
- low speed test products; the installation of elements in the armature is practically limited by the speed of the element per minute.

The first step taken to eliminate the deficiencies noted was the improvement of the test armature for pure crystals, which should be accurately characterized. One solution is to use modular reinforcement, in which the crystals are mounted on inserts of the same length, placed in layers between two end elements, each consisting of a segment of the transmission line and a connector. The two end elements can be fully characterized by VAC at the end points of transmission lines by calibrating the SOLT (Short-Open-Load-Thru) or TRL (Thru-Reflect-Line) type calibration; the conductors are welded by thermocompression welding to the installed inserts or between two end elements connected end-to-end. Then the crystal is installed (fixed) in the insert, mounted between the end elements and connected with wires to the transmission lines. Such a method is one of the main ones for precise characterization and extraction of a model.

The main achievements of these improvements were the elimination of the location factor, the assembly, the output contacts and the reinforcement as sources of variability, the reduction of the cost of integration, and the calibration process. The remaining drawbacks are variations of fastening loops and destruction (products, plates), as well as the duration of the method, which prevent its use in mass applications, such as obtaining statistical models or acceptance tests.

11.8.2 Coplanar Probes

The solution for accurate and mass microwave MMIC testing came from Cascade Microtech, the first company to make commercially available RF and microwave probes together with comprehensive (hardware-software) support for their use.

For measurements on semiconductor wafers with direct current and for digital applications, for a comparatively long time, some types of probes with a large number of leads placed on metal or ceramic cloths with excessive mounting were available. Despite improvements in the design of RF probes by several companies for their own (internal) use, they focused on short-circuited standard DC probes, which made it difficult to characterize and use them at microwave frequencies.

The idea of Cascade Microtech using coplanar structures in the probe with a GSG (ground-signal-ground) configuration (ground-signal-ground) up to its tip made it possible to obtain reproducible 50-Ohm matching with the device under test, thereby providing the possibility of well-reproducible non-destructive microwave measurements at the level of semiconductor wafers. The coaxial transmission line of traditional RF probes is converted to a coplanar signal and probe earth points. This allows you to measure the signal coplanar or microstrip circuit relative to the earth. At the same time there are no intermediate connections, delay times and uncertainties, as well as the cost of the device itself.

Such probes are usually characterized by a GS (ground-signal) and/or GSG (ground-signal-ground) configuration and are widely used for accurate high-frequency measurements. Probes with a GSG configuration provide enhanced performance in the frequency bands above 12 GHz and can be used up to 100 GHz with appropriate construction performance. Probe distance between the signal and the ground determines the distance (pitch) between outputs. In a usual probe, a pitch is equal to 0.006 inches (0.152 mm). Due to small size, RF parameters and physical strength of the probe is significantly influenced by choice of material.

Many companies, including Cascade Microtech and PicoProbe, specialize in RF probes.

Figure 11.9 shows several designs of modern co-planar probes by Cascad Microtech.

These probes include:

- probe head (Fig. 11.9a), which is an air/co-planar (2 RF-contacts) combined with eight DC contacts (probes) for provision of multi-contact applications with MMIC or for modular probing; it is powered with direct current, up to 65 GHz; possible combinations of RF contacts—GSG, GS, or SG; ensures low-low inserted losses at the RF;
- wideband WPH probe (Fig. 11.9b); standard contact configuration—GSG; standard reflection losses ≥20 dB up to 65 HHz; ensures 50-Ohm matching for testing of multi-port MMIC on semiconductor wafers;
- ACP 40 probe (Fig. 11.9c), ensures precise measurements on semiconductor waves of low-impedance powerful integral SHF devices; ensures probing within the frequency range of 800 MHz–100 GHz with alteration of characteristics of all S-parameters; GSG contact configuration;
- new air/co-planar probe ACP 110 (Fig. 11.9d), ensures high-temperature (−55 °C–200 °C) probing in the DC range of up to 110 GHz with 1 mm connector produced by Hewlett-Packard; relevant calibration standards have been designed for the probe; ensures low insertion and reflection losses; GSG contact configuration.

11.8.3 Automated Complexes of Testing of ICs on Semiconductor Wafers

Automatic installations for testing of semiconductor wafers with devices (connectors) for mounting of probe manipulators working with probes Cascade Microtechs appeared approximately at the same time as co-planar probes. Aligent (known as HP at the time) introduced the HP 8510 vector network analyzer, which ensured much faster and easier calibration method for microwave testing equipment and started using 50-Ohm matched microwave MMIC. Combination of the results of these events and their realization in the corresponding automated installations

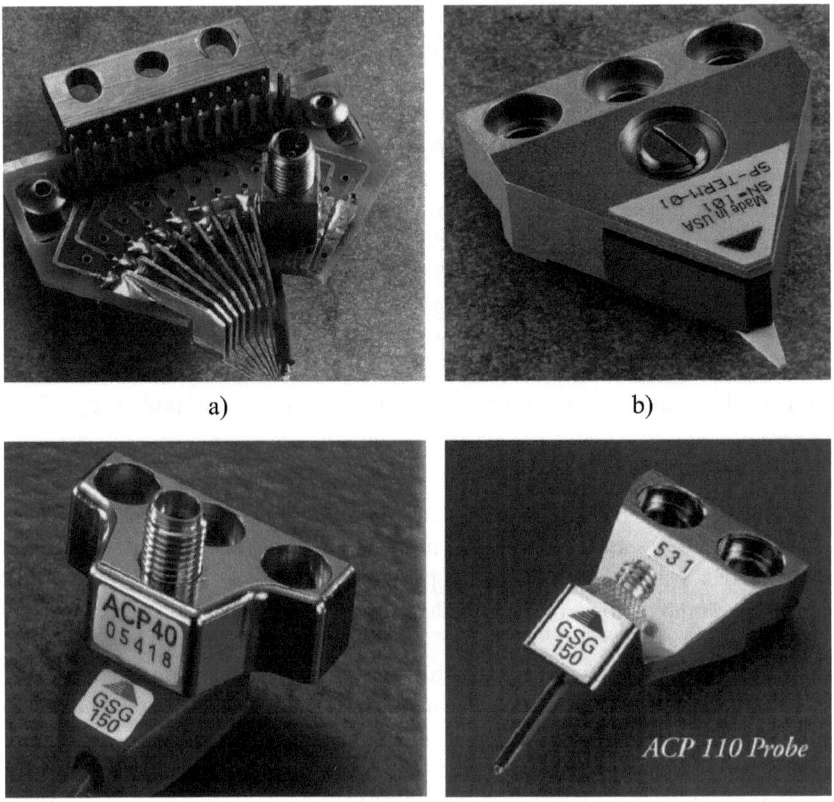

a) b) c) d)

Fig. 11.9 Several type of co-planar probes produced by Cascade Microtech (USA)

for IC testing on semiconductor wafers allowed to fully change the processes of IC characterization and selection during serial production. As early as in late 80s, many MMIC manufacturers were already estimating characteristics of semiconductor wafers based on the results of RF testing of standard transistor cells in the process of monitored control and provided their customer with known-good chips.

Typical automated system for testing of semi-conductor wafers includes a measurement system, a probe installation, RF probes, and DC probes. An example of such installation is shown in Fig. 11.10.

The selection of the used measuring equipment is based on targeted parameters of the specific tested devices, required frequency range, and precision of measurement. Tested parameters can include power gain, reflection coefficients and isolation, noise characteristics (noise coefficients), etc. Depending on the measured parameters, composition of the measuring equipment can include signal sources, including multiplier-based, scalar and vector network analyzers, spectrum analyzers, pulse power testers, noise coefficient testers, etc., as well as corresponding switching devices.

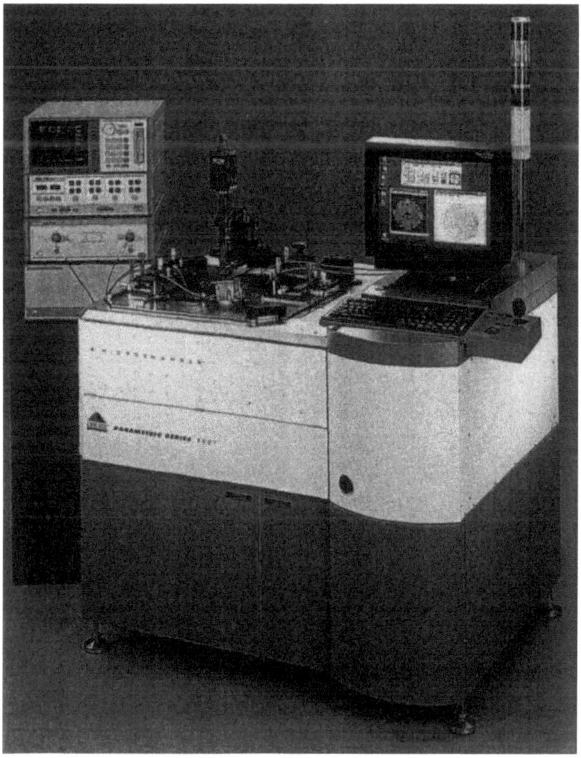

Fig. 11.10 Automated probe installation PS21 RF by Cascade Microtech with HP measurement equipment

Figure 11.11 shows a photo of a typical working probe installation. This testing installation has cassette supply and configuration of the test head type and ensures automatic matching (orientation and alignment of wafers).

Test head links up with the test interface that is installed in the hole in the left side of the installation. Probe holder used a ring-type probe plate as shown in Fig. 11.12.

Upper plate of the probe installation can be customized depending on the type of the probes used with ring frames or micromanipulators. Standard probes are installed on the top plate of the probe installation with the help of micromanipulators located in quadrants (circle quarters), which gives access to each of the four sides of the integral circuit (Fig. 11.13).

Manufacturers of probe installations for testing of semiconductor wafers offer various types of top plates depending on the applied structures of probes.

Table 11.2 gives data on typical characteristics of probes.

Specialized vector circuit analyzer Hewlett-Packard of type HP 85109C, which allows performing fully calibrated measurement on semiconductor wafers within the frequency range of 45 MHz–110 GHz, is a quite effective measurement system, used with nearly all probe installation for testing of ICs on semiconductor wafers within the millimeter wave range. HP 85109C can be configured differently, depending on the optimization of the used frequency range or measurement precision. In case with wideband measurements, the combined mode allows performing measurement

Fig. 11.11 Running probe installation for RF testing

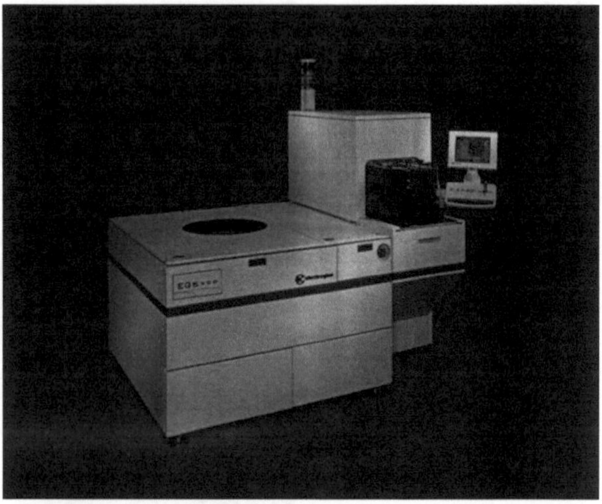

Fig. 11.12 Ring-type plate of a RF probe

within the range of 45 MHz–65 GHz with one connection of the tested device. In order to ensure higher precision of measurements, HP85109C is used with 2.4 mm coaxial transmission line within the frequency range of 45 MHz–50 GHz, with V-band waveguide during measurements within 50–75 Hz and W-band waveguide of 75–110 GHz.

HP 85109C uses the structure of the new 1.0 m coaxial connector to solve the problem of connection of the system waveguide ports with waveguide inputs of the probes (Fig. 11.14). Before invention of such adapters, sections of rigid waveguides were used, which restricted movement of probes due to their rigidity and could cause difficulties in protection of the probe and could cause difficulties in

Fig. 11.13 Standard probes on manipulators

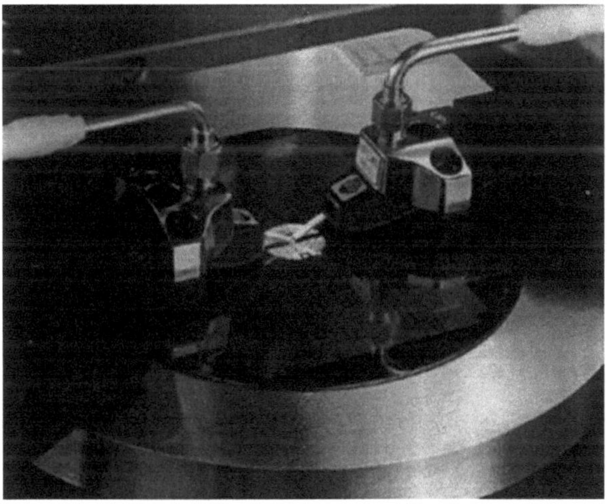

Table 11.2 Data on typical characteristics of probes

Type of the probe	Frequency range, GHz	Insertion loss, dB	Losses reflection, dB	Mutual losses, dB	Contact repeatability, dB
WPH-205	DC–50	2.5	−12	−50	−50
WPH-405	DC–65	3.0	−12	−50	−50
WPH-505	50–75	3.5	−10	−40	−50
WPH 805	75–110	4.0	−10	−40	−50

Fig. 11.14 Waveguide-coaxial (1.0 mm) adaptors

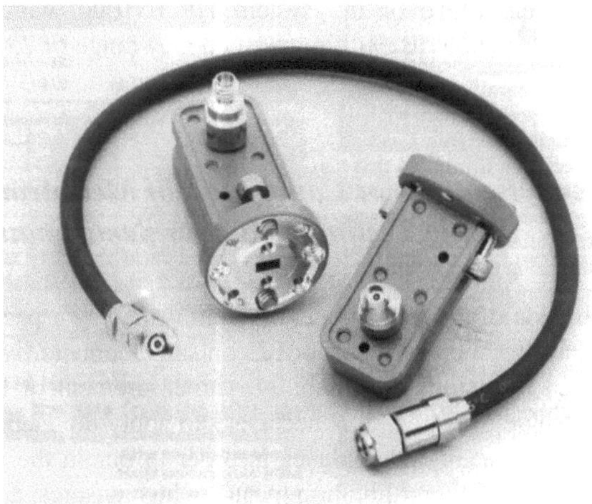

protection of the probe plane. In HP 85109C, short sections of coaxial (1.0 mm) cable are used on each port together with 1 mm coaxial-waveguide adaptors on probe installation and probes. Such cables make it possible to move the probes without connectors voltage, thus ensuring flexible interface required for such applications.

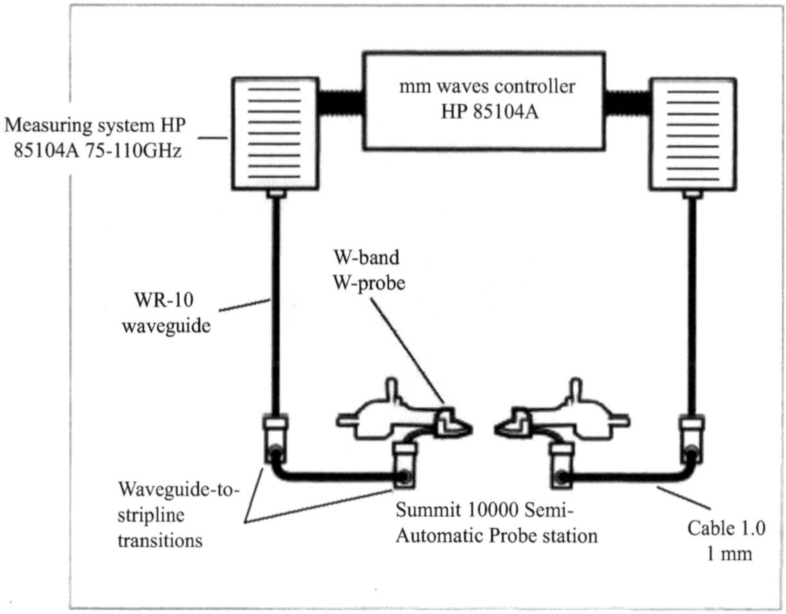

Fig. 11.15 Configuration of the measurement system of the range 75–110 GHz

Configuration of the measurement system HP 85109 for probing on semiconductor wafers in the W band (frequency range 75–110 Hz, waveguide WR-10) is shown in Fig. 11.15.

Dynamic range of the system HP 85109C within the frequency range of 50–75 GHz is 65 dB, within the frequency range of 75–110 GHz—about 60 dB.

11.8.4 Features of Calibration of Measuring Systems during Testing of ICs on Semiconductor Wafers

Calibration of the measurement system is the key aspect that determines the quality of parameter measurement for ICs on semiconductor wafers. In order to ensure the required precision and reliability of measurement of IC parameters on the tested semiconductor wafers, it is necessary to apply the method of calibration of devices placed next to interface planes in order to obtain the most precise testing results. The most effective calibration means, which ensures, among other things, complete vector correction of RF probe terminals, is a vector network analyzer. Application of the relevant methods of calibration of measuring circuits including vector network analyzer allows using them for testing of the millimeter range MMIC.

Recovery of parameters of circuits or devices containing coplanar, strip, microstrip or another line requires application of ideal loads with precisely measured parameters. Recovery of true parameters of devices in the millimeter wave range is usually performed through the procedure of TRL-calibration, which uses the transition line which connects two transducers (T), a reflecting element, e.g. short-circuiter (R) and the line section (L).

Thus, removal of the probe (elimination of its effect on measurement results) requires calibration standards (examples, measures). Thus, vector measurements require calibration loads in the idle run and short circuit mode and loading measures. Probe suppliers normally offer calibration standards designed specifically for their probes. As a rule, calibrating measures are proposed (SS, NL, matched load, passed line, pin or stub in the line), executed on ceramic substrates or made on actually tested semiconductor wafers that ensure standard calibration of probe tips.

Figure 11.16 shows certain calibration standards suggested by Cascade Microtech.

Electrical parameters, physical geometry, and substrate material are determined by the customer (user).

As a rule, the used configuration of contacts is GS, SG, GSG, or the one required by the customer. The used frequency range is up to 110 GHz. The interval of possible steps is 50–3000 μm. Standards are supported by calibration methods of types SOLT, TRL, LRM/LRRM.

In case with scalar measurements or if composite probe assemblies are used, it is possible to use alternative calibration standards with reduced measurement precision. Alternative calibration standards may be custom test structures printed on ceramic substrate or test structure of a semiconductor wafer.

When selecting the probing technology and the type of probe for any applications, it is necessary to consider the way of calibration, maximum used frequency, as well as the price of probes and cost of calibration schemes.

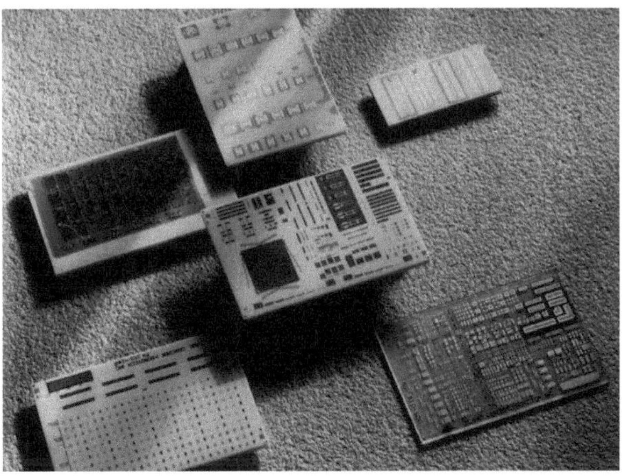

Fig. 11.16 Calibration standards of Cascade Microtech

11.8.5 On Advantages on RF and Microwave Testing of ICs on Semiconductor Wafers

Wide application of RF and SHF testing of ICs on semiconductor wafers in industry can be explained by the following factors:

- precision of RF testing depends on calibration performed at the tip of the probe in the point of contact with the tested product. The applied calibration methods are trustworthy, supported by elaborate and carefully worded calibration standards and easily implemented with the help of the relevant software; this leads to precise description (precise models) of the designed and serially produced devices and increased percentage of output of non-defective products during industrial production;
- reproducibility of measurement results with stable impedance of the probe (regardless of whether it is 50 Ohm or a customer value) and automatic adjustment of the probe position in relation to the contact area ensured by modern probe installations. Location of the probe installation on contact areas during testing and calibration is critical, especially in the range above 10 GHz and for tested products with narrowband matching;
- non-destructive tests of the measured devices allowing to supply known-good chips to customers. This ability is a key for multi-chip modules or in case of using reverse chips on the same board. Correlation between the results of testing of assembled semiconductor wafer and devices (chips) is perfect if grounding MMIC is properly implemented, and circuits with DC shift are similar.
- short clock cycle for testing or statistical characterization and removal of models of beans which allow to carry out modeling and prediction for successful production of ICs.

High capacity (efficiency) with complete automation of testing and probing efficiency and low price reducing with the coefficient 10 over 10 years.

Further development of microelectronic technologies, modernization and development of new types of microwave probes, expansion of the list of ICs, movement up the frequency range, increase in reliability and precision of measurements—all these determine the prospectivity of further works for implementation of vector network analyzers as the most effective means of measurement used in automated stations of IC testing on semiconductor substrates. This is related to improvement of the applied error models during adjustment of IC parameters, as well as development of new means and ways of calibration of VNAs designed for work with new measuring media and in new frequency ranges.

11.9 Vector Analyzers of Millimeter Wave Networks. Classification and Building Principles

11.9.1 Types and Classifications of Vector Network Analyzers

Since S-parameters of SHF devices are complex values, all possible ways of their measurement can be determined as *amplitude-phase*. Information and measurement systems or devices designed for measurement of complex S-parameters of SHF devices are called *testers of S-parameters or vector network analyzers*, since any complex parameter can be represented by the corresponding vector characterized by its amplitude and phase depending on the measurement frequency.

When building vector network analyzers, it is necessary to take the following key questions into account [11–13].

1. In order to obtain information about S-parameters, *the tested device shall be switched into the measuring path of the device* and *supplied* with a *SHF signal modulated on the basis of a certain parameter (amplitude, frequency, or phase).* The signal that passes from the tested space or reflected from it carries useful information about its properties, removal of which requires comparison *of this measuring signal with a source (reference) signal.* Therefore, *network analyzer circuit shall contain measurement and reference channels and information signal converter.* At the same time, implementation of any method of measurement of S-parameters requires comparison of the unknown phase of the measuring signal (reflected from the tested device or passing through it) with the phase of a coherent reference signal. Such a comparison can be made by interfering and converting the measurement and reference signals on a non-linear element—the primary converter of the SHF signals. Therefore, all possible modifications of vector network analyzers are in fact interferometers, SHF measuring paths of which are based on a bridge circuit with reference and measurement channels.
2. A possible and frequently used method of obtaining information about the measured amplitude-phase characteristics of S-parameters is *the generation of quadrature, i.e. phase-shifted by $\pi/2$, signals*, processing of makes it possible to extract information on both amplitude–frequency and phase-frequency dependencies of the examined S-parameters.
3. Determination of characteristics of various studied SHF devices within their complete frequency ranges requires the vector network analyzer to allow for measurements in sufficiently wide and adjusted strips of bands with indication of information on the monitor screen, i.e. perform panoramic measurements with the possibility to adjust width of the frequency band of display and alteration of the targeted characteristics.
4. Vector analyzers of SHF circuits, which are measurement devices, shall guarantee the required measurement precision of the tested parameters of tested products; i.e., they shall be metrologically provided with the relevant methods and means of calibration and adjustment.

5. Modern requirements for information and measurement systems and devices, including vector network analyzers, also include requirements for automation of measurements and application of digital methods of processing, storage, and presentation of the received measuring information.

Specific view of the circuit of a SHF measurement path also largely determines the structure of the vector network analyzer in general. At the same time, the applied circuits of SHF measuring tracts (waveguide, coaxial, etc.) can be differentiated based on the requirements for VNA characteristics regardless of the applied methods of measurement of *S*-parameters.

A universal classification attribute of VNA SHF paths and VNAs themselves is the number of poles (number of inputs/outputs) of the SHF path circuit. This number is first determined by the nomenclature of test objects that can only include two-pole devices or devices represented by equivalent two-poles or four-poles, which, in their turn, can be either unilateral or bilateral, or be represented by equivalent multi-poles in the common case. If a test object is a multi-pole, it can be transformed into two-pole or four-pole by means of connecting it to corresponding outputs of matched loads. Second, the number of poles is determined by the requirements for the volume of the source measurement information, i.e. the number of output signals of primary converters, on which the functional possibilities and metrological characteristics of the VNA depend.

According to the VNA classification based on the number of poles, two-poles are allocated to the VNA input at the measurement signal source, which is usually represented by sweep oscillators, two-poles—to the VNA output for two-pole connection and two more poles—for four-pole connection. The remaining number of poles is determined by the number and type of primary converters of SHF signals.

Most known VNA circuits have measuring paths characterized by a number of poles from 6 up to 14.

The following can be used as other classification attributes of VNA:

- the character of output signals of the used primary measurement VNA converters;
- means of modulation of measurement and reference signals;
- methods of generation of quadrature signals of measurement information, etc.

Let us consider the first of the listed attributes—the character of output signals of primary converters. This attribute allows us to subdivide all known ways of measurement of *S*-parameters (and, subsequently, VNA structures) into two large groups:

- methods with detection of the interference SHF signal initially received from one signal source (one sweep oscillator) and further split in the SHF path of the analyzer into reference and measurement signals—these methods form the basis for creation of *homodyne VNAs*;
- methods with conversion of frequency of measurement and reference signals with various frequencies received from various sources. Source of the measuring signal is generally referred to as measuring generator, source of the reference signal (of other frequency) is known as heterodyne. Primary VNA analyzers implementing these methods are not detectors but mixers, with the help of which

the information about values of the measured *S*-parameters is removed from SHF range to the range of intermediate (low) frequencies. These methods are also called frequency transition methods; they form the basis for creation of *heterodyne VNAs*.

It is necessary to point out the following circumstance. In course of analysis of operation of homodyne and heterodyne network analyzers, output signals of primary converters are determined (calculated) based on standard methods of the classic theory of detection and conversion of signal frequencies. In this, it does not matter how the reference signal was obtained during frequency conversion—from a single heterodyne or by means of modulation of reference and measurement signals received from the same source and set at a frequency interval equal to the intermediate frequency from each other due to modulation. Therefore, certain modifications of homodyne VNAs are in fact hybrid network analyzers: on the one hand, these are single-generator, i.e. homodyne VNAs; on the other hand, based on the principle of receipt of output signals of primary converters, they can be qualified as network analyzers with frequency shift. The primary converters themselves in such circuits of analyzers are in fact mixers that are often traditionally called homodyne detectors. Characteristics of such hybrid network analyzers are close to characteristics of heterodyne VNAs; at the same time, they are cheaper and simpler in terms of design.

Circuits of homodyne analyzers of SHF networks can employ various types of modulation of reference and measurement signals (amplitude, frequency, phase modulation, or manipulation) and various types of modulation of the reference and measurement signals at the same time; there are designs where modulation of reference and measurement signals is completely absent, while the information of the measured *S*-parameters is performed using direct current. Next, we are going to take a closer look at VNA designs with separate types of modulation of reference and measurement signals and with various ways of formation of quadrature signals of the measurement information (hardware-based, circuit-based, software-based, etc.).

11.9.2 Heterodyne VNAs

Heterodyne vector analyzers of SHF networks are testers of *S*-parameters that use frequency conversion of course SHF reference and measurement signals received from different sources into informational low-frequency signals, measurement of amplitude and phase ratios of which allows obtaining the required information about complex *S*-parameters of the studied device.

Block diagrams of a two-channel heterodyne converter of SHF signals and a heterodyne VNA are shown in Figs. 11.17 and 11.18.

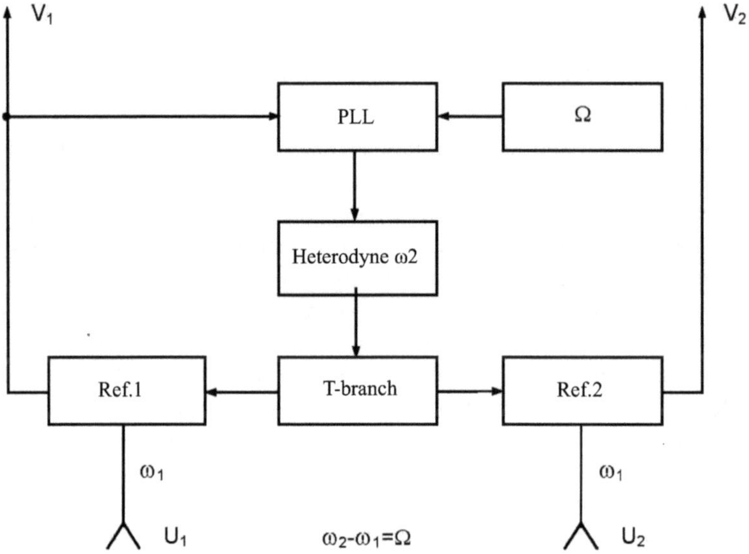

Fig. 11.17 Block diagram of a two-channel heterodyne converter of SHF signals

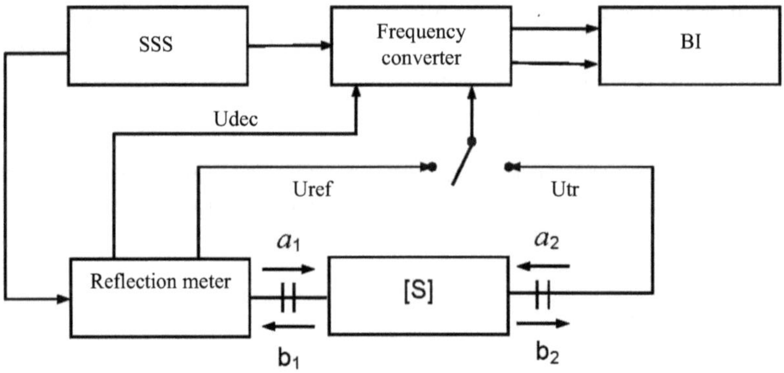

Fig. 11.18 Block diagram of a heterodyne VNA

In such analyzers, incoming informational SHF signals of frequency ω_1 are

$$\begin{cases} U_1 = V_0 \cos(\omega_1 t), \\ U_2 = V_0 \rho \cos(\omega_1 t + \varphi), \end{cases} \tag{11.57}$$

where ρ and φ are the modulus and phase of the measured parameter.

Heterodyne converter transforms them (in a linear manner) into the voltage of low intermediate frequency Ω:

$$\begin{cases} V_1 = m_1 V_0 \cos \Omega t, \\ V_2 = m_2 V_0 \cos (\Omega t + \varphi). \end{cases} \tag{11.58}$$

At the same time, an important requirement for heterodyne VNAs is the permanence of frequency Ω of the output voltages, which is ensured with the help of the phase-locked loop system (PLL) which tunes the heterodyne to the frequency of input signal of the reference channel U_1.

Advantages of heterodyne VNAs include high sensitivity in terms of the level of input signals (about 10-12-10-14 W) and selectivity in terms of the frequency spectrum of input signals, high limits of measured depletions (up to -80 to -100 dB), wide functional possibilities (ensure measurements of complex values of *S*-parameters, VSWR, depletions, full resistance and conductivity of linear and non-linear devices, group delay measurements, measurements of narrowband and wideband devices, etc.), good metrological characteristics, including high resolution in amplitude (0.001 dB) and phase (0.01°).

However, these advantages of heterodyne VNAs are ensured due to greater complexity of the circuit, while the amplifiers themselves are very expensive units (for example, costly frequency synthesizers are used to obtain the required characteristics as both source and heterodyne, etc.). The main disadvantage of heterodyne converters of measuring SHF signals is the complexity of creation of wideband heterodynes, two-channel mixers, and PLL systems ensuring compliance with the condition

$$\omega_1 - \omega_2 = \text{const.} \tag{11.59}$$

Downsides of such testers also include their relatively low action speed, which does not allow using them in a time mode close to real time.

Heterodyne VNAs in the millimeter wave range can be exemplified by systems of type 8510XF produced by Hewlett-Packard and working within the range of up to 110 GHz.

High cost and complexity of the VNAs based on the method of frequency translation, deterioration of parameters in case of attempts to simplify the circuit, absence of stable sources of SHF signal, and wideband frequency converters—all these predetermine the interest to homodyne VNAs, especially in solving measurement tasks within the shortwave part of the millimeter wave range (3–2 mm and shorter).

11.9.3 Homodyne VNAs. Measurement Methods Depending on the Modulation of the Reference and Measuring Signals

Circuit with Absence of Signal Modulation

In designing circuits of homodyne VNAs, various types of modulation of measurement and reference signals can be used—first of all, an option is possible where modulation of SHF signals (reference and measurement) can be totally absent, while processing of information about the measured *S*-parameters is performed using direct current. This way is used in development of simple and cheap homodyne VNAs that, however, cannot compare to heterodyne VNAs in terms of their characteristics.

The main downside of such analyzers is the complexity of calibration and measurement algorithms that do not allow implementation of panoramic measurement mode, low limits of measured depletions (not higher than 40 dB), and the need to use a generator with increased output power (more than 20 MW).

Since this variant is not promising, it is not considered further.

Circuit with Amplitude Modulation of the SHF Signal in the Source Itself

The next way of designing a homodyne VNA circuit is based on using amplitude modulation of the SHF signal in the source itself. In used types of measuring generators, amplitude modulation is usually performed by modulating voltage of squarewave type. This is quite an effective way to increase sensitivity and expand the dynamic range of a VNA; however, it has its own principal limitations due to the presence of the component proportional to the power of the reference signal. This component prevents measurement of the level of effective interference signal that is lower than the reference signal. Therefore, even though the dynamic range of the analyzer is increased, it does not exceed 40 dB.

Output signals of primary converters in case of such modulation contain non-linear members that significantly complicate algorithms of calibration and processing of the measurement data. This version of design of a homodyne VNA circuit is not a prospective one either in terms of achievable parameters.

Other ways of measurement of *S*-parameters with detection of SHF-signals suggest modulation of SHF signals in channels (tracts) of network analyzers. This modulation can be both amplitude-based (AM) and phase-based (PM).

11.9.4 Circuit with Amplitude Modulation in the Measuring Channel

11.9.4.1 Schaffer–Kohn Method

Schaffer–Kohn method suggests modulation based on amplitude of the SHF signal only in the measurement channel, the signal of which is known as modulated carrier, unlike non-modulated subcarrier in the reference channel. Due to this reason, this method is sometimes referred to as *method with modulated subcarrier*.

The main advantage of the method is elimination of the component in the detected signal that is proportional only to the power of the reference signal and limits the dynamic range of the analyzer with modulation of the source (generator) signal. This component during modulation in the channel is actually the direct current voltage; it is filtered by selective amplifier set to the modulation frequency Ω. Moreover, if the balance AM of the measurement signal is implemented, non-linear member in the interference signal also becomes negligible due to suppression of the carrier. The expression for the detected voltage determining the modulus $|S_x|$ and the argument φ_x of the measurement *S*-parameter becomes extremely simple

$$U_1 = 2\alpha K_1 K_2 E_0^2 |S_X| \cos(\varphi_X + \varphi_H) \cos \Omega t, \tag{11.60}$$

where K_1, K_2 are the moduli of the aggregated coefficient of transmission of propagation circuits (channels) of reference and measurement signals from the data take-off point in the path to the detector with a transmission coefficient of α; φ_H is the phase shift taking into account non-identity of phase-frequency characteristics (PFC) of these circuits.

Voltage (11.60) is denoted by U_1, since it is proportional only to the real part of the measured parameter S_X. In order to determine the modulus $|S_X|$ and phase of φ_x, it is necessary to additionally have an expression of the form

$$U_2 = |S_X| \sin(\varphi_X + \varphi_H) \cos \Omega t, \tag{11.61}$$

i.e. it is necessary to have additional quadrature (shifted by π/2 in relation to U_1) signal. The detector with the output voltage determined by the expression (11.60) or (11.61) is known as scalar homodyne detector (SHD).

There are two ways of creation of hybrid VNAs during implementation of the balanced AM of the measurement signal.

The first way is single-band modulation of the measurement signal with suppression of the carrier implemented by means of double-balanced modulators or single-sideband modulators (SSM). Such modulators are very common in SHF equipment. Their advantage lies in relatively high speed of action, since they are based on Schottky diodes or p-i-n diodes. The main disadvantage of such modulators with single sideband is insufficient suppression of the second sideband (up to 20 dB with the necessity to have at least 40 dB) due to non-identity of frequency characteristics of the modulator channels. This disadvantage is noticeably manifested in wideband VNAs of the millimeter range. One possible way of overcoming this disadvantage of modulators is automatic correction of non-identity of the frequency characteristics of modulator channels by software and hardware means.

Second, in case with balanced AM of the measurement signal it is possible to use double-balanced mixers with suppression of image channel signal as primary converters of SHF signals. Such mixers are also well known in SHF equipment. In terms of structure and resulting effect they are equivalent to single-sideband modulators; however, they have the same downsides. In this case, they are manifested in insufficient suppression of the image channel signal.

Devices of the type of double-balanced mixers with suppression of the image channel signal, unlike scalar homodyne detectors (SHD), are known as *vector homodyne detectors (VHD)*. The aim of various circuit realizations of VHDs is to replace separately formed voltages (11.60) and (11.61) with one voltage of the form

$$U = \alpha K_1 K_2 \, |S_X| \, E_0^2 \cos \, [\Omega t + (\varphi_X + \varphi_H)]. \tag{11.62}$$

Voltage (11.62) is equal to the output voltage of the heterodyne network analyzer mixer. In terms of its information value, it is equivalent to output voltages (11.60) and (11.61) received with the help of two SHDs; this is why the VHD notion was introduced.

Historically, development of the Schaffer–Kohn method took the way of reduction of the initial level of measurement signal as compared to reference signal. In this case, non-linear member in the SHD output voltages can be neglected without noticeable increase in the error of measurement of $|S_X|$ even in case with non-balanced AM. This method, suggested by Kohn, limits the dynamic range of the circuit analyzer and, subsequently, limits the measurement of $|S_X|$. In order to avoid this limitation, it was suggested to use double amplitude modulation of the measurement signal with various modulation frequencies and process measurement information at differential modulation frequency. As a result, measurement limits could be extended to 70 dB.

An interesting modification of the Schaffer–Kohn method was suggested by I.I. Chuprov. Since it is impossible to eliminate non-linear member in SHD output voltages with normal AM without significant (by at least 20 dB) decrease in the initial level of the measuring signal in relation to the reference signal, it was suggested to account for the signal using additional voltage of the frequency F_M proportional to the value $|S_X|^2$. It is ensured with the help of an additional SHF signal converter installed in the measurement channel. Source system of non-linear equations is linearized with consideration of this voltage and solved in a much easier way as a result. However, non-identity of amplitude and phase characteristics (APC) of SHDs and additional conversion bring its own errors, and as a result the limits of measurement of $|S_X|$ can only be expanded to 60 dB.

11.9.4.2 Measuring Path of the Homodyne VNA with Modulation of the Measuring Signal

Generalized block diagram of the measurement path of homodyne network analyzer with modulation of the measurement signal is shown in Fig. 11.19.

If the SHF signal source (SO) produced amplitude-modulated signal with modulation frequency of Ω, output voltage of the SHD is described with the following expression:

$$U = \alpha E_0^2 K_1^2 [2\, K_1 K_2\, |S_X| \cos(\varphi_X + \varphi_H) + K_2^2 |S_X|^2] \cos \Omega t, \tag{11.63}$$

From this expression, it can be seen that the presence of the first member complicates significantly the possibilities of measurement of small information signals determined by second and third members, while the presence of the third member complicates significantly the algorithm of measurement data processing, as the necessity to solve a system of non-linear equations appears. Therefore, there is a natural desire to get rid of first and third members in the Eq. (11.63).

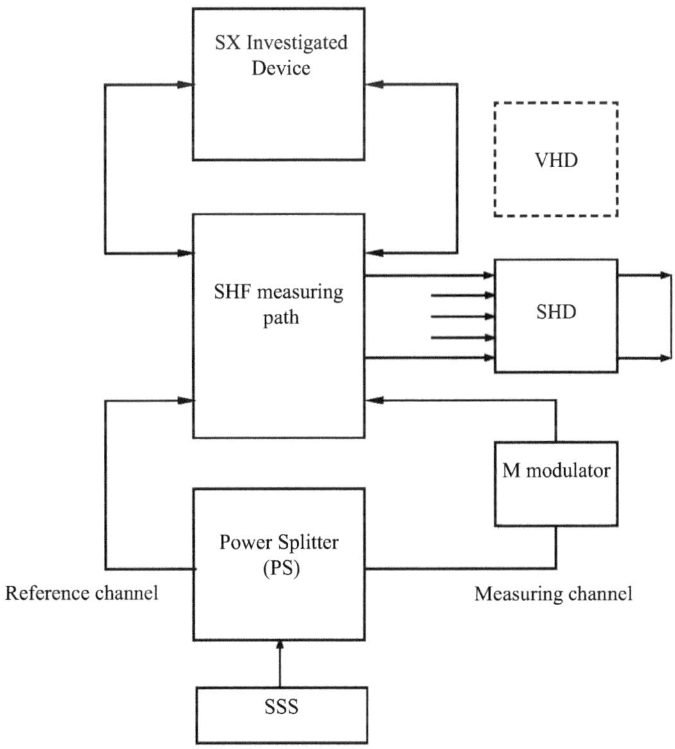

Fig. 11.19 Block diagram of the measuring SHF path with measuring signal modulation

In the event of subcarrier modulation in the measuring path, the interference signal is determined by the expression

$$U = \alpha M E_0^2 \left[2\, K_1 K_2 |S_X| \cos\left(\varphi_X + \varphi_H\right) + K_2^2 |S_X|^2\right] \cos \Omega t, \tag{11.64}$$

where M is the amplitude modulation coefficient.

The detected output signal of the reference signal determined by the first member of the expression (11.64) is actually DC voltage; it is filtered out by the selective amplifier set to the frequency Ω.

The second step was also made by Kohn; it consists in using balanced amplitude modulation of the measuring signal, with which the third member in the expression (11.63) becomes negligible.

Let us write down the main expressions for this case.

As we know, measurement signal U_m presenting AM signal with balanced modulation is described by the expression

$$U_m = K_2 E_0 \frac{M}{2} \times |S_X| \left[\cos\left(\omega_0 t + \Omega t + \varphi_X + \varphi_m\right) + \cos\left(\omega_0 t - \Omega t + \varphi_X + \varphi_m\right)\right], \tag{11.65}$$

where φ_m is the measurement signal phase.

In measurement channel, where there is no modulation, the signal U_r is described by the expression

$$U_r = K_1 E_0 \cos\left(\omega_0 t + \varphi_r\right), \tag{11.66}$$

where φ_r is the phase of the reference signal.

In this case, the resulting interference signal with balanced amplitude modulation has the form

$$U = \alpha M K_1 K_2 |S_X| \cos\left(\varphi_X + \varphi_H\right) \cos \Omega t, \tag{11.67}$$

where

$$\varphi_H = \varphi_m - \varphi_r. \tag{11.68}$$

Signal of the form (11.67) can be obtained from the expression (11.63) differently, by using balance mixer instead of a usual SHD. In this case, expression (11.67) will only differ by presence of the relevant proportionality ratio.

11.9.4.3 Circuits with Linear Phase Modulation and Phase Shift Keying of the Measuring Signal

Along with amplitude modulation of the measurement signal, one of the first classic methods applied in homodyne circuit analyzers was the method with application of *phase modulation* (PM) or *phase manipulation* (PMP) of the measurement signal. Widely known is application *of linear phase modulation* of the measuring signal within 0–2π performed with the help of electromechanical Fox phasers or their analogs—ferrite polarization rotators (for example, Fox phasers are used in units to measure depletion and calibrate attenuators of the millimeter wave range of type DK1–15 within the frequency range of 37.5–78.3 GHz, and electrical phasers, linearly modulated according to the "saw" law are used in installations of type DK1–19 and DK1–21 in frequency ranges of 78.33–118.1 GHz and 129.2–142.8 GHz respectively). Periodic change in the phase of the measuring signal within 0–2 π according to the linear law is known to be equivalent to shift in the frequency of this signal by the value equal to the phase change rate. In fact, this is a single-sideband modulation of the measuring signal, which is necessary for creation of homodyne VNAs. Such network analyzers can be implemented without serious problems in the shortwave part of the millimeter wave range. However, it is necessary to point out (as a disadvantage of the Fox phasers) the fact that these phasers do not ensure high speeds of phase change and, subsequently, great shifts of the measuring signal frequency—tens and thousands of cycles per second.

The generally accepted value of the measurement signal frequency is 100 kHz, at which measurement information is processed in modern network analyzers. The frequency of 100 kHz is optimal, since it allows to obtain high speed of data processing ensuring real-time scale during the study of characteristics of *S*-parameters; at the same time, the influence of *1/f* noises is sufficiently minimized.

Unlike Fox electromechanical phasers, modern electrical ferrite phasers ensure the possibility of linear modulation with frequencies of several dozen kHz (up to 100 kHz) within frequency ranges of up to 180 kHz and can be used for creation of VNA circuits. Such phase modulators are used, in particular, in VNAs R4206, R4204, R4203, and R4202 produced by Elmika (Lithuania). We shall only point out that production of such ferrite phasers is a challenging technological task.

The other method suggests *discrete phase manipulation* of the measuring signal *based on the algorithm 0/π*. This phase manipulation is known to be equivalent to balance AM with the output signal spectrum equal to

$$E(t) = \frac{2}{\pi}E_0\left[\cos\left(\omega_0 + \Omega\right)t - \cos\left(\omega_0 - \Omega\right)t + \frac{1}{3}\cos\left(\omega_0 + 3\Omega\right)t - \frac{1}{3}\cos\left(\omega_0 - 3\Omega\right)t + \ldots\right], \quad (11.69)$$

where $\Omega = 2\pi F$ corresponds to the frequency F of the measuring signal phase switching with amplitude E_0; in addition, switching voltage determining the spectrum (11.69) shall be a squarewave.

As in case with balanced AM, the spectrum contains no carrier, due to suppression of which the non-linear member becomes negligible in the interference signal.

Comparison of the expression (11.69) with the expression describing the signal spectrum with balanced AM demonstrates that, first of all, even with the amplitude modulation coefficient $M = 1$, the ratio of the amplitudes of the first (useful) sidebands of signals is $4/\pi \approx 1.27$; due to this, dynamic range of the network analyzer with phase modulation according to the algorithm 0-π is expanded.

Second, deviation of the switched phase shift from 180° has less effect on the carrier suppression level than non-identity of characteristics of balanced amplitude modulator arms.

It should be noted that there are currently binary phase rotators based on pin diodes within the range of 110–120 GHz. They have quite high speed, reliability, and good operational parameters (insensitive to static electricity, do not require additional supply voltage and have much smaller dimensions than balance modulators).

When using phase manipulation according to algorithm 0-π, calibration and adjustment algorithm are similar to the case with balanced AM.

Everything above allows us to make a conclusion about full competitive ability of this modulation method as compared to balanced AM and prospectiveness of its use in circuits of millimeter waverange VNA circuits.

It is also possible to use discrete phase manipulation according to the algorithm 0-π/2 in VNA circuits. This case corresponds to a non-balanced AM, while the output signal spectrum is described by the expression

$$E(t) = \frac{\sqrt{2}}{\pi} E_0 \left[\frac{1}{2} \cos \omega_0 t - \cos (\omega_0 + \Omega) t + \cos (\omega_0 - \Omega) t - \frac{1}{3} \cos (\omega_0 + 3\Omega) t + \frac{1}{3} \cos (\omega_0 - 3\Omega) t \ldots \right]. \tag{11.70}$$

As can be seen from (11.70), the level of the carrier in the output signal does not exceed $0.707E_0$, while in case with non-balanced AM it would amount to E_0, which also indicates the advantage of the phase manipulation.

11.9.4.4 Circuit with Reference Signal Modulation

Among various ways of signal modulation in homodyne network analyzers, worth noting is the case when only the reference signal is modulated. Modulation can be performed both by amplitude and by phase of this signal.

During modulation of the reference signal, the detected signal contains a constant component proportional to $|S_X|$, which is filtered out by the selective amplifier. However, the detected signal also contains the component proportional to the power of the reference signal.

This unwanted component can be excluded by using the voltage of the sweep oscillator WKS system sensor, which is also proportional to the power of the reference signal, as compensating voltage. WKS system sensor is plugged into the reference channel, thus solving two tasks at once. It is also possible to include a special two-channel programmed amplifier with subtraction circuit that excludes non-identity of the AFC of WKS system sensor and SHD from the measurement results. In practice, it is impossible to fully suppress or consider the algorithmically undesired component; therefore, the normal measurement limit of $|S_X|$ does not exceed 40 dB.

Circuit of homodyne analyzer employing linear phase modulation (according to the "saw" law) of the reference signal in the millimeter wave range is applied in installation of depletion measurement and calibration of attenuators of types DK1–15, DK1–15/1, DK1–19, and DK1–21 (=37.5–142.8 GHz).

The principle of action of such units is based on transformation of the millimeter range signal into a signal of fixed intermediate frequency and measurement of depletion of the tested four-pole by means of replacing depletion of the four-pole at working frequency with depletion of the reference attenuator at intermediate fixed frequency.

Block diagram of the unit is shown in Fig. 11.20.

Dynamic range of depletions measured by such units is 0–80 dB. Precision of the transmission ratios measurement is 0.15–1.22 dB.

The units make it possible to measure phase shifts within 0–360° with an error of 0.6–3.1 °.

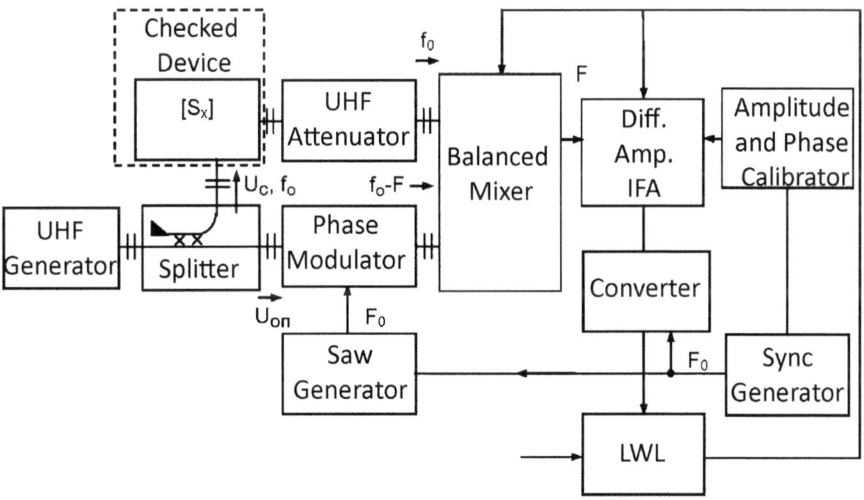

Fig. 11.20 Block diagram of a homodyne depletion tester with phase modulation of the reference signal

11.10 Homodyne Circuits with Simultaneous Modulation of Reference and Support Signals

Circuits with simultaneous modulation of both reference and measurement signals give a possibility to combine the advantages that are characteristic of methods of modulation of only measurement or only reference signals. The method allows to obtain the resulting signal of the form (11.62), which is free from all non-linear members. Such solution allows to obtain the resulting signal characteristic of heterodyne VNAs by using relatively simple means. This method is universal in relation to the used means of modulation. The most important thing is that the modulation has to be separate, and frequencies of modulating voltages shall be different. Since processing of the measurement information is carried out at differential frequency, the method is equally effective when using PPM and AM. The advantage of the method is its feasibility in nearly entire range of millimeter waves, including its shortwave portion.

Block diagram of a homodyne network analyzer with separate modulation of measurement and reference signals is shown in Fig. 11.21.

Let us consider an instance of separate amplitude modulation requiring application of autonomous amplitude modulators for reference (M1) and measurement (M2) channels. Significant is the difference in modulation frequencies; the modulating signal can take any form. Let us consider it a sinewave. Useful signal extraction and processing of measurement information are performed at differential or aggregated modulation frequency.

It should be noted that similar measurement signal can also be obtained due to double modulation of the measurement signal; however, separate modulation is much more easy to carry out than double modulation. Therefore, it is more preferable and equally fitting for the modulation type and frequency range of VNAs.

Let us demonstrate that even in case with regular AM of the reference signal with the frequency of Ω_1 and of the measurement signal with the frequency of Ω_1 output voltages of amplitude-phase detectors (APD) are described by expressions (11.60) and (11.61) in case of approximation of sweep oscillator diodes by second-level polynomial. In spectral composition of output voltages with these initial conditions, we get:

- DC voltage

$$U_1^{\mathrm{I}} = \alpha_1 E_0^2 \left[K_1^2 + 2K_1K_2\,|S_X| \cos(\varphi_X + \varphi_H) + K_2^2 |S_X|^2\right], \tag{11.71}$$

- voltage of the frequency Ω_1:

$$U_1^{\mathrm{II}} = \alpha_1 M_1 E_0^2 \left[K_1^2 + 2K_1K_2\,|S_X| \cos(\varphi_X + \varphi_H)\right]\, \cos\Omega_1 t, \tag{11.72}$$

where M_1 is the value of the AM coefficient in the reference channel;

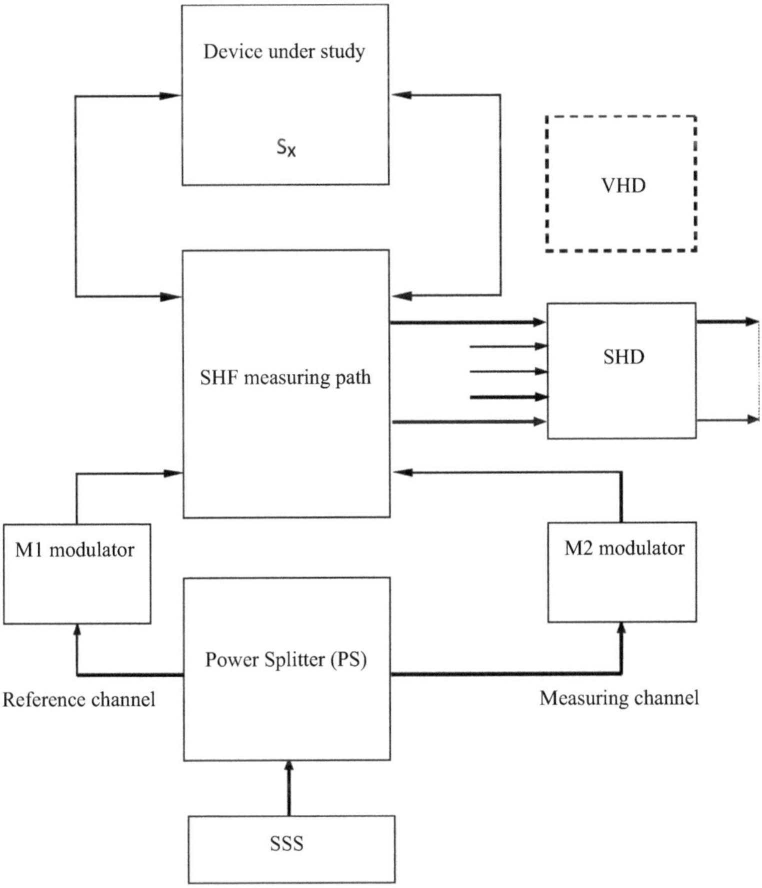

Fig. 11.21 Block diagram of a measuring CHF path with separate modulation of reference and measurement signals

– voltage of the frequency Ω_2:

$$U_1^{\mathrm{III}} = \alpha_1 M_2 E_0^2 \Big[2K_1 K_2 \, |S_X| \cos\left(\varphi_X + \varphi_H\right) + K_2^2 |S_X|^2 \Big] \cos \Omega_2 t, \tag{11.73}$$

where M_2 is the value of the AM coefficient in the measurement channel;

– voltage of the frequency $2\Omega_1$:

$$U_1^{\mathrm{IV}} = \frac{1}{2} \alpha_1 M_1^2 K_1^2 E_0^2 \cos \Omega_1 t, \tag{11.74}$$

– voltage of the frequency $2\Omega_2$:

$$U_1^{\mathrm{V}} = \frac{1}{2}\alpha_1 M_2^2 K_2^2 E_0^2 |S_X|^2 \cos 2\Omega_2 t, \tag{11.75}$$

– voltages of the frequencies ($\Omega_1 + \Omega_2$) and ($\Omega_1 - \Omega_2$):

$$\begin{aligned} U_1^{\mathrm{VI}} = \alpha_1 M_1 M_2 K_1 K_2 E_0^2 |S_X| \cos\left(\varphi_X + \varphi_H\right) \\ \times \left[\cos\left(\Omega_1 + \Omega_2\right)t + \cos\left(\Omega_1 - \Omega_2\right)t\right]. \end{aligned} \tag{11.76}$$

The obtained expressions indicate that if voltage of the frequency ($\Omega_1 + \Omega_2$) determined by the first member in (11.76) or voltage of the frequency ($\Omega_1 - \Omega_2$) determined by the second member of the expression (11.76) is selected as the measurement data signal, each of the expression will be free from non-linear members. Moreover, each of these voltages is in fact not the product of direct detection of amplitude-modulated SHF signal, but the product of conversion of the frequency of SHF signals presented by side bands in the spectra of reference ($\omega_0 \mp \Omega_1$) and measurement ($\Omega_0 \pm \Omega_2$) signals.

The analysis of the obtained expressions shows that this method is a generalization of the methods considered above. It included DC voltage (11.71) characteristic of the twelve-pole method, voltage of the frequency Ω_1 used in circuits with input signal modulation, and finally, voltage of the frequency Ω_2 used in the Schaffer–Kohn method. At the same time, components of the frequency voltage ($\Omega_1 \pm \Omega_2$) are present at the APD output, and each of them is completely free from non-linear members according to (11.73).

It should be noted that the dynamic range of such analyzer is significantly affected by values of modulation coefficient of reference and measurement channels M_1 and M_2 as a result of non-optimal energy ratios occurring in case of amplitude modulation. Transition from amplitude modulation to phase modulation according to the algorithm 0-π eliminates this downside and expands the dynamic range of VNAs.

In order to prove similarity between characteristics of hybrid homodyne and heterodyne network analyzers, let us present voltages (11.69) and (11.70) not as a result of direct detection of AM signals, but as a result of conversion of the frequency of the signals that are sidestrips in spectra of reference and measurement signals. Accepting modulating voltages of frequencies Ω_1 and Ω_2 as sinewaves, let us graphically illustrate the frequency conversion process (Fig. 11.22).

It can be seen that output of each SHD in fact contains two frequency signals ($\Omega_1 - \Omega_2$) and ($\Omega_1 + \Omega_2$). They interfere with each other, and it is possible to isolate resulting voltages of the frequencies ($\Omega_1 - \Omega_2$) and ($\Omega_1 + \Omega_2$) at the outputs of the corresponding selective amplifiers, amplitudes of which are determined by ratios (11.72) and (11.73); their initial phases do not depend on φ_{x}.

Interference effect and corresponding transfer of information on the value of φ_x from initial phases of partial voltages to the resulting voltage amplitude are the difference between SHDs and VHDs. Mathematically, it is reflected in the form of expressions for voltages (11.60) and (11.64). In practice, it only affects the algorithm

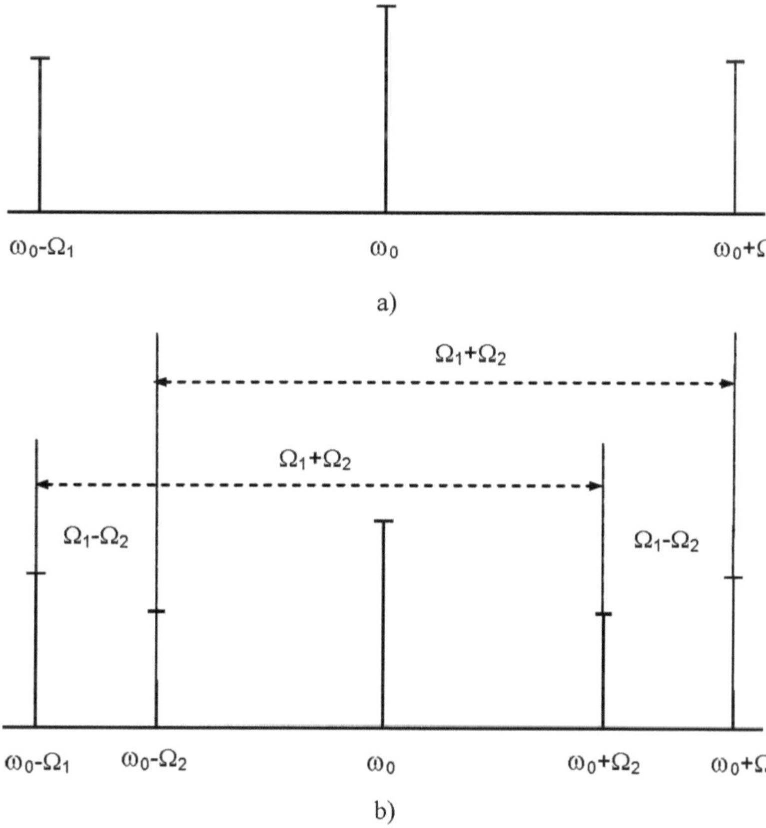

Fig. 11.22 Signal spectrum during modulation of the reference (**a**) and measurement (**b**) signals

of processing of measurement information. As for the essence of SHD operation, hybrid network analyzers employ conversion of frequencies of reference and measurement signals instead of their direct detection.

Thus, the method realizing separate modulation of reference and measurement signals turns out to be universal in relation to the type of modulation itself and is successfully implemented even in modulation of the simplest types.

For this method, we can write down

$$U_1 = \alpha_1 M_1 M_2 K_1' K_2' E_0^2 \left|S_X\right| \cos(\varphi_X + \varphi_H) \cos(\Omega_1 - \Omega_2)t, \qquad (11.77)$$

$$U_2 = \alpha_1 M_1 M_2 K_1'' K_2'' E_0^2 \left|S_X\right| \sin(\varphi_X + \varphi_H + \delta) \cos(\Omega_1 - \Omega_2)t. \qquad (11.78)$$

Let us point out another crucial circumstance.

As can be seen from (11.74), voltage with a frequency of $2\Omega_1$ is proportional to E_0^2, i.e. output power of the sweep oscillator, and does not depend on $|S_X|$ and φ_X. Therefore, it can be used for functioning of the sweep oscillator WKS system, and there is no need for a special sensor of the WKS system. Indeed, for functioning of the WKS system it is only necessary to isolate the voltage (11.74) with the help of selective amplifier set to the frequency of $2\Omega_1$. It should be noted that in case with APD, we have not one, but two voltages of type (11.74):

$$U_1 = \frac{1}{2}\alpha_1 M_1^2 K_1'^2 E_0^2 \cos 2\Omega_1 t, \tag{11.79}$$

$$U_2 = \frac{1}{2}\alpha_2 M_1^2 K_1''^2 E_0^2 \cos 2\Omega_1 t, \tag{11.80}$$

In practice, it is more correct for WKS from the methodical point of view to use voltage

$$U_{\mathrm{APM}} = \frac{1}{2}\left(U_1^{\mathrm{IV}} + U_2^{\mathrm{IV}}\right), \tag{11.81}$$

averaging the characteristic spread of α_1 and α_2, $K_1{}'$ and $K_1{}''$.

It is possible to use another way and use U_{WKS} not for stabilization of the sweep oscillator output power, but for standardization of the information signals (11.77) and (11.78). Such standardization, performed as a division operation, can be implemented by analog (using the voltage relation) or digital (when processing information with a computer) methods. This includes the effect of the WKS system on characteristics of the sweep oscillator, first of all—on the stability of the oscillator operation.

Advantages of the method with separate modulation of measuring and reference signals described above predetermine the feasibility of its use and popularity in creating waveguide vector network analyzer in the millimeter wave range. Most developed homodyne VNAs for sections of waveguide paths 11×5.5 mm, 7.2×3.4 mm, 5.2×2.6 mm, 3.6×1.8 mm, 2.4×1.2 mm, and 1.6×0.8 mm are based on this circuit with application of various types of modulation (amplitude, linear phase, phase manipulation). Circuits and operating principle of various types of such homodyne VNAs are detailed below.

11.11 Block Diagrams of Automatic Analyzers of Homodyne Networks

11.11.1 Block Diagrams of Network Analyzers with 14-Pole Measuring SHF Path

Block diagram of a VNA with fourteen-pole measuring path and phase manipulation in the measuring channel is shown in Fig. 11.23; the corresponding diagram with implementation of separate modulation in reference and measurement channels is shown in Fig. 11.24 [12, 22].

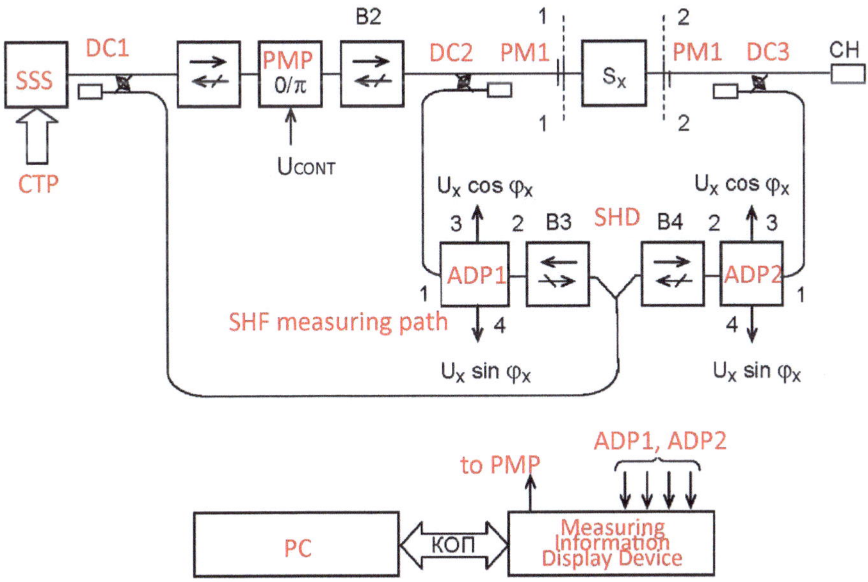

Fig. 11.23 Structural diagram of a fourteen-pole homodyne VAN with PM in the measurement channel

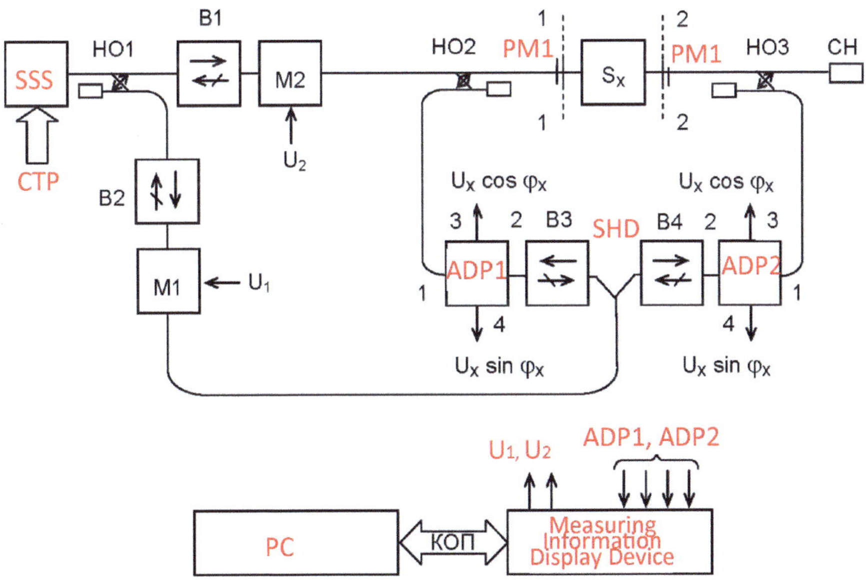

Fig. 11.24 Block diagram of a fourteen-pole homodyne VNA with separate modulation of reference and measurement signals

In the diagram shown in Fig. 11.23, the signal from the sweep oscillator output is supplied to the input of the directional coupler DC1, the outputs of which form measurement and reference channels. In order to reduce spurs between reference and measurement channels, it is necessary to use couplers with the highest possible directionality (not less than 35–40 dB). The measuring channel includes a phase manipulator operating in accordance with the algorithm 0-π. Since PM outputs are characterized by a relatively high VSWR level (of order 2–3), reduction of its effect is ensured by ferrite valves V1 and V2 implemented in the circuit.

The signal is supplied from the PM output through the directional coupler DC2 to the tested device characterized by the dispersion matrix $[S_X]$ and then to the directional coupler DC3. Coupler DC2 is oriented to the signal reflected from the tested device, coupler DC3—to the signal passing through the tested device. If possible, parameters of couplers DC2 and DC 3 shall be equivalent to those of the coupler DC1, since they are also interconnected by compromise relations between useful signal and spur. In addition, for the directional coupler DC2 oriented to the reflected signal, high directionality is a more crucial requirement, since it determines the lower limit of the value of the measured reflection coefficient and the error in its measurement. For example, if the DC2 directionality is 30 dB, the calculated limit of the measured reflection coefficient cannot be lower than 0.001. Introduction of the coupler DC3 into the analyzer circuit requires certain explanation, since it at first seems only to introduce additional depletion, which reduces the measurement limits by the value of the transitional depletion. However, first of all, introduction of DC2 with matched load ML at its output ensures better matching in the path, which is especially important in the millimeter range, and reduce the measurement error for parameters S_{11} and S_{22}. Second, its introduction allows for additional improvement of the APD2 output with the output of the tested device. Third, it allows for increased isolation of measurement and reference channels.

On the diagram in Fig. 11.23, APDs are represented by quadrature bridges detailed further. Each of the APDs is connected to the output of the secondary channels of couplers DC2 and DC3 and ensures simultaneous presentation of the measurement data on reference and transmitted signals. The reference signal is supplied to the inputs of 2 APD1 and APD2 through the power splitter PS with transitional depletion of 3 dB and isolation between outputs of at least 12 dB. Application of the power splitter DM is due to the need to ensure zero phase progression of division of the reference signal between APD1 and APD2. The circuit is fitted with valves V3 and V4 as an additional measure for increasing isolation between reference and measurement channels, as well as between the channels of reflected and passed waves.

The difference of the block diagram in Fig. 11.24 from the diagram in Fig. 11.23 lies in the use of separate modulators in reference and measurement channels. As noted before, these modulators can be both phase and amplitude. As compared to PPMs, modulators have better matching; therefore, during their use it is not required to additionally install isolating valves at the input and output of each of them. However, there is another problem associated with the necessity to exclude the possibility of penetration of the modulating signal from the reference channel to the measurement channel, and vice versa. For this purpose, the circuit includes valves V1 and V2 that in combination with the directional coupler DC1 ensure isolation of at least 75 dB.

Figures 11.23 and 11.24 show sections of waveguides FM1 that are used in calibration and serve as phase measures (PM). Their length is equal to a quarter of the mean wavelength in the measurement path waveguide $\frac{\lambda_{g0}}{4}$. In the shortwave part of the millimeter wave length, the section of the waveguide with such length is represented by a thin washer. As a rule, two of such washers are inserted in the measurement channel and taken into account during equation of electric lengths of channels. Therefore, they are shown as parts of the measuring path on VNA diagrams. In addition to PM1, there are also PM2, the length of which exceeds the length of PM1 by $\frac{\lambda_{g0}}{4}$. Therefore, if PM1 washers are replaced with PM2 washers, it will change the reflection coefficient phase at the medium frequency of the VNA working range by $2\varphi_0 = 180°$. Frequency dependence of the altered phase can be calculated according to the known formula

$$2\varphi(f) = \frac{4\pi\Delta l}{\lambda_g} = \frac{4\pi\Delta l}{c} = \frac{4\pi\Delta l}{c} f\sqrt{1 - \left(\frac{c}{2\alpha f}\right)^2}, \tag{11.82}$$

where Δl is the difference between geometrical lengths of MPs; s is the speed of light; f is the measurement frequency; α is the size of the wide wall of the MP waveguide.

Calculation of the frequency pitch $2\varphi(f)$ for the frequency range of a standard waveguide demonstrates that it does not exceed 180 ± 60 °.

11.11.2 Block Diagram of the Network Analyzer with 10-Pole Measuring HF Path

Block diagram of a ten-pole homodyne VNA with separate modulation of measurement and reference signals is shown in Fig. 11.25. This diagram differs from the ones shown in Figs. 11.23 and 11.24 in simplification of the SHF path and implementation of additional possibilities in switching of SHF signal propagation directions, which consist in the method of separate modulation of reference and measurement signals.

The main components of the VNA SHF path in Fig. 11.25 are the measuring bridge (MB), power splitter (PS), AFD, and modulators M1–M3. Measuring base of the MB is the ring bridge included into the balance bridge. Properties of such in-phase bridge with two inputs and two in-phase outputs are determined by the possibilities of waveguide power splitters. The measurement bridge is formed from the ring bridge by means of rupture of the channel connecting inputs 2 and 3 of the MB in Fig. 11.25. Two matched loads (not shown in Fig. 11.25) that ensure proper

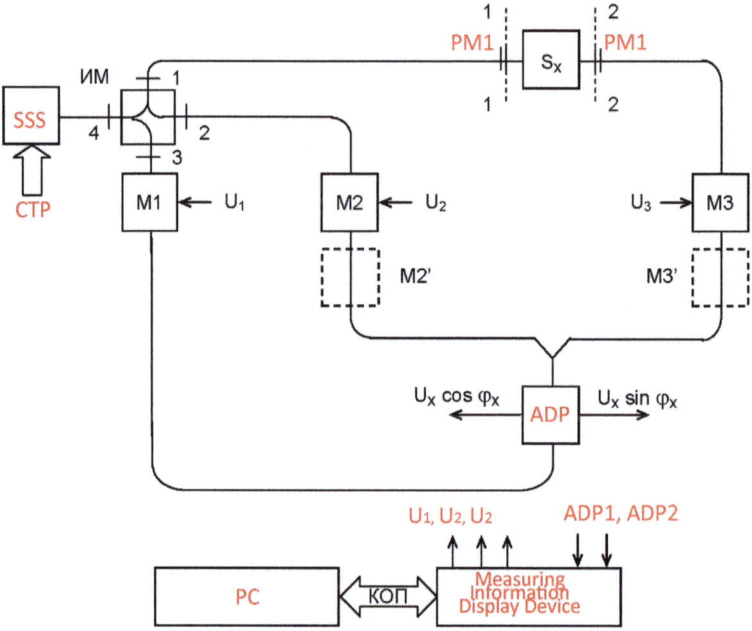

Fig. 11.25 Block diagram of a homodyne VNA with separate modulation of reference and measurement signals

matching of the MB at the side of outputs 2 and 3 are inserted into the rupture. The signal from the sweep oscillator output is transferred to the MB input and split into two equal parts that are transmitted to outputs 1 and 3 in phase. Output 1 is the beginning of the measuring channel, to which tested two-poles and four-poles are connected; output 3 is the beginning of the reference channel. The signal reflected from two-pole or four-pole input returns to output 1 and is also spilt into two even parts, one of which is absorbed in the output device of the generator, while the other is supplied to the output 2. It is possible to include an isolating valve between sweep oscillator and MB if the generator output VSWR exceeds the valve VSWR.

Circuit of supply of the reflected signal from the modulator M2 starts from output 2. Due to higher isolation (over 35 dB), direct transmission of the generator signal to the APD measuring input between MB input and output 2 is prevented. Spurious coupling between the reference channel and the reflected wave channel (between outputs 2 and 3) is completely excluded due to inclusion of two matched loads between these outputs.

Thus, properties of the measurement bridge allow us to exclude two traditional directional couplers included into SHF path of *S*-parameter testers: directional coupler forming reference and measurement channels and directional coupler of the reflected wave.

Due to regularity of waveguides of all channels of the measuring bridge, the use of the third coupler of the passed wave is no longer required. The signal that has passed through the tested four-pole is transmitted directly through the modulator M3 to the measuring input of APD. In this version of the circuit, structure of the SHF path is significantly simplified due to replacement of three directional couplers with one.

When separate modulation of reference and measurement channels is used, it is possible to increase the level of automation of measurements and exclude mechanical SHF switches in addition to general simplification of the SHF path. This is achieved through installation of two modulators in the measurement channel (unlike Fig. 11.24, where one modulator is used). Modulator M2 is plugged into the reflected wave channel; M3—into the passed wave channel. They work in the following manner.

If it is necessary, for instance, to isolate a signal reflected from the two-pole, signal with a frequency F_2 is supplied to the modulator M2; permanent voltage is supplied to the modulator M3 and locks it. This prevents penetration of any signal through the power splitter to the measurement input of the quadrature bridge and also increases isolation of the channel of the reflected and passed waves.

In case of measurement of the signal passed through the four-pole, the modulator M2 is locked, and the signal with a frequency of F_2 is supplied to the modulator M3. In order to prevent parameter variety of serial amplitude modulators and ensure necessary isolation between the reflected and passed signals, it is possible to install additional switch modulators M1 and M2. Conditions requiring installation of modulators M3I and M3I depend on the specific isolation of the path.

11.11.3 Block Diagram of the Vector Network Analyzer with 8-Pole Measuring SHF Path

Block diagram of an eight-pole homodyne VNA with separate modulation of measurement and reference signals is shown in Fig. 11.26.

As can be seen from the figure, block diagram of such network analyzer is generalizing for the diagrams shown in Figs. 11.24 and 11.25. SHF path is indeed based on directional couplers, like in the diagram shown in Fig. 11.24. Switching and modulation of signals proportional to the reflected and passed waves are performed with the help of the modulators M2 and M3 like in the circuit shown in Fig. 11.25. The difference lies in the absence of APD and introduction of phaser into the reference channel in order to obtain quadrature signals at the SHD output, which reduces the number of poles.

The signal from the sweep oscillator output (Fig. 11.26) is split into the reference and measurement signals with the help of the directional coupler DC1 with transitional depletion of about 10 dB. The reference signal is supplied to the modulator M1 from the output of the secondary channel of DC1. The implemented circuit, which

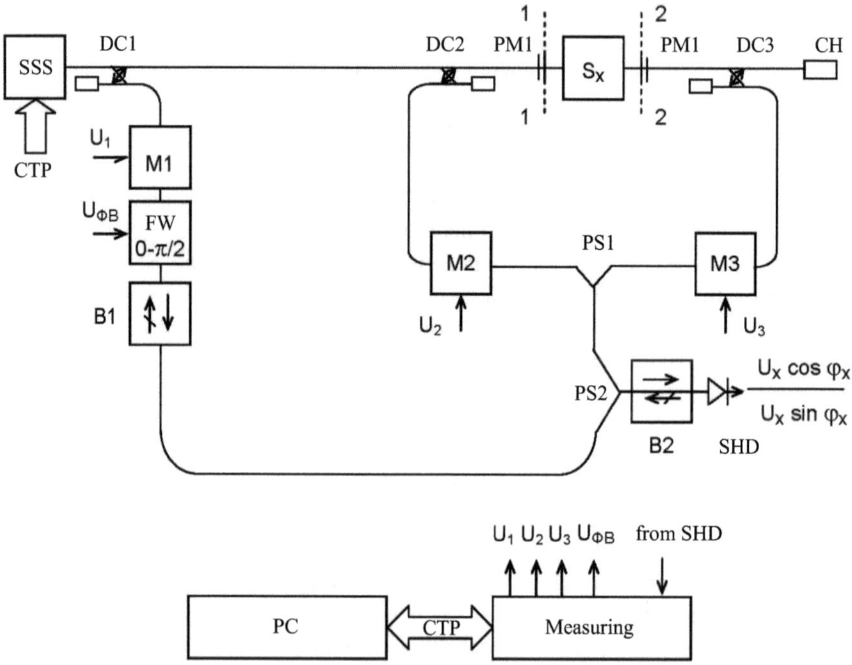

Fig. 11.26 Block diagram of an eight-pole homodyne VNA with separate modulation of reference and measurement channels

was designed for operation in waveguide path of section 1.6×0.8 mm, employs serial amplitude modulators, in which input and output VSWR did not exceed 1.15–1.20. Such matching makes it possible to use modulators M1–M3 without additional isolating valves. The signal from the output of modulator M1 is supplied to the phaser, in which periodical change of its phase 0–π/2 occurs. VSWR values of the phaser inputs are within the same limits as the modulator M1. However, it is plausible to install the isolating valve V1 in the reference channel after the phaser, as the installed power splitters PS1 and PS2 have insufficient isolation between inputs 1 and 2 (about 12 dB).

In the measurement channel after the coupler DC1, the signal passes through the coupler DC2, reflects from the tested device, and is supplied to the modulator M2. The signal passed after the tested signal is transmitted to the modulator M3 through the coupler DC3. Since modulators M2 and M3 work in terms (similar to the circuit Fig. 11.25), while couplers DC2 and DC3 have high directionality, no additional means are required to ensure isolation between the channels of reflected and passed waves.

11.12 Network Analyzers by Keysight

11.12.1 PNA-X Series Network Analyzers

In addition to mass components requiring testing, today's developers and engineers often have to deal with extremely complex high-speed, SHF and microwave radioelectronic systems, such as onboard and land radar stations, ultra-wideband communication systems (including mobile, cell, satellite and space), etc. [30]. Modern onboard and land antenna systems [31, 32] in addition to antennas contain a number of built-in SHF and microwave devices—waveguides and coaxial lines, transmission lines, filters, pre-selectors and mixers, heterodynes, high-sensitivity radio receivers, signal switches, etc. (Fig. 11.27).

Very recently, testing of such systems required application of dozens of complex and expensive devices—multimeters, measuring generators, oscilloscopes, spectrum analyzers, testers of gain, phase response and *S*-parameters, and many more. Their connection to the tested equipment and calibration required a lot of time, while the abundance of devices forced creation of specialized and extremely expensive measuring laboratories.

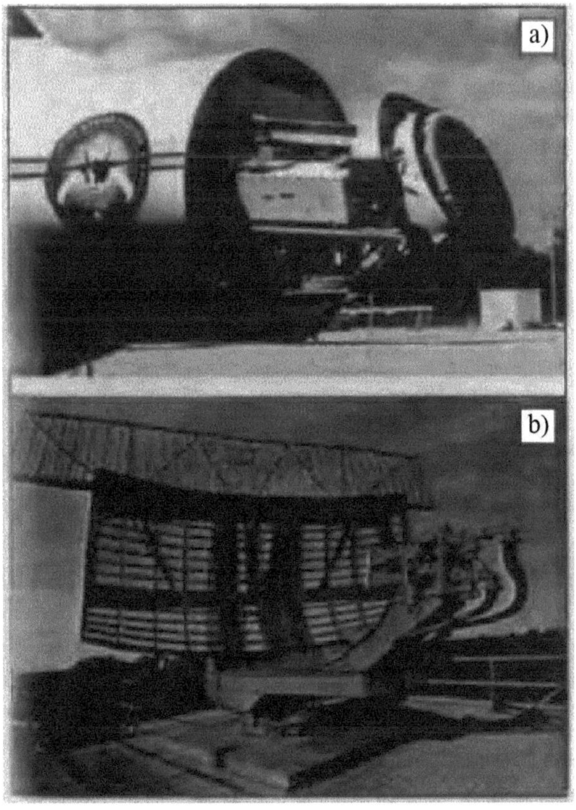

Fig. 11.27 Modern antenna systems: (**a**) onboard; (**b**) land

The greatest achievement in the development of multi-purpose measuring devices of such type was the creation of vector network analyzers that allow measuring dozens of parameters of the tested equipment with minimum number of connections and create measuring systems for especially complex and special applications (Fig. 11.28) [32].

Open architecture with a built-in PC operated by Windows system ensures extended possibilities of connection of network analyzers to other devices and maximum flexibility in their use. Low limit of the network analyzers covers the region of radio frequencies, which extends the field of application of network analyzers. The devices are fulled with means of computer math and allow easy analysis of both linear and non-linear circuits and devices both in stationary mode and in pulse mode.

Far-field region antenna testing (Fig. 11.28) employs vector analyzer of networks manufactured by Keysight technologies of PNA-X series (Fig. 11.29). This is a fairly complex tabletop analyzer—a typical example of such devices. Its characteristic attribute is the presence of ports to which the tested equipment is connected. In this case, four ports are provided: two of them are outputs of measurement generators, the other two—outputs of measurement receivers. In general, analyzers can have two, four, or more ports. Outside ports are high-quality coaxial connectors.

Simplified functional diagram of a 4-port network analyzer is shown in Fig. 11.30. From the figure, it can be seen that the analyzer contains two measuring generators with digital synthesis of frequency and its quick tuning. Generators and other units (options) can be configured differently with the help of mechanical and

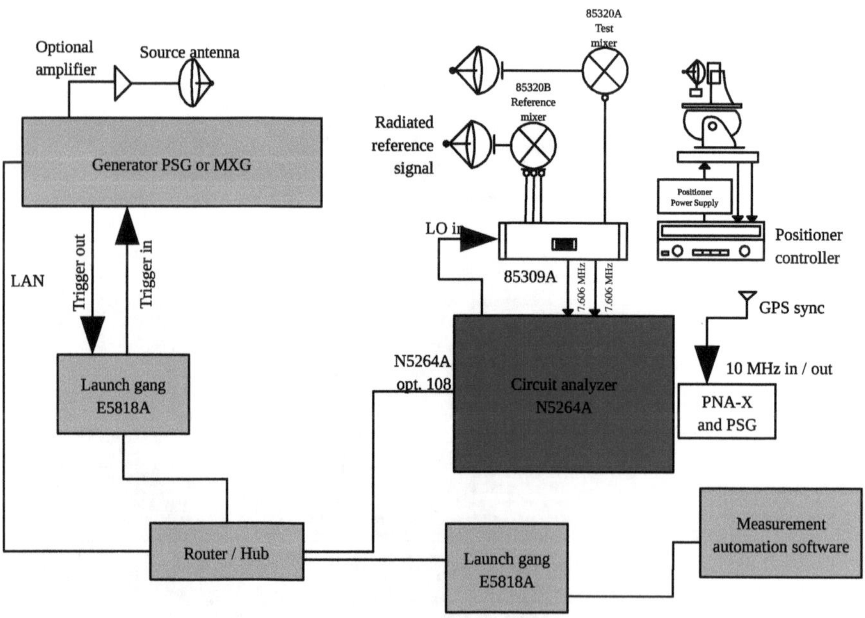

Fig. 11.28 System for testing of antennas in far-field region with receiver based on vector network analyzer by Keysight

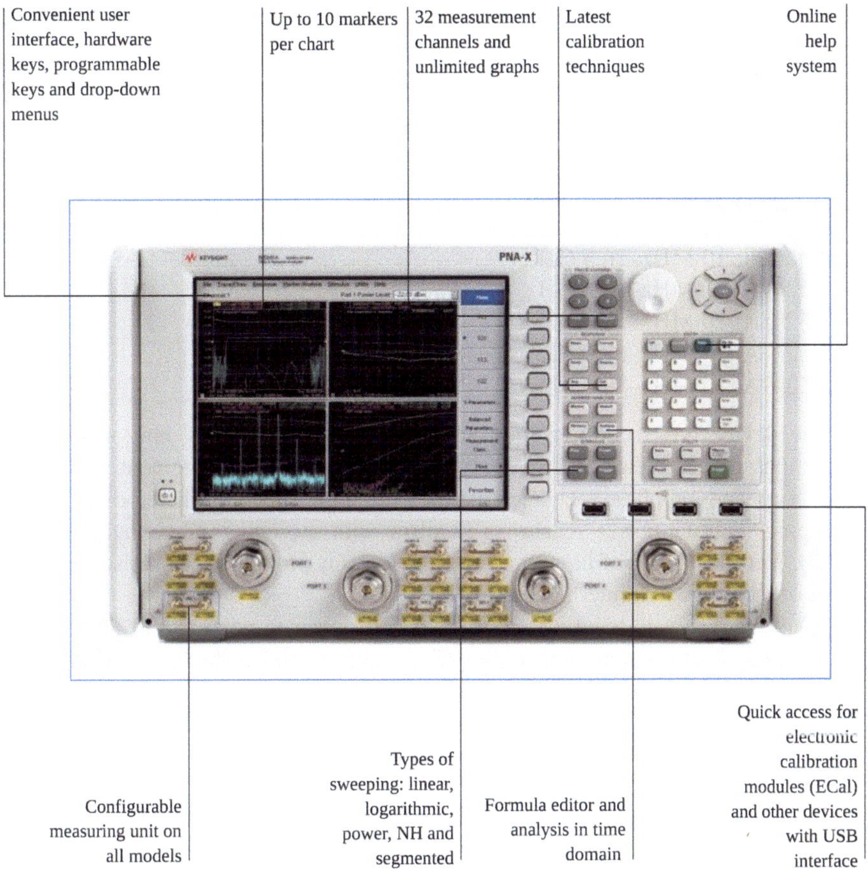

Fig. 11.29 Vector network analyzer of series PNA-X, by Keysight

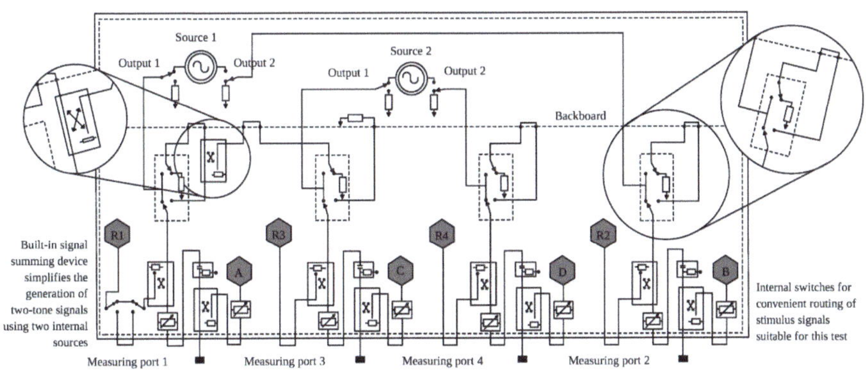

Fig. 11.30 Block diagram of the vector network analyzer of series PNA-X by Keysight

electronic switches. For example, one generator can be configured as the source of the input signal, the other one—as the heterodyne of the tested mixer or receiving module. Configuration of units and options of the circuit analyzer is performed with the help of high-precision matched paths with ultra-wide bandpass and low losses.

Generation of sinewave signals is ensured in the widest frequency range (depending on the specific model):

- N5241 A: 10 MHz–13.5 GHz;
- N5242A: 10 MHz–26.5 GHz;
- N5244A: 10 MHz–43.5 GHz;
- N5245A: 10 MHz–50 GHz;
- N5247A: 10 MHz–67 GHz;
- N5249A: 10 MHz–8.5 GHz.

The device has second internal source for measurement of inter-modulation distortions of parameter S22 in the working mode and measurement of various frequency and amplitude characteristics with high speed of heterodyne signal sweeping. Output power reaches +22 dBm with sweeping range by power of 48 dB. Very low level of harmonics (≤ -60 dBs) is provided for precise measurement of harmonic and intermodulation distortions.

Analyzer receivers have a very low level of intrinsic noise -130 dBm in the 10 Hz band with compression points of 0.1 dB at the level of +13 dBm. High sensitivity is ensured during measurements of linear *S*-parameters in pulse modes. In standard configuration, the maximum IF band reaches 5 MHz, and the receiver can be used for measurement of short radio pulses.

Network analyzer series PNA-X by Keysight present the most powerful and multi-purpose SHF analyzers in the market. They ensure a breakthrough in the field of SHF and microwaves. The devices have the following properties:

- A complex of measurements is performed during one connection.
- Frequency range is from 10 MHz to 8.5/13.5/26.5/43.5/50 GHz.
- 2-port and 4-port models.
- Two built-in signal sources with high characteristics.
- Single-block technical solution for measurement of circuit parameters in the pulse mode.
- Inner device for signal integration and mechanical switches.
- Calibration methods in line with the latest achievements in this field.
- 10.4-inch display with sensor screen function.

It should be noted that network analyzers PNA-X by Keysight employ a number of innovative and patented solutions that significantly increase the main technical characteristics of analyzers. For example, in normal analyzers of spectra and networks, it is necessary to use extremely narrowband permit filters, due to which the time of spectral analysis increases greatly. The patented method of spectral zeroing, in which the zero frequencies of the spectrum coincide with unwanted spectral lines, allows to bring the level of unwanted frequencies to -120 dB as compared to—(70–80) dB with usual spectral analysis.

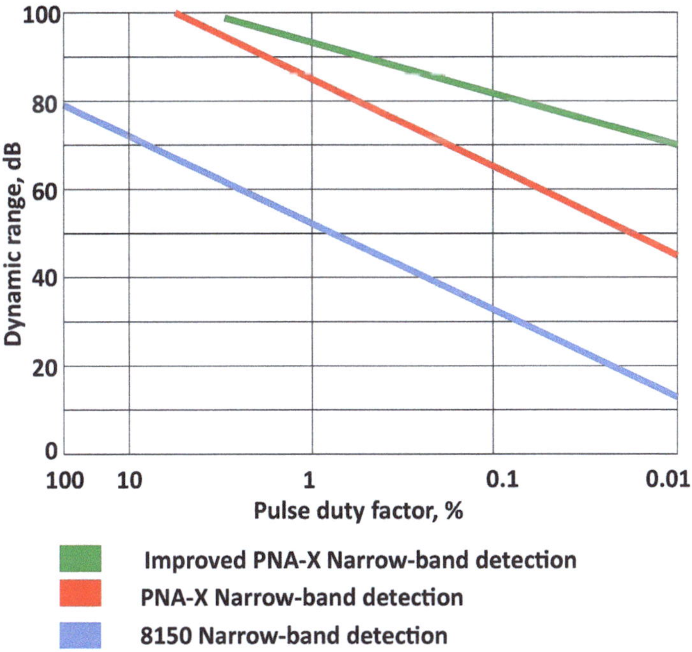

Fig. 11.31 Dynamic range of network analyzers PNA-X

As a result, it was possible to expand maximum IF band in analyzers PNA-X to 15 MHz and measure linear *S*-parameters in pulse modes during wideband detection of short radio pulses (pulse duration >100 ns). PF bands exceeding 15 MHz have no practical value since an increase in the IF band also leads to an increase in the level of the receiver's own noise, which impairs the signal-to-noise ratio and increases the measurement error. As can be seen from Fig. 11.31, dynamic range of network analyzers PNA-X has increased dramatically, especially in case with low coefficient of radio pulse filling.

In a number of applications (for instance, in radar location), spectrum and circuit analyzers work in the strobing mode. Along with hardware strobing, which is used in PNA series devices, analyzers PNA-X also employ the software strobing method patented by Keysight (Fig. 11.32).

11.12.2 Features of Testing of Amplifiers with PNA-X Network Analyzers

The use of network analyzer PNA-X allows for simple implementation of the principle "one connection - a complex of measurements." After calibration and connection to the tested device (TD), users can measure nearly all necessary parameters of the TD. For examples, during testing of amplifiers (Fig. 11.33) it is

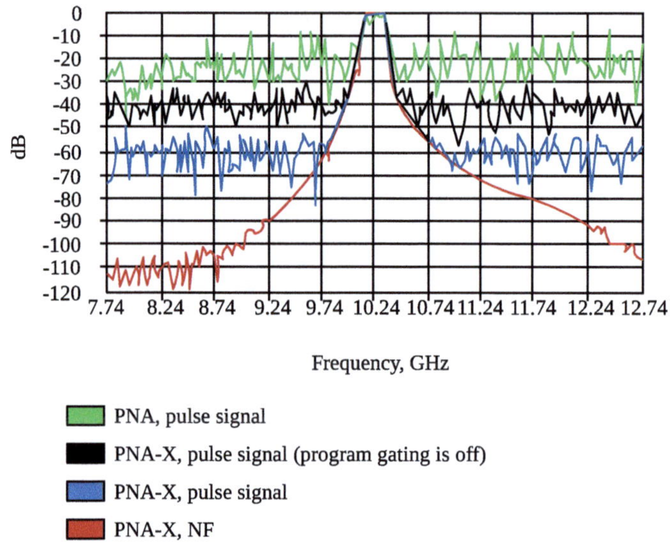

Fig. 11.32 Comparison of PNA/PNA-X for measurement of network parameters in the pulse mode

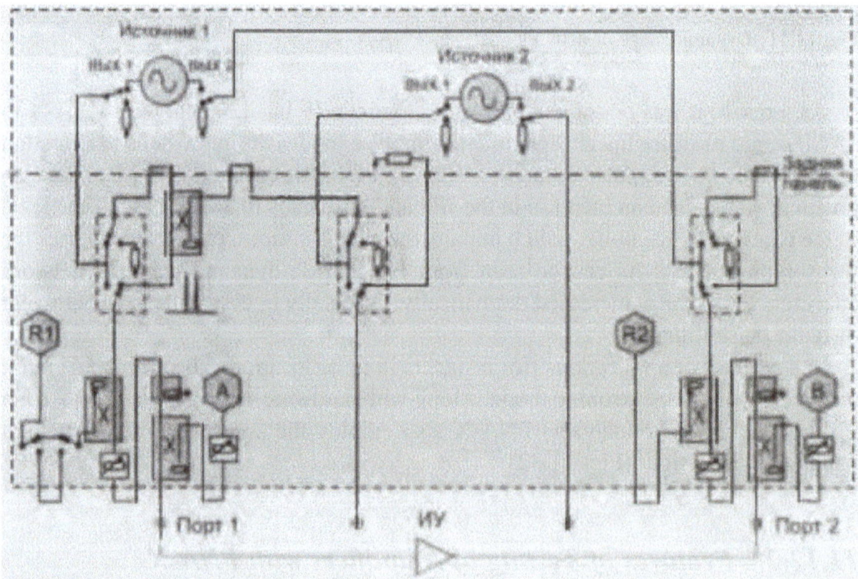

Fig. 11.33 Connection of the amplifier to network analyzer

possible to simultaneously measure matching by input and output, amplification coefficient, harmonics of the output signals, compression point at 1 dB, coefficient of transformation of amplitude modulation into phase modulation, level of the third-order intersection point (TOI), and the noise coefficient. Network analyzers of series PNA-X are perfectly suited for measurement of parameters of active components, such as power amplifiers, low-noise amplifiers, frequency converters, transmission/receipt modules, and antennas.

Properties of technical solutions of PNA-X during testing of power amplifiers and low-noise amplifiers [32]:

- One connection, complex of measurements at frequencies up to 67 GHz (with microwave options up to 1.1 THz).
- Measurement of the noise coefficient with the help of the cold source using improved matching error correction in the source in order to ensure high measurement precision.
- Support of measurement of amplifiers with high amplification ratio (>30 dB).
- Bidimensional sweeping modes during measurement of the dependence of amplification coefficient compression from frequency (Fig. 11.34).
- Alteration of intermodulation distortions in frequency sweeping and power sweeping modes.
- Measurement of harmonic distortions in sweeping mode.
- Measurement of the added power coefficient of efficiency.
- Alteration of the S22 parameter in the working mode.
- Single-block technical solution for measurement of S-parameters in the pulse mode.

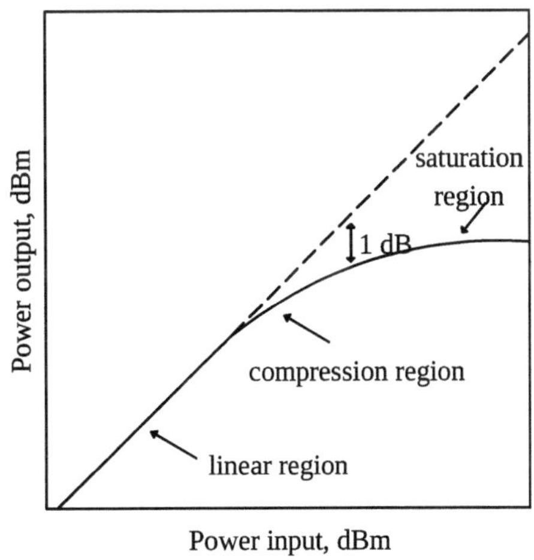

Fig. 11.34 Dependence of the output amplifier power on the input power

- Measurements using true differential stimulus signals.
- Expanded calibration of frequency unevenness for measurement of power amplifiers.
- Maximum range of power sweeping: ≥40 dB.
- Two signal sources with extremely low level of harmonics (≤−60 dBs).
- High value of the compression point at 0.1 dB (>+13 dBm).
- Expansion of possibilities due to connection of vector signal generators and analyzers to HF access points at the rear panel of the analyzer for measurement of EVM, ACPR, CCDF, and other parameters.

11.12.3 Features of Testing of Non-linear Devices with PNA-X Network Analyzers

Many radioelectronic devices (real amplifiers, mixers, detectors, etc.) are non-linear in principle. First types of circuit analyzers were only used for testing of linear devices. However, PNA series analyzers can also analyze non-linear circuits and devices. Testing of non-linear devices with network analyzers PNA-X provides the following possibilities [32]:

- Measurements of non-linear X parameters and plotting of their graphs and diagrams.
- Measurements of frequency converters with built-in heterodyne: precise measurement of the absolute phase and group delay (GD)—Fig. 11.35.
- Measurements of the amplification coefficient compression, amplification coefficient in the field region and in the compression point, input and output power in the compression point, and APC linearity.
- First ever single-unit technical solution for measurement of network parameters in the impulse mode based on network analyzer with internal pulse modulators and pulse generators.
- Measurements using true differential stimulus signals.
- Measurements using scalar calibration of the mixer: transfer of precision of the power tester to measurements of the amplification coefficient/conversion losses.
- Measurements using vector calibration of the mixer.
- Measurement of the absolute group delay (GD) for mixers and frequency converters.
- Measurements of intermodulation and harmonic distortions: measurement of distortions in frequency sweeping and power sweeping modes without the use of external components.

Figure 11.35 shows connection of a converter with built-in heterodyne to the network analyzer PNA-X. Converter heterodyne frequency measurement is performed using new technical solution (option 084) based on application of vector calibration of the mixer. At first, actual value of the IF signal frequency is measured.

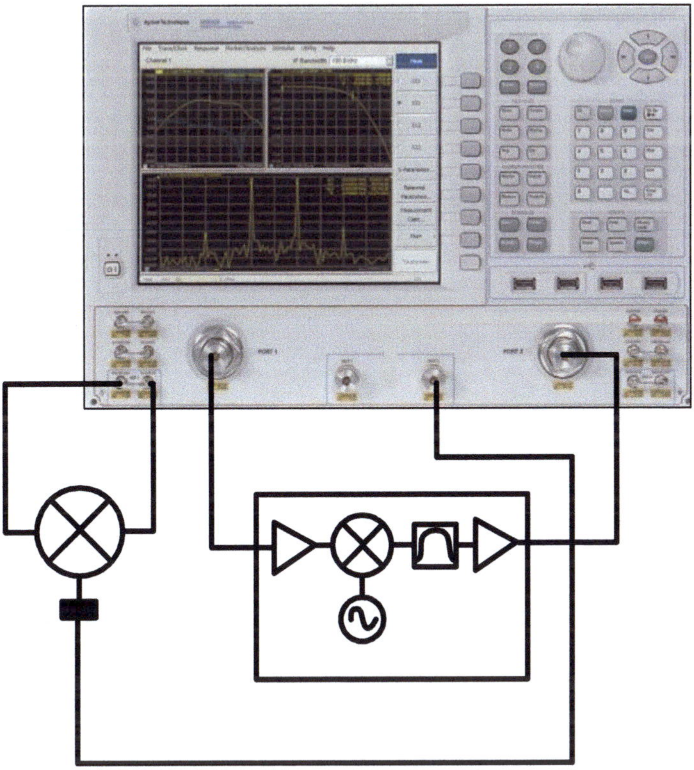

Fig. 11.35 Measurements of characteristics of a converter with built-in heterodyne

Using the result of this measurement and knowing the frequency of the incoming HF signal, it is possible to calculate the heterodyne frequency.

It requires two sweeping cycles of the network analyzer. The first sweeping cycle is wideband and allows approximate measurement of the built-in heterodyne frequency in the selected frequency range (Fig. 11.36). In this example, nominal frequency value is 5 Hz, and the maximum deviation of the heterodyne frequency in accordance with the specifications shall not exceed ±500 kHz. Thus, central frequency of the receiver is set to 5 GHz, and the frequency range—to 1 MHz. Since 200 points are used for this measurement, frequency resolution is 1 MHz/ 200 = 5 kHz.

However, the analyzer requires more precise determination of the heterodyne frequency for group delay measurement. The second sweeping cycle ensures the necessary precision of the analyzer. During this cycle, measurements of the phase P in time are performed (Fig. 11.37). After that, it is possible to determine the frequency deviation: $\Delta F = -\Delta P/(360 \times \Delta T)$. As a result of wideband and precision sweeping cycles, working frequency of the heterodyne can be determined precisely enough for performance of reliable group delay measurements.

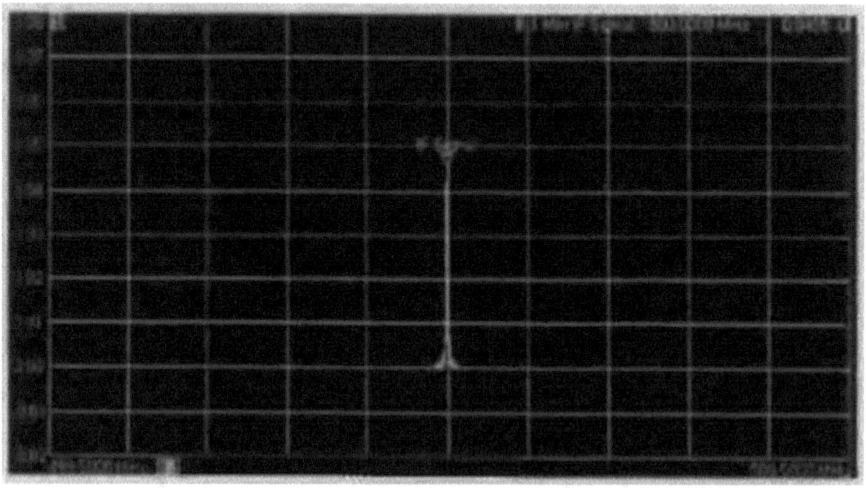

Fig. 11.36 Spectrum during wideband cycle of the converter testing

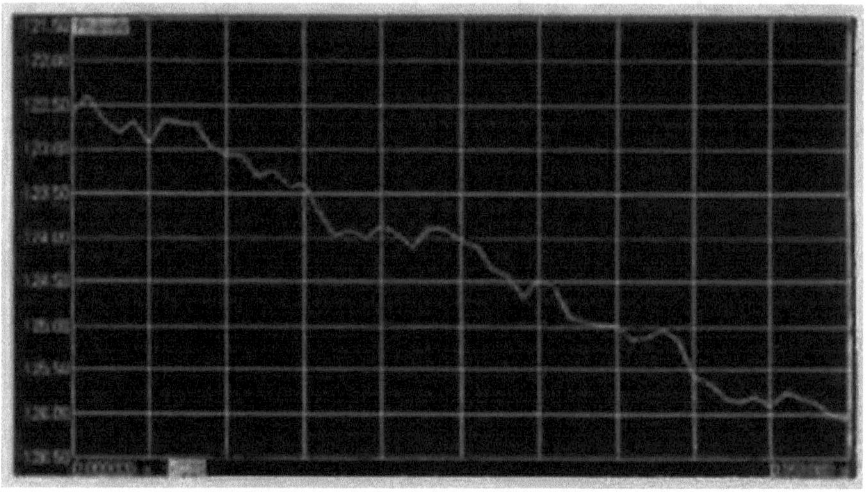

Fig. 11.37 Oscillogram of narrowband cycle of converter testing

Precision of this method can also be demonstrated by comparing measurements of the frequency converter with heterodyne, which is either synchronized with the common source of the reference signal or not depending on the case, as shown in Fig. 11.38. Since the results are nearly identical, it is evident that this technical solution is capable of effectively solving the problem of measurement of the absolute group delay of frequency converters with one built-in heterodyne.

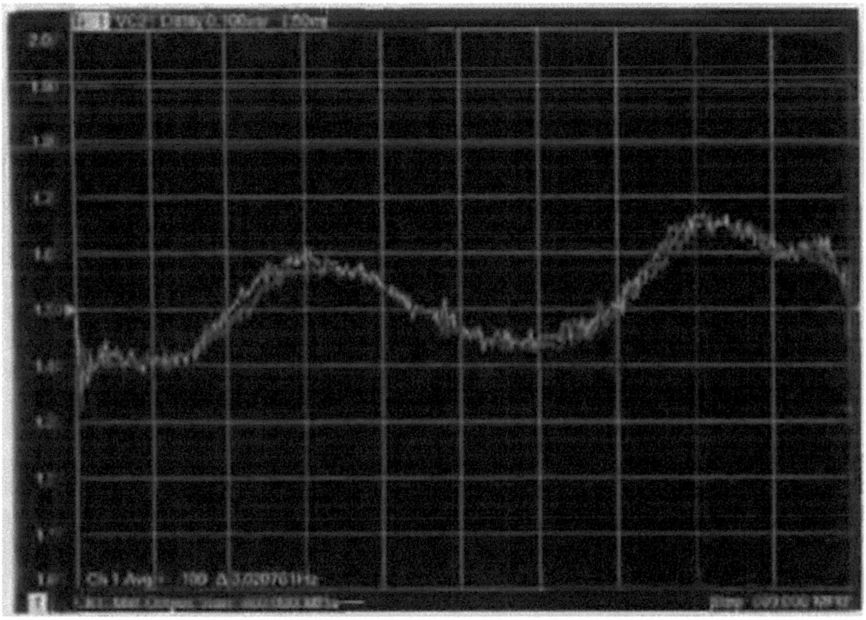

Fig. 11.38 Comparison of the results of measurement of absolute group delay for cases when the heterodyne signal source is either synchronized with the common reference signal or not

This is also proven by Fig. 11.39, which shows comparison of the results of group delay measurement on different circuit analyzers with different calibration methods. It indicates the effectiveness of the technical solution applied in PNA-X analyzers.

A two-frequency signal is usually used for measurement of intermodulation distortions. Availability of two generators in analyzers simplifies the testing procedure. Measurement results can be presented as graphs of the spectrum and frequency dependences of parameters (Fig. 11.40). They clearly demonstrate the essence of such distortions—appearance of differential and summation intermodulation frequency.

Network analyzers PNA-X can test devices with differential inputs and outputs (Fig. 11.41). Such devices are often used in high-speed systems, ensuring high protection from interference at low signal level.

Figure 11.42 shows the comparison of results of testing of devices with non-symmetrical and differential signals.

11.12.4 Means of Calibration of Network Analyzers

In development of network analyzers, Keysight pays great attention to creation of the means of calibration for the analyzers. The problem is that regular calibration

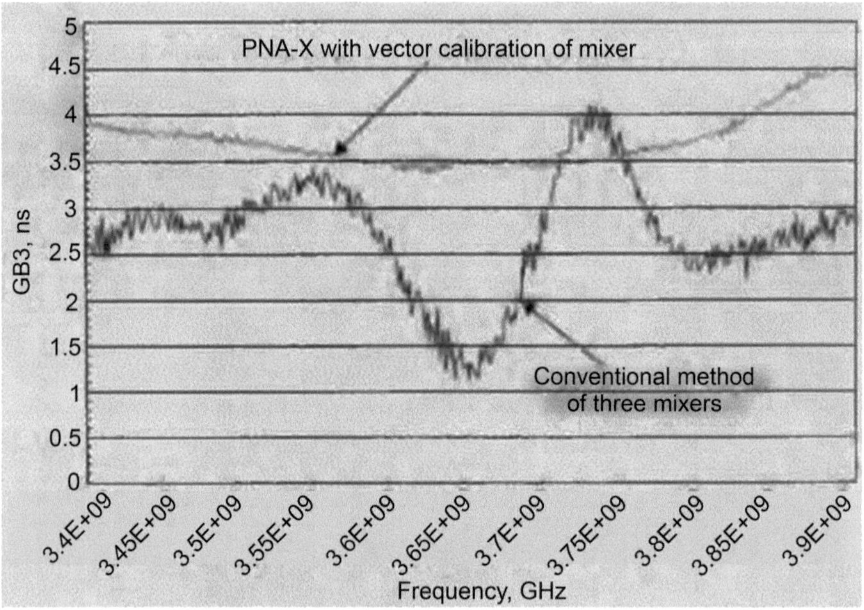

Fig. 11.39 Comparison of group delays for different analyzers of Keysight network analyzers with different calibration methods

means are only partially suitable for network analyzers. In terms of stability of amplitude and other parameters the latter are often on par with regular calibration devices; therefore, they are not applicable. Moreover, network analyzers are multi-port devices, and all ports, as well as certain combination of ports, need separate calibration means. As a result, the number of calibration operations and the time spent for this operation sometimes increase by more than an order of magnitude. For example, 4-port analyzers require 15 steps of calibration, 8-port—31 steps.

Methods of calibration during short circuits, idle run, introduction of an unknown jumper, etc. are widely used. These methods are included in the innovative method of quick calibration Quik SOLT. For 8-port calibration, the number of steps is reduced to 10. Calibration procedures are detailed in analyzer operation manuals.

Special attention is paid to obtaining even frequency characteristics of various parameters of measuring generators and receivers. For instance, Fig. 11.43 shows frequency dependencies of the level of the second harmonic and output power of generator PNA-X characteristic of a calibrated analyzer. Their high evenness indicates proper operation of the calibration system and the system of automatic adjustment of parameters of the generator output signal.

After calibration, draft is observed in the time of the measurement parameters. It also greatly depends on the frequency, which can reach extremely high values for network analyzers. Typical character of the amplitude and phase drift over 24 h and within the frequency range of up to 20 GHz of the measured sinewave signal is shown in Fig. 11.44.

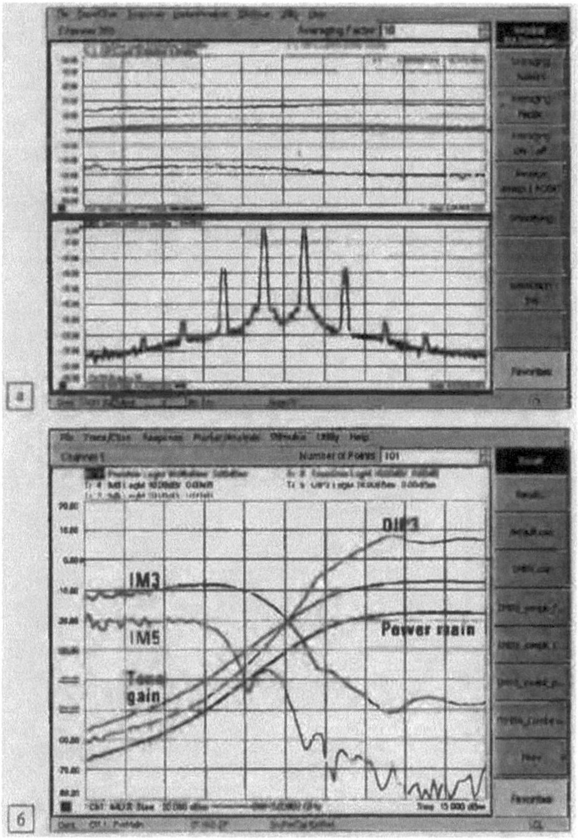

Fig. 11.40 Spectral estimation of intermodulation distortions: (**a**) spectrum; (**b**) graphical dependencies of parameters

Calibration modules ECal (Fig. 11.45) created for certain ways of measurement and significantly simplifying calibration procedure are widely used for calibration of network analyzers produced by Keysight.

11.12.5 Measuring Systems Based on Network Analyzers

Even though network analyzers are extremely powerful and multi-purpose measuring devices, sometimes their possibilities are not enough to solve ever-increasing in difficulty tasks of research and testing of radioelectronic components, devices, and systems—e.g. mass systems of mobile radio communication and satellite systems using high-speed digital modulation and manipulation. In such cases, other measuring devices can be connected to network analyzers (Fig. 11.46).

Figure 11.47 shows the block diagram of interaction between system devices and the network analyzer. MXG signal generator and MXA spectrum analyzer are connected to the connectors on the rear panel of PNA-X. They can be activated for measures of parameters requiring digital modulation or demodulation.

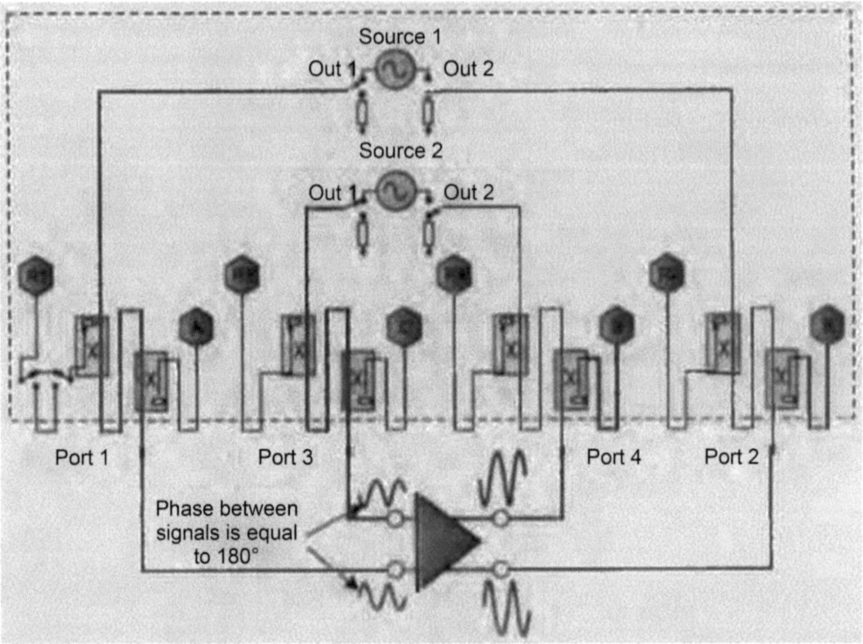

Fig. 11.41 Configuration of the analyzer for testing of devices with differential input and output

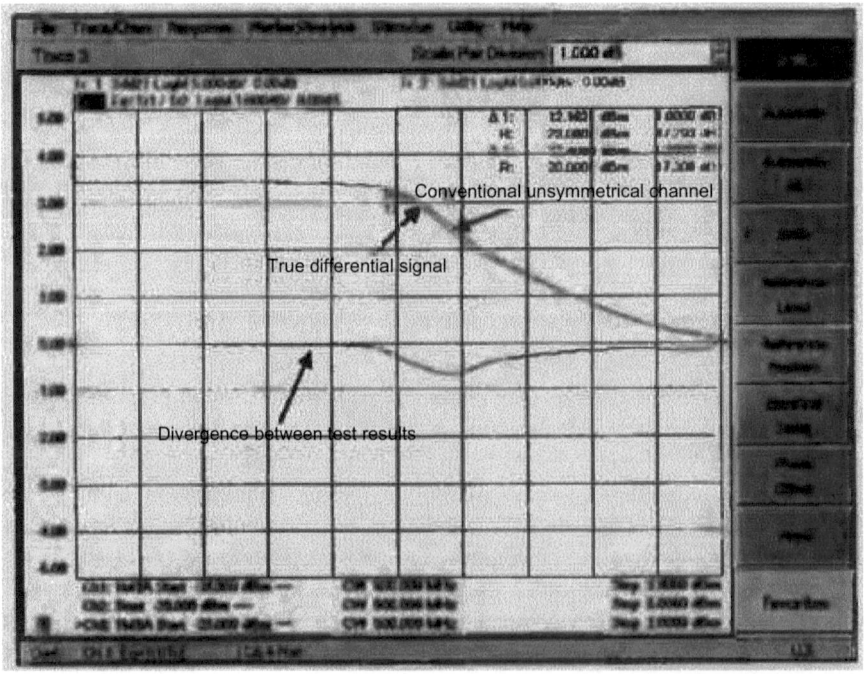

Fig. 11.42 Comparison of results of testing of devices with non-symmetrical and differential signals

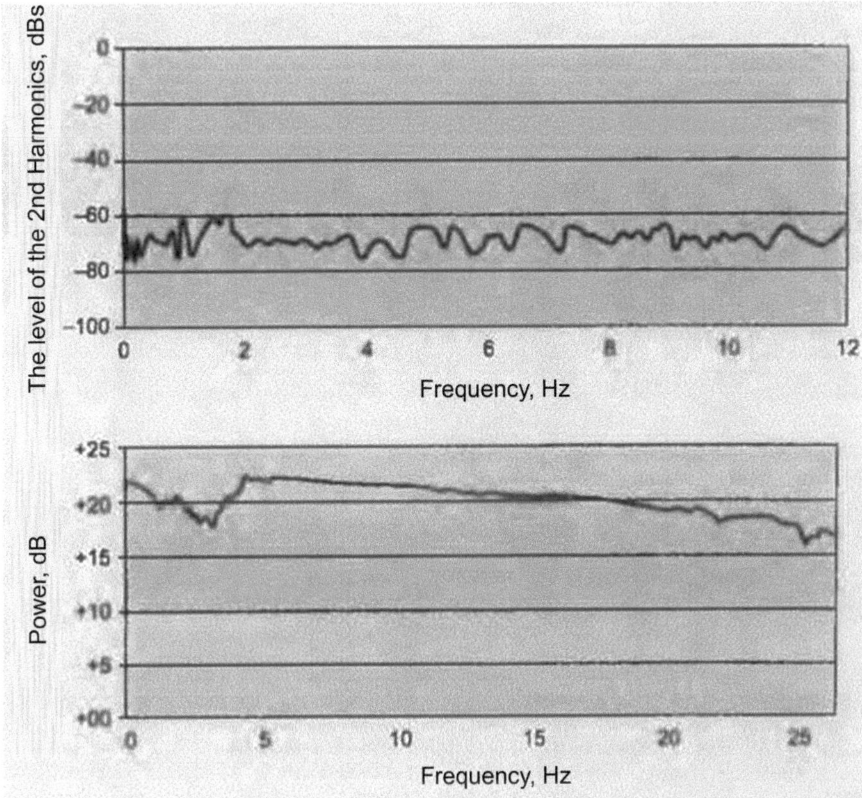

Fig. 11.43 Dependence of the second-level harmonic and output power of generators of network analyzers PNA-X on frequency

Keysight produces multiple options for network analyzers, application and replacement of which allows to adjust these devices to solve certain classes of measurement tasks. Therefore, it is possible to create systems for research and testing of the millimeter wave techniques (Fig. 11.48). Here, the pair of analyzer ports is expanded to eight ports, to which four units of millimeter waves are connected. The second pair of ports remains free for use.

There is a wide choice of modules (options) for expansion of the frequency range to the field of millimeter wavelengths from 50 GHz to 1.1 THz for satisfaction of specific needs:

- N5262AW15 from 50 up to 75 GHz;
- N5262AW12 from 60 up to 90 GHz;
- N5262AW10 from 75 up to 110 GHz;
- N5262AW08 from 90 up to 140 GHz;
- N5262AW06 from 110 up to 170 GHz;
- N5262AW05 from 140 up to 220 GHz;
- N5262AW03 from 220 up to 325 GHz;

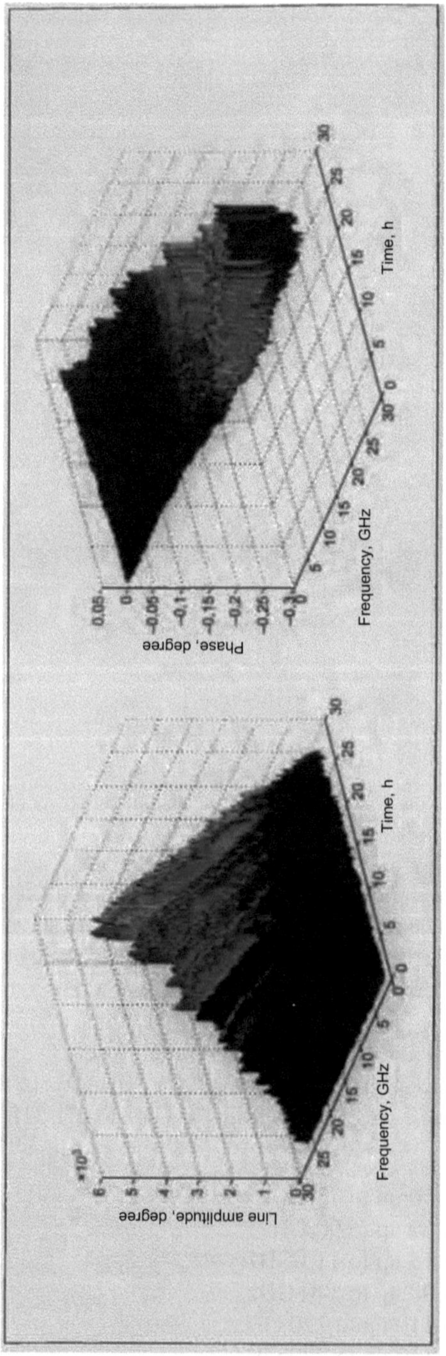

Fig. 11.44 Drift of generator signal amplitude and phase after calibration

Fig. 11.45 Calibration modules ECal: (**a**) variety of ECal modules; (**b**) ECal module with connectors of various types; (**c**) use of the user characterization function for creation of a custom ECal module with the required connectors

- N5262AW02 from 325 up to 500 GHz;
- N5256AW01 from 500 up to 750 GHz;
- N5262AW01 from 750 to 1.1 GHz.

11.12.6 PNA-L Network Analyzers

Network analyzers of series PNA-L (Fig. 11.49) overlap the frequency range of 300 kHz–8.5 or 13.5 GHz, as well as 10 MHz–20, 43.5, and 50 GHz and possess the functions that make it possible to significantly increase the efficiency of developments and tests. There are also 4-port models with a frequency range from 300 kHz to 13.5 or 20 GHz (Fig. 11.48).

Simplified block diagram of a 4-port analyzer PNA-L is shown in Fig. 11.50. The analyzer has two generators and four measurement ports.

Main possibilities of PNA-L analyzers:

- Measurement speed—not more than 4–9 μs per point.
- Superior characteristics and expanded possibilities in terms of analyzer connection to other devices as compared to network analyzers of series 872x.
- Second built-in source (only for 4-port models) (optional).

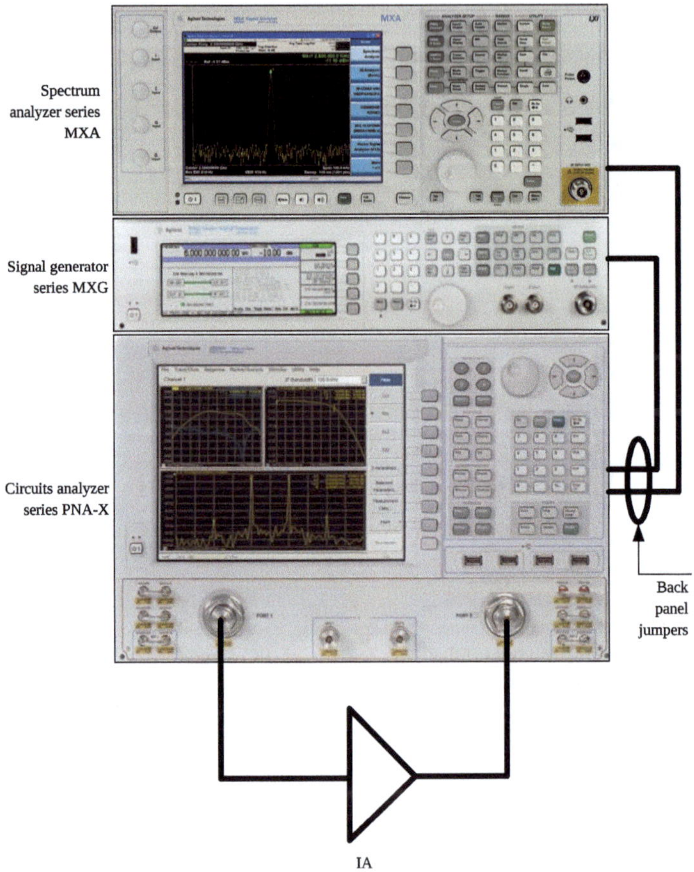

Fig. 11.46 Connection of additional devices (signal generator with digital modulation and spectrum analyzer) to the network analyzer

- Analysis in the temporal region and measurements with frequency displacement (optional).
- Electronic calibration (ECal) modules supplied by additional order allow calibration to be performed ten times faster as compared to mechanical calibration.
- Up to 32 independent measurement channels.

11.12.7 PNA Series Network Analyzers

Analyzers for SHF range circuits of series PNA (without additional letters) cover frequency ranges from 10 MHz up to 20, 40, 50, or 67 GHz, at the same time demonstrating high precision [33]. They are perfect for measurement of parameters

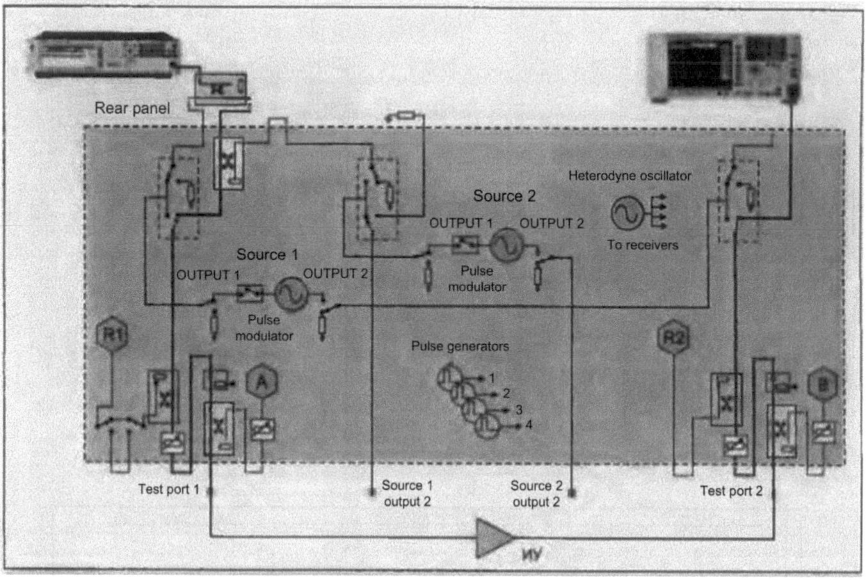

Fig. 11.47 Block diagram showing interaction between system devices and the network analyzer

Fig. 11.48 Multi-port measurement system of millimeter waves based on network analyzer

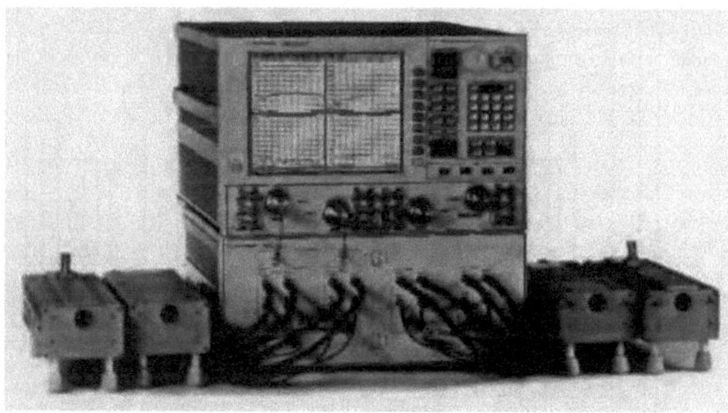

of precision devices of SHF range, such as components for mobile and satellite communication. Appearance of a two-port analyzer of the PNA series is shown in Fig. 11.51. The analyzers have a dynamic measurement range of 123 dB at measuring ports and 136 dB on condition of direct access to the receiver inputs. It also ensures the possibility to carry out TRL/LRM calibration of objects in the connection devices or during probe measurements at wafers. Moreover, architecture of the receiver provides the mode with frequency shift for measurement of parameters of mixers and frequency converters. The possibility to alter configuration of the measuring unit allows easy connection of external units of reflectometers and perform precise measurements of multi-pole devices.

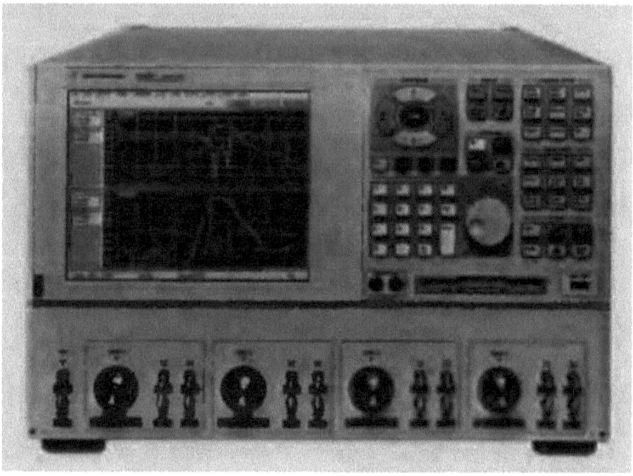

Fig. 11.49 Network analyzer of PNA-L series

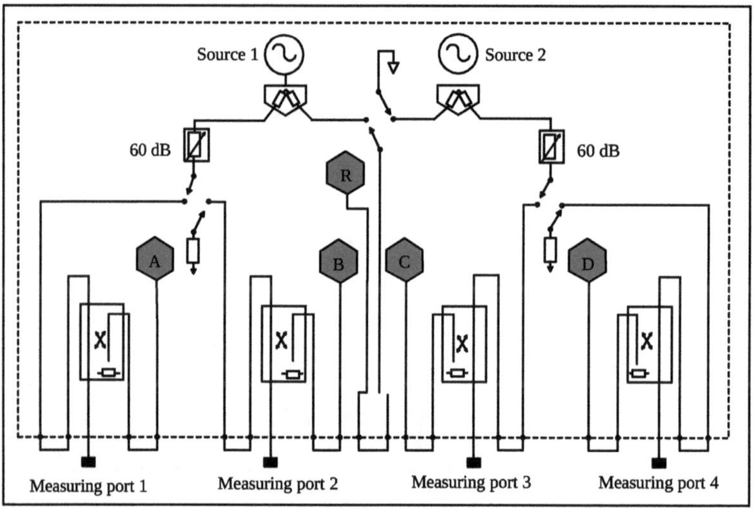

Fig. 11.50 Simplified block diagram of a 4-port PNA network analyzer

Functional properties of PNA analyzers:

- Dynamic measurement range: 136 dB on condition of direct access to the receiver.
- Extended functional possibilities due to the use of various measurement units (reflectometer units).
- Extended possibilities of adjustment of mixer and frequency converter parameters (optional).
- Electronic calibration (ECal) modules supplied by additional order allow calibration to be performed ten times faster as compared to mechanical calibration.

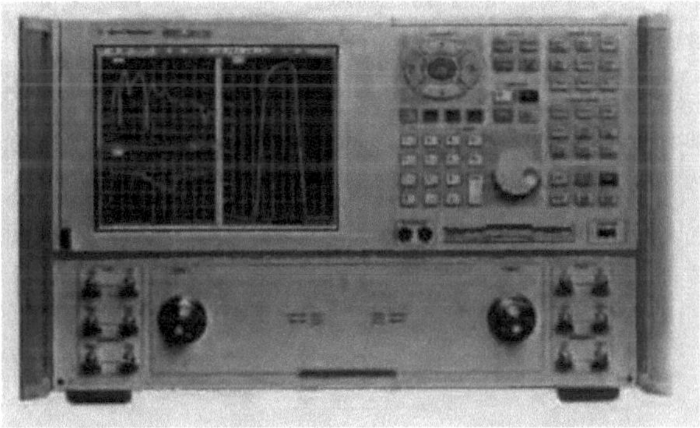

Fig. 11.51 Two-port analyzer of SHF networks PNA by Keysight

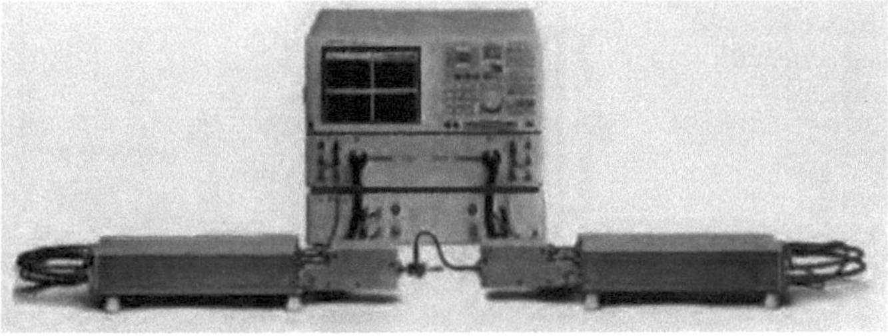

Fig. 11.52 Network analyzer of millimeter range N5251A by Keysight

- Antenna and impulse measurements (optional).

Device N5251A by Keysight is a network analyzer of series PNA (Fig. 11.52) designed for millimeter range of wavelengths. This model combines unrivaled high characteristics, measurement speed, and rich possibilities in terms of connection for solution of parameter measurement tasks in devices with coaxial connectors and on wafers up to 110 GHz. Moreover, within certain frequency bands solutions are available that ensure overlapping a wider range.

Possibilities of analyzers N5251A:

- Single sweeping cycle from 10 MHz to 110 GHz.
- Small-sized test heads and two built-in synthesizers.
- 29 settings of the IF band, 32 independent measuring channels, 64 graphs, and 16,001 measurement points.
- Code conversion utility for transition from Aligent 8510 network analyzers to PNA analyzers.

Changes in the parameters SHF and microwave radiation of different substances often make it possible to make conclusions about subtle physical properties of substances. Keysight presents state-of-the-art methods for measurement of dielectric and magnetic properties of materials. For example, new radial two-layer cylindrical dielectric resonator with frequency up to 10 GHz (Fig. 11.53) measures complex

Fig. 11.53 Connection of the radial two-layer cylindrical dielectric resonator B5072A to the network analyzer

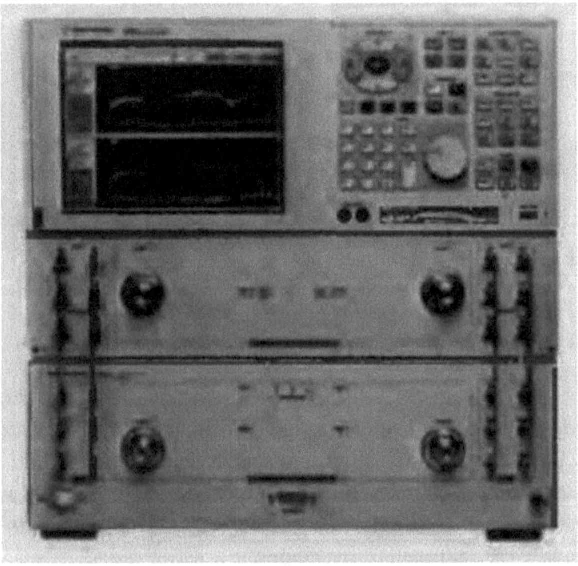

Fig. 11.54 Network analyzer-based system for microwave physical research

parameters of dielectric constant, dielectric loss tangent of thin films, uncoated substrates, and other low-loss sheet materials as part of a complete technical solution for testing for compliance with IPC-TM-650 2.5.5.13. Software for measurement of material parameters automates the use of various methods in a wide range of frequencies and different media including transmission lines, free space, NRL arch, and volumetric resonators.

The set of dielectric probes 85070D contains software and hardware means for measurement of complex parameters of dielectric constant of liquids and matched solids from 200 MHz up to 50 GHz. Measurement of electromagnetic properties of materials is important during all stages of life of products (design, input control, production process control, and quality assurance). Keysight sets measurement standards in this field, using its 20 years of experience in development of innovative products (Fig. 11.54).

11.12.8 Non-linear Vector Network Analyzers (NVNA)

Research and testing of non-linear devices and systems with moderate non-linearity is performed using non-linear vector circuit analyzers (NVNA) [34, 35]. They employ a system of non-linear X-parameters that are extensions of S-parameters reduced to them in case of very low non-linearity. X-parameters are linear approximation of coefficients of the Fourier series of non-linear function with low change in the signal amplitude in the conditions of work with high signal level. NVNA include analyzers Keysight PNA-X with options for operation with non-linear networks (Fig. 11.55).

Figure 11.56 illustrates the appearance of combination components of the output signal of non-linear tested DUT device for two-tone input signal.

The system (Fig. 11.55) easily measures such distortions and reproduces the spectrum of output signal frequencies on the network analyzer screen (Fig. 11.57) even if the signal is a multi-tone one.

Comparison of measurements based on application of non-linear X-parameters and results of modeling of physical devices with the help of the electromagnetic 3D modeling software (Fig. 11.58) indicates fairly high degree of correspondence and

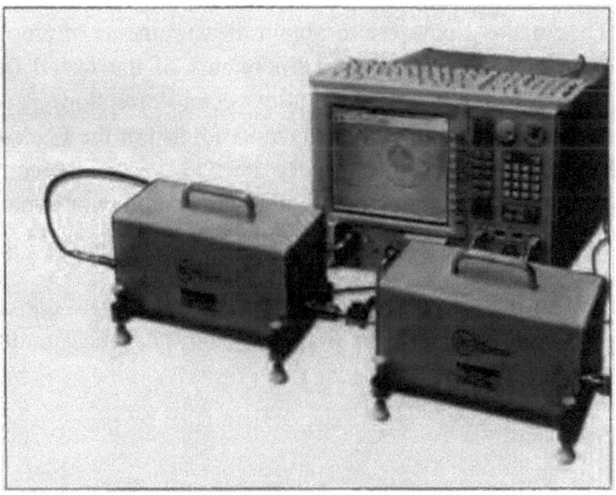

Fig. 11.55 Installation for measurement of parameters of non-linear devices based on network analyzer Keysight PNA-X

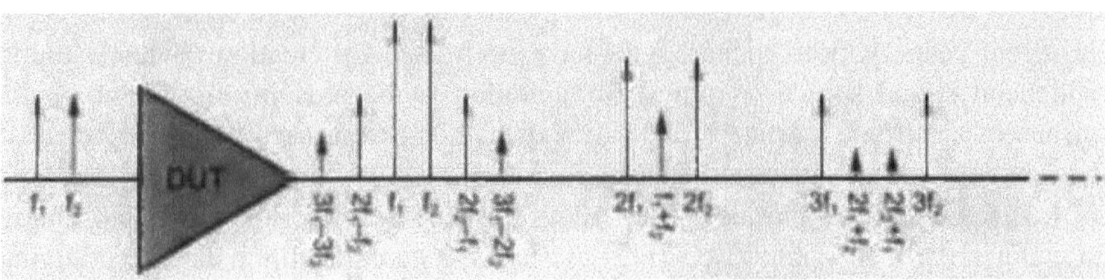

Fig. 11.56 Combination distortions of a non-linear device at two-frequency signal

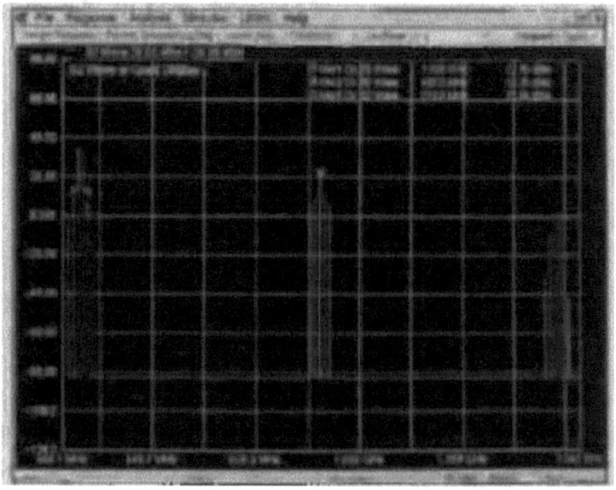

Fig. 11.57 Installation screen demonstrating combination components of a multi-tone signal

great possibilities of non-linear vector network analyzers in research and testing of actual physical devices.

Measurements considering non-linearity in the field of the pulse envelope ensure deeper understanding of time-dependent memory effects in active non-linear devices that are caused, for instance, self-heating or the effect of displacement circuits. They make it possible to obtain measurements of amplitude and phase of the main frequency and harmonic components of the tested device pulse envelope in the temporary realm calibrated after vector correction.

Long list of network analyzers (not just of the PNA series) produced by Keysight complicates their selection. Reviews [32–36] at the company's website [31] contain manuals on selection and configuration of network analyzers. Let us confine ourself to presentation of possibilities of network analyzers provided by Keysight in the industrial sphere (Table 11.3).

The data given below demonstrate that network analyzers of PNA series are not only the most high frequency, but also most multi-purpose among devices of this class.

11.12.9 Signal Generator Keysight UXG

In recent years, various technologies have been used for creation of quick-tuning wideband signal sources required for imitation of hazards in adjustment of the means of electronic warfare (EW). Similar methods were used in development of heterodynes and generators of RF pulses that can quickly tune within a wide range of frequencies and amplitudes. These ideas formed a basis for the new rapid-tuning signal generator Keysight UXG (Fig. 11.59). UXG architecture using direct digital synthesis (DDS) and new switching technology made it possible to create a generator that was perfect for tasks where quick heterodyne tuning or highly realistic

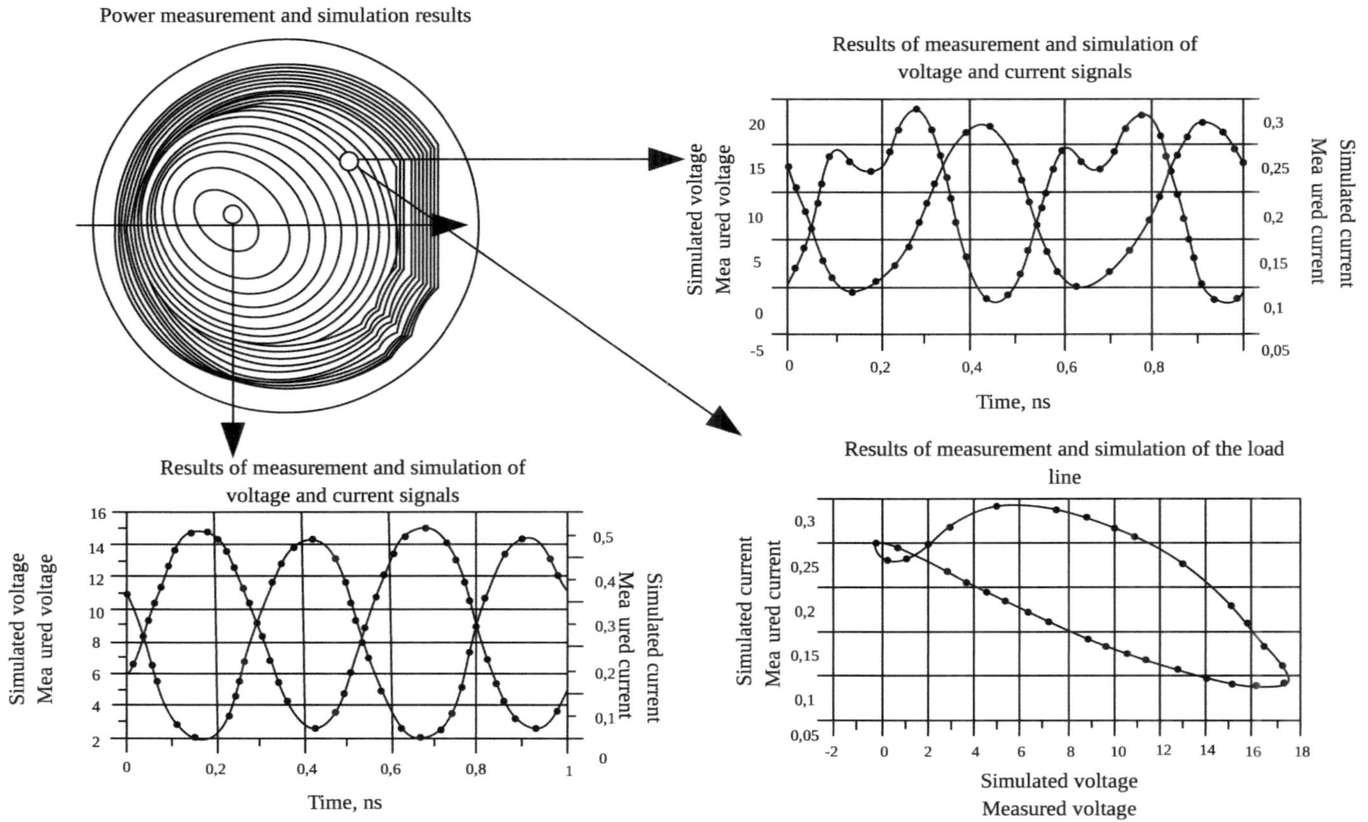

Fig. 11.58 Comparison of results obtained from actual measurements with the results obtained during modeling in the environment of ADS modeling program

Table 11.3 Comparison of network analyzers based on their possibilities in the industrial sphere

Production tests	E5100 series	Combined analyzer series	ENA-L series	ENA series	PNA-L series	PNA-X series	PNA-(SHF) series	PNA series (millimeter range)	5757 series
Programming									
Possibility of internal programming	•	•	•	•	•	•	•	•	
Quick data transmission			•	•	•	•	•	•	
Calibration									
Calibration with elimination of the adapter influence				•	•	•	•	•	
Calibration of SOLR (by unknown jumper) using mechanical means of calibration or e-calibration modules ECa				•	•	•	•	•	
E-calibration			•	•	•	•	•	•	
Measurement and analysis									
Segmented sweeping	•	•	•	•	•	•	•	•	
Pre-start tests	•	•	•	•	•	•	•	•	•
Addition and removal of circuits				•	•	•	•	•	
Signal shape analysis command	•		•	•					
Interfaces									
Interface of local network			•	•	•	•	•	•	
In/out port	•	•	•	•	•	•	•	•	
Manipulator control interface	•	•	•	•	•	•	•	•	
VGA output	•	•	•	•	•	•	•	•	•

The bullet symbol indicates that the device in the column has the relevant feature from the first column

Fig. 11.59 Rapid-tuning signal generator Keysight UXG (at the top) is a powerful basis for creation of reliable heterodynes and expandable hazard imitators

imitation of EW environment is required. The DDS technology is implemented on the basis of digital-to-analog converter (DAC) designed by Keysight, which expands the dynamic range and reduces the phase noise significantly [37] (www.keysight.com/find/UXG).

UXG characteristics and functions are based on three innovative technologies:

- Signal generation with the help of DDS allowing to obtain exclusive signal quality within the band from 10 MHz up to 20 or 40 GHz.
- Electronic switches that change output levels over only 600 ns within a wide range of amplitudes with precise synchronization over time.
- Direct support of the pulse description words (PDW)—native language of EW engineers—makes it possible to create complex and precise signals.

The block diagram shown in Fig. 11.60 illustrates architecture of the UXG generator based on DDS. The heart of the DDS is a quick-action DAC operating at high frequency that has a sufficiently wideband for minimization of the number of cascades of step-up frequency converter. An array of doublers and filters is used to multiply the signal to the maximum frequency of 40 GHz.

DAC covers the octave with central frequency of 1 GHz, significantly expands the dynamic range, and reduces the phase noise. As a result, we obtain the possibility to create extremely clean signals (Fig. 11.61).

These characteristics are combined with the best commutation speed in the industry and extremely small delay of frequency change commands. For example, frequency can be changed over just 100 ns. Depending on the selected operation mode, delay from the moment of receipt of an external command until frequency change can be just 250 ns. Moreover, UXG is capable of generating wide pulses with linear frequency modulation with frequency of 10–25% of the carrier.

Comprehensive hazard imitation in EW application requires rapid change of the amplitude that would correspond to the frequency change speed of the radiation source. This requires matched switching of electronic attenuators; this task is solved

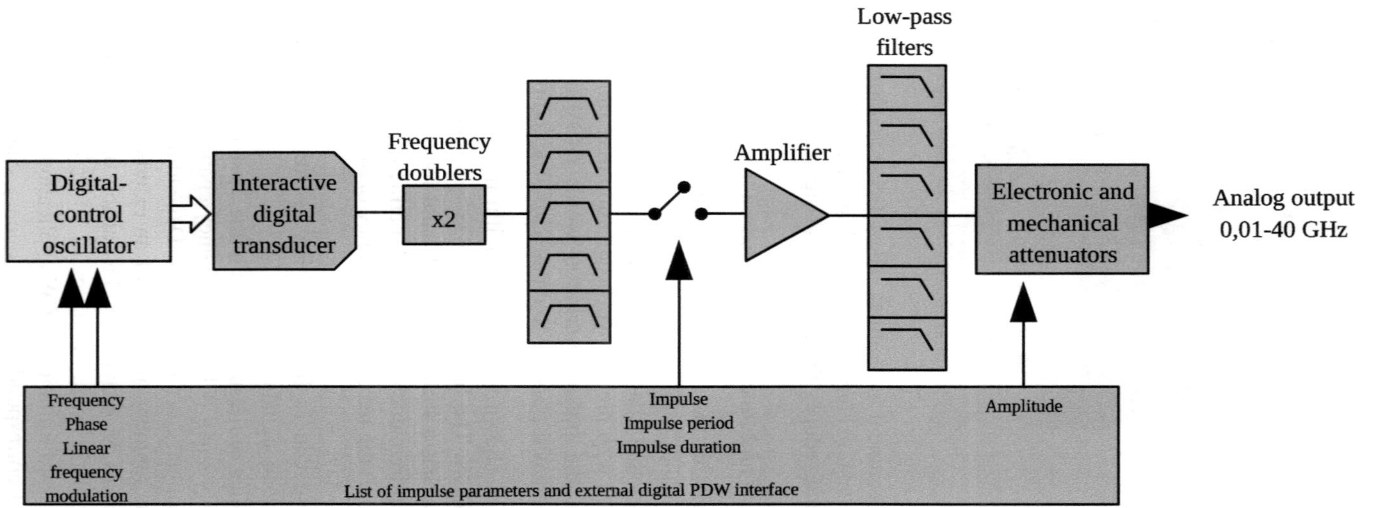

Fig. 11.60 UXG architecture makes it possible to model complex signal scenarios precisely due to quick switching, phase reproducibility, and wide possibilities of the pulse modulation

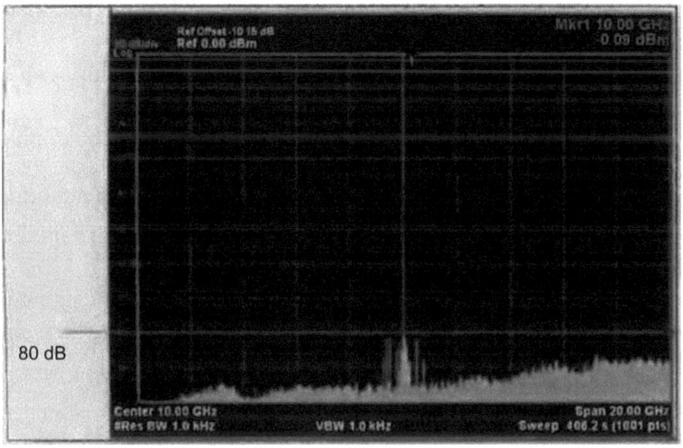

Fig. 11.61 In this measurement of the spectrum with a swath of 20 GHz, the modulated 10 GHz signal of a quick-tuning UXG generator demonstrates a wide dynamic range, low level of intrinsic noise, and minimum level of spurs

in UXG with the help of integral switches based on field SHF nanotransistors designed and manufactured by Keysight. Such new switches are designed for SHF and millimeter range frequencies and are quick enough to minimize distortions of the pulse shape. As a result, quick amplitude change within 80 dB can be used within the range of output levels of 0–120 dBm, which makes it possible to work with various tuning scenarios. In order to optimize cost and functionality of the device for various applications, quick-action electronic attenuator is supplied as optional.

In certain scenarios, for imitation of hazards it is necessary to preserve certain phase and frequency ratios between pulses, frequency hopping, and consequently interleaved signals. UXG is capable of generating signals with any phase or frequency change algorithms without limiting the speed of frequency or amplitude tuning. Moreover, it supports any set ratios in case of change in the frequency, amplitude, and pulse characteristics or with positive or negative edge. It is also possible to unite several UXGs based on their clock frequency and start-up, which ensures coherence of several sources and makes it possible to imitate returnable directional antennas or generate scenarios with a given fall angle for evaluation of DF receivers.

In EW sphere, there is a constant race between development and implementation of improved systems, as well as between means of analysis and countermeasures applied to counter these systems. DAC architecture used in UX and based on DDS suggests competitive offers for EW applications, be it the demand for quick heterodyne tuning or imitation of the EW environment using PDW:

- Signal quality. UXG generates signals with low level of spurs and phase noise and high detuning from the carrier. Such quality is typical of devices using PLL instead of DDS.

- Rapid tuning of frequency, amplitude, and phase. All three parameters of the output signal can change over just 100 ns.
- Support of several signals and complex scenarios. Sampling sequences can imitate multiple emitters with one signal source.
- Phase repeatability. Calculations for signal generation include phase drives programmed to any desired phase ratio.
- Good adaptation to improving hazards. UXG tuning speed and flexibility make it possible to precisely imitate equivalent characteristics of the modern hazard and adapt to them as they transform.
- Simple way from given pulse characteristics to the real output signal. UXG can directly reproduce complex dynamic and pulse signals, creating them on the basis of mathematical description and given pulse characteristics.

The last point is very important: UXG understands the PDW language directly. Each word describes all parameters of a separate pulse: frequency, duration, amplitude, linear FM change rate, and much more. PDW charts can be transferred to UXG memory, and complex sequences can be reproduced quickly. Separate pulses can have duration of only 5 ns, rise/fall time of 1 ns, and suppression during the pause equal to 80 dB. Markers are provided to synchronize PDW performance with other devices; IDW sequences can be synchronized and controlled by external start-up signals supplied to UXG.

Due to these possibilities, the generator UXG is designed for all-round and realistic testing during early stages of development and allows the engineers to check system characteristics before implementation of costly, potentially long and poorly replicable field tests. Moreover, the device significantly reduces the time between collection of data on the new signal and creation of realistic modeled hazards.

In conditions where quick alteration of tactical data is possible, better testing ensures deeper confidence in correct operation of the EW system. Quick-tuning signal generator Keysight UXG helps EW engineers to create complex scenarios whenever they are needed.

UXG generator is a powerful ready-made solution to be used, for example, as a reliable heterodyne or as a part of scaled hazard imitator. Since UXG supports the programming language of EW systems, it ensures quicker integration of newly obtained intelligence information into current signal scenarios. Due to unrivaled characteristics, such as switching speed and phase noise, UXG allows to create more and more complex models that correspond precisely to the reality.

References

1. Abubakirov, B. A., Gudkov, K. G., & Nechaev, A. V. (1984). *Measurement of parameters of radiotechnical circuits*. M.: Radio i Svyaz (in Russian).
2. Andreev, V. S. (1982). *Theory of nonlinear electrical circuits*. M.: Radio i Svyaz (in Russian).

3. Andrianov, A. V., & Shpak I. I. (1987). *Digital information processing in measuring devices and systems*. Minsk: Vysh. Shkola (in Russian).
4. Bondarenko, I. K., Deinega, G. A., & Magrachev, Z. V. (1969). *Computer-aided measurements of parameters of microwave paths*. M.: Sovetskoye Radio (in Russian).
5. Valitov, R. A. (1963). *Radiotechnical measurements*. M.: Sovetskoye Radio (in Russian).
6. Valitov, R. A., et al. (1969). *Submillimeter-wave engineering*. M.: Sov. Radio (in Russian).
7. Vikulov I. K. (1977). Millimeter waves abroad. In *Reviews of electronic engineering. Series 1, Microwave electronics* (Vol. 11, No. 481) (in Russian).
8. Glebovich, G. V., et al. (1984). *Studies of objects with picosecond pulses*. M.: Radio i Svyaz (in Russian).
9. Goldenberg, L. M., Matyushkin, B. D., & Polyak, M. N. (1985). *Digital signal processing: Reference book*. M.: Radio i Svyaz (in Russian).
10. Gonorovsky, I. S. (1986). *Radio engineering circuits and signals*. M.: Radio i Svyaz (in Russian).
11. Gusinsky, A. V., Sharov, G. A., & Kostrikin, A. M. (2009). *Vector analyzers of millimeter-wave circuits*. Monograph in 3 parts. Minsk: BSUIR (in Russian).
12. Gusinsky, A. V. (1990). Automatic homodyne analyzer of the millimeter-wave networks. In *Prospects for the development and application of automated radio-measuring devices: Proceedings of the Republican scientific-technical conference*. Minsk: MNIPI (in Russian).
13. Gusinsky, A. V. (1991). *Automatic homodyne analyzers of the millimeter-wave networks with modulation of the measurement and reference signals: Thesis—C.Sc.* Minsk: MRTI (in Russian).
14. Elizarov, A. S. (1978). *Computer-aided measurements of parameters of linear non-reciprocal microwave two-port networks*. M.: Sovetskoye Radio (in Russian).
15. Efimov, I. E., & Shermina, G. A. (1979). Waveguide transmission lines. M.: Svyaz (in Russian).
16. Max, J. (1983). *Methods and signal processing equipment for physical measurements*. M.: Mir (in Russian).
17. Kostrikin, A. M. (1987). *Automated S-parameter measurements of microwave devices based on the switching method: Thesis—C.Sc.* Minsk: MRTI (in Russian).
18. Kostrikin, A. M. (1982). Automatic testers for microwave switch circuits with a ring-type test path. In *XXXIV all-union scientific session NTORES n.a. A. S. Popov: Abstracts*, M. (in Russian).
19. Lanne, A. A. (1969). *Optimal synthesis of linear electric circuits*. M.: Svyaz (in Russian).
20. Petrovsky, A. A. (1988). *Methods and microprocessor means of real-time processing of broadband and fast processes*. G.V. Rimsky (Ed.). Minsk: Nauka i Tehnika (in Russian).
21. Hasanova, L. G. (Ed.) (1988). *Solid-state microwave devices in communication technology*. M.: Radio i Svyaz (in Russian).
22. *EA-10/12: EA guidelines on the evaluation of vector network analyzers (VNA)*. May 2000. [Electronic source]. Access mode: http://www.european- accreditation.org
23. Vanderstein, G., & Rolain, Y. (1997). An improved siiding-load calibration procedure using a semiparametric circle-fitting procedure. *IEEE Transactions on Microwave Theory and Techniques*, 45(7), 1027–1033.
24. Gartner, U. F., Schneider, I., & Schick, B. (1989). *Homodyne network analysis - improving the accuracy and sensitivity by redundant phase coding and double modulation*. AEU, Band H3, Heft 1.
25. Belous, A. I., Yemelyanov, A. V., & Chigir, G. (2008). *Test structures in the IC quality control systems* (p. 208). Minsk: Integral Poligraf (in Russian).
26. Belous, A. I., Yemelyanov, A. V., Turtsevich, A. S., & Chigir, G. G. (2011). *The methods to enhance the IC reliability on the basis of test structures* (p. 240). Ministry of Education of the Republic of Belarus, Gomel State University n.a. F. Skaryna - Gomel GTU n.a. F. Skaryna (in Russian).
27. Belous, A. I., Yemelyanov, A. V., & Chigir, G. G. (2007). An express-method for predicting the reliability of integrated circuits. In *Proceedings of the 8th international conference "Modern Information and Electronic Technologies"*, Odessa, p. 281 (in Russian).

28. Belous, A. I., Emelianov, V. A. & Chyhir, R. R. (2009, June 23–26). Express control of safe dielectric reliability in the process of the submicron microcircuit production. In *Proceedings of the 6th international conference Meet 2009*, Zakopane, Lublin, p. 16.
29. Belous, A. I., & Syakersky, V. S. (2006). Effective methods to control the quality of integrated circuits. In *Proceedings of the 1st International conference "Global information systems. Problems and development trends"*, Tuapse, Kharkov, pp. 87–88 (in Russian).
30. Dyakonov V.P. Ultrafast solid-state electronics. In 2 volumes. M.: DMK-Press, 2013 (in Russian).
31. *Website of Keysight Technologies*. http://www.keysight.com/
32. Agilent PNA-X Series. (2009). *The premier-performance microwave network analyzers*. Agilent Technologies (in Russian).
33. *Agilent PNA-Series microwave network analyzers*. Procurement Manual. Agilent Technologies, 2009 (in Russian).
34. *Agilent. Network analyzers*. Selection Guide. Agilent Technologies. 2007 (in Russian).
35. *Agilent Nonlinear Vector Network Analyzer (NVNA)*. Agilent Technologies, 2011.
36. *Agilent. Network analyzer basics*. Agilent Technologies. http://cp.literature.agilent.com/litweb/pdf/5965-7917E.pdf
37. Kalinin, O. (2015). Application of new methods of signal generation for advanced modeling of EW environment. *Komponenty i Tehnologiya*, 2, 114–116 (in Russian).

Index

A
Active devices, 598
Active electronically scanned array (AESA), 363
Active Phased Array Radars (APAR), 363
Adaptive algorithms, 20
Adaptive Coding and Modulation (ACM) algorithms, 335
Address signals and command signals, 104
ADIsimPower, 752, 779, 792
AL-102B diode, 737
Altai system, 153
Amplitude and phase characteristics (APC), 868
Amplitude–frequency characteristic (AFC), 214, 452
Amplitude–frequency response, 596
Amplitude modulation (AM)
  linear phase modulation and phase shift keying, 871, 872
  measuring path, homodyne VNA, 869, 870
  reference signal modulation, 872, 873
  Schaffer–Kohn method, 867, 868
Amplitude-phase detectors (APD), 874
Analog-to-digital converters (ADC), 238, 488, 499, 737
Antenna-feeder devices (AFD), 513
Application layer, 21
Application level software, 17
Application-specific integrated circuits (ASIC), 741
Arbitration code signals, 85
Associated Power Technologies (APT), 810
Asymmetric interference, 721
Asynchronous method, 10
Asynchronous transmission, 10, 11
Attenuation curves, 729
Attenuators, 873
Automatic gain control (AGC) systems, 74, 242, 282

B
Back drilling, 639
Bandwidth, 11
Base transistor designs, 149
Baseband system, 31
Belarusian State University of Informatics and Radioelectronics (BSUIR Minsk), 470
Beryllium and copper seal, 707
Bidirectional data bus, 102
Bismaleimide triazine (BT), 676
Bit error rate (BER), 292, 295, 614
Bits per second (BPS), 11
Blind vias, 640
Block-by-block transfer, 84
Bounded media, 37
Bridges, 13, 55–57
Broadband transmission, 32
Bulk acoustic waves (BAW), 204

C
Cable assemblies
  angular velocity, 248
  appearance, 245
  assembly electrical length, 249
  customer specifications, 245
  electrical parameters, 252

A. Belous, V. Saladukha, *High-Speed Digital System Design*,
https://doi.org/10.1007/978-3-030-25409-4

Cable assemblies (*cont.*)
geometric length, 245, 246
inner and outer conductors, coaxial cables, 245
maximum frequency, 247
microwave devices, 244
parameters, 246
phase stability, 245, 249
PHASEFLEX radio frequency cable, 244
PTF, 250
radiation resistance, 245, 252
RF microwave, 252
semi-rigid, 245
shielding attenuation, 248
temperature coefficient, 249
time-consuming process, 252
velocity of propagation, 248
wavelengths, 249
Calculating Analytical Machines (CAM), 14
Capacitive crosstalk, 493–496
Carrier aggregation (CA), 221
Cascade Microtech, 853, 854
Cavendish Kinetics, 221
CDMA2000, 154
Ceramic Column Grid Array (CCGA), 656
Ceramic condensers, 523–525
Channel access method, 49
Chip manufacturers, 604
Chip-scale Packages (CSP), 656, 666, 667, 695
Chuka converter circuit, 305
Circuit switching, 51
Clock and data recovery (CDR), 295–297
Clock regenerator, 599
Close interaction field, 705
Cloud computing, 162
CMOS
chip scaling, 148
MEMS (*see* Microelectromechanical systems (MEMS))
technology, 195, 196
Coaxial cable, 8
Coaxial low-pass filters, 525, 526, 530
Coefficients of thermal expansion (CTE), 693, 694
Combined correction circuit, 100
Common-mode interference, 727
Communication channels, 145, 163, 168, 169
Communication line immunity, 90
Communication lines, 81, 87, 91
Compensation methods, 597
Competitive access method, 49
Compression force, 708
Computer aided design (CAD), 235
Conductive interference, 721
Conductive surfaces, 706
Conductive/inductive interference, 454
Conductor shielding test, 607
Continuous data transmission technology, 154
Conventional equivalent frequency dip circuit, 66
Coplanar waveguides, 566
Copper foil etching profiles, 558
Corrosion, 708
COSMOS project, 425, 427–430
Cross Polarization Interference Canceller (XPIC), 335
Crosstalk, 491
Crosstalk disturbance, 594, 604
CSD technology, 153
"Cutting and stitching" method, 614

D
Daisy chain, 85
Data circuit-terminating equipment, 36
Data compression, 11
Data link, 19
Data reception (DR) signals, 84
Data transmission channels, 5, 80, 111
Data transmission modes and media
bounded and wireless, 37
data terminal equipment, 35
DCE, 36
duplex transmission, 36
FC connectors, 42
functions, 36
modern high-speed networks, 37
SC-connector, 41
transmission media, 37
UTP, 39
Data transmission rate, 340
DCS-control topology
chips of, 314–316
DC/DC converters, 326, 329, 330, 332, 333
energy-saving mode, 317, 318, 320
seamless transition, 320, 326
DDR3 interface, 611
Dedicated lines, 8
Defense Advanced Research Projects Agency (DARPA), 168
Delay line (DL), 203
Department of Defense (DoD), 15
Deserializers, 293
Design rule checkers (DRC), 606
Designing high-speed digital circuits, 569
Differential signal buffer, 95

Digital down converter (DDC), 241
Digital front-ends (DFE), 488
Digital microchips, 817
Digital signal processors (DSP), 238, 488, 490
Digitally controlled potentiometer (DCP), 99
Digital-to-analog converters (DAC), 488
    499, 911
Direct Chip Attach (DCA), 667
Direct digital synthesis (DDS), 908
Directional coupler (DC), 182, 183
Directional waveguide power divider, 182
Directive 336EC 89, 721
Discontinuous conduction mode (DCM), 317
Distributed MEMS transmission line
    (DMTL), 225
Distributed stream computing platforms
    (DSCPs), 165
Diverse Accessible Heterogeneous Integration
    Program (DAHI) project, 431–433
DOCSIS 3.0 and 3.1 specifications, 71, 72
DOCSIS network operation principle, 71
Double-balanced modulators, 867
Double data rate (DDR)
  DDR3 modifications, 104
  design, 101
  DQS and CLK signals, 104
  DQS signal, 102
  DRAM chips, 102–104
  DRAM module, 103
  IBIS model, 105
  recording mode, 102
  and SDRAM, 101

E
e2v products, 61
e2v semiconductors, 62
Earth leakage currents, 717
Earth's magnetic field, 735
Efficient Power Conversion company (EPC),
    412, 413, 416, 419, 422, 423
eGaN® FET, 412, 413, 416, 417, 420–426
EGLE wideband magnetometer, 731
Electric migration (EM), 695, 697
Electrical capacitance, 87
Electrical fast transient (EFT), 714
Electrical rule checks (ERC), 612
Electrode potential, 708
Electroless nickel immersion gold (ENIG), 578
Electromagnetic compatibility (EMC), 600,
    606, 720, 802
  aspects of, 449
  characteristics, 485, 486
  electronic devices, 447
  high-speed electronic devices and
    modules, 448
  issues, 491
  microcontrollers, 478, 479, 481–484
  printed circuit boards, 473–476, 478
  radioelectronic systems, 452
  regulations and standards, 456–459, 461,
    463, 465
  systems, components and basic concepts,
    488, 489
  telecommunication systems, 487, 488
Electromagnetic environment (EME), 453
Electromagnetic fields (EMF), 731, 732
Electromagnetic interference (EMI), 447–449,
    452, 453, 466, 468, 473, 483, 484,
    486, 703, 704, 802, 803
  classification, 455, 456
  in telecommunications systems
    ceramic condensers, 523–525
    coaxial low-pass filters, 525, 526, 530
    surface mount filters, 530–532
Electromagnetic noise (EMN), 606
Electromagnetic protection, 704
Electromagnetic radiation
  EMI, 709
  high-speed electronic devices, 706
  metallization, 708
  microwave element, 704
  modern telecommunication systems, 709
  PCB circuitry techniques, 704
  protective device, 704
  protective screen, 705
Electromagnetic screens, 730–738
Electromagnetic waves (EMW), 732
Electromigration, 693, 695–697, 699
Electron-beam lithography, 205
Electronic calibration (ECal), 902
Electronic devices
  high-speed, 659, 665, 668
  miniature, 658
  packaging, 671
  physical dimensions, 655
  spacecraft, 661, 682, 684–686, 688, 690,
    692
Electronic equipment, 802–804, 807, 812
Electronic Numerical Integrator
    and Computer, 14
Electronic warfare (EW), 3, 908
Electronics Industries Association
    (EIA), 23, 340, 347
Electrostatic discharges (ESD), 448, 486
Elementary telecommunication system, 4

Embedded component packaging (ECP), 584
Embedded component technology, 583, 584
Embedded Die Technology (EDT), 658
EMI NE filters, 716
Emitted interference, 726
Encoders
  AMT31 series, 189, 191
  asymmetric quadrature signal, 190
  capacitive, 185, 188–190
  differential signal, 190
  digital functions, 189
  digital quadrature signals, 186
  magnetic, 185, 188
  non-capacitive, 190
  non-programmable, 190
  optical, 185, 187
  oscilloscope, 189
  output switching signals, 186
  radioelectronic systems, 185
  strips/lines, 188
  system controller, 186
  vector control, 186
Enhanced Data rates for GSM Evolution (EDGE), 154
Enhanced mobile broadband networks (eMBB), 166
Equalizers, 98, 99
Equivalent electrical circuit, 618
Equivalent series inductance (ESL), 797
Equivalent series resistance (ESR), 221 303, 797
Ethernet-type local networks, 8
European Committee for Electrical Standardization (CENELEC), 457
European Committee for Standardization (CEN), 457
European Space Agency (ESA), 269
European Telecommunications Standards Institute (ETSI), 23, 457
European Union's EMPOWER project, 584
EV10AQ190 converter, 73
EV12AS200A, 76
Experimental installation, 737
External influencing factors (EIF), 534, 537
External memory interfaces, 81
External modem, 7
External T-shaped matching chains, 141
External π-shaped matching chains, 140

F

Fan packaging at wafer level (FOWLP), 658
Fast Fourier transform (FFT), 216
Fiber-optic cables, 9, 40
Fiber-optic communication, 589
Fiber-optic communication lines, 9
Fiber-optic connectors, 41
Fiber-optic technologies
  active part, 277
  arbitrary format signals, 282
  characteristics, 275
  data acquisition systems, 285
  electromagnetic interference, 282
  electromagnetic protection, 285
  electronic technology, 273
  functional diagram, 281, 282
  IFPD, 280
  information collection systems, 275
  LD, 280
  local area networks, 285
  multimode optical cables, 279
  non-linear optics, 279
  OMRD-04, 283, 284
  OMRD-06, 282–284
  optical radiation, 278, 280
  output and input electrical signals, 280
  passive part, 277
  physical values, 275
  properties, 280
  threshold current, 281
  unbalanced digital sequences, 278
  VCSEL laser structures, 279
  watt-ampere curve, 281
Field-effect transistors, 218
Field-oriented control (FOC), 186
Field-programmable gate array (FPGA), 126
Fifth-generation (5G)
  cloud computing, 162
  cloud services, 159
  electromagnetic radiation, 164
  element base, 164
  file sharing networks, 159
  forecasts, 158
  frequency domains, 163
  heterogeneous network principle, 163
  higher productivity, 157
  implementation
    energy efficiency and sustainability, 161
    frequency spectra, 161
    low-power M2M devices, 160
    MBBA, 159
    proximal communication, 161
    super-reliable communications, 161
    ultra-high bandwidth and network capacity, 160
  IoT (*see* Internet of Things (IoT))

MBAN, 158
mesh-networks, 163
methods, 162
METIS project, 161
millimeter-wave communication systems, 164
mm-band antenna systems, 163
network connectivity, 158
one-for-all technology approach, 157
phased antenna arrays, 168, 169
radio access technologies, 158, 159
radio-relay systems, 164
remote video surveillance, 159
single transmission channel, 162
transmission delay time, 162
unrestricted network access, 157
video streaming, 159
virtual and augmented reality, 159
wireless access networks, 161
Finishing coating
data transmission line, 579
microstrip, 578
SMT boards, 578
First generation (1G), 153, 156
Flexible printed circuit boards, 564
Flip-chip technology, 668, 681
Flyback topology, 783
Forward Error Correction (FEC) technology, 295, 335
Fourier transducers (FT), 216
Fourth generation (4G), 155, 156
Fox phasers, 871
Frames, 11
Frequency controlled drives, 716
Frequency division duplex (FDD), 334
Frequency division multiplexing (FDM), 32
Frequency ranges and sensitivity, 75
Frequency spectrum, 74
Frequency-time transformations, 847
Frequency transition methods, 863
Full-duplex serial channel, 96
Full-duplex transmission systems, 12

G
Gallium nitride transistors, 407, 411, 412, 414–416, 419, 423
Galvanic metallization, 553
GaN-based technology, 193
GaN-on-SiC technology, 193
GaN technology, 194
Gas discharge protective devices, 514, 515
Gas discharge protectors (GDT), 713
Gateway, 13, 59
General Packet Radio Service (GPRS), 154
Geostationary orbit (GSO), 687
Global network Internet, 8
Global networks, 19
GPON networking, 73
Graphical user interface (GUI), 299, 335, 405, 406, 410, 749
Graphs indicate interference, 725
Ground instability, 498, 499
Grounding conductors, 601
Grounding systems
earthing system, 806
electrical power systems, 806
electromagnetic impulse, 806
electronic devices, 807–809, 812
electronic equipment, 802–804, 807
electronic mechanisms, 805
equipotential surfaces, 804, 805
floating-type, 809–811
lightning discharge, 806
NE EMI, 807
zero potential, 804, 805
Ground-signal-ground (GSG) configuration, 852
Group delay time (GDT), 822

H
Half-duplex transceiver, 121
Half-duplex transmission, 12
Hamming function, 214
Heat management, 706
Heavy charged particles (HCP), 684
Hermeticity, 672
Heterodyne, 169
Heterodyne VNAs, 863–865
Heterogeneous integration technology
COSMOS project, 425, 427–430
DAHI project, 431–433
NEXT project, 430, 431
SMART-LEES project, 433–435
High electron mobility transistors (HEMT), 218
High frequency (HF)
absorbing filters, 181, 182
adapters, 181
attenuators, 181, 182
coaxial communication lines, 179, 180
DC, 182, 183
harmonic distortion, 178

High frequency (HF) (*cont.*)
high-power telecommunication systems, 178
maximum permissible values, 179
microwave radiation, 178
passive intermodulation, 178
power level restrictions, 179
PSs, 183, 184
terminal equipment, 181
waveguides, 179, 180
High-frequency conducted interference, 721
High-speed applications, 642
High-speed battery-powered electronic devices
ADP1111
asynchronous converters, 788
characteristics, 783
external power transistor, 786, 787
functional diagram, 784
output voltage, 786
PWM modulator, 785
switch-on circuit, 787
wiring schemes, 786, 788
ADP1870/71 IC, 790
ADP187x controller, 792, 793
DC/DC converters, 783, 784
inverting DC/DC converters, 783, 784
price-quality criterion, 789
synchronous step-up and step-down converters, 791, 794
ZETA topology, 789
High-speed channels, 8, 49–51
High-speed data networks
CAM, 14
computer groups, 15
EDS AS, 14
EDVAC, 14
ENIAC, 14
global network, 16
organizations, 16
personal computers, 15
High-speed data transmission, 23, 38, 43
High-speed data transmission systems, 1
bus topology, 48
directly connected topology, 46
hybrid topologies, 48
mixed topology, 46
modern switches, 48
Ring topology, 48
High-speed electronic devices, 447, 448, 532, 533, 633
circuit level, 704
designers, 543
electromagnetic radiation, 703
EMC (*see* Electromagnetic compatibility (EMC))
EMI (*see* Electromagnetic interference (EMI))
IDL (*see* Intermodulation distortion level (IDL))
MEMS (*see* Microelectromechanical systems (MEMS))
protective equipment, 704
protective screen, 705, 706
technological features, 543
High-speed network applications, 32
High-speed printed circuit boards, 616–633
High-speed signal wires, 608
High-speed telecommunications devices, 314
Hittite Microwave company, 377, 388, 391, 400, 404, 406
Homodyne VNAs
amplitude-modulated SHF signal, 876
amplitude modulation, 866
APD, 874
differential/aggregated modulation frequency, 874
eight-pole measuring path, 883, 884
fourteen-pole measuring path, 878–881
frequency conversion process, 876
interference effect, 876
non-optimal energy ratios, 876
reference signals, 874
signal modulation, 866
signal spectrum, 877
sweep oscillator WKS system, 877
ten-pole measuring path, 881–883
waveguide paths, 878
Hughes Network Systems, 223

I

I/O interfaces, 81, 86
IBIS-AMI behavioral models, 614
iCoupler technology, 777, 779
iDCS-Control, 333
Immersion gold plating, 580
Immersion process, 579
Improved Mobile Telephone Service (IMTS), 153
IMT-Advanced, 155, 156
Inductive crosstalk, 492, 494–496
Information and communication technologies (ICT), 4, 149, 151
open information sources, 2
organization and functioning, 3
R&D areas, 3

Information technology equipment (ITE), 722
Infrared (IR) communication systems, 9, 10
Infrared data transmission, 42
Insertion losses (IL), 203
Institute of Electrical and Electronics Engineers (IEEE), 23
Integrated circuits, 295, 340, 347, 356, 357, 364, 368, 369, 428, 432
Integrated circuits electromagnetic emission model (ICEM), 485
Integrated radiation protection (IRP), 684
Interchange circuit, 80
Interdigital transducers (IDTs), 204, 209
Interface bus exchange protocols, 84
Interface Buses
- asynchronous protocol, 84
- AT-96 and industrial PC buses, 82
- component-level buses, 83
- data transmission, 83
- digital devices, 82
- DS and DR signals, 84
- ISA bus, 82
- Multibus II bus, 82
- multiprocessor system, 85
- PCI architecture, 82
- PICMG, 82
- skew, 83
- STD bus, 82
- synchronous protocol, 83
- VL-Bus, 82
- VME bus, 82

Interface communication lines, 88
Interference filtering, 718
Intermediate Bus Architecture (IBA), 741
"Intermediate" generation
- 2.5G, 153, 156
- 2.75G, 154, 156
- 3.5G, 155
- 3.75G, 155

Intermodulation distortion level (IDL)
- amplitude-phase conversion, 502
- electric circuits, 501
- electronic assembly, 502
- frequency-division multiplex, 501
- GSM-900 system, 503
- IMI components, 502
- non-linear node, 502
- PIM, 501

Intermodulation distortions, 897
Intermodulation products, 184
Internal bus, 81
Internal feedback photodiode (IFPD), 280
Internal modem, 7
International Electrotechnical Commission (IEC), 244
International Radio Consultative Committee, 453
International Space Station (ISS), 730
International Standardization Organization (ISO), 16
International Telecommunication Union (ITU), 22, 154, 166
Internet calibration systems, 847
Internet Engineering Task Force, 23
Internet of Everything, 150
Internet of Things (IoT), 150, 151
- AMPS mobile phone networks, 167
- concept of, 165
- e-call technology, 167
- eMBB, 166, 167
- 5G, 165
- LTE-M, 167
- wireless communications, 166

Internet research group, 23
Internet Research Task Force (IRTF), 23
Intolerable electromagnetic interference (IEI), 452, 454
Intra-Cavity Interconnection (ICIC), 674, 675
Invalid reverse path occurrence, 607
Inverse SEPIC, 783
iOA series, 510
iQA series, 511
iTA series, 510

J

JESD204B standard, 600
JESD204B system, 599

K

Kaelus company, 510
Keysight equipment, 171
Keysight network analyzers
- calibration, 895–897
- measuring systems based, 897, 899
- NVNA, 907, 908
- PNA series, 902, 904–906
- PNA-L series, 901, 902
- PNA-X series (*see* PNA-X series network analyzers)
- UXG signal generator, 908, 911 913, 914

Kirchhoff's law, 136
Known Good Die (KGD), 667
Kotelnikov theorem, 65

L
Laser diode (LD), 280
Laser transmission, 42
LDMOS technology, 193, 194
Leadframe packages, 484, 485
Leading edge pulse time (LEPT), 644, 650
Leadless connector, 698
LeaXield module, 718, 719
Line matching, 90
Linear phase modulation, 871
Linear Technology, 307
Liquid photoimageable solder mask (LPISM), 579
Logical encoding, 29
  redundant codes, 30
  scrambling, 30
Long-term evolution (LTE), 155
Loop far-end crosstalk generation, 621
Loop near-end crosstalk generation, 621
Loop output signal, 622–625
Loop-type delay lines, 633
LoRa, 172
Low-noise amplifier (LNA), 67
Low-power wide-area network (LPWAN), 171–173
Low-voltage transmission systems, 122
LVDS-interface
  and DVI, 349, 352, 354, 355
  building, 347–349
LVPECL driver, 142
LVPECL standard
  DC coupling, 136
  drivers, 136
  and HCSL, 135
  matching, 135
  matching chain, 137
  PNP transistors, 135
  receivers, 136
  standard matching chain, 136
  transmission lines, 135
  valve, 135

M
Machine-to-machine (M2M) communication, 161
Machine-to-machine (M2M) connections 165, 172
Manchester code, 28
Manchester coding principle, 28
Mask compatibility, 616
Massachusetts Institute of Technology (MIT), 165
Massive machine-type communications (mMTC), 167
Mathematical expressions, 617
Maxim-Dallas, 307
Maxwell's equations, 606, 644
Measurements and analysis of microwave devices
  characteristics, 817
  coplanar testers, 817
  hardware and software, 817
  millimeter wave band (*see* Millimeter wave band)
  semiconductor wafers, 817
  vector analyzers, 817, 818
Mechanical deformations (MD), 695
Meggitt Safety Systems, 263, 265, 266
Message switching, 52
Message-switched network, 52
Metal oxide semiconductor field-effect transistor (MOSFET), 301, 302, 411, 414–416
Metal-dielectric waveguides (MDWG), 842
Metallized seals, 707
Metal-semiconductor field-effect transistor (MESFET)
  alloying level, 201
  amplification mode, 201
  channel and buffer layer parameters, 198
  channel layer–buffer layer, 199
  chemical vapor deposition, 197
  electrophysical modeling, 199
  epitaxial growth method, 198
  epitaxial layer parameters, 199
  ionizing radiation, 198
  Lely method, 197
  parasitic capacitances, 199, 202
  Schottky junction, 198
  semi-insulating substrate, 198
  threshold voltage, 199
  two-dimensional model, 200
Method with modulated subcarrier, 867
Micro-Cap software package, 310
Microcircuit assembly, 657, 666
Microcircuit chip, 666, 667, 677, 678
Micro-Coax UFA210B cable, 250
Microcontrollers, 478, 479, 481–484
Microelectromechanical systems (MEMS), 177
  actuation voltage, 221
  broadband antenna tuning, 220
  cantilever beam, 222
  capacitance values, 222
  capacitors, 226, 227
  Cavendish's radio frequency, 222

characteristics, 219
communication protocols, 218
components, 219
configurable source identifiers, 220
death grip problem, 219
ESR, 221, 222
frequency tuning, 222
installation technology, 233–235, 238
LGA type package, 220
low temperature processing, 218
non-volatile memory circuit, 221
radio frequency, 220, 221
resonators, 228, 229
RF systems, 218
SHF devices
inductance coils, 230
micromechanical resonator circuits, 232
micro-processed cavity resonators, 232
microwave technology, 230
switch-based circuits, 231
transmission line, 232
varactors, 231
switches
actuation electrodes, 224
advantages, 223
BEOL-metallization layers, 223–225
electrostatic actuation, 224
oscillator frequency, 226
RIT operations, 223
thermal actuator, 223
waveguide, 224
thermally controlled, 223
tunable capacitors, 222
Microelectronic devices, 146
Microelectronic integrated circuit (MIC), 466
Microelectronic microwave devices, 530
Microelectronic products, 658, 659
Microelectronics technologies, 145, 581, 850
Microminiaturization, 556
Microprocessor relay protection device (MRPD), 714
Microprocessor-based relay protection devices (MRPDs), 802, 804
Microprocessor system, 87
MicroSiP modules, 332
Microstrip configuration, 579
Microstrip cross section, 648
Microstrip lines, 478, 495, 496, 501, 649
Microwave antennas, 715
Microwave data transmission systems, 42
Microwave devices
mechanical components, 537
reliability and durability, 536
stricter requirements, 537
Microwave monolithic circuits (MMIC), 671–674
Microwave systems, 9, 711
Milandr company, 347, 352
Millimeter wave band
absolute errors, 820
attenuation measurement limit, 819
calibration algorithm, 820
equivalent multi-terminals, 818
measurement tasks, 819
modem, 333, 335
phase errors, 819
radioelectronics, 818, 819
structural simplification, 819
transceiver chips, 337–340
transmission lines, 818, 819
VAC, 819
waveguide channels, 820
waveguide flange connections, 820
waveguide model measurement, 821
MIL-STD-188-125, 712, 714
Minimum frequency manipulation (MSK), 337, 338
Mir orbital station, 731
Missile and air defense systems (MD/AD), 2
Mobile broadband access (MBBA), 158, 159
Mobile communication system, 153
Modem, 7
Modern computers, 17
Modern fiber-optic cables, 41
Modern high-power commercial telecommunication systems, 716
Modern high-speed electronic systems, 85
Modern radio equipment, 709
Modern satellite telecommunications systems, 68
Modern telecommunication systems, 3
Modulated carrier, 867
Modulated vector network analyzer (MVAC), 847–849
Molecular beam epitaxy (MBE), 672
Molex, 253, 254, 256, 258
Monolithic integrated circuits (MIS), 164
Monolithic microwave integrated circuits (MMIC), 851
Moore's Law, 146, 147
MORNSUN company, 328, 340, 342, 343, 345
MOSFET transistors, 752
Multi Chip Modules (MCM), 657, 667, 678
Multi Level Transmission—3, 29
Multichannel resonator method, 205
Multilayer film screens (MFS), 730
Multilayer PCB typical design, 549
Multilayer printed circuit, 548

Multilayer printed circuit boards, 548
Multimode fibers (MMF), 40, 279, 292
Multiplexing, 32
Multi-port measurement system, 903
Multiprocessor system, 86
Mu-metal, 732

N

National Semiconductor, 307, 309
Near-end crosstalk voltage expression, 620
Near-end crosstalk voltage waveform, 620
Network adapters, 12, 30
Network equipment, 12
Network processes architecture, 18
Network segment, 54
Network technology, 152
Neural networks, 146
NEXT project, 430, 431
Nonadaptive algorithms, 20
Non-functional pads, 640
Non-linear microwave circuits, 849, 850
Non-linear vector circuit analyzers (NVNA), 907, 908
Northrop Grumman Space Technology (NGST), 671, 673
Nyquist frequency, 636
Nyquist limit, 25
Nyquist zones, 67

O

Offset strip line design sketch, 565
100 ohm differential pair, 568
Ohm's law, 89
Ohmic losses, 593
OMRD-04
- photodetector, 284
- receiving module, 283
- ST-type optical connectors, 283
- transmitting module, 283, 284

OMRD-06
- electro-optical parameters, 284
- FC type optical connector, 283
- functional diagram, 283
- photodetector circuit, 282
- receiving module, 282

OnStar technology, 167
Open Shortest Path First (OSPF), 59
Optical multiplexers, 34
Optical transport networks (OTN), 292
Oscillogram, 894
Overmolded Plastic Array Carrier (OMPAC), 676

P

Packaging technology
- BGA chip assembly technology, 676, 677
- Chip-on-Board, 666, 667
- development, 655–658
- high-speed electronic devices, 658, 659, 661, 663
- MCM and printed boards, 678, 680, 681
- 3D products, 681, 682
- TSV, 688–692
- WLP technology, 668, 671–673, 675

Packet switching, 52, 54
Passive intermodulation (PIM)
- band, coaxial and waveguide transmission lines, 504, 505
- directional couplers, frequency duplexers and transformers, 505, 506
- evaluation methods, 507–509
- external sources, 506, 507
- IDL, 501
- iTA series, 510
- level, 510–512
- and non-linear phenomena, 503
- printed circuit board material, 504
- quantity of connect–disconnect cycles, 504
- radio frequency connectors, 503, 504

Passive intermodulation interference (PII), 184
Passive optical network (PON), 73
PCB transmission line base (PCB TL), 569
PCS-7448 and PC8548 microprocessor, 76
PF402 cable, 275
Phase manipulation (PMP), 871
Phase modulation (PM), 871
Phase shifter, 826
Phase temperature factor (PTF), 249
Phased antenna arrays (PAA), 234, 826
Phase-frequency characteristics (PFC), 867
Phase-frequency response (PFR), 217
Phase-locked loop system (PLL), 96, 865
Phase-stable SHF cables
- creation, 253
- Huber + Suhner SucoPearl RF cables, 269, 271
- manufacturing technology, 253
- phase track RF cables, 264, 266
- SiO_2 insulation, 260, 262–264
- Temp-Flex RF cables, 253, 254, 256, 258, 260

Photomultiplier tubes (PMT), 730
Physical and chemical electroplating processes, 554
Physical and logical data coding
- AMI code, 27
- ASCII, 24

4B/5B code, 30
4B/5B logic code, 30
code table, 24
data transmission, 26
digital transmission, 25
ECL and CMOS logic, 24
logical encoding, 29
MLT-3 code, 29
modern computer networks, 24
NRZ code, 26
NRZI, 27
signals, 25
Plastic Ball Grid Array (PBGA), 656, 658, 662
Plastic fiber-optic cables, 41
Plastic Pin Grid Array (PPGA), 656
Plated Through-hole (PTH), 656
PNA series network analyzers, 902, 904–906
PNA-L series, 904
PNA-L series network analyzers, 901, 902
PNA-X series network analyzers
amplifiers TD, 889, 891, 892
antenna systems, 885
inter-modulation distortions, 888
non-linear devices, 892–895
4-port, 886
properties, 888
radio frequencies, 886
radio pulse filling, 889
SHF and microwave radioelectronic systems, 885
sinewave signals, 888
spectral zeroing, 888
strobing method, 889
technical characteristics, 888
VANs, 886
Point of load (POL), 742
Point of Load Alliance (POLA), 744
Polling access, 50
Polymorphism, 699
Power distribution network (PDN), 613, 794, 800
Power distribution systems, 741
Power source interference, 499–501
Power splitters (PSs), 183, 184
Power supply rejection ratio (PSRR), 499, 500
Power supply sequence, 742
Power supply systems
control signals, 766, 767, 772, 773
controllers and drivers, 749, 751–753
current reader amplifiers, 779
electronic equipment, 741
high-speed electronic devices
advantages, Z-One system, 746
analog power management, 745
characteristics, power systems, 747
distributed power systems, 744
electromagnetic interference, 743
intermediate power bus, 743
microcontrollers, 742
monitoring service, 744
oscillogram, 743
POL, 742
POL analog control circuit, 745
process synchronization, 743
programmed FPGA data, 742
supply voltage levels, 741
switching sequence, 742
transient processes, 742, 743
voltage converters, 742
Z-One bus, 748
Z-POL converters, 748
Z-Series digital power controller, 746
hot plugging controllers, 769, 774–777
interference suppression, 797–801
LDO-stabilizers, 757, 758, 761, 782
load switches, 777, 779–781
multi-level supervisors, 768
power consumption, digital monitors, 778
pulse voltage stabilizers
ADP1614, 755
ADP2323/ADP2325, 754
ADP2370/ADP2371, 754
ADP2384/ADP2386, 753
ADP2441/ADP2442, 755, 759
parameters, 760
supervisor chips, 761, 762, 765–767
telecommunication devices, 749
wiring diagram, 792, 794–797
Power synchronization system, 742
Power system monitoring, 746
Precision-guided munition (PGM), 2
Preliminary analysis, 611
Preliminary topological diagram, 610
Pre-preg technology, 547
Presentation layer, 21
Printed circuit board (PCB), 679, 693, 697, 796
classification, 551, 583
computer-aided design systems, 606
conductive pattern, 545
conductors, 557
copper nanoparticles, 586
creation and development, 544
density, 555
design fragments, 547
design methodology, 609
design tools, 606, 641

Printed circuit board (PCB) (*cont.*)
digital devices, 547
double-sided, 546
electromagnetic incompatibility, 608
electronic components, 556
embedding active and passive components, 583
etching technology, 545
flexibility, 551
flexible PCB appearance, 550
heat removal process, 558
high heat dissipation, 558
introduction, 557
layout design, 603
LED scrolling signs, 551
LED-generated heat, 552
manufacturing, 557
manufacturing techniques, 546, 553
matching element, 641
materials and substrates, 563
microstrip lines, 564
multilayer printed circuit board, 554
paraffin-impregnated paper, 544
predecessor, 546
preliminary analysis, 610
properties, 563
radioelectronic devices, 556
reflectometer, 573
rough surface, 573
SI, 605
signal transmission, 572
signal transmission lines, 563
signal transmission time, 571
structural element, 544
surface effect, 574
through-holes, 608
transition, 544
transmission lines, 571
zigzag lines, 616
Printed circuit boards (PCBs), 544, 553, 568
Printing plate, 546
Protective coatings, 578
Protective seals, 706
Protective shutdown devices (PSD), 717
Protocol data unit (PDU), 18
Protocols, 18
Pseudo-random bit stream (PRBS), 639
Pseudo-random frequency search (PRFS), 216
PT210 cable, 271
PTLS600 cable, 277
Public *dial-up* telephone communication, 8
Pulsar Research and Development Institute, 203
Pulse description words (PDW), 911
Pure steel, 732
PTSRB141 cable, 273

Q
Quasilinear tuning, 228
Quasistationary magnetic fields
multilayer screens, 732
mu-metal, 732
nonmagnetic layer, 733
PMT cases, 734
shielding, 731
shielding efficiency, 733

R
Radiated interference, 454
Radiation protection screens (RPS), 683, 684, 687, 688
Radio and electronic equipment, 720
Radio frequency chips, 435, 436, 438, 439, 441–443
Radio frequency resource (RFR), 453
Radio frequency systems
field-effect transistor, 195
frequency spectrum, 196
GaAs, 193
GaN, 193, 194
high-power SHF MESFET transistors (*see* Metal-semiconductor field-effect transistor (MESFET))
LDMOS, 193
power amplifiers, 192
power summation efficiency, 195
product creation, 197
silicon-based power amplifiers, 195
silicon-based technologies, 193
stationary communication systems, 193
virtual heterojunction bipolar transistor, 195
Radio transmission, 42
Radioelectronic systems (RES), 452, 453 456, 467
Radiostation, 454
Reference clock signal, 96
Reflection, 593
Reflective gratings (RG), 203
Reflectometers, 837, 839
Relaxation oscillation (ringing), 496, 498
Reliability
description, 447
mechanical components, 532–537
quiet transistors, 516
Resonance effect analysis, 651

Resonant frequency, 635, 636, 652
Reverse-treated foil (RTF), 575
RFHIC company, 357–360
RG resonators, 204
Rite Aid pharmacy network, 10
RJ-45 connector, 38, 243
Rolled copper, 575
Rosenberger Company, 511
Routers, 13, 59
Routing Internet Protocol (RIP), 59
RS-485 bus, 116
RS-485 interface
  transceivers, 340, 342
  use, 342, 345–347
RS-485 standard, 121
Russian Federation laws, 23
Russian GOST R 50397-92, 720

S
Satellite communication systems, 9
Satellite signal processing system, 69
Scalar homodyne detectors (SHD), 867, 868
Schaffer–Kohn method, 867, 868, 876
Schottky barrier field transistor, 197
Schottky diode, 308, 309
Schottky transistor–transistor logic elements, 470, 472
Second generation (2G), 153, 156
Semi-additive process, 586
Semiconductor protective devices, 516, 517, 519–522
Semiconductor wafers
  coplanar probes, 852–854
  IC testing
    automated probe installation, 854, 855
    calibration measurement system, 858, 859
    HP 8510 vector network analyzer, 853
    HP 85109C, 856, 858
    RF and microwave, 860
    ring frames/micromanipulators, 855
    ring-type plate, 855, 856
    running probe installation, 855, 856
    typical characteristics of probes, 855, 857
    vector circuit analyzer, 855
    waveguide-coaxial adaptors, 857
  SHF integral circuits, 850–852
Semtech SerDes series chips, 293
Sensitive circuits, 606
Sentaurus TCAD, 199
SEPIC topology
  bipolar voltage source, 304, 306
  components of the circuit, 308, 309
  converter chip, 306, 307
  modeling process, 310–312
  parameters, 299–304
SerDes devices, 614
SerDes series
  basic standard IEEE 802.3, 292, 293
  by Semtech, 293–296
  structure and operation principle, 296–299
SerDes transceiver, 95
SerDes type, 79
SerDes type transceivers, 94, 97
  CDR, 95
  classical correction methods, 98
  devices, 97
  equalizer, 100
  parallel bit stream, 96
  PLL circuit, 96
  principle of operation, 94
  uncoded data, 95
Serial data buses, 94
Serial data line analysis, 615
Serializers, 293
Session layer, 20
Shannon–Hartley theorem, 595
SHF chips
  application, 356, 357
  by Hittite Microwave
    analog voltage control attenuation, 372, 373
    attenuators, 369–371
    digital control, 374, 375
    frequency synthesizers, 387, 388, 391, 393, 397–399, 403, 404, 406
    ultra-wideband amplifiers with fixed gain, 375, 377, 379, 383–385, 387
  GaN-Based SHF-amplifiers, 359, 361, 363, 364, 367–369
  wireless communication systems, 357, 358
Shielding, 731
Short-Open-Load-Thru (SOLT), 852
Short-Short-Load-Load-Thru (SSLLT) Calibration Technique, 846
Short text messages (SMS), 153
SigFox, 172
Signal damping, 12
Signal distortions, 93
Signal integrity (SI), 490, 491, 592, 595–597, 603, 605
Signal transmission lines, 643
Signal waveforms, 646
Signal-to-noise ratio (SNR), 595

Silicon carbide (SiC), 193, 197
Silicon dioxide (SiO_2) insulation
cable design, 262
inorganic dielectric, 260
laser welding, 263
operating temperature, 262, 263
oxygen-free copper, 262
phase temperature dependence, 264
plunger pressing, 262
radiation resistance, 262
sealed SMA connector, 263
specifications, 264
Silicon on Insulator (SOI), 692
Silver coating, 579
Simplex transmission systems, 12
Simultaneously switching noises (SSN), 482, 486
Single-Ended Primary-Inductance Converter (SEPIC), 783
Single-mode fibers (SMF), 40, 279, 280
Single-mode optical fiber (SMF), 292, 293
Single-sideband modulators (SSM), 867
Single-sided printed circuit board, 546
Slightly larger than a microcircuit chip (SLICC), 666, 667
Small Electronic Accounting Machine (SEAM), 14
SMART-LEES project, 433–435
SMV 11 selective microvoltmeter, 724
Software defined radio systems (SDR), 67
Software-defined networks (SDN), 162
Software-defined radio systems (SDR)
active USB-8P8C converters, 243
ADC, 238
antenna-feeder path, 241
block diagram, 240
communication equipment, 238
data transmission channels, 243
digital radio receivers, 238
digital-controlled quadrature heterodyne, 242
DSP, 238
DVB-T digital television receiver, 242
evolution stages, 240
intersymbol interference, 239
market monopolization, 239
microelectronic technologies, 238
multichannel signal processing algorithms, 243
R820T chip, 242
reception path, 239
reference oscillator, 240
RTL2832U chip, 242
signal demodulation, 239
single radio interface, 243
structure, 240
system operating parameters, 239
telecommunication technologies, 238
transmission path, 240
universal modems, 239
wireless services, 238
Solder mask defined (SMD), 696, 697
Solder mask over bare copper procedure (SMOBC), 574
SP304_FEP cable, 278
SP306_FEP cable, 278
Space communication systems, 68
Spanning Tree Algorithm, 60
Spectrum, 894
Spice-modeling, 105
Spontaneous electromagnetic interference (SEI), 452–454
SPRUT experimental equipment, 730
Stacked die, 690
Standard fin-line design, 567
Standard frequency spectrum, 74
Standard microstrip line design, 564
Standard MIL-STD-188-125, 711
Standard modem, 7
Standard separation capacitor ratings, 142
Standard unshielded twisted pair, 39
Standing-wave ratio (SWR), 181
Strip and microstrip lines, 623
Strip transmission lines, 564
Structured cable systems
administrative subsystem, 44
cables, 43
data transmission systems, 43
EIA/TIA 568, 45
EIA/TIA 569, 46
EIA/TIA 606, 46
EIA/TIA 607, 46
EIA/TIA standards, 43
horizontal subsystem, 44
TIA/EIA 568, 43
TIA/EIA 568A standard, 44
TIA-568A/B standard, 47
transmission medium, 45
trunk segment, 44
workplace subsystem, 44
Super high frequency (SHF)
cable assemblies (*see* Cable assemblies)
computing facilities (PCs), 827
increase measurement accuracy, 827
massive and inexpensive vector analyzers, 827

measured circuit parameters, 821–823
measurement tasks
frequency-selective devices, 824
integrated circuits, 826
linear passive devices, 823
microwave semiconductor microassemblies, 826
non-linear devices, 824
PAA, 826
parameters and characteristics, microwave devices, 825
time domain, 825
ultra-wideband devices, 824
MEMS (*see* Microelectromechanical systems (MEMS))
passive electronic components, 185
PII, 184
probe methods
four-probe method, 834, 836
frequency sweeping, 833, 834
measuring line errors, 831–833
measuring waveguide line, 828–831
millimeter waves, 837, 838
multi-probe measuring line, 834–837
reflectometric method, 837, 839
Surface acoustic waves (SAW), 177
acoustic synchronism, 211
antenna connector, 214
antiresonance, 213
applications, 212
audio frequency signal, 212
bridge-type impedance filters, 211
broadband filter, 217
broadcast television systems, 214
channel filter, 215
differential bridge circuit, 214
DL, 203
electromechanical coupling coefficient, 213
expansion inductances, 217
frequency response, 215
G-circuits, 217
hardware functions, 216
IDT, 204, 205, 215
image distortion, 212, 213
impedance filters, 207, 209
LC filters, 207, 209, 210
lithium niobate, 213, 214
microwave oscillators, 202, 203, 207
optimal matching, 206
out-of-band suppression, 214
PFR, 217
piezoelectric substrate, 211
resonators, 203, 204
self-oscillators, 205, 206, 216
single-input resonator, 204, 205
spectral density, 205–207
symmetric transformer, 211, 212
TCS transceiver devices, 216
TCSs, 208
T-type bandpass impedance, 209
T-type notch impedance, 210
two-input resonator, 205
ultra-narrow differential bridge filter, 216
Surface effect, 594
Surface Laminar Circuit (SLC), 680
Surface mount filters, 530–532
Surface-mounting technology (SMT), 656
Suspended slow-wave (SSW), 225
Suspended strip line, 565
Suspended strip line design sketch, 566
Swept-signal sources (SSSs), 819
Switching technologies, 51
packet switching, 52
Symmetric reflecting couplers (SCR), 33
Symmetrical interference, 721
Synchronization, 94
Synchronous communication, 11
Synchronous multiplexers, 33
Synchronous packet data transmission, 38
System bus, 81
System in Package (SiP), 663, 665, 666, 691
System on chip (SoC), 291, 333–340, 665, 666

T

Tape BGA (TBGA), 658
Tape-automated Bonding (TAB), 666, 676, 678
TD1211D485H transceiver, 345
Telecommunication Industry Association (TIA), 23
Telecommunication systems (TCS)
advantages and disadvantages, 8
analog signal, 6
communication channels, 7
communication lines, 6
components, 5
data transmission channels, 5
digital signal, 6
effectiveness and capabilities, 10
EIA, 23
electronic signals, 6
elementary modem, 7
flow diagram, 4
functions, 6
GSM, 23
hardware and software components, 6

Telecommunication systems (TCS) (*cont.*)
IAB, 23
IEEE, 23
IETF, 23
internal modem, 7
ISO, 22
ITU-T developments, 22
modem, 7
narrowband and wideband, 1
network protocols, 6
network software, 6
organization, 1
OSI, 22
performance, 1
protocol, 6
protocols and standards, 24
segments, 1
standards, 21
structure, 5
subscribers, 8
technical solutions, 1
TIA, 23
transmission channels, 1
types, 5
universal standards, 6
Telecommunication systems (TCSs), 207, 208
Telecommunication technologies, 1
Telecommunications Industry Association (TIA), 340, 347
Telecommunications software, 13
Telephone lines, 8
Temp Flex radio frequency cables
141-1701, 258, 261
141SC-1901, 254, 257
cordless-tubular insulation, 258
high-flexibility, 258, 262
MULTIFLEX_141, 254, 257
parameters, 254, 255
PFA insulation, 261
phase change dependence, 254, 256
solid PFA insulator, 258
Teflon knee effect, 253
twisted PFA strands, 256
UFB142A, 258, 261
ultra high-speed coaxial cable, 256
Ultra Low Loss Microwave Coax Series, 258, 259
Temperature coefficients of linear expansion (TCLE), 661
TERRABUS project, 590, 591
Tested device (TD), 889
Texas Instruments, 307, 314, 316, 318, 322, 326, 333, 419
TF4 technology, 266
Thermal conductivity, 561
Thermal effects, 706
Thermocoax, 264, 268
Thermomechanical fatigue (TMF), 694, 695
Thermomigration (TM), 693, 696, 698
Thevenin equivalent circuit, 137
disadvantages, 137, 138
VCC0, 139
π- and T-shaped matching chains, 139
Third generation (3G), 154, 156
Third-Generation Partnership Project (3GPP), 172
Third-order intersection point (TOI), 891
Three-dimensional array, 613
3D-assembly technologies, 659, 668
Three-dimensional mesh spacing, 614
Through Silicon Vias (TSV), 688–690, 692
Thru-Reflect-Line (TRL), 852
Thru-vias, 637, 639
Time division duplex (TDD), 334
Time division multiplexing (TDM), 32
Times Microwave Systems, 264, 267
Token data transmission method, 50
Token Ring network, 48
Traditional technologies, 732
Transceivers, 337–340, 342, 344, 346, 347
Transmission line model (TLM), 478
Transmission speed, 11
Transmitting high-frequency signals, 89
Trunk-modular organization, 80
T-shaped matching chain, 140
Twisted pair, 38
Twisted pair cable, 9
Two-dimensional electron gas (2DEG), 412, 414

U

UFA210B cable, 251
UFA210V cable, 250
UltraCMOS 10 technology, 195, 196
Ultra-reliable machine type communication (uMTC), 167
UMTS technology, 155
Unit interval (UI), 597
Universal Internet, 161
Universal wireless test system, 76
University of California at San Diego (UCSD), 168

Unshielded twisted pair (UTP), 112
cables, 38
disadvantage, 39
and fiber optics, 39
transmission medium, 44

V
Vacuum protective devices, 522, 523
Van Allen radiation belts, 683
Vector analyzers of circuits (VAC), 819
Vector homodyne detectors (VHD), 868
Vector network analyzers (VNAs)
AM (*see* Amplitude modulation (AM))
amplitude-phase, 861
anti-transverse flanges, 846
calibration methods, 846
classification, 862
degree of reference, 845
development, 840, 841
factors, 844
flange connections, 845
geometric tolerances, 845
heterodyne, 863–865
homodyne, 862, 863, 866
installation torque values, 846
new frequency ranges and measurement media, 841–844
testers of *S*-parameters, 861, 862
two-pole devices, 862
uniformity of measurement, 846
Video transmission, 98
Viewpoint software, 190, 191
Voltage controlled oscillators (VCOs), 230
Voltage regulator module (VRM), 794
Voltage stabilization module, 742
Voltage standing-wave ratio (VSWR), 247
Voltage-controlled oscillator (VCO), 298, 391, 402, 403, 407

W
Wafer level packaging (WLP) technology, 668, 671–673, 675
Wafer-Scale Assembly (WSA), 674
WALOPACK materials, 687
Wave impedance, 593
Wavelength division multiplexing (WDM), 33, 35
Weightless, 172
Whiskers, 693, 697–699
Wide area network (WAN), 45
WiMAX, 155
Wired communication circuit, 87
Wired communication lines, 86, 94
matching, 92
parameters, 93
signal, 92
transmitter chip, 92
Wireless communication channels, 9
Wireless media, 38
Wireless networks, 42
Wireless telecommunication systems
components, 145
high-speed electronic devices, 152
information and communication technologies, 151
information transmission lines, 152
long-term and short-term development, 150
microcircuits, 148
power consumption, 152
road map, 146, 147
silicon processors, 146
stages, 149
3D integration, 146
transistors, 148
WiSpry, 219, 221
WKS system sensor, 873
World Wide Network, 149
World Wide Web, 150
WS1041 circuit, 221
WSI042 circuit, 221

X
Xilinx company, 334, 339, 340

Z
Zero potential, 27